Assessing the Safety of Existing Embankment Dams

Many existing embankment dams were designed and constructed to standards below current requirements, potentially posing risks to downstream populations. Dam owners must assess whether these structures provide adequate safety or require upgrades to reduce risks to acceptable levels. This book provides in-depth analysis of dam failure mechanics alongside relevant engineering assessment methods.

The book provides comprehensive methods for assessing embankment dam safety across four critical failure modes: overtopping, slope instability, deformations under seismic loading, and internal erosion and piping. Assessment approaches include deterministic (standards-based), risk enhanced deterministic, quantitative risk assessment (QRA) and simplified QRA methodologies. These approaches represent current industry practice, drawing from the authors' extensive experience in dam safety assessment and upgrade design across more than 200 dams in Australia, New Zealand, Canada, the USA and Panama.

The book is designed to help dam owners, their consultant engineers and engineering geologists develop defensible safety assessments and upgrade designs for existing dams.

Assessing the Safety of Existing Embankment Dams

Robin Fell, Mark Foster, Gavan Hunter and Chi Fai Wan

CRC Press
Taylor & Francis Group
Boca Raton London

CRC Press is an imprint of the
Taylor & Francis Group, an **informa** business

Designed cover image: Hume Dam, Australia showing Embankment No. 1 and the Spillway. Courtesy of Murray Darling Basin Authority and Water NSW

First edition published 2027
by CRC Press
4 Park Square, Milton Park, Abingdon, Oxon, OX14 4RN

and by CRC Press
2385 NW Executive Center Drive, Suite 320, Boca Raton FL 33431

CRC Press is an imprint of Informa UK Limited

British Library Cataloguing-in-Publication Data
A catalogue record for this book is available from the British Library

ISBN: 9781032889047 (hbk)
ISBN: 9781032892641 (pbk)
ISBN: 9781003542025 (ebk)

DOI: 10.1201/9781003542025

Typeset in Times
by Apex CoVantage, LLC

Contents

Author Biographies

Robin Fell has 58 years of experience in dam engineering, geomechanics, landslides and slope stability, applied to a wide range of projects and in several countries. This has involved work in Papua New Guinea, Fiji, Solomon Islands, Hong Kong, Thailand, Tanzania, New Zealand, Canada, Panama and the USA, as well as throughout Australia. He has specialised in embankment dam engineering, geotechnical engineering of dams and associated structures, and risk assessment for dams and landslides. He has worked on over 150 dams and a further 100 dams in Portfolio Risk Assessments. He worked for 10 years with the Australian government on hydropower projects, and 10 years with consulting geotechnical engineers before being appointed Professor of Geotechnical Engineering at UNSW in 1985. There he supervised research into internal erosion and piping of embankment dams and risk assessment of dams and landslides. He has consulted on peer review panels for dams throughout Australia, and in Canada, the USA, Panama and New Zealand. He is acknowledged internationally for his expertise in internal erosion and piping of dams and risk assessment for dams and landslides. He was the lead author of the book *Geotechnical Engineering of Dams*, published by Taylor & Francis, and ICOLD Bulletin 164 on Internal Erosion of existing dams, levees and dikes and their foundations.

Mark Foster has 30 years of experience in geotechnical and dam engineering, with specialist expertise in dam safety upgrade design, embankment dams and risk assessment. He has conducted independent review of dam safety assessments for existing dams and new dams. He is an active member of the Australian National Committee on Large Dams (ANCOLD) and was on the working group that completed updating of the guidelines on risk assessment for dams. He completed his doctorate in 1999 under the supervision of Professor Fell at UNSW, titled "Methods for Estimating the Probability of Failure of Embankment Dams". He is a co-author of the widely referenced book *Geotechnical Engineering of Dams* published by Taylor & Francis.

Gavan Hunter has more than 30 years of experience in geotechnical engineering for a range of water, civil and mining projects. He specialises in geotechnical dam engineering and has been involved with the assessment of performance, investigation, design and construction of upgrades for numerous dam projects in the past 20 years. These projects have covered a broad range of geotechnical engineering application including soft ground engineering, slope stability assessment, risk assessment, foundation design, assessment of performance (deformation, seepage, pore pressure), cutoff wall design, liquefaction assessment, earthworks control, and design and construction advice. Embankment types have included water storage and tailings dams, and range in size from small dams up to very large dams, including some of Australia's most significant dams. For the past five years, he has provided independent technical review to water authorities and mining companies on a range of dam projects, mostly in Australia but also internationally. He completed his doctorate in 2003 under the supervision of Professor Robin Fell, titled "Pre and Post Failure Deformation Behaviour of Landslides in Soil Slopes". He represents ANCOLD on the ICOLD Embankment Dams Committee (Committee E).

Chi Fai Wan has over 30 years of post-graduate experience, principally in the field of water and dams engineering. He has specialist skills regarding internal erosion of embankments, having researched aspects of initiation of piping erosion and soil erosion rate. He has published more than 30 technical papers and reports on soil erosion testing, dam design, risk assessment, probabilistic design and analysis, numerical modelling, and internal erosion in dams, and was a guest lecturer in dams engineering courses at the Sydney University and the University of New South Wales. He has worked as a numerical modeller, design engineer, risk analyst and specialist in internal erosion problems for dam consultants in Australia, Hong Kong, and the USA. He has served as design manager for dam raising and safety upgrade projects, and as technical lead in numerous dam safety reviews using both traditional and risk assessment methods.

Introduction

1

Many existing embankment dams were designed and constructed to lesser standards than modern dams and potentially pose a risk to the population downstream of the dam. Dam owners need to assess whether these dams are adequately safe or need upgrade works to reduce the risks to tolerable limits.

The book is intended to guide owners and their consultants to reach defensible assessments of the safety of the dams they own or are considering.

Methods are presented for assessing the safety of existing embankment dams for failure by the four important Potential Failure Modes: overtopping, slope instability, deformations under seismic loading and internal erosion and piping. This may be done in the following four ways.

1.1 DETERMINISTIC ASSESSMENT

The deterministic method is called the "standards based" method by some.

- *Overtopping failure.* The annual probability of the Inflow Design Flood (IDF) which the dam should be able to pass without failure by overtopping is related to acceptable flood criteria in national or international guidelines. These criteria are related to the Consequence Classification of the dam, which is defined in terms of Potential Life Loss, or Population at Risk if the dam were to fail. Lower Annual Exceedance Probability (AEP) IDFs are required for dams with large consequences of failure and not for dams with small consequences of failure.

  The dam is expected to have freeboard above the maximum reservoir level resulting from the IDF so the likelihood of failure from waves overtopping the embankment is negligible.
- *Slope instability.* The factors of safety for normal operation and flood are compared to commonly adopted acceptable factors of safety, taking into account the settlement and deformation behaviour of the dam.
- *Deformations under seismic loads.* Assessment of the settlement of the crest under the loading from the Earthquake Design Ground Motion (EDGM). This is compared to the freeboard between the dam crest and normal operating reservoir level to assess whether the settlement would result in overtopping.
- *Internal erosion and piping in the embankment, foundation and from embankment to foundation.* Comparison of the as designed and constructed embankment zoning, filter properties, foundation conditions (soil or rock) and cutoff design and treatment with modern design criteria.

DOI: 10.1201/9781003542025-1

1.2 RISK ENHANCED DETERMINISTIC ASSESSMENT

This approach is based on deterministic assessment but adds database methods to estimate annual probability of failure to aid in the dam safety assessment.

1.3 QUANTITATIVE RISK ASSESSMENT (QRA)

Quantitative Risk Assessment (QRA) is a process in which the annual probability of failure of the dam is estimated for all Potential Failure Modes (PFMs), Potential Life Loss (PLL) is estimated and the risk is compared to tolerable societal and individual risk criteria.

There is an event tree for each PFM and conditional probabilities estimated for each reservoir load partition or seismic load partition. Methods are described to estimate the conditional probabilities for initiation, continuation, progression, detection and intervention, and breach.

It is assumed that those using the QRA approach will have an understanding of risk analysis principles. Guidance on these can be obtained in ANCOLD (2022) *Guidelines on Risk Assessment* and USACE/Reclamation (2019) *Best Practices in Dam and Levee Safety Risk Analysis*.

1.4 SIMPLIFIED QRA

For the Simplified QRA, the concept is to provide a method based on the QRA but assessing the annual probability of failure for only the most important PFMs. For these PFMs, fewer branches of the event tree and fewer reservoir/seismic load partitions are modelled and, for some PFMs, simplified methods are used to estimate conditional probabilities of failure. The annual probability of breach for individual PFMs and the related PLL are estimated and the outcome considered in the semi-quantitative tolerable risk matrix. The semi-quantitative tolerable risk matrix is calibrated to the quantitative risk criteria.

QRA and Simplified QRA also take account of the outcomes of deterministic assessments.

Deterministic methods are the most familiar to most owners and consultants but are not well suited for assessing the safety of a dam where modern design criteria are not met. It is expected that many existing dams will not satisfy deterministic criteria.

While quantitative risk assessment methods are well suited to assessing the safety of such dams, fewer owners and consultants are familiar with quantitative risk assessments and many countries do not have guidelines on tolerable risk criteria. The risk enhanced deterministic and Simplified QRA methods are intended to fill the gap between deterministic and QRA methods.

The book provides in-depth detail on the mechanics of failure of dams by overtopping, slope instability, deformations under seismic loads and internal erosion piping, and relevant engineering assessment methods. These represent the state of practice in practical terms and are intended to inform consultants and owners so they can better assess the safety of their dams.

The book covers most types of embankment dams. It does not cover tailings dams or hydraulic fill dams because the PFMs are particular to each dam, and the dominant PFMs are often different to other embankment dams.

Levees and dikes are not specifically discussed but the principles presented could be applied to them.

The structure of the book is:

Chapter 2. Terminology
Chapter 3. Description of and Assessment of Potential Failure Modes and Data Gaps
Chapter 4. Geotechnical Investigations of Existing Dams and Their Foundations
Chapter 5. Hydrologic and Seismic Hazard Assessment
Chapter 6. Estimating the Likelihood of Failure by Overtopping or Throughflow of the Embankment
Chapter 7. Slope Instability and Deformations Leading to Loss of Freeboard and Overtopping
Chapter 8. Deformations of Embankment Dams under Seismic Loading and Assessment of the Likelihood of Loss of Freeboard
Chapter 9. Internal Erosion and Piping—Concentrated Leak Erosion
Chapter 10. Internal Erosion and Piping—Backward Erosion Piping, Suffusion and Global Backward Erosion
Chapter 11. Contact Erosion and Other Mechanisms of Erosion into Soil and Rock Foundations
Chapter 12. Internal Erosion and Piping—Filters and Continuation of Internal Erosion
Chapter 13. Internal Erosion and Piping—Progression, Detection, Intervention and Repair, and Breach
Chapter 14. Quantitative Risk and Simplified QRA Approaches to Assessing the Safety of a Dam for Internal Erosion and Piping
Chapter 15. Historical Database Screening Method for Assessing the Annual Probability of Failure Resulting from Internal Erosion and Piping
Chapter 16. Assessment of the Safety of Concrete Face Rockfill and Asphalt Face Embankment Dams
Chapter 17. Assessing the Safety of Concrete and Asphaltic Concrete Core Embankment Dams
Chapter 18. Dam Safety Assessment
Chapter 19. Measures for Upgrading Dams to Reduce Risks
Appendix A. Methods for Estimating Conditional Probabilities of Transverse Cracking Leading to Concentrated Leak Erosion
Appendix B. Guidelines for Backfilling Investigation Drill Holes, Holes Resulting from Cone Penetration Tests, and Test Pits (ANCOLD, 2020)
Appendix C. Development of the Historical Database Screening Method for Assessing the Annual Probability of Failure Resulting from Internal Erosion and Piping

Terminology 2

2.1 TYPES OF EMBANKMENT DAMS

The terms shown in Figure 2.1 have been adopted.

2.2 DAM ENGINEERING TERMINOLOGY

2.2.1 General Dam Engineering

Accident: An incident characterised by malfunction or abnormality outside the design assumptions and parameters that adversely affects a dam's primary function of impounding water. Accidents can progressively lead to or heighten the risk of failure but are normally amenable to corrective action.

Three categories of accidents are listed by ICOLD (1974):

Accident Type 1 (A1): An accident to a dam which has been in use for some time, but which has been prevented from becoming a failure by immediate remedial measures, including possibly drawing down the water.

Accident Type 2 (A2): An accident to a dam which has been observed during the initial filling of the reservoir, and which has been prevented from becoming a failure by immediate remedial measures, including possibly drawing down the water.

Accident Type 3 (A3): An accident to a dam during construction, i.e. settlement of foundations, slumping of side slopes etc., which have been noted before any water was impounded and where the essential remedial measures have been carried out, and the reservoir safely filled thereafter.

Classification: A system of assigning dams to categories, usually on the basis of the consequences of failure, so that the appropriate dam safety standards can be applied.

Consequences of failure: Impacts on the downstream area of a dam as a result of failure of the dam or its appurtenances. The term consequence refers to life loss or environmental, cultural and economic losses above and beyond that which would have occurred in the same event or conditions had the dam not failed.

Dam crest flood: The reservoir inflow flood which causes the reservoir level to rise to the dam crest level.

Dam crest level: The level of the upper surface of the embankment including any road pavement. For dams with a parapet or wave wall the dam crest level is the top of a parapet wall or wave wall provided the wall is designed to withstand the load from the reservoir up to the top of the wall.

DOI: 10.1201/9781003542025-2

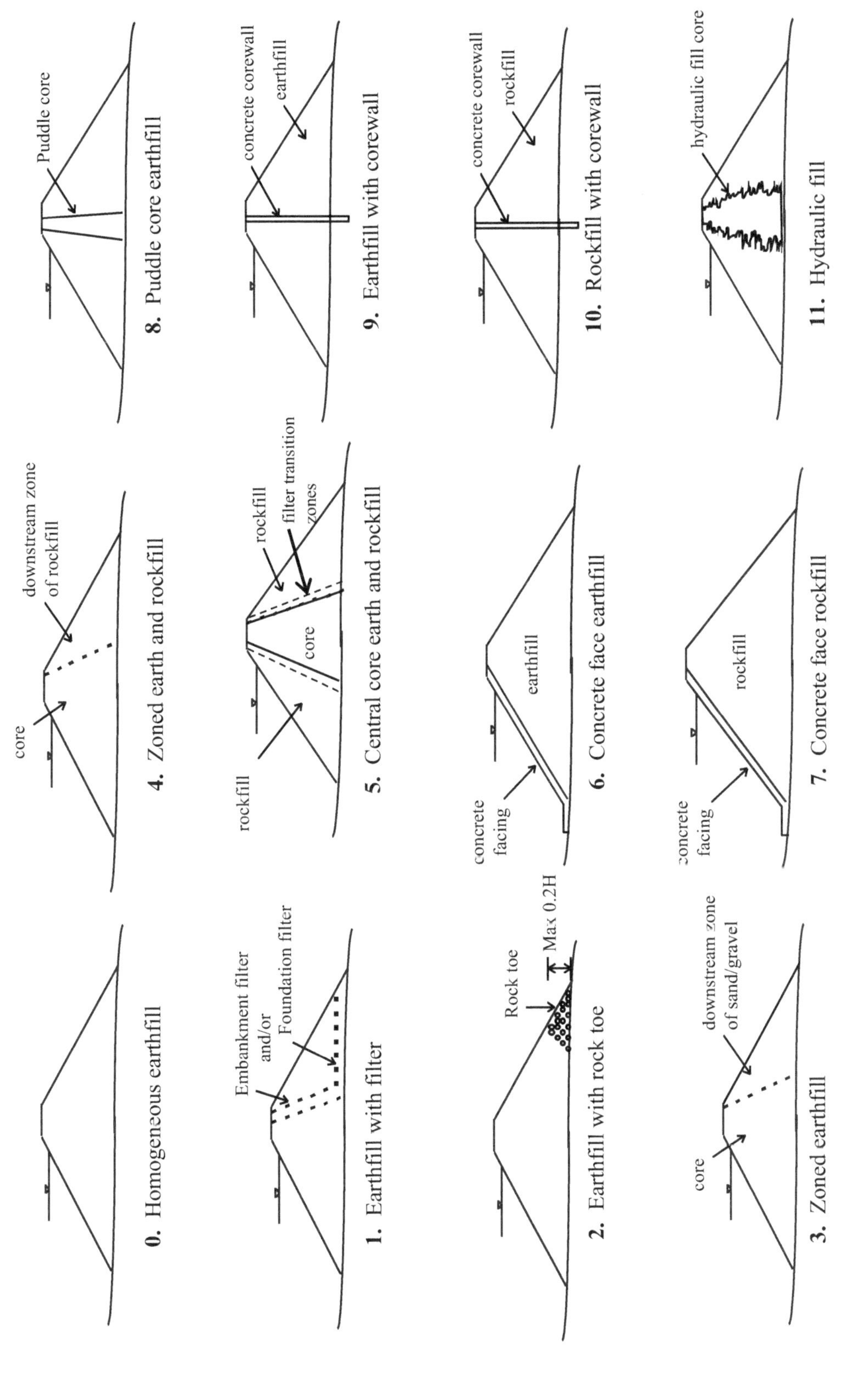

FIGURE 2.1 Dam zoning categories.

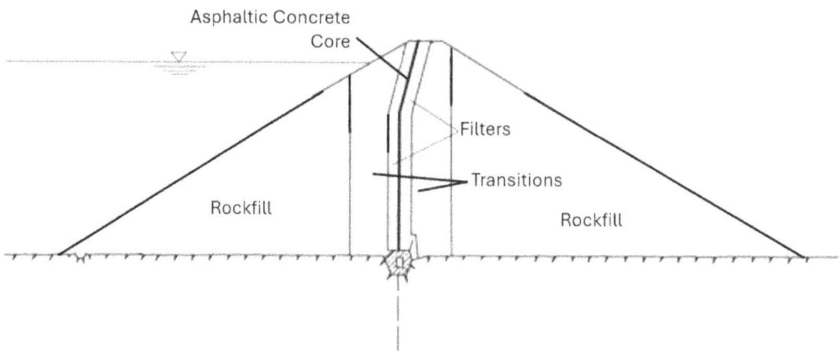

FIGURE 2.1 (Continued)

It is important to use actual levels. These may vary from the design drawings because of built in camber, post-construction settlement and/or reprofiling of the crest.

Failure of dam: Collapse or movement of part of a dam or its foundation, so that the dam cannot retain water. In general, a failure results in the release of large quantities of water, imposing risks on the people or property downstream (ICOLD, 1995). Incidents to dams during construction are only considered as failures when large amounts of water were involuntarily released downstream.

Two categories of failures are listed by ICOLD (1974):

Failure Type 1 (F1): A major failure involving the complete abandonment of the dam.

Failure Type 2 (F2): A failure which at the time may have been severe but has permitted the extent of damage to be successfully repaired and the dam brought into use again.

Freeboard: The freeboard for a dam is the vertical distance between a specified reservoir water surface level and the crest of the dam, without allowance for camber of the crest of the dam.

- *Normal freeboard* is the vertical distance between the crest of the dam without allowance for camber, and the normal reservoir Full Supply Level.
- *Minimum freeboard* is the vertical distance between the crest of the dam without allowance for camber, and the maximum reservoir water surface (MRWS) that results from routing the Inflow Design Flood (IDF) through the reservoir.

Full Supply Level (FSL): The maximum normal operating water surface level of a reservoir.

Incident: Either a failure or accident, requiring major repair.

Large dam: A dam which is more than 15 metres in height (measured from the lowest point in the general foundations to the crest of the dam), or any dam between 10 and 15 metres in height which meets one of the following conditions:

a. the crest length is not less than 500 metres
b. the capacity of the reservoir formed by the dam is not less than one million cubic metres

c. the maximum flood discharge dealt with by the dam is not less than 2000 cubic metres per second
d. the dam is of unusual design.

(ICOLD, 1974, 2019)

Reservoir: The body of water that is impounded by a dam.

2.2.2 Internal Erosion and Piping

The following is from Fell et al. (2024) and the terminology in ICOLD *Bulletin 164, Internal Erosion of Existing Dams, Levees and Dikes and Their Foundations* (ICOLD, 2017).

Backward Erosion Piping (BEP): Backward Erosion Piping involves the detachment of soil particles when the seepage exits to a free unfiltered surface. The detached particles are carried away by the seepage flow, and the process gradually works its way towards the upstream side of the embankment or its foundation until a continuous pipe is formed. For BEP to occur, the roof of the pipe must be formed by a plastic soil layer of the foundation or embankment, and the pipe is essentially horizontal (Figure 10.1).

Breach: Breach is the final phase of the internal erosion process resulting in uncontrolled release of the contents of the reservoir. It may occur by one of the following four phenomena (listed in order of their observed frequency of occurrence).

- Gross enlargement of the pipe.
- Slope instability of the downstream slope.
- Unravelling of the downstream face.
- Overtopping (e.g. due to settlement of the crest from suffusion and/or due to the formation of a sinkhole from a pipe in the embankment).

Concentrated leak erosion: Erosion in a concentrated leak may occur in a crack in an embankment or its foundation caused by differential settlement, desiccation, freezing and thawing; by hydraulic fracture; or by a gap adjacent to a structure, such as an outlet conduit or spillway wall; or it may occur in a continuous zone containing poorly compacted materials which form an interconnecting voids system or gap. The concentration of flow causes erosion of the walls of the crack or interconnected voids.

Contact erosion in soil: Soil contact erosion is a form of internal erosion which involves selective erosion of fine particles from the contact with a coarser layer, caused by the flow passing through the coarser layer. For instance, along the contact between layers of silt and gravel-sized particles. It relates only to conditions where the flow in the coarser layer is parallel to the interface between the coarse and fine layer (Figure 2.2). Selective erosion of fine particles from the contact with a coarser layer caused by the flow through the fine layer with flow normal to the contact surface is considered by another phase of the erosion process: filtration or continuation.

Contact erosion into open joints in a rock foundation: Contact erosion may occur at the interface of embankment fill and open joints in a rock foundation. The erosion is caused by water flowing in the open joints. Contact erosion may also occur into open fissures or cracks in soil in the foundation.

Continuation (filtration): Continuation is the phase of the erosion process where the relationship of the particle size distribution between the base (core) material and the filter controls whether erosion will continue. Foster and Fell (2001) and Foster (1999) define four levels of severity of continuation: No Erosion, Some Erosion, Excessive Erosion, and Continuing Erosion.

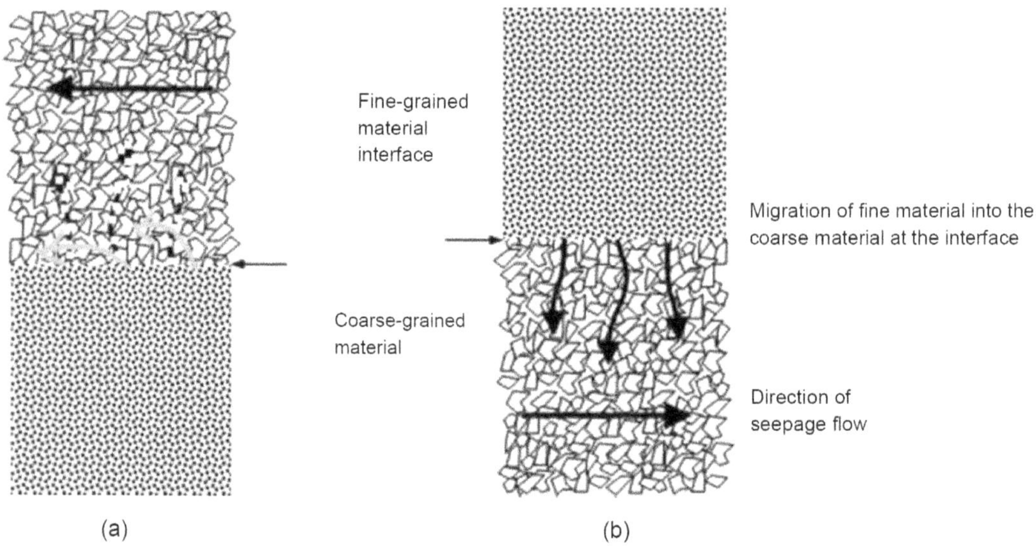

FIGURE 2.2 Diagram describing soil contact erosion.

- **Continuing Erosion**: The filter is coarser than the Continuing Erosion criteria and is too coarse to allow the eroded base materials to seal the filter resulting in unrestricted erosion of the base soil.
- **Excessive Erosion** defines conditions where erosion of the base soil will be excessive before it seals. Filters between the Excessive Erosion boundary and the Continuing Erosion boundary will eventually seal but only after significant erosion of the base soil. In dams there may be large leakage flows before the filter seals by clogging of the surface of the filter by eroded base soil.
- **Some Erosion**: The filter is between the No Erosion and Excessive Erosion boundaries. The filter quickly seals after particles of the base material clog the surface of the filter.
- **No Erosion**: The filter is finer than the No Erosion criterion and seals with no or practically No Erosion of the base material. Filters designed and constructed according to modern filter design criteria will satisfy No Erosion criteria.

Detachment: Detachment is the first stage of the erosion process. Particle detachment occurs by the hydraulic shear forces developed by the seepage flow velocity. The mechanics are determined by whether the soil is plastic or non-plastic.

Detection: Detection is the ability to detect the internal erosion process during one of the erosion phases. Internal erosion is most easily detected in the progression or breach phases.

Dilative and contractive soil: Dilative or contractive describes the initial tendency of the soil on the surface of rupture to increase (dilate) or decrease (contract) in volume under drained shear (or to develop negative or positive pore pressures in undrained shear in a saturated condition).

Dispersion is a physico-chemical phenomenon contributing to concentrated leak erosion. Soils in which the clay particles will detach from each other and from the soil structure and go into suspension without a flow of water are termed dispersive clays.

The dispersivity of a soil is directly related to its clay mineralogy. In particular, soils with a high exchangeable sodium percentage such as montmorillonite tend to be dispersive, while kaolinite and related minerals (e.g. halloysite) are non-dispersive. Soils with Illite present tend to be moderately dispersive.

The dispersivity depends also on the pore water chemistry. Low pore water salt concentrations lead to greater dispersivity, and high salt concentrations can suppress dispersion in susceptible soils.

Erosion of a soil particle is the detachment and transport of that particle by forces caused by seepage flow (drag and lift forces, which are related to viscosity and turbulent shear stresses). Detachment is a first condition expressing that drag and lift forces are larger than resisting forces. Transport is the second condition expressing first a hydro-mechanical criterion that the drag forces are larger than the rolling or sliding resistance of the buoyancy weight of the particle and second a geometric criterion: voids or cracks exist which are large enough for eroded particles to pass through. This last criterion makes the difference between internal and external erosion.

Flaw: A flaw is a physical condition that allows seepage concentration. Examples include (i) a continuous crack or (ii) a gap in a poorly compacted layer, in which a concentrated leak may form.

Global Backward Erosion (GBE): Global Backward Erosion occurs in broadly graded silt sand gravel soils in which the voids are over-filled, so the coarser particles are "floating" in the finer soil matrix. Finer particles move within the soil matrix and may be eroded from the soil. Some use the term suffosion to describe what is termed GBE in this document.

Heave (also known as ***"blow out"***) occurs in non-plastic soils which are confined by an overlying lower permeability stratum when seepage pore pressures are such that the effective stress becomes zero (pore water pressure ≥ total stress). "Heave" may often be followed by Backward Erosion Piping of the non-plastic soil.

Hydraulic fracture occurs in the core or foundation of embankment dams when the pore pressure exceeds the minor principal stress and the effective stress in the core or foundation becomes zero (or even slightly negative if the soil can withstand tensile stresses). The pressure of the water seeping through the core from the reservoir exceeds the remaining compressive stress and forms a crack or further opens an existing crack in which internal erosion by concentrated leak may initiate.

Initiation: Initiation is the first phase of the internal erosion process, when one of the phenomena of detachment of particles occurs. Five initiation phenomena are defined: concentrated leak, Backward Erosion Piping, Suffusion, Global Backward Erosion and contact erosion (each of which are defined separately in this section).

Internal erosion: Occurs when soil particles within an embankment dam or its foundation are carried downstream by seepage flow. Internal erosion can initiate by concentrated leak erosion, backward erosion, suffusion and soil contact erosion.

Internal instability: In soils subject to internal instability, finer particles in the soil are able to move within the soil voids under the forces imposed on the particles by seepage flow. The phenomenon does not require a crack within the soil in which erosion may occur as is required in concentrated leak erosion, or a free surface from which particles detach and a roof of plastic soil as is required for Backward Erosion Piping.

In this document internal instability includes Suffusion, Global Backward Erosion and where particles are able to move within the soil voids without erosion of the particles from the soil.

Location of initiation of erosion: This is the physical initiating point of internal erosion in the structure such as in the embankment, foundation, along a conduit or along a spillway interface.

Piping: Piping is a potential progression phase of internal erosion which results in the formation of a continuous tunnel called a "pipe" between the upstream and the downstream side of the embankment or its foundation. Internal erosion is commonly described as "internal erosion and piping" but piping is actually the culmination of a process of erosion in which a number of phases must occur and be sustained in order that a "pipe" develops through the dam or its foundation and allows the passage of considerable quantities of water, which may lead to a breach.

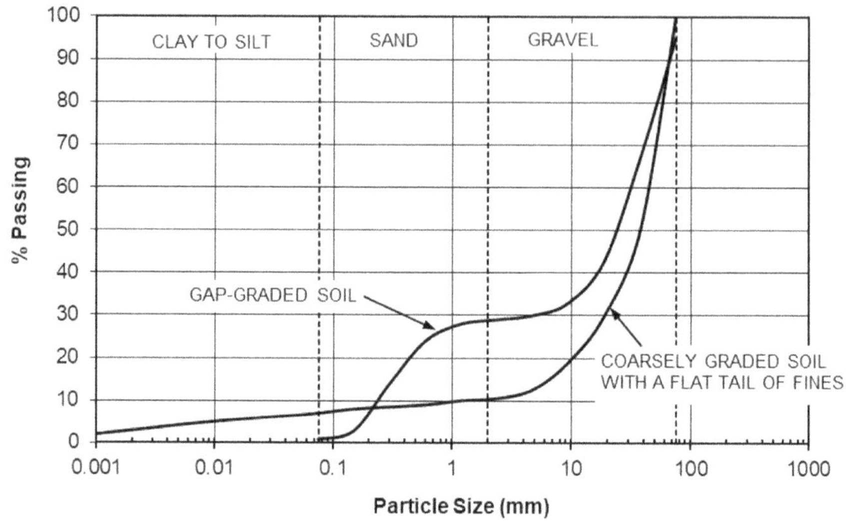

FIGURE 2.3 Soil types which are potentially subject to suffusion.

Progression: Progression is the phase of internal erosion where hydraulic shear stresses within the eroding soil may or may not lead to the erosion process being on-going, and in the case of Backward Erosion Piping and concentrated leak erosion to formation of a pipe. The main issues are whether the pipe will collapse or whether upstream zones may control the erosion process by flow limitation.

Self-filtering: In soils which self-filter, the coarse particles prevent the internal erosion of the medium particles, which in turn prevent erosion of the fine particles. Soils which self-filter are internally stable. Soils which potentially will not self-filter include soils which are susceptible to suffusion and very broadly graded soils which are subject to Global Backward Erosion.

Sinkhole: A sinkhole is a cavity formed in the core of a dam or in a dike which may form as a result of erosion initiated by a concentrated leak, contact erosion or suffusion, or erosion into open joints in a foundation or conduit. The sinkhole cavity is usually near vertical and forms above the zone or point where erosion initiates. The progression of development of sinkholes is often a slow process with soil falling from the top of the sinkhole being carried away by the initiating erosion mechanism. Sinkholes may take years to manifest themselves at the crest of the dam, levee or dike.

Suffusion: This is a form of internal erosion of internally unstable soils which involves selective erosion of finer particles from the matrix of coarser particles, in such a manner that the finer particles are removed through the constrictions between the larger particles by seepage flow, leaving behind a soil skeleton formed by the coarser particles. The volume of finer particles is such that they fit within the voids formed by the coarser particles, that is, the voids are under-filled.

Coarsely graded and gap-graded soils, such as those shown schematically in Figure 2.3, are susceptible to suffusion.

2.2.3 Earthquakes

The following terminology is from Fell et al. (2015) based on ANCOLD (2019) and ICOLD (2012).

Active fault: A fault, reasonably identified and located, known to have produced historical earthquakes or showing evidence of movements in Holocene time (i.e. in the last 11,000 years) and large faults that have moved in latest Pleistocene time (i.e. between 11,000 and 35,000 years ago).

Operating Basis Earthquake (OBE): The OBE is that level of ground motion which will cause only minor and acceptable damage to the dam and appurtenant structures. The dam, appurtenant structures and equipment should remain functional and damage from the occurrence of earthquake shaking not exceeding the OBE should be easily repairable.

***Safety Evaluation Earthquake (SEE)*:** The SEE is the maximum level of ground motion for which the dam should be designed or analysed. Damage can be accepted but there should be no uncontrolled release of water from the reservoir. The term *Safety Evaluation Earthquake* replaces the term *Maximum Design Earthquake* (MDE) in ICOLD and ANCOLD.

Maximum Credible Earthquake (MCE): The MCE is the largest reasonably conceivable earthquake magnitude that is considered possible along a recognised fault or within a geographically defined tectonic province, under the presently known or presumed tectonic framework.

Maximum credible earthquake ground motion: This is the most severe ground motion affecting a dam site due to an MCE.

The OBE and SEE are expressed in terms relevant to the design of the dam or appurtenant structure. It may be in terms of peak ground acceleration and Moment Magnitude Mw, peak ground velocity or response spectra.

2.2.4 Liquefaction and Related Phenomenon

The following definitions are from Fell et al. (2015).

Liquefaction: All phenomena giving rise to a reduction in shearing resistance and stiffness and development of large strains as a result of increase in pore pressure under cyclic or monotonic (static) loading of contractive soils.

Cyclic liquefaction is a form of temporary liquefaction, where the cyclic loading causes shear stress reversal and an initial liquefaction (zero effective stress) condition develops temporarily.

Cyclic mobility is a form of temporary liquefaction where the shear stresses are always greater than zero.

Cyclic softening is a term used to describe the reduction of shear strength and stiffness of clays and plastic silts under cyclic loading.

Initial liquefaction is the condition when effective stress is momentarily zero during cyclic loading.

Flow liquefaction is the condition where there is a strain weakening response in undrained loading and the *in situ* shear stresses are greater than the steady state undrained shear strength.

Temporary liquefaction is the condition where there is a limited strain weakening response in undrained loading; at larger strain the behaviour is strain hardening.

These phenomena apply to monotonic (static) as well as cyclic loading and are apparent in contractive soils, both non-plastic and those with some clay content.

2.3 SOME STATISTICAL TERMINOLOGY

The following terminology is largely based on ANCOLD *Guidelines on Risk Assessment* (ANCOLD, 2022).

ALARP (as Low As Reasonably Practicable) principle: That principle which states that risks, lower than the limit of tolerability, are tolerable only if risk reduction is impracticable or if its cost is grossly disproportionate (depending on the level of risk) to the improvement gained.

Aleatory uncertainty: Uncertainty that arises from the inherent randomness of natural events and properties.

Annual Exceedance Probability (AEP): The probability that an event of specified magnitude will be equalled or exceeded in any year.

Conditional probability: The probability of an outcome, given the occurrence of some event. For example, given that a flood has reached the crest of an embankment dam, the probability of the dam failing is a conditional probability.

Consequences: In relation to risk analysis, the outcome or result of a risk being realised. Includes flood impacts in the downstream as well as upstream areas of the dam resulting from failure of the dam or its appurtenances, as well as indirect impacts over an indefinitely large area.

Dam break analysis: An analysis that provides an estimation of downstream flooding effects resulting from dam failure. The analysis includes a Dam Breach analysis and the routing of the dam-break hydrograph through the downstream floodplain.

Deterministic: Describing a process with an outcome that is always the same for a given set of inputs. Thus, the outcome is "determined" by the input. Deterministic contrasts with random, which describes a process with an outcome that can vary even though the inputs are the same.

Deterministic dam safety assessment: Uses established standards and rules to evaluate a dam's safety, focusing on specific conditions and potential failures.

Epistemic uncertainty: Uncertainty that arises from the limits to knowledge.

Except in exceptional circumstance: In relation to tolerable risk criteria, circumstances in which governments allow otherwise unacceptable risks to continue in the wider interests of society.

Failure Mode: A way that failure can occur, described by the means by which element or component failures must occur to cause loss of the sub-system or system function.

Failure Modes and Effects Analysis (FMEA): An inductive method of analysis where particular faults or initiating conditions are postulated and the analysis reveals the full range of effects of the fault or the initiating condition on the system. FMEA can be extended to perform what is called failure modes, effects and criticality analysis (FMECA). In a FMECA, each failure mode identified is ranked according to the combined influence of its likelihood of occurrence and the severity of its consequences.

Frequency: A measure of likelihood expressed as the number of occurrences of an event in a given time or in a given number of trials (see also likelihood and probability).

Hazard: Threat; condition which may result from either an external cause (e.g. earthquake, flood or human agency) or an internal vulnerability, with the potential to initiate a failure mode. A source of potential harm or a situation with a potential to cause loss.

Individual risk: The increment of risk imposed on a particular individual by the existence of a hazardous facility. This increment of risk is an addition to the background risk to life, which the person would live with on a daily basis if the facility did not exist.

Intervention: Intervention is the ability to stop or delay the progression of a Potential Failure Mode prior to the commencement of a breach. Intervention can also lower risk by providing enough advance warning to minimise downstream consequences should the Dam Breach.

Likelihood: A qualitative description of probability and frequency.

Load: The main loads exerted on a dam are hydrostatic loads (which are related to the reservoir level) and seismic loads (which may be expressed in terms of earthquake magnitude and peak ground acceleration [PGA]). The likelihood of a given reservoir level load is expressed as the annual probability or frequency that the maximum reservoir level is within a certain range. The likelihood of a given seismic load is expressed as the annual probability or frequency of a loading parameter (e.g. peak ground acceleration) being in a certain range.

Load condition: A load condition is a theoretical combination of different loads exerted on a dam at a given time for analysis purposes. The loading condition is the first node of the internal erosion assessment or failure from other Potential Failure Modes. The loading condition controls whether certain failure processes are initiated, progress, can be detected and intervention can be successful.

Population at Risk (PAR): All those persons who would be directly exposed to floodwaters within the dam-break affected zone if they took no action to evacuate.

Portfolio Risk Assessment (PRA): A particular form of risk assessment or analysis, which aims to make a comparative estimation of risks over all of, or many of, the dams of a single owner or single regulatory or other jurisdiction.

Potential Failure Modes (PFM): Specific chains of events leading to failure of a dam.

Potential Failure Modes Analysis (PFMA): Analysis to identify specific chains of events leading to failure of a dam.

Potential Loss of Life (PLL): The number of people expected to lose their life as a result of uncontrolled release of all or part of the stored water in the reservoir.

Probability: A measure of the degree of confidence in a prediction, as dictated by the evidence, concerning the nature of an uncertain quantity or the occurrence of an uncertain future event. It is an estimate of the likelihood of the magnitude of the uncertain quantity, or the likelihood of the occurrence of the uncertain future event. This measure has a value between zero (impossibility) and 1.0 (certainty).

There are two main interpretations:

Statistical—frequency or fraction: The outcome of a repetitive experiment of some kind like flipping coins. It includes also the idea of population variability. Such a number is called an "objective" probability because it exists in the real world and is in principle measurable by doing the experiment.

Subjective probability (degree of belief): Quantified measure of belief, judgement or confidence in the likelihood of an outcome, obtained by considering all available information honestly, fairly and with a minimum of bias. Subjective probability is affected by the state of understanding of a process, judgement regarding an evaluation or the quality and quantity of information. It may change over time as the state of knowledge changes.

Qualitative risk analysis: An analysis which uses word form, descriptive or numeric rating scales to describe the magnitude of potential consequences and the likelihood that those consequences will occur.

Quantitative risk analysis: An analysis based on numerical values of the potential consequences and likelihood, the intention being that such values are a valid representation of the actual magnitude of the consequences and the probability of the various scenarios which are examined.

Quantitative Risk Assessment (QRA): An assessment based on quantitative risk analysis.

Residual risk: The remaining level of risk at any specified time before, during and after a program of risk mitigation measures has been taken.

Risk: Measure of the probability and severity of an adverse effect to life, health, property, or the environment.

In the general case, risk is estimated by the combined impact of all triplets of scenario, probability of occurrence and the associated consequence. As a special case, average (annualised) risk can be estimated by the mathematical expectation of the consequences of an adverse event occurring (that is, the product of the probability of occurrence and the consequence, combined over all scenarios).

Risk analysis: The use of available information to estimate the risk to individuals or populations, property or the environment, from hazards. Risk analyses generally contain the following steps: scope definition, hazard identification and risk estimation.

Risk assessment: The process of reaching a decision recommendation on whether existing risks are tolerable and present risk control measures are adequate, and if not, whether alternative risk control measures are justified or will be implemented. Risk assessment incorporates, as inputs, the outputs from the risk analysis and risk evaluation phases.

Risk enhanced deterministic safety assessment: A deterministic dam safety assessment enhanced with quantitative risk from analyses of databases of dam failures and accidents.

Risk-Informed Decision Making (RIDM) is a method of dam safety evaluation that uses the likelihood of loading, system response given the loading and consequences of failure to estimate risk. This risk estimate can be used to inform decisions regarding dam safety investments.

Risk-Informed Safety Assessment (RISA) is a method of dam safety evaluation that uses the results of QRA or Simplified QRA along with engineering assessments and deterministic acceptance criteria.

Simplified QRA is based on the QRA but assesses the annual probability of failure for only the most important PFMs. The annual probability of failure for an individual PFM and the related PLL are estimated and the outcome considered in the semi-quantitative tolerable risk matrix.

Societal risk: The risk of multiple fatalities resulting from a failure.

Standards Based Approach: The traditional approach to dams engineering, in which risks are controlled by following established rules as to design events and loads, structural capacity, safety coefficients and defensive design measures.

Tolerable risk: A risk within a range that society can live with so as to secure certain net benefits. It is a range of risk that we do not regard as negligible or as something we might ignore, but rather as something we need to keep under review and reduce still further if and as we can. In the context of these guidelines, tolerable risk means that risks are reduced to such levels that ALARP conditions are continually met.

Unacceptable risk: A risk so high that it must be reduced regardless of the sacrifice involved in its reduction in order to provide equity. Sometimes called intolerable risk by others.

Uncertainty: Any situations without sureness, whether or not described by a probability distribution. In the context of dam safety, uncertainty can be attributed to (i) inherent variability in natural properties and events and (ii) incomplete knowledge of parameters and the relationships between input and output values. Previously used by some to refer to situations where the likelihood of potential outcomes could not be described by objectively known probability density functions, but this usage appears to be declining.

Description of and Assessment of Potential Failure Modes and Data Gaps

3

3.1 THE MOST IMPORTANT POTENTIAL FAILURE MODES FOR EMBANKMENT DAMS

3.1.1 Introduction

A Potential Failure Mode is a sequence of events starting from an initiating mechanism, such as internal erosion in a defect, flaw or seepage path in the dam or its foundation, progression of erosion and breach which may lead to an uncontrolled release of the reservoir.

As shown in Table 3.1, Foster (1999) and Foster et al. (2000 a, 2000b) found that for dams up to 1986 the most important failure modes leading to uncontrolled release of the water stored in the reservoir were overtopping and internal erosion and piping, with downstream and upstream slides significantly less. There were no failures resulting only from sliding, that is, not associated with piping or overtopping.

3.1.2 Failure Resulting from Throughflow, Overtopping or Wave Overtopping of the Crest of the Dam

As shown in Figure 3.1 there are three hydraulic scenarios:

1. Throughflow occurs when the reservoir level is above the top of the core and water flows through a rockfill zone above the core.
2. Wave overtopping occurs when wave action on the reservoir results in intermittent flow of water over the crest.
3. Overtopping of an embankment occurs when the reservoir level is above embankment crest level.

The consequences of overtopping and throughflow depend on the embankment zoning.

DOI: 10.1201/9781003542025-3

TABLE 3.1 Overall Statistics of Failure of Large Embankment Dams up to 1986 (Foster, 1999; Foster et al., 2000a, 2000b)

MODE OF FAILURE	*% TOTAL FAILURES* [(a,b)] *(WHERE MODE OF FAILURE KNOWN)*
Overtopping	34%
Spillway or gate failure resulting in uncontrolled flows that erode the toe of the embankment or result in overtopping failure	13%
Piping through embankment	32%
Piping from embankment into foundation	2%
Piping through foundation	15%
Downstream slide	3%
Upstream slide	1%
Earthquake	2%
Totals[(c)]	102%
Total overtopping	47%
Total piping	49%
Total slides	4%

Notes:
(a) Percentages based on the % of cases where the mode of failure is known. They have been rounded to the nearest whole number.
(b) Percentages are for failures of embankment dams in operation only, i.e. excluding failures during construction.
(c) Percentages do not necessarily sum to 100% as some dams were classified as multiple modes of failure. For example, development of piping through the embankment may cause saturation of the downstream slope which then initiates a downstream slide. In these cases, all the modes of failure that were involved in the dam incident are assigned to the mode of incident in the database.

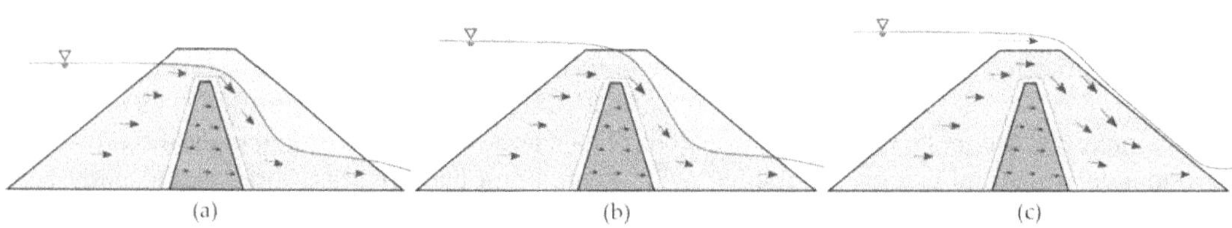

FIGURE 3.1 (a) Throughflow above the core; (b) partial overtopping of the crest with throughflow; (c) overtopping of the crest with throughflow and potential overflow of the downstream rockfill (Dezert et al., 2022).

3.1.2.1 Earthfill Embankments

The process of failure resulting from overtopping of the dam crest may be broadly broken into three phases:

- Initiation of erosion on the downstream slope or at the downstream toe.
- Progression of erosion back through the dam crest, usually in the form of head-cut erosion.
- Breach commences. This occurs when erosion progresses back through the full width of the dam crest and reaches the reservoir.

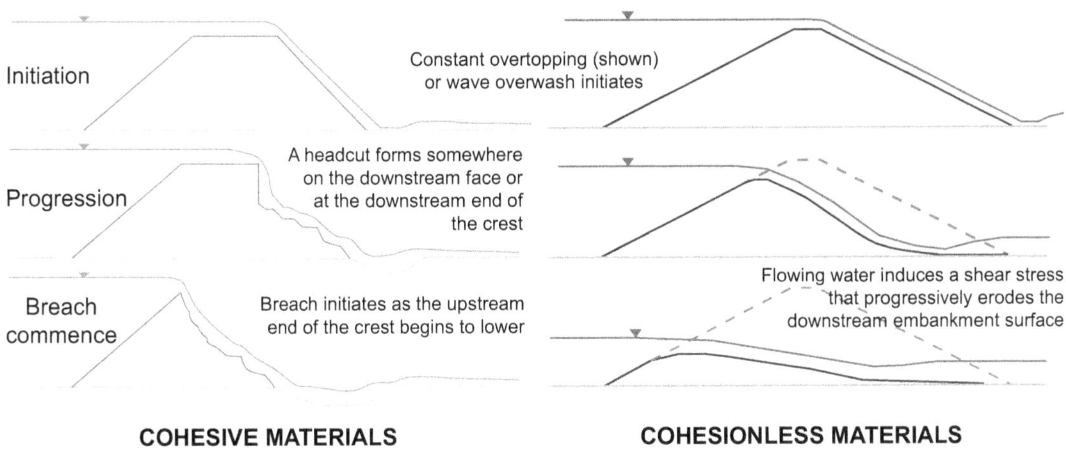

FIGURE 3.2 Initiation, progression and breach commence, resulting from overtopping failure of an earthfill embankment dam. (From Powerpoint presentation, USACE/Reclamation 2019, Chapter D3.)

As shown in Figure 3.2, in plastic (cohesive) soils, erosion initiates on the downstream slope and usually progresses in the form of a head-cutting in an upstream direction of a deep erosion channel that can eventually reach the reservoir and breach the dam. For embankment downstream slopes of non-plastic (cohesionless) soil, a head-cut may form, or erosion may progress more uniformly.

3.1.2.2 Earthfill Embankments Which Have Rip-Rap Protection on the Downstream Slope

The process of failure resulting from overtopping of the dam crest may be broadly broken into three phases:

- Initiation and progression of erosion of the rip-rap to expose the earthfill.
- Initiation of erosion on the earthfill on the downstream slope.
- Progression of erosion back through the dam crest, usually in the form of head-cut erosion.
- Breach commences. This occurs when erosion progresses back through the full width of the dam crest and reaches the reservoir.

3.1.2.3 Earth and Rockfill Embankments

The process of failure resulting from overtopping or throughflow of the dam crest may be broadly broken into three phases:

- Initiation. As shown in Figure 3.1b and c, the flow over the crest and in the rockfill on top of the core enters the rockfill and emerges lower down towards the toe of the embankment. Initiation of erosion is in the form of unravelling.

  For throughflow (Figure 3.1a) the mechanism of initiation is similar if the downstream rockfill is sufficiently permeable. If it is not, flow may emerge high up the downstream slope and initiation may include unravelling, slope instability or internal erosion and piping.
- Progression. Unravelling advances upstream and may stabilise. If flows are sufficient unravelling advances further upstream. If the rockfill becomes saturated, it may become unstable, and the downstream slope begins to slide.
- Commencement of breach. If the sliding and unravelling processes progress to encompass the crest of the embankment and the upstream slope to the reservoir, then breach commences.

3.1.2.4 Concrete Face Rockfill Dam (CFRD), Asphalt Face Rockfill Dam (AFRD)

The Potential Failure Modes from overtopping are similar to that for earth and rockfill embankments. However, there are differences that need to be considered that include the following:

- The concrete or asphaltic concrete face may provide a resistive element in the progression phase and require significant erosion on the downstream side of the face before it ultimately fails.
- A common detail of CFRD is to have a reinforced vertical extension of the face or a parapet wall at the crest. For these dams, during the progression phase the overtopping flow is directed onto the rockfill crest resulting in erosion, formation of void on the downstream side and then structural collapse of the concrete element. The detailing of the connection to the face slab is a factor in the collapse mechanism.
- Some AFRDs have a concrete parapet wall at the crest and some CFRD have been raised by addition of a concrete parapet wall. Failure of one or more parapet wall segments prior to overtopping may occur. This could be due to uplift (from seepage at poor joint detailing or performance) or bearing capacity.

3.1.2.5 Concrete Corewall Earthfill

The Potential Failure Modes from overtopping are similar to that for earthfill embankments. The concrete corewall may provide a resistive element in the progression phase and may require significant erosion on the downstream side of the corewall before it ultimately fails. The level of the top of the corewall is also a consideration in the failure mechanism as this may be Full Supply Level in some dams and up to dam crest level in others.

3.1.2.6 Concrete Core Rockfill Embankments (CCRD) and Asphalt Core Rockfill Embankments (ACRD)

The Potential Failure Modes from overtopping are similar to that for earth and rockfill dams. As for concrete core earthfill embankments, the core may provide a resistive element in the progression phase.

For earthfill embankments a sealed road pavement on the crest may have some effect on progression but not initiation of erosion on the downstream slope. The effect is likely to be minor.

For earth and rockfill embankments a pavement may have some effect on flow into the rockfill during overtopping and therefore on initiation and progression.

If the embankment has a concrete parapet or wave wall on the dam crest, then erosion may initiate where the flows spilling over the top of the crest wall impinges onto the dam crest leading to loss of support to the wall, wall collapse and subsequent erosion of the embankment.

3.1.3 Slope Instability and Deformations Leading to Loss of Freeboard and Failure

This PFM relates to failure by deep-seated slope instability of the downstream slope of the embankment, or the embankment and foundation during operation of the reservoir, either during normal operations or flood loading.

For failure to occur the slope must reach a factor of safety ≤ 1.0 and the resulting deformations sufficient to result in the lowering of the dam crest to give loss of freeboard and subsequent overtopping of the dam crest (Figure 3.3).

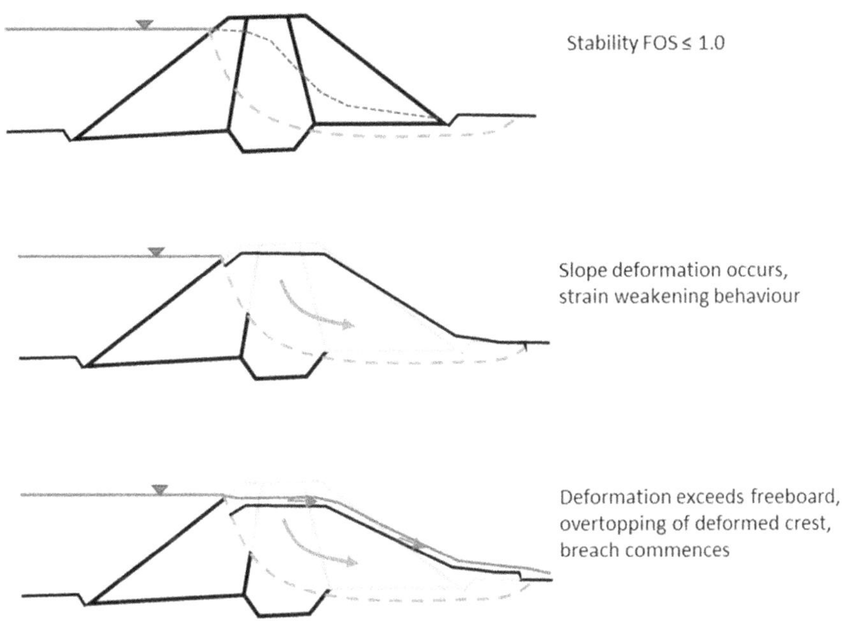

FIGURE 3.3 Progression of downstream slope instability for an embankment dam.

Figure 3.4 shows a situation where breach by slope instability is more likely:

- There is a small freeboard compared to the dam height and the dam crest is narrow.
- The soils in the embankment and foundation are strain weakening including:
    - High plasticity or high clay size content clays.
    - Loss of shear strength in a sandy gravel if a zero effective stress, or heave condition is reached.

Other situations where breach by slope instability is more likely include:

- The soils are contractive on shearing, such as saturated poorly compacted plastic soil or loose non-plastic soil such as sand or silty sand.
- There is low discharge capacity in the downstream zone making a rise of pore pressure likely where the initial displacement exposes the failure surface to the reservoir.
- The soil in the foundation is subject to liquefaction and large loss of strength under seismic loads.

Figure 3.5 shows conditions where breach by slope instability is less likely:

- There is a large freeboard compared to the dam height, and the dam crest is wide.
- The soil in the embankment and foundation is non-strain weakening.
- There is a large discharge capacity in the downstream rockfill zone making it unlikely pore pressures will rise even if the initial displacement exposes the failure surface to the reservoir.

Slope instability during construction when the reservoir is yet to fill, or of the upstream slope under drawdown of the reservoir, does not result in uncontrolled release of the water in the reservoir so is not considered to be a dam failure.

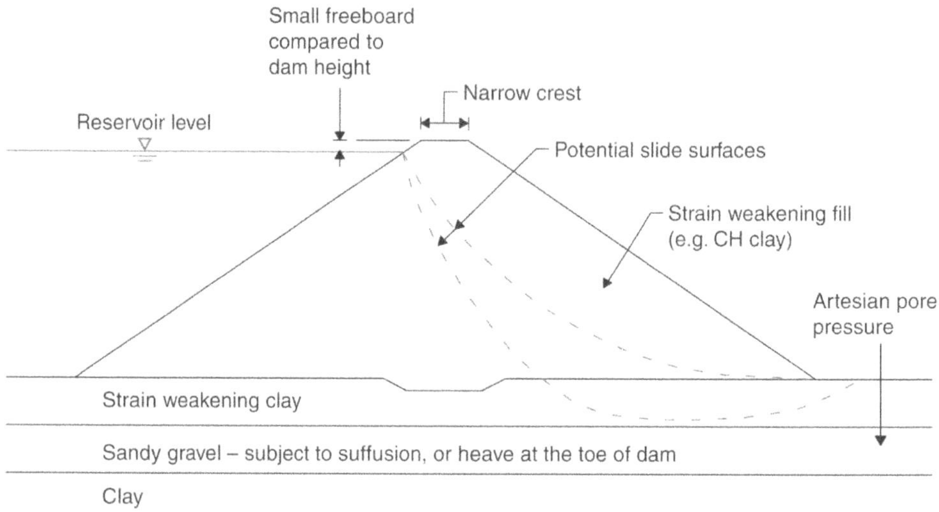

FIGURE 3.4 Conditions in which breach by slope instability are more likely (Fell et al., 2015).

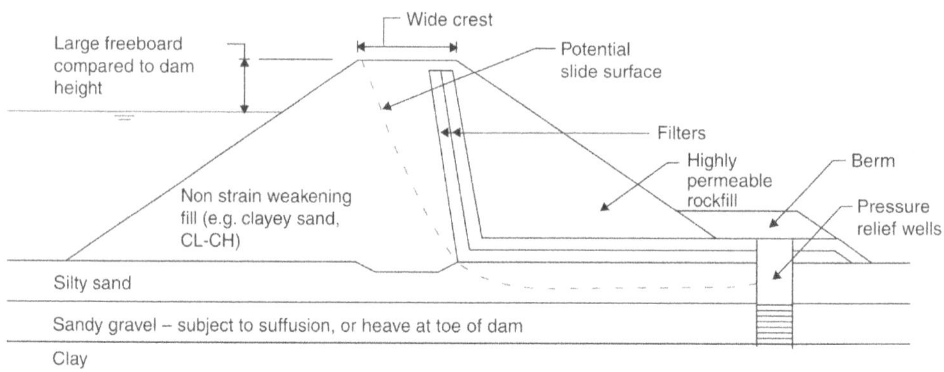

FIGURE 3.5 Conditions in which breach by slope instability are less likely (Fell et al., 2015).

While drawdown induced slides of the upstream slope of embankment dams are fairly common (and have the potential to present the owner with a business risk in this damaged condition), the authors are not aware of a dam failure (breach) resulting from a drawdown induced slide of the upstream slope.

Slope instability may occur in the progression of internal erosion and piping and is covered under internal erosion and piping PFMs. In this case, slope instability occurs due to increases in seepage and/or pore pressures within the embankment caused by internal erosion processes. Slope instability may take the form of deep-seated instability or sloughing or unravelling of the downstream slope.

3.1.4 Failure Resulting from Internal Erosion and Piping

Internal erosion and piping may occur within the embankment, in the foundation, around conduits through the embankment or adjacent to spillway or other retaining walls, or from embankment to foundation.

The process of internal erosion and piping may be broadly broken into four phases:

- Initiation of erosion.
- Continuation of erosion.
- Progression to form a pipe.
- Commencement of a breach.

This is shown in Figures 3.6(a) and (b) for internal erosion through the embankment. Similar processes apply for internal erosion through the foundation, and from the embankment to the foundation and are shown in Figure 3.6(c)–(e).

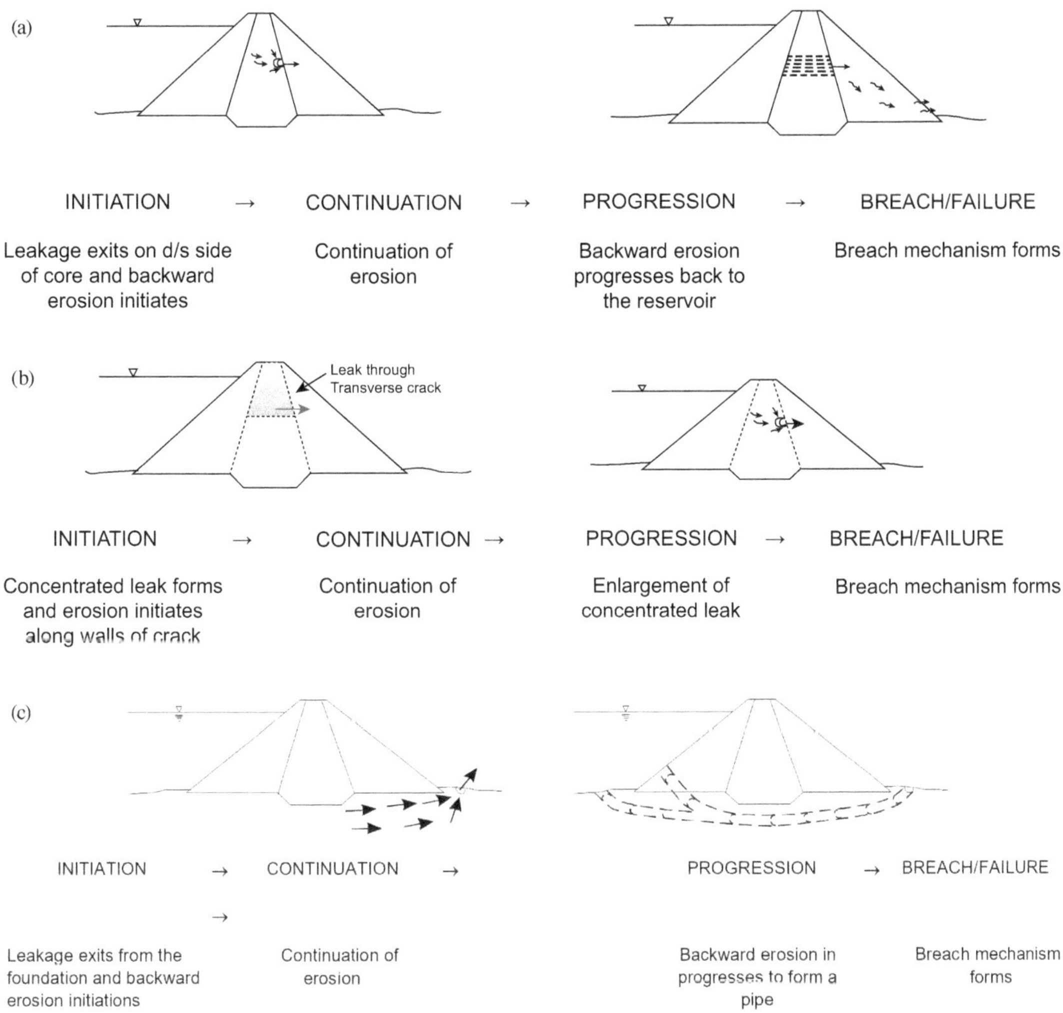

FIGURE 3.6. Models for the development of failure by internal erosion (Fell et al., 2024, adapted from Foster and Fell, 1999b). (a) Internal erosion in the embankment initiated by Global Backward Erosion or Suffusion. (b) Internal erosion in the embankment initiated by erosion in a concentrated leak. (c) Internal erosion in the foundation initiated by Backward Erosion Piping. (d) Internal erosion from the embankment to foundation initiated by Global Backward Erosion in a non-plastic core. (e) Internal erosion by contact erosion at the embankment—foundation contact.

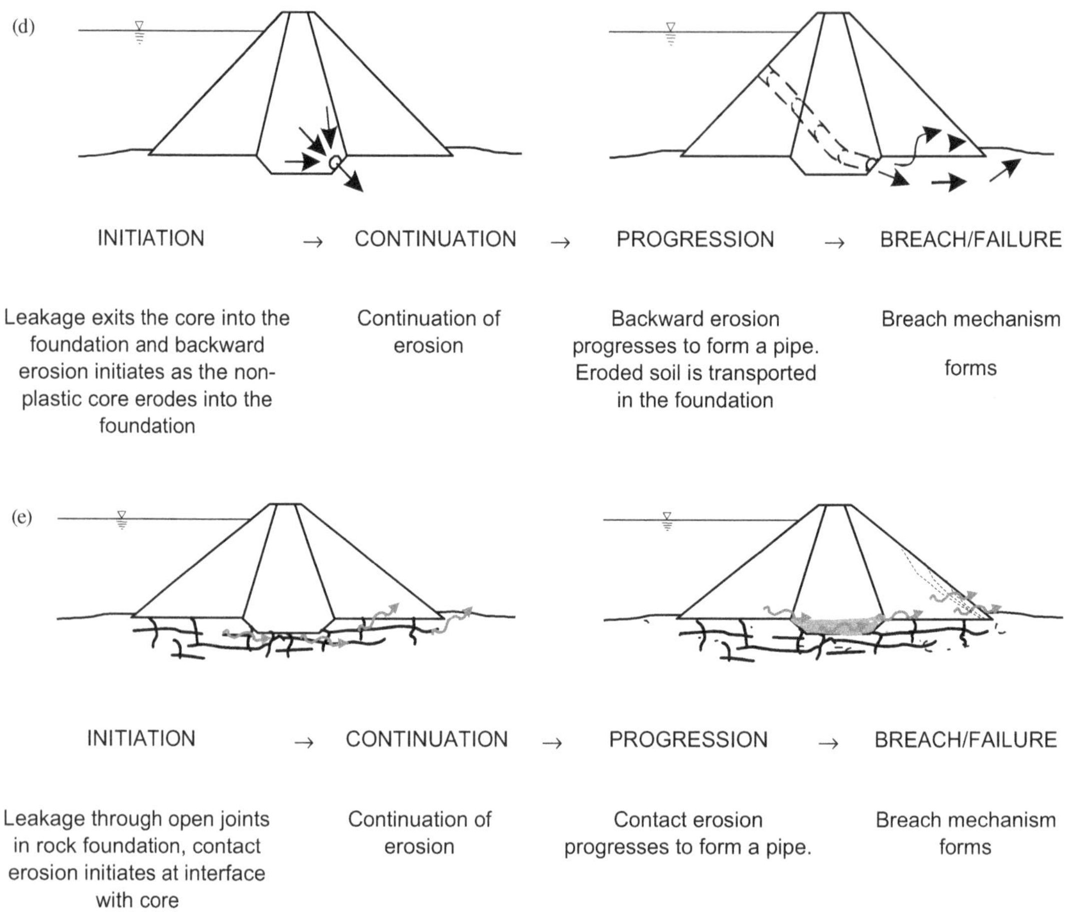

FIGURE 3.6. (Continued)

Initiation is the first phase of the process and considers the existence of a flaw or continuous zone of non-plastic soil in which internal erosion may initiate. There are several mechanisms by which erosion can initiate in the embankment or foundation:

- Concentrated leak erosion.
- Backward Erosion Piping.
- Suffusion.
- Global Backward Erosion.
- Contact erosion at or into open joints in a rock foundation.
- Contact erosion in soil.

These processes and detailed discussion of PFM are described in Chapters 9 to 15.

The nature of the soil in the dam or foundation determines its mechanism of initiation of internal erosion:

- Non-plastic soils such as silts, sands and silty sands and silt, sand and gravel mixtures are subject to Backward Erosion Piping, contact erosion, suffusion or Global Backward Erosion depending on their particle size distribution.

- Plastic soils, such as clays, clayey sands, and clayey sandy gravels are subject to concentrated leak erosion and contact erosion. Backward erosion and suffusion cannot occur in these soils under the gradients normally experienced in dams and their foundations but may occur if local gradients are very high.
- Silt sand gravel soils with very low plasticity (PI ≤ 7) are potentially subject to suffusion or Global Backward Erosion but most likely subject only to concentrated leak erosion and contact erosion.
- Dispersive plastic soils are soils in which, because of their clay mineralogy and the water chemistry, erosion will initiate in cracks or concentrated leaks under very low hydraulic stresses and gradients.

Section 10.2.3 provides more details.

Continuation of internal erosion relates to whether there are filters or a transition zone downstream of the core, and whether these have been designed and constructed so they will act to arrest internal erosion should it initiate. Foster and Fell (1999b, 2001) and Foster (1999) define four levels of severity of continuation: No Erosion (NE), Some Erosion (SE), Excessive Erosion (EE) and Continuing Erosion (CE).

The authors' experience is that many dams constructed prior to the 1980s had filters which were too coarse to satisfy No Erosion filter criteria, or had excessive non-plastic or plastic fines so are susceptible to holding a crack and therefore not being effective as filters. The fines content also reduces the drainage capacity of the filters.

It is also our experience that dams with filters constructed of crushed rock are susceptible to being gap graded, having excessive fines passing 0.075 mm and susceptible to segregation.

Progression is the third phase of the internal erosion process:

For *concentrated leak erosion and contact erosion at open joints in a rock foundation*, progression depends on whether hydraulic shear stresses within the eroding flaw may or may not lead to the enlargement of the pipe. Whether this occurs depends on whether the pipe will collapse and whether upstream zones may control the erosion process by flow limitation or crack filling.

For *Backward Erosion Piping*, progression depends on whether conditions are present such that a roof can be sustained and the hydraulic gradient in the soil beneath the dam or levee in which the erosion is occurring is high enough. If not, the progression will reach an equilibrium condition.

For *suffusion and Global Backward Erosion and contact erosion in soil*, progression depends on the seepage gradient or velocity in the soil. This may be influenced by flow limitation by upstream zones.

For all initiating mechanisms, the degree to which erosion progresses is dependent on whether the filters are No, Some, Excessive or Continuing Erosion.

Dam Breach is the final phase of internal erosion. It may occur by one of the following four phenomena (listed below in order of their observed frequency of occurrence).

- Gross enlargement of the pipe.
- Development of a sinkhole from the pipe to the crest of the embankment resulting in settlement of the crest and overtopping.
- Slope instability of the downstream slope resulting in loss of freeboard.
- Unravelling of the downstream face resulting in loss of freeboard.
- Overtopping due to settlement of the crest from suffusion and/or collapse of an eroding pipe in the embankment).

In this book we discuss commencement of a breach but not the development of the breach leading to uncontrolled release of the reservoir.

Figure 3.7 shows examples of possible locations of initiation of internal erosion. Other locations where concentrated leak erosion may initiate are over small steps on the foundation profile and poorly compacted soil at the embankment-foundation contact.

Internal erosion and piping may also initiate through transverse cracks in the embankment formed under seismic loads.

Figure 3.8 shows examples of crest details which may result in relatively high likelihood of internal erosion and piping because the reservoir may rise above the top of the core and filters (Figure 3.8a) or above cracks in the upper part of the core above the filters (Figure 3.8b). These details are common in Australian dams constructed in the 1960s and 1970s and we would expect in dams elsewhere.

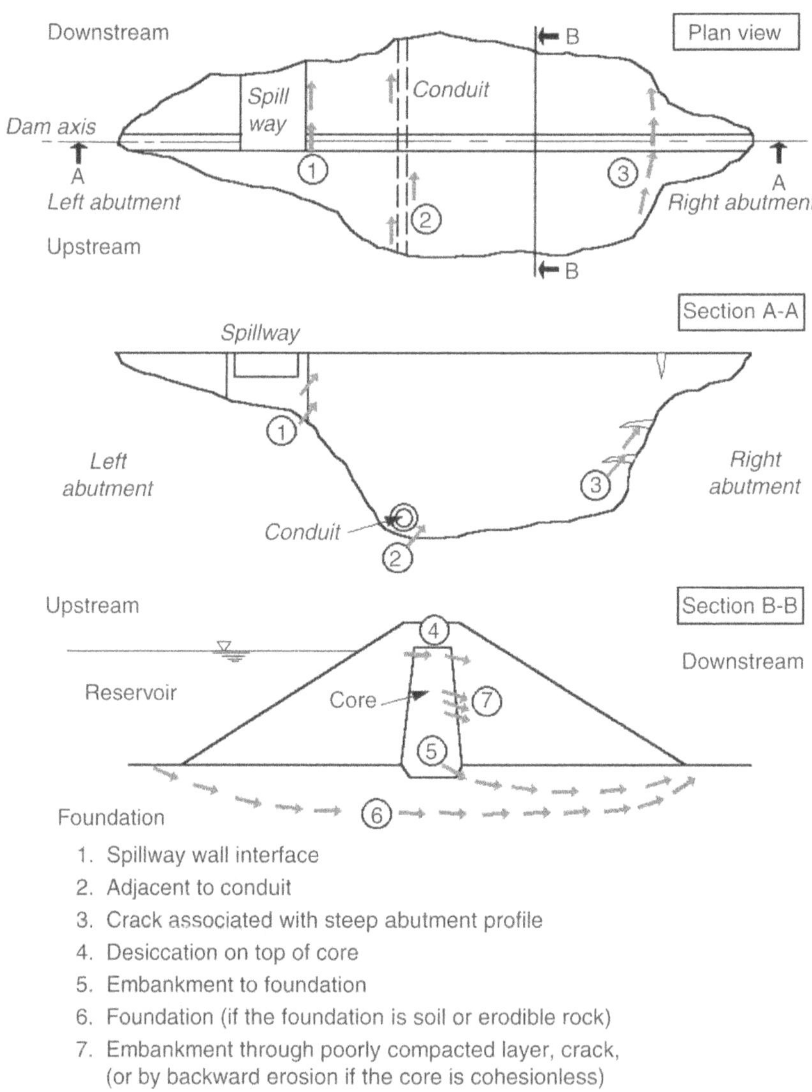

FIGURE 3.7 Examples of possible locations of initiation of internal erosion (Fell and Fry, 2007; Fell et al., 2015).

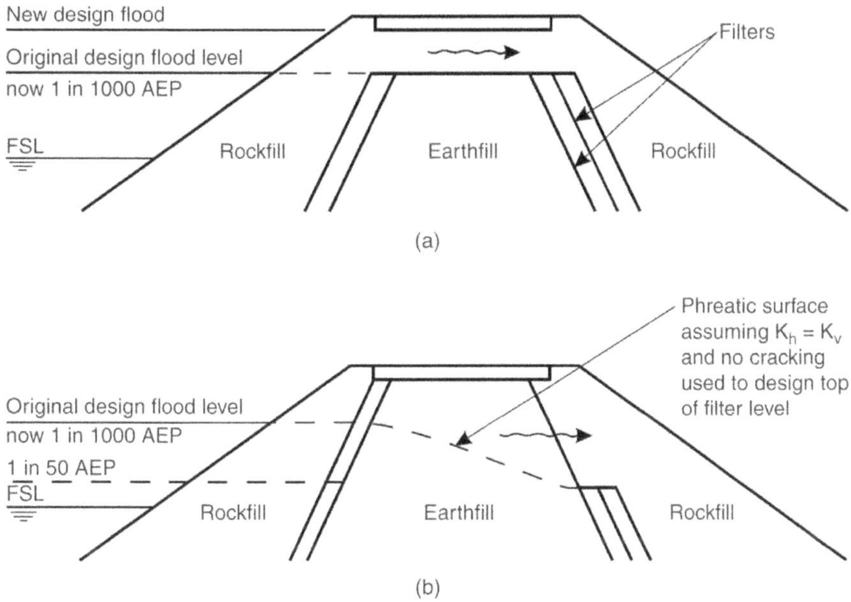

FIGURE 3.8 Examples of embankment crest details which may result in relatively high likelihood of internal erosion (Fell et al., 2008, 2015).

3.1.5 Leakage PFMs for Concrete Face Rockfill Dams and Asphaltic Concrete Face Rockfill Dams (ACFD)

Figure 3.9 shows the potential failure path for CFRD due to leakage through the face slab.

As can be seen the PFMs are somewhat different to other embankment dams and are:

- Opening of joints in the face slab or damage to the face slab resulting from settlements during construction, in service or under seismic loading, resulting in leakage into the embankment fill. Leakage into the embankment can also occur if internal erosion develops in the foundation below the plinth.
- Leakage flows not being controlled by Zone 2D either because it was constructed very coarse (as was the practice for early CFRD) or internal erosion of Zone 2D is not controlled by Zone 2E and Zone 3A.
- Saturation of the downstream slope resulting in slope instability, or large leakage flows resulting in unravelling of the downstream slope. If the rockfill is relatively fine and there are no high permeability zones in the embankment, this is more likely.

For asphaltic concrete face dams, similar principles apply to evaluating CFRD, although initiating mechanisms are different and zoning may differ too. See Chapter 16 for details.

3.1.6 Concrete and Asphalt Core Embankment Dams

See Chapter 17 for a discussion of PFM for these dams.

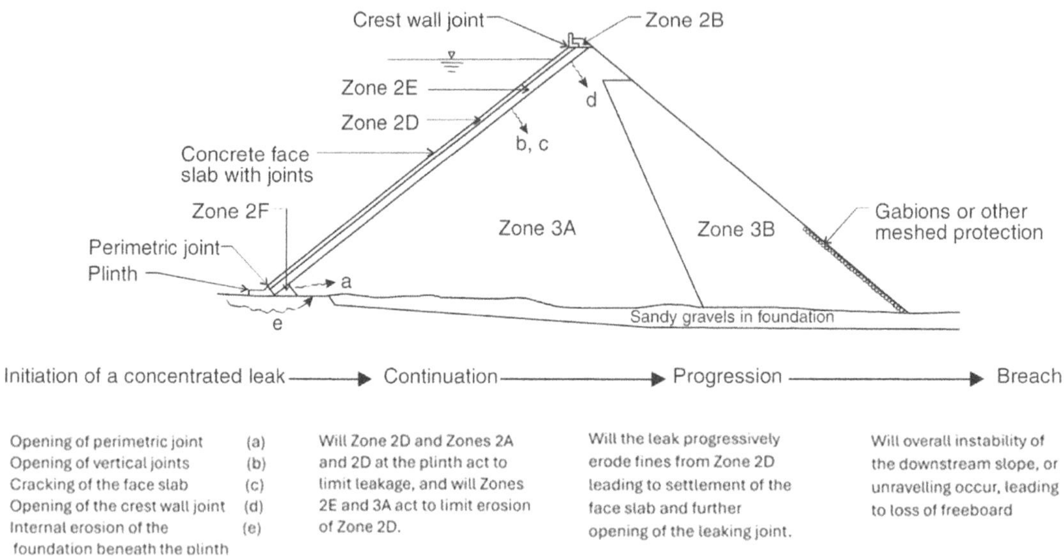

FIGURE 3.9 Failure paths for CFRD (adapted from Fell et al., 2015).

3.1.7 Slope Instability and Pre-Existing Landslides at Dam Sites

It is not uncommon for the slopes of the site of the dam and appurtenant structures to be subject to slope instability as the excavations for the structures are carried out. Many sites have pre-existing landslides which are likely to be re-activated as the structures are constructed. At some sites the landslides may be so large that they jeopardise the feasibility of the project.

Fell et al. (2005, 2015) include examples of landslides affecting embankment dams.

3.1.8 Reservoir Rim and Catchment Area Slopes

There are four key PFMs associated with reservoir rims that may need to be considered:

(a) *Landslides into the reservoirs* that may either generate waves leading to the overtopping of the dam or creating a debris dam that then breaches leading to a cascade failure of the embankment dam.
(b) *Leakage out of the reservoir* through the reservoir rim to the extent that the functionality of the project to store water is compromised. This is usually related to the presence of karst limestone in the reservoir rim but may relate to the presence of alluvial soils or other zones of high permeability forming the reservoir rim.
(c) *Potential landslides (including debris flows) in the catchment area* which may be triggered by rainfall events, bushfire denuding the slopes or seismic loading. These may create landslide dams in the stream feeding into the reservoir affecting inflow flood magnitudes to be considered, silt loads affecting stability or operation of the dam, or extreme turbidity make the stored water unfit for purpose.
(d) *Potential landslides (including debris flows and rockfalls)* that may lead to complete or partial blockage of spillways or inoperability of spillway gates. This can impact on the ability

of the spillway to safely discharge flood inflows, which can lead to the development of other failure modes (such as overtopping of the dam crest).

The PFMs for (b) and (c) represent potential operational risks for the dam operations, whereas PFMs for (a) and (d) represent a potential dam safety risk.

3.2 ASSEMBLY AND ASSESSMENT OF AVAILABLE DATA

The quality and credibility of a dam safety assessment will suffer unless the analyst team is fully aware of all pertinent information relevant to the Potential Failure Modes Analysis (PFMA). To the maximum extent practicable, existing data from investigations and design, construction records, performance history, and post-construction investigations need to be reviewed.

At screening and preliminary stages of a dam safety assessment, existing data supplemented by engineering judgement usually provide a sufficient basis for evaluation. If significant risk, including consideration of uncertainty, for the project is indicated, field investigations, laboratory testing, and additional analyses may be warranted, particularly for detailed and very detailed risk assessments.

ANCOLD (2022) provides guidance on the engineering and risk analysis methods required for screening, preliminary, detailed and very detailed risk assessments. These are described in Section 3.6.

At the outset of a study, assemble all available records that are relevant to the assessment of the PFMA. Commonly there is not a listing of such material and, with changes in organisations and personnel over time, an exhaustive search process is often needed to discover the available information. Catalogue and list the sources of data in the study report.

The following are typical of the documents which should be obtained:

- Hydrology reports.
- Spillway rating curves and estimates of AEP of flood levels and for dam crest flood.
- Geological reports.
- Site investigation reports.
- Materials investigation reports.
- Design reports.
- Construction reports.
- Construction photographs.
- Geological record reports.
- Geological mapping of foundations.
- Foundation grouting records.
- "As built" drawings.
- Full details of any modifications to the dam.
- Safety inspection reports (routine, annual and comprehensive).
- Instrumentation monitoring records and plots, including detailed review of the records and embankment performance.
- Safety review reports including results of stability analyses.
- Past performance and incident reports.
- Seismic hazard study reports including any specific studies of identified faults in proximity to the dam.

Discussions with the original dam designers and construction personnel are almost always valuable. However, whilst the recollections of designers and construction people are to be respected, memory is sometimes unreliable after many years. Independent validation of recollections should be sought where practicable.

"As built" or "work as executed" drawings are often not correct, and investigations may be necessary to confirm critical details such as zoning in the upper part of the embankment.

Documented information may be sparse for older dams. Considerable time and effort may be needed to locate scattered sources of information and to establish the credibility of the needed details.

The study documentation needs to record all sources of information. Others who refer to the study in the future need to know the sources on which it was based.

3.3 SITE INSPECTION

The authors recommend that as part of the Potential Failure Modes Analysis (PFMA) the dam and appurtenant structures be inspected by the risk analysis team and independent reviewers accompanied by operating and surveillance personnel.

It is not sufficient to rely on design and construction drawings, specifications and reports, or on past dam safety or annual inspections. Those inspections are not directed towards PFMA and are not necessarily undertaken by persons with expertise in PFMA.

In regard to slope instability and internal erosion and piping assessments, site inspection allows:

- An assessment of such details as local settlements and deformations in embankments which may be indicative of developing instability and/or internal erosion; openings between the embankment and spillway walls (or between earthfill and concrete dam sections) which may become pathways for initiation of internal erosion.
- Observation of seepage on the embankment or in the foundation which may indicate concentrated leak(s) including observation of any material being transported by the seepage. Seepage may also indicate high pore pressures which may affect slope stability.
- Observation of cracking and undue deformations in spillway walls and slabs which may indicate marginal stability and cracks or gaps which may form pathways for concentrated leak or contact erosion.
- Observation of maintenance of slopes can reveal animal burrows and other issues caused by poor maintenance.
- Inspection of exposures of the rock upon which the dam is founded which provides information relevant to assessment of the potential for contact erosion and erosion from the embankment to foundation.
- Potential identification of slope instability of the abutments affecting the dam and appurtenant structures and may create open pathways for internal erosion and piping in the foundation.
- Discussions with the personnel who maintain and operate the dam, spillway gates and outlet structures.

Site inspection also allows pertinent observations of characteristics for the assessment of dam crest overtopping failure modes, including potential presence and location of low points in the dam crest and at the abutments, condition of grass or rockfill surface layer, and conditions down the downstream abutment groin where flows are likely to concentrate under an overtopping condition.

Personnel should think *failure modes* during the inspection. Experience is that such matters can have a significant effect on proper identification of Potential Failure Modes and on estimated probabilities of failure.

As detailed in ANCOLD (2022), a PFMA and risk analysis carried out without a site inspection should be regarded as potentially flawed.

3.4 POTENTIAL FAILURE MODES ANALYSIS

3.4.1 Introduction

The authors recommend that those assessing the safety of any existing dam carry out a PFMA regardless of whether the safety is being assessed by quantitative or semi-quantitative risk assessment or deterministic assessment.

The PFMA process should include:

- Assembly and review of available information about the dam and its foundations.
- A site inspection of the dam.
- A workshop where those carrying out the safety assessment meet with the dam operators, surveillance staff and owner's representatives. The workshop participants should discuss all Potential Failure Modes and develop a thorough understanding of the sequence of events which may lead to failure of the dam. For example, for internal erosion PFMA, this would involve assessment of the potential location of seepage and erosion paths through the embankment and foundation in which erosion might initiate, consideration of progression and the potential for detection and intervention, and breach mechanisms.
- The sequence of events from initiation to breach should be documented by annotating cross sections and longitudinal sections of the embankment and its foundation to help visualise the failure mode. An example is given in Figure 3.10.

The possibility of overlooking potentially important failure modes is reduced by considering the particular details of the dam and its appurtenant structures, such as details of walls retaining the embankment and conduits through the embankment, and by assembling construction photographs and reports and inspecting the dam as part of the failure modes assessment. It is also reduced by having the failure modes assessment done by a team which includes the engineer and engineering geologist most familiar with the dam and dam operating and surveillance staff and facilitated by a person experienced in failure modes analysis.

USACE/Reclamation (2019) Chapter A3 describes good practice for PFMA.

3.4.2 Example of Description of the Potential Failure Mode

Figure 3.10 depicts an example of the description of a failure mode for an initiating event involving concentrated leak erosion for seepage path A in a transverse crack in Zone 3.

3.4.2.1 *Failure Mode Description*

"The reservoir rises to elevation 102 m which is 1 m above the historic high reservoir elevation and 2 m above the base of a transverse crack resulting from cross valley differential settlement. Hydraulic shear stresses on the sides of the crack exceed the critical shear stress of the soil so erosion initiates and

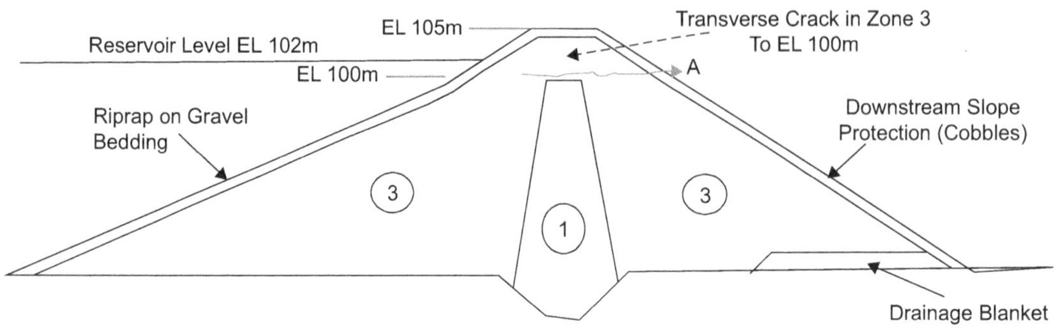

FIGURE 3.10 Failure mode for concentrated leak erosion for seepage path A, as described in Section 3.4.2.1.

begins a concentrated leak erosion process. The downstream cobble layer (slope protection) does not filter the Zone 3 earthfill material, and/or the cobble layer does not have sufficient overburden pressure on the emerging crack, and the slope protection layer blows off from reservoir pressure reaching the layer.

"Erosion progresses as a roof forms in the developing pipe in the Zone 3 material.

"The rip-rap bedding layer does not function as a crack stopping material because it does not have sufficient volume due to its limited thickness and/or it is not of the proper gradation to be filtered by the downstream slope protection material. There is no upstream zone of material to limit flows.

"Wet spots and flowing water appear on the downstream face of the embankment, but these seepage expressions are not seen by either the public or project personnel. If the seepage expressions are seen and reported, intervention efforts are unsuccessful because the efforts simply do not stop the erosion process, or the erosion progresses too rapidly to allow meaningful intervention efforts to be implemented.

"The crack widens, which leads to increased seepage velocities and hydraulic shear stress and more erosion of the Zone 3 material to form an enlarging pipe. Eventually this process of increasing seepage velocities and erosion progresses to full breach. The breach occurs by mechanisms typical for this type of embankment (i.e. gross enlargement of the developing pipe, slope instability or loss of freeboard as the pipe collapses)."

It should be noted that the description covers the initiation through to commencement of breach. Terms such as "piping in the embankment," "piping in the foundation," "concentrated leak erosion" or "concentrated leak erosion in a transverse crack resulting from cross valley differential settlement" are not adequate descriptions of the complete failure mode.

3.5 ASSESSMENT OF DATA GAPS AND ADDITIONAL SITE INVESTIGATIONS

3.5.1 A Checklist of Information to Inform the Safety Assessment

The following is a checklist of information which is needed to assess the safety of an existing dam.

3.5.1.1 Hydrology and Dam Break Reports

- AEP of reservoir levels from Full Supply Level (FSL) to dam crest level and above dam crest level if the floods result in overtopping of the dam.
- AEP of the dam crest flood and the historic high reservoir level.
- Historic reservoir levels as a percentage time versus level.
- Flood outflow hydrographs and information on duration of floods.
- For dams with gated spillways, gate operating rules and reliability.
- Dam break studies showing the extent of inundation in the event the dam breaches, the Population at Risk (PAR) and if available, estimated Potential Life Loss (PLL).

3.5.1.2 Seismic Hazard

- AEP of the Earthquake Design Ground Motions (EDGM). For some PFMs this may be the AEP of Peak Ground Acceleration (PGA) at the dam site and the Magnitude contributions to these PGAs.
- For more detailed assessments if required, time-history data for representative earthquakes or specific fault studies where these are in proximity to the dam in question and the fault is active.

3.5.1.3 Engineering Geology

Engineering Geological Model of the dam site and reservoir area. This includes:

Soil Foundations

- The stratigraphy, continuity and geological origins of the foundation soils.
- The classification, particle size distribution, and Atterberg limits of each soil.
- The dispersion properties of plastic soils.
- The shear strength, compressibility and permeability of each soil.
- The presence of low strength defects or seams in soil foundations (including fissuring, relict defects from weathering, pre-sheared surfaces from landslides), and if present, their shear strength or influence on shear strength of the soil mass.
- The presence of potentially liquefiable soils and their Standard Penetration Test (SPT) and Cone Penetration Test (CPT) and Piezocone (CPTU) properties which allow assessment of the likelihood they will liquefy and their liquefied strength.
- The presence of acid sulphate soils, and halite and gypsum rich soils.
- The level to which the core and the shoulders of the embankment were taken into the foundation.
- Foundation preparation beneath filters on the foundation, e.g. rolling or trafficking by construction equipment of the foundation prior to placing the filter will inhibit seepage from flowing into the filter and it may bypass the filter.
- Piezometer data where available and the relation to reservoir and tailwater levels.
- Seepage measurement data and relationship to reservoir level and rainfall.
- Photographs taken during construction.

Rock Foundations

- The stratigraphy, continuity and geological origins of the foundation rock.
- Where applicable, bedding thickness and orientation.
- The orientation, persistence and nature of rock defects, i.e. joints, shears and faults.

- The presence of low strength defects or seams in the foundations (including bedding crushed seams and shears resulting from valley stress relief and pre-sheared surfaces from landslides, and if present, their location, orientation persistence and shear strength).
- The degree of weathering insofar as this affects other properties.
- The level to which the core and the shoulders of the embankment were taken into the foundation.
- Foundation preparation beneath filters on the foundation, e.g. trafficking of the foundation by construction equipment for weathered rock foundations prior to placing the filter will inhibit seepage from flowing into the filter and it may bypass the filter.
- The permeability of rock in the foundation as determined by water pressure tests, i.e. Lugeon tests.
- The depth and spacing of grouting and grout take, and closure criteria.
- Piezometer data where available and the relation to reservoir and tailwater levels.
- Seepage measurement data and relationship to reservoir level and rainfall.
- Photographs taken during construction.

These factors are often not uniform across the valley in which the dam is sited. Abutments may have different engineering geological conditions to the river section, and soil conditions almost always vary within the river section.

Dam Embankment

- The actual embankment zoning particularly in the upper parts of the embankment.
- The source and processing of all zones within the embankment.
- The sequence and timing of construction, construction methods for placement of materials, level of site supervision and quality control records where these are available.
- The classification, particle size distribution, and Atterberg limits of the soil in each zone.
- The dispersion properties of plastic soils.
- The specified and actual layer thickness, degree of compaction and compaction water content relative to standard optimum water content of all zones.
- Pore pressure distributions within the embankment and how these have changed over time since construction and their relation to reservoir level.
- Post-construction crest settlement and deformation records, and if available other deformation data (including from during construction). Where available, cross arm settlement data.
- The presence of cracking or gaps adjacent to structures.
- Observations of seepage or cracking in the embankment.
- Records of any incidents or unusual observations since construction.
- Records of any modifications undertaken to the dam since original construction.

The following should be noted when assessing this information:

- For many low consequences of failure embankment dams, not all these data will be available or required to allow a satisfactory safety assessment. For higher consequence of failure dams, assessments may have to be made without all the data or, as discussed later, consideration given to obtaining critical data by site investigations.
- The quality of the data available will be variable, but the authors' experience is that even old data not up to modern standards can be valuable and should generally not be discarded.
- In many cases the zoning of the dam as constructed is not as designed, and even not as shown on "as-constructed" drawings. This is particularly so in the upper parts of the embankment

where construction working room was limited and borrow areas were becoming worked out. Common examples are cores and filter zones that are finished at lower levels in the embankment than indicated on the design drawings. This may also occur resulting from large settlements after construction.

- The actual gradations of materials in all zones are often not as specified or recorded in construction reports and technical papers. A common feature of dams constructed in the 1960s and 1970s is that filters were constructed with large fines content, e.g. as much as 10% to 20%, and with plastic fines, meaning the filter material may hold a crack and may not be effective as a filter. Segregation of coarsely graded filters is also potentially an issue.

  Older dams may also display more variability because placement methods were adjusted during construction and between seasons as different borrow areas and machinery became available.

3.5.2 Assessment of Data Gaps

Based on the assessed PFM and the available information, those assessing the safety of the dam should consider whether the data available is sufficient to make a defensible assessment of the dam safety.

This may be done by first assessing safety using the available data and then identifying the factors which most affect the outcome.

If the uncertainty in these factors significantly affects the outcome, consideration should be given to whether site investigations should be carried out to obtain the missing data.

Examples of such data gaps include:

- No SPT or CPTU test data in the foundation or zones in the embankment which are potentially subject to liquefaction.
- Limited or no particle size distributions and plasticity data for a foundation potentially subject to Backward Erosion Piping.
- No as-constructed particle size distributions for filters. As noted earlier, it should not be assumed that filters were constructed as specified. Filters were often constructed with high fines content and may hold a crack and be ineffective as a filter.
- Strength testing of weak seams in the foundation of embankment dams founded on geological environments where adversely orientated, weak geological features are known to exist.
- Pore pressures in such seams.
- Uncertainty in the constructed details particularly of zones important to the PFMs and risk assessment (e.g. crest detailing, top of core, top of filters, construction sequencing in the upper embankment, etc.).
- Pore pressure conditions within the embankment or its foundation where these are important in the PFMs being considered (e.g. pore pressures within sandy zones in the foundation for assessing liquefaction or BEP).

3.6 ANCOLD (2022) LEVELS OF RISK ANALYSIS

ANCOLD (2022) provides Tables describing levels of risk assessment and the engineering inputs for each level.

TABLE 3.2 Levels of Risk Assessment (ANCOLD, 2022)

LEVEL	*TYPE*	*ENGINEERING INPUTS*	*ESTIMATION OF PROBABILITIES OF FAILURE*	*ESTIMATION OF CONSEQUENCES*	*RISK EVALUATION METHOD*
Screening	Qualitative or Quantitative	Basic	Screening to preliminary	Basic to moderate	Basic
Preliminary	Quantitative	Moderate to basic	Preliminary	Moderate	Moderate to basic
Detailed	Quantitative	Advanced to moderate	Detailed	Advanced to moderate	Advanced to moderate(4)
Very detailed	Quantitative	Advanced to very advanced	Very detailed	Advanced to very advanced	Advanced or very advanced

Notes:
(1) *Risk assessment* is always done in parallel with *Engineering Assessment*.
(2) Only *detailed* or *very detailed* risk assessments should be used, in conjunction, where applicable with traditional standards-based assessments, for concluding whether a dam meets safety requirements or requires upgrade measures. *Screening* and *preliminary* studies should only be used to rank risks, or to get early identification of dam safety issues. Many *detailed* studies will also only be used to rank risks.
(3) Portfolio Risk Assessment (PRA) can be carried out at screening to detailed level, the level determined by the proposed use for the assessment. However, note (2) also applies to PRAs.
(4) Advanced or very advanced risk evaluation methods should be used in studies to conclude if a dam meets safety requirements.

TABLE 3.3 Levels of Engineering Inputs to Risk Assessment (ANCOLD, 2022)

LEVEL	*GENERAL DESCRIPTION OF TYPICAL ISSUES*
Basic	• Assemble readily available design, construction, monitoring and surveillance data and reports. • Use and adapt existing flood and earthquake studies. • Use existing analysis of embankment stability or use judgement. • Assess filters and internal erosion and piping based on existing data. • Assess liquefaction by presence of liquefiable materials and judgements using existing data. • Assess concrete dam stability using judgement, existing calculations or basic calculations. • Assess gates, valves by operational history and judgement.

(Continued)

TABLE 3.3 (Continued)

LEVEL	*GENERAL DESCRIPTION OF TYPICAL ISSUES*
Moderate	• Detailed search for and assembly of design, construction, monitoring and surveillance data and reports. • Flood studies to modern standards, earthquake hazard report specific to the dam. • Reassess embankment stability using existing data. • Assess filters and internal erosion and piping in detail using existing data or minor additional data. • Assess liquefaction by methods detailed in ANCOLD (2019) to give AEP of liquefaction using existing data or limited additional data. • Assess concrete dam stability under flood loading using conventional analysis with estimated properties for dam and foundation based on actual geology, mapping and laboratory testing. For earthquake use spectral analysis pseudo-dynamic. • Assess gates and valves condition and reliability by inspection.
Advanced	• Detailed search for and assembly of design, construction, monitoring and surveillance data and reports. • Flood studies to modern standards, seismic hazard report specific to the dam. • Assess embankment stability and potential post-failure deformations in detail, with new investigations of conditions if needed. • Assess filters and internal erosion and piping in detail using existing data supplemented by sampling and testing of as-constructed materials as necessary. Carry out erosion/filter testing if needed. • Assess liquefaction by methods detailed in ANCOLD (2019) to give AEP of liquefaction and post-liquefaction deformations, using existing and additional data as necessary. • Assess concrete dam stability under flood loading using conventional analysis with measured or estimated uplift pressures, and foundation properties investigated in detail. For earthquake, use rigid block/Newmark analysis as detailed in ANCOLD (2019). • Assess gates and valves condition and reliability by detailed inspection and test operation, historic maintenance and performance data and basic reliability analysis as needed. Broadly covering the complete system including power, controls and access.
Very Advanced	• Detailed search for and assembly of design, construction, monitoring and surveillance data and reports. • Flood studies to modern standards, earthquake hazard report specific to the dam. • Assess embankment stability and potential post-failure deformations in detail, with new investigations of conditions if needed. • Assess filters and piping in detail using existing data supplemented by sampling and testing of as-constructed materials as necessary, carry out erosion/filter testing if needed. • Assess liquefaction by methods detailed in ANCOLD (2019) to give AEP of liquefaction and post-liquefaction deformations, using existing and additional data as necessary. Post-failure deformations may be estimated numerically, or dynamic numerical analysis carried out. • Assess concrete dam stability under flood loading using conventional analysis with measured or estimated uplift pressures, and foundation properties investigated in detail. For earthquake, use rigid block/Newmark analysis, modelling the uncertainty in the parameters. If critical, use linear or nonlinear dynamic numerical analyses as detailed in ANCOLD (2019). • Assess gates and valves condition and reliability by detailed inspection and test operation, historic maintenance and performance data and detailed reliability analysis as needed. Cover the complete system including power, controls and access.

Note: These levels are a guide only. If the results are not sensitive to the engineering inputs, a lower level of engineering input may be acceptable.

Geotechnical Investigations of Existing Dams and Their Foundations

4

4.1 INTRODUCTION

Geotechnical investigations are the activities undertaken to assess and evaluate the geotechnical characteristics of the foundations of a dam and appurtenant structures, sources of earthfill, filters and rockfill for embankment dams, and concrete aggregates for concrete dams and appurtenant structures and reservoir rim and abutment stability. For existing dams, it includes assessing the embankment zoning, the materials used to construct the dam and their ability to perform as required, foundation treatment, and the water pressures within the dam and its foundations.

Geotechnical investigations involve the disciplines of engineering geology, hydrogeology, soil mechanics, rock mechanics and civil engineering.

Figure 4.1 shows the stages in the geotechnical investigations and development of the geological and geotechnical models. While this figure is directed towards investigations of foundations for new dams, the same principles apply to investigating the foundations and embankment of an existing dam.

The investigations are often targeted to Potential Failure Modes (PFM), particularly those for which risk analyses or dam safety assessments have shown to be the most important and for which there are gaps in data.

It is important to note that the process is often carried out iteratively in more than one stage so that as information is gathered, the geological and geotechnical models are refined, gaps in knowledge are identified or reassessed and if required, further investigations planned and carried out.

Geotechnical investigations are always required for the design of new dams and appurtenant structures, no matter how small the dam may be. They are often required for existing dams when there are vital gaps in available information from the design and construction records for the dam and appurtenant structures which may be required for dam safety assessments or upgrades. The extent of the investigations will vary commensurate with the complexity of conditions and the consequence of failure classification of the dam.

The methods described here are suitable for both new dams and for existing dams. For the former data are limited but the process is simpler. For the latter there may be a large amount of data but frequently this is of variable quality and may use various logging or testing standards, often requiring conversion or re-calibration. As discussed in Section 3.2 significant effort is required to review this data

 DOI: 10.1201/9781003542025-4

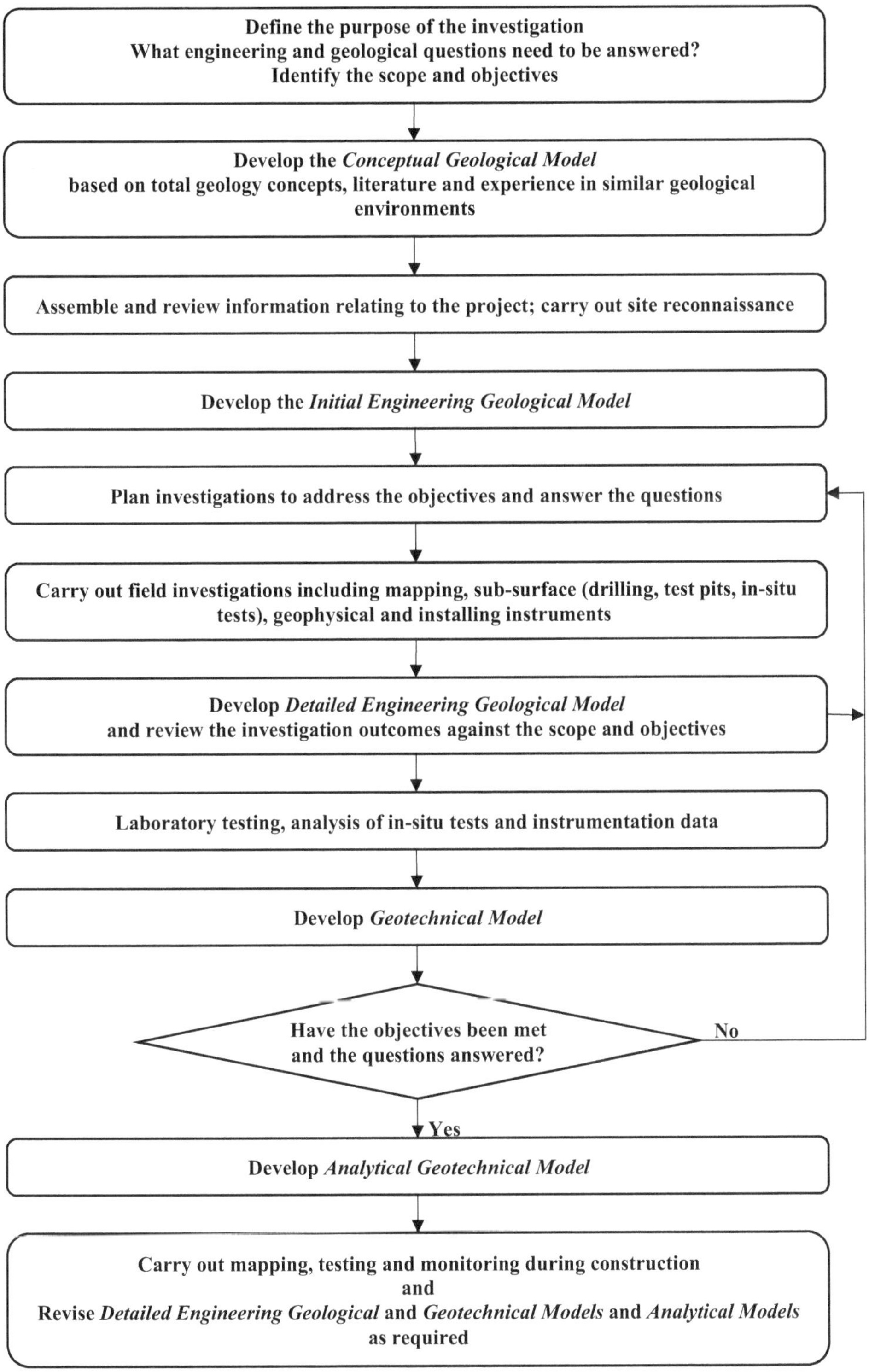

FIGURE 4.1 Stages in the geotechnical investigations and development of the geological and geotechnical models (ANCOLD, 2020).

and establish its relevance. In most cases this should be done after an Initial Engineering Geological Model has been prepared and an initial site reconnaissance undertaken.

4.2 GEOLOGICAL, ENGINEERING GEOLOGICAL AND GEOTECHNICAL MODELS

An important part of geotechnical investigations is the development of geological, engineering geological and geotechnical models.

4.2.1 Foundations of an Existing Dam

For the foundations of an existing dam, the models may be summarised as:

A Conceptual Geological Model is based on the tectonic setting and regional geological data with input from information published by geologists from, for example, state or country geological surveys, academics and practitioners. It takes into account the geological conditions that can be anticipated as described, for example, in Chapter 3 in Fell et al. (2005, 2015).

The Initial and Detailed Engineering Geological Models emphasise the geological aspects that are significant for the project and presents them in a form that engineers will understand.

The Geotechnical Model builds on the Engineering Geological Model by characterising the geotechnical properties of critical components of it based on the results of field and laboratory work. It is used by civil and geotechnical engineers and others as the basis for analysis of stability, groundwater flow, erosion potential etc.

The Analytical Geotechnical Model is a simplified version of the Geotechnical Model that will allow engineering analysis to take place. The Analytical Geotechnical Model only contains the essential elements required to meet the objectives of the investigation. In practice it will probably consist of a number of simplified plans and sections with quantified engineering parameters.

The Conceptual Geological Model will cover the whole of the dam site and surrounding valley slopes and some distance upstream and downstream of the site.

The Engineering Geological Models will also cover the whole of the dam site and surrounding valley slopes and some distance upstream and downstream of the site. They should not focus only on the dam footprint.

The Geotechnical Models are likely to be structure specific, e.g. embankment, or spillway, but should extend sufficiently beyond the footprint to allow adequate understanding of the overall site characteristics.

Geological and Engineering Geological Models are presented as a set of plans and sections, with accompanying text. For most dams two-dimensional (2D) models illustrated by maps and cross sections are sufficient to illustrate the geology, but for dams with complex geological conditions three-dimensional (3D) digital models may be required.

If 3D digital models are developed, the engineering geological and geotechnical models should also be presented in conventional geological drawings (e.g. geological maps, 2D cross sections and block diagrams) as it is often too difficult to apply 3D models directly to analyses.

These models should incorporate all relevant data including geological and geomorphological mapping carried out in the original site investigations for the dam, mapping and observations during construction and in operation.

The Engineering Geological and Geotechnical Models should include foundation levels and treatment for the core and shoulders of the embankment and details of grouting. For embankments on a soil foundation, there may have been a cutoff constructed.

For many dams these models will already be available in some form but may require supplementation and updating as additional site investigations are carried out.

4.2.2 Existing Dam Embankment

4.2.2.1 Initial and Detailed Embankment Engineering Models

For most dams there will be sufficient information from the original design documentation to develop an initial engineering model of the embankment including the geometry, zoning, description of the materials used for construction, and specifications for each zone of the embankment.

There may be some older dams for which not all this data is available, in which case those carrying out the safety assessment may have to rely on limited data and their experience of what was done for similar dams to develop the initial model.

A detailed Embankment Engineering Model will be developed as additional site investigations are completed.

4.2.2.2 Embankment Geotechnical Model

The Embankment Geotechnical Model assigns geotechnical and hydrogeological properties for each zone of the embankment and the foundation. It is developed by assembling and assessing field and laboratory testing from original investigations prior to and during construction, results of quality control testing during construction, photographs during construction and monitoring of deformations and pore pressures. Where there have been site investigations carried out after construction, these results should be incorporated into the model.

4.2.2.3 Analytical Geotechnical Model

As discussed previously, the Analytical Geotechnical Model is a simplified version of the Geotechnical Model that will allow engineering analysis to take place.

In practice the embankment and foundation models are integrated to provide the overall model of the dam and its foundations.

4.3 PLANNING THE INVESTIGATIONS

The steps in planning the investigations are:

(a) Develop initial Engineering Geological Models and Embankment Engineering Models from existing information.
(b) Identify PFM for the dam as described in Section 3.4 and identify gaps in data as described in Section 3.5.

(c) Define the objectives of the investigations and assess the most appropriate methods for investigations and the location and the number of investigations.
(d) Assess whether investigations may potentially increase the risks to the dam. Plan measures to minimise the risks and ensure they are tolerable.
(e) Amend the Dam Safety Emergency Plan (DSEP) and Emergency Action Plan (EAP) as required. If risk is to be managed, there may also be a need to involve the regulator.

4.4 GEOTECHNICAL SITE INVESTIGATION METHODS AND THEIR APPLICATION

4.4.1 Geological and Geomorphological Mapping

4.4.1.1 Geological Mapping

For existing dams there may be mapping existing from the original site investigations and during construction. Where necessary, these should be supplemented by mapping exposures around the dam.

The maps should show the following types of factual information and observations:

- Ground surface contours.
- Geomorphic features, e.g. slope changes, areas of hummocky ground, karstic terrain, erosion, surface processes.
- Geological surface features, e.g. areas of rock outcrops, scree, boulders and soil, superficial deposits.
- Features of the *in situ* rock, e.g. rock types and their boundaries, attitudes of bedding and foliation, the nature, location, persistence and orientation of important geological defects such as joints, sheared or crushed zones and faults.
- Groundwater features, e.g. springs, seepage, areas of swamp and vegetation indicating moist or wet ground.
- The location of tracks, roads and other infrastructure and exposures in them which have been mapped.

The most important end product of geological mapping is the understanding of the geology and geological history of an area which should be illustrated by the map and geological sections. This is part of the development of geological and geotechnical models.

4.4.1.2 Geomorphological Mapping

Geomorphological mapping should be completed concurrent with the geological mapping, and the mapping combined onto one set of maps.

The shape of the ground surface reflects both the underlying geology and the geomorphological development of the area. Geomorphological mapping of surface features can provide an indication of the distribution of subsurface materials, their structure and areas of possible mass movement, e.g. landslides.

Geomorphological mapping is developed from existing contour plans and ground based observations. The authors' experience is that contours developed from Light detection and ranging (Lidar) surveys are superior to other contours in showing slope changes and other details. The cost of doing Lidar surveys is often offset by the reduced effort required in the field.

It is recommended that geomorphological mapping be carried out for existing dams, as the authors' experience is that many dams were constructed in sites which have pre-existing landslides that were not recognised by the designers, or in some cases by those who carried out earlier dam safety assessments. Geomorphological mapping may assist in identifying such landslides.

The geomorphological mapping should include the dam footprint and the areas upstream and upslope of the footprint. The mapping should be based on pre-construction contours, aerial photographs, geological mapping and by mapping the area surrounding the dam from contours and on the site.

Figure 4.2 shows an example of geomorphological mapping for an existing dam with one abutment formed of landslide debris from a large ancient landslide high in the right abutment.

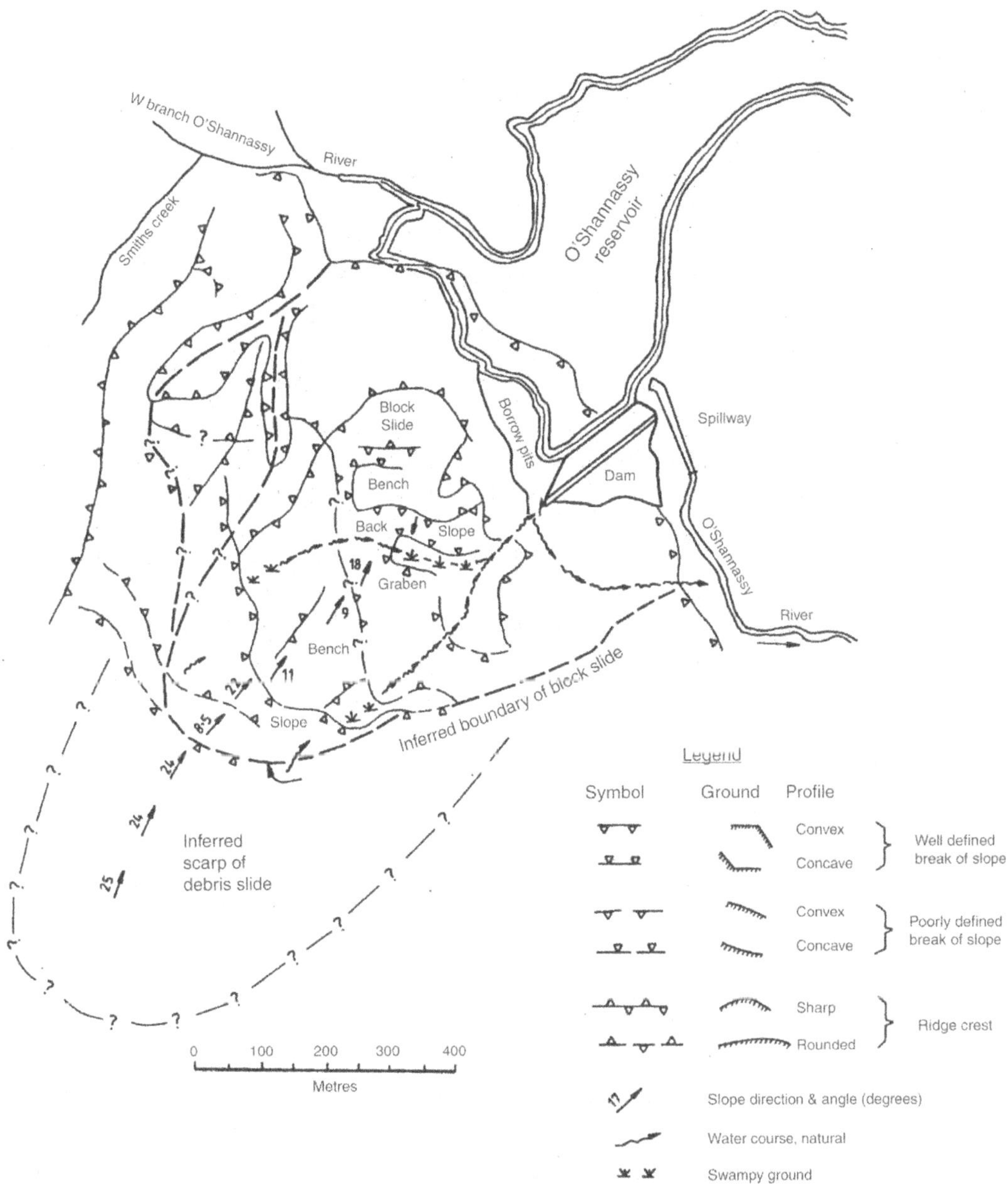

FIGURE 4.2 Large-scale geomorphological plan of landslide area at O'Shannassy Dam, near Melbourne, Australia (Fell et al., 2015).

Geological and geomorphological mapping should be carried out consistent with AS 1726–2017 or equivalent standards and guidelines.

Anon (1972), Anon (1981), Dearman (1991), Fell et al. (2005, 2015), Hutchison (2001), Varnes (1978) and Brunsden et al. (1975) describe mapping methods.

It is important that the mapping is carried out by engineering geologists who have experience in such mapping, understand the objectives of the investigation, and recognise and concentrate on the important geological features.

4.4.2 Test Pits and Trenches, Logging of Exposures

Test pits excavated by a rubber tyred backhoe or tracked excavator are effective in providing information on subsurface conditions in dam foundations or existing dam embankments.

In foundations in soil, they allow inspection of the structure of soil including layering of sediments, presence of rootholes and other features which control mass permeability, and fissuring which potentially affects soil strength. The absence or presence of groundwater is indicated, and the sources of inflows can usually be observed and their flow rates recorded.

In foundations in rock, they allow logging of stratigraphy and the orientation, nature and persistence of bedding and defects such as joints, and other structures which are not always observed in the drill core.

In embankment dams, they allow determination of the actual zone geometry and sampling of the core, filters and rockfill.

Test pits have often shown that the zoning of embankments was not as shown in design or as-constructed drawings. They allow assessment of segregation and contamination of filters, and sampling of sufficient size to allow particle size distributions to be obtained. Figures 4.3 to 4.6 show an example of where filters were not constructed to the lines required by construction drawings and segregation of the filter which had a wide range of particle sizes. The filter had been placed by spreader boxes. The dam was designed and construction supervised by organisations with extensive experience in dam engineering. Zhong (2023) provides details.

In planning embankment test pits, allowance should be made for trench support, managing dam safety during the investigations, and backfilling to engineered fill standards.

Photographs should always be taken of logged exposures, test pit excavations and spoil piles.

Some limitations of test pits are the following:

- Pit depth is limited by the reach of the machine and its ability to dig the material. The maximum depth of a pit dug by backhoe is 4 m, but the machine commonly refuses on very weak rock. Larger excavators can reach to depths of more than 6 m and, depending on the nature of the rock defects, can excavate into weak rock.
- Local laws on health and safety support requirements of excavations must be followed. Collapse of pit walls occurs commonly, particularly in wet ground. The effective use of steel or timber support or benching can enable safe access to pits for logging and sampling.
- Groundwater inflows may cause excessive collapse and limit effective excavation, even with supports.
- Test pits can result in defects within the embankment or foundation if the backfill soils are not backfilled with appropriate materials and properly compacted. In practice, pits in the crest of embankments should not cross the whole of the core or crest.

A logical extension of the use of test pits is the excavation of trenches, using either a tracked excavator or bulldozer and ripper. Such trenches provide virtually continuous exposures of the subsurface materials at sites where there is little natural outcrop. Careful cleanup by small backhoe or hand tools

FIGURE 4.3 Photograph of test pit in a filter with specified gradation as shown in Figure 4.6 showing segregation. (Photo courtesy of AECOM and Melbourne Water.)

FIGURE 4.4 Close-up photograph showing segregation. (Photo courtesy of AECOM and Melbourne Water.)

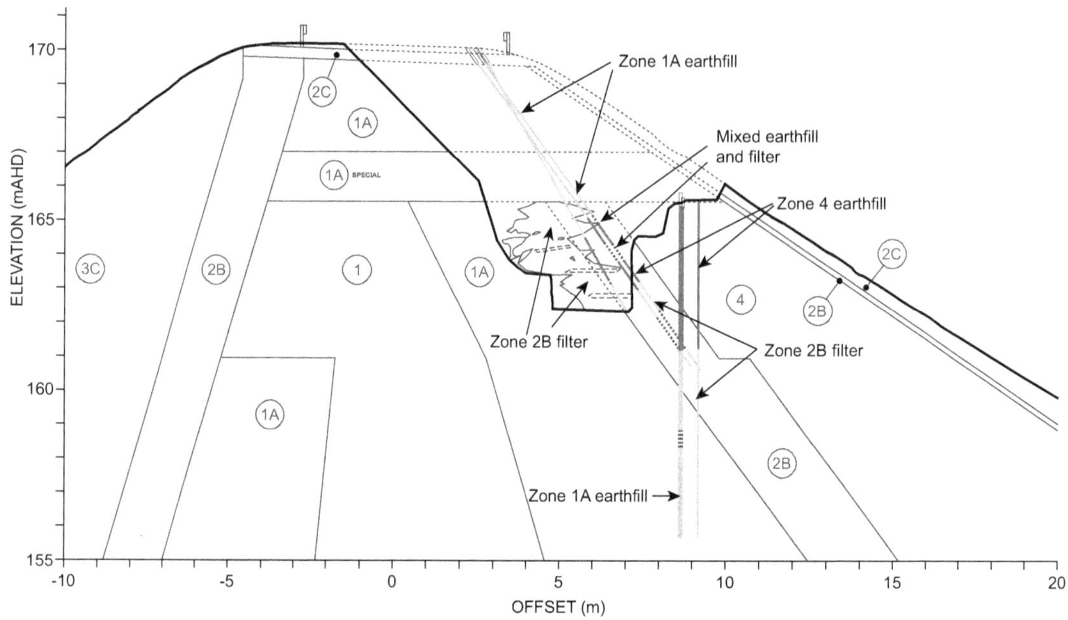

FIGURE 4.5 (A) Cross section of dam showing design zoning boundaries and actual boundaries as determined in a test pit and sonic drill holes. (Courtesy of AECOM and Melbourne Water.)

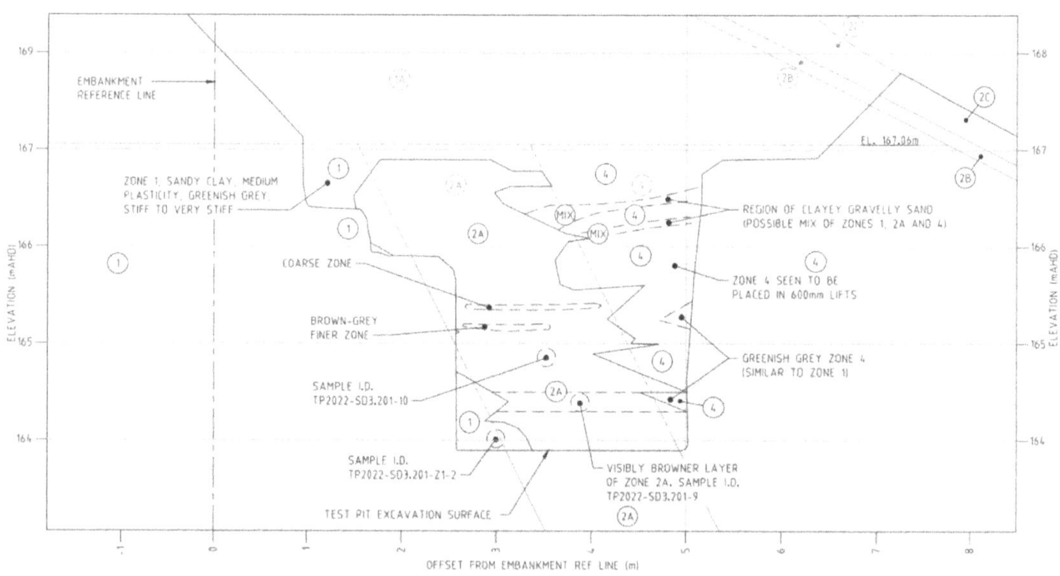

FIGURE 4.5 (B) Detailed cross section of dam showing design zoning boundaries and actual boundaries as determined in a test pit and sonic drill holes. (Courtesy of AECOM and Melbourne W ater.)

is usually required to expose defects, particularly in rocky materials. Trench exposures are logged in a similar manner to test pits.

At sites for large structures in geologically complex areas with little natural exposure, trenching can be the most practical method of providing the answer to vital safety assessment and remedial

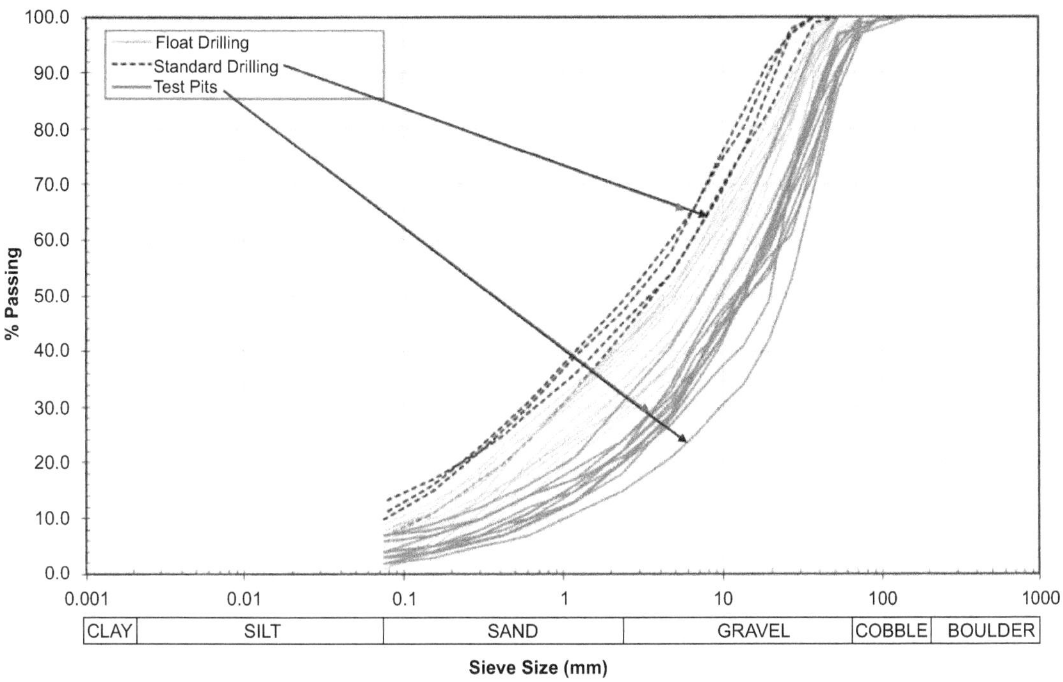

FIGURE 4.6 Specified gradation of the filter and gradations on samples taken from test pits and sonic drill holes. Note that the gradations for samples from drill holes are affected by drilling and are finer than the actual gradations which are shown by tests on samples from test pits. (Courtesy of AECOM and Melbourne Water.)

design questions. A disadvantage of trenching is that it can cause much disturbance of the site but, as with test pits, trenches can be backfilled and revegetated. Trenches are more likely to be subject to instability than pits because there is no benefit from 3D effects.

It is important to note that much valuable information can be obtained by cleaning up existing road cuttings or other exposures and logging the exposed soil and rock. Photogrammetric analysis methods can be used to assist in logging and interpretation of rock exposures in cuttings.

4.4.3 Drilling and Sampling Methods

Tables 4.1a and 4.1b summarise drilling methods, their applicability, advantages and disadvantages, and precautions regarding their use. Details are provided in Section 4.4.4.

The most commonly used drilling, *in situ* testing and sampling methods are the following:

4.4.3.1 Plastic Soil and Very Low Strength Rock Foundations

For investigating the foundation for new embankments in stiff plastic soils or very low strength rock several options are available, and a combination of methods is usually required.

Drilling by hollow flight auger or wash boring with drilling fluid support and sampling with thin walled tubes is most common and gives the least disturbed samples for laboratory testing. Good results can also be achieved by core drilling, with the core being less disturbed than with sonic drilling. Logging of the tube samples or core provides important information on fissuring and defects.

TABLE 4.1A Drilling Techniques and Their Applicability (ANCOLD, 2020, adapted from Fell et al., 2015)

DRILLING METHOD	*HOLE DIAMETER*	*SUPPORT*	*APPLICABILITY*	*ADVANTAGES*	*DISADVANTAGES AND PRECAUTIONS*
Auger drilling—Solid flight	Usually 100 mm	Self-support, i.e. the hole must support itself.	Clayey soils, moist silty sand and sand above the water table.	Ease of setting up, rapid drilling, continuous disturbed sample recovered.	Limited depth—30 m normal maximum. Not applicable for sandy soils below water table or soft clays because hole needs support.
Auger drilling—Hollow flight	Usually 100, 150 or 200 mm, up to 300 mm available	Self-support above water table if augers used as for solid flight augers. Augers used to support below water table.	As above, and for sandy soils and soft clays below the water table by using the augers for support and sampling through centre of augers.	As above, can be used below water table if flights are filled with water or mud. Easy sampling. Can be used in dam cores.	Hollow flights are not normally filled with water or mud so "blowing" or disturbance at the base of the hole may occur in granular soils and soft clays when withdrawing the plug in the base of the augers for sampling or testing.
Rotary drilling—Noncoring in soil and rock (wash boring)	75–100 mm common	Self-support in clayey soils. Casing and/or drilling mud in granular soils and soft clays. Usually no support in rock, but need mud if rock is weak or highly fractured.	All soil and rock types.	Allows full range of sampling and testing techniques. Rapid drilling in rock.	Poor identification of soil types in drill cuttings. Virtually impossible to identify thin beds of sand in clay. Proper identification requires tube or SPT samples. Unable to penetrate gravels efficiently. Must case or fill hole with water or mud to prevent blowing in loose granular soils and disturbance of soft clays. Note that even stiff clays can fail by shear at 30 m depth if not supported by mud. Rock samples difficult to identify in cuttings. May induce hydraulic fracture in dam cores and should not be used for that purpose.

(*Continued*)

TABLE 4.1A (Continued)

DRILLING METHOD	*HOLE DIAMETER*	*SUPPORT*	*APPLICABILITY*	*ADVANTAGES*	*DISADVANTAGES AND PRECAUTIONS*
Rotary drilling—Coring in soil and rock	"N" hole 76 mm, "H" hole 96 mm. Use "H" or "Q" in soil and poor rock. Core diameters NQ 48 mm HQ 63 mm PQ 85 mm	Hole is supported by casing through soil overlying rock. No support in rock. Water, mud or foam used as drill fluid.	All rock types. With care good recovery can be achieved in soils with any cohesion.	Continuous core of rock. Continuous "undisturbed" sample allows identification of stratification in alluvial soils. Allows for orientation of cores and optical logging.	Poor recovery in weak, friable and closely jointed rock. No recovery in clean sand and silt. Soil and weathered rock samples are usually disturbed by stress relief and swell due to water used from drilling. "H" and "Q" size cores give better recovery but are heavier and more expensive, but the incremental cost is small and usually worth spending. NMLC can give better recovery than NQ, the wireline equivalent and sometimes better than HQ. Sometimes core size is based on packer testing equipment.
Sonic drilling	75 mm to 250 mm. Common available 150 mm	Outer casing of sonic equipment.	All soil and low strength rock. Can penetrate stronger rock but not preferred method. Can drill up to 150 m with large drill rigs. If dry drilling methods are required, they are limited to depths up to approx. 50 m.	Rapid drilling. Recovery of a disturbed "core" of the soil or rock. This includes non-plastic soils such as sand, silts and sand gravel cobble soils. Does not require mud support and only limited water so may be used where hydraulic fracture is a problem, such as in dam cores.	Less commonly available. More expensive than conventional rotary drilling per day, but this may be offset by higher productivity and better sample recovery. The "core" is disturbed by the drilling and by poor handling procedures. This can be reduced by using a plastic inner liner in the core barrel. Potential for sonic vibrations to cause densification of loose sands below the drill bit and adversely affect SPT N values, which is particularly critical for liquefaction assessments.

TABLE 4.1B Drilling Techniques and Their Applicability (ANCOLD, 2020, adapted from Fell et al., 2015)

DRILLING METHOD	*HOLE DIAMETER*	*SUPPORT*	*APPLICABILITY*	*ADVANTAGES*	*DISADVANTAGES AND PRECAUTIONS*
Reverse circulation (rotary)	Usually BQ Duo-tube	Outer casing of reverse equipment.	Sands, silty sand, sand with fine gravel.	Rapid drilling. Complete recovery of soil drilled (disturbed). Useful for investigation of sand sources.	Less commonly available. More expensive equipment. Mixes strata. No testing or undisturbed sampling possible. More common in mining.
Cable tool percussion bit and bailer	Up to 750 mm	Casing which may or may not be recovered.	Sands, gravel, clay, rock (fractured).	Can penetrate gravels. Can drill large diameter holes. Simple equipment. May give continuous but very disturbed recovery in clay soils.	Slow drilling. Results in strata mixing, breakage of rock fragments and loss of fines which can give misleading particle size distribution. Very heavy tools for transport and handling. Seldom used for site investigations as other methods are better.
Cable tool tube	Up to 200 mm	Casing usually recovered.	All soils and weak rocks.	Provides continuous samples and useful groundwater data.	Slow drilling. Very heavy tools for transport and handling. Seldom used for site investigations as other methods are better.
Hammer drill (impact drill, down the hole hammer)	Usually 50–100 mm	Self-supporting in rock. Needs casing in soils.	Gravel, rockfill, boulders, rock (to just drill a hole).	Very rapid penetration rate (e.g. 200 m in 24 hours). Can penetrate rockfill and soil with gravel and cobbles.	Samples limited to rock chips. May jam and lose equipment in loose fractured rock, rockfill or boulders. Penetration in clay is limited due to blockage of air passages.
Bucket auger, casing and auger, casing and clam shell	900–1200 mm	Casing which may or may not be recovered.	Sands and gravels but applicable to all soils.	Drills through gravelly soils without breaking up into smaller sizes. Suitable for investigation of sand/gravel filter sources.	Costly. Sub-sampling and testing also expensive because of large sample size.
Rotary drilling with rig on all-terrain vehicle	As for other rotary drilling.	As for other rotary drilling.	Drilling in swamps, water up to 0.5 m deep.	Allows access where truck mounted rig would not enter because of soft ground.	Smaller rig mounted. Applicable to 30 m depth.
Rotary drilling with rig on small jack-up barge	As for other rotary drilling.	As for other rotary drilling.	Drilling in water up to 6 m deep.	Float rig into position and set up on bottom, unaffected by tide or waves.	Costly to establish. Smaller rig mounted. Applicable to 30 m depth.

Test pitting or trenching provides the clearest visual picture of the foundation in the upper 3 to 4 m (and deeper) with suitable battering or trench support and is recommended for logging soil and rock structure, defects and potential shear zones.

For very low strength plastic soils, a combination of wash boring with drill fluid support or hollow flight augering, with thin walled tube sampling, and CPTU and vane shear is most commonly used.

4.4.3.2 Non-Plastic Soil Foundations Including Investigation for Liquefaction

For investigating non-plastic soil foundations such as sand and silty sand, several options are available, and a combination of methods is usually required.

Wash boring with drill fluid support provides good results, provided care is taken to avoid blow in and disturbance to the base of the hole, and is the preferred method.

If hollow flight augers are used for drilling into foundations of loose silty sands and sands below the water table, there is a high likelihood that the soil will flow into the auger as the plug at the bottom is removed. This will result in spuriously low SPT blow counts. If hollow augers must be used, then drill with drilling fluid and remove the plug at the base of the augers slowly.

Sonic drilling has application for investigation of foundations with gravels or boulders (e.g. glacial till or fluvial soils). Whilst there is breakdown of the gravel and coarser fractions, there is generally good recovery of the finer matrix and SPT testing can be performed in silty sand and sand layers or matrix, although the blow counts may be affected by disturbance from the drilling process so must be used with caution.

Test pitting or trenching provides the clearest picture of the structure and stratigraphy of the soil profile but pits in non-plastic soils are subject to collapse, particularly below the water table.

A combination of CPT or CPTU and wash bore drilling with SPTs is recommended for investigation and assessment for potentially liquefiable sandy soils. Seismic Piezocone (SCPTU) may be used to provide the shear wave velocity, which can be used as an additional method for assessing liquefaction potential.

4.4.3.3 Rock Foundations

Rotary core drilling methods are recommended for drilling in rock foundations. When coring in rock it is important that the drillers understand that the objective is to recover core, not drill the hole, and to achieve this it may be necessary to use short drill runs, and therefore slower drilling rates. Because of this, the contracts for coring should be on an hourly or daily rate, not per metre. The experience of the operator and their capability to change drill bits and drilling technique to improve core recovery is important for the success of investigation programs.

Sonic drilling is not recommended because the drilling often pulverises the rock and disturbs what is recovered.

4.4.3.4 Earthfill Core of Embankment Dams

Drilling with air or water must not be used because of the risk of hydraulic fracture, and this precludes rotary wash boring and coring and hammer drilling methods. See Section 4.4.5 for details.

Test pitting is often useful in the upper 2 to 3 m (and in practice up to 6 m) with suitable battering or trench support.

CPTU provides a continuous record that is useful for characterising strength and identifying soft layers, voids and pore pressure response, which is an indication of contractive and dilative behaviour. It allows identification of soil classification, but experience shows it requires verification by other methods that allow sampling of the soil for laboratory testing.

A combination of these methods is usually required for investigating earthfill cores.

4.4.3.5 Filters in Embankment Dams

Sonic methods are best suited for drilling through and recovery of filters in embankment dams. However, some breakdown of the coarser particles is likely to occur, creating additional fines and this may give a misleading idea of the fines content. Experience is that the amount of fines passing 0.075 mm generated by drilling is not necessarily large (2% to 4% if good drilling practices are followed) but is significant. The breakdown of coarser particles also results in underestimation of the % coarse particles.

Test pits are the best method for exposure of and sampling of filters but are restricted in the practicable depth they can be excavated to.

4.4.3.6 Rockfill in Embankment Dams

Sonic drilling within rockfill generally results in large core loss particularly for clean rockfill with high voids content, as the sample is lost into the voids of the rockfill. Sonic drilling with larger diameter sample barrel and using catchers designed to retain the sample may improve core recovery, but the recovered core is not suitable for characterisation of the rockfill.

Hammer drill and sonic drill are best suited to drilling holes in rockfill where no recovery of the material is required, e.g. to investigate earthfill beneath rockfill.

Excavation of large test pits is the best method for investigation and characterisation of rockfill in embankment dams but is limited by the depth achievable.

4.4.3.7 Installing Instruments in Embankment Dams and Their Foundations

The best drilling methods for installation of instruments in embankment dams are sonic drilling and hollow flight augering as these methods provide a casing to support the hole, and the backfilling can be done gradually as the casing is extracted. Sonic drilling has the benefit of vibration of the casing on extraction to avoid sand arching and grabbing the instrument. Grabbing is where the sand backfill is caught between the casing and instrument such that further casing extraction will pull the instrument or its cable, leading to damage and likely breakage.

4.4.3.8 Backfilling Investigation and Instrumentation Boreholes

It is important at the planning stage of investigations to consider the backfilling methods. The key objectives for backfilling investigation holes within existing embankments include:

- Backfilling to a condition similar to the existing condition prior to investigations. This includes material types, internal zoning and compaction.
- The dam safety risks for the condition after backfilling should be equal to or, where possible, less than the existing condition prior to investigations.

When backfilling the hole, it is important to avoid leaving voids and the potential for connecting pathways within the core. It is common practice to use a non-shrink bentonite cement grout for backfilling boreholes. Backfilling with grout should be done in short lengths to avoid applying high fluid pressures that can hydraulically fracture the core.

See Appendix B for guidance on backfilling site investigations and instrumentation boreholes.

4.4.4 Details of Drilling Methods

The following provides more details of drilling methods and their application and limitations in regard to investigations within embankment dams and their foundations.

4.4.4.1 Auger Drilling

Auger drilling provides a cost-effective drilling method for investigation of earthfill zones within embankments and in soil and weak rock foundations.

There are several auger drilling methods available, and each has its advantages and disadvantages. Auger methods include:

- Hollow flight augering, typically ranging in diameter from 125 mm up to 250 mm. Larger sizes up to 300 mm diameter are available in the USA.
- Solid stem continuous flight augering, typically ranging in diameter from 75 mm up to 250 mm.
- Bucket augers, which can drill larger diameter (up to 1.2 m diameter) and within earth-rockfill zones and weak rock foundations.
- Hand augering.

Hollow flight augering is the preferred method of auger drilling within the earthfill core of embankments as it presents the safest method for maintaining the integrity of the core. The casing of the auger prevents caving or sloughing in boreholes and also supports holes against squeezing when drilling in soft to firm clay earthfill (puddle cores, for example). Sampling and testing can be undertaken through the base of the auger. These are the main advantages over solid flight augers which are removed after each run, so the hole has to be self-supporting. Other advantages include:

- The ability to install instruments (e.g. piezometers or inclinometers). Consideration needs to be given to the method of installation and backfill around the instrument in the planning phase as this will impact the size of augers required.
- Tremie pipes for grouting boreholes can be installed through the hollow flight. In caving conditions, the grouting should be done progressively with the extraction of the augers.
- There are hollow flight systems available that can recover a continuous soil core within an inner sample barrel.

Limitations of hollow flight augering methods include the following:

- Whilst hollow flight augering can be used below the water table, care needs to be taken to avoid "blow back" when the base is opened for sampling. This requires the use of a fluid (water or mud) within the hollow stem to equalise the pressure head. Failure to do this can result in disturbance of the soil below the base of the hole, potentially blocking the bit leading to termination of the hole. This has consequential problems with backfilling the hole.

FIGURE 4.7 Large diameter hollow auger drill rig (Fell et al., 2015; courtesy CMEC).

- The drilling depth is limited to approximately 40 to 50 m. However, achieving this depth requires a reasonably powerful rig. The depth achievable with smaller rigs may be limited to 15 to 20 m depth.
- Large diameter (300 mm) equipment as shown in Figure 4.7 has been used to drill up to 80 m depth holes in United States Bureau of Reclamation (USBR) dams (Fell et al., 2015).
- They are not suitable for drilling in dominantly granular soils, gravels or rockfill.

Solid flight augers do have application for investigation of earthfill cores, but it is recommended that the depth be limited (to about 10 m) because of the risk of collapse or sloughing blocking the borehole as the solid auger is withdrawn as the hole is progressed. Where solid flight augering is contemplated, the drilling contractor should always have available back-up hollow flight augers in case the holes collapse.

Solid flight augers should not be used in the following applications:

- For drilling in soft to firm earthfill cores (puddle cores, for example) where squeezing of the hole will be an issue.
- Drilling below the water table or phreatic surface due to the likelihood of hole collapse (this includes in layered earthfill once seepage is encountered). Persisting with auger drilling in these conditions can result in damage to the embankment due to formation of voids and the collapse of the hole leads to mixing of the soils being drilled.

- Drilling in sands or through embankment filter zones. Solid flight auger drilling through filter zones should not be contemplated because of the risks to the dam through collapse and loss of filter integrity.
- Drilling in predominantly granular soils, gravels or rockfill.

Bucket type augers are generally used for constructing piles. They can have application for investigations, mainly for investigation of borrow areas and of the earth-rockfill shoulders of an embankment. The larger diameter and greater power of these rigs allow them to drill within gravelly materials, and disturbed samples can be recovered. They are not recommended for drilling within the earthfill core because of the difficulties in backfilling such a large diameter borehole. In addition, they are not suitable for drilling in soft to firm earthfill and drilling below the water table because the hole will be subject to collapse.

4.4.4.2 Non-Core Rotary Drilling ("Wash Boring")—Figure 4.8

Non-core rotary drilling methods (or wash boring) use water-based drilling fluids. These methods should not be used for investigation of earthfill cores or foundations within the embankment footprint because of the risk of hydraulic fracture of the core during drilling.

The methods are suitable for investigation of soil foundations for new dams and outside the embankment footprint for existing dams. They are the preferred method for investigating alluvial foundations in which SPT are required. The methods do have limitations as highlighted in Table 4.1. They are useful for installing instrumentation in foundations as the casing provides the support of the hole.

Rotary methods using compressed air (and/or foam) as the drilling fluid also fall into this category. Under no circumstance should they be used for drilling within the core or foundation of embankments because of the high risk of hydraulic fracture.

4.4.4.3 Rotary Core Drilling—Figure 4.9

Rotary core drilling methods use water-based drilling fluids to transport cuttings to the surface and cool the drilling bit. Air with foam can also be used but is far less common. They are used for drilling through rock and recovering a continuous core sample. This method can also be used for drilling in very stiff cohesive soils with good recovery achieved. The use of water (or foam or air) precludes the use of core drilling within earthfill cores because of the risk of hydraulic fracture during drilling.

See Section 4.4.5 regarding core drilling in rock in the foundation beneath the embankment footprint.

Standard sizes for core drilling are "N" and "H". "N" sized equipment drills a 76 mm diameter hole and recovers a 52 to 54 mm diameter core (NMLC or NW barrel type) or 46 to 48 mm diameter core for wireline drilling (NQ). "H" sized equipment drills a 96 mm diameter hole and recovers a 76 mm diameter core (HW barrel type) or 60 to 64 mm diameter core for wireline drilling (HQ). Larger "P" sized equipment is also available and drills a 123 mm diameter hole and recovers an 83 to 85 mm diameter core for wireline drilling.

Several considerations for rotary core drilling are as follows:

(a) Discussions with the drilling company in the planning phase on the expected conditions and purpose of the drilling will be of benefit particularly with selection of barrel and bit types best suited to the likely conditions and the use of additives to the drilling fluid for bit lubrication and improving circulation.
(b) The drilling rig and drilling equipment should be complete and in good condition. If not, damage to the core and poor core recovery can result.
(c) The experience of the drilling rig operator is important and can significantly impact the quality of the drilling results.

(d) Bit selection is very important in core drilling. Bit choice is where operator experience can significantly impact the quality of core recovered. The operator should be equipped with a range of drilling bits and spare bits for drilling in hard rock. Allowance for bit wear should be in tender documentation.
(e) Supervising engineering geologists and geotechnical engineers and should be experienced or well supported by experienced persons. They should be able to have an informed discussion with drill rig operator on drilling methods, equipment and core recovery.
(f) Drilling without additives to the water will reduce the quality of core recovery due to poor bit lubrication, erosion of soils and weak rocks under the high flow velocities and also removal of cuttings (additives assist with cutting removal and reducing fluid loss). The use of additives (and core recovery) needs to be balanced against the impacts on rock mass permeability where water pressure testing is undertaken. Additives also need to be compatible with water quality and turbidity considerations for the particular project, and this needs to be discussed with the drilling company in the planning stage.
(g) Core recovery and core quality generally improves with increasing core diameter, but drilling rates may decrease and costs will be somewhat higher. This is particularly important for recovery of weak seams or joints in the rock mass (which are important in the assessment of dam foundations) and drilling soils and weaker rocks. As a result, it is common to use "H" sized equipment.
(h) Refer to Section 4.4.6 for discussion on core orientation and Section 4.7 for downhole imaging within boreholes.

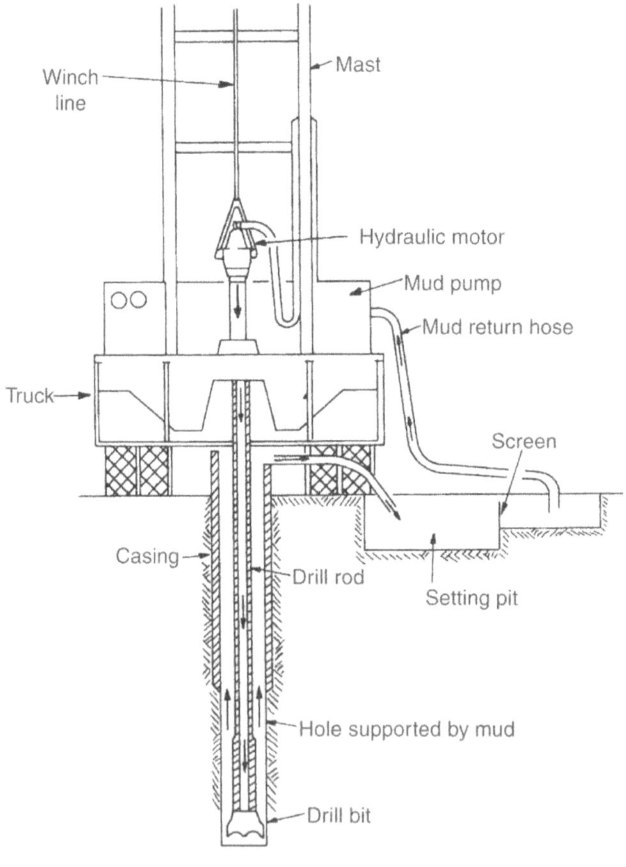

FIGURE 4.8 Rotary drilling rig—non-coring, also known as wash boring (Fell et al., 2015).

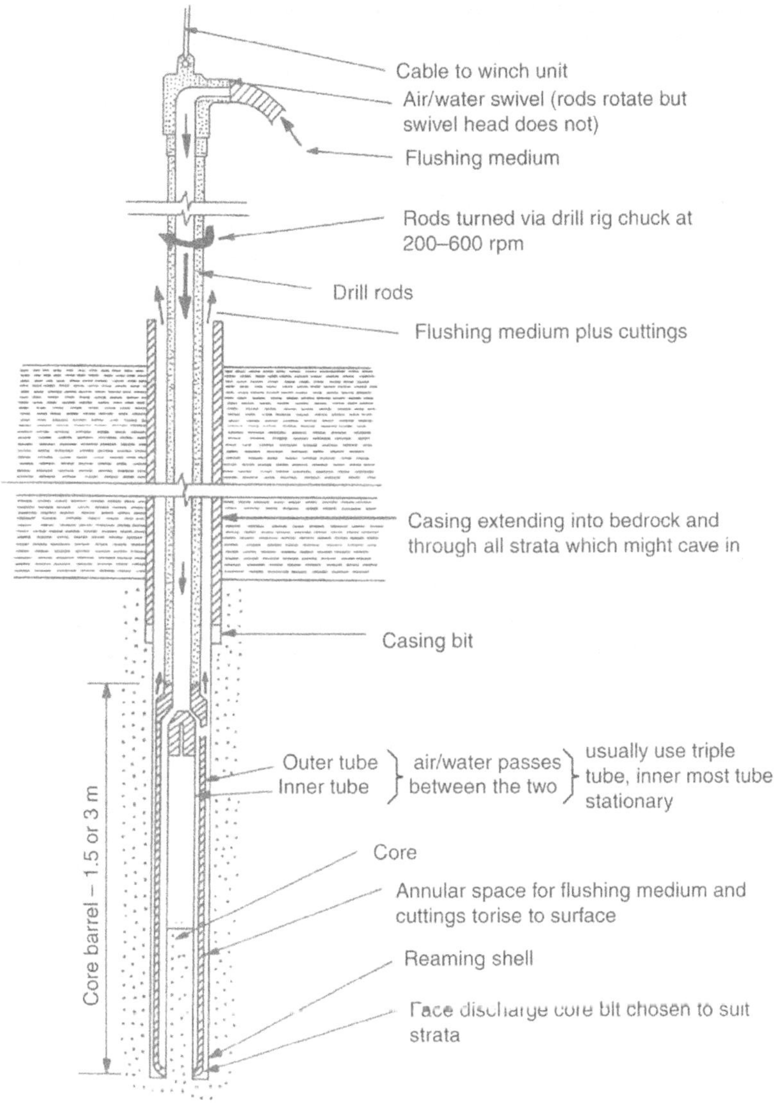

FIGURE 4.9 Rotary drilling rig—coring (Fell et al., 2015).

(i) There are a range of barrel types available (with double and triple tube), and those that reduce fluid exposure on the drilled core will aid core recovery, especially in soils and weak rocks. For these barrels the lifter casing extends into the bit shank and fluid contact with the core is limited to the crown of the drilling bit. There are also face discharge bits designed to further reduce the fluid exposure on the drilled core.

(j) Triple tube core barrels generally provide the best protection to core during drilling (thereby reducing core damage) and during handling of core (with the split liners). Single tube barrels should not be used.

(k) Wireline systems do have advantages with the drilling of deeper holes and in providing a casing against hole collapse or fall in (during drilling and water pressure testing) and are lower cost than non-wireline drilling. However, the core is smaller so core quality may be poorer, particularly for NQ equipment.

(l) Fluid pressure should be monitored during drilling to avoid the use of high pressures that may hydraulically fracture and damage the foundation. High fluid pressures also increase erosion of the core.

4.4.4.4 Sonic Drilling—Figures 4.10 and 4.11

In recent years there has been a trend towards using sonic drilling for investigation into existing embankments and their foundations due to the benefits of not requiring fluid (water or air) for drilling, recovery of a relatively continuous sample, the increased production rates over other drilling methods and savings in environmental clean-up.

However sonic drilling has not always been successful because of breakdown of rock and gravel resulting from the drilling action and inevitable sample disturbance in other soils. There is also potential for hydraulic fracture if water is used to aid the drilling.

In view of this it is recommended that sonic drilling is only used where appropriate. This includes:

(a) Drilling with dry methods in well compacted embankment cores to detect softened and wetter layers and to detect potential shear zones or other potential defect pathways, and to install instruments (where these are essential).
(b) Drilling through rockfill where a drill hole is required to install instruments.
(c) Drilling in alluvial soils in dam foundations to determine the stratigraphy, and recover silt, sand and gravel soils for laboratory testing for gradation, as for example is required for liquefaction assessments.

Sonic drilling should not be used to core rock because it pulverises the rock.

Penetration of the formation by sonic drilling involves fracturing, shearing and/or displacement of the strata. As a result, the wall and base of the borehole will be disturbed. In addition, the high frequency vibration can locally cause liquefaction and loosening of susceptible strata (e.g. saturated sands) or densification (e.g. loose sands or gravels) which may extend some distance from the sonic head.

In view of this, sonic drilling generally should not be used where SPT, *in situ* permeability and pressuremeter tests are to be carried out.

Consideration should also be given to the reliability of laboratory testing undertaken on sonic drilled samples, given the samples are likely to be disturbed.

Further details are provided in ANCOLD (2020).

4.4.4.5 Hammer Drilling

Hammer drilling methods use compressed air to flush the cuttings from the hole. These methods include downhole hammer (more common for geotechnical investigations) and surface hammer or air-track (more common for mining drill and blast).

These methods have application for drilling through rockfill in embankment dams, particularly where further investigation of the foundation below the rockfill is required. The method has limited application for characterisation of rockfill because of the fragmentation that occurs. The use of a casing advancing system is required for drilling through rockfill to maintain circulation and therefore retrieving cuttings. The use of foam (as a surfactant to the air stream) can improve cutting retrieval. Borehole diameters range from 100 to 400 mm.

The ODEX system is one system that is designed to advance the casing during drilling. The system has an eccentric bit that can be retrieved leaving the casing in place. Once the rockfill has been drilled through, the casing can be grouted in and the foundation of underlying strata then drilled through using other drilling methods.

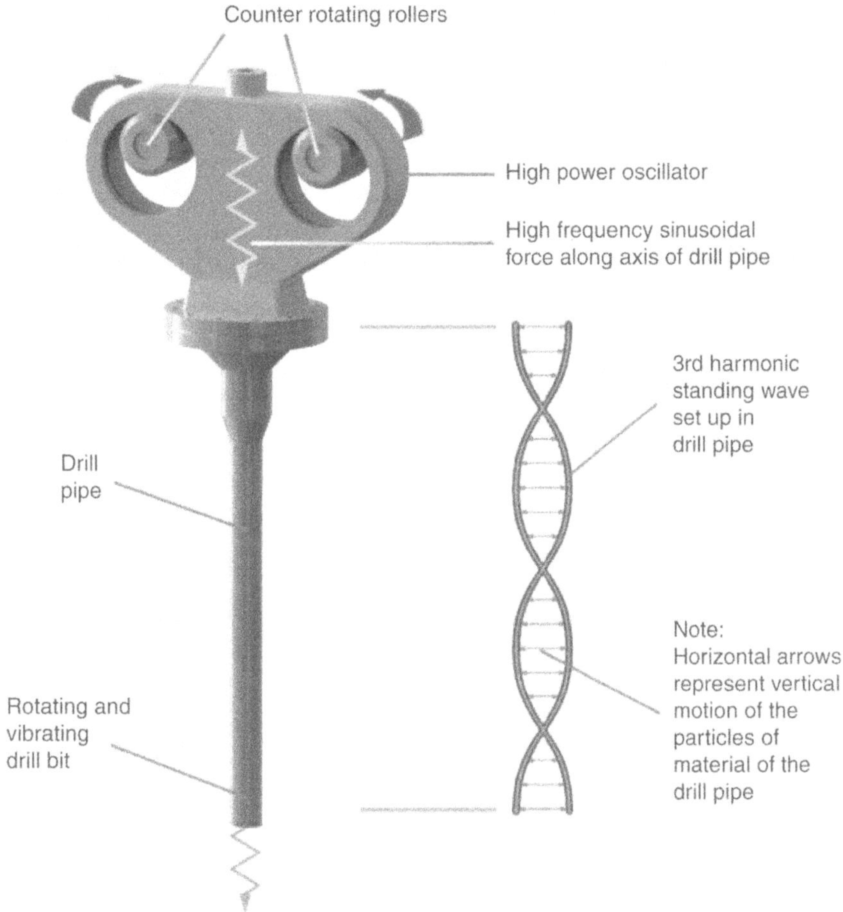

FIGURE 4.10 Principles of the sonic vibration action (Sonic Drill Corporation).

4.4.5 Drilling in the Core of Embankment Dams

Drilling into the core of embankment dams should be avoided because of the risk of hydraulic fracture of the core, which may lead to initiation of internal erosion and piping.

If holes *must* be drilled into the core of an existing embankment dam, it is recommended that:

(a) The holes should be drilled with a dry drilling method such as solid or hollow augers, or dry sonic drilling.
(b) Never use drilling fluids (water, bentonite and synthetic muds) or high pressure air when drilling because these have the potential to hydraulically fracture the core.
(c) Avoid drilling methods that can result in caving conditions within the earthfill core or foundation below the core; for example, solid auger drilling in silty sand to clayey sand cores below the water table or in soft to firm puddle core dams when squeezing conditions are encountered.
(d) Use methods to backfill the drill hole which avoid leaving voids or which might hydraulic fracture the core.

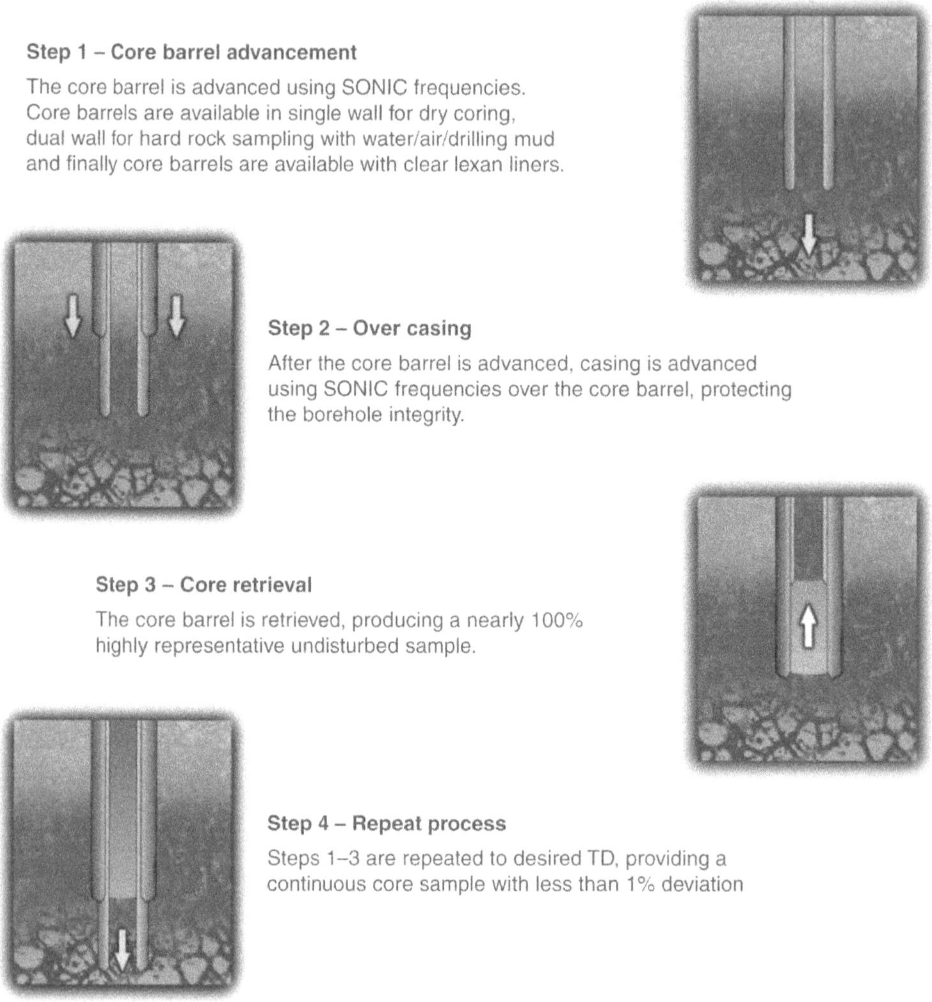

FIGURE 4.11 Steps in the sonic drilling process (Sonic Drill Corporation).

For drilling through the embankment to sample the foundation, use dry drilling methods within the earthfill core and soil layers within the foundation, then install and seal a steel casing socketed into the rock foundation (allowing time for the sealing grout to set) before continuing with core drilling of the foundation with water flush methods. Even taking these precautions there is some chance drill fluid pressure will find its way into the core, and unless drilling and sampling of the foundation below the core is essential it should be avoided.

The United States Army Corps of Engineers (USACE) and Federal Energy Regulatory Commission (FERC) provide guidelines for drilling in such conditions:

- USACE (2014) ER 1110-1-1807 Drilling in Earth Embankments.
- FERC (2016) Guidelines for Drilling in and Near Embankment Dams and Their Foundations Version 3.1.
- If using sonic drilling in the core of embankment dams, use dry drilling methods as described in ASTM D6914–04.

4.4.6 Core Orientation

The methods available to orient core include the following:

4.4.6.1 Core Orientation: Downhole

A number of core orientation devices exist in the industry broadly separated into mechanical and electronic systems.

Mechanical core orientation systems typically employ a retractable prong impression device to "map" the shape of the rock core stub prior to the commencement of each core run. A rotational weight, free to find its position at the bottom of the tool, is locked in place as the impression is made. After the rock core is recovered, the retractable prongs are matched to the top of the rock core and the locked rotational weight is used to identify the bottom side of the inclined borehole. It is important to note that any soil or rock fragments sitting on the top of the rock core stub can result in a poor impression, making rock core orientation difficult or sometimes impossible using this method. If the orientation is successful, a "reference line" is drawn onto the rock core from top to bottom, indicating the bottom of hole.

Electronic core orientation systems are available that use accelerometers to record "time stamped" barrel orientation data throughout the drilling process. The orientation of the barrel at the completion of each core run is used to orient the end of the rock core. A "reference line" is drawn onto the rock core from bottom to top (i.e. in the opposite direction to that of the mechanical systems), indicating the bottom of hole.

For both systems, rotation of the core within the barrel prior to marking the reference line can result in errors. It is essential that potential errors are checked, by comparing the reference line between adjacent core runs.

Davis (2015) presents a review of the available devices.

Selection of the most appropriate orientation system for field investigations will largely be driven by the investigating engineering geologist's experience with core orientation methods. Once orientation systems are ranked, selection will be based on availability (as the equipment is often rented by the consultant or drilling contractor), with consideration also afforded to the compatibility with the driller's existing equipment, cost and required training of the driller and the geologist supervising the drilling.

The cost of downhole core orientation is low as a proportion of the investigation cost. There is also a small additional drilling time to include orientation.

4.4.6.2 Core Orientation: Geological Logging

For vertical boreholes the defect dip angle is measured as the acute angle from horizontal. Rock core extracted from vertical boreholes is generally not oriented, so the dip direction cannot be directly measured. For both vertical and inclined boreholes, the defect dip direction can sometimes be estimated by measurement of the structure relative to well defined site bedding or metamorphic layering. Marjoribanks (2011) provides more details.

4.5 IN SITU TESTING

4.5.1 General Applicability

Table 4.2 summarises *in situ* test methods and their applicability. For investigations for dams, the most commonly used methods are:

(a) SPT, CPTU and CPT and SCPTU for investigating potentially liquefiable soils in dam foundations and their liquefied shear strength.
(b) CPTU and CPT and SPT for determining the stratigraphy, relative density, stiffness, and drained shear strength of non-plastic soils, and undrained shear strength and stiffness of plastic soils in alluvial foundations and mine tailings.
(c) CPTU and CPT for locating weak layers within the cores of dams and estimating the undrained shear strength and stiffness of the soils generally.
(d) Vane shear for estimating the undrained shear strength of soft and very soft soils (Cu less than about 50 kPa) in foundations.

4.5.2 Cone Penetration Tests (CPTU, CPT, SCPTU)

There are three types of Cone Penetration Tests, the electric cone penetrometer (CPT), piezocone (CPTU) and seismic piezocone (SCPTU).

The CPTU is a CPT with a piezometer in the cone to measure dynamic pore pressures generated as the cone is pushed into the ground. For most applications the CPTU is preferred to CPT because the additional information assists in identifying contractive non-plastic or very low plasticity soils, which are potentially liquefiable, and contractive plastic clays, the strength of which is controlled by undrained rather than drained strengths. The CPTU also is better suited to detecting thin layers of sand in clayey estuarine or alluvial sediments.

A CPTU fitted with equipment to measure the shear wave velocity is also available. It is termed the seismic CPTU and designated SCPTU. The shear wave velocity can be useful for liquefaction assessments and determination of the dynamic shear modulus for numerical modelling embankment dams and their foundations under seismic loads.

The CPTU can be used to carry out *in situ* permeability tests in the form of dissipation tests when the cone penetration is paused and the rate of pore pressure dissipation monitored.

Specifications for these tests should require compliance with Australian Standard AS 1289, 6.5.1 or ASTM Standard D578–12, Standard Test Method for Electric Friction Cone Penetrometer Testing of Soils.

4.5.2.1 Saturation of CPTU Piezometer

The good condition and saturation of piezometer ceramic filter is essential to the correct operation of the CPTU. If the filter is broken, fouled or contains air, the pore pressures reported will be anomalous. Good practice requires the development of a vacuum around the cone to saturate the filter prior to insertion into the ground. Typically, the vacuum is drawn on the cone whilst the filter is submerged in a water soluble medium, such as glycerine. This ensures that once a vacuum has been achieved and air is removed from the filter, glycerine is drawn into the filter when the vacuum is removed, rather than air. This assists in acquiring accurate pore pressures.

Caution should be used when interpreting measured pore pressures if the CPTU penetrates through heavily overconsolidated clays or compacted clay above the groundwater level. Penetration through these materials can cause the filter to become desaturated.

4.5.2.2 Interpretation of Soil Properties from CPT and CPTU

There are software packages available to interpret CPTU and CPT data. While these are very useful to estimate soil classification, strength (undrained for plastic soils, effective friction angle for non-plastic soils), stiffness, and relative density (of non-plastic soils), some packages provide interpreted data which

TABLE 4.2 *In Situ* Test Methods and Their Applicability (Fell et al., 2015 from Lunne et al., 1997)

		SOIL PARAMETERS													*GROUND TYPE*						
		SOIL													*HARD*	*SOFT*					
GROUP	*DEVICE*	*TYPE*	*PROFILE*	u	$*\phi'$	S_u	I_D	m_v	c_v	k	G_o	σ_h	*OCR*	$\sigma\text{-}\varepsilon$	*ROCK*	*ROCK*	*GRAVEL*	*SAND*	*SILT*	*CLAY*	*PEAT*
Penetrometers	Dynamic	C	B	-	C'	C	C	-	-	-	C	-	C	-	-	C	B	A	B	B	B
	Mechanical	B	A/B	-	C	C	B	C	-	-	C	C	C	-	-	C	C	A	A	A	A
	Electric (CPT)	B	A	-	C	B	A/B	C	-	-	B	B/C	B	-	-	C	C	A	A	A	A
	Piezocone (CPTU)	A	A	A	B	B	A/B	B	A/B	B	B	B/C	B	C	-	C	-	A	A	A	A
	Seismic (SCPT/SCPTU)	A	A	A	B	A/B	A/B	B	A/B	B	A	B	B	B	-	C	-	A	A	A	A
	Flat dilatometer (DMT)	B	A	C	B	B	C	B	-	-	B	B	B	C	C	C	-	A	A	A	A
	Standard Penetration Test (SPT)	A	B	-	C	C	B	-	-	-	C	-	C	-	-	C	B	A	A	A	A
	Resistivity probe	B	B	-	B	C	A	C	-	-	-	-	-	-	C	C	-	A	A	A	A
Pressuremeter	Pre-bored (PBP)	B	B	-	C	B	C	B	C	-	B	C	C	C	A	A	B	B	B	A	B
	Self boring (SBP)	B	B	A[1]	B	B	B	B	A[1]	B	A[2]	A/B	B	A/B[2]	-	B	-	B	B	A	B
	Full displacement (FDP)	B	B	-	C	B	C	C	C	-	A[2]	C	C	C	-	C	-	B	B	A	A
Others	Vane	B	C	-	-	A	-	-	-	-	-	-	B/C	B	-	-	-	-	-	A	B
	Plate load	C	-	-	C	B	B	B	C	C	A	C	B	B	B	A	B	B	A	A	A
	Screw plate	C	C	-	C	B	B	B	C	C	A	C	B	-	-	-	-	A	A	A	A
	Borehole permeability	C	-	A	-	-	-	-	B	A	-	-	-	-	A	A	A	A	A	A	B
	Hydraulic fracture	-	-	B	-	-	-	-	C	C	-	B	-	-	B	B	-	-	C	A	C
	Crosshole/downhole/surface seismic	C	C	-	-	-	-	-	-	-	A	-	B	-	A	A	A	A	A	A	A

Applicability: A = high; B = moderate; C = low, - none; $*\phi'$ = Will depend on soil type; [1] = Only when pore pressure sensor fitted; [2] = Only when displacement sensor fitted. Soil parameter definitions: u = *in situ* static pore pressure; ϕ' = effective internal friction angle; S_u = undrained shear strength; m_v = constrained modulus; c_v = coefficient of consolidation; k = coefficient of permeability; G_o = shear modulus at small strains; σ_h = horizontal stress; OCR = overconsolidation ratio; $\sigma\text{-}\varepsilon$ = stress-strain relationship; I_D = density index cannot in fact be determined with any reliability using CPTU and CPT. This includes estimates of permeability, shear wave velocity, *in situ* stress ratio, overconsolidation ratio and sensitivity (peak/remoulded undrained strength). These estimates are determined using empirical relationships built into the packages which are often not referenced or widely used in practice because they are only very approximate.

It should also be noted that the interpretations of soil classification from CPT and CPTU are not always correct, and they should be checked and calibrated for the site from laboratory test on samples from boreholes, and soil classification from borehole logs. This potential discrepancy can be significant, particularly for liquefaction assessments.

Estimates of undrained shear strength from CPT/CPTU should be correlated against *in situ* vane shear testing and on important projects, laboratory tests to obtain the empirical cone factor (N_{KT}) which relates cone resistance to undrained shear strength. See Section 7.2.8.

4.5.3 Standard Penetration Test (SPT)

SPT should be carried out according to AS 1289 6.3.1 or ASTM D1586–11. For investigations for liquefaction, it is essential that the SPT hammer energy is determined on site using the drill equipment and SPT hammer used for the investigations. ASTM D4633–16 "Standard Test Method for Energy Measurement for Dynamic Penetrometers" describes how this should be done.

4.5.4 Vane Shear Test

Vane shear tests should be carried out according to AS 1289 6.2.1 or ASTM D2573. The test is suited to measure the undrained strength of plastic soils with an undrained shear strength up to about 50 kPa.

The test should be carried out to determine the peak and large strain undrained strength. The ratio of the peak and large-scale strengths is known as the sensitivity of the soil. For reliable results the largest vane that measures the strength for the soil should be used. Using smaller vanes tends to potentially underestimate the strength because of disturbance as the vane is pushed into position.

4.5.5 Dilatometer Test (DMT)

A dilatometer test is a push-in test with a wedge shaped device. It has a small pressure cell which is used to measure pressure versus displacement at given intervals as it is inserted. The test should be carried out according to ASTM D6635.

The DMT has no particular advantage to the CPT or CPTU and, as there are more widely used methods for estimating properties using CPT/CPTU, generally those tests are preferred.

4.5.6 In Situ Permeability Tests of Soils

The permeability of soil in a dam foundation may be of importance if the cutoff and/or shoulder of the embankment is founded in the soil, as in dams on alluvial, colluvial and glacial deposits and on some deeply weathered residual or lateritised soils.

In general, the structure of the soil controls water flow, for example:

- sandy layers in alluvial soils,
- root holes and fissures in residual and alluvial soils,

- worm burrows in alluvial and residual soils,
- leached zones, infilled root holes and relict joints in lateritised soil and extremely weathered rocks.

Hence it is seldom possible to obtain a realistic estimate of soil mass permeability from laboratory testing.

Because the soil structure tends to be blocked or smeared by the drilling action, pump-in type permeability tests in boreholes can also give quite misleading results, with measured permeability values one or two orders of magnitude lower than the actual soil mass permeability.

Where possible, pump out tests should be conducted with a pump well and observation wells. Figure 4.12 shows a typical arrangement.

In a pump out test the inflow of water removes the soil blocking or smearing the hole side, and realistic estimates of permeability can be obtained.

The design of the pump well, observation wells and the test program analysis need to recognise the effect of different vertical and horizontal permeabilities in the soil. The results are often difficult to assess, and it is recommended that a groundwater hydrologist be employed in the planning, execution and analysis of such tests.

In the event that the groundwater level is below the depth of interest, it is necessary to carry out pump-in tests. Even when considerable care is taken to clean the hole, it should be understood that the results of such tests give lower bound estimates and that the actual permeability may be up to one or two orders of magnitude greater than indicated by the test result.

4.5.7 In Situ Permeability Tests of Rock

The permeability of the rock mass can be determined by either constant head or falling head tests. As shown in Figures 4.13 and 4.14, a section of a drill hole is isolated using a sealing packer and water is added to maintain a constant head, or the rate of fall in water level is measured after a slug of water is added to the hole. Both methods suffer from the potential effects of smear and clogging of defects, but by careful flushing of the hole before testing, reasonable values can be obtained. In falling head

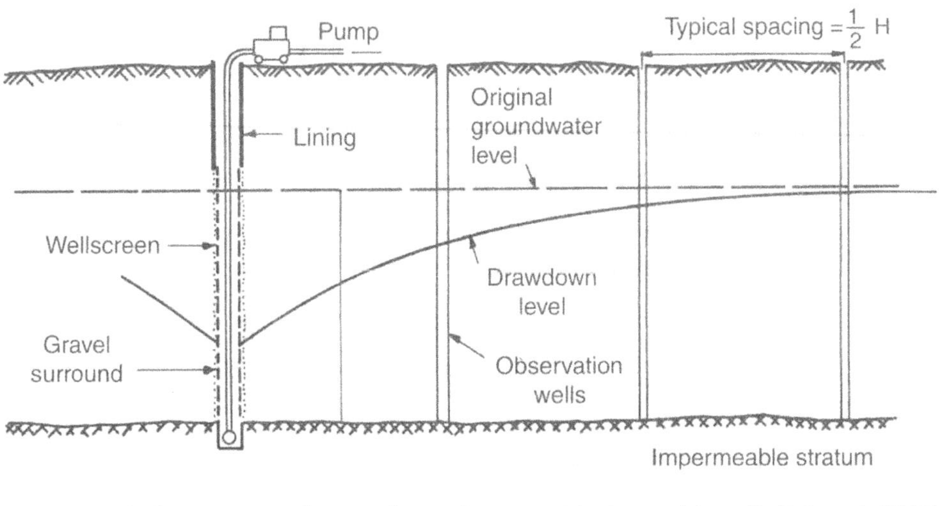

FIGURE 4.12 Typical arrangement for a well pumping test with observation wells (Fell et al., 2015).

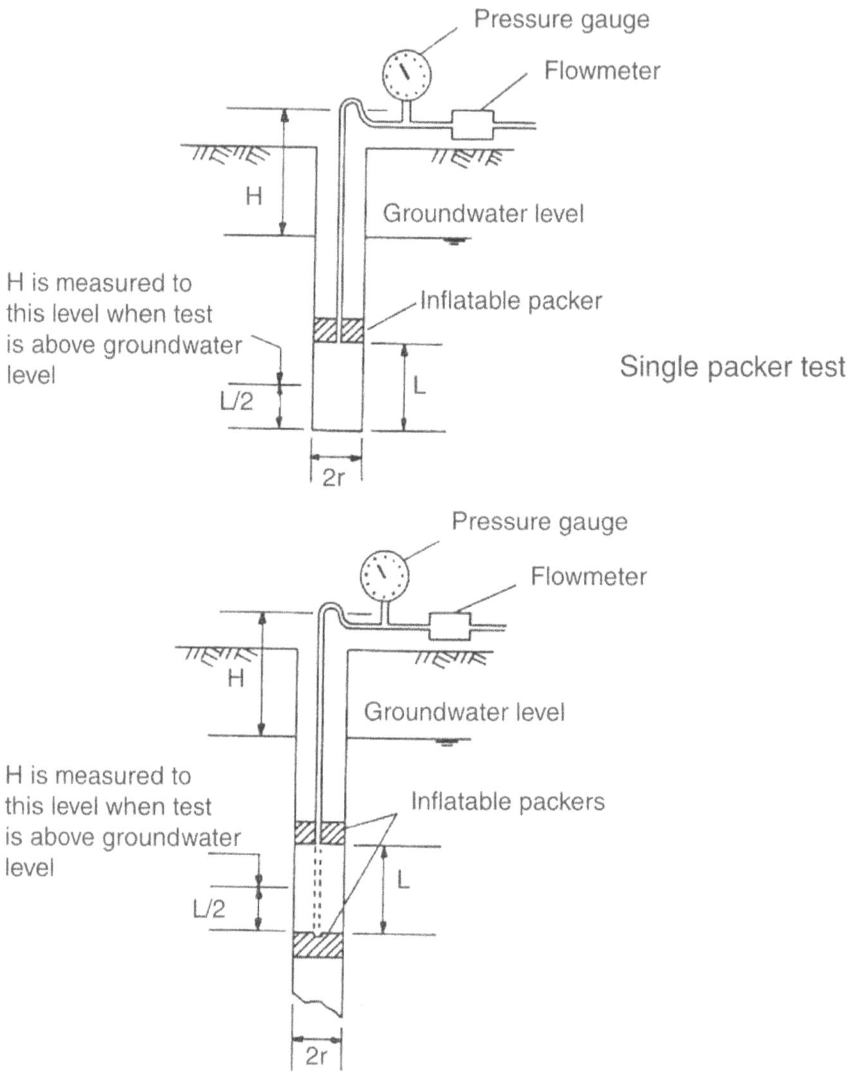

FIGURE 4.13 Packer permeability testing. Single packer and double packer methods (Fell et al., 2015).

tests the additional pressure that can be added to the test section is limited by the level of the test section and the practicability of extending the pipe imposing the head above ground level.

The most common and effective method of measuring rock mass permeability is the water pressure test (also known as the Lugeon or "packer" test).

The test consists of isolating a section of drill hole and pumping water under pressure into that section until the flow rate for any given pressure is constant (i.e. it is a constant head test). The use of successive rising and falling test pressures as discussed later establishes the relationship between the volume of water accepted into the section and the pressure, to provide an estimation of permeability and indicate water flow mechanisms.

As rock substance is generally almost impermeable, the permeability determined in this test represents an indication of the number, continuity and opening of the rock defects which intersect the wall of the borehole in the test section.

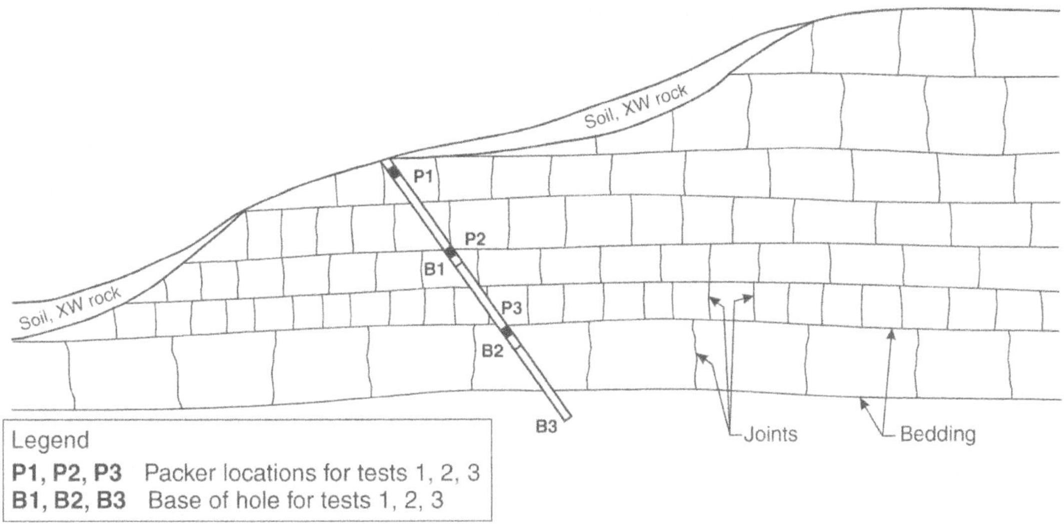

FIGURE 4.14 Borehole inclined to intersect joints in the rock (Fell et al., 2015).

Results are expressed in Lugeon (uL) units. A Lugeon is defined as the water loss of 1 litre per minute per metre length of test section at an effective pressure of 1 MPa.

Indicative rock permeabilities are:

LUGEON	*RANGE*	*CONDITION*
< 1	Low	Joints tight
1–5	Low/Mod.	Small joint openings
5–50	Mod./High	Some open joints
> 50	High	Many open joints

There is no unique relationship between Lugeon value and equivalent rock mass permeability (ke). Moye (1967) recommended use of the equation:

$$ke = \frac{QC}{LH} \tag{4.1}$$

where ke = the equivalent coefficient of permeability (m/sec)

Q = the flow rate (m^3/sec)

L = the length of the test section (m)

H = net head above the static water table at the centre of the test section (m)

$$C = \frac{1 + \ln(L/2r)}{2\pi} \tag{4.2}$$

r = the radius of the hole (m).

This is based on the assumption of radial laminar flow in a homogeneous isotropic rock mass, a condition seldom, if ever, achieved, but it is an approximation commonly used.

Hoek and Bray (1981) suggest the use of the equation:

$$ke = \frac{Q \ln(2mL / D)}{2\pi LH} \tag{4.3}$$

where $m = (ke/kp)^{1/2}$:

kp = equivalent permeability parallel to the hole
ke = equivalent permeability normal to the hole
D = diameter of hole.

They suggest that for most applications ke/kp is about 10^6. This implies no fractures parallel to the hole and in most rocks would not be a reasonable approximation.

If it is assumed that ke/kp = 10, and water pressure testing in 5 m lengths of NMLC hole (75 mm diameter), then 1 Lugeon is equivalent to ke = 1.6×10^{-7}m/sec.

For ke/kp = 1, i.e. homogeneous, isotropic conditions, m = 1 and 1 Lugeon is equivalent to ke = 1.3×10^{-7}m/sec. The Moye (1967) formula gives similar results to this homogeneous isotropic case.

4.5.7.1 Sequence of Testing in a Cored Borehole

In situ permeability tests should be carried out progressively as the borehole is cored/advanced using a single packer arrangement. Commonly a 3 metre or 6 metre length (i.e. one or two standard core runs) of hole is tested. For 6 metre test sections, if significant water takes are measured the packer can be relocated to test the lower 3 metres so the location of the high take can be better defined.

It is not good practice to drill the complete hole and then do water pressure tests using double packers or testing the complete hole in one test. With double packers, water may leak past the lower packer and not be observed. For a single test in a hole, the water losses are averaged over the complete hole so higher and lower permeability zones are not identified. The required progressive single packer method should be specified in the site investigations brief, as it is more time consuming and hence costly.

In recent years some consultants (sometimes at the request of owners) have opted to reduce the cost of investigations by doing either no *in situ* permeability testing or testing only sections of boreholes which they judge to be higher permeability based on the drill core. This is poor practice and should not be followed. It is not possible to tell which sections are high permeability from drill core and in any case knowledge of the full permeability profile is required to assist in selecting the foundation levels and depth of grouting for the dam.

4.5.7.2 Test Procedure

The test involves:

- Measurement of groundwater level.
- Washing out of drill hole. Circulation of drilling water should be continued for at least 15 minutes after the water appears clear.
- Installation of packer at the selected level.
- Connection of the water supply system.
- Application of the test pressures and measurement of water loss.
- Removal of equipment.

4.5.7.3 Test Pressures

Tests should be carried out at five pressures, e.g. rising from 33%, 66% and 100% of maximum pressure, then reducing to 66% and 33%. If this is done, jacking/hydraulic fracture can be detected and plotting these data allows suitable interpretation. Failure to adopt this method can result in significant errors in interpretation of results (e.g. a single point test that causes hydraulic fracture will significantly over-state the permeability).

The existing groundwater level should be recorded. For interpretation methods which use individual pressure data to estimate permeability, the groundwater level must be accounted for. For the 5-point method described earlier, it is not specifically required for interpretation of the results as detailed in Figure 4.15.

It is important to also calibrate the head loss through the packer at the start of the project. Failure to do so can significantly underestimate the Lugeon value where there are high flow rates, and in the case of very high flow rates it could well be the packer that is controlling the pressure and not the ground.

4.5.7.4 Maximum Test Pressures

Field practice suggests that to avoid hydraulic fracture (also known as "jacking"), the maximum test pressures should not exceed 22 kPa per metre at the test section. This is equivalent to 12 kPa per metre depth measured at surface level to account for the hydraulic pressure of the water column. The gauge height must be factored into the test pressure calculation at the test section, as does the topography (e.g. steep valley slopes) and inclination of the hole.

It is important to recognise that the horizontal stress in the sides of a valley may be less than the vertical stress calculated from the depth of the test section and the unit weight of the rock, so jacking may occur at pressures less than discussed earlier. This can be detected by a sudden increase in inflow rates with test pressure.

4.5.7.5 Presentation and Interpretation of Results

The results are best plotted with the effective test pressure at the centre of the test section against the flow rate. The effective test pressure is the gauge pressure corrected for the elevation difference between gauge and water table and for friction losses in the system. Where the water level is not known or suspected to be locally elevated due to the effects of the testing, the results are plotted using gauge pressure only. Typical results plotted in this way are shown in Figure 4.15. Note that the data from the increasing pressures do not plot identically to those on the decreasing pressures. The plot is judged to be parallel to the 6 Lugeon line.

Fell et al. (2005, 2015) provide guidance on how to best analyse the data and to check for the potential for jacking/fracturing of the rock mass as a result of the testing process.

4.5.8 Pressuremeters

Both the self-boring (SBPM) and Menard type pressuremeters which are inserted into a pre-drilled borehole have application in dam engineering, particularly in the assessment of the modulus of soil and weathered rock foundations. These pressuremeters give the most accurate estimate of modulus provided that the test method and interpretation are correct.

SBPM test results can also be used to estimate the *in situ* horizontal stress in the ground.

It should be recognised that when testing in rock the pressuremeter only tests a relatively small sample of the rock mass, so if the defects controlling the rock mass stiffness are widely spaced the test may give results which are higher than the stiffness of the rock mass.

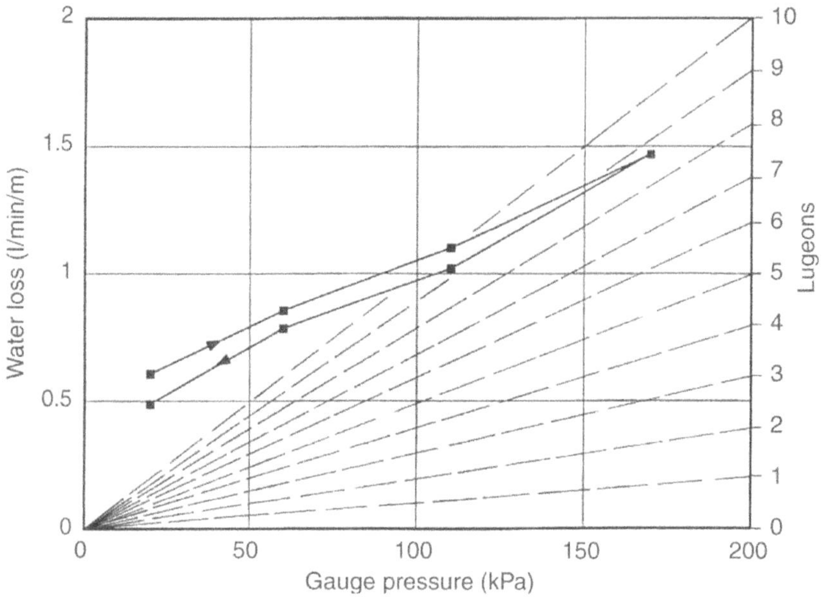

FIGURE 4.15 Typical results of packer test (Fell et al., 2015).

The tests are relatively expensive (in comparison to CPT/CPTU and SPT) but affordable on most large dam site investigations where the project requirements and site conditions warrant it.

4.6 GEOPHYSICAL METHODS

Table 4.3 shows the application and limitations of geophysical techniques.

The most commonly used surface techniques are Multi-Channel Analysis of Surface Waves (MASW), electrical resistivity and microgravity, and to a lesser extent Time Domain Electromagnetic Conductivity profiling (TDEM), Seismic Refraction Tomography (SRT) and seismic reflection. In the past seismic refraction was used extensively for investigation of dam foundations, and that data is still useful.

Of the techniques available, Ground Penetrating Radar (GPR) has been the most abused, i.e. used or suggested for use where it was not the most suited method. It should only be used where the physics of the situation are such that the technique may work.

It is recommended that advice be obtained from an engineering geophysicist familiar with the available techniques and their limitations on which methods (if any) should be used for the site investigation.

All geophysical methods require "ground truthing" or calibration of site conditions with borehole or test pit data. The interpretation of the geophysical testing should be carried out with that data available. Too often interpretation is done in isolation, and because of this the interpretations are not reliable.

TABLE 4.3 Application Geophysical Methods (ANCOLD, 2020)

TYPE OF SURVEY	*APPLICATIONS*	*LIMITATIONS*
Seismic refraction	Determines compression "P" wave and shear wave "S" wave velocities depending on the energy source used. Mostly "P" wave. From these can be interpreted depth to rock and properties of the overlying soil and weathered rock.	May be unreliable unless strata are thicker than a minimum thickness. Velocities must increase with depth; velocity inversions cannot be detected. Difficult to interpret if rock profiles are irregular or stepped. Information represents average values. "P" wave velocity of all saturated soils are the same, and to differentiate, "S" wave data is required.
Seismic refraction tomography (SRT)	As above but uses source and receptors in boreholes.	Allow detection of low velocity zones below high velocity zones.
Multi-channel analysis of surface waves (MASW)	Determines shear wave velocities based on dispersion of Rayleigh waves to depths up to 30 metres. Not constrained by requiring velocities to increase with depth.	Relatively sophisticated analysis which may produce complex results that sometimes cannot be explained by other subsurface data.
Seismic reflection	Produces geological stratigraphic and structural data using seismic waves similar to sound waves on shore or acoustic air blast off shore.	Difficult to distinguish the arrivals of the reflected waves from refracted and direct surface waves for shallow investigations. Mostly applicable where bedrock depth is > 30 m. Seldom useful for dam investigations as a result.
Self-potential (SP) profiling	Uses the current generated by flow of water (the streaming potential) in soils or into open cracks to locate seepage areas and concentrated flows, e.g. into open cracks in dams or reservoir. The only technique that directly indicates subsurface water flow.	Accuracy is affected by presence of ferrous objects. May need to be accompanied by a magnetic survey.
Electrical resistivity	Measure the resistivity of soil and fractured rocks to locate boundaries between materials with contrasting resistivity, e.g. between clean granular and clay strata, fractured and non-fractured rock, degree of saturation, position of water table and soil-rock interface. Resistivity imaging may locate cavities.	Sometimes it is difficult to interpret and subject to correctness of the assumed subsurface conditions; does not provide any indication of engineering strength properties. Accuracy affected by presence of ferrous objects. Correct selection of electrode array required to ensure reliable and robust results.

(*Continued*)

TABLE 4.3 (Continued)

TYPE OF SURVEY	*APPLICATIONS*	*LIMITATIONS*
Electromagnetic conductivity profiling (TDEM)	Locates boundaries between clean granular and clay strata; assessment of degree of saturation, position of water table and soil-rock interface. More rapid reconnaissance than electrical resistivity.	Sometimes difficult to interpret and subject to correctness of the assumed subsurface conditions; reliant on resistivity contrast; does not provide any indication of engineering strength properties. Accuracy affected by presence of ferrous objects.
Magnetic	Locates ferrous objects. Used in combination with self-potential and resistivity surveys.	Does not locate other metals.
Microgravity	May be used to delineate the boundary between alluvial soils and rock, including paleo-channels. Where seismic refraction methods cannot be used, e.g. in densely populated urban areas, locates small volumes of low density materials and larger cavities.	Use of expensive and sensitive instruments in rugged terrain may be impractical. Requires precise levelling.
Ground-penetrating radar	Provides a subsurface profile and may locate cavities, buried objects, boulders and soil-rock interface.	Success is site specific and depends on having a contrast in dielectric properties. It has limited penetration particularly in clay materials and laterites.

4.7 BOREHOLE IMAGING

Downhole imaging in cored boreholes allows the remote logging and determination of actual defect dip and dip direction, including bedding, joints, open defects, and the presence and orientation of crushed or sheared seams without the need for oriented core recovery. The drilling action may wash away crushed seams and shears and open defects, so they may not be detected in the drill core and may otherwise only be inferred indirectly through drill water loss or high takes in *in situ* permeability tests.

Table 4.4 summarises the methods, their application and limitations.

It is recommended that appropriate imaging should be specified for all important foundation investigations in rock. The added cost will be offset manyfold by the added and better quality data.

Optical and acoustic versions of televiewer are available, and in some cases, the data needed for analysis warrants both tools being run. Depending on the lithology, one method may be better suited than the other.

Optical methods (OTV) are best utilised in clean, cored holes where flushing of the drilling fluids has been completed.

Acoustic televiewer (ATV) is typically the most utilised. It relies on an image, or synthetic optical image, created by the response of the rock to a sonic input.

The tool can be used for defect detection and logging and geotechnical rock mass data collection. Acoustic imaging requires a liquid medium between the probe and the borehole wall so the technique will not work in dry holes.

For determination of the orientation of rock defects, the best results are achieved when televiewers are used in conjunction with downhole core orientation techniques. This allows for correlation of significant defects and provides a level of redundancy should one orientation method fail.

It should be noted that the interpretation of the imaging requires experienced persons, and there should be interaction between those interpreting the downhole logs and the engineering geologists logging the drill core from the holes. Engineering geologists should not rely on the geophysical contractor or geophysicists to recognise the origin of defects.

The use of televiewers is not a substitute for detailed logging of rock core, as it is still necessary to record geotechnical characteristics of natural defects (e.g. shape, roughness, infill, etc.). It is not possible to identify the geological origin and defect type confidently using the ATV and OTV alone. For example, it is often not possible to distinguish crushed seams (geological faults) from extremely weathered seams and infilled seams without carefully pulling them apart during the logging process, and it is not possible to distinguish joints from sheared surfaces without looking at the defect surface.

Experience shows that in fact the best results are achieved when the televiewer logs are compared directly to the drill core. Televiewer results are usually not available until after the initial logging of the core, and it is essential that the core is thoroughly reviewed when the televiewer results are available and that consultants and owners allow enough time for this review to be carried out.

It is strongly suggested that auto-pick algorithms (algorithms that process the image automatically looking for linear or sinuous features [i.e. defects or bedding]) are not utilised without checking against the drill core as the algorithms can produce a significant population of acoustic artefacts.

4.8 INSTALLATION OF PIEZOMETERS AND OTHER INSTRUMENTS INTO EXISTING EMBANKMENTS

Piezometers and other instruments such as inclinometers should only be installed in existing embankment dam cores if there is a known deficiency in the performance of the dam which requires investigation.

TABLE 4.4 Applications and Limitations of Borehole Logging Methods (ANCOLD, 2020)

METHOD	*PARAMETER MEASURED*	*APPLICATIONS*	*LIMITATIONS*
Ultra-sonic log	Determination of sonic velocities, attenuation of shear velocity in shear zones, or imaging the sides of the borehole wall	Shows fractures, other discontinuities.	Requires uncased hole. Image less clear than borehole camera.
Thermal profile	Temperature	Zones of inflow (lower/higher temperature water).	Open hole not necessarily required.
Caliper log	Borehole diameter	Infers lithology by borehole enlargement by erosion during drilling by drill fluid.	Requires an uncased hole.
Induction log	Electrical conductivity	Infers lithology by clay content, permeability, degree of fracturing.	Lower resolution than resistivity log, but can evaluate unsaturated zone and PVC cased boreholes.
Resistivity	Electrical resistivity	Infers lithology (particularly between sand and clay).	Applicable only in saturated zone; requires open hole.
Self-potential	Electrical potential from mineral reaction and groundwater flow	Infers lithology, oxidation/reduction zones, subsurface flows.	Applicable only in saturated zone; requires open hole. Data difficult to interpret.
Natural gamma	Natural gamma radiation	Infers presence of clay and shale.	Mud coating on borehole wall may affect results.
Gamma gamma	Natural density	Provides log of density, from which lithology may be inferred.	Provides only density. Health and safety issues.
Neutron	Moisture content (above water table). Porosity (below water table).	Log of moisture content, from which lithology or wetter zones may be inferred.	Provides only water contents. Health and safety issues.
Borehole camera or video, e.g. RAAX	Visual image of fractures and structure	Can identify structure. May be able to interpret dip of fractures and opening of joints, and bedding partings	Requires uncased hole. Images affected by water quality, smearing of sides of hole.
Acoustic reflectivity	Acoustic reflections	Can identify lithology and structure. May be able to interpret dip of fractures and orient cores.	Requires uncased hole.

This is because there are significant risks in installing instruments into the core of embankment dams because of the possibility of hydraulic fracture of the core during installation. If they must be installed use the drilling and hole backfill methods described in Sections 4.4.5 and Appendix B.

4.9 LABORATORY TESTING OF SOIL AND ROCK

4.9.1 Testing Soil

Laboratory tests of soil should be carried out according to Australian Standard AS 1289 or by ASTM standards or standards applicable in the country in which the dam is sited.

4.9.2 Testing Rock

Laboratory testing of rock should be carried out according to the relevant International Society for Rock Mechanics (ISRM) or ASTM standards or standards applicable in the country in which the dam is sited.

4.9.3 Hole Erosion Tests

For testing the erosion properties of soils for assessment of internal erosion and piping by concentrated leak erosion, it is recommended to follow the procedures described in Wan and Fell (2004a, 2004b). Corrections should be applied for entry and exit losses as described in Bonelli et al. (2013). Testing should include specific testing for critical shear stress by varying the head. The critical shear stress should not be estimated only from correlations with the erosion rate index.

Details are provided in Section 9.13.6.

4.9.4 All Laboratory Tests

It is important to use laboratories which have the required equipment and technical expertise to do the tests with experience of testing for dam investigations. Many laboratories do not have that expertise, particularly for rock testing.

Refer to Chapter 6 of Fell et al. (2015) for a discussion on soil testing methods, their application and common problems.

Hydrologic and Seismic Hazard Assessment

5

5.1 INTRODUCTION

To assess the safety of an existing dam it is necessary to know the hydrologic and seismic loads the dam is subject to.

5.1.1 Hydrologic and Seismic Loads for Quantitative Risk-Based Assessments

Hydrologic load: Annual Exceedance Probability (AEP) of reservoir levels under normal operating and flood conditions allowing for the reliability of gated operation where present.

Seismic Load: AEP of the ground motions at the dam site.

For risk-based assessments, loads for the full range of AEPs are considered.

5.1.2 Hydrologic and Seismic Loads for Deterministic Assessments

5.1.2.1 Dam Classification

For deterministic assessments, the hydrologic and earthquake design loads are related to the dam classification, which is determined by the consequences of failure of the dam. These terms are defined as follows:

Classification: A system of assigning dams to categories, usually on the basis of the consequences of failure, so that the appropriate dam safety standards can be applied.

Consequences of failure: Impacts on the downstream area of a dam as a result of failure of the dam or its appurtenances. The term consequence refers to life loss or environmental, cultural and economic losses above and beyond that which would have occurred in the same event or conditions had the dam not failed.

The CDA (2013) dam classification is shown in Table 5.1. We have adopted this classification throughout this book.

DOI: 10.1201/9781003542025-5

TABLE 5.1 CDA (2013) Dam Classification (Published with the permission of the Canadian Dam Association)

DAM CLASS	*POPULATION AT RISK [NOTE 1]*	*INCREMENTAL LOSSES*		
		LOSS OF LIFE [NOTE 2]	*ENVIRONMENTAL AND CULTURAL VALUES*	*INFRASTRUCTURE AND ECONOMICS*
Low	None	0	Minimal short-term loss No long-term loss	Low economic losses; area contains limited infrastructure or services
Significant	Temporary only	Unspecified	No significant loss or deterioration of fish or wildlife habitat Loss of marginal habitat only Restoration or compensation in kind highly possible	Losses to recreational facilities, seasonal workplaces and infrequently used transportation routes
High	Permanent	10 or fewer	Significant loss or deterioration of *important* fish or wildlife habitat Restoration or compensation in kind highly possible	High economic losses affecting infrastructure, public transportation and commercial facilities
Very High	Permanent		Significant loss or deterioration of *critical* fish or wildlife habitat Restoration or compensation in kind possible but impractical	Very high economic losses affecting important infrastructure or services (e.g. highway, industrial facility, storage facilities for dangerous substances)
Extreme	Permanent	More than 100	Major loss of *critical* fish or wildlife habitat Restoration or compensation in kind impossible	Extreme losses affecting critical infrastructure or services (e.g. hospital, major industrial complex, major storage facilities for dangerous substances)

Note 1: ***Definitions for Population at Risk:***

None: There is no identifiable Population at Risk, so there is no possibility of loss of life other than through unforeseeable misadventure.

Temporary: People are only temporarily in the dam-breach inundation zone (e.g. seasonal cottage use, passing through on transportation routes, participating in recreational activities).

Permanent: The Population at Risk is ordinarily located in the dam-breach inundation zone (e.g. as permanent residents); three consequence classes (high, very high, extreme) are proposed to allow for more detailed estimates of potential loss of life (to assist in decision making if the appropriate analysis is carried out).

Note 2: ***Implications for Loss of Life:***

Unspecified: The appropriate level of safety required at a dam where people are temporarily at risk depends on the number of people, the exposure time, the nature of their activity, and other conditions. A higher class could be appropriate, depending on the requirements. However, the design flood requirement, for example, might not be higher if the temporary population is not likely to be present during the flood season.

5.1.2.2 Hydrologic Load

The selection of the Inflow Design Flood (IDF) is usually related to the dam classification. These are typically based on national, state or regional guidelines or regulations. Table 5.2 shows an example prepared by the Canadian Dam Association (CDA, 2013).

TABLE 5.2 An example of Inflow Design Flood and Earthquake Design Ground Motion for Deterministic Assessments (CDA, 2013; published with the permission of the Canadian Dam Association)

DAM CLASS [NOTE 1]	*ANNUAL EXCEEDANCE PROBABILITY—FLOODS [NOTE 2]*	*ANNUAL EXCEEDANCE PROBABILITY—EARTHQUAKES [NOTE 3]*
Low	1/100	1/100
Significant	Between 1/100 and 1/1000 [note 4]	Between 1/100 and 1/1000
High	1/3 between 1/1000 and PMF [note 5]	1/2475 [note 6]
Very High	2/3 between 1/1000 and PMF [note 5]	1/2 between 1/2475 and 1/10,000 or MCE [note 5]
Extreme	PMF [note 5]	1/10,000 or MCE [note 5]

This table addresses two major natural hazards only and does not consider the many other types of hazard that must be considered in dam safety assessments.

Acronyms: PMF, Probable Maximum Flood; AEP: annual exceedance probability; MCE: maximum credible earthquake
Note 1: As defined in Table 2.1, Dam Classification (Section 2.5.4; Table 5.1 in this book).
Note 2: Simple extrapolation of flood statistics beyond 10 AEP is not acceptable.
Note 3: Mean values of the estimated range in AEP levels for earthquakes should be used. The earthquake(s) with the AEP as defined in Table 6.1B (Table 5.2 in this book) is then input as the contributory earthquake(s) to develop the EDGM parameters as described in Section 6.5 of these (CDA) guidelines.
Note 4: Selected on basis of incremental flood analysis, exposure, and consequences of failure.
Note 5: PMF and MCE have no associated AEP.
Note 6: This level has been selected for consistency with seismic design levels given in the National Building Code of Canada.

5.1.2.3 *Seismic Load*

The Earthquake Design Ground Motion (EDGM) is related to the dam classification. Table 5.2 shows an example also by CDA.

The dam safety authorities in the country or state in which the dam is located may have such deterministic criteria.

The authors favour risk-based approaches to evaluating dam safety. These may be Simplified QRA or QRA. For both it is required that the reservoir levels and seismic loads are partitioned as described in the following sections.

CDA Dam Safety Technical Bulletin—Hydrotechnical Consideration for Dam Safety (2007) explains how to do the interpolation for high and very high classification dams as follows:

> "For the Low and Significant classes, if the reservoir is small, the flood peak is the only characteristic required. However, for medium and large reservoirs, it is necessary to determine the peak, volume and hydrograph for the IDF.
>
> *For the High and Very High classes, the methodology recommended to compute the IDF hydrograph is as follows:*
>
> - Compute the peak, volume and hydrograph for the 1/1000 year flood.
> - Compute the PMF.
> - Plot both floods on one graph, adjusting the time origins of the floods to both peaks coincide; For each time step compute the IDF discharge with equation (1).

Equation (1)

$$Q_{IDF} = Q_{1/1000} + C^*(Q_{PMF} - Q_{1/1000})$$

Where:

Q_{IDF} *is the discharge of the IDF for the selected time step.*
$Q_{1/1000}$ *is the discharge of the 1/000 year flood for the selected time step.*
Q_{PMF} *is the discharge of the PMF for the selected time step.*
C equals 1/3 for High Consequence class and 2/3 for Very High Consequence class."*

5.2 ASSESSMENT OF HYDROLOGIC AND SEISMIC HAZARD

In many cases there will be existing reports on hydrologic and seismic hazards. These should be reviewed to determine if they are adequate for the safety assessment. The methods for assessing hydrologic and seismic hazard are evolving with time, and older assessments often have used outdated methods or data so may need updating.

The review and any updated hazard assessments should be carried out by hydrological and seismic specialists with local experience.

5.3 HYDROLOGIC LOAD AND PARTITIONING OF THE RESERVOIR LEVELS

The hydrologic load on the embankment is obtained by the routing of several floods ranging from low to high annual probability of occurrence through the spillway to obtain the AEP of reservoir levels. Figure 5.1 shows an example.

If the floods result in overtopping of the embankment, it is preferable that the flows over the embankment are modelled because they can be significant and affect the AEP of levels reached. If flow over the embankment crest is modelled, the analyst should check the actual crest levels based on survey as there may be localised low points in the crest level or camber remaining.

When the dam has spillway gates, the flood routing analyses should consider the various combinations of spillway gates failing to operate on demand and include this within the risk analysis based on a gate reliability assessment.

For QRA of internal erosion and piping and slope instability, the reservoir levels should be partitioned to coincide with the following:

- Full Supply level (FSL) or the top of active conservation level.
- Pool of Record (POR) level, which is the maximum level the reservoir has reached during its operation.
- Changes in embankment design such as top of core elevation. For example, at 1 in 1000 Annual Exceedance Probability (AEP) flood in Figure 3.8a; and at 1 in 50 AEP flood in Figure 3.8b.
- Embankment crest level.
- Geological features which occur above a particular level in the foundation (e.g. a highly permeable gravel layer).

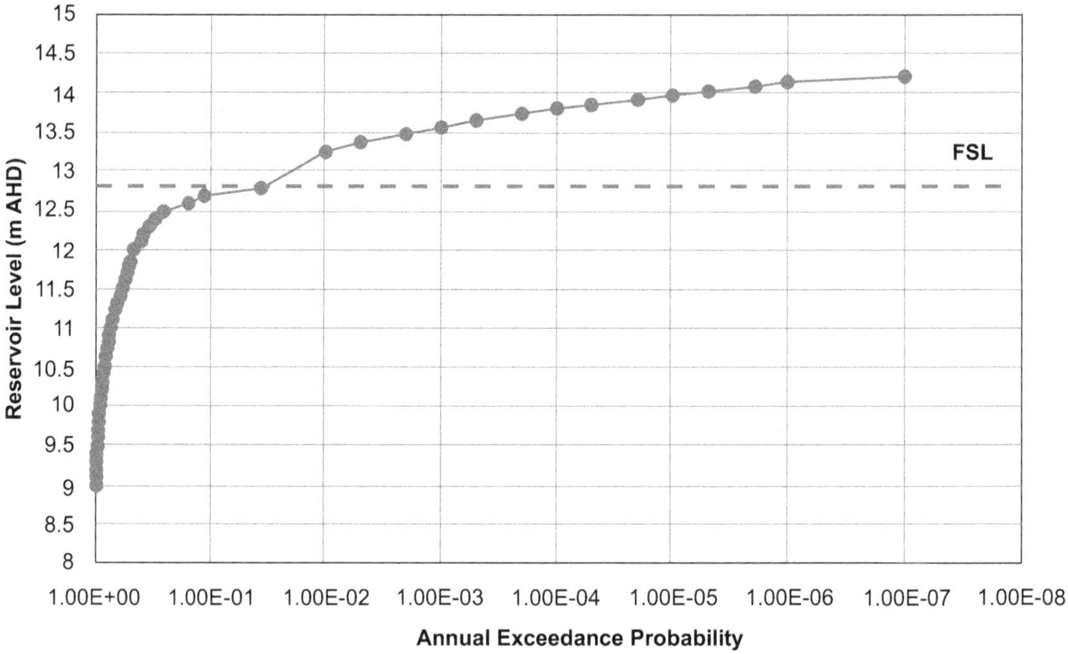

FIGURE 5.1 An example of reservoir level versus annual exceedance probability.

- Topographic features such as major changes in foundation profile if these are above the pool of record.
- Reservoir level where stability analyses indicate marginally low FOS values are predicted to occur.

For flood control dams it will probably be necessary to partition at, for example, 1 in 100 AEP and 1 in 1000 AEP flood levels even if these do not coincide with design features. For flood control dams which have not yet experienced a flood reaching the reservoir spillway crest level this may be used as a partition boundary.

Table 5.3 shows an example of reservoir level partitioning.

For assessing the likelihood of failure resulting from overtopping of the embankment, the data requirements will be:

- AEP of the dam crest flood.
- AEP of reservoir levels above dam crest level, for example 0.15 m, 0.3 m, 0.5 m and 1.0 m above dam crest level.
- Hydrographs of overtopping flows for each depth above dam crest level, or as a minimum, the duration of overtopping.

These partitions and those for earthquake loads are used by some to develop system response curves.

TABLE 5.3 Example of Reservoir Level Partitions

RESERVOIR LEVEL	*SIGNIFICANCE*	*ANNUAL EXCEEDANCE PROBABILITY OF THE RESERVOIR LEVEL*	*RESERVOIR LEVEL PARTITION RANGE*	*PROBABILITY/ANNUM THE RESERVOIR LEVEL IS IN THIS RANGE*
R_{MIN}	Minimum pool level of record (or "minimum pool of record")	1.0		
			R_{MIN}–R_{TOC}	0.5
R_{TOC}	Top of active conservation level (or "full pool")	0.5		
			R_{TOC}–R_{POR}	0.48
R_{POR}	Pool of record level (or "historic high reservoir level")	0.02		
			R_{POR}–R_{ECL}	0.0199
R_{ECL}	Embankment crest level (or "dam crest flood")	0.0001		
			$> D_{ECL}$	0.0001
TOTAL				1.0

5.4 EARTHQUAKE LOAD PARTITIONING

Evaluate the earthquake ground motion, e.g. peak ground acceleration and earthquake magnitude contribution versus annual exceedance probability for the site.

Figures 5.2 and 5.3 are examples.

The seismic hazard should be evaluated taking into account historic earthquakes and the potential influence of active faults in the vicinity of the dam, and the shear wave velocity of the rock at the dam site. The hazard should be evaluated for earthquakes with a $M_W \geq 5.0$. Smaller magnitude earthquakes will cause minimal damage to an embankment dam.

The potential amplification or de-amplification of ground motions by soils overlying the rock in the foundation should be assessed separately, not by including these in the shear wave assessment.

ANCOLD (2019) provides details of guidance of these issues. National, state or regional guidelines in which the dam is situated may apply.

Partition the loads to form a table as for the example shown in Table 5.4.

The earthquake peak ground acceleration partition values may be selected to coincide with the damage class contours given in Figure 8.2 for earth dams and Figure 8.3 for earthfill and rockfill dams.

If the foundation of the dam is subject to liquefaction the partitions may be selected to coincide with the PGAs for no or very little liquefaction; an intermediate extent of liquefaction; and extensive liquefaction.

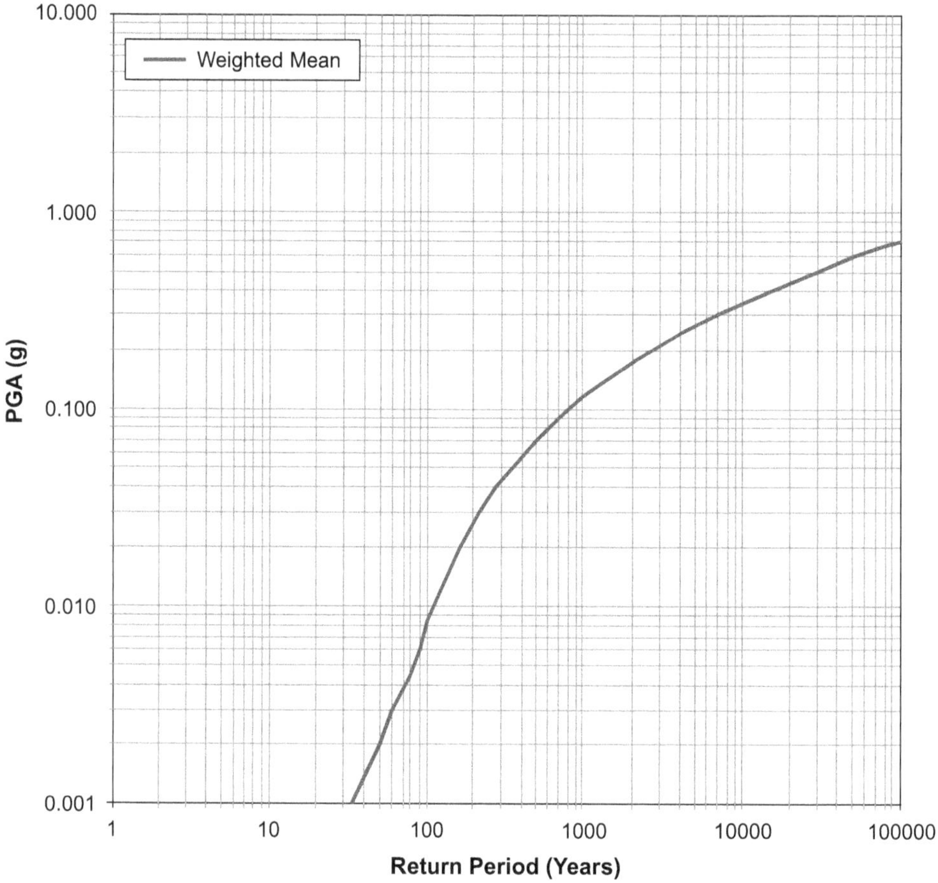

FIGURE 5.2 Example of peak ground acceleration versus AEP.

For more detailed analyses of deformations under seismic loads, earthquake time history accelerograms will be required. These should be selected by the seismologist in consultation with the risk analyst team.

5.5 CONSIDERATION OF EARTHQUAKE AFTERSHOCKS

Consider whether the dam, or appurtenant structures critical for the operation of the dam post-earthquake, are likely to be sufficiently damaged by the main earthquake ground motion, so as to be more vulnerable to further damage by earthquake aftershocks. In such a case, seek advice from the seismologist on the likely magnitude, focal location and seismic ground motions at the dam site and analyse the consequent effects on the dam and appurtenant structures.

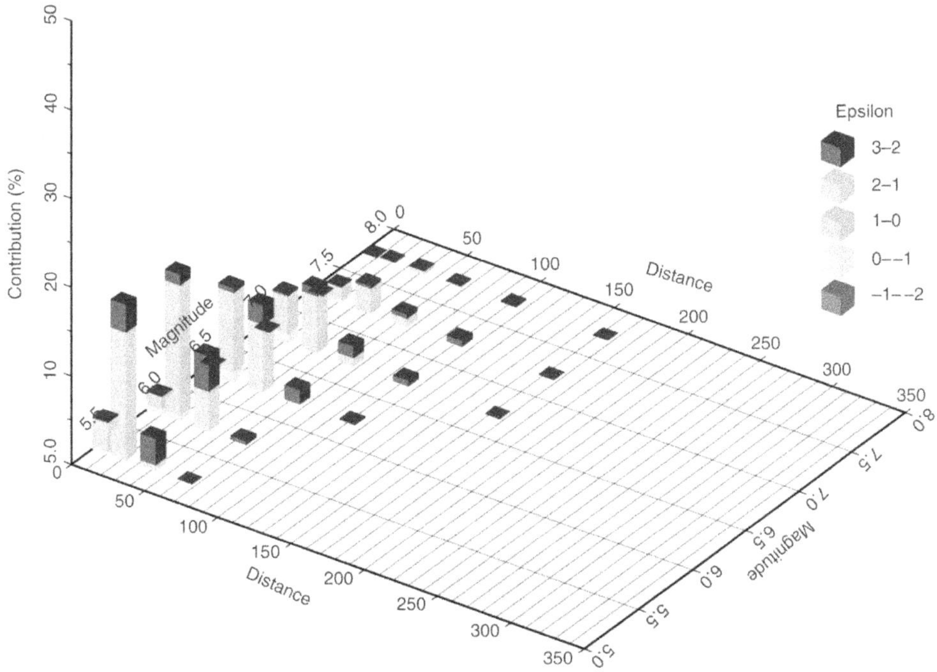

FIGURE 5.3 Example of magnitude-distance contributions to the seismic hazard.

TABLE 5.4 Example of Earthquake Load Partitions

EARTHQUAKE PEAK GROUND ACCELERATION	*REPRESENTATIVE EARTHQUAKE MAGNITUDE M_W*	*ANNUAL EXCEEDANCE PROBABILITY OF THE EARTHQUAKE LOADING*	*PGA IN PARTITION RANGE*	*PROBABILITY/ANNUM THE LOADING IS IN THIS STAGE*
			0 to 0.10 g	0.99
≤ 0.10 g	6.0	0.01		
			0.01 g to 0.20 g	0.009
0.20 g	6.0	0.001		
			0.20 g to 0.30 g	0.0009
0.30 g	7.0	0.0001		
			> 0.30 g	0.0001
TOTAL				1.0

It should also be considered how the safety of the dam will be managed in the period through the aftershocks until repairs can be completed. This may require interim repairs and increased surveillance.

An example of a dam that might be susceptible to aftershock seismic ground motions is a concrete face, asphalt face or other membrane face rockfill dam that has poor drainage capacity. Thus, the rockfill could get saturated from the leakage through the damaged face slab or membrane following the main earthquake, leaving it with a low static factor of safety and susceptible to much larger deformations and damage during the aftershock.

Concrete dams and appurtenant structures, such as walls retaining the embankment that may experience significant displacement during the primary ground motion (which are sufficient to compromise the foundation drainage or passive or post-tensioned anchors), should be considered for aftershocks.

Estimating the Likelihood of Failure by Overtopping or Throughflow of the Embankment

6

6.1 INTRODUCTION

There are three hydraulic scenarios:

H1. Overtopping of an embankment occurs when the reservoir level is above embankment crest level.
H2. Wave overtopping occurs when wave action on the reservoir results in intermittent flow of water over the crest.
H3. Throughflow occurs when the reservoir level is above the top of the core and water flows through a rockfill zone above the core.

Figures 6.1 and 6.2 show examples of overtopping and throughflow.

When an embankment dam is overtopped, the water flowing over the crest and downstream slope potentially initiates erosion, which may progress upstream to erode the embankment and result in uncontrolled release of the reservoir, i.e. breach.

Figure 6.3 shows typical hydraulic conditions during embankment overtopping. The maximum or "terminal" velocity is reached within the supercritical flow area and is often within a few metres of the crest.

The mechanism of erosion and breach formation depends on the embankment zoning:

6.1.1 Earthfill Embankments

Initiation: Erosion is initiated by surface erosion on the downstream slope near the crest.

Progression: Progression occurs by head cutting. The head-cuts advance upstream.

Commencement of breach: If the flow depth and duration are sufficient, the head cutting reaches the reservoir and breach initiates.

DOI: 10.1201/9781003542025-6

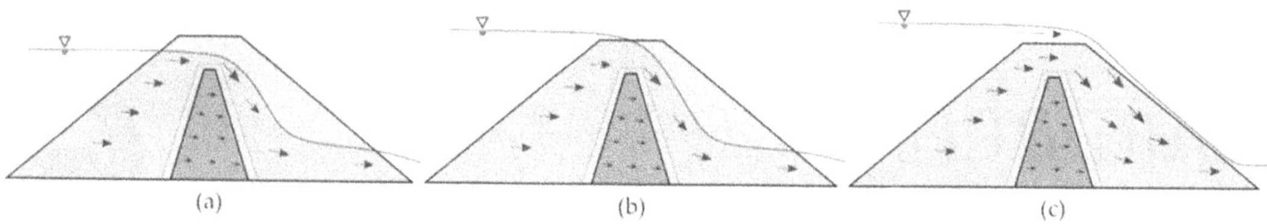

FIGURE 6.1 (a) Throughflow above the core; (b) partial overtopping of the crest with throughflow; (c) overtopping of the crest with throughflow and potential overflow of the rockfill (Dezert et al., 2022).

During test 1-03, overtopping.

During test 1-03, overtopping. Incipient failure.

FIGURE 6.2 Photographs of field scale tests on 6 m high embankments. (a) Throughflow in Test 4-02 Homogeneous uniform rockfill 300 mm to 400 mm compacted in 1 m layers. (b) Overtopping in Test 1-03 Rockfill with a moraine core. Rockfill well graded 0 to 500 mm compacted in 1 m layers (EBL, 2005).

Figure 6.4 shows this process in a test embankment subject to overtopping. The critical stages for defining failure by overtopping are Stages 1 and 2 (head-cut formation and head-cut advance). Once the head-cut progresses to the upstream face (Stage 3), the breach process initiates and progresses to breach (Stage 4).

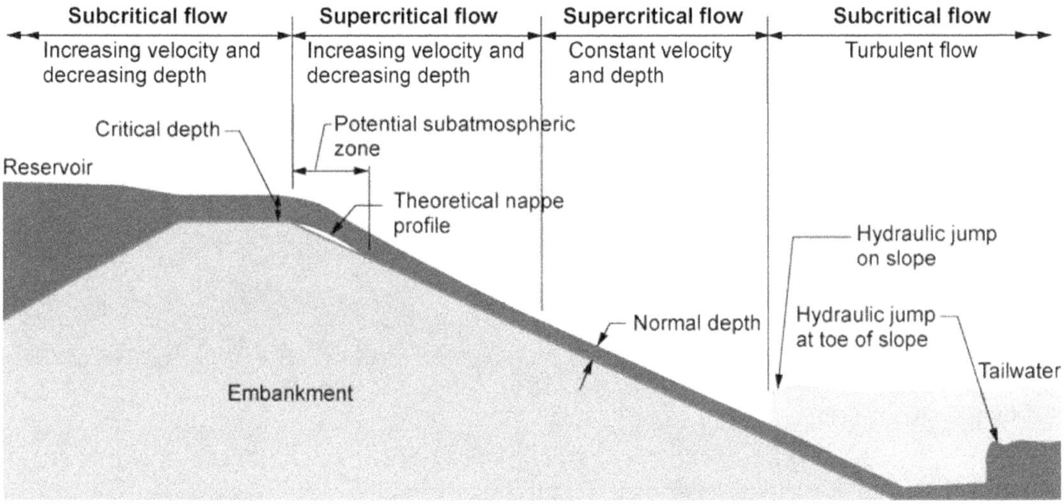

FIGURE 6.3 Typical hydraulic conditions during overtopping of an earthfill embankment (FEMA, 2014).

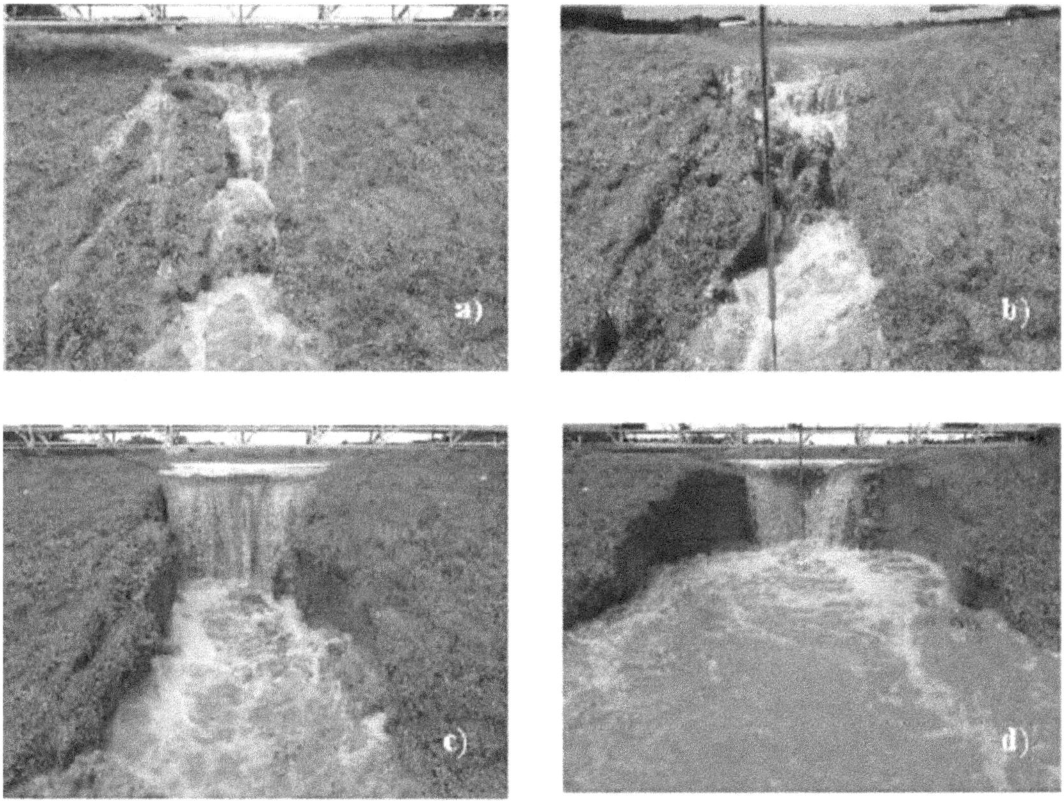

FIGURE 6.4 Stages of head-cut progression in an earthfill embankment subject to overtopping. (a) Rills and cascade of small overfalls during Stage 1; (b) consolidation of small overfalls during Stage I; (c) head-cut at downstream crest, transition from Stage 1 to Stage 2; (d) head-cut at upstream crest; (e) and (f) transition from Stage 3 to Stage 4 at breach formation (Hanson et al., 2003).

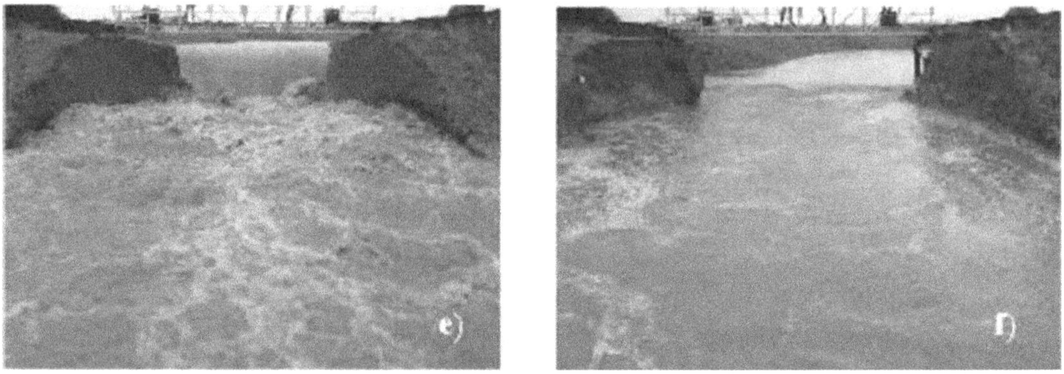

FIGURE 6.4 (Continued)

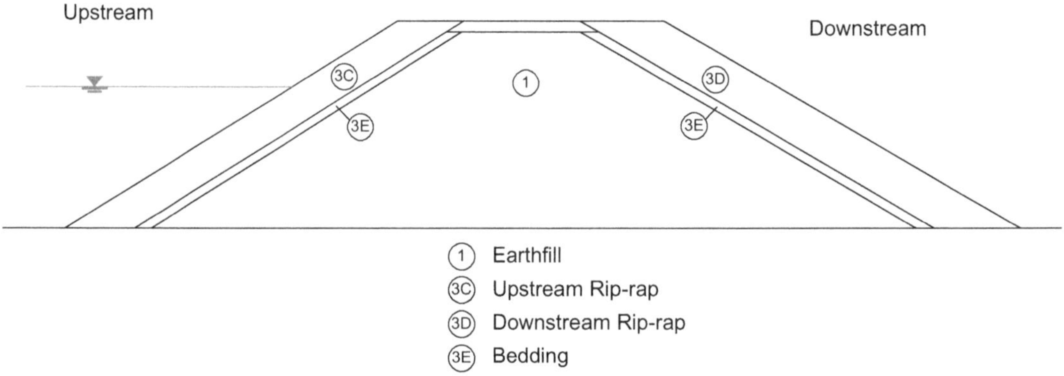

FIGURE 6.5 Earthfill embankment with downstream rip-rap slope protection.

6.1.2 Earthfill Embankments Which Have Rip-Rap Protection on the Downstream Slope

Figure 6.5 shows an earthfill embankment with downstream rip-rap protection.

Initiation: The rip-rap protects the earthfill from erosion unless flows over the crest are sufficient to result in removal or dislodgement of enough rocks to expose the bedding material under the rip-rap. If the flow is sufficient to fill the voids in the rip-rap, it will slide off the slope.

Progression: The bedding material being smaller size than the rip-rap erodes to expose the earthfill. Then initiation of erosion and head cutting of the earthfill progresses as described previously.

Commencement of breach: If the flow depth and duration are sufficient, the head cutting reaches the reservoir and breach initiates.

6.1.3 Earth and Rockfill Embankments

Initiation: As shown in Figure 6.1b and c, the flow over the crest and in the rockfill layer on top of the core enters the downstream rockfill zone and emerges lower down towards the toe of the embankment. Depending on the flow, size of the rockfill and the downstream slope of the embankment individual rocks are destabilised.

If the rockfill vertical permeability is insufficient to allow all the flow over the crest to enter the rockfill, there will be flow over the downstream slope.

In both cases initiation of erosion is in the form of unravelling.

For throughflow, Figure 6.1a, the mechanism of initiation is similar if the downstream rockfill is sufficiently permeable. If it is not, then flow may emerge high up the downstream slope and initiation may include unravelling, slope instability or internal erosion and piping.

Progression: Unravelling advances upstream and may stabilise if the downstream face reaches a stable slope to resist unravelling. If flows are sufficient, unravelling advances further upstream. If the rockfill becomes saturated, it becomes unstable, and the downstream slope begins to slide. Figure 6.2b shows an example.

Commencement of breach: If the sliding and unravelling progresses to encompass the crest of the embankment and the upstream slope to the reservoir, then breach commences.

6.1.4 Structure of the Chapter

Section 6.2 details the overtopping scenarios for all embankment types.

Section 6.3 describes the method for estimating wind-generated wave heights and run-up.

Section 6.4 describes methods for how flows resulting from overtopping and wave overtopping are estimated.

Sections 6.5, 6.6, 6.7 and 6.8 describe the engineering analyses methods for each type of embankment and how to assess the safety of the embankment using deterministic Simplified QRA and Quantitative Risk Assessment methods.

Section 6.9 discusses overtopping and wave overtopping of parapet and wave walls.

Section 6.10 discusses the estimation of the likelihood of detection and intervention in QRA methods.

The development of the breach and resulting dam-break hydrograph are not covered in this book. Readers should refer to USACE (2017b), Reclamation (2017) and Bowles et al. (2014) for guidance on those aspects.

6.2 OVERTOPPING SCENARIOS

Several factors affect the methods to be used for assessing the likelihood of failure of embankment dams by overtopping:

6.2.1 Hydraulic Loading

H1, H2 or H3 as discussed earlier.

6.2.2 Embankment Zoning

(Z1) Earthfill embankments with no rip-rap downstream slope protection (Types 0, 1, 2, 3, 6 and 8 in Figure 2.1).

(Z2) Earthfill embankments with rip-rap downstream slope protection (Types 0, 1, 2, 3, 6 and 8 in Figure 2.1).

(Z3) Zoned earth and rockfill and central core earth and rockfill (Types 4 and 5 in Figure 2.1).
(Z4) Earthfill with core wall (Type 9 in Figure 2.1).
(Z5) Concrete face rockfill (Type 7 in Figure 2.1).
(Z6) Rockfill with core wall (Type 10 in Figure 2.1) and asphalt core rockfill (Type 11, Figure 2.1).

For overtopping, Z4 may be considered as a variant of Z1 and Z2, and Z6 as a variant of Z3.

6.2.3 Crest Level

The level of the crest of the embankment is as follows:

(a) The level of the upper surface of the embankment including any road pavement.
(b) The top of a parapet wall or wave wall provided the wall is designed to withstand the load from the reservoir up to the top of the wall. If the wall is not so designed, then it is as for (a).

It is important to use actual crest levels. These may vary from the design drawings because of built in camber, post-construction settlement and/or reprofiling of the crest.

6.3 CHARACTERISTICS OF WIND-GENERATED WAVES ON THE RESERVOIR, AND ESTIMATION OF WAVE RUN-UP

6.3.1 Introduction

The wave overtopping failure mode (hydraulic scenario H3) requires an engineering assessment of the size of wind-generated waves and wave run-up that could occur during the design wind event.

For a deterministic assessment, a design wind event is selected based on state or national guidelines or codes, and this event is assumed to occur at the maximum reservoir level for the Inflow Design Flood event. For a risk-based safety assessment, a range of wind events should be evaluated, and the probability of the wind event occurring in an adverse direction and at the same time as the maximum reservoir level is estimated and included in the failure probability estimate.

The following outlines the method described in USBR (2012). The method is applicable to dam reservoirs and is the preferred approach for assessing whether a dam has sufficient freeboard to prevent significant overtopping flows. The procedures are too lengthy to include here other than in summary form, and for detailed design or detailed assessment of important structures readers should refer to the original text.

6.3.2 Fetch

In reservoirs, fetches are limited by the land surrounding the body of water. The shorelines are irregular, and an effective fetch is calculated by constructing nine radials from the point of interest on the dam, at 3 degree intervals as shown in Figure 6.6. The length of each radial is measured and arithmetically averaged as in Equation 6.1.

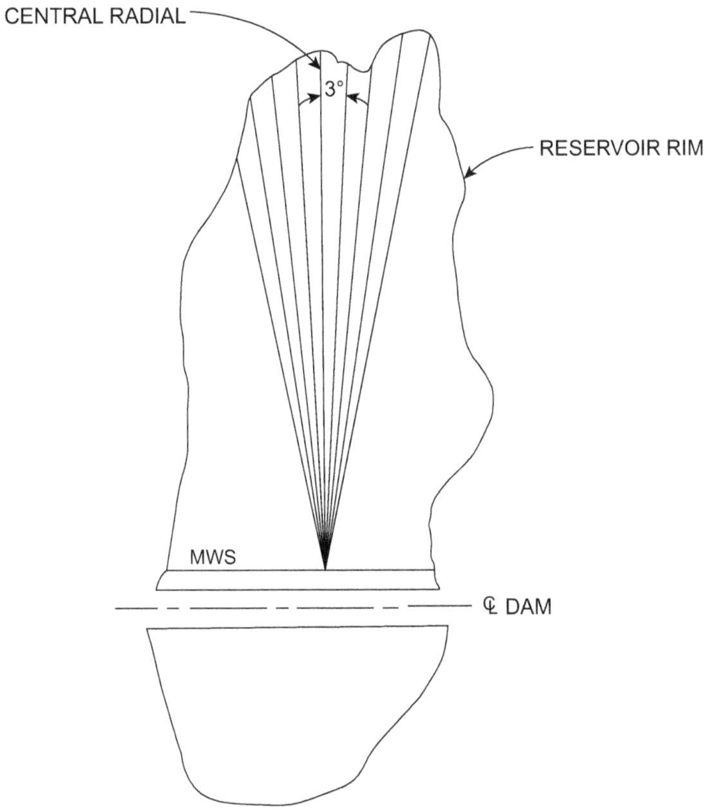

FIGURE 6.6 Method for calculating effective fetch (Fell et al., from USBR, 2012).

This calculation should be performed for directions where the central radial is normal to the dam axis and where the total spread results in the longest possible set of radials. For each fetch calculated, the angle of the central radial in respect to the dam axis should be determined. This will be used to adjust the wave run-up.

$$\text{Fetch} = \sum_{1}^{n} \left(\text{length of radials}\right) / \left(\text{number of radials}\right) \tag{6.1}$$

6.3.3 Design Wind

Design wind velocity estimates should be obtained from the Bureau of Meteorology or equivalent organisation. This may be done for a range of wind durations and annual likelihood of occurrence.

The critical wind duration depends on the fetch; longer durations are critical for longer fetches. The estimates should allow for local topographic effects and, if known, the likely wind direction for high wind velocities.

Generally, design velocities are controlled by the fetch. The duration needed for a given wind velocity to generate the highest waves for a given fetch is designated the minimum duration which can be estimated from:

$$t_{min} = 1.6\left[\left(F\right)^{0.67} / \left(V_W\right)^{0.34}\right] \tag{6.2}$$

where:

t_{min} = minimum wind duration to generate maximum wave height in hours
F = fetch in km
V_W = wind velocity over water in km/h.

The estimates should be adjusted from the over land wind velocity to over water wind velocity by applying the correction factors shown in Figure 6.7. In Figure 6.7 U_L is the wind velocity over land and U_W is the wind velocity over water.

USBR (2012) describes how these data may be used to estimate the annual probability of over water wind velocities being exceeded.

In some cases, it may be reasonable to consider the predominant wind directions relative to the fetch direction when deciding upon the appropriate wind velocity to use for the determination of whether the crest will be overtopped. Factors to think about here would include the size of the storage and the expected timing of the peak water level during a flood. In small storages, where thunderstorms would be the likely reason for the inflow flood and the peak water level would be reached very quickly, the critical wind should be assumed as coming from any direction.

6.3.4 Wave Height

For the assessment of the adequacy of freeboard, i.e. the difference between dam crest level and reservoir level, the significant wave height in metres (Hs), which is the average of the highest one-third of the waves in the wave spectrum, should be used. Hs can be estimated from Equation 6.3.

$$Hs = 0.00366\,(F)^{0.5}\,V_W\,(1.1 + 0.01\,V_W)^{0.5} \quad (6.3)$$

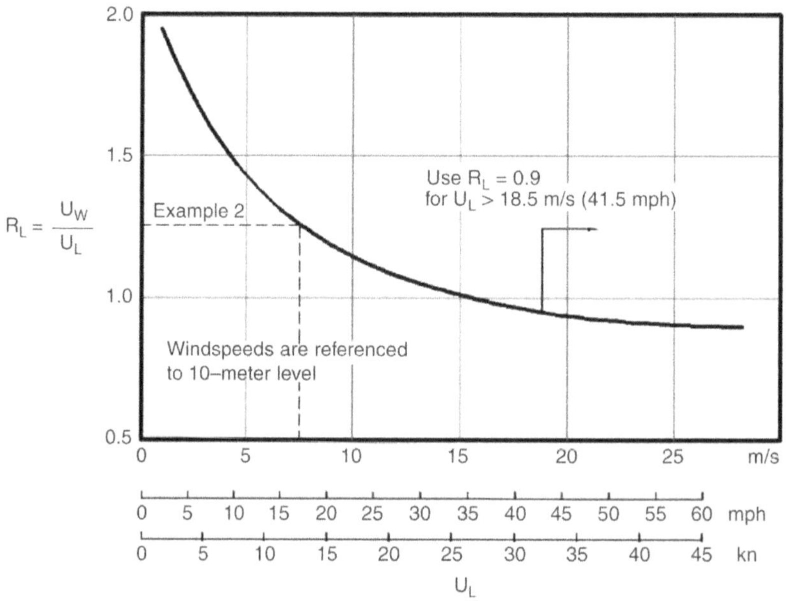

FIGURE 6.7 Ratio of wind velocity over water (U_W) to wind velocity over land (U_L) as a function of wind velocity over land (Fell et al., from USBR, 2012).

where:

Hs = significant wave height in metres (= H_{m0} in Section 6.3.6)
F = fetch in km
V_W = wind velocity over water in km/h.

This equation is valid for most reservoirs where the reservoir is relatively deep compared to the wind-generated wave length (i.e. water depth > 0.5 L, where L is the deep water wave length calculated in Equation 6.4).

For shallow reservoirs it is suggested that USACE (2002, 2011) Part VI be used.

Eurotop (2018) is used in Sections 6.3.6 and 6.5 for estimating wave run-up and overtopping flows. In those sections the symbol H_{m0} is used for the significant wave height Hs.

6.3.5 Wave Length and Wave Period

USBR (2012) indicate that the deep water wave length L (in metres) can be computed from Equation 6.4:

$$L = 1.56\ T^2 \tag{6.4}$$

where T is the wave period determined from:

$$T = 0.335\ F^{0.33}\ V_W^{0.33}\ (1.1 + 0.01\ V_W)^{0.167} \tag{6.5}$$

Note: In Sections 6.3.6 and 6.5, $L = L_{m0}$ and $T = T_{m-1.0}$ in Section 6.3.6.

6.3.6 Wave Run-Up from Eurotop (2018)

Eurotop (2018) provides detailed methods and supporting information for estimating wave run-up and overtopping flows. What is presented here and in Section 6.4.2 is a summary. Readers should refer to the original report which is freely available to download.

Figure 6.8 shows wave breaking, run-up and overtopping on an embankment.

Figure 6.9 shows the definitions of run-up and freeboard.

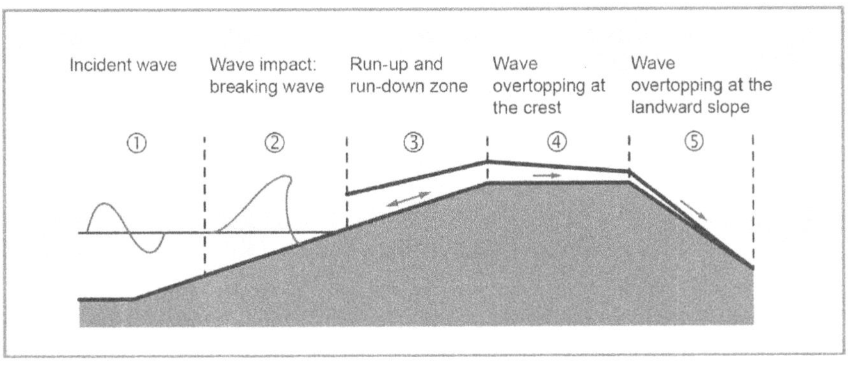

FIGURE 6.8 Wave breaking, run-up and overtopping on an embankment (Eurotop, 2018).

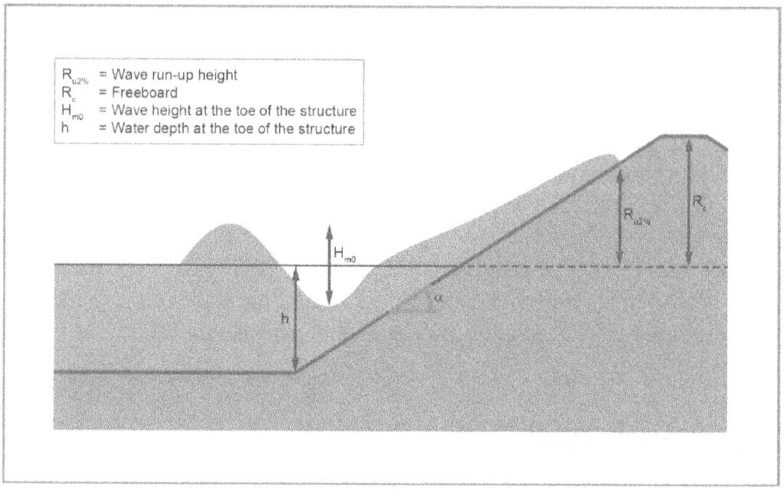

FIGURE 6.9 Definition of wave run-up ($R_{u2\%}$), freeboard (R_c) and upstream slope angle (α) (Eurotop, 2018).

6.3.6.1 Run-Up for Smooth Upstream Slopes

Eurotop (2018) indicates that for upstream slopes 2H:1V and flatter, the mean value wave run-up for which no more than 2% of the waves will be exceeded is given by:

$$\frac{R_{u2\%}}{H_{m0}} = 1.65.\gamma_b.\gamma_f.\gamma_\beta.\xi_{m-1.0} \tag{6.6}$$

with a maximum of:

$$\frac{R_{u2\%}}{H_{m0}} = 1.0.\gamma_f.\gamma_\beta\left(4 - \frac{1.5}{\sqrt{\left(\gamma_b.\xi_{m-1,0}\right)}}\right) \tag{6.7}$$

where:

$R_{u2\%}$ = wave run-up height exceeded by 2% of the waves (m)
H_{m0} = significant wave height (m)
γ_b = influence factor for a berm, = 1.0 if no berm
γ_f = influence factor for roughness of the upstream slope
γ_β = influence factor for oblique wave attack
$\xi_{m-1.0}$ = breaker parameter or surf similarity factor

And

$$\xi_{m-1.0} = \tan\alpha/(H_{m0}/L_{m-1.0})^{0.5} \tag{6.8}$$

where:

$L_{m-1.0}$ = deep water wave length (m)

$$= g\,(T_{m-1.0})^2/2\pi = 1.56\,(T_{m-1.0})^2 \tag{6.9}$$

and $T_{m-1.0}$ = spectral wave period (seconds)

g = acceleration due to gravity = 9.81 m/s^2.

6.3.6.2 Run-Up on Upstream Slopes Steeper than 2H:1V Including Slopes with Rip-Rap Slope Protection

For upstream slopes steeper than 2H:1V Eurotop (2018) indicates that wave steepness has no influence, and the mean run-up is given by:

$R_{u2\%}/H_{m0} = 0.8 \cot\alpha + 1.6$ with a minimum of 1.8 and a maximum of 3.0.

They recommend that for design and assessment the run-up be assessed from:

$R_{u2\%}/H_{m0} = 0.86 \cot\alpha + 1.71$ with a minimum of 1.93 and a maximum of 3.21.

This is the mean plus one standard deviation.

Eurotop (2018) and USBR (2012) provide details of influence factors. These are summarised in Table 6.1:

6.3.6.3 Influence Factor for Roughness of the Upstream Slope (γ_f)

TABLE 6.1 Values of Influence Factor γ_f for Structures (Data from Eurotop, 2018, and USBR, 2012)

UPSTREAM SLOPE SURFACE	*INFLUENCE FACTOR FOR ROUGHNESS OF THE UPSTREAM SLOPE* γ_F
Smooth impermeable surface e.g. concrete, asphalt	1.0
Smooth blocks	1.0
Grass (3 cm in length)	0.9–1.0
Rocks (1 layer, impermeable core)	0.6
Rocks (1 layer, permeable core)	0.45
Rocks (2 layers, impermeable core)	0.55
Rocks (2 layers, permeable core)	0.40

6.3.6.4 Influence Factor for Oblique Wave Attack (γ_β)

As shown in Figure 6.10, the angle of wave attack is defined as the angle between the direction of propagation of waves and the axis perpendicular to the structure. For perpendicular wave attack $\beta = 0°$.

For Smooth Impermeable Upstream Surfaces

For wave run-up:

$$\gamma_\beta = 1 - 0.0022\beta \text{ for } 0 \le \beta \le 80° \qquad (6.10)$$
$$\gamma_\beta = 0.824 \text{ for } \beta > 80°$$

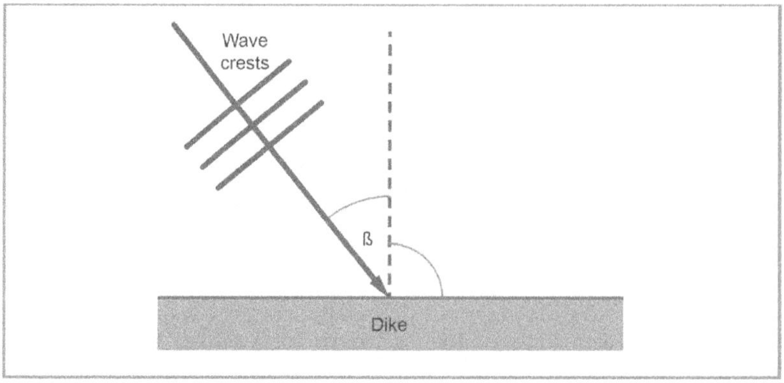

FIGURE 6.10 Definition of the angle of wave attack (β) (Eurotop, 2018).

For wave overtopping:

$$\gamma_\beta = 1 - 0.0033\beta \text{ for } 0 \leq \beta \leq 80° \quad (6.11)$$

$$\gamma_\beta = 0.736 \text{ for } \beta > 80$$

For rough upstream surfaces

$$\gamma_\beta = 1 - 0.0063\beta \text{ for } 0 \leq \beta \leq 80° \quad (6.12)$$

For $\beta > 80°$ the result for $\beta = 80°$ can be applied.

6.3.6.5 Influence Factor for Presence of an Upstream Berm (γ_b)

See Eurotop (2018). If there is no berm, $\gamma_b = 1.0$.

6.4 ESTIMATION OF THE FLOWS RESULTING FROM OVERTOPPING, WAVE OVERTOPPING AND THROUGHFLOW

6.4.1 Flows Resulting from Overtopping

Figure 6.11 shows the variables used in the calculation of flows, depths and velocities.

The overtopping flow, Q, can be estimated using the broad crested weir equation:

$$Q = CLH^{1.5} \quad (6.13)$$

The rate of overtopping flow per unit length of dam, q, is given by:

$$q = \frac{Q}{L} = CH^{1.5} \quad (6.14)$$

FIGURE 6.11 Variables used in the calculation of flows, depths and flow velocities (Ostrovsky and Wan, 2023).

Critical flow depth, Y_C, on the embankment crest is:

$$Y_C = \left(\frac{q^2}{g}\right)^{\frac{1}{3}} \tag{6.15}$$

Flow depth (brink depth), Y_1, and velocity of flow, V_1, at the downstream edge of the embankment crest are:

$$Y_1 = \frac{Y_c}{1.4} \tag{6.16}$$

$$V_1 = \frac{q}{Y_1} \tag{6.17}$$

Flow depth Y_2 and terminal velocity of flow, V_2 , on the downstream slope are:

$$Y_2 = \left(\frac{qn}{S^{1/2}}\right)^{3/5} \tag{6.18}$$

$$V_2 = \frac{q}{Y_2} \tag{6.19}$$

where in the prior equations:

C = weir coefficient
H = overtopping head (total) above embankment crest (m)
L = embankment length (m)
g = gravitational acceleration (m/s^2), $g = 9.81$ m/s^2
S = downstream slope of embankment
n = Manning's roughness factor.

The weir coefficient C = 1.7 based on Eurotop (2018) and Hewlett et al. (1987).

Hewlett et al. (1987) recommend use of Figure 6.12 which relates Manning's roughness factor to the downstream slope.

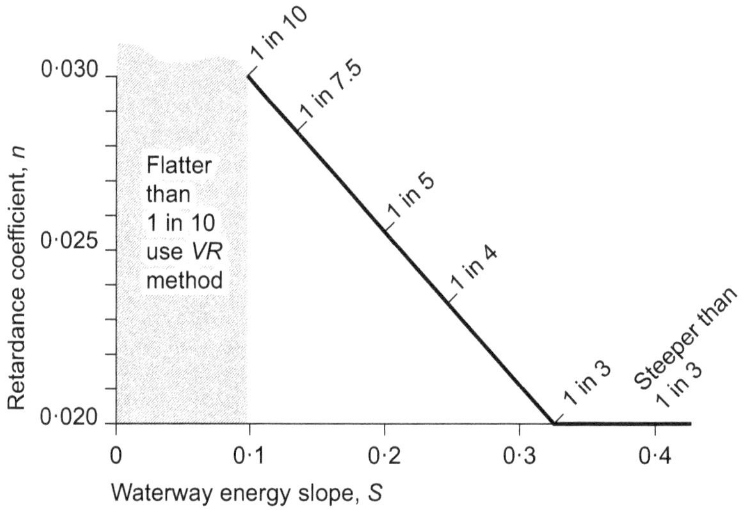

FIGURE 6.12 Recommended Manning's *n* values for grassed slopes steeper than 1 in 10. (Hewlett et al., 1987; reproduced with permission of CIRIA through PLSclear.)

6.4.2 Flows Resulting from Overtopping by Waves

If the assessment in Sections 6.3 and 6.4.1 indicate that waves are likely to overtop the embankment, assess the overtopping flow using the procedure detailed next, which is a summary of the method described in Eurotop (2018).

Readers should refer to Eurotop (2018) for details of the method.

The mean value approach method for estimating the average overtopping flow is:

$$\frac{q}{\sqrt{\left(g \cdot H_{m0}^{3}\right)}} = \frac{0.023}{\sqrt{tan \propto}} \gamma_b \cdot \xi_{m-1.0} \cdot \exp\left[-(2.7 \frac{R_c}{(\xi_{m-1.0} \cdot H_{m0} \cdot \gamma_b \cdot \gamma_f \cdot \gamma_\beta \cdot \gamma_\nu)})^{1.3}\right] \tag{6.20}$$

with a maximum of

$$\frac{q}{\sqrt{\left(g \cdot H_{m0}^{3}\right)}} = 0.09 \cdot \exp[-[1.5 \frac{R_c}{(\xi_{m-1.0} \cdot H_{m0} \cdot \gamma_f \cdot \gamma_\beta \cdot \gamma^{*})})^{1.3}] \tag{6.21}$$

where:

q = mean overtopping discharge per metre structure width (m^3/s per m)
g = gravitational acceleration = 9.81 m/s^2
α = upstream slope angle (degrees)
R_c = freeboard (m)
H_{m0} = wave height (m)
γ_b = influence factor for a berm, = 1.0 if no berm
γ_f = influence factor for roughness of the upstream slope
γ_β = influence factor for oblique wave attack
γ_v = influence factor for a vertical wall on the upstream crest or slope

γ^* = influence factor for non-breaking waves on steep slopes where there is a wall on the upstream slope or crest

$\xi_{m-1.0}$ = breaker parameter or surf similarity factor.

See Section 6.3.6 for guidance on estimation of the influence factors. See Eurotop (2018) for estimating γ_v and γ^*. Both = 1.0 if there are no walls.

If $\xi_{m-1.0} < 2$, waves are considered to be breaking, and if > 2, non-breaking.

Eurotop (2018) recommends that for design or deterministic assessment the following be used to allow for uncertainty in the estimates:

$$\frac{q}{\sqrt{\left(g \cdot H_{m0}^{3}\right)}} = \frac{0.026}{\sqrt{tan \propto}} \gamma_b \cdot \xi_{m-1.0} \cdot \exp\left[-(2.5 \frac{R_c}{(\xi_{m-1.0} \cdot H_{m0} \cdot \gamma_b \cdot \gamma_f \cdot \gamma_\beta \cdot \gamma_v)})^{1.3}\right] \tag{6.22}$$

with a maximum of

$$\frac{q}{\sqrt{\left(g \cdot H_{m0}^{3}\right)}} = 0.1035 \cdot \exp\left[-(1.35 \frac{R_c}{(\xi_{m-1.0} \cdot H_{m0} \cdot \gamma_f \cdot \gamma_\beta \cdot \gamma^*)})^{1.3}\right] \tag{6.23}$$

6.4.3 Flows Resulting from Throughflow in Rockfill Over the Top of the Embankment Core

This is scenario H3 shown in Figure 6.1(a). Throughflow may also occur in earthfill dams which have a layer of rockfill above the core as shown in Figure 6.13(a) or earth and rockfill dams which have low permeability rockfill in the downstream shoulder as shown in Figure 6.13(b).

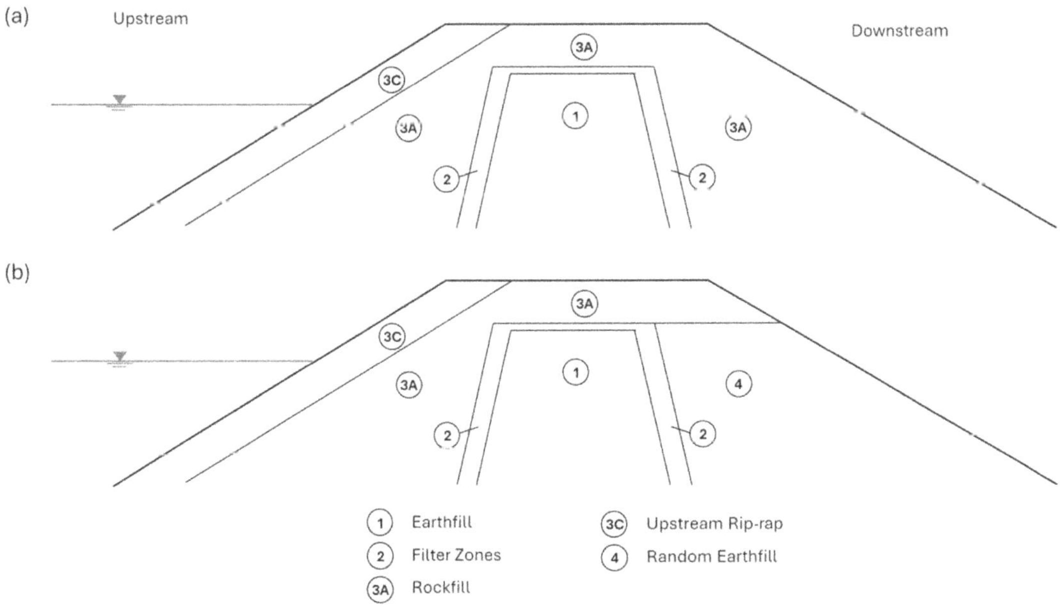

FIGURE 6.13 (a) Earth and rockfill dam with rockfill over the core and rockfill downstream shoulder. (b) Earth and rockfill dam with rockfill over the core and with low permeability downstream shoulder.

Flows and seepage flow nets should be modelled by carrying out seepage analysis of the embankment using suitable parameters.

The estimates of throughflow rate will be dependent on the geometry and zoning of the crest, the permeability and the permeability ratio k_H/k_V of the rockfill above the core, and the permeability and permeability ratio k_H/k_V of the downstream shoulder.

6.4.3.1 Permeability of Rockfill

There are several approaches to estimating the permeability of the rockfill:

6.4.3.2 EBL (2005) Method to Estimate Rockfill Permeability

EBL (2005) indicates that flow in rockfill may be turbulent and discusses how the turbulent permeability of rockfill may be estimated:

$$k_t = \left(n^3 \cdot g \cdot d_t\right) / \left[\beta_0 \cdot (1-n)\right] \tag{6.24}$$

where:

k_t = turbulent permeability (m^2/sec^2)
n = porosity
g = gravitational acceleration = 9.81 m/s^2
β_0 = particle shape factor = 3.6 for quarried or crushed rock
d_t = characteristic dimension of the rockfill = 1.7 d_{10} where d_{10} = particle size of rockfill for which 10% is finer (m).

EBL (2005) back-analysed measured flow in a 6 m high field scale model for turbulent flow using CFD program FLUENT and for laminar flow using PLAXIS. For the model constructed of rockfill 30 mm to 500 mm size, the matching permeability coefficients were 6.5 to 8.1 × 10^{-3} m^2/sec^2 for turbulent and 1.0 to 1.5 × 10^{-1} m/sec for laminar flow. They concluded that the relationship between these permeabilities fitted reasonably well using the relationship:

$$v = q / a = k_{lam} \cdot i = \left(k_t \cdot i\right)^{0.5} \tag{6.25}$$

where:

v = mean velocity (m)
q = flow (m^3/ sec)
a = area (m^2)
k_{lam} = laminar permeability (m/sec)
k_t = turbulent permeability (m^2/sec^2)
i = hydraulic gradient.

EBL (2005) found that the two models gave similar piezometric surfaces. As most rockfill will have lower permeability than the EBL (2005) example, the potential errors in use of conventional seepage analysis methods are likely to be within accepted values. If estimates of Reynolds number indicate turbulent flow conditions, then CFD modelling using the turbulent permeability might be used.

From the previous it is suggested that the laminar permeability of coarse rockfill with small finer fraction may be estimated using Equation 6.24 to estimate the turbulent permeability and Equation 6.25 to convert that estimate to an equivalent laminar permeability. To do this the seepage gradient relevant to the analysis should be used.

6.4.3.3 Measured Permeabilities in Dam Rockfill

Table 6.2 presents the results of permeability tests in rockfill dams. Most are from concrete face rockfill dams where there were concerns regarding the drainage capacity, so they represent a somewhat biased sample. These may be used as a guide when estimating permeabilities.

TABLE 6.2 Permeabilities Measured in Rockfill Dams

ROCK TYPE	*LAYER THICKNESS (M)*	*HORIZONTAL PERMEABILITY (M/SEC)*	*UNCONFINED COMPRESSIVE STRENGTH OF ROCK (UCS) (MPA)*
Mangrove Creek Dam Field Trials			
Fresh Siltstone	0.6	6×10^{-4} to 2.0×10^{-3} m/sec	25
Fresh Siltstone Minus 150 mm size	0.3	$1.x\ 10^{-5}$ m/sec	25
Fresh Sandstone[a]	0.6	1×10^{-7} m/sec	22
Weathered Siltstone	0.6	$3.x\ 10^{-4}$ m/sec	
Mangrove Creek Dam During Construction			
Zone 1B Intermixed Fresh and Weathered	0.6	Max 1.2×10^{-4}, Min 2.3×10^{-9}, Mean 1×10^{-5} m/sec	
Zone 1B Drainage Zone Fresh Siltstone	0.6	Max 2×10^{-4}, Min 3.9×10^{-6} Mean 3×10^{-5} m/sec	45 to 64
Winneke Dam			
Weathered Metamorphics	0.5 to 0.6	1×10^{-5} m/sec	4 to 24
Fresh Metamorphics	0.7 to 0.9	5×10^{-3} m/sec	70 to 100
Salvajina Dam			
Siltstone and Sandstone	0.9	3×10^{-4} m/sec	n/a
Shuibuya Dam			
Limestone	0.8	$> 1 \times 10^{-3}$ m/sec	55 to 75
Rossvatn Test Dam 1A-01[b]			
Fresh Metavolcanics	1.0	1.3×10^{-1} m/sec	
Kangaroo Creek Dam[c]			
Slightly Weathered Gneiss and Schist	1.2	3.7×10^{-4} m/sec	20 to 35
Clarrie Hall Dam Field Trials [d]			
Zone 1—MW Rhyolite (D_{90}= 50 to 120 mm, D_{10} = 1 to 2 mm)	0.4 to 0.8	1 to 5×10^{-3} m/sec	40 to 50 (dry)
Zone 2—MW Rhyolite (D_{90} 150 to 600 mm, D_{10} 3 to 7 mm)	1.0 to 1.3	4×10^{-3} to 2×10^{-2} m/sec	40 to 50 (dry)
Zone 3—SW to FR Rhyolite (D_{90} 700 mm, D_{10} > 30 mm)	0.7 to 1.2	2×10^{-2} to 1×10^{-1} m/sec	120 to 130 (dry)

Notes:

(a) Fresh sandstone broke down to silty sand under rolling.

(b) Estimated from measured throughflow and by back-analysis (EBL, 2005).

(c) Estimated from particle size distributions using the EBL (2005) method. Breakdown occurred on the surface of each layer resulting in 0.05 to 0.15 m of silt sand material with low permeability.

(d) Zone 1 compacted with 4 to 6 passes of an 8 tonne smooth drum vibrating roller (SDVR), Zones 2 and 3 compacted with 4 to 6 passes of a 16 tonne SDVR, 0% to 30% water added (by volume).

6.4.3.4 Ratio of Horizontal and Vertical Permeability in the Rockfill

Rockfill is usually placed in the dam by dumping from haul trucks, spread by bulldozers into layers 0.6 m to 1.2 m thick, and compacted by vibratory steel drum rollers. This may result in segregation of coarser particle towards the bottom of each layer. As shown in Figure 6.14, this also results in a layer of finer crushed rock on the surface of each layer.

The thickness and permeability of this crushed layer depend on the strength of the rock and the susceptibility to break down during compaction. This is often related to the rock type, UCS and the degree of weathering. Slightly weathered and fresh rock is usually less susceptible to crushing than more weathered rock.

The authors' experience is the following:

- Granitic rock and dolerite are least susceptible to crushing.
- Volcanics such as basalt, andesite, dacite, trachyte and rhyolite are more variable. If the rock is high strength with widely spaced joints, such as columnar basalt, its susceptibility to crushing is low. If lower strength and more closely jointed, it is moderately susceptible.
- Pyroclastic deposits are characterised by extreme variability in engineering properties over short distances laterally and vertically. They range from extremely low density "collapsing" type soils to extremely strong rocks. Their susceptibility to crushing is directly related to this.
- Schistose rocks including slate, phyllite and schist are highly susceptible to crushing.
- Mudrocks, claystone, siltstone, mudstone and shale are generally highly susceptible to crushing. Some high strength siltstones are less susceptible.
- Sandstone, greywacke and conglomerate are generally highly susceptible to crushing. Some of high strength and/or subject to metamorphism are less susceptible.
- Limestone varies from extremely susceptible to crushing to moderately susceptible depending on the strength and/or degree of metamorphism.
- The UCS strength is a significant factor—generally:
  UCS > 70 to 100 MPa: minimal breakdown on surface.
  UCS < 50 to 60 MPa: do get some breakdown and formation of an upper layer with excess fines.
  UCS < 20 to 30 MPa: significant breakdown during compaction forming a soil like surface layer.

Photographs taken during construction of the dam are often a useful guide to performance.

As a guide, the crushed layer thickness is likely to be about 0.05 m for the least susceptible rocks, and up to 0.15 m for the most susceptible.

The permeability of the crushed layer is likely to be about 0.1 k_H for least susceptible rocks, and 0.01 to 0.001 k_H for the most susceptible.

The suggested approach to estimating the equivalent permeability ratio k_{HE}/k_{VE} of rockfill is the following:

(a) Estimate the thickness of the crushed layer ($T_{Crushed}$)
(b) Estimate the likely ratio of the permeability of the crushed layer to the horizontal permeability of the layer ($k_{Crushed}/k_H$) and from this $k_{Crushed}$
(c) Calculate the equivalent horizontal and vertical permeabilities, k_{HE} and k_{VE} from Equations 6.26a and 6.26b
(d) Calculate the permeability ratio of the layer of rockfill k_{HE}/k_{VE}

FIGURE 6.14 Placement of rockfill in Kenyir Dam showing segregation and formation of layer of crushed rock on the surface of each layer. (Courtesy of SMEC.)

$$k_{HE} = [(T_{Crushed}) \times (K_{Crushed}) + (T_{Layer} — T_{Crushed}) \times (K_{Horizontal})] / T_{Layer} \tag{6.26a}$$

$$k_{VE} = [T_{Layer}]/[(T_{Layer} — T_{Crushed})/(K_{Horizontal})]+[(T_{Crushed})/(K_{Crushed})] \tag{6.26b}$$

For example, for a layer thickness = 1.0 m, crushed thickness = 0.1 m, and crushed permeability 1/10th of the equivalent horizontal permeability, $k_{HE} = 0.91\ k_H$ and $k_{VE} = 0.52\ k_H$ so $k_{HE}/k_{VE} = 1.75$.

If the crushed permeability = 1/100th of the horizontal permeability, $k_{HE}/k_{VE} = 9.9$, say 10.

6.4.3.5 *Sherard et al. (1984) Formula for Estimating Permeability*

Sherard et al. (1984) carried out laboratory permeability tests on 15 different filter sands and sandy gravels with D_{15} in the range from 0.1 to 10 mm. The materials were mostly broadly graded and with little or no fines.

From these tests they concluded that the permeability could be estimated from:

$$k = C\ (D_{15})^2 \tag{6.27}$$

where: k = permeability in m/sec, D_{15} = particle size for which 15% is finer in mm, C = 0.0035 with a range of median test values of C = 0.002 to C = 0.006 for densely compacted filters (70% relative density).

Sherard et al. (1984a) found that the permeability of the same materials placed very loose were higher by a ratio in the range 1.1 to 3.0 but up to 13.7 for a very broadly graded silt sand gravel soil.

This approach may be used within the limits of the tests it was developed on, i.e. sand-gravel materials with $D_{15} \le 10$ mm.

6.5 FAILURE BY OVERTOPPING AND WAVE OVERTOPPING OF TYPE Z1 EARTHFILL EMBANKMENTS

6.5.1 Introduction

As described in Sections 3.1.2 and 6.1 and shown in Figure 3.2, when an earthfill dam is overtopped, erosion initiates on the upper part of the downstream slope and progresses by head cutting, and breach initiates when the head-cut reaches the reservoir.

There are several approaches to estimating the likelihood of failure by overtopping:

Database, based on analysis of case data. These methods incorporate all three stages of the process.
Simplified QRA, using simplified methods to model each of the three stages.
Quantitative, using computer simulation of the three stages.

There are a number of published methods available for each of these. What follows are what the authors feel are the most appropriate.

6.5.2 Database Method

The database method is based on the method reported in Jacobs (2016) and Maslin and Rodd (2016).

For that study a total of 31 cases of earthfill embankment dams, which had been subject to overtopping and had either breached or not breached but experienced minor or major erosion on the downstream slope, were assessed. Much of the data was from Powledge et al. (1989). The table of estimated probability of breach was developed in a workshop using expert elicitation taking account of the case data and the mechanics of erosion.

For this book, additional cases from Gilbert and Miller (1991) and Druyts and Van Den Berg (1988) have been added, giving a total of 40. Added detail for the cases has been obtained from Gilbert and Miller (1991), Flynn et al. (2021) and Ostrovsky and Wan (2023) and the data in the original database corrected as required.

Table 6.3 summarises the case data. It can be seen that:

- Breach occurred with overtopping durations as low as 2 h, even 1 h.
- The database is all clay soils except that Spitskop had semi-pervious shoulder and Tygerpoort was clay soil but with poor compaction which explains why it breached despite only 0.3 m maximum depth and a short duration of overtopping.
- There is a high probability of breach if water depth > 0.5 m. Ten of 15 breach cases had this. There are only three "No breaches" and those had overtopping durations ≤ 2 h.
- A water depth 0.3 to 0.5 m, say 0.4 m, is more likely to give moderate damage than minor if breach does not occur.

There is insufficient data to assess the effect of the downstream slope, the plasticity of the soil and whether there is grass cover.

TABLE 6.3 Data on Earthfill Embankments Which Have Been Subject to Overtopping of the Crest

EMBANKMENT	HEIGHT (M)	DOWN-STREAM SLOPE	MATERIAL	MAXIMUM OVERTOPPING DEPTH (M)	DURATION[c] (HR)	EXTENT OF DAMAGE	ESTIMATED EROSION RATE Kd[f] CM³/N.SEC
R. D. Bailey cofferdam, WV	18	1.5H:1V	Well-graded sand and sandstone Rolled rockfill	2.0	4[b]	Breach	0.5
Bloomington cofferdam, WV	9	2.5H:1V	Clayey sand gravel	1.5	10	Breach	0.25
Rainbow Lake, MI	14	4H:1V	Sandy silty clay	0.45 to 0.6	14	Breach	0.6
Oros, Brazil[e]	30	?	Core CL	0.8	12–18	Breach	0.35
Armando de Salles, Brazil[e]	33	2H:1V	Homogeneous earthfill	1.2	0.3[i]	Breach	25
Euclides de Cumba[e]	61	?	Homogeneous earthfill	1.3	7.5	Breach	0.4
Clarence Canyon Cofferdam	14	3H:1V	Stiff clay	1.5	20	Breach	0.15
McCarty, TX	16	2H:1V	Sandy clay	?	Few to 24	Breach	?
Little Blue River Levee, MO	4.6	3H:1V	CL-CH (vegetated)	0.25	20[d]	Breach	0.9
Jacksonport Levee, AR	4.6	?	Clay	0.3–0.6	1 week	Breach	< 0.1
Kaskasia Island Levee[g]	7	3H:1V	Clay with silt (CL?)	0.3 to 0.6	6 to 12	Breach	1
Chaq-Chaq[g] Iraq	15	2H:1V	Earthfill	0.5 to 0.6	1.0 to 1.5	Breach	7
Cheaha Crk Watershed No. 6[g]	7	2.5H:1V	Earthfill	1.5	3	Breach	0.9
Macchu II[g] India	26	2H:1V	Earthfill	0.6	2	Breach	4
Ogayarindotameike[g] Japan	19	2H:1V	Earthfill	0.76	4	Breach	1.5
Spitskop[g] South Africa	17	2H:1V	Semi-pervious d/s	0.3	4	Breach	3.5
Tygerpoort[g] South Africa	19	2H:1V	Earthfill, poor comp	0.3	1	Breach	20[j]
Hebgen[h]	37	2H:1V	Earthfill	0.9	0.17	No breach	> 50[h]
McLean Bottom Levee[g]	6	3H:1V	Silt clay (CL?)	0.3 to 0.6	12–24	No breach	< 0.6
Steinbachtal[g] Germany	17	2.4H:1V	Earthfill	0.3	3	No breach	< 5
Test Embankment 1, soil type 3	2.3		Medium plasticity clay	0.45	20	Moderate	< 0.5
Test Embankment 2, soil type 3	1.5		Medium plasticity clay	0.3	70	Moderate	< 0.2
Awonga Saddle Dam	3		Clayey gravel	0.4	12	Moderate	< 0.8
Irene Court	1.5	2H:1V?	Clayey gravel	?	?	Moderate	
Rocky Run Dam, IA	6		Loess	0.5	5	Moderate	< 2
Rocky Run Protection Levee, IA	2.7		Loess	0.6	6	Moderate	< 1

(Continued)

TABLE 6.3 (Continued)

EMBANKMENT	*HEIGHT (M)*	*DOWN-STREAM SLOPE*	*MATERIAL*	*MAXIMUM OVERTOPPING DEPTH (M)*	*DURATION[c] (HR)*	*EXTENT OF DAMAGE*	*ESTIMATED EROSION RATE Kd[l] CM^3/N.SEC*
Elm Fork, Tex. Site 9	11	2H:1V	Clay (vegetated)	0.12	3	Moderate	< 10
Colorada Ret. Structure W-1	15			0.9	1–2	Moderate	< 3
B. Everett Jordan cofferdam, NC	9	2H:1V	Silty clay	0.3	< 24	Moderate[f]	< 0.6
Little Blue River Levee, MO	4.6	3H:1V	Clay (vegetated)	0.3	12	Minor	< 1
Colorada Ret. Structure M-1	12	2.5H:1V	SC-CL (vegetated)	0.4	2–4	Minor	< 3
Colorado Ret. Structure S-1	7	2.5H:1V		0.25	2–4	Minor	< 6
Upper Elk River Nos. 37, 41, 42 and Big	12–14	2.5H:1V	Clay and gravel	0.09–0.8	1–4	Minor	< 5
W. Fork Point Remove Creek Site 2	11		Lean clay, gravel, and silty sand	0.4	< 2	Minor	< 6
W. Fork Point Remove Creek Site 4	15		Lean clay gravel, and silty sand	0.3	< 2	Minor	< 7
East Willow Creek W/S Str. D-2, MN	8		Lean clay	0.45–0.6	1.0+	Minor	< 10
Bogue Creek Dam No. Y-30–57, MS	6		Lean clay	0.2	< 1	Minor	< 20
Deep Creek W/S Dam 30A, NC	12		Lean clay, silt, silty	0–0.2	< 1	Minor	< 25
Randall RC&D CA-191-BH, SD	7		Lean clay	0.3	< 12	Minor	< 1
Klinkner (Westfork Kickappo W), WI	8		Silt-lean clay	0.6	2	Minor	< 4

Notes:
(a) Data mainly from Jacobs (2016) and Maslin and Rodd (2016).
(b) Breaching would probably have occurred more quickly if crest width had not been approximately 122 m.
(c) Elapsed overtopping time until breaching.
(d) During 1977 overtopping of 0.3 m, 12 h, no breaching; during 1982 overtopping, breaching after 20 h.
(e) Brazilian dams.
(f) Relief provided by "cutting crude spillway".
(g) Data from Flynn et al. (2021).
(h) Hegben was seiche wave resulting from earthquake.
(i) From description in Gilbert and Miller (1991), this dam probably did not fail by general overtopping.
(j) High Kd a result of poor compaction.
(k) Embankment slope data from table in Ostrovsky and Wan (2023) checked against Gilbert and Miller (1991) where available.
(l) Kd values have been estimated from Figure 6.17 using the flood depth and duration for each case.

TABLE 6.4 Database Method for Estimating the Probability of Commencement of Breach (P_{BFS}) Resulting from Overtopping of Embankments Constructed of Plastic Soils (Adapted from Jacobs, 2016; Maslin and Rodd, 2016)

MAXIMUM DEPTH OF OVERTOPPING	*PROBABILITY OF BREACH DURATION OVERTOPPING ≤ 2 HOURS*	*PROBABILITY OF BREACH DURATION OVERTOPPING 3 TO 6 HOURS*	*PROBABILITY OF BREACH DURATION OVERTOPPING > 6 HOURS*	*PROBABILITY OF BREACH DURATION OVERTOPPING ≤ 2 HOURS*	*PROBABILITY OF BREACH DURATION OVERTOPPING 3 TO 6 HOURS*	*PROBABILITY OF BREACH DURATION OVERTOPPING >6 HOURS*
	DOWNSTREAM SLOPE 3H:1V			*DOWNSTREAM SLOPE 2H:1V*		
0–0.15 m	0.01	0.05	0.1	0.02	0.1	0.2
> 0.15–0.3 m	0.05	0.1	0.3	0.1	0.3	0.5
> 0.3–0.5 m	0.15	0.3	0.5	0.3	0.5	0.8
> 0.5–1 m	0.4	0.5	0.7	0.5	0.7	1
> 1 m	0.8	1	1	0.9	1	1

Probability of breach was estimated for each case in Table 6.3 using the Jacobs (2016) Maslin and Rodd (2016) method. For "Breach" cases the estimates were correspondingly high but for "Minor Damage" cases the estimates were too high. Many of those cases have short duration of overtopping.

As a result, and taking account of the additional case data, an added category for overtopping durations ≤ 2 h has been added to the Jacobs (2016), Maslin and Rodd (2016) table. Table 6.4 is the updated table.

For the database method the Inflow Design Flood (IDF) is routed through the reservoir. If the embankment is overtopped, the duration of overtopping is to be estimated from the flood hydrograph. For large reservoirs this will require routing the flood through the reservoir. For small reservoirs the outflow and inflow hydrographs will be similar.

For overtopping durations > 20 hours, a higher probability of commencement of breach than shown in Table 6.4 may be appropriate.

The annual probability of commencement of breach = (Annual probability of the IDF causing the overtopping) × (P_{BFS}).

Table 6.4 is applicable to embankments constructed of a plastic soil. Where the embankment material is non-plastic, it is expected that the probability of breach would be higher.

It should be recognised that there is significant uncertainty in these estimates because of the limited number of case data available and because whether breach commences is dependent on more factors than just overtopping depth and duration and the downstream slope. These include presence and condition for grass cover on the downstream slope, the erosion properties of the soil in the embankment (i.e. critical shear stress and erosion rate), crest width and shape of the flood hydrograph.

There is, however, also uncertainty in the Simplified QRA and quantitative methods described next, and it is useful to compare the results of those methods with this method as a check.

6.5.3 Simplified QRA Method

The suggested method is:

(a) Define the geometry and zoning of the embankment.
(b) Determine the hydrographs of overtopping flows for a range of AEP of floods likely to control the risk.
(c) Estimate the terminal velocity V_2 on the downstream slope for the maximum reservoir level for each flood partition as described in Section 6.4.1.
(d) Estimate the time for initiation (T_{INIT}) of erosion on the downstream slope using Figure 6.15. This is from CIRIA (1987) and Hewlett et al. (1987).
(e) Estimate the time for head cutting (T_{HC}) using plots of head-cut progression time versus erosion coefficient (Kd) shown in Figure 6.16.

Morris et al. (2012) indicate that the CIRIA (1987) method in Figure 6.16 has a factor of safety compared to the original version of this plot in CIRIA Technical Note 71. Comparing the two shows that the factor of safety varies from about 1.33 for 1 h, 1.2 for 2 h, 1.08 for 5 h and 1.00 for 50 h. The adoption of Figure 6.16 means that there is some conservatism. This could be overcome by applying these factors of safety, but given the uncertainty in the method that may be unnecessary.

This method should not be applied to embankments with rock rip-rap or other forms of downstream slope protection. See Section 6.6 for such embankments.

Figure 6.16 has been developed using the WinDAM C breach modelling program to allow estimation of the time for head-cut to progress through the crest (T_{HC}).

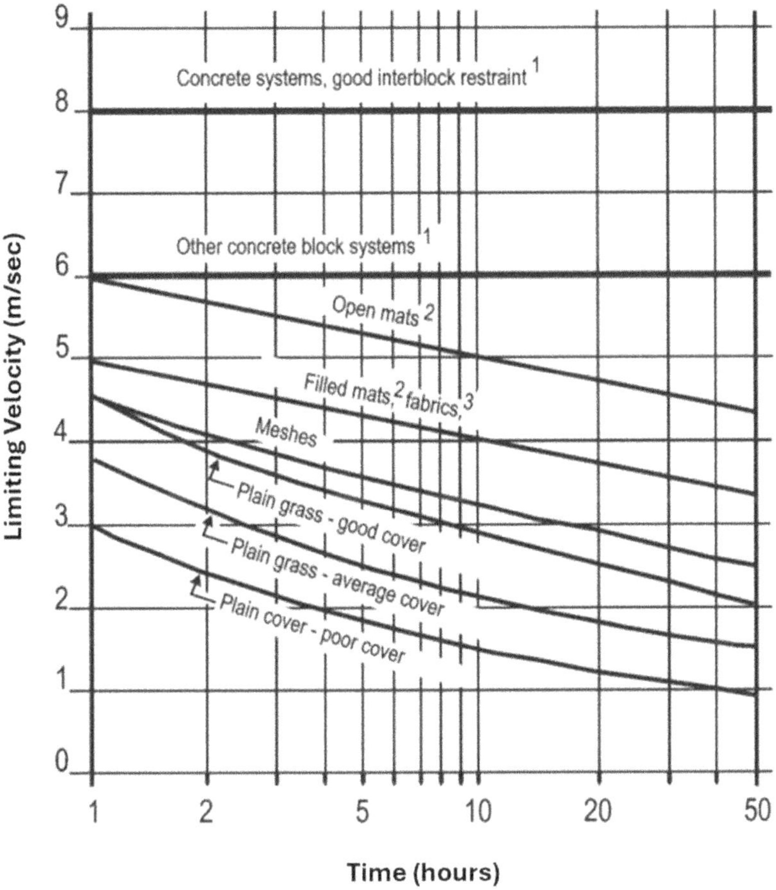

LEGEND:

Plain grass—good cover. Dense tightly knit turf established for at least two growing seasons.

Plain grass—average cover. Not defined but between good cover and poor cover.

Plain cover—poor cover. Uneven tussocky grass growth with bare ground exposed or significant proportion of non-grass weed species.

FIGURE 6.15 Recommended limiting velocity for erosion of grass cover. (CIRIA, 1987; Hewlett et al.,1987; reproduced with permission of CIRIA through PLSclear.)

The modelling completed and assumptions are:

- Modelling carried out for a generic homogenous earthfill dam with 15 m high, 6 m crest width and 3H:1V upstream and downstream slopes (also analysed 2H:1V slopes).
- "Bare earth" slopes (no grass cover, although the model software can include grass or rip-rap cover).
- Earthfill material parameters:
    - Erosion rate parameter Kd—varies 0.2–50 $cm^3/N.s$.
    - Undrained shear strength (Cu) = 50 kPa. This effects the height of the head-cut before the head-cut scarp face becomes unstable.
    - Critical shear stress = 0 Pa
    - Total unit weight = 20 kN/m^3
- Simple hydrographs to give a constant uniform overtopping flow over the dam crest of depths of 0.15 m, 0.3 m, 0.5 m, 1 m and 2 m above the dam crest level.

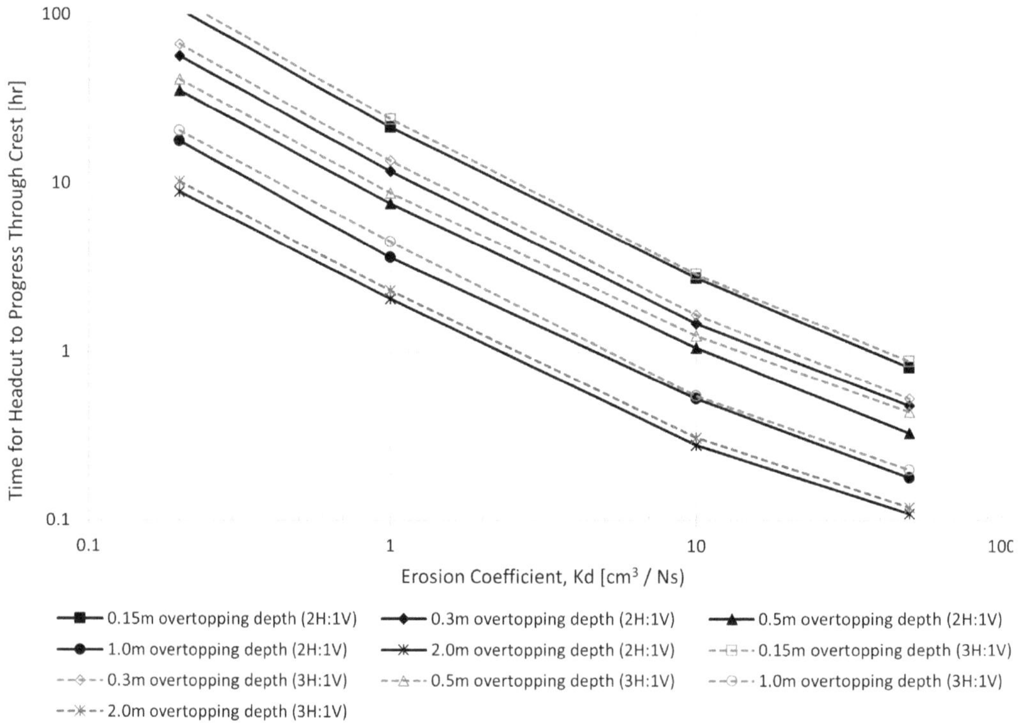

FIGURE 6.16 Time for head-cut to progress through the crest (hour) for generic homogeneous earthfill dam.

- The software provides two options for head-cut assessment methods—Hanson and Robinson stress model method, and Temple and Hanson energy model method. Both methods give similar results.

The time for the head-cut process to start and then progress through the crest to the upstream face was extracted for each model run to produce Figure 6.16.

6.5.4 Physically Based Numerical Modelling

There are two models most widely used:

WinDam C. Developed by the Agricultural Research Service and Natural Resources Conservation Service, USA. It can model breach resulting from overtopping and internal erosion for homogeneous embankments.

EMBREA. Developed at H. R. Wallingford UK. It can model breach from overtopping for homogeneous and zoned embankments constructed of plastic or non-plastic soil. It can model downstream surface protection layers such as grass cover. Monte Carlo simulations may be computed to produce a probability distribution of results.

For each model, the erosion process leading to commencement of breach as follows:

Phase 1. Failure of the vegetation cover on the downstream slope if present and initiation of erosion.

Phase 2. Downward and downstream erosion that leads to formation of a near-vertical head-cut in the vicinity of the initial failure.
Phase 3. Upstream advance and deepening of the head-cut resulting from flow over the vertical or near vertical face until the reservoir is reached.
Phase 4. Widening and deepening of the breach.

The main purpose of the models is to model the breach widening process and resulting flood hydrograph (i.e. for dam break studies), but as part of the program the initiation of erosion and head cutting leading to the commencement of breach is modelled.

The methods model the hydraulic shear stress imposed on the downstream slope and the rate of erosion based on that stress, the critical shear stress (τ_c) and an erosion rate coefficient Kd. Head cutting is influenced by the undrained shear strength (Cu) of the soil which controls the height of the vertical head-cut before it fails, allowing progression of the erosion process.

Table 6.5 which relates Kd and critical shear stress to soil classification has been developed from Briaud et al. (2017), Robbins and Corcoran (2022), USACE (2019), Fell et al. (2008, 2024) and the back analysis case data in Table 6.3. This may be used as a guide.

For detailed assessments, Jet Erosion Tests (JET) on the soil from the embankment is the preferred approach to estimate Kd. However, from published data the JET seems to give higher critical shear stresses than obtained in Hole Erosion Tests (HET), so high as to not be reasonable, and it is suggested the JET critical shear stresses not be relied upon and the values in Table 6.5 adopted or Hole Erosion Tests carried out on the soil.

6.5.5 Suggested Methods for Assessing Dam Safety for Overtopping Type Z1 Earthfill Embankments

6.5.5.1 *Deterministic Assessment*

In the deterministic approach to assessing dam safety against overtopping failure of the embankment, the Inflow Design Flood which the dam should be able to pass without failure by overtopping is related to the consequences of failure. Lower AEP IDFs are required for dams with large consequences of failure than for a dam with a small consequence of failure.

The dam is expected to have sufficient freeboard above the maximum reservoir level resulting from the IDF that the likelihood of failure from waves overtopping the embankment is negligible.

The CDA (2013) guideline on freeboard for embankment dams is as follows:

The crest level should be set so the embankment is protected against the most critical of the following cases:

- No overtopping by 95% of the waves caused by the most critical wind with a frequency of 0.001/annum when the reservoir is at maximum normal elevation.
- No overtopping by 95% of the waves caused by the most critical wind when the reservoir is at its maximum level during passage of the IDF.

CDA (2013) suggest the AEP of wind frequency used for the freeboard calculation for the IDF load condition are:

Low consequence dam AEP = 0.5
Significant consequence dam AEP = 0.1
High, very high or extreme consequence dam AEP = 0.01

TABLE 6.5 Erosion Rate Coefficient (Kd, $cm^3/N.s.$) for well-compacted plastic soils. For poorly compacted soil, i.e. no formal compaction as in older dams, adopt 10× these values.

			EROSION RATE COEFFICIENT Kd $CM^3/N.S$				
SOIL CLASSIFICATION	*BRIAUD EROSION CLASS (a)*	*EROSION RATE INDEX (b)*	*USACE BEST PRACTICES D1*	*BRIAUD (2017)*	*BACK ANALYSIS (c)*	*SUGGESTED (d)*	*CRITICAL SHEAR STRESS PA (e)*
Clay, Silty Clay CL	High II	3 to 4	0.2 to 2	0.1 to 1	0.1 to 5 (0.5)	0.1 to 5 (0.5)	5
Clay, Silty Clay, Sandy Clay, CL-CH	Medium to High III to II	4	0.05 to 0.5	0.05 to 0.5	0.1 to 5	0.05 to 2 (1)	25
Clay, Silty Clay, Sandy Clay, CH	Medium to Low III to IV	4 to 5	0.005 to 0.25	0.005 to 0.05		0.01 to 0.5 (0.1)	25–60
Clayey Sand SC	High to Medium II to III	3	0.01 to 0.5	0.05 to 0.5		0.05 to 2 (0.5)	4
Silty Sand SM	High to Medium II to III	2 to 3	0.05 to 0.5	0.05 to 0.5		0.5 to 20 (5)	2
Sand fine to medium SP	Very High to High I to II	1?	1 to 5 or more	0.5 to 5		5 to 100 (50)	1

Notes:
(a) Briaud (2008).
(b) Fell et al. (2008, 2024).
(c) Back analyses data in Table 6.3.
(d) Suggested range and best estimate.
(e) Fell et al. (2008, 2024). The values shown are for non-dispersive soils. For dispersive soils the critical shear stress is 1 to 2 Pa.

If waves are assessed as discussed in Section 6.3 and the freeboard equals or exceeds these requirements, there will be at most very minor overtopping flows and the likelihood of failure by overtopping may be assumed to be negligible.

Based on this the recommended Deterministic Approach is as follows:

Determine the maximum reservoir level for the IDF by routing the flood through the spillway. Assume that the reservoir is at FSL or normal operating level at the beginning of the flood.

(a) If the maximum reservoir level exceeds the dam crest level, the dam fails to satisfy deterministic safety criteria.
(b) If the maximum reservoir level is below dam crest level but the estimated wave run-up for the 2% waves ($R_{u2\%}$) from Section 6.3.6 exceeds the freeboard, the dam fails to satisfy deterministic safety criteria.
(c) If the freeboard exceeds the wave run-up for the 2% waves, there will be at most very minor overtopping flows and the likelihood of failure by overtopping may be assumed to be negligible and the dam satisfies deterministic safety criteria.

Use the actual crest level allowing for settlement since construction, residual camber and repairs to the crest.

6.5.5.2 Risk Enhanced Deterministic Assessment

Determine the maximum reservoir level for the IDF by routing the flood through the spillway.

Then assess as described for Deterministic Assessment, but if the IDF results in overtopping of the crest use the database method described in Section 6.5.2 to provide additional information on the likelihood of failure to assist in making the dam safety assessment. Flow concentration effects would not usually be allowed for in the risk enhanced assessment.

6.5.5.3 Simplified QRA Assessment

Determine the maximum reservoir level for each flood load partition by routing the floods through the spillway. Assume that the reservoir is at FSL or normal operating level at the time of the flood.

(a) If the maximum reservoir level exceeds the dam crest level, estimate the resultant maximum flow depth and velocity and duration over the crest for the maximum reservoir level as described in Section 6.4.1.
(b) Estimate the time for initiation of erosion (T_{INIT}) and time for head cutting (T_{HC}) as described in Section 6.5.3.
(c) Use engineering judgement to estimate the probability of commencement of breach given the overtopping (P_{BFSQ}) by comparing (T_{INIT} + T_{HC}) with the duration of the overtopping hydrograph.
(d) If the maximum reservoir level for the flood load partition is below dam crest level but the estimated wave run-up for the 2% waves ($R_{u2\%}$) exceeds the freeboard, estimate the mean estimate maximum flow rate from Equation 6.21 as the flow rate and calculate as detailed in (b) and (c).
(e) If the freeboard exceeds the run-up for the 2% waves, there will be at most very minor overtopping flows and the likelihood of failure by overtopping may be assumed to be negligible.
(f) It is assumed that once breach initiates it will progress. No allowance is made for detection and intervention.

The annual probability of commencement of breach

$$= \Sigma\ [(\text{Annual probability of the flood}) \times (P_{BFSQ})] \text{ for all the flood load partitions.} \quad (6.28)$$

Compare the result of this approach with the database method. If the results are greatly different, determine the reasons, and if necessary, revise the Simplified QRA approach.

6.5.5.4 Quantitative Risk Assessment

For most Quantitative Risk Assessments (QRA) use the same approach as for Simplified QRA but allow for detection and intervention as detailed in Section 6.10.

For very detailed QRA use physically based numerical modelling as detailed in Section 6.5.4.

The approach for using these models is as follows:

(a) Define the geometry and zoning of the embankment.
(b) Determine the hydrographs of overtopping flows for the range of AEP of floods.
(c) Determine the critical shear stress, erosion rate coefficient and undrained shear strength of the earthfill in the embankment. Guidance on assessing these is provided in Section 6.5.4. As these parameters are uncertain, assign a best estimate and range to each parameter.
(d) Run the model for each of the overtopping hydrographs. Assess which lead to loss of protective grass cover, head-cut erosion progresses and commencement of breach.
(e) Make further runs to check the sensitivity to the input parameters. If the program can model Monte Carlo analyses, do that.
(f) Based on these results, assign conditional probabilities for the loss of the surface grass protection layer and head-cut erosion progresses across the dam crest for each flood load partition.
(g) Consider detection, intervention and breach as described in Section 6.10.
(h) Use event tree methods to determine the total annual probability of commencement of breach. Figure 6.17 shows a typical event tree.

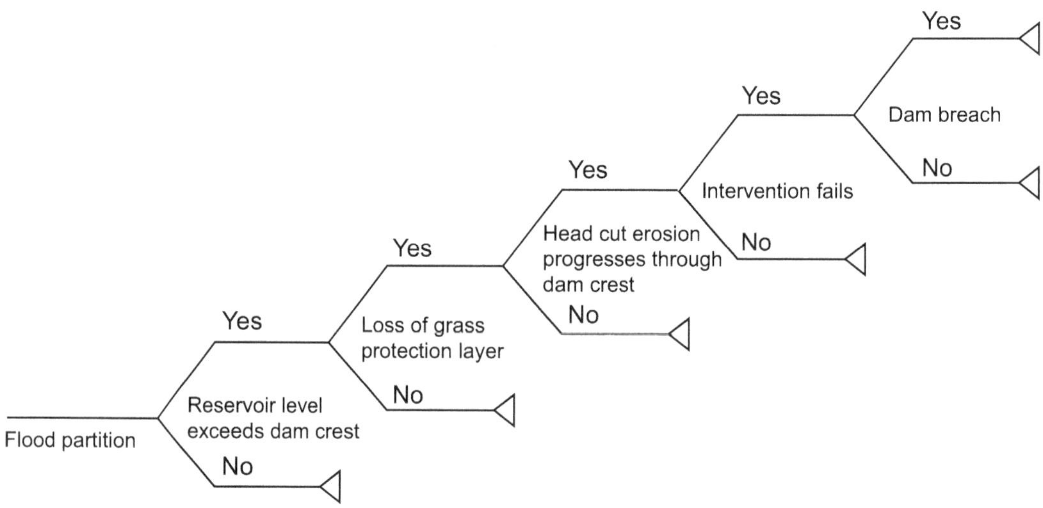

FIGURE 6.17 Event tree for modelling the annual probability of failure by overtopping of an earthfill embankment.

6.6 FAILURE BY OVERTOPPING OR WAVE OVERTOPPING OF TYPE Z2 EARTHFILL EMBANKMENTS WITH RIP-RAP DOWNSTREAM PROTECTION

6.6.1 Introduction

The following discussion applies to earthfill embankments with a layer of rock rip-rap on the downstream slope for erosion protection.

Methods for design of the rip-rap are summarised in Abt et al. (2013). They gathered test data from the several laboratory studies which have been carried out and used these to check the performance of 21 available design methods.

The test data covered a wide range of sizes of rip-rap, with D_{50} sizes from 15 mm to 655 mm. Much of the data was for $D_{50} < 60$ mm, so relatively small size. The slopes of the rip-rap ranged from 0.2% (i.e. almost flat) to 50%, with only about 12 of the 96 cases with slopes ≥ 30%, i.e. ≥ 3H:1V, which is what most dams and levees are.

They concluded that the methods of Khan and Ahmad (2011), Chang (1998) and Hartung and Scheuerlein (1970) gave the best predictions over the full range of data. The Mishra (1998) and Frizell et al. (1998) methods performed well for the coarser rip-rap ($D_{50} \geq 250$ mm).

The Frizell et al. (1998) method was developed based on large-scale tests by the Bureau of Reclamation and is considered by the current authors most applicable to embankments with rip-rap of that size. The Khan and Ahmad (2011) method is considered applicable to the full range of sizes but more particularly for rip-rap $D_{50} < 200$mm.

The following sections describe the Khan and Ahmad (2011) and Frizell et al. (1998) methods.

For both methods the rate of overtopping flow per unit length of dam, q (m^3/m/sec) is given by:

$$q = \frac{Q}{L} = CH^{1.5} \tag{6.29}$$

where:

C = Weir coefficient
H = overtopping head (total) above weir crest (m).

As discussed in Section 6.4.1, C may be assumed to be 1.7. Frizell et al. (1998) assumed C = 1.57.

6.6.2 Khan and Ahmad (2011)

The Khan and Ahmad (2011) relationship is:

$$D_{50} = 0.66t^{0.58}S^{0.22}C_u^{-0.45}q_f^{0.22} \tag{6.30}$$

where:

D_{50} = size of rock in the rip-rap for which 50% is finer (m)
t = thickness of rip-rap layer (m)
S = embankment slope relative to the horizontal (ratio of vertical to horizontal)
C_u = Uniformity coefficient of rip-rap (D_{60}/D_{10})
q_f = overtopping discharge/unit length of levee or dam (m^3/m/sec).

Failure is defined by Khan and Ahmad (2011) as when the bedding layer or geofabric under the rip-rap is exposed. This equation has no factor of safety. It was based on a multivariate regression analysis of published data.

Khan and Ahmad (2011) do not provide guidance on what factor of safety to apply for design.

This method was developed using data within the following limits:

D_{50} 25 to 200 mm
t 75 to 300 mm
S 0.1V:1.0H to 0.25V:1.0H
C_u 1.25 to 2

It is likely to be less reliable if applied outside those limits.

6.6.3 Frizell et al. (1998)

Frizell et al. (1998) describe the results of large-scale laboratory experiments carried out by USBR on rip-rap subject to overtopping flow. The test facility was 3 m wide, 15 m vertical drop on a slope of 2H:1V. Maximum discharge was 4.5 m^3/sec.

They tested rip-rap with the following characteristics:

- D_{50} size 386 mm, 0.6 m thick over a 203 mm thick gravel bedding.
- D_{50} size 655 mm, 0.6 m thick placed over the rip-rap and bedding layer from the first test.
- D_{50} size 271 mm, 0.53 m thick, over gravel bedding.

Frizell et al. (1998) note the following:

(a) Failure of the rip-rap was defined as removal or dislodgement of enough material to expose the bedding material.
(b) Failure occurred with the measured water depth still below the surface of the rip-rap. A corollary is that if the flow exceeds the interstitial flow capacity of the rip-rap, the rip-rap will fail. For design, the interstitial flow capacity must exceed the overtopping flow.
(c) Failure occurred within the slope. Tests were carried out to demonstrate this and that the weak point is not the toe of the slope.
(d) The design curves in Figure 6.18 combine empirical data with accepted sediment transport equations and are not simply a best fit of the data. No details are provided in the paper. The Frizell et al. (1998) design method was developed for angular rip-rap supported upon a bedding layer and applies only for flows which are carried in the interstitial flow of the rip-rap. No factor of safety applies.

The Frizell et al. (1998) design method involves:

- Make a first estimate of the required D_{50} size (m) of the rip-rap for the design overtopping discharge/unit length of levee or dam, q (m^3/m/sec) and an assumed uniformity coefficient C_u and embankment slope S using Figure 6.20.
- Determine the required rip-rap layer thickness t (m) to transmit the overtopping flow within the interstitial voids in the rip-rap using:

$$t = q/(v_i n_p) \quad (6.31)$$

where:

t = rip-rap layer thickness (m)
q = overtopping discharge/m of crest length (m^3/m/sec)
v_i = interstitial velocity (m/sec)
n_p = porosity of the rip-rap

and

$$v_i = 2.4S^{0.58}\, C_u^{-2.22} \sqrt{gD_{50}} \tag{6.32}$$

where:

S = embankment slope relative to the horizontal (ratio of vertical to horizontal)
C_u = uniformity coefficient of rip-rap (D_{60}/D_{10})
g = gravitational constant (9.81 m/sec^2).

If the calculated thickness is greater than the minimum thickness of 2 D_{50}, increase the D_{50} or the thickness so all flow is within the interstitial voids. Re-do the calculations until a solution is reached.

Those wanting to use this method should reference the paper which includes a worked example. Frizell et al. (1998) say,

> A factor of safety may be applied by the designer, as necessary. For example, if the design is for the probable maximum flood, no factor of safety may be required. However, if the design is for the 1 in 100 year flood over the service spillway, a factor of safety may be required based on agency policy or experience or judgement of the designer.

NCHRP (2006) provide more information on the Frizell et al. (1998) design method.

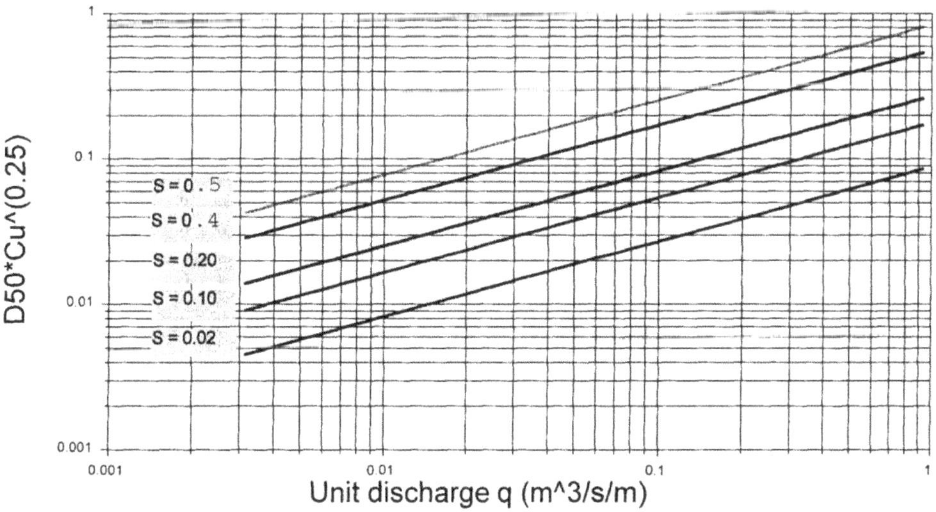

FIGURE 6.18 Design curves to size rip-rap protection on embankments of various slopes. The curves represent the point of incipient failure. No safety factor has been included (Frizell et al., 1998).

6.6.4 Peirson et al. (2008)

Peirson et al. (2008) carried out flume tests in a flume 900 mm wide, and length 4.2 m for 0.4V:1H slope, and 8.4 m for slopes ≤ 0.2V:1H. Sandstone rock rip-rap with D_{50} 76 mm and 109 mm, and basalt rock with D_{50} 94 mm was tested. It was placed in two layers on a base consisting of two layers of 40 mm gravel. Failure was taken as being when the rock rip-rap was displaced to expose the base gravel.

Peirson et al. (2008) found that the overtopping flow/unit width to initiate rip-rap failure (exposure of the base layer) $q_{over.fail}$ is given by:

$$q_{over,fail}\sqrt{\rho/\left(g\left(\rho_S-\rho\right)d_{50}^3\right)} = B10^{-1.37\tan\theta} \tag{6.33}$$

where:

ρ = density of water (tonnes/m^3)
ρ_s = rock density (tonnes/m^3)
g = gravitational acceleration (9.81 m/sec^2)
d_{50} = size of rip-rap with 50% finer by mass (m)
θ = slope angle (degrees)
B = 1.60 for rock density 2.3 tonnes/m^3 and 1.76 for rock density 2.7 tonnes/m^3.

This method is for dumped rockfill. Peirson et al. (2008) found that selectively placing rock to minimise porosity gave larger failure flows by approximately 30%. This would be applicable to some older dams which have downstream slopes paved with hand placed rock, but the rock is not mortared into place.

6.6.5 Importance of Bedding Layer

These methods for design of a protective rip-rap layer assume there is a granular bedding layer between the rip-rap and the embankment fill. If there is no granular bedding layer, then the flow that occurs through the rip-rap layer may erode the embankment fill material, causing the rip-rap material to be undermined and displaced such that it exposes the fill directly to the overtopping flows. In this case, the rip-rap layer is unlikely to be effective.

The authors' experience with geotextile placed beneath the rip-rap rather than a granular bedding layer is that the geotextile is readily damaged and/or water flows beneath the geotextile concentrating flows which may initiate erosion of the earthfill. It is recommended therefore that if geotextile is present, it be considered as for no bedding layer.

6.6.6 Suggested Methods for Assessing Dam Safety for Overtopping of Type Z2 Earthfill Embankments with Rip-Rap Downstream Protection

6.6.6.1 Deterministic Assessment

As for Type Z1 embankments. See Section 6.5.5.

6.6.6.2 Risk Enhanced Deterministic Assessment

Determine the maximum reservoir level for the IDF by routing the flood through the spillway.

Then assess as described for Deterministic Assessment, but if the IDF results in overtopping of the crest use the Khan and Ahmad (2011) or Frizell et al. (1998) method described in Sections 6.6.2 and 6.6.3 to assess whether the rip-rap would fail. Do not consider erosion of the earthfill beneath the rip-rap.

6.6.6.3 Simplified QRA and Quantitative Risk Assessments

For failure to occur, first the rip-rap layer must fail, and then erosion in the earthfill initiates and progresses to breach the embankment. Figure 6.19 shows an example of an event tree.

(A) Determine the maximum reservoir level for each flood load partition by routing the floods through the spillway.

(a) If the maximum reservoir level exceeds the dam crest level, estimate the unit discharge (m^3/m/sec) (q) over the crest for the reservoir level being considered using Equation 6.29. Include consideration of flow concentration effects (e.g. due to uneven crest level, or flow concentration along the abutment groin).
(b) Estimate the mean and range of downstream slope angle D_{50} and uniformity coefficient Cu of the rip-rap from samples and gradations and/or from observations on the downstream slope. Note that the failure of surface erosion protection measures is dependent on the consistency of the rip-rap gradation and gradient of the slope. If there are areas on the downstream slope which are deficient in the larger particle sizes or locally steeper (i.e. due to poor construction or degradation), then those areas may preferentially erode exposing the embankment fill and causing displacement of the rip-rap further up slope.
(c) From these, estimate the mean D_{50} and C_u combination, and reasonable upper bound and lower bound combinations.

(B) Assess the likelihood of failure of the rip-rap:

Rip-rap with a D50 coarser than 250 mm.

It is recommended that the Frizell et al. (1998) method be applied for rip-rap with a D_{50} coarser than 250 mm.

The steps are:

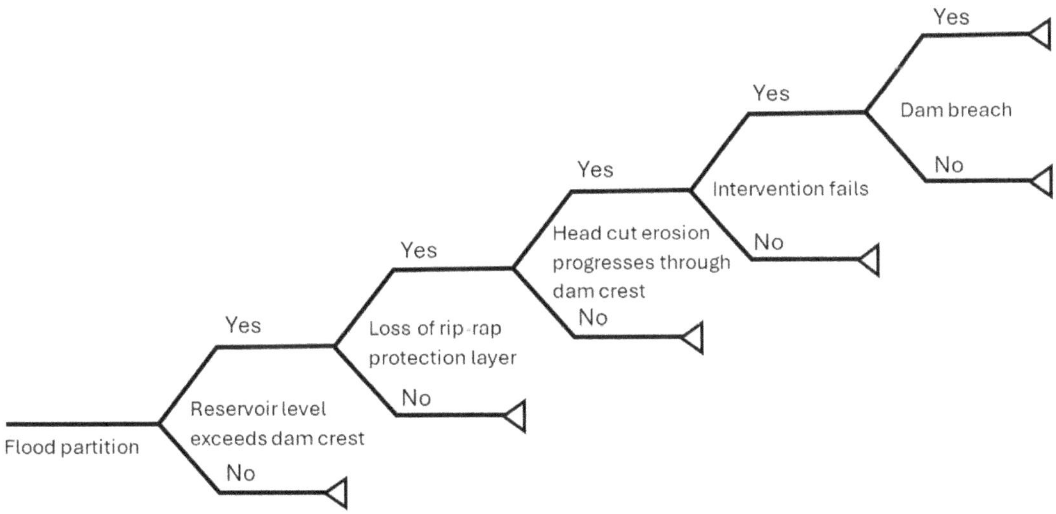

FIGURE 6.19 Event tree for modelling the annual probability of failure by overtopping of an earthfill embankment with rip-rap slope protection.

(a) Estimate the unit discharge resulting in incipient failure (q_{Fmean}) of the rip-rap for the most likely D_{50} and Cu combination for the embankment downstream slope (S) using the design curves in Figure 6.20. Note these have no factor of safety.
(b) Allow for the model uncertainty by also estimating $q_{Fmin} = 0.8x\ q_{Fmean}$ and $q_{Fmax} = 1.2x\ q_{Fmean}$ from Figure 6.20.
(c) Compare these unit discharges with the estimated overtopping unit discharge for this reservoir stage.
(d) Use expert elicitation to estimate the probability of failure of the actual rip-rap ($P_{Failriprap}$) from these results.

As a guide to estimating these probabilities:

- If $q > q_{Fmax}$ then $P_{Failriprap} \approx 0.9$
- If $q = q_{Fmax}$ then $P_{Failriprap} \approx 0.5$
- If $q < 0.8\ q_{Fmin}$ then $P_{Failriprap} \approx 0.1$
- If $q < 0.3\ q_{Fmean}$ then $P_{Failriprap} < 0.01$

(e) If the maximum reservoir level for the flood load partition is below dam crest level but the estimated wave run-up for the 2% waves ($R_{u2\%}$) exceeds the freeboard, estimate the mean estimate maximum flow rate from Equation 6.21 as the flow rate and calculate as detailed in (d) to (g).
(f) If the freeboard exceeds the run-up for the 2% waves, there will be at most very minor overtopping flows and the likelihood of failure by overtopping may be assumed to be negligible.
(g) For Simplified QRA, it is assumed that once breach initiates it will progress. No allowance is made for detection and intervention.
(h) For QRA consider detection, intervention and breach as described in Section 6.10.

For rip-rap which is selectively placed rock, adjust the unit discharge resulting in failure upwards by 30% as recommended by Peirson et al. (2008).

If proprietary concrete blocks or other form of slope protection has been constructed on the crest and downstream slope, refer to the manufacturer design manuals to assess the ability of the slope protection to pass the floods without failure.

For Rip-Rap with a D_{50} Less than 200 mm

For rip-rap with a D_{50} less than 200 mm, use the procedure described earlier, but substitute Equation 6.30 from the Khan and Ahmad (2011) method in step (d).

For Rip-Rap with a D_{50} between 200 mm and 250 mm

For rip-rap with D_{50} between 200 mm and 250 mm, it is suggested that both the Frizell et al. (1998) and Khan and Ahmad (2011) methods be used and judgement applied to assessing the likelihood of failure.

(C) Estimate the likelihood of failure of the exposed earthfill below the rip-rap:

For cases where the rip-rap is assessed as likely to fail, the earthfill will be exposed. Then estimate the probability of commencement of breach (P_{Breach}) given the rip-rap has failed using the methods described in Section 6.5.

The probability of commencement of breach for the reservoir level being considered = ($P_{Failriprap}$) × (P_{Breach}).

$$\text{Annual probability of breach for the flood stage} = \Sigma\,[(P_{FS}) \times (P_{Failriprap}) \times (P_{Breach})] \tag{6.34}$$

where:

P_{FS} = annual probability of the flood stage.

Repeat for all reservoir stages being considered in the risk analysis.

(D) Assess the likelihood of successful detection and intervention to prevent breach as described in Section 6.10.

(E) Use event tree methods to determine the total annual probability of commencement of breach.

6.7 FAILURE BY OVERTOPPING OR WAVE OVERTOPPING OF TYPES Z3, Z5 AND Z6 EMBANKMENTS WITH DOWNSTREAM ROCKFILL

6.7.1 Introduction

The methods described in this section are applicable to dams with a downstream rockfill zone and include (Z3) Zoned Earth and Rockfill and Central Core Earth and Rockfill (Types 4 and 5); (Z5) Concrete Face Rockfill (Type 7); (Z6) Rockfill with Core Wall (Type 10); and Asphalt Core Rockfill (Type 11).

Depending on the rate of flow overtopping the crest, the zoning of the embankment and permeability of the rockfill, the following may occur:

(a) All overtopping flow enters the rockfill and emerges lower down on the downstream slope of the rockfill.
(b) Overtopping flow partly enters the rockfill as in (a) and the remainder flows down the surface of the downstream slope.
(c) Almost all the overtopping flow flows down the surface of the downstream slope.

For an embankment the conditions may be (a) for small depths of overtopping, becoming (b) for greater depths of overtopping. Condition (c) is likely to occur where the rockfill has low vertical permeability. See Section 6.4.3 for information on rockfill permeability.

6.7.2 Deterministic Assessment for Types Z3, Z5 and Z6 Embankments

As for Type Z1 embankments. See Section 6.5.5.

There is no risk enhanced method for these types of dams.

6.7.3 Assessment of Flow Conditions

To assess which flow condition applies and the flows per unit width of embankment do the following:

(a) Estimate the rate of overtopping flow passing over the dam crest per unit length of dam, q, $m^3/s/m$ using Equation 6.29.

(b) Estimate the rate of inflow per unit length of dam crest into the downstream rockfill (q_{inflow}) by the following approximation:

$q_{inflow} = k_v.i.A$ = vertical permeability of rockfill × seepage gradient × 10 m length of downstream slope. The seepage gradient is taken as 1.0 for vertical seepage flow under gravity. The 10 m is a length judged reasonable by the authors.

Care should be taken to allow for reduced vertical permeability of the rockfill resulting from breakdown on the surface of each layer caused by compaction equipment as discussed in Section 6.4.3.

The assessment should allow for throughflow that occurs through rockfill layers above the top of the core zone if such layers are present ($q_{through\_crest}$).

(c) Estimate the downstream slope overtopping flow rate per unit length of dam.

$$q_{overflow} = q + q_{through\_crest} - q_{inflow} \tag{6.35}$$

As can be seen from Table 6.6, for all the flow to enter the downstream slope the vertical rockfill permeability would need to be high. If the permeabilities in Table 6.6 are compared to the permeabilities in Table 6.2, it will be apparent that for many or most dams almost all the overtopping flow will flow over the downstream slope.

6.7.4 Condition (A) All Overtopping Flow Enters the Rockfill and Emerges Lower Down on the Downstream Slope of the Rockfill

6.7.4.1 Engineering Assessment

EBL (2005) reviewed the literature regarding methods for assessing the capability of a dam to withstand throughflow and overtopping. These are summarised in Figure 6.20. With the exception of the Martins (1981) tests they are based on flume tests, so not modelling throughflow.

TABLE 6.6 Estimated Rate of Inflow per Unit Length of Dam into a Downstream Rockfill Zone

DAM CREST OVERTOPPING DEPTH	*ESTIMATED FLOW RATE OVER THE CREST (M³/S PER M LENGTH)*	*VERTICAL PERMEABILITY OF ROCKFILL FOR ALL FLOW TO PASS INTO UPPER 10 M OF ROCKFILL (M/S)*
0.15m	0.1	> 0.01
0.25m	0.2	> 0.02
0.5m	0.6	> 0.06
1.0m	1.7	> 0.17

EBL (2005) compared those methods with the results of their own throughflow field tests 1-01, 4-02, and 2-02, examples of which are shown in Figure 6.20.

The data in Figure 6.20 shows that methods based on flume tests are conservative for throughflow. EBL (2005) point out that the Solvik (1981) method is for threshold flow, with individual rocks beginning to move, whereas the other data is for collapse flow or incipient failure. The Solvik (1981) method has a factor of safety of 1.8 on incipient failure and should not be compared to the other methods or the case data.

EBL (2005) proposed the "recommended eq." curve in Figure 6.20 to fit the field and laboratory throughflow test data. This is given by:

$$d_{50} = 0.43 S_0^{0.43}\, q^{0.78} \tag{6.36}$$

where:

d_{50} = rock size for which 50% by mass is smaller (metres)
S_0 = slope of the rockfill (V/H), e.g. 1:1.5
q = unit discharge in m^3/m/sec.

Note that Figure 6.20 is for downstream slope of 1V:1.5H so equation 6.36 should be used rather than Figure 6.20 for other slopes.

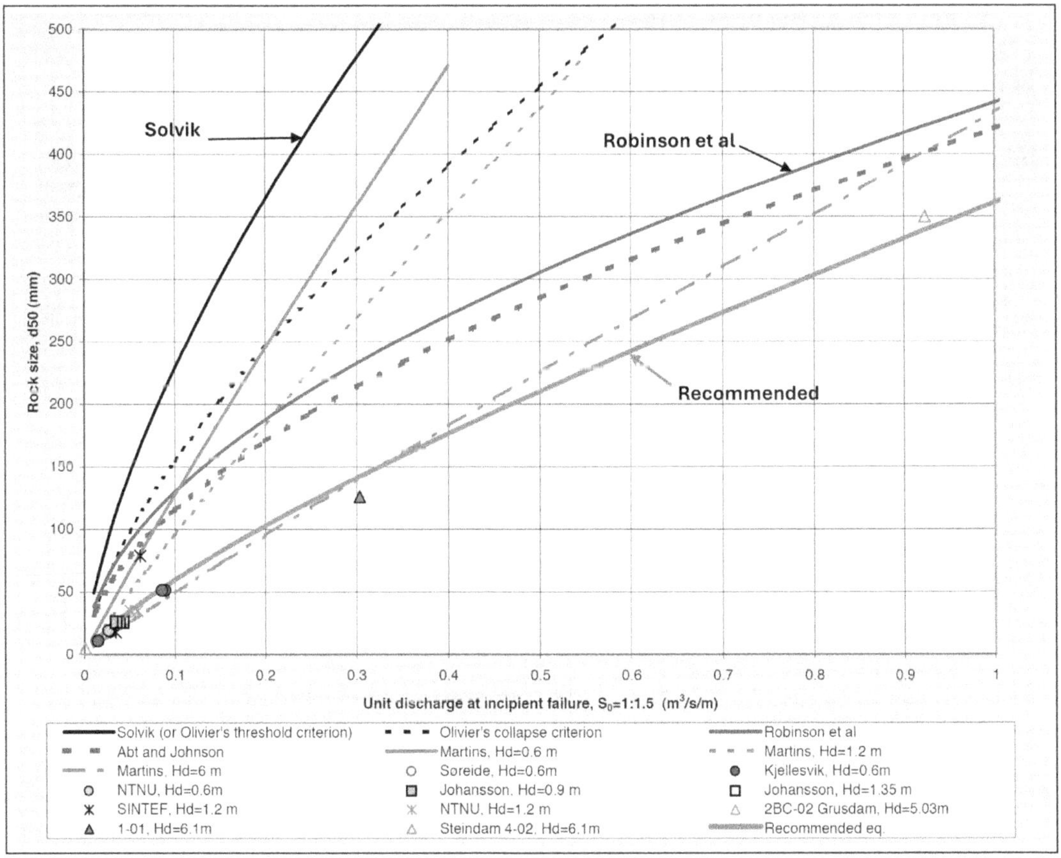

FIGURE 6.20 Rock size at incipient failure for a downstream slope of 1.5H:1V (EBL, 2005).

6.7.4.2 Assessment of Dam Safety

Simplified QRA and Quantitative Risk Assessment

The steps to estimate the probability of breach ($P_{BreachTF}$) are:

(a) Estimate the unit discharge (flow/m length) (q) over the crest for the reservoir level being considered as described in Section 6.7.2.

(b) Consider the potential for the flow through the downstream rockfill zone to concentrate and exit out at the downstream toe due to the valley profile. Concentration of flow at the toe of the dam will be most pronounced for embankments in V-shaped valleys.

(c) Estimate the mean and range of D_{50} of the rockfill from samples and gradations and/or from observations in test pits. Do not include outlier values.

(d) From these, estimate the mean D_{50} and reasonable upper bound and lower bound values.

(e) Estimate the unit discharge (q_{IF}) at incipient failure for these D_{50} values for the embankment downstream slope (S) using the EBL (2005) method as shown in Equation 6.36.

(f) Use expert elicitation to estimate the probability of unravelling initiating ($P_{UnravelTF}$) by comparing q to $q_{IF.}$ For design EBL (2005) adopt a factor of safety of 1.5 on D_{50}.This is equivalent to a factor of safety of 1.68 on q, so the probability of failure should be relatively low for a factor of safety of 1.68.

As a guide to estimating these probabilities:

- If $q > 1.2\ q_{IF}$ then $P_{BreachTE} \approx 0.9$
- If $q = q_{IF}$ then $P_{BreachTE} \approx 0.5$
- If $q < 0.8\ q_{IF}$ then $P_{BreachTE} \approx 0.1$
- If $q < 0.6\ q_{IF}$ then $P_{Failriprap} \leq 0.01$

(g) If the maximum reservoir level for the flood load partition is below dam crest level but the estimated wave run-up for the 2% waves ($R_{u2\%}$) exceeds the freeboard, estimate the mean estimate maximum flow rate from Equation 6.21 as the flow rate and calculate as detailed in (b) to (f).

(h) If the freeboard exceeds the run-up for the 2% waves, there will be at most very minor overtopping flows and the likelihood of failure by overtopping may be assumed to be negligible.

(i) For Simplified QRA, assume that once unravelling commences it will progress ($P_{BreachTF} = P_{UnravelTF}$). No allowance is made for progression or detection and intervention.

(j) For Quantitative Risk Assessments, use expert elicitation to estimate the probability of unravelling progressing ($P_{ProgressTF}$) by considering the following factors:

- Evaluate what slope angle the displaced rockfill material would need to attain for unravelling to cease (from Equation 6.36) and assess whether this slope angle could develop without losing the dam crest.
- Assess whether there is sufficient time for the unravelling process to develop completely through the downstream rockfill and to erode or destabilise the core zone (where present).

(k) For Quantitative Risk Assessments, consider progression of unravelling as described in (j), and detection, intervention and breach as described in Section 6.10.

Repeat for all reservoir stages in the risk analysis.

For QRA estimate the annual probability of failure using event tree methods. Figure 6.21 shows an example of the event tree.

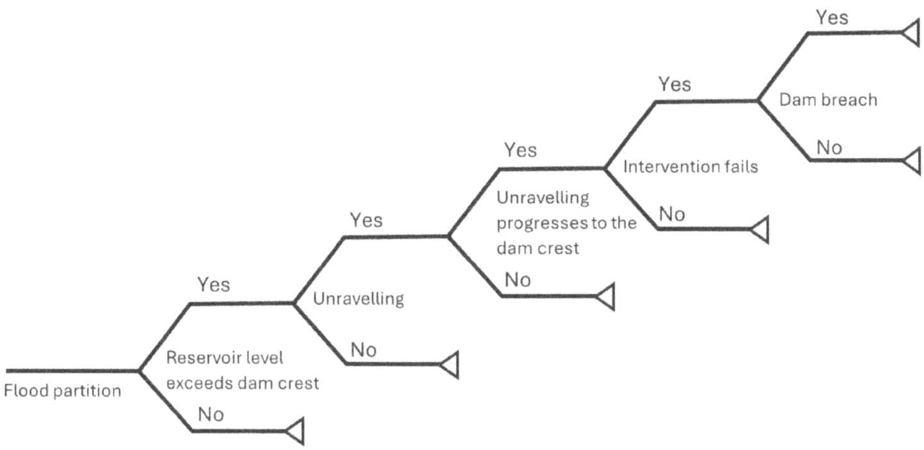

FIGURE 6.21 Event tree for modelling the annual probability of failure by overtopping of a rockfill embankment for overtopping flows entering the rockfill.

6.7.5 Condition (B) Overtopping Flow Partly Enters the Rockfill as in (A) and the Remainder Flows down the Surface of the Downstream Slope

Simplified QRA and Quantitative Assessments

The suggested method is:

(a) Estimate the rate of flow per unit length of dam into the downstream rockfill (q_{inflow}) and assess the probability of breach resulting from this flow ($P_{BreachTF}$) as detailed in Section 6.7.4.
(b) Estimate the overtopping flow rate per unit length of dam $q_{overflow} = q - q_{inflow}$ and estimate the probability of breach resulting from overtopping ($P_{BreachOF}$) as detailed in Section 6.7.6.
(c) The overall probability of breach for any reservoir stage $= 1 - (1 - P_{BreachTF}) \times (1 - P_{BreachOF})$.
(d) If the freeboard exceeds the run-up for the 2% waves there will be at most very minor overtopping flows and the likelihood of failure by overtopping may be assumed to be negligible.
(e) For Simplified QRA assume that once breach commences it will progress. No allowance is made for detection and intervention.
(f) For QRA consider detection, intervention and breach as described in Section 6.10

Repeat for all reservoir stages in the risk analysis.

6.7.6 Condition (C) All Flow down the Surface of the Downstream Slope

Simplified QRA and Quantitative Assessments

(a) Estimate the overtopping flow rate per unit length of embankment for each reservoir stage.

(b) Use the method in Section 6.6.6 to assess the likelihood of failure of the outer surface of rockfill. When making this assessment use the gradation of the rockfill on the outer slope. This is often coarser than the rest of the rockfill because the larger rocks are often pushed to the outer slope during construction.
(c) If the overtopping continues the surface exposed will be the next to fail until the crest is breached.
(d) For Simplified QRA assume that once failure of the outer surface of rockfill occurs then it will progress. No allowance is made for detection and intervention.
(e) For QRA, use expert elicitation to estimate the probability of erosion progressing ($P_{ProgressTF}$) by considering the following factors:
  - Evaluate what slope angle the displaced rockfill material would need to attain for further loss of rockfill to cease and assess whether this slope angle could develop without losing the dam crest.
  - Assess whether there is sufficient time for the erosion process to develop completely through the downstream rockfill and to erode or destabilise the core zone (where present).

(f) There may be an opportunity to detect and intervene because the process is progressive.

6.8 FAILURE BY THROUGHFLOW IN ROCKFILL OVER THE TOP OF THE EMBANKMENT CORE

Examples of this scenario are shown in Figure 6.13. The throughflow emerges at the downstream slope within the rockfill.

6.8.1 Engineering Assessment

(a) Estimate the horizontal permeability of the rockfill using the methods described in Section 6.4.3.
(b) Estimate the throughflow/m of embankment length (q_{THF}) (m^3/m/sec) based on the embankment zoning and geometry, and the horizontal permeability.
(c) Estimate the mean and range of D_{50} of the rockfill from samples and gradations and/or from observations in test pits. Do not include outlier values.
(d) From these, estimate the mean D_{50} and reasonable upper bound and lower bound values.
(e) Estimate the unit discharge (q_{IF}) at incipient failure for these D_{50} values for the embankment downstream slope (S) using the EBL (2005) method as shown in Equation 6.36.

6.8.2 Deterministic Assessment

Calculate the factor of safety against incipient failure of rockfill at the downstream slope = (q_{THF})/(q_{IF}). If the factor of safety is ≤ 1.68, then deterministic criteria are not met.

6.8.3 Simplified QRA and Quantitative Risk Assessment

(f) Use expert elicitation to estimate the probability of failure ($P_{BreachTF}$) by comparing q to $q_{IF.}$ For design EBL (2005) adopt a factor of safety of 1.5 on D_{50}.This is equivalent to a factor of safety of 1.68 on q so the probability of failure should be relatively low for a factor of safety of 1.68.

As a guide to estimating these probabilities:

- If $q > 1.2\ q_{IF}$ then $P_{BreachTE} \approx 0.9$
- If $q = q_{IF}$ then $P_{BreachTE} \approx 0.5$
- If $q < 0.8\ q_{IF}$ then $P_{BreachTE} \approx 0.1$
- If $q < 0.3\ q_{IF}$ then $P_{BreachTE} \leq 0.01$

(g) Repeat for all reservoir stages in the risk analysis where the reservoir level is above the base of the rockfill layer.

(h) For Simplified QRA, assume that once breach commences it will progress. No allowance is made for detection and intervention.

(i) For QRA, consider detection, intervention and breach as described in Section 6.10.

6.9 OVERTOPPING AND WAVE OVERTOPPING OF PARAPET AND WAVE WALLS

If an embankment has a parapet or wave wall and it is overtopped or overtopped by wave action, as shown in Figure 6.22, failure may occur by:

(a) Structural failure of the parapet wall.
(b) Failure of the parapet wall by sliding or overturning.
(c) Scour of the downstream toe of the parapet wall by plunging flow leading to failure by sliding or overturning.
(d) Failure from overtopping or wave overtopping by the failure modes discussed in earlier sections.

Figure 6.22 shows these schematically.

Guiding principles for assessment of parapet walls due to overtopping flows and/or wave action:

- **Structural failure of the wall**: Apply structural standards or codes of practice to confirm whether the parapet wall sections have sufficient structural capacity to resist forces from

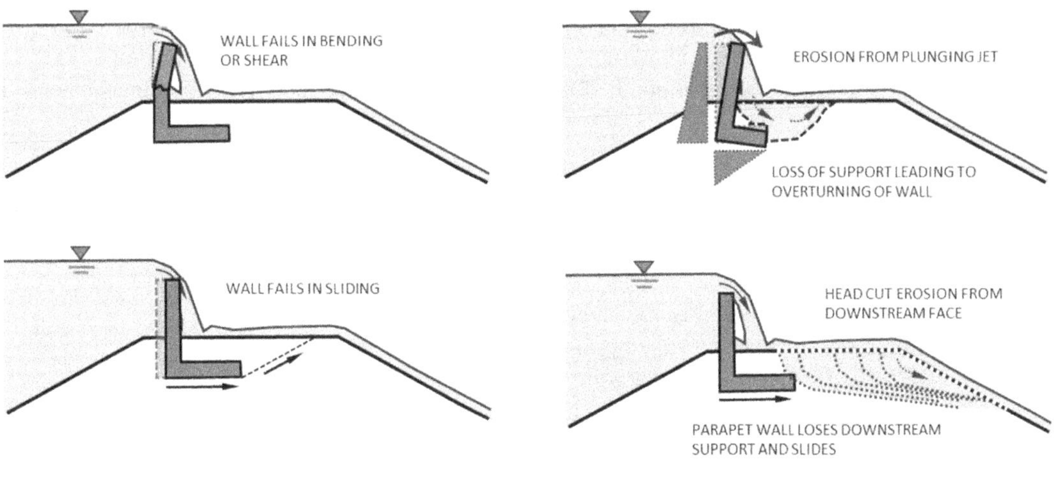

FIGURE 6.22 Potential failure modes for parapet and wave walls.

water loads, including from wave action. Guidance on estimating forces on vertical walls due to wave action is provided in Eurotop (2018).

- **Global instability (without erosion)**: Perform stability checks on the parapet wall to assess what height of reservoir level would be required for the parapet wall section to become unstable (either by sliding, overturning or bearing capacity).
- **Crest erosion leading to global instability of the wall**: Perform stability checks on the parapet wall to assess how much loss of wall backfill would be required for the parapet wall section to become unstable (either by sliding or overturning). Assess the potential for erosion by overtopping flows to remove the backfill for the wall to become unstable. Guidance on the depth of erosion of earthfill by plunging flows is provided in Mason and Arumugam (1985), Hoffmans (2009), Bombardelli et al. (2018) and Di Nardi et al. (2021).
- **Seepage and potential for uplift**: For dams where a parapet wall has been added after original construction, evaluate the joint details and potential for differential movement between the added parapet and original dam, and consider the potential for seepage resulting in uplift below the base slab of the parapet wall.

There are a wide range of designs of parapet and wave walls, so no general guidance is practicable. Often minor details can be important. Chapter 19 provides additional information.

6.10 CONSIDERATION OF DETECTION, INTERVENTION AND BREACH IN EVENT TREE ANALYSES

Event tree analyses for QRA should consider the likelihood of successful detection and intervention of the developing overtopping failure.

Some factors affecting the likelihood of detection include whether:

- Reservoir levels are monitored and the levels transmitted to dam safety personnel.
- There are dam operating personnel at the dam at the time of the flood able to observe reservoir levels.
- Flood forecasting is available from the Bureau of Meteorology.
- The rate at which the reservoir level rises in the flood.

Some factors which affect the likelihood of successful intervention to prevent commencement of breach are:

- The depth and duration of overtopping. Embankments may survive if overtopped by small depths and short duration.
- The depth and duration of wave overtopping if the maximum reservoir level is below the crest level.
- The length of the embankment crest being overtopped.
- The availability of measures to temporarily raise the crest level, e.g. sand bags.
- The ability to draw down the reservoir.
- The safety of personnel attempting to intervene.

Each dam will have particular characteristics, so these conditional probabilities will need to be estimated by expert elicitation.

It is expected that for most dams the likelihood of successful detection and intervention will be low.

For Simplified QRA, it will be sufficient to assume that detection and intervention are unsuccessful.

Once breach from overtopping commences it is highly likely it will progress, and the reservoir will be lost. If particular circumstances indicate that may not be the case (for example, if the reservoir level drops below the invert before full breach develops), then the conditional probability of breach may be included in the event tree analysis.

Slope Instability and Deformations Leading to Loss of Freeboard and Overtopping

7

7.1 DATA ASSEMBLY AND ASSESSMENT

7.1.1 Data Required to Assess Factors of Safety and Deformations

The following is a check list of information which will be required to assess factors of safety and deformations.

7.1.1.1 Hydrology

(a) Annual Exceedance Probability (AEP) of reservoir levels from Full Supply Level (FSL) to dam crest level.
(b) AEP of the historic high reservoir level.
(c) Historic reservoir levels as a percentage time versus level.
(d) Outlet capacity to draw down the reservoir if required.

7.1.1.2 Dam Embankment

(a) The actual embankment zoning, particularly in the upper parts of the embankment.
(b) The classification, particle size distribution and Atterberg limits of the material in each zone including consideration of segregation (for filter and rockfill materials) and sub-layering in rockfill where susceptible to breakdown.
(c) The specified and actual layer thickness, degree of compaction and compaction moisture content relative to Standard Optimum Moisture Content of all earthfill zones.
(d) The shear strength, compressibility and permeability of each earthfill, filter and rockfill zone.

DOI: 10.1201/9781003542025-7

(e) The specified and actual layer thickness and degree of compaction of filters.
(f) The specified and actual layer thickness, roller details and number of passes for rockfill zones.
(g) Records of quality control testing during construction to support (b) to (f).
(h) Results of site investigations and laboratory testing carried out since construction of the dam.
(i) Pore pressure distributions within the embankment and how these have changed over time since construction and their relation to reservoir level.
(j) Post-construction crest settlement and deformation records, and if available other deformation data. Where available, cross arm settlement data.

7.1.1.3 Soil Foundations

(a) The stratigraphy, continuity and geological origins of the foundation soils.
(b) The classification, particle size distribution and Atterberg limits of each soil.
(c) The shear strength, compressibility and permeability of each soil.
(d) The presence of low strength defects or seams in soil foundations (including fissuring, relict defects from weathering, pre-sheared surfaces from landslides), and if present, their shear strength or influence on shear strength of the soil mass.
(e) The presence of non-plastic soils and their SPT and CPTU properties which allow assessment of the shear strength and compressibility of the soils.
(f) The level to which the core and the shoulders of the embankment were taken into the foundation and the criteria which defined these levels.
(g) Piezometer data where available and the relation to reservoir and tailwater levels.
(h) Photographs taken during construction.

7.1.1.4 Rock Foundations

(a) The stratigraphy, continuity and geological origins of the foundation rock.
(b) Where applicable, bedding thickness and orientation.
(c) The orientation, persistence and nature of rock defects, i.e. joints, shears and faults.
(d) The presence of low strength defects or seams in the foundations (including bedding crushed seams and shears resulting from valley stress relief and pre-sheared surfaces from landslides, and if present, their location, orientation persistence and shear strength.
(e) The degree of weathering insofar as this affects other properties.
(f) The level to which the core and the shoulders of the embankment were taken into the foundation and the criteria which defined these levels.
(g) Piezometer data where available and the relation to reservoir and tailwater levels.
(h) Photographs taken during construction.

The foundations conditions are often not uniform across the valley in which the dam is sited. Abutments may have different engineering geological conditions to the river section, and soil conditions almost always vary within the river section.

7.1.2 Engineering Embankment and Engineering Geology Models

Develop an engineering embankment model and engineering geology model of the foundation, geotechnical models of the embankment and foundation, and from these the analytical geotechnical model.

See Sections 4.2.1 and 4.2.2 for details. In most cases these models will be developed progressively, initially on available data, and updated as additional site investigation data becomes available.

Select representative cross sections of the embankment and its foundation for stability analyses.

7.2 SHEAR STRENGTH OF CLAY SOILS

7.2.1 Drained Strength—Definitions

When soil is sheared slowly in a drained condition (so that there is sufficient time for dissipation of pore pressures induced by shearing), the stress displacement curves will take the general form shown in Figure 7.1.

The behaviour is dependent on whether the soil has a high or low clay fraction (finer than 0.002 mm) content and on whether it is normally consolidated (NC) or overconsolidated (OC), but has the following common features:

- A peak strength is obtained at a small displacement.
- A reduction of strength to the critical state or fully softened strength then occurs with further displacement. For overconsolidated soils this is due to an increase in moisture content with dilation of the soil as it is sheared. The fully softened strength corresponds to the critical state (Skempton, 1985), i.e. when continuing displacement occurs without further change in

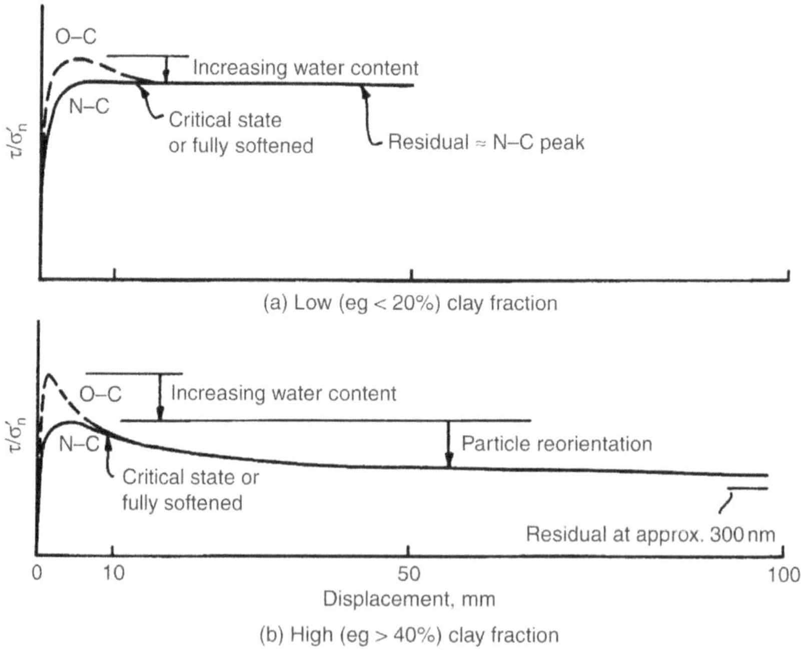

FIGURE 7.1 Diagrammatic stress displacement curves at constant normal stress σ'_n (Skempton, 1985). (Used with permission of Emerald Publishing conveyed through Copyright Clearance Center Inc.)

volume or moisture content. Note that normally consolidated and overconsolidated samples of the same soil will tend to achieve the same critical state or fully softened condition.

- With continuing displacement of soils with a high clay fraction content, particle reorientation occurs, resulting in a further reduction in shear strength. The minimum value of shear strength achieved at large displacements is the "residual" strength.
- The residual and fully softened strengths are significantly different for high clay fraction content soils, but not for soils with low clay fraction content (see Figure 7.1).
- Strength envelopes for these cases are defined by:
    - peak strength c' ϕ'
    - softened strength c'_s ϕ'_s
    - residual strength c'_R ϕ'_R

Figure 7.2 shows the relationship between these parameters. Note that often $c'_s \approx c'_R \approx 0$, and $\phi'_s \approx \phi'$ and $c' = 0$ if curvature of the strength envelope is allowed for.

An undisturbed sample of soil, e.g. from the core of a dam, may behave in an overconsolidated manner at low normal stress and in a normally consolidated manner at high normal stresses (in excess of the pre-consolidation pressure). This affects the pore pressure response of the soil during shear (see Section 7.2.3) and also the drained load-deformation behaviour as shown in Figure 7.3.

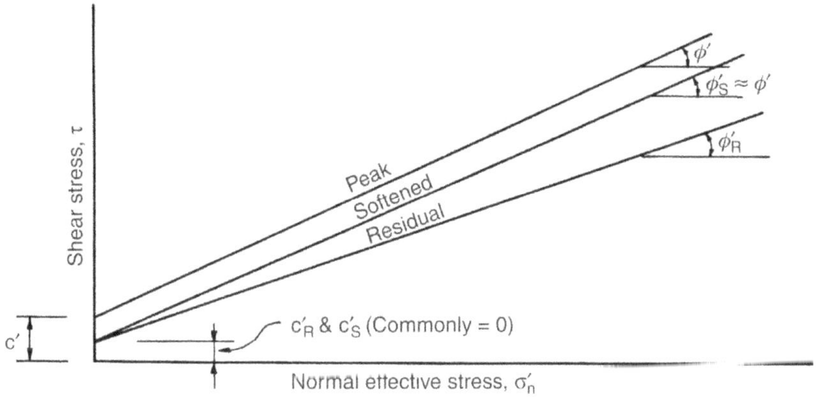

FIGURE 7.2 Relationship between peak, softened and residual strengths (Fell et al., 2015).

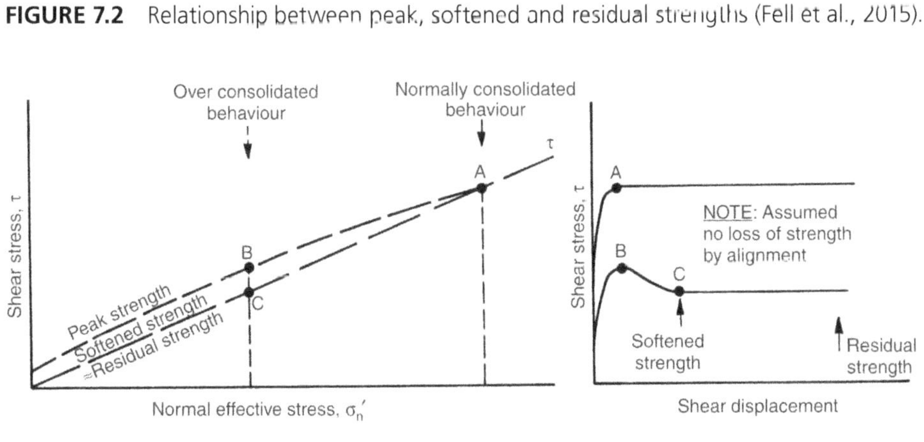

FIGURE 7.3 Diagrammatic stress displacement behaviour (Fell et al., 2015).

7.2.2 Development of Drained Residual Strength ϕ'_R

Lupini et al. (1981) carried out a series of ring shear tests on sand-bentonite mixtures and suggested that there were different mechanisms of residual plane development depending on the clay fraction percentage present. The mechanisms are turbulent or rolling shear, where the presence of rounded silt size particles prevents alignment of clay particles on the shear surface; sliding shear when the effects of the silt is overridden by the predominance of clay particles; and a transition between these two conditions. These results were summarised by Skempton (1985) and are reproduced in Figure 7.4.

Skempton (1985) presented results of field residual (strengths back-analysed from slope failures) and ring shear tests on a range of soils. These are reproduced in Figure 7.5.

These figures show that "sliding shear" with complete reorientation of the clay particles is likely to occur only where the clay fraction (finer than 0.002 mm) exceeds 50% of the total soil, and that for less than about 25% clay fraction, "turbulent" or "rolling" shear occurs without the influence of clay particle alignment. Note that Figure 7.5 applies to clays with a Plasticity Index/Clay Fraction (PI/CF) ratio of 0.5 to 0.9.

It should be noted that the rather small range of residual strength for each clay size fraction in Figure 7.5 is an artefact of the few data points used in the plot. As shown in Fell et al. (2015), the actual range is larger, e.g. for high clay fraction soils the range is about 8° to 14°.

Apart from the clay fraction, the mineralogy of the clay also has an effect on residual strength. This is particularly so when the clay fraction is large. This reflects the fact that the different clay minerals have different particle shape and different inter-particle bonding. Most clay minerals, e.g. Kaolinite, Illite, Chlorite and Montmorillonite, are platy structures and are therefore subject to alignment when sheared. Montmorillonite has a particularly thin plate structure and weak interplate bonds and leads to the lowest residual strength value with ϕ'_R as low as 5°. Kaolinite has a ϕ'_R of approximately 15° and Illite approximately 10°.

Some clay minerals do not have a plate structure, e.g. halloysite has a tubular structure, attapulgite a needle like structure and some amorphous clay sized minerals such as gibbsite, haematite and bauxite

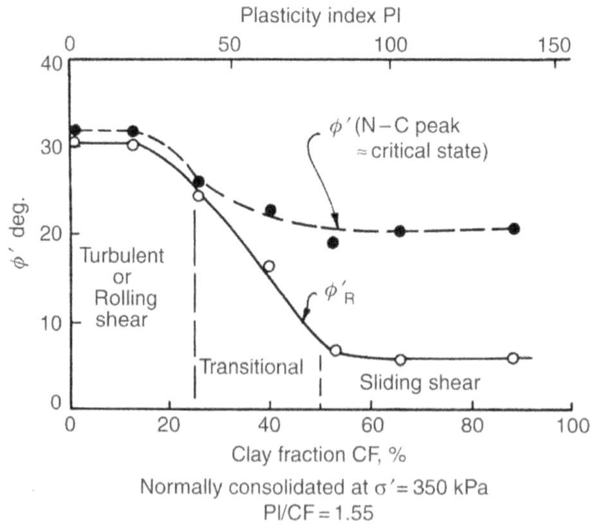

FIGURE 7.4 Ring shear tests on sand-bentonite mixtures (from Skempton, 1985 after Lupini et al., 1981). (Used with permission of Emerald Publishing conveyed through Copyright Clearance Center Inc.)

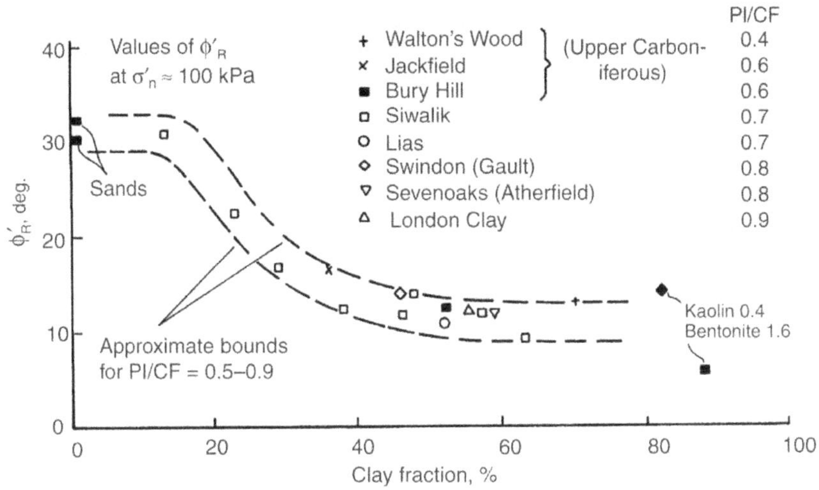

FIGURE 7.5 Field residual and ring shear tests on sands, kaolin and bentonite (Skempton, 1985). (Used with permission of Emerald Publishing conveyed through Copyright Clearance Center Inc.)

have essentially granular structures. This leads to much higher residual friction angles, commonly greater than 25°.

7.2.3 Undrained Strength Conditions

Undrained strength conditions occur when the loading time and the properties of the soil in the slope are such that there is insufficient time for pore pressures generated during the loading period to dissipate. Duncan (1996) suggests this can be assessed from the dimensionless time factor, T, which is expressed as:

$$T = \frac{Cv\, t}{D^2} \tag{7.1}$$

where c_v = coefficient of consolidation (m²/year); t = loading time (years); D = length of the drainage path (metres).

Duncan indicates that, if the value of T exceeds 3.0, it is reasonable to treat the material as drained, and, if T is less than 0.01, undrained. If the value of T is between 3.0 and 0.01, both undrained and drained strengths should be considered. The undrained strength of a soil may be less than or greater than the drained strength depending on whether the soil is contractive or dilatant when it shears.

It is important to note that the strength available in the case of undrained shear of the contractive soil (Su) is less than that for drained loading and less than assumed for conventional effective stress analysis, because of the positive pore pressures generated during shearing. As discussed later, this can happen in coarse grained as well as fine grained soils.

Cooper et al. (1997) discuss an example where the use of conventional effective stress analysis overestimated by about 50% the factor of safety of an earth dam constructed with poorly compacted clay which, if saturated, is contractive on shearing and would generate excess pore water pressure.

7.2.4 Triaxial Testing for Drained Strength Parameters, Some Common Errors

It is the authors' experience that many organisations, at least in the past, failed to follow correct test procedures despite the fact that they have been long established. This needs to be considered when reviewing laboratory testing for existing dams.

There are several common sources of error in triaxial testing, which often lead to an overestimation of effective cohesion, usually with a reduction of effective friction angle. This may lead to an overestimation of strength in the working stress range. The sources of error are as follows:

(i) *Testing at a too high stress range*. The Mohr's circle envelope for most overconsolidated soils is curved in the manner shown in Figure 7.6. If laboratory tests are carried out at higher stresses than will be encountered in the field, this will result in an overestimation of shear strength in the field. If the field stresses were between 0 and 100 kPa and testing had been carried out at confining stresses between 100 and 400 kPa, the strengths would be overestimated by 10% to 300%. This potential problem can be overcome by specifying the correct stress range for the triaxial tests and by recognising possible curvature effects when selecting design parameters from the p′-q plot. It is also important to plot the individual Mohr's circle plots for a sample, so the curvature can be identified.

(ii) *Not saturating the sample adequately*. Soil samples (even if taken from below the water table) may not be saturated before testing in the laboratory. Compacted soils even if compacted wet of optimum moisture content, as is commonly required in dam construction, are usually only 90% to 95% saturated.

If triaxial tests are carried out correctly, the samples are back pressure saturated and the pore pressure response to added confining stress checked to confirm the samples are saturated. Some earlier triaxial tests were not saturated.

The effects of partial saturation are:

- Pore pressure changes during shearing are less than for a saturated sample. However, the trend (whether +ve or −ve changes) is not affected (Lade, 1986).

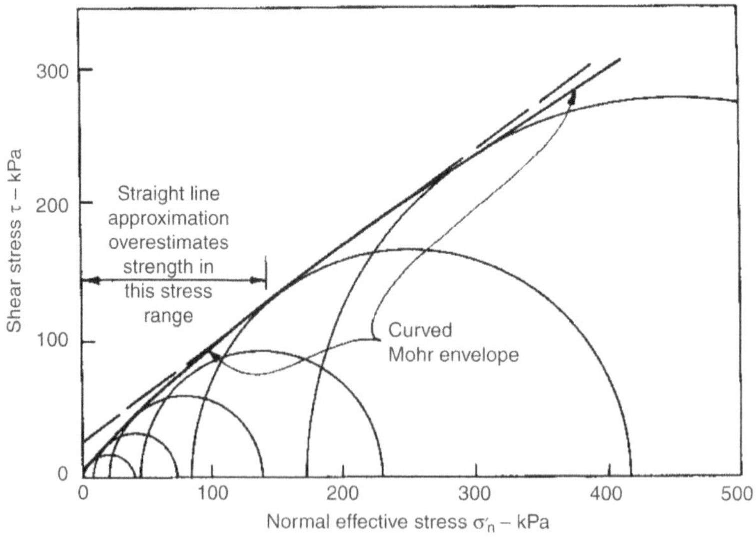

FIGURE 7.6 Typical Mohr's envelope plot (Fell et al., 2015).

- Because pore pressure changes are less than for a saturated soil, the undrained strength will be affected. For overconsolidated clays, Δu is −ve, and lower than for a saturated soil, so the undrained strength is reduced; for normally consolidated clays, Δu is +ve and lower than for a saturated soil, so the undrained strength is increased.
- The magnitude of actually measured pore pressures may be incorrect, usually underestimated, because of the presence of air in the pore system.

Provided the pore pressures are measured correctly, the effective strength envelope will not be greatly affected even though the total stress envelope is markedly affected.

These effects cannot be quantified when reviewing test data, but the potential effects should be considered.

(iii) *Testing at too high a strain rate.* Consolidated Drained (CD) and Consolidated Undrained with Pore Pressure measurement (CUDPP) triaxial tests must be sheared at a sufficiently slow rate to allow dissipation of pore pressure in the CD test and equalisation of pore pressure throughout the sample in the CUDPP test.

The effect of testing at too high a strain is illustrated in Figure 7.7 for an overconsolidated clay. If a CUDPP test is sheared too quickly, the pore pressure changes measured at the ends of the sample are less than the actual changes at the centre of the sample where shearing is occurring, because of the restraint effects of the top and bottom of the sample. Since there is a reduction in pore pressure during shearing of an overconsolidated clay, the Mohr circle plots based on measured pore pressures are displaced to the left of where they should be.

The negative pore pressure response to shearing is less for the sample at higher confining stress where the soil behaves closer to a normally consolidated soil, so the displacement of the Mohr's circle for the higher confining stress is less than that for the lower confining stress.

This yields overall an increase in c′, and a decrease in ϕ′. The effect is similar for CD tests, where negative pore pressure will remain in the centre of the sample if the strain rate is too high, giving an apparent increase in failure stress (σ'_1), which again will be greater for samples at lower confining stresses than for higher confining stresses.

(iv) *Staged testing.* In staged testing, the one soil sample is usually saturated, consolidated and sheared at the lowest selected confining stress, then consolidated to the second confining stress and sheared, then further consolidated to the third confining stress and sheared. This reduces the amount of sample preparation time, and hence the cost of testing. The procedure often gives acceptable results but may tend to give lower strengths for the second and particularly third stage, when testing strain weakening or compacted dilative soils due to sample dilation and softening and loss of strength due to displacement on the shear plane.

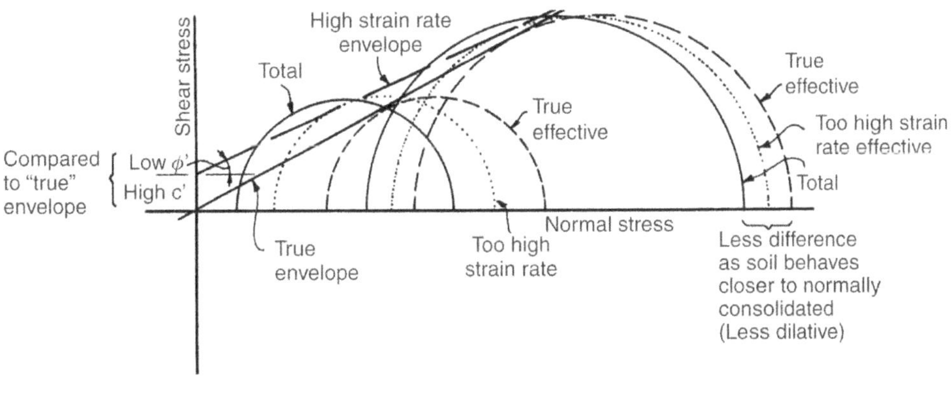

FIGURE 7.7 The effect of high strain rate on CUDPP triaxial tests (Fell et al., 2015).

Where staged testing gives high effective cohesion, it is suggested that the second and third stages be given less weight in interpreting the strengths and assume they are underestimating the strength.

7.2.5 Direct Shear Test

Direct shear tests are used to obtain the peak and fully softened or critical state (constant volume) strengths of granular soils (sand, silt, gravel) and plastic soils. They are also used to determine the residual strength of plastic soils, by using the multiple reversal technique, which is described later, and the strength on existing surfaces of weakness in soil, e.g. shear surfaces, slide surfaces and fissures.

These tests are all done in drained conditions. Those for plastic soils are saturated prior to testing. The test is seldom used to determine undrained strengths and in the opinion of the authors should not be.

It is the authors' experience that when direct shear and triaxial test data are both available, the peak strengths derived from direct shear tests are sometimes higher than from the triaxial tests. This is mainly due to the inability to saturate the soil properly, along with the problem that the direct shear forces failure on a predetermined plane, which is not coincident with the principal stress direction at failure. This means that care should be taken in using the direct shear results and the test should not be used as the primary means to obtain peak strengths for plastic soils. The direct shear is a suitable way to obtain the residual strength provided the tests are carried out correctly.

True laboratory residual strength of clay soils only develops with significant displacement. Skempton (1985) gives the "typical displacement values" reproduced in Table 7.1.

Most laboratory shear box equipment is 75 mm × 75 mm or 60 mm × 60 mm, and the maximum practical displacement is of the order of 6 mm to 10 mm, sufficient only to measure the peak and possibly the softened strength.

Residual strength is therefore obtained with a direct shear machine by repeatedly shearing the sample until the strength is not further reduced by further shearing (the "multiple reversal" method). Typical load displacement curves for "turbulent" and "sliding" shear are shown in Figure 7.8 for tests which are progressing well. Most tests are more irregular.

The repeated shearing is achieved in several ways:

- Shear to limit of travel, unload (partly), rewind rapidly to start, then reload and re-shear etc. until a minimum is reached, or

TABLE 7.1 Typical Displacements at Various Stages of Shear in Clays Having Clay Fraction > 30% (Skempton, 1985)

	DISPLACEMENT MM[(2)]		
STAGE	*OVERCONSOLIDATED*		*NORMALLY CONSOLIDATED*
Peak	0.5–3		3–6
Rate of volume change approx. zero[(1)]		4–10	
At residual + 1° (ϕ'_R + 1°)		30–200	
At residual (ϕ'_R)		100–500	

Notes:
(1) i.e. fully softened strength.
(2) For σ_n' < 600 kPa.
(3) Clay fraction = % finer than 0.002 mm.

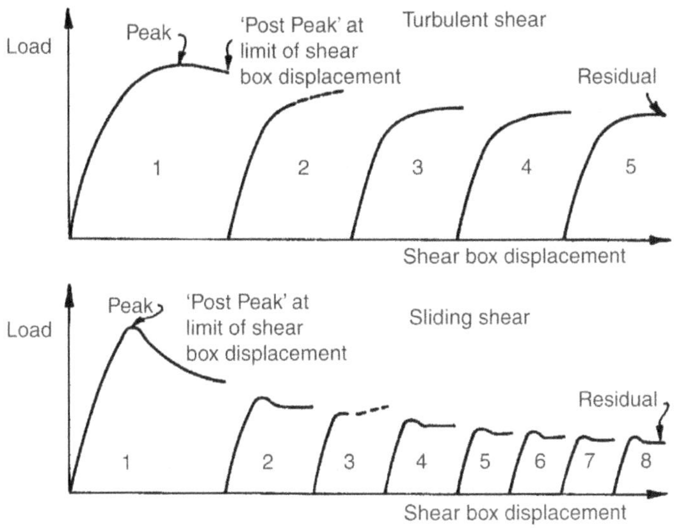

FIGURE 7.8 Typical load displacement curves for direct shear tests (Moon, 1984).

- Shear to peak, then unload partly, and wind back and forth rapidly until a total travel of say 50 mm is achieved, reload and re-shear etc. until a minimum is achieved, or
- Shear back and forth under load at a slow strain rate until a minimum is achieved.

Further acceleration of achievement of the residual strength is sometimes accomplished by cutting a slide plane into the sample prior to testing. The cut surfaces are sometimes polished on a glass plate prior to testing.

Problems which arise in the use of the direct shear to assess peak strength include the following:

- *It is difficult to ensure that the sample is properly saturated.* Saturation is achieved by soaking under a confining stress until there is no further volume change. Unlike triaxial testing, pore pressure cannot be measured, so it is necessary to rely on monitoring the consolidation deformation of the sample during saturation and consolidation. The best approach is to saturate under a low confining pressure until swell (or consolidation) ceases, then load to the first test confining pressure, and monitor until settlement ceases.
- *Testing at a too high strain rate* leads to overestimation of shear strength in a similar way to triaxial testing, i.e. c′ is overestimated, ϕ' underestimated. This can be overcome by monitoring the consolidation to obtain t_{50}, t_{90} or c_v value, and then using formulae developed by Akroyd (1975), Bowles (1978) or Head (1985) to estimate the strain rate. Skempton (1985) showed, however, that the residual strength is not greatly sensitive to strain rate provided reasonably low strain rates are used. This fact may be used to speed up test rates, provided that when a constant shear stress has been reached the strain rate is reduced, and the resulting shear stress is not significantly lower than the previous minimum (see Figure 7.8).
- Testing at a too high confining stress range and staged testing can lead to an overestimation of c′ and underestimation of ϕ' in the same way as for triaxial testing.

Problems which arise in the use of the direct shear to assess residual strength include the following:

- Repeated reversal of the direction of shearing is necessary, and this may partially destroy the alignment of particles on the shear plane, preventing a true residual strength being achieved.

- Repeated reversal of the shear box often leads to soil squeezing out between the two halves of the box. If this happens, or if the box tilts, interference can occur between the two halves of the box leading to erroneous results.

The authors' experience is that one of the biggest problems is that laboratory personnel (and the engineers supervising them) do not understand the definition of residual strength. Figure 7.9 shows some of the all too common misconceptions. All are likely to result in overestimation of the actual residual strength.

If direct shear test data is available for an existing dam, the laboratory test sheets should be reviewed to see if the results are reliable, taking account of the preceding comments.

7.2.6 Ring Shear Test

In the ring shear, an annular ring shaped specimen (Figure 7.10) is subject to a constant normal stress σ'_n confined laterally and caused to shear on a plane of relative rotary motion at a constant rate of rotation.

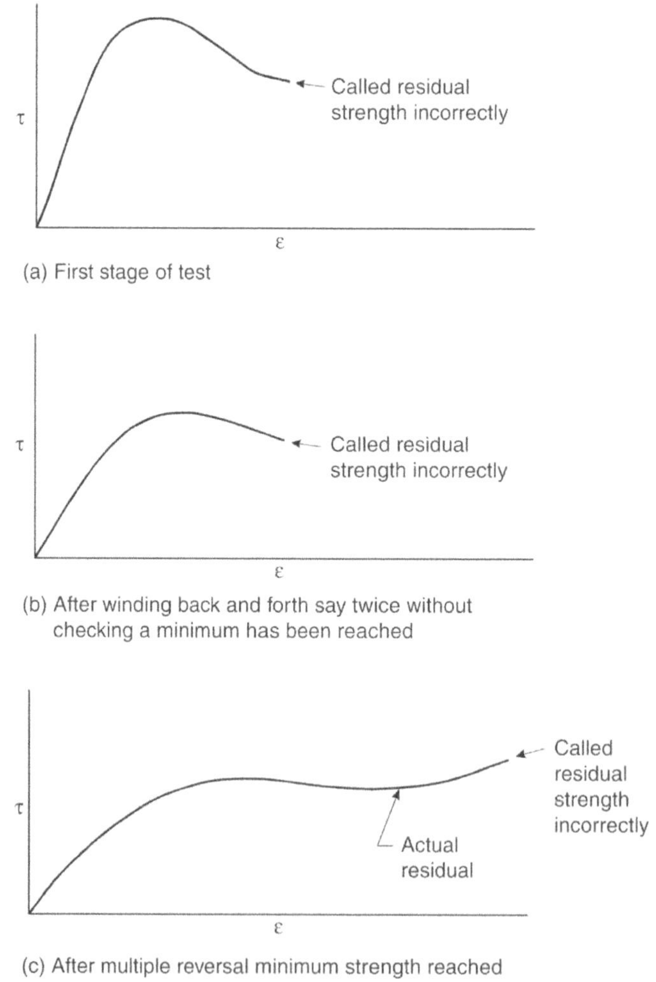

FIGURE 7.9 Some all too common misconceptions in interpreting residual strength from direct shear tests (Fell et al., 2015).

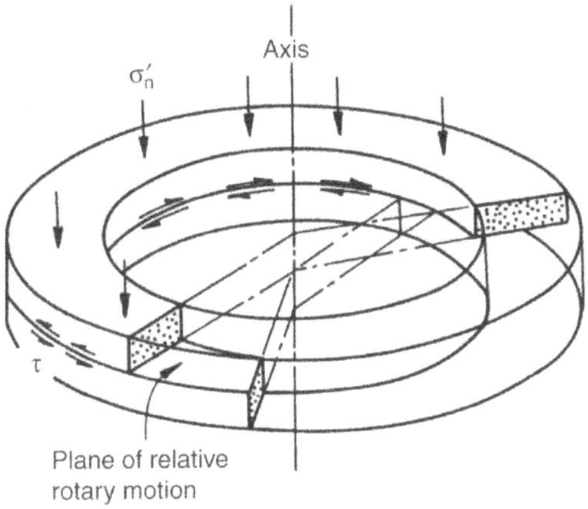

FIGURE 7.10 Ring shear test sample (Fell et al., 2015).

The test procedure is detailed in BS1377–7:1990 and ASTM D6467–21e1. The sample in the Bromhead machine is outside diameter 100 mm, inside diameter 70 mm and only 5 mm thick.

The ring shear devices have essentially unlimited "strain" availability, overcoming the major objection to the reversing direct shear test. However, there are limitations:

- Only remoulded samples can be tested.
- Only the residual strength can be obtained. The peak strength is affected by remoulding and non-uniform stress conditions.
- The sample may tend to squeeze out the sides of the ring and lead to the top and bottom parts of the ring interfering with each other, giving erroneously high strengths. This is usually apparent from observations of the tests and the test data.
- Saturation and testing at a strain rate appropriate to give the required drained conditions is necessary as for the direct shear test.
- Any particles of sand in the soil can form a groove on the shear surface and give erroneously high strengths. The authors have overcome this by sieving the dried sample on a 150 micron sieve and testing the minus 150 micron soil placed in the ring shear at about the plastic limit.
- As summarised in Fell et al. (2015), there is literature to indicate that ring shear tests may underestimate the actual residual strength by 1° to 2°.

If ring shear test data is available for an existing dam, the laboratory test sheets and results should be reviewed to see if the results are reliable, taking account of the preceding comments.

7.2.7 Laboratory Testing for Undrained Strength

The undrained strength of plastic soils can be obtained from several laboratory tests:

(a) Unconfined compression (UC)
(b) Unconsolidated undrained (UU)
(c) Consolidated undrained (CU)

It must be recognised that the undrained strength is affected by:

- The degree of disturbance of the soil sample. Disturbance can occur because of stress relief on sampling, the sampling technique (disturbance as the thin walled tube or other sample is pushed into the soil) and handling (e.g. drying or disturbance during transportation, extrusion of the sample from the sampling tube, or trimming in the laboratory).
- The moisture content of testing, e.g. soils allowed to dry from their field moisture content, will be stronger than those tested at the field moisture content. Thin walled tubes should be sealed as soon as they are taken to avoid drying.
- The stress path, in particular whether tested in compression, extension, or direct simple shear (see Figure 7.11).
- The test procedure, e.g. UC and UU tests do not simulate the *in situ* test stresses.

Sample disturbance for soft clays, such as may be present in foundations of small dams, puddle cores or saturated, poorly compacted clay fill, and the sensitivity of undrained strength to moisture content for stiff clays, make UC and UU tests totally unreliable, and they should not be used for design purposes. In particular for soft clays, they may seriously underestimate the real strength—but you never know by how much.

Tests should therefore be done by CU methods, and they should, so far as practicable, simulate the *in situ* stress conditions, and the stress path. More details are given in Fell et al. (2015)

7.2.8 Undrained Strength from Cone Penetration (CPT) and Piezocone (CPTU) Tests

The undrained strength of clays can be estimated from the cone resistance q_c of the static cone penetrometer test (CPT).

The undrained strength Su is related to q_c by

$$Su = \frac{q_c - \sigma_{vo}}{N_K} \tag{7.2}$$

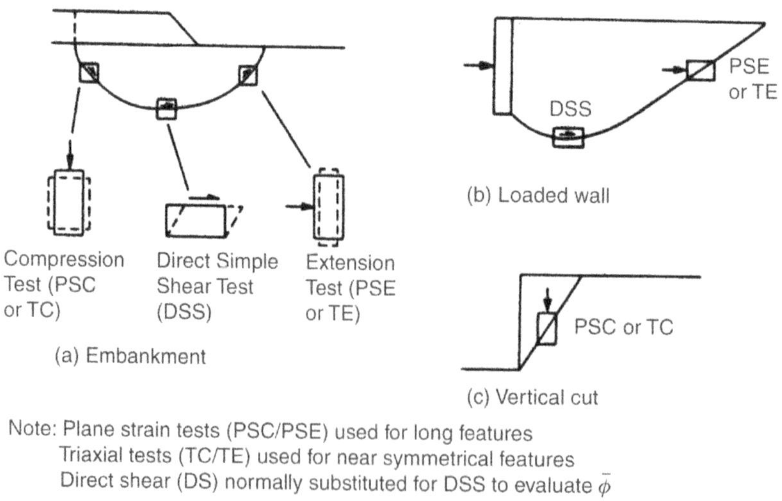

FIGURE 7.11 Relevance of laboratory shear tests to shear strength in the field (Kulhawy, 1992). (Used with permission of ASCE conveyed through Copyright Clearance Center Inc.)

where:

q_c = cone resistance

σ_{vo} = *in situ* total overburden stress

N_K = empirical cone factor.

Figure 7.12 shows the relation between N_K and Plasticity Index of the soil.

If piezocone (CPTU) tests are carried out, the corrected cone resistance q_r is calculated from:

$$Su = \frac{q_T - \sigma_{VO}}{N_{KT}} \tag{7.3}$$

where N_{KT} is the empirical cone factor related to q_T.

Aas et al. (1986) present cone factors N_{KT} for several sites in Norway. These are reproduced in Figure 7.13.

N_{KT} varies from about 12 to 22 for normally consolidated soils. N_K and N_{KT} values show a much larger range for overconsolidated clays.

Given that the value of the correlation factor N_K or N_{KT} is uncertain, strengths cannot be estimated with a high degree of accuracy. The potential error is higher in overconsolidated clays. This uncertainty should be taken into account when reviewing data available for a dam safety assessment.

The correlations between undrained strength and cone resistance are developed for saturated soil conditions. When the CPT or CPTU is used in dams, the soils may be partially saturated above the phreatic surface. In view of this, where the undrained strength is needed with some accuracy, the CPT or CPTU strengths should be calibrated against laboratory tests on samples from the dam and *in situ* shear vane testing where this is practical to do so.

The CPT and CPTU have the advantage of providing a continuous profile of q_c and hence a continuous profile of undrained strength. Lower undrained strength layers may be detected in embankment

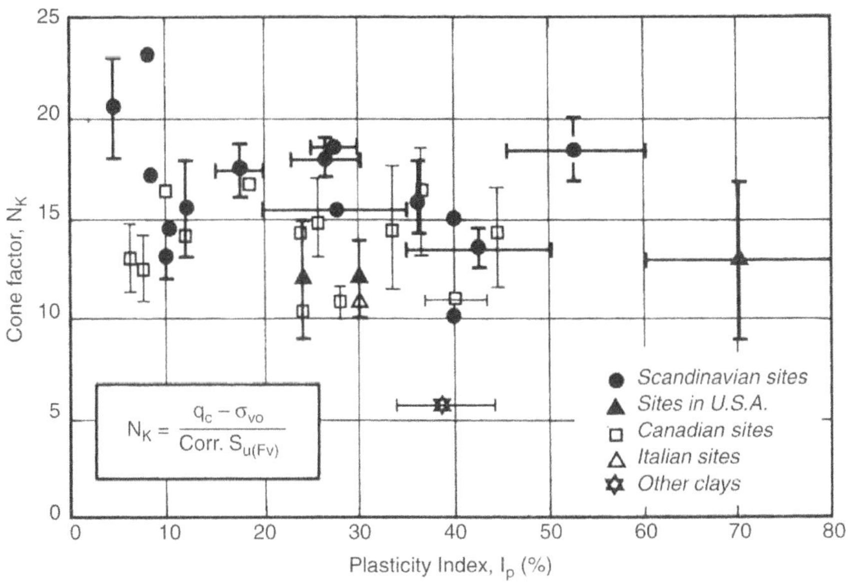

FIGURE 7.12 Empirical cone correction factor N_K for CPT based on corrected vane shear strengths (Aas et al., 1986). (Used with permission of ASCE conveyed through Copyright Clearance Center Inc.)

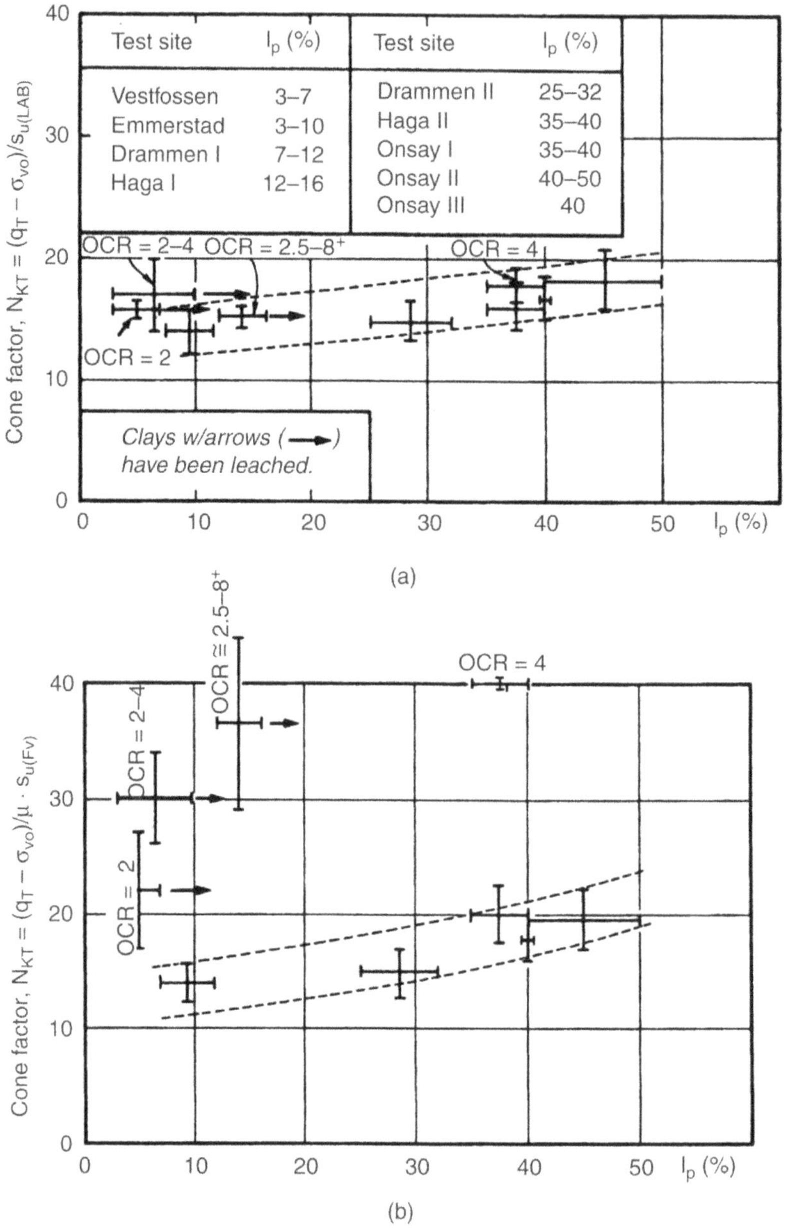

FIGURE 7.13 Cone factor N_{KT} for CPTU in normally consolidated and overconsolidated clays (Aas et al., 1986). (Used with permission of ASCE conveyed through Copyright Clearance Center Inc.)

cores or foundations, and the CPT and CPTU plots should be inspected to see if such lower strength zones are present.

The CPT or CPTU may not be able to penetrate well compacted earthfill. They will refuse on gravelly soils.

CPT or CPTU probes through dam cores should be backfilled as detailed in Appendix C. Care should be taken to avoid hydraulic fracture or not fully backfilling the probe hole and, in some situations, it may be judged too risky to probe through the core.

7.2.9 Undrained Shear Strengths from Vane Shear Tests

The vane shear may be used to measure the undrained strength of very soft to firm (Su less than about 60 kPa) plastic soils. The strength measured by the vane should be corrected to account for the soil plasticity, anisotropy and rate of shearing.

This may be done using the Azzouz et al. (1983) correction using the equation $Su = \mu S_{uv}$ where Su = design undrained shear strength; S_{uv} = undrained strength measured by the vane; μ = correction factor from Figure 7.14.

7.2.10 Estimation of the Undrained Strength of Overconsolidated Soils from OCR

Jamiolkowski et al. (1985) showed that the normalised shear strength Su/σ'_{vc} versus OCR graphs for some soils were close to linear on a log-log plot. Hence:

$$(Su/\sigma'_{vc})_{OC} = (Su/\sigma'_{vc})_{NC}\,(OCR)^{m} \tag{7.4}$$

where σ'_{vc} = effective vertical stress in the test.

For non-cemented clays with a Plasticity Index I_P less than 60%, $m \approx 0.8$ and $(Su/\sigma'_{vc})_{NC} = 0.23 \pm 0.04$. So:

$$(Su/\sigma'_{vc})_{OC} = (0.23 \pm 0.4)\,OCR^{0.8} \tag{7.5}$$

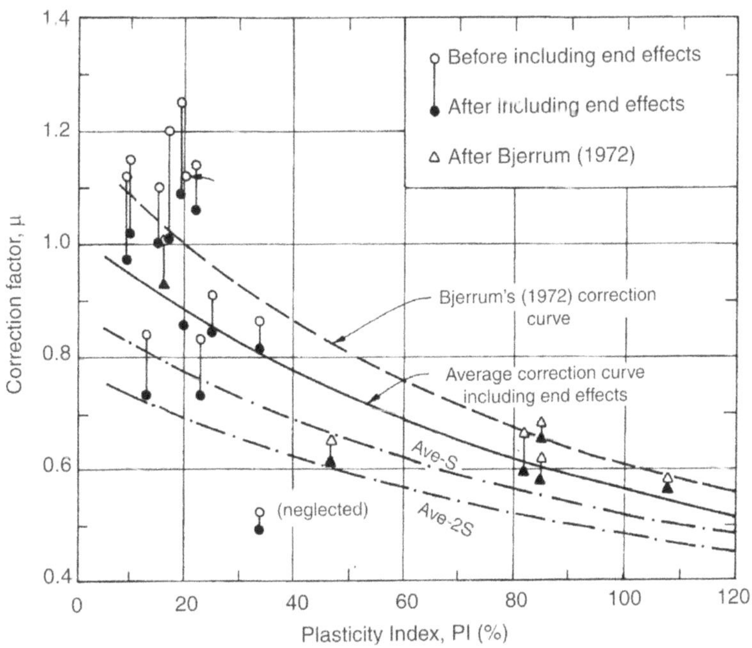

FIGURE 7.14 Vane shear correction factors (Azzouz et al., 1983). (Used with permission of ASCE conveyed through Copyright Clearance Center Inc.)

This method can be used as an initial estimate of strength, or as a check on other methods.

7.2.11 Estimation of Undrained Strength from Effective Stress Shear Parameters

Wroth and Houlsby (1985) give for normally consolidated soils:

$$\left(Su/\sigma'_{vo}\right)_{NC} = 0.5743\,\frac{3\sin\phi'_{tc}}{3-\sin\phi'_{tc}} \tag{7.6}$$

where subscript tc indicates triaxial compression tests with isotropic consolidation.

Thorne (1984) gave the relationship:

$$Su_{tc} = \frac{c'\cos\phi' - Uo\sin\phi'}{1-\left(1-2A_f\right)\sin\phi'} \tag{7.7}$$

where $U_o = \dfrac{-(1+2Ko)\sigma'_{vo}}{3} = \dfrac{-\left(\sigma_{vo'} + 2\sigma'_{ho}\right)}{3}$;

A_f = Skempton "A" at failure, c′ = effective cohesion, ϕ′ = effective friction angle, Ko = earth pressure coefficient at rest.

This relationship is applicable to both normally consolidated and overconsolidated clay, with A_f and Ko dependent on the degree of overconsolidation.

7.3 ASSESSMENT OF DRAINED, EFFECTIVE STRESS SHEAR STRENGTHS OF CLAY SOILS

7.3.1 Peak, Residual or Fully Softened Strength?

Whether peak, softened or residual strengths are used in the analysis of slope stability depends on the presence or absence of existing slide surfaces or fissuring. The following guidelines are given:

- Where there is an existing slide surface, e.g. in the foundation of a dam or where a dam has failed by sliding, the field residual strength should be used for the slide surface. This applies regardless of how long it has been since sliding last occurred. The exception could be where there is definite evidence of re-cementing.
- Bedding surface shears formed by folding of rock strata have strengths approaching residual and, unless extensive tests on the surfaces show otherwise, residual strengths should be used. Allowance should be made for waviness of the shear surface and for parts of the surface which may have rock-to-rock contact.
- Compacted soils and soils which have no fissuring or cracks should be assigned peak strength parameters c′ and ϕ′ with ϕ′ determined from saturated remoulded samples

compacted to the same density ratio and moisture content the soil would be in the dam or from undisturbed samples from the embankment.

- The majority of dam embankment design and assessment of the stability of existing dams is based on peak strength parameters. If the dam has experienced significant deformations of the upstream slope under drawdown conditions, it is likely that the strains have localised to a degree and the soil dilated sufficiently to where fully softened strengths may be more applicable.
- The upper parts of the embankment above the phreatic surface will be partially saturated and have negative pore pressures. The soil in the core is also potentially overconsolidated from the rolling during construction. These effects should be allowed for when assessing the stability of the embankment as they have a significant effect on the factors of safety of the upper parts of the embankment.
- Fissured soils have between peak strength and residual strength depending on the nature of fissuring, orientation, continuity and spacing of the fissures. Such soils do occasionally occur in dam foundations, e.g. Ross River Dam and Prospect Dam. This is discussed in more detail in Section 7.3.4. Soils with relict joints, such as extremely weathered basalt, behave in a similar manner.

The effective cohesion (c′) should be zero or very small. The value of c′ adopted is particularly critical in the analysis of the stability of smaller dams and the upper parts of larger dams because it has a major effect on calculated factors of safety when failure surfaces are shallow. It is recommended that the effective cohesion adopted for design should be between 1 kPa and 5 kPa for peak strength and should be zero or say 1 kPa for residual and softened strengths. The adoption of 1 kPa helps avoid obtaining low factors of safety for very shallow surfaces parallel to the downstream slope.

In situations where the strength envelope is markedly curved or gives a higher effective cohesion, a bi-linear envelope such as that shown in Figure 7.15 should be used.

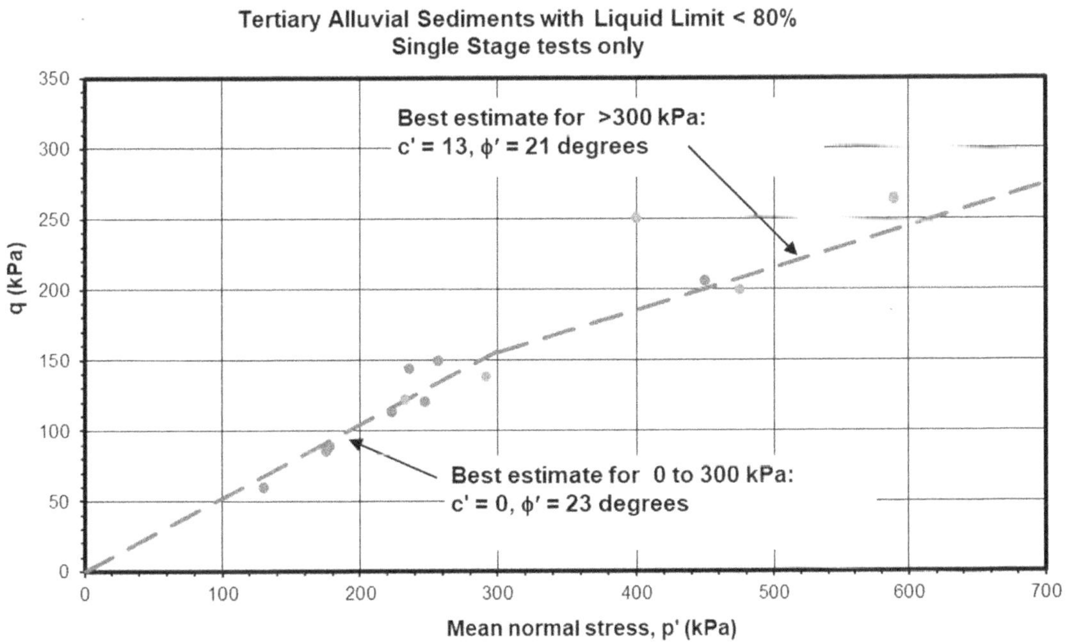

FIGURE 7.15 Example of p′-q plot of triaxial shear tests which exhibit bi-linear characteristics.

7.3.2 Review of Laboratory Test Results

Laboratory triaxial and direct shear tests should be reviewed taking account of the discussion in Sections 7.2.4 and 7.2.5.

Tests with high c′ and low ϕ′, for example, c′= 100 kPa, ϕ′ = 14°, are almost certainly erroneous because as shown in Figure 7.16 soils do not have peak ϕ′ < 20° and most ϕ′ are not < 25°. Test data which are clear outliers to the rest of the tests should be reviewed and the reason for the outlier determined. In many cases the result should be discarded.

7.3.3 Selection of Effective Stress Design Parameters in Clay Soils

7.3.3.1 Dams for Which There Are No Laboratory Strength Tests

For dams where no triaxial or direct shear tests are available, an initial assessment of stability may be carried out by estimating ϕ′ from Figure 7.16. Those carrying out the analyses may have strength data from other projects on soils with the same classification and plasticity, and that data may be used along with Figure 7.16.

7.3.3.2 Dams for Which There Are a Number of Triaxial Tests Available

When several triaxial tests have been carried out on the one soil, it is recommended that the design shear strength parameters are obtained from a p′-q plot of the test results, rather than by, say, averaging the individual c′,ϕ′ values from each test. It is important to use results from the effective stress range applicable to the field problem, particularly for soils with bi-linear strength envelopes.

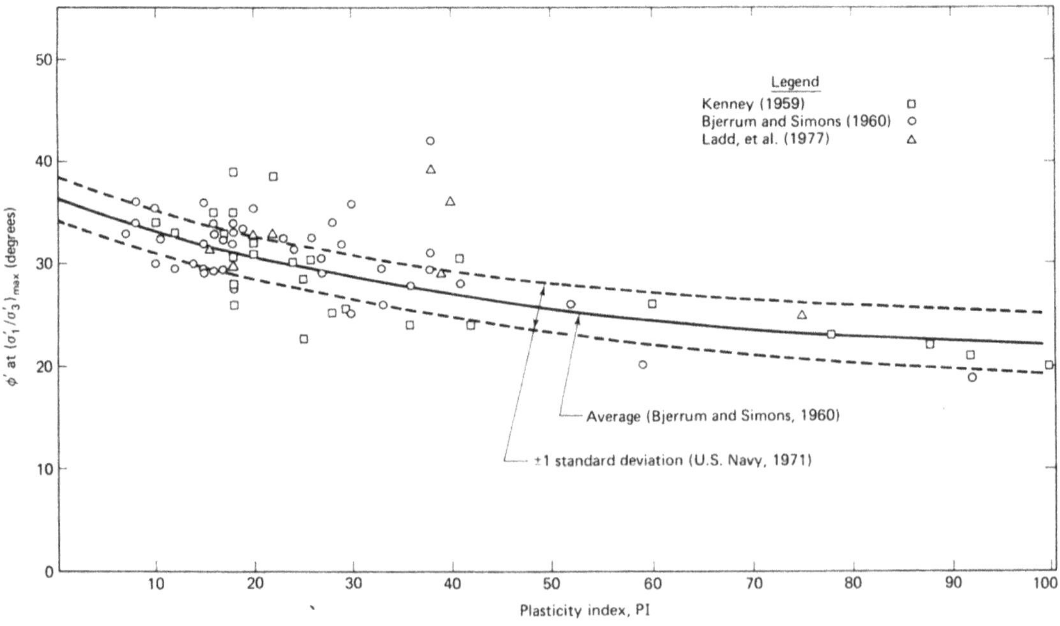

FIGURE 7.16 Correlation between ϕ′ and Plasticity Index from triaxial compression tests on clay soils (Holtz and Kovacs, 1981).

The p′-q plot in Figure 7.18 is a graph of the apex points of the Mohr's circles from the test results as shown in Figure 7.17. It can be shown that $\phi' = \sin^{-1}(\tan\phi')$; $c' = a/\cos\phi'$ where a and α are obtained from the p′-q plot as shown on Figure 7.18.

When selecting the design parameters for design of dam embankments, it is common to bias towards the conservative by selecting a line with, say, 75% of the test points above and 25% below (i.e. a lower quartile line). This is after culling erroneous results and clear outliers as discussed in Section 7.3.2. In Section 7.12 the authors refer to this strength as the "reasonable but conservative" strength.

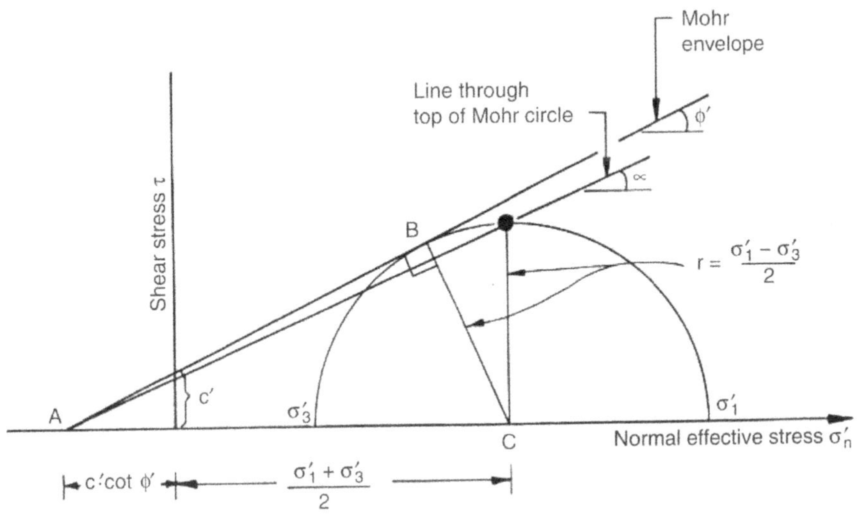

FIGURE 7.17 Basis of p′-q plot where $p' = (\sigma'_1 + \sigma'_3)/2$, $q = (\sigma'_1 - \sigma'_3)/2$ (Fell et al., 2015).

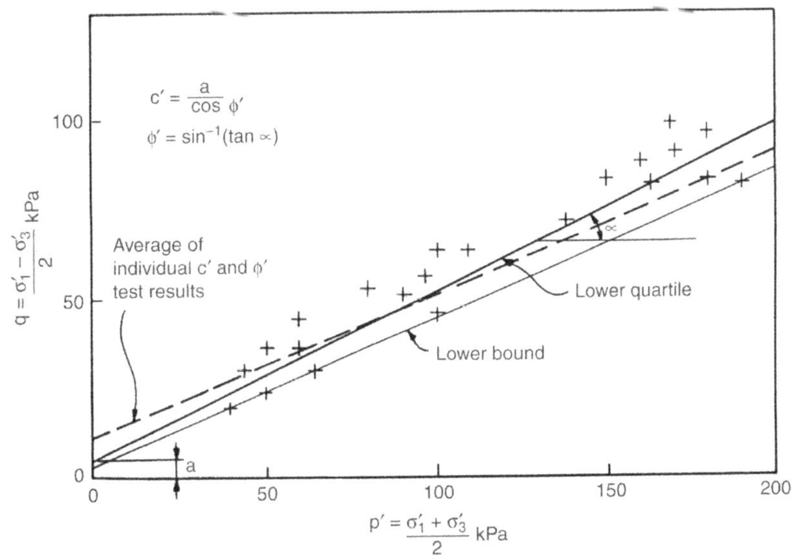

FIGURE 7.18 Typical p′-q plot of triaxial test results (Fell et al., 2015).

For the safety assessment of an existing dam, it is suggested that:

(a) Lower quartile strength be assessed as described earlier. This can be used in deterministic assessments.
(b) Median strengths also be assessed since these are used in QRA approaches as described in Section 7.14.
(c) Lower bound strengths be assessed for use in sensitivity analyses.

It is essential to use judgement in selecting the design line, rather than using say a least squares regression analysis, as judgement enables allowance to be made for:

- Poor or less reliable individual test results.
- The general trend for the second and particularly third stage of staged tests to give a lower strength than the true one because of excessive deformation and some loss of strength from the peak due to displacement on the shear surface.
- General curvature of the Mohr's circle envelope—think bi-linear strengths.
- Adoption of low effective cohesion (c′) which correctly models the behaviour of most soils.

The average of the c′ and ϕ′ values from which the p′-q plot was derived (from laboratory test results), is also shown in Figure 7.18. Using averages tends to give a larger c′ and lower ϕ′ than using the p′-q diagram. This is generally non-conservative for smaller dams, shallow failure surfaces in larger dams and landslide stability, because the strength is overestimated in the working stress range.

If direct shear tests have been used to determine strength parameters, the results should be plotted on a graph of normal stress vs shear stress and design parameters selected by a similar procedure.

7.3.4 Shear Strength of Fissured Soils

Fissures are discontinuities in soils which have similar implications to the strength and permeability that joints have in rock masses—they reduce the soil mass strength and may increase the soil mass permeability. Fissured soils are sometimes present in the foundations of dams—either left in place by design or because they were not removed by those who built the dam, not realising the effect on the shear strength.

Fissures in soil have a wide range of characteristics relating to the properties of the soil mass, the depositional and weathering history and climate (when the fissures were formed and now). It is useful to log and classify the fissures systematically.

Walker et al. (1987) provide a fissure classification system which is simple and practical to use. The main components are shown in Figure 7.19. Figure 7.20 shows an example of fissured soils.

Fissures are caused by one or more of the following:

(a) Shrink swell (desiccation) due to seasonal moisture content changes.
(b) Differential consolidation and settlement in sedimentary soils.
(c) Freeze thaw.
(d) Stress relief due to erosion or retreat of glaciation.

It is important to recognise that the fissures are often formed in a different geological and climatic condition from the present—e.g. in periods of glacial retreat, or when the ground surface was

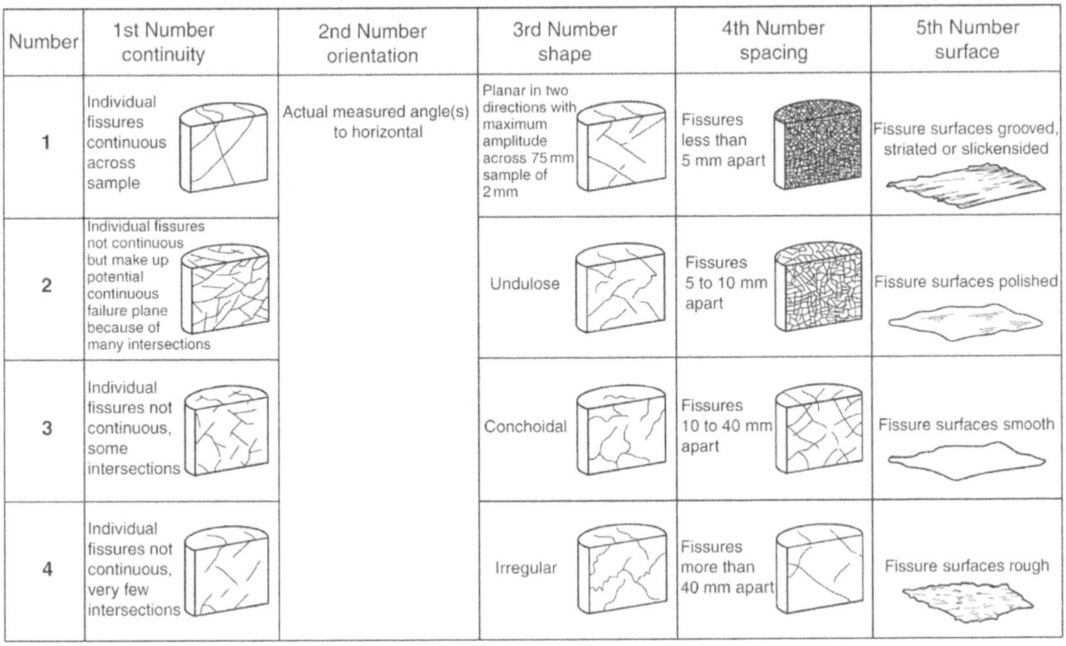

FIGURE 7.19 Fissure Classification Explanation Sheet (Walker et al., 1987). (Reproduced by permission of Taylor & Francis Group.)

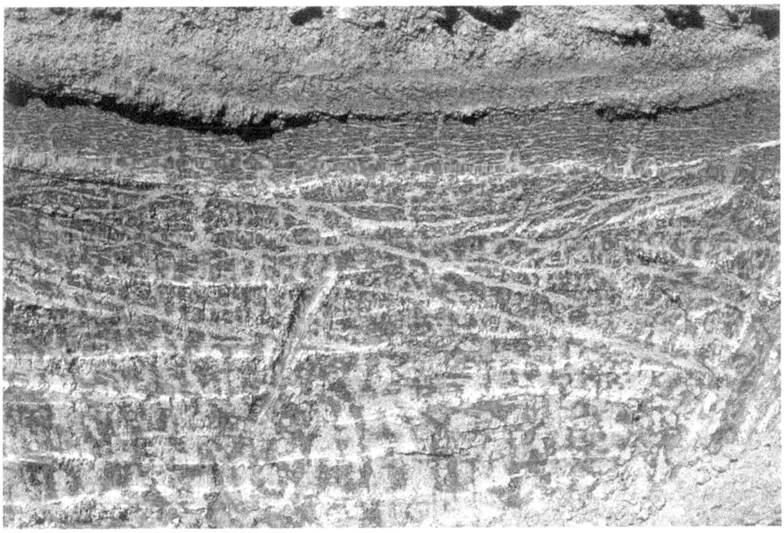

FIGURE 7.20 Weathered Tertiary clay, which exhibits shrink-swell (desiccation) induced fissures, the major sets being inclined at about 20° to the horizontal, consistent with passive failure on swelling (Fell et al., 2015).

(a) View of side of test pit. Pit is about 1.5 m deep. Fissures are the inclined light coloured lineations.

FIGURE 7.20 (Continued)

(b) View of pieces of the fissured soil. The large piece is about 300 mm across.

lower than now. The fissured soils may in some cases be overlain by more recent sediments which are not fissured.

The shear strength of fissured soils is dependent on:

- The peak, fully softened and residual strength of the soil substance.
- The continuity, orientation, shape and spacing of the fissures and the nature of the fissure surfaces.
- These are in turn related to the origin of the fissures.

For example, smooth, planar fissures may only reduce the soil mass strength to the fully softened strength, but polished and slickensided fissures will reduce the strength in the direction of the fissures to near the residual strength.

It is recommended that the following procedure be followed to assess the shear strength of fissured clays:

(a) Map the fissures for orientation, continuity and surface characteristics as detailed in Figure 7.19.
(b) Analyse the orientation using stereo projection methods as one does for rock joints.
(c) Carry out laboratory tests to determine the peak, fully softened and residual strengths.
(d) Assess the strength of the soil mass parallel to the fissure surfaces (φ_f') accounting for the continuity and nature of the fissure surfaces, e.g. polished slickensided fissures would be assigned residual strength, and if they were 50% continuous, the strength determined as approximating to the average of residual and fully softened strength. Normal to the fissures, and for rough fissure surfaces, fully softened strength would apply.

This procedure is highly judgemental, and should where practicable be backed up with back analyses of failures in the same geological materials.

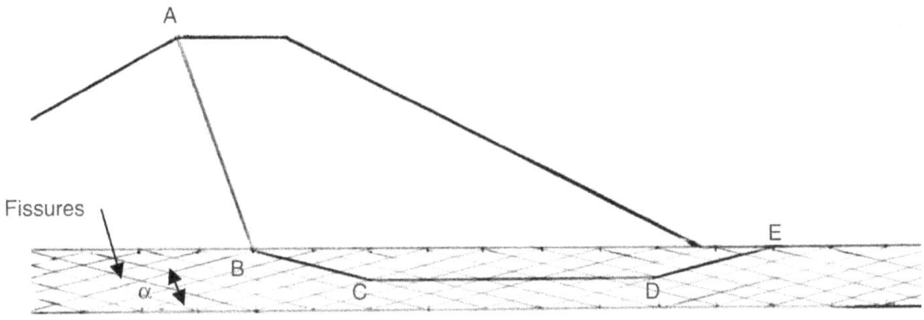

FIGURE 7.21 Method for estimating the shear strength of fissures soils based on fissure orientation (Fell et al., 2015).

For failure surfaces which are not parallel to the fissure direction but inclined at an angle α to the fissure orientation as shown in Figure 7.21, the shear strength along the failure surface C–D can be taken as $\varphi_f' + \alpha$ in a similar manner to how the shear strength on rock defects is estimated as the basic friction angle plus angle of dilation. The limiting strength will be the fully softened strength of the soil because the soil strength will control where shearing is at a steep angle to the fissure orientation.

For failure surfaces B–C and D–E the fissure strength φ_f' applies.

As most fissures formed by desiccation are oriented at relatively steep angles to the horizontal, if detailed investigation can demonstrate there are no low angle fissures present, the fissures will have little effect on the shear strength of failure surfaces, which are predominantly horizontal.

The undrained strength is also affected by the presence of fissures. This is best assessed by using the procedures described in Section 7.2.11 to estimate the strength from the effective stress strengths (which have been assessed allowing for the fissures).

CPT, CPTU and vane shear tests do not allow the soil to fail along the fissures, so they will generally measure the soil substance strength, not the fissured mass strength, and hence will potentially overestimate the strength.

7.4 UNDRAINED, TOTAL STRESS PARAMETERS FOR CLAY SOILS

7.4.1 Triaxial Compression, Extension or Direct Simple Shear Strength

As discussed in Section 7.2.3 and shown in Figure 7.11, it may be necessary to allow for the stress path in selecting the undrained strength. The undrained strength estimated only from triaxial compression tests may significantly overestimate the strength.

For most analyses of stability of embankments, the shape of potential failure surfaces is such that triaxial compression tests give a reasonable representation of the stress path.

If the embankment is founded on vert soft to firm clay soils, the shape of the potential failure surface may be similar to that in Figure 7.11, and using only triaxial compression test data will potentially overestimate the average strength along the failure surface in the foundation.

Note that vane shear strengths corrected as detailed in Figure 7.14 allow for strength anisotropy.

7.4.2 Selection of Undrained Strength Design Parameters

7.4.2.1 Very Soft to Firm Clays in Dam Foundations

The undrained strength of natural very soft to firm clay deposits, e.g. marine clays, usually shows a considerable scatter, which is due to natural stratification of the deposit and measurement errors. Figures 7.21 and 7.23 show typical soft ground vane shear strength profiles with an overconsolidated crust above the near normally consolidated soil.

In these circumstances it is recommended that, where the data plots as in Figure 7.22, the median strength be used after discarding outlying results. Where there is considerable scatter as in Figure 7.23, a more conservative, e.g. lower quartile, strength be adopted to allow for many of the higher strengths being a result of the vane strength being influenced by shells and roots.

In the overconsolidated crust, Ferkh and Fell (1994) adopted an empirical approach, using the median strength line which uses a strength equal to the median strength up to where the liquidity index, LI = (moisture content – plasticity index)/(liquid limit – plastic limit) is less than 1.0.

Those guidelines were developed for soft clay deposits, which are not usually found where dams are being built, but the principles are applicable. If the soil in the overconsolidated crust is fissured, assess the fissured strength for the overconsolidated crust, as detailed in Section 7.3.4.

For dams founded on stiff or stronger residual, there will be no vane shear data because the soil is too strong for vane shear tests. The approach for assessing the undrained strength detailed later for embankments may be used.

7.4.2.2 For Embankments

(a) The undrained strength varies within and between the layers of the compacted fill, reflecting varying degrees of softening as the soil wets up with seepage from the reservoir or weaker layers on winter shut down surfaces. This is demonstrated in Figure 7.24 with the variation in CPT cone resistance. Note that $Su = (q_c - \sigma_{vo})/N_K$.

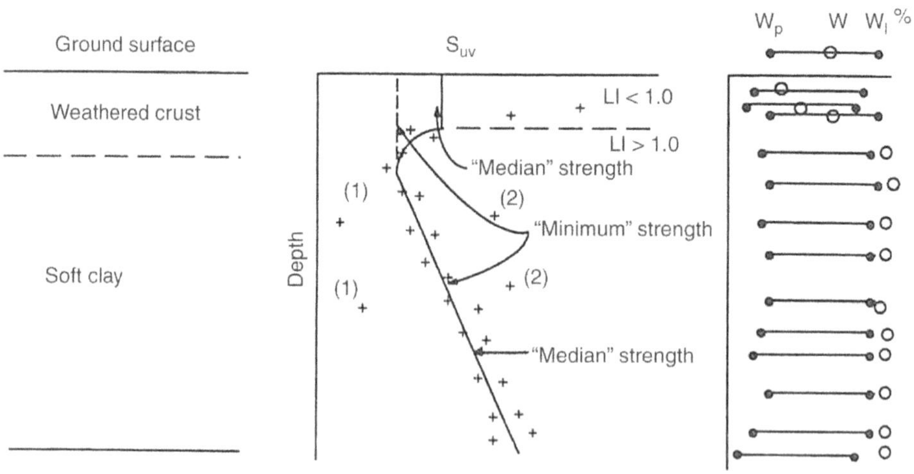

FIGURE 7.22 Selection of foundation strength—Normal case. Notes: (1) Discard, probably disturbed. (2) Discard, probably affected by shells, roots etc. (Ferkh and Fell, 1994).

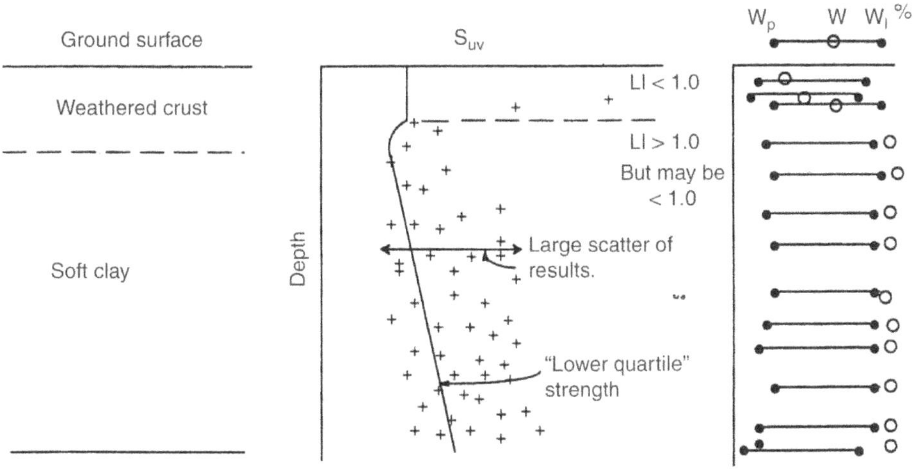

FIGURE 7.23 Selection of foundation strength—influenced by shells, roots etc. and large scatter of data (Ferkh and Fell, 1994).

(b) The earthfill has significant undrained strength at the surface, because the soil is over-consolidated from compaction and is in a partially saturated condition. The selection of undrained strength and stability analysis should allow for these features and may include:
- Allowance for cracks, or softened zones around old cracks with little or no shear strength. These are unlikely to persist below FSL.
- Allowance for water pressures in these cracks.
- Averaging of the strength for failure surfaces which will pass through many layers of the compacted soil, i.e. steeply sloping failure surfaces.
- Adoption of the strength of lower strength layers for failure surfaces which are horizontal. However, care needs to be taken in this, since weak layers may not persist across the zone, unless they are due to desiccation or freezing in a prolonged shutdown during construction or a particularly poor period in compaction control.

7.4.2.3 For Foundations and Embankments

There will sometimes be data from more than one type of undrained shear strength test available. These should be each taken into account and judgement used in assessing the strength for stability analyses, taking account of the quality and inherent uncertainty of each data set.

7.4.3 Shear Strengths for Analysis of Stability under Reservoir Drawdown Conditions

7.4.3.1 General Approach

For the analysis of the upstream slope under reservoir drawdown, it is recommended that the following procedure, which is based on USBR (1987) and Duncan et al. (1990), be used.

(a) Assess for each zone in the dam whether drainage will occur during drawdown, using the approach suggested by Duncan (1992). Estimate the dimensionless time factor T from:

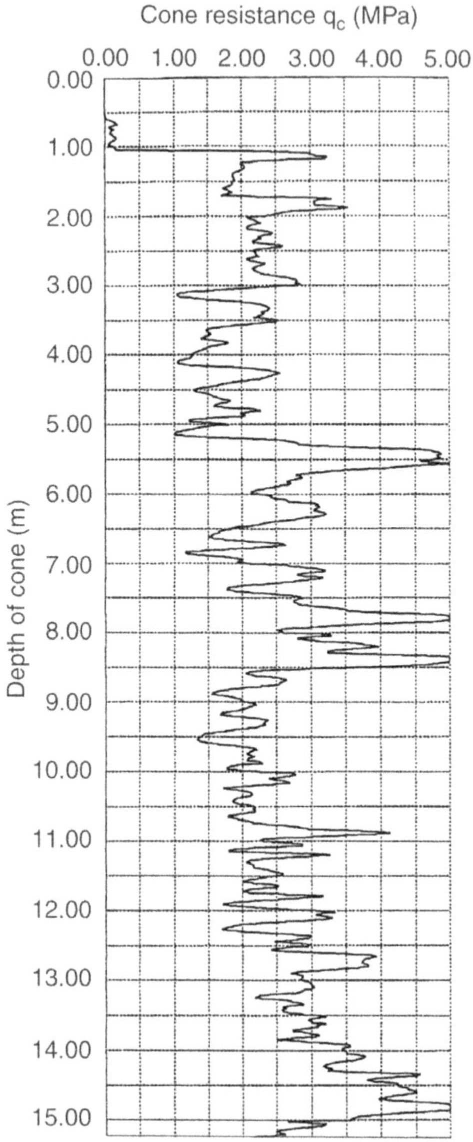

FIGURE 7.24 Cone Resistance q_c determined from Cone Penetration Test, versus depth below dam crest for Eppalock Dam. (Courtesy Goulburn Murray Water and Snowy Mountains Engineering Corporation.)

$$T = \frac{C_v t}{D^2} \tag{7.8}$$

where C_v = coefficient of consolidation (m²/year); t = time for drawdown (years); D = drainage path length (m).

If $T > 3.0$, the zone will be free draining, so effective stress parameters will apply.

If $T < 0.01$, the zone will be undrained, so undrained strengths should be used.

If > 0.01 and < 3, the zone will be partially drained. For drawdown analysis assume the zone will be undrained.

(b) For the zones which are below normal maximum reservoir operating level, e.g. below Full Supply Level, develop effective stress and total stress strength envelope from conventional isotropically consolidated undrained (IC-U) triaxial tests, as shown in Figure 7.25.

Note that the effective stress Mohr circle will be to the left or right of the total stress circle, depending on whether the pore pressures generated during shear are positive or negative.

(c) For these zones, develop the composite shear strength diagram shown in Figure 7.26, from the effective stress and undrained strength envelopes in Figure 7.25.

(d) For undrained or partially drained zones below the normal maximum operating level, determine the strength from the composite envelope, i.e. the minimum of effective stress and undrained strengths. The normal stress used to select the shear strength from the composite envelope should be the effective consolidation stress acting on the base of the slice after drawdown, i.e. the total normal stress minus the pore pressure taken from the drawdown flow net (or from measured pore pressures), after allowing for the factors discussed in Section 7.9.

(e) For the zones which are above normal maximum operating level, e.g. above FSL, estimate the undrained strengths from the undrained strength envelope, allowing for any known cracking or softening.

This procedure has the following characteristics:

(i) It uses undrained strengths, where these are less than effective stress strengths, to allow for the effects of pore pressures developed during the undrained loading.

(ii) It uses effective stress strengths, where these are lower than undrained strengths, in areas normally submerged by the reservoir. This is as recommended by USBR (1987), US Corps of Engineers (1970) and Duncan et al. (1990) and is to allow for possible drainage (the drainage paths may be short in these low stressed areas) and softening under the low confining stresses. It will also allow for the cracking, dilation and softening of the earthfill which occurs because of settlements during construction and under repeated drawdown events.

(iii) It uses the effective stress after drawdown to estimate the strengths as recommended by Duncan et al. (1990). They demonstrated that the use of the effective stresses prior to drawdown as recommended by US Corps of Engineers (1970) (and USBR, 1987) is conservative. However, those authors caution that these effective stresses should be estimated conservatively, and the default position would be to use the effective stress prior to drawdown which is more readily estimated.

(iv) It uses undrained strengths above the normal maximum operating level (after allowing for cracking and softening) because effective stress strengths in this area, where confining stresses are low, greatly underestimate the real strength of the partially saturated soil and, if used, could give misleadingly low strengths in the upper part of the dam. Dams which have experienced larger than normal settlements should be assumed to have cracking and softening present.

7.4.4 Shear Strength of Partially Saturated Clay Soils

In general, unsaturated soils have a shear strength that is larger than that of the same soil saturated because the negative pore water pressure within the soil (known as the matric suction) increases the effective stress.

The most widely used approach for modelling the shear strength of partially saturated soils in slope stability programs is that of Fredlund et al. (1978). They proposed that the shear strength could be expressed as a linear combination of the net normal stress $(\sigma_n - u_a)$ and of the matric suction $(u_a - u_w)$

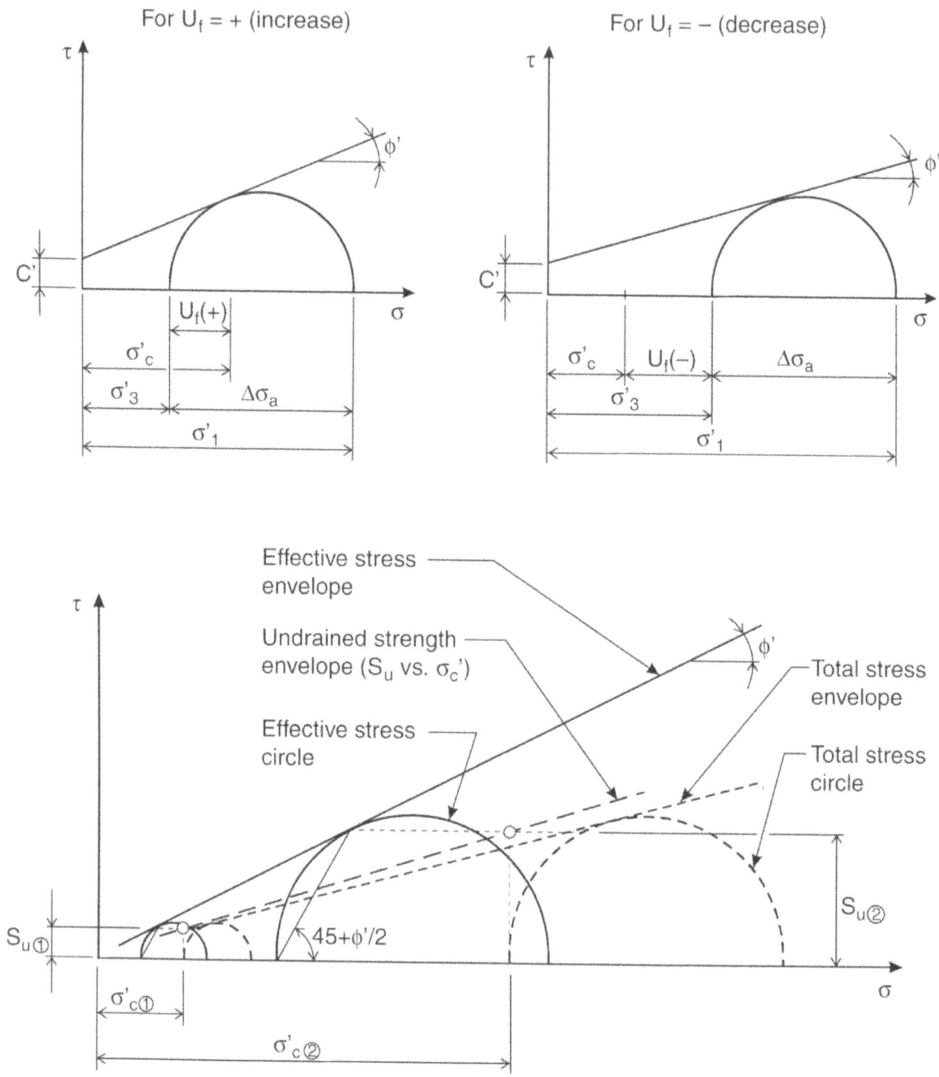

FIGURE 7.25 Shear strength envelopes for isotropically consolidated undrained (IC-U) strength tests (Fell et al., 2015).

$$\tau_f = c' + (u_a - u_w)\, tan\phi^b + (\sigma_n - u_a)\, tan\phi' \tag{7.9}$$

where:

c' = effective cohesion (the same as for the saturated soil)
ϕ' = effective angle of internal friction (the same as for the saturated soil)
ϕ_b = an angle defining the increase in shear strength for an increase in matric suction
σ_n = total normal stress
u_w = pore water pressure
u_a = pore air pressure

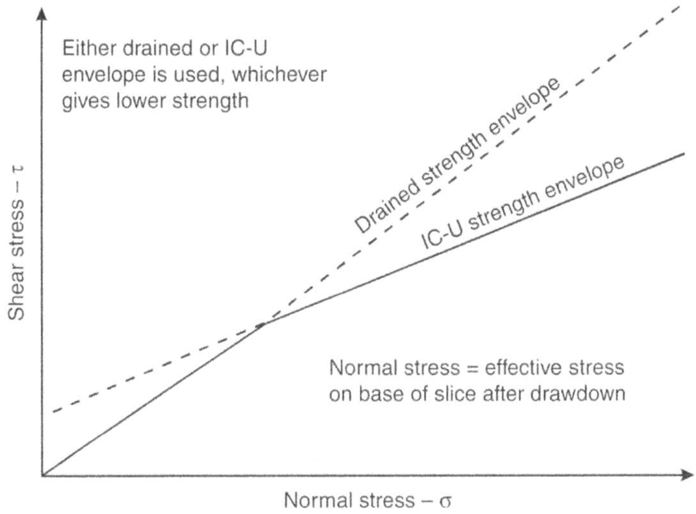

FIGURE 7.26 Composite shear strength envelope (Fell et al., 2015).

$(\sigma_n - u_a)$ = net normal stress
$(u_a - u_w)$ = matric suction.

As discussed in Fredlund et al. (1978), the partially saturated shear strength envelope is nonlinear, with ϕ_b approaching ϕ' at high degrees of saturation. Fredlund et al. (1978) and Fredlund and Rahardjo (1993) describe how the strengths can be determined in laboratory tests.

Fredlund and Rahardjo (1993) present some test data from several authors which showed a range of ϕ_b from 13 to 22 degrees and $(\phi' - \phi_b)$ from 5 to 21 degrees with larger values applying to soils with high ϕ' values. The SLOPE W manual suggests most ϕ_b values are between 15 and 20 degrees.

7.5 SHEAR STRENGTH OF FILTERS

The shear strength of compacted sand and sandy gravel filters may be estimated from the relative density specified or measured during construction and particle size as shown in Figure 7.27.

Well compacted granular filters in dams, e.g. compacted with four passes of a 10 tonne smooth drum vibratory roller, in layers 500 mm compacted thickness, should give a relative density of at least 70%, so the effective friction angle should be around 38° for fine sand and 40° for well graded filter sands and sandy gravels.

In reality, at low confining stresses, e.g. less than 200 kPa, ϕ' is likely to be 2° to 5° higher than these figures, and this should be allowed for.

The effective friction angle of coarseg rained granular soils (gravelly sands and gravels) can be estimated from Figure 7.28. Note that for this plot the gravel % is based on plus 4.75 mm size.

For dams for which there is no specified compaction, or it is known that filters were placed and compacted in thick layers without formal compaction, assume that the relative density is around 30%.

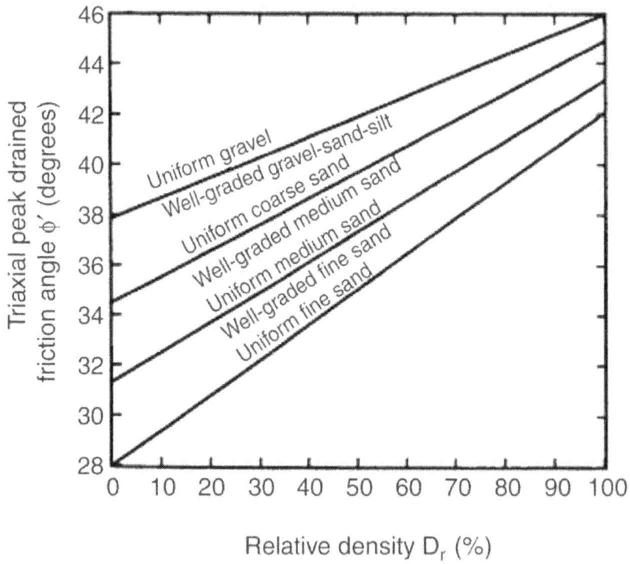

FIGURE 7.27 Peak effective friction angle of sands as a function of relative density and grain size (Schmertmann, 1978; modified from Burmister, Donald M. *The Importance and Practical Use of Relative Density in Soil Mechanics*, ASTM Proc. Vol. 48. 1948.)

7.6 SHEAR STRENGTH OF ROCKFILL

The shear strength of rockfill is dependent on the effective normal stress, and to a lesser extent on the bulk dry density, void ratio or degree of compaction, the unconfined compressive strength of the rock, the uniformity coefficient of the grading, maximum grain size, fines content and particle shape (Marsal, 1973, Douglas, 2003).

Douglas (2003) carried out an extensive analysis of the factors affecting the shear strength of rockfill. This involved establishing a large database from published and unpublished triaxial testing, and statistical analysis.

Figure 7.30 shows the variation of secant friction angle ϕ'_{sec} with normal stress σ_n for dense rockfill. The strong dependence of ϕ'_{sec} on σ_n is evident, as is the wide variation in ϕ'_{sec}.

Douglas (2003) carried out nonlinear statistical analyses on the database. An analysis of the form shown next was found to be the most effective for relating ϕ'_{sec} to σ'_n.

$$\phi'_{sec} = a + b\sigma'^{c}_{n} \tag{7.10}$$

The analysis resulted in the following constants:

$$a = 36.43 - 0.267ANG - 0.172FINES + 0.756(c_c - 2) + 0.0459(UCS - 150) \tag{7.11}$$

$$b = 69.51 + 10.27ANG + 0.549FINES - 5.105(c_c - 2) - 0.408(UCS - 150) - 0.408 \tag{7.12}$$

$$c = -0.3974 \tag{7.13}$$

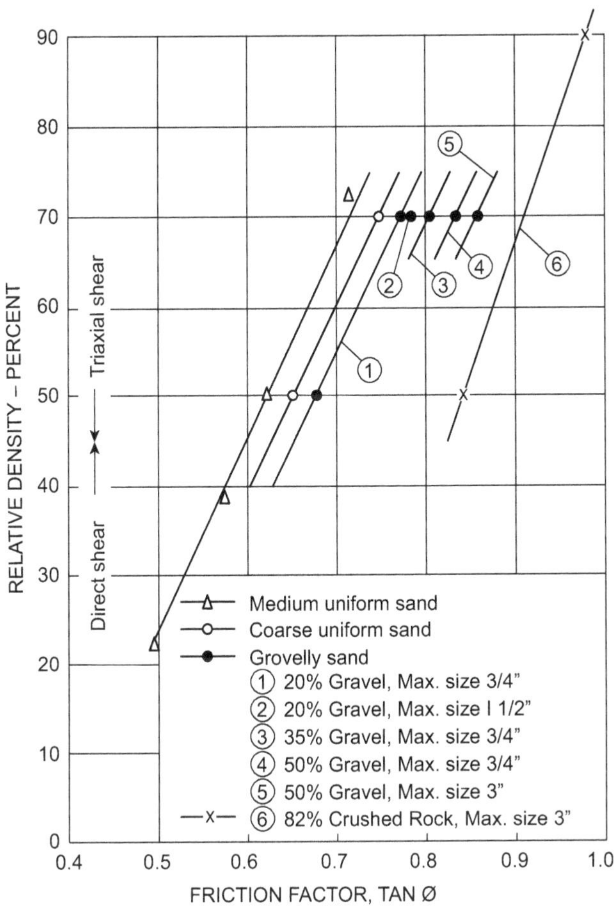

FIGURE 7.28 Peak friction angle for coarse grained soils as a function of relative density and grain size (USBR, 1998).

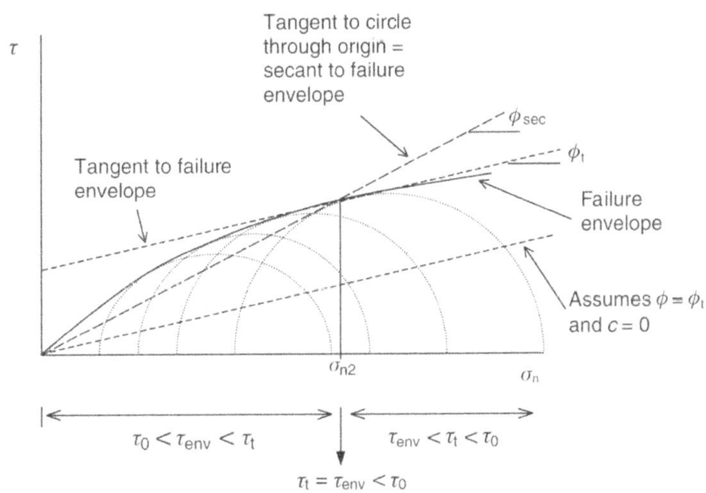

FIGURE 7.29 Methods for representing the shear strength of rockfill (Douglas, 2003).

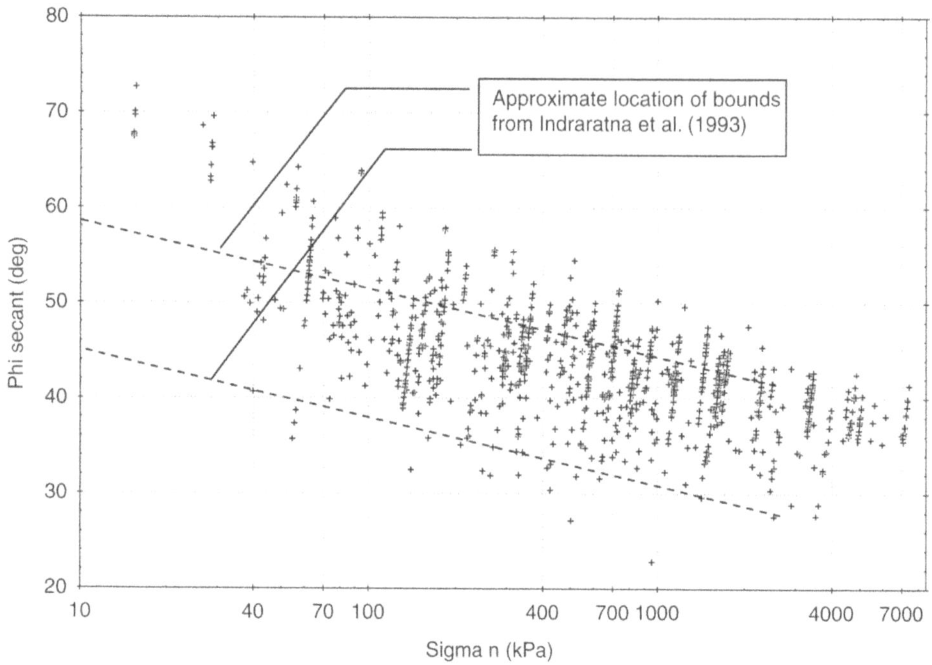

FIGURE 7.30 Variation of secant friction angle ϕ'_{sec} with normal stress σ'_n (Douglas, 2003).

where:

ANG = (angularity rating −5) for angularity rating > 5.5 otherwise 0, rounded particles have an angularity rating of 0, crushed angular rock a rating of 8. The rating is subjective (Douglas, 2003).

FINES = percentage of fines passing 0.075 mm (%).

c_c = coefficient of curvature = $\frac{d_{30}^2}{d_{10}d_{60}}$

UCS = unconfined compressive strength of the rock substance (MPa).

For many analyses of stability, it will be sufficient to use the data in Figure 7.30 and Equations 7.10 to 7.13 to select the strength of well compacted rockfills. A bi-linear strength envelope should be used to better represent the curved Mohr–Coulomb envelope. This can be plotted by calculating ϕ'_{sec} from Equations 7.10 to 7.13 at varying σ_n applicable to the stability analysis.

An approximate approach is to use Figure 7.30 to estimate ϕ'_{sec} versus σ_n. For well compacted rockfill consisting of high UCS low fines content quarried rock such as granite, use the median of the data in Figure 7.30. For medium UCS high fines content quarried rockfill which breaks down on compaction, e.g. schist, use towards the lower 10% of the data in Figure 7.30. Using a bi-linear strength envelope which models the high friction angle at low confining stresses is important. Doing so rather than some average strength avoids obtaining misleadingly low factors of safety for shallow failure surfaces in the upper parts of dams which have rockfill shoulders.

Loose dumped rockfill at low σ_n will have ϕ'_{sec} about 4 degrees lower than calculated using this approach. At high σ'_n the difference will be negligible.

Douglas (2003) also carried out an analysis of the database using the principal stresses σ_1' and σ_3'. Details are summarised in Fell et al. (2015). This may be used for cases where the shear strength of rockfill is critical to stability.

7.7 SHEAR STRENGTH OF NON-PLASTIC SOILS IN THE FOUNDATION

7.7.1 Methods Usually Adopted

The peak shear strength (ϕ') of silts, sands and gravels in the dam foundation or within the embankment can be assessed by *in situ* testing, such as SPT, CPT or CPTU.

7.7.1.1 Standard Penetration Test (SPT)

The effective friction angle ϕ' can be estimated by using the SPT to estimate the relative density and then estimating the strength from the relative density and general grain size characteristics. Schmertmann (1978) developed such a method, which is shown in Figure 7.27.

Lunne et al. (1997) indicate the laboratory tests on which Figure 7.27 is based were at a confining stress of 150 kPa, and that lower values of ϕ' should be used at high confining stresses.

The relative density is best estimated using the method suggested by Skempton (1986). He reviewed the available information and proposed a method where the SPT value is corrected for the energy of the hammer, and to an effective overburden pressure of 100 kPa by:

$$N_{60} = \text{Nm ERm} \tag{7.14}$$

where:

N_m = measured SPT "N" value
N_{60} = "N" value corrected to a hammer of 60% efficiency
ERm = Energy Rating of the SPT hammer

and

$$(N_1)_{60} = C_N\, N_{60} \tag{7.15}$$

where: C_N = factor to correct N_{60} to 100 kPa effective overburden stress

$$= \frac{200}{100 + \sigma'_{vo}} \text{ for fine, normally consolidated sand} \tag{7.16}$$

$$= \frac{300}{200 + \sigma'_{vo}} \text{ for coarse, normally consolidated sand} \tag{7.17}$$

$$= \frac{170}{70 + \sigma'_{vo}} \text{ for overconsolidated sand} \tag{7.18}$$

where:

σ'_{vo} = effective overburden stress at the level of the SPT test (in kN/m^2).

Skempton (1986) then uses the $(N_1)_{60}$ value to estimate relative density from Table 7.2. For $D_R \geq 35\%$, this can be approximated by

$$(N_1)_{60}/D_R^{\,2} = 60 \tag{7.19}$$

The method requires measurement of the energy rating of the SPT equipment. This should be done on the site being investigated as the energy transmitted to the rods and the SPT varies depending on the type of equipment and the condition of drill rods etc. This is particularly important where SPT is being used to assess liquefaction. See Section 8.7.4.

7.7.2 Cone Penetration (CPT) and Piezocone (CPTU) Tests

The effective friction angle ϕ' can be determined directly from the CPT and CPTU using charts which have been developed by several authors. Lunne et al. (1997) reviewed the methods and concluded that the Robertson and Campanella (1984) and two other methods were to be preferred. The Robertson and Campanella (1984) method is simple to use and is shown in Figure 7.31.

Robertson and Campanella (1984) indicate that:

- The method applies to normally consolidated, uncemented, moderately incompressible (grains), predominantly quartz sands, but can be used for silty sand, sandy silt, sand and gravelly sand.
- For highly overconsolidated sands ϕ' may be up to 2° lower than predicted from Figure 7.31.
- For highly compressible sand (e.g. sand with many shells, or mica pieces), the method predicts conservatively low friction angles. For the sands included in Robertson and Campanella's analysis the effect could be up to 3°.

7.8 SHEAR STRENGTH OF BEDDING CRUSHED SEAMS AND POTENTIALLY LOW STRENGTH STRATA IN THE FOUNDATION

When a river valley is formed the *in situ* stresses in the earth's surface are relieved, resulting in horizontal movement of the valley slopes towards the river, and heave or "bulging" of the valley floor and slopes adjacent to valley. Figure 7.32 shows the effects for sub-horizontally bedded rocks.

TABLE 7.2 Method for Estimating Relative Density (D_R) from $(N_1)_{60}$ Value Skempton (1986)

RELATIVE DENSITY %	*CONDITION*	*$(N_1)_{60}$*
0		0
	Very loose	
15		3
	Loose	
35		8
	Medium dense	
65		25
	Dense	
85		42
	Very dense	
100		58

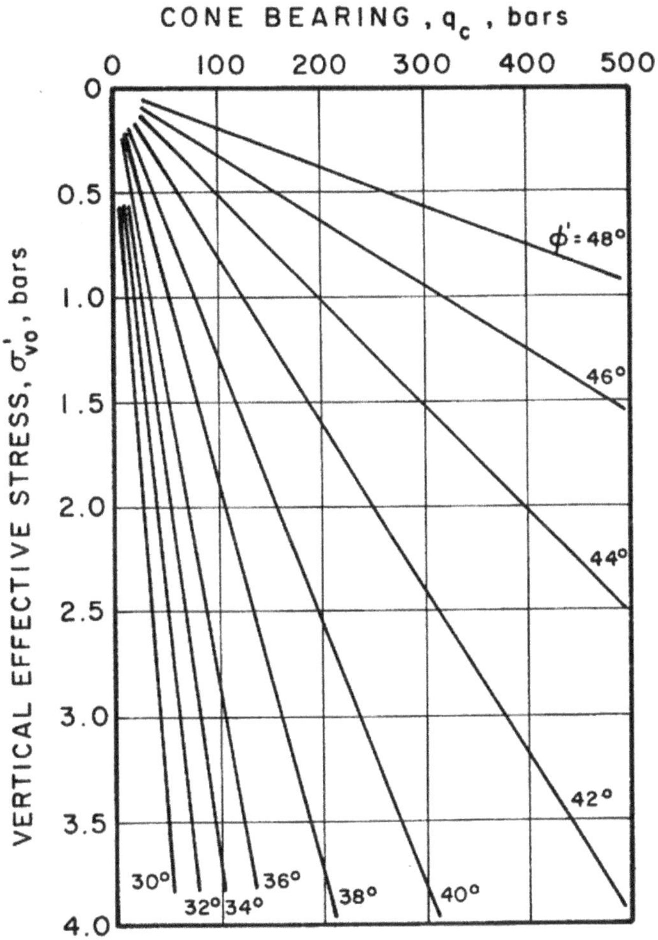

FIGURE 7.31 Robertson and Campanella's (1984) method for estimating ϕ' for cohesionless soils from the Cone Penetration Test (CPT). (Reproduced courtesy of P.K. Robertson.)

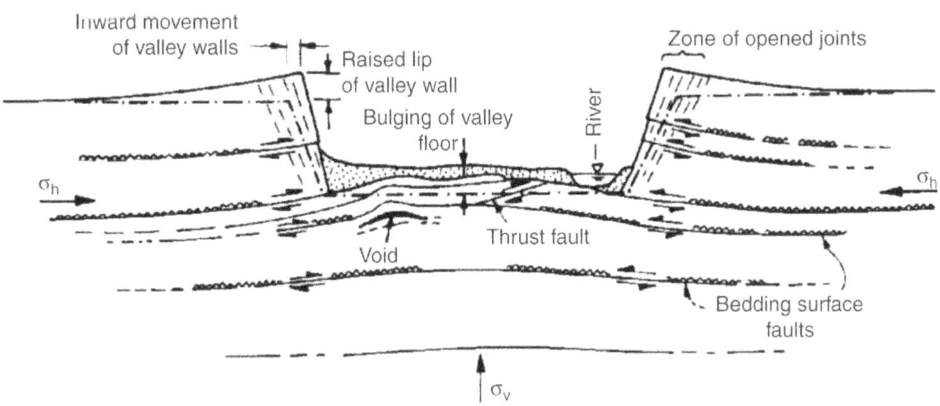

FIGURE 7.32 Complex valley structures related to stress release in sub-horizontal bedded rocks (Fell et al., 2015, based on Patton and Hendron, 1972).

The *in situ* horizontal stresses σ_t which would have been present before the valley was formed are reduced to zero at the sides of the valley. The result is movement of the valley wall inwards toward the valley. This results in opening of the sub-vertical joints adjacent to valley wall. If the rock is bedded, for example sandstone and siltstone, these will have different rock mass stiffness, so the amount of movement is not uniform, and differential movement occurs on the boundaries of the beds. This forms bedding surface faults or shears, or crushed seams which develop in the weaker of the beds, so for example usually in the siltstone or shale if interbedded with sandstone.

The relative displacements can be quite large so often the bedding faults, shears or crushed seams may be at or near residual strength or have soil strength. Because of the mechanism of formation these defects tend to be persistent over tens or hundreds of metres.

The *in situ* stresses are concentrated beneath the valley floor as a result of load transfer as the excavation of the valley removed lateral support from the rock layers above the floor and vertical load from the rock beneath the floor. This results in gentle anticlines, valley heave and bulge. This in itself results in differential movement between beds and bedding faults and can also result in thrust faults in the valley floor and even voids due to buckling of the beds in the floor. These voids may become infilled with alluvium.

Bedding surface faults, shears and crushed seams are also formed by folding of interbedded sedimentary rocks. The mechanism is shown in Figure 7.33. As the rocks are folded, differential movements occur between the beds. These localise on the boundary of the finer grained lower strength rock such as claystone, shale and siltstone.

It should be noted that these features form even in gently folded rocks where there may be little discernible variation in dip of the beds across a site. These tend to combine with defects formed by valley stress relief so the exact origins of the defects may be a combination effect.

These features will almost invariably be present and should be anticipated as part of the Conceptual Geological Model.

Typically, bedding surface faults and shears will have lengths which have sheared or crushed rock with strength controlled by the strength of the crushed seam, and lengths which have clean rock to rock contact with strength controlled by the basic friction angle of the rock and large and small scale waviness.

7.8.1 Shear Strength of Clean Rock to Rock Contact

The shear strength of clean, rough (or smooth) discontinuities may be assessed using the approach of Patton (1966), who recommended use of the form of Equation 7.20 or 7.21.

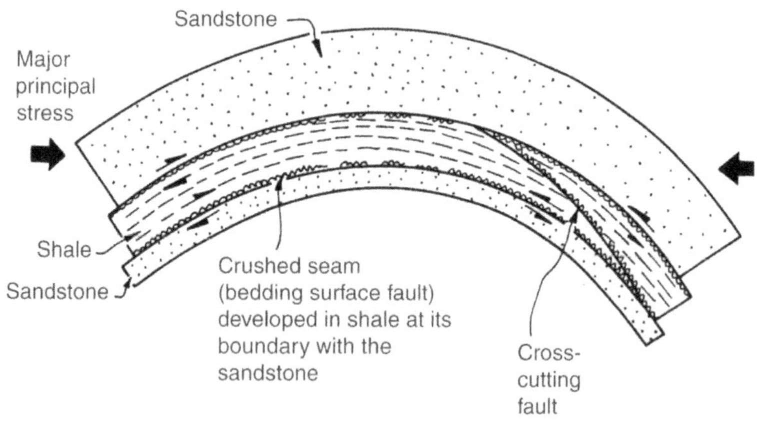

FIGURE 7.33 Usual way in which bedding surface faults are formed in mudrocks (Fell et al., 2005, 2015).

$$\tau = \sigma'_n \tan(\phi'_b + i) \qquad \text{if } \sigma'_n < \sigma'_{ns} \tag{7.20}$$

or

$$\tau = c'_a + \sigma'_n \tan\phi'_b \qquad \text{if } \sigma'_n > \sigma'_{ns} \tag{7.21}$$

where $\sigma'_n = \sigma_n - u$ = effective normal stress; σ'_{ns} = effective normal stress at which the strength of asperities is exceeded; u = internal water pressure (pore pressure) within the discontinuity; ϕ'_b = basic friction angle (for wet surfaces) of the discontinuity wall rock; σ'_n = effective normal stress across discontinuity; i = average roughness angle; c'_a = effective apparent cohesion.

As shown in Figure 7.34 at low normal stresses the mechanism of shear follows the roughness features on the shear surface. At high normal stresses the mechanism is shearing of the intact rock.

The basic friction angle (ϕ'_b) is derived from direct shear tests on sawn and lightly ground surfaces of the discontinuity or alternatively by measuring both vertical and horizontal displacements of a defect specimen during laboratory direct shear testing and calculating the roughness angle i, which is then subtracted from the measured friction angle to determine ϕ'_b.

The roughness or waviness angle (i) is generally determined by field observations of surface roughness, at a scale appropriate to the dam foundation, in the direction of potential displacement (i.e. the first order roughness in Figure 7.35).

The Patton (1966) method does not allow for the second order roughness.

The bi-linear Patton (1966) model can be replaced with the curvilinear approach of Barton and Bandis (1982, 1991). Their method measures the small-scale roughness at laboratory scale and has corrections to apply these to approximate the first order roughness or waviness at field scale.

The authors have some reservations about how this is done, particularly the method advocated by Barton (1982) relating amplitude of waviness to length of the "wave" because it gives excessively high strengths in some situations, and suggest the Barton (1982) method not be used. Hencher and Richards (2014) allowed for large-scale or first order waviness as described earlier and added the lower bound of dilation from the laboratory tests to account for small-scale or second order waviness. The authors are of the opinion that it is reasonable, at least in some situations, to add the effects of small-scale roughness provided that both the large-scale waviness and small-scale roughness are conservatively estimated.

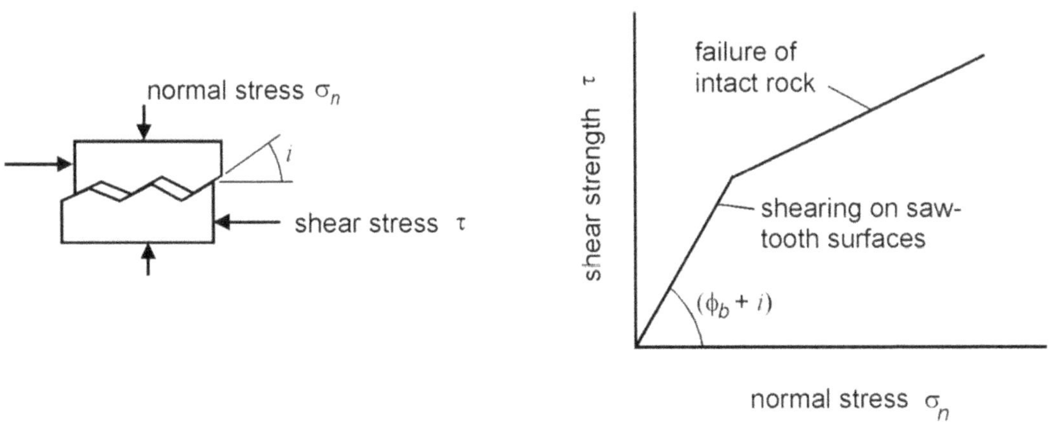

FIGURE 7.34 Shearing on a rock surface with basic friction angle ϕ'_b and first order roughness angle i.

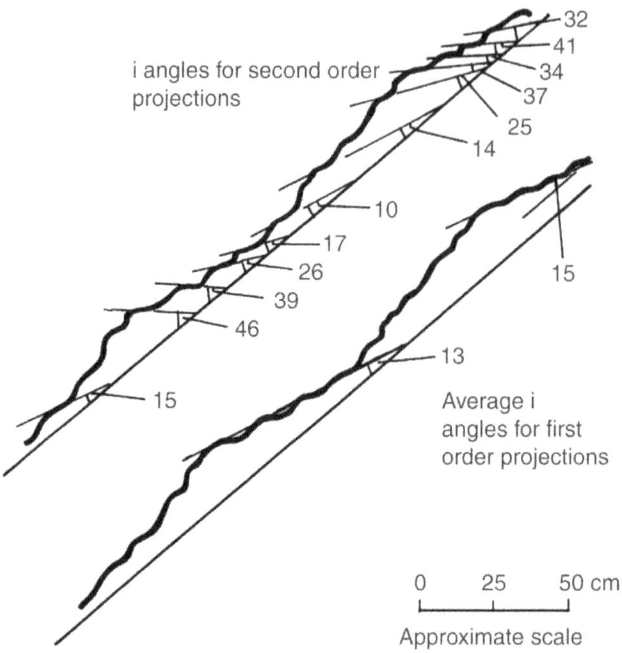

FIGURE 7.35 Measurement of roughness angles (i) for first and second order roughness on a rock surface (Fell et al., 2015 adapted from Patton, 1996).

It is suggested that both Barton and Bandis (1982, 1991) and Hencher and Richards (2014) are used and considered when assigning the shear strength.

7.8.2 Shear Strength of Infilled Joints and Crushed Seams Showing Evidence of Previous Displacement

The shear strength of previously displaced features (e.g. sheared or crushed seams) is determined from:

$$\tau = \sigma'_n \tan (\phi'_r + i) \tag{7.22}$$

where ϕ_r' = effective residual friction angle of the infill, sheared or crushed material; i = average first order roughness angle for those parts of the shear surface with infilled joints or crushed seams.

The residual frictional strength of the feature is generally determined by direct shear tests on undisturbed samples measured at sufficient displacement to ensure that the residual strength of the feature has been reached. For thick gouge (crushed rock) zones or, if undisturbed samples cannot be obtained, remoulded, normally consolidated samples of the material are tested for their residual strength. Gouge is often layered so care should be exercised to ensure that the weakest layer of material is selected for testing.

In many cases the crushed seam is too thin to obtain sufficient material, or the seam has too many rock fragments in it to test in direct shear. In these cases, the shear strength may be estimated based on experience of shear strength of similar materials, taking account of the particle size distribution and plasticity of the crushed seam.

7.8.3 Shear Strength of Intact Rock Bridges

The shear strength of intact rock bridges within a potential failure surface is approximated by the rock mass strength and may be estimated using the method of Hoek et al. (2002) or Hoek and Brown (1997)

7.8.4 Weighted Average Shear Strength

The shear strength of potential failure surfaces with lengths of rock to rock contact and crushed seams should be determined as a weighted average, allowing for the % of each and their shear strength.

If intact rock "bridges" form part of the potential shear surface, the analyses should first assume that all the load is taken on the rock bridges. This allows for the rock bridges being stiffer than the remainder of the potential failure surfaces. If the bridges are unable to take the load, assign the crushed seam strength to those parts of the potential failure surface and re-analyse.

7.8.5 The Need for Specialist Rock Mechanics and Engineering Geology Input

Where bedding crushed seams or faults are known to be or may be present, experienced specialist engineering geologist and rock mechanics advice should be obtained.

7.9 ESTIMATION OF STEADY STATE PORE PRESSURES

7.9.1 Estimation from Piezometer Data

For existing dams, there is sometimes piezometer data on which to assess the pore pressures. The data need to be considered carefully, particularly in regard to whether equilibrium has been reached and whether there is sufficient instrumentation to define the pore pressures over the whole of the dam. The reliability of the existing pore pressure data should also be carefully considered. The implication of the high k_H/k_V on pressures should be considered.

In cases where the piezometer data is limited or insufficient to allow a reliable estimate of pore pressures, seepage analyses as described next should be carried out to supplement the piezometer data.

7.9.2 Estimation from Numerical Seepage Analyses

7.9.2.1 Steady State Seepage Condition at Full Supply Level

Steady state seepage conditions are usually assumed for the assessment of the long term stability of the downstream slope of the dam. These are usually based on the reservoir being at Full Supply Level.

It may take many years to reach steady state conditions in the dam core. The authors' experience is that large dams may not have reached equilibrium pore pressures even 20–30 years after construction, with construction pore pressures dominating in the lower part of the dam and fully saturated conditions not established in the upper parts. It is however good practice to design or assess an existing dam for the steady state conditions, making appropriately conservative assumptions to estimate the pore pressures.

Pore pressures for the steady state seepage condition are estimated by calculating the flow net for the embankment section by finite element methods.

The following issues are important when calculating the flow net for embankment dams.

(a) *Zoning of the embankment*. Embankment zoning clearly has a vital role in determining the pore pressures in the embankment. All zones should be modelled with their permeability estimated.
(b) *Anisotropic permeability of embankment earthfill*. Earthfill in dam embankments is compacted in layers and the action of rolling, possible drying and cracking of the surface of each layer and greater compaction of the upper part of the layers compared to the lower part will almost invariably lead to the horizontal permeability (k_H) being greater than the vertical permeability (k_V). It would not be unusual for $k_H/k_V = 15$ and even as high as $k_H/k_V \approx 100$. This has a marked effect in pore pressures in an embankment, particularly if no seepage control

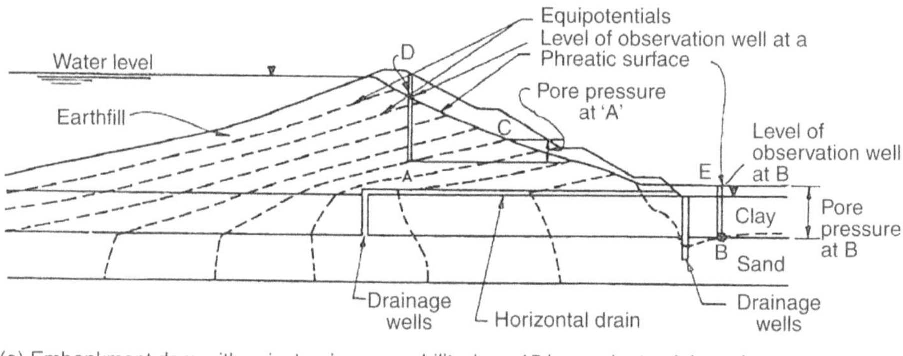

(a) Embankment dam with anisotropic permeability, $k_h = 15\ k_v$; equipotentials and pore pressures. (Fell, 1987)

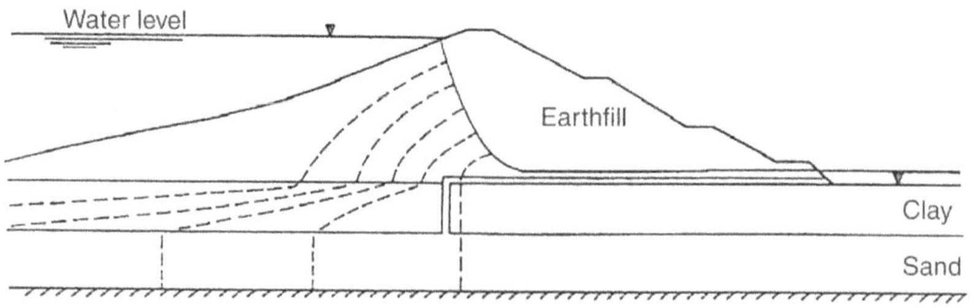

(b) Embankment dam with isotropic permeability; equipotentials.

FIGURE 7.36 Effect of anisotropy of permeability on seepage flow net for Mardi Dam (Fell et al., 2015).

measures such as a vertical drain are incorporated in the design. The effect of anisotropy on pore pressures is evident in Figure 7.36. Note in the figure that the seepage flow net is also influenced by the high permeability horizontal drain.

(c) *Foundation permeability.* The permeability of the foundation (compared to that of the embankment) has an important effect on the seepage flow net and the foundation must be modelled along with the embankment. No dam foundation is "impermeable" and in the majority of cases, a rock foundation will have a permeability greater than that of compacted earthfill. Most rock masses have a permeability of between 1 and 10 lugeons (10^{-7} to 10^{-6} m/sec) compared to compacted clay earthfill with a $k_V \approx 10^{-7}$ to 10^{-10} m/sec. Soils in a dam foundation (which will usually be affected by fissures, root holes or layering during deposition) will almost invariably have a permeability far greater than compacted earthfill, e.g. in the range 10^{-5} to 10^{-6} m/sec.

The authors have too often seen seepage modelling which does not include the foundation. Such modelling is incorrect and potentially misleading and should not be relied upon.

It is difficult to accurately predict steady seepage pore pressures. There are several mechanisms which can lead to pressures equivalent to those where a high k_H/k_V ratio is assumed, i.e. reservoir head persists through the core. The authors' advice is that in the absence of piezometer data, design and assess the safety of all embankments on the assumption that k_H/k_V is ≥ 15.

The authors' own practice for an earth and rockfill embankment with a vertical drain is to ignore the flow net effect, i.e. assume the flow lines are horizontal in the core to the vertical drain. This is potentially conservative as it implies $k_H/k_V = \infty$ but only affects the factor of safety marginally over that if $k_H/k_V = 15$ were used and allows one to be confident that pore pressures have been estimated conservatively.

7.9.3 Pore Pressures under Flood Conditions

Pore pressures in the low permeability zones in a dam under flood conditions are very difficult to predict. There are two broad issues:

- Will pore pressures in the body of the dam, for example in the core, rise with the flood water level?
- Will pore pressures in the vicinity of the dam crest above normal Full Supply Level rise with the flood water level?

Considering these in turn:

(a) *Pore pressures in the body of the dam.* There are some general observations:
   (i) Whether the pore pressures will respond rapidly to the reservoir rise depends on the degree of drainage which can be assessed as described in Section 7.4.3. If a zone is assessed to be free draining (Time Factor $T > 3$) or in the intermediate zone (T 0.01 to 3), then it should be assumed the pore pressures will respond as the reservoir level rises in the flood.
   (ii) For non-draining zones ($T < 0.01$) the pore pressures will respond slowly, so in most flood conditions there will be little or no rise in pore pressures.
   (iii) Where monitoring of piezometers in the dam shows a rapid response to reservoir level, regardless of the prior assessments, assume the pore pressures will respond to the reservoir level.

(b) *Pore pressures near the crest of the dam.* The pore pressures near the crest of the dam will depend on whether the zones are free draining, as detailed earlier, but also on whether the permeability of the core is affected by cracking due to differential settlement and desiccation. In many dams the low permeability zones have not been taken to dam crest end and high permeability zones, e.g. rockfill, are used, so these will saturate under the flood.

The authors are reluctant to assume that low permeability zones will not develop pore pressure during floods due to the possibility of higher permeability of the core resulting from cracking and would normally check the case where the pore pressures above full Supply Level rise with the reservoir level.

7.10 DRAWDOWN PORE PRESSURES AND THE ANALYSIS OF STABILITY UNDER DRAWDOWN CONDITIONS

7.10.1 Some General Issues

When the reservoir level behind an embankment dam is lowered, the stabilising influence of the water pressure on the upstream slope is lost. If the reservoir level is dropped sufficiently quickly that the pore pressures in the slope do not have time to reach equilibrium with the new reservoir water level, the slope is less stable.

Many failures of dam slopes (and natural slopes around reservoirs) have occurred during drawdown. Duncan et al. (1990) give some examples.

In view of the difficulty in estimating the pore pressures under drawdown conditions, particularly those developed on shearing in low permeability, clayey soils, it is common practice to use the procedure detailed in Section 7.4.3.

For free draining rockfill or gravels, the pore pressures will dissipate as the reservoir level drops, so there are no residual pore pressures.

For well compacted, medium permeability soils, e.g. silty sands, pore pressures can be estimated using flow nets. In these cases, any shearing induced pore pressures will be negative, due to dilation of the dense soil, so they can be conservatively ignored. If the silty sands are loose and drawdown is in fact rapid, undrained strengths should be used.

7.10.2 Estimation of Drawdown Pore Pressures, Excluding the Effects of Shear-Induced Pore Pressures

Pore pressures in the upstream slope of a dam under drawdown conditions are determined by the change in water level, geometry and zoning of the slope, the relative permeabilities of the zones and, importantly, layers within each zone and the relative permeability of the foundation.

Some of these points are illustrated in Figure 7.37 and discussed as follows:

(i) In most cases the assumption is made that drawdown is "instantaneous", and hence the flow net changes from the steady state case to the drawdown case instantaneously. This is somewhat conservative but sufficient for most analysis.

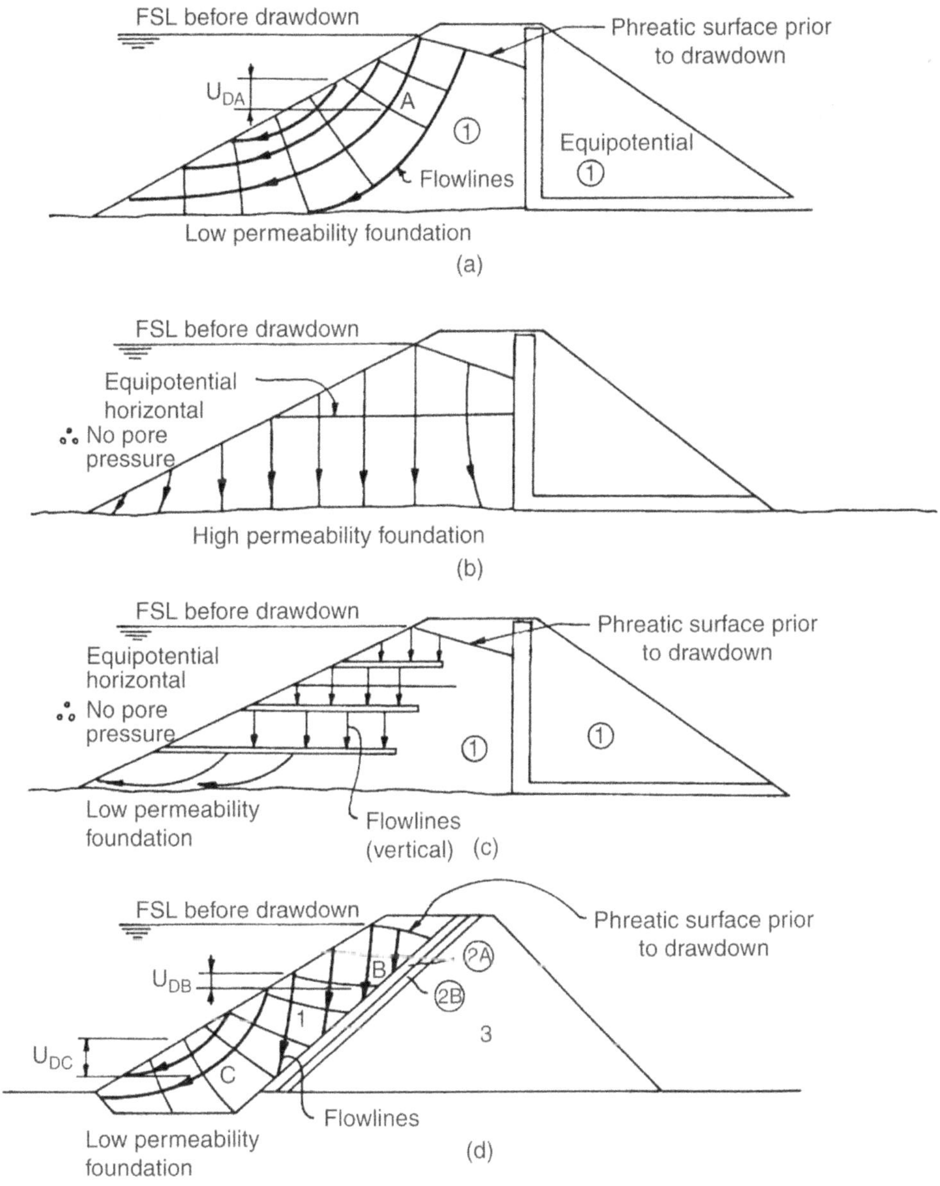

FIGURE 7.37 Drawdown flow nets (a) earthfill with chimney drain, low permeability foundation; (b) earthfill with chimney drain, high permeability foundation; (c) earthfill with horizontal drainage blankets; (d) sloping upstream core, low permeability foundation. Note that flow nets assume isotropic permeability and $k_H = k_V$ and the reservoir is completely drawn down (Fell et al., 2005, 2015).

(ii) If there is piezometer data under reservoir drawdown, it may be used to provide a better and less conservative estimate of the pore pressures.

(iii) The pore pressures in the embankment are immediately reduced on drawdown due to the unloading effect of removing the water. Hence pore pressures on the surface of earthfill in Figure 7.37(a)–(d) are all zero.

(iv) In Figure 7.37(a), the foundation permeability is low compared to that of the earthfill and flow will be towards the upstream face. This gives relatively high pore pressures compared to Figure 7.37(b). In the latter case, the foundation is more permeable than the earthfill, so flow lines will be near vertical and equipotentials near horizontal, giving pore pressures which should be close to zero.

(v) In Figure 7.37(c), horizontal drains are incorporated in the upstream slope. These cause flow lines to be near vertical on drawdown (except near the low permeability foundation) and should result in low pore pressures.

(vi) In Figure 7.37(d), the flow lines are influenced by the presence of the filter zones and rockfill, giving lower pore pressures than for Figure 7.38(a,) at least in the upper part of the slope.

The diagrams in Figure 7.37 are sketched with the implicit assumption that $k_H/k_V = 1$ and permeability is isotropic (uniform). In reality, if there are lower permeability layers within the zones, perched water tables may result, completely altering the pore pressures. Hence great care must be taken in the analyses, and in the absence of piezometer data it would be more common to assume that the pore pressures are those which exist prior to the drawdown, adjusted for the unloading effect of removing the water.

Drawdown flow nets and pore pressures can be obtained by finite element analysis. Factors which must be considered in the analyses are:

- Permeability anisotropy in the earthfill, i.e. allowing for $k_H > k_V$ and low permeability layers.
- The relative permeability of the foundation and the embankment zones. In practice this means that the foundation should be modelled with the embankment.
- The effect of removal of buoyancy from the upstream rockfill zone (if present) should be allowed for. This will induce increased pore pressures.

For embankments such as those shown in Figure 7.37, the analysis is sensitive to the assumptions made. For a central core earth and rockfill dam, the analysis may control the slope but will not be so sensitive to assumptions.

7.10.3 Allowance for the Effects of Shear-Induced Pore Pressures

Shear-induced positive pore pressures in low permeability zones are allowed for as described in Section 7.4.3 by adopting undrained strengths at low normal stresses.

In well compacted soil shear-induced pore pressures will be negative and are usually ignored, which is conservative.

For loose saturated non-plastic soils, e.g. sand, consider the potential for collapse settlement and generation of positive pore pressures as described in Section 7.15.

7.11 SOME DETAILS OF THE ANALYSIS OF FACTORS OF SAFETY FOR NORMAL OPERATION, FLOOD AND DRAWDOWN

7.11.1 Introduction

For most dams it is appropriate to assess the stability using Limit Equilibrium Analyses (LEA) using commercial computer programs.

It is not common to require Finite Element Analyses (FEA) unless the dam is showing signs of excessive deformation, in which case FEA will also be carried out and will be useful to assist in understanding the mechanics of deformation.

It is essential that the computer model include the foundation except where the foundation is rock with no weak seams and has a much higher shear strength than the embankment materials.

7.11.2 Some Common Problems

Some common problems in LEA (which are often simply omissions and/or occur through lack of proper care) include:

(a) Use of circular rather than non-circular analysis, where failure surfaces will clearly follow weak surfaces—e.g. bedding surface shears. Figure 7.38 shows examples where non-circular analysis must be used.
(b) Omitting to model cracks in the ground surface and the water pressure which will often develop in the cracks due to rainfall or surface water flowing into the crack. This can have an effect on calculated factors of safety, particularly for smaller failure surfaces near a dam crest.
(c) Not modelling anisotropy of strengths, e.g. in fissured clays or stratified soil and rock.
(d) Not clearly showing on the drawing of the cross section and failure surface the properties used for that particular analysis. If the two are separated, confusion is likely to occur.
(e) Quoting the minimum factors of safety for shallow failure surfaces which will not form viable failures resulting in breach of the dam. These can be avoided by specifying 1 kPa effective cohesion, properly modelling the nonlinearity of soil and rock shear strengths, modelling partial saturation effects and only considering potential failure surfaces which encompass most if not all of the dam crest.

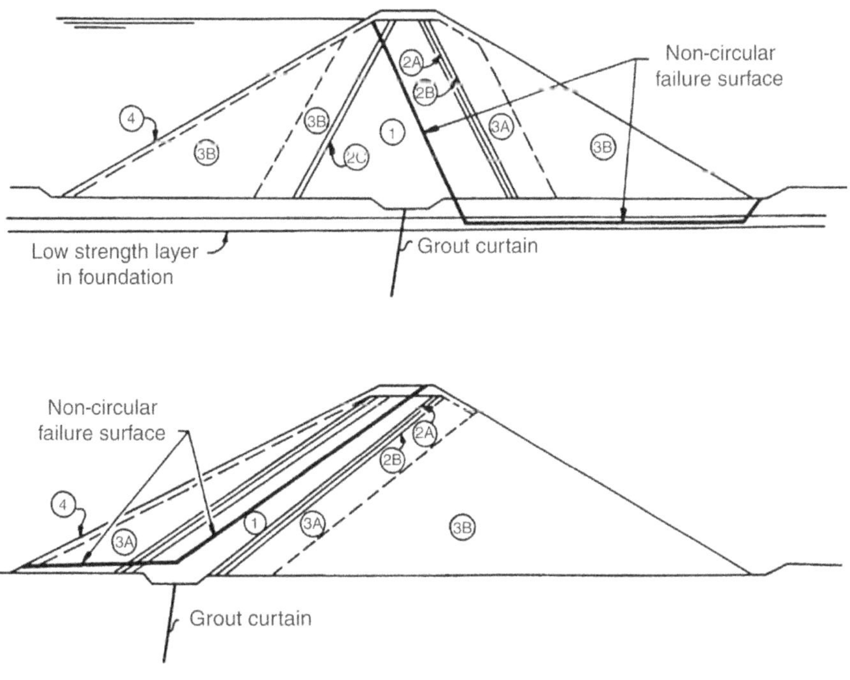

FIGURE 7.38 Examples where non-circular failure surfaces must be analysed (Fell et al., 2015).

7.11.3 Some Detailed Issues for Drawdown Analyses

(a) When analysing the stability of the upstream slope of a dam, it is important to consider shallow and deep failure surfaces. Shallow surfaces are more likely to drain during the reservoir drawdown so drained parameters may be most appropriate. Deeper surfaces are less likely to drain so undrained strengths may apply. In most cases shallow surfaces will have lower factors of safety, but depending on the embankment geometry and freeboard may be unlikely to breach the dam. Deeper surfaces may be more likely to result in loss of freeboard at the crest, although in most cases the potential for this to lead to overtopping and breach will be very low, as the sliding occurs when the reservoir level is drawn down.

(b) Consideration needs to be given to the potential for the surfaces to cause cracking of the core and displacement of filter zones and the potential consequential development of the piping and internal erosion. Note that while upstream slope failure surfaces may not result in a breach of the dam, they could still mean that the dam and reservoir must be taken out of service for a considerable time so that repair works can be done. Such an outcome should be taken into account when deciding if shallow failures could be acceptable.

(c) Multiple reservoir drawdowns often result in increased deformation of the upstream slope. This deformation is likely to result in some strain softening, giving, in effective stress terms, weakening from peak to fully softened strength as a minimum, and potentially towards residual strength. It is also likely to result in longitudinal cracking in the crest of the dam, and this cracking should be modelled in the stability analyses. It should be assumed that the cracks will fill with water under rain events following the drawdown. The slope failure at San Luis Dam in California is such a case (Stark and Duncan, 1987, 1991).

7.12 ASSESSMENT OF ACCEPTABLE FACTORS OF SAFETY

Most dams are designed, and the safety of existing dams are assessed using factors of safety as acceptance criteria. The factors of safety adopted are reasonably universal, with for example similar values used by the US Bureau of Reclamation (USBR, 1987, 2013; USACE, 2003), Building Research Establishment Guide to the Safety of Embankment Dams in the United Kingdom (BRE, 1990) and the Norwegian Geotechnical Institute (1992).

Table 7.3 is based on these and lists what can be regarded as baseline minimum factors of safety applicable to design of new dams and assessment of existing dams with high consequence of failure, with the qualifications listed in the notes.

Table 7.4 details factors which should influence the selection of a minimum acceptable factor of safety and their extent of influence. These are not used by the organisations listed earlier, but the authors recommend applying them when assessing the safety of an existing dam.

Tables 7.3 and 7.4 are based on the following:

(a) The strengths used in the analysis should be "reasonable but conservative", as discussed in Section 7.3.3, and pore pressures estimated as discussed in Section 7.9 with a high k_H/k_V in compacted earthfill.

TABLE 7.3 Baseline Recommended Minimum Acceptable Factors of Safety and Load Conditions (Fell et al., 2015)

SLOPE	*LOAD CONDITION*	*RESERVOIR CHARACTERISTIC*	*MINIMUM FACTOR OF SAFETY*
Upstream and Downstream	End of construction	Reservoir empty	1.3
Downstream	Steady state seepage	Reservoir at normal maximum operating level (Full Supply Level)	1.5
Downstream	Maximum flood	Reservoir at maximum flood level	1.5 free draining crest zones,1.3 otherwise
Upstream	Drawdown	Rapid drawdown to critical level	1.3

Notes:

(1) These factors of safety apply to design of new high consequence of failure dams, on high strength foundations, with low permeability zones constructed of soil which is not strain weakening, using reasonable but conservative shear strengths and pore pressures developed from extensive geotechnical investigations of borrow areas, laboratory testing and analysis of the results and using the methods of analysis detailed previously. It is assumed there will be monitoring of deformations by surface settlement points during construction and during operation of the dams.

(2) Pore pressures should be estimated as described in Section 7.9 with a suitably conservative k_H/k_V.

(3) "Free draining crest zones" means the pore pressures in zones near the crest will respond to reservoir level as it rises. For dams with a low permeability earthfill core taken to crest level, the pore pressures will not respond to reservoir rise and lower factors of safety may be acceptable.

TABLE 7.4 Factors Which Influence the Selection of Factor of Safety and Their Effect on the Baseline Minimum Factor of Safety (Fell et al., 2015)

FACTOR	*DESCRIPTION*	*RECOMMENDED CHANGE TO THE BASELINE MINIMUM FACTOR OF SAFETY*
Existing (vs new) dam	A lower factor of safety may be adopted for an existing dam which is well monitored and performing well.	0 to −0.1
Soil or weak rock foundation	A higher factor of safety may be needed to account for the greater uncertainty of the strength.	0 to +0.2 for effective stress +0.1 to +0.3 for undrained strength analyses
Strain weakening soils in the embankment or foundation	A higher factor of safety may be needed to account for progressive failure, and greater displacements if failure occurs.	0 to +0.2
Limited (little or no good quality) strength investigation and testing, particularly of soil and weak rock foundations	A higher factor of safety should be used to account for the lack of knowledge. Detailed investigations will be required to confirm conditions.	+0.1 to +0.3 for effective stress analyses, +0.3 to +0.5 for undrained strength analyses
Contractive soils in the embankment or foundation	A higher factor of safety may be needed to account for the greater uncertainty in the undrained strength.	+0.1 to +0.3 for undrained strength

Note: These figures are given for general guidance only. Experienced geotechnical professionals should use their own judgement, but note the principles involved in this table.

(b) The greater uncertainty in undrained strengths than in effective stress strengths means that higher factors of safety should be used where undrained strengths control stability.
(c) The lower minimum factors of safety for the construction condition are based on the consequences of failure usually being lower than for a dam under steady state seepage. However, if stability is controlled by undrained strengths, the uncertainties in the shear strength are often high, so unless there is good monitoring of pore pressures and displacements, high factors of safety should be used.
(d) The lower factors of safety for rapid drawdown are again predicated on the assumption that the dam is unlikely to breach under a drawdown failure. It should be recognised that by adopting lower factors of safety for the construction and drawdown conditions there is an acceptance of a higher probability of instability.
(e) Factors of safety during construction of dam upgrades should be as for the steady state seepage case because the consequences of failure are potentially high.
(f) The factors of safety apply to non-trivial failure surfaces. Failure surfaces passing only through free draining rockfill may acceptably have lower factors of safety, because provided the slope can be constructed; there is no possible change in conditions which will cause instability. If realistic (high) effective friction angles are used for rockfill at low confining stresses, this problem will generally not arise.

If the estimated factor of safety ≥ the adjusted baseline factor of safety and if the crest deformations are monitored and show normal behaviour as discussed in Section 7.13, the likelihood of failure leading to loss of freeboard may be assumed to be extremely low or negligible. This has been the authors' approach for assessing the safety of all but extreme consequence of failure dams. For extreme consequence of failure dams, risk based approaches are used.

7.13 REVIEW AND ASSESSMENT OF DEFORMATIONS OF THE DAM DURING OPERATION

Many dams have crest survey marks which are surveyed beginning when or soon after construction was completed. For well monitored dams the survey includes settlement and horizontal displacement parallel and normal to the crest centreline, but some only have settlement survey.

These data can be useful in identifying abnormal behaviour which may or may not be related to slope instability.

Hunter (2003) and Hunter and Fell (2003c) gathered data from a large number of earth and earth and rockfill dams, where the core has been generally compacted to a minimum density of 98% of standard maximum dry density, with a moisture content in the range 2% dry to 1% wet of Standard Optimum Moisture Content. Most of the cores were "dry placed", with a moisture content 0.5% to 1% dry of Standard Optimum Moisture Content.

Figures 7.39 and 7.40 show post-construction crest settlements 10 years and 20 to 25 years after the end of construction for zoned earthfill and earth and rockfill embankments.

Figure 7.41 shows long term post-construction crest settlement rates for these dams.

Tables 7.5 and 7.6 show typical ranges of post-construction settlement and long term settlement rates for different core materials, core widths and compaction moisture content. These exclude a small number of "outliers" which had significantly larger settlements. Hunter (2003) and Hunter and Fell (2003c) have details.

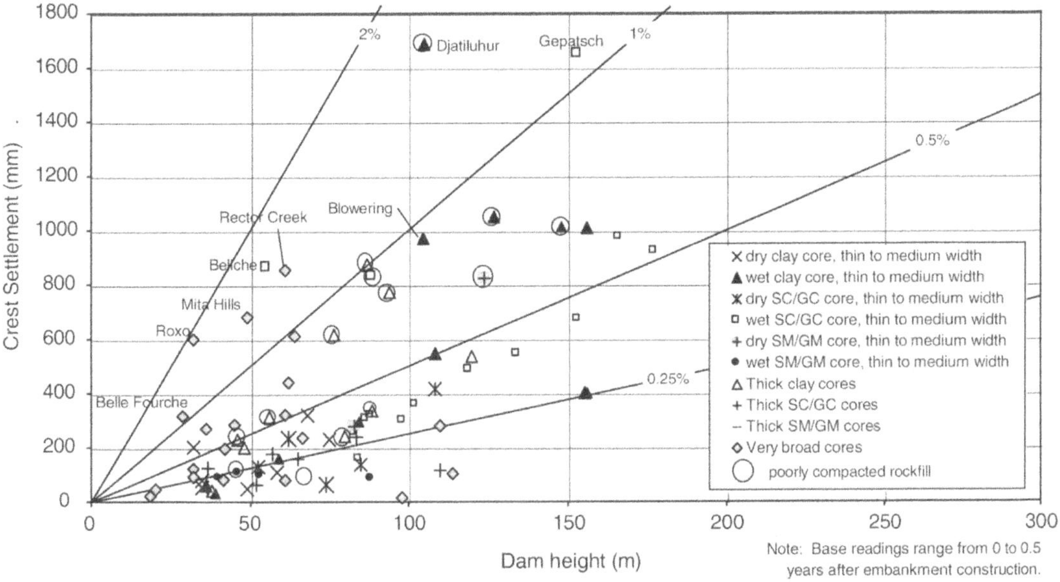

FIGURE 7.39 Post-construction crest settlement at 10 years after end of construction (Hunter, 2003; Hunter and Fell, 2003c).

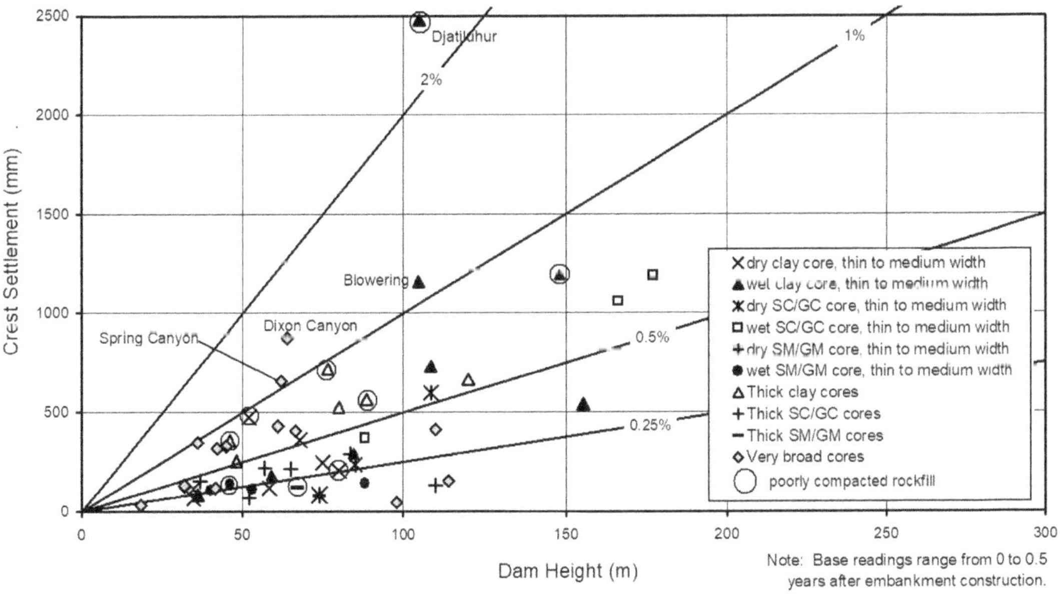

FIGURE 7.40 Post-construction crest settlement at 20 to 25 years after end of construction (Hunter, 2003; Hunter and Fell, 2003c).

In Table 7.5 the core width is classified as:

Thin width < 0.5 dam height.
Medium width ≥ 0.5 and ≤ 1.0 dam height.
Wide width > 1.0 dam height.

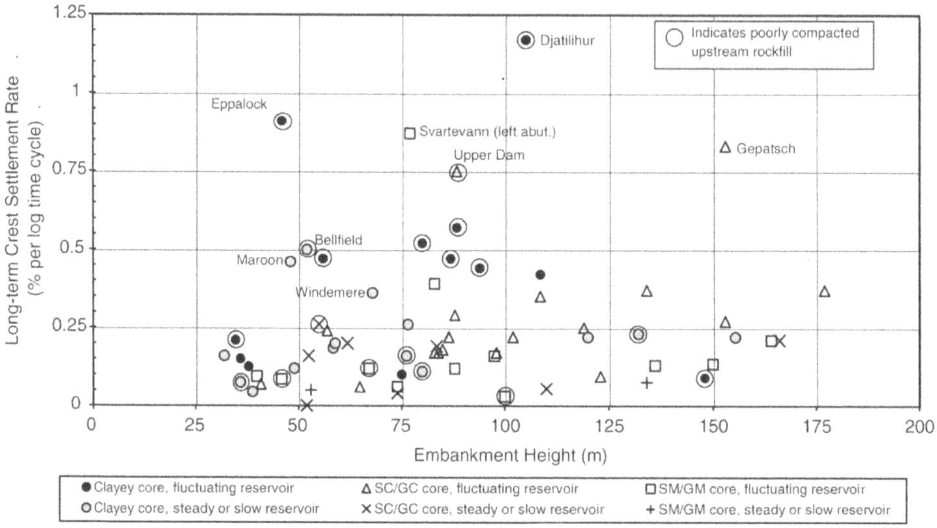

FIGURE 7.41 Long-term post-construction crest settlement rate for zoned earthfill and earth-rockfill embankments of thin to thick core widths (Hunter, 2003; Hunter and Fell, 2003c).

Hunter (2003) and Hunter and Fell (2003c) conclude:

- Nearly all dams experience less than 1% crest settlement post construction for periods up to 20 to 25 years and longer after construction.
- Most experience less than 0.5% in the first 3 years and less than 0.75% after 20 to 25 years.
- Smaller magnitude settlements are observed for dry placed clayey sands and clayey gravels, and dry to wet placed silty sands and silty gravels.
- A broader range of settlement magnitude is shown for clay cores, wet-placed clayey sand to clayey gravel cores, and embankments with very broad core widths.
- For zoned earth and rockfill dams, poor compaction of the rockfill is over-represented for case studies at the larger end of the range of crest settlement.

The authors' experience is that even very high dams with cores constructed of well compacted broadly graded silt sand gravel, e.g. glacial moraine or alluvial "gravels", have very small post-construction settlements, less than 0.1%.

If a dam exhibits larger than usual post-construction settlements or long term settlement rates or accelerated periods of settlement, the reason(s) for this should be assessed.

Outliers to the general trend of post-construction crest settlement are generally evident in at least two of the plot types, usually the settlement versus time plot and then either or both of the total settlement and settlement rate plots. The settlement versus time plots are the most comprehensive for assessment of potential outliers. Figures 7.42 and 7.43 show an example.

It can be seen in Figure 7.42 that accelerated rates of settlement coincided with major drawdown of the reservoir. Lateral displacement also reacted to the reservoir drawdown as shown in Figure 7.43.

For Upper Yarra Dam, the crest settlement and lateral deformation data along with that from internal settlement gauges showed that a wedge of the clay core and "soft rock" shoulders had formed an internal shear surface as shown in Figure 7.44. This affected about half the length of the embankment.

TABLE 7.5 Typical Range of Post-Construction Crest Settlement (Hunter, 2003; Hunter and Fell, 2003c)

	CORE PROPERTIES			*CREST SETTLEMENT (% OF DAM HEIGHT)* (1)		
CORE WIDTH	*CLASS'N*	*MOISTURE CONTENT*	*NO. CASES*	*3 YEARS*	*10 YEARS*	*20 TO 25 YEARS*
Thin to medium	CL/CH	Dry	9	0.05 to 0.55	0.10 to 0.65	0.20 to 0.95
		Wet	11	0.04 to 0.75	0.08 to 0.95	0.20 to 1.10
	SC/GC	Dry	5	0.10 to 0.25	0.10 to 0.40	< 0.5
		Wet	18	0.15 to 0.80	0.20 to 1.10	< 1.1
Thin to thick	SM/GM	All	16	0.06 to 0.30	0.10 to 0.65	< 0.5 to 0.7
Thick	CL/CH	All (most dry)	12	0.02 to 0.75	0.10 to 1.0	0.5 to 1.0
	SC/GC	All (most dry)	5	0.05 to 0.20	0.10 to 0.35	0.10 to 0.45
Very broad	all	All (most dry)	18	0.0 to 0.60	0.0 to 0.80	0.05 to 0.76

Note: (1) Excludes possible outliers.

TABLE 7.6 Summary of Long Term Crest Settlement Rates (% per log cycle of time) (Hunter, 2003; Hunter and Fell, 2003c)

CORE WIDTH	*CORE CLASSIFICATION*	*RESERVOIR OPERATION*	*NO. CASES*	*LONG TERM SETTLEMENT RATE (% PER LOG CYCLE OF TIME)*	*COMMENTS*
Thin to thick	CL/CH	Fluctuating	14	0.09 to 0.57	Higher rates at Djatiluhur (1.17%) and Eppalock (0.91%).
		Steady or slow	15	0.04 to 0.50 (most < 0.26)	Very high at Ataturk (3.1%). Higher rates at Maroon (0.46%), Windemere (0.36%) and Bellfield (0.50%).
	SC/GC	Fluctuating	18	0.06 to 0.37	Higher rates at Upper Dam (0.75%) and Gepatsch (0.83%).
		Steady or slow	8	0 to 0.26	
	SM/GM	Fluctuating	12	0.03 to 0.21	Higher rates at Svartevann (0.87%) and Andong (0.39%).
		Steady or slow	2	< 0.10	Only two cases, Round Butte and Parangana
Very broad	All types (most CL)	Fluctuating	17	0.07 to 0.70 (most < 0.35)	Very high rates (> 1.8%) at Belle Fourche and Roxo (near buttress). High rates at Rector Creek (1.65%), Dixon Canyon (1.38%) and Spring Canyon (1.31%).
		Steady or slow	2	0.08 and 0.44	Only two cases, Eucembene (0.08%) and Enders (0.44%).

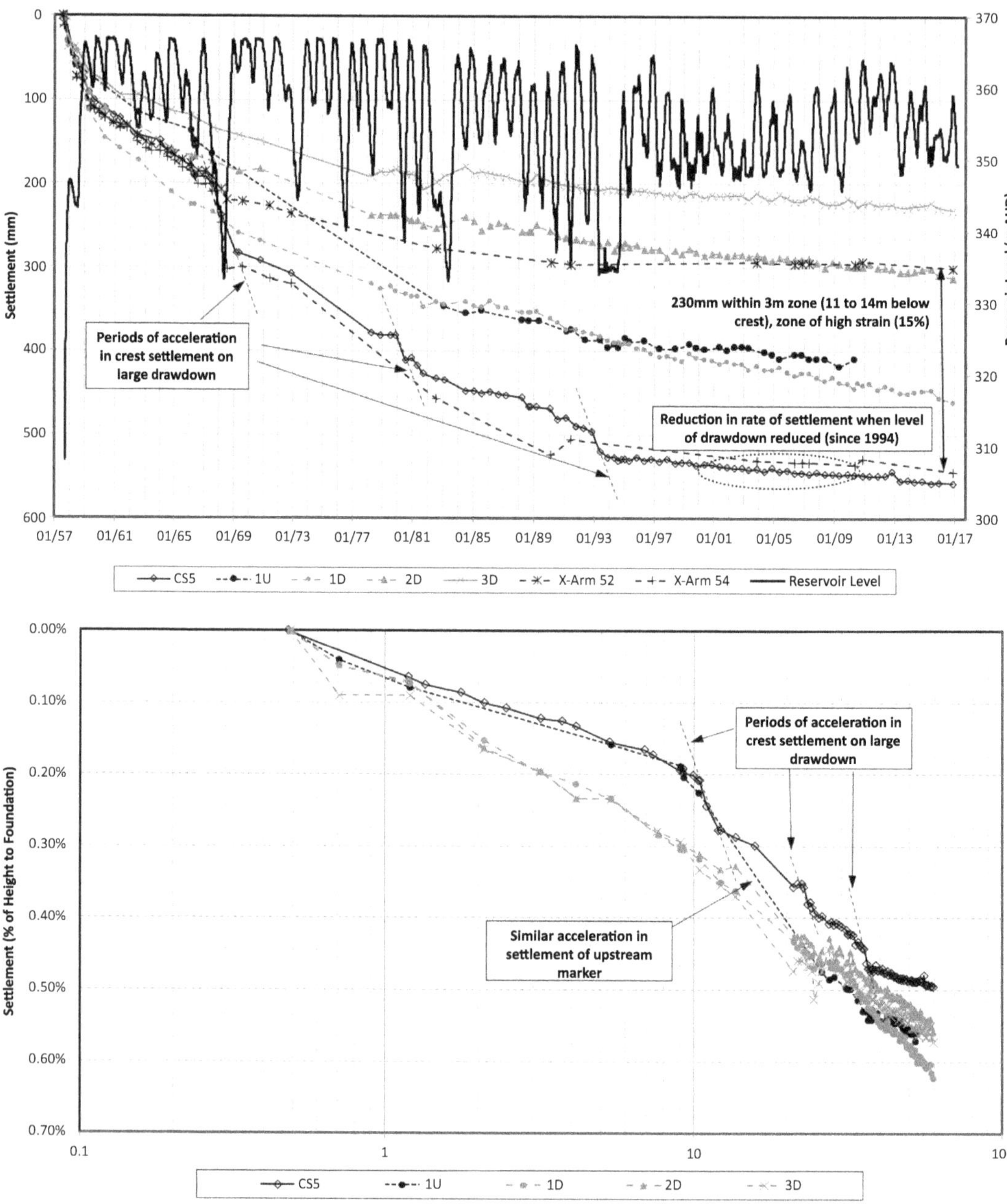

FIGURE 7.42 Post-construction settlement versus time at the maximum height section. (Courtesy of Melbourne Water and AECOM.)

The outer rockfill shoulders were not so well compacted and were less stiff than the core, so the wedge of core and shoulder had insufficient support under drawdown conditions. The overall factor of safety for a potential failure surface including the postulated shear was only marginally below 1.3, i.e. the commonly accepted satisfactory factor of safety, so the abnormal behaviour was not a threat to dam safety by loss of freeboard. Similar behaviour to this has been observed in central core rockfill dams

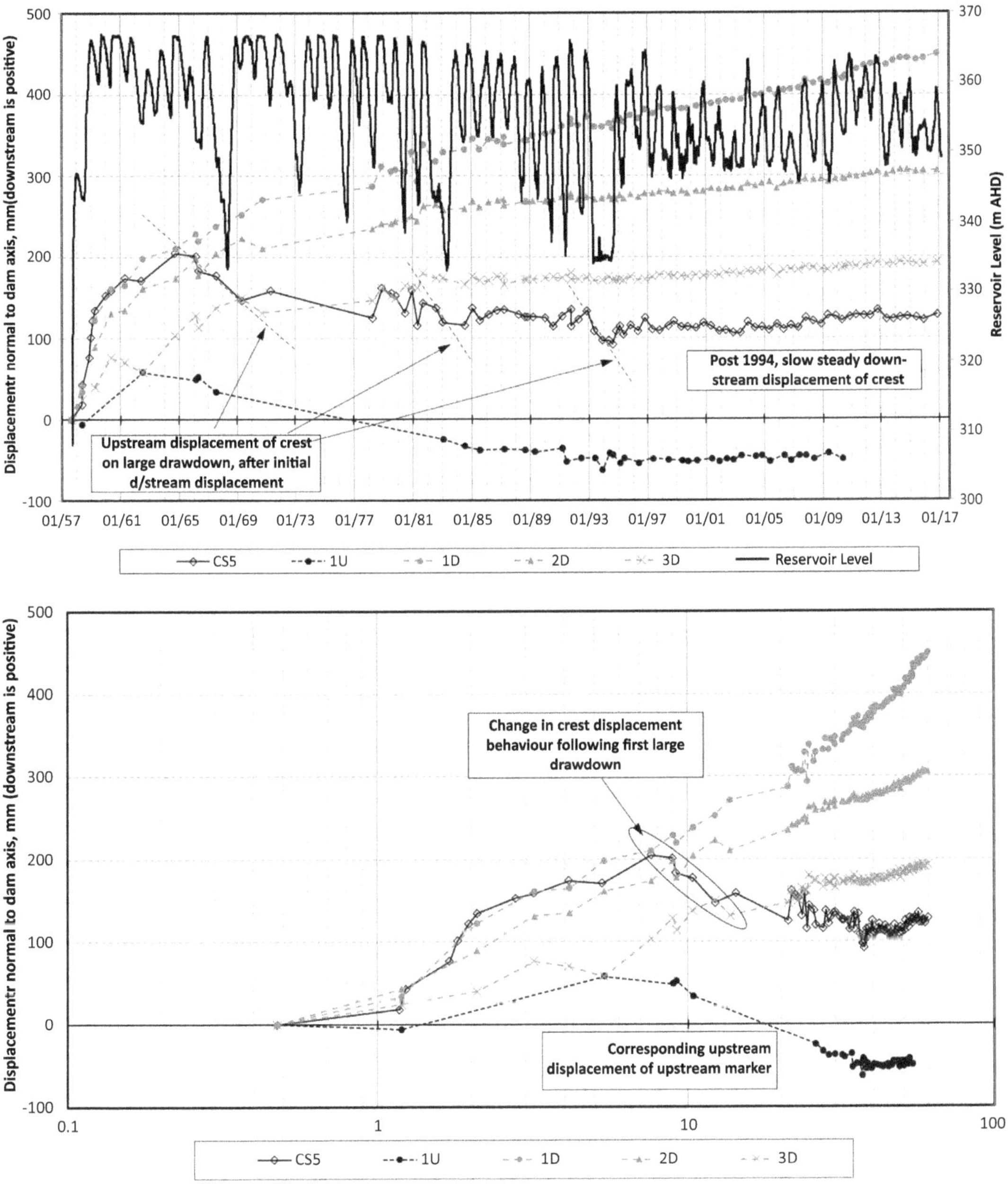

FIGURE 7.43 Post-construction displacement normal to the dam axis versus time at the maximum height section. (Courtesy of Melbourne Water and AECOM.)

with the rockfill not being compacted and placed by bulldozer in layers several metres thick, e.g. Eppalock Dam (Davidson et al., 2000).

It is important to note that use of the term "abnormal" in relation to the deformation behaviour of an embankment does not necessarily equate with potential instability. In nearly all cases identified as "abnormal" or potentially "abnormal", the overall stability of the embankment is not in question.

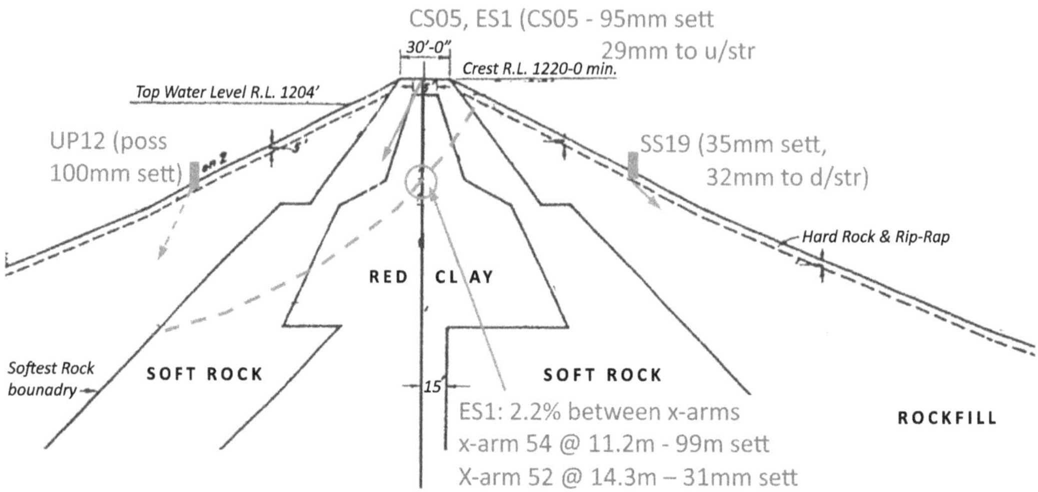

FIGURE 7.44 Estimated deformation of Upper Yarra Dam on 1967/1968 reservoir drawdown and postulated internal shear mechanism. (Courtesy of Melbourne Water and AECOM.)

7.14 ASSESSING THE PROBABILITY OF FAILURE RESULTING FROM DOWNSTREAM SLOPE INSTABILITY

7.14.1 Introduction

This section applies to slope instability under normal operating and flood conditions. For deformations and potential loss of freeboard under seismic loads see Chapter 8.

Instability of the upstream slope under reservoir drawdown is considered unlikely to result in loss of freeboard so is not considered here.

For a dam to fail by loss of freeboard resulting from downstream slope instability requires:

(a) Initiation of slope instability—factor of safety reduces below 1.0 and movement of the unstable slope initiates.
(b) Progression of slope instability—slope movement progresses and results in loss of freeboard such that there is continuous overtopping of the dam crest.
(c) Detection and intervention fail to stop the process.
(d) Breach—once continuous overtopping occurs, the probability of breach of the embankment can be assumed to be virtually a certainty.

In reality there are two opportunities to detect and intervene:

- Before the factor of safety falls below 1.0, e.g. when it is around 1.2 to 1.1 when most slopes begin to deform but a complete failure surface has yet to form.
- After the factor of safety falls below 1.0.

The likelihood of successful detection and intervention depends on whether the failure surface intercepts above or below the reservoir. Figure 7.45 shows these scenarios.

Figure 7.46 shows an event tree modelling the sequence of events from initiation of instability to breach.

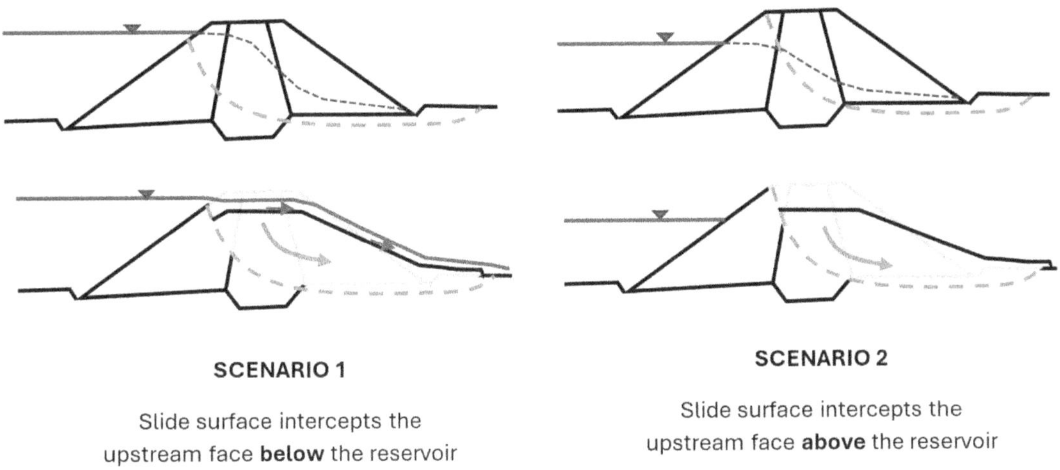

FIGURE 7.45 Scenarios for downstream slope instability.

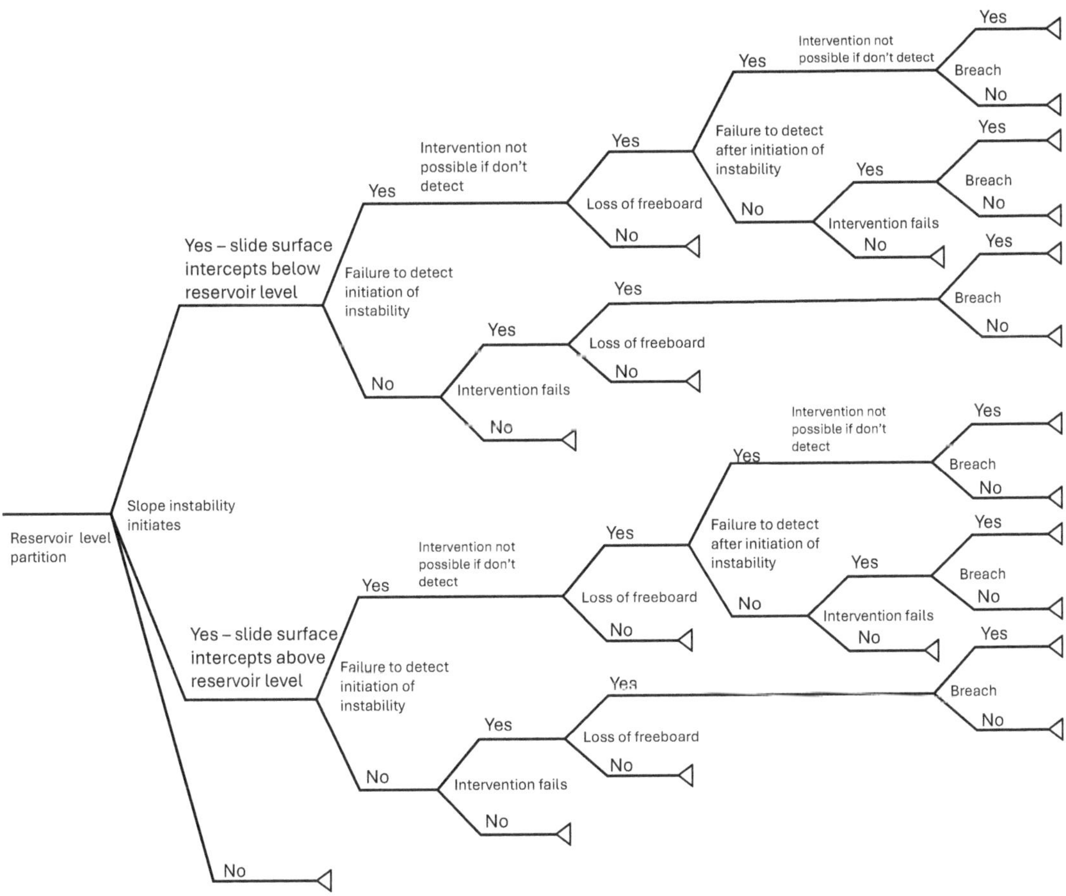

FIGURE 7.46 Event tree for modelling the annual probability of failure of an embankment from slope instability.

In a deterministic assessment approach, the likelihood of failure should be assessed for the reservoir at Full Supply Level or normal operating level for gated spillways, and for the reservoir level corresponding to the Inflow Design Flood.

For a risk based approach, the annual probability of failure (P_{AF}) is determined from:

$$P_{AF} = \Sigma\,(P_{AL} \times P_I \times P_{LFI} \times [(P_{DF} \times P_B) + ((1 - P_{DF}) \times P_{IF} \times P_B)] \tag{7.23}$$

where:

P_{AL} = annual probability of reservoir level partition
P_I = probability of initiation of slope instability
P_{LFI} = probability of loss of freeboard given initiation of slope instability
P_{DF} = probability of failure to detect instability
P_{IF} = probability intervention fails (if instability is detected)
P_B = probability of breach.

This should be assessed for all reservoir partitions as described in Chapter 5 following the event tree structure in Figure 7.46.

7.14.2 Estimating the Probability of Initiation of Downstream Slope Instability (P_I)

See Section 7.12 for a discussion on acceptable factors of safety.

7.14.2.1 Dams with Factors of Safety ≥ the Adjusted Baseline Factor of Safety

If the estimated factor of safety ≥ the adjusted baseline factor of safety from Tables 7.4 and 7.5 and if the crest deformations are monitored and show normal behaviour as discussed in Section 7.13, then the likelihood of failure leading to loss of freeboard may be assumed to be extremely low or negligible.

The strengths used in the analysis should be "reasonable but conservative", as discussed in Section 7.3.3. and pore pressures estimated as discussed in Section 7.9 with a high k_H/k_V ratio in compacted earthfill.

7.14.2.2 Dams with Factors of Safety < the Adjusted Baseline Factor of Safety, and Extreme Consequence Dams

It is recommended that the modified Silva et al. (2008, 2010,2025) method as described next be used to estimate the conditional probability of failure for each reservoir level being considered.

7.14.2.3 Silva et al. (2008, 2010) Method

Silva et al. (2008, 2010) present Figure 7.47, which relates annual probability of failure of a slope to the factor of safety and the level of engineering applied to the slope.

The levels of engineering described by Silva et al. (2008, 2010) are:

Category I—facilities designed, built and operated with state of the practice engineering. Generally, these facilities have high failure consequences.

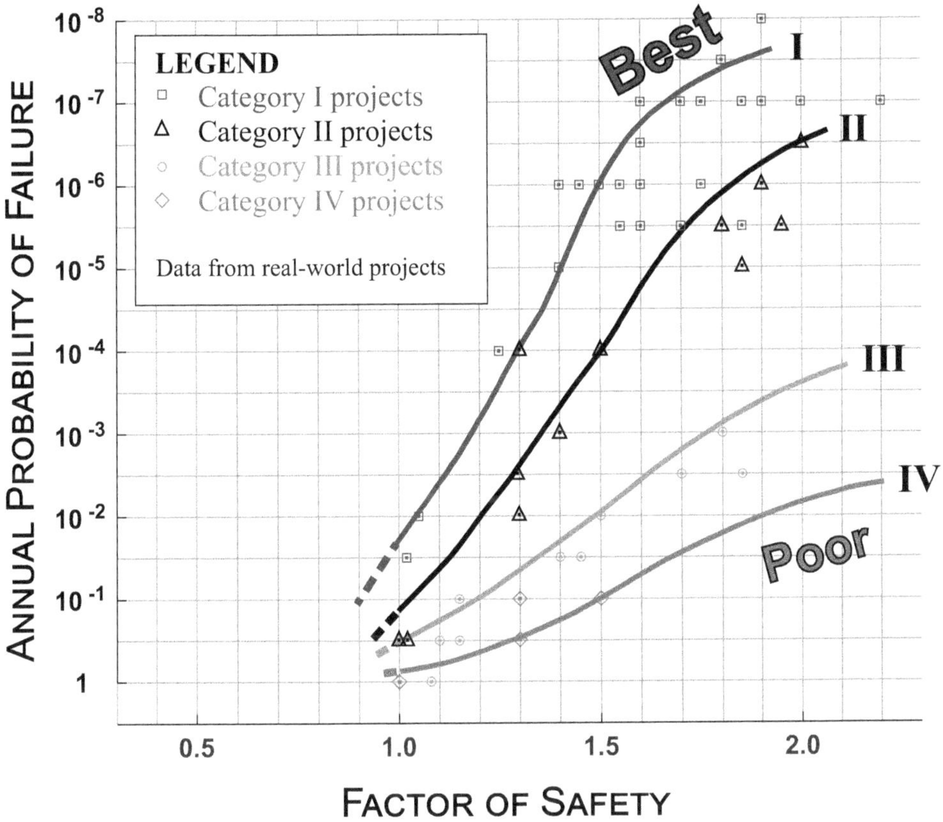

FIGURE 7.47 Factor of safety versus annual probability of failure (Silva et al., 2008, 2010). (Used with permission of ASCE conveyed through Copyright Clearance Center Inc.)

Category II—facilities designed, built and operated using standard engineering practice. Many ordinary facilities fall into this category.

Category III—facilities without site-specific design and sub-standard construction or operation. Temporary facilities and those with low failure consequences often fall into this category.

Category IV—facilities with little or no engineering.

Table 7.7 provides more detail on investigation, testing, analyses and documentation. Construction and operation and monitoring related to each category.

Silva et al. (2008, 2010) indicate that they interpolate between curves by assigning partial values for the different characteristics. For example, if a particular structure meets most of the Category I criteria but only had part time supervision by a qualified engineer during construction (a Category II attribute), they use the weighting values in parenthesis in Table 7.7 to compute the interpolated value as shown in Table 7.8.

The annual probability of failure is then obtained by linear interpolation in Figure 7.47.

Silva et al. (2008, 2010) indicate that the factor of safety should be estimated from:

- Representative geometry of the slope.
- Best estimates of shear strength parameters.

TABLE 7.7 The Characteristics of the Engineering Categories (Silva et al., 2008; used with permission of ASCE conveyed through Copyright Clearance Center Inc.)

	DESIGN				
LEVEL OF ENGINEERING	*INVESTIGATION*	*TESTING*	*ANALYSES AND DOCUMENTATION*	*CONSTRUCTION*	*OPERATION AND MONITORING*
I (Best) Facilities with high failure consequences	• Evaluate design and performance of nearby structures	• Run lab tests on undisturbed specimens at field conditions	• Determine FS using effective stress parameters based on measured data (geometry, strength, pore pressure) for site determination	• Full time supervision by qualified engineer	• Complete performance program including comparison between predicted and measured performance (e.g. pore pressure, strength, deformations)
	• Analyse historic aerial photographs	• Run strength test along field effective and total stress paths	• Consider field stress path in stability	• Construction control tests by qualified engineers and technicians	• No malfunctions (slides, cracks, artesian heads)
	• Locate all nonuniformities (soft, wet, loose, high, or low permeability zones)	• Run index field tests (e.g. field vane, cone penetrometer) to detect all soft, wet, loose, high, or low permeability zones	• Prepare flow net for instrumented sections	• No errors or omissions	• Continuous maintenance by trained crews
	• Determine site geologic history	• Calibrate equipment and sensors prior to testing program	• Predict pore pressure and other relevant performance parameters (e.g. stress, deformation, flow rates) for instrumented section	• Construction report clearly documents construction activities	
	• Determine subsoil profile using continuous sampling		• Have design report clearly document parameters and analyses used for design		
	• Obtain undisturbed samples for lab testing of foundation soils		• No errors or omissions		
	• Determine field pore pressures		• Peer review		
	(0.20)	(0.20)	(0.20)	(0.20)	(0.20)

II (Above average) Ordinary facilities	• Evaluate design and performance of nearby structures • Exploration program tailored to project conditions by qualified engineer (0.40)	• Run standard lab tests on undisturbed specimens • Measure pore pressure in strength tests • Evaluate differences between laboratory test conditions and field conditions (0.40)	• Determine FS using effective stress parameters and pore pressures • Adjust for significant differences between field stress paths and stress path implied in analysis that could affect design (0.40)	• Part-time supervision by qualified engineer No errors or omissions (0.40)	• Periodic inspection by qualified engineer • No uncorrected malfunctions • Selected field measurements • Routine maintenance (0.40)
III (Average) Unimportant or temporary facilities with low failure consequences	• Evaluate performance of nearby structures • Estimate subsoil profile from existing data and borings (0.60)	• Index tests on samples from site (0.60)	• Rational analyses using parameters inferred from index tests (0.60)	• Informal construction supervision (0.60)	• Annual inspection by qualified engineer • No field measurements • Maintenance limited to emergency repairs (0.60)
IV (Poor) Little or no engineering	• No field investigation (0.80)	• No laboratory tests on samples obtained at the site (0.80)	• Approximate analyses using assumed parameters (0.80)	• No construction supervision by qualified engineer • No construction control tests (0.80)	• Occasional inspection by non-qualified person • No field measurements (0.80)

TABLE 7.8 Example of Interpolation of Category (Silva et al., 2008)

Design	
Investigation	0.2
Testing	0.2
Analysis and documentation	0.2
Construction	0.4
Operation	0.2
Interpolated Category	1.2

From this it is implied that the best estimate of pore pressures should be used.

The discussion indicates that effective stress strengths would be used or stress paths to failure if using strengths in terms of total strength.

The method was developed by subjective judgement of the Silva et al. (2008) authors and experienced engineers, and the probability estimates anchored by:

- Factor of safety = 1.5, annual probability of failure = 0.0001. This is based on historical performance of earth dams designed and constructed with conservative engineering practice as described in Baecher et al. (1980), Whitman (1984) and Christian et al. (1992).
- Factor of safety = 1.0, annual probability of failure = 0.5 based on probabilistic theory.

Data from 75 projects was used to develop Figure 7.47. The projects included zoned and homogeneous earth dams, tailings dams, natural and cut slopes and earth retaining structures. No details are available.

7.14.2.4 Discussion

The Silva et al. (2008) paper includes Figure 7.48 which shows the curves in Figure 7.47 plotted with log-normal curves and the anchor points from Whitman (1984).

We note that:

(a) The log-normal distributions are conditional probabilities, not annual probabilities. Silva et al. (2008) do not provide the coefficient of variation, V_F (defined as the ratio standard deviation/mean), for these curves.

(b) The "average earth dam design" annual probability of failure is 10^{-4}/annum as Whitman (1984) reports it, but this is for all failure modes including overtopping, internal erosion and piping and slope instability.

(c) Foster et al. (1998, 2000a) analysed case data for dam failures and accidents and found that downstream slope instability resulted in a small percentage of failures. Dams which had modest or good compaction control have a historical frequency of about 10^{-3} for downstream slope instability accidents, i.e. not resulting in breach of the reservoir. Assuming $V_F = 0.15$ as suggested by Whitman (1984) is applicable for dams and a factor of safety = 1.4, this plots between Categories II and III so is a reasonable alternative anchor point.

(d) The anchor point of probability of failure = 0.5 for factor of safety = 1.0 is from reliability theory so is also not an annual probability.

Silva (Silva et al., 2025) was consulted by the authors and indicated that he and others use Figure 7.47 routinely to estimate system response probabilities (conditional probabilities) or annual exceedance probabilities. He indicates:

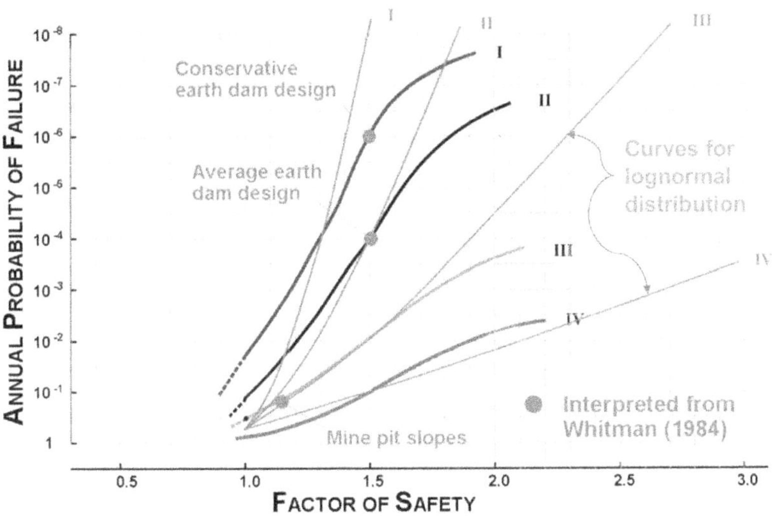

FIGURE 7.48 Factor of safety versus annual probability of failure with log-normal distributions, and data from Whitman (1984); (Silva et al., 2008). (Used with permission of ASCE conveyed through Copyright Clearance Center Inc.)

> The simplest way to explain this is that the plots provide a snapshot relationship between the Factor of Safety for a specific field condition (involving geometry, strength, pore pressures) and the corresponding failure probability. Changing conditions outside the boundaries of the mathematical models used to compute the Safety Factor (e.g. hydraulic loads and other external loads, intervention activities) provide the annualization.

He recommended using an updated version of Figure 7.47, which is shown in Figure 7.49.

Figure 7.50 shows the original Silva et al. (2008), revised Silva et al. (2025), the Silva et al. (2008) log-normal curves, and log-normal plots for $V_F = 0.07, 0.1, 0.15, 0.2$ and 0.3.

It can be seen that the Silva et al. (2025) revised Category II, Category III and Category IV curves follow log-normal curves for $V_F = 0.10, 0.18$ approx., and 0.3 up to factor of safety = 1.5.

The Silva et al. (2025) curve for Category I has not been changed from Silva et al. (2008). Given that Category I is the best defined factor of safety, the current authors believe the probability of failure should = 0.5 for a factor of safety = 1.0 and that curve should follow $V_F = 0.07$ up to factor of safety = 1.3.

We have adopted the revised Silva et al. (2025) method with this modification, and assuming that the vertical axis is conditional probability of slope instability initiating given a factor of safety, not an annual probability.

7.14.2.5 Recommended Modified Silva et al. (2008, 2025) Method

The recommended method is to use Figure 7.51 to estimate the conditional probability of initiation of instability versus factor of safety.

Category Descriptions

The category descriptions from Silva et al. (2008, 2025) have been modified to better apply to embankment dams.

Category I—Dams investigated, designed, built and operated with state of the practice engineering. Monitoring of pore pressures and crest deformations available since end of construction show normal behaviour.

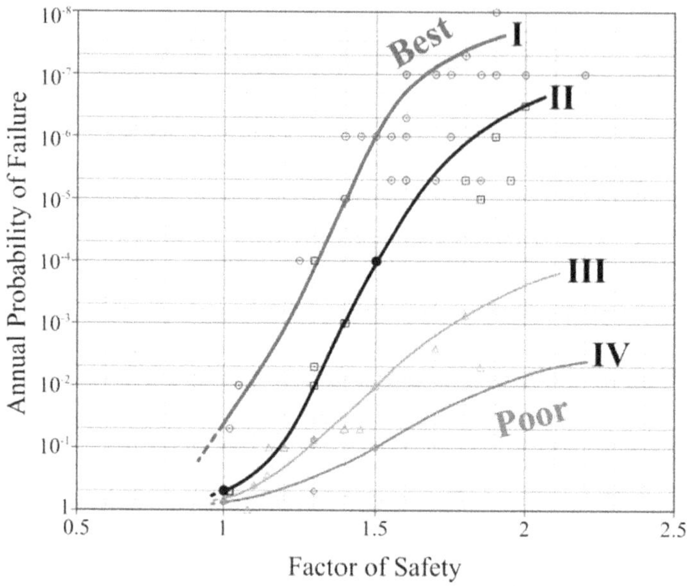

FIGURE 7.49 Revised plot of factor of safety versus annual probability of failure (Silva et al., 2025). (Used with permission of F. Silva.)

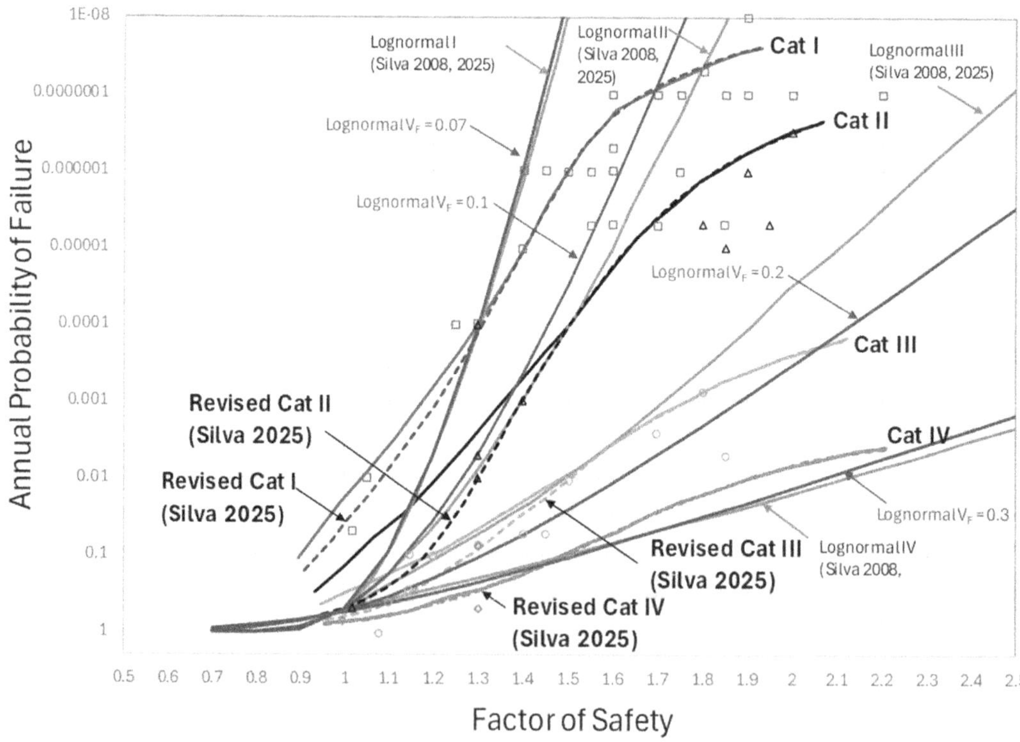

FIGURE 7.50 Annual probability of failure versus factor of safety with plots of Silva et al. (2008), Silva et al. (2025) with Silva et al. (2008, 2025) and additional lognormal curves by the current authors.

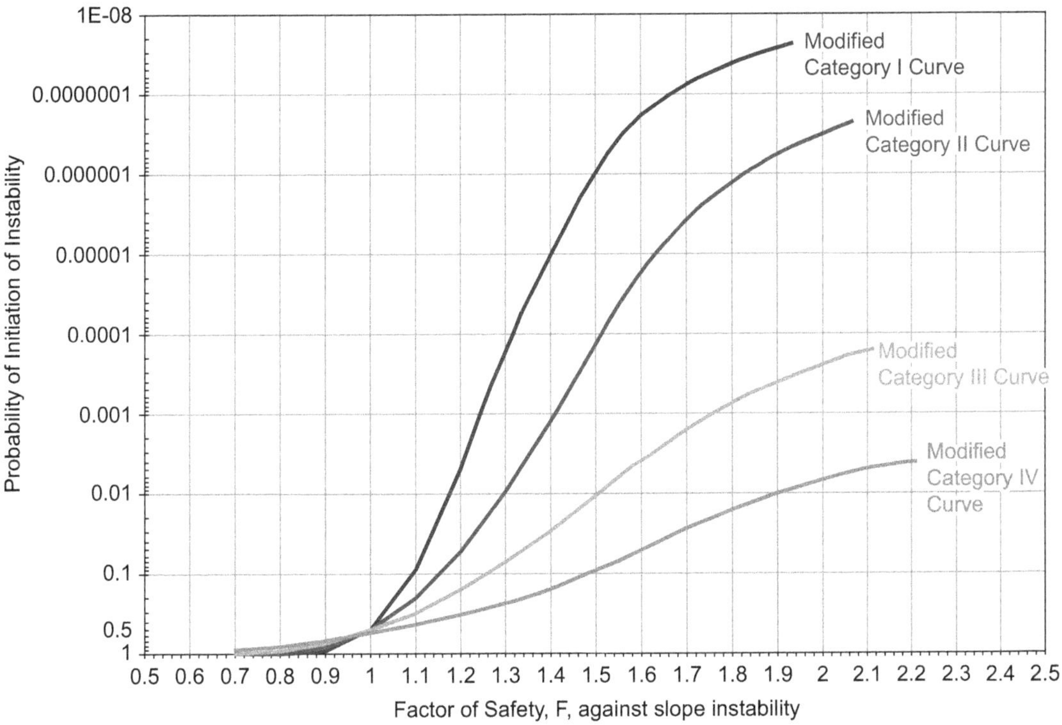

FIGURE 7.51 Modified Silva et al. (2008, 2025) method for estimating the conditional probability of initiation of slope instability from the factor of safety using best estimate shear strengths and pore pressures.

Only dams with complete documentation of investigations, design and construction supervision and quality control testing, and with extensive monitoring will satisfy this category.

> *Category II—Dams investigated, designed, built and operated using standard engineering practice.* Monitoring of crest deformations available since end of construction shows normal behaviour. Monitoring of pore pressures would be a requirement for dams where the factor of safety is sensitive to pore pressures.

Documentation of quality control during construction may be incomplete but there is confidence that the fill was compacted according to the specification.

If deformation monitoring is not continuous or sufficiently extensive to be assured of normal behaviour, the category may lie between Categories II and III.

If monitoring shows abnormal behaviour, the category should lie between Categories II and III with the actual category depending on details.

Many dams designed and constructed since the 1960s and particularly the 1970s by organisations experienced in dams and with good construction control will fall into this category.

> *Category III—Dams designed and built without site specific investigations and design.* Fill likely to have been compacted by rollers but with little or no construction control. No monitoring of crest deformations but no abnormal deformations of the crest apparent.

If crest settlements have been monitored for several years and show normal behaviour, the category may lie between Categories III and II with the actual category depending on details.

If the crest shows abnormal behaviour, the category should lie between Categories III and IV with the actual category depending on details.

Many dams constructed between 1930s and 1960s will fall into this category, but so will some dams constructed up to current times.

Category IV—Dams with little or no engineering. Fill not compacted with rollers. No crest deformation monitoring, and crest surface appears irregular or has been reconstructed to repair excessive deformations or settlement.

Many pre-1930s and some pre-1950s dams fall into this category

The category may change if there have been extensive recent site investigations of the embankment and foundations.

Hunter (2003) analysed slope failures of embankment dams during and after construction. Table 7.9 summarises the failures during construction.

It is notable that failures were all controlled by undrained strengths in poorly compacted fill or weak layers in the foundation. Contractive behaviour occurs in saturated poorly compacted plastic soils (e.g. ≤ 90% standard MDD). A similar trend applies to post-construction downstream slope instability, but there are fewer failures.

Whitman (1984) indicates that $V_F = 0.3$ for slopes controlled by undrained strengths determined by vane shear tests. That is equivalent to Category IV. The current authors suggest that if the undrained strength is known more reliably, e.g. from a combination of estimates from CPTU and pore pressure response in triaxial tests, then a lower category would be appropriate.

For dams with contractive soils in the embankment and/or foundation adopt Category IV if there is limited undrained strength data and no monitoring. Adopt Category III or between Category III and Category II if there is good quality undrained strength data (e.g. from triaxial tests and CPTU) and recent monitoring shows no unusual behaviour.

Contractive behaviour also occurs in foundations of low strength plastic soil, such as marine clays.

Table 7.7 from Silva et al. (2008) provides more detail on relating categories to investigation, testing, analyses and documentation, construction, operation and monitoring related to each category. Readers should use Table 7.7 to assist in selecting a category. Interpolation should be carried out as described earlier.

Normal and abnormal deformation behaviour is discussed in Section 7.13.

TABLE 7.9 Mechanics of Slope Failure for Failures in Embankment Dams during Construction (Hunter, 2003)

FAILURE TYPE	*NO. CASES*	*COMMENTS*
Embankment only, undrained failure in wet placed fill	5	Mainly homogeneous earthfill embankments with materials placed well wet of OMC (average greater than 4% wet, but material dependent).
Embankment only, undrained failure controlled by wet layers in fill	6	Undrained instability within wet, low undrained strength layers in the embankment. Mainly homogeneous earthfill embankments. Has been observed in one puddle core dam. Failure was during construction on a winter shutdown surface.
Embankment and foundation, undrained failure in low strength foundation layer	10	Failure controlled by weak layers in foundation. Either pre-sheared seams at residual strength or low undrained strength layers.
Embankment and foundation, progressive failure of foundation and undrained failure in the embankment	1	Only one case, Carsington. Progressive failure within the overconsolidated clay foundation layer (with solifluction shears) and undrained failure within the embankment.

Estimating the Factor of Safety (F of S)

The factor of safety for input into the modified Silva et al. method should be estimated from:

- Representative geometry of the slope.
- Best estimates of shear strength parameters.
- Best estimate of pore pressures.
- Potential failure surfaces which include the dam crest or a large part thereof, not shallow failure surfaces on the downstream slope.

7.14.3 Estimation of the Likelihood of Loss of Freeboard Given Slope Instability Has Initiated (P_{LFI})

Estimate the probability of loss of freeboard given initiation of slope instability (P_{LFI}) from Table 7.10 and Figure 7.52.

As detailed in Table 7.10, strain weakening may occur from three mechanisms:

(a) Drained shear in high clay size content soils. See Section 7.3. These soils are subject to progressive failure.
(b) Undrained shear in saturated poorly compacted plastic soils. See Section 7.4.
(c) Undrained shear in loose saturated silt sand gravel soils. See Section 7.15.3.

The method just described is based on the engineering judgement of the authors and is approximate. For important decisions and if F of S < 1.0, carry out numerical deformation analyses because this gives a better modelling of the potential failure surface than limit equilibrium stability analyses do. See Section 8.4 for details.

7.14.4 Estimation of the Probability of Not Detecting Slope Instability

There are three opportunities to detect slope instability:

(a) Before slope instability has fully initiated. Many slopes which fail display cracking, local deformations and local accelerated displacement in the hours, days or weeks before instability has initiated forming a failure surface. These may be observed by inspection and/or by crest survey monitoring.
(b) After slope instability has initiated and is progressing. This will be observable by inspection.
(c) After progression, provided that the failure surface does not encompass the whole of the dam crest.

7.14.4.1 Detection before Slope Instability Has Fully Initiated

The likelihood of not detecting the initiation of instability is dependent on:

(a) The qualifications and experience of the persons carrying out inspections and the person interpreting the inspection and monitoring data. In most cases the routine inspections are carried out by the persons operating the dam. Interpretation is carried out by the dam safety manager. For successful detection both need to be qualified and experienced for their role.

TABLE 7.10 Factors Which Influence the Likelihood of Loss of Freeboard Given Initiation of Slope Instability

FACTOR	*RELATIVE IMPORTANCE FACTOR (RF)*	*LIKELIHOOD FACTOR (LF)*			
		LESS LIKELY (1)	*NEUTRAL (2)*	*MORE LIKELY (3)*	*MUCH MORE LIKELY (4)*
Presence of strain weakening soil in the embankment or foundation	(3)	***Soil Classification*** Low to medium plasticity, clay size content < 20%.	Clays, sandy clays, clay size content 20% to 40%.	Clays, sandy clays, high plasticity; clay size content (% passing 0.002 mm) 40% to 50%.	As for More Likely, but with clay size content > 50%.
		Degree of Compaction: Plastic Soils All very well compacted to e.g. ≥ 98% standard MDD, moisture content 1% dry of optimum to 1% wet of standard OMC.	Well compacted to e.g. 95%–98% standard MDD, moisture content 2% dry of optimum to 1% wet of standard OMC.	Compacted in layers too thick for compaction equipment < 93% standard MDD, 2% to 3% dry of standard OMC.	No formal compaction, density ratio < 90% standard MDD, ≥ 4% dry of standard OMC.
		Non-Plastic Soil Saturated medium dense to dense dilative sand, gravel or rockfill ***and*** Unsaturated soil	Saturated medium dense sand and gravel.	Saturated, very loose sand or loose silty sand contractive on shearing.	Saturated very loose silt and sand.
Freeboard compared to dam height for reservoir stage.	(2)	> 7%	≈ 5%	< 3%	< 1%
Crest width	(1)	> 9 m	≈ 6 m	< 4 m	≤ 3 m

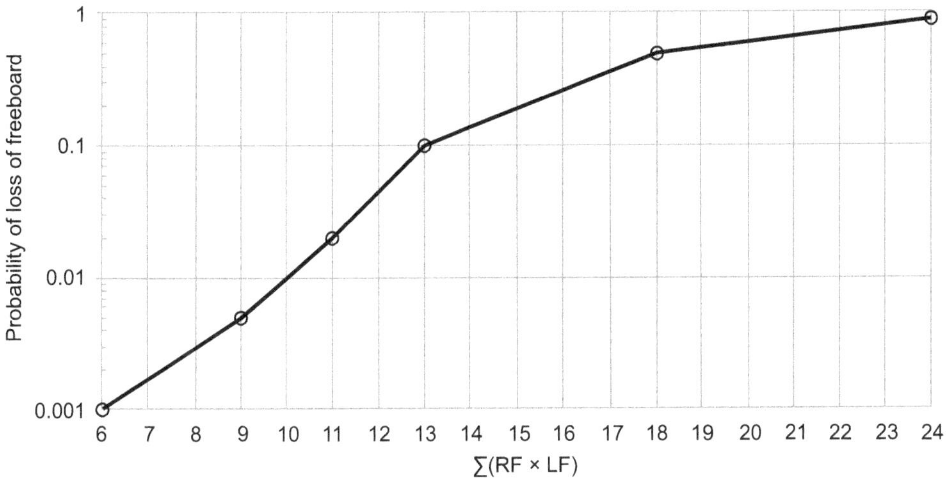

FIGURE 7.52 Estimation of the probability of loss of freeboard given initiation of slope instability (P_{LFI}) versus Σ (Relative importance factor [RF]) × (Likelihood factor [LF]).

(b) The inspection regime for the dam. Instability of dams with, for example, daily inspection is more likely to be detected than those with weekly or monthly inspection.

(c) Crest survey monitoring of the dam:

- The onset of slope instability is sometimes apparent in settlement and displacements of the crest. If survey is annual or longer this can only assist in detecting failures which are in slow developing non-strain weakening soils.
- Crest survey marks are often at 30 m spacing or more, which may be too wide a spacing to detect abnormal deformations. Crest survey marks 10 m apart are preferred but are usually only present when there has been some evidence of excessive deformation and the marks are installed to monitor this.
- If there is evidence of crest reconstruction to bring the crest back to design levels, this may be an indicator of marginal stability.

Tables 7.11 and Figure 7.53 may be used to estimate the probability of not detecting the initiation of slope instability.

7.14.4.2 Detection after Slope Instability Has Initiated and Is Progressing

Whether slope instability is detected given it has initiated, and whilst the movement is occurring, is dependent on the frequency of inspection of the dam, and the post initiation of instability velocity of movement.

Hunter (2003) analysed a number of case histories of dams which failed by slope instability during construction. The results are shown in Figures 7.54 and 7.55.

From these data the common range of maximum velocity is:

- High plasticity clays (CH) 0.1 m to 10 m/day.
- Medium to low plasticity clays (CL-CH and CL) 0.01 m to 10m/day.
- Clayey sand (SC) and sandy clay 0.1 m to 1 m/day.

TABLE 7.11 Factors Influencing the Likelihood of Not Detecting the Initiation of Slope Instability before Initiation Is Fully Developed

	RELATIVE IMPORTANCE FACTOR	*LIKELIHOOD FACTOR (LF)*			
FACTOR	*(RF)*	*LESS LIKELY (1)*	*NEUTRAL (2)*	*MORE LIKELY (3)*	*MUCH MORE LIKELY (4)*
Qualifications and experience of the person doing and interpreting the inspection and interpreting the monitoring data	(3)	Geotechnical professional experienced in slope instability and embankment dam deformation behaviour.	Civil engineer experienced in embankment dams and their monitoring.	Engineer or other professional with little experience in dams or slope instability.	Person with no qualifications in engineering, little or no experience in dams, and no experience in slope instability.
Frequency of inspection	(2)	Daily	Weekly	Monthly	Annually
Crest survey monitoring	(1)	Annual crest survey; survey marks 10 m apart or closer.	Annual crest survey; survey marks 30 m apart or more.	Five yearly survey; survey marks 30 m apart or more.	No crest survey.

Notes:
(a) Daily and weekly inspections are usually carried out by the reservoir manager. He/she needs to be trained in what to observe and to recognise signs of instability.
(b) If there is evidence of crest reconstruction to bring the crest back to design levels, this may be an indicator of initiation of instability. Assess the reasons for the reconstruction and if due to instability act to address the issue.

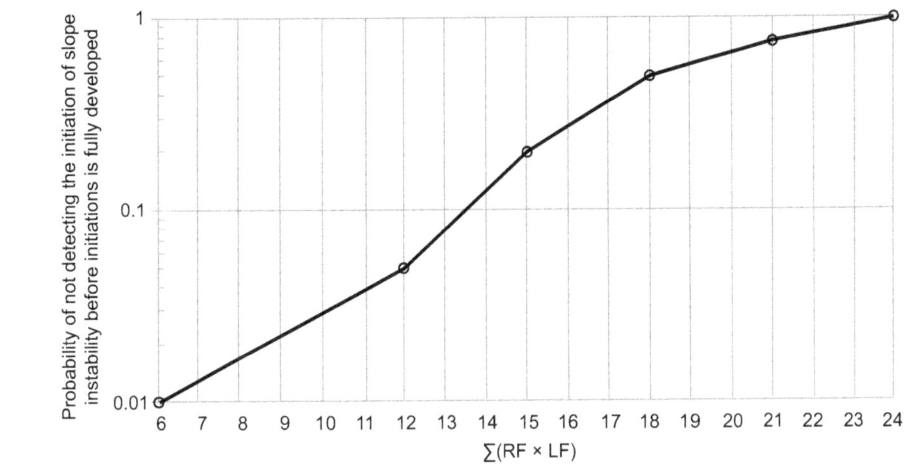

FIGURE 7.53 Estimation of the probability of not detecting the initiation of slope instability before initiation is fully developed (P_{NDI}) versus Σ (Relative importance factor [RF]) × (Likelihood factor [LF]).

TABLE 7.12 Probability of Not Detecting Slope Instability after Instability Has Initiated and Is Progressing (P_{NDP})

	STRAIN WEAKENING CLASS IN TABLE 7.10			
FREQUENCY OF INSPECTION	*1*	*2*	*3*	*4*
Daily	0.1	0.25	0.5	1.0
Weekly	0.5	0.75	0.9	1.0
Monthly	1.0	1.0	1.0	1.0
Annual	1.0	1.0	1.0	1.0

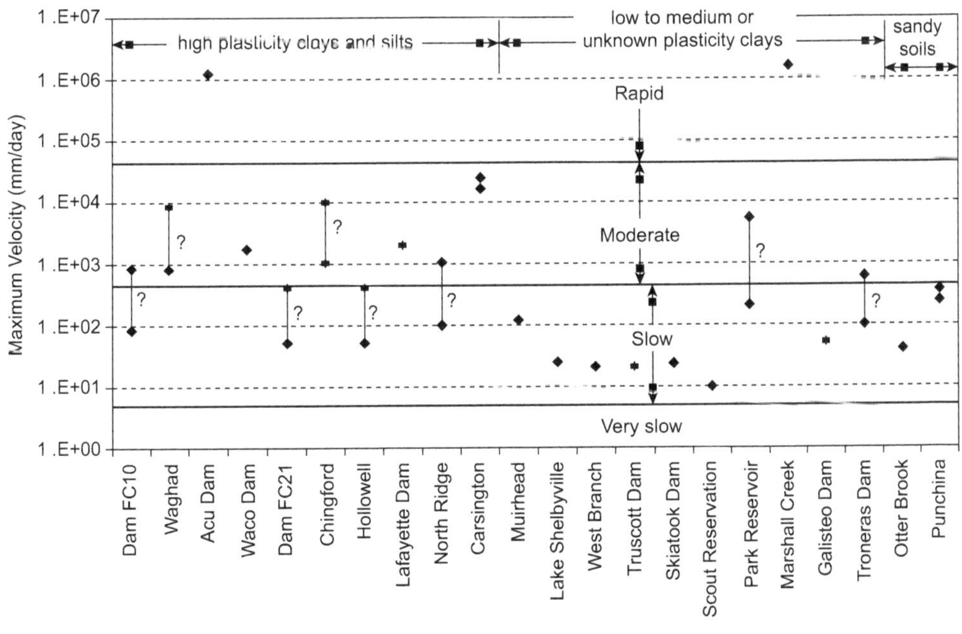

FIGURE 7.54 Maximum post-failure velocity of slides in embankment dams during construction (Hunter, 2003).

Non-plastic soils which are potentially subject to static liquefaction are likely to have maximum velocities from 1 m to 10 m/day.

These velocities can be related to the presence of strain weakening and progressive failure in the soils in Table 7.10. Given that the velocities are high for most soils, the likelihood of detection whilst the movement is occurring is low. Table 7.12 has some suggested values.

7.14.4.3 Detection after Progression Is Complete

The Modified Silva method described in Section 7.14.2 requires consideration of failure surfaces which encompass the dam crest or a large part thereof.

For failure surfaces which encompass the dam crest and the back scarp is below the reservoir level, the probability of unsuccessful intervention will be 1.0 and the probability of detection is not relevant.

For failure surfaces with the head scarp above reservoir level, detection is likely provided that the inspection frequency is sufficiently close.

Some factors to consider are as follows:

- Where the critical failure surfaces from the stability analyses are located, and whether they result in a narrow or wide remaining section of embankment. Narrow remaining embankment may fail in hours or days. Wide remaining embankment may survive for days or weeks.
- Whether the back scarp is expected to be steep and subject to retrogressive failure. Experience in landslides is that the retrogression takes place in hours or days after the initial failure for steep back scarps.
- The nature of the embankment materials remaining.

If the inspection regime is daily, the likelihood of successful detection is probably around 0.5 to 0.9.

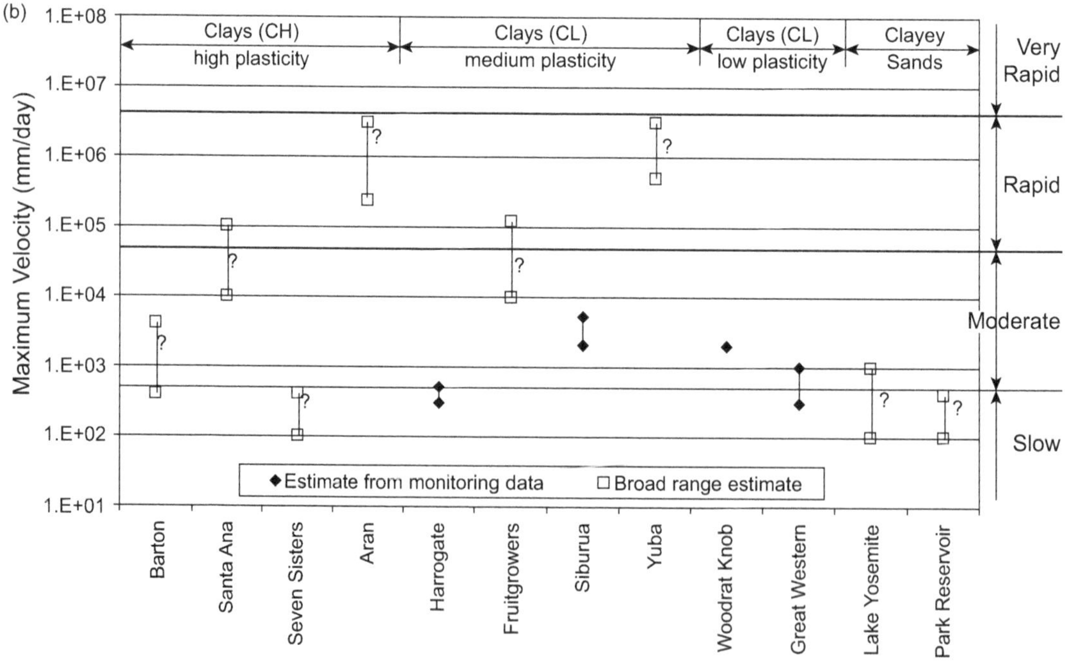

FIGURE 7.55 Maximum post-failure velocity in the downstream shoulder of embankment dams post-construction (Hunter, 2003).

If it is weekly, the likelihood of detection is probably around 0.1.

If it is monthly, the likelihood of detection is probably negligible.

Take account of whether the public have access to the dam as members of the public are likely to recognise a failure and warn the relevant authority.

7.14.5 Estimation of the Likelihood of Unsuccessful Intervention and Repair

General

Intervention and repair to prevent the progression of slope instability and breach can take several forms, including the following:

(a) Drawing down the reservoir level using spillway gates or outlet valves.
(b) Sandbagging or other methods to raise the crest in the affected area.
(c) Placement of sandbags or rip-rap to protect the head scarp from erosion.
(d) Building a weighting berm to reduce the likelihood of slope instability or unravelling.

In many larger dams the only practical intervention will be reservoir drawdown.

It should be recognised that there may be reluctance on the part of the reservoir owner or operator to release water given the lost revenue that may result, or if release of reservoir water is likely to result in property damage and loss of life, for example if levee banks downstream of the dam are likely to be overtopped by the flood resulting from release of the water.

Also, it should be recognised that health and safety considerations of the personnel may affect their ability to access or work near a damaged or marginally stable dam. The highest priority for dam safety personnel and emergency agencies is likely to be the safety of their personnel.

7.14.5.1 Estimation of the Likelihood of Unsuccessful Intervention before Initiation of Instability Is Fully Developed

This will depend on the details of the dam, whether the potential failure scarp leaves part of the crest above reservoir level and whether that part of the crest is capable of retaining water, the time available to intervene and resources available to intervene. Tables 7.13 and Figure 7.56 are provided as a general guide. Those involved in the dam safety assessment should make their own estimate.

TABLE 7.13 Factors Which Influence the Likelihood of the Probability of Unsuccessful Intervention before Initiation of Instability Is Fully Developed Given Instability Is Detected

		LIKELIHOOD FACTOR (LF)			
FACTOR	*RELATIVE IMPORTANCE FACTOR (RF)*	*LESS LIKELY (1)*	*NEUTRAL (2)*	*MORE LIKELY (3)*	*MUCH MORE LIKELY (4)*
Strain weakening class in Table 7.10	(3)	1	2	3	4
Ability to draw down the reservoir	(2)	Drawdown 1m/day	Drawdown 0.3 m/day	Drawdown 0.1 m/week	Drawdown < 0.1 m/month
Freeboard compared to dam height at the time of incident	(1)	> 7%	≈ 5%	< 3%	< 1%

7.14.5.2 Estimation of the Likelihood of Unsuccessful Intervention after Slope Instability Has Initiated and Is Progressing

This will depend on the details of the dam, the time available to intervene and resources available to intervene.

From Figures 7.54 and 7.55 the time available will be very short, potentially hours or a day or two. Table 7.13 and Figure 7.57 are provided as a general guide for situations where the dam operators have

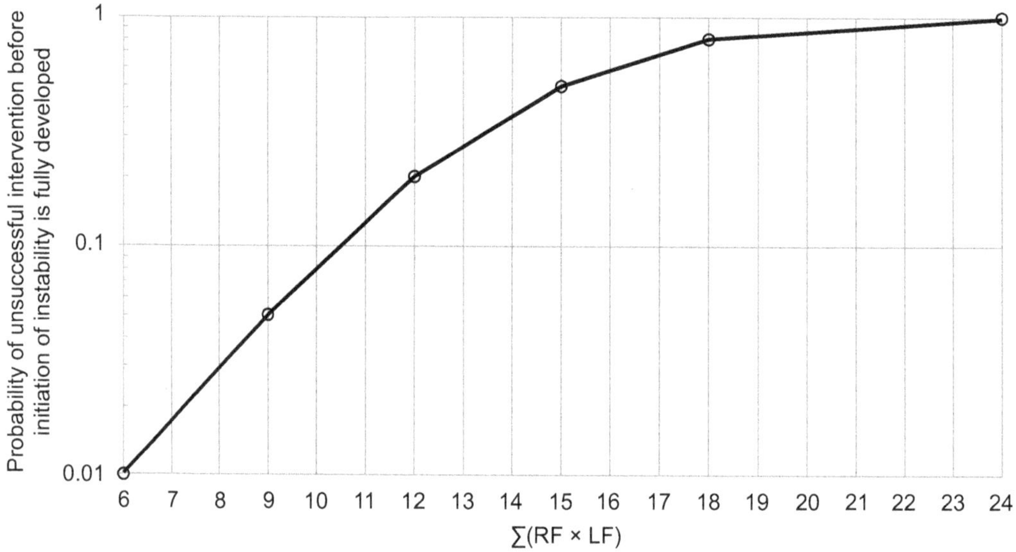

FIGURE 7.56 An indicative guide to estimating the probability of unsuccessful intervention before initiation of instability is fully developed (P_{IFI}) given instability is detected versus Σ (Relative importance factor [RF]) × (Likelihood factor [LF]).

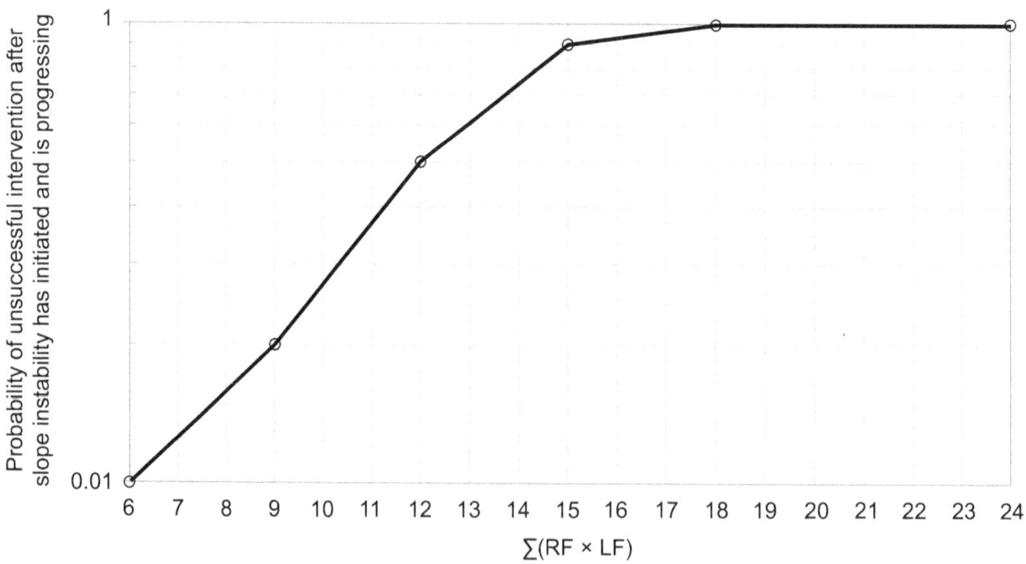

FIGURE 7.57 An indicative guide to estimating the probability of unsuccessful intervention after slope instability has initiated and is progressing (P_{IFP}) given instability is detected versus Σ (Relative importance factor [RF]) × (Likelihood factor [LF]).

good emergency action plans and good resources to intervene. They should not be relied upon. Those involved in the dam safety assessment should make their own estimate. In many cases there will be a high likelihood that intervention will not be practicable.

7.14.6 Estimation of the Probability of Breach

Sections 7.14.2 to 7.14.5 give the probability of slope instability leading to loss of freeboard and detection and intervention is unsuccessful for each reservoir level.

The authors consider that to further refine the estimate to account for the likelihood of breach for overtopping as described in Chapter 6 is beyond the accuracy of the methods and suggest that it is reasonable to assume that breach is a certainty, i.e. $P_B = 1.0$.

7.15 STATIC LIQUEFACTION AND ITS POTENTIAL TO LEAD TO LARGE DEFORMATIONS AND LOSS OF FREEBOARD

7.15.1 What Is Static Liquefaction?

Static liquefaction is a phenomenon in which saturated, loose sand, silty sand or silt loses strength under sustained (static) shear loading. This is a result of generating high pore water pressures in the soil mass and results in very low undrained strength. The disturbance causing liquefaction is by monotonic loading and not associated with dynamic events such as earthquakes.

The forms of change in effective stress which can trigger static liquefaction and flow sliding include rising piezometric pressures, loss of soil suction on wetting of partially saturated soils and steepening or heightening of cut or fill slopes.

The stress-strain behaviour is usually brittle, and the steady state undrained strength is much lower than the static shear stresses, creating a large force imbalance resulting in acceleration, increased velocity and flow sliding.

Static liquefaction may occur in a number of situations:

- Poorly compacted constructed fills for roads and buildings e.g. historically in Hong Kong (HKIE, 1998; Hunter and Fell, 2003c).
- Mine waste dumps, e.g. from coal washeries, Aberfan (Bishop et al., 1969; Hutchison, 1988) and Canadian mine waste dumps (Dawson, 1994; Dawson et al., 1998).
- Hydraulic fill dams, e.g. Fort Peck Dam (U.S. Army Corps of Engineers, 1939).
- Dams or levees constructed on loose sand or silty sand foundation soils, or with poorly compacted sand or silty sand fill, e.g. fill placed in thick layers by bulldozer without roller compaction, e.g. Edenville Dam, USA (France, 2024; France et al., 2022).
- Hydraulically placed mine and industrial tailings, e.g. Feijão Tailings Dam, Brazil (Arroyo and Gens, 2022; Robertson et al., 2019).
- Excavations into loose saturated naturally occurring sand and silty sand, or by natural erosion of marine slopes (Sladen and Hewitt, 1989; Kramer, 1988; Koppejan et al., 1948; Cornforth et al., 1974) e.g. Inskip Point Australia (ABC news items).

France et al. (2022) indicate that static liquefaction flow failure is clearly illustrated in video recordings of the Edenville Dam USA (2020) and Feijão Tailings Dam, Brazil (2019) failures. According to the authors, the two videos illustrate that static liquefaction shows little to no visible signs of distress until just before failure and rapid failure typically occurring in seconds or tens of seconds suggesting sudden development of a large force imbalance that creates acceleration of the failure mass.

7.15.2 The Mechanics of Undrained Shear of Saturated Non-Plastic Sand and Silt Sand Soils

Figure 7.58 shows schematically the undrained behaviour of saturated sand in triaxial compression. A soil which has an initial void ratio higher than the ultimate state (steady state, critical state/line) will tend to contract (become denser) in drained loading, reaching a void ratio equal to that at the ultimate state (critical state) line at that mean effective stress. In saturated undrained loading the soil is unable to contract, so positive pore pressures are developed, which will give strain weakening behaviour, reaching the steady state undrained strength S_{us} termed S_u, ultimate undrained strength in Figure 7.58. If the static gravitational shear stress q_{st} in a slope is significantly greater than S_{us}, rapid sliding in what is called flow liquefaction will result.

A soil which has an initial void ratio just above the ultimate state line may show limited strain softening, with the soil wanting to contract on initial shear and to dilate at larger strains, so developing positive and then negative pore pressures in undrained loading. A soil with an initial void ratio below the ultimate state line will tend to dilate, or develop negative pore pressures in undrained loading, giving a strain "hardening" or strengthening response.

The susceptibility of a soil to flow liquefaction is dependent on the particle size distribution, plasticity, void ratio (which may be characterised by the relative density), the initial stress conditions and the stress path of loading, e.g. triaxial compression or extension.

France (2024) summarises the mechanics as follows:

The research results cited above (Arroyo and Gens, 2022; Castro, 1969; Kramer, 1996; Olsen et al. 2000; Olsen and Stark, 2003; Sasitharan et al., 1993; Robertson et al., 2019; Zhu et al., 2024), together with the work of numerous other researchers, have led to a concept of static liquefaction triggering illustrated in Figure 7.59. In this concept, a loose sand or silt exists with a static shear stress, as would occur beneath an embankment slope. Under this condition, an undrained static liquefaction flow failure could be triggered either by (a) a rapid undrained stress change, as illustrated by the "blue line" in the figure; (b) a drained increase in pore water pressure, as illustrated by the "orange line" in the figure; or (c) a combination of the two. Some researchers have also suggested that creep can trigger static liquefaction. In any event, the stress path moves to intersect the instability line, the dashed line in the figure, and the rapid undrained static liquefaction is triggered. The instability line shown in Figure 7.59 has been given other names by other researchers (e.g., collapse line, flow liquefaction surface), but the concept is the same.

A number of researchers have investigated the instability line with the results indicating mobilized friction angles on collapse lines ranging from about 14° to 22° for sands with drained friction angles ranging from 27° to 32°. Kramer (1996) suggested that for "isotropic initial conditions, the slope of the flow liquefaction surface (FLS) (φ_L') is about two-thirds the slope of the drained failure envelope for clean sand." As noted above, the FLS is the same as the collapse line. Olson and Stark (2003) suggest that collapse can initiate with ratios of undrained strength to pre-failure effective vertical stress ($Su/\sigma'v$) ranging between 0.23 and 0.31, which correspond to mobilized friction angles of about 13° to 17°. Olson and Stark (2003) also suggested equations relating trigger Su/σ_V' to SPT and CPT values.

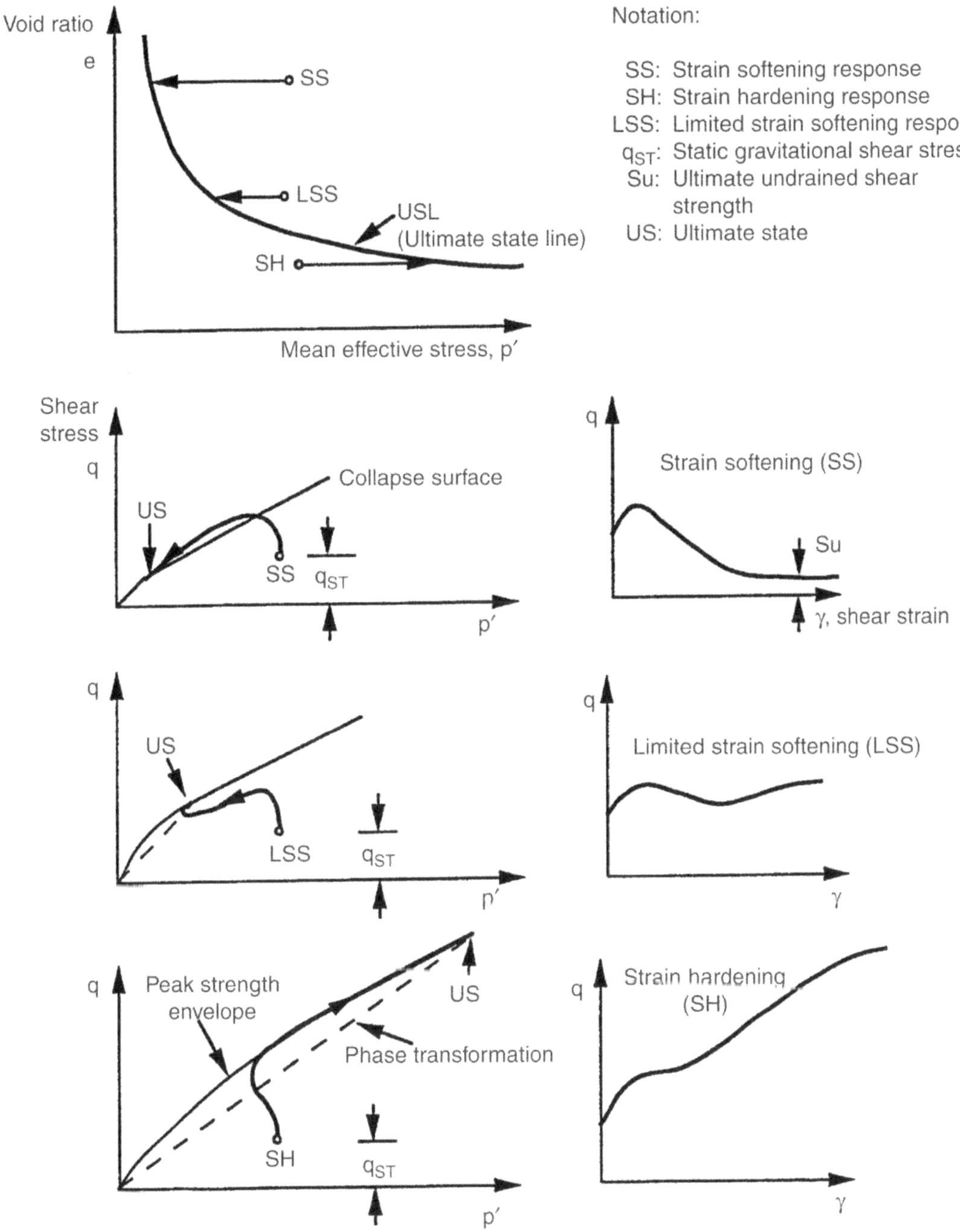

FIGURE 7.58 Schematic of undrained monotonic behaviour of sand in triaxial compression (Robertson, 1994). (Reproduced with permission of P.K. Robertson.)

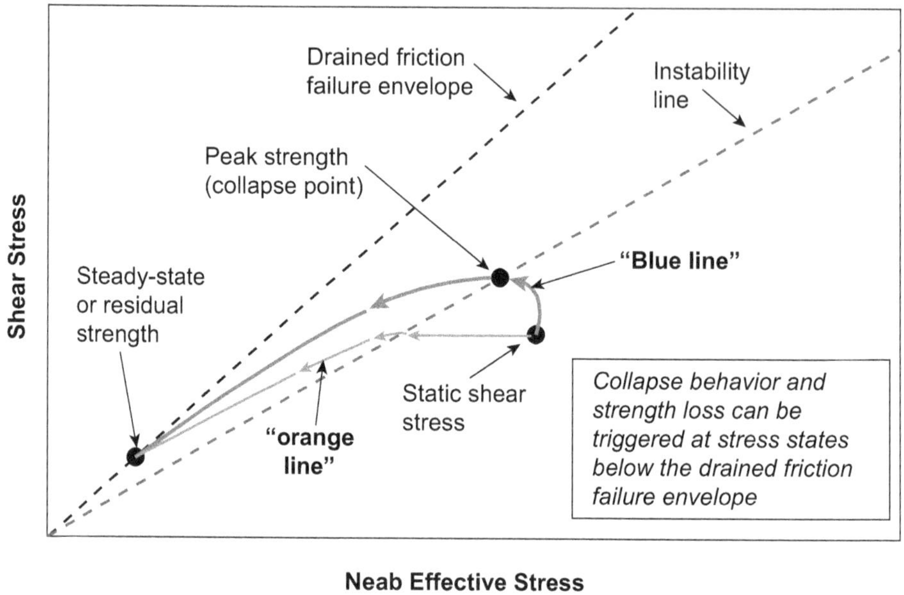

FIGURE 7.59 Concept of static liquefaction triggering (France, 2024). (Reproduced with permission of J. France.)

7.15.3 The Classification of Soils Which Are Subject to Static Liquefaction and Flow Sliding

Based on an analysis of 350 case studies and an understanding of the mechanics of contractive soils, Hunter and Fell (2003c) determined that the approximate bounds of particle size distribution of soils susceptible to static liquefaction and flow sliding are as shown in Figure 7.60. The coarse boundary, representative of sandy gravels with a trace of mostly silt fines (from coal mine waste dumps), is probably an upper bound particle size distribution due to permeability constraints. Materials with coarser particle size distributions are likely to be sufficiently permeable that pore water pressures developed on contraction can be dissipate. Most of the data for Figure 7.60 is from constructed slopes, but it would be expected that natural soils with particle sizes within the boundaries shown in Figure 7.60 would also be susceptible to flow liquefaction.

A guide to the limits of plasticity of soils which can be subject to flow liquefaction can be obtained from the earthquake literature. Non-plastic soils which fall within the bounds shown in Figure 7.60 are potentially liquefiable. For soils with some plasticity the authors have relied upon the Seed et al. (2003) method, which is shown in Figure 8.5 and discussed in Section 8.6.1. These underlying mechanics of seismically induced and static liquefaction are similar, so the Seed et al. (2003) method should be applicable to potential for static liquefaction under static loading.

7.15.4 Relative Density or Degree of Compaction of Soils Susceptible to Static Liquefaction and Flow Sliding

Based on their case data, Hunter and Fell (2003c) found examples of soils which it could be assessed from their post-failure behaviour as having been susceptible to static liquefaction and flow sliding, included the following:

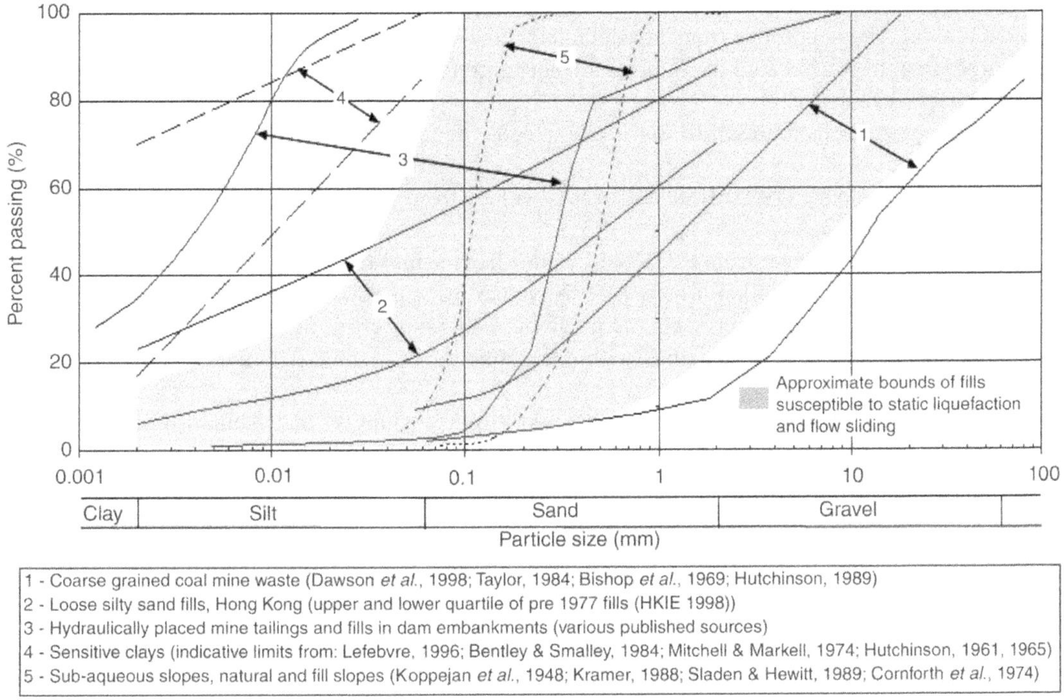

FIGURE 7.60 Particle size distributions of material types susceptible to liquefaction and flow sliding (Hunter and Fell, 2003c).

(i) Clean sands with relative density less than about 15%–30% (i.e. very loose to loose). This is likely to vary depending on the particle size distribution and the particle shape (Cubrinovski and Ishihara, 2000).

(ii) Silty sands with relative densities up to 45%–60% (i.e. very loose to medium dense).

(iii) Silty sands with clay contents less than about 10%–20%, such as those derived from decomposed granite with density ratios below 85%–90% of standard maximum dry density (HKIE, 1998); flow slides in these materials are typically of shallow depth (up to 3–4 m) and on slopes steeper than about 30°to 40°.

(iv) Sandy gravels and gravelly sands (in coal waste spoil piles and coking coal stockpiles) with trace to some silt fines (less than 5% to 10% finer than 0.075 mm) at void ratios greater about 0.3 (Dawson, 1994; Eckersley, 1986); field studies in these material types (Eckersley, 1990; Dawson et al., 1998) show that the moisture content at placement and the method of placement have a significant effect on the initial void ratio and therefore the potential for flow sliding.

A useful and practical guide to whether sandy and silty sandy soils are likely to be susceptible to flow liquefaction can be obtained from Standard Penetration Test "N" values as shown in Figure 7.61. This figure is applicable to slopes which fail under static (monotonic) loading and cyclic loading from earthquakes.

The figure compares *in situ* methods that have been developed to evaluate flow liquefaction potential of sandy soils (sands and silty sands)—this is flow liquefaction triggered by earthquake and static loading conditions. The lines are liquefaction flow/no-flow boundaries for the various methods—to the left are flow conditions and to the right no-flow.

This separates sands and silty sands where contraction on shearing results in large loss in strength and potential for flow conditions from "quasi liquefaction" (develop high pore pressures but not necessarily large strength loss because the samples are not contractive at larger shear strains) and those sands that are denser and dilative.

The various methods represented in the plot include the following:

- Baziar and Dobry (1995) from SPT $(N_1)_{60}$ values from field cases of earthquake-induced flow liquefaction.
- Ishihara (1993) is also from SPT $(N_1)_{60}$ values from actual flow failure sites of earthquake-induced flow liquefaction where SPT N values were known from testing prior to the earthquake. The zone represents the range of results—if left of the shaded zone then high likelihood of flow liquefaction; if within the zone then possibly; if to the right then in the no-flow region.
- The e_{max}–e_{min} relationships are from the work by Cubrinovski and Ishihara (2000) from laboratory testing. They tested a number of different sands and silty sands (with up to 30% non-plastic fines content) and developed relationships for flow/no-flow boundary based on vertical effective stress differentiating the characteristics of the sand using e_{max}–e_{min}. Hence each line represents a flow/no-flow boundary and depends on e_{max}–e_{min} and K_o.

What is apparent from the plot is reasonable correlation of Ishihara (1993) and Baziar and Dobry particularly at 75 to 125 kPa effective vertical stress. These are also comparable to the Cubrinovski and Ishihara (2000) flow/no-flow boundaries for e_{max}—e_{min} of 0.5 to 0.6 (but not so for lower e_{max}—e_{min} values). In practice the authors suggest that the Ishihara (1993) right side boundary be used rather than the other less conservative methods for low overburden stresses and the Baziar and Dobry (1995) boundary at high overburden stresses.

The method should be used with an adequate degree of caution and in recognition of the uncertainty associated with several of the empirical correlations used to derive the flow liquefaction/no-flow boundaries.

Additional information to assist in assessing if static liquefaction and flow sliding will occur is to consider the liquefied undrained strength. For static liquefaction and flow sliding to occur, the liquefied undrained strength should be small. This is discussed in Section 8.8.3. Figures 8.19 and 8.21(a) and (b) show the Idriss and Boulanger (2008) methods for estimating this strength from SPT and CPTU data.

The major change in behaviour is at SPT $(N_1)_{60cs} \approx 12$ and CPTU $(q_{c1N})_{cs} \approx 90$, so these may be taken as a guide to separate static liquefaction and flow sliding from non-flow conditions.

The state parameter Ψ is the difference between the *in situ* void ratio e_0 and the void ratio at the critical state e_{cs} at the same mean effective stress p′. Figure 7.62 shows the definition.

Davies et al. (2002) suggest that for mine tailings, undrained shear phenomena will not be of concern, i.e. the soil will be dilatant, if the state parameter $\Psi < -0.1$. They provide Table 7.14 as a guide to behaviour. Davies et al. (2002) point out that Been (1999) suggest $\Psi < -0.08$ and < -0.10 respectively as their practical limits of minimum state to ensure satisfactory engineering performance provided the drained strength of material is sufficient for all loading conditions.

Winckler et al. (2014) quote Jefferies and Been (2006) as suggesting that for $\Psi > -0.05$ strain softening and loss of strength in undrained shear can be expected. They provide Figure 7.63 which relates contours of Ψ to the CPTU data.

From these references it is suggested that $\Psi > -0.05$ may be used as a guide to soils which may be subject to static liquefaction and flow liquefaction, and that CPTU testing may be used to assess Ψ.

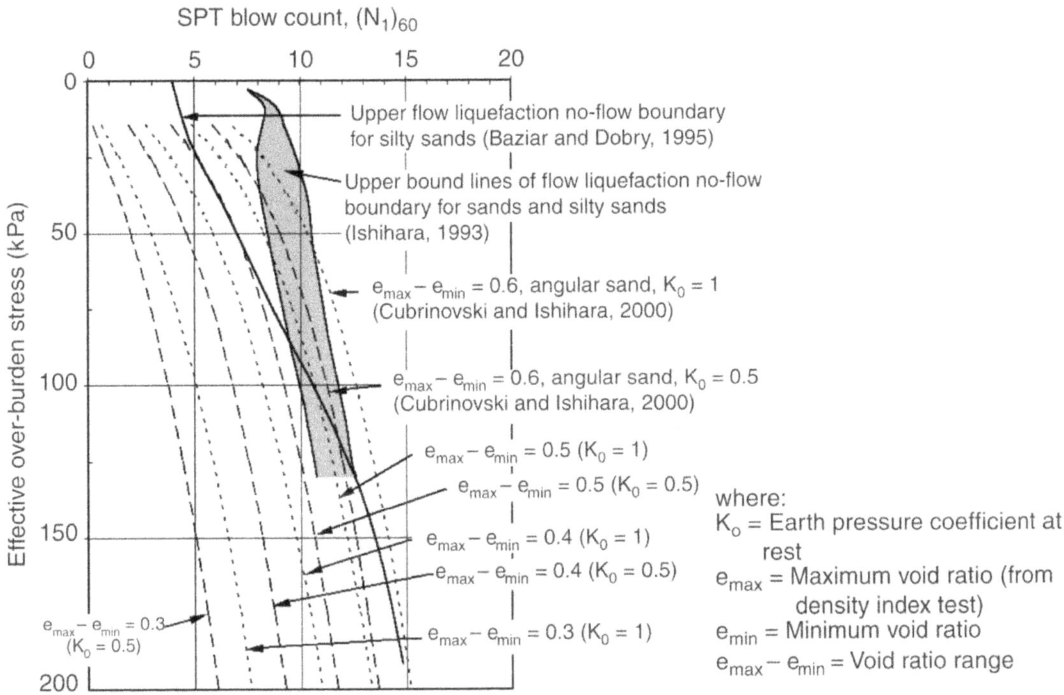

FIGURE 7.61 Comparison of flow liquefaction no-flow boundaries in terms of SPT $(N_1)_{60}$ for sands and silty sands from monotonic laboratory undrained tests and earthquake-triggered field cases e_{max}–e_{min} void ratio range; K_o earth pressure coefficient at rest (Hunter and Fell, 2003c after Cubrinovski and Ishihara, 2000).

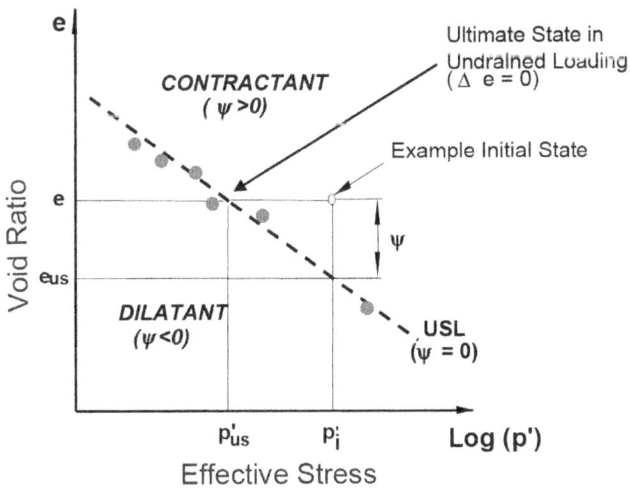

FIGURE 7.62 Typical results of steady state strength testing on cohesionless materials (Davies et al. (2002).

TABLE 7.14 Static Liquefaction General Design Guidelines (Davies et al., 2002)

$(N_1)_{60}$	Ψ_{**}	$S_U/(P')$	SHEAR STRAIN γ_F (%)	
0–4	+0.12	0.05–0.10	25–50	> 100
4–10	+0.05	0.10–0.20	10–25	30–100
10–15	0	0.15–0.4	8–15	20–35
15–20	−0.08	0.3–0.5	5–10	15–25
> 20	< −0.10	> 0.5 (< ~ 0.68)*	< 5	< 15

* Dilatency ignored regardless of state.
** Approximate only, depends on steady state compressibility of tailings, λ.

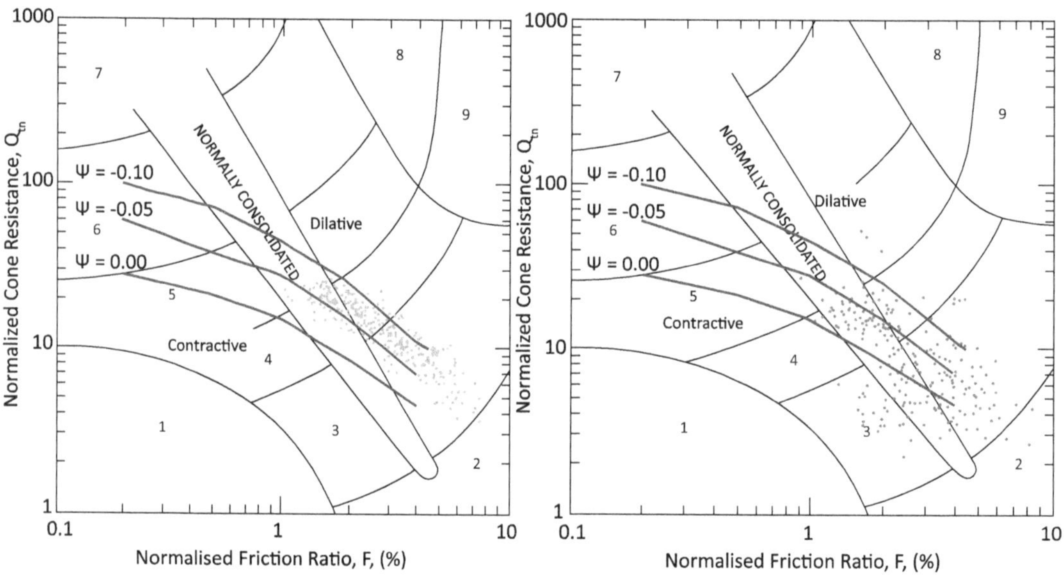

FIGURE 7.63 Normalised cone resistance versus normalised friction ratio showing contours of state parameter Ψ. The data points are for dilative behaviour (left graph) and contractive behaviour (right graph) (Winckler et al., 2014). (Reproduced with permission of C. Winckler.)

7.15.5 Estimation of the Large Strain Undrained Strength S_{us} in Soils Which Have Liquefied

The most practical way to estimate the large strain undrained strength S_{us} in soils subject to liquefaction is to use the results of standard penetration or Piezocone Penetration Tests and Figures 8.19 and 8.21a and b.

It should be noted that for most of these soils the liquefied strength is equivalent to an equivalent undrained friction angle of between 3 and 6 degrees. This is for soils which have a drained friction angle of more than 25 degrees, so there is a large reduction in strength.

7.15.6 Assessing the Likelihood a Dam May Fail as a Result of Static Liquefaction

The likelihood a dam may fail as a result of static liquefaction of soil in the foundation or in the embankment may be estimated as follows:

(a) Determine the as constructed zoning of the embankment and the classification, particle size distribution and Atterberg limits of the material in each zone.
(b) Determine the stratigraphy and the classification, particle size distribution and Atterberg limits of each of the strata in the foundation.
(c) Assess which zones in the embankment and strata in the foundation may be subject to static liquefaction using the data in Section 7.15.3.
(d) If there are no potentially liquefiable soils in the embankment and foundation, the probability of static liquefaction = zero. Many dams with a core constructed of plastic soils such as clays, sandy clays, clayey sands, or earth and rockfill dams and founded on rock or on plastic soils will satisfy this requirement.
(e) For embankment zones and foundation strata potentially liquefiable as described in (c):
    - Estimate the relative density and/or degree of compaction of the embankment zones from construction records of layer thickness and compaction equipment, and if available, the compaction specification and construction records. Zones which have been compacted to 95% Standard Maximum Dry Density (SMDD) or 75% relative density will have a probability of static liquefaction = zero.
    - Assemble site investigation SPT and CPTU data for the strata and the foundation.
(f) Using this data, assess the likelihood that zones or strata are potentially subject to static liquefaction using the methods described in Section 17.5.4 and engineering judgement. Use qualitative or quantitative methods to do this, taking account that the result may be different for the several methods in Section 17.5.4.
(g) For those zones or strata assessed as potentially subject to static liquefaction, estimate the liquefied undrained strength ratio from Figures 8.19 to 8.21(a) and (b).
(h) Assess the drained and where applicable undrained shear strength of the other zones and strata and estimate the pore pressure in these zones and strata.
(i) As a first guide to performance, analyse the stability assuming the strengths in (h) but for zones potentially susceptible to static liquefaction assign a strength = 2/3 the drained strength as an approximation of the instability line/collapse line. This approximates to the collapse surface as suggested by France (2024) and summarised in Section 7.15.2.
(j) For static liquefaction to be triggered, the factor of safety for this condition needs to approach 1.0. To assign an annual probability to the trigger condition, carry out analysis for pore pressure conditions representing:
    - Normal operating conditions, i.e. with an annual probability = 1, and for estimated pore pressures.
    - Maximum recorded pore pressures with the estimated AEP of that loading, e.g. reservoir level.
    - Estimated pore pressures for say 1 in 100 AEP, and 1 in 1000 AEP reservoir level.

    Use the results of these stability analyses and subjective judgement to estimate the annual probability for the loading condition that triggers static liquefaction.

(k) If the factor of safety on the instability/collapse line is < 1.0, or better < 1.3 to allow for the approximation, proceed to (l). If the factor of safety > 1.3, the likelihood of loss of freeboard and failure will be low.

(l) Carry out limit equilibrium stability analyses using the liquefied undrained strength strengths from (g) and other strengths from (h). Do this for normal operation and under the relevant reservoir level and estimated pore pressure conditions. This analyses the slope given static liquefaction has occurred and the liquefied undrained strength condition is reached in those zones.

(m) For cases where the post-liquefaction stability factor of safety in (l) is < 1.0, estimate the conditional probability that freeboard may be lost using the method described in Section 7.14.3.

(n) Estimate the overall annual probability the static liquefaction may result in loss of freeboard and the dam fail taking account of the estimated probabilities in (j) to (m).

Figure 7.64 shows the event tree to model static liquefaction leading to loss of freeboard and breach.

These methods are approximate. For high consequence of failure dams which are assessed as potentially subject to static liquefaction and the estimated resulting deformations are large, a dam or tailings dam engineer expert in static liquefaction should be consulted to assist in the assessment. Most difficult will be to assign an annual probability to the trigger event.

For modern conventional dams the likelihood of failure as a result of static liquefaction is likely to be negligible for many dams as described in (d) and (e).

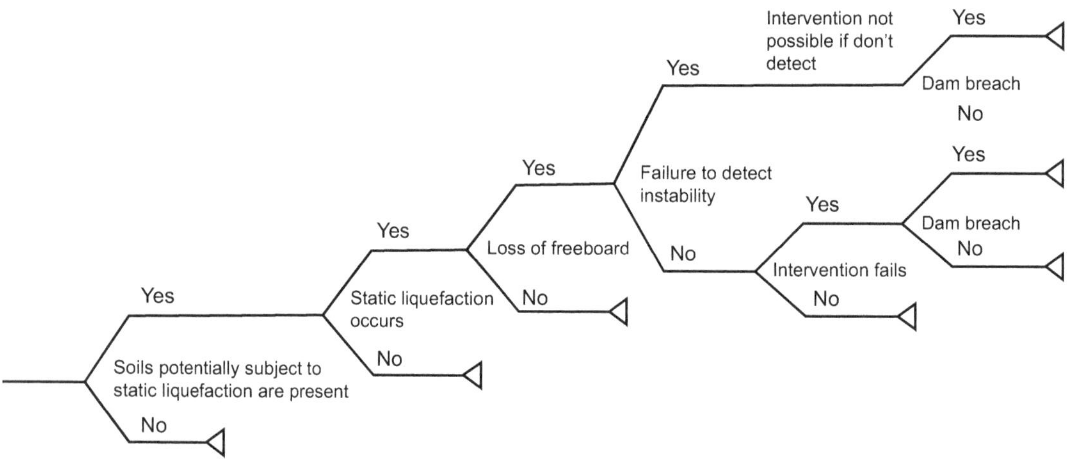

FIGURE 7.64 Event tree for static liquefaction leading to loss of freeboard and breach.

8 Deformations of Embankment Dams under Seismic Loading and Assessment of the Likelihood of Loss of Freeboard

8.1 INTRODUCTION

Most embankments will, under large seismic loads, yield during part of the loading cycle, resulting in permanent deformations. However, that does not mean the dam has "failed", provided the deformations are tolerable. Hence the emphasis in this chapter is on deformations, not stability.

It is only dams which experience liquefaction or are founded on or constructed of strain weakening soils resulting in post-earthquake factors of safety below 1.0 that have a potential stability issue.

Section 8.2 discusses the historic performance of embankment dams during earthquakes.
Sections 8.3 and 8.4 describe the methods for estimating deformations under seismic loads for embankments not subject to liquefaction.
Section 8.5 discusses the pseudo-static method of analysis.
Sections 8.6 to 8.10 describe the assessment of liquefaction of soils in the embankment and foundation, and post-earthquake strengths.
Section 8.11 discusses post-earthquake stability and deformation analyses for embankments subject to liquefaction.
Section 8.12 discusses application of the results of analyses to assessing the safety of the dam.

DOI: 10.1201/9781003542025-8

8.2 PERFORMANCE OF EMBANKMENT DAMS DURING EARTHQUAKES

Major reviews of past performance have been conducted by Sherard (1967), Sherard et al. (1974), Seed (1979, 1983) and Seed et al. (1978, 1985). These studies show that based on the data available at that time:

- Even at short distances from the epicentres, there have been no complete failures of embankments built of clay soils, but several dams have come close to failure.
- Well constructed dams of clay soils on clay or rock foundation, not susceptible to strain weakening, can withstand extremely strong shaking resulting from earthquakes of up to M8.25 with peak ground accelerations ranging from 0.35 g to 0.8 g. They suffer deformations and cracking but few, if any, have failed catastrophically.
- Dams which have suffered complete failure resulting from deformation leading to loss of freeboard as a result of earthquake shaking have been constructed primarily with saturated sandy materials or on saturated sand foundations. Liquefaction is a major contributory factor in these failures. The dams most susceptible to failure under earthquake loading are hydraulic fill dams and tailings dams constructed using upstream methods, because they are susceptible to liquefaction if the fill or tailings are granular and saturated.
- In dams constructed of saturated cohesionless soils, the primary cause of damage or failure is the build-up of pore water pressure (liquefaction) under the earthquake loading.
- There are very few cases of dam failures during the earthquake shaking. Most of the failures occur from a few minutes up to 24 hours after the earthquake. However, cracking and displacements do occur during the earthquake.

Experience world-wide since these reviews confirms those conclusions.

The following suggested approach to assessing performance under earthquake loading is based on the authors' experience and relies on risk based logic. This is considered applicable because the earthquake loadings are often expressed in probabilistic terms and, even if they are expressed in deterministic terms, there is a significant uncertainty to those loads. Also supporting a risk based logic approach is that the methods for estimating deformations are often approximate, and even the most sophisticated methods have a range of possible outcomes.

8.3 DATABASE AND SIMPLIFIED METHODS FOR ESTIMATING DEFORMATIONS UNDER EARTHQUAKE LOADING FOR EMBANKMENTS NOT SUBJECT TO LIQUEFACTION

8.3.1 Introduction

There are several methods available for estimating the deformations which may occur in embankments and their foundations during seismic loading. These can be summarised as:

Empirical database methods. These are applicable to embankments and their foundations which do not liquefy or experience significant loss of strength due either to build-up of pore pressure or strain weakening. These should only be applied if the criteria listed in Section 8.3.2 are satisfied and should only be relied upon if the estimated deformations are much less than tolerable, e.g. crest settlements are much less than the available freeboard.

Simplified methods for estimating deformations during earthquakes. The methods described in Section 8.3.3 are also only applicable to embankments and their foundations which do not liquefy or experience significant loss of strength due either to build-up of pore pressure or strain weakening. They assume the post-earthquake deformations are negligible and the deformations during the earthquake are due to the action of the horizontal inertia forces induced by the earthquake.

Numerical methods. Refer to Section 8.4 for methods. These methods are used for estimating deformations during and post-earthquake for non-liquefied and liquefied conditions.

It should not be assumed that the database and simplified methods are conservative, and where estimated deformations are near to the tolerable deformations, more advanced methods of analysis should be used or measures taken to reduce deformations.

Where significant strain weakening of overconsolidated high clay content soils within the dam foundation or in the embankment may occur due to displacements induced by the earthquake, post-earthquake stability and deformation analyses should be undertaken taking account of the strain weakening.

The authors also emphasise that the post-earthquake internal erosion and piping risk must not be ignored as it may be an important Potential Failure Mode. This is discussed in Section 9.9.

8.3.2 Empirical Database Methods for Estimating Deformations under Earthquake Loading for Dams Not Subject to Liquefaction

Empirical database methods are based on recorded deformations, dam geometry and earthquake loading. The following summarises the methods.

8.3.2.1 Swaisgood (1998, 2003) Empirical Method for Estimating Crest Settlements

Swaisgood (1998, 2003) gathered data on crest settlement, dam height, type, depth of alluvium in the foundation, earthquake magnitude and peak ground acceleration and the focal distance of the dam to the earthquake.

Figure 8.1 shows relative crest settlement (settlement/{dam height + thickness of alluvium in the foundation}) versus peak ground acceleration (bedrock).

This method should only be used to give an approximate estimate of settlements, where there is no potential for liquefaction or significant strain weakening. Account should be taken of the scatter in the data as shown in Figure 8.1.

8.3.2.2 Pells and Fell (2002, 2003) Empirical Method for Estimating Settlement, Damage and Cracking

Pells and Fell (2002, 2003) gathered data from 305 dams, 95 of which reported cracking, and classified these for damage according to the system shown in Table 8.1.

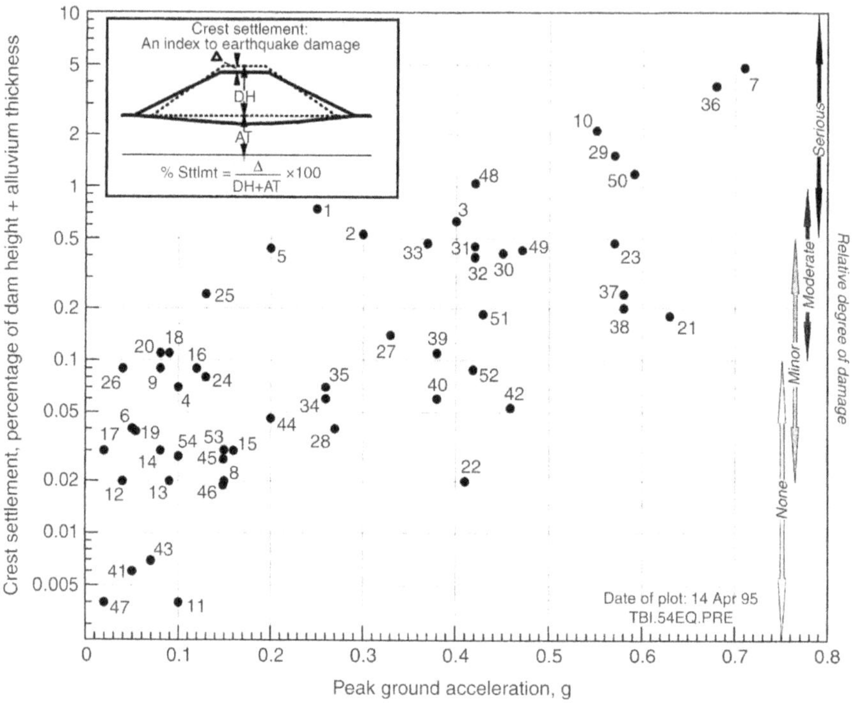

FIGURE 8.1 Relative crest settlement versus peak ground acceleration (Swaisgood, 2003). (Reproduced with permission of NZSEE.)

TABLE 8.1 Damage Classification System for Embankment Dams under Earthquake Loading (Pells and Fell, 2002, 2003)

DAMAGE CLASS		*MAXIMUM LONGITUDINAL CRACK WIDTH*[(1)] *(MM)*	*MAXIMUM RELATIVE CREST SETTLEMENT*[(2)] *(%)*
NUMBER	*DESCRIPTION*		
0	No or slight	< 10	< 0.03
1	Minor	10–30	0.03–0.2
2	Moderate	30–80	0.2–0.5
3	Major	80–150	0.5–1.5
4	Severe	150–500	1.5–5
5	Collapse	> 500	> 5

(1) Maximum crack width is taken as the maximum width, in millimetres, of any longitudinal cracking that occurs.
(2) Maximum relative crest settlement is expressed as a percentage of the maximum dam height (from general foundation to the dam crest).

Figures 8.2 and 8.3 show plots of damage contours versus earthquake magnitude and peak ground acceleration for earthfill and earth and rockfill dams with respect to crest settlement and cracking. These figures can be used to estimate the magnitude of crest settlement based on the range of relative crest settlements given for each damage class given in Table 8.1.

In terms of the potential for earthquake-induced cracking, Pells and Fell (2002, 2003) concluded that earthquakes cause settlement, lateral spreading and cracking of embankment dams. Longitudinal

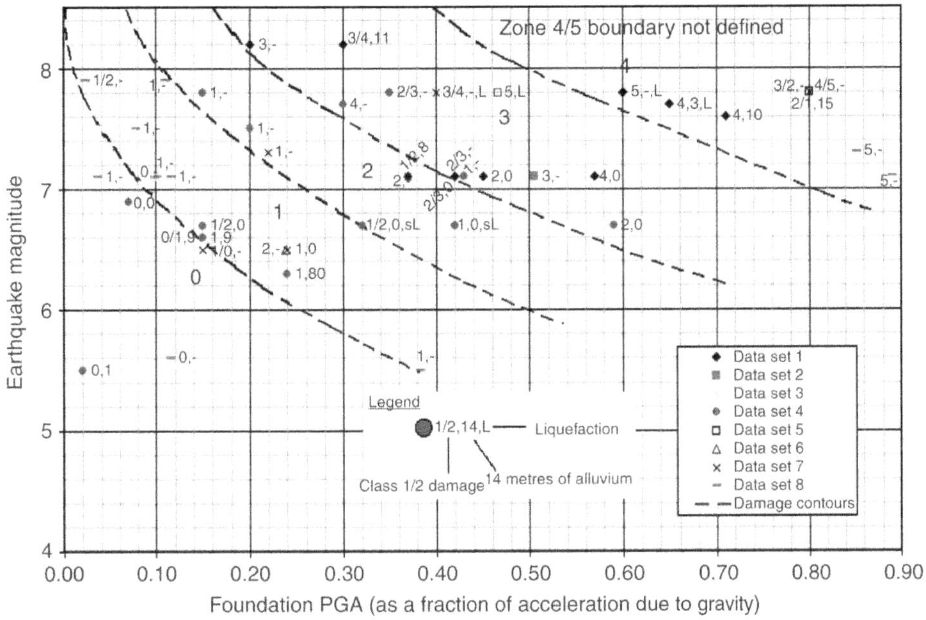

FIGURE 8.2 Earthfill dams—contours of damage class versus earthquake magnitude and peak ground acceleration. Contours drawn without consideration for cases that had liquefaction (Pells and Fell, 2002, 2003).

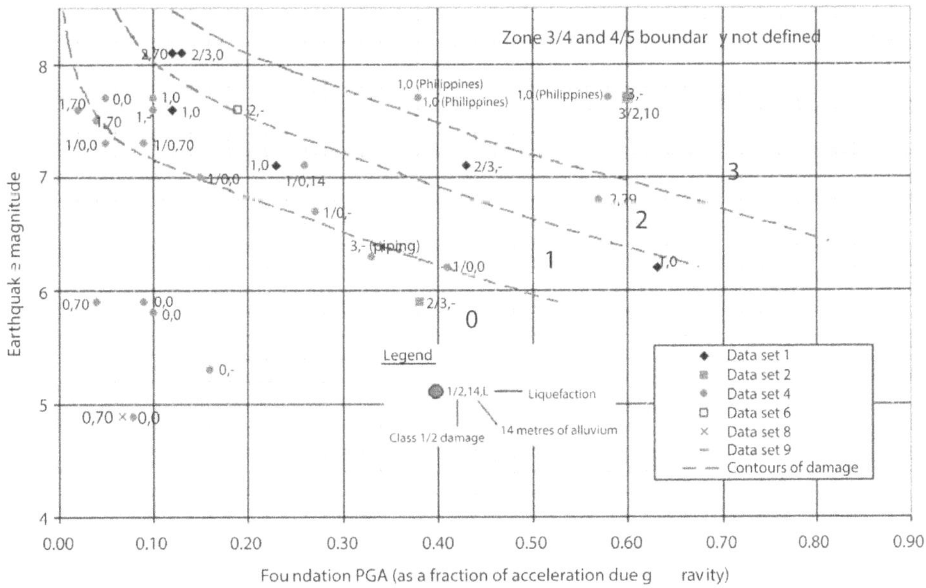

FIGURE 8.3 Earth and rockfill dams—contours of damage class versus earthquake magnitude and peak ground acceleration. Contours drawn without consideration for cases that had liquefaction. Boundaries shown assume rockfill is well compacted. For dumped or poorly compacted rockfill use Figure 8.2 to estimate the damage (Pells and Fell, 2002, 2003).

cracks are more common than transverse cracks and are mostly in the upper part of the dam, more likely on the upstream face than the downstream. Section 9.9 describes the earthquake PGA and magnitude likely to result in transverse cracking extending below the reservoir level which may result in initiation of concentrated leak erosion.

8.3.3 Simplified Methods of Deformation Analysis for Dams Where Liquefaction and Significant Strain Weakening Does Not Occur

8.3.3.1 General Principles

These methods are all based on the Newmark (1965) approach. In this, the sliding of a soil mass along a failure surface was likened to slipping of a block on an inclined plane. Newmark (1965) envisaged that failure would initiate and movements would begin when the inertia forces exceed the yield resistance and that movements would stop when the inertia forces were reversed. Movement was assumed to happen only in the downslope direction. He proposed that once the yield acceleration and the acceleration time history of a slipping mass are determined, the permanent displacements can be calculated by double integrating the acceleration history above the yield acceleration as shown in Figure 8.4.

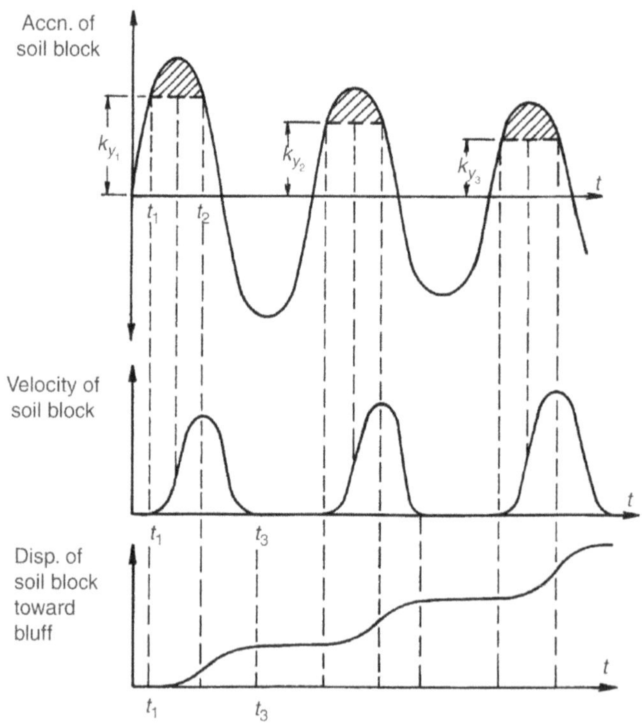

FIGURE 8.4 Double integration method for determination of the permanent deformation period of an embankment (Martin, 1988).

According to this approach, the permanent deformation in a sliding mass is a function of:

- The amplitude of the average acceleration time history of the sliding mass, which in turn is a function of the base motion, the amplifying factor of the embankment and the location of the sliding mass within the embankment.
- The duration of the earthquake, which is a function of the magnitude of the earthquake.
- The acceleration of the potential sliding mass.

It is stressed that Newmark's approach is limited in application to compacted clayey embankments and dry or dense non-plastic soils and rockfill that experience very little reduction in strength due to cyclic loading. The approach should not be applied where embankments or their foundations are susceptible to liquefaction or strain weakening because it will significantly underestimate displacements. They may also underestimate deformations for dams with poorly compacted rockfill subject to collapse settlement.

8.3.3.2 Makdisi and Seed (1978) Method

The Makdisi and Seed (1978) approach is based on Newmark's method but modified to allow for the dynamic response of the embankment. The approach was developed from a series of deformation analyses performed on a large number of embankments subjected to earthquake loading.

Bray and Travasarou (2007) used this method as well as their own (described later), and for the dams they analysed the Makdisi and Seed (1978) method generally underestimated the displacements estimated using nonlinear analyses.

The authors' recent experience is that most use the Bray and Travasarou (2007) method rather than the Makdisi and Seed (1978) method.

8.3.3.3 Bray and Travasarou (2007) Method

A simplified procedure was proposed by Bray and Travasarou (2007) for estimating earthquake-induced permanent displacements in earth dams using a Newmark type model. This procedure is based on the results from a set of simplified nonlinear analyses using 688 recorded ground motions from 41 earthquakes. The flexibility of the dam system and the interaction between yielding and seismic loading were considered by using a nonlinear coupled stick-slip deformable sliding model. The flexibility of the dam structure is captured through an estimate of the initial fundamental period *Ts*.

Key parameters of this procedure are the yield acceleration *ky* (in g), the initial fundamental period of the embankment, *Ts*, the value of spectral acceleration for a damping of 5% and a degraded response period equal to 1.5 *Ts*.

The authors' experience is that the Bray and Travasarou (2007) method usually but not always gives larger estimates of settlement than the Pells and Fell (2002, 2003) method. In some cases, the Bray and Travasarou (2007) estimates seem excessive.

8.3.3.4 Bray, Macedo and Travasarou (2018) Method

Bray et al. (2018) used the Bray and Travasarou (2007) approach to model embankments subject to seismic time histories from subduction zone earthquakes. They indicate that the displacement estimates were similar to those for the crustal earthquake motions used by Bray and Travasarou (2007) but that that method was conservative when used to estimate seismic displacements for subduction zone earthquakes.

The authors have limited experience with this method so cannot provide guidance on its use other than to suggest that it be used for subduction zone sourced earthquakes.

8.4 NUMERICAL METHODS FOR ESTIMATING DEFORMATIONS DURING AND POST-EARTHQUAKE FOR NON-LIQUEFIED AND LIQUEFIED CONDITIONS

Dynamic numerical codes used in practice may be divided into two main categories: total stress codes and effective stress codes. A brief discussion of some of the more frequently used codes within each category is provided in the following sub-sections.

8.4.1 Total Stress Codes

The total stress codes, as can be inferred from the classification, are based on the total stress concept and do not take account of pore pressures in the analysis. They should only be used in situations where the seismically induced pore pressures are negligible. The total stress codes may be divided into two main categories: (1) codes based on the equivalent linear (EQL) method of analysis, and (2) fully nonlinear codes.

8.4.1.1 Equivalent Linear Analysis (ELA)

ELA is essentially an elastic analysis and was developed for approximating nonlinear behaviour of soils under cyclic loading. Typical of the EQL codes used in practice are SHAKE, QUAD-4, QUAD4M and FLUSH. SHAKE is a one-dimensional wave propagation program and is used primarily for site response analysis. QUAD-4, QUAD4M and FLUSH are two-dimensional versions of SHAKE and are used for seismic response analysis of dams and embankments. The programs EERA (Equivalent-linear Earthquake site Response Analyses of Layered Soil Deposits) by Bardet et al. (2000) and DEEPSOIL are also available for these analyses. Given the elastic nature of the ELA analysis, these codes cannot take account of material yielding and material degradation under cyclic loading. Therefore, they tend to predict a stronger response than actually occurs. Also, they cannot predict the permanent deformations.

The authors' experience is that for analyses of potential liquefaction of dam foundations the one-dimensional SHAKE analyses are conservative, and it is better to use two-dimensional analyses such as QUAD-4. However, 1D analyses do not account for amplification of the edges of the crest of the dam and tend to underestimate the stresses.

8.4.1.2 Fully Nonlinear Analysis

More accurate and reliable predictions of permanent deformations can be obtained using the elasto-plastic nonlinear codes. Typical of the elasto-plastic nonlinear codes used in the analysis of embankments are DIANA, ANSYS and FLAC. The constitutive models used in these codes vary from simple hysteretic nonlinear models to more complex elasto-kinematic hardening plasticity models. Compared to the ELA codes, the elasto-plastic nonlinear codes are more complex and put heavier demand on computing time. However, they provide more realistic analyses of embankments under earthquake loading, especially under strong shaking.

8.4.2 Effective Stress Codes

Effective stress codes allow modelling pore pressure generation and dissipation in materials susceptible to build-up of excess pore pressure and liquefaction and thus are suited to obtain better estimates of permanent deformations under seismic loading. The effective stress codes may be divided into two main categories: fully coupled and semi-coupled.

8.4.2.1 *Fully Coupled Codes*

In the fully coupled codes, the soil is treated as a two-phase medium, consisting of soil and water phases. Two types of pore pressures are considered, transient and residual. The transient pore pressures are related to recoverable (elastic) deformations and the residual pore pressures are related to non-recoverable (plastic) deformations. A major challenge in fully coupled codes is to predict residual pore pressures. The residual pore pressures, unlike the transient pore pressures, are persistent and cumulative and thus exert a major influence on the strength and stiffness of the soil skeleton. The transient pore pressures are cyclic in nature, and their net effect within one loading cycle is often equal to zero. An accurate prediction of residual pore pressures requires an accurate prediction of plastic volumetric deformations. In the fully coupled codes this is often achieved by utilising elasto-plastic models based on kinematic hardening theory of plasticity (utilising multi-yield surfaces) or boundary surface theory with a hardening law.

Generally speaking, fully coupled prediction of pore pressures under cyclic loading is very complex and difficult. The validation studies performed on a number of these codes suggest that the quality of response predictions is strongly path dependent. When the loading paths are similar to the stress paths used in calibrating the models, the predictions are good. As the loading path deviates from the calibration path, the predictions become less reliable. Apart from the numerical difficulties, part of this unreliability is also due to the poor or less than satisfactory characterisation of the soil properties required in the models. For instance, because of sampling problems, it is often very difficult to accurately determine volume change characteristics of loose sands as required by these models. In general, the accuracy of pore pressure predictions in fully coupled modes is highly dependent upon the quality of the input data.

Fully coupled codes include PLAXIS, ABAQUS and FLAC2D and 3D.

FLAC contains a number of general purpose constitutive models including Mohr–Coulomb but also provides for the use of user-defined constitutive models. These include a bounding surface hypoplasticity model for modelling cyclic liquefaction and pore water pressure generation during earthquake loading, e.g. Wang and Makdisi (1999) by AMEC Geomatrix Inc.; FLAC-UBCSAND (Beaty and Byrne, 2011), PM4Sand (Dafalias and Manzari, 2004; Boulanger, 2010; Boulanger and Ziotopoulou, 2012, 2013; Ziotopoulou and Boulanger, 2012, 2013) and PM4Silt (Boulanger and Ziotopoulou, 2018, 2023).

8.4.2.2 *Semi-Coupled Codes*

Compared to the fully coupled codes, the semi-coupled codes are more robust and less susceptible to numerical difficulties. However, they are theoretically less rigorous. In these codes empirical relationships are used to relate cyclic shear strains/stresses to pore pressures. The empirical nature of the pore pressure generation in these codes generally puts less restriction on the type of plasticity models used in the codes. The parameters they require are often routinely obtained in the laboratory or in the field.

8.4.3 Discussion

USSD (2022) provides a summary of the available methods and their application, as well as cautions regarding their use.

The authors' experience is that more advanced dynamic methods are complex and potentially expensive and should only be done by very experienced persons who understand the limitations of the analysis and the need to use well-considered properties and good understanding of the geotechnical model and mechanisms. The analysis methods are controlled by the quality of data put into them, the limitations of the methods themselves and particularly of those doing the analysis.

They are in most cases only to be contemplated for dams and foundations subject to liquefaction and where the extra refinement of the analyses is warranted, e.g. where the simplified methods are not really applicable and/or are resulting in marginal assessments on whether remedial works are required.

Any nonlinear deformation analysis needs to be documented in sufficient detail that it can be reasonably scrutinised. All input parameters, all numerical details (boundary conditions, damping parameters etc.) and enough output plots to evaluate reasonableness of initial conditions and dynamic responses would be required. These should be reviewed by persons expert in the practice.

It is a big step up from nonlinear analyses to use fully coupled analyses, as the time and cost involved is substantial and results are subject to the details of assumptions of input parameters, and assumptions are difficult to review. In one case the authors were involved in, the added refinement resulted in outcomes difficult to reconcile with less sophisticated analyses despite those carrying out the analyses being experienced in the use of the programs.

It is important to recognise the uncertainties in the results of seismic deformation analyses of embankment dams. USSD (2022) states the following:

> *"Case histories of back analysis of dam performance during past earthquakes (e.g. Mejia et al., 1992; Boulanger et al., 1995; Zeghal and Abdel-Ghafar, 1992; Naesgard and Byrne, 2007; Beaty and Byrne, 2008; Mejia and Dawson, 2012; Hadidi et al., 2014; Chowdhury, 2019; and Boulanger, 2019) provide a rough measure of overall uncertainty in estimated embankment seismic response and deformations under circumstances of dam and foundation characterization typical, or perhaps at the higher end, of the range encountered in U.S. practice. Such case histories suggest that overall uncertainty in the simulated seismic response and deformations of embankment dams in practice is considerable. For example, calculated permanent seismic displacements at the dam crest, from available case histories, are commonly a factor of two or three off from the observed displacements. Accordingly, it is critically important to consider such uncertainty in the interpretation of the results of seismic deformation analyses of embankment dams and in the evaluation of their seismic safety".*

8.5 PSEUDO-STATIC METHOD

Up until the 1970s, the pseudo-static analysis was the standard method of stability assessment for embankment dams under earthquake loading. The approach involved a conventional limit equilibrium stability analysis, incorporating a horizontal inertia force to represent the effects of earthquake loading. The inertia force was often expressed as a product of a seismic coefficient "k" and the weight of the sliding mass W. The larger the inertia force, the smaller the safety factor under the seismic conditions. In this approach a factor of safety (FOS) of < 1 implies failure, whereas FOS > 1 represents seismically safe conditions.

The seismic coefficients used in this approach were typically less than 0.2 and were related to the relative seismic activity in the areas to which they apply.

The pseudo-static method of analysis, despite its earlier popularity, was based on several restrictive assumptions. For instance, it assumed that the seismic coefficient acting on the potential unstable mass is permanent and in one direction only. In reality, earthquake accelerations are cyclic, with direction reversals. Also, the concept of failure used in the approach was influenced by that used in static problems where a factor of safety of less than one cannot be permitted, as the stresses producing this state will exist until large deformations change the geometry of the structure.

However, under seismic conditions, it may be possible to allow the FOS to drop below one, as this state exists only for a short time. During this time, earthquake-induced inertia forces cause the potentially unstable masses to move down the slope. However, before significant movement takes place, the direction of the earthquake loading is reversed and the movement of the soil masses stop as, once again, the FOS rises above one. In fact, experience shows that a slope is likely to remain stable post-earthquake despite having a calculated pseudo-static FOS less than 1.0.

In view of these limitations, the pseudo-static method should not be used in assessing dam safety.

Some limit equilibrium stability analysis programs allow pseudo-static analyses, and some consultants still use that facility. The results are quite meaningless.

8.6 SOILS SUSCEPTIBLE TO LIQUEFACTION

8.6.1 Methods Based on Soil Classification and in Situ Moisture Content

It has long been recognised that saturated sands, silty sands and gravelly sands are susceptible to liquefaction. As described by Seed et al. (2003), as more data has been gathered it has been recognised that coarser gravelly soils and silty and silty clay soils may also be subject to liquefaction.

Hunter and Fell (2003a, 2003b) gathered data from case studies where flow liquefaction had occurred, mainly failures of fills and cut slopes under static loading conditions. These data would also be reasonably applicable to earthquake loading. Figure 7.60 shows the data separated into classes of slope, with a suggested range of soils susceptible to flow liquefaction.

Seed et al. (2003) note that gravelly soils are often more permeable so can dissipate cyclic generated pore pressures unless they are contained within finer strata which inhibit drainage during cyclic loading.

It should be recognised that particle size distribution alone is not able to identify soils which are susceptible to liquefaction.

There are a number of methods available to identify soils which are susceptible to liquefaction based on the particle size gradation, plasticity and *in situ* moisture content of the soil. These include:

8.6.1.1 Seed et al. (2003) Criteria

Seed et al. (2003) gathered a large database of field performance. Based on this data they suggested that Figure 8.5 be used.

Seed et al. (2003) indicate that:

- Soils within Zone A represent those which are potentially susceptible to "classic" cyclic liquefaction.
- Soils within Zone B may be liquefiable.
- Soils within Zone C (soils not plotting in Zones A or B) are not generally susceptible to "classic" cyclic liquefaction but should be checked for potential sensitivity (loss of strength with remoulding or monotonic accumulation of shear deformation).

They define "classic" cyclic liquefaction as referring to soils which experience significant loss of strength and stiffness due to cyclic pore pressure generation.

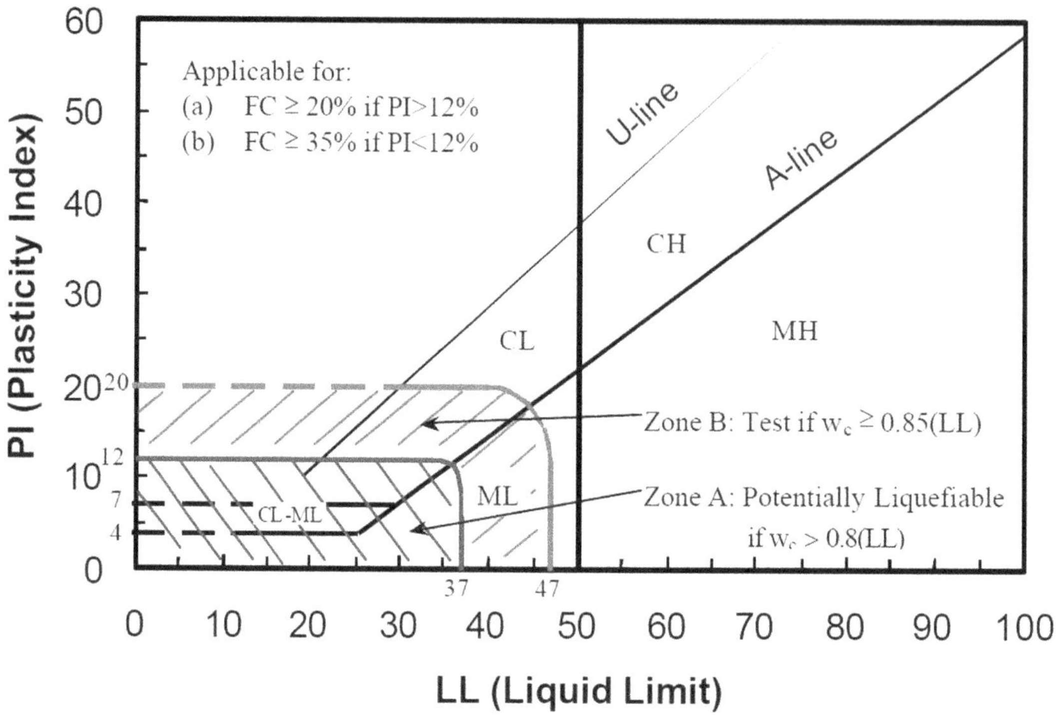

FIGURE 8.5 Seed et al. (2003) method for assessing liquefiable soil types.

Note that the method requires the *in situ* moisture content, liquid limit and fines content of the soil. This requirement is sometimes overlooked by consultants.

8.6.1.2 Bray and Sancio (2006, 2008)

Bray and Sancio (2006, 2008) developed the method shown in Figure 8.6 based on field data and cyclic testing. It has similar variables as the Seed et al. (2003) method arranged in a different manner and probably will give a similar outcome for most soils.

8.6.1.3 Boulanger and Idriss (2006)

Boulanger and Idriss (2006) and Idriss and Boulanger (2008) present Figures 8.7 and 8.8 to distinguish between soils which exhibit sand-like behaviour and liquefy, and soils which exhibit clay-like behaviour and cyclic softening. These are based on cyclic laboratory tests. There is an intermediate zone which is not clearly defined, but they conclude that soils with a PI $\geq$ 7 can reasonably be expected to exhibit clay-like behaviour. Soils with PI < 7 should be considered soils that probably exhibit sand-like behaviour (that is they are potentially liquefiable) unless shown to be otherwise through detailed *in situ* or laboratory testing. They indicate that these criteria are intended only to provide guidance in choosing engineering procedures.

In PEER (2023), Professor Boulanger said this method has the objective of guiding the analysis method, i.e. sand-like or clay-like. He preferred for it to be interpreted as a cyclic strength evaluation procedure criterion.

For application to dams and their foundations, the method remains useful for assessing the soils in the embankment and foundation because it identifies soils which will be sand-like and hence potentially susceptible to liquefaction and large loss of strength and stiffness.

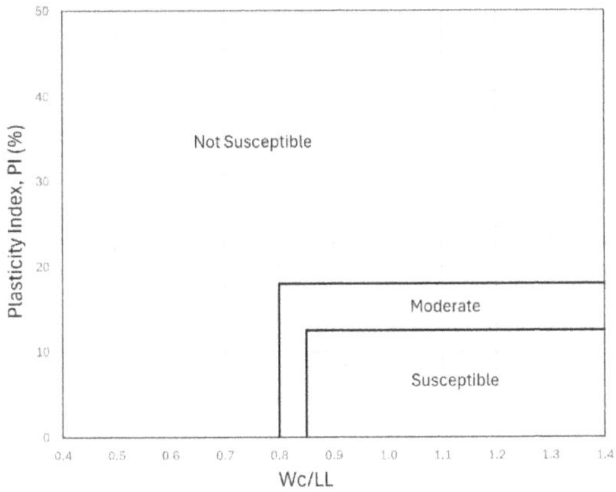

FIGURE 8.6 Figure summarising the Bray and Sancio (2006, 2008) method for assessing the susceptibility of soils to liquefaction.

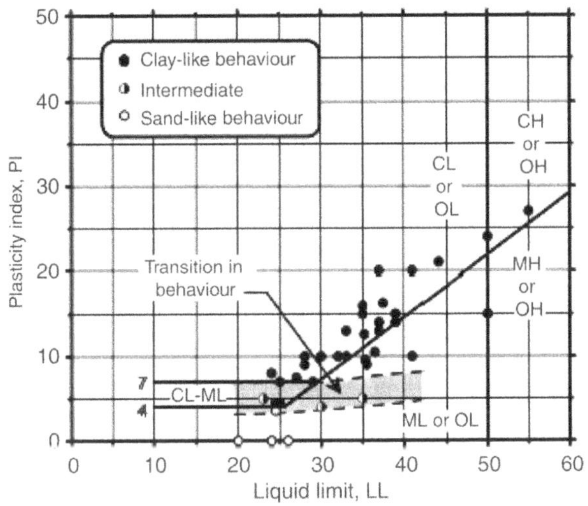

FIGURE 8.7 Atterberg limits chart showing representative values for soils which exhibited clay-like, sand-like and intermediate behaviour in cyclic loading (Boulanger and Idriss, 2006). (Used with permission of ASCE conveyed through Copyright Clearance Center Inc.)

8.6.2 Methods Based on CPT/CPTU Indices

8.6.2.1 Robertson and Wride (1998) Soil Behaviour Type Index I_c for Cone Penetration Tests

This method described in Youd et al. (2001) is from Robertson and Wride (1998) and calculates the Soil Behaviour Type index I_c as follows:

$$I_c = [(3.47 - \log Q)^2 + (1.22 + \log F^2)]^{0.5} \quad (8.1)$$

where:

$$Q = [(q_c—\sigma_{vo})/P_a][P_a/\sigma'_{vo}]^n = [(q_c—\sigma_{vo})/\ \sigma'_{vo}]^n \quad (8.2)$$

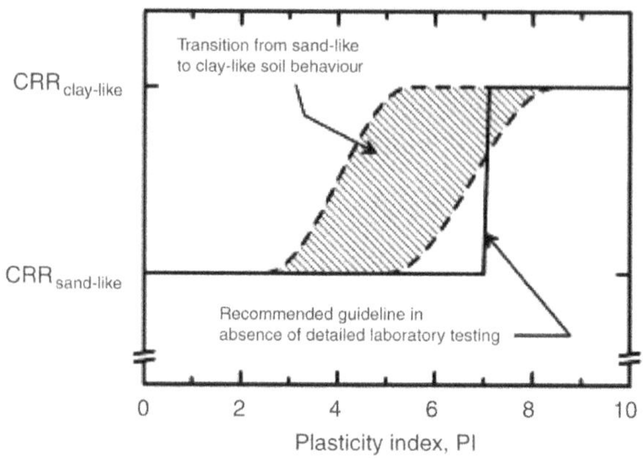

FIGURE 8.8 Schematic illustration of the transition from sand-like to clay-like behaviour with increasing Plasticity Index and the recommended guideline for practice (Boulanger and Idriss, 2006). (Used with permission of ASCE conveyed through Copyright Clearance Center Inc.)

and the normalised friction ratio F from:

$$F = [fs/(q_c—\sigma_{vo})]100\% \quad (8.3)$$

where fs = CPT sleeve resistance, (kPa); σ_{vo} = total vertical stress at test level, (kPa) $\sigma_{vo'}$ = effective vertical stress at test level; q_c is the cone resistance (kPa) and assume n = 1.0 for the initial calculation.

If $I_c > 2.6$, the soil is classified as clayey, so is unlikely to liquefy.

If $I_c < 2.6$, the soil is likely to be granular, so recalculate Q, and I_c using n = 0.5.

If the recalculated $I_c < 2.6$, the soil is non-plastic and granular.

If the recalculated $I_c > 2.6$, the soil is likely to be silty, possible plastic, so I_c should be recalculated using n = 0.7. Such soils are marginal and Youd et al. (2001) suggest they be sampled and tested to verify behaviour and/or checked using other criteria.

The first author has been involved in a project where soils with $I_c > 2.6$ but < 2.9 were considered as potentially liquefiable by the consultant and peer reviewer, both of whom were experts in the field. This approach was also independently taken by the authors on other projects where the other criteria indicated the soils might be liquefiable.

8.6.2.2 Robertson (2016) for Piezocone Penetration Tests

Robertson (2016) presents Figure 8.9 to be used to identify contractive soils which may be susceptible to flow liquefaction and cyclic softening.

The Robertson (2016) method attempts to distinguish the difference between the behaviour of ideal soils that are either "loose" or "dense" of critical state in conjunction with soils being "sand-like", "transitional" or "clay-like".

The boundary shown in Figure 8.9 represented by $I_B = 32$ represents the lower boundary for most sand-like ideal soils. The region represented by $22 > I_B < 32$ is defined as "transitional soils", which are soils that can have a behaviour somewhere between that of either sand-like or clay-like ideal soil.

The CD = 70 boundary shown in Figure 8.9 applies only to ideal soils with little or no microstructure. Aged and/or cemented soils can be contractive at large strains and inherently produce relatively high values of Q_{tn} due to increased stiffness and strength. The boundary represented by $I_B = 22$ represents the upper boundary for most clay-like ideal soils.

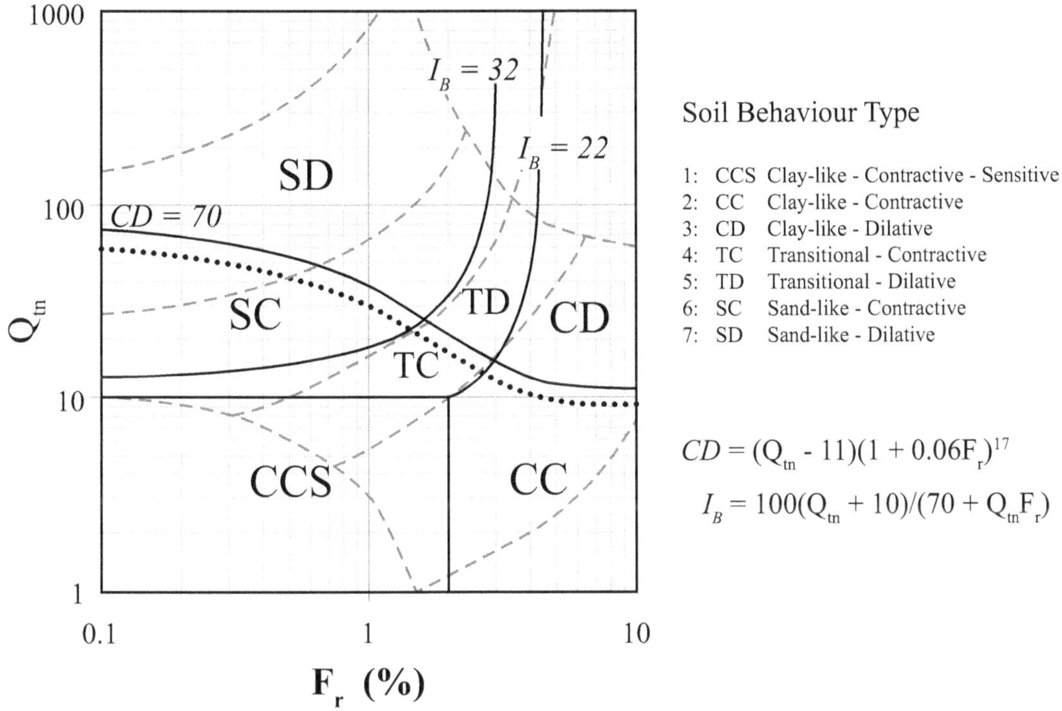

FIGURE 8.9 Soil Behaviour Type (SB_{Tn}) chart based on Q_{tn} and F_r %. Solid lines show Soil Behaviour Type boundaries and dashed lines show boundaries suggested by Robertson (1990) (Robertson, 2016). (Used with permission of Canadian Geotechnical Journal conveyed through Copyright Clearance Center Inc.)

8.6.3 Methods Based on Geology and Age of the Deposit

Idriss and Boulanger (2008) refer to Youd and Perkins (1978) and indicate that the soils which are most susceptible to liquefaction are non-plastic or very low plasticity fill, alluvial, fluvial, marine and deltaic soils. They indicate soils are most susceptible when they are recently deposited, becoming more resistant to liquefaction as they become older. The most susceptible soils are Holocene or younger soils; Pleistocene soils are low or very low susceptibility except for loess.

Youd et al. (2001) discuss the effects of the age of the deposit and conclude that for human-made structures such as fills and embankment dams, aging effects are minimal and corrections for age should not be applied in calculating liquefaction resistance.

8.6.4 Discussion and Recommended Approach

It is emphasised that the methods discussed previously are to determine whether a soil is potentially susceptible to liquefaction given a sufficiently large earthquake ground motion. Just because a soil is susceptible does not mean it will liquefy. That depends on the earthquake loading.

The authors have experience that for a dam where foundation liquefaction was thoroughly assessed using SPT and CPT/CPTU data, the Robertson and Wride (1998) and Robertson (2016) methods identified a number of strata as non-liquefiable, whereas the Seed et al. (2003) method indicated the soil was potentially liquefiable. The Seed et al. (2003) result was consistent with the soil properties from SPT sample recovery. In view of this it seems unwise to rely upon CPT/CPTU data and the Robertson (2016) method alone, and the authors recommend both SPT and CPTU tests be carried out.

We suggest the following:

(a) Use the method of Seed et al. (2003) or Bray and Sancio (2006) to assess whether a soil is potentially liquefiable. The methods should be used with all required data inputs including *in situ* moisture content, liquid limit and % fines.
(b) Use the Boulanger and Idriss (2006) method to assess whether the soil will have sand-like or clay-like behaviour.
(c) Where CPT data is available (which is preferable), also use the method of Robertson and Wride (1998) and/or Robertson (2016).
(d) Use this data together to assess the likelihood the soils are potentially liquefiable.

Consider the results as shown in Table 8.2.

Where soils plot in Zone B of the Seed et al. (2003) method, they should be considered as potentially susceptible to liquefaction and/or cyclic softening. If their behaviour becomes critical to the assessment of the safety of the dam, either the soil should be assumed to be liquefiable or undisturbed samples should be subject to cyclic loading tests rather than relying on these semi-empirical methods to predict behaviour.

8.7 THE "SIMPLIFIED METHOD" FOR ASSESSING LIQUEFACTION RESISTANCE OF A SOIL

8.7.1 Background to the Simplified Method

The most widely accepted, simplest and most practical method of assessing whether there is a potential for liquefaction for horizontal ground conditions was developed originally by Professor H.B. Seed and his co-workers.

The method is semi-empirical and based on the maximum acceleration induced by the earthquake a_{max}, the SPT "N" value corrected for the SPT hammer energy and for overburden pressure $(N_1)_{60}$, earthquake magnitude (M) and fines content of the soil (% passing 0.075 mm). It is based on recorded cases of liquefaction during earthquakes in the United States, Japan and China.

TABLE 8.2 Assessing the Likelihood That a Soil Is Potentially Liquefiable

RESULT OF ASSESSMENT	*LIKELIHOOD THESE STRATA ARE POTENTIALLY LIQUEFIABLE*
Methods (a), (b), and (c) indicate liquefiable and sand-like	Highly likely to certain
Methods (a) and (b) indicate liquefiable and sand-like	Likely
Only Seed et al. (2003) or Bray and Sancio (2006) indicate liquefiable, others do not	Likely to possible Depending on if Zone A or Zone or susceptible to moderate
Only Robertson and Wride (1998) or Robertson (2016) indicates liquefiable, others do not	Unlikely
Methods (a), (b), and (c) indicate non-liquefiable and clay-like	Liquefaction can be discounted

There has been a gradual development of refinements of the method, including use of updated databases of case histories, greater application of the Cone Penetration Test, modification to earthquake magnitude corrections and to the corrections for high overburden stress, static shear stresses and age of the soil deposit.

The authors have mostly used the method described in Idriss and Boulanger (2008) and subsequent refinements in Boulanger and Idriss (2012, 2015a, 2015b), Idriss and Boulanger (2015), Montgomery et al. (2014) and Boulanger et al. (2012, 2013).

The following sections detail the simplified method based mainly on the Idriss and Boulanger (2008) approach as updated in the later references.

Those carrying out liquefaction assessments should refer to the original references, in particular Idriss and Boulanger (2008), so they understand the details of how the method was developed.

8.7.2 The Simplified Method—Outline

The method requires the calculation or estimation of two variables for evaluation of liquefaction resistance of soils:

(a) The Cyclic Stress Ratio, CSR, which is a measure of the cyclic load applied to the soil by the earthquake.
(b) The cycle resistance ratio, CRR, which is the capacity of the soil to resist liquefaction.

The CRR is estimated from Standard Penetration Tests (SPT), Cone Penetration Tests (CPT) or, less frequently, the shear wave velocity. Figure 8.10 shows the CRR based on SPT as proposed by Idriss and Boulanger (2008) and Boulanger and Idriss (2012).

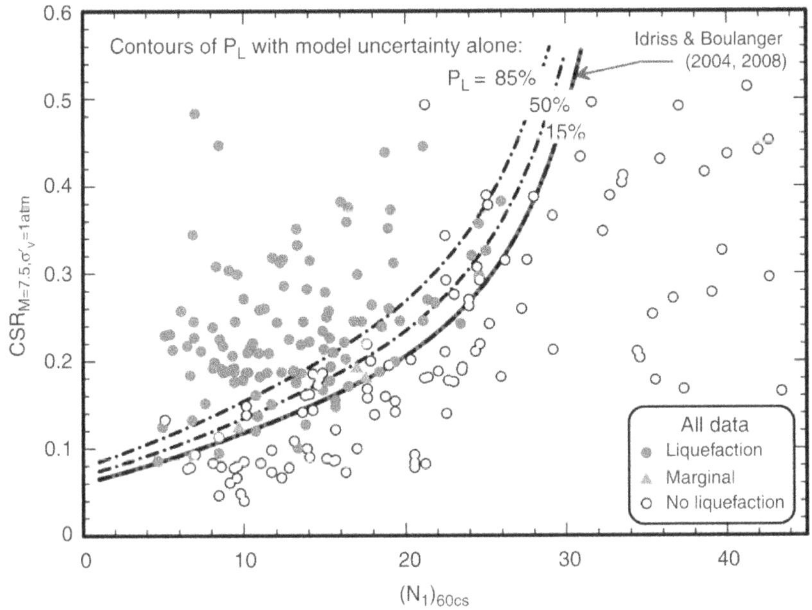

FIGURE 8.10 SPT clean-sand base Cyclic Resistance Ratio (CRR) curves for Magnitude 7.5 earthquakes, and an effective overburden stress of 100 kPa showing probabilities of liquefaction of 15%, 50% and 85% (Boulanger and Idriss, 2012). (Used with permission of ASCE conveyed through Copyright Clearance Center Inc.)

If the CSR is greater than the CRR, liquefaction is likely to occur.

For most purposes a simple comparison of CSR with CRR with engineering judgement will be sufficient. However, the methods available allow for the uncertainty in the boundary between liquefiable and non-liquefiable soils. Idriss and Boulanger (2008), Boulanger and Idriss (2012) indicate that the base curve in Figure 8.10 represents a probability of liquefaction of 15%. That is, to have a CRR greater than CSR is not a guarantee of no liquefaction. These uncertainties are only the model uncertainties and do not include the uncertainty relating to ground conditions.

It is emphasised that the methods need to be used with reasonable engineering judgement and should not be regarded as giving precise outcomes.

The methods are applicable to soils which are judged to be potentially liquefiable and sand-like, as discussed and summarised in Section 8.6.

To avoid bias in the analysis from short duration, high-frequency acceleration estimated from near-field small magnitude earthquakes, the peak horizontal acceleration should be based on the influence of earthquakes M5 or greater. Smaller magnitude earthquakes have too few cycles of motion to induce cyclic liquefaction.

8.7.3 Evaluation of Cyclic Stress Ratio (CSR)

The CSR is calculated from:

$$CSR = (\tau_{av}/\sigma'_{vo}) = 0.65(a_{max}/g)(\sigma_{vo}/\sigma'_{vo})r_d \tag{8.4}$$

Where a_{max} = peak horizontal acceleration at the ground surface generated by the earthquake; g = acceleration of gravity; σ_{vo} and σ'_{vo} are total and effective vertical overburden stresses, respectively, and r_d = stress reduction coefficient, which Idriss and Boulanger (2008), Boulanger and Idriss (2012) indicate can be calculated from:

$$r_d = \exp\left(\alpha(z) + \beta(z)M_W\right) \tag{8.5}$$

$$\alpha(z) = -1.012 - 1.126\sin\left(\frac{z}{11.73} + 5.133\right) \tag{8.6}$$

$$\beta(z) = 0.106 + 0.118\sin\left(\frac{z}{11.28} + 5.142\right) \tag{8.7}$$

Where z = depth below ground surface in metres and the arguments inside the sine terms are in radians.

It should be recognised that there is some uncertainty in r_d. This uncertainty is greater at depth. Boulanger and Idriss (2012) discuss this and recommend that the CSR (or equivalent r_d values) at depths greater than about 10 m should be based on dynamic response studies. These require sufficient subsurface site characterisation of the site and should allow for the variability in possible input motions.

The peak horizontal acceleration at the ground surface should be calculated allowing for amplification or de-amplification of the peak horizontal bedrock acceleration estimated in the absence of increased pore pressure or the onset of liquefaction.

If there is an existing embankment, the peak horizontal acceleration should be estimated at the crest of the embankment using the same assumptions and from this and the ground surface value, an acceleration representative of what would apply to the liquefiable layer estimated.

Alternatively, and better, estimate the cyclic stresses directly at the level of the potentially liquefiable layer using 2D dynamic response analyses (e.g. QUAD-4) of the embankment. The authors recommend this be done for all studies where important decisions on dam safety are to be made. Such dynamic response analyses must include a sufficient number of input acceleration time series and adequate site characterisation details to be reasonably robust.

8.7.4 Evaluation of Cyclic Resistance Ratio for M7.5 Earthquakes ($CRR_{7.5}$) from the Standard Penetration Tests Using the Boulanger and Idriss (2012), Idriss and Boulanger (2008) Method

The CRR for M7.5 earthquakes can be estimated from the curves in Figure 8.10. This plot requires the $(N_1)_{60cs}$ which is the $(N_1)_{60}$ with correction for fines content as detailed below.

Idriss and Boulanger derived an expression for the $CRR_{7.5}$ which fits the 15% probability curve. Some may wish to use this in spreadsheet calculations but if this is done it should be recognised it represents the 15% probability curve and is not the absolute boundary between liquefiable and non-liquefiable soils.

The equation is:

$$CRR_{7.5} = exp\left(\frac{(N_1)_{60CS}}{14.1} + \left[\frac{(N_1)_{60CS}}{126}\right]^2 - \left[\frac{(N_1)_{60CS}}{23.6}\right]^3 + \left[\frac{(N_1)_{60CS}}{25.4}\right]^4 - 2.8\right) \tag{8.8}$$

The $(N_1)_{60}$ is the SPT blow count normalised to an effective overburden pressure of 100 kPa, hammer energy of 60%, borehole diameter, rod length and sampling method. $(N_1)_{60}$ is given by:

$$(N_1)_{60} = N_m C_N C_E C_B C_R C_S \tag{8.9}$$

where N_m = measured standard penetration resistance; C_N = factor to normalise N_m to a common reference effective overburden stress; C_E = correction for hammer energy ratio (ER); C_B = correction factor for borehole diameter; C_R = correction factor for rod length; C_S = correction for samplers with or without liners.

Boulanger and Idriss (, 2014) indicate that C_N should be calculated from:

$$C_N = \left(\frac{p_a}{\sigma'_{vo}}\right)^{0.784-0.0768\sqrt{(N_1)_{60}}} \le 1.7 \tag{8.10}$$

with $(N_1)_{60}$ limited to values ≤ 46

where σ'_{vo} = effective vertical stress at the time of doing the SPT test (kPa); P_a = atmospheric pressure (100 kPa).

The equation for CPT is:

$$C_N = \left(\frac{P_a}{\sigma'_{vo}}\right)^{1.338-0.249(q_{c1N})^{0.264}} \leq 1.7 \qquad (8.11)$$

Figure 8.11 shows plots of the relationships.

The hammer energy ratio C_E varies between hammer types and for individual hammers and drill rig set ups. Tests on Australian free fall SPT hammers at Hume Dam in 1990s gave low C_E of about 0.6. Tested efficiencies at Hume Dam in 2015 range from 80% to 90% and in 2020, 66% to 72% for one drill rig, and 80% to 83% for a second.

It is clear that the efficiency for free fall hammers is equipment dependent and should be measured for each project. If it is not, conservative assumptions will have to be made which may prove costly in overly conservative assessments of liquefaction potential.

Correction for borehole diameter (C_B) is seldom needed. C_B = 1.0 for holes 65–125 mm diameter, C_B = 1.05 for holes 150 mm diameter and C_B = 1.15 for holes of 200 mm diameter. Correction for rod length, C_R is necessary for rod length < 3 m, where C_R = 0.75. No other corrections are needed as they were not applied to develop the database.

SPT tests should be done with the split inner tubes in place, in which case C_S = 1.0. If the inner tube was left out, C_S = 1.1 to 1.3.

Idriss and Boulanger (2008), Boulanger and Idriss (2012) indicate that the correction for fines content, FC (% passing 0.075 mm sieve), should be calculated from:

$$(N_1)_{60CS} = (N_1)_{60} + \Delta(N_1)_{60} \qquad (8.12)$$

$$\Delta(N_1)_{60} = \exp\left(1.63 + \frac{9.7}{FC+0.01} - \left(\frac{15.7}{FC+0.01}\right)^2\right) \qquad (8.13)$$

Boulanger and Idriss (2012) indicate that the $(N_1)_{60CS}$ values used in liquefaction assessments should be the mean values of the critical stratum, which is consistent with their treatment in analysing the case data. Figure 8.12 shows the relationship for SPT and CPT.

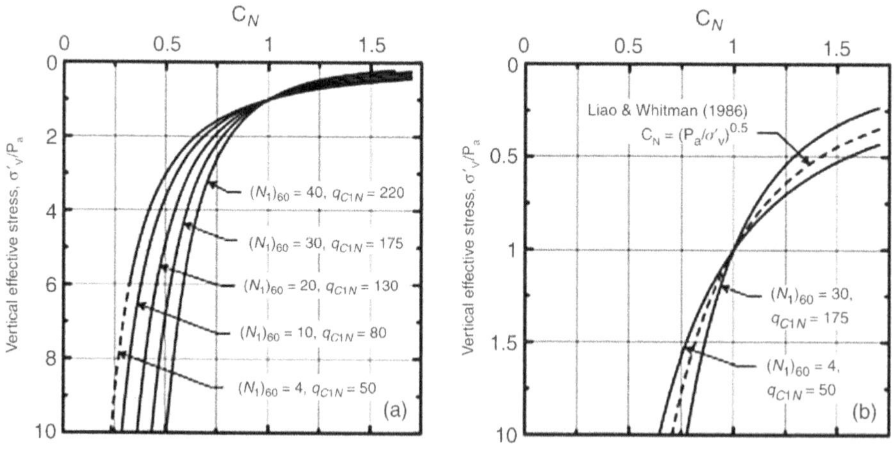

FIGURE 8.11 Overburden correction factor C_N for (a) vertical effective stresses up to 1000 kPa and (b) vertical effective stresses up to 200 kPa (Boulanger and Idriss, 2014).

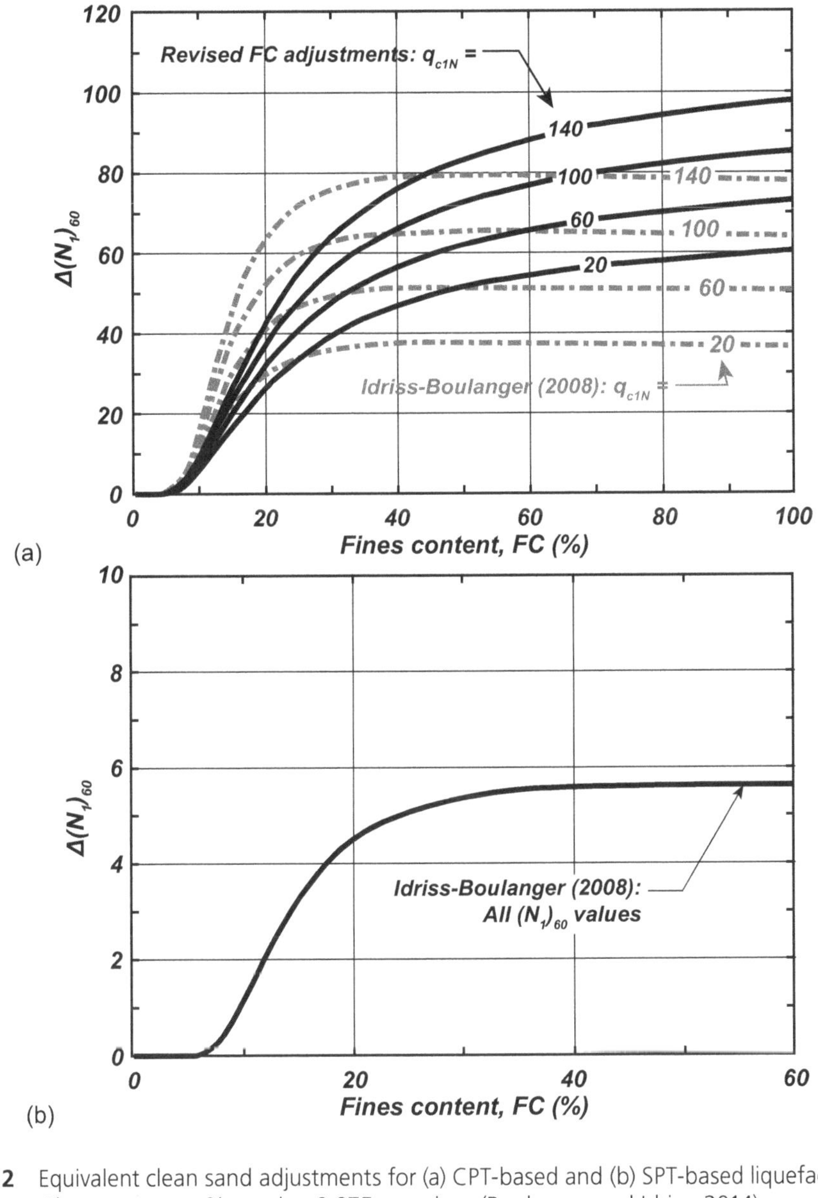

FIGURE 8.12 Equivalent clean sand adjustments for (a) CPT-based and (b) SPT-based liquefaction triggering procedures. Fines content = % passing 0.075 mm sieve (Boulanger and Idriss, 2014).

8.7.5 Evaluation of the Cyclic Resistance Ratio for M7.5 Earthquake ($CRR_{7.5}$) from Cone Penetration Tests Using the Boulanger and Idriss (2015a) Method

The CRR from M7.5 earthquakes can be estimated from the curve in Figure 8.13.

In these figures q_{c1N} is the normalised dimensionless cone penetration resistance calculated from:

$$q_{c1N} = C_Q(q_c/P_a) \tag{8.14a}$$

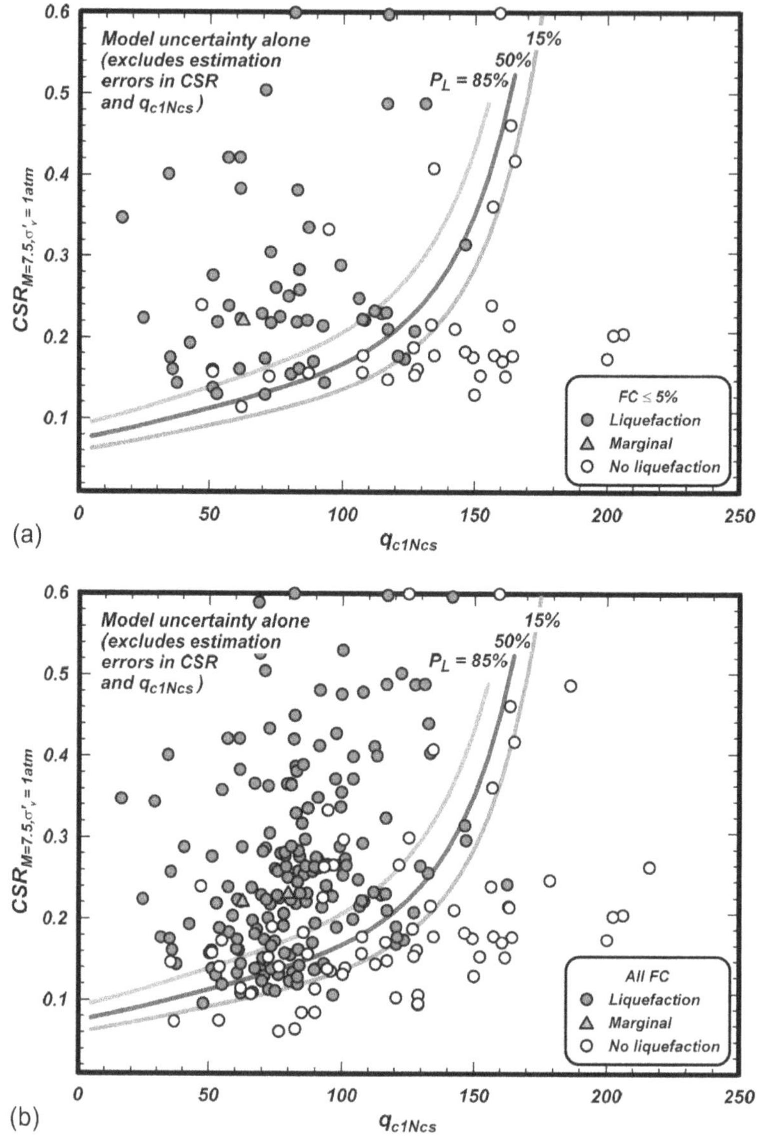

FIGURE 8.13 Curves of Cyclic Resistance Ratio for Magnitude 7.5 earthquakes and an effective overburden stress of 100 kPa (a) clean sands; (b) all sands, with data from compiled cases (Boulanger and Idriss, 2014).

where:

$$C_Q = (P_a / \sigma'_{vo})^n \tag{8.14b}$$

where C_Q = normalising factor for cone penetration resistance; σ'_{vo} = effective vertical stress at the time of doing the CPT test (kPa); P_a = atmospheric pressure (100 kPa); n = exponent that varies with soil type; and q_c = field cone penetration resistance measured at the tip in kPa.

The exponent n varies between 0.5 and 1.0 and can be estimated using the Soil Behaviour Type index I_c as described in Section 8.6.2.

The correction to an equivalent clean sand value $(q_{c1N})_{cs}$ is done by:

$$(q_{c1N})_{cs} = q_{c1N} + \Delta q_{c1N} \tag{8.15}$$

where:

$$\Delta q_{c1N} = \left(11.9 + \frac{q_{c1N}}{14.6}\right) exp\left[1.63 - \frac{9.7}{FC+2} - \left(\frac{15.7}{FC+2}\right)^2\right] \tag{8.16}$$

Figure 8.13(a) shows this relationship for a number of q_{c1N}.

Idriss and Boulanger (2015) provide equations for the curves in Figure 8.13(b).

Theoretical as well as laboratory studies indicate that CPT tip resistance is influenced by softer soil layers above or below the cone tip. As a result, measured CPT tip resistance is smaller in thin layers of granular soils sandwiched between softer layers than in thicker layers of the same granular soil. The amount of the reduction of penetration resistance in soft layers is a function of the thickness of the softer layer and the stiffness of the stiffer layers.

Youd et al. (2001) recommend the correction factor shown in Figure 8.14.

Idriss and Boulanger (2008) have added the equation to model the Youd et al. (2001) correction factors shown in Figure 8.14.

8.7.6 Evaluation of Cyclic Resistance Ratio for M7.5 Earthquake ($CRR_{7.5}$) from Shear Wave Velocity Using the Andrus and Stokoe (2000) and Rollins et al. (2022) Methods

Shear wave velocity Vs, obtained from surface to down hole or by cross-hole testing, can be used to assess the CRR. This is particularly useful where the potentially liquefiable soils are gravelly, and the gravel particles affect the SPT "N" value and prevent the penetration of a CPT. However, the test is likely to have difficulty locating thin, low velocity strata interbedded within the gravelly soils.

Idriss and Boulanger (2008) discuss the ability of shear wave methods to identify liquefiable soils and conclude that it is the least sensitive of the methods compared to SPT and CPT. They conclude that the method is best used to identify conditions in which the likelihood of liquefaction is high, low or uncertain.

The authors' experience is also that there is often no clear discrimination using this test, and it is recommended that it be used only in conjunction with SPT and/or CPT data.

Nevertheless, the authors appreciate that there may be conditions such as sites with significant coarse gravels present that make the use of SPT or CPT or both impracticable and that shear wave velocity testing may be the only choice available.

Consideration may be given to using the Becker Penetration Test (BBT) for investigations in gravelly soils if the equipment is available.

The CRR for M7.5 earthquake can be estimated from the curve in Figure 8.15.

In this figure Vs is corrected to an effective overburden stress of 100 kPa using:

$$V_{s1} = V_s \left(\frac{P_a}{\sigma'_{vo}}\right)^{0.25} \tag{8.17}$$

where V_{s1} = overburden stress corrected shear wave velocity, m/sec, P_a = atmospheric pressure, 100 kPa and σ'_{vo} = initial effective vertical stress in kPa.

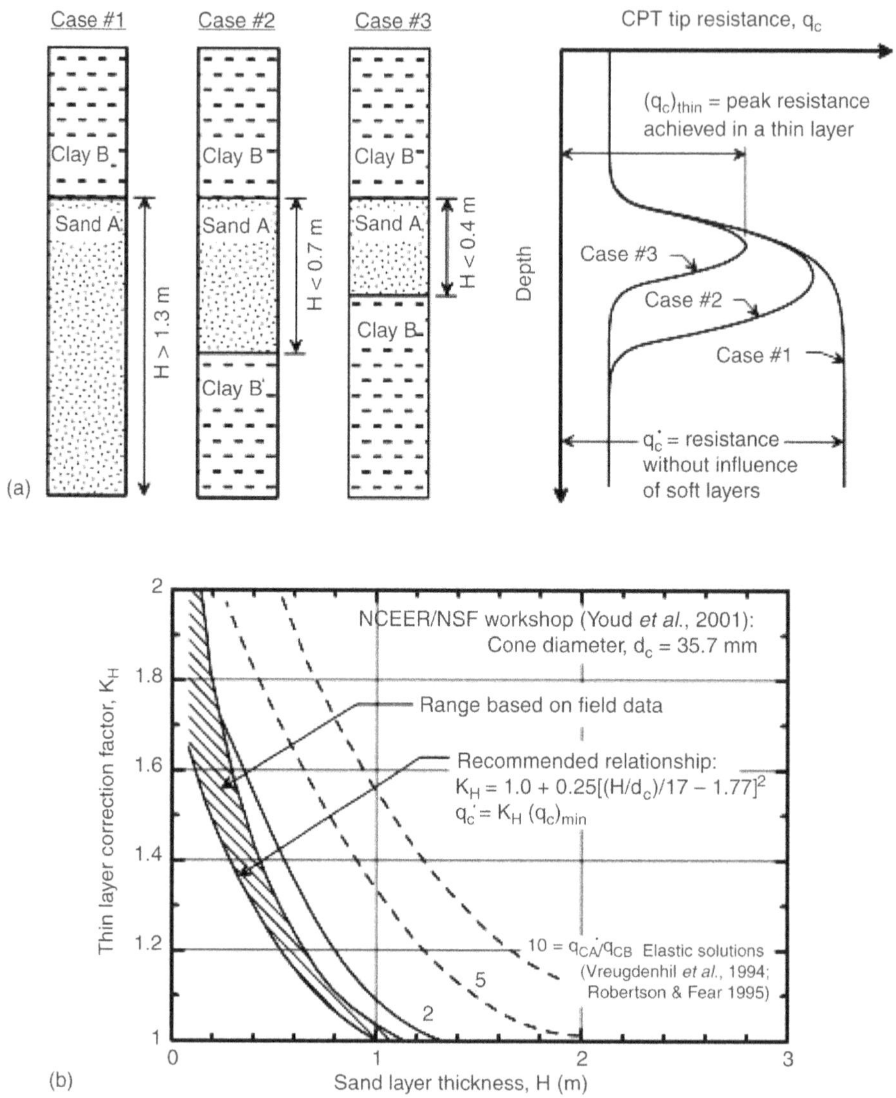

FIGURE 8.14 Thin-layer correction factor K_H for determination of equivalent thick layer CPT resistance: (a) a schematic of thin layer effects and (b) the relationship recommended by Youd et al. (2001). H = thickness of the interbedded layer in mm, q_{cA} and q_{cB} = cone resistances of the stiff and soft layers, respectively, and d_c = diameter of the cone in mm (Idriss and Boulanger, 2008).

The recommended lines in Figure 8.15 are dashed above $CRR_{7.5}$ of 0.35, to indicate field data is limited in this range, and do not extend below V_{s1} of 100 m/sec because there is a lack of data.

Rollins et al. (2022) developed Figure 8.16 from a larger database. This shows considerable uncertainty in the boundary between liquefiable and non-liquefiable data.

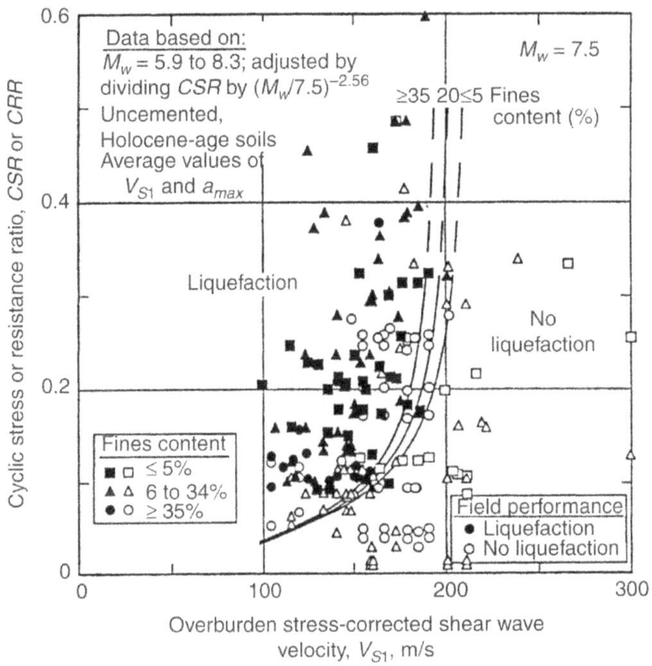

FIGURE 8.15 Shear wave clean sand base Cyclic Resistance Ratio curve for Magnitude 7.5 earthquake and an effective overburden stress of 100 kPa with data from compiled case histories (Youd et al., 2001, from Andrus and Stokoe, 2000). (Used with permission of ASCE conveyed through Copyright Clearance Center Inc.)

8.7.7 Earthquake Magnitude Scaling Factors and Factor of Safety Against Liquefaction

The procedures outlined earlier give $CRR_{7.5}$, Cyclic Resistance Ratios for M7.5 earthquakes.

Smaller magnitude earthquakes giving the same peak horizontal acceleration are less likely to initiate liquefaction, because the earthquake will have fewer cycles of motion. Larger magnitude earthquakes are more likely to initiate liquefaction. This is allowed for by using a Magnitude Scaling Factor (MSF) in the equation for Factor of Safety (FS) against liquefaction:

$$FS = (CRR_{7.5}/CSR)MSF \tag{8.18}$$

where:

CSR = calculated Cyclic Stress Ratio generated by the earthquake shaking and
$CRR_{7.5}$ = Cyclic Resistance Ratio for Magnitude7.5 earthquakes.

Boulanger and Idriss (2015b) discuss MSF and recommend use of the equations:

$$MSF = 1 + \left(MSF_{max} - 1\right)\left[8.64\exp\left(\frac{-M}{4}\right) - 1.325\right] \tag{8.19}$$

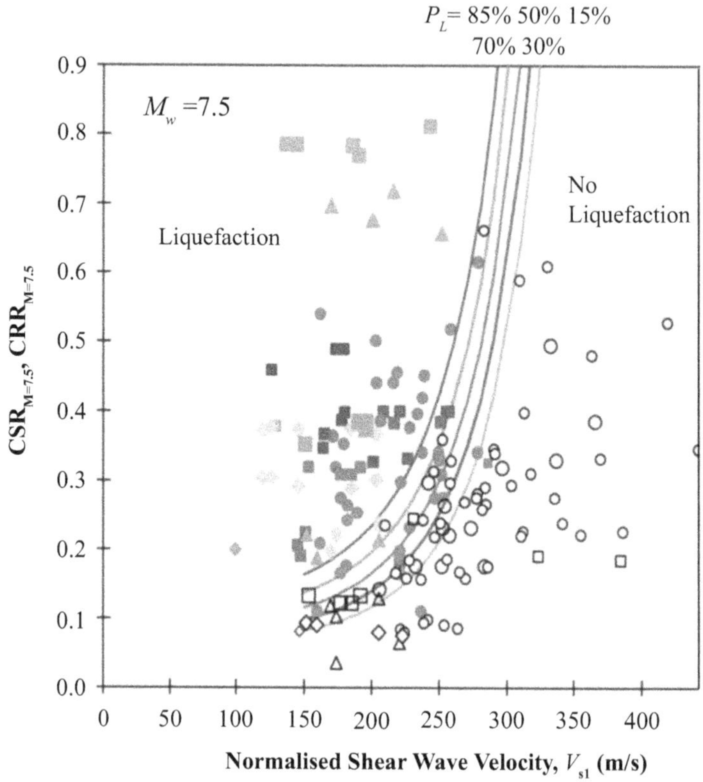

FIGURE 8.16 Cyclic Stress Ratio (CSR) and Cyclic Resistance Ratio (CRR) versus shear wave velocity normalised to effective overburden stress of 100 kPa showing proposed curves of probability of liquefaction (P_L). Solid markers denote liquefaction sites. Open markers denote non-liquefaction sites (Rollins et al., 2022). (Used with permission of ASCE conveyed through Copyright Clearance Center Inc.)

where MSF_{max} was related to q_{c1Ncs} values as:

$$MSF_{max} = 1.09 + \left(\frac{q_{c1Ncs}}{180}\right)^3 \leq 2.2 \tag{8.20}$$

These are shown in Figure 8.17 for SPT and CPT.

8.7.8 Corrections for Overburden Stress and Static Shear Stress

The method of assessment of liquefaction potential described earlier is for horizontal or gently sloping ground and for depths less than about 15 m.

For assessments of liquefaction for embankment dams, the confining stresses may be higher than this and the dam will impose static shear stresses, which may alter the liquefaction potential.

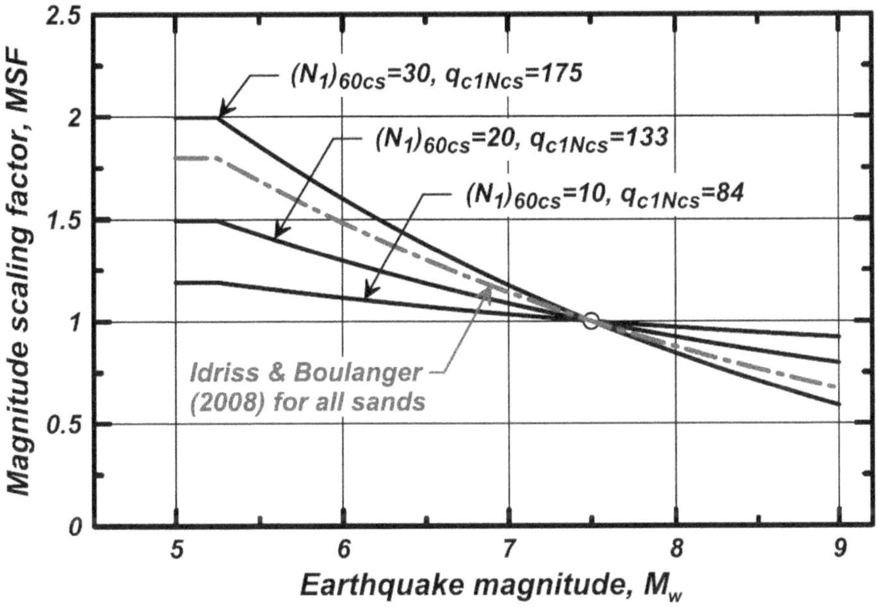

FIGURE 8.17 Magnitude Scaling Factors (MSF) proposed by Boulanger and Idriss (2015b). (*Used with permission of Journal of Soil Dynamics and Earthquake Engineering* conveyed through Copyright Clearance Center Inc.)

The factors K_σ and K_α may be applied to allow for the high overburden stresses and static shear stress respectively. They are applied by extending Equation 8.18 to:

$$FS = \left(CRR_{7.5} / CSR\right) MSF \cdot K_\sigma \cdot K_\alpha \tag{8.21}$$

K_σ and K_α can be estimated as follows:

8.7.8.1 Overburden Stress Factor K_σ

Youd et al. (2001), Seed et al. (2003) and Idriss and Boulanger (2008) use different methods to account for the overburden stress and to calculate K_σ.

(a) Youd et al. (2001) recommend the method of Hynes and Olsen (1999), which is based on laboratory data.
(b) Seed et al. (2003) and Cetin et al. (2004) used statistical regression analyses to develop criteria for effective stresses from 30 kPa to 175 kPa for SPT-based assessments. For higher stresses they recommended the Hynes and Olsen (1999) method.
(c) Idriss and Boulanger's (2008, 2010) recommendeded method is based on laboratory tests data and theoretical considerations.

Figure 8.18 shows the three methods.

It can be seen that Idriss and Boulanger (2008, 2010) recommend somewhat larger values of K_σ than the other authors, particularly at higher stresses. It is one of the important distinctions between the methods.

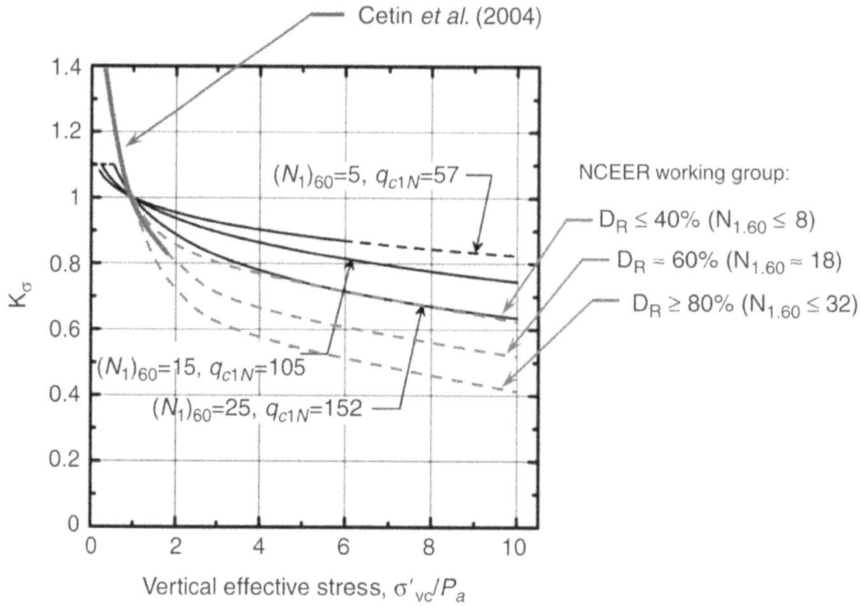

FIGURE 8.18 K_σ values recommended by NCEER (Youd et al., 2001); Idriss and Boulanger (2008); Cetin et al. (2004); (Seed, 2010).

The Idriss and Boulanger (2008, 2010) method is modelled by the following equations:

$$K_\sigma = 1 - C_\sigma \ln\left(\frac{\sigma'_v}{P_a}\right) \leq 1.1 \tag{8.22}$$

where the coefficient C_σ is given by:

$$C_\sigma = \frac{1}{18.9 - 2.55\sqrt{(N_1)_{60cs}}} \leq 0.3 \tag{8.23a}$$

$$C_\sigma = \frac{1}{37.3 - 8.27(qc1N)^{(0.264)}} \leq 0.3 \tag{8.23b}$$

Idriss and Boulanger (2010) discuss the Cetin et al. (2004) and Moss et al. (2006) methods and argue they are inconsistent with each other and inconsistent with available experimental data.

Montgomery et al. (2014) concluded that the Idriss and Boulanger (2008, 2010) method gave reasonably good fit to data for clean sands and was slightly unconservative for sands with fines contents 7% to 35%. They suggested that a value intermediate between the Idriss and Boulanger (2008, 2010) method and Youd et al. (2001) might be used for silty sands, but caution that the C_N value needs to be considered if that is done. See Montgomery et al. (2014) for more detail.

8.7.9 Sloping Ground Correction Factor K_α

The CRR is affected by the presence of static shear stresses, such as exist within slopes or embankment dams. Idriss and Boulanger (2008) indicate that the case history data are insufficient to empirically determine this effect. They refer to the method of Harder and Boulanger (1997), which is based on laboratory tests, and say the validity of these relationships for *in situ* sands is difficult to assess. Figure 8.19 presents that data.

Harder and Boulanger (1997) suggested the use of Figure 8.19 for soils with an effective confining stress less than 300 kPa. They point out that at significantly higher confining stresses, sandy soils will be more contractive and lower K_α may be appropriate.

The static shear stresses should be calculated from numerical analyses of the dam and its foundation.

Boulanger and Idriss (2014) indicate that they have not included K_α in that report because the database is dominated by level or near level ground conditions.

It is clear there is considerable uncertainty in the estimation of K_α. It seems that K_α will be less than 1.0 (meaning liquefaction is more likely) for cases where there are significant shear stresses from the slope and relative densities lower than about 50%, so, marginal cases should be assumed liquefiable if there are significant existing shear stresses.

8.8 LIQUEFIED UNDRAINED SHEAR STRENGTH

8.8.1 Some General Principles

As shown in Figure 7.53, undrained triaxial tests on saturated sands which display strain weakening (softening) behaviour reach a steady state undrained strength, S_{us}.

In field situations, such soils may experience some pore pressure redistribution, and the equivalent term is "liquefied shear strength $S_{u\,(LIQ)}$". This is also referred to in the literature as the mobilised undrained critical strength and undrained residual strength.

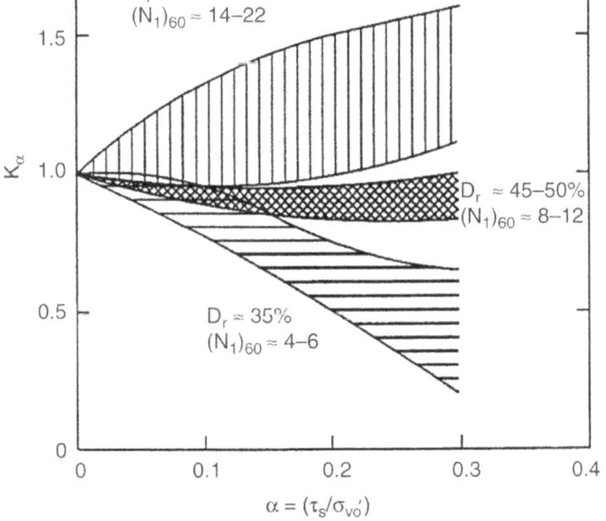

FIGURE 8.19 Sloping ground correction factor K_α suggested by Harder and Boulanger (1997). τ_s = static shear stress, σ'_{vo} = effective vertical stress (Idriss and Boulanger, 2008).

To check the stability of a dam after the earthquake motion has ceased, a limit equilibrium stability analysis is carried out with zones which have been assessed as liquefied under the earthquake loading assigned the liquefied shear strength, $S_{u\,(LIQ)}$.

Saturated zones of the dam or foundation which are subject to limited strain softening (temporary liquefaction), or even strain hardening, will experience pore pressure build-up in the earthquake which should be modelled in this "post-earthquake stability analysis" as described in Sections 8.10.2 and 8.11.

If the factor of safety is greater than 1.0 in these conditions, the slope will have experienced deformations during the earthquake, but no further deformation should occur after the earthquake motion ceases apart from those resulting from dissipation of pore pressures. If, however, the factor of safety is less than 1.0, deformation will occur under gravity loading. In soils which are sufficiently loose and contractive, a flow slide may occur, with large, rapid deformations.

There is significant uncertainty in estimate of $S_{u\,(LIQ)}$ which should be taken into account as detailed in Section 8.12.2.

8.8.2 Background to the Assessment of the Liquefied Shear Strength $S_{u(LIQ)}$

The steady state undrained strength, S_{us}, and liquefied shear strength $S_{u(LIQ)}$ of a soil is dependent on the void ratio, confining stresses, stress path of loading and soil fabric.

As discussed in Idriss and Boulanger (2008), Seed et al. (2003) and Seed (2010), the shear strength that the liquefied soil mobilises in the field is also affected by void redistribution, particle intermixing and other field mechanisms which are not replicated in the laboratory.

As a result, those authors concluded that the only practical methods for estimating the liquefied shear strength $S_{u(LIQ)}$ is by back analysis of case studies and relating these to SPT and CPT data. There are not many cases available and often for these the SPT and CPT data is variable and sometimes of limited quality and/or the geometry of the embankment and the stratigraphy of the foundation is complicated. As a result, considerable judgement has been exercised by authors in developing the methods, and the results are uncertain, with a wide range of strengths possible.

The issue is further complicated because there are two schools of thought:

- Those such as Seed et al. (2003), Seed (2010) who believe that the liquefied shear strength $S_{u(LIQ)}$ should be related to the SPT $(N_1)_{60CS}$ value directly with no allowance for the effective overburden stress. This is termed a "critical state" based method.
- Those such as Olsen and Stark (2002) who believe the normalised strength ratio $S_{u(LIQ)}/\sigma'_{vc}$ should be used and related to the SPT $(N_1)_{60CS}$ or $(N_1)_{60}$ values. That is, the residual strength is controlled by the effective overburden stress. Idriss and Boulanger (2008) and Boulanger and Idriss (2014) and Robertson (2010, 2022) have also developed such methods.

Idriss and Boulanger (2008) provide both methods without guidance on which to adopt but say they believe the residual strength ratio method provides a better representation of the potential effects of void redistribution than the liquefied residual strength, while it is recognised that neither correlation fully accounts for the numerous factors that influence void redistribution processes.

The authors understand that current USA practice is mostly to use the residual strength ratio method and that has been the authors' approach, but also check outcomes using the critical state method.

Some seem to be relying more on the results of laboratory tests on "undisturbed" samples of the potentially liquefiable soil. The authors are reluctant to do this because disturbance of the soil during sampling and subsequent handling in transit and in the laboratory of natural soils and sandy dam fills is almost certain to occur, affecting the test results.

8.8.3 Some Methods for Assessing the Strength of Liquefied Soils in the Embankment and Foundation

8.8.3.1 "Critical State" Based Methods

Figure 8.20 shows the Idriss and Boulanger (2008) method which is based on case studies from Seed (1987), Seed and Harder (1990) and Olsen and Stark (2002).

This requires the median *(N1)*60cs-Sr value for the liquefied strata. The corrections for fines content for calculating the *(N1)*60cs-Sr value is based on Seed (1987) and this is different to the fines content corrections used for the liquefaction triggering correlations discussed in Section 8.7.4. The correction for fines content for estimating $S_{u\,(LIQ)}$ is based on:

$$(N_1)_{60cs-Sr} = (N_1)_{60} + \Delta(N_1)_{60-Sr} \tag{8.24}$$

where Δ*(N1)*60-Sr are based on the values given in Table 8.3. Boulanger and Idriss (2014) indicate that these corrections continue to appear reasonable.

8.8.3.2 Normalised Strength Ratio Methods

Figures 8.21 and 8.22 show the Olsen and Stark (2002) and Idriss and Boulanger (2008) methods for SPT and for CPT data. σ'_{vo} is the pre-failure vertical effective stress for CPT and SPT.

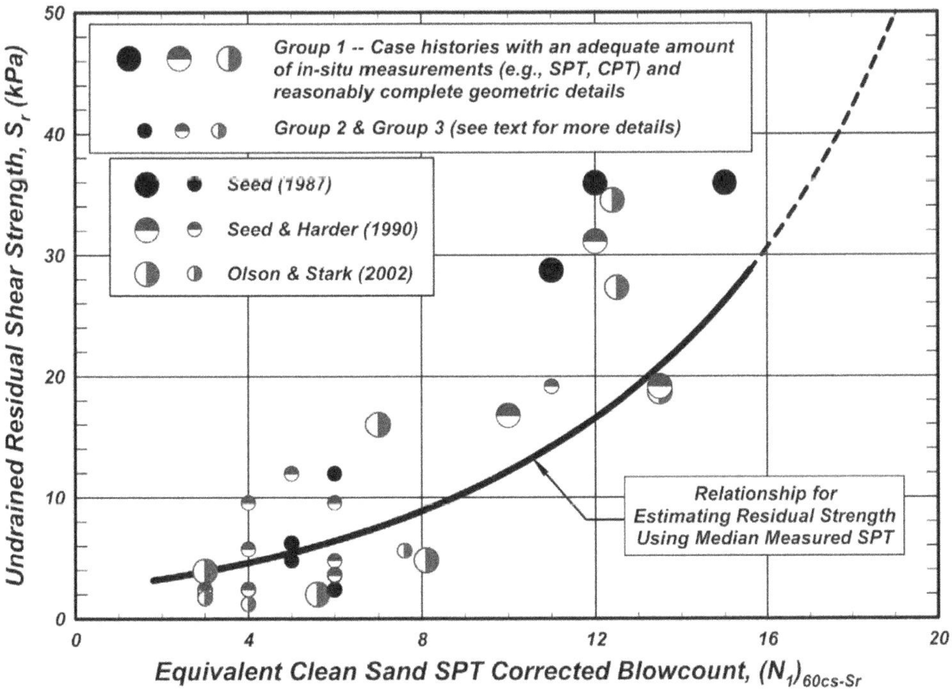

FIGURE 8.20 Liquefied shear strength $S_{u\,(LIQ)}$ versus SPT $(N_1)_{60CS}$ value according to Idriss and Boulanger (2008); Boulanger and Idriss (2014).

TABLE 8.3 Values of $\Delta(N_1)_{60\text{-SR}}$ for Fines Correction for Estimating $S_{u\,(LIQ)}$ from Seed (1987) and Idriss and Boulanger (2008)

FINES CONTENT (% PASSING 0.075 MM)	$\Delta(N_1)_{60\text{-}Sr}$
10	1
25	2
50	4
75	5

Olsen and Stark (2002) suggest the central dashed lines may be used for design. These are described by:

$$S_{u(LIQ)} / \sigma'_{v0} = 0.03 + 0.0143(q_{ci}) \pm 0.03 \quad for \ q_{ci} < 6.5\ MPa \tag{8.25}$$

and

$$S_{u(LIQ)} / \sigma'_{v0} = 0.03 + 0.0075((N_1)_{60}) \pm 0.03 \quad for \ (N_1)_{60} < 12 \tag{8.26}$$

Note the Olsen and Stark (2002) correlations for SPT and CPT use the $(N_1)_{60}$ and q_{ci} values which are not corrected for the fines content of the soil, whereas the Idriss and Boulanger (2008) correlations use the equivalent clean sand SPT and CPT values $(N_1)_{60\text{-}sr}$ and $(q_{c1})_{Ncr\text{-}sr}$.

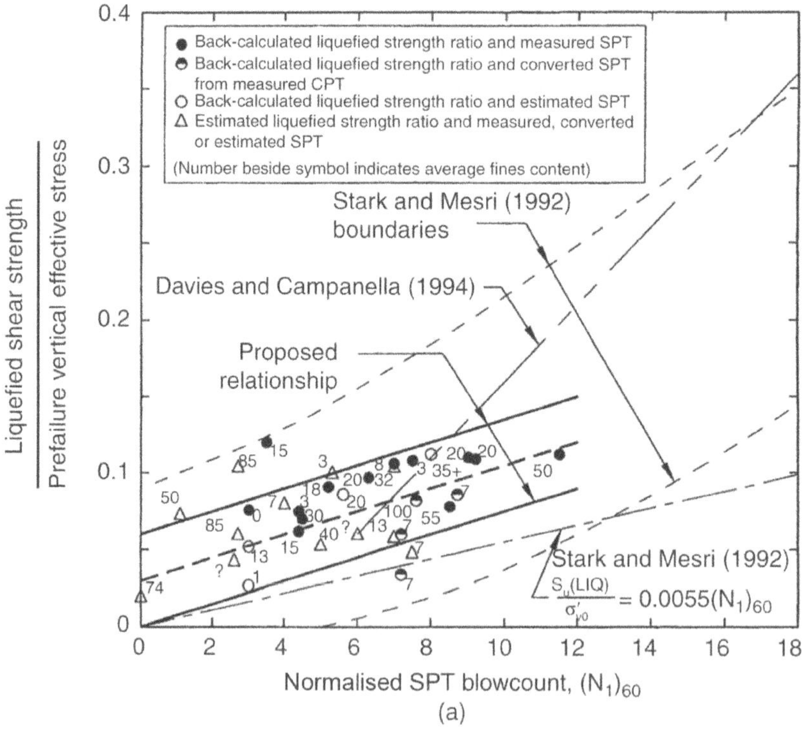

FIGURE 8.21 (A) Liquefied undrained strength ratio $S_{u\,(LIQ)}/\sigma'_{vo}$ versus SPT $(N_1)_{60}$ value (Olsen and Stark, 2002). (*Used with permission of Canadian Geotechnical Journal* conveyed through Copyright Clearance Center Inc.)

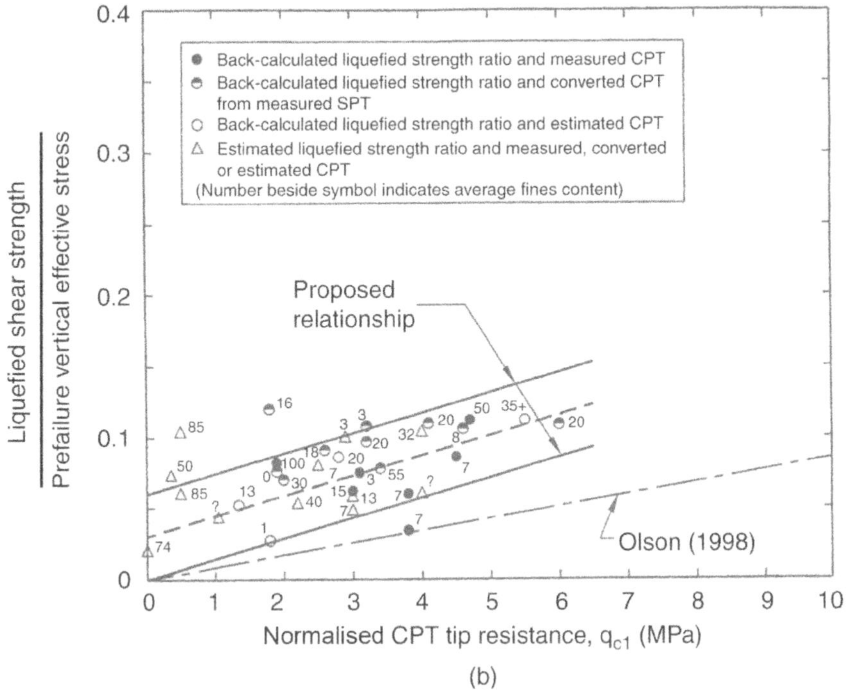

FIGURE 8.21 (B) Liquefied undrained strength ratio $S_{u\,(LIQ)}/\sigma'_{vo}$ versus CPT q_{c1} value (Olsen and Stark, 2002).

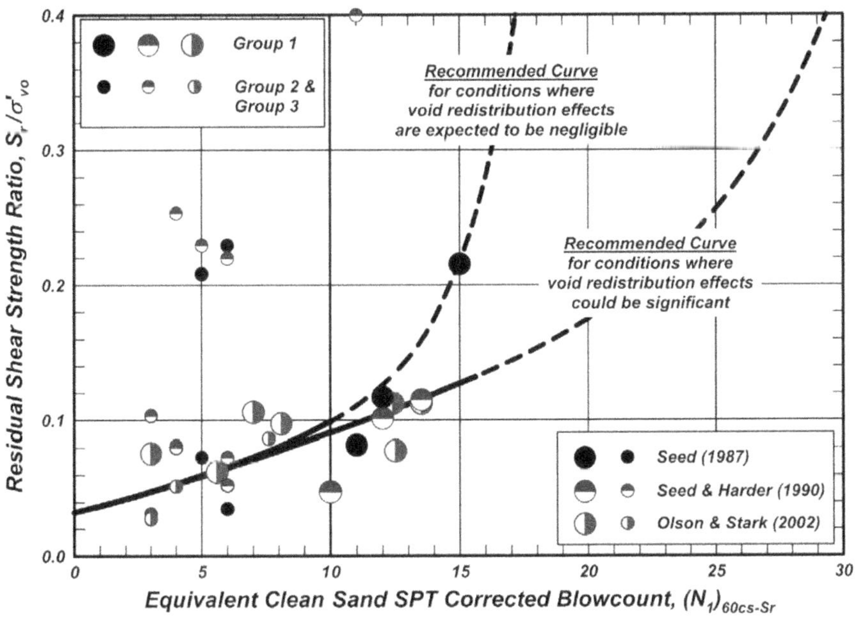

FIGURE 8.22 (A) Liquefied undrained strength ratio $S_{u\,(LIQ)}/\sigma'_{vc}$ versus SPT $(N_1)_{60CS}$ value (Idriss and Boulanger, 2008, 2015). (Used with permission of *Journal of Soil Dynamics and Earthquake Engineering* conveyed through Copyright Clearance Center Inc.)

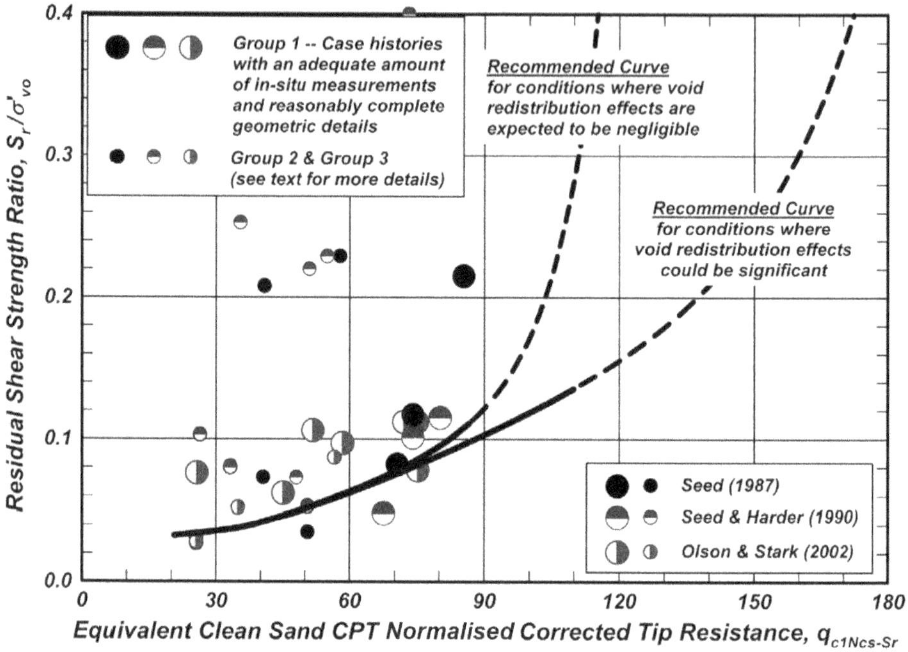

FIGURE 8.22 (B) Liquefied undrained strength ratio $S_{u\,(LIQ)}/\sigma'_{vc}$ versus CPT $(q_{c1N})_{cs}$ value (Idriss and Boulanger, 2008, 2015). (Used with permission of *Journal of Soil Dynamics and Earthquake Engineering* conveyed through Copyright Clearance Center Inc.)

TABLE 8.4 Values $\Delta\,(q_{c1Ncs})$ for Fines Correction for Estimating $S_{u\,(LIQ)}$ Idriss and Boulanger (2008, 2015)

FINES CONTENT (% PASSING 0.075 MM)	$\Delta\,(q_{c1Ncs})$
10	10
25	25
50	45
75	55

Table 8.4 shows the fines content correction for CPTU tests.

The Idriss and Boulanger (2008) line for $< (N_1)_{60-Sr} < 15$ plots a little lower, although the plots are not directly comparable because Olsen and Stark (2002) use $(N_1)_{60}$ and Idriss and Boulanger (2008), $(N_1)_{60-Sr}$.

Robertson (2022) Method

Robertson (2022) presents Figure 8.23 which is based on case histories including additional case data since the Robertson (2010) paper and adopting the drained strength for $Q_{1n,cs} \geq 80$ where the soils would be dilative. Robertson (2022) suggests using a minimum strength ratio of 0.02 and/or 1 kPa minimum liquefied strength.

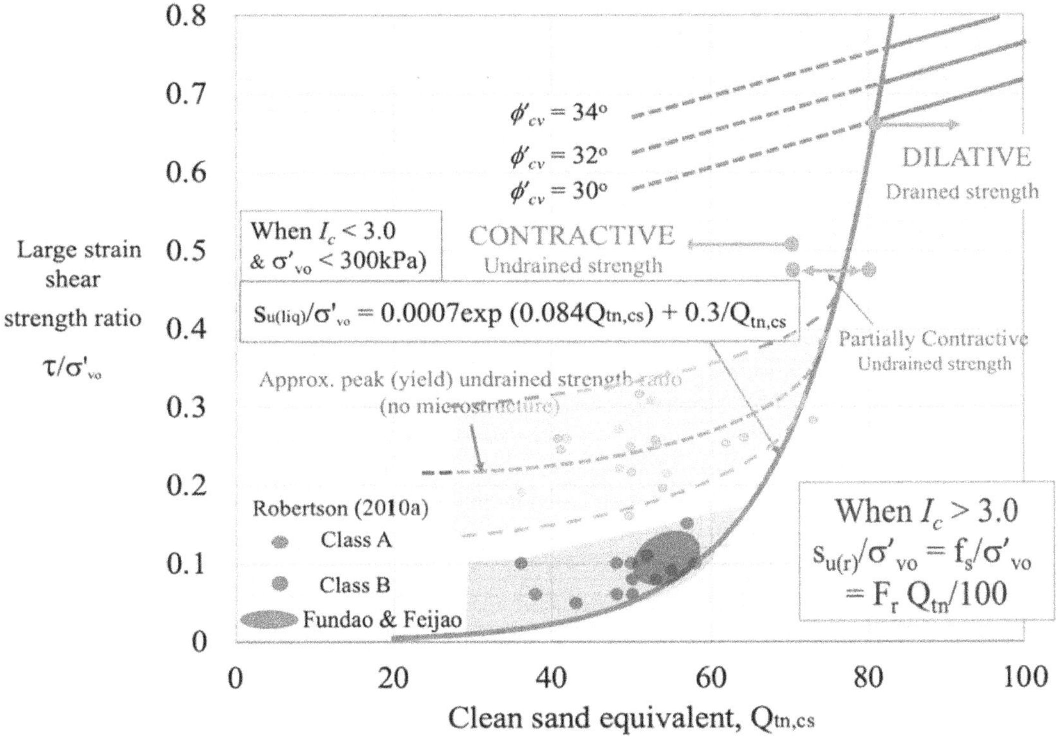

FIGURE 8.23 Relationship between large strain shear strength ratio, or liquefied strength ratio ($S_{u(LIQ)}/\sigma'_{vc}$) versus CPT clean sand equivalent CPT $(q_{c1N})_{cs}$ for I_c < 3.0 (Robertson, 2022). Class A and B refer to the quality of data points and are described in Robertson (2010). The light grey shading represents the likely range of strength ratios. (Used with permission of *Canadian Geotechnical Journal* conveyed through Copyright Clearance Center Inc.)

Robertson (2022) corrects CPT data to clean sand equivalent by:

$$Q_{\text{tn,cs}} = Q_{\text{tn}} K_{\text{c}} \tag{8.27}$$

where:

$$K_{\text{c}} \approx 15 - \frac{14}{1 + (I_{\text{c}}/2.95)^{11}} \quad \text{for } I_{\text{c}} \leq 3.0 \tag{8.28}$$

when $I_{\text{c}} < 1.7, K_{\text{c}} = 1.0$

8.8.3.3 Discussion

The authors have experience of the Robertson (2022) method in saturated sand fill placed in thick layers by bulldozer with no formal compaction. It gave liquefied strength ratios more than double those from Idriss and Boulanger (2008) for CPT and SPT and considered too high for the very loose sand. We therefore caution on the use of the method, and recommend it not be used alone.

Some adopt a minimum liquefied shear strength of 1 kPa when using the normalised strength ratio method to avoid numerical convergence issues. This is not mentioned in Idriss and Boulanger (2008, 2010). As discussed earlier, Robertson (2022) does suggest a minimum of 1 kPa. The authors advise caution in adopting this because our experience is that it can have a significant effect on the estimation of deformations using numerical methods.

8.8.3.4 Some Other Factors to Consider

Olsen and Stark (2002) indicate that, in their view, at least for silty sands with > 12% fines, it is reasonable to allow for the increase in $S_{u\ (LIQ)}$ which would be indicated by the increase in σ'_{vo} from the construction of a berm. For high consequence of failure projects, they suggest laboratory consolidation tests be carried out to confirm that the strength is parallel to the steady state line, which is implicit in the assumption that $(S_{u\ (LIQ)}/\sigma'_{vc})$ is constant.

Idriss and Boulanger (2008) favour the stress ratio approach but do not give any clear direction on what to do in regards designing berms for remedial works.

Seed et al. (2003) favour the critical state based methods which unless consolidation occurs under the loads of a berm would result in no strength increase.

This is particularly important if it is planned to use a berm to improve the post-earthquake stability of a dam. Based on the preceding points, it would appear unwise to allow for the strength increase from the increased effective stress of a berm to be constructed for remedial works for clean sands and silty sands with ≤ 12% fines, and the $S_{u(LIQ)}$ obtained without the berm should be assumed to apply after the berm is built, unless sampling and laboratory testing indicates otherwise.

If strength increase was to be allowed for example in silty sands with fines content > 12%, it would also be wise to confirm the strength increase by carrying out SPT or CPT through the berm after it is constructed.

8.8.3.5 Recommended Approach to Assessing the Liquefied Undrained Strength Soils in the Embankment and Foundation

Based on the preceding, the following is recommended:

(a) Use both critical state and normalised strength ratio methods. Apply a greater weighting to the normalised strength ratio method.
(b) Either the Idriss or Boulanger (2008, 2015) or Olsen and Stark (2002) methods may be used. These may be compared to the Robertson (2022) method but should be relied upon more than the Robertson (2022) method.
(c) Use SPT and CPT tests wherever practicable. Give equal weighting to each or weight towards the test giving the lower strength.
(d) Do not allow for any increase in the liquefied undrained strength beneath a berm which is to be added to improve post-earthquake stability unless the requirements described earlier are met. If a berm has already been constructed, carry out SPT and CPT in the liquefiable soils below the berm and use these data to assess the liquefied undrained strength.
(e) Take account of the fact that there are significant uncertainties in these methods and use strengths between the best estimate and lower bound strengths.

8.9 METHODS FOR ASSESSING THE POST-EARTHQUAKE STRENGTH OF NON-LIQUEFIED SOILS IN THE EMBANKMENT AND FOUNDATION

8.9.1 Saturated Potentially Liquefiable Non-Plastic Soils

Saturated non-plastic soils (e.g. sand, silty sand), which are potentially liquefiable but which have not been subject to sufficient cyclic shear stress to reach a liquefaction condition, will develop positive pore pressures during the cyclic loading of the earthquake. The amount of positive pore pressure will depend on many factors relating to the soil and the cyclic loading.

Idriss and Boulanger (2008) refer to Marcuson et al. (1990) for level ground conditions but warn that pore pressures will be higher for situations where static shear stresses are present. The Marcuson et al. (1990) data is shown in Figure 8.24 and relates the residual excess pore pressure ratio R_u, at the end of the earthquake loading to the factor of safety against liquefaction, FS_L, calculated from Equation 8.21. Idriss and Boulanger (2008) also refer to other references and conclude that if the factor of safety against liquefaction is greater than 1.2 to 1.3, the pore pressure buildup will be < 50% and deformations should be small.

It is suggested that if the factor of safety against liquefaction of non-liquefied soils is ≥ 1.3 the pore pressures may be ignored is post-earthquake stability analyses, but if the factor of safety against liquefaction is < 1.3, the pore pressures should be modelled, and Figure 8.23 may be used as an approximate estimate.

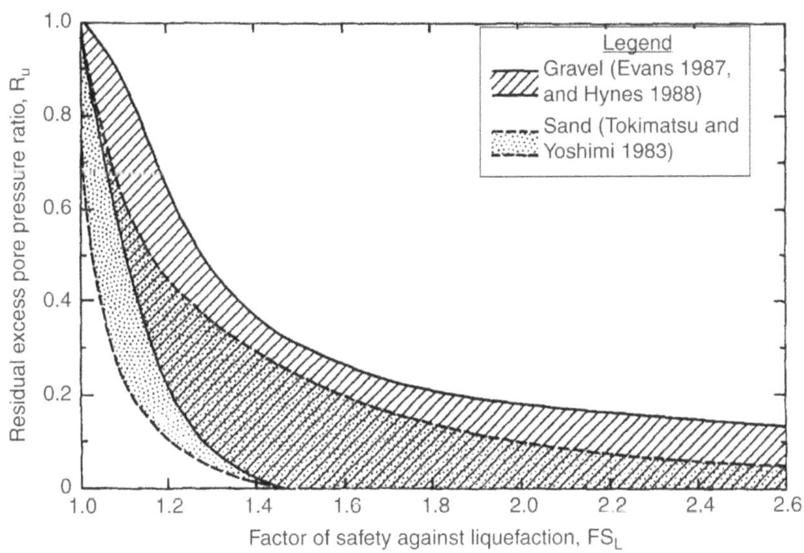

FIGURE 8.24 Method for approximate estimation of the residual excess pore pressure R_u in non-liquefied zones from the factor of safety against liquefaction FS_L (Marcuson et al., 1990). (Reproduced with permission of Sage Publications.)

If these pore pressures become critical, and the project is sufficiently important, seek expert advice. For more critical projects, carefully planned and executed cyclic laboratory tests should be used.

8.9.2 Cyclic Softening in Clays and Plastic Silts

Boulanger and Idriss (2006) and Idriss and Boulanger (2008) describe how to estimate the effects of cyclic softening of clays and plastic silts. The discussion relates to saturated soils with overconsolidation ratios of 1 to 4. Such soils sometimes occur in association with liquefiable soils but are not common. Readers should refer to Boulanger and Idriss (2006) and Idriss and Boulanger (2008) and later literature for details.

The boundary between which this behaviour and liquefaction occur is discussed in Section 8.6.1.

8.9.3 Plastic Soils in the Embankment and in the Foundation

For post-earthquake stability analyses, the potential loss of strength from pre-earthquake strengths will depend on the amount of cracking and deformations of the soil. These are likely to be greatest in the upper part of the embankment and least in the foundation. The following are suggestions on how this may be done.

For compacted clay soils in the embankment and heavily overconsolidated soils in the foundation not subject to cyclic softening:

(a) For embankments with expected extensive longitudinal cracking and deformation, e.g. Pells and Fell (2002, 2003) Class 3 Damage, adopt the composite undrained shear strength/effective stress envelope as described in Section 7.4.3.
(b) For embankments with expected minor damage, e.g. Pells and Fell (2002, 2003) Damage Class 1, adopt the pre-earthquake strengths.
(c) For embankments with expected Pells and Fell (2002, 2003) Damage Class 2, adopt pre-earthquake strengths with 10% loss of strength to allow for disturbance.

For dams where the embankment clay is not well compacted as in some older dams, and/or the foundation clays are contractive in nature:

(d) For Damage Class 3, adopt composite strength as described in Section 7.4.3 for soils in the embankment. For the foundations allow for strain weakening taking account of the sensitivity of the soil.
(e) For Damage Class 1, adopt the pre-earthquake strengths.
(f) For Damage Class 2, for soils in the embankment adopt pre-earthquake strengths with 10% loss of strength to allow for disturbance. For soils in the foundation allow for partial strain weakening which will be less than for Damage Class 3.

8.9.4 Compacted Non-Plastic Soils

Well compacted free draining rockfill, filters and dense sands and gravel fill will dilate on shearing. Adopt the effective stress strengths c', ϕ', with no change in the pore pressures. If large deformations are expected in the earthquake, the dense granular materials may have loosened and will have a strength approaching the critical state strength rather than the peak strength. In practice this can be accommodated by a small reduction from the expected peak strengths.

8.10 DATA ASSESSMENT AND ASSESSMENT OF LIQUEFACTION POTENTIAL

8.10.1 Site Investigations Requirements and Development of a Geotechnical Model of the Foundation

The following are some recommendations regarding site investigations of foundations of dams for assessing the likelihood and consequences of liquefaction.

8.10.1.1 Review of Available Data

Gather together all available data on the foundations, including for existing dams, site investigations and laboratory testing carried out prior to construction, geological mapping, reports and photographs taken during construction, monitoring data and site investigations and laboratory testing carried out since the dam was constructed.

8.10.1.2 Preliminary Geological and Geotechnical Models and Planning Site Investigations

Draw plans and sections of the available data, taking account of the depositional environment of the foundation soils.

From this develop interpretive geological and geotechnical models of the strata in the foundation and summarise the properties in a form useful for liquefaction assessment.

Assess the adequacy of the available data and plan site investigations to supplement this data.

8.10.1.3 Site Investigation Methods

Potentially Liquefiable Soils in the Foundation and Embankment

(a) Non-plastic soils in the foundation and embankment such as horizontal filters should be investigated by boreholes in which SPT tests are carried out in potentially liquefiable soils, and undisturbed thin walled tube samples taken of other strata, and by CPT and CPTU.

(b) Use of both SPT and CPT, CPTU is necessary because CPT, CPTU are best able to detect layering in the strata, but SPT and laboratory tests are required to provide data for some parameters and to make available two methods for assessing the liquefaction potential and post-earthquake liquefied strengths. Some of these (SPT and CPT) should be done adjacent to each other for correlation purposes.

(c) Boreholes should where practicable be drilled using wash boring methods with casing and drilling mud support. If hollow flight augers are used (as may be necessary for holes drilled through an existing dam), take measures to prevent "blowing" of the strata at the base of the augers with subsequent reduction in SPT blow counts.

(d) Caution should be used in interpretation of SPT data where sonic drilling methods are used as the sonic vibrations could lead to densification of loose sand materials below the drill bit. The authors prefer to confirm SPT data in sonic drill holes with SPT in adjacent boreholes drilled using conventional methods and drilling mud, subject to it being acceptable to drill using wash boring without risks of hydraulic fracture of the core of the embankment.

(e) Take samples from all SPT tests in potentially liquefiable strata and test for moisture content and fines content (% passing 0.075 mm sieve). For representative samples test for Atterberg limits. In stratified strata, ensure different soil types that are collected in the SPT samples are carefully separated and tested.
(f) The energy efficiency of the SPT hammer being used in the investigations should be measured. If it is not measured, assume 60% efficiency even for free fall drop hammers.
(g) Record the level of the water table during drilling and preferably install piezometers to monitor groundwater pressures at the time of the investigations and as they fluctuate with river flows and possibly reservoir level and rainfall.
(h) For SPT tests in gravelly soils take blow counts for each 75 mm as an aid to detecting interference by gravel particles and allow adjustment to allow for this. For gravelly soils consider the use of shear wave velocity in addition to boreholes with SPT or Becker Penetration Test.
(i) For CPT and CPTU tests in layered soils adjust the data for the effects of soft and stiff layers as detailed in Figure 8.14. CPTU is preferred over CPT as the measured dynamic pore pressures provide an indication of the contractive vs dilative behaviour of the soil under dynamic loading conditions.

Other Embankment Materials

Depending on what information is available from site investigations before the dam was constructed, construction control testing and investigations since construction, it may be necessary to drill boreholes and/or excavate test pits to sample the other materials in the embankment for laboratory tests on particle size distribution, plasticity, stiffness and drained and undrained shear strengths. These should be done taking account of the guidance in Chapter 4.

Shear Wave Velocity of the Foundations and Embankment Materials

A required property for the assessment of amplification using 1D SHAKE or 2D QUAD-4 analyses is the dynamic shear modulus which can be estimated from the shear wave velocity V_S of the soil overlying rock in the foundation and the embankment materials. Site specific geophysical testing should be carried out to determine these.

If that is not possible, the shear wave velocity can be estimated from CPT/CPTU and SPT data using the methods described in Wair et al. (2012). One project the authors were involved in gave shear wave velocities significantly lower than velocities estimated by geophysical methods.

8.10.1.4 Geological and Geotechnical Models

Draw plans and sections of the embankment and foundations using all available data including the data from the new investigations and taking account of the depositional environment of the foundation soils.

Of particular importance is the persistence of potentially liquefiable strata in the foundation and embankment. Take account of the depositional environment of soils in the foundation and how embankment fill was placed and compacted.

Analyse the SPT and CPT data to characterise the properties of the potentially liquefiable strata and other strata in the foundation. Analyse the classification, stiffness and strength data of the embankment materials.

From this develop an updated geotechnical interpretive model.

8.10.2 Analysis of Liquefaction Potential and Post-Earthquake Strengths

The analysis of liquefaction potential is best carried out as follows:

- Assess the seismic hazard for the dam site. For deterministic assessments use the Earthquake Design Ground Motion for the AEP relevant to the dam consequence of failure category. For risk based assessments select ground motion partitions as described in Section 5.4.
- Assemble the SPT and CPT/CPTU data for potentially liquefiable strata into spreadsheets. In so doing assess the data for quality and inconsistencies and remove or "tag" as unreliable poor quality or inconsistent data.
- Where using spreadsheets, for SPT include all "N" results in the spreadsheet. For CPT/CPTU enter data at 50 mm or smaller intervals as recorded.
- Analyse the data to determine if the soils are potentially liquefiable using the criteria summarised in Section 8.6. Exclude from further considerations soils which are not liquefiable, or which are above the water table and will not be saturated under any condition.
- Estimate the Cyclic Stress Ratio using the simplified method allowing for amplification effects of the ground motions. This can be done using one-dimensional elastic dynamic analysis such as SHAKE using appropriate parameters. Do this for each ground motion partition.
  Experience is that such analyses can be conservative, and it is better to carry out two-dimensional dynamic elastic analyses such as QUAD-4 to estimate the cyclic shear stresses. Do this for a selection of representative earthquake time histories.
- Estimate the Cyclic Resistance Ratio using the simplified method.
- For the potentially liquefiable strata, assess the factor of safety against liquefaction triggering for a number of peak ground accelerations within the AEP range to be assessed for the site. Take account of the representative earthquake magnitude for each AEP in making this assessment.
- Assess the extent of potentially liquefiable strata assuming that strata with a factor of safety against liquefaction triggering ≤ 1.1 will liquefy. Allow for uncertainty in the CSR as shown in Figure 8.10 for SPT and Figure 8.13 for CPT/CPTU.
- From these data, plot plans and sections showing contours of the AEP to trigger liquefaction of the liquefiable zones.
- For major decisions on dam safety and the need for upgrades, it may be necessary to carry out fully coupled numerical analyses using programs as described in Section 8.4.2. These should only be carried out by organisations expert in such analyses.

The post-earthquake strength of liquefied strata and other strata should be estimated as follows:

- For the strata predicted to liquefy (i.e. FS for liquefaction < 1.1), estimate the liquefied undrained strength (Section 8.8). There are often several different liquefiable strata with different properties.
- For saturated non-liquefied non-plastic strata, estimate the residual pore pressure built up during the earthquake based on the FS for liquefaction as described in Section 8.9.1, Figure 8.24. Include these pore pressures in the post-earthquake stability analyses if the factor of safety against liquefaction of that stratum is < 1.3. For higher factors of safety against liquefaction, these pore pressures will be small and may be ignored.
- For saturated plastic strata which are subject to cyclic softening, estimate the effects of cyclic softening (Section 8.9.2).
- For compacted fills allow for the effects of cracking and deformation. These will be earthquake loading dependent (Section 8.9.3).
- If there is strain weakening overconsolidated clay in the foundation, estimate the strain weakened strength allowing for the earthquake load being considered and potential deformations in the strain weakening strata (Section 8.9.3).

8.11 POST-EARTHQUAKE STABILITY AND DEFORMATION ANALYSES FOR EMBANKMENTS SUBJECT TO LIQUEFACTION

8.11.1 Post-Earthquake Stability Analysis

The post-earthquake stability analysis is done with conventional limit equilibrium analysis methods. No loading from the earthquake is applied, since this is a post-earthquake analysis.

The following is guidance regarding inputs to the analysis:

(a) The geometry of the embankment and foundations is as constructed.
(b) Piezometric surface is as before the earthquake.
(c) The shear strengths of liquefied zones are modelled as a strength ratio, but with some analyses carried out with liquefied strengths as a check.
(d) Residual pore pressure build-up is modelled in non-plastic zones which are not liquefied as described in Section 8.9.1.
(e) Pore pressures in well compacted filters and rockfill which will behave in a dilative manner do not change from the pre-earthquake pore pressures.
(f) Separate analyses will be required for seismic loads ranging from those where liquefaction is just initiated to those where the maximum seismic load being considered and the extent of liquefaction will be greater.
(g) It will be essential to model non-circular failure surfaces if the liquefied zones are relatively thin.

The sensitivity of the factors of safety to the extent and continuity of liquefied zones and the uncertainty in the strength inputs should be assessed.

8.11.2 Deformation Analyses

8.11.2.1 *Khalili et al. (1996) Approximate Method*

The potential post-failure deformation of the dam may be assessed using the simplified method of Khalili et al. (1996).

The Khalili et al. (1996) method considers two modes of failure defining the upper and lower bounds of post-failure deformation:

- A "rapid model" based on the principle of conservation of energy, where the potential energy of the slide mass is resisted by the frictional forces along the surface of rupture and approximated using the residual shear strength.
- A "slow model" based on equations of equilibrium between the driving force of the slide mass and resisting forces along the surface of rupture and approximated using the residual shear strength, i.e. assuming the rate of movement would be so slow that the effect of inertia forces of the sliding mass is negligible.

The failure model is shown in Figure 8.25 and based on the assumptions of a circular slide surface, the failure is plane strain, the failing mass moves as a rigid body and energy losses during failure are only due to the frictional forces acting along the slip surface. Approximate solutions for the rapid model

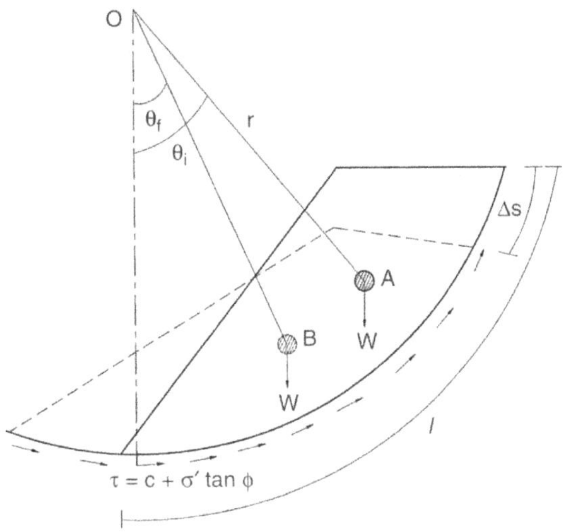

FIGURE 8.25 Failure model for post-failure deformation (Khalili et al., 1996). (Reproduced by permission of Taylor & Francis Group.)

are calculated using Equation 8.29 and for the slow model using Equation 8.30 where $FS_{residual}$ is the factor of safety calculated using residual strengths along the surface of rupture, and θ_i and θ_f are the initial and final positions of the centre of gravity as defined in Figure 8.24:

$$\text{Rapid model } \theta_f = 2\theta_i(FS_{residual} - 0.5) \tag{8.29}$$

$$\text{Slow model } \sin\theta_f = FS_{residual} \sin\theta_i \tag{8.30}$$

Khalili et al. (1996) found that for most cases analysed the rapid model gave the best estimate of deformations.

This method is approximate and should only be used as a guide to behaviour. It does not satisfactorily model failure surfaces controlled by thin liquefiable strata which are non-circular.

8.11.2.2 Static Post-Earthquake Numerical Analyses

Post earthquake deformations of cases where the post-earthquake stability factor of safety is < 1.0 can be estimated by a static analysis of the embankment and foundations with liquefied shear strengths in liquefied zones and other strengths as described in Section 8.8.

Programs such as FLAC are best able to do this as they remain computationally stable at large deformations. The analyses should include sensitivity to input parameters.

The deformations which may occur during the earthquake motion are not included in this method.

8.11.2.3 Dynamic Earthquake Analyses

Fully coupled dynamic analyses may be used to estimate deformations during and post-earthquake. During the earthquake, excess pore pressures are generated from dynamic analysis and deformations are modelled. After shaking ceases, a consolidation analysis is required to model the dissipation of these pressures and the resulting deformations. Post-seismic deformations are assessed through static analysis.

These are a major step up in complexity and cost and should only be required for major dam safety assessments where the outcomes from other approaches are marginal.

8.12 ASSESSMENT OF DAM SAFETY RESULTING FROM SEISMIC LOADING

8.12.1 Embankments Not Susceptible to Liquefaction

8.12.1.1 Deterministic Approach

(a) Use one or more of the database methods to estimate the deformation for the Earthquake Design Ground Motion. If the estimates of crest settlements are less than 10% of the available freeboard before the earthquake with reservoir level at FSL or normal operating level, accept that the likelihood or failure by overtopping is negligible for that loading and the embankment satisfies deterministic criteria.
(b) If the estimated crest settlements by the database methods are greater than this, use one or more of the simplified methods to estimate deformations. If the estimated deformations from both database and simplified methods are less than 20% the available freeboard, accept that the likelihood or failure by overtopping is negligible for that loading and the embankment satisfies deterministic criteria.
(c) If fully coupled numerical analyses are carried out the potential accuracy is ±50% (see 8.4.3), it may be concluded that deterministic criteria are not satisfied if estimated deformations using best estimate parameters are > 40% of available freeboard.

If the crest settlements are greater than these the dam does not satisfy deterministic criteria.

8.12.1.2 Quantitative Risk Assessment Approach

(a) Estimate crest settlements using database and simplified methods for each seismic load partition.
(b) Estimate the conditional probability of loss of freeboard (P_{LF}) based on these estimates taking account of the uncertainty in the estimated settlements and variation in reservoir level.
(c) Estimate the likelihood of unsuccessful detection (P_{UD}), intervention and repair (P_{UIR}), and probability of breach (P_B) using Sections 7.14.4, 7.14.5 and 7.14.6. Allow for potential earthquake damage to communications and infrastructure in making the assessments.

The annual probability of breach for the earthquake load partition = $[(P_{EDM}) \times (P_{LF}) \times (P_{UD}) \times (P_{UIR}) \times (P_B)]$

where:

P_{EDM} = annual probability of the earthquake load partition.

In most cases if freeboard is predicted to be lost then there will be a very low or negligible likelihood of successful detection, intervention and repair, and breach will occur.

8.12.2 Embankments and/or Foundations Which Are Subject to Liquefaction

8.12.2.1 Deterministic Approach

Estimate the post-earthquake factor of safety for the Earthquake Design Ground Motion (EDGM) using:

- Representative geometry of the slope and foundations. Pay particular attention to the extent and continuity of liquefied strata for the EDGM.
- Lower bound estimates of shear strength parameter for liquefied zones, reasonable but conservative shear strengths for non-liquefied zones.
- Conservative estimates of pore pressure build-up and strain weakening in other zones.
- Model the whole dam and foundation and obtain factors of safety for downstream and upstream failure surfaces. Take account of the guidance in Section 7.12.

If the post-earthquake factor of safety is greater than 1.3, accept that the earthquake deformations will likely be minor and so the likelihood of failure will also be small and probably tolerable subject to the quality and extent of information available.

Deterministic criteria may then be satisfied for low and significant consequence of failure classification. For high, very high and extreme classification dams, use risk based methods to assess dam safety.

8.12.2.2 Risk Enhanced Deterministic Approach

(a) Estimate the post-earthquake factor of safety for the Earthquake Design Ground Motion using:
   - Representative geometry of the slope and foundations. Pay particular attention to the extent and continuity of liquefied strata.
   - Best estimates of shear strength parameters including liquefied shear strength.
   - Best estimates of pore pressure build-up and strain weakening in other zones.
   - Model the whole dam and foundation and obtain factors of safety for downstream and upstream failure surfaces. Take account of the guidance in Section 7.12.

(b) Use the Modified Silva et al. (2008) method described in Section 7.14.2 to estimate the conditional probability of initiation of instability as discussed for static loading. Assume Category III to allow for the uncertainty in liquefied strength and undrained strength of other plastic soils.

(c) Estimate the likelihood of loss of freeboard given instability initiates using Tables 7.10 and Figure 7.52. Assume that the reservoir is at Full Supply Level.

Use engineering judgement to assess the dam safety taking account of these results and the dam consequence of failure classification. If in doubt use the QRA method.

8.12.2.3 Quantitative Risk Assessment Approach

Earthquake partitions are developed up to and beyond the design earthquake. A threshold earthquake ground motion below which earthquake-induced deformations are assessed to be negligibly low is usually identified. For each partition of earthquake ground motion (with annual probability P_{EGM}):

(a) Estimate the post-earthquake factor of safety using:
   - Representative geometry of the slope and foundations. Pay particular attention to the extent and continuity of liquefied strata.
   - Best estimates and reasonable lower bound shear strength parameters including liquefied shear strength.
   - The best estimate and conservative estimate of pre-earthquake pore pressures.
   - Model the whole dam and obtain factors of safety for downstream and upstream failure surfaces. Take account of the guidance in Section 7.12.

(b) Carry out post-earthquake static numerical deformation analyses for:
   - The parameters giving the lowest post-earthquake factor of safety.
   - The best estimate parameters.
   - The reasonable but conservative parameters between these.

(c) From these analyses evaluate the estimated crest settlement, lateral displacements upstream and downstream.

(d) Compare these to the freeboard and use expert elicitation to estimate the likelihood of loss of freeboard (P_{LF}).

Take account of the results of the analyses but also the potential uncertainty in the numerical method and inputs used. Analyses may be carried out for Full Supply Level and other reservoir levels to account for the seasonal nature of reservoir levels. Assess the probability of the reservoir level being within a particular range at the time of the earthquake—this is based on the proportion of time the reservoir level is within that range (P_{RL}).

(e) Estimate the likelihood of unsuccessful detection (P_{UD}), intervention and repair (P_{UIR}), and probability of breach (P_B) using Sections 7.14.4, 7.14.5 and 7.14.6. Allow for potential earthquake damage to communications and infrastructure in making the assessments.

The annual probability of breach for the earthquake load partition = $(P_{EGM}) \times \Sigma[(P_{RL}) \times (P_{LF}) \times \{(P_{UD}) \times (P_B) + (1\text{-}P_{UD}) \times (P_{UIR}) \times (P_B)\}]$
where:

P_{EGM} = annual probability of the earthquake load partition.

Use event tree methods to estimate the annual probability of breach for all earthquake load partitions. Figure 8.26 shows a typical event tree.

In most cases if freeboard is predicted to be lost then there will be a very low or negligible likelihood of successful detection, intervention and repair, and breach will occur.

For cases where the estimated likelihood of failure is intolerable, either design remedial works so the residual likelihoods of failure are tolerable or use advanced numerical methods to make more refined estimates of deformations. Allow for the range of liquefied strengths and other properties. Use these to assess the likelihood of failure taking account of the still approximate nature of these estimates.

The extent to which an embankment is analysed for deformations should be consistent with the consequence of failure and size of the dam. In many cases it will be better to design and construct remedial measures than continuing to do more and more sophisticated analyses.

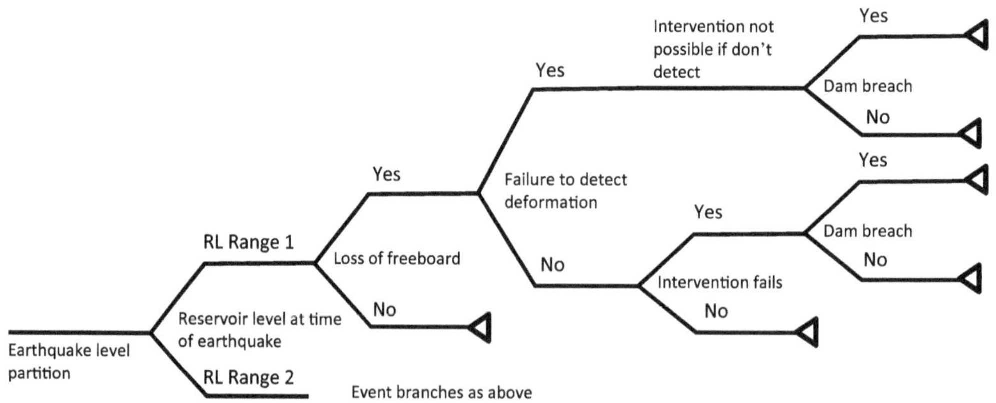

FIGURE 8.26 Event tree for modelling the annual probability of failure of an embankment from loss of freeboard under seismic loading.

Internal Erosion and Piping—Concentrated Leak Erosion

9

9.1 INTRODUCTION

Concentrated leak erosion may occur in any soil which may hold a crack, including plastic soils such as clay, sandy clay, clayey sand, and non-plastic soils with high fines contents such as silt, sandy silt and silty sands.

Cracks or gaps may form in the embankment as a result of:

- Cross valley differential settlements in the embankment and foundation.
- Settlement of the embankment adjacent to spillway walls or other structures.
- Around conduits or culverts resulting from poor compaction of fill around the conduit.
- Collapse settlement of poorly compacted soil in the embankment.
- Desiccation and freezing.
- Hydraulic fracture resulting from cross valley arching and arching of the core onto the shoulders of the embankment.
- Hydraulic fracture over small-scale irregularities in the foundation profile.

Concentrated leak erosion may also occur in cracks in foundation soil and weathered rock which has soil-like properties.

Whether erosion will initiate in a crack in an embankment or its foundation depends on:

- Whether the crack or gap persists from the reservoir side to downstream.
- Whether the water flowing through the crack will cause the soil on the sides of the crack to swell, closing the crack or reducing the width so the stresses imposed on the sides of the crack by the water flowing through the crack are insufficient for erosion to progress.
- Given there is a crack, whether the hydraulic stresses imposed on the sides of the crack by the water flowing through the crack are sufficient to initiate erosion.
- Whether the flow velocity is sufficient to transport the detached particles through the crack (Figure 9.1).

DOI: 10.1201/9781003542025-9

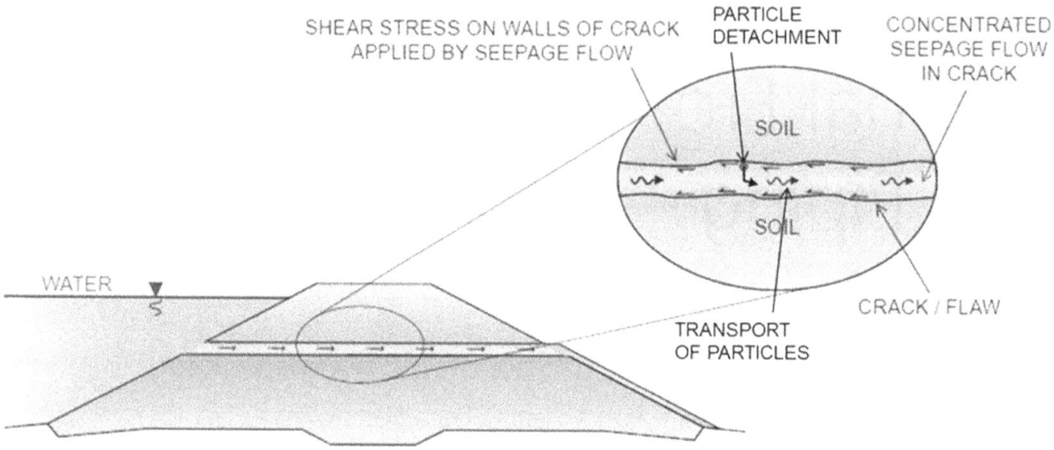

FIGURE 9.1 Detachment and transport of particles in concentrated leak erosion (Fell et al., 2015).

Whether in the absence of filters or where filters are too coarse, the erosion will progress to form a pipe depending on:

- Whether the soil will hold open the pipe ("will it hold a roof?").
- Whether upstream zones will limit the erosion process or crack filling stops the erosion process.

Also important is the rate of erosion and enlargement of the pipe because this influences whether the leak can be detected and intervention taken to lower the reservoir or stop the erosion process before breach occurs.

Hydraulic fracture can occur in the embankment or foundation when the water pressure ($\gamma_w.h$) at a particular location exceeds the sum of the minor principal total stress (σ_3) and the tensile strength (σ_t):

$$\gamma_w.h \geq \sigma_3 + \sigma_t \tag{9.1}$$

Figure 9.2 shows an example. An initial opening forms on the upstream side of the core due to a defect or crack. Water pressure in the initial water filled crack increases, for example due to increasing reservoir level. If the water pressure in the crack exceeds the sum of the minor principal stress and tensile capacity of the soil at the crack tip, then the crack is jacked open, and it extends further into the core. The process progresses from upstream to downstream.

Hydraulic fracture can also occur due to use of high pressures when drilling with water or air, or by pressure grouting.

There are several Potential Failure Modes and initiating mechanisms for concentrated leak erosion. These are listed in Table 9.1. This numbering system has been used throughout to facilitate use of the methods in Appendix A and for consistency with Fell et al. (2024).

The section which discusses the initiating mechanism is shown. Appendix A provides details of methods to quantify the likelihood of cracking or persistent gaps for those wanting to use quantified risk assessment approach to assessing dam safety. Those wanting to use deterministic or engineering assessment approach will find the methods in Appendix A methods useful as aids to judgement.

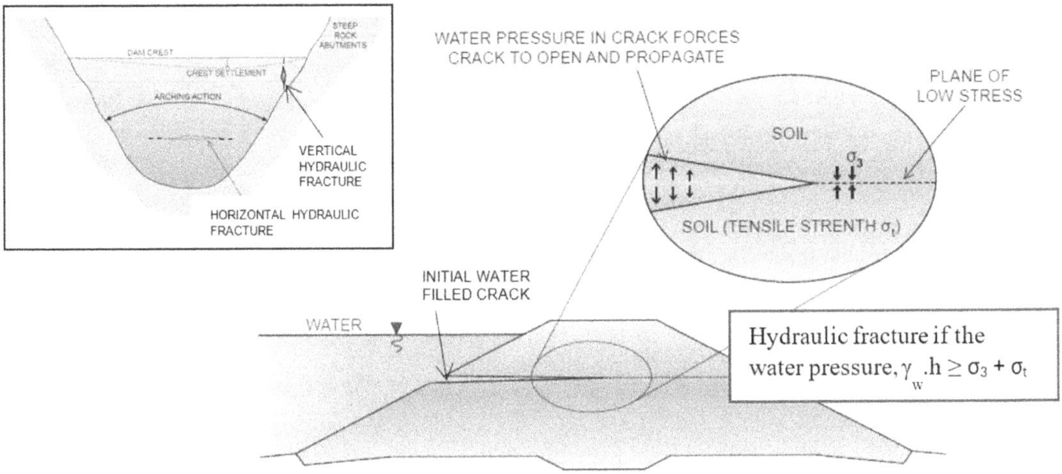

FIGURE 9.2 Mechanism of formation of hydraulic fractures in embankment dams (Fell et al., 2015).

TABLE 9.1 Internal Erosion and Piping. Concentrated Leak Erosion Initiating Mechanisms

INITIATING MECHANISM	*SECTION*
IM 1 Transverse cracking in the upper embankment resulting from cross valley differential settlement	9.2.1
IM 2 Transverse crack or gap adjacent to a cliff at the top of the embankment formed by differential settlement	9.2.9
IM 3 Transverse hydraulic fracture or cracking resulting from cross valley arching	9.2.2
IM 4 Transverse crack resulting from cross section differential settlement due to poorly compacted shoulders	9.2.6
IM 5 Transverse cracking in the upper or lower embankment resulting from differential settlements in soil in the foundation under the core	9.2.3
IM 6 Transverse cracking resulting from differential settlements due to embankment staging	9.2.4
IM 7 Desiccation cracking in the crest of the dam	9.8.1
IM 8 Desiccation cracking on seasonal shutdown layers and on the surface of staged embankments	9.8.1
IM 9 Transverse cracking in the lower/middle embankment resulting from cross valley differential settlements	Covered in IM 1
IM 10 Hydraulic fracture resulting from arching of the core onto the shoulders of the embankment	9.2.7
IM 11 Transverse cracking in the lower embankment resulting from differential settlement in soil in the foundations under the core	Covered in IM 5
IM 12 Transverse cracking or hydraulic fracture at the foundation contact due to small-scale irregularities in the foundation profile under the core	9.2.5
IM13 Transverse cracking in the upper part of the embankment caused by earthquake	9.9
IM 14 Transverse crack or continuous gap resulting from poorly compacted zone within the core	9.5

(Continued)

Table 9.1 (Continued)

INITIATING MECHANISM	*SECTION*
IM 15 Transverse cracking or continuous flaw resulting from poorly compacted layer on the core–foundation contact	9.5.5
IM 16 Transverse cracking or continuous flaw resulting from freezing at the crest of the embankment	9.8.2
IM 17 Transverse cracking or continuous flaw resulting from freezing on seasonal shutdown layers and on the surface of staged embankments	9.8.2
IM 18 Continuous crack or gap associated with a conduit through the embankment.	9.6.1
IM 19A Erosion into a (non-pressurised) conduit, and	9.7
IM 19B Leakage out of a (non-pressurised) conduit leading to erosion along the conduit	9.7
IM 20 Transverse crack or gap resulting from a poorly compacted zone associated with a spillway or abutment wall	9.6.2
IM 21 Transverse crack or gap adjacent to a spillway or abutment wall resulting from differential settlement adjacent the wall due to wall profile or foundation profile adjacent the wall	9.2.8
IM 22 Transverse crack or gap adjacent to a spillway or abutment wall resulting from differential settlement adjacent the wall	9.2.8
IM 23 Persistent crack or gap at the interface of wrap-around connecting embankment to concrete gravity dam	9.2.10
IM 26 Persistent crack in plastic soil in the foundation	9.10

9.2 TRANSVERSE CRACKING IN THE EMBANKMENT RESULTING FROM DIFFERENTIAL SETTLEMENTS

9.2.1 IM 1 Cracking Resulting from Cross Valley Differential Settlement

As an embankment dam is constructed, the partially saturated compacted soil in the embankment consolidates and settlement occurs. Where the valley sides are steep and/or have steps in the profile, such as shown in Figure 9.3, differential settlements occur due to the variations in the height of the embankment, and these can lead to tensile or low stress zones in which cracks may form.

It should be recognised that these stresses are inevitable even in well-designed and constructed dams and are not caused by poor design or construction practice.

Figures 9.4 and 9.5 show transverse cracking which occurred over a bench in the foundation of Buffalo Dam, Australia. When investigated in 2001, 35 years after construction, the crack was open at the top of the core (about 40 mm wide) and then transitioned with depth into a softened zone containing cracks. Probing indicated this crack probably extended several metres to below reservoir Full Supply Level. The dam had post-construction settlement of 1%. Longitudinal cracking was also observed that was the result of lack of support for the core from the dumped rockfill shoulders. The cracking of Buffalo Dam probably occurred soon after construction but was not observed until years later when the road pavement was removed for site investigations.

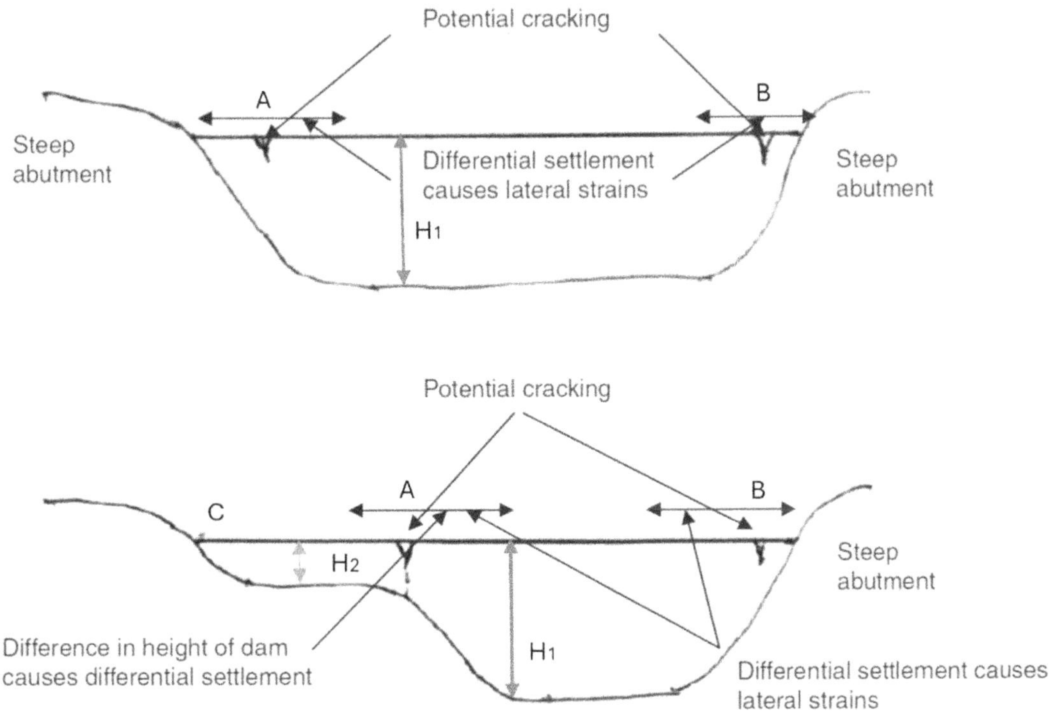

FIGURE 9.3 Situations in which cross valley differential settlement may lead to cracking or lateral strain and low stress zones subject to cracking (Fell et al., 2024).

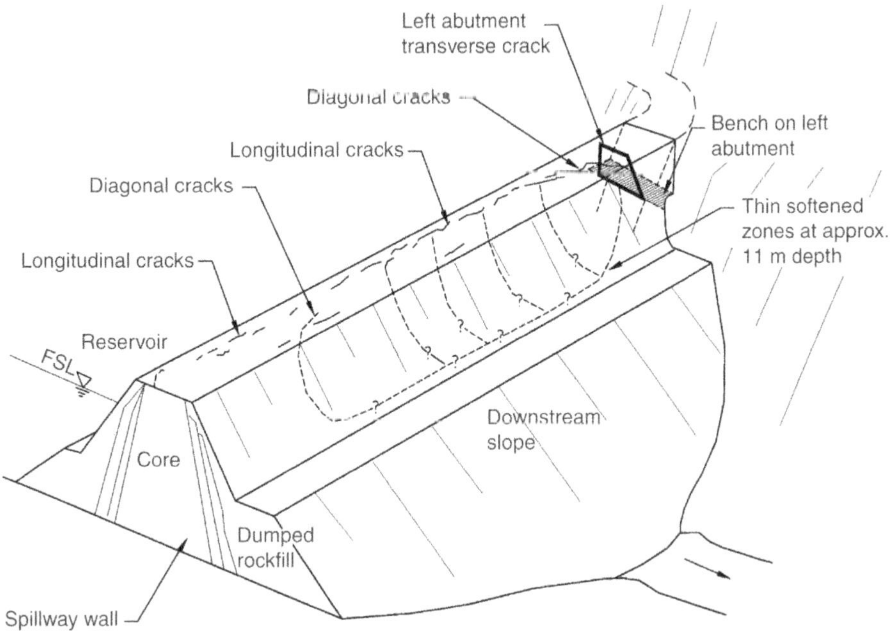

FIGURE 9.4 Sketch showing transverse and longitudinal cracking in Buffalo Dam (Newman and Foster, 2006).

(a) (b)

FIGURE 9.5 Buffalo Dam. (a) Bench in the abutment over which transverse cracking occurred. (b) Transverse crack. (Courtesy of Goulburn-Murray Water and AECOM.)

Figure 9.6 shows transverse cracking of Matahina Dam, New Zealand (Gillon, 2007). The cracking occurred over steps in the slope correction concrete. These resulted in erosion of the core and sinkhole development.

Figures 9.7 and 9.8 show cracking at Tullaroop Dam, Australia. As described in Hunter et al. (2007) and Hunter et al. (2017), a 60 mm wide diagonal crack (refer to Figures 9.7 and 9.8) developed in the downstream shoulder of the main embankment over the left abutment. The crack developed over a two week period in April 2004 (55 years after construction). The crack was greater than 2 m depth and extended from the crest to the toe, terminating at the downstream edge of the crest, but did not extend across the crest. Narrow transverse cracks (less than 2 mm wide) had developed years earlier on the crest over the left abutment.

Investigation and analysis were undertaken to understand the mechanics of the crack formation. Significant factors contributing to the crack formation were the presence of pre-existing shear planes in the foundation not removed as part of the embankment construction, and extensional strains resulting from the sharp change in foundation slope on the left abutment from excavation of the temporary diversion channel. The timing of the crack in 2004 was assessed as due to a combination of sustained drawdown and the extended drought period (resulting in drying profile in the downstream shoulder of the earthfill).

During construction of a filter berm on the downstream slope in 2015 (Hunter et al., 2017), additional fine transverse cracks were observed in the location shown in Figure 9.8. These were a result of extensional strains over the change in foundation slope on the mid-right abutment. This is discussed further in Section 19.6.1.

As shown by the performance of Tullaroop Dam, cracking may occur many years after construction of the dam as the built-in stresses are relieved; in that case by record low reservoir levels and prolonged drought.

Based on the cracking observed at Buffalo Dam (Newman and Foster, 2006) and Matahina Dam (Gillon, 2007), it is considered wise to assume cracking or low stress zones conducive to hydraulic fracture may persist down to the step in the foundation profile.

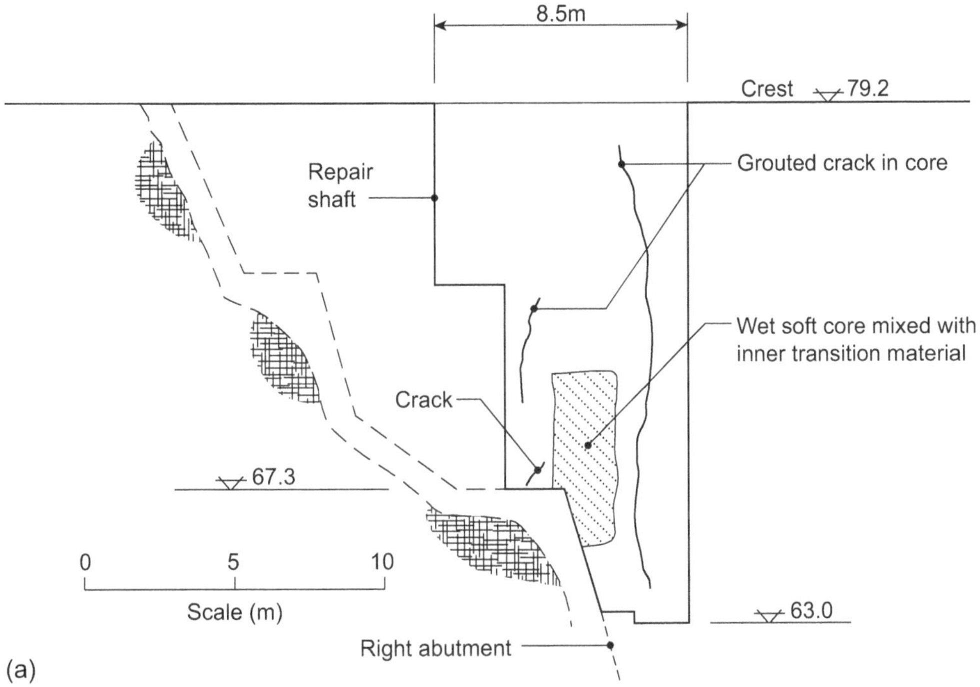

(a)

(b)

FIGURE 9.6 Cracking of the core on Matahina Dam (Gillon, 2007).
(a) Section of core over slope correction concrete showing cracking and softened zones.
(b) Photograph of abutment showing slope correction concrete.

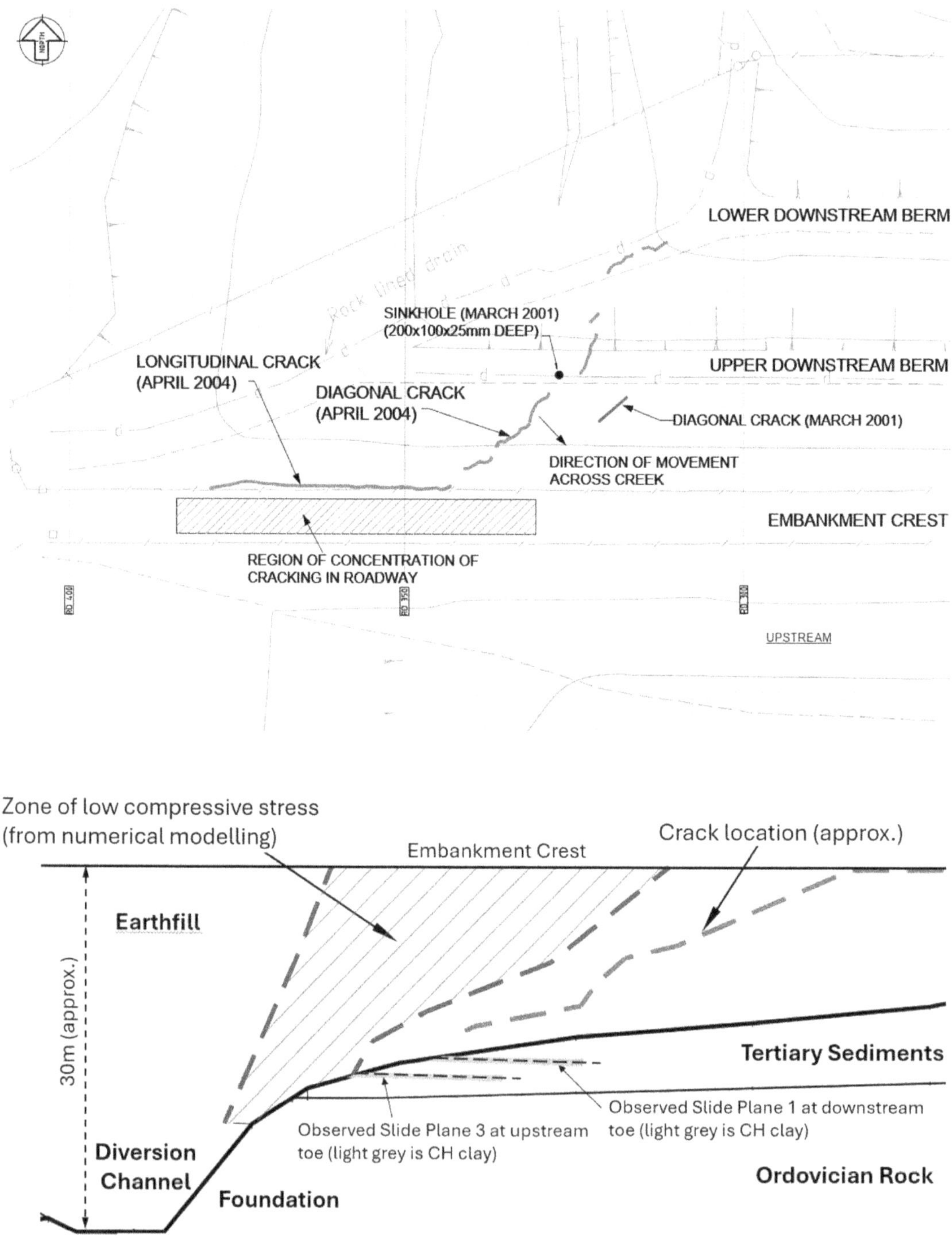

FIGURE 9.7 Tullaroop Dam. Crack locations on left abutment, main embankment (Hunter et al., 2007) and section of the left abutment showing the location of shear planes within the foundation.

Bui et al. (2004, 2005) modelled sharp breaks in slope and smoothed curved shape and showed that the shapes of steps in the abutment profile, i.e. rounded or sharp break in slope, is a secondary factor to the overall geometry.

For most dams, about 80% to 90% of the total settlement occurs during construction (Hunter, 2003; Hunter and Fell, 2003c), and as shown by Bui et al. (2004, 2005), the stresses set up in the construction phase influence the likelihood of low stress zones and cracking.

He (2016), He et al. (2017, 2019, 2021) carried out extensive numerical analyses of stresses and strains which develop in embankment dams resulting from differential settlements during and post construction. This has shown that post-construction settlements rather than those during construction dominate the horizontal tensile stresses and strains and the potential for developing cracking.

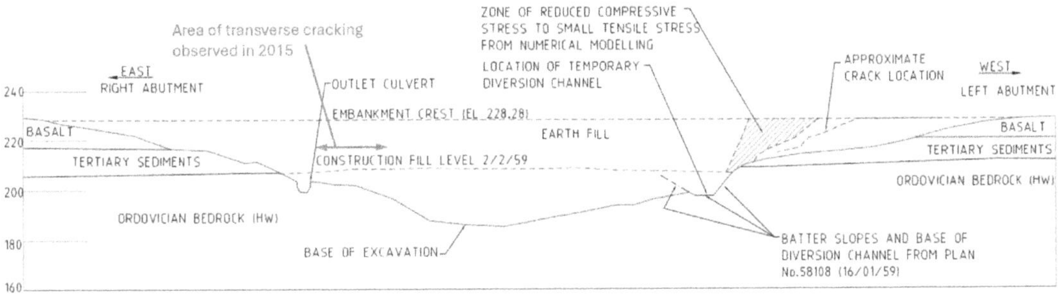

FIGURE 9.8 Tullaroop Dam. Longitudinal section of main embankment along the dam axis (Hunter et al., 2017) showing the location of the diagonal crack in 2004 over the left abutment and the area of transverse cracking observed in 2015.

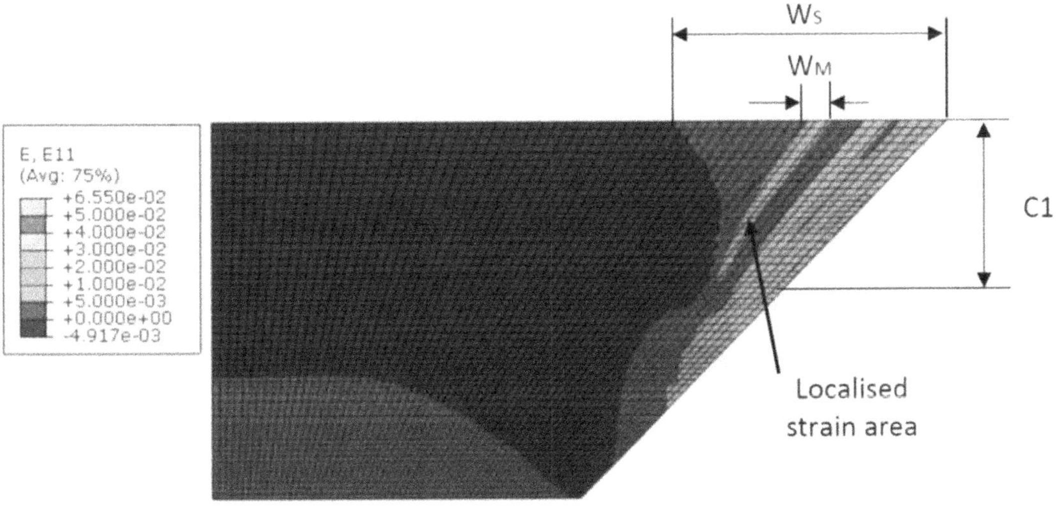

FIGURE 9.9 Elasto-plastic analyses—valley profile without a step showing localisation of horizontal strains. WS is the distance from the abutment to zero tensile strain. WM is the width of the localised strain zone with greatest strains. C1 is the height from crest to 2% strain boundary in the localised strain area with largest strains (He et al., 2017).

Figure 9.9 shows the localisation of horizontal strains which occur near the abutments of dams with steep abutment slopes resulting from post-construction settlements.

Figure 9.10 shows the localisation of horizontal strains which occur over a step in the foundation profile resulting from post-construction settlements.

Based on the analyses by He et al. (2017), Fell et al. (2024) developed Figures 9.11 and 9.12 to relate localised tensile strains to post-construction crest settlement and the slope of the abutment. Figure 9.13 defines the slope geometry terms. Figure 9.12 is for valley profiles with steps. Figure 9.11 is for valley profiles without a step. Field data from crest survey markers reported in Purss (2022) and Purss and Hunter (2022) confirm strain concentration does occur.

These figures can be used to estimate the localised tensile strain which can be compared to the strains likely to result in cracking as described in Section 9.3.

The methods require as input the maximum total post-construction settlement of the embankment crest expressed as a percentage of the maximum height of the dam above the core foundation. Figure 9.3 shows the height (H1) to be used for assessing potential cracking at locations A and B. For potential cracking at a location such as C which is sufficiently far from the highest part of the dam, the height is H2 and the settlement is the settlement at that location.

The crest settlement can be determined by the following:

(a) *Directly from monitoring data.* If crest settlements have been measured from the end of construction to the present or close to the present, these data can be used directly.

(b) *Estimation from incomplete monitoring data.*
For many dams, monitoring of crest settlement only began a number of years after construction so the data is incomplete. Fell et al. (2024) describe how to use this data to estimate total post-construction settlement.

(c) *Estimation from surveyed crest levels relative to the design or as constructed levels.*
The amount of crest settlement can be estimated by comparing the current levels with design or as constructed crest levels.

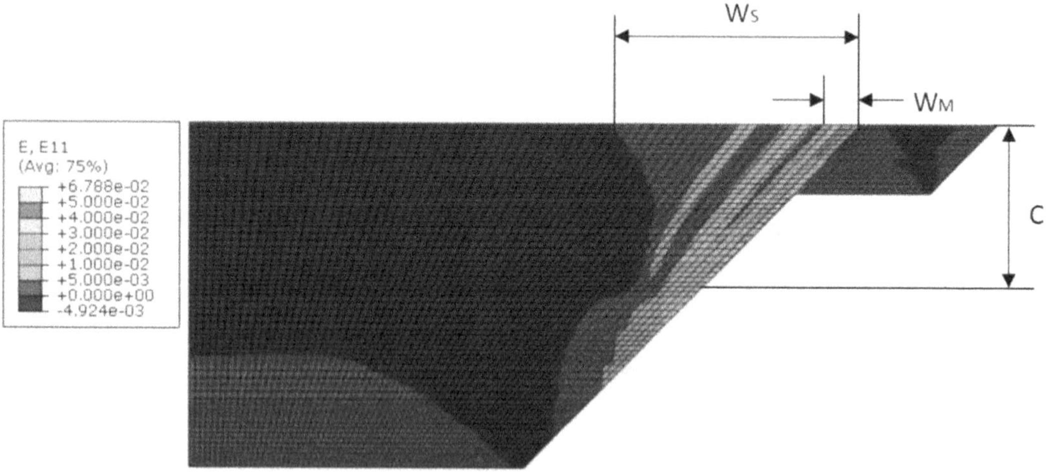

FIGURE 9.10 Elasto-plastic analyses–valley profile with a step showing localisation of horizontal strains. WS is the distance from the abutment to zero tensile strain. WM is the width of the localised strain zone with greatest strains. C1 is the height from crest to 2% strain boundary in the localised strain area with largest strains (He et al., 2017).

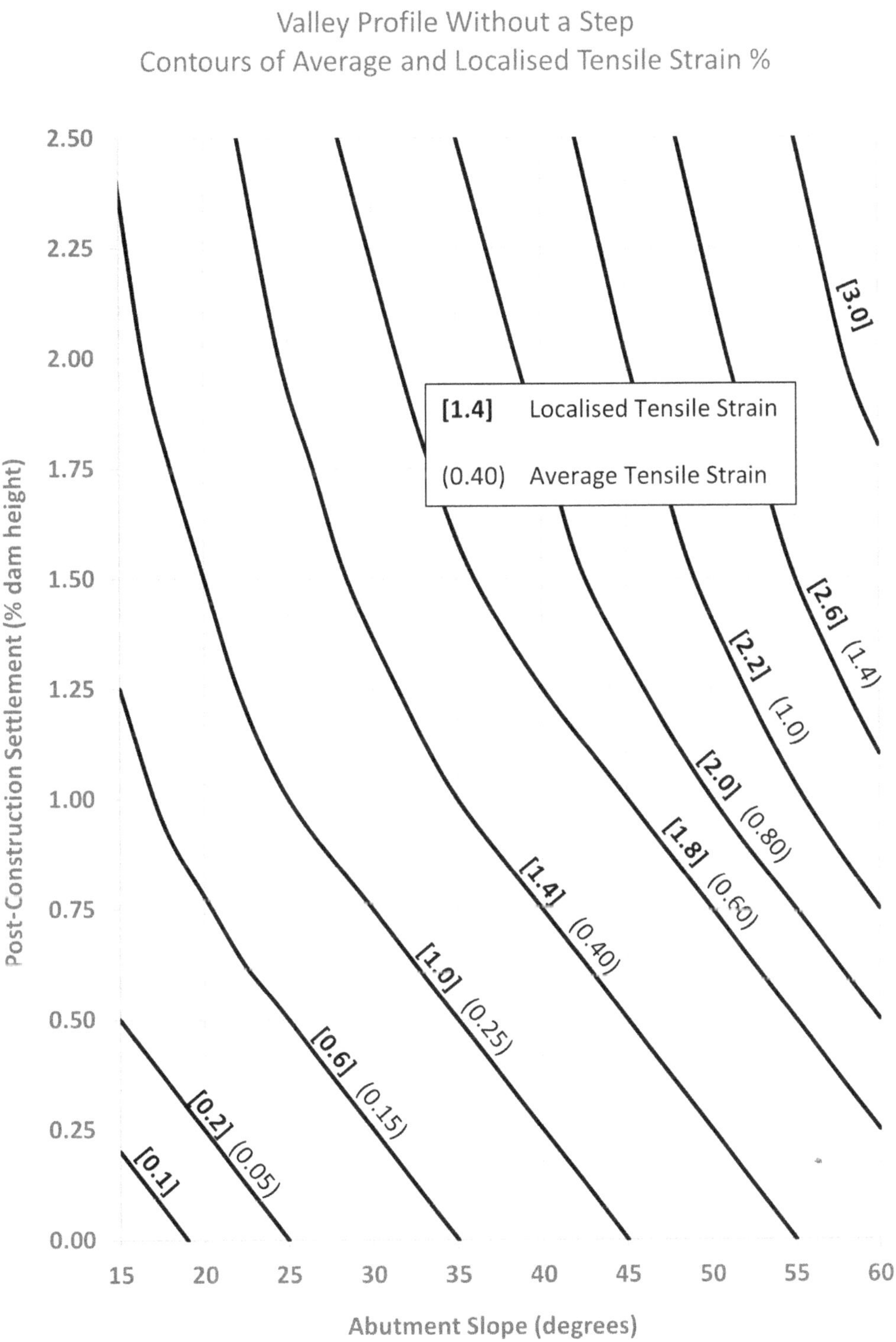

FIGURE 9.11 Embankment with no step in the abutment. Post-construction settlement (% of embankment height) versus abutment slope in degrees, with contours of average and localised % strain in the tensile zone at the crest of the embankment (Fell et al., 2024).

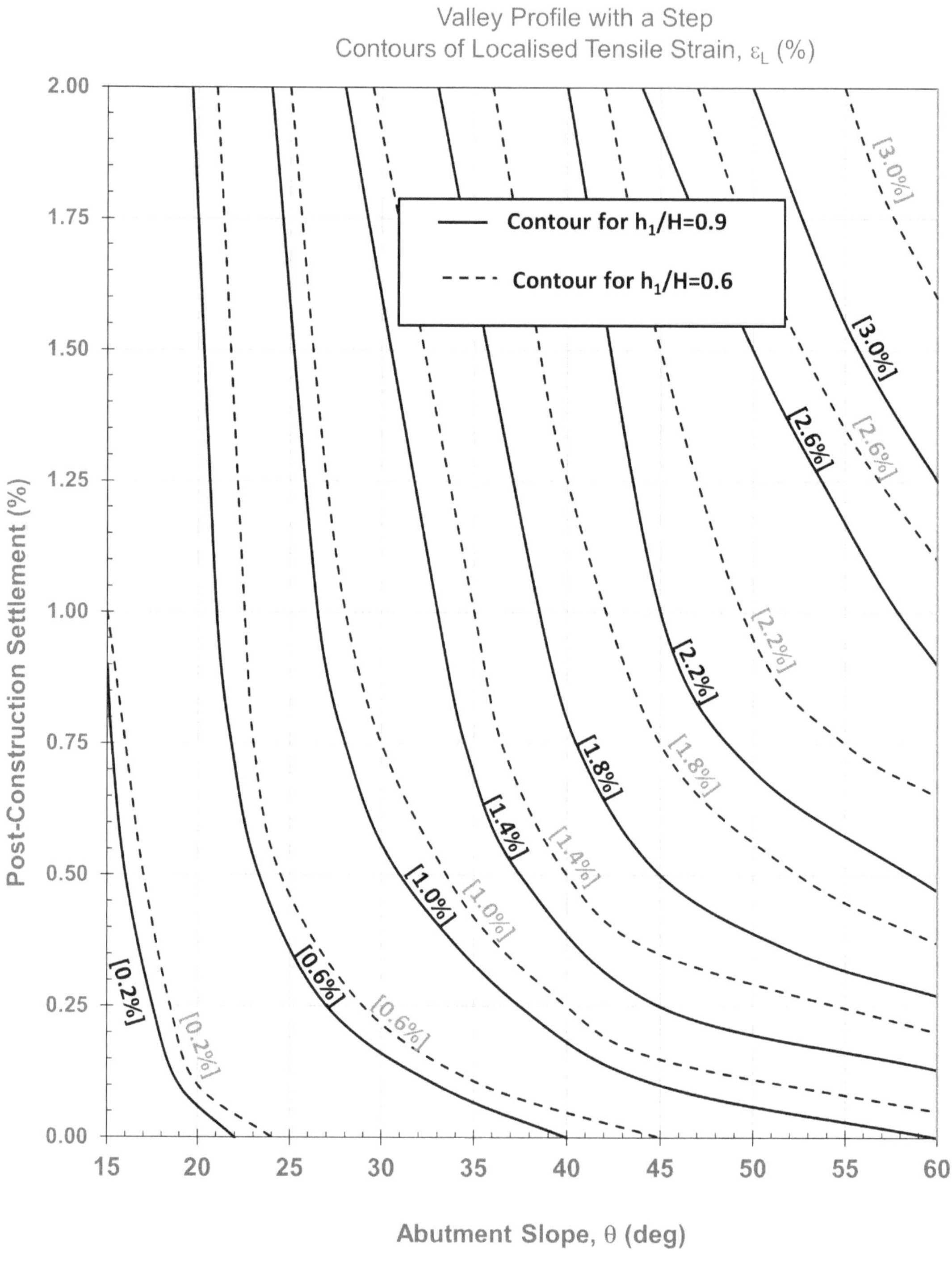

FIGURE 9.12 Valley profile with a step. Contours of localised strain relative to abutment slope (degrees), post-construction settlement (%) and position of the step (h_1/H) (Fell et al., 2024).

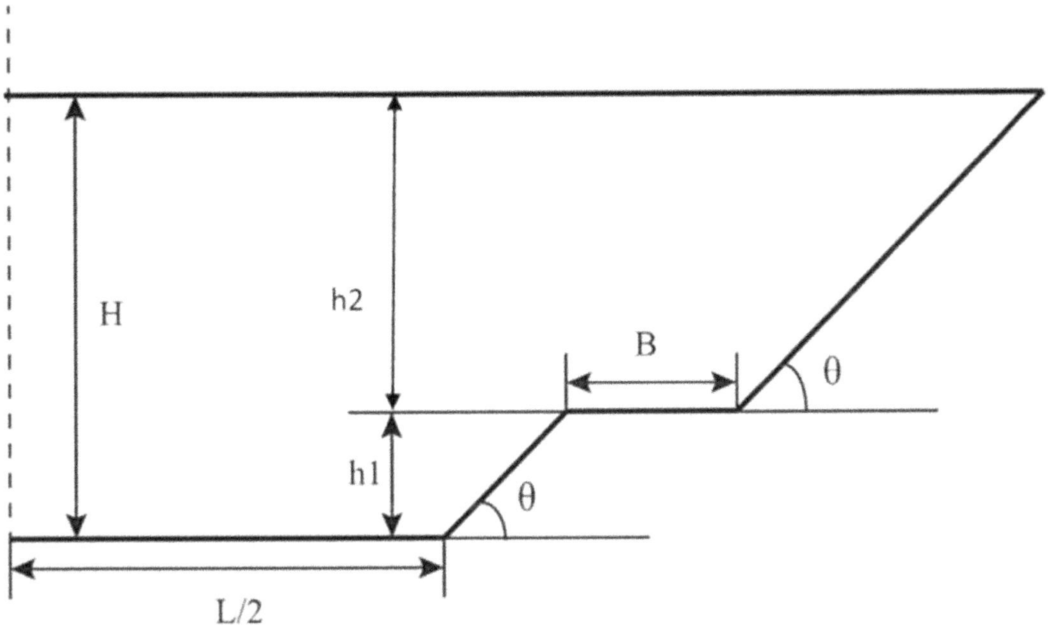

FIGURE 9.13 Cross valley section model showing definition of terms (Fell et al., 2024).

However, there are several potential pitfalls in doing this:

- The dam will often have been constructed with camber to allow for post-construction settlement, so this must be accounted for.
- The upper part of embankments constructed in the 1960s and 1970s was often not built as shown on the design drawings.
- As constructed drawings have often been found to be inaccurate.
- Reprofiling of the embankment crest may have been done sometime after construction.

Experience has shown that the level of the top of the core relative to design level is a more reliable guide to the amount of settlement, probably because those constructing the dam were more careful in regard to the level of the top of the core and the road may have been reprofiled. This sometimes requires test pits in the crest to determine the level of the top of the core.

(d) *Estimation from settlement data for similar dams.*
Hunter and Fell (2003c) present information on the deformation behaviour of embankment dams. This can be used to provide a guide to the settlement a dam may have experienced post construction. Figures 7.39, 7.40 and 7.41, and Tables 7.5 and 7.6 summarise the data.

Purss (2021) compared strains measured in extensometers in four dams with measuring points spaced at 3 m, with strains measured with crest survey marks, standardised to 50 m spacing. He also had data from four dams with crest survey marks spaced at closer intervals. The dams with extensometer data had relatively uniform slopes, whereas the others had steps. These data are shown in Figure 9.14.

The data for the dams with uniform slopes show ratios of localised strain to strain measured over 50 m up to 2.5, and those for stepped foundation profiles up to about 4.

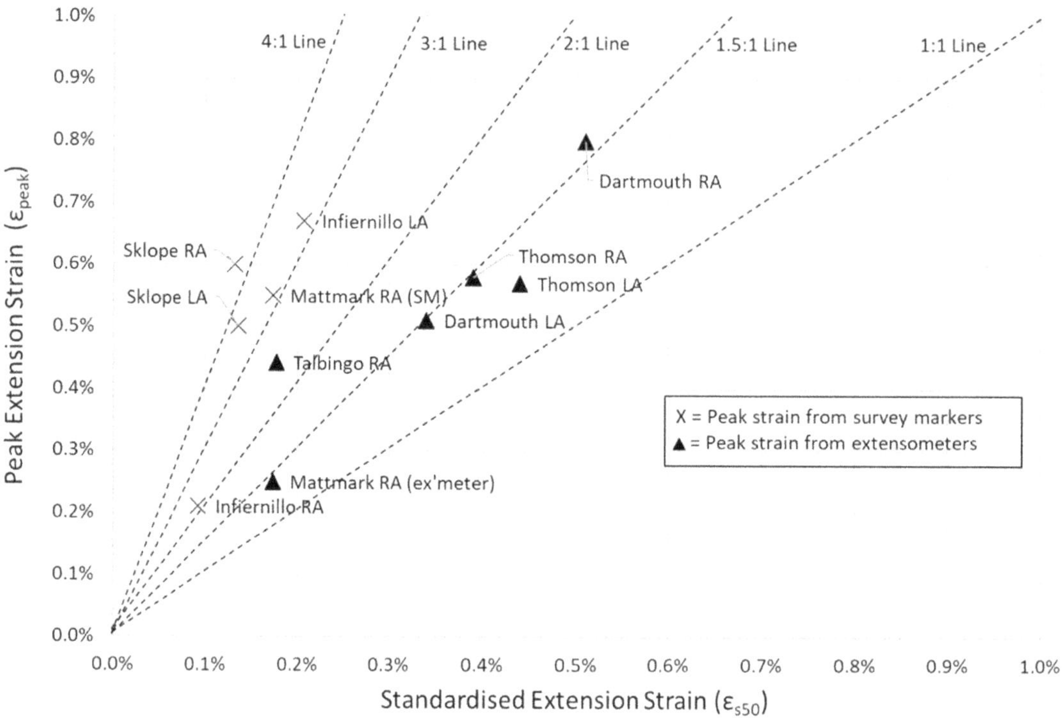

FIGURE 9.14 Comparison of "standardised extension strains" measured over 50 m to "peak extension strains" (Purss, 2021; Purss and Hunter, 2022).

It should be noted that the ratios are related to the widths of the strain values. For Figure 9.11 the localised strains from numerical analyses are mostly measured over 1 to 5 m and the average strains measured over 40 to 60 m. This results in the ratio between localised and average strains varying from 1.85 to 4, greater than the data from Purss (2021) where localised strains are averaged over greater width.

The contours in Figure 9.11 are developed from the numerical analysis average strains, so the ratios of localised to average strain are also based on the numerical analyses.

Purss (2021) prepared Figure 9.15 from extensional strains and post-construction settlements from survey monitoring of a number of dams. This supports the assumption in Figures 9.11 and 9.12 that the extensional strains are larger for steep abutment slopes than for flatter slopes.

9.2.2 IM 3 Cracking and Hydraulic Fracture Resulting from Cross Valley Arching

If the valley in which the dam is constructed is narrow and steep, cross valley arching can occur and the vertical stresses are shed onto the sides of the valley. This can lead to a situation where hydraulic fracture can occur. Fell et al. (2008, 2024) suggest that cross valley arching is most likely to occur if the width of the valley base is less than a quarter of the dam height and the valley sides are steeper than about 60 degrees. It is unlikely to be an issue if the width of the valley base is greater than three quarters the dam height and the valley sides are flatter than 45 degrees.

9.2.2.1 Mud Mountain Dam in Washington State, USA

Information on the dam is taken from Graybeal (1988), Graybeal and Levallois (1991), Eckerlin (1992, 1993) and USCOLD (1988).

The dam, which is 130 m high, was constructed in 1941. Figure 9.16 shows a cross section of the dam, and Figure 9.17 a section across the valley. The core is glacial till consisting of sand, gravel and

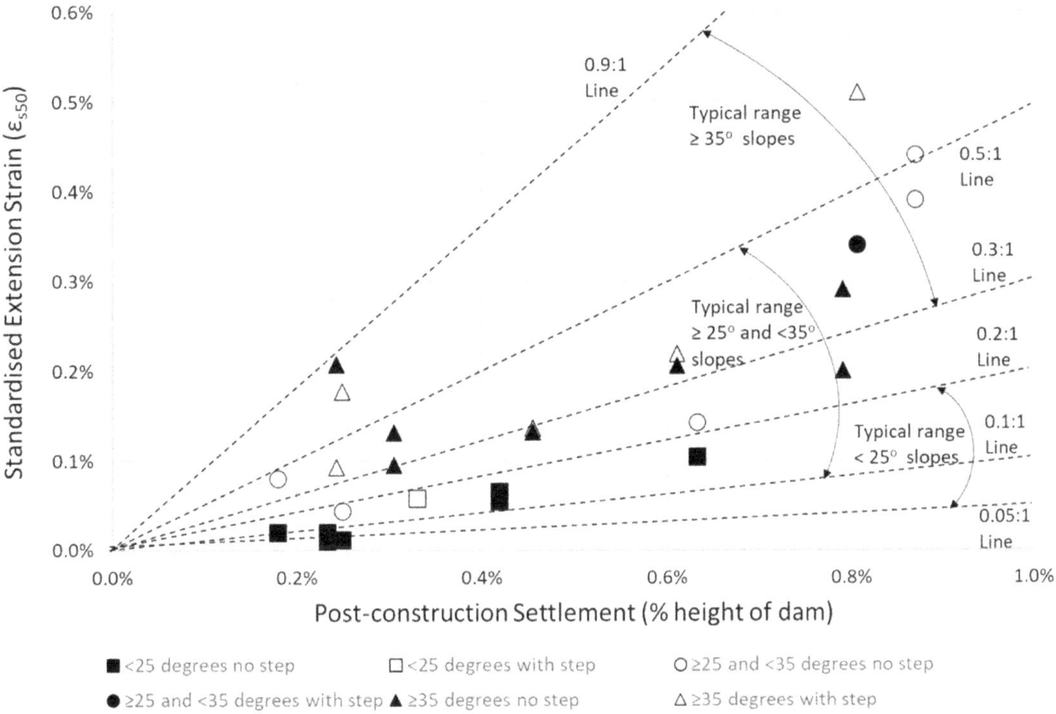

FIGURE 9.15 Standardised extension strains versus post-construction settlement (Purss, 2021; Purss and Hunter, 2022).

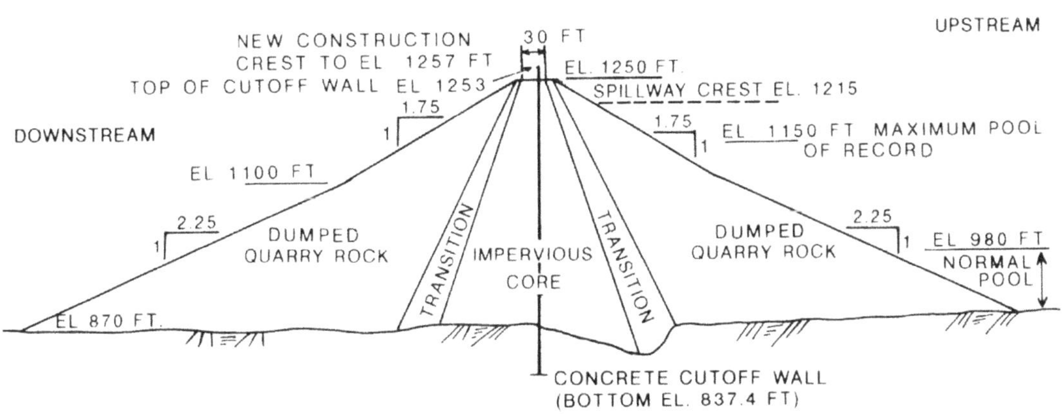

FIGURE 9.16 Mud Mountain Dam cross section (Eckerlin, 1992). (Used with permission of AEG.)

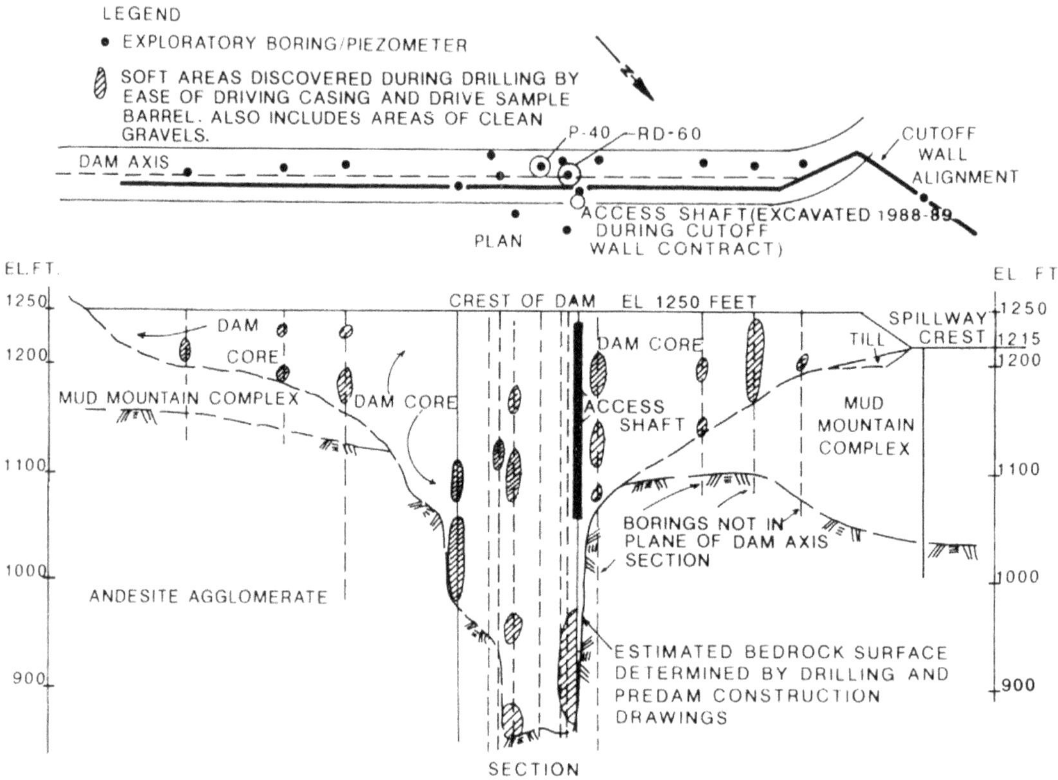

FIGURE 9.17 Mud Mountain Dam valley section showing loose, soft and low density zones (Eckerlin, 1992). (Used with permission of AEG.)

silt. It was compacted in 150 mm lifts by sheepsfoot rollers. The transition zone is crushed rock placed in lifts and tracked by bulldozer. The rockfill was dumped and sluiced. The foundation is a low compressibility rock.

The maximum reservoir level reached in 1988 was RL 350 m (RL 1148 feet), 31 m below crest level. The reservoir normally fluctuated below RL 300 m (below RL 985 feet).

Monitoring of pore pressures showed the core was deteriorating with time. Investigations carried out in 1985 showed extensive loose or soft zones, low density zones and zones with little or no fines. These zones and where drill mud was lost are shown in Figure 9.17.

It was concluded that (Graybeal, 1988) settlement and arching of the dam core within the deep narrow gorge had produced loose zones, voids and/or cracks in the upper part of the core. Suffusion had removed fines from the coarser fraction of the core.

A concrete cutoff wall was constructed in 1988–1989. Details are given in Eckerlin (1992) and Graybeal and Levallois (1991).

Figure 9.18 shows the results of cross valley numerical modelling by Bui et al. (2005) of the dam.

The modelling estimates minor principal stresses and vertical stresses much lower than the stresses calculated from the depth of fill, and lower than the pore pressures which develop as the reservoir fills, so hydraulic fracture is predicted, which is consistent with the field behaviour.

There are no gradations of the core available to the authors, so we are not able to ascertain whether the erosion mechanism was a concentrated leak in hydraulic fractures or whether suffusion and/or Global Backward Erosion was involved. It seems likely it may have been a combination given that the erosion of the core took place over a number of years.

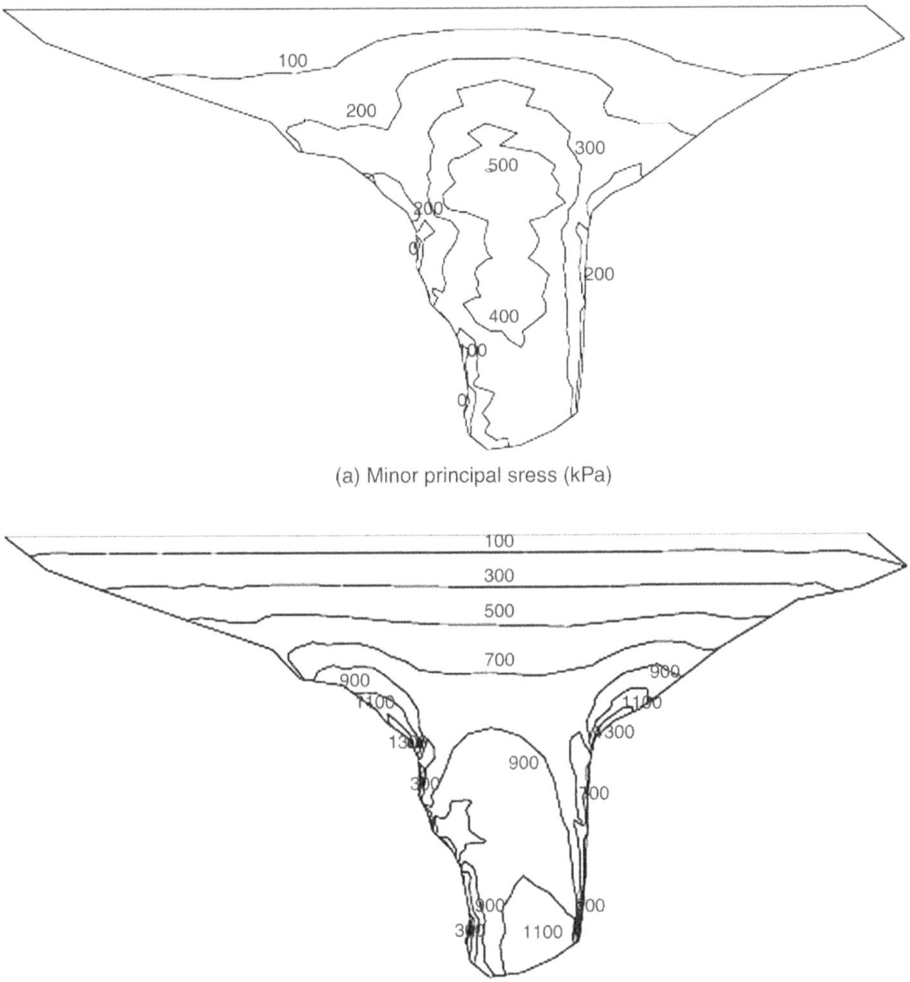

(a) Minor principal sress (kPa)

(b) Vertical stress (kPa)

FIGURE 9.18 Mud Mountain Dam—minor principal stress and vertical stress distributions (kPa) (Bui et al., 2005).

9.2.3 IM 5 Cracking in the Upper and Lower Embankment Resulting from Differential Settlements in the Soil in the Foundation under the Core

Figure 9.19 shows foundation conditions which are likely to lead to differential settlement and cracking or formation of low stress zones conducive to hydraulic fracture.

He et al. (2017) carried out numerical analyses to model these conditions. Figure 9.20 is an example of the results showing strain localisation. Fell et al. (2024) used the He et al. (2017) analyses to develop Figure 9.21 which relates post-construction settlement to the slope of the soil rock interface. Use Table 9.2 to adjust total settlements for the settlement during construction.

These can be used to estimate the localised tensile strain which can be compared to the strains likely to result in cracking described in Section 9.3.

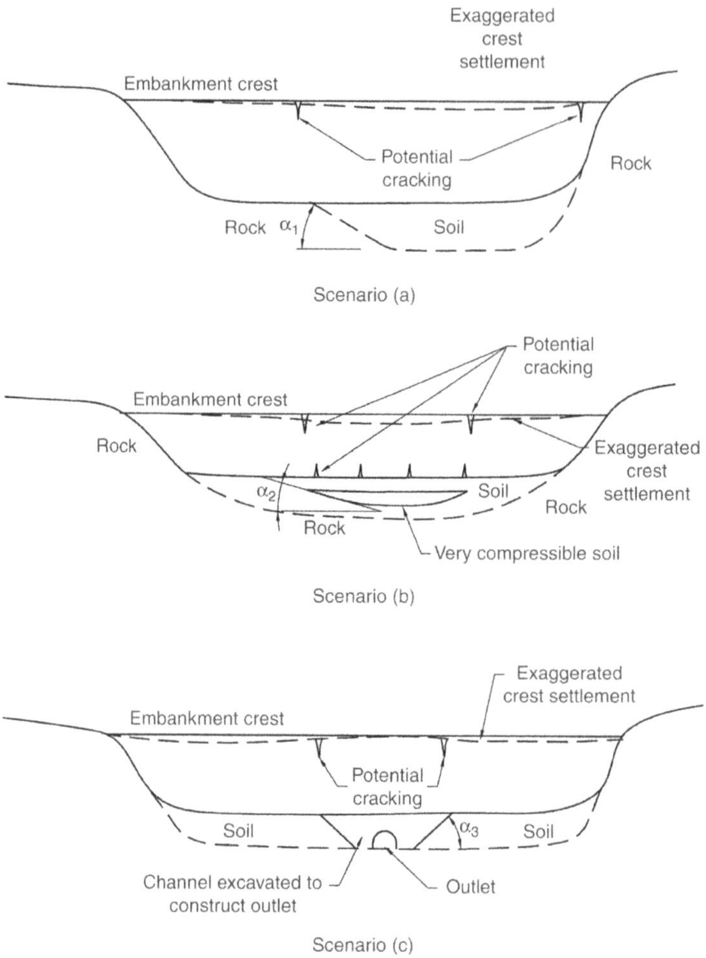

FIGURE 9.19 Situations which may cause differential settlement in the foundation of dams leading to cracking, lateral strains and low stress zones subject to hydraulic fracture (Fell et al., 2008, 2024).

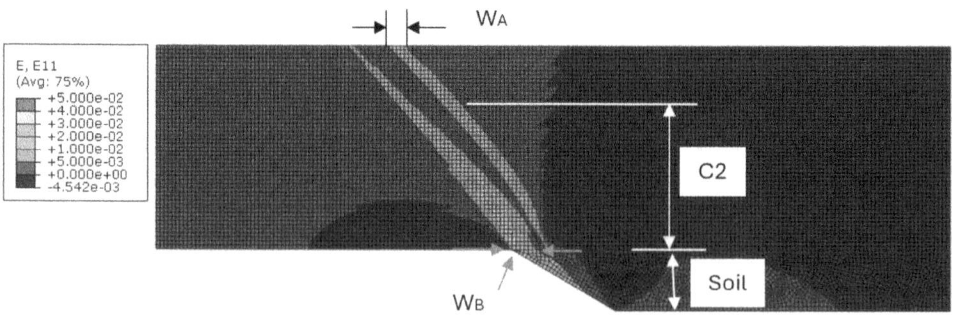

FIGURE 9.20 Example of results of elasto-plastic analyses for embankment founded partly on a soil foundation—horizontal strains during construction, valley profile with a step showing dimensions. Notes: (a) Tensile strains widths W_A and W_B measured between zero strain contours. (b) If tensile strains persist to the crest W_A is to be measured. (c) C2 is the vertical height of the 2% tensile strain above the step unless noted in the table Fell et al., 2024, from He et al. (2017).

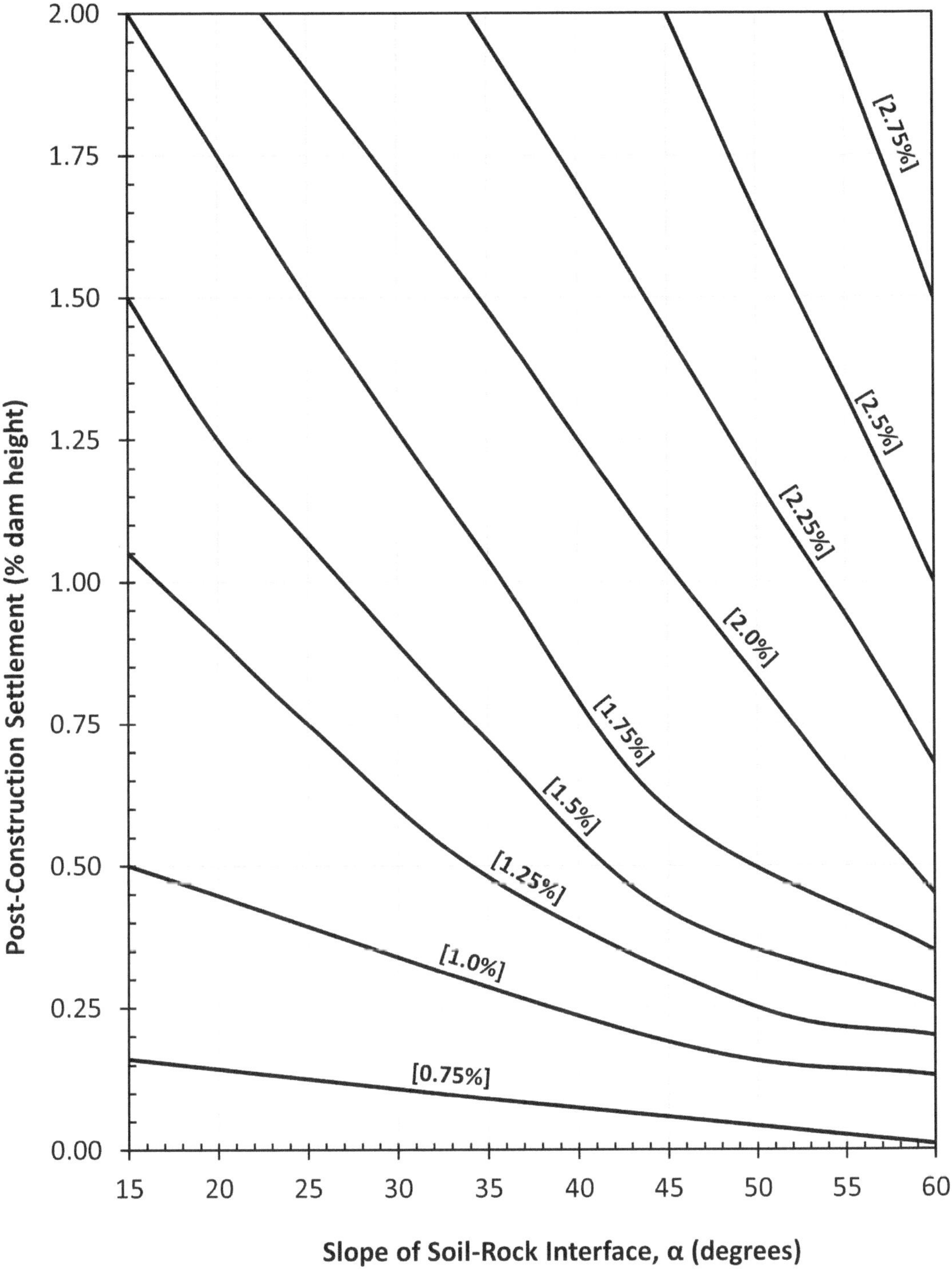

FIGURE 9.21 Embankment with soil in part of the foundation. Post-construction settlement in the soil (% of embankment height) versus slope of the soil/rock interface (α) in degrees, with contours of localised % strain in the tensile zone at the crest of the embankment (Fell et al., 2024).

9.2.4 IM 6 Likelihood of Transverse Cracking Due to Differential Settlements Resulting from Embankment Staging

This situation is shown in Figure 9.22 and arises where the embankment has been staged during construction. There is a potential for differential settlement to occur if the "existing" (first stage) embankment is a significantly higher modulus than the remainder of the embankment. If this is the situation, use the method described in Section 9.2.1 for IM 1 to assess the likelihood a crack or hydraulic fracture will result. If, however, there is no or little difference in the modulus, the probability of cracking will be negligible. In most cases the latter will apply.

TABLE 9.2 Adjustment to Localised Strains from Figure 9.21 to Allow for Estimated Settlements in the Soil During Construction (Fell et al., 2024)

ESTIMATED SETTLEMENT DURING CONSTRUCTION AS A % OF EMBANKMENT HEIGHT	*SLOPE OF THE SOIL-ROCK INTERFACE α (DEGREES)*	*ADJUSTMENT TO FIGURE 9.20 LOCALISED STRAINS (% LOCALISED STRAIN)*
1	15	Minus 0.3%
3	15	0
5	15	Plus 0.3%
1	30	Minus 0.5%
3	30	0
5	30	Plus 0.5%
1	45	Minus 0.5%
3	45	0
5	45	Plus 0.5%
1	60	Minus 0.5%
3	60	0
5	60	Plus 0.5%

Note: See Figure 9.19 for definition of α.

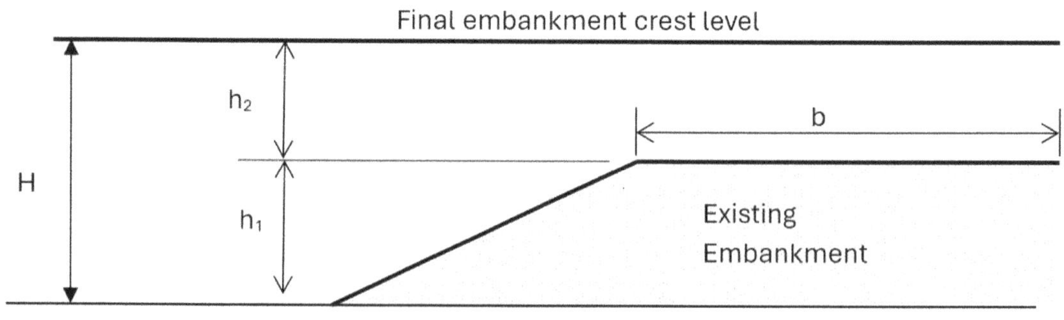

FIGURE 9.22 Longitudinal section through staged embankment (Fell et al., 2024).

9.2.5 IM 12 Cracking and Hydraulic Fracture at the Foundation Contact Due to Small-Scale Irregularities in the Foundation Profile under the Core

Small-scale irregularities in the foundation profile under the core may comprise steps, benches or depressions in the foundation rock, or irregularities formed in constructing diversion channels or haul roads. For cracking or hydraulic fracture to occur, the irregularities in the foundation profile must persist across the full width or nearly the full width of the embankment core.

These irregularities are usually relatively small compared to the embankment height (e.g. < 10% of the height at that location). Larger-scale features which may result in cracking near the dam crest are discussed in Section 9.2.1.

Figure 9.23 shows one of the mechanics of the process where low stresses are formed at the base of such steps.

Hydraulic fracture may also occur in low stress zones above irregularities in the foundation profile such as that shown in Figure 9.24.

There are two approaches to assessing the likelihood of transverse cracking or hydraulic fracture over such features:

Empirical method. This method is described in Appendix A and is calibrated against historic data of failures and accidents. It is approximate and reasonable to use for initial evaluation and screening but should not be relied upon for important decisions.

Method based on numerical analyses and hydraulic fracture mechanics. This method should be used where conditions are identified as potentially conducive to formation of low stress zones and where important decisions are being made relating to this Potential Failure Mode.

For both methods the important first steps are as follows:

(a) Determine the profile of the foundation along the centreline of the core by assembling contours of the stripped core foundation during construction, other survey such as along the grout curtain, and photographs of the foundation of the core of the dam upon which the core

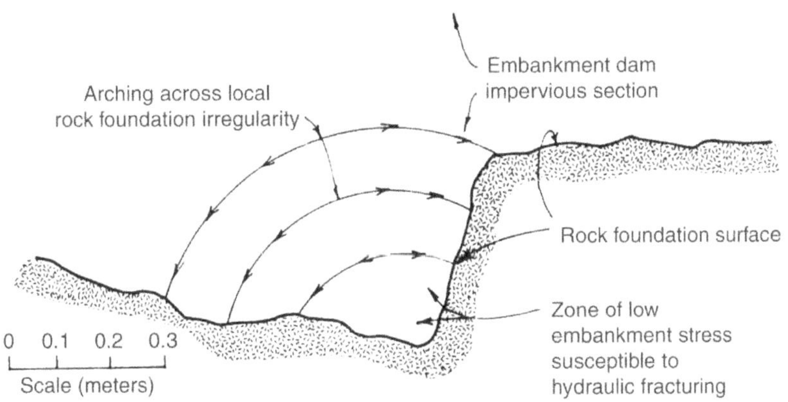

FIGURE 9.23 Zones of low embankment stress caused by localised arching adjacent to small-scale irregularities in the foundation rock allowing hydraulic fracture (Sherard, 1986). (Used with permission of ASCE conveyed through Copyright Clearance Center Inc.)

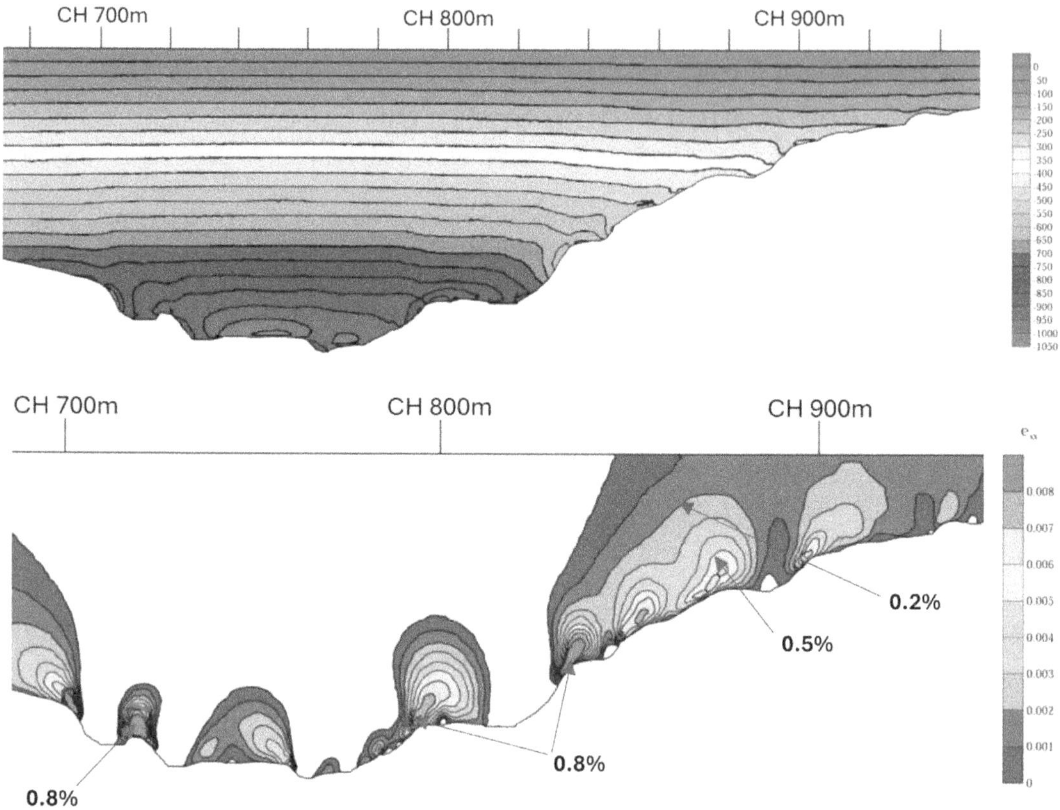

FIGURE 9.24 Example of low horizontal stress and extensional strains (%) in the dam over irregularities in the foundation profile (AECOM, 2023). (Permission to reproduce provided by AECOM.)

was constructed. Allow for slope correction treatment where steep steps in the foundation have been reduced in slope by trimming or adding concrete.

(b) From these select foundation irregularities which persist across the full width or close to the full width of the core and their geometry is conducive to developing low stress zones.

For the numerical method, the approach to assessing the likelihood of hydraulic fracture in such low stress zones is as follows:

(c) Carry out cross valley 2D elasto-plastic FEA of the dam and its foundation. Model the conditions at the end of construction and to the present allowing for post-construction settlement in the embankment, and for dams constructed on a compressible foundation, settlements in the foundation during and post construction. For irregularities that change in shape across the foundation below the core, several sections may need to be analysed to evaluate the varying stress conditions.
(d) From these analyses estimate the horizontal stresses over each low stress zone. For FEA using total unit weights, the resulting stresses will be the minor principal total stresses.
(e) Estimate the reservoir level which could result in the hydraulic pressure causing hydraulic fracture. For this use Equation 9.2 which includes allowance for the tensile strength of the soil.

(f) Compare this reservoir level with the reservoir stage under consideration and estimate the likelihood of hydraulic fracture occurring, taking into account the persistence of the irregularity across the foundation below the core, the factors of safety against hydraulic fracture and the performance of the dam as shown by seepage measurements.

The equation for estimating the critical hydraulic fracture pressure (U_F) given the minor principal total stress (σ_{PT}) is:

$$U_F = m\,\sigma_{PT} + n \qquad (9.2)$$

where m is the ratio of the fracture pressure to confining pressure and n is the apparent tensile strength obtained from laboratory fracture tests. Hydraulic fracture tests require specialised equipment which are generally not available. Therefore, in the absence of such tests, data is provided in Fell et al. (2024) Supporting Document which may be used to estimate the "m" parameter, taking account of the classification of the soil in the core, the compaction of the core and degree of saturation. It is usually between 1.0 and 1.15. Section 9.3 provides data on the tensile strength of soils.

9.2.6 IM 4 Cracking Resulting from Cross Section Differential Settlement Due to Poorly Compacted Shoulders

This mechanism is best described by the Eppalock Dam and Upper Yarra Dam case data. This is summarised as follows. For more information refer to Fell et al. (2024) Supporting Document.

9.2.6.1 Eppalock Dam

Eppalock Dam is a 47 m high central core earth and rockfill dam. It was constructed in 1962. A shear developed through the upper part of the core of Eppalock Dam because of reduced support for the wedge of core shown in Figure 9.25 resulting from collapse settlement of the upstream dumped rockfill. The wedge of core moved upstream as determined from crest survey vectors, localised strains in the internal settlement gauge and approximate location of shearing estimated from CPTU. Figure 9.26 shows further displacement which occurred during construction of upgrade works. Davidson et al. (2000) describe the dam and discuss the mechanism involved.

The shear potentially forms a through-going defect in the core from upstream to downstream in which concentrated leak erosion may initiate.

The following summary is taken from Hunter and Fell (2003c).

A number of other notable findings, detailed by Woodward Clyde (1999), relating to the deformation behaviour at Eppalock dam are:

Longitudinal cracking on the crest of the embankment. Cracking was initially observed as early as 1973 and since then has persisted. In 1985 the crest was repaired on account of the cracking, but cracking was again observed in 1986. Test pitting in the late 1990s indicated the longitudinal cracks were open (5 to 10 mm width) to about 1.2 m depth with softening around the crack. Below 1.2 m the crack persisted as a softened zone up to 100 mm in width to depths in excess of 3.8 m. The dip of the discontinuity was 80 to 90 degrees to upstream. Woodward Clyde (1999) commented that road base levels either side of a crack showed a vertical differential to upstream, but that it was not conclusive if it was due to a shear type displacement, because it could be a result of the crest works in 1985.

A series of softened zones at angles of 45 to 70 degrees to the horizontal were observed in the upper 4 m of the core in tube samples recovered from boreholes drilled in the 1990s. The moisture content in the softened zones was about 3% greater than the "normal" moisture content in the embankment.

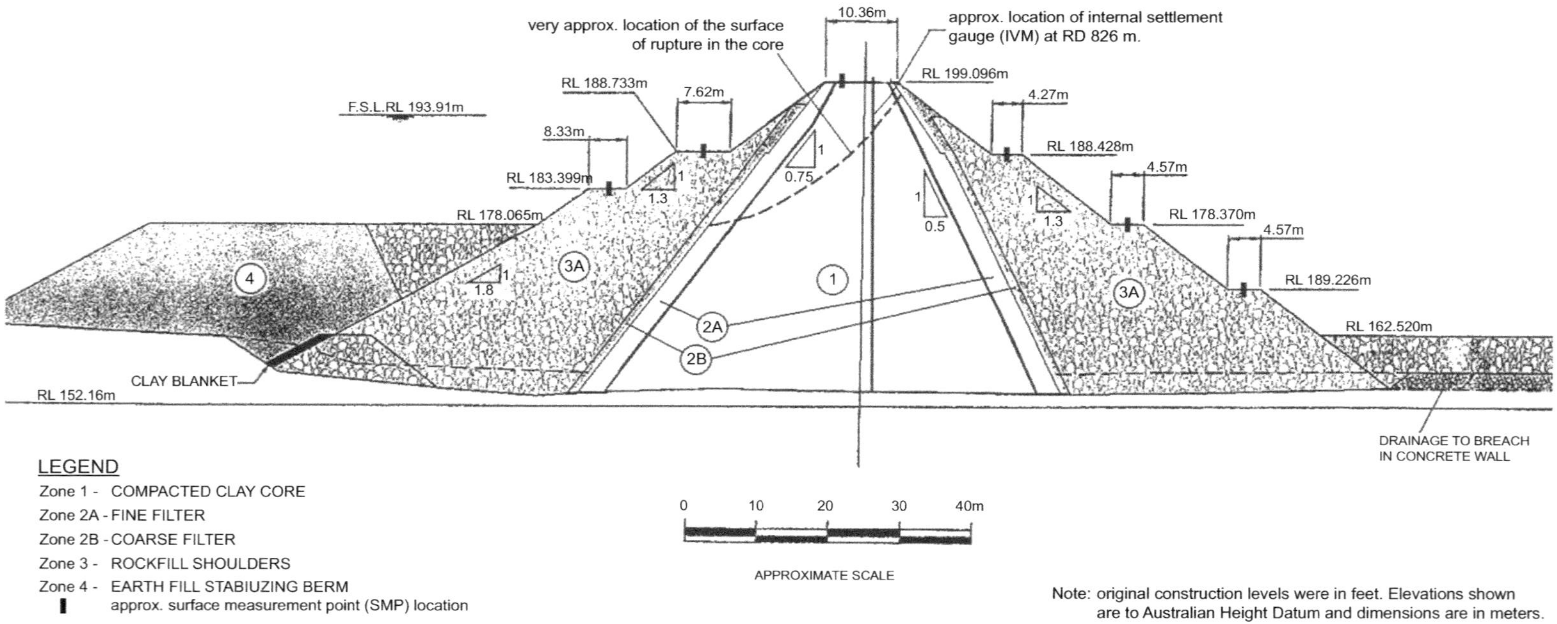

FIGURE 9.25 Eppalock Dam, original design at maximum section showing wedge of core which displaced upstream (Hunter and Fell, 2003c).

FIGURE 9.26 Eppalock Dam. Photograph of the downstream end of displacement of the wedge of core which occurred during construction of upgrade works. In the photograph the base of the wedge is about 1 m above the filter. (Permission to publish photograph approved by Goulburn-Murray Water.)

Piezocone testing (in the 1990s) indicated softening in the upper 5 to 6 m of the core. This depth of softening was also confirmed from tube sampling and pressuremeter testing in boreholes.

Test pitting in the rockfill indicated it was "dirty", containing up to 20% to 25% finer than 25 mm. The finer fractions predominantly filled the voids of the larger basalt rock pieces.

Inclinometers were installed in the crest in 1997. On drawdown in 1998 (35 to 36 years) a localised shear type displacement of 1 to 2 mm was observed at 11 m depth below crest level in IN2, located next to SMP CS1. The localised displacement occurred between mid March and mid April 1998 when the reservoir was drawn down below 186.8 m AHD to a low at 186.4 m AHD. Further localised shear type displacement at this depth and also at 4 m depth were detected during (and following) placement of additional rockfill on the upper berm of the upstream slope at the time of the remedial works in 1999.

Woodward Clyde (1999) concluded from their analysis that two mechanisms were occurring:

A down drag on the clay core and resultant tensile cracking caused by the significantly greater settlement of the rockfill shoulders post construction.

Conditions approaching limiting equilibrium of the upper portion of the upstream slope. Limit equilibrium analysis showed that under a reservoir drawdown condition, rainwater penetrating the existing cracks and softening of the undrained strength of the clay core, the factor of safety of the upstream slope was approaching 1. They envisaged that under a drawn down reservoir, heavy rainfall could trigger shear type deformation. For the deeper slide surface, as indicated by the shear type displacements at 11 m depth in IN2, the factor of safety against sliding is around 1.3.

Remedial works were undertaken in 1999 to improve dam safety of the embankment, including improvement of the stability of both the upstream and downstream slopes and placement of a new multi-zone filter in the upper part of the downstream filter. Design and construction of the remedial works are summarised by Davidson et al. (2000) and Fox and Fyfe (2001). During the construction works shear type deformations (Davidson et al., 2000) were observed along existing longitudinal cracks in the core. The most significant surface expression of the shear type deformation was on the downstream batter of the exposed core (Figure 9.25). Shear type deformations were also observed in inclinometers as described earlier.

9.2.6.2 *Upper Yarra Dam*

Upper Yarra Dam, shown in Figure 9.27, is an example of where a localised internal shear affecting the upper part of the dam has occurred during large drawdowns of the reservoir. Permanent deformations including accelerated settlement and upstream displacement of the crest and upper upstream slope occurred on large drawdowns. The shear was evident as large, localised strains in vertical settlement gauges, and these zones were where loss of bentonite drilling fluid occurred in boreholes drilled from the crest in earlier investigations. The dam has since been upgraded with the upgrade addressing the potential for internal erosion and piping initiating along the shear surface and other internal erosion mechanisms. Details are given in Section 19.6.1.

Both Eppalock and Upper Yarra Dams displayed accelerated vertical crest settlements and movement upstream during large reservoir drawdowns.

Factors which indicate that a through-going shear in the core may be present are:

(a) Larger than usual crest settlements and displacement upstream.
(b) Large settlements and/or lateral displacements of the crest occurring over a short period, usually associated with large drawdown of the reservoir.
(c) Poorly compacted upstream zones such as rockfill dumped in layers several metres thick with no roller compaction, rockfill or gravel fill compacted by dozer tracking or by small rollers in thick layers or weathered rockfill subject to collapse settlement placed in layers greater than 1.2 metres thick.
(d) Localised strain within the core identified in cross arm settlement devices.

The presence of shears may be confirmed by installing and monitoring inclinometers, localised very low CPTU cone resistance in otherwise cone high resistance core, and vectors of displacement consistent with movement along an upstream dipping shear surface.

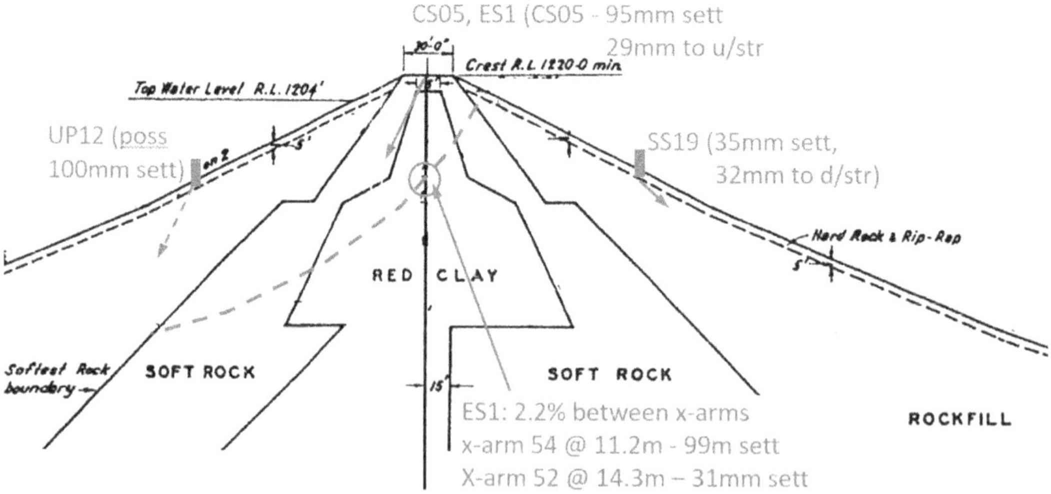

FIGURE 9.27 Postulated internal shear within the earthfill core of Upper Yarra Dam; shown are estimated deformations in the 1967/1968 drawdown. (Courtesy of AECOM and Melbourne Water Corporation.)

Whether such shears potentially form a defect in which concentrated leak erosion may initiate depends on the geometry of the surface and its location relative to reservoir level and can only be assessed on a case by case basis.

9.2.7 IM 10 Hydraulic Fracture or Gap Resulting from Arching of the Core onto the Shoulders of the Embankment

He (2016) and Bui et al. (2004) carried out parametric studies of 50 m high embankment cross sections shown in Figure 9.28. Vertical stresses are shed from the core onto the stiffer filters and shoulders in all models. The effect is most marked for the narrow core.

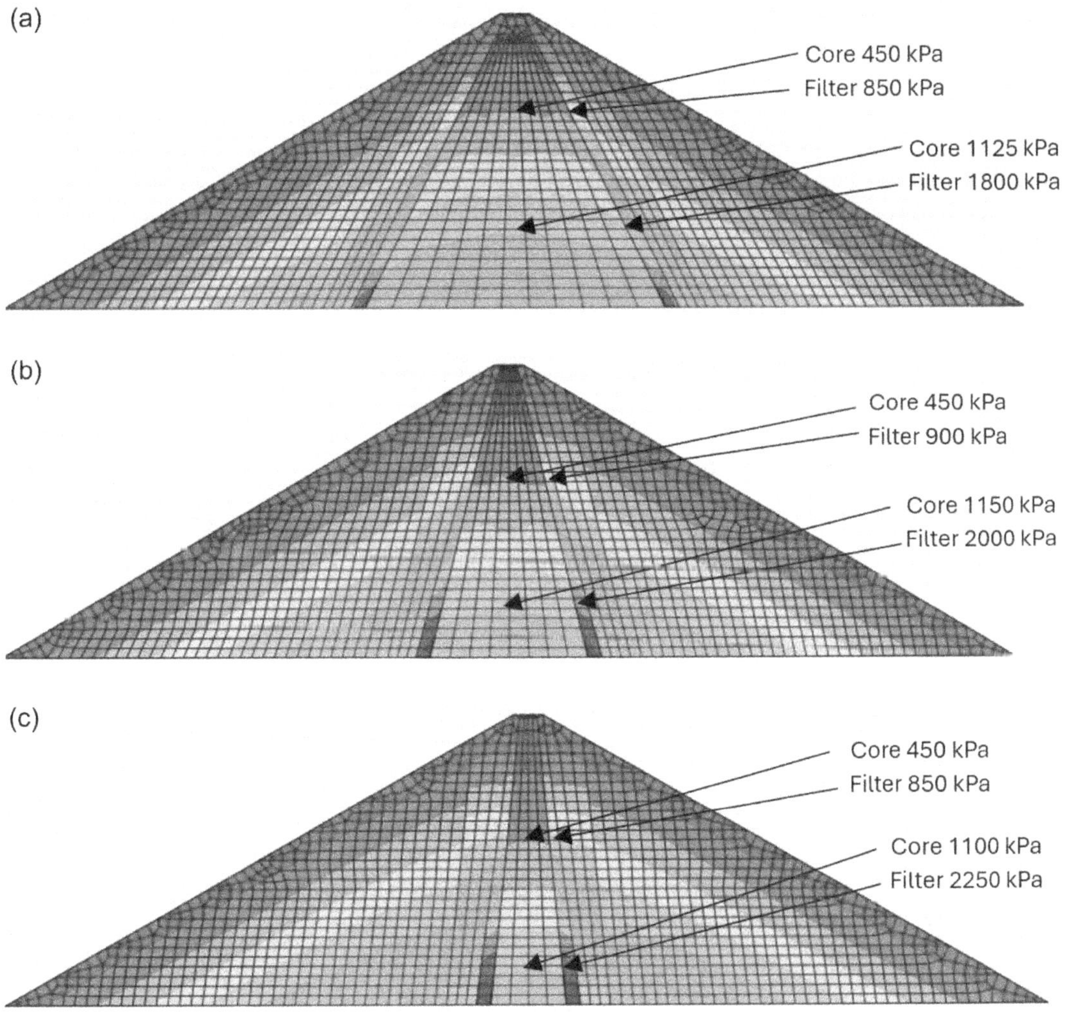

FIGURE 9.28 Effect of the slope of the core on the vertical stress distribution (He et al., 2017).
(a) Model D1—Slope of core = 0.5H:1V.
(b) Model D2—Slope of core = 0.25H:1V.
(c) Model D3—Slope of core = 0.1H:1V.

He (2016) modelled the stresses and deformations of Dartmouth Dam for which good monitoring of internal deformations and internal strains were available. The dam is 175 m high. Figure 9.29 shows the cross section. It can be seen that some arching effect is present even though the core width is about 0.75 × height. Analyses showed that the reduction is not sufficient to result in hydraulic fracture.

Bui et al. (2004, 2005) modelled the effects of softening of the core on saturation by reducing the modulus of the core from 40 MPa to 30 MPa, 20 MPa and 10 MPa. Figure 9.30 shows the vertical stresses in a narrow core at the end of construction and the stresses when the modulus of the core is reduced to 10 MPa. The result is to shed load from the core onto the filters and shoulders. This approach approximates to modelling the effects of collapse settlement of the core. Settlement may also occur if the core is constructed from silt sand gravel soils subject to Global Backward Erosion or Suffusion.

Skempton (1990) provides relevant information on the historical performance of UK puddle dams. This indicates that a hydraulic gradient across the puddle core of ≥ 6 is critical for hydraulic fracture. This is approximately equivalent to W/H ≤ 0.17. Rather than the gradient being critical it is the narrow core resulting in arching onto the shoulders and low vertical stresses in the core.

The authors have seen evidence of a gap and very soft zone through which water had flowed within the core of a puddle core dam just below the phreatic surface. This was seen in a test pit in the abutment and was evident as near zero CPT cone resistance elsewhere in the core. The gap is formed by collapse settlement of the core below the phreatic surface and arching of the unsaturated core above the phreatic surface.

A similar mechanism may result at the top of a narrow core with the fill above the core bridging across the top of the core.

To summarise, as detailed in Bui et al. (2004, 2005) and Fell et al. (2008, 2024), cracking and hydraulic fracture is most likely to be a problem for cores which are very narrow (core width less than 0.25 of the embankment height) and for soils subject to collapse compression on saturation (poorly compacted soil placed dry of optimum moisture content). It is unlikely to be a problem for cores which are wider than 0.5 to 1.0 of the embankment height and the core is well compacted at around optimum moisture content. It is not sufficient to cause hydraulic fracture to just have stiff shoulders and a less stiff core. There needs to be differential settlement of the core after construction.

9.2.8 IM 21 and IM 22 Crack or Gap Adjacent to a Spillway or Abutment Walls

There are two scenarios which may result in a crack or gap forming adjacent to the wall. The first scenario (IM 21) is where a gap may form due to the wall type and profile of the wall as shown in Figure 9.31. A gap may also form if the retaining wall is slender and has marginal factor of safety against stability so deflects under the pressure of the fill. Many walls were designed for active earth pressure, whereas modern design is for at rest pressures which are higher.

The second scenario (IM 22) is where a gap forms due to settlement of the fill adjacent to the wall, as depicted in Figure 9.32.

He et al. (2017, 2019) carried out a limited amount of numerical analysis of the fill settlement scenario as depicted in Figure 9.32. Figure 9.33 shows the geometry of the model and Table 9.3 the results.

It is assumed that all the post-construction settlements occur in S (the soil below the base of the wall).

The analyses model the soil in S as viscoelastic and the soil in W is modelled as elastic. The Young's modulus of the soil in S is varied to give the required % settlement of the crest of the embankment away from the wall.

The interface between the wall and soil is modelled with and without a contact interface element, which allows opening of the gap between the wall and soil. The friction coefficient between wall and dam is taken as tan (φ′) where φ′ = 30°.

For those analyses without contact elements, tensile stresses develop as shown in Figure 9.34. For analyses with contact elements, a gap is modelled as shown in Figure 9.35.

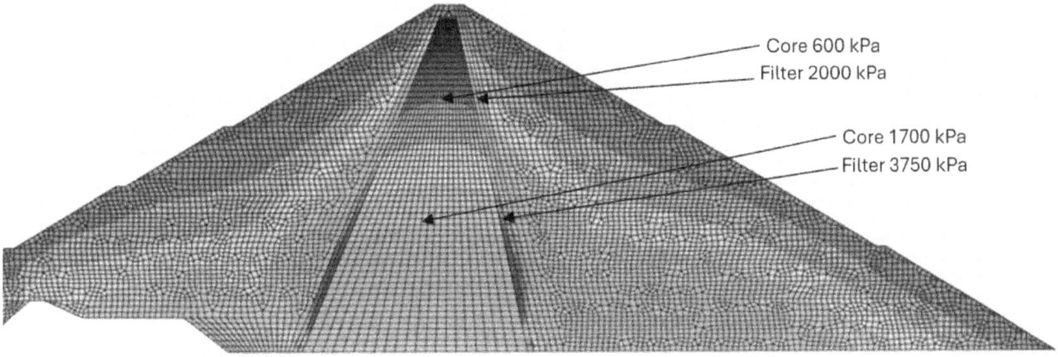

FIGURE 9.29 Vertical stress distribution of Dartmouth Dam cross section at end of construction (He, 2016).

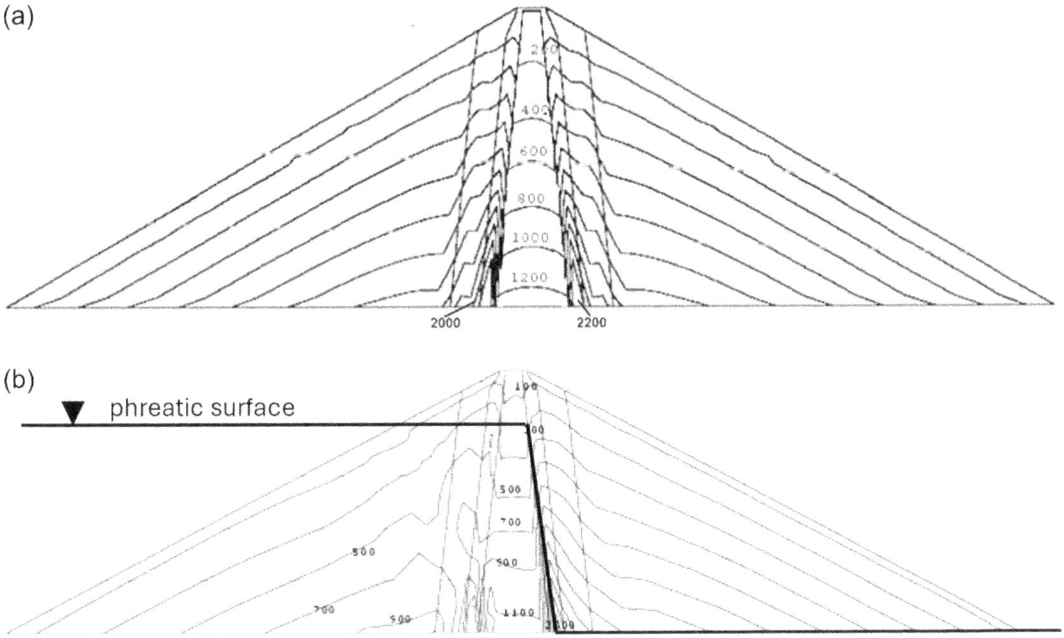

FIGURE 9.30 Vertical stresses in narrow central core earth and rockfill dams (Bui et al., 2004, 2005).
(a) Vertical stress during construction (kPa).
(b) Vertical stress (kPa) allowing for softening of the core on saturation from 40 MPa to 10 MPa.

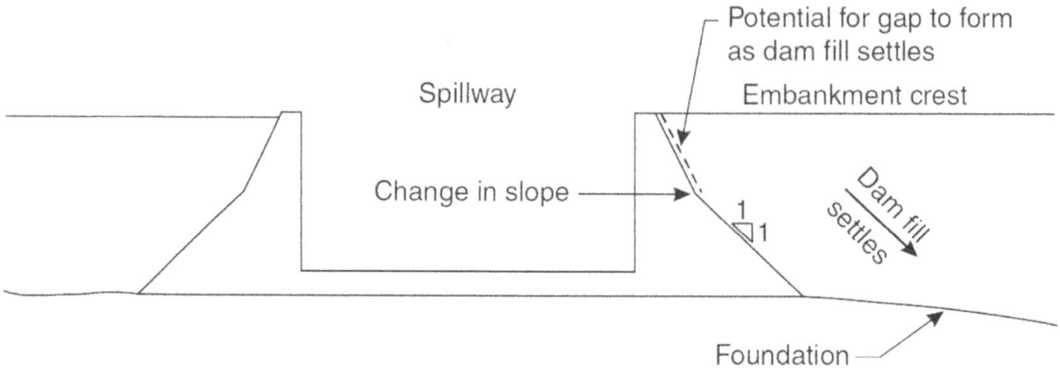

FIGURE 9.31 A situation where a gap may form between the dam fill and spillway wall resulting from the change in slope of the retaining wall, IM 21 (Fell et al., 2024).

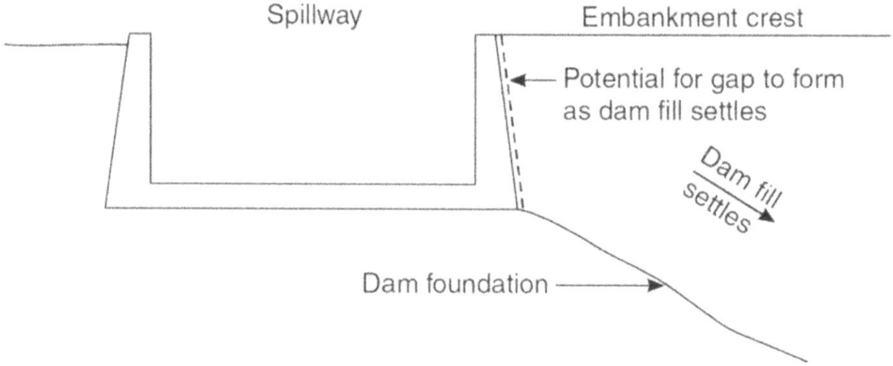

FIGURE 9.32 Situation where a gap may form between the fill and a wall, IM 22 (Fell et al., 2024).

Tables 9.4 and 9.5 summarise the results of the analyses. Appendix B of He et al. (2017) has the detailed results.

The following should be noted:

(a) In the models with no contact elements, D_W is controlled numerically by the thickness of the layer in the numerical model. The modelled width and depth of a tensile zone formed during construction are small because as each layer is added to the numerical model, the layers below go back into compression.
(b) The models with contact elements are a better guide to the potential depth and width of a gap.
(c) The % post-construction settlement is a major factor in whether a gap is likely to form, and the width and depth of the gap.
(d) For elastic analyses the width of the gap should be scale dependent. The models are for wall height 15 m. For most models the depth of embankment below the base of the wall is also 15 m.

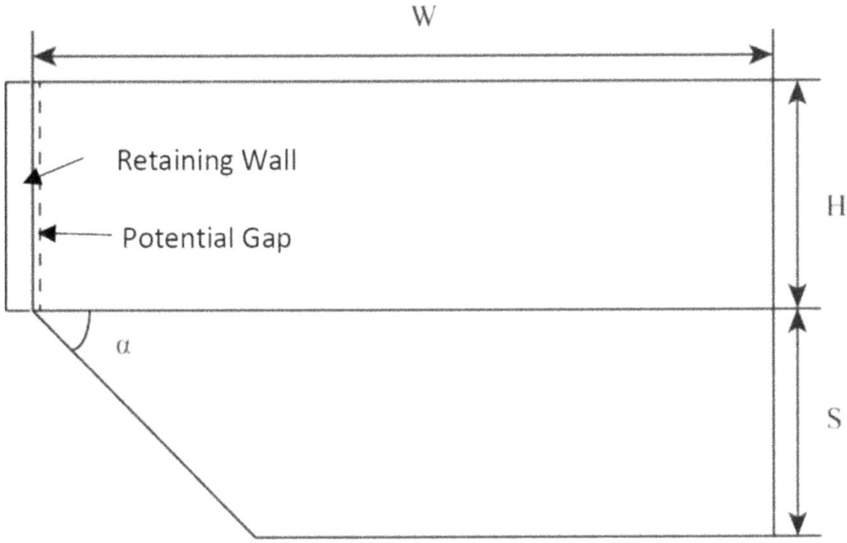

FIGURE 9.33 Retaining wall at the top of a slope in the foundation containing the earthfill embankment showing definition of the parameters in He et al. (2017) numerical analyses.

TABLE 9.3 Geometry and Material Properties—Retaining Wall at the Top of a Slope (He et al., 2017, 2019)

	GEOMETRY						
MODEL	*H (M)*	*S (M)*	*W (M)*	*Θ (°)*	*α (°)*	*SOIL PROPERTIES*	*POST-CONSTRUCTION SETTLEMENTS (% OF DAM HEIGHT)*
W1	15	15	50	90	45	E = 30 MPa, v = 0.3	0
W2	15	15	50	90	30	E = 30 MPa, v = 0.3	0
W3	15	30	50	90	45	E = 30 MPa, v = 0.3	0
W4	15	15	50	90	45	E = 30 MPa, v = 0.3, φ = 0	0
W1 (0.1%)	As for W1						0.10%
W1 (0.25%)	As for W1						0.25%
W1 (0.5%)	As for W1						0.50%
W1 (1%)	As for W1						1%

Notes:

(a) Post-construction settlements are assumed to occur in S.

(b) In this table, Θ = slope of retaining wall.

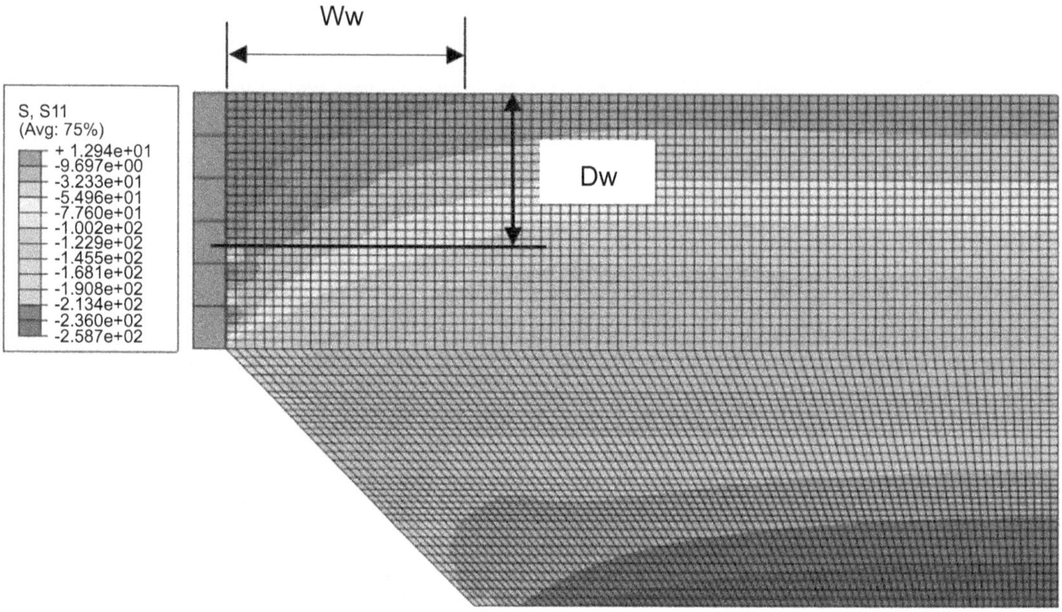

FIGURE 9.34 Example of results of analyses with soil–wall contact remaining intact. Horizontal stresses. W_W and D_W define the boundary of zero horizontal stress (He et al., 2017).

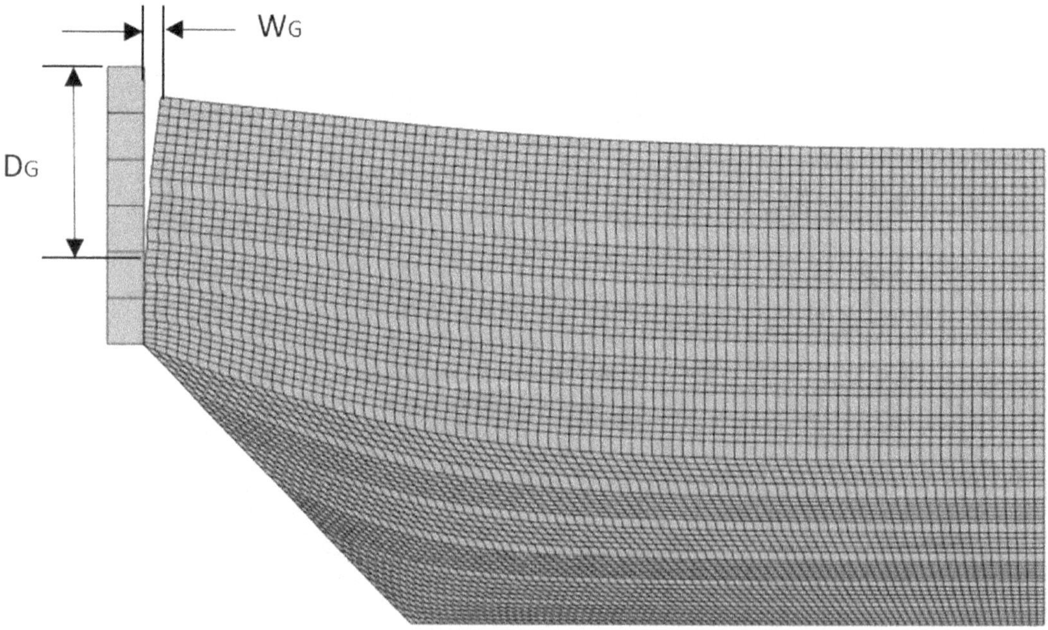

FIGURE 9.35 Example of results of analyses with soil–wall contact remaining intact. Horizontal stresses. W_G and D_G define the geometry of the gap formed at the wall (He et al., 2017).

TABLE 9.4 Results of Analyses—Retaining Wall at the Top of a Slope (He et al., 2017)

MODEL	GEOMETRY H (M)	S (M)	W (M)	Θ (°)	α (°)	MODEL WITH NO CONTACT ELEMENTS ZERO TENSILE STRESS W_W (M)	D_W (M)	MODEL WITH CONTACT ELEMENTS GAP FORMED AT WALL W_G (MM)	D_G (M)
CONSTRUCTION CONDITION									
W1	15	15	50	90	45	12.64	1.5	2.10	2
W2	15	15	50	90	30	15.50	1.5	2.09	2
W3	15	30	50	90	45	18.10	1.5	3.50	2
W4	15	15	50	90	45	14.37	1.5	1.10	1.5
POST-CONSTRUCTION CONDITION									
W1 (0.1%)	As for W1					12.07	1.5	9.75	7
W1 (0.25%)	As for W1					9.77	1.5	25.80	9.75
W1 (0.5%)	As for W1					7.47	1.5	56.00	14
W1 (1%)	As for W1					5.74	1.5	111.00	14.5

TABLE 9.5 Retaining Wall at the Top of a Slope (Effects of post-construction settlements within the embankment on the gap between the wall and the embankment; modelling with contact elements; He et al., 2017)

DEPTH OF EMBANKMENT BELOW RETAINING WALL % WALL HEIGHT	SLOPE FOUNDATION BELOW THE WALL α (°)	POST-CONSTRUCTION SETTLEMENT (% HEIGHT)	WIDTH OF GAP FORMED BETWEEN WALL AND EMBANKMENT (MM)	DEPTH OF GAP (M)	DEPTH OF GAP (% WALL HEIGHT)
EFFECT OF POST-CONSTRUCTION SETTLEMENTS					
100	45	0	2	2	15
100	45	0.1	10	7	50
100	45	0.25	25	9.75	65
100	45	0.5	55	14	90
100	45	1.0	110	14.5	100
CONSTRUCTION CONDITION MODELS					
100	45	0	2	2	15
100	30	0	2	2	15
200	45	0	3.5	2	15
100	45	0	1	1.5	10

These analyses show that a gap is not likely to form during construction, but even quite small post-construction settlements can lead to a gap. The depth and width of the gap increase with increased post-construction % settlement.

The assessment should take account of field observations. Signs of a gap potentially having formed include separation of the crest road pavement from the wall.

For important decisions on dam safety or potential upgrades, test pits at this location will provide more definitive assessments.

9.2.9 IM 2 Crack or Gap Adjacent to a Cliff at the Upper Part of the Embankment Formed by Differential Settlement

Settlement of the embankment adjacent to a cliff (or vertical excavation, e.g. following a vertical joint) in the upper abutment of a dam may lead to a crack or gap forming between the embankment and the cliff. Figure 9.36 shows the situation where this may occur.

For a gap to form, the cliff needs to be close to the sloping part of the foundation profile and the slope of the foundation relatively steep, i.e. both $W_b/H_w \leq 2.5$ and $\beta_1 \geq 15°$.

See Appendix A for details of a method to estimate the likelihood of a crack or gap being present.

The assessment should take account of field observations. Signs of a gap potentially having formed include separation of the crest road pavement from the cliff.

For important decisions on dam safety or potential upgrades, test pits at this location will provide more definitive assessments.

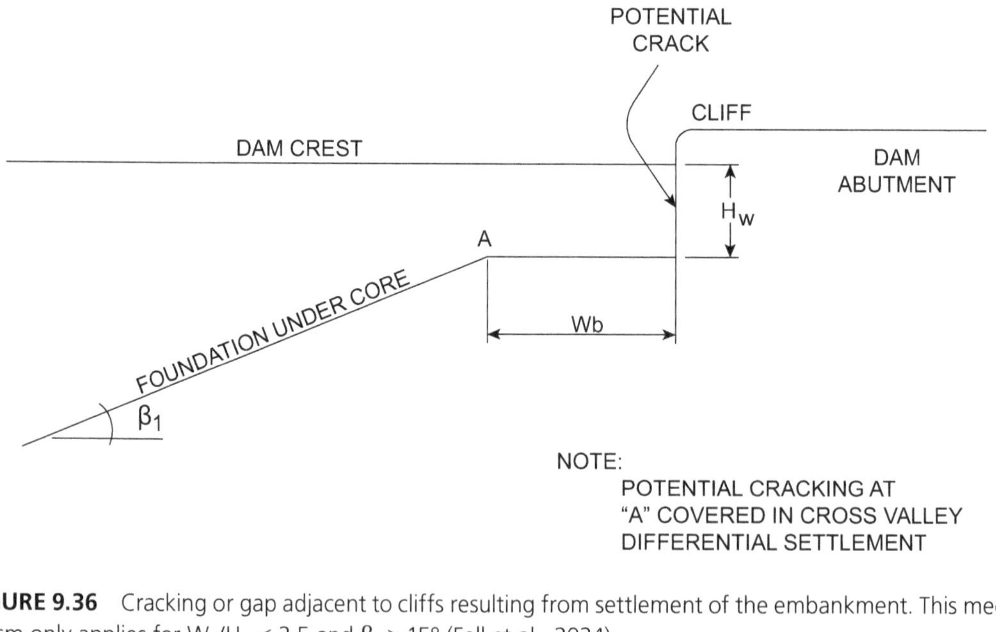

FIGURE 9.36 Cracking or gap adjacent to cliffs resulting from settlement of the embankment. This mechanism only applies for $W_b/H_w \leq 2.5$ and $\beta_1 \geq 15°$ (Fell et al., 2024).

9.2.10 IM 23 Crack or Gap at the Interface of Wrap-Around Connecting Embankment Dam to a Concrete Gravity Dam

Where there is a wrap-around junction between a concrete gravity dam and an embankment dam as shown in Figure 9.37, differential settlements similar to those described in Section 9.2.8 may occur and may result in gaps between the core and the concrete. These may be assessed as described in Section 9.2.8.

9.3 LIMITING STRESSES AND STRAINS LIKELY TO RESULT IN TRANSVERSE CRACKING

As discussed in He et al. (2017, 2019), numerical analyses show that for differential settlement initiating mechanisms, strains tend to localise, so it is important to be able to predict what localised strains are likely to result in cracking. There are several sources of information on what localised tensile strains may result in cracking.

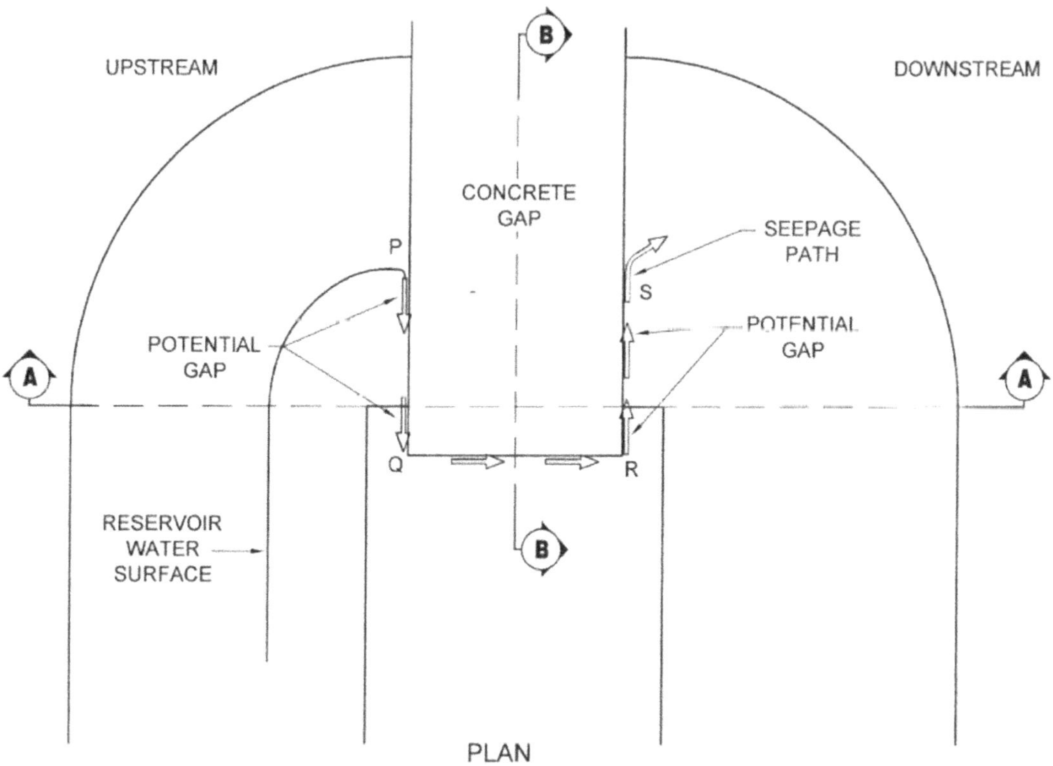

FIGURE 9.37 Wrap-around details for connection of embankment dam to concrete gravity dam (Fell et al., 2008, 2024).

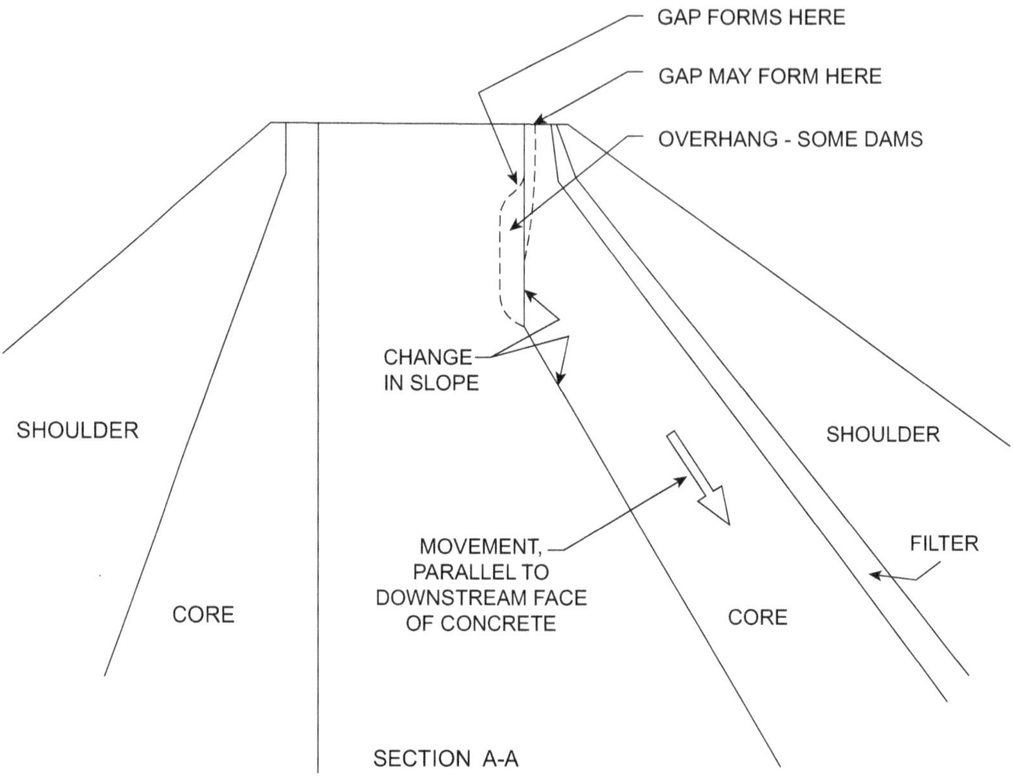

FIGURE 9.37 (Continued)

9.3.1 Laboratory Tension Tests on Soils

9.3.1.1 Direct Tension Tests by Win (2006)

Figures 9.38 and 9.39 show the results of direct tension tests carried out by Win (2006). The soils tested were Soil C silty sandy clay of low plasticity CL; Soil D sandy silty clay of low plasticity CL; and Soil F clayey silty sand of low plasticity SC.

The tensile strengths for these soils ranged from about 10 kPa to 65 kPa with the lower strengths for soils compacted wet of Standard Optimum Moisture Content, and the highest strength for those compacted dry of optimum. The tensile strain at failure were mostly < 0.002 (< 0.2%), in the range 0.05% to 0.2% strain.

These tests were all relatively short duration tests. Attempts to do longer term tests were unsuccessful because the samples dried out despite being coated in wax.

9.3.1.2 Flexural Bending Tests by Leonards and Narain (1963)

Leonards and Narain (1963) carried out flexural bending tests on six soils. The classification properties of the soils are given in Table 9.6. For each soil "short term" (two days) and "long-term" (four weeks) tests were carried out. One soil was also tested over six months. The results of the tests are summarised in Table 9.7. More details are given in the paper.

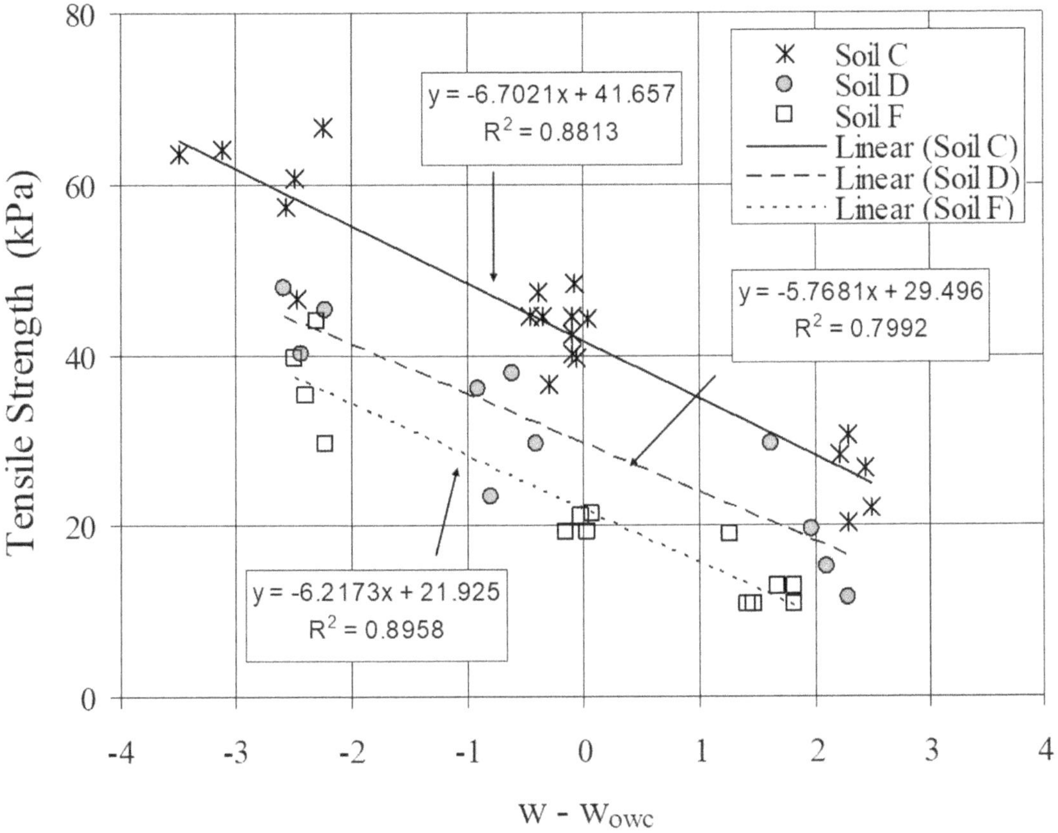

FIGURE 9.38 Effect of compaction moisture content on tensile strength (Win, 2006).

It can be seen that:

(a) The tensile stresses and strains at failure are similar to those from Win (2006).
(b) The long term (four week) tensile strains at failure are larger than the short term (two days) but not by a lot. The six month test was also not much higher strain than the short term test.
(c) The high plasticity soil (limestone clay) had significantly larger strains to failure than the others, particularly the non-plastic soil (see the Shell Oil Co. Dam shown in Table 9.7).
(d) The tensile strengths are generally higher than for the soils tested by Win (2006) but not by a large amount. The limestone clay plastic soil has much higher tensile strength. Note the bottom four tests of that soil were compacted to Modified Proctor energy.

9.3.2 Numerical Analyses of Buffalo and Red Willow Dams

He (2016) carried out numerical modelling of Buffalo Dam and Red Willow Dam. Both dams had experienced transverse cracking detected a number of years after construction. Details are given in He (2016) and He et al. (2017).

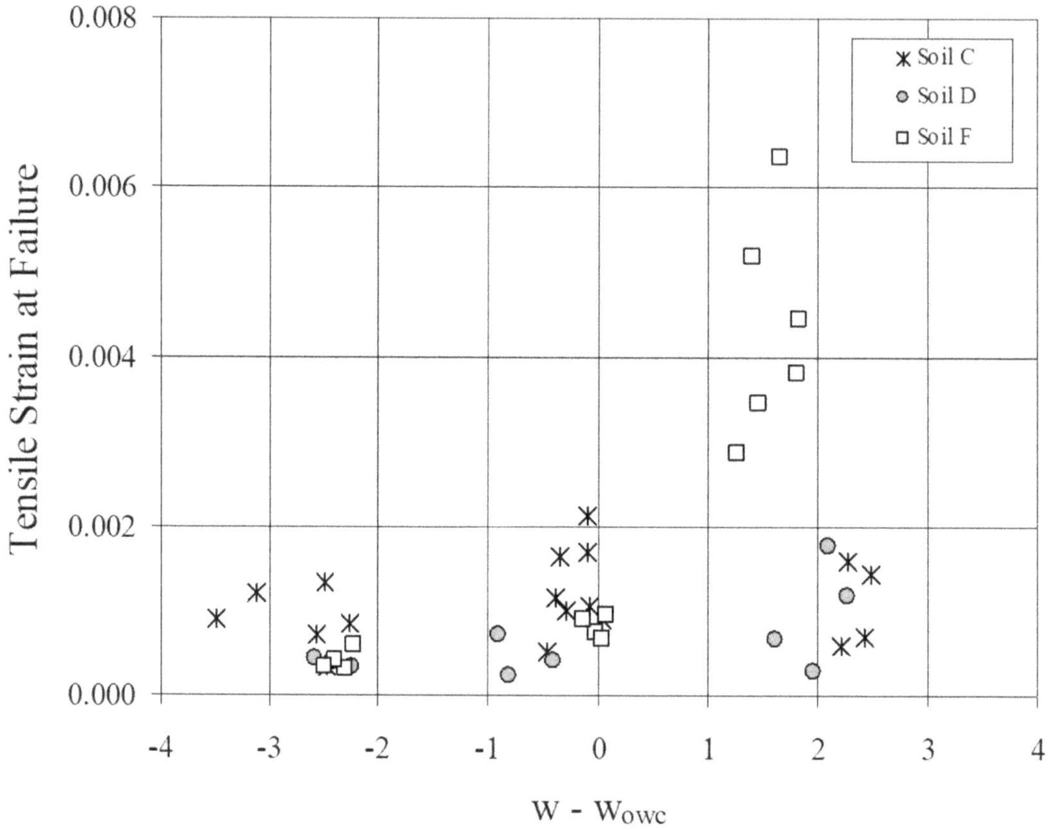

FIGURE 9.39 Effect of compaction moisture content on tensile strain at failure (Win, 2006).

TABLE 9.6 Summary of Geotechnical Properties of Soils Tested by Leonards and Narain (1963)

SOIL	*LIQUID LIMIT %*	*PLASTICITY INDEX %*	*SPECIFIC GRAVITY*	*% FINER THAN 0.075 MM*	*OPTIMUM MOISTURE CONTENT %*	*STANDARD MAXIMUM DRY DENSITY (T/M3)*
Portland Dam	29	8	2.74	25	16.3	1.79
Rector Creek Dam	38	16	2.60	11	19.8	1.64
Woodcrest Dam	Non-plastic		2.64	21	10.2	2.04
Shell Oil Co. Dam	Non-plastic		2.65	22	11.2	1.92
Willard Dam Test Embankment	31	11	2.72	28	16.4	1.76
Limestone Clay	72	45	2.75	82	25.9	1.54

TABLE 9.7 Summary of Flexural Beam Tests Carried Out by Leonards and Narain (1963)

	COMPACTION CONDITION		TENSILE STRAIN AT FAILURE		TENSILE STRESS AT FAILURE	
SOIL	DRY DENSITY % SMDD	MOISTURE CONTENT (%)	SHORT TERM (%)	LONG TERM (%)	SHORT TERM (KPA)	LONG TERM (KPA)
Portland Dam	100	OWC	0.17	0.19	56	56
	99	OWC −2.8	0.12	0.16	88	74
Rector Creek Dam	100	OWC	0.16	0.17	65	52
	93	OWC− 1.0	0.12	0.13	47	43
Woodcrest Dam	100	OWC	0.18	0.24	80	77
	98	OWC−3.0	0.17	0.20	52	48
Shell Oil Co. Dam	100	OWC	0.07	n/a	40	n/a
	88	OWC+1.0	0.06	n/a	30	n/a
Willard Test Embankment	100	OWC	0.20	0.24	46	38
	94	OWC−3.9	0.18	0.24	71	67
Limestone Clay	100	OWC	0.26	0.29(a)	192	166(b)
	97	OWC−2.0	0.17	0.32	167	115
	97	OWC+2.0	0.24		107	
	93	OWC+3.6	0.28		82	
	115	OWC−8.4	0.12		265	
	111	OWC−10.4	0.08		230	
	111	OWC−6.4	0.105		178	
	107	OWC−2.9	0.14		160	

Notes: (a) Very long term 0.33%. (b) Very long term 165 kPa.

9.3.2.1 Buffalo Dam, Australia

Figure 9.40 shows the valley profile and the constitutive models used for the analysis. Also shown is the upper left abutment area including a step in the profile above which cracking was observed in 2000, 35 years after construction. As described in Section 9.2.1, the cracking was 40 mm wide at the top of the core and extended to the foundation. There was also extensive longitudinal cracking in the core. It is likely the cracking had been present from soon after construction was completed but not apparent because the core was covered by a layer of rockfill.

The 35 m high central core earth and rockfill embankment experienced 1% settlement post construction. The rockfill was not compacted and this contributed to the relatively large settlement.

The numerical modelling showed that the tensile strains in the elasto-plastic analyses localise over the step and for the analyses with drained shear strengths reach about 6%, with a total relative displacement of about 33 mm. For modelling with undrained strengths in the upper 10 m of the embankment, the localised tensile strain above the step is about 3% and total accumulated relative displacement of 16 mm. These better model the actual likely soil properties.

9.3.2.2 Red Willow Dam, USA

Red Willow Dam (US Bureau of Reclamation, 2013, 2014) was constructed by USBR in 1961. It was a homogeneous earthfill dam about 30 m (100 ft) high founded on 15 m of alluvium. Construction settlements were about 2.5% to 3.5% of embankment height and post construction about 0.2% of embankment height. Most of the settlement was in the alluvium.

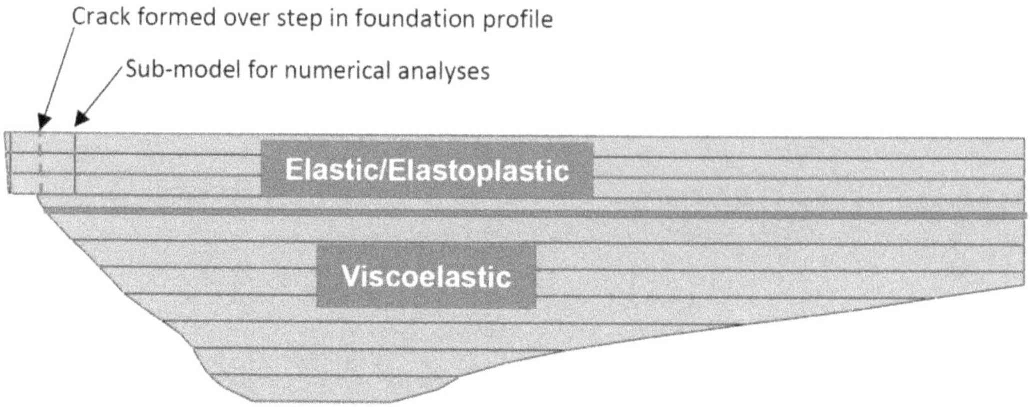

FIGURE 9.40 Longitudinal section of Buffalo Dam showing foundation profile and location of crack (He, 2016; He et al., 2017).

Cracking was observed in 2009 over the outlet conduit and over the old creek channel where the depth of alluvium over the rock foundation was less. The cracking was on the downstream slope about 6 m (20 ft) above the top of the conduit and a similar height above the general top of alluvium at the old creek location. Cracking did not extend to the embankment crest.

At the location of cracking the post-construction settlements would be about 1% of the height of the embankment assuming 0.2 m post-construction settlement at that location.

Figure 9.41 shows the results of elasto-plastic numerical modelling with modelled tensile strains over the culvert and old creek channel.

The end of construction tensile strains using elasto-plastic analysis with drained strengths are 0.25% at the creek and 2% at the conduit. With post-construction settlements the strains increase to 0.3% (0.45%) and 3% (4%) respectively for 0.2% (1%) settlement. The 1% settlement data is considered to be a better representation.

The relative displacements in these localised strain areas were 11 mm and 51 mm respectively. Measured maximum crack widths were 25 mm and 50 mm.

He (2016) modelled this with crack propagation modelling and was able to replicate the measured widths. This showed that modelling the crack alters the stress distribution and gives crack depths 35% deeper than the elastic continuum modelling.

Over the conduit localised computed strains were large (3% to 4%), but it is notable that cracking also occurred over the creek where computed strains were only 0.3% to 0.45%.

9.3.2.3 Upper Yarra Dam, Australia

Upper Yarra Dam is a 90 m high zoned earth and rockfill dam. Figure 9.27 shows the zoning at the top of the embankment. Construction was completed in 1957. The dam experienced 0.5% crest settlement from 1957 to 2016.

As part of investigations and analysis for design of a dam upgrade elasto-plastic numerical analysis was carried out by AECOM and reported in AECOM (2017). Crack propagation analyses were also carried out by Dr Ke He, reported in Pells Sullivan Meynink (2017). The results of the analyses are shown in Figure 9.42 and summarised in Table 9.8. Chia et al. (2022) summarise the analyses.

The analyses estimated extension strains of 0.4% to 0.5% at location R1 in the upper part of the right abutment, and 0.3% at location L1 in the upper part of the left abutment. The crack propagation analyses estimated a crack 19 mm wide at R1 and no open crack at L1.

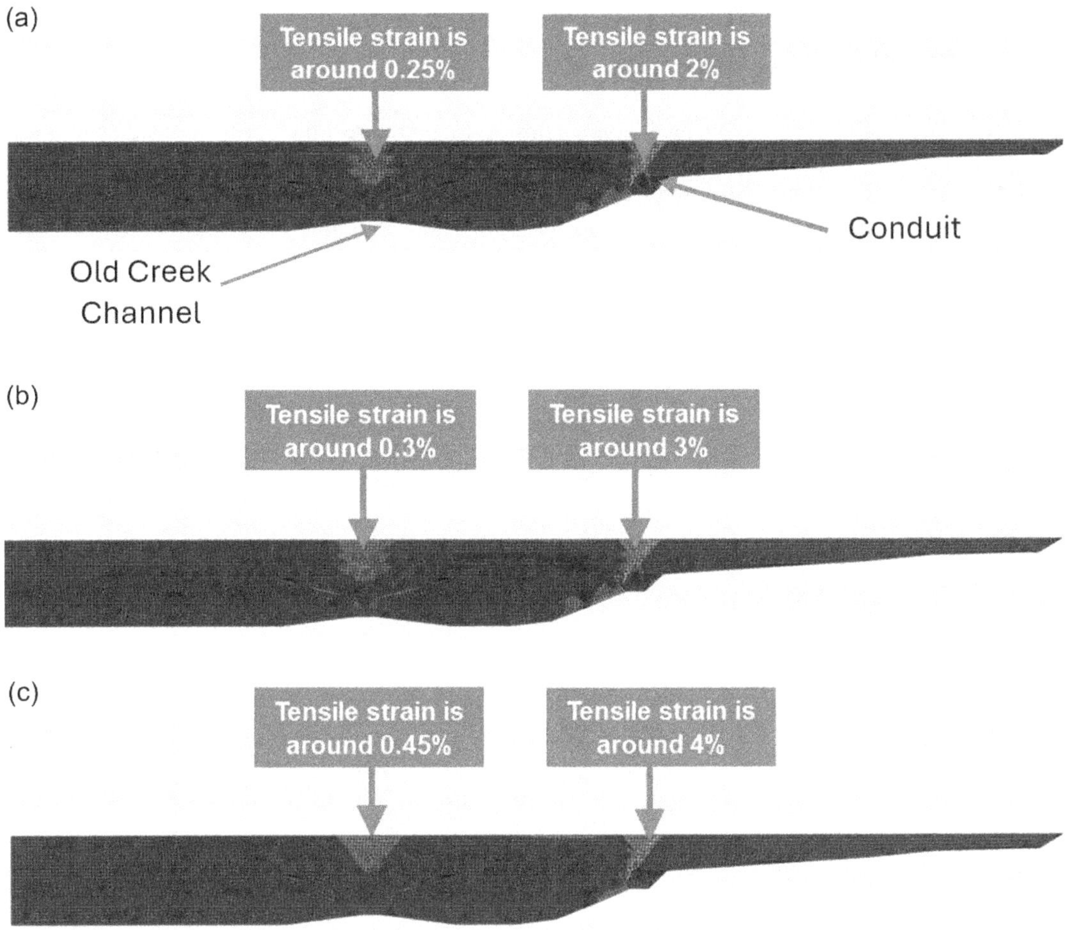

FIGURE 9.41 Figure 9.41 Red Willow Dam. Development of tensile strain zones during and after construction; elasto-plastic drained analysis, section along cracking (He, 2016; He et al., 2017).
(a) End of construction.
(b) Post-construction settlement = 0.2% of dam height.
(c) Post-construction settlement = 1% of dam height.

The downstream face of the upper part of the core was exposed during construction of the upgrade. This showed:

(a) No cracking other than desiccation cracks in the left abutment.
(b) Multiple vertical cracks at about 2 m average spacing over a length of 20 m in the right abutment between Ch 430 m and Ch 450 m in Figure 9.42; i.e. to the right of R2, not in the modelled area of largest strains. The cracks were hairline (< 1 mm) on excavation and opened up at the surface to 5 mm to 10 mm after desiccation. Post-construction average strains measured from surface survey were 0.09% to 0.14% in this area. It is interpreted that these are extensional strain cracks and that the strains have resulted in multiple cracks rather than a single crack as modelled numerically.

Upper Yarra Dam has been upgraded as detailed in Chia et al. (2022).

9.3.3 Case Studies from the Literature and Unpublished Reports

There are a number of well documented case studies in the literature which provide insights into the conditions in which cracks have been observed in embankment dams. A number of these are discussed in the PhD thesis of Dr James Sherard (Sherard, 1953) and in Leonards and Narain (1963). Table 9.9 lists the cases and the source(s) of information. The embankment zoning categories are as shown in

TABLE 9.8 Summary of Crack Propagation Analyses for Upper Yarra Dam (AECOM, 2017; Pells Sullivan Meynink, 2017)

POSTULATED CRACK	*CHAINAGE (M)*	*EXISTING CONDITION (CRACK WIDTH/ DEPTH)*	*1 IN 10,000 AEP EARTHQUAKE (CRACK WIDTH/ DEPTH)*	*1 IN 50,000 AEP EARTHQUAKE (CRACK WIDTH/ DEPTH)*
RIGHT ABUTMENT				
R1	410	19 mm/11.9 m	Open to foundation	
R2	430	4 mm/4.9 m	15 mm/15.3 m	40 mm/23.8 m
R3	480	Closed	Closed	Closed
R4	435	Closed	Closed	Closed
LEFT ABUTMENT				
L1	935	Closed	Closed	Closed
L2	965	7 mm/8.5 m	25 mm/13.5 m	59 mm/13.5 m
L3	950	Closed	Closed	33 mm/19.3 m
L4	935	Closed	Closed	Closed

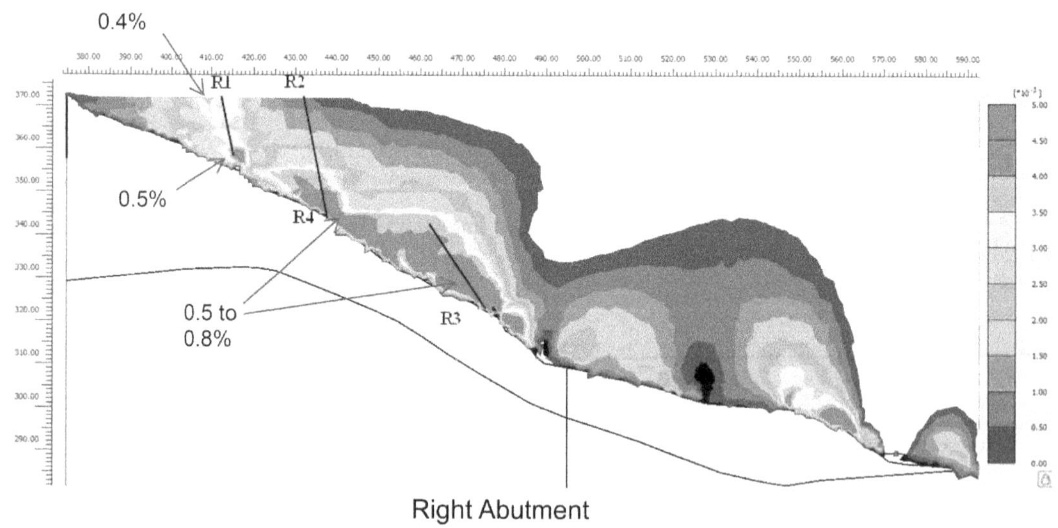

FIGURE 9.42 Post-construction extension strains in Upper Yarra Dam (AECOM, 2017). (Permission to reproduce provided by AECOM.)

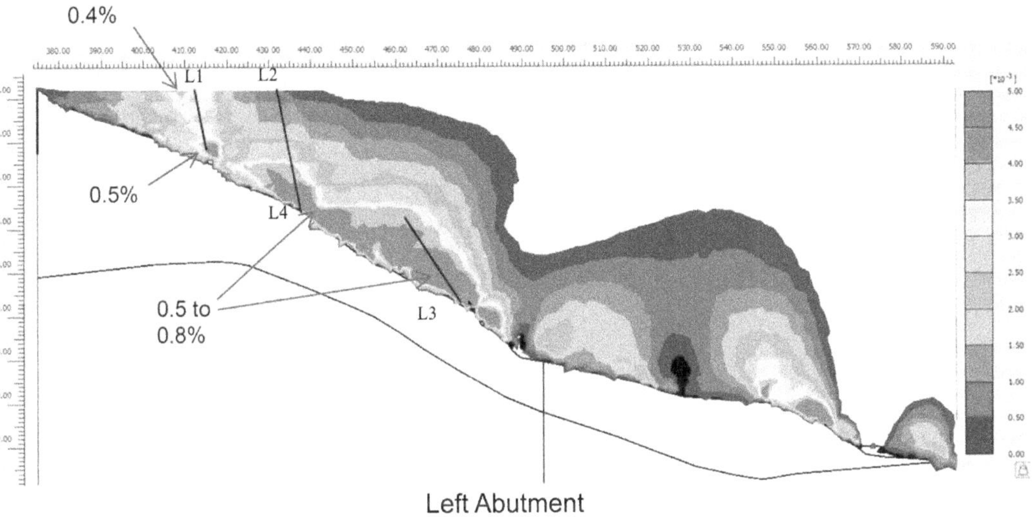

FIGURE 9.42 (Continued)

Figure 2.1. Table 9.10 summarises the cracking observed in these dams, and some of the factors which influence whether cracking may occur.

The following are some general observations from the data in Table 9.10:

(a) A number of the embankments in which cracking was observed experienced 1% to 2% post-construction settlement. However, cracking was also observed where post-construction settlements were significantly smaller, e.g. Matahina, which had very steep abutments with a step; Red Willow with 0.2% but 3.5% including settlements during construction; and Tullaroop with 0.23%. Figure 7.39 shows post-construction settlements for a number of dams. It can be seen that most dams settle less than 0.5% in the 10 years after construction.

(b) Embankments with a step in the abutment profile appear susceptible to cracking over or adjacent to the step, e.g. Matahina, Buffalo, Round Butte, East Branch.

(c) Many of the embankments cracked on first filling. A number of these were embankments which experienced collapse settlement of the embankment fill compacted very dry of Standard Optimum Moisture Content. Cracking of others was not detected until many years after construction, e.g. Buffalo, Matahina, Red Willow. However, it is possible that the cracks in Buffalo and Matahina formed soon after construction but had gone undetected.

(d) Seismic loads result in settlement, and this further settlement may result in transverse cracking.

(e) The depth of cracking may be substantial and greater than 100× crack width assumed in Fell et al. (2008). If the cracking is over a step in the abutment, it is likely to penetrate to the step, e.g. Buffalo.

(f) The tensile strains measured between survey marks on the crest of Rector Creek and Woodcrest dams were around 0.25% to 0.35%. Both dams experienced cracking, so this may be a guide to the average tensile strains resulting in cracking.

To get a feel for the maximum localised strains which might have occurred in these dams, the parametric model(s) by He et al. (2017), which most closely models the valley profile, and the post-construction settlement have been used to estimate the localised strains and accumulated strains at the

TABLE 9.9 Case Studies in Which Transverse Cracking in Embankment Dams Is Described (Fell et al., 2024)

DAM	*COUNTRY*	*YEAR CONSTRUCTED*	*EMBANKMENT TYPE*	*REFERENCE(S)*
Rector Creek	USA	1946	Homogeneous earthfill	Sherard (1953), Leonards and Narain (1963)
Apishapa	USA	1919–20	Earthfill with rock toe	Sherard (1953)
Scofield	USA	1925–26	Earth and rockfill	Sherard (1953)
Lake Francis	USA	1899	Homogeneous earthfill	Sherard (1953)
Masterton	USA	1950	Earthfill with rock toe	Sherard (1953)
Kern	USA	1948	Earthfill with rock toe	Sherard (1953)
Portland	USA	1956	Homogeneous earthfill	Leonards and Narain (1963)
Woodcrest	USA	1953	Zoned earthfill	Leonards and Narain (1963)
Shell Oil	USA	1944–45	Homogeneous earthfill	Leonards and Narain (1963)
Stockton Creek	USA	1950	Earthfill with rock toe	Sherard (1973)
Boggy Creek	USA	1970	Homogeneous earthfill	Sherard et al. (1972)
Wister	USA	1946–48	Homogeneous earthfill	Sherard (1973)
East Branch	USA	1952	Zoned earthfill	Sherard (1973)
Virginia Ranch	USA	1961–63	Zoned earthfill	Sherard (1973)
Round Butte	USA	1962?–64	Central core earth and rockfill	Sherard (1973)
Duncan	Canada	1964–67	Zoned earthfill	Gordon and Duguid (1970)
Matahina	New Zealand	1964–66	Central core earth and rockfill	Gillon (2007)
Tullaroop	Australia	1958–59	Homogeneous upper, zoned earthfill lower	Hunter et al. (2015)
Buffalo	Australia	1964–65	Central core earth and rockfill	Newman and Foster (2006)
Red Willow	USA	1961	Homogeneous earthfill	US Bureau of Reclamation (2013, 2014).
Austrian	USA	1948–50	Zoned earthfill	Harder et al. (1991); Pells and Fell, (2002, 2003), Forster and MacDonald (1998).

TABLE 9.10 Part 1: Summary of Cracking Observed in Embankment Dams and Factors Which Influence Whether Cracking May Occur (Fell et al., 2024)

			VALLEY PROFILE			MAXIMUM OBSERVED SETTLEMENT		OBSERVED CRACKING			
DAM	*HEIGHT (M)*	*CORE FOUNDATION*	*UPPER SLOPE (DEGREES)*	*SLOPE CHANGE LEVEL (% HEIGHT)*	*LOWER SLOPE (DEGREES)*	*% HEIGHT*	*CAUSE*	*LOCATION RELATIVE TO ABUTMENT (M)*	*WIDTH (MM)*	*DEPTH (M)*	*WHEN CRACKING WAS OBSERVED*
Rector Creek	46	Rock									
Left abutment			35	None	n/a	1.3	Construction.	20	19 (12a)	>7	First fill
						1.3	Collapse	23	12	?	3 years
Right abutment			25	None	n/a	1.3	Settlement	65	12	?	3 years
Apishapa	34	Rock abut.									
Left abutment		9 m alluv.	23	33	83	2.2	Collapse	24	50–75(b)	?	First fill
Right abutment			23	33	63	2.2	Settlement	27	50–75(b)	9 m(c)	First fill
Scofield	19	Rock abut. 4.5 m alluv.	"Very steep"	n/a	n/a	2.4	Collapse settlement	Adjacent each abutment	37	n/a	First fill
Lake Francis	16	Rock abut.									
Left abutment		1.5m alluv.	7	55	8	n/a	Collapse	No crack			First fill
Right abutment			25	n/a	n/a	n/a	settlement	"near abutment"	"large"	10 m (d)	First fill
Masterton	18	Rock	"Steep"	n/a	n/a	6.7	Collapse settlement	"Over abutment"	125 (75–100(e)	>5	First fill
Kern	16	Rock abut. 1.8m alluv.	"Near vertical"	n/a	n/a	4.8 (7.6)(f)	Collapse settlement	"Near abutments"	"Several" Width n/a	>6 (g)	First fill
Portland	14	Rock					Collapse				
Left abutment			16	55	52	1.7	settlement	No crack			
Right abutment			18	44	35	1.7		30, 36	"cracks"	n/a	First fill

TABLE 9.10 Part 2: Summary of Cracking Observed in Embankment Dams and Factors Which Influence Whether Cracking May Occur (Fell et al., 2024)

			VALLEY PROFILE			MAXIMUM OBSERVED SETTLEMENT		OBSERVED CRACKING			
DAM	*HEIGHT (M)*	*CORE FOUNDATION*	*UPPER SLOPE (DEGREES)*	*SLOPE CHANGE LEVEL (% HEIGHT)*	*LOWER SLOPE (DEGREES)*	*% HEIGHT*	*CAUSE*	*LOCATION RELATIVE TO ABUTMENT (M)*	*WIDTH (MM)*	*DEPTH (M)*	*WHEN CRACKING WAS OBSERVED*
Woodcrest	13	Rock Abut. 11 m alluv.									
Left abutment			3	76	26	1.5	Foundation settlement	77 (h) 82 (i)	1 to 3 37	n/a 9	1 year 6 years
Right abutment			12	None	n/a	1.5		None			
Shell Oil	23	Soil?									
Left abutment			24	None	n/a	0.6	Collapse settlement	11–24 (i)	4 cracks "cracks"	n/a n/a	First fill 4 years
Right abutment			24	None	n/a	1.6		9–30 (i)	4 cracks	n/a	First fill
Stockton Creek	24	Rock	35	70	70	n/a		None (j)	n/a	n/a	First fill
Boggy Creek	9	Rock Abut. 14m alluv.	14	0	43	n/a	Foundation settlement	None (k)	n/a	n/a	First fill
Wister	30	Rock River Channel 10 m alluv.	15 River bank	100	n/a	n/a	Differential settlement	None (l)	n/a	n/a	First fill
East Branch	56	Rock	40	60	? Construction road cut	n/a	Differential settlement	None (m)	n/a	n/a	4.5 years
Virginia Ranch abutment	10–13	Rock	Minus 13	n/a	15	1.9	Differential settlement	5 m from wall	25 (6n)	2.7	3 years
Round Butte	134	Rock	10	90	70	n/a	Differential settlement	22 (o)	25	>5	< 1 year?

TABLE 9.10 Part 3: Summary of Cracking Observed in Embankment Dams and Some Factors Which Influence Whether Cracking May Occur (Fell et al., 2024)

			VALLEY PROFILE			MAXIMUM OBSERVED SETTLEMENT		OBSERVED CRACKING			
DAM	*HEIGHT (M)*	*CORE FOUNDATION*	*UPPER SLOPE (DEGREES)*	*SLOPE CHANGE LEVEL (% HEIGHT)*	*LOWER SLOPE (DEGREES)*	*% HEIGHT*	*CAUSE*	*LOCATION RELATIVE TO ABUTMENT (M)*	*WIDTH (MM)*	*DEPTH (M)*	*WHEN CRACKING WAS OBSERVED*
Duncan	30 (15)(p)	Rock abut. 380 m alluv. and lacustrine	15	n/a	n/a	15	Foundation settlement	25 m from edge of alluvium in foundation	25 to 80	>10.5	During constr'n
Matahina	85	Rock									
Left abutment			70	80 (2 m step)	70	0.1 (0.3q) 0.05	Embankment (q)	9 m from abutment	2.5 m soft zone	3 m to 15 m	21 years
Right abutment			50	86 (3 m step)	75	(0.3q)		16 m from abutment	50 widest	1.5 m to 16 m	First fill
Tullaroop	42	Rock and tertiary soils									
Left abutment			7	55	50	0.23	Embankment	15 m–45 m left slope change	60 (15r)	> 3 m	45 years
Buffalo	35	Rock	80	76 (3 m step)	47	1.0	Embankment	Over edge of step	40	8 (to step)	36 years
Red Willow	30	Rock abut. 15 m alluv.									
Outlet conduit			0	0	90	0.2 (3.5s)	Foundation settlement	Over conduit	50 (5t)	> 2.7	48 years
Creek bed			15	Base creek	−15	0.2 (3.5s)		Over creek	25 (19t)	> 3.3	48 years
Austrian	56	Rock									
Left abutment			40 (u)	n/a	n/a	1.4(1.8v)	Seismic load	0–15 m(w) At abut.	200 450	3.1 8	39 years 39 years
Right abutment			40 (u)	n/a	n/a	1.4(1.8v)		0–7 m(x)	250	7	39 years (Y)

Notes:
(a) Crack width at 7 m depth = 12 mm. Note that strains measured on crest survey marks 100 ft apart were < 0.25%; 0.36% and 0.23% over the cracked areas.
(b) Estimated by Sherard based on other cases.
(c) As measured on upstream slope.

(*Continued*)

TABLE 9.10 (Continued)

(d) From depth from which water flowed in the crack in downstream slope.
(e) 125 mm maximum at crest, 75 to 100 mm on upstream slope.
(f) 4.8% without allowance for camber, 7.6% allowing for camber.
(g) Measured to water level.
(h) 1954 data. Note that the strains measured on the crest survey marks 100 ft apart were $< 0.09\%$ and 0.37%.
(i) 16 m–21 m valley side of slope change.
(j) Dam failed by concentrated leak erosion. Sherard concluded that this was due to hydraulic fracture over steep cut in the abutment.
(k) Dam failed by concentrated leak erosion. Sherard concluded that this was due to hydraulic fracture resulting from differential settlement and desiccation.
(l) Dam experienced concentrated leak erosion. Sherard concluded that this was due to differential settlement of fill in river channel compared to fill on alluvium.
(m) Sherard concluded concentrated leak erosion occurred in crack over cut for construction road.
(n) Differential settlement occurred over a convex slope adjacent retaining wall. Fill was relatively poorly compacted. Crack 25 mm wide at crest, 6 mm at 1.8 m.
(o) Crack formed over a step in the abutment.
(p) Height of dam when cracks were mapped in shaft.
(q) Crest settled 50 mm on first filling, 30 mm in earthquake. No data on settlement in between. The foundation settled 200 mm during construction and figure in brackets includes that. Abutment cracking observed in 1988 but probably occurred during construction.
(r) Crack was diagonal on downstream slope. Crack widths 5 mm to 15 mm at 2.5 m to 3 m below surface.
(s) 2.5% to 3.5% end of construction, additional 0.2% post construction.
(t) Figures in brackets are crack widths at base of test pit.
(u) Estimated from contours. No longitudinal section available.
(v) First figures are settlement from seismic loading. Figure in brackets is total settlement since construction including seismic loading.
(w) Several cracks from abutment to 15 m from abutment.
(x) Separation of embankment and spillway wall.
(y) After Loma Prieta earthquake.

crest. These results are presented in He et al. (2017) and summarised in Table 9.11. They are approximate and only intended to provide a feel for the order of magnitude of the strains.

9.3.4 Summary of the Data

Table 9.11 summarises the information presented in Sections 9.3.1, 9.3.2 and 9.3.3.

From this it is concluded that:

(a) Post-construction settlement is a guide to whether cracking may occur. Post-construction settlements > 1% and in particular > 2% are most common in case data where cracking has been observed. However, post-construction settlements as low as 0.1% have resulted in

TABLE 9.11 Summary of Information on What Post-Construction Settlements and Localised Tensile Strains May Result in Cracking (He et al., 2017)

NATURE OF THE INFORMATION	*DETAILS*	*POST-CONSTRUCTION SETTLEMENTS*	*LIMITING LOCALISED TENSILE STRAINS*
Laboratory direct tension tests	Win (2006) Low plasticity soils	Not applicable	0.05% to 0.2%, independent of compaction moisture content relative to optimum.
	Leonards and Narain (1963)		
	Non-plastic soils		0.06% to 0.18%
	Low plasticity soils.		0.12% to 0.20%
	High plasticity soils		0.29% to 0.32%
Numerical modelling of dams in which cracking was observed or was not present	He (2016) and He et al. (2017, 2019, 2021)		
	Buffalo Dam	1%	Modelled 6% tensile strain over a step in valley profile using drained strength parameters, and 3% tensile strain using undrained strengths.
	Red Willow Dam	0.2%, 3.5% during construction	Modelled tensile strain over conduit 3% and 0.3% over creek bed.
	AECOM, 2017, Upper Yarra Dam	0.4% to 0.5% Post construction	Several cracks < 1 mm open.
Case studies of cracking from the literature and unpublished reports	He et al. (2017, 2019)		
	From Table 9.10 Majority of dams	1% to 2%,	Generally not available. Average tensile strains at crest for Rector Creek and Woodcrest Dams 0.25% to 0.35%. Based on data in Fell et al. (2024) (Figure 9.12), localised strains may be 4x average strain so 1% to 1.4% localised strain.
	Exceptions: Matahina	0.1%, very steep abutment.	
	Tullaroop	0.23%, affected by desiccation	
	From parametric modelling	n/a	Seven cases 1.5% to 2%, five 1.0% to 1.5%, two 0.5% to 1%, one < 0.5%. Note these are very approximate.

cracking if the abutment slope is very steep, and there are also examples with post-construction settlements around 0.2%. Hence it is concluded that post-construction settlement alone is not a reliable guide but is a useful screening tool.

(b) From laboratory tests, localised tensile strains < 0.05% are unlikely to result in cracking.

(c) From analyses of case data, localised tensile strains > 1.5% and particularly > 2% are likely to result in cracking. There are however cracked examples where estimated localised strains are as low as 0.5%. Estimated localised strains for Upper Yarra were 0.4% to 0.5% but there were no cracks > 1 mm open. This supports the possibility that 0.5% localised strain represents a low likelihood of wide-open cracks.

(d) Only three of the 20 cases in Table 9.10 Parts 1–3 resulted in failure by internal erosion and piping leading to breach. Four had no cracking observed, so only three out of 16 or about 20% would have had through-going transverse cracks. This lack of persistence needs to be taken account of when assigning the probability of transverse cracking which can lead to initiation of concentrated leak erosion.

(e) For cracks to be open they must be in tensile stress zones.

9.4 THE USE OF NUMERICAL MODELLING TO ASSESS THE LIKELIHOOD OF TRANSVERSE CRACKING

For important decisions in dam safety assessments and risk assessments and design of upgrades, numerical modelling has been used for several Australian dams to determine the location and depth of potential tension zones, tensile strains and strain localisation. Crack propagation modelling is then carried out to determine the potential location and depth of cracking. This is done for construction, post construction, and settlements induced by seismic loading. Chia et al. (2022) describe this process which was used as an input to the design of the upgrade of Upper Yarra Dam.

Figure 9.43 shows the valley cross section and the areas of the abutments for which crack propagation modelling was done. Figure 9.44 is an example of the outputs of the FEA and Figure 9.45 the output from the crack propagation modelling. Crack propagation modelling usually results in deeper cracks than indicated by tensile stress zones from the FEA analyses.

These analyses have been used on several major dams in Australia, and the authors suggest that cross valley numerical modelling should be regarded as a requirement for major dam safety assessments, risk assessments and design of upgrade works. Crack propagation modelling is valuable but not widely available.

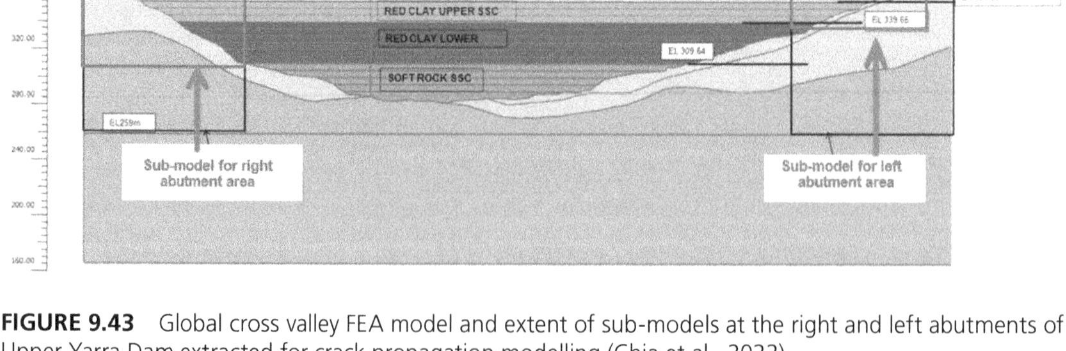

FIGURE 9.43 Global cross valley FEA model and extent of sub-models at the right and left abutments of Upper Yarra Dam extracted for crack propagation modelling (Chia et al., 2022).

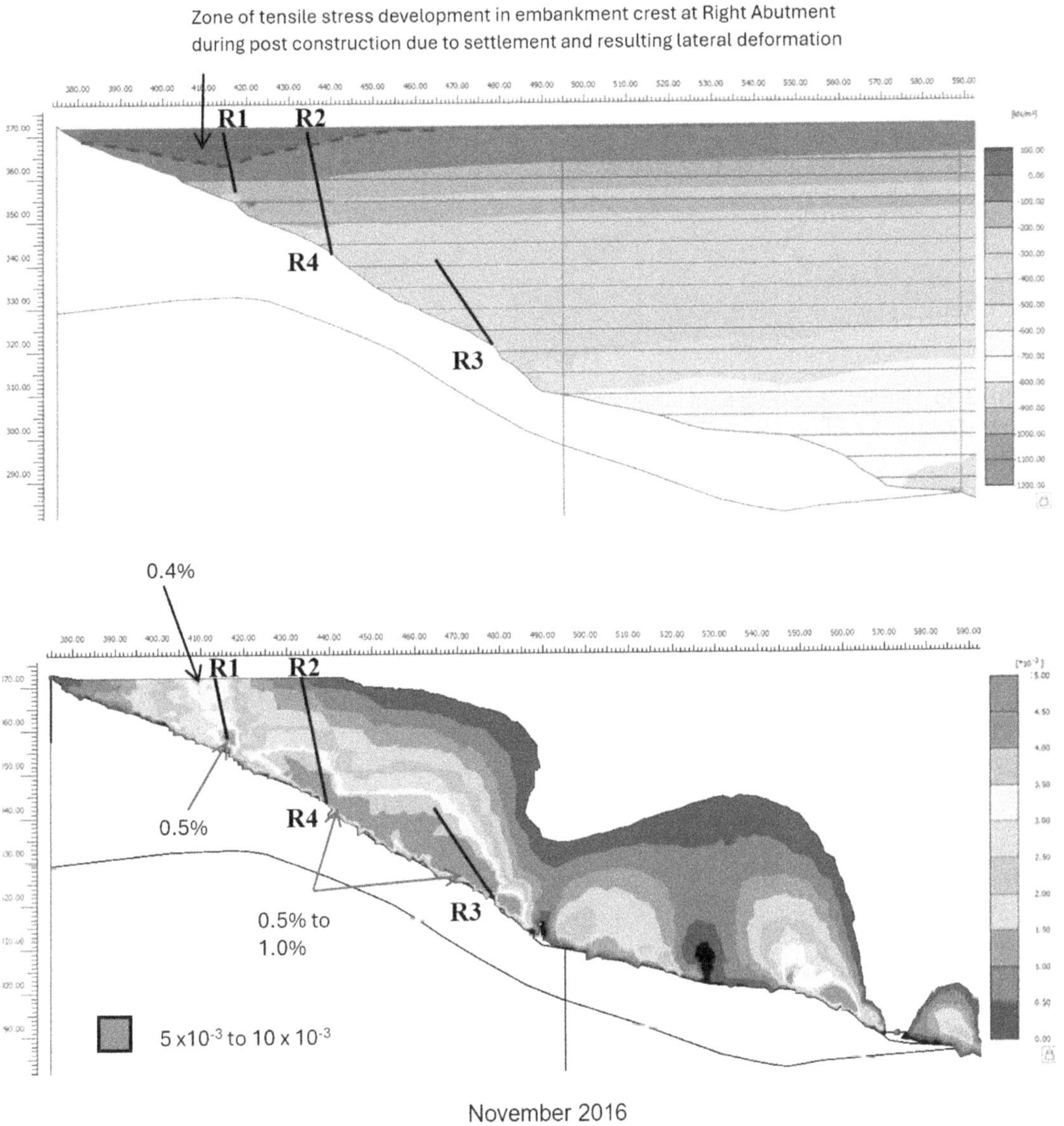

FIGURE 9.44 Upper Yarra Dam right abutment: cross valley analyses results 59 years after construction: total horizontal stresses (top) and cumulative longitudinal extension strain (bottom) (Chia et al., 2022).

These methods are applicable to the following initiating mechanisms:

IM 1 Cracking resulting from cross valley differential settlement.
IM 5 Cracking in the upper and lower embankment resulting from differential settlements in the soil in the foundation.
IM 12 Cracking and hydraulic fracture at the foundation contact due to small-scale irregularities in the foundation profile.
IM 21 and 22 Crack or gap adjacent to a spillway or abutment wall.

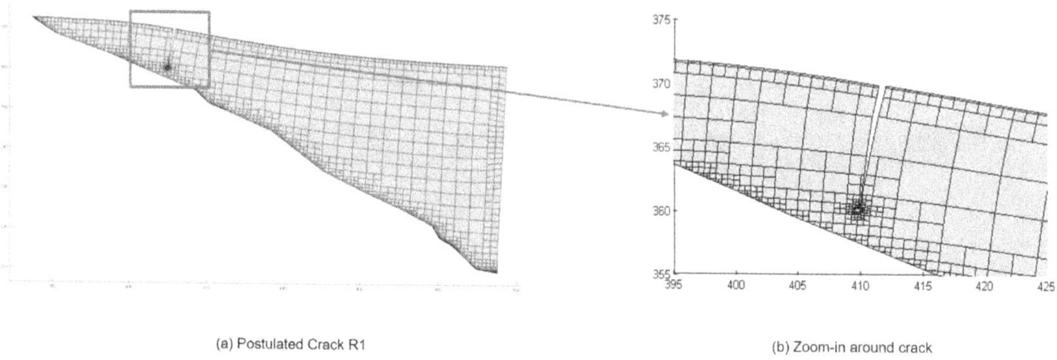

FIGURE 9.45 Upper Yarra Dam right abutment. Example of the results of crack propagation modelling for postulated crack R1 (Chia et al., 2022).

9.5 IM 14 TRANSVERSE CRACKING OR GAP RESULTING FROM POORLY COMPACTED ZONE WITHIN THE CORE

9.5.1 Introduction

It is well documented (e.g. Sherard, 1973; Foster et al., 2000a) that internal erosion and piping occurs in poorly compacted cohesive soils. This is particularly so for dispersive soils. The mechanism is potentially of two types as shown in Figure 9.46.

- The soil behaves as a series of clods with openings between the clods through which water passes.
- The soil collapses on saturation forming a gap (open pathway) through which the water can flow with relative ease.

This is most likely where there is poorly compacted soil against a conduit or a wall but is possible within layers of soil. Sherard (1973) gives examples of this.

The authors are aware of cases where a gap has been seen in test pits and/or zero CPT cone resistance value occurs just below the phreatic surface in the core of the dam. This has been caused by settlement of the soil in the saturated zone and arching of the unsaturated soil over the saturated material. In these cases, the gap may be a pathway for water at reservoir levels greater than the historic high.

The large number of piping incidents experienced in Swedish central core earth and rock fill dams (Nilsson, 2007a, 2007b) constructed with broadly graded glacial till cores may have been initiated by collapse settlement of the core which was compacted by small rollers in 800 mm thick layers.

Figure 9.47 shows concentrated leaks formed in Tunbridge Dam, Tasmania. Given the multiple leaks at the same level, it appears likely that these formed by collapse settlement of a poorly compacted layer or on a construction shut down surface which was not properly compacted.

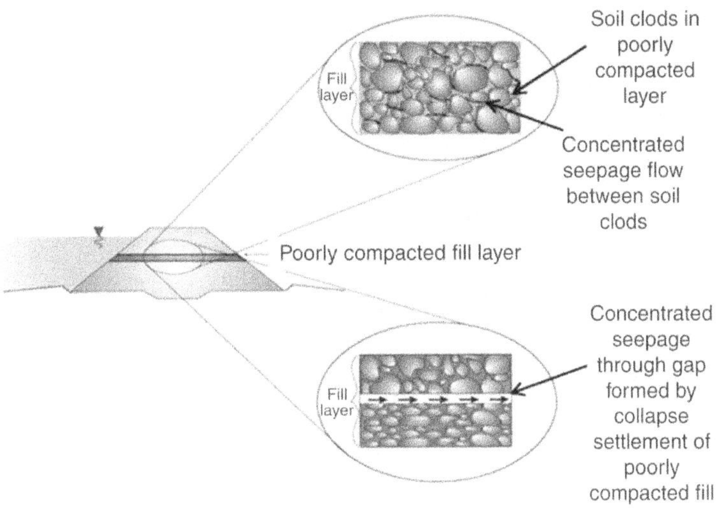

FIGURE 9.46 Mechanism of formation of concentrated seepage or gap in poorly compacted layer in the embankment (Fell et al., 2015).

FIGURE 9.47 Concentrated leaks in Tunbridge Dam possibly related to collapse settlement of a poorly compacted layer.

9.5.2 The Mechanics of Collapse Settlement of Poorly Compacted Soil

9.5.2.1 *Conditions Required for Collapse Settlement to Occur*

Collapse settlement occurs when poorly compacted partially saturated soils are wetted. Most of the effect will occur on first wetting of the poorly compacted layer, so it will occur on first filling, or in the upper parts of the dam when reservoir levels equal or exceed the pool of record. However, there is some chance a layer will still experience some further collapse settlement on second and subsequent reservoir level rises which allow water to reach the layer.

Lawton et al. (1989, 1992) quote Barden et al. (1973) and Mitchell (1976) as saying that four conditions are necessary for collapse to occur in a soil:

(a) An open, partially unstable, partially saturated fabric.
(b) A high enough total stress so that the structure is unstable.
(c) A sufficiently large soil suction or the presence of bonding or cementing agent that stabilises the soil in the partially saturated condition.
(d) The addition of water to the soil which reduces the soil suction or softens or destroys the bonding agent, thereby causing shear failures at the inter-aggregate or intergranular contacts.

Lawton et al. (1989) conclude from a review of the literature:

(e) For any set of conditions, the amount of collapse generally decreases with increasing pre-collapse moisture content, increasing pre-collapse dry density and decreasing overburden pressure.
(f) For any soil there are combinations of initial dry density, moulding moisture content and overburden pressure at which no volume change will occur when the soil is inundated.
(g) For a given soil, there appears to be a critical moulding moisture content at soaking above which no collapse will occur. For some soils the critical moisture content is above Standard Proctor optimum moisture content.
(h) There is a critical degree of saturation for a given soil above which negligible collapse will occur regardless of the magnitude of the pre-wetting overburden pressure.

9.5.2.2 Published Data on Collapse Settlement of Soils

Lawton et al. (1989) carried out oedometer collapse settlement tests on a soil which was taken from a deep fill in Southern California. It seems likely the soil was of alluvial origin. El Howayek et al. (2011) carried out tests on two loess soils from Indiana.

Charles and Skinner (2001) carried out oedometer collapse settlement tests on four soils. Two were coal mine colliery spoil which were silty sand gravels, so not typical of dam core material. The other soils were a mudstone typical of open cast mine backfill in the United Kingdom and boulder clay.

Table 9.12 summarises the properties of the soils. These tests are oedometer consolidation tests first carried out at as-compacted moisture content and second, soaked. The difference in sample thickness is the collapse settlement.

Figures 9.48, 9.49 and 9.50 summarise the results of the Lawton et al. (1989) tests.

It can be seen that for the soil Lawton et al. (1989) tested:

- The amount of collapse settlement is dependent on the relative compaction and moisture content relative to optimum moisture content.
- The greatest collapse settlement is for soils compacted well dry of optimum moisture content and low relative % relative compaction.
- For overburden pressures less than about 100 kPa the soil compacted to 90% Standard Proctor relative compaction (i.e. very poor compaction) swelled on saturation. For the soil

TABLE 9.12 Summary of Properties of Soils Tested by Lawton et al. (1989), Charles and Skinner (2001) and El Howayek et al. (2011)

TEST	*LAWTON ET AL. (1989)*	*EL HOWAYEK ET AL. (2011)*		*CHARLES AND SKINNER (2001)*	
		SOIL A	*SOIL B*	*MUDSTONE*	*BOULDER CLAY*
ASTM classification	SC	CL	CL	CL	CL-CH
Liquid Limit	34	37%	39%	39	47
Plasticity Index	15	18%	15%	19	27
% Gravel	0	0	0	31	9
% Sand	62	5	14	26	22
% Silt	23	70	72	43	69
% Clay	15	25	14		
Standard compaction Maximum dry density kN/m^3	17.7	16.6	15.4	16.7	16.3
Standard compaction optimum moisture content %	15	18	21	12	14

Note: Clay size is taken as minus 0.002 mm.

compacted to 95% relative compaction, the soil swelled at overburden pressure less than about 200 kPa.

- For the soil compacted to 90% Standard Proctor relative compaction some settlement occurred for the sample compacted 1% wetter than OMC. For the soil compacted to 95% relative compaction no settlement occurred. For the soils compacted at moisture content 4% and 7% above OMC no settlement occurred.

The soils tested by El Howayek et al. (2011) behaved in a similar manner to the soil tested by Lawton et al. (1989). Figure 9.51 shows an example of the results of El Howayek et al. (2011) tests.

Table 9.13 has been developed by Fell et al. (2024) from these data which are for plastic soils. In the absence of better information, it may be applied to non-plastic soils. However, caution should be applied in doing this and, for important decisions, laboratory tests on the soil should be carried out.

9.5.3 Mechanisms Which May Form a Continuous Gap or Crack

Poorly compacted zones within the embankment core can result in four mechanisms which may form a continuous gap or crack in which concentrated leak erosion may initiate:

Mechanism A. The soil behaves as a series of clods with openings between the clods through which water passes.

Mechanism B. The soil surrounding a conduit or adjacent to a wall collapses on saturation forming a gap at the contact of the soil with a conduit or wall.

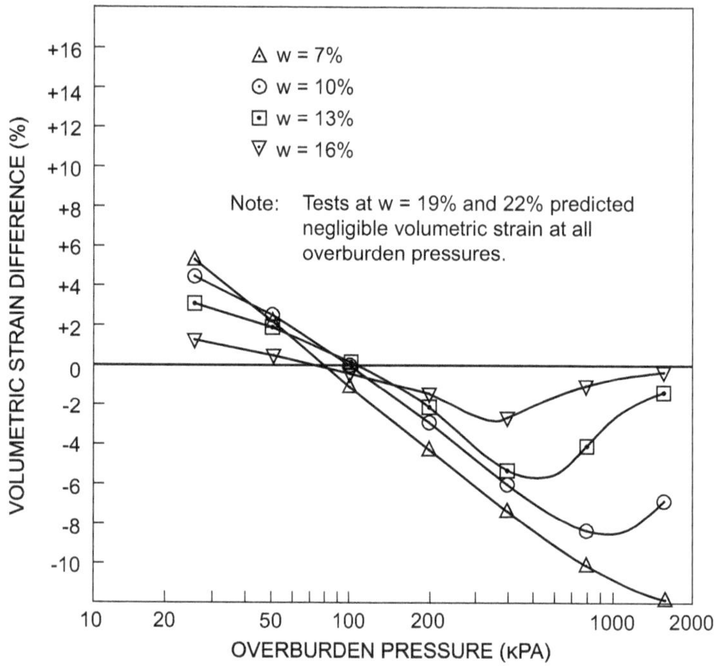

FIGURE 9.48 Effect of moulding moisture content on volumetric strain difference versus overburden pressure curve at relative compaction 90% of Standard Proctor dry density. Standard Optimum Moisture Content is 15% Lawton et al. (1989). (Used with permission of ASCE conveyed through Copyright Clearance Center Inc.)

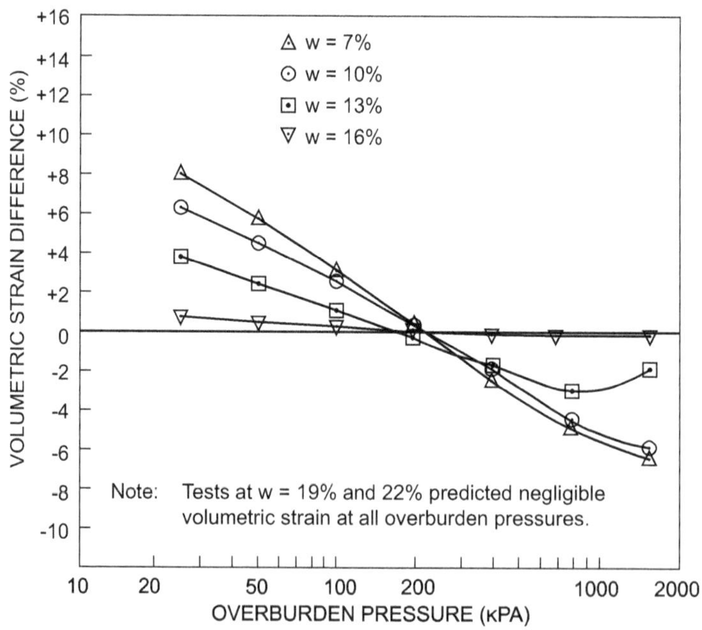

FIGURE 9.49 Effect of moulding moisture content on volumetric strain difference versus overburden pressure curve at relative compaction 95% of Standard Proctor dry density. Standard Optimum Moisture Content is 15% Lawton et al. (1989). (Used with permission of ASCE conveyed through Copyright Clearance Center Inc.)

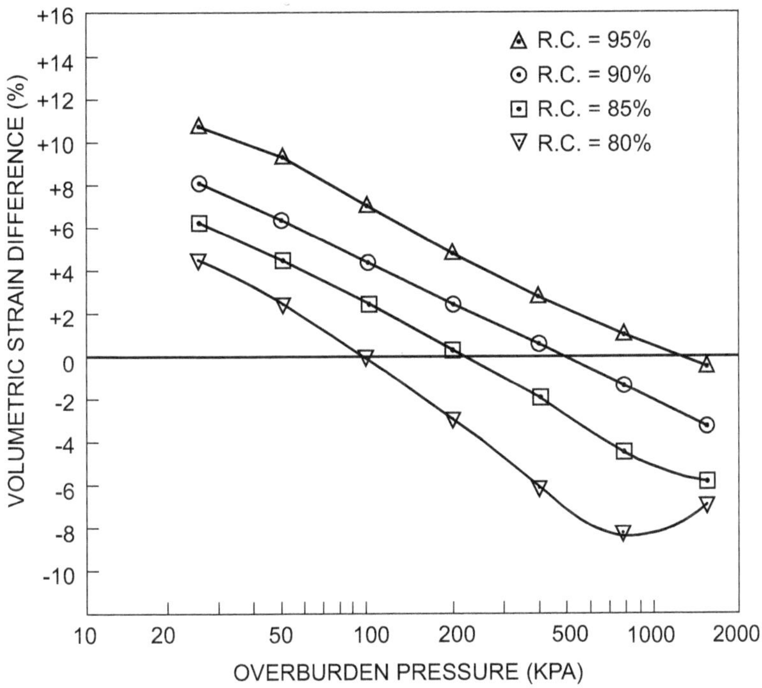

FIGURE 9.50 Effect of relative compaction on volumetric strain difference versus overburden pressure curve at moulding moisture content = 10%, i.e. 5% dry of Standard Proctor optimum moisture content Lawton et al. (1989). Relative compactions shown are for Modified Proctor. The equivalent Standard Proctor relative compactions are 90%, 95%, 100% and 106%. (Used with permission of ASCE conveyed through Copyright Clearance Center Inc.)

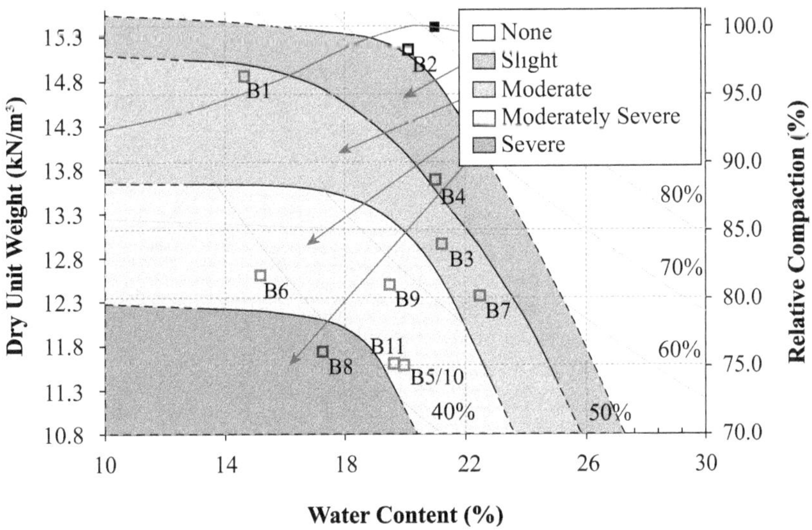

FIGURE 9.51 Contours of collapse potential for Soil B. Slight = 0.1% to 2%; Moderate = 2.1% to 6%; Moderately Severe = 6.1%–10%; Severe >10%. Data is for Standard Proctor compaction and 200 kPa overburden stress. OMC = 21% (El Howayek et al., 2011).

TABLE 9.13 Amount of Collapse Settlement Which May Occur on Saturation versus Compaction Properties (Fell et al., 2024)

DESCRIPTION OF THE METHOD AND DEGREE OF COMPACTION OF THE CORE		*AMOUNT OF COLLAPSE SETTLEMENT AS A PERCENTAGE OF THE LAYER THICKNESS (C_S) (a)*		
QUALITATIVE DESCRIPTION	*QUANTITATIVE DESCRIPTION*	*OVERBURDEN TOTAL STRESS 30 KPA*	*OVERBURDEN TOTAL STRESS 200 KPA*	*OVERBURDEN TOTAL STRESS 400 KPA*
Soil placed with no formal compaction, e.g. by horse and cart in old dams, or by pushing into place by excavator or bulldozer in very thick layers > 1 m thick Well dry of OWC.	Layers very poorly compacted, dry of Standard Optimum Moisture Content e.g. < 85% standard dry density ratio, 5% dry of standard OWC.	3%	6%	12%
Soil placed with no formal compaction, e.g. by horse and cart in old dams, or by pushing into place by excavator or bulldozer in thick layers 0.6 m to 1 m thick. Well dry of optimum moisture content.	Layers very poorly compacted, dry of Standard Optimum Moisture Content e.g. < 90% standard dry density ratio, 3% dry of standard OWC.	1.5%	3%	6%
Soil placed and compacted by bulldozer, no compaction by rollers, or rolled in thick layers beyond the capability of the roller. Dry of optimum moisture content.	Layers poorly compacted, dry of Standard Optimum Moisture Content e.g. < 93% standard dry density ratio, 2% to 3% dry of standard OWC.	1%	1.5%	2%
Soil rolled in layers near the limit of the capability of the rollers, at moisture contents dry of standard OWC.	Compacted to e.g. 95% standard dry density ratio, moisture content 2% to 3% dry of standard OWC.	Will not collapse	Will not collapse	1%
Soil compacted by suitable rollers in suitable layer thickness. Around optimum moisture content.	Well compacted to e.g. 98% standard dry density ratio, moisture content 1% dry of optimum to 1% wet of standard OWC.	Will not collapse	Will not collapse	Will not collapse
As above but with good documentation and records.	All very well compacted to e.g. ≥ 98% standard dry density ratio, moisture content 2% dry of optimum to 1% wet of standard OWC.	Will not collapse	Will not collapse	Will not collapse.

Note: (a) Estimated settlements are for first filling and reservoir levels above pool of record.
For reservoir levels below POR assume settlement = 0.1 C.

Mechanism C. The soil collapses on saturation causing settlement of the embankment above the poorly compacted layer. The settlement results in arching of the core onto the filters and/or shoulders, resulting in a gap at the top of saturated soil, most likely at the phreatic surface.

Mechanism D. Additional settlement of the embankment resulting from collapse settlement in the poorly compacted layer causes transverse cracking as shown in Figure 9.52. This is unlikely to result in transverse cracking unless the poorly compacted layer is thick and very poorly compacted.

For valley cross sections such as 1–1 in Figure 9.52, the effect of collapse settlement will be to result in tensile strains over the ends of the poorly compacted layer. That is at A and B.

For valley cross sections as 2–2 in Figure 9.52, the effect of collapse settlement will be to result in tensile strains over the ends of the poorly compacted layer. That is at C and D.

Saturation and potential collapse settlement are most likely to occur on first filling of the reservoir or when the reservoir rises above the historic high level. Collapse settlement of the upstream shoulder

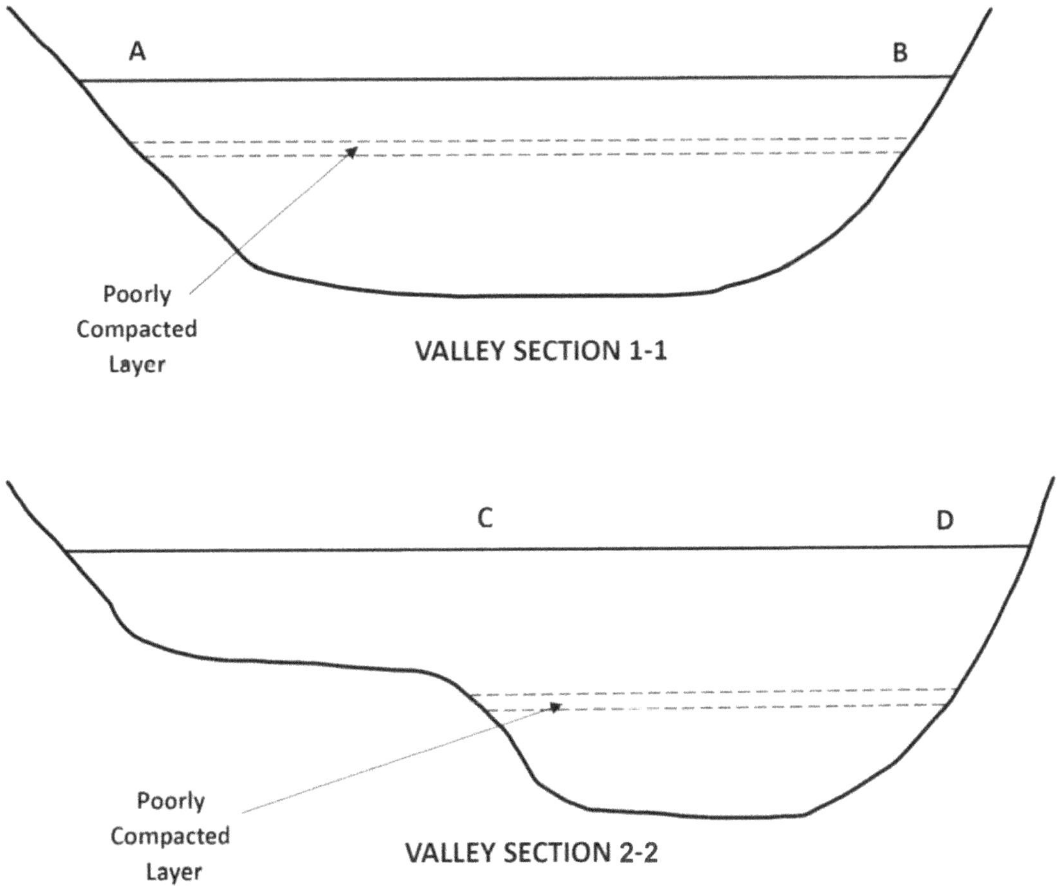

FIGURE 9.52 Locations of tensile strains resulting from collapse settlement in poorly compacted layers (Fell et al., 2024).

as the reservoir fills or of the downstream shoulder due to high tailwater levels can also result in increased settlement of the core.

In Fell et al. (2008, 2015) it was assumed that the weight of the soil above the layer subject to collapse settlement would be supported by non-collapsed soil adjacent. It was assumed that a gap would form in the collapsed soil layer equal to the estimated collapse settlement.

However, as shown for example in Figure 9.49, for the soil to collapse it must have an overburden stress on it of at least 100 kPa or 200 kPa. The soil cannot collapse with no overburden stress which is implicit in assuming the load is supported by adjacent non-collapsed soil. Instead, it swells.

It is therefore considered that what was assumed in Fell et al. (2008, 2015) was conservative and the updated procedure is described in Section 9.5.4.

9.5.4 Estimation of the Likelihood and Thickness of a Continuous Gap or Crack Formed by Collapse Settlement

9.5.4.1 Overall Approach

The overall approach to estimating the probability of initiation of internal erosion in a gap or crack formed in the core resulting from collapse settlement is as follows:

(a) Estimate the probability of a continuous poorly compacted layer or layers and the location, thickness, degree of compaction and compaction moisture content relative to Standard Optimum Moisture Content.
(b) Estimate the total settlement which may occur on saturation of the layer(s).
(c) Estimate, given this settlement, the likelihood of a continuous gap or crack through the embankment core for each of the four mechanisms.
(d) Estimate the thickness of that gap or width of crack for each mechanism.

Care should be taken not to double count the effects of collapse settlement with the effects of observed settlement in an existing dam already considered in other failure modes.

9.5.4.2 Probability of a Continuous Poorly Compacted Layer

A poorly compacted layer within the core may be present as a result of one or more of the following:

(a) Layers of the soil being too thick for the compaction equipment.
(b) Compaction moisture content being significantly dry of optimum moisture content.
(c) Use of incorrect compaction equipment, e.g. use of a steel drum roller rather than a tamping feet roller for plastic soil.
(d) Poor construction supervision leading to (a) and/or (b) and (c).

To result in initiation of internal erosion the poorly compacted layer must persist across the full width of the core. This is more likely to occur where the core is narrow and as a result more likely in the upper parts of an embankment.

The likelihood of a continuous poorly compacted layer is estimated based on the available information from construction records and investigations of the core using engineering judgement.

9.5.4.3 Settlement Which May Occur on Saturation of the Poorly Compacted Layer

This is calculated from the layer thickness and likely collapse settlement from Table 9.13.

9.5.4.4 The Likelihood of a Continuous Gap or Crack through the Embankment Core

Mechanism A
Estimate the probability of the soil within the poorly compacted layer behaving as a series of clods using Table 9.14. Take account of any observations from site investigations into the poorly compacted layer(s) and of the soil in undisturbed bulk samples in the laboratory.

The probability of a continuous gap for this mechanism is equal to (Probability of a poorly compacted layer being present) × (Probability of a continuous pathway within the poorly compacted layer), from Table 9.14.

Poorly compacted plastic soils may behave as a series of clods with openings between the clods. Non-plastic soils are unlikely to form clods but may settle locally on saturation resulting in continuous pathways for water to flow. In the absence of other information Table 9.14 may be applied to non-plastic soils.

TABLE 9.14 Method for estimating the probability that the soil in the poorly compacted layer behaves as a series of clods with openings between the clods forming a continuous pathway through the core (Fell et al., 2024)

	PROBABILITY OF CONTINUOUS PATHWAY WITHIN A POORLY COMPACTED LAYER	
COMPACTION METHOD	*ABOVE POR*	*BELOW POR*
Soil placed with no formal compaction (e.g. by horse and cart in old dams, or by pushing into place by excavator or bulldozer in very thick layers). Well dry of OWC.	0.2	0.01
Soil placed with no formal compaction (e.g. by horse and cart in old dams, or by pushing into place by excavator or bulldozer in very thick layers). No control on layer thickness. Well dry of optimum moisture content.	0.05	0.0025
Soil placed and compacted by bulldozer, no compaction by rollers, or rolled in thick layers beyond the capability of the roller. Dry of optimum moisture content.	0.001	0.0001
Soil rolled in layers near the limit of the capability of the rollers, at moisture content ≥ 4% dry of standard OWC.	0.0001	Negligible
Soil compacted by suitable rollers in suitable layer thickness. Around optimum moisture content.	Negligible	Negligible
As above but with good documentation and records.	Negligible	Negligible

Note: See Table 9.13 for additional description of compaction conditions.

Mechanism B
This is covered in Section 9.6.1 for conduits and 9.6.2 for walls.

Mechanism C
This mechanism is covered in Section 9.2.6. Consider the effect of poorly compacted layers when evaluating the likelihood of hydraulic fracture.

Mechanism D
For tensile stresses and strains at A and B, the collapse settlement in the poorly compacted layer is assumed to be additional post-construction settlement unless it is already included in the settlement assessment for cross valley differential settlement (Section 9.2.1).

For tensile stresses and strains at C and D, the collapse settlement in the poorly compacted layer is assumed to be additional post-construction settlement in the foundation (Section 9.2.3).

9.5.5 IM 15 Transverse Cracking or Continuous Flaw Resulting from Poorly Compacted Layer on the Core Foundation Contact

A poorly compacted layer at the core foundation contact may be present as a result of one or more of the following:

(a) Layers of the soil being too thick for the compaction equipment.
(b) Compaction moisture content being significantly dry of optimum moisture content.
(c) Irregular surface of the soil or rock foundation making compaction difficult.
(d) Use of incorrect compaction equipment, e.g. use of tamping feet roller rather than rubber tyre equipment to compact the contact soil on a rock foundation.
(e) Poor construction supervision leading to (a) and/or (b) and (d).

To result in initiation of internal erosion the poorly compacted layer must persist across the full width of the core. This is more likely to occur where the core is narrow in the upper parts of an embankment.

The mechanics of development of a crack or gap are as for IM 14 described in Section 9.5. A method for estimating the likelihood of a poorly compacted layer at the core foundation contact is described in Appendix A.

9.6 IM 18 AND IM 20 CONTINUOUS CRACK OR GAP ASSOCIATED WITH CONDUITS EMBEDDED IN THE EMBANKMENT AND WALLS

9.6.1 IM 18 Transverse Crack or Gap Associated with Conduits

A continuous crack or gap around a conduit through the embankment may result from the following mechanisms:

Mechanism AC. Poor compaction of the soil around the conduit resulting from:

(a) Round conduits not being provided with concrete surround making compaction below the conduit difficult or impossible. See Figure 9.53(b).

(b) Conduits constructed in deep narrow trenches such as in Figure 9.54(b). The narrow trench makes compaction difficult and any settlement results in arching of the backfill onto the sides of the trench.

(c) Conduits constructed in trenches just the width of the conduit without enough room provided for heavy compaction equipment. Figure 9.55 shows examples of good and poor practice.

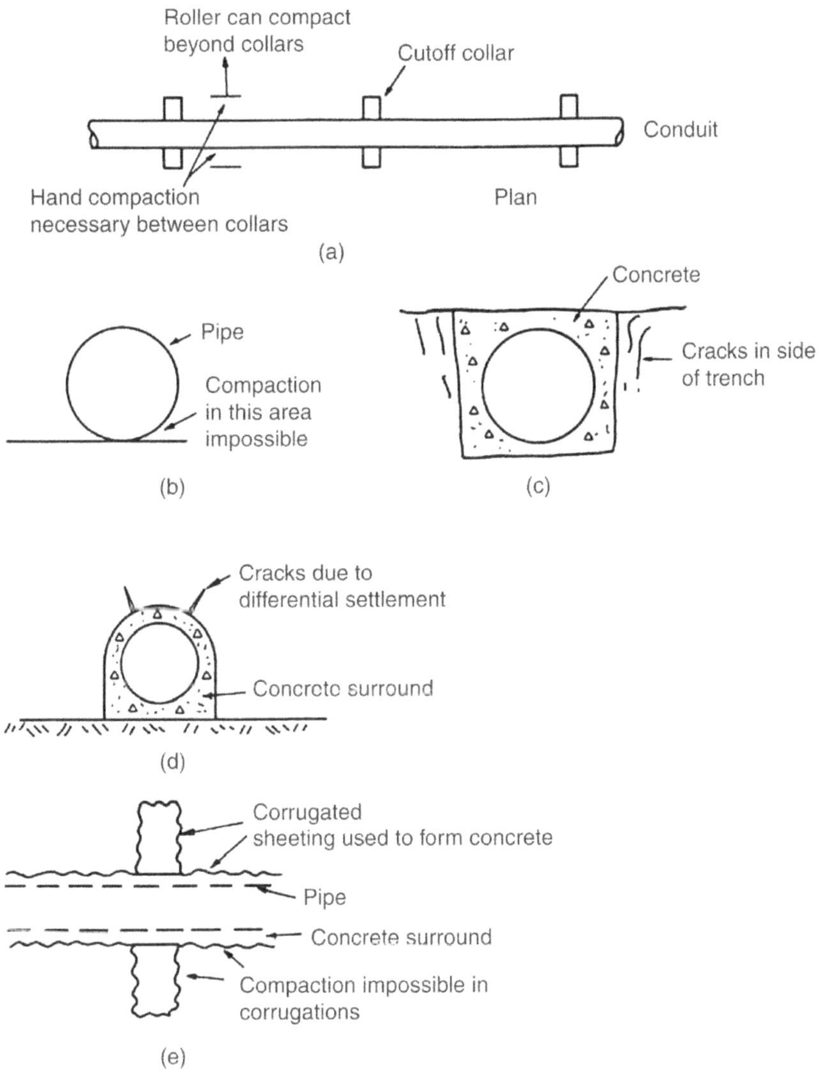

FIGURE 9.53 Some causes of internal erosion around conduits: (a) inadequate compaction due to the presence of cutoff collars; (b) inadequate compaction under pipe; (c) cracking in soil or extremely weathered rock in the sides of a trench; (d) cracks due to differential settlement; (e) use of corrugations or other roughening of the surface of cutoff collars or concrete surround (Fell et al., 2015).

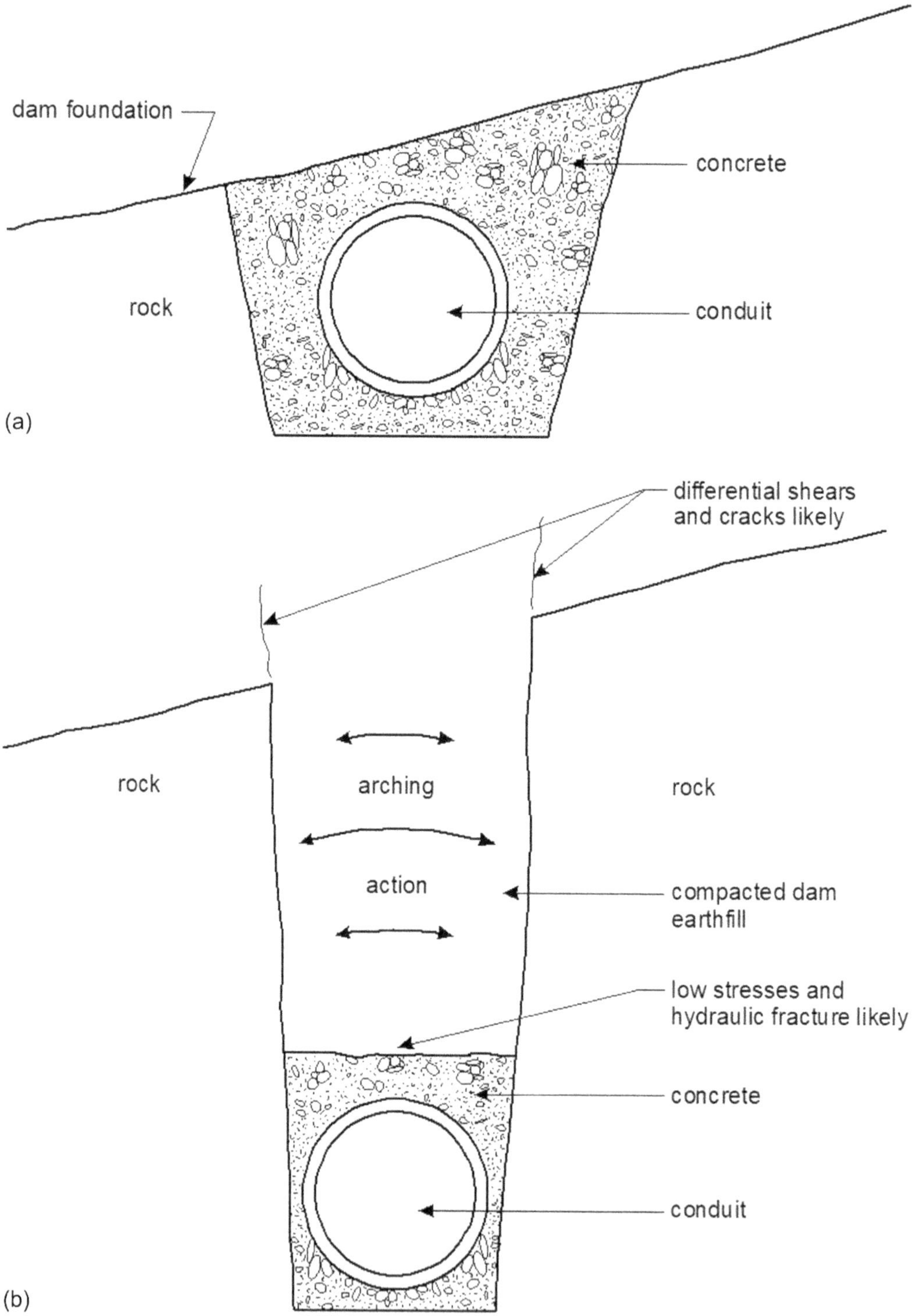

FIGURE 9.54 Conduit embedded in rock foundation: (a) good practice, (b) poor practice resulting in the potential for poor compaction and collapse settlement and hydraulic fracture (Fell et al., 2008, 2024).

(d) The use of cutoff collars which make compaction of the soil by compaction equipment such as rubber-tyred rollers difficult or impossible, so backfill compaction is done by hand. Figure 9.53(a) shows an example.
(e) Use of corrugated metal pipe as the conduit, or corrugated metal as formwork for the concrete as shown in Figure 9.53(e). The corrugations make compaction against the pipe difficult.

These scenarios are susceptible to collapse settlement of the poorly compacted fill on saturation forming a gap or creating conditions susceptible to hydraulic fracture.

Mechanism BC. Cracking of soils in the sides of the trench in which the conduit is constructed by desiccation and/or stress relief during construction. Figure 9.53(c) shows an example.

Mechanism CC. Differential settlement of the soil around the conduit causes low stress zones subject to hydraulic fracture [Figure 9.53(d)]. This mechanism should be assessed using the approach detailed in Section 9.2.5, IM 12, with the conduit geometry considered as part of the foundation profile.

The approach to estimating the likelihood of a gap or crack formed around a conduit is:

9.6.1.1 Mechanism AC

(a) Estimate the probability of a continuous poorly compacted layer using the method described in Appendix A. The method includes all the scenarios in *Mechanism AC*.

Estimate the thickness of the gap or width of crack which may result using the method described in Appendix A.

9.6.1.2 Mechanism BC

Use engineering judgement/expert elicitation to estimate the probability of a continuous open desiccation crack or stress relief crack in soil in the sides of the trench and the width of the crack based upon the available data.

The potential for swell of the soil closing the gap or crack should not be included if the cracking mechanism is assessed to be due to a hydraulic fracture mechanism.

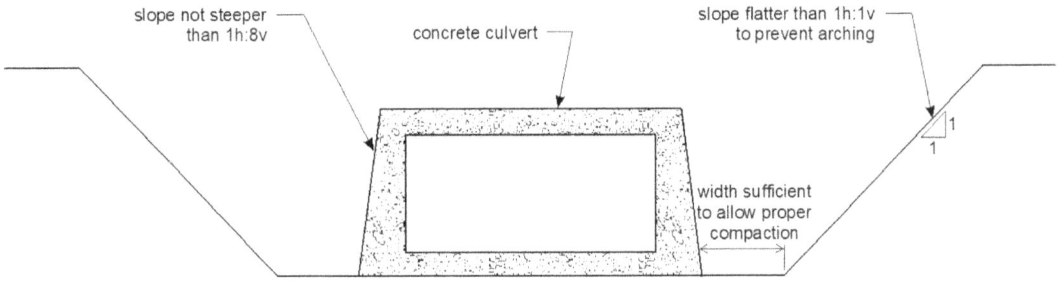

FIGURE 9.55 Trench and culvert details for concrete cast in *in situ* or pre-cast rectangular concrete culvert. Good practice is as shown. Poor practice has little or no width between the culvert and slope, near vertical slope and vertical side slopes for the culvert which preclude use of rollers to compact the fill against the culvert (Fell et al., 2024).

FEMA (2005b) "Conduits through Embankment Dams. Best practices for design, construction, problem identification and evaluation, inspection, maintenance, renovation and repair" provides a comprehensive coverage of the issues relating to conduits in embankment dams.

9.6.2 IM 20 Transverse Crack or Gap Associated with Poor Compaction Adjacent to a Spillway or Abutment Wall

9.6.2.1 Overall Approach

A continuous layer of poorly compacted soil adjacent to a spillway or abutment wall may result in formation of a transverse crack or gap by the following mechanisms:

Mechanism AW. The soil behaves as a series of clods with openings between the clods in which water passes.

Mechanism BW. The soil collapses on saturation causing settlement of the fill. This results in a gap and/or low vertical stress zone within the poorly compacted layer or adjacent to the wall.

The overall approach to estimating the probability of a gap or crack formed at the wall is:

(a) Estimate the probability of a continuous poorly compacted layer or layers, and the location, degree of compaction and compaction moisture content relative to Standard Optimum Moisture Content.
(b) Estimate the total settlement which may occur on saturation of the layer(s).
(c) Estimate, given this settlement, the probability of a continuous gap or crack through the embankment core for each of the two mechanisms.
(d) Estimate the thickness of that gap or width of crack for each mechanism.

Care should be taken not to double count the effects of the poorly compacted layer with the effects of development of a crack or gap resulting from differential settlement covered in IM 21 and IM 22.

Appendix A describes a method for estimating the likelihood of the presence of a crack or gap formed by poor compaction next to a wall.

9.7 IM 19A EROSION INTO A NON-PRESSURISED CONDUIT AND IM 19B LEAKAGE OUT OF A PRESSURISED CONDUIT LEADING TO EROSION ALONG THE CONDUIT

9.7.1 IM 19A Erosion into a Non-Pressurised Conduit

As shown in Figure 9.56, erosion of the soil surrounding a crack or open joint (a "defect") in a conduit may result in a sinkhole developing above the defect in the conduit. The mechanism for development of the sinkhole depends on the properties of the soil surrounding the conduit. This may be:

- Global Backward Erosion or Suffusion if the soil is silt/sand/gravel.
- Backward Erosion Piping if the soil is uniform fine to medium sand, e.g. as backfill around the conduit.
- Concentrated leak erosion if the soil is plastic.

Depending on the location of the defect, a sinkhole may form below or above reservoir level. If it forms to exit into the reservoir the PFM may progress by hydraulic fracture along the exterior of the conduit resulting in piping failure along the conduit.

9.7.2 IM 19B Erosion of Soil Resulting from Leakage out of a Pressurised Conduit

As shown in Figure 9.57, if a conduit is flowing under pressure, leakage may occur out of cracks or open joints (defects) in the conduit. This may result in:

- Backward Erosion Piping of the soil around the conduit downstream of the defect if the soil is uniform fine to medium sand.
- Hydraulic fracture along the exterior of the conduit resulting in piping along the conduit downstream of the defect.

If the leakage out of the conduit occurs downstream of the core zone, then there would need to be an interconnecting defect through the core zone or for the pipe to head-cut upstream to form a continuous concentrated leak through the embankment.

A method for assessing the likelihood of initiation of IM 19A and IM 19B is presented in Appendix A.

9.8 CRACKING RESULTING FROM DESICCATION AND FREEZING

9.8.1 IM 7 and IM 8 Cracking Due to Desiccation

9.8.1.1 IM 7 Desiccation Cracking in the Crest of the Dam

Desiccation cracking occurs when the soil in the core of the embankment is dried from its original moisture content. Whether cracking occurs and the potential depth and width of cracking are dependent on:

(a) The climate, best considered as the average annual net evaporation which is (average annual pan evaporation) – (average annual rainfall). Dams sited in areas with large net evaporations are more likely to have desiccation cracking, and the depth and width are likely to be greater than for a dam located where there is small net evaporation.
(b) Whether the core is covered by a sealed or gravel road or a layer of rockfill. This is because the road pavement, being silty sandy gravel, cannot sustain large suctions, so drying of the core below the pavement is inhibited or in many cases prevented.

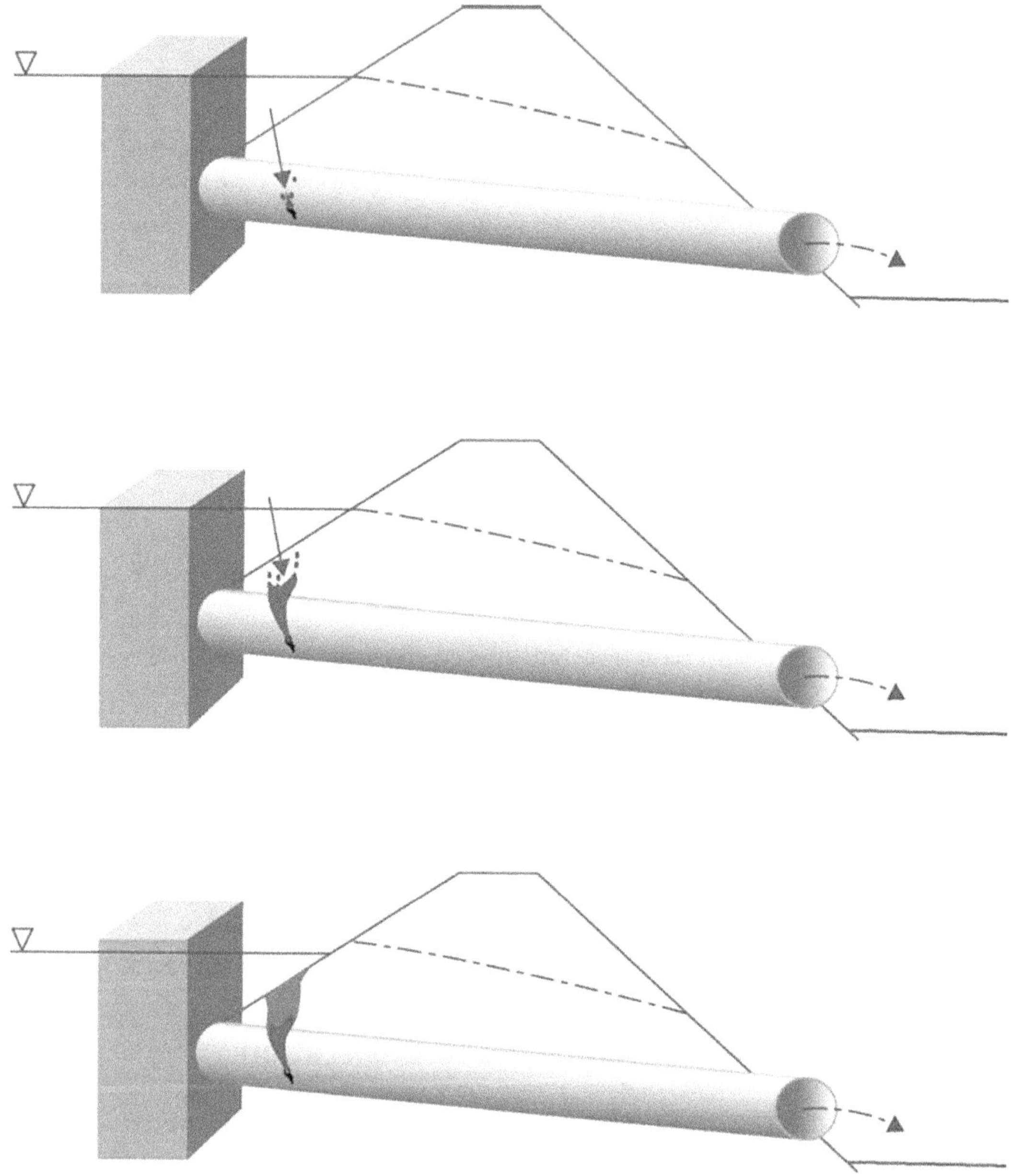

FIGURE 9.56 Erosion of the soil into a crack or open joint in a non-pressurised conduit resulting in a sinkhole (FEMA, 2005b).

(c) The Unified Soil Classification of the core material is a factor. The more plastic the soil the greater the likelihood of deep and wide desiccation cracking.
(d) Whether the soil was compacted wet or dry of optimum moisture content. Wet of optimum may result in somewhat greater shrinkage than dry of optimum.

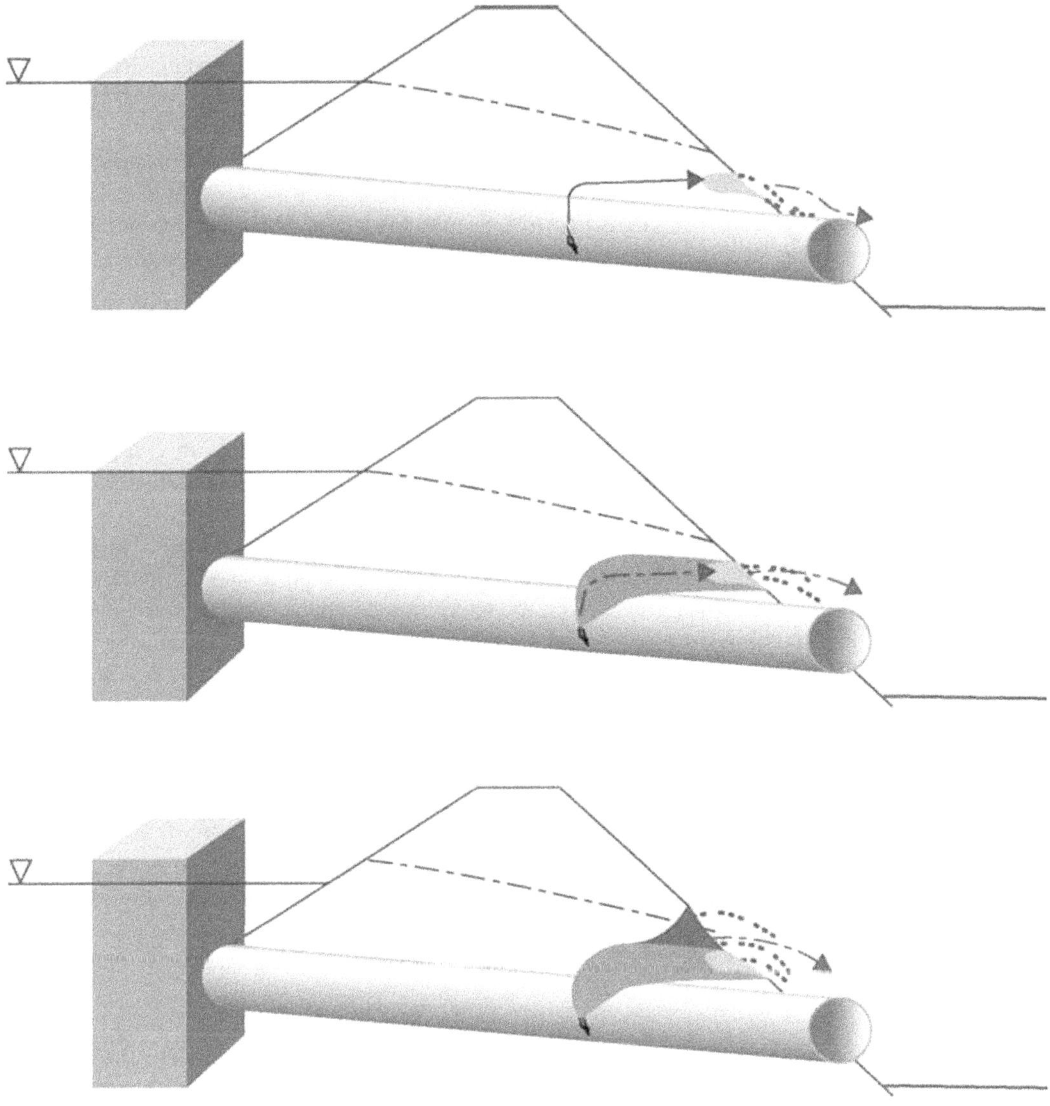

FIGURE 9.57 Leakage out of a pressurised conduit resulting in erosion of the soil around the conduit (FEMA, 2005b).

Based on a review of the literature, desiccation cracking can extend to significant depths. For example:

(i) Talbot (1994) reported desiccation cracking to a depth of 5 m in arid regions of the USA even in soils not associated with high shrinkage and swelling. The cracks were spaced up to 7 m to 10 m and were a few millimetres wide.

(ii) Lawrence (2002), summarising the outcome of the "Workshop on cracking of homogeneous earth dams in semi-arid environments", indicated that a number of dams in Maricopa Country, Arizona, had experienced extensive desiccation cracking. Some details are:
- The 22 dams are up to 32 m high, but most are less than 13 m high, with a combined crest length of 19 km. The dams were 26 to 48 years old.
- The average rainfall is 175 mm to 250 mm, with an annual net evapo-transpiration of 1700–1900 mm.
- The soils in the embankments classify as SM, SC, CC-ML, CL and ML, and were compacted to 100% standard maximum dry density (SMDD) at or near optimum moisture content.
- The soils had dried by about 6% to 7% since construction.
- Desiccation cracks extended to a depth of 6.4 m, with an average spacing of 3.6 m. The cracks were up to 75 mm wide.

(iii) As part of a study for Melbourne Water Corporation of the likelihood of concentrated leak erosion in desiccation cracks in flood retention basin embankments, data was gathered from consultants to Melbourne Water and from owners of dams in Queensland, Victoria, Tasmania and South Australia. This data is recorded in Fell et al. (2024). This showed that:
- For earthfill not covered by a road pavement, cracking extended to around 1.7 m for net evaporation of up to 425 mm and to around 3.5 m for net evaporation 900 mm.
- For earthfill covered by a gravel road, the maximum depth of cracking was 1.0 m and most had no cracking.

The lesser cracking beneath road pavements is to be expected because the equilibrium moisture content beneath paved surfaces or concrete slabs is around the Standard Optimum Moisture Content, so if fills were compacted close to Standard Optimum Moisture Content, as most dam cores will be, there should be little or no moisture change and as a result little or no cracking.

Based on this case data, Fell et al. (2024) developed Table 9.15 which gives maximum likely depth and width of desiccation cracking. If the dam has significant freeboard, desiccation cracking may not extend below FSL and may only be reached by the reservoir under flood conditions.

Apart from the data on Lawrence (2002), there is no case data to provide guidance for net evaporations greater than 1500 mm.

Fell et al. (2024) used the case data to develop Tables in Appendix A for estimating the likelihood of cracking being present, and if so whether it would persist across the crest of the embankment. The probability of desiccation cracking persisting across the dam core has been assumed to be high for wide cracks and low for small cracks. This is based on observations of desiccation cracking.

For important dam safety decisions, test pit excavations at the crest are recommended to be carried out to help evaluate the potential for the desiccation cracking to be present and to evaluate the characteristics of the cracks if present (width, depth and persistence across the crest). In assessing this data take account of the rainfall and evaporation leading up to when the test pits were excavated.

9.8.1.2 IM 8 Desiccation Cracking on Seasonal Shutdown Layers and on the Surface of Staged Embankments

Desiccation cracking may also occur on seasonal shut down surfaces during construction or on the surface of the first stage of dams which are built in two stages. Good construction practice would be to remove the desiccated soil, but this was not always done.

TABLE 9.15 Part A: Suggested Widths and Depths of Desiccation Cracking for Assessing Concentrated Leak Erosion, Net Evaporation 175 mm to 425 mm (Fell et al., 2024)

	UNIFIED CLASSIFICATION OF THE SOIL IN THE EMBANKMENT				
CREST CONDITIONS	*SILTY SAND, CLAYEY SAND, CLAYEY SILT SM-SC-ML*	*SILTY CLAY AND SANDY CLAY CL LL < 35%*	*CLAY AND SANDY CLAY CL-CH LL 35% TO 50% AND MH*	*CLAY CH LL 50% TO 65%*	*CLAY CH LL > 65%*
No road or pathway, or < 0.05 m thick gravel	1 to 2 mm ≤ 0.5 m	1 to 3 mm 0.5 to 1 m	5 to 10 mm at surface 2 to 5 mm at 1.0 m 1 mm at 1.5 m	10 to 20 mm at surface 3 to 10 mm at 1.5 m 1 mm at 2 m	20 to 30 mm at surface 5 to 10 mm at 2 m 1 mm at 3 m
Gravel road or pathway ≥ 3 m wide and 0.05 m to 0.2 m thick OR Sealed pathway < 2 m wide	No cracking expected	1 mm< 0.5 m	Estimate by interpolation between rows above and below allowing for road or pathway conditions	Estimate by interpolation between rows above and below allowing for road or pathway conditions	Estimate by interpolation between rows above and below allowing for road or pathway conditions
Gravel road ≥ 0.2 m thick ≥ 3 m wide OR Concrete or sealed pathway ≥ 2 m wide	No cracking expected	No cracking expected	No cracking expected	1 to 3 mm at surface ≤ 0.5 m	2 to 5 mm at surface 0.5 m to 1.0 m
Sealed road or concrete pathway ≥ 3 m wide	No cracking expected	No cracking expected	No cracking expected	No cracking expected	No cracking expected

9.8.2 IM 16 Cracking or Continuous Flaw Due to Freezing and IM 17 Cracking or Flaw Resulting from Freezing on a Seasonal Shutdown Layer

9.8.2.1 IM 16 Transverse Cracking or Continuous Flaw Resulting from Freezing at the Crest of the Embankment

The effects of frost on embankment dams are described in Vuola et al. (2007). These effects cause extra water to be drawn into the soil by capillarity action. This causes ice lenses to form with associated heave and potentially cracking and/or loosening of the soil. The cracking is most likely to be longitudinal but may also be transverse at the crest of the dam. Figures 9.58, 9.59 and 9.60 show the potential effects.

Vuola et al. (2007) provide an equation for estimating the depth of frost penetration.

TABLE 9.15 Part B: Suggested Widths and Depths of Desiccation Cracking for Assessing Concentrated Leak Erosion, Net Evaporation 450 mm to 900 mm (Fell et al., 2024)

	UNIFIED CLASSIFICATION OF THE SOIL IN THE EMBANKMENT				
CREST CONDITIONS	*SILTY SAND, CLAYEY SAND, CLAYEY SILT SM-SC-ML*	*SILTY CLAY AND SANDY CLAY CL LL < 35%*	*CLAY AND SANDY CLAY CL-CH LL 35% TO 50% AND MH*	*CLAY CH LL 50% TO 65%*	*CLAY CH LL > 65%*
No road or pathway, or < 0.05 m thick gravel	1 to 2 mm ≤ 0.5m	2 to 5 mm 0.5 to 1.5 m	10 to 20 mm at surface 2 to 5 mm at 1.5 m 1 mm at 2 m	15 to 30 mm at surface 5 to 10 mm at 2 m 1 mm at 3 m	20 to 40 mm at surface 5 to 10 mm at 2.5 m 1 mm at 4m
Gravel road or pathway ≥3 m wide and 0.05 m to 0.2 m thick OR Sealed pathway < 2 m wide	No cracking expected	2 to 5 mm ≤ 0.5 m	Estimate by interpolation between rows above and below allowing for road or pathway conditions	Estimate by interpolation between rows above and below allowing for road or pathway conditions	Estimate by interpolation between rows above and below allowing for road or pathway conditions
Gravel road ≥ 0.2 m thick ≥3 m wide OR Concrete or sealed pathway ≥ 2 m wide	No cracking expected	No cracking expected	No cracking expected	2 to 5 mm at surface 0.5 m to 1 m	5 to 10 mm at surface 1 m to 1.5 m
Sealed road or concrete pathway ≥ 3 m wide	No cracking expected	No cracking expected	No cracking expected	No cracking expected	No cracking expected

There are several references which provide information on the susceptibility of soils to frost penetration. These include Wallace (1987), Holtz and Kovacs (1981), Vuola et al. (2007) and USACE (1984) which gives the most specific data. Core materials which are most susceptible to freezing and ice lens formation include silts, clayey silts, silty sands, silty gravels and clayey sands and gravels with a Plasticity Index < 12.

9.8.2.2 IM 17 Transverse Cracking or Continuous Flaw Resulting from Freezing on Seasonal Shutdown Layers and on the Surface of Staged Embankments

Cracking or development of a flaw resulting from freezing on a shutdown surface only applies where there has been a seasonal shutdown during construction, or the embankment has been staged with the stage being subject to freezing conditions during the shutdown.

TABLE 9.15 Part C: Suggested Widths and Depths of Desiccation Cracking for Assessing Concentrated Leak Erosion, Net Evaporation 925 mm to 1500 mm (Fell et al., 2024)

	UNIFIED CLASSIFICATION OF THE SOIL IN THE EMBANKMENT				
CREST CONDITIONS	*SILTY SAND, CLAYEY SAND, CLAYEY SILT SM-SC-ML*	*SILTY CLAY AND SANDY CLAY CL LL < 35%*	*CLAY AND SANDY CLAY CL-CH LL 35% TO 50% AND MH*	*CLAY CH LL 50% TO 65%*	*CLAY CH LL > 65%*
No road or pathway, or < 0.05 m thick gravel	2 mm to 5 mm 0.5 to 1 m	5 mm to 10 mm 1 m to 2 m	10 to 20 mm at surface 2 to 5 mm at 2.0 m 1 mm at 3 m	20 to 40 mm at surface 5 to 10 mm at 2.5 m 1 mm at 4 m	20 to 50 mm at surface 5 to 10 mm at 3 m 1 mm at 5 m
Gravel road or pathway ≥ 3 m wide and 0.05 m to 0.2 m thick OR Sealed pathway < 2 m wide	No cracking expected	2–5 mm 0.5 m to 1 m	Estimate by interpolation between rows above and below allowing for road or pathway conditions	Estimate by interpolation between rows above and below allowing for road or pathway conditions	Estimate by interpolation between rows above and below allowing for road or pathway conditions
Gravel road ≥ 0.2 m thick ≥ 3 m wide OR Concrete or sealed pathway ≥ 2 m wide	No cracking expected	No cracking expected	No cracking expected	2 to 5 mm at surface 0.5 m to 1 m	5 to 10 mm at surface 1.5 m to 2 m
Sealed road or concrete pathway ≥ 3 m wide	No cracking expected	No cracking expected	No cracking expected	No cracking expected	No cracking expected

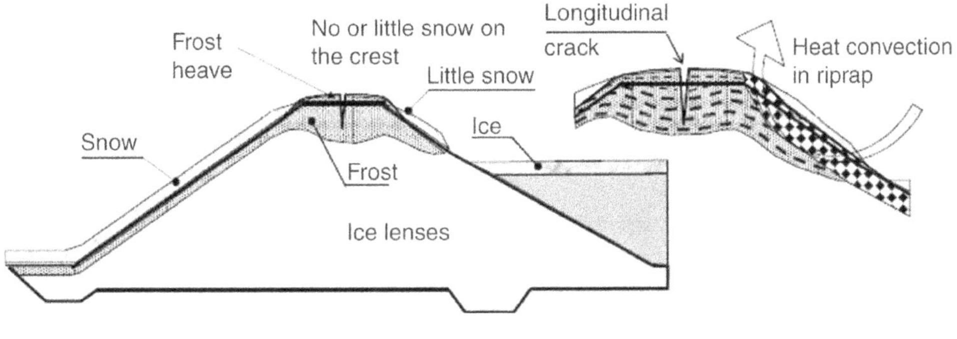

FIGURE 9.58 Effects of frost on an embankment dam (Vuola et al., 2007).

Fell et al. (2008, 2024) developed methods for estimating the likelihood and depth of cracking or gaps resulting from freezing for both IM 16 and IM 17. These are presented in Appendix A.

9.9 IM 13 TRANSVERSE CRACKING IN THE UPPER PART OF THE EMBANKMENT CAUSED BY EARTHQUAKE

When embankment dams experience a large earthquake, they often settle and spread in the upstream–downstream direction as shown in Figure 9.61. Many exhibit longitudinal cracking, and some, transverse cracking. An example is Austrian Dam shown in Figure 9.62 (Harder et al., 1991; Forster and MacDonald, 1998; Vrymoed and Lam, 2006).

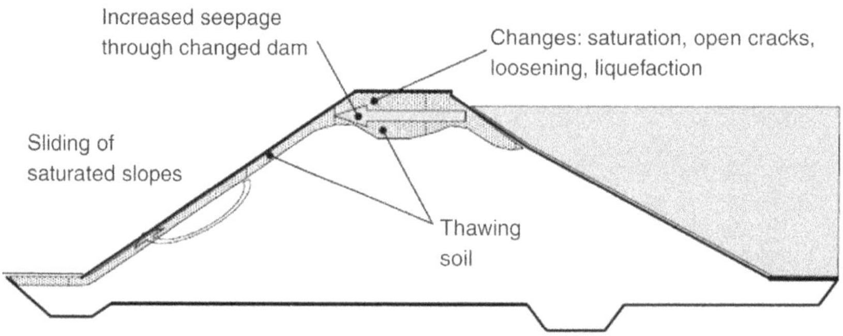

FIGURE 9.59 Damage phenomena occurring in embankment dams from frost thaw (Vuola et al., 2007).

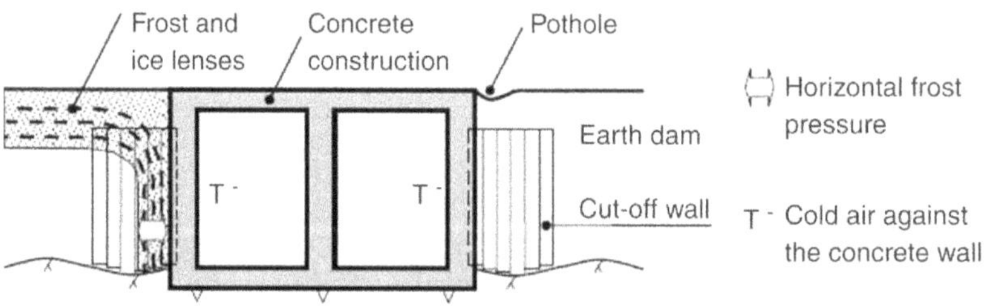

FIGURE 9.60 Effect of frost against a spillway structure (Vuola et al., 2007).

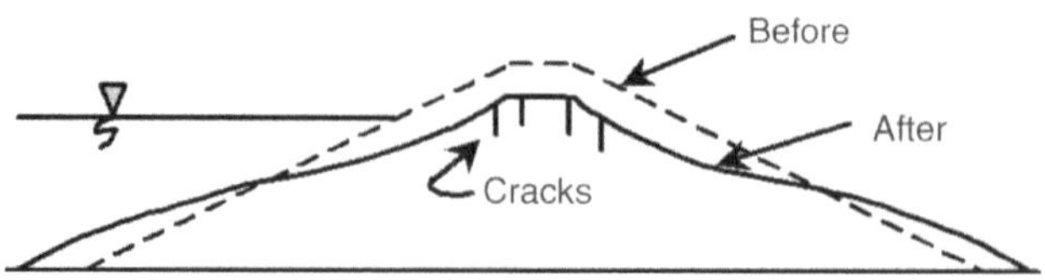

FIGURE 9.61 Settlement, lateral spreading, and longitudinal cracking of embankment dams in earthquakes.

Austrian Dam only settled about 0.5% but developed wide deep transverse cracks near the spillway wall as shown in Figure 9.62b.

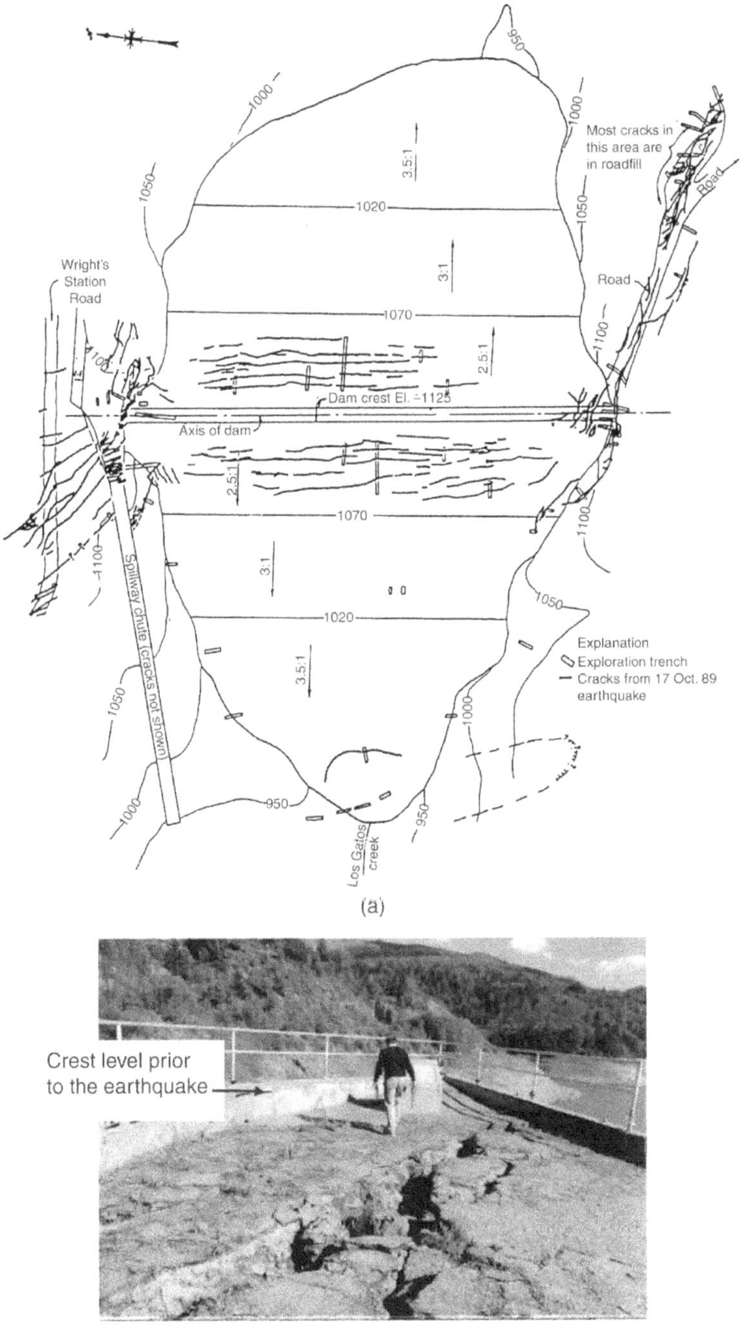

FIGURE 9.62 Transverse and longitudinal cracking of Austrian Dam during the Loma Prieta earthquake. (a) Plan of the cracking. (b) Cracking adjacent to the spillway (Forster and MacDonald, 1998). (Photograph courtesy of S. Todaro.)

Transverse cracks are important because concentrated leak erosion may initiate in them if the reservoir is above the base of cracking and the cracks are persistent.

Pells and Fell (2002, 2003) analysed case data for dams not subject to liquefaction. These are presented in Figures 8.2 and 8.3 and Table 8.1. If liquefaction occurs the deformations are likely to be large and the likelihood of cracking greater.

Pells and Fell (2002, 2003) concluded:

- Earthquakes cause settlement, lateral spreading and cracking of embankment dams. Slope instability may occur, but it is not common. Longitudinal cracks are more common than transverse cracks and are mostly in the upper part of the dam, more likely on the upstream face than the downstream.
- There are a number of mechanisms that can lead to formation of either of these types of cracks. Many of these mechanisms are common to both seismic loading and normal operating loading. There are seen to be more mechanisms that may lead to the formation of longitudinal cracks than transverse cracks, largely because lateral displacement is more readily achieved in the upstream–downstream directions.
- For visible longitudinal cracks to occur the dam needs to experience a Magnitude 6.5 or greater earthquake, and a peak ground acceleration greater than about 0.15 g for earthfill dams and 0.3 g for earth and rockfill dams. Alternatively, it needs an earthquake of M7 or greater, and a PGA of 0.05 g for earthfill and, say, 0.15 g for earth and rockfill dams. For hydraulic fill dams, visible cracking may occur for M6 earthquakes and 0.05 PGA.
- Dams which experience damage of Class 2 (relative settlement of 0.2% to 0.5% and/or longitudinal cracks 30–80 mm wide) or greater are highly likely to experience transverse cracking. However, transverse cracking has been observed at under M6.0 to M6.5 earthquakes, at PGA as low as 0.1 to 0.15 g.
- There is evidence to show that higher seismic loads will result in the formation of larger and deeper cracks and are more likely to cause transverse cracking due to the greater differential settlements across the valley.
- At low seismic loading, there is evidence to show that only one type of cracking is likely to develop. This may be either longitudinal (more common) or transverse, depending on factors within the embankment that make it predisposed to a particular form. It is the nature of the embankment, zoning and the foundation geometry and presence of compressible materials, and not the seismic loading, that differentiate between which form of cracking develops and in particular determine whether transverse cracking occurs at low seismic intensities.
- The susceptibility of an embankment to a particular type of crack might not be related to dam type, but a weak relationship has been developed with dam shape—dams in steeper valleys tend to be more susceptible to transverse cracking near the abutments.

The settlement resulting from earthquake loading may result in transverse cracking associated with differential settlements in the foundation (IM 5) and adjacent to walls (IM 21 and IM 22).

IM 13 includes these mechanisms.

The suggested procedure to estimate the likelihood of cracks is:

First: Determine the earthquake hazard for the site as discussed in Chapter 5. The PGA from the hazard studies are usually "bedrock" values. For dams on rock foundations use these as the peak ground acceleration at the dam foundation level. For dams on deep weathered foundations or alluvium, use SHAKE or equivalent analyses to determine the PGA at foundation level allowing for amplification or de-amplification.

Second: Estimate the Pells and Fell (2002, 2003) damage class which the embankment may experience as a result of the earthquake using Figure 8.2 for earthfill dams or Figure 8.3 for earth and rockfill dams.

Third: Estimate the crest settlement as a % of embankment height above cutoff level from the Pells and Fell (2002, 2003) damage class classification in Table 8.1.

Fourth: Estimate the crest settlement for the earthquake loading using the Bray and Travasarou (2007) or Bray et al. (2018) methods. These are described in Section 8.3.

Fifth: Determine the likely crest settlement taking into account the results of Pells and Fell (2002, 2003), Bray and Travasarou (2007) or Bray et al. (2018) methods. These methods assume there is no severe loss of strength, e.g. from liquefaction.

If the shoulders of the embankment are poorly compacted rockfill, the settlements are likely to be at the upper bound of the Pells and Fell (2002, 2003) method. The Bray and Travasarou (2007) and Bray et al. (2018) methods do not allow for this and may underestimate deformations.

Sixth: Add this estimated crest settlement to that experienced by the dam post construction.

Seventh: Use that total % settlement to estimate the likelihood of transverse cracking using the methods described in Section 9.2.1 (for cross valley differential settlement), Section 9.2.3 (for differential settlement in the foundation) and Section 9.2.8 (for cracking or gap adjacent walls). Use Section 9.11 to estimate the width and depth of potential transverse cracking.

Do this for each earthquake load partition.

This procedure applies to situations where liquefaction does not occur in the dam or its foundations. If flow liquefaction is predicted, assume as an approximation the damage class in one greater than from Figures 8.2 and 8.3.

For important decisions on dam safety and upgrades where liquefaction of the foundations is predicted, use numerical modelling to estimate the post-earthquake settlement. These are described in Section 8.4.

9.10 IM 26 PERSISTENT CRACK IN PLASTIC SOIL IN THE FOUNDATION

9.10.1 Overall Approach

Using engineering judgement and/or expert elicitation, estimate the probability (P_d) there is soil or erodible extremely weathered rock beneath the embankment in which an open persistent single crack or interconnected pattern of cracks may exist from upstream of the core to downstream of the core.

Cracking may be the result of relict defects in the foundation soil or weathered rock, differential settlement in the foundation or desiccation cracks in the foundation soil which was not stripped from the foundation.

The question is best put in terms of the width of the crack. For example: "what is the probability of an open persistent crack or pattern of cracks 5 mm wide?"

It will be useful to consider the question for 2 mm, 5 mm, and say 10 mm wide cracks, as there may be a higher likelihood of narrow cracks and lower likelihood of wider cracks.

If exposures of the foundation or in adjacent exposures adjacent to the embankment show no open cracks or only a few cracks of limited persistence, it is sometimes useful to ask, "Given the lack of evidence of persistent open cracks, what mechanism could result in persistent open cracks the full width of the core foundation?"

9.10.2 Some Factors to Consider in Estimating the Likelihood of the Presence of a Persistent Crack

9.10.2.1 Factors Which Apply to All Causes of Cracking

(a) The mode of internal erosion is concentrated leak erosion. For erosion to initiate the crack must be open for the full distance from upstream of the core to downstream of the core.
(b) The length of the contact between the core and the foundation is dependent on the height of the dam at that location and the core geometry. The length will vary from the upper abutment to maximum section. It may be tens of metres at the maximum section. The likelihood of persistent open cracks is likely to be less for long contact lengths than shorter contact lengths.
(c) Existing cracks, particularly narrow cracks, in the foundation are potentially closed by the pressure exerted on the foundation by the embankment.
(d) A guide to the likelihood of open persistent cracks is to map exposures of the soil and extremely weathered rock adjacent to and in the vicinity of the embankment. These exposures are often in road cuttings. For important dam safety decisions, exploratory pits are recommended to be excavated at the downstream toe to expose the foundation soils to allow careful inspection of the soil structure, and to confirm the depth of stripping below the embankment.
(e) The criteria adopted for the cutoff foundation and the degree to which they were followed during construction. Well designed and constructed dams have cutoffs taken below foundation soils and erodible rock with open defects.
(f) Monitoring of foundation seepage flows and how these vary with reservoir level. If there is a postulated persistent open defect below current reservoir operating levels, then there will be seepage flow along the defect. The flow can be estimated using the method described in Section 9.13.5. For example, from Table 9.16 for a seepage gradient of 0.25, a 2 mm crack 2 m deep could result in 2.5 l/sec, a 5 mm crack 12 l/sec and a 10 mm crack 40 l/sec. If these flows are not observed, the postulated wide defects are not there and their presence can be assigned a negligible probability.
(g) The fact that there is seepage through the foundation of a dam does not mean there are persistent open defects in erodible soil or rock. Seepage may be a result of water flowing in open defects in non-erodible rock.

TABLE 9.16 Flow Rates in Defects in a 2 m Deep Defect in a Rock Foundation versus Average Gradient in the Defect (Fell et al., 2024)

WIDTH OF DEFECT (MM)	*FLOW RATE FOR 2 M DEEP DEFECT (L/SEC)*		
	GRADIENT = 0.15	*GRADIENT = 0.25*	*GRADIENT = 0.4*
2 mm	2	2.5	3.5
5 mm	10	12	16
10 mm	35	40	60
25 mm	160	180	250
50 mm	300	340	475

Note: Flow rate is proportional to the depth of the open defect.

9.10.2.2 Cracking Resulting from Relict Defects in the Soil and Extremely Weathered Rock

Relict defects include joints, bedding partings, shears and faults. The orientation, persistence and nature of the defects should be determined by engineering geological mapping of exposures and from mapping carried out during construction of the dam. Mapping must include observations of whether the defects are open, and if so the width and persistence of the open defect.

Only defects with orientations such that they potentially persist from upstream to downstream are relevant.

It is useful to consider first if there are potentially persistent defects, and second if they are open. For example, a fault or shear zone may be persistent, over many metres, but it may be seen in all mapping that the defects are closed. By comparison, joints may be open but not persistent.

9.10.2.3 Cracking Due to Differential Settlement

The situations which are likely to result in cracking in the foundation soils are essentially the same as those causing cracking low in the embankment due to differential settlement in the foundation. An example is shown in Figure 9.19(b). Given this, it is suggested that if in using Section 9.2.3 to assess the probability of cracking in the embankment due to differential settlement in the foundation, the presence of cracking in the embankment is likely to be on the foundation/embankment contact, then this probability can also be assigned to the probability of cracking in the plastic soil in the foundation beneath the embankment. Whether erosion initiates in the embankment core or in the foundation will depend on which has the lower critical shear stress. Only one mode should be carried forward.

If the assessment of cracking in Section 9.2.3 is for features such as those in Figure 9.19(a) and (c) which will cause cracking near the crest of the dam, then cracking in the foundation due to differential settlement may be assumed to be negligible.

9.10.2.4 Cracking Due to Desiccation

Some factors to consider in assessing the likelihood of persistent or interconnected desiccation cracking in the foundation include:

- Whether the soils in the foundation on which the dam was built are susceptible to desiccation in the climatic conditions at the site. This can be assessed using Section 9.8.1 and considering the foundation soil as the "core material".
- Whether desiccation cracking is evident in the soil surrounding the dam and if so the depth, width and persistence of the cracks.
- Whether the cracking is likely to persist below the level of the cutoff for the dam. This can be assessed using observed desiccation crack depths for the site and/or using the data in Section 9.8.1 to estimate the depth and width of cracking for conditions pre-construction and knowing the depth of the cutoff below the ground surface.
- Soil in which desiccation cracking is observed will likely wet up under the embankment from seepage and groundwater flows and reach an equilibrium moisture content at which the cracks are likely to close.
- The weight of the embankment is likely to close cracks.

9.10.2.5 Open Defects Due to Other Processes

Interconnected open defects can also occur in soil foundations due to soil structure, including fissured soils, relict defects in lateritic soil profiles, holes left from decomposed roots, etc. Some factors to consider in assessing the likelihood of persistent or interconnected defects under these scenarios include:

(a) Lateritic soil profiles commonly contain soil structure leading to high vertical permeability. The structure is a result of leaching and tree roots rotting with time and the resulting holes remaining open or being infilled with sand. Exposures of lateritic profiles seen by the authors indicate that the features are mainly vertical and are not likely to be persistent horizontally.

(b) Fissured soils. There are two types of fissuring:
- Blocky fissuring caused by desiccation of the upper soil profile. The cracks formed are often infilled with sand eroded into the fissures. The overall effect is to result in higher mass permeability from the soil substance permeability. Whether they form a continuous open defect in which erosion can initiate may be assessed by excavating test pits and mapping the fissures.
- Fissures formed by shrink and swell of high plasticity clays. Figure 7.20 shows an example. These fissures are commonly closed and do not form persistent open defects.

9.10.3 Critical Shear Stress of Foundation Soils and Weathered Rock

Where foundation conditions indicate that persistent open defects may occur in erodible foundation soils, the critical shear stress of soil in the foundation should be determined by laboratory Hole Erosion Tests (HET) on representative soils at the *in situ* density and moisture content as it would be beneath the embankment.

For potential erosion in relict defects, HET can be carried out on undisturbed or disturbed samples of the soil or completely weathered rock at the *in situ* density and moisture content. However, the results of such tests are likely to significantly underestimate the critical shear stress because the defects often have coatings of iron and other oxides. The authors do not know of a way to test the erosion properties of such surfaces, so judgement needs to be applied in estimating the critical shear stress.

Defects in highly and moderately weathered and fresher rock will have large critical shear stresses and will not be susceptible to erosion in narrow cracks at the seepage gradients beneath a dam.

9.11 ESTIMATION OF THE WIDTH AND DEPTH OF CRACKING

9.11.1 Cracking Resulting from Cross Valley Differential Settlements. IM 1, IM 5, IM 6 and IM 13

9.11.1.1 Crack Widths

Fell et al. (2024) assembled case data on crack widths and depths and the results of numerical analyses by He (2016) and He et al. (2017). The data on crack widths is plotted in Figure 9.63.

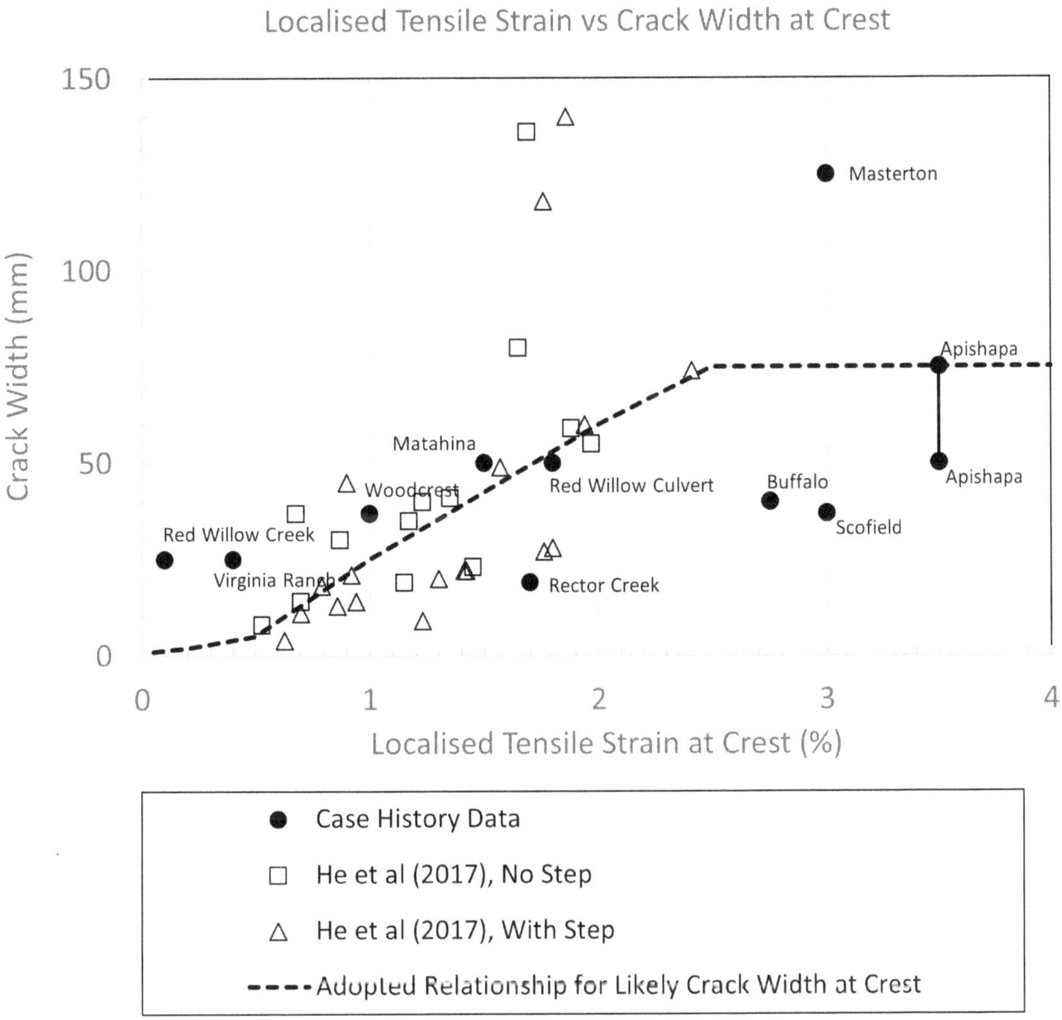

FIGURE 9.63 Crack widths estimated from numerical analyses. Valley profiles with and without step (Fell et al., 2024).

For the case history data, the crack widths, depths and post-construction settlement and abutment slope are taken from Table 9.10. The localised strains are estimated from Figures 9.11 and 9.12.

For the He et al. (2017) data, the widths are from numerical analyses and calculated from the modelled width of the localised strain and the average localised strain.

From these tables and Figure 9.63 it can be seen that:

(a) The widths generally are related to the localised tensile strain which is to be expected.
(b) The widths for with and without step are similar so they are treated as one database.
(c) There is a significant scatter in the data which should be considered when using it.
(d) Most of the actual observed widths are for post-construction settlements ≥ 1%.

The dashed line in Figure 9.63 is the "best estimate" assessed by the authors.

It should be noted that these are likely crack widths at the dam crest which might occur if strains are localised in one position. In practice the strains may be distributed over a number of cracks. This will be more likely for strains which accumulate slowly over many years, and less likely for strains which occur rapidly such as in earthquake-induced settlement, or settlement induced by collapse settlement on first filling. It is also considered more likely strains will result in a single crack for large, localised strains than small, localised strains. We have insufficient case data to provide more specific guidance on this.

9.11.1.2 Crack Depths

Figure 9.64 plots the ratio of crack depth to crack width versus crack width.

For the case history data, the crack widths and depths are taken from Table 9.13.

For the He et al. (2017) data, the widths are from numerical analyses and calculated from the modelled width of the localised strain and the average localised strain.

Also shown are case data from Bolton et al. (2006) for seven dams. There is data in that paper on cracking but not the other data detailed in Table 9.10.

From this there are two ways of summarising the data:

(a) A stepped relationship between the ratio of crack depth to crack width as shown in the dashed line.
(b) A continuously varying relationship as shown by the solid line.

The continuing varying relationship fits the data better and avoids discontinuities, so it is recommended to be used.

For the adopted line the ratio of crack depth (D) to crack width (W) is:

$$D/W = (250 - 1.33W) \tag{9.3}$$

So the crack depth is:

$$D\ (\text{metres}) = W\ (250 - 1.33W)/1000 \text{ where W is in mm.} \tag{9.4}$$

Observation of cracking in test pits at Buffalo Dam, Tullaroop Dam and Red Willow Dam is that cracks tend to reduce in width at depth but not necessarily as much as would be expected by a uniform taper. Crack propagation analyses at Upper Yarra Dam (Figure 9.45) showed that cracks remain relatively wide at depth rather than uniformly tapered. There is insufficient case data to specify actual crack profiles, so it is suggested that when estimating the hydraulic shear stresses in the crack both triangular and rectangular crack shapes are considered.

9.11.2 Widths of Cracking and Hydraulic Fracture Resulting from Arching. IM 3, IM 10, IM 12 and IM 30

IM 3, IM 10, IM 12 and IM 30 crack widths should be estimated as shown in Appendix A. These have been estimated by Fell et al. (2024) from case data which are in the Fell et al. (2024) Supporting Document.

The widths are approximate and given only as a guide.

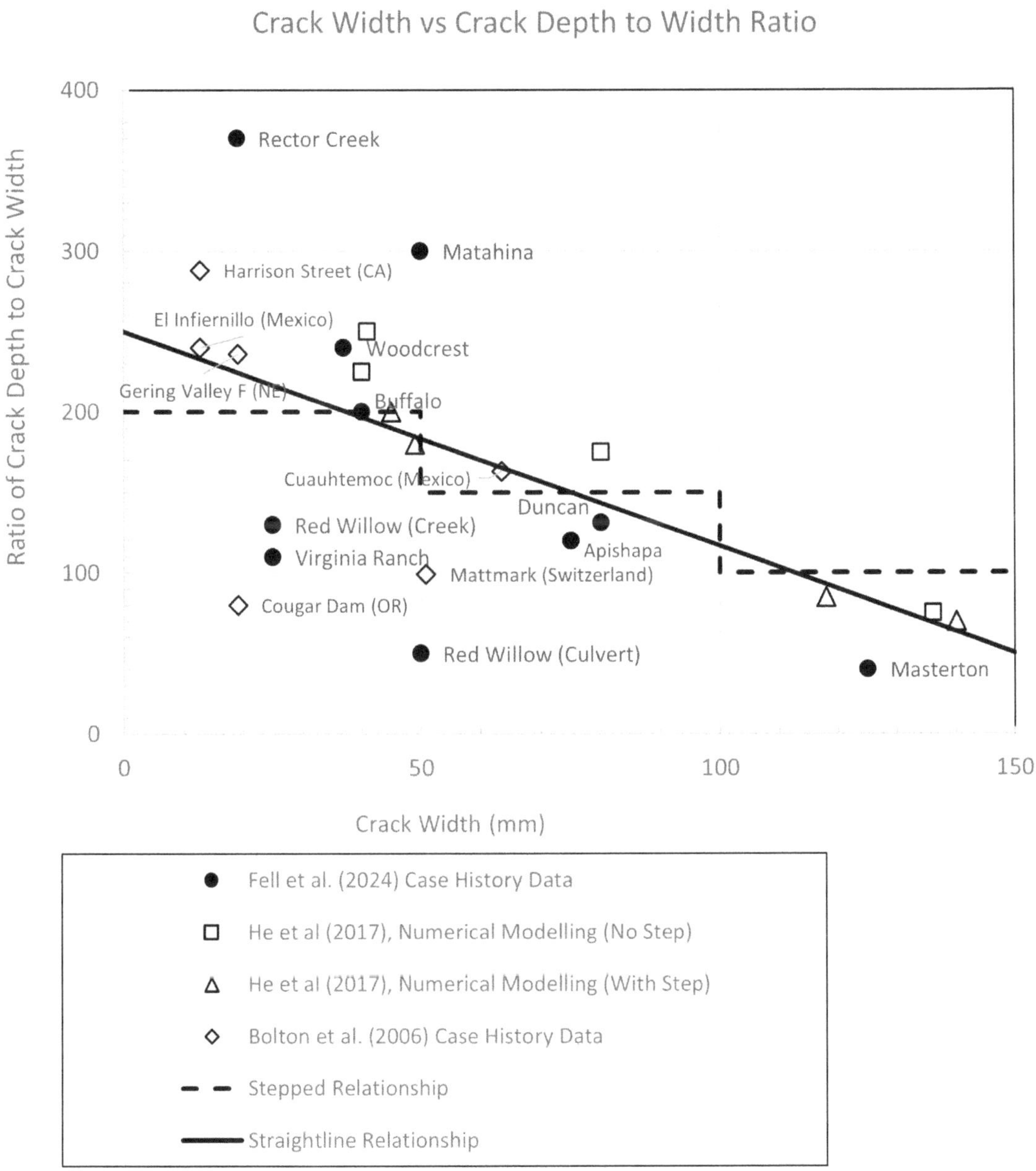

FIGURE 9.64 Ratio of crack depth to crack width versus crack width (Fell et al., 2024).

9.11.3 Estimation of the Width of a Crack or Flaw Resulting from Poor Compaction in the Core. IM 14

Mechanism A. The soil behaves like a series of clods with openings between the clods through which water passes.

The actual width is uncertain. Estimate the width based on the understanding of the conditions within the poorly compacted layer. As a fall-back approach adopt the gap sizes shown in Table 9.17 (which are based on judgements made by the Fell et al., 2024, authors).

TABLE 9.17 Estimated "Opening Gap Size" for Poorly Compacted Soils That Behave as a Series of Clods with Openings between the Clods Forming a Continuous Pathway through the Core (Fell et al., 2024)

COMPACTION METHOD	*ESTIMATED EFFECTIVE "OPENING GAP SIZE" BETWEEN SOIL CLODS*
Soil placed with no formal compaction, e.g. by horse and cart in old dams, or by pushing into place by excavator or bulldozer in very thick layers > 1 m thick. Well dry of OWC.	3–6 mm
Soil placed with no formal compaction, e.g. by horse and cart in old dams, or by pushing into place by excavator or bulldozer in thick layers 0.6 m to 1 m thick. Well dry of optimum moisture content.	2–5 mm (for the soil within the bottom of the poorly compacted layer)
Soil placed and compacted by bulldozer, no compaction by rollers or rolled in thick layers beyond the capability of the roller. Dry of optimum moisture content.	2–3 mm (for the soil within the bottom of the poorly compacted layer)
Soil compacted by suitable rollers in suitable layer thickness. Around optimum moisture content.	1–2 mm (for the soil within the bottom of the poorly compacted layer)

Mechanism B. The soil surrounding a conduit or adjacent to a wall collapses on saturation forming a gap at the contact of the soil with a conduit or wall.

For IM 18 for conduits, and IM 20 for walls.

The suggested procedure is:

- Assess the thickness of the layer of soil which is poorly compacted (T). There may be a single layer or several layers. To do this take account of the dimensions of the conduit and the trench in which it is placed.
- Estimate the amount by which the layer may collapse (C_S) using Table 9.13 as a guide. Table 9.13 requires an overburden stress to select C_S. For this mechanism, use the reservoir pressure at the level of the conduit as the stress causing collapse.
- Estimate the width or height of the gap (G) which is formed from G = (T) × (C_S).

Mechanism C. The soil collapses on saturation causing settlement of the embankment core which results in arching of the core onto the filters and/or shoulders, resulting in a gap at the top of the saturated soil.

See Section 9.11.2.

Mechanism D. Additional settlement of the embankment resulting from collapse settlement in the poorly compacted layer causes transverse cracking.

Use the method in Section 9.5.

9.11.4 Estimation of the Width and Depth of Gaps Formed at Walls and Cliffs. IM 2, 21 and 22

IM 21 Transverse crack or gap adjacent to a spillway or abutment wall resulting from differential settlement adjacent the wall due to wall profile or foundation profile adjacent to the wall.

Estimate the potential width and depth of a potential gap taking into account:

(a) The height and slope of the wall, in particular whether it has a flatter lower part and steeper upper part as shown in Figure 9.31 and their dimensions.
(b) Lateral movements from crest survey points.

(c) The stiffness of the wall and its factor of safety for the load applied by the embankment.
(d) Signs and amount of deformation such as movement relative to adjacent parts of the wall, gaps between road kerbing and the wall.

IM 2 Transverse crack or gap adjacent to a cliff at the top of the embankment formed by differential settlement

and

IM 22 Transverse crack or gap adjacent to a spillway or abutment wall resulting from differential settlement adjacent to the wall.

The procedure is:

(a) Calculate the width of the gap at the top of the embankment where it abuts the wall or cliff using Table 9.18.
(b) Calculate the depth of the crack using Tables 9.19 and 9.20.
(c) Assume that the gap varies in width linearly from the maximum width at the surface to zero at maximum depth. For estimates of hydraulic shear stresses in the crack check assuming the crack is rectangular shape.

The parameters are defined in Figure 9.36. Figure 9.65 shows how the effective wall heights in Table 9.20 were estimated.

The method may be conservative for walls higher than about 10 m because the soil adjacent to the crack may not be able to support a deeper vertical face.

TABLE 9.18 Gap Widths in mm at the Top of a Wall or Cliff versus W_b/H_w and % Settlement for Foundation Slope β_1 = 45° (Fell et al., 2024)

POST-CONSTRUCTION SETTLEMENT, S/H (%)	*W_b/H_w*					
	0	*0.25*	*0.5*	*1.0*	*2.0*	*2.5*
0	1	1	< 1	< 1	< 1	No gap
0.1	5	4	2.5	1	< 1	No gap
0.25	12	10	6	2.5	< 1	No gap
0.5	25	20	12	5	1	No gap
1.0	50	40	25	10	2.5	No gap
2.0	100	80	50	20	5	No gap

Note: For β_1 = 15° multiply gap width by 0.25; for β_1 = 30° multiply gap width by 0.75; for β_1 = 60° adopt same gap width values as for β_1 = 45°.

TABLE 9.19 Gap Depth (Fell et al., 2024)

POST-CONSTRUCTION SETTLEMENT S/H (%)	*TOTAL DEPTH OF GAP (% EFFECTIVE WALL HEIGHT H_E)*
0	15
0.1	50
0.25	65
0.5	90
1.0	100
2.0	100

TABLE 9.20 Effective Wall Height (H_E) for Calculation of Depth of Gap at Walls or Cliff (Fell et al., 2024)

FOUNDATION SLOPE (β_1) DEGREES	*W_B/H_W*				
	0	*0.25*	*0.5*	*1.0*	*2.0*
15	H_w	$0.93H_w$	$0.87H_w$	$0.73H_w$	$0.46H_w$
25	H_w	$0.88H_w$	$0.77H_w$	$0.54H_w$	$0.08H_w$
45	H_w	$0.75H_w$	$0.50H_w$	$(0.05H_w)$	$(0.05H_w)$
60	H_w	$0.57H_w$	$0.14H_w$	$(0.05H_w)$	$(0.05H_w)$

Note: Values in brackets are nominal minimums.

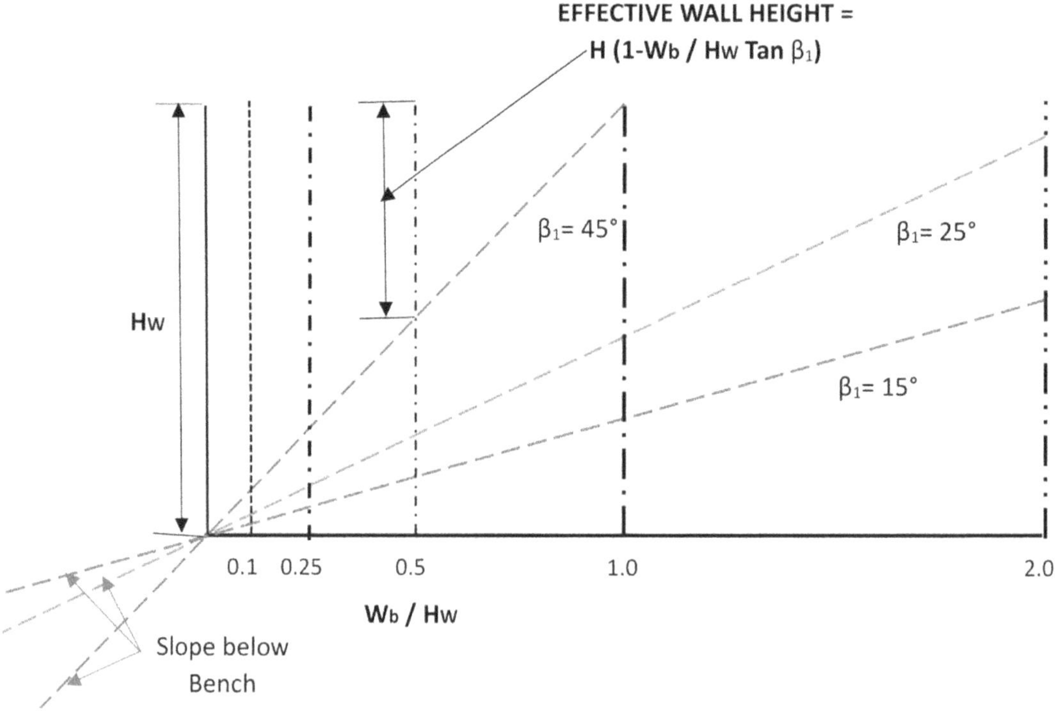

FIGURE 9.65 Definition of effective wall height for estimation of gap depth (Fell et al., 2024).

9.12 ASSESSMENT OF WHETHER A CRACK WILL CLOSE RESULTING FROM SWELL OF THE SOIL ON THE SIDES OF THE CRACK

9.12.1 General Principles

The general principles are:

(a) The soil in the sides of a crack may swell when water flows in the crack.
(b) How much swell occurs depends on the swell properties of the soil, the degree of compaction, and moisture content relative to Standard Optimum Moisture Content and whether the soil is at as-compacted moisture content or has dried since compaction.
(c) The swell process takes time, and during that time, the water flow in the crack may erode the sides of the crack, countering the swell process.
(d) Whether a crack closes or remains open depends on the initial crack width, the amount and rate of swell, and the rate of erosion and crack widening.

As discussed in Section 9.13, the hydraulic shear stress imposed on the sides of a crack is greater at the surface of the water in the crack than lower in the crack, and greater at the downstream end of the crack than at the upstream end. The hydraulic shear stresses at the upstream end of a crack with a free exit or with a filtered exit are small and for some soils may be less than the critical shear stress of the soil, so erosion will not occur there or be slow. This should be taken into account when estimating the likelihood the swell will close the crack.

9.12.2 Recommended Method

The recommended method is:

(a) Estimate the width of the crack at the reservoir stage being considered using the method described in Section 9.11.
(b) Carry out laboratory surface swell tests on representative samples of the soil in the core, and for the *in situ* density ratio and moisture content relative to Standard Optimum Moisture Content (SOMC). Measure the total swell and rate as described in Savage et al. (2019). In this method the soil is compacted into a standard 100 mm compaction mould at the required relative compaction and moisture content. The mould with soil in it is submerged in water and the surface swell measured at suitable time intervals. The results are in mm of swell.
(c) If using a quantitative risk based method, use this data to assign a probability that the crack will close by swelling by expert elicitation in a workshop. This probability should be assigned to a branch in the event tree with the question, "Does the crack close by swell of the soil in the sides of the crack?" The other branch is for when "Swell does not close the crack" (designated $P_{IC\_Not\_Close}$).
(d) If using an engineering based assessment, use the guiding principles listed next to assess whether swelling is sufficient to close the crack.

Guiding principles for estimating probabilities by expert elicitation or an engineering assessment approach:

- Swelling of the walls of the crack will occur on both faces of the crack where the crack passes through soil, but only one face if a gap develops adjacent to a concrete structure (e.g. spillway wall or conduit structure).
- If the total swell amount exceeds the estimated crack width and this occurs relatively quickly compared to the rate of reservoir rise, then it is likely the crack would close prior to erosion commencing.
- If the rate of erosion (determined from the erosion rate index, IHET) in the upstream part of the crack is low and less than the total swell, then it is likely the crack will close.

- Soil swell effects are likely to be more important where the predicted crack widths are narrow (1 mm or up to 5 mm).
- Figure 9.66 shows the time for an open pipe to erode from an initial 1 mm diameter to 5 mm diameter depending on the erosion rate index (I_{HET}) and hydraulic gradient within the pipe. This figure can be used as a guide as to whether there is sufficient time for a crack to swell shut prior to erosion progressing back from the downstream end of the crack where hydraulic shear stresses are highest.
- The gradient in the upstream parts of cracks will typically be very low (< 0.1), so for example for a soil with $I_{HET} = 4$ and hydraulic gradient of 0.1 it takes > 18 hours to enlarge from 1 mm to 5 mm. However, for cracks with a free exit (no filter or transition), erosion initiates at the downstream end and may progress upstream more rapidly.
- If the total swell amount is much less than the estimated crack width, or the loading conditions involve a rapid rise in reservoir level relative to the rate of swell, then it is unlikely the crack would close before erosion commences.
- If the mechanism of formation of the crack is a rapid process (e.g. under an earthquake-induced crack scenario) or hydraulic fracture, then swelling is unlikely to occur before erosion starts and should be ignored.

If no samples of the soil used in the core are available for laboratory testing, estimate the Seed et al. (1962) swelling potential as described later. Use the data in Figure 9.67 to get a feel for the likely swell of the sides of the crack. Such estimates will be approximate and should not be relied upon for important decisions on dam safety or upgrades if the amount and rate of swell is an important factor. For those cases, consideration should be given to obtaining samples of the core or from borrow areas used for construction of the core and carrying out laboratory swell tests.

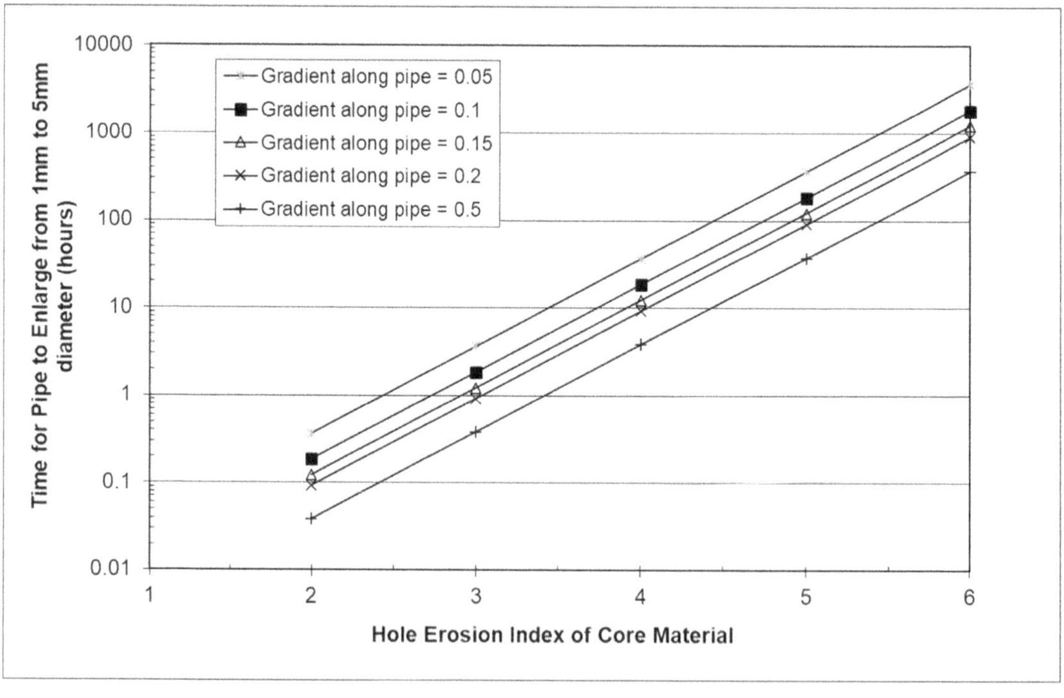

FIGURE 9.66 Approximate time for pipe to enlarge from 1 mm to 5 mm diameter (Fell et al., 2024).

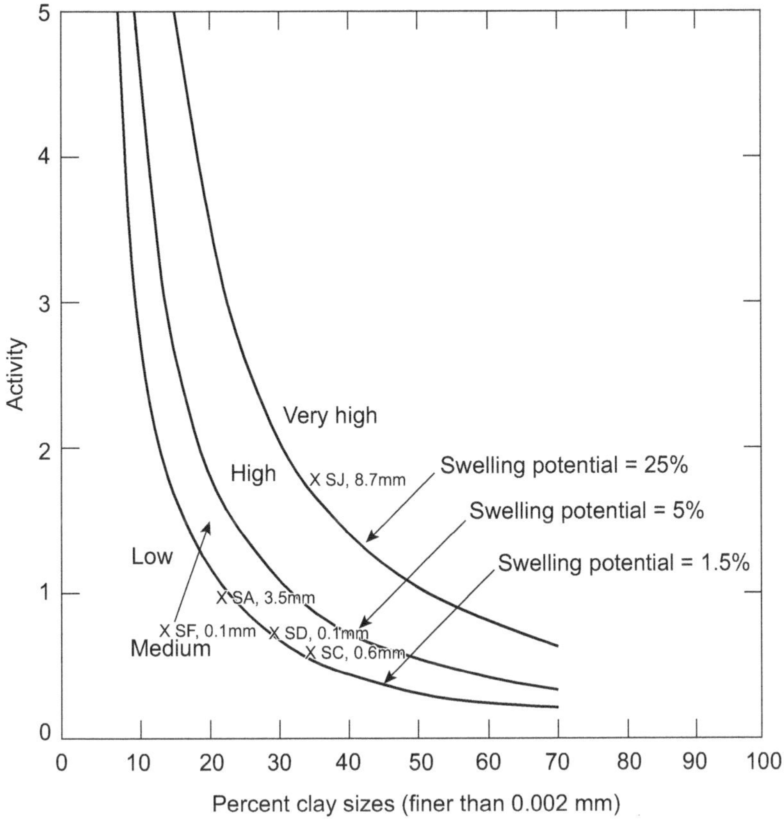

FIGURE 9.67 Classification chart for swelling potential (Holtz and Kovacs, 1981, after Seed et al., 1962), with Savage et al. (2019) data on swell in mm for samples compacted to SMDD at SOMC.

Seed et al. (1962) developed the relationship shown in Figure 9.67 based upon laboratory tests where soils were compacted to 100% standard compaction at Standard Optimum Moisture Content in a 25 mm thick consolidation ring. A surcharge of 1 psi, or 6.9 kPa, was applied and the sample flooded.

$$\text{The swell potential} = (\text{increase in thickness})/(\text{initial thickness}) \times 100\ \% \qquad (9.5)$$

A swell potential of 25% is equivalent to a swell of 6 mm and 5% equivalent to 2.5 mm.

The activity of the soil is defined as:

$$\text{Activity} = (\text{Plasticity Index})/(\%\ \text{passing}\ 2\ \mu\text{m} - 5\%) \qquad (9.6)$$

and

clay size is taken as minus 0.002 mm or 2 μm.

Figure 9.67 shows the Seed et al. (1962) plot with data from Savage et al. (2019) tests at SMDD and standard OMC.

The Savage et al. (2019) data is limited but follows the general trend of the Seed et al. (1962) plot. Soil SD, Buffalo Dam, has very low swell, which is related to its clay mineralogy and has no Smectite or Illite.

The effect of swelling will be that the soil moisture content is increased with only a small reduction in unit weight. As a result, the critical shear stress will likely to be increased and erosion rate reduced because the degree of saturation increases. In most cases it will be sufficient to assume that swell has no effect on the critical shear stress.

9.13 THE MECHANICS OF EROSION IN CONCENTRATED LEAKS

9.13.1 The Procedure for Assessing Whether Erosion Will Initiate

The procedure for assessing whether erosion will initiate in a crack or flaw is as follows:

(a) For cracking resulting from cross valley differential settlement, estimate the depth and width of the vertical crack or gap at the reservoir surface level for the reservoir stage being considered using the methods described in Section 9.2 (IM 1 to IM 5, IM 21, IM 22, IM 23), Section 9.9 (IM 13) and Section 9.6 (IM 18, IM 20).

(b) For cracking resulting from desiccation and freezing, estimate the depth and width of vertical crack or gap at the reservoir level stage being considered using methods described in Section 9.8 (IM 7, IM 8, IM 16, IM 17).

(c) For horizontal cracking or hydraulic fracture resulting from arching, estimate the location and width of the crack or fracture using the methods described in Sections 9.2.7 (IM 10) and 9.2.5 (IM 12).

(d) For cracking, gap or flaw resulting from poorly compacted zones, estimate the location and width of the crack or gap using methods described in Section 9.5 (IM 14, IM 15) and Section 9.6 (IM 18).

(e) Assess whether the crack may close resulting from swell of the sides of the crack as described in Section 9.12. If swell will result in closure of the crack, then erosion will not initiate. If using a quantitative risk based approach, assign a probability that the crack will close based upon the quality of the swell data and whether the amount of swell is more than that required to close the crack, or only marginally sufficient.

(f) For cases where swell is insufficient to close the crack, ignore the effects of swell and carry forward the original estimates of crack width. This is potentially conservative and if there is good quality data on swell the risk analyst may opt to allow for swell and carry forward crack widths adjusted for swell.

(g) For failure modes giving vertical cracks ([a] and [b] earlier) use the Peirson et al. (2022) and Peirson and Fell (2024, 2025) methods described in Sections 9.13.3 and 9.13.4 to estimate the hydraulic shear stresses at the surface of the water flowing in the crack at the downstream and upstream end of the crack. For cracks with no downstream filter or transition, use the method in Section 9.13.3. For cracks with a downstream filter or transition, use the method in Section 9.13.4.

(h) For failure modes giving horizontal cracks or gaps below the reservoir level, use the method described in Section 9.13.2 to estimate the average hydraulic shear stress in the crack or gap.
(i) These estimates are for each reservoir stage being considered taking account of the geometry of the core of the embankment and the assumed crack dimensions and location relative to the reservoir surface.
(j) Compare the hydraulic shear stress to the critical shear stress which will initiate erosion for the soil in the core of the embankment or foundation (τ_c) at the degree of saturation of the soil on the sides of the crack. In doing this take account of the dispersion properties of the soil and the chemistry of the seepage water.

There is always some uncertainty regarding the input parameters, and this should be allowed for in estimating whether erosion will initiate. There should be checks on the sensitivity of the outcome to the assumptions made.

9.13.2 Estimation of Hydraulic Shear Stresses in Cylinders and Horizontal Gaps below Reservoir Level

Wan (2006), Wan and Fell (2002, 2004a, 2004b) and Fell et al. (2008) give details on how to estimate the hydraulic shear stresses in cracks and cylinders as shown in Figure 9.68. These apply to confined flow conditions only and do not apply to vertical open cracks that extend above the reservoir level.

They give the following equations for estimating the hydraulic shear stress on the surface of a cylindrical pipe, or parallel sided transverse crack in an embankment. The assumptions are:

- Applies to confined flow conditions only (i.e. where the entrance of the pipe or crack/gap is completely submerged).
- Linear head loss from upstream to downstream.
- Steady uniform flow along the crack.
- Zero pressure head at the downstream end.
- Uniform frictional resistance along the surface of the crack or cylindrical pipe.
- Driving force = frictional resistance.

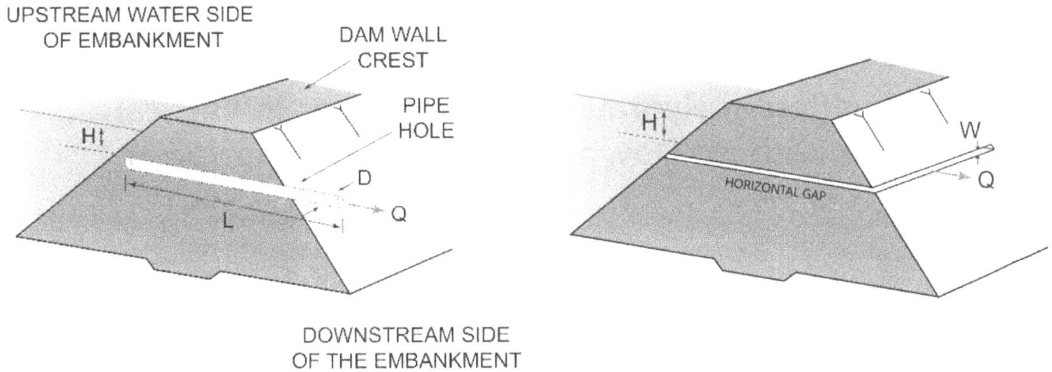

FIGURE 9.68 Geometry of pipes and cracks for estimation of hydraulic shear stresses.

(a) Cylindrical pipe:

$$\tau = \rho_w \frac{gH_f d}{4L} \tag{9.7}$$

(b) Horizontal transverse crack

$$\tau = \frac{\rho_w gH_f^{\,2} W}{2(H_f + W)L} \tag{9.8}$$

where:

τ = hydraulic shear stress in N/m^2
ρ_w = density of water in kg/m^3
g = acceleration due to gravity = 9.8 m/s^2
H_f = head loss in pipe or crack due to friction in metres
L = length of pipe or crack base in metres
D = diameter of the pipe in metres
W = width of crack in metres.

The rate of erosion ($\acute{\varepsilon}$) is given by:

$$\acute{\varepsilon} = C_e (\tau - \tau_c) \tag{9.9}$$

where:

C_e = coefficient of soil erosion (kg/sec/m^2/pa)
τ = hydraulic shear stress (N/m^2)
τ_c = critical shear stress (N/m^2).

And

$$I_{HET} = -\log (C_e) \tag{9.10}$$

Where I_{HET} is the erosion rate index which ranges from 1 to 6, with $I_{HET} \leq 1$ extremely rapid erosion and $I_{HET} \geq 6$ extremely slow erosion.

9.13.3 Estimation of Hydraulic Shear Stresses in Vertical Cracks with No Downstream Filter or Transition

Peirson et al. (2022) and Peirson and Fell (2024) developed a method for estimating the hydraulic shear stresses within vertical transverse cracks in embankment dams. This was based upon numerical modelling and laboratory tests which showed that the flow is controlled by the exit conditions at the downstream end of a crack. It also showed that the hydraulic shear stresses are highest at the downstream and much lower at the upstream end. The highest stresses occur at the surface of the water profile.

Figure 9.69 shows the scenario being modelled by Peirson et al. (2022), Peirson and Fell (2024). Figure 9.70 shows the dimensions and coordinate system used.

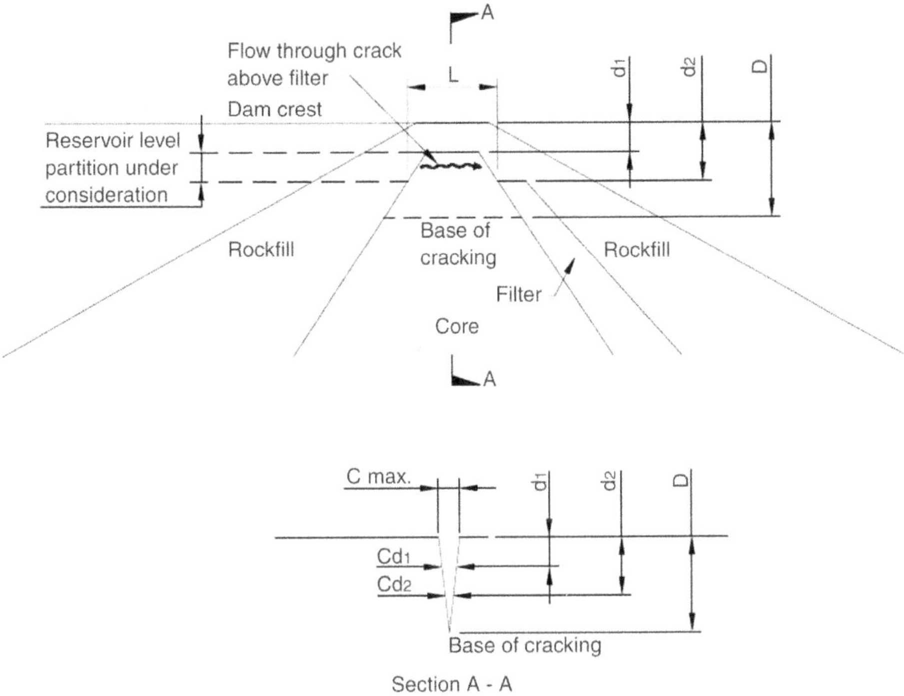

FIGURE 9.69 Detailed geometry of vertical crack in the top of the core for estimation of the hydraulic shear stress in a triangular crack (Fell et al., 2008, 2015).

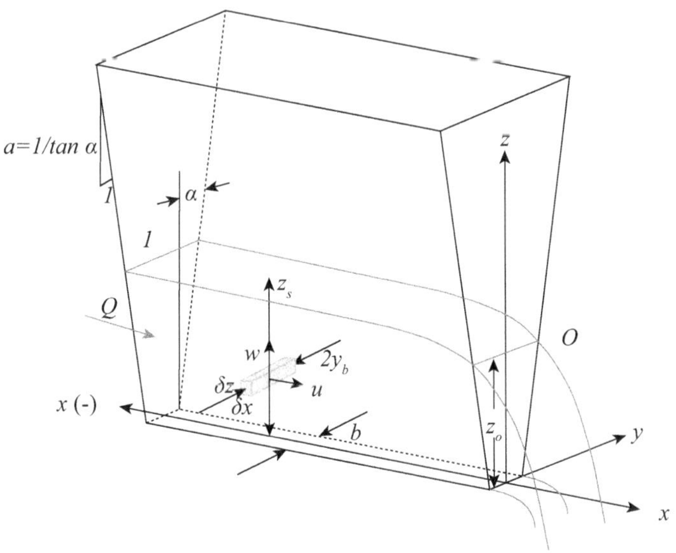

FIGURE 9.70 Dimensions and coordinate system for Peirson et al. (2022), Peirson and Fell (2024) method. (Copyright © 2025 Taylor & Francis Group, reprinted by permission of Informa UK Limited)

Figures 9.71 and 9.72 show normalised profiles of water surface elevation along cracks up to 30 m in length and the upstream water level in a 30 m long crack as a function of crack geometry and water depth at free overflow.

Figures 9.73 and 9.74 provide the flow rate through the crack and the maximum hydraulic shear stress at the downstream end for the assumed crack geometry.

Figure 9.75 allows the hydraulic shear stress distribution along the crack to be estimated so that the region of potential erosion can be determined.

As a practical example, assume a core 20 m wide with a water level 10 m above the base of the crack at its upstream side. Assume also a rectangular crack geometry with a width of 8 mm.

The upper panel of Figure 9.71 yields an estimate of $\left(z_x - z_0\right)/\left(z_{30} - z_0\right) = 0.83$.

Assuming that $z_0 = 0$, as an initial estimate, gives $z_{30} = 12.05$ m which, with Figure 9.72, gives $z_0 = 2.1$.

Recalculating gives $z_{30} = 11.6$ which, with Figure 9.72, gives $z_0 = 2.0$ and $z_{30} = 11.7$ (converged).

Figure 9.73 yields an estimate of 120 L/s of flow through the crack.

Figure 9.74 gives a maximum hydraulic shear stress of 100 Pa at the downstream end.

Figure 9.75 gives the hydraulic shear stress at the upstream end (x = 20 m) = 0.06 τ_{max} = 12 Pa.

Peirson et al. (2022), Peirson and Fell (2024) indicate that the maximum hydraulic shear stress is expected to remain within 10% of its maximum value over a distance of 5 m from its downstream end.

The Peirson et al. (2022), Peirson and Fell (2024) method is iterative, and for screening and preliminary risk analyses approximate estimates of the hydraulic shear stresses may be obtained from Tables 9.21 and 9.22.

The following can be seen from these tables:

(a) The hydraulic shear stresses are largest at the downstream end of cracks. The analyses show that the highest stresses at any location along the crack will occur at the surface of the flow in the crack. Lower down in the crack at the exit and elsewhere along the crack the shear stresses will be smaller than at the surface.
(b) The minimum hydraulic shear stresses occur at the upstream end of the crack.
(c) The hydraulic shear stresses at the mid-length of the crack are only marginally higher than the minimum.
(d) The shear stresses are somewhat smaller for rectangular cracks than for the same depth triangular crack.

The shear stresses estimated using average gradient (τ_{av}) method in Section 9.13.2 are significantly lower than the estimated maximum shear stresses at the downstream end of cracks. They are, however, larger than the minimum shear stresses at the upstream end of cracks.

Peirson indicates that the flow in vertical cracks is downstream controlled, so the greatest hydraulic shear stress is always at the downstream end of the crack and near the surface of the water in the crack. As this erosion occurs, the crack at that location widens so in that part of the crack the water surface drops. The new control becomes the non-eroded crack just upstream of the eroded part. This process will progress upstream and not cease unless the crack closes by swell at the upstream end or filters arrest erosion.

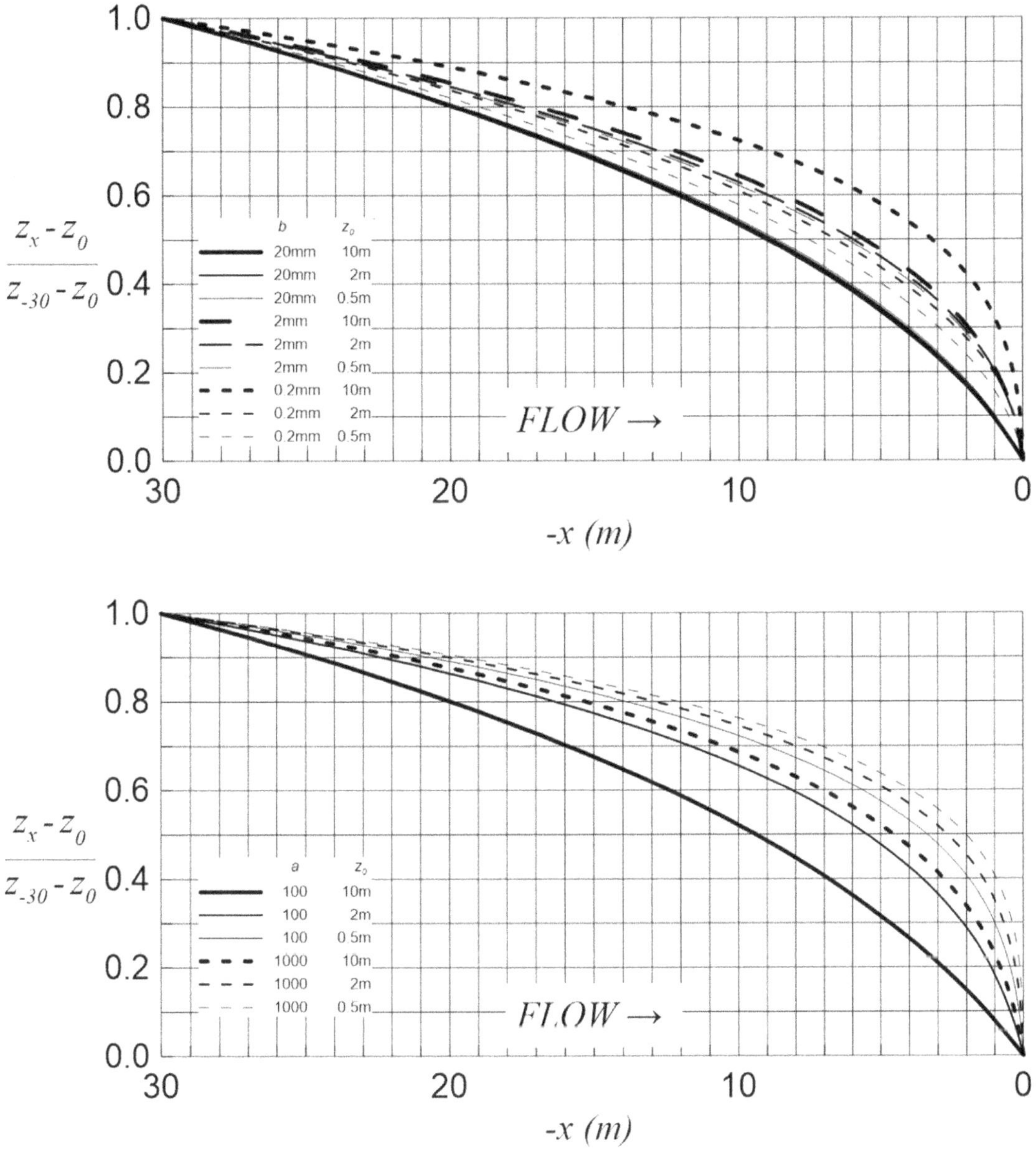

FIGURE 9.71 Normalised profiles of water surface elevation along vertical cracks up to 30m in length. Light lines indicate $z_0 = 0.5m$, medium $z_0 = 2m$ and heavy lines indicate $z_0 = 10m$. Upper panel, rectangular cracks with line styles: short dashes $b = 0.4mm$; long dashes $b = 2mm$; continuous $b = 20mm$. Lower panel, triangular cracks with line styles: continuous $a = 100$; short dashes $a = 1000$. Note in this normalisation, the $b = 20mm$ cases are indistinguishable (Peirson et al., 2022; Peirson and Fell, 2024). (Copyright © 2025 Taylor & Francis Group, reprinted by permission of Informa UK Limited, trading as Taylor & Francis Ltd, https://www.tandfonline.com.)

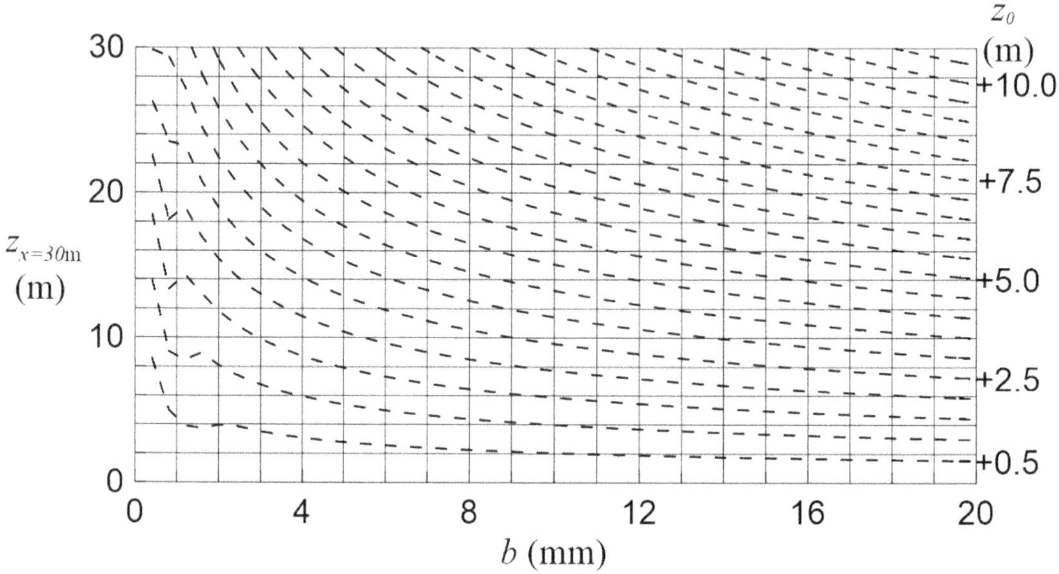

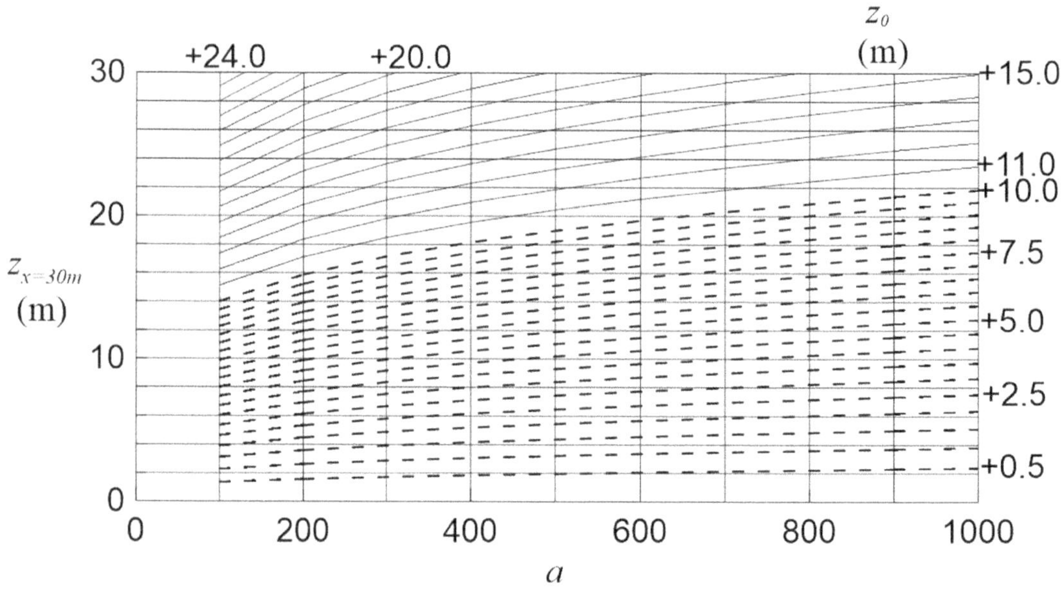

FIGURE 9.72 The upstream water level in a 30 m long vertical crack as a function of crack geometry and water depth at free overflow. Upper panel, rectangular cracks; lower panel, triangular cracks (Peirson et al., 2022; Peirson and Fell, 2024). (Copyright © 2025 Taylor & Francis Group, reprinted by permission of Informa UK Limited, trading as Taylor & Francis Ltd, https://www.tandfonline.com.)

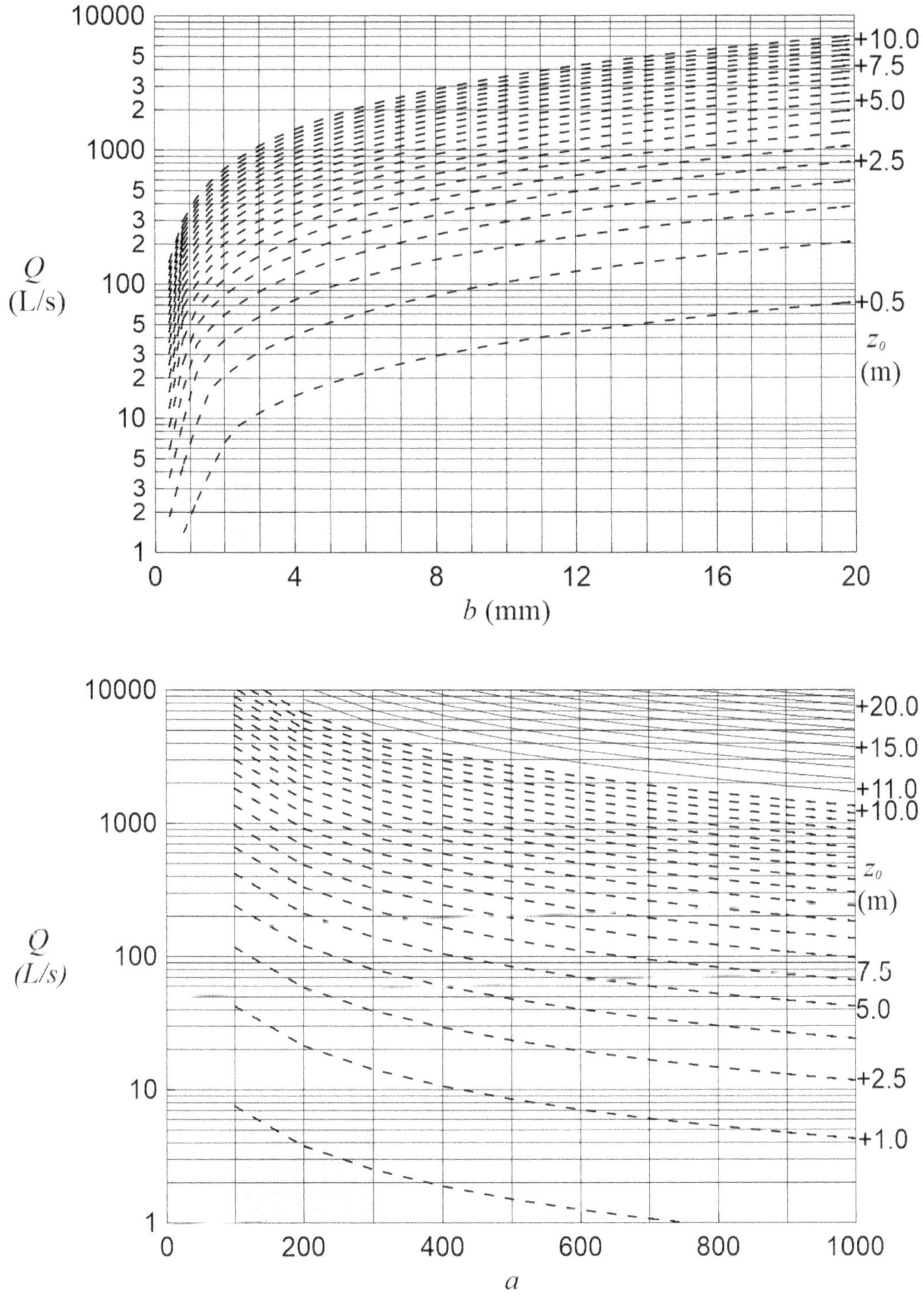

FIGURE 9.73 Flow through vertical cracks as a function of crack geometry and water depth at free overflow. Upper panel, rectangular cracks; lower panel, triangular cracks (Peirson et al., 2022; Peirson and Fell, 2024). (Copyright © 2025 Taylor & Francis Group, reprinted by permission of Informa UK Limited, trading as Taylor & Francis Ltd, https://www.tandfonline.com.)

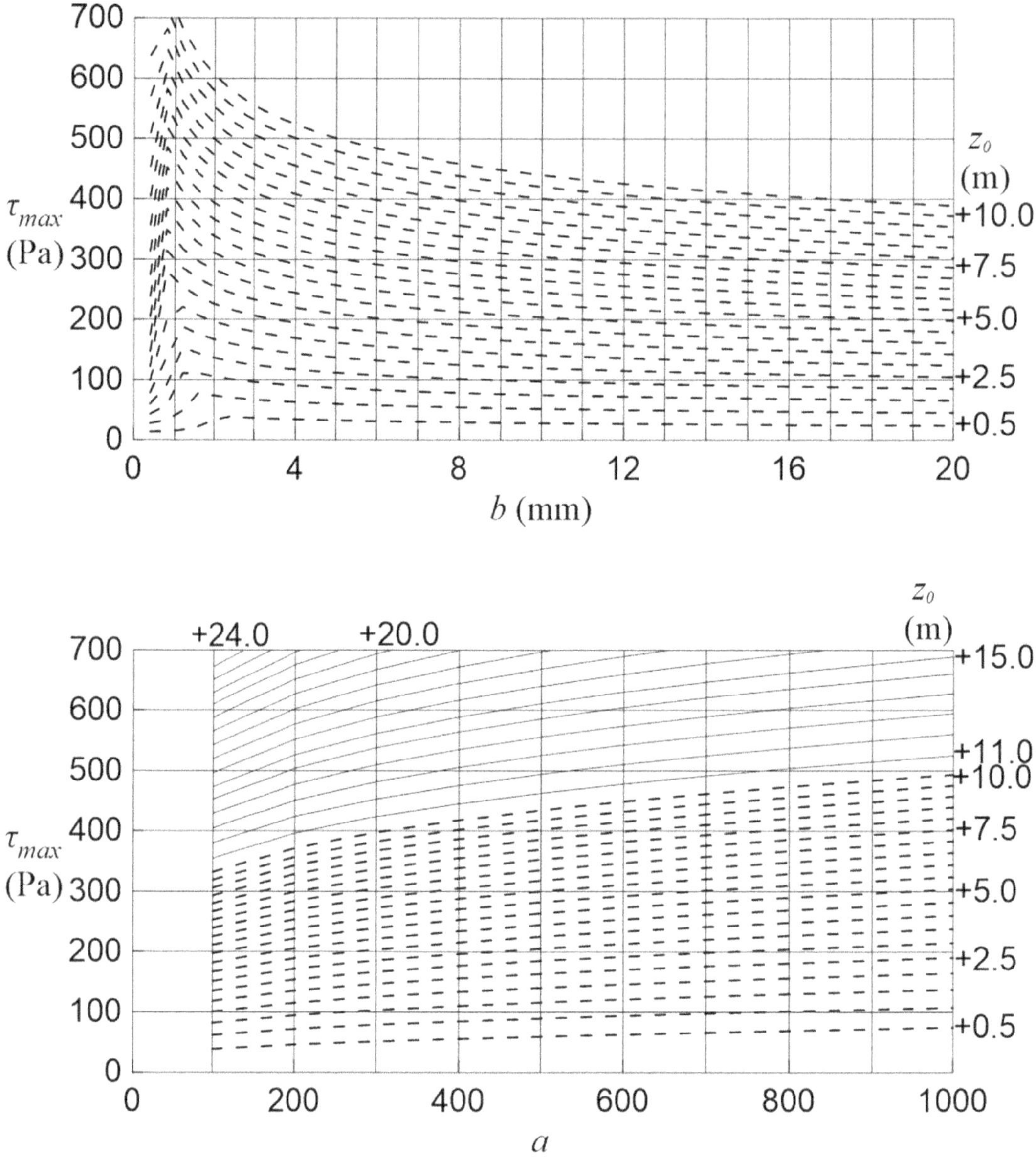

FIGURE 9.74 The maximum hydraulic wall stress occurring in a vertical crack as a function of crack geometry and water depth at the free overflow. Upper panel, rectangular cracks; lower panel, triangular cracks (Peirson et al., 2022; Peirson and Fell, 2024). (Copyright © 2025 Taylor & Francis Group, reprinted by permission of Informa UK Limited, trading as Taylor & Francis Ltd, https://www.tandfonline.com.)

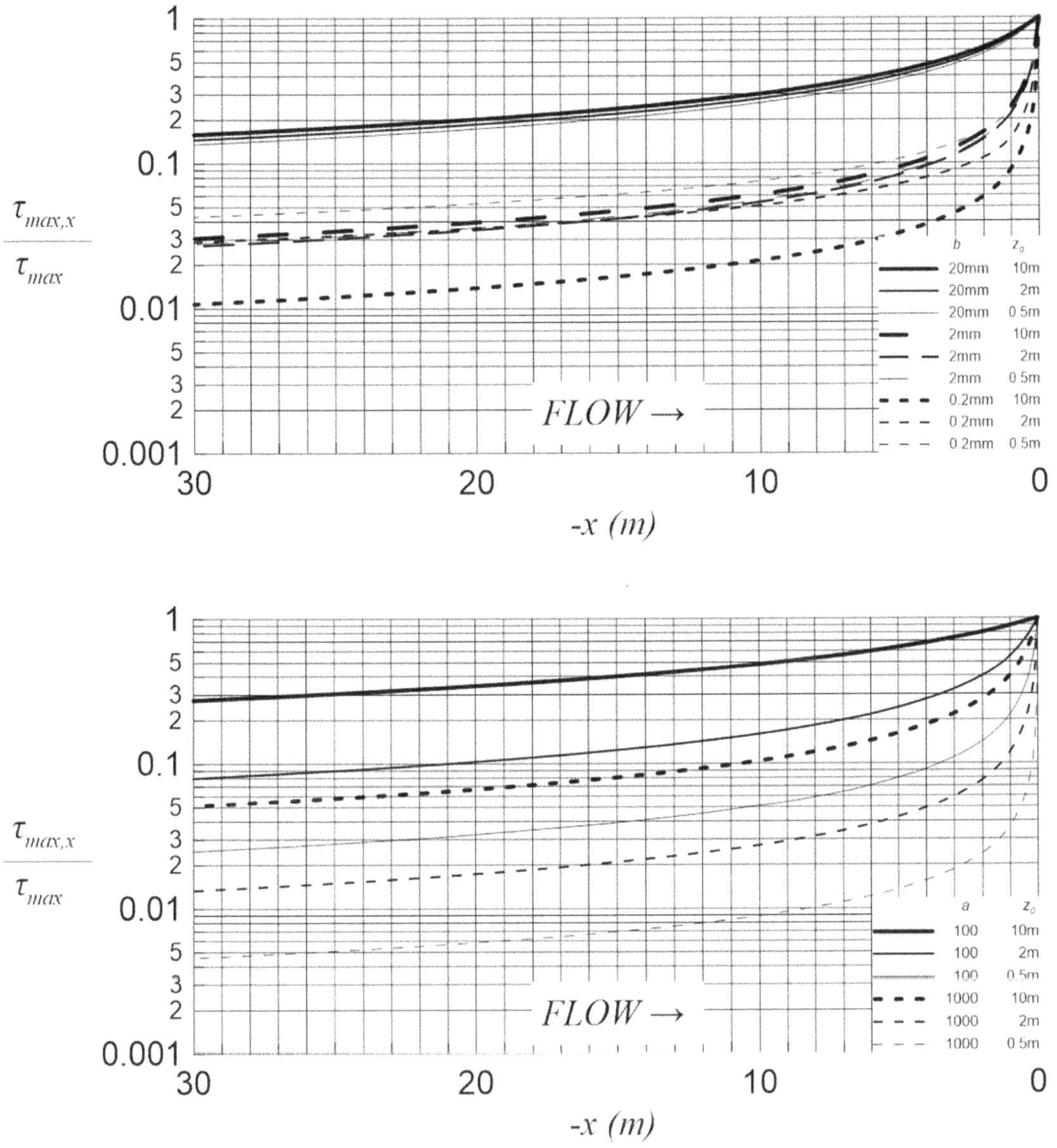

FIGURE 9.75 Maximum hydraulic wall stresses along cracks up to 30 m in length normalised by the maximum hydraulic wall stress at the downstream end. Light lines indicate $z_0 = 0.5m$, medium $z_0 = 2m$ and heavy lines indicate $z_0 = 10m$. Upper panel, rectangular cracks with line styles: short dashes $b = 0.4mm$; long dashes $b = 2mm$; continuous $b = 20mm$. Lower panel, triangular cracks with line styles: continuous $a = 100$; short dashes $a = 1000$ (Peirson et al., 2022; Peirson and Fell, 2024). (Copyright © 2025 Taylor & Francis Group, reprinted by permission of Informa UK Limited, trading as Taylor & Francis Ltd, https://www.tandfonline.com.)

TABLE 9.21 Hydraulic Shear Stresses (N/m² or Pa) at the Downstream End (τ_{max}), Upstream End (τ_{min}) and Mid-Point (τ_{mid}) of Triangular Cracks, versus Depth of the Base of a Crack (D) below Reservoir Level at ($Z_{x=30}$) and Crack Depth-Width Aspect Ratio (a) (Also shown are the hydraulic shear stresses estimated using average gradients; Fell et al., 2024)

WATER DEPTH (D) (M) AT THE UPSTREAM END OF THE CRACK, I.E AT	*Z_0 = WATER DEPTH AT THE DOWNSTREAM END OF THE CRACK (M)*			*HYDRAULIC SHEAR STRESS AT THE DOWNSTREAM END τ_{MAX} (PA)*			*HYDRAULIC SHEAR STRESS AT THE UPSTREAM END τ_{MIN} (PA)*			*HYDRAULIC SHEAR STRESS AT THE MID-POINT OF THE CRACK τ_{MID} (PA)*			*HYDRAULIC SHEAR STRESS ESTIMATED USING AVERAGE GRADIENT (PA)*		
$Z_{X=30}$ (M)	*a=100*	*a=200*	*a=400*	*a=100*	*a=200*	*a=400*	*a=100*	*a=200*	*a=400*	*a=100*	*a=200*	*a=400*	*a=100*	*a=200*	*a=400*
2 m	0.9	0.7	0.55	55	60	60	2	1	0.5	2.5	2	1	8	4	2
4 m	2.0	1.7	1.4	100	100	100	8	4	2	11	7	2	24	12	6
6 m	3.5	3.0	2.5	150	130	145	18	9	4.5	25	13	6	60	30	15
8 m	5.0	4.5	3.5	190	170	185	32	15	7	44	25	11	100	50	25
10 m	6.5	5.5	4.5	230	200	225	46	24	11	92	36	16	160	80	40

Note: See Figure 9.70 for definition of a, the taper of the crack. Crack depth/crack width = 0.5a. b = 0 for this table.

TABLE 9.22 Hydraulic Shear Stresses (N/m^2 or Pa) at the Downstream End (τ_{max}), Upstream End (τ_{min}) and Mid-Point ($\tau_{x=15}$) of Rectangular Cracks, versus Depth of the Base of a Crack (D) below Reservoir Level at ($Z_{x=30}$) (Also shown are the hydraulic shear stresses estimated using average gradients (τ_{av}); Fell et al., 2024)

WATER DEPTH (D) (M) AT THE UPSTREAM END OF THE CRACK, I.E. AT	*Z_0 = WATER DEPTH AT THE DOWNSTREAM END OF THE CRACK (M)*			*HYDRAULIC SHEAR STRESS AT THE DOWNSTREAM END τ_{MAX} (PA)*			*HYDRAULIC SHEAR STRESS AT THE UPSTREAM END τ_{MIN} (PA)*			*HYDRAULIC SHEAR STRESS AT THE MID-POINT OF THE CRACK τ_{MID} (PA)*			*HYDRAULIC SHEAR STRESS ESTIMATED USING AVERAGE GRADIENT τ_{AV} (PA)*		
$Z_{X=30}$ (M)	*b = 2 MM*	*b = 5 MM*	*b = 10 MM*	*b = 2 MM*	*b = 5 MM*	*b = 10 MM*	*b = 2 MM*	*b = 5 MM*	*b = 10 MM*	*b = 2 MM*	*b = 5 MM*	*b = 10 MM*	*b = 2 MM*	*b = 5 MM*	*b = 10 MM*
2 m	0.25	0.25	0.4	10	20	25	0.4	0.8	1.25	0.5	0.9	1.8	0.4	2	4
4 m	0.5	0.5	0.75	15	35	45	0.5	1.4	2.25	0.75	1.5	3.5	0.6	3	7
6 m	0.65	0.8	1.15	25	55	70	0.7	2.2	3.5	1.25	2.5	5.25	2	5	10
8 m	0.9	1.1	1.5	35	80	90	1.2	3.2	4.5	1.75	3.5	6.75	2.5	7	13
10 m	1.1	1.45	1.9	45	95	105	1.5	4	5.25	2.25	4.25	7.5	3.25	8	16

Note: b = width of crack with no taper; i.e. $a = 0$. Values are listed to the second decimal to maintain relativity. In reality they are not so precise.

9.13.4 Estimation of Hydraulic Shear Stresses in Vertical Cracks with a Downstream Filter or Transition

9.13.4.1 Description of the Peirson and Fell (2025) Modelling

Professor Peirson, reported in Peirson and Fell (2025), modelled flow in vertical cracks which exit into a filter at the downstream end of the crack. Figure 9.76 shows the conceptualisation of the crack-filter interface.

Since many filters are coarser than required for no-erosion performance, Peirson and Fell (2025) modelled filter permeabilities representative of filters which will allow Some or Excessive Erosion as defined by Foster and Fell (2001) before sealing.

Using the Sherard et al. (1984) method the permeability is estimated from:

$$k_v = C(D_{15})^2 \tag{9.11}$$

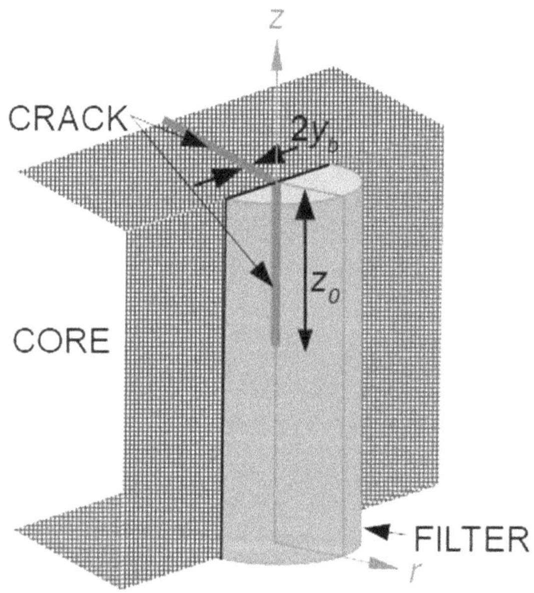

FIGURE 9.76 Conceptualisation of the crack-filter interface (Peirson and Fell, 2025). (Copyright © 2025 Taylor & Francis Group, reprinted by permission of Informa UK Limited, trading as Taylor & Francis Ltd, https://www.tandfonline.com.)

where D_{15} in mm is the particle size for which 15% is finer and C is a coefficient equal to 0.0035, based on a range of median test values of $C = 0.002$ to $C = 0.006$.

The Sherard and Dunnigan (1989) No Erosion filter limiting D_{15} size for sandy clays with 15% to 85% passing 0.075 mm is 0.7 mm, so these materials would have a permeability = 1.7×10^{-3} m/sec. Using Foster and Fell's (2001) criteria of Some Erosion, $D_{15} = 9D_{85}$ and so is likely to be about 2 mm with a permeability = 1.4×10^{-2} m/sec. Similarly, Foster and Fell's (2001) criteria of Excessive Erosion, $D_{15} = 9D_{95}$ and so is likely to be about 5 mm with a permeability = 8.7×10^{-2} m/sec.

Given this, the Peirson and Fell (2025) modelling used permeabilities 1×10^{-2} m/sec, 3×10^{-2} m/sec and 1×10^{-1} m/sec approximating to No-erosion, Some Erosion and Excessive Erosion filters for Type 2 soils, i.e. soils with between 40% and 85% finer than 0.075 mm.

The crack width/water level/discharge/hydraulic shear stress behaviours of the combined crack-filter model are summarised for these three permeabilities in the Supplemental Information to Peirson and Fell (2025). This paper is open access and available free of charge at https://doi.org/10.1080/19648189.2025.2547252.

For reference, the results of Peirson et al. (2022) for flow with a free-flow exit are also shown because Peirson and Fell (2025) found that the influence of the crack hydraulics on the water level/flow/shear stress behaviour may dominate, especially for filters of high permeability.

Figures 9.77 and 9.78 present the water level-flow behaviour at the downstream end, and 30 m upstream of a crack with a filter of permeability of 1×10^{-1} m/sec. Figure 9.79 presents the shear stresses at the downstream end where they will be at a maximum. These figures are representative of the cases modelled and provided in the Supplemental Information of Peirson and Fell (2025). From these Peirson and Fell (2025) indicate it can be seen that:

(a) The upper panel of Figure 9.77, which is for flow depths at the downstream end (Zo), shows that for crack widths of 1, 2 and 4 mm, the discharge from cracks without a filter downstream is predicted to be lower than that where a filter with hydraulic conductivity of 1×10^{-1} m/sec is present. Consequently, flow is controlled by the crack geometry rather than the filter, and the presence of the filter makes no difference to the flow emerging from the crack. The flow should always be determined by whichever curve gives the lowest flow. In contrast, for crack widths of 20 and 40 mm, the flow with the filter is lower than that from a crack with no filter. Therefore, the flow is determined by the filter curve.

Similar behaviour can be observed in the lower panel of Figure 9.77 except that the filter curve with $a = 500$ crosses the no filter curve at approximately $z_0 = 5.5$ m. This indicates that for downstream water depths less than 5.5 m, flow is controlled by the crack geometry. Above water depth of 5.5 m, flow is controlled by the filter.

(b) In Figure 9.78 which is for flow depths at the upstream end of a 30 m long crack (Z30) the effect of the resistance to flow in the crack can be observed for the high permeability filter. In the upper panel, there is no difference between the filter and no filter curves. In spite of the differences between the filter and no filter curves in the upper panel of Figure 9.78, the effect of the resistance to flow in the crack smooths out these differences

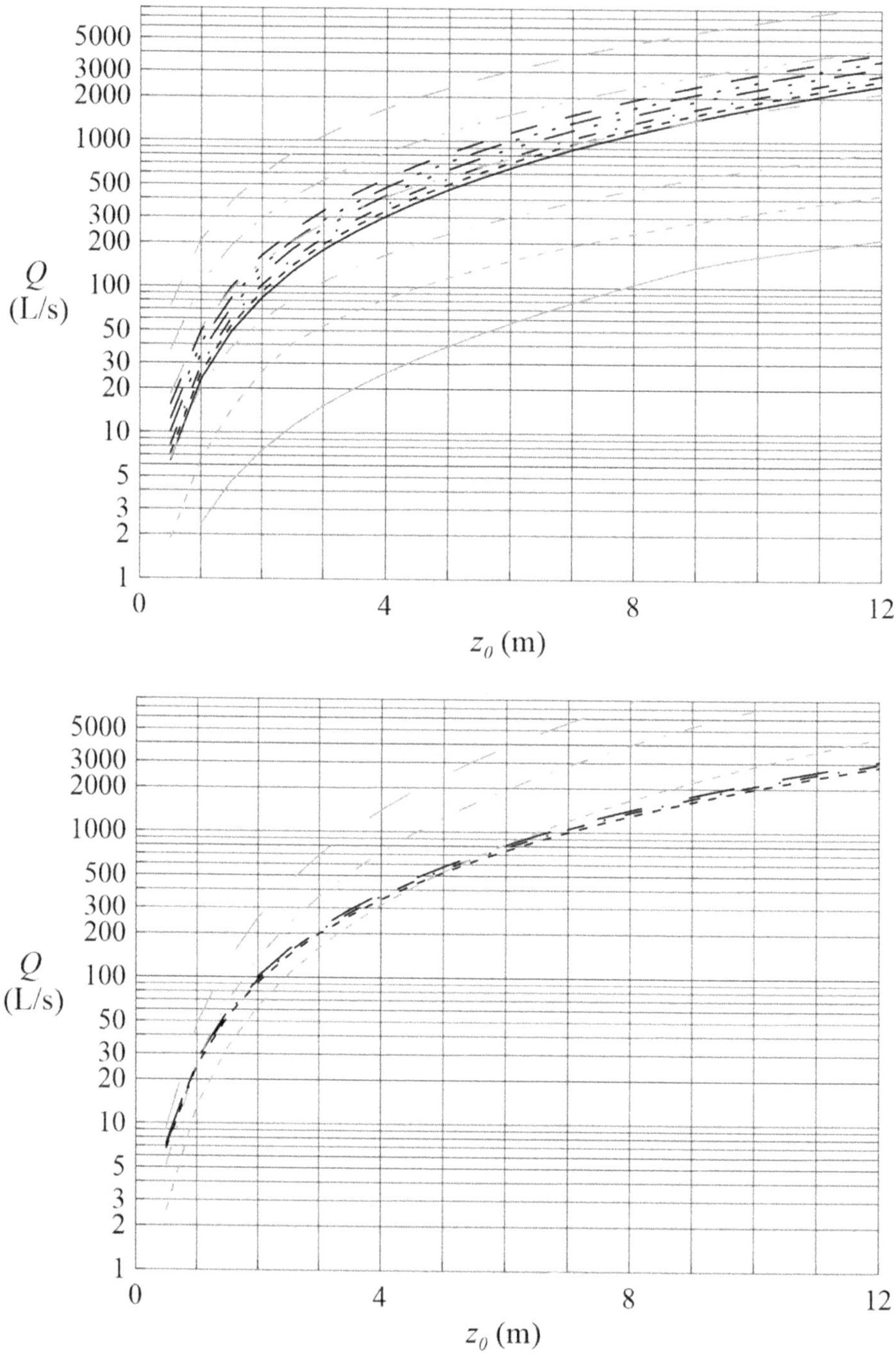

FIGURE 9.77 Flow-depth relationships at the downstream end of cracks. Grey lines—no filter downstream (Peirson et al., 2022). Black lines—with a filter of 1×10^{-1} m/sec. Upper panel (Supplemental A.R.1E–1.Q.0): Rectangular cracks, where line styles indicate crack width *b*: solid line 1 mm; short dashes 2 mm; dash-dot line 4 mm; dash-double dot 10 mm; dash-triple dot 20 mm; long dashes 40 mm. Lower panel (A.T.1E–1.Q.0): Triangular cracks with base crack width *b* = 1 mm and line styles: short dashes *a* = 500; dash-dot line *a* = 200; long dashes *a* = 100 (Peirson and Fell, 2025). (Copyright © 2025 Taylor & Francis Group, reprinted by permission of Informa UK Limited, trading as Taylor & Francis Ltd, https://www.tandfonline.com.)

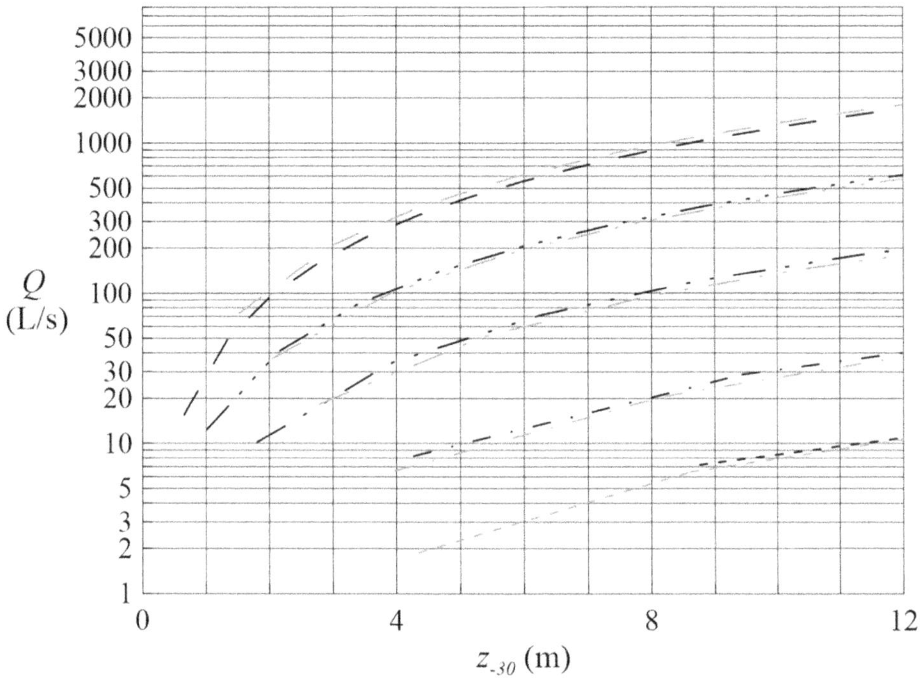

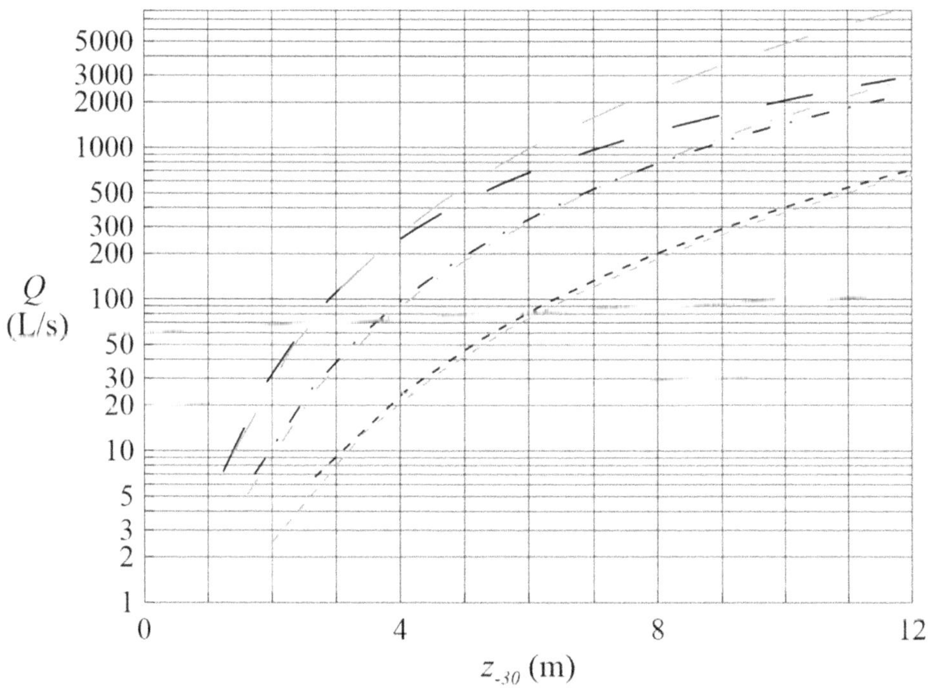

FIGURE 9.78 Computed water levels 30 m upstream of the crack outflow with no filter downstream and with a filter of 1×10^{-1} m/sec. Upper panel: rectangular cracks (as per Supplemental Figure A.R.1E–1.Q.30), lower panel: triangular cracks (A.T.1E–1.Q.30). The line styles are identical to those used in Figure 9.77 (Peirson and Fell, 2025). (Copyright © 2025 Taylor & Francis Group, reprinted by permission of Informa UK Limited, trading as Taylor & Francis Ltd, https://www.tandfonline.com.)

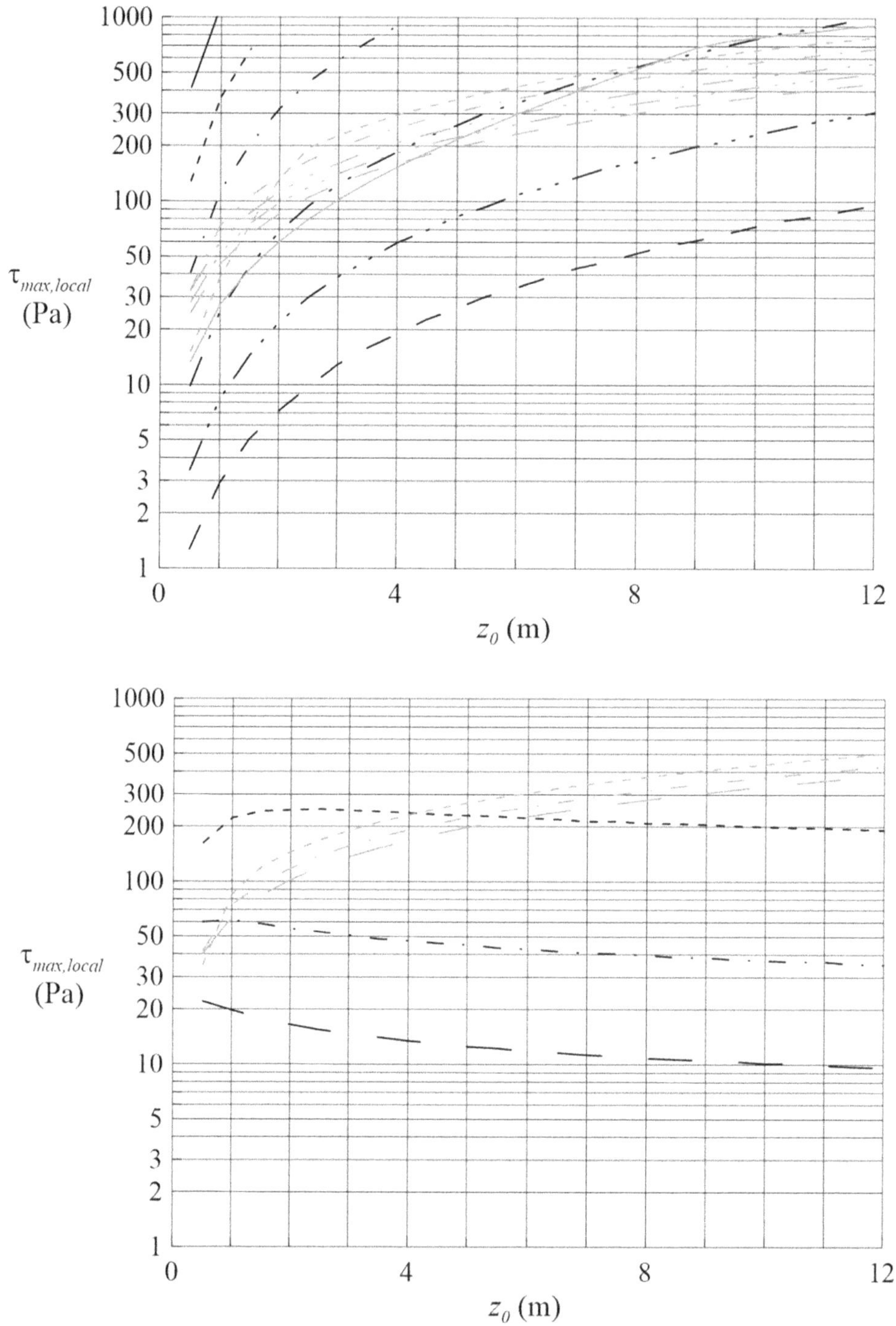

FIGURE 9.79 Computed maximum wall stresses within the cracks at the outflow with no filter downstream and with a filter of 1 × 10⁻¹ m/sec. Upper panel (as per Supplemental Figure A.R.1E–1.T.0): rectangular cracks, lower panel (A.T.1E–1.T.0): triangular cracks. The line styles are identical to those used in Figure 9.78 (Peirson and Fell, 2025). (Copyright © 2025 Taylor & Francis Group, reprinted by permission of Informa UK Limited, trading as Taylor & Francis Ltd, https://www.tandfonline.com.)

and the filter has no influence on water levels 30 m upstream of the crack exit. Similar behaviour can be observed in the lower panel of Figure 9.78 except that, for cracks with a = 100 and 200, the filter can be observed to affect water levels above 4 and 9 m respectively 30 m upstream of crack exits.

(c) Corresponding behaviour can be observed in the hydraulic stress graphs (Figure 9.79, as an example) to that in the flow graphs. Again, where a lower shear stress is predicted in the absence of a filter, it is the crack walls which are dominating the energy losses in the system and no filter value is applicable rather than that with the filter. Careful inspection of Figure 9.79 in comparison with Figure 9.77, for example, shows that the crossover points between filter-dominated and crack-dominated behaviour, for a given geometry, are identical. These crossover points will change according to the permeability of the filter.

9.13.4.2 Estimating Discharges, Hydraulic Shear Stresses and Surface Profiles along Vertical Cracks

Peirson and Fell (2025) describe the following procedure for estimating discharges, hydraulic shear stresses and surface profiles along vertical cracks with and without a downstream filter. The procedure has been edited to be compatible with the remaining text in this book.

1. Determine the range and configuration of possible crack geometries and the maximum water levels in the reservoir upstream of the dam as described in earlier sections.
2. Use the flow graphs presented in Peirson and Fell (2025) Supplemental to determine the flow through a given crack from the upstream water level and the length of the crack. The flow graphs are presented for a sequence of crack lengths. Flow can be obtained by interpolation between crack lengths that bracket the crack length under investigation. This process may have to be undertaken iteratively.

   There will be slopes in both the upstream and downstream faces of the core. The crack length is the horizontal distance between the surface point of entry in the reservoir and the core surface point of the emergent flow downstream as shown in Figure 9.80 and will vary according to the flow and downstream filter properties.
3. Once the flow is determined, check that the crack length assumed in step 2 is consistent with the corresponding depth of flow emerging from the crack at the downstream point using the graphs in Peirson and Fell (2025) Supplemental. If not, this step will need to be repeated. Once consistent crack length, flow and downstream depths have been determined, the Peirson and Fell (2025) Supplemental graphs can be used to determine water levels at specified points along the crack. Discharge-water level and shear stress-water level graphs for 0, 2, 5, 10, 20 and 30 metres upstream from the downstream end are provided. From the local water level, the local maximum hydraulic shear stress can be directly determined.
4. Peirson and Fell (2025) have assumed that almost all of the dams being assessed will have a core width at the level of potential cracking less than 30 m. For crack lengths greater than 30 m, a conservative result will be obtained if the crack length is assumed as 30 m.
5. The graphs in Peirson and Fell (2025) Supplemental provide sufficient information to enable the shear stress distribution of a freely outflowing crack to be compared with that with a filter downstream of a given permeability.

As a practical example, refer to the plots in Peirson and Fell (2025) Supplemental. Assume a core 20 m wide with a water level 10 m above the base of the crack at its upstream side. Assume also a rectangular crack geometry with a width of 8 mm.

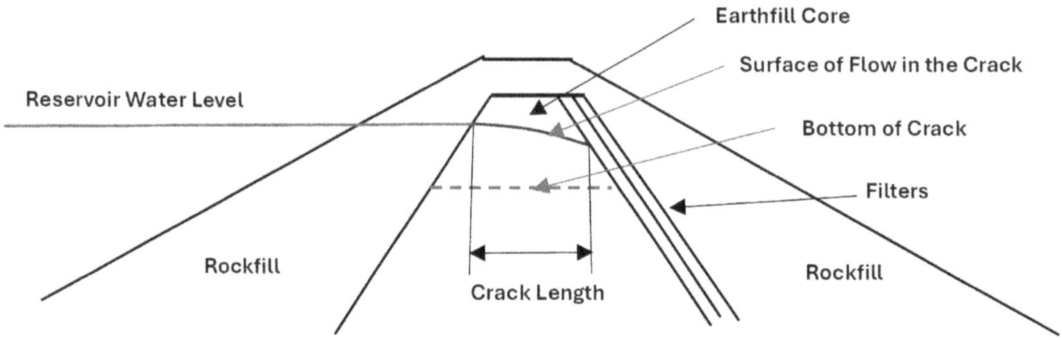

FIGURE 9.80 Definition of crack length for use in Peirson and Fell (2025) modelling of flows on cracks with a downstream filter and Peirson and Fell (2024) modelling of flow in cracks with free outflow.

If that crack is free flowing at its downstream end, Figure A.R.1E-2.Q.20 yields a flow rate through the crack of 120 l/sec. Figure A.R.1E-2.T.20 gives a local maximum shear stress at the upstream end of 6.5 Pa. Figure A.R.1E-2.Q.0 shows that the water level at the downstream end is 2.1 m. The maximum shear stress, from A.R.1E-2.T.0, is estimated to be 120 Pa.

A water level and shear stress profile can be developed from the intervening figures. For example, Figure A.R.1E-2.Q.2 shows that the water level 2 m upstream of the downstream end is 4.5 m and Figure A.R.1E-2.T.2 gives a maximum shear stress of 28 Pa. Five metres upstream of the downstream end the water level is 7.0 m and Figure A.R.1E-2.T.5 gives a maximum shear stress of 19 Pa.

Repeating these calculations with an assumed filter of permeability of 1×10^{-2} m/sec, Figure A.R.1E-2.Q.20 tells us that the flow rate is slightly reduced to 90 l/sec. Figure A.R.1E-2.T.20 gives a local maximum shear stress at the upstream end is less than 4.5 Pa. Figure A.R.1E-2.Q.0 shows that the water level at the downstream end is 6.1 m. The maximum shear stress, from A.R.1E-2.T.0, is estimated to be 8 Pa.

For an identical geometry, now assume that the core width is 7 m rather than 20 m and the water level in the reservoir is 10 m above the base of the crack with the filter in place at the downstream end. Figure A.R.1E-2.Q.5 indicates that the flow would be 140 l/sec if the depth was 10 m at a location 5 m upstream of the downstream end. Similarly, Figure A.R.1E-2.Q.10 yields 120 l/sec for a 10 m depth at 10 m upstream. Linear interpolation gives 128 l/sec at 7 m, as required. Water levels and stresses can then be derived along the crack. For example, Figure A.R.1E-2.Q.0 and Figure A.R.1E-2.T.0 indicate a depth and maximum shear stress at the downstream end of 7.7 m and 12 Pa.

9.13.4.3 Tables with Typical Results

Tables 9.23 and 9.24 present crack flows and hydraulic shear stresses for a range of crack widths and water depths for a 30 m long crack. The minor differences for the values for where there is no filter compared to those in Tables 9.21 and 9.22 are due to variations in interpolation and extrapolation.

These are presented to show the effect of the filters on the flows and hydraulic shear stresses. It can be seen that the filter has little effect for 2 mm cracks but has a large effect on downstream shear stresses for wide cracks. These tables should not be applied to crack lengths less than 30 m because the results are dependent on the crack length. Use the tables in Peirson and Fell (2025) Supplemental which cover a range of lengths.

9.13.4.4 Selection of Filter Permeability for Analyses

As discussed in Chapter 12, for many existing dams the filter and soil in the core have a range of particle size distributions. There may be scenarios where the filter is a No Erosion filter to the core, others where it is a Some Erosion, ExcessiveEerosion or Continuing Erosion.

The suggested procedure for selecting the filter permeability for use in estimating flow and hydraulic shear stresses is:

- No Erosion filter. Use filter permeability for the filter grading at the coarse boundary of no-erosion filter.
- Some Erosion filter. Use filter permeability for the filter grading at the coarse boundary of some-erosion filter.
- Excessive Erosion filter. Use filter permeability for the filter grading at the coarse boundary of Excessive Erosion filter.
- Continuing Erosion filter. Estimate assuming there is no filter, i.e. free outflow condition as described in Section 9.3.3.

As shown in Figure 9.81 the variability in permeability can be considered as being in layers in the filter. The flow conditions will not be as modelled in Figure 9.76, which assumes a homogeneous isotropic permeability. The flow where the crack exits into a coarse more permeable layer of Zone 2A filter will discharge through that layer into the Zone 2B filter and then flow down that filter. There may be somewhat higher resistance to discharge from the crack into the filter than assumed in the Peirson and Fell (2025) analyses but not by a significant amount.

If there is only a single filter in an earthfill dam with an embankment filter drain as in Figure 2.1, flow in the filter may be controlled by the permeability of the less permeable layers of filter. However, flow from the crack will spread horizontally and laterally within the more permeable layers of filter. It is considered therefore that the previous assumptions should be applied and will be somewhat on the conservative side in regard to hydraulic shear stresses in the crack.

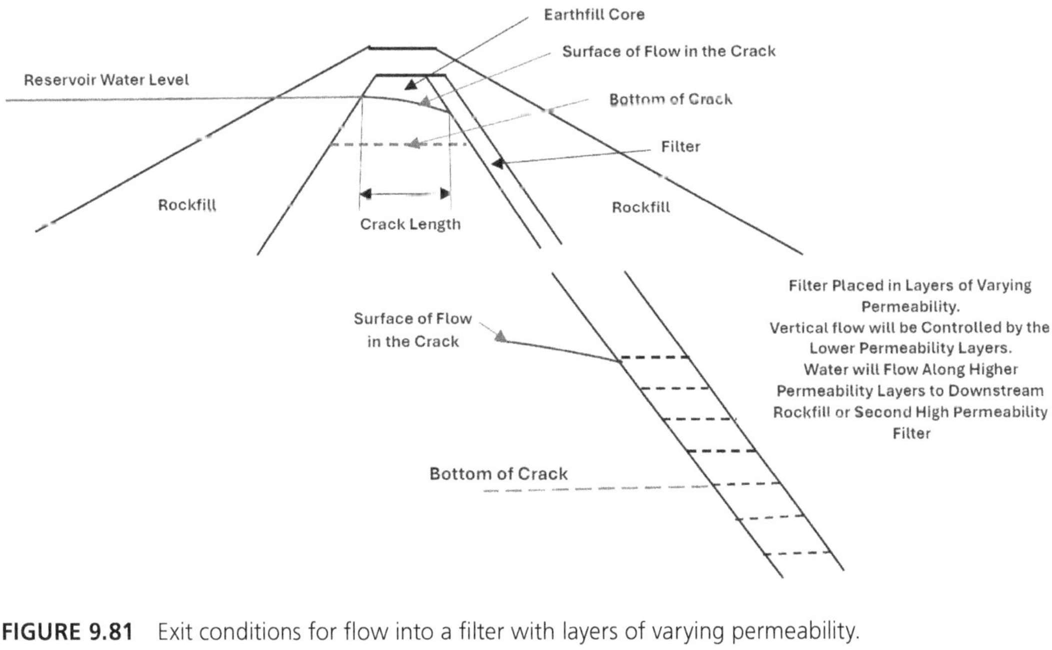

FIGURE 9.81 Exit conditions for flow into a filter with layers of varying permeability.

TABLE 9.23 Estimated Flow (Q, l/sec) and Hydraulic Shear Stresses (Pa) at the Downstream End (τ_{max}) and Upstream End (τ_{min}) in Rectangular Vertical Cracks with and without a Filter Downstream (kv is the permeability of the filter where the crack exits [m/sec]. Data is for a 30 m long crack. Do not apply to other crack lengths.)

		CRACK WIDTH																			
		2 MM				4 MM				10 MM				20 MM				40 MM			
Z_{30} M	k_v M/S	Q L/S	Z_0 M	τ_{MIN}	τ_{MAX}	Q L/S	Z_0 M	τ_{MIN}	τ_{MAX}	Q L/S	Z_0 M	τ_{MIN}	τ_{MAX}	Q L/S	Z_0 M	τ_{MIN}	τ_{MAX}	Q L/S	Z_0 M	τ_{MIN}	τ_{MAX}
1	No filter	<1	<0.25	<<1	8	2	0.3	<1	20	7	0.1	1	10	20	0.3	1.8	20	50	0.4	2.5	23
1	0.1	<1	<.3	<<1	7	1.8	0.3	<1	18	7	0.4	<1	8	12	0.5	1.6	3.4	28	0.6	1.5	1.4
1	0.03	<1	<.3	<<1	7	1.4	0.4	<1	4	3	0.5	<1	1.2	8	0.9	<1	1	12	0.9	<1	<1
1	0.01	<1	<0.5	<<1	<1	1.3	0.5	<1	1	2.5	0.8	<1	<1	4	1.1	<1	<<1	5	1	<<1	<<1
2	No filter	1	0.3	<1	8	3	0.3	<1	25	15	0.4	1.2	27	35	0.5	2.3	29	100	0.7	4.1	35
2	0.1	1	0.3	<1	8	3	0.3	<1	25	11	0.6	1.1	12	35	1	2	8.5	90	1.5	2.8	5
2	0.03	1	0.4	<1	8	2	0.5	<1	5	10	1	1.4	3	28	1.6	1.2	1.8	45	1.9	<1	<1
2	0.01	1	0.7	<1	3.5	2	0.8	<1	<1	8	1.9	<1	1.1	12	0.5	<1	<1	17	2	<1	<1
5	No filter	2.2	0.6	<1	18	8.8	0.6	<1	37	45	1.1	2.8	65	150	1.3	5.5	65	450	1.7	11	75
5	0.1	2.2	0.6	<1	17	8.8	0.6	<1	40	45	1.3	2.8	30	140	2	5	21	400	3.4	7.5	14
5	0.03	2.2	0.6	<1	16	8.8	1.1	1.2	16	45	2.3	2.5	10	120	4.1	3.3	7	210	4.8	2.2	2.9
5	0.01	2.2	0.9	<1	5	8.8	1.9	1.1	4.8	35	3.8	1.7	2.9	65	4.8	1	1.2	80	5	<1	<1
10	No filter	8	1.3	<1	60	28	1.3	<2	90	140	2.2	6.1	120	450	2.7	11	270	1300	3.5	21	130
10	0.1	8	1.3	<1	55	28	1.3	<2	90	140	2.2	6.1	75	450	3.9	11	58	1200	6.5	17	40
10	0.03	8	1	<2	45	28	1.9	2.2	35	140	4.2	5.1	23	390	6.9	7.5	16	750	9.3	5.9	7.5
10	0.01	8	1.9	1.3	16	28	3.5	2.1	11	110	7.5	3.7	7.5	210	9.1	2.7	3.1	290	10	1.1	1.1

TABLE 9.24 Estimated Flow (Q, l/sec) and Hydraulic Shear Stresses (Pa) at the Downstream End (τ_{max}) and Upstream End (τ_{min}) in Triangular Vertical Cracks with and without a Filter Downstream. (kv is the permeability of the filter where the crack exits [m/sec]. Data is for a 30 long crack. Do not apply to other crack lengths.)

		CRACK SLOPE a = CRACK DEPTH/CRACK WIDTH											
		a = 100				a = 200				a = 500			
Z_{30}	k_v M/S	Q L/SEC	Z_0 M	τ_{MIN}	τ_{MAX}	Q L/SEC	Z_0 M		τ_{MIN}	τ_{MAX}	Z_0 M	τ_{MIN}	τ_{MAX}
1	No filter	4	0.3	<1	25	2	0.4	<1	36	<1	<.3	<1	<20
1	0.1	4	0.4	<1	22	2	0.4	<1	35	<1	<.3	<1	<20
1	0.03	3	0.7	<1	2.5	1.4	0.4	<1	6.1	<1	<.3	<1	12
1	0.01	2.5	0.9	<1	<1	1.4	0.7	<1	1.2	<1	<.5	<1	<3.5
2	No filter	30	0.8	2.1	50	10	0.7	1.3	50	2.5	0.4	<1	28
2	0.1	30	1.1	2.1	19	10	0.7	1.3	50	2.5	0.4	<1	28
2	0.03	26	1.9	1.2	1.9	10	1.2	1.3	6	2.5	0.6	<1	22
2	0.01	10	2	<1	<1	6.5	1.7	<1	<1	2.5	1	<1	4
5	No filter	550	2.7	12	120	180	2	7	110	40	1.3	3	110
5	0.1	450	4.4	3	12	180	3.8	7	47	40	1.3	3	110
5	0.03	180	5	1.3	1.3	140	4.8	3.5	5	40	2.5	3	29
5	0.01	60	5	<1	<1	55	5	<1	<1	35	4	2	3.8
10	No filter	5000	6.6	50	220	1800	5.8	28	240	380	3	11	190
10	0.1	2000	9.5	9	10	1200	8.4	17	40	380	4	11	210
10	0.03	650	9.9	1.1	1.3	600	9.8	3.9	4.1	350	7.8	3.8	22
10	0.01	210	9.9	<1	<1	200	9.8	<1	<1	180	9.5	2.8	3.1

9.13.4.5 *Application of the Results of Analyses*

The hydraulic shear stresses at the downstream end and upstream end of the crack are to be compared to the estimated critical shear stress of the soil to assess whether erosion will initiate. There are three scenarios:

(a) If the hydraulic shear stresses at both downstream and upstream ends are less than the critical shear stress, erosion will not initiate.
(b) If the hydraulic shear stresses exceed the critical shear stress at the downstream end but not the upstream, check if the upstream part of the crack will close by swell of the sides of the crack using the method described in Section 9.12. If it does close, erosion will initiate but not progress.
(c) If the hydraulic shear stresses at both downstream and upstream ends are greater than the critical shear stress and the upstream crack does not close by swell, erosion will initiate throughout the crack.

Take account of the uncertainty in the estimates of hydraulic shear stresses and critical shear stress when making this assessment.

9.13.4.6 *Filters Backed by Low Permeability Downstream Shoulder*

The prior method assumes that the filter thickness is greater than crack flow plume radius. Commonly constructed filters may have a width of 1 or 1.5 m. Analyses by Peirson and Fell (2025) indicate that for filters that width, the plume may be restricted by the presence of the downstream shoulder if it has a low permeability. The effect of an impermeable layer will be to increase the water levels immediately upstream of the filter leading to estimates of higher flows but lower shear stress in the crack. Ignoring any impermeable filter backing layer will lead to a somewhat but not greatly conservative result.

9.13.4.7 *Filters Backed by a Second and Higher Hydraulic Conductivity Filter*

Peirson and Fell (2025) describe a method for modelling this scenario. In most cases the effect will be small and unlikely to significantly affect outcomes.

9.13.4.8 *Non-Vertical Filters*

The Peirson and Fell (2025) analyses assumed that the filter lies in a vertical column. Where there is a slope in the downstream face of the core, the plume emerging below the crack will spread laterally and then flow in a thin layer down the core face. However, in the immediate vicinity of the crack-filter interface, flow behaviour will be dominated by the pressures at the downstream end of the crack. Consequently, they recommended that the effects of the downstream core slope be ignored when determining the crack flows and shear stresses within the crack.

9.13.5 Flows in Narrow Cracks in the Core or Foundation

Peirson and Fell (2024) present Figure 9.82 for estimating flow rate for confined flow conditions in narrow cracks. This gives the same or similar flows to the prior equations.

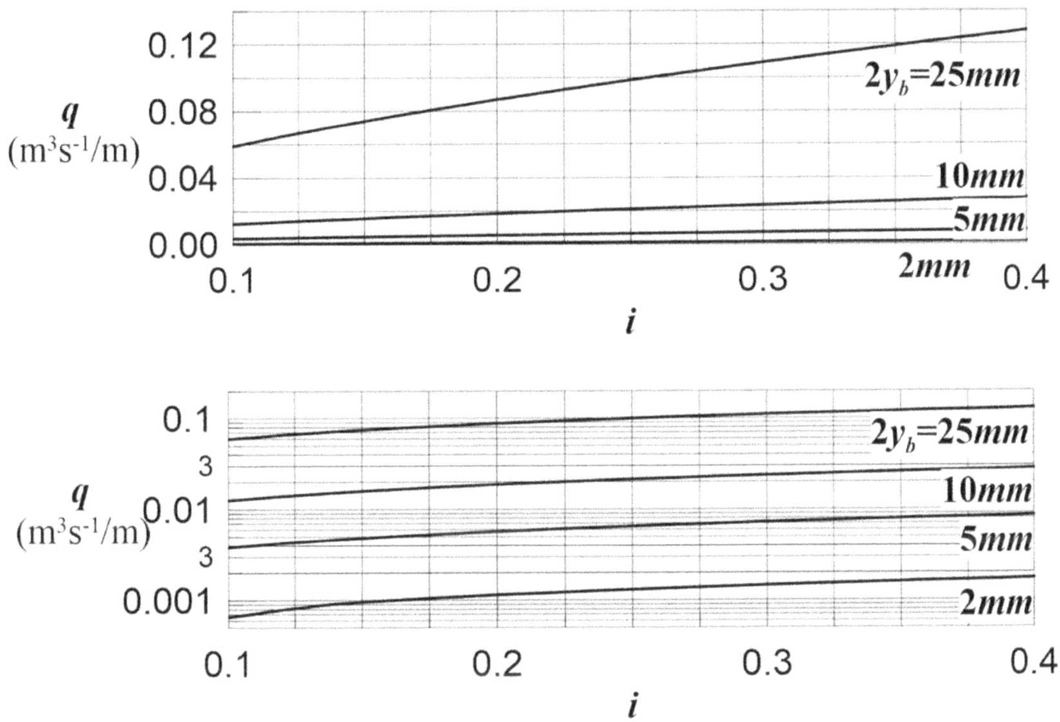

FIGURE 9.82 Flow per unit length of crack normal to direction of flow generated in submerged cracks of width $2y_b$ subject to a hydraulic gradient i. Upper panel: linear flow axis. Lower panel: logarithmic flow axis (Peirson and Fell, 2024). (Copyright © 2025 Taylor & Francis Group, reprinted by permission of Informa UK Limited, trading as Taylor & Francis Ltd, https://www.tandfonline.com.)

9.13.6 Erosion Properties of Soils in the Core of Embankment Dams

9.13.6.1 Basic Principles

As discussed in Section 9.13.2, the resistance to initiation of erosion of the core to concentrated leak erosion is characterised by the critical shear stress (τ_c) and the rate of enlargement of the pipe in the progression phase is characterised by the erosion coefficient (I_{HET}).

The erosion properties of soils for concentrated leak erosion can be determined by:

- Rotating cylinder tests, e.g. Arulanandan and Perry (1983); Chapius (1986a, 1986b); Chapius and Gatien (1986); Lim (2006); Lim and Khalili (2010).
- Slot Erosion Tests (SET) and Hole Erosion Tests (HET), e.g. Wan (2006); Wan and Fell (2002, 2004a, 2004b); Bonelli et al. (2007).
- Jet Erosion Tests (JET), e.g. Hanson (1990, 1991); Hanson and Cook (2004).

These tests allow determination of the critical shear stress for initiation of erosion and the erosion rate. At this time the most widely used tests are the HET and the JET. Figure 9.83 shows the laboratory test set up for the HET. Figure 9.84 shows a plot of rate of erosion (ε) versus hydraulic shear stress (τ) showing

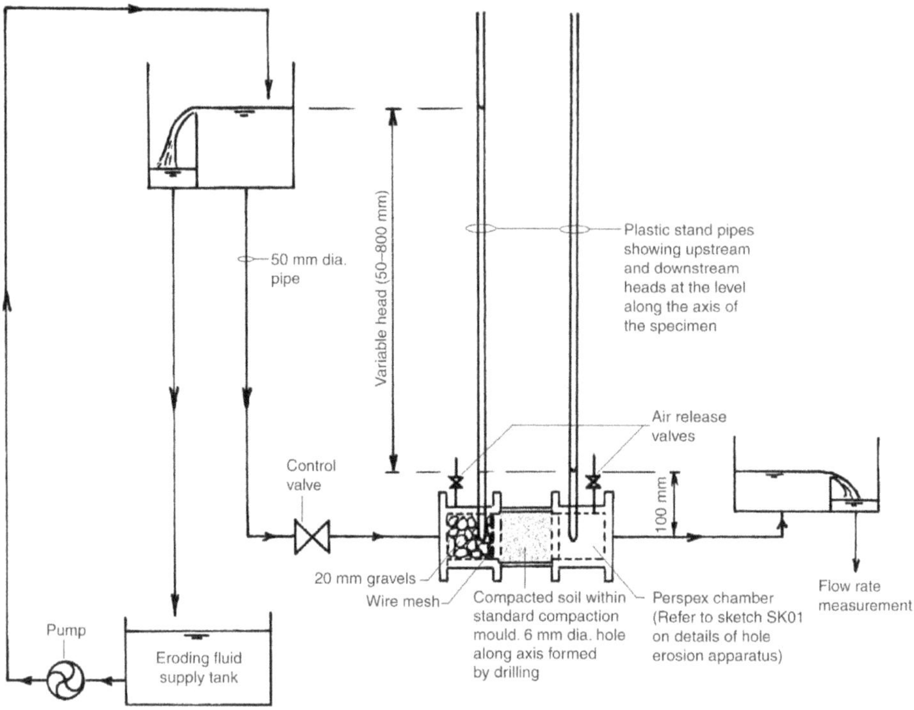

FIGURE 9.83 Hole Erosion Test equipment (Wan and Fell, 2002, 2004a, 2004b).

extrapolation to determine the critical shear stress (τ_c) at zero rate of erosion and the coefficient of soil erosion (C_e). In practice the data is seldom of such good quality as shown in Figure 9.84 and extrapolation give spurious results including negative values, so the authors recommend determination of the critical shear stress by varying the gradient in the test until a sufficiently low gradient is reached that no erosion is initiated.

In practice it can be difficult to interpret the initiation of erosion as the head is incrementally increased. In addition to observing erosion, it is necessary to monitor the flow rate. The initiation of erosion usually is apparent by an increased flow rate for a given head as shown in Figure 9.85.

Table 9.25 shows typical representative critical shear stress related to the soil Unified Soil Classification. This is for soils compacted to 95% SMDD at optimum moisture content.

The Representative Erosion Rate Index $\tilde{I}_{HET}$ is the hole erosion index I_{HET} for soil compacted to a density ratio of 95% of standard maximum dry density at optimum moisture content.

Soils can be classified into six groups according to their Representative Erosion Rate Index, $\tilde{I}_{HET}$. The six groups are as shown in Table 9.26. Note that this is a logarithmic scale and the rate of erosion of soils varies by up to five orders of magnitude.

In the absence of laboratory test values, the $\tilde{I}_{HET}$ can be related approximately to soil properties. Table 9.27 has been developed from test data to give a first approximation to the likely range of $\tilde{I}_{HET}$ for different classifications of non-dispersive soils.

It is emphasised that it is better to carry out a series of Hole Erosion Tests at varying heads or to use the method of Bonelli et al. (2007), Bonelli and Brivois (2008) to define the critical shear stress (τ_c) than to rely on these relationships. The tests should be carried out at the relative compaction and moisture content representative of those in the embankment because as discussed later, the properties are dependent on the degree of saturation.

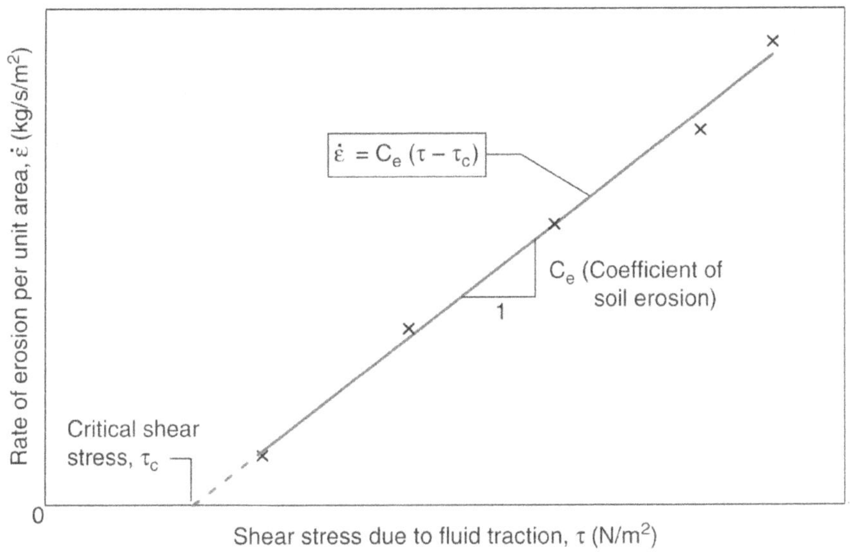

FIGURE 9.84 Plot of rate of erosion (ε) versus hydraulic shear stress (τ) showing extrapolation to determine critical shear stress (τ_c) at zero rate of erosion and coefficient of soil erosion (C_e) (Wan and Fell, 2002, 2004a, 2004b).

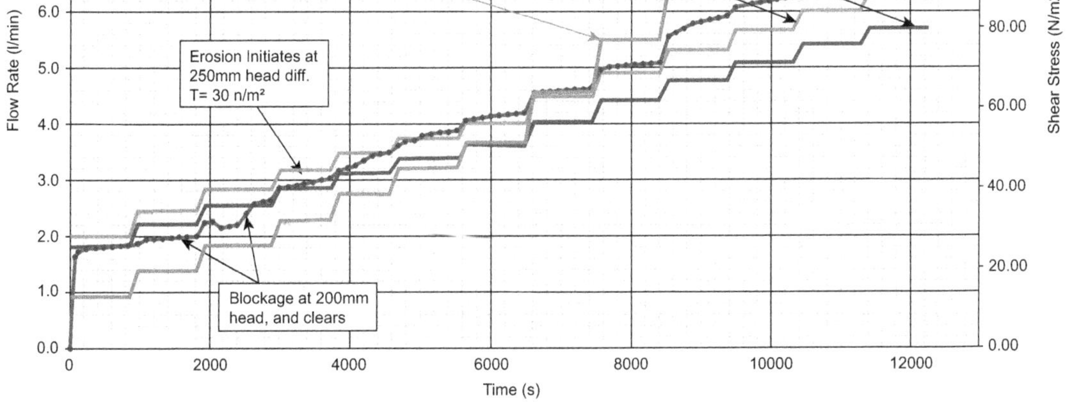

FIGURE 9.85 Flow rate and pressure differential versus time for a Hole Erosion Test. The interpreted critical shear stress is 30 Pa where measured flow rate exceeds the flow rate for the initial hole diameter and length. (Courtesy AECOM and Melbourne Water.)

TABLE 9.25 Typical Values and Likely Range of Representative Critical Shear Stress (τ_c) versus Unified Soil Classification (Fell et al., 2024)

	REPRESENTATIVE CRITICAL SHEAR STRESS (τ_C) PA[(a)]			
	NON-DISPERSIVE SOIL BEHAVIOUR		DISPERSIVE SOIL BEHAVIOUR	
UNIFIED SOIL CLASSIFICATION	BEST ESTIMATE	LIKELY RANGE	BEST ESTIMATE	LIKELY RANGE
SM with < 30% fines	2	1 to 5	1	0.5 to 2
SM with ≥ 30% fines	2	1 to 5	1	0.5 to 2
SC with < 30% fines	2	1 to 5	1	0.5 to 2
ML	2	1 to 5	1	0.5 to 2
SC with ≥ 30% fines	4	2 to 10	2	1 to 5
CL-ML	4	2 to 10	2	1 to 5
CL	5	2 to 20	2	1 to 5
MH	5	2 to 20	2	1 to 5
CL-CH	25	10 to 50	5	2 to 10
CH with Liquid Limit < 65%	25	10 to 50	5	2 to 10
CH with Liquid Limit ≥ 65%	60	25 to 100	5	2 to 10

Note: (a) To be used with caution. For important decisions carry out HET to determine the critical shear stress (τ_c).

TABLE 9.26 Descriptors for Erosion Rates of Soils (Wan, 2006)

GROUP NO.	EROSION INDEX	RATE DESCRIPTION
1	< 2	Extremely rapid
2	2–3	Very rapid
3	3–4	Moderately rapid
4	4–5	Moderately slow
5	5–6	Very slow
6	> 6	Extremely slow

Three other points to note are the effect of degree of saturation of the soil and the effect of dispersion.

9.13.6.2 Effect of Degree of Saturation of the Soil

The previous discussion relates to the properties of soils at the representative compaction conditions. That is at 95% standard maximum dry density and Standard Optimum Moisture Content. Wan and Fell (2002, 2004a, 2004b) and Lim (2006) have found that the erosion rate index for the hole erosion (I_{HET}) and rotating cylinder (I_{RCT}) tests are related to the degree of saturation of the soil.

Figure 9.86 shows results of testing three soils. Soil S is a residual granitic soil from Serpentine Dam, Western Australia, classifying as MH; Soil B is a sandy clay of low plasticity from Fairbairn Dam, Queensland; Soil F is a clay of low plasticity from Boggy Creek Dam in Oklahoma, USA; and "50% of kaolin" is an artificial soil consisting of 50% kaolin and the remainder fine sand size particles.

It will be seen that there is a significant dependence of the erosion rate on degree of saturation for the two clay soils. There was, however, less dependence on the degree of saturation for the silty soil. Wan and Fell (2002) noted a similar trend with two other residual granitic soils showing little change in the erosion rate with increased degree of saturation.

TABLE 9.27 Representative Erosion Rate Index ($\tilde{I}_{HET}$) versus Soil Classification for Non-Dispersive Soils Based on Wan and Fell (2002)

UNIFIED SOIL CLASSIFICATION	*EROSION RATE INDEX ($\tilde{I}_{HET}$)*		
	LIKELY MINIMUM	*BEST ESTIMATE*	*LIKELY MAXIMUM*
SM with < 30% fines	1	< 2	2.5
SM with > 30% fines	< 2	2 to 3	3.5
SC with < 30% fines	< 2	2 to 3	3.5
SC with > 30% fines	2	3	4
ML	2	2 to 3	3
CL-ML	2	3	4
CL	3	3 to 4	4.5
CL-CH	3	4	5
MH	3	3 to 4	4.5
CH with Liquid Limit < 65%	3	4	5
CH with Liquid Limit > 65%	4	5	6

Notes:

(1) Use best estimate value for best estimate probabilities. Check sensitivity if the outcome is strongly dependent on the results.

(2) For important decisions carry out Hole Erosion Tests, rather than relying on this table, which is approximate.

That the critical shear stress and erosion rate depend on the degree of saturation of the soil is an important finding because it means that once the core of a dam constructed of clay soil is saturated, it will have a slower rate of erosion and a higher initial shear stress. Just as important, this does not apply to silty sand cores such as decomposed and residual granites.

9.13.6.3 Effect of the Testing Method on the Critical Shear Stress to Initiate Erosion (τ_c) and the Erosion Rate Index

Critical Shear Stress

As discussed earlier, the critical shear stress to initiate erosion (τ_c) (or for the HET the initial shear stress, τ_0) is a difficult property to measure. Lim (2006), Lim and Khalili (2010) have found the form of the hydraulic shear stress versus erosion rate plot varies for different soils, with some plotting as bi-linear plots, others with the intercept on the "X" and "Y" axes. Wan and Fell (2002) had similar problems and this is the reason they adopted the procedure of varying heads in the HET to define the initial shear stress.

Bonelli and colleagues at CEMAGREF have refined the HET and developed an improved method for determination of the critical shear stress. Details are given in Bonelli et al. (2007), Bonelli and Brivois (2008), and Benahmed and Bonelli (2012).

Erosion Rate Index

Wan and Fell (2002, 2004a, 2004b) obtained a strong correlation between the erosion rate index from the Hole Erosion Test and the Slot Erosion Test. In the Hole Erosion Test erosion is from the sides of an initially 6 mm diameter hole drilled into a 100 mm long by 100 mm diameter cylindrical sample oriented horizontally. In the Slot Erosion Test erosion is from the sides of an initially 10 mm by 2.2 mm cross section by 1 metre long slot oriented horizontally.

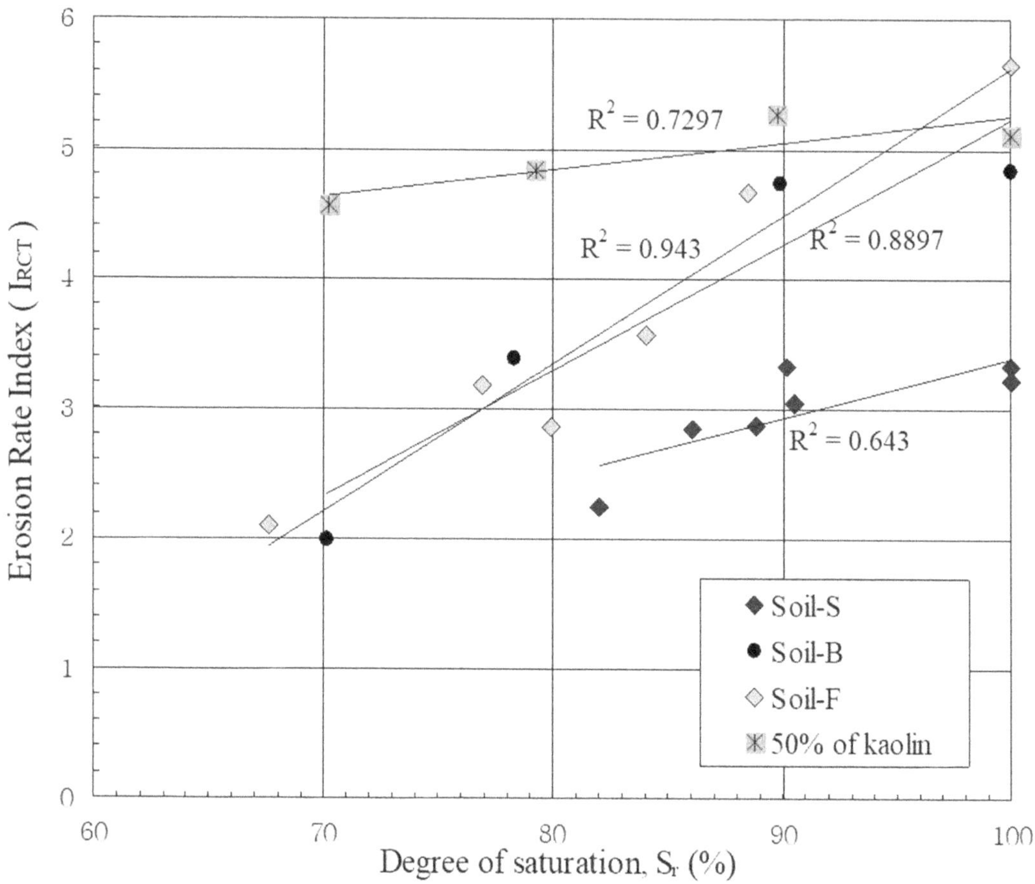

FIGURE 9.86 Relationship between degree of saturation and Erosion Rate Index (I_{RCT}) for non-dispersive soils (Lim, 2006).

Lim (2006), Lim and Khalili (2010 investigated the erosion properties of soils using rotating cylinder and Hole Erosion Tests. In the rotating cylinder test the erosion is from the vertical sides of a 100 mm diameter by 100 mm high sample. Figure 9.87 shows a comparison between the erosion rate indices for these tests. For dispersive soils there is good correlation with the HET giving slightly larger index (slower rate of erosion). However, there is a large difference for the non-dispersive soils with the rotating cylinder test giving rates 10 to 15 times higher than those in the HET. Lim (2006) attributes this to the vertical face of the RCT, and to surface and/or body slaking occurring in the RCT which could not occur in the HET because of the relatively large size of the slaking particles and the small hole in the HET.

Wahl et al. (2008) carried out tests where identically prepared remoulded soil samples were tested by the Hole Eosion and Jet Jrosion tests. They found that both methods ranked the soils similar for their relative erodibility, but the JET method indicated a higher rate of erosion, up to one or more orders of magnitude, and lower critical shear stresses by up to two or more orders of magnitude. They speculated that differences between the erosion rates observed in the two tests were related to several factors, including:

- Differences in the nature of the hydraulic attack upon the eroding surface in each test.
- Differences in the way that the flow exploits different weaknesses in the soil structure.
- Differences in the geometry of the exposed soil surface.

- Inaccurate or incomplete analytical models for the stress distributions produced at the soil surface.
- Use of a linear erosion model for analysis, when erosion rate versus applied stress may actually be nonlinear.
- Performing tests in different stress ranges and with an opposite progression of applied stress (high to low in JET, and typically well above the critical stress; low to high in HET, beginning near the critical stress level).

The range of soil types and compaction conditions for the tests by Wahl et al. (2008) was limited, but the greatest differences seemed to occur in samples with a coarse or non-uniform soil structure (generally

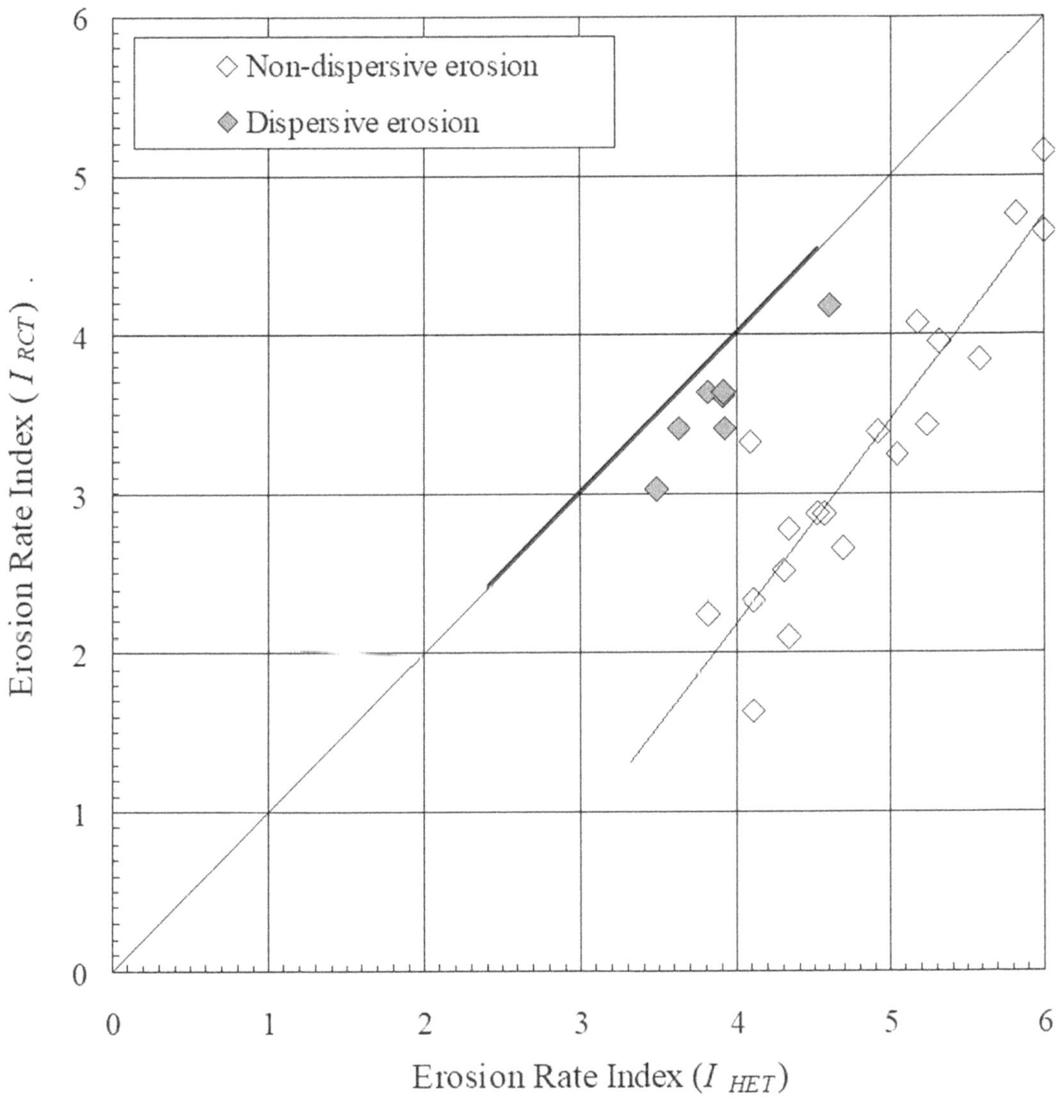

FIGURE 9.87 Correlation between erosion rate index from Rotating Cylinder and Hole Erosion Tests by Lim (2006).

samples containing more clay and compacted significantly dry of optimum moisture content). This suggested that the JET is more sensitive to soil fabric, perhaps because the stress environment produced by the impinging jet is more readily able to exploit weak zones in a non-uniform soil structure. This may also be related to the different geometry of the exposed soil surface in each test: a planar surface in the JET versus a small, confined hole for the HET. Aggregates of soil particles that can be dislodged and removed in the JET environment may be held in place by the surrounding soil mass in the HET.

It is considered that the HET better models erosion conditions in a crack and pipe in the early stages of development than does the rotating cylinder. However, it could be that the HET underestimates the rate of erosion as the pipe develops and the size allows the mechanisms modelled in the rotating cylinder to take effect. This could be allowed for by using a varying erosion rate index when calculating the rate of enlargement of a pipe.

9.13.6.4 Effect of Dispersion, Slaking, Soil Structure and Shear Strength on Erosion Properties

Dispersion

As discussed earlier, soils which show dispersive behaviour, soils classifying as Emerson Crumb Class 1 or 2 and Pinhole Dispersion D1 and D2, will have a very low critical shear stress if the eroding fluid does not inhibit the dispersion. That is, if the eroding water has low salts content.

However, Lim (2006) showed that for rotating cylinder tests the Erosion Rate Index is not greatly affected by whether the soil is dispersive after the initially rapid part of the erosion process.

Slaking

The term "slaking" or "soil slaking" is defined as "disintegration of unconfined soil after exposure to the air and subsequent immersion in a fluid, usually water; no external confining pressure is assumed to act over the soil prior to immersion" (Moriwaki and Mitchell, 1977).

Lim (2006) showed that slaking is the most important process involved in erosion in the RCT. They proved a strong correlation between the rates of slaking from a sample held statically in water to the erosion rate index from the RCT (I_{RCT}). He also showed that the slaking process was correlated strongly to the degree of saturation of the soil, with the slaking rate being up to 30 to 50 times lower between soils at 70% degree of saturation and those at 100% degree of saturation. This corresponds with the behaviour of the erosion rate index for clay soils obtained by HET.

This is a very important finding as it helps explain the actual mechanics of the erosion process in clay soils being strongly linked to slaking.

Soil Structure

Benahmed and Bonelli (2012), Lim (2006) and Wahl et al. (2008) have all noted that soil structure has an important effect on the erosion properties. They found that erosion rates are significantly higher for the same soil if the soil is compacted dry of optimum moisture content and the soil forms aggregated particles and/or micro-cracks. These allow erosion of blocks of the soil rather than of individual particles. This is one of the reasons higher erosion rates are measured in JET than HET, as the HET test is stopped with a relatively small hole diameter not allowing the "blocks" of soil to dislodge from the sides of the hole. This behaviour was also noted in rotating cylinder tests by Lim (2006).

Shear Strength of the Soil

Contrary to what many have assumed, the erosion rate and critical shear stress are unrelated to the effective stress or undrained shear strength of the soil. Soils with similar undrained strengths can have an erosion rate 100 or 1000 times different and critical shear stresses ranging over 1 Pa to 100 Pa depending on dispersion properties, clay mineralogy and degree of saturation.

Internal Erosion and Piping—Backward Erosion Piping, Suffusion and Global Backward Erosion

10

10.1 DESCRIPTION OF BACKWARD EROSION PIPING, SUFFUSION AND GLOBAL BACKWARD EROSION

10.1.1 Backward Erosion Piping (BEP)

As discussed in Section 10.2.1 Backward Erosion Piping (BEP) occurs in poorly graded fine to medium grained sands (SP) and silty sands (SM).

As shown in Figures 10.1 and 10.2, the BEP process begins at a free surface on the downstream side of an embankment. This free surface may be in a ditch or other excavation penetrating into the eroding non-plastic soil as shown in Figure 10.2 or, as in Figure 10.1, might form by first heaving and cracking the plastic strata overlying the non-plastic soil.

Backward erosion piping initiates where critically high hydraulic gradients at the toe of a dam erodes particles upwards or erodes particles horizontally in the case where the non-plastic soil layer daylights in the side of a ditch.

The process progresses beneath the levee or dam embankment in small erosion channels in which the flow velocity can transport the eroded particles downstream. For this to occur, the embankment, or the plastic soil strata overlying the non-plastic soil, must form a roof for the eroding channels to remain open. The presence of backward erosion piping is often exhibited by the presence of sand boils at the downstream side of the embankment.

If the seepage gradient beneath the embankment is sufficiently high, the eroding channels will progress to the upstream side of the embankment to connect with the reservoir. If that occurs the channels are rapidly eroded to form a pipe. The pipe enlarges rapidly and leads to breach of the embankment by slope instability or loss of freeboard if the pipe collapses to leave the top of the embankment below reservoir level.

DOI: 10.1201/9781003542025-10

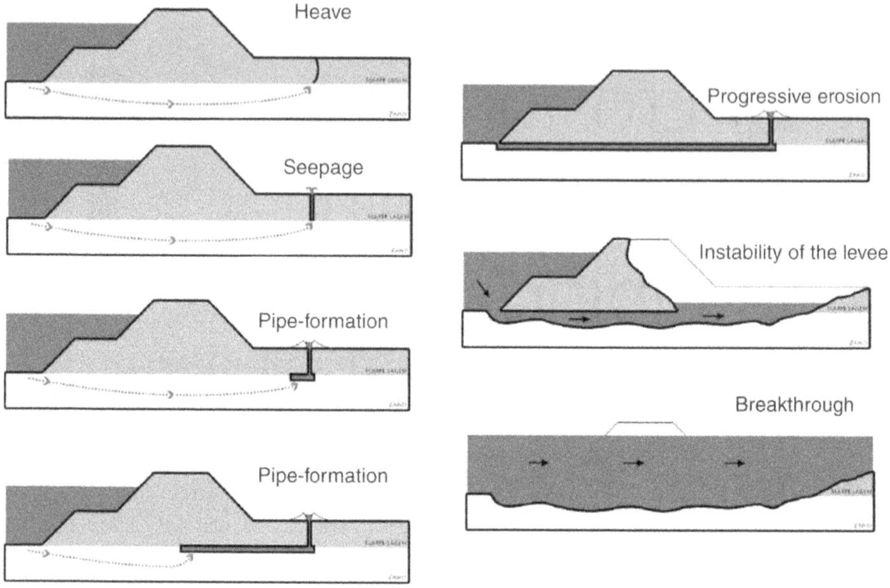

FIGURE 10.1 Backward Erosion Piping process; adapted from Sellmeijer et al. (2012). (Copyright © 2025 Taylor & Francis Group, reprinted by permission of Informa UK Limited, trading as Taylor & Francis Ltd, https://www.tandfonline.com.)

The BEP process is described in more detail in Robbins and Van Beek (2015), from which the following description has been adapted.

The first requirement for BEP to occur is the presence of seepage through the foundation (shown as process 1 in Figure 10.2). For backward erosion to initiate, a free, unfiltered exit must exist at the downstream side of an embankment. This free exit can either be due to a pre-existing defect in the confining layer (e.g. root hole, fence post hole or animal burrow) or to a crack resulting from heave (uplift) of the plastic confining layer. It is also possible to have a large, unfiltered exit area due to the presence of ditches, borrow pits or simply lack of a natural confining layer in certain geologic conditions.

Given the presence of a defect and an unfiltered exit, the seepage velocities at the exit must be high enough to locally fluidise the sand bed and begin lifting particles of the non-plastic foundation sand up, through the defect, to the ground surface (process 3 in Figure 10.2).

Once particle transport has initiated, erosion channels begin to be formed at the top of the non-plastic strata. In large-scale laboratory experiments such as those shown in Figures 10.3 and 10.4, the channels are seen to be narrow, meandering and braiding. Similar behaviour is observed in medium and smaller scale laboratory experiments. Hoffmans and van Rijn (2018) indicate that the channel width lies in the range of 10–30 times the d_{50} particle size of the sand and the channel height ranges from 5 to 10 times the d_{50} particle size. The channel height is often less than 2 mm.

For the channels to propagate horizontally beneath the confining layer, sufficient horizontal hydraulic forces (gradients) must exist for soil to continuously be released from the edge of the void and be transported to the defect location (processes 3 and 4, Figure 10.2). The magnitude of the horizontal gradients in the vicinity of the defect is reliant on the amount of discharge through the defect. As the defect increases in size, more flow will pass through it, creating a larger horizontal gradient towards the opening.

The shear stress imparted on the soil grains by the viscous, laminar flow must exceed the critical shear stress required for particle transport on the bottom of the pipe (process 6, Figure 10.2). If the

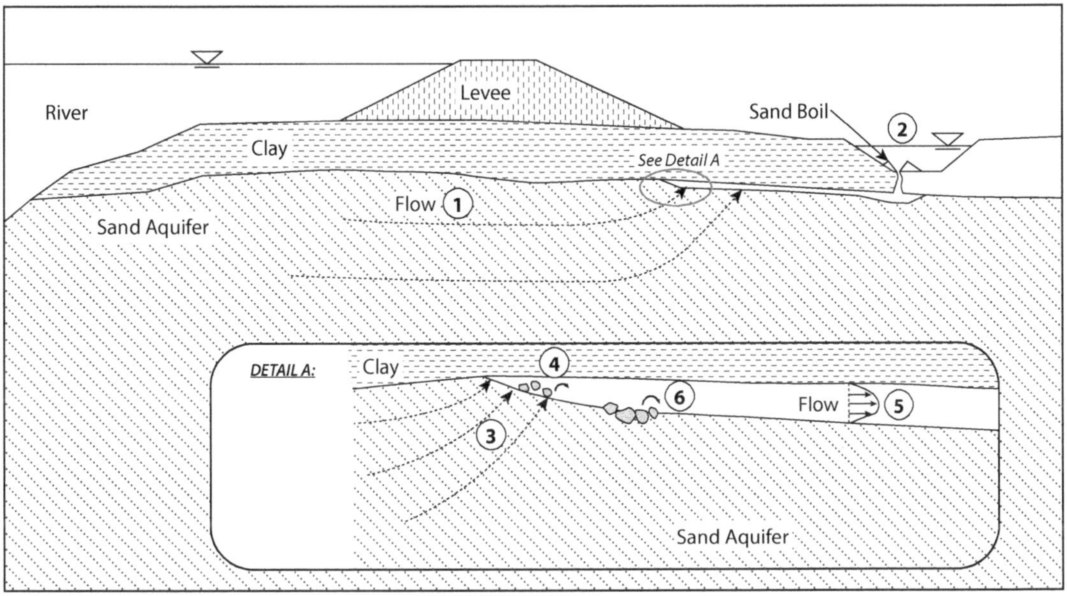

FIGURE 10.2 Illustration of the physics of backward erosion piping (Robbins and Van Beek, 2015).

The process includes (1) Darcy flow; (2) exit-related hydraulic conditions such as orifice flow, pipe flow or constant head boundary conditions; (3) liquefaction or fluidisation at the pipe head due to concentrated flow leading to (4) occasional bursts of high suspension solids into the pipe; (5) laminar flow conditions in the open pipe which cause (6) sediment transport along the bottom of the pipe. Items (1)–(6) must all be occurring simultaneously for piping to progress to failure.

flow in the pipe is insufficient to transport the particles, the pipe will become blocked and backward erosion will stop.

If backward erosion in the channels progresses beneath the structure to the river or reservoir, forming a continuous open channel, the erosion process changes to that of concentrated leak erosion. The erosion rate is rapid because the non-plastic sand soil has a low erosion rate index. The shallow channels enlarge to form a long, shallow pipe directly beneath the top stratum.

The individual physical processes in BEP interact as an overall system. During the progression phase head is lost in the sand aquifer, in the channels and in the exit area, e.g. a sand boil.

The exit condition has the ability to limit the exit flow, thereby limiting the amount of flow in the pipe. This is why sandbagging sand boils during flood fighting efforts slows or may halt the progression. The increased water level within the sandbag ring decreases the pressure head difference across the structure, thus limiting flow.

An interruption in the sediment transport in the pipe will increase the resistance in the pipe, potentially stopping the erosion at the pipe head.

10.1.2 Suffusion

Suffusion occurs when water flows through widely graded or gap-graded non-plastic soils such as river alluvium, colluvium in the bed of rivers in mountainous areas, embankment cores constructed of glacial origin soils, and in filters which have very wide or gap gradings or an excessive fines content.

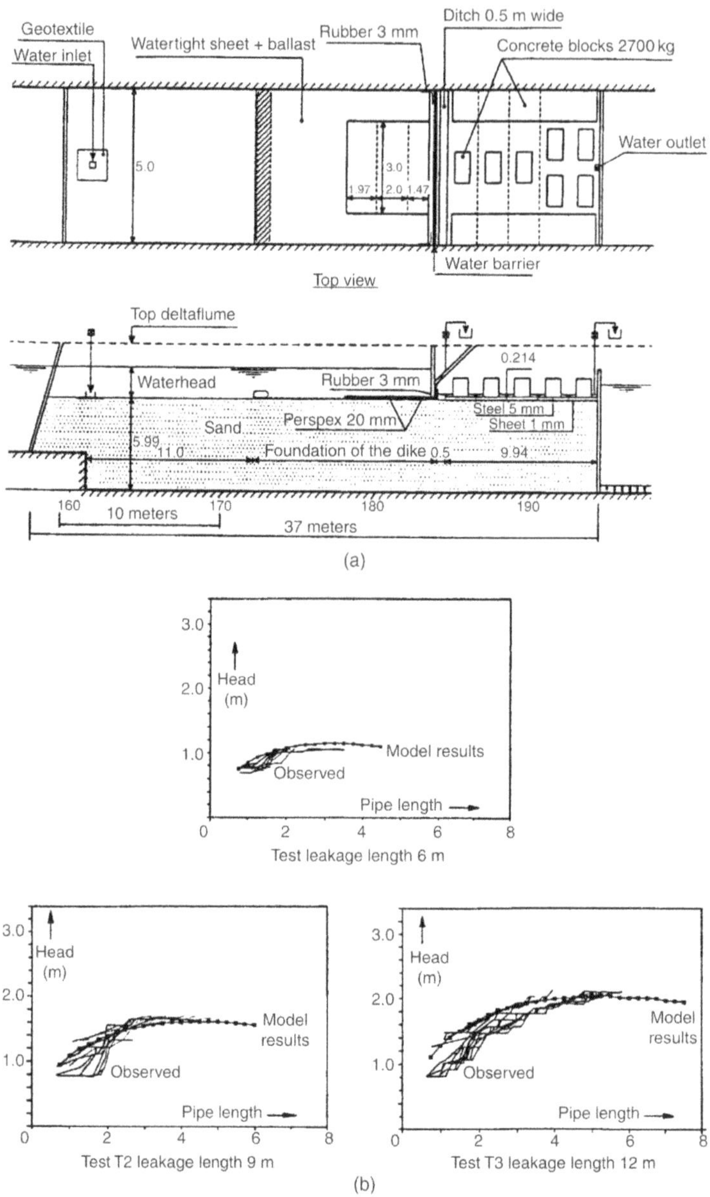

FIGURE 10.3 Delft Hydraulics large-scale laboratory piping flume test model (a) test equipment, (b) typical test results showing the development of the length of the channel (pipe) as the head increases (Weijers and Sellmeijer, 1993).

Soils, such as those shown schematically in Figure 10.5, are potentially susceptible to suffusion. Examples of the gradation of soils subject to suffusion are given in Section 10.2.

The phenomenon does not require a crack within the soil in which erosion may initiate, as is required in concentrated leak erosion, or a roof of plastic soil as is required for backward erosion piping.

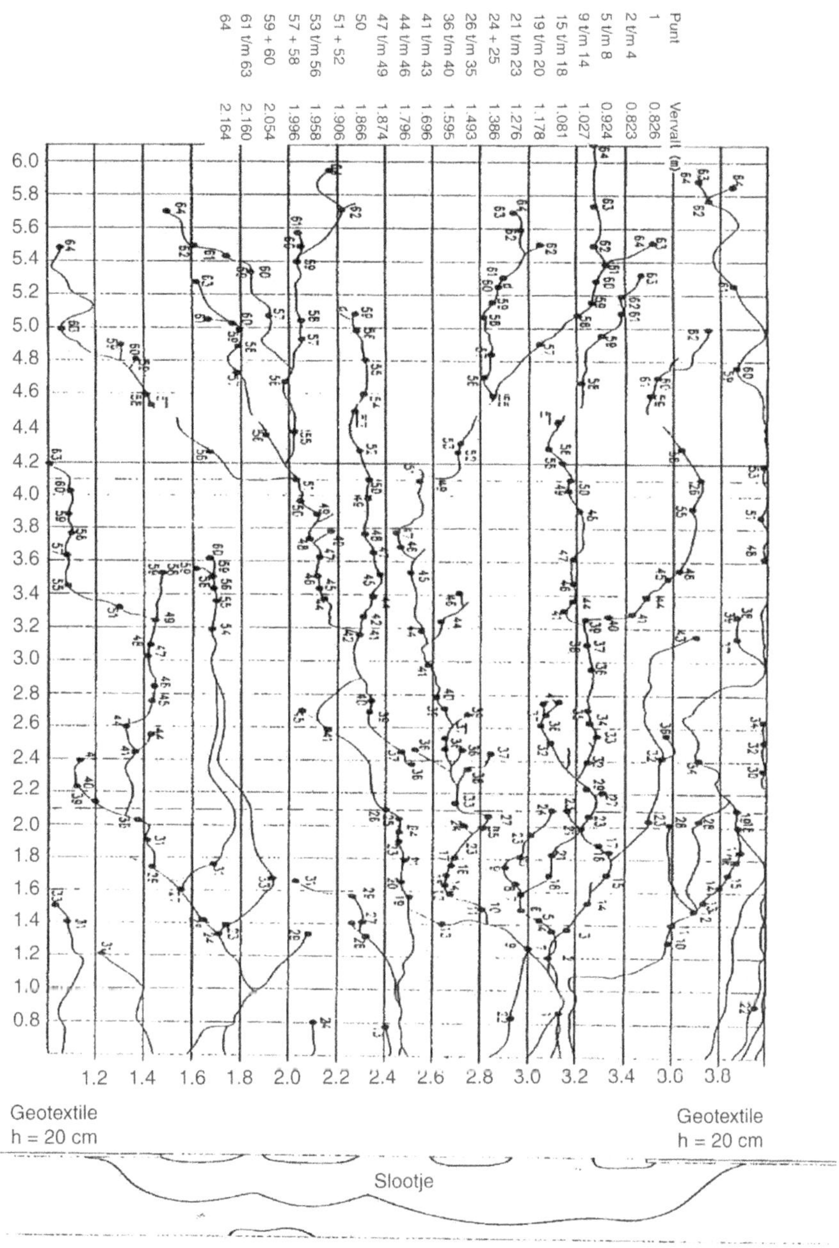

FIGURE 10.4 Plan view and section of an example of the development of piping channels in the Delft Experiments. Dimensions in metres. The numbers are the time in minutes from when erosion initiated (Silvis, 1991).

The small particles of soil are transported throughout the soil by the seepage flow through the pores of the coarser particles. The coarser particles are not transported, and the effective stresses on the soil are largely transferred through the matrix of the coarser particles.

For suffusion to occur, the following three criteria, geometric criterion, stress criterion and hydraulic criterion, must be satisfied (Wan and Fell, 2004c, 2008):

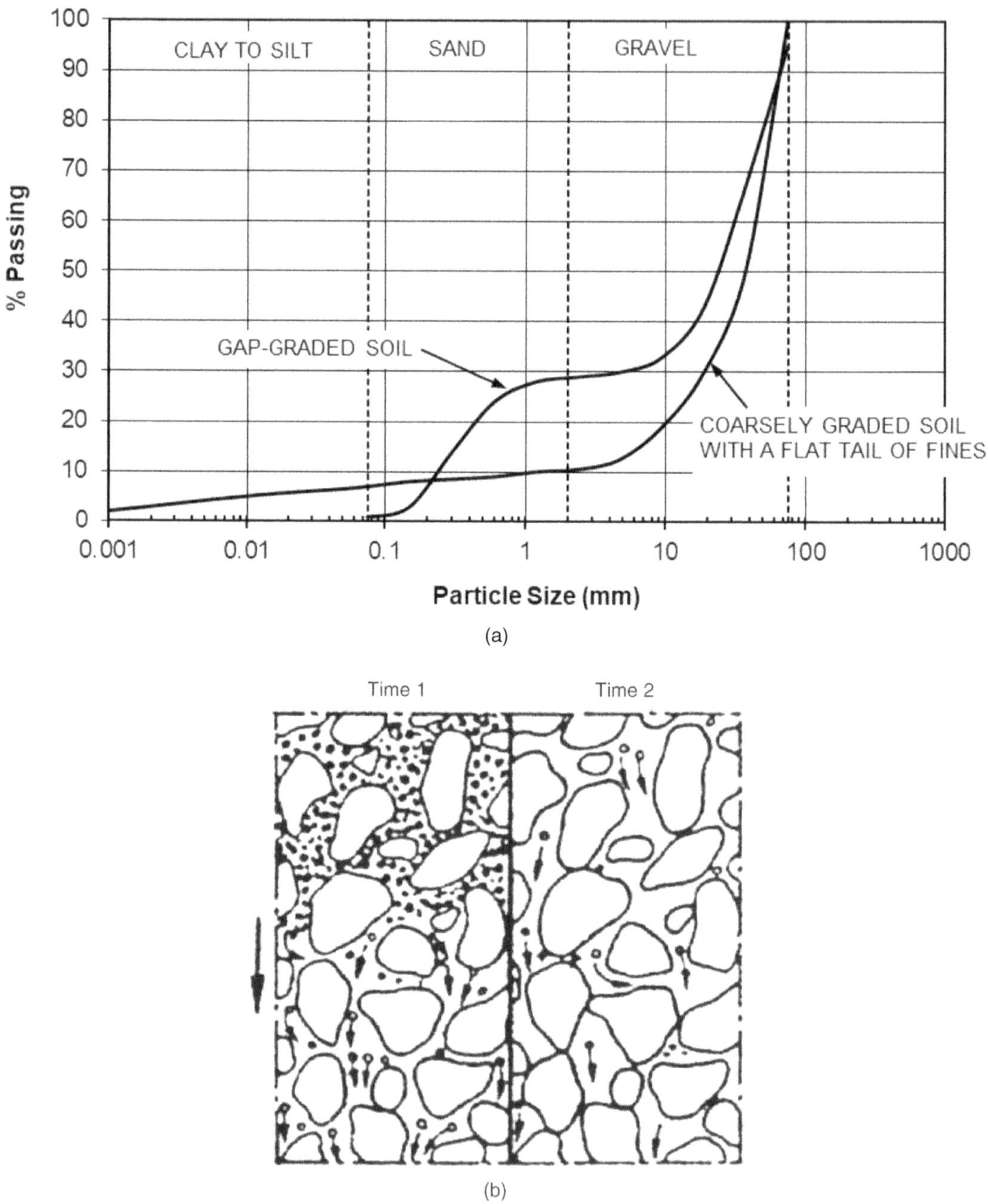

FIGURE 10.5 (a) Schematic of soil gradation types which are potentially susceptible to suffusion; (b) schematic of suffusion process (Fell et al., 2015).

Criterion 1: The size of the finer soil particles must be smaller than the size of the constrictions between the coarser particles, which form the basic skeleton of the soil.

Criterion 2: The amount of finer soil particles must be less than enough to fill the voids of the basic skeleton formed by the coarser particles. That is the voids must be "under-filled". If there are more than enough finer soil particles for void filling, the coarser particles will be "floating" in the matrix of fine soil particles, instead of forming the basic soil skeleton. This is termed "over-filled" voids.

Criterion 3: The velocity of flow through the soil matrix must impose a high enough stress to overcome the stresses imposed on the particles by the surrounding soil and to move the finer soil particles through the constrictions between the larger soil particles.

Suffusion results in an increase in permeability, greater seepage velocities and potentially higher hydraulic gradients, possibly accelerating the rate of suffusion. Stable situations may be reached with some of the finer fraction remaining in equilibrium with the seepage stresses. Suffusion may re-commence during cycling periods of water loads or during higher reservoir or river water level.

Suffusion occurring within an embankment core or the foundation of a dam may also lead to some settlement of the embankment.

Suffusion in filters is likely to lead to the D_{15} size of the filter increasing, making it less likely to act as a no-erosion filter.

10.1.3 Global Backward Erosion

Global Backward Erosion (GBE) occurs in widely graded silt sand gravel soils in which the voids are over-filled, so the coarser particles are "floating" in the finer soil matrix. Finer particles move within the soil matrix and may be eroded from the soil if there is a free exit surface, or the filter or transition is too coarse to arrest erosion such as shown in Figure 10.6. GBE may also occur in silt sand gravel alluvium or fluvioglacial soils in the foundation. Examples of gradations of soils subject to GBE are shown in Section 10.2.

GBE potentially results in an increase in permeability if finer fractions are eroded from the soil. However, internally stable situations may be reached with some of the finer fraction remaining in equilibrium with the seepage stresses, in which case, lower permeabilities may result in that part of the soil. GBE may re-commence during cycling periods of water loads or during higher reservoir or river water level. GBE occurring within an embankment core or the foundation of a dam may also lead to some settlement of the embankment.

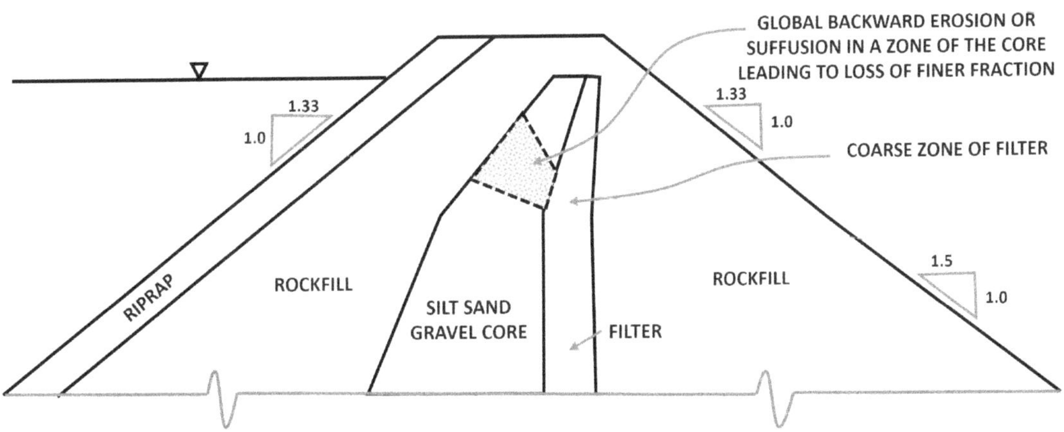

FIGURE 10.6 An example of where Global Backward Erosion or Suffusion may occur in the core of a dam.

10.2 SCREENING TEST OF SUSCEPTIBILITY TO BEP, SUFFUSION AND GBE BASED ON SOIL CLASSIFICATION

10.2.1 Typical Gradings for Soils Subject to Backward Erosion Piping (BEP), Suffusion and Global Backward Erosion (GBE)

The following are some typical gradings of soils subject to BEP, Suffusion and GBE. These are provided for information. They do not represent the total range of susceptible soil gradations.

10.2.1.1 *Backward Erosion Piping*

The data in Figures 10.7(a) and (b) from the Lower Mississippi and Sacramento Rivers are gradations of foundation soils in which BEP has occurred. Those for the Netherlands in Figure 10.7(c) are gradations which were tested in laboratory experiments. Van Beek (personal communication) indicates some of those soils as tested have less fines content than occur in the field. This adjustment to the field gradations was done to facilitate laboratory testing. All soils tested displayed BEP in the laboratory experiments.

Table 10.1 summarises the gradation data from these three sources. The soils are non-plastic poorly graded fine to medium grained sands (SP) and silty sands (SM). Most have a uniformity coefficient Cu

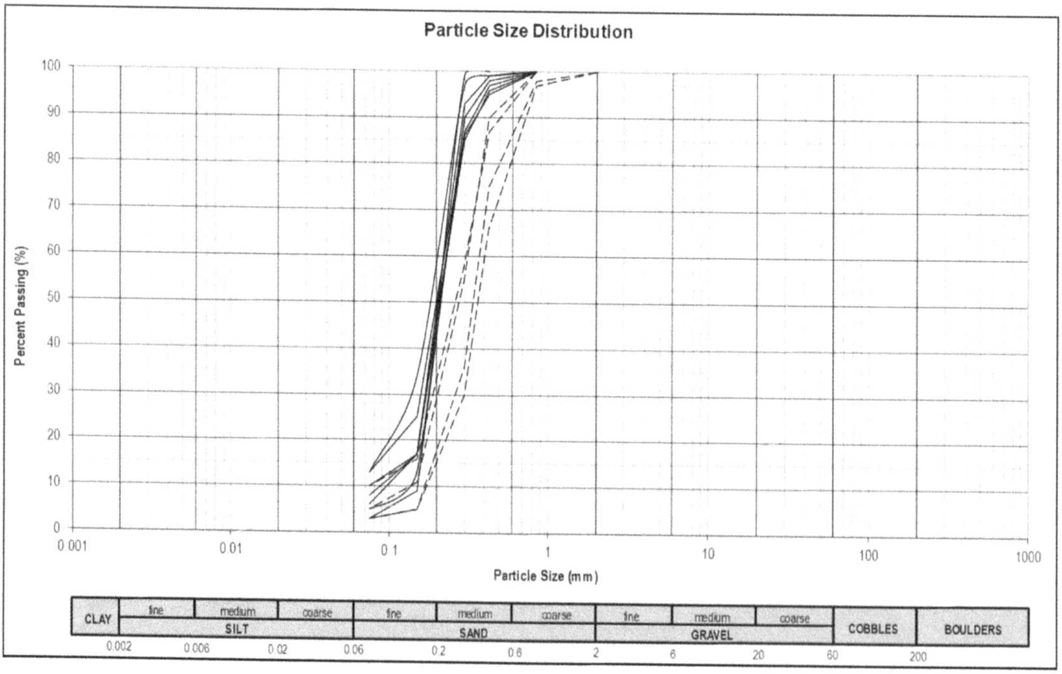

FIGURE 10.7(A) Gradations for foundation sands where sand boils have been observed in levee foundations along the Lower Mississippi River. (Plotted from data in USACE, 1956. Dashed data is from Hole in the Wall, Louisiana; solid lines from Cotton Bayou, Louisiana (Fell et al., 2024.)

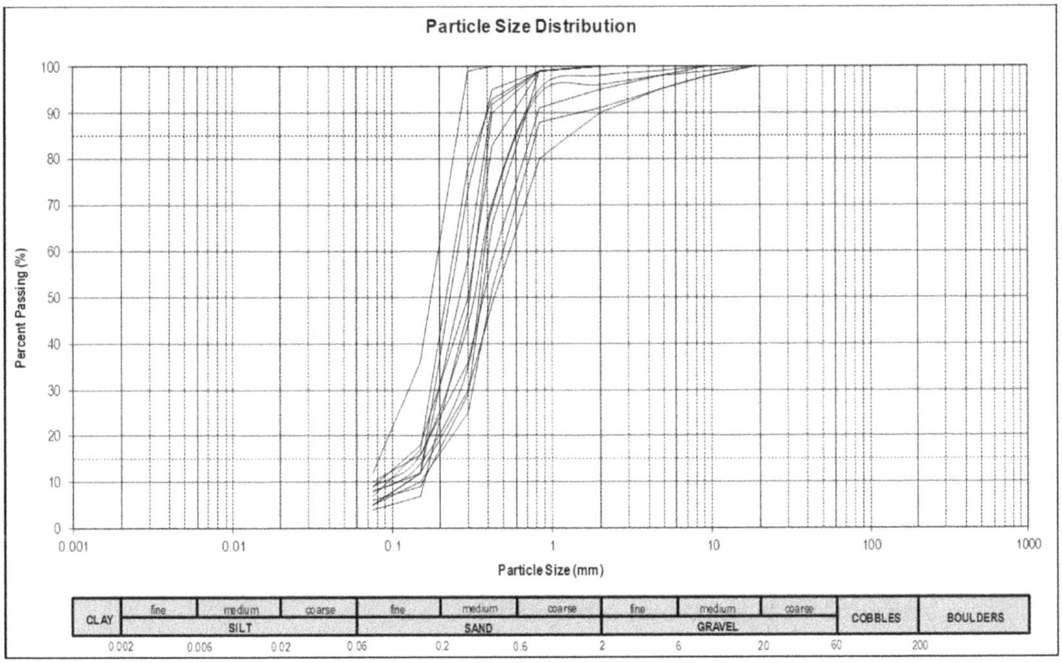

FIGURE 10.7(B) Gradations for foundation sands where sand boils have been observed in levee foundations in Sacramento, CA, USA. (Data courtesy of Mark Stanley, HDR Consultants.) (Fell et al., 2024.)

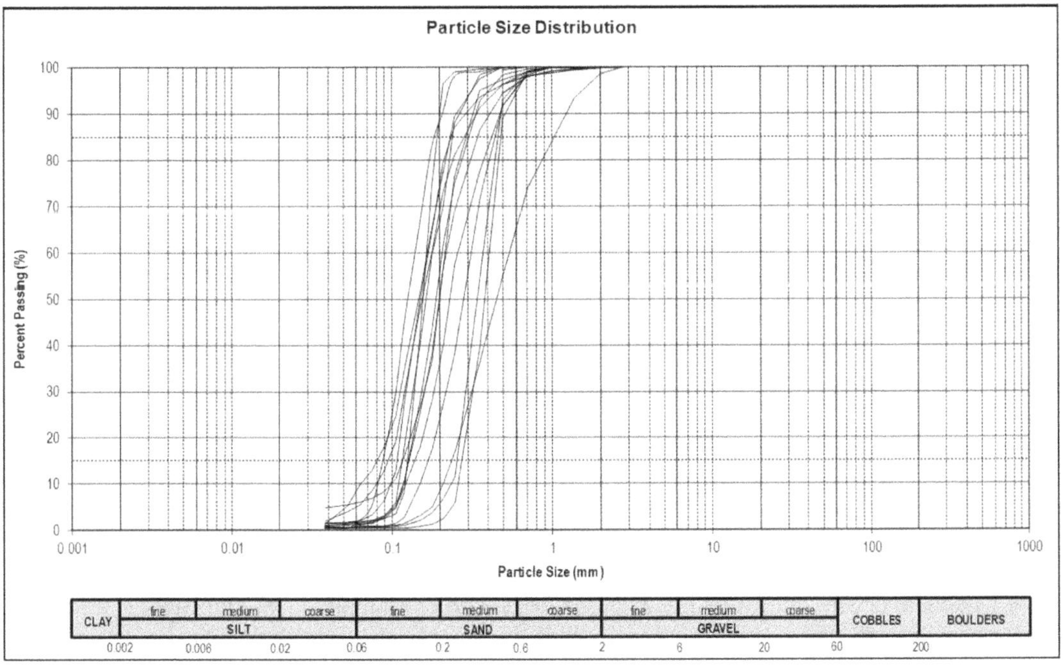

FIGURE 10.7(C) Particle size distributions for sands used in Deltares and IJkdjik Netherlands BEP laboratory experiments. (Data taken from Van Beek et al., 2011, and Van Beek, personal communication, 2021). (Fell et al., 2024.)

TABLE 10.1 Particle Size Characteristics of Soils Subject to Backward Erosion Piping from the Sacramento area, Lower Mississippi, and Netherlands (Fell et al., 2024)

LOCATION	D_{60} *(MM)*	D_{10} *(MM)*	*% PASSING 0.075 MM*	*UNIFORMITY COEFFICIENT* D_{60}/D_{10}
Sacramento area	0.23 to 0.63	0.06 to 0.20	3 to 10	1.8 to 7.2; most 1.8 to 4.1
Lower Mississippi Hole in the Wall	0.23 to 0.46	0.08 to 0.21	3 to 9	1.9 to 3.5
Lower Mississippi Cotton Bayou	0.17 to 0.27	0.06 to 0.15	3 to 13	1.5 to 3.5
Netherlands Laboratory testing	0.15 to 0.50	0.06 to 0.27	0 to 13	1.3 to 2.6

< 4. Many of the soils have appreciable silt content. Most fall in the range 0% to 10% but some have up to 13% silt.

Koelewijn and Van Beek (2024) reported that sand boils had been observed in fine to medium sands in the Netherlands with fines contents up to 10% to 15%.

It may be concluded that if the soils in the foundation of the dam, dike or levee have gradations similar to these soils, then BEP is possible. If soils have greatly different gradations, it may be unlikely that they will be subject to BEP.

10.2.1.2 Global Backward Erosion and Suffusion

Figures 10.8, 10.9 and 10.10 show gradations of soils subject to suffusion and GBE. These are all laboratory test gradations. The authors have found few data for gradations of embankment and foundation soils subject to suffusion and GBE. ICOLD (2017) Volume 2 has one example, Jonas Dike, which is a gap-graded soil. The other examples in ICOLD (2017) of Laguna Dam and Saint Pardoux Dam are not suffusion as defined in this book but concentrated leak erosion in holes in lateritic foundation soils and in residual granitic soil respectively.

Testing by Douglas et al. (2016, 2019) showed that soils with very minor or no erosion from the sample showed changes of pore pressures during the test and changes in gradation within the test sample showing they were internally unstable; that is, particles were moving within the sample.

10.2.2 Limits on the Plasticity of Soils Subject to BEP, Suffusion and GBE

There are several sources of information:

(a) The participants at the Aussois Workshop (Fell and Fry, 2007) considered that at gradients likely to occur within a dam foundation, backward erosion piping was probably restricted to non-plastic soils or soils with only limited plasticity. However, the limits of plasticity at which backward erosion could occur were not known.

(b) For practical purposes Fell and Fry (2013) concluded that based on the data available, soils with Plasticity Index > 7, should be considered not subject to suffusion at the gradients

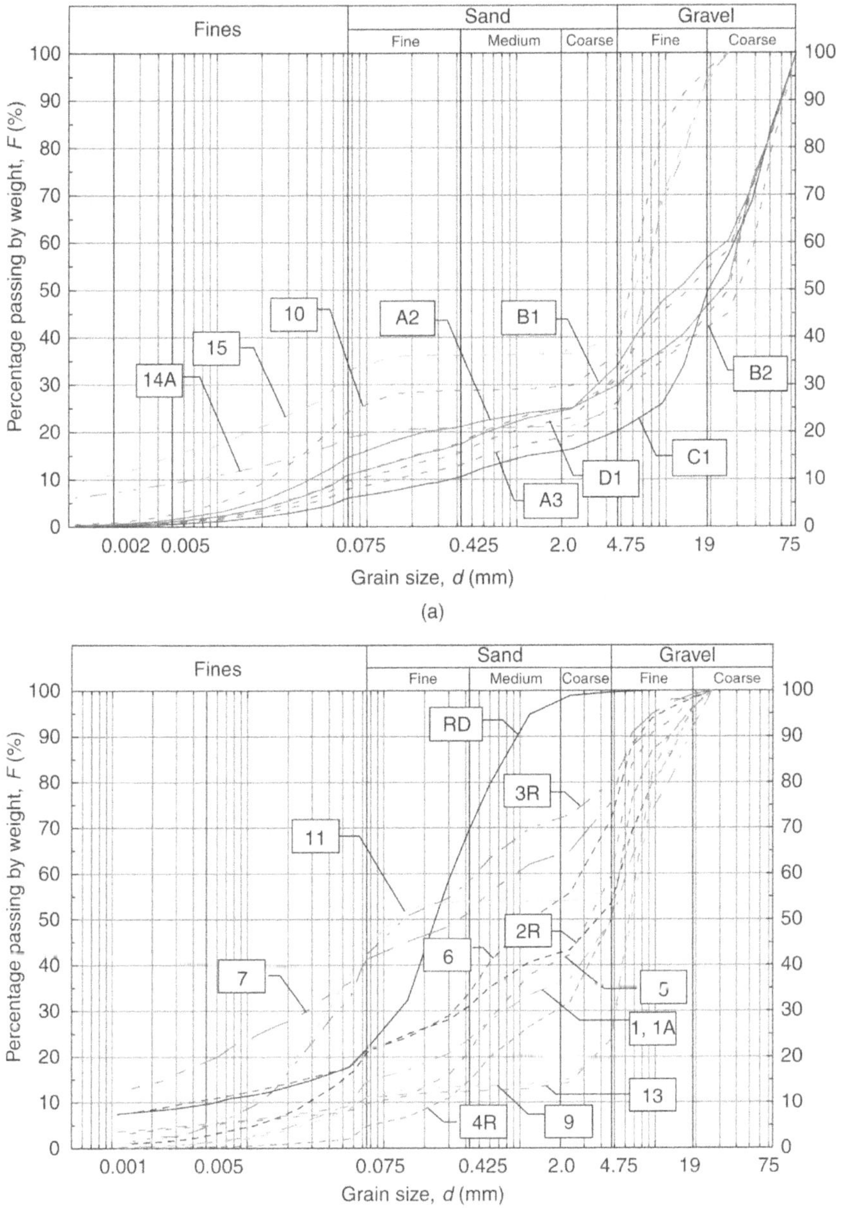

FIGURE 10.8 Soil samples tested as being suffusive by Wan and Fell (2004c, 2008).

usually experienced in dams and their foundations. If for some particular reason the gradient is higher than about 4, then soils with a Plasticity Index ≤ 12 should be considered for suffusion.

(c) Some authors have used the terms suffusion and suffosion to describe internal erosion in global backward erosion laboratory experiments on internally stable non-plastic soils and

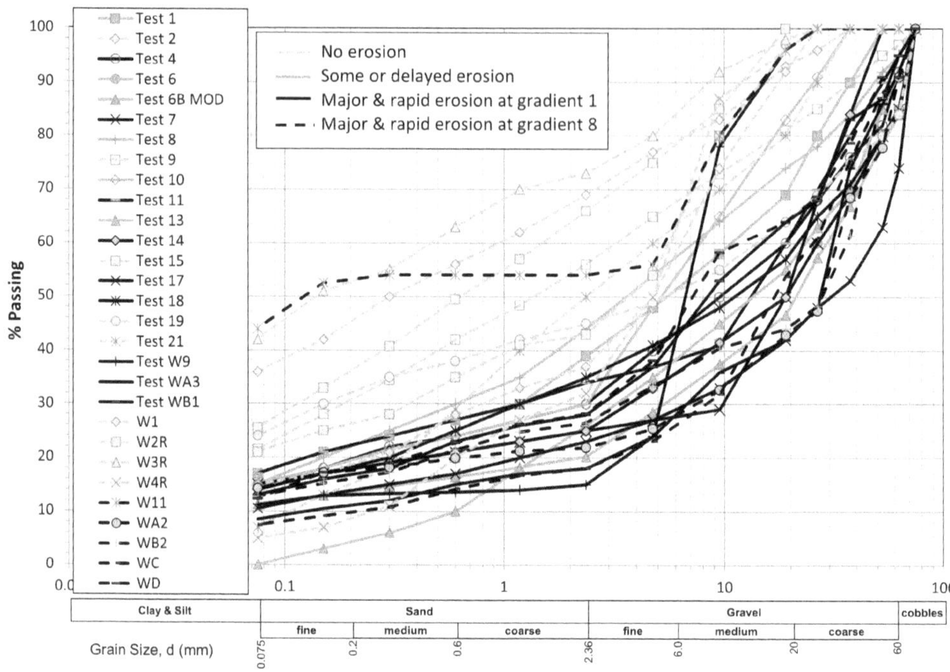

FIGURE 10.9 Gradations of soils exhibiting major and rapid erosion at gradient 1, representing suffusive soils; some or delayed erosion representing GBE, and no or very minor erosion representing internal instability (Douglas et al., 2016, 2019).

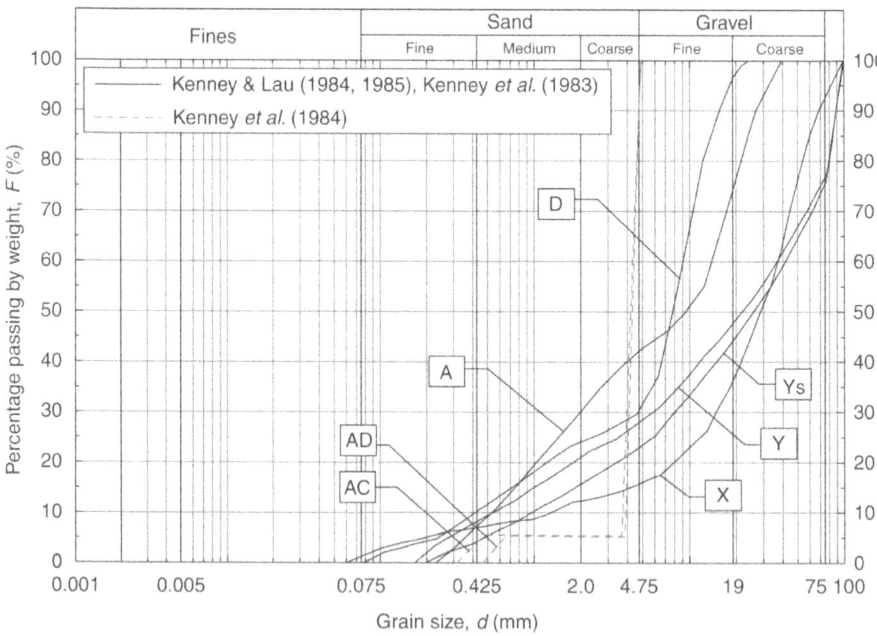

FIGURE 10.10 Soil samples tested as being internally unstable (suffusive) by Kenney and Lau (1985, 1986) and Kenney et al. (1985).

plastic soils under very high gradients, typically > 15. This includes the tests by Moffat and Fannin (2011) and Moffat et al. (2011), Marot et al. (2007), Bendahmane et al. (2008) and Sail et al. (2011). These soils have over-filled voids between the coarser particles, so the finer fraction is not free to move under the gradients normally experienced in dams. The phenomena observed are not suffusion as defined in this book. It may be concentrated leak erosion due to hydraulic fracture or a form of GBE.

(d) Wan and Fell (2004c) tested some soils with plastic fines consisting of 5% to 22% kaolin. The Plasticity Index ranged from 9% to 13%. Of the six samples tested only one, a gap-graded soil (soil 15 in Figure 10.8) experienced internal erosion, described by Wan and Fell (2004c) as suffusion. Those tests were at a gradient of 8.

(e) USBR/USACE Best Practices document (USBR/USACE (2019) indicates, "BEP occurs in non-plastic soils or those with a low plasticity index (PI)". It does not specify a PI limit. In Fell et al. (2008) an upper limit of PI ≤ 4 was adopted with the consensus of those developing that document from USBR, USACE, URS and UNSW.

(f) Dr Vera van Beek, Deltares; Dr Bryant Robbins, USACE; and Mr Noah Vroman, Director of the Dam and Levee Safety Production Center, USACE, all have indicated they do not know cases of BEP occurring in plastic soils. The authors have seen no reported field data to suggest that BEP may occur in plastic soils. All case data is for non-plastic soils.

It is concluded that it is most likely that BEP may only occur in non-plastic soils. However, in the absence of definitive data, the authors are reluctant to rule out soils with a PI up to 4 being subject to BEP and have adopted that limit. This is consistent with the limit for sand-like behaviour in liquefaction assessment. See Section 8.6.

For suffusion and GBE there is laboratory evidence that soils, particularly gap-graded soils, with some plasticity experience erosion under higher gradients as described earlier.

The Unified Soil Classification separates GM from GP soils and SP from SM soils at a Plasticity Index of 4, and GP from GC soils, and SM from SC soils at a Plasticity Index of 7. It is therefore considered that for the purposes of this document non-plastic soils should be defined as having a PI ≤ 4, and plastic soils a PI ≥ 7.

This results in the recommendations on limits to plasticity detailed in Section 10.2.3.

10.2.3 Conclusions

Is the soil plastic or non-plastic and what is the soil classification?

If the soil is non-plastic or has a Plasticity Index ≤ 4, then follow the procedures in Section 10.3 to assess the probability of initiation of backward erosion piping for SP, SM and ML soils, and Section 10.6 for initiation of suffusion and global backwards erosion for GW, GP and GM and SW soils.

If the soil is plastic with a Plasticity Index > 7, then the likelihood of BEP and suffusion and GBE is negligible under the seepage gradients which occur in foundations and the embankment of earth and earth and rockfill dams. For such soils, assess if concentrated leak erosion may occur through a crack in the soil using the procedures detailed in Sections 9, 11, 12 and 13.

If the soil is low plasticity with a Plasticity Index between 4 and 7, the likelihood of BEP is negligible. Suffusion or GBE may occur in gap-graded GW, GP and GM and SW soils including for seepage gradients ≤ 4.

10.3 IM 24—ASSESSMENT OF THE PROBABILITY OF INITIATION OF BACKWARD EROSION PIPING IN A LAYER OF NON-PLASTIC SOIL OR SOIL WITH PLASTICITY INDEX ≤ 4 IN THE FOUNDATION

10.3.1 Description of Method

The following has been prepared using quantitative risk-based terminology. Those wishing to use deterministic or risk enhanced approaches should find the approach a useful guide to engineering judgement.

The steps to be followed are:

(a) Estimate the probability (P_{cl}) there is a continuous layer of non-plastic soil or soil with PI $\leq$ 4 classifying as SP, SM and ML from upstream of the embankment, to downstream of the embankment.

If there is no continuous layer such as shown in Figure 10.11, then $P_{cl} = 0$, BEP is not possible, and the probability of BEP may be taken as negligible. In some situations, there will be some uncertainty and P_{cl} should be assessed from the geotechnical borehole data, understanding of the depositional conditions in which the soils were deposited, seepage and piezometer data, including response of the piezometers to changes in reservoir level.

(b) Determine the potential exit conditions for seepage flow taking into consideration the stratigraphy and geometry of the foundation soils, the presence of drainage ditches or other excavations such as borrow areas and building foundations/basements which may fully or partially penetrate the soils overlying the soil potentially subject to BEP.

(c) Assess the probability (P_H) of reaching a "heave" or critical gradient (i_{cr}) condition which are zero effective stress conditions, at the toe of the embankment, or into ditches for the reservoir stage under consideration using the method described in Section 10.3.2.

(d) Calculate the probability of initiation of backward erosion (P_I) from the results of the prior assessments.

$$(P_I) = (P_{cl}) \times (P_H) \tag{10.1}$$

Rather than this calculation, this may be modelled as two separate nodes in the event tree.

Note that the "heave" condition may not need to result in actual heave of the ground if the overlying strata has through-going root holes or other penetrations which may act as exits for seepage flow in which BEP may initiate.

For situations where sand boils have been observed, the probability of initiation of backward erosion (P_I) for the reservoir or river level at which sand boils have been observed and above is assumed to be 1.0.

To allow for the observation in the field Sills and Vroman (2007), Wolff (2002) and Glynn and Kusmaul (2004), and laboratory tests by Allan (2018), that sand boils re-occur at lower river levels and therefore lower average gradients beneath the levee in successive floods, assume that the probability of initiation of backward erosion (P_I) = 1.0 for subsequent floods resulting in an average gradient beneath the embankment 0.95x that when boils were observed.

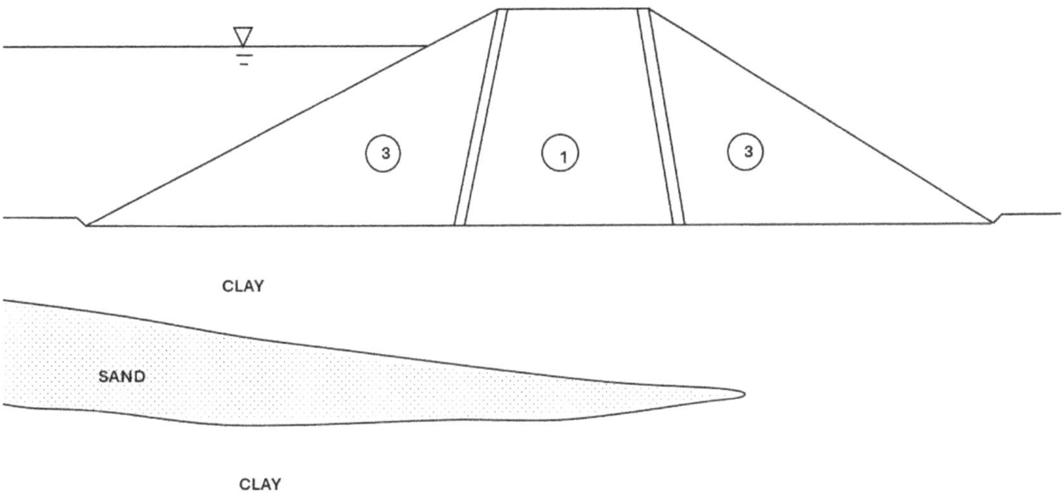

FIGURE 10.11 An example of a situation where there is no continuous layer of non-plastic soil in the foundation and backward erosion cannot occur (Fell et al., 2024).

Take account of the comments at the end of Section 10.4.2 regarding the effect of repeated sand boil activity damaging the foundation and potentially making seepage analyses using undamaged parameters unreliable.

10.3.2 Estimation of the Probability of Heave or Reaching the Critical Gradient (P_H) from Piezometer Data and/or Seepage Flow Net Models

(a) Develop a seepage flow net of the embankment and the foundation based upon all available data and, so far as practicable, calibrated to piezometric data and in consideration of seepage observations. Repeat the analysis for each reservoir level partition.

(b) Calculate the factor of safety against heave or reaching the critical gradient from the following equations:

Factor of safety against heave.

$$F_{UT} = \frac{\sigma_v}{u} \tag{10.2}$$

where:

σ_v = total vertical stress at any point in the foundation—kN/m^2
u = pore pressure at the same point—kN/m^2

or for the conditions shown in Figure 10.12:

$$F_{UT} = \frac{h\gamma_{sat}}{h_p\gamma_w} \tag{10.3}$$

where:

γ_{sat} = unit weight of saturated foundation soil—kN/m^3
γ_w = unit weight of water—kN/m^3
h_p = piezometric head at point X—metres
h = depth below ground surface of point X—metres.

Factor of safety on critical gradient

The alternative method for estimating the factor of safety for conditions such as those in Figure 10.12 where there is no overlying plastic strata is to consider the gradient of the flow net. If the gradient approaches the critical vertical gradient, liquefaction and heave can be expected to occur.

$$F_{UC} = \frac{\text{critical gradient}}{\text{actual gradient}} \tag{10.4}$$

For vertically up seepage, the critical gradient i_{cr} can be calculated from (Terzaghi and Peck, 1948, 1967):

$$i_{cr} = (G - 1) / (1 + e) \tag{10.5}$$

where:

G = soil particle density (specific gravity) in tonnes/m^3
e = void ratio

or from:

$$i_{cr} = (\gamma_{sat} - \gamma_w) / \gamma_w \tag{10.6}$$

Where there is a layer of plastic or other low permeability soil overlying the soil which may be subject to backward erosion, the most critical heave condition usually occurs from the base of the lower permeability layer to the ground surface.

(c) Estimate the probability of heave or reaching the critical gradient from F_{UT} and/or F_{UC} from Table 10.2.

The selected probability should take account of the quality of the data on which the factors of safety are based.

Table 10.2 has been developed by the authors of Fell et al. (2024) based on their experience and taking account of the following information.

Figure 10.13 relates the severity of seepage observed at 16 levee sites in the Lower Mississippi during the 1950 flood. As discussed in USACE (2005), the data would appear to indicate that the gradient

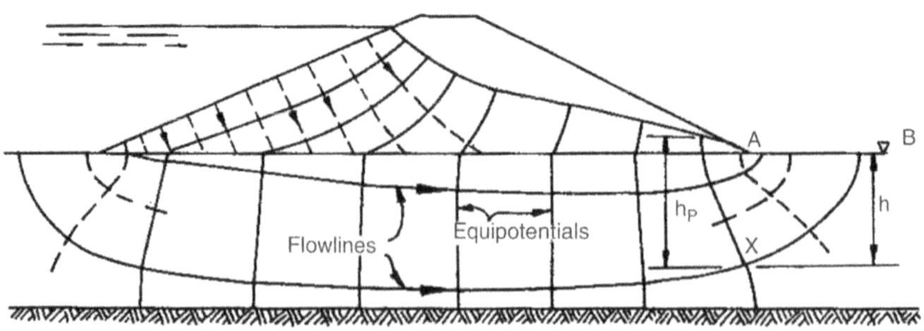

FIGURE 10.12 Cross section of an embankment and dam foundation showing seepage flow net and definition of terms (Fell et al., 2005, 2015). Factor of Safety Against Critical Gradient

TABLE 10.2 Estimation of the Probability of Heave or Reaching the Critical Gradient (P_H) from the Estimated Factor of Safety (Fell et al., 2024)

ESTIMATED FACTOR OF SAFETY		*PROBABILITY OF HEAVE OR REACHING THE CRITICAL GRADIENT (P_H)*
F_{UT}	F_{UC}	
1.6		Negligible[(a)]
1.3	2.0	0.005 to 0.05
1.23	1.7	0.02 to 0.2
1.12	1.3	0.05 to 0.5
1.05	1.15	0.1 to 0.9
1.0	1.0	0.5 to 0.99
0.92	0.85	0.99
0.80	0.66	0.999

Note: (a) Negligible or extremely low probability depending on the quality of data available. Based on F of S ≥ 1.6 being the required factor of safety for USACE (2022) design.

required to cause sand boils varies considerably, and sand boils might even occur at an exit gradient of 0.5. However, USACE (1956) report explains these data as follows:

> the gradient required to cause sand boils varies considerably at the different sites, and relatively low gradients were recorded near some sand boil areas. This may be due to the fact that at sites where sand boils developed previous to the 1950 high water, only fairly low excess head may have been required to reactivate boils in 1950 and, as a relief of pressure occurs at the boil, readings of piezometers near the boil may be somewhat lower than those farther from the boil.

The authors have assumed that a similar uncertainty applies to sites where piezometers may not be positioned in the critical location and/or that seepage analyses will not model seepage pressures accurately.

USACE (1956) and Wolff (2002) show that local geology has an important influence on the occurrence of sand boils. Sand boils are more likely to occur where swales from point bar deposits cross the levee at an angle and concentrate seepage at the toe.

USACE (2022) indicates that the required factor of safety for design of a levee is $F_{UT} \geq 1.6$ at the landward toe of the levee. Lower factors of safety are allowed at locations landside of the levee toe.

Sills and Vroman (2007), Wolff (2002) and Glynn and Kusmaul (2004) indicate that field observations show that initiation and progression of backwards erosion as evidenced by the appearance of sand boils with eroding soil may occur at successively lower river levels on levee banks. That is at successively lower overall gradients.

Allan (2018) carried out laboratory BEP tests in a 1.3 m flume, and circular exit in which the head was cycled a number of times to represent successive floods, for a medium uniform sand, Cu = 1.3, d_{50} = 0.3 mm, the head to re-initiate pipe tip progression was on average only about 2% less per cycle. For a fine to medium silty sand Cu = 2.6, d_{50} = 0.19 mm and 7% passing 0.075 mm the head to re-initiate pipe tip progression was on average about 5% less per cycle for one test and 13% for another.

Allan (2018) also noted when sand boiling at the exit began after each cycle. The overall trend of these were some increases in head for the first soil boiling, and a reduction for the second.

The second soil tested by Allan (2018) is closer to the grading of the soils experiencing sand boils in the Lower Mississippi levee foundations shown in Figure 10.7(a), so the laboratory tests support the field observations.

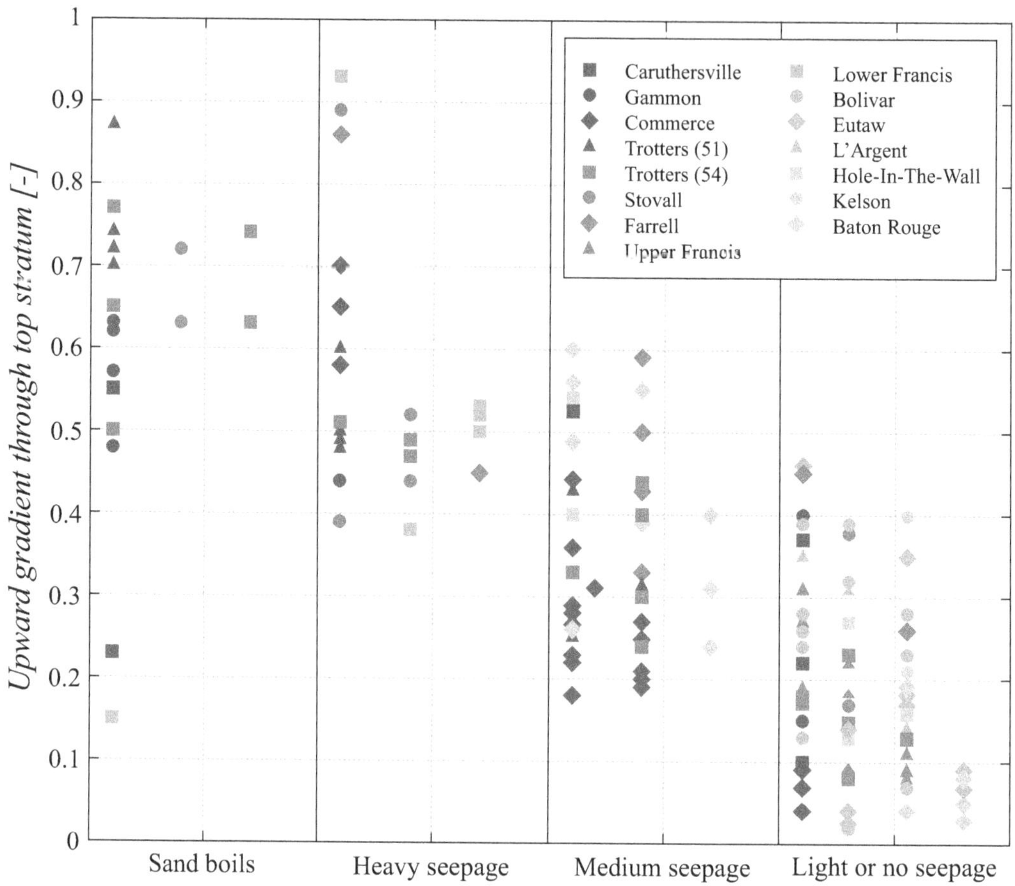

FIGURE 10.13 Severity of seepage related to upward gradient through the top stratum (Robbins and Van Beek, 2015, plotting data from USACE, 1956).

In view of this data, it is concluded that allowance should be made for the likelihood that sand boils can be expected to recur at successively lower reservoir levels. From the data available it has been assumed that this is 5% per cycle.

10.4 IM 24 ESTIMATING THE PROBABILITY OF PROGRESSION OF BACKWARD EROSION PIPING, GIVEN INITIATION OF BACKWARD EROSION

10.4.1 Estimating the Probability (P_{roof}) of a "Roof" above the Soil Subject to Backward Erosion Piping

Estimate the probability (P_{roof}) that there is a soil layer overlaying the non-plastic soil which will form a roof for the backward erosion channels. This layer may be naturally occurring plastic or high silt content sandy silt or silty sand, or plastic soil embankment fill. It may also be a concrete structure such as a spillway.

Take account of the site investigations data and depositional environment in making this assessment. Also take into consideration the guidance on the ability of soils to hold a roof in Section 13.2.2.

10.4.2 Estimating the Probability of Progression of Backward Erosion Where Heave or Reaching the Critical Gradient Is Predicted, or Sand Boils Have Been Observed (P_P) and There Is a Roof for the Pipe

10.4.2.1 Introduction

There are currently three methods most widely used for estimating the critical average gradient beneath the embankment at which backward erosion will progress from the downstream toe to upstream of the embankment. These are:

Sellmeijer and Co-Workers at Deltares Method
Described in Van Beek et al. (2012a, 2013) and Sellmeijer et al. (2012).

Hoffmans Darcy Shields Method
Described in Hoffmans (2014), Hoffmans and van Rijn (2018) and Van Beek and Hoffmans (2017).

Schmertmann Method
Described in Schmertmann (2000), Robbins and Sharp (2016) and Allan (2018).

All methods compare the critical gradients to the average seepage gradient (i_{avf}) in the non-plastic foundation layer beneath the dam at the reservoir stage under consideration, or to point gradients obtained from seepage flow nets.

Fell et al. (2024) found that at the time (2023) none of the three methods could be relied upon to confidently predict field behaviour. Trial calculations by the authors of Fell et al. (2024) for a range of foundation geometries and soil gradations showed that:

(a) The three methods give different answers, sometimes very different.
(b) The *Schmertmann* and *Sellmeijer* methods gave reasonably similar results for soils with Cu 1.33 to 1.8. This includes for a wide range of geometries (L and D/L as defined later).
(c) There is clearly a correlation between Cu and critical gradient, quite a significant one at the laboratory scale. Sellmeijer does not include test data with Cu > 2.6 in his calibration and may underestimate the effect of Cu as a result.
(d) The *Hoffmans* method is insensitive to the geometry (D/L) and scale (L). This is not consistent with what Hoffmans (2014), Hoffmans and van Rijn (2018) and Van Beek and Hoffmans (2017) say. The method assumes that the critical gradient is controlled by transport of the sediment in the channels, which is controlled by Shields coefficient, which in turn is controlled by the d_{50} size. However, as flow into the channels is affected by the thickness of the sand layer with thin layers (small D/L) having less inflow than thick layers D/L, and probably L should have more effect. Both Sellmeijer and Schmertmann allow for this.
(e) Given the large differences (both higher and lower) between the Hoffmans method and the Schmertmann and Sellmeijer methods; the lack of dependence on geometry and scale; and it being less widely used, it is recommended that the Hoffmans method not be relied upon

at this stage, but those doing dam safety assessments may choose to use it to check the outcomes from the other methods. This approach is endorsed by the current authors.

It is recommended that:

- For soils with Cu ≤ 2.5 use the Schmertmann and Sellmeijer methods and give each equal weighting.
- For soils with Cu ≥ 3, use the Schmertmann method.
- For soils with Cu between 2.5 and 3, use the Schmertmann and Sellmeijer methods with a greater weighting to the Schmertmann method.

In these methods the likelihood that backward erosion will progress is related to the seepage gradients in the non-plastic strata under the embankment. For practical purposes in most cases this can be taken as the average seepage gradient (i_{avf}) in the non-plastic foundation layer beneath the dam at the reservoir stage under consideration.

The average seepage gradient (i_{avf}) is defined as the hydraulic head difference divided by the seepage path length. As shown in Figure 10.14, if there is upstream blanketing with lower permeability soil much of the head may be lost in seepage through this "blanket". In practical terms this gradient may be the estimated average seepage gradient (i_{avf}) = (H2 – H_3) / L. This allows for the head losses through the upstream clay layer.

The head difference should be determined with piezometers and/or estimated from the finite element seepage analysis used to assess the likelihood of heave or critical gradient. If in doubt of the effectiveness of the upstream blanket, neglect the blanketing effect.

If there is a sheet pile, soil bentonite or concrete cutoff wall, the average gradient of interest is the average gradient in the soil subject to backward erosion from downstream of the cutoff wall to the toe of the embankment allowing for the effectiveness of the cutoff wall. Figure 10.15 shows this scenario. The average seepage gradient downstream of the cutoff wall (i_{avf}) = (H2 – H_3) / L where the dimensions are as shown in Figure 10.15.

If this gradient is sufficient to progress BEP to the cutoff wall, the risk analysis team will need to consider whether the cutoff and depth is of sufficient quality to stop progression. This will need to be done using engineering judgement backed up by seepage analyses.

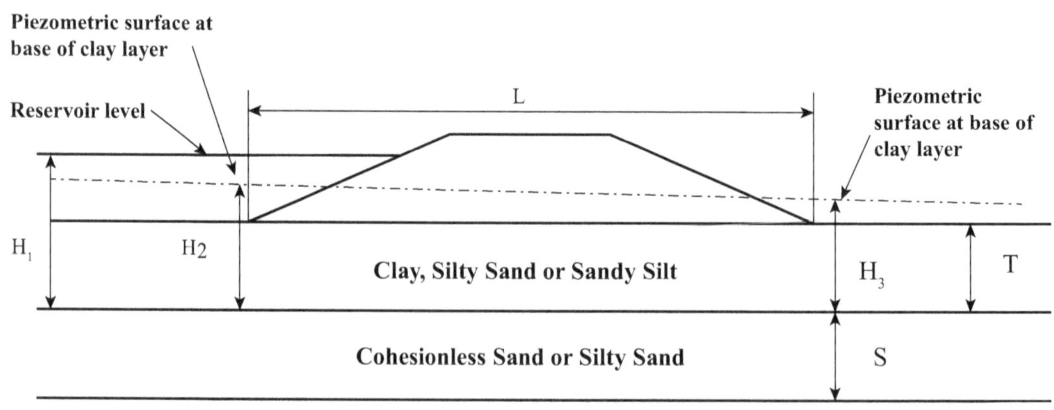

FIGURE 10.14 Section through embankment and foundation showing definition of terms to estimate the average gradient in the foundation sand (Fell et al., 2024).

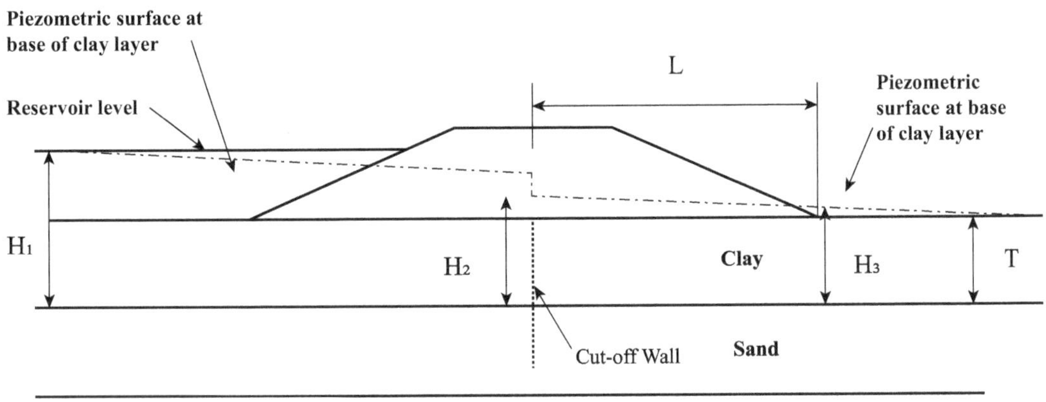

FIGURE 10.15 Section through embankment and foundation showing definition of terms to estimate the average gradient in the foundation sand for a foundation with a cutoff wall (Fell et al., 2024).

The other factors which affect the likelihood of progression of backward erosion are the particle size characteristics and permeability of the non-plastic sand and the geometry of the embankment and foundation strata.

The exit conditions for seepage also affect the critical average gradient. That is whether the seepage and potential backward erosion exit into a single sand boil or into a ditch or other excavation which exposes a significant length of the non-plastic soil strata. This is a result of 3D effects. The critical average gradient is least for a single sand boil. This is discussed further in Section 10.5.4.

10.4.2.2 Recommended Method for Estimating the Probability of Progression of BEP Given Initiation of Erosion (P_H) and a Roof for the Pipe (Proof)

The approach to be used to make this assessment is as follows:

(a) Estimate the average seepage gradient (i_{avf}) through the non-plastic soil layer in the foundation beneath the dam as described earlier for the level of the reservoir stage under consideration.

(b) Estimate the critical average seepage gradient to result in progression of backward erosion using the Schmertmann method as described in Section 10.5.2 and modified as described in Section 10.5.3, and the Sellmeijer method as described in Section 10.5.1. As these methods are calibrated against models with plane and slope exit conditions, they are termed i_{cps}.

(c) For soils with relative density < 50%, apply the correction factor C_{rd} shown in Table 10.3 to the average critical gradients to allow for the fact that low relative density soils test with lower critical gradients than high relative density soils.

(d) Calculate the critical average gradients corrected for the relative density and exit conditions which are assessed as most likely to apply using:

$$(i_{pmt})_{Corrected} = C_{rd} \times C_{ec} \times i_{cps} \tag{10.7}$$

where C_{ec} is determined from Table 10.4.

TABLE 10.3 Correction Factors for Critical Average Gradient (C_{rd}) Related to Relative Density of the Non-Plastic Soil Layer (Fell et al., 2024)

RELATIVE DENSITY %	*CORRECTION FACTOR* C_{rd}
< 40	0.6
≥ 40 and < 50	0.8
≥ 50	1.0

TABLE 10.4 Correction Factors (C_{ec}) for Critical Average Gradient Depending on Exit Conditions (Fell et al., 2024)

EXIT CONDITIONS	*CORRECTION FACTOR* C_{ec}
Plane	1.0
Slope/Beach	1.0
Ditch/Slot	0.8
Hole/ Single Boil	0.7

Section 10.5.4 describes the basis for assigning these correction factors. Where there is uncertainty regarding which exit condition might apply, it is recommended it be assumed that the single sand boil condition is applied, as this seems to be the most common situation in practice.

(e) Estimate the probability of progression (P_{PIR}) using (i_{avf}) and $(i_{pmt})_{Corrected}$ as inputs to calculate a factor of safety, and Table 10.5 as a guide to judgement. Do this for both the Sellmeijer and Schmertmann methods. In doing so, take account of the uncertainty in the input parameters to the Sellmeijer and Schmertmann methods. Table 10.5 has been developed by the current authors to allow for the inherent uncertainty in the two methods. Users should use their judgement, taking account of all the factors involved.

(f) From these assess the probability of progression of BEP given there is a roof for the pipe, taking account of the weighting of the two methods as discussed earlier. If equal weighting is given to the two methods, average the probabilities obtained from Table 10.5 for the factors of safety for each method. Do not average the factors of safety before using Table 10.5.

The probability of progression with uncertainty allows for model uncertainty and test data variability. This is discussed in Section 10.5.5.

Monte Carlo simulation methods can also be applied to estimate the probability of progression using the Sellmeijer and Schmertmann methods based on Robbins and Sharp (2016).

10.4.2.3 Estimation of the Likelihood of Progression for Dams and Levees Which Have Experienced Sand Boil Activity

The methods described previously are applicable to dam and levee foundations which have not experienced sand boil activity in the past.

TABLE 10.5 A Guide to Estimation of the Probability of Progression of BEP in the Foundation Given Initiation Is Predicted (P_{PIR}) and a Roof for the Pipe

FACTOR OF SAFETY $= (i_{pmt})_{Corrected}/(i_{avf})$	*PROBABILITY OF PROGRESSION* (P_{PIR}) *ALLOWING FOR UNCERTAINTY*
4.0	0.001
3.0	0.005
2.0	0.05
1.5	0.30
1.0	0.50
0.66	0.70
0.33	0.80

Sills and Harder (2016) describe case studies where repeated sand boil activity has resulted in damage to the foundation from the BEP channels, so modelling with undamaged soil parameters as the prior methods rely upon is incorrect and likely to be unconservative. They suggest that the following approach be taken to identify situations where progression has extended well upstream and is likely to lead to breach, so urgent action is required to intervene:

(a) Determine from observations the river stage at which sand boils are becoming active.
(b) Run a seepage model of the embankment and its foundation at that river stage using the undamaged parameters and calculate the seepage gradient in the sand at the toe of the levee. Call this the apparent gradient.
(c) If the apparent gradient is 0.3 or less, seepage distress is occurring at a very low apparent gradient and immediate remediation measures should be taken.

That is, if the foundation is damaged, and the apparent gradient F_{UC} is less than 0.3, the probability of progression should be taken as approaching 1.0.

10.4.3 Estimation of Parameters for Use in Estimation of the Critical Gradient for Progression of Backward Erosion Piping

Each method for prediction of progression of backward erosion has a number of parameters as inputs to the analyses. Some guidance on estimation of the parameters is provided in Section 10.5.5.

10.4.4 Estimation of the Probability of Progression Given Initiation (P_{IP}) for This Reservoir Stage

Estimate the probability of progression of backward erosion given initiation (P_{IP}) from the results of the prior assessments.

$$(P_{IP}) = (P_{roof}) \times (P_{PIR}) \tag{10.8}$$

The mean or best estimate value should be used for assessing the safety of the dam and estimating risks, but it is suggested that the approximate 16% and 84% ranges should also be reported and may be used for the safety assessment.

10.5 METHODS FOR ESTIMATING THE LIKELIHOOD OF PROGRESSION OF BACKWARDS EROSION PIPING

10.5.1 Sellmeijer and Co-Workers Model

In the 1980s Sellmeijer and co-workers from Delft Hydraulics and Delft Geotechnics Laboratories in the Netherlands carried out laboratory backward erosion piping tests in flumes. These are reported in De Wit et al. (1981), De Wit (1984) and Weijers and Sellmeijer (1993). The tests were mostly on fine to medium sands, with a few tests on medium to coarse sands. The sands were relatively uniform (Cu = 1.58 to 3.53). Early tests were on small-scale models (base length 0.8 m), but later tests were on very large models as shown in Figure 10.3.

The tests showed that critical head and resultant average gradient occurs when the length of the channel (l) is about 0.4 to 0.5 of the flow path length L. As the piping channels extends further, the head required for backward erosion to progress is less. Figure 10.4 shows the development of the piping channels in the large experiments. Figure 10.19 defines l and L.

Deltares carried out testing on small and medium-scale laboratory experiments. These are summarised in Van Beek et al. (2010, 2011). The test set-ups are shown in Figures 10.16 and 10.17. Full scale tests were carried out at IJkdijk. Figure 10.18 shows a schematic cross section for those tests.

Sellmeijer (1988), Sellmeijer and Koenders (1991), Koenders and Sellmeijer (1992) developed a mathematical model for backward erosion piping based on the experiments carried out at Delft Laboratories in the 1980s.

This model was refined using the results of the Deltares and IJkdijk testing and is presented in Van Beek et al. (2012a, 2013) and Sellmeijer et al. (2011).

The critical gradient is determined as a product of three contributions: resistance factor, scale factor and geometrical shape factor. The equations for the critical gradient at which backward erosion will progress are:

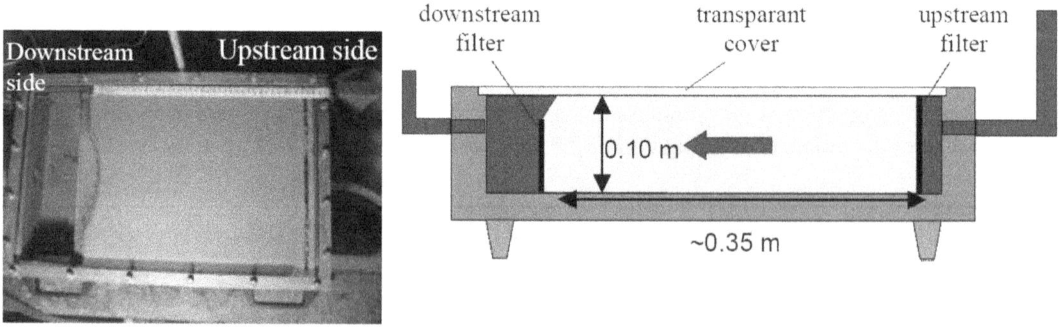

FIGURE 10.16 Set-up of Deltares small scale laboratory tests (Van Beek et al., 2011). (Copyright © 2025 Taylor & Francis Group, reprinted by permission of Informa UK Limited, trading as Taylor & Francis Ltd, https://www.tandfonline.com.)

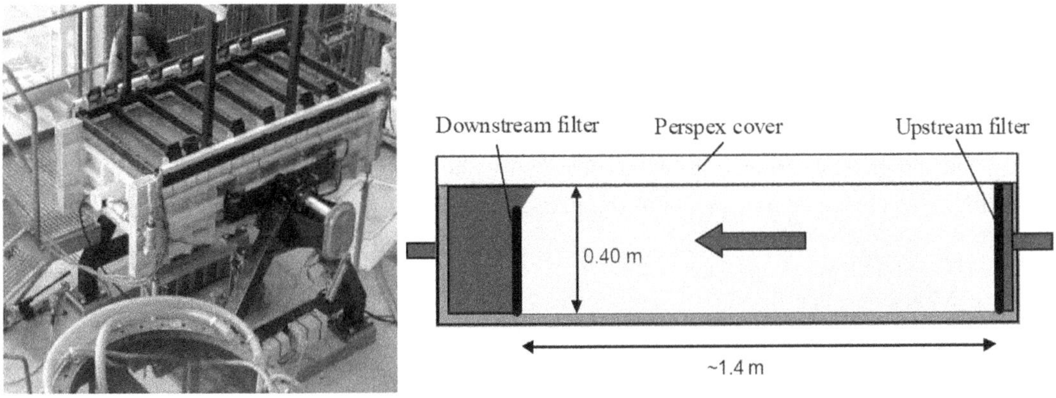

FIGURE 10.17 Set-up of Deltares medium-scale laboratory tests (Van Beek et al., 2011). (Copyright © 2025 Taylor & Francis Group, reprinted by permission of Informa UK Limited, trading as Taylor & Francis Ltd, https://www.tandfonline.com.)

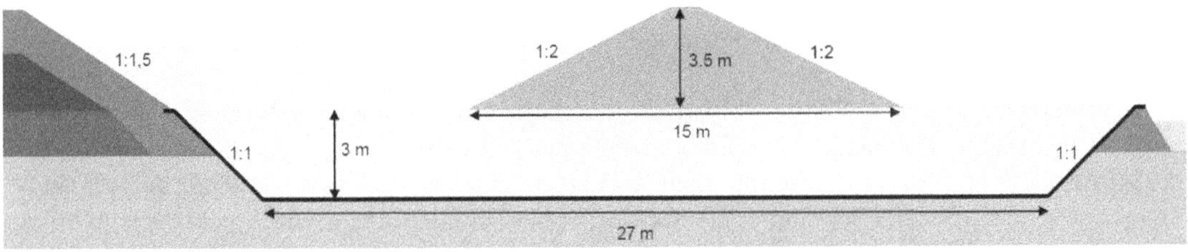

FIGURE 10.18 Schematic cross section of the IJkdijk full-scale tests (Van Beek et al., 2011). (Copyright © 2025 Taylor & Francis Group, reprinted by permission of Informa UK Limited, trading as Taylor & Francis Ltd, https://www.tandfonline.com.)

$$
\begin{aligned}
&\frac{H}{L} = \frac{1}{c} = F_{\mathrm{R}}\, F_{\mathrm{S}}\, F_{\mathrm{G}} \\
&F_{\mathrm{R}} = \eta\, \frac{\gamma_{\mathrm{p}}'}{\gamma_{\mathrm{w}}} \tan\vartheta \left(\frac{RD}{RD_{\mathrm{m}}}\right)^{0.35} \left(\frac{U}{U_{\mathrm{m}}}\right)^{0.13} \left(\frac{KAS}{KAS_{\mathrm{m}}}\right)^{-0.02} \\
&F_{\mathrm{S}} = \frac{d_{70}}{\sqrt[3]{\kappa\, L}} \left(\frac{d_{70\mathrm{m}}}{d_{70}}\right)^{0.6} \\
&F_{\mathrm{G}} = 0.91 \left(\frac{D}{L}\right)^{\frac{0.28}{\left(\frac{D}{L}\right)^{2.8} - 1} + 0.04}
\end{aligned}
\qquad (10.9 \text{ to } 10.12)
$$

where:

H [m]: hydraulic head across structure
L [m]: seepage length (= base length of the embankment)
D [m]: thickness of sand layer under the embankment
c [-]: erosion coefficient
F_{R} [-]: resistance factor

F_S [-]: scale factor
F_G [-]: geometrical shape factor
RD [%]: relative density
U [-]: uniformity coefficient $Cu = d_{60} / d_{10}$
KAS [%]: roundness
d_{70} [m]: soil particle diameter for which 70% by weight of the soil is finer
γ_p'[kN/m³]: submerged unit weight of soil particles 9.8(G–1)
G [t/m³]: soil particle density
γ_w [kN/m³]: unit weight of water
η [-]: Whites drag coefficient
ϑ [deg]: bedding angle (angle of repose) of sand
K [m²]: intrinsic permeability.

$$\text{Where: } K = \frac{\upsilon}{g}k \tag{10.13}$$

υ [m²/sec]: kinematic viscosity
g [m/sec²]: gravity
k [m/sec]: hydraulic coefficient of permeability, also termed hydraulic conductivity.

H, D and L are defined in Figure 10.19. In these equations the variables are normalised by the mean values in the data set. The mean values are as detailed in Table 10.6.

Sellmeijer et al. (2011) indicate that their data set assumes a bedding angle of 37 degrees and a Whites coefficient of 0.25. For water at 20° Celsius, $K = 1.02 \times 10^{-7}$ (k) where K is m², k is in m/sec. They indicate that they ignore the roundness and uniformity coefficient terms as they do not contribute significantly.

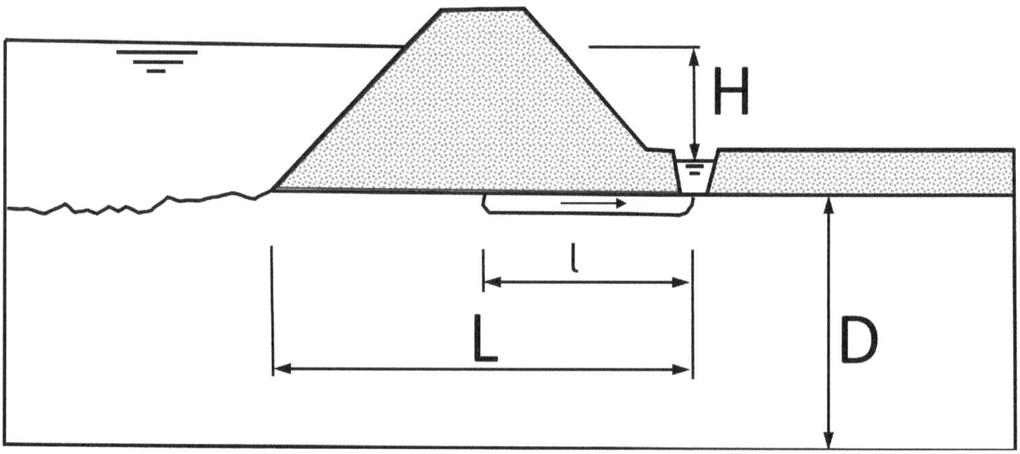

FIGURE 10.19 Geometry of Delft backward erosion piping model (figure adapted from Koenders and Sellmeijer, 1992). (Used with permission of ASCE conveyed through Copyright Clearance Center Inc.)

Sellmeijer et al. (2011) indicate that the equations should only apply to soils falling within the limits of the parameters during the laboratory testing. These limits are given in Table 10.6.

They found that the equations predicted the large-scale IJkdijk experiment behaviour quite well for the fine-grained soil ($d_{70} = 180$ μm), but that it was not so accurate for the soil with a $d_{70} = 260$ μm. For that test the model predicted a critical head of 2.25 m, whereas the experiment resulted in 1.74 m so the critical gradient estimated by their method was overestimated by about 30%.

There are very significant scale effects. The smaller the structure, the higher the critical gradient. In the tests the value of the scale factor F_S ranged from 0.134 for the large-scale IJkdijk tests to 0.192 for the medium-scale tests and 0.421 for the small-scale tests.

Sellmeijer et al. (2012) emphasise that the prior equations do not include any margin of safety and that for design rules they are likely to apply factors of safety.

Van Beek et al. (2012) consider the case where the soil subject to backward erosion piping is underlain by coarser sand strata as shown in Figure 10.20. They indicate that Equations 10.9 to 10.13 apply provided that:

The permeability k, is replaced by $k_{h,avg}$

where:

$$k_{h,avg} = \sum_{m=1}^{n} \frac{k_{h,m} D_m}{D_{tot}} \tag{10.14}$$

TABLE 10.6 Properties of Soils for Which the Sellmeijer et al. (2011) Method Is Applicable

PARAMETER	*MINIMUM*	*MAXIMUM*	*MEAN*
RD	50%	100%	72.5%
U	1.3	2.6	1.81
KAS	35%	70%	49.8%
d_{70}	150 μm or 1.5E-04 m	430 μm or 4.3E-04 m	208 μm or 2.08E-04 m

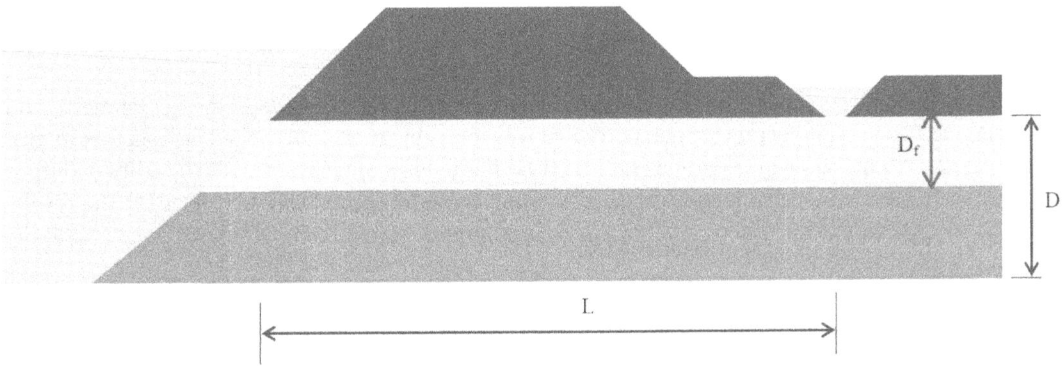

FIGURE 10.20 Geometry of layered foundation model (Van Beek et al., 2012b).

$k_{h,m}$ = horizontal permeability of layer m [m/sec]
D_m = thickness of layer m[m]
D_{tot} = total thickness of m layers [m]
provided D/L < 0.3 and k_c/ k_f < 10

where:
k_c= horizontal permeability of the coarser layer
k_f = horizontal permeability of the finer layer.

The presence of a higher permeability layer facilitates seepage flow into the backward eroding channels so the critical gradient for progressions is reduced compared to when there is only the layer subject to BEP.

10.5.2 Schmertmann Method

Schmertmann (2000) developed a method for predicting the progression of backward erosion piping which is based on tests in flumes at the University of Florida. These were carried out in a number of sets of tests, including those by Townsend et al. (1988) who carried out 15 tests. Those tests were carried out on a range of soils from fine to medium sands, up to coarse sand/fine gravel mixes. The test arrangement is shown in Figure 10.21.

The tests were different to the Delft tests in that a starter pipe was pre-formed at the downstream end. The head differential was increased progressively to keep the piping channel progressing upstream until the critical head was reached, causing the channel to form a pipe to the upstream source.

Most of the tests (32 out of 39) which Schmertmann (2000) relied upon were carried out on sands with a uniformity coefficient less than 3.2. The other seven tests were carried out on gap-graded soils and a well-graded soil with uniformity coefficients around 6.

Schmertmann (2000) found that the critical gradient $\bar{i}_{pmt}$ was related to the uniformity coefficient Cu (d_{60}/d_{10}) of the soils tested. He also plotted the Delft tests from de Wit et al. (1981), and Weijers and Sellmeijer (1993) and found a similar correlation.

There were few tests on soils with a Cu > 3. It is notable that two of the three data sets (tests 5 and 6 in Figure 10.22) with Cu around 6 were gap graded. The test descriptions in Townsend et al. (1988) indicate fine sand did erode in these tests, so the interpretation of the tests is somewhat complicated and should not be given much weight in assessing the overall behaviour. These tests and two tests on well-graded sand with a d_{60} of 1.6 mm withstood a gradient of greater than 1 without developing a pipe.

The test geometries used at the University of Florida and Delft were not the same, so Schmertmann (2000) applied correction factors for geometry and was then able to plot all the results together to give Figure 10.22 with his recommended design line if no flume tests are carried out. The exit conditions for these tests were a slope or plane. Table 10.7 shows the reference values for adjusting the results from the other laboratories. These must be used for reference to the field parameters when applying the method.

Schmertmann (2000) proposed a design method based around the results of the flume tests in Figure 10.22, which applies corrections for a number of factors to relate the laboratory tests to field conditions. Readers who intend to use the method should read the paper so they can better understand the background to the method.

The method calculates a factor of safety against piping initiating and progressing to form a complete pipe. It does this at various points along the flow path along which piping may progress,

TABLE 10.7 Reference Values Robbins and Sharp (2016) from Schmertmann (2000)

VARIABLE	*REFERENCE VALUE*
Seepage Length	1.52 m
Layer Depth	0.30 m
Layer Width	0.30 m
Grain Size (d_{10})	0.20 mm
K_h/K_v	1.5
Relative Density	60%
Seepage Path Inclination	0 degrees

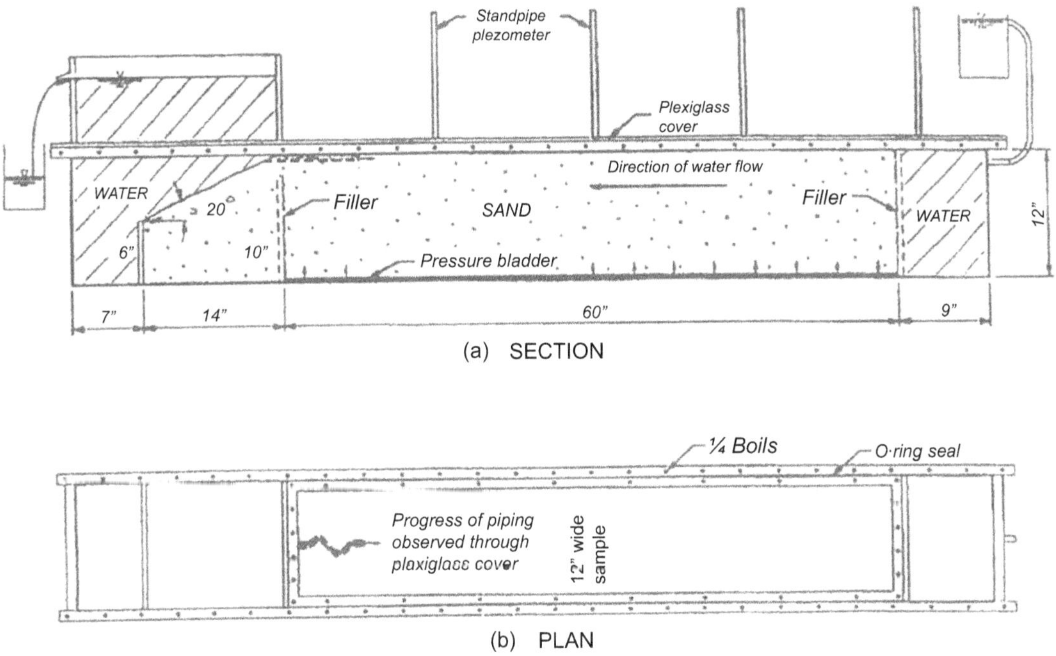

FIGURE 10.21 Geometry of the University of Florida backward erosion flume (Note: dimensions are in inches, one inch = 25.4 mm) (Schmertmann, 2000). (Used with permission of ASCE conveyed through Copyright Clearance Center Inc.)

beginning at the downstream toe of the embankment. This point factor of safety at point x is calculated from:

$$F_{px} = \frac{(C_D\, C_L\, C_S\, C_K\, C_z\, C_\gamma)\, i_{pmt}\, (C_\alpha)}{C_R\, i_{fx}} \tag{10.15}$$

where:

i_{pmt} = maximum point seepage gradient needed for complete piping in the flume test based on the soil coefficient of uniformity C_u (from flume tests, or Figure 10.22)
i_{fx} = hydraulic seepage gradient at point x parallel to the flow path
C_D = correction factor for (D/L)
C_L = correction factor for total pipe length L
C_R = correction factor for dam axis curvature
C_S = correction factor for grain size
C_z = correction factor for high permeability under layer
C_α = adjustment for pipe inclination
C_γ = correction factor for density
D = depth of piping sand layer, in direction perpendicular to α (m)
L = direct (not meandered) length between ends of a completed pipe path, from downstream to upstream exit, measured along the pipe path (m)
C_K = correction factor, for permeability anisotropy

Schmertmann provides information on the correction factors. The following is a summary:

C_D = depth/length (D/L) factor is determined from Figure 10.23. In this, D, the depth of the piping sand layer, is measured in a direction perpendicular to the flow lines. For a vertical flow path, D/L = infinity, and C_D = 0.715. For horizontal flow paths, D = vertical depth. For steeply inclined flow paths interpolate between these limits.
C_L = length factor is calculated from:

$$C_L = (L_t/L_f)^{0.2} \tag{10.16}$$

where L_f = length L, transformed in relation to k_H/k_v in the piping layer:

$$= (L)/\sqrt{k_h/k_v} \tag{10.17}$$

L_t = flume model length
= 1.52 m (5 feet) if Figure 10.24 is being used to estimate i_{pmt}.
C_S = grain size factor is calculated from:

$$C_S = (d_{10f}/0.20)^{0.2} \tag{10.18}$$

where d_{10f} = size in millimetres of the 10% finer than by weight size in the piping soil at point x.
C_K = anisotropic permeability factor is calculated from:

$$C_K = (1.5/R_{kf})^{0.5} \tag{10.19}$$

where:

R_{kf} = (the permeability in one direction, usually horizontal) / (permeability in a perpendicular direction, usually vertical)
= k_h/k_v of the piping soil
C_Z = under-layer factor. If the layer susceptible to piping is underlain by a more permeable "under-layer" of thickness d meters, and permeability ku (m/sec), C_Z is calculated from Figure 10.24. In this figure k_p is the permeability of the piping layer (m/sec) and r is the equivalent radius

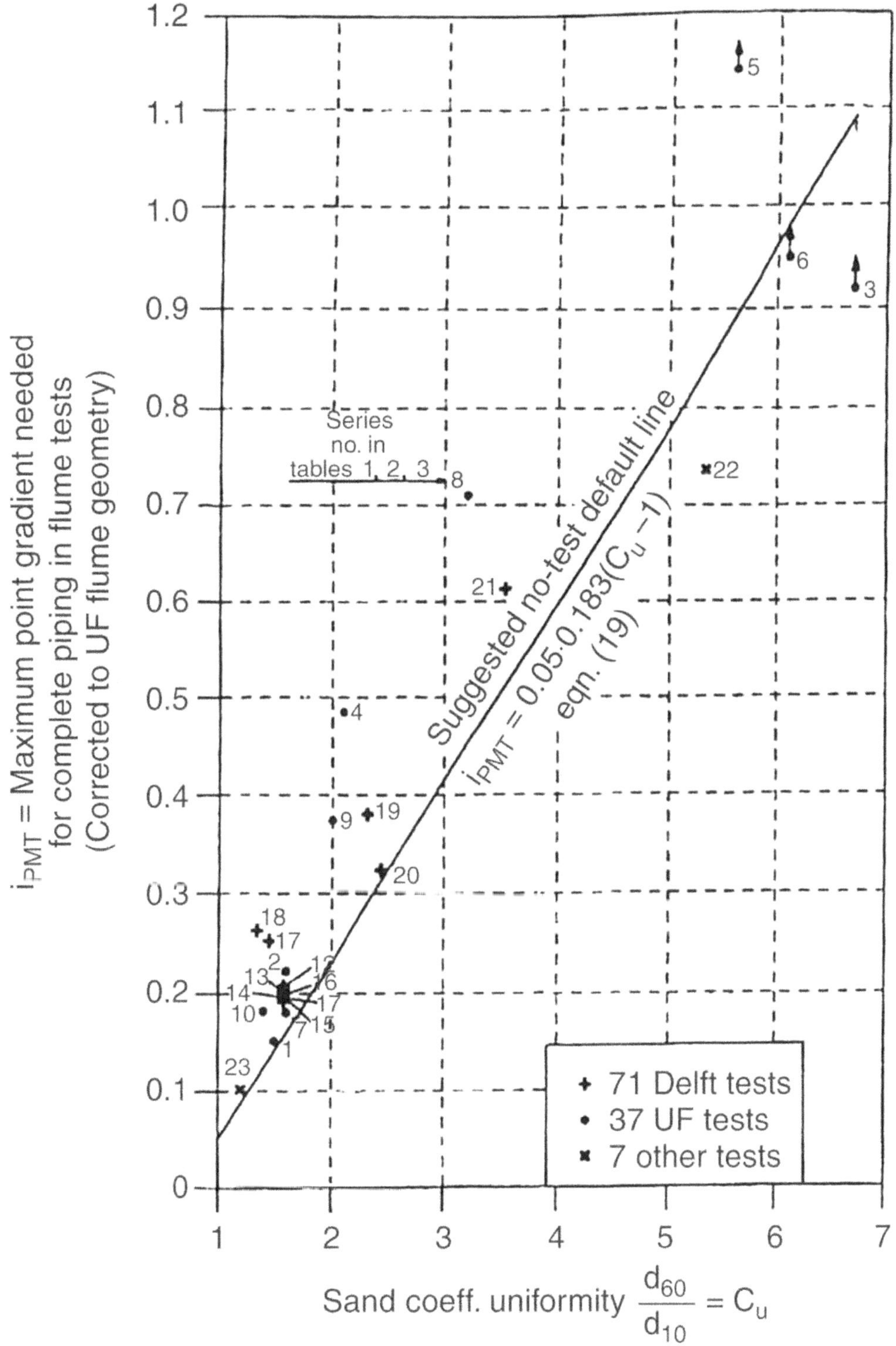

FIGURE 10.22 Maximum point gradient, i_{pmt}, needed for complete piping in the University of Florida (UF), Delft and other flume tests versus uniformity coefficient of soil. (Note: UF tests L = 5.0 ft [1.52 m] and D/L = 0.20) (Schmertmann, 2000). (Used with permission of ASCE conveyed through Copyright Clearance Center Inc.)

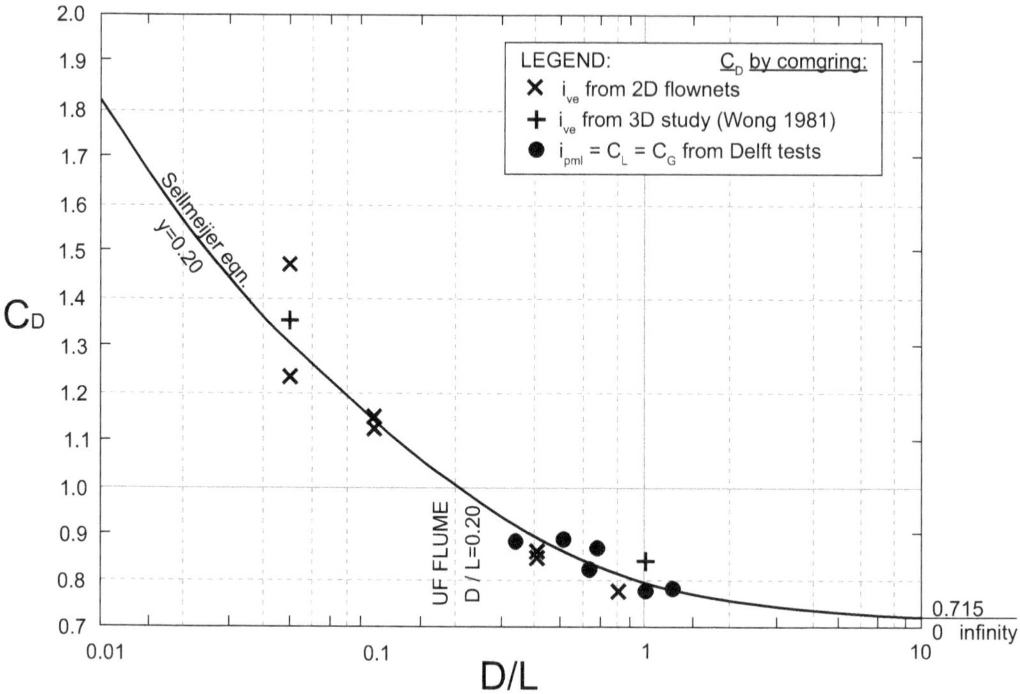

FIGURE 10.23 C_D—depth/length factor versus depth/length (Schmertmann, 2000). (Used with permission of ASCE conveyed through Copyright Clearance Center Inc.)

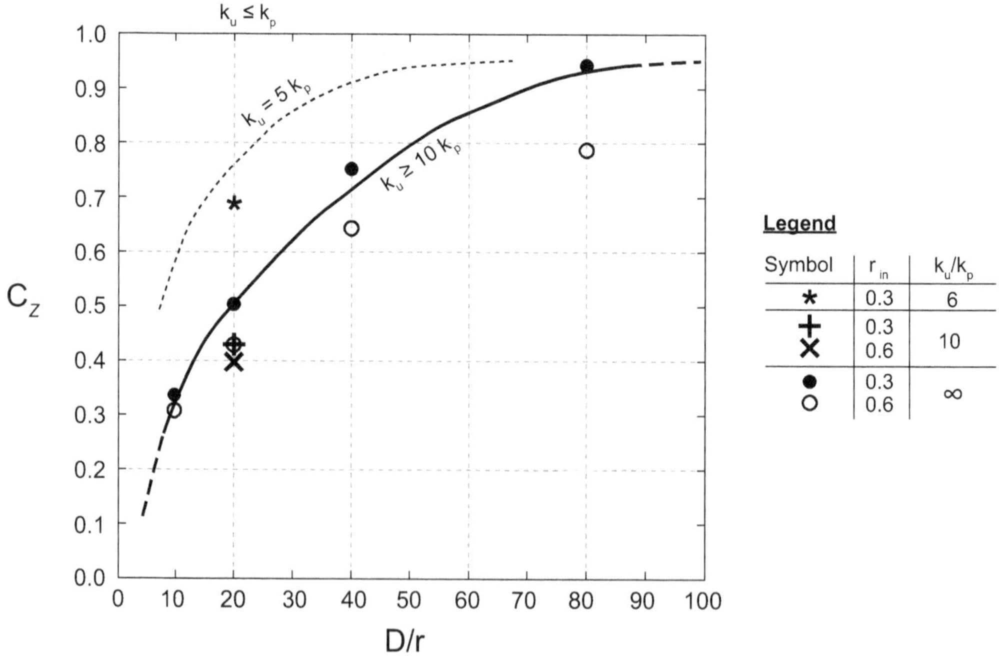

FIGURE 10.24 Under layer factor C_Z versus layer depth (D)/pipe radius(r) (Schmertmann, 2000). (Used with permission of ASCE conveyed through Copyright Clearance Center Inc.)

of the tunnel developing as piping progresses (meters). Note that Schmertmann uses small tunnel radii in his examples (2.5 and 6 mm).

C_γ = density factor is calculated from:

$$C_\gamma = 1 + 0.4(D_{rf}/100 - 0.6) \tag{10.20}$$

where:

D_{rf} = relative density of the piping layer (%)

C_α = pipe inclination adjustment

$= (i_{po}) / (i_{p\alpha})$

This is based on the inclination α of the piping channel (the flow line) estimated on the true (not transformed) section. The inputs to Figure 10.25 are α and $i_{p\alpha}$.

In Figure 10.27, $i_{p\alpha}$ is obtained by making all the other corrections to i_{pmt} using Equation 10.21.

$$i_{p\alpha} = [(C_D\, C_L\, C_S\, C_K\, C_Z\, C_\gamma) / (C_R)]\, i_{pmt} \tag{10.21}$$

Where α is zero (flow line horizontal), there is no correction.

C_R = convergent or divergent flow factor.

If the axis of the embankment is curved in plan, this factor is applied. If it is a straight line, C_R = 1. For curved embankments, C_R is estimated from:

$$C_R = \frac{R_1 + R_0}{2R} \tag{10.22}$$

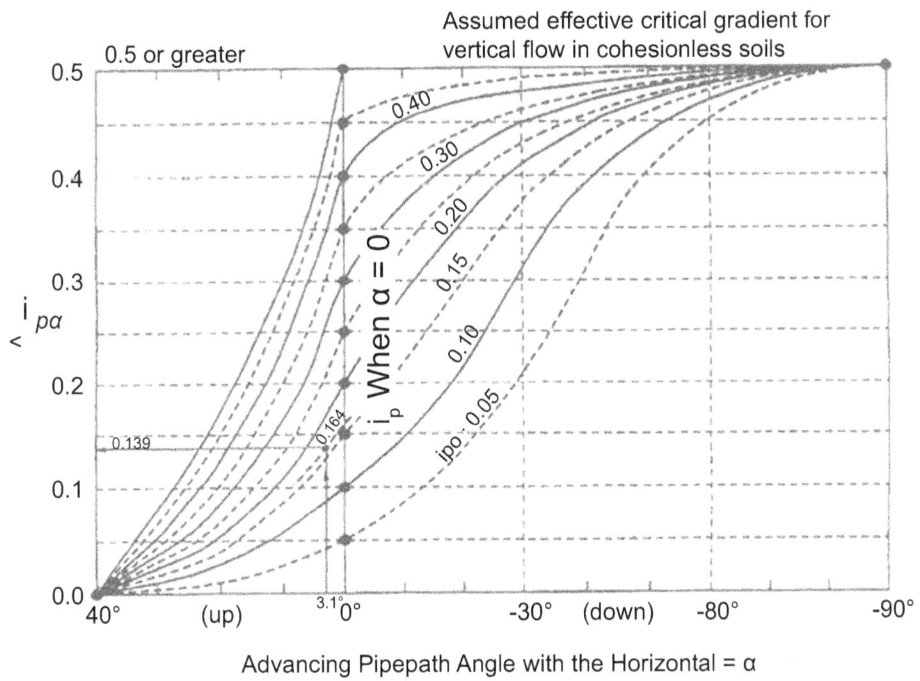

FIGURE 10.25 Graph to obtain i_{po} from $i_{p\alpha}$ and a (Schmertmann, 2000). (Used with permission of ASCE conveyed through Copyright Clearance Center Inc.)

where:

R = radius to point on the pipe path in a dam with curved axis (m)
R_1 = shortest radius to an end of completed pipe path (m)
R_0 = longest radius to an end of completed pipe path.

Schmertmann (2000) recommends the following procedure:

10.5.2.1 Point Method

(a) Draw a section of the proposed or actual dam, with both the dam and foundation soils indicated. Choose a section representative of a length of the dam and with the likely lowest F_p profile for that length of dam.
(b) Construct a flow net, based on tests or best estimates of the permeability and anisotropy of each of the soils involved. For the investigation of an existing dam, construct the flow net with the aid of piezometer data.
(c) Select a potential trial pipe path through the dam section that provides the necessary roof to prevent pipe collapse and also provides a downstream discharge to keep the pipe open. The pipe path should follow a flow path or flow line.
(d) Obtain i_{pmt} for each of the soils in the trial pipe path. Either test the sands involved in a standard laboratory piping flume test, such as the University of Florida test, or use a conservative default value based on a sand's C_u (Figure 10.22) for i_{pmt} in the absence of tests.
(e) At any and all points "x" along the trial pipe path, correct the i_{pmt} to match the field conditions by using the various correction factors discussed earlier. Apply these corrections to obtain the gradient i_{px} needed in the field to move the pipe path through each point "x" and calculated F_{px}.
(f) Determine enough F_{px} points to get an F_p profile along the trial pipe path. By inspection of this profile the engineer needs to decide if the trial pipe path has an acceptable F_p variation. If "yes", then one needs to investigate other trial pipe paths in the present or additional dam sections to determine the most critical situations. If "no", then the dam requires redesign until one gets a "yes" answer for all creditable pipe paths. The engineer can ignore a "no" finding but then must include proper filter protection in the design and accept the lack of redundancy.

Readers should refer to the Schmertmann (2000) paper for the details of this procedure.

10.5.2.2 Average Method

An average F_p can be calculated using estimated average values for the correction factors, and the average gradient (H2 – H3) / L in Figure 10.14. Note that in many cases, important detail may be lost if the average method is used. However, the authors' experience is that the average method is preferred by most practitioners.

10.5.2.3 Uncertainty in the Method

Schmertmann (2000) discusses the uncertainty associated with estimate of the factor of safety F_{px}. He assesses the coefficients of variation for each correction factor, including a correction factor C_G which he has used to convert the global gradients $\bar{i}_{pmt}$ from the laboratory tests to i_{pmt} in Figure 10.22. Despite reservations about the lack of substantive information to support the uncertainty analysis, the authors have recommended it be considered as discussed later in Section 10.5.4.

10.5.3 Modifications to the Schmertmann (2000) Method

10.5.3.1 Robbins and Sharp (2016)

Robbins and Sharp (2016) took the laboratory test results from Schmertmann (1995), Townsend et al. (1988), de Wit (1984), Weijers and Sellmeijer (1993), Kohno et al. (1987) and Muller-Kirchenbauer et al. (1993), corrected them using the reference values listed in Table 10.7 and plotted the individual results as shown in Figure 10.26. Figure 10.27 shows that data but with the source documents detailed superimposed on Figure 10.22. The data points in Figure 10.22 use average results for each soil.

Robbins and Sharp (2016) analysed these data statistically to develop quantiles at 10% intervals shown in Figure 10.26.

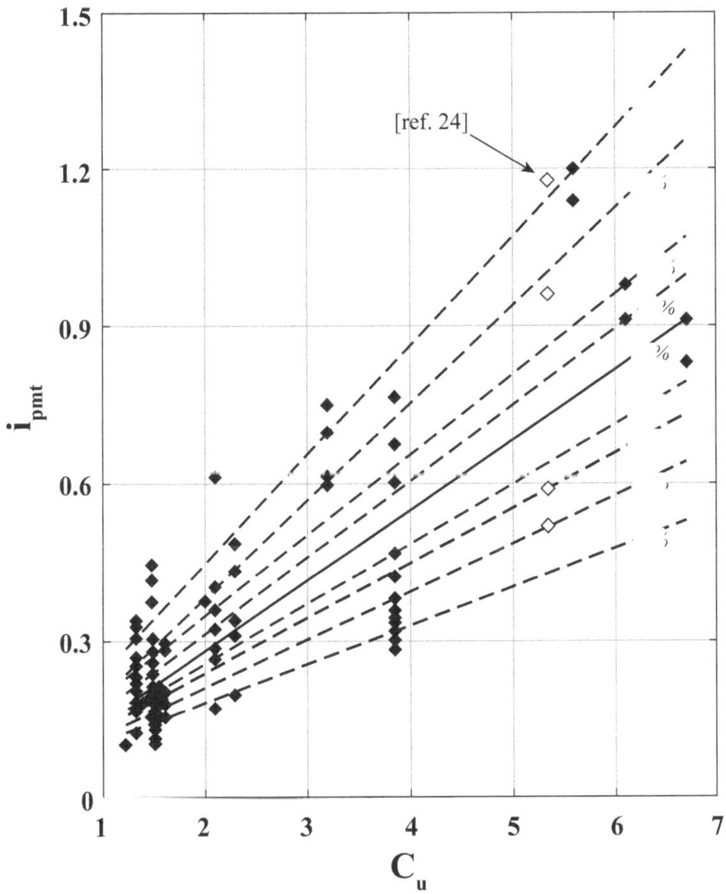

FIGURE 10.26 Critical point gradients from individual laboratory tests and best fit quantile regression lines for the 10th to 90th quantiles. Open points are from different testing configurations of Kohno et al. (1987). (Robbins and Sharp, 2016.)

There are a lot of data for soils with a Cu ≤ 2.3 which are the soils most likely to be subject to BEP. There are a number of tests at Cu = 3.85. These are by de Wit (1984) and carried out in 0.80 m and 2.4 m flumes with a plane exit so can be relied upon. However as discussed later they are for low relative density soils. The tests at Cu = 5.6 and Cu = 6.1 are by Townsend et al. (1988) and as discussed earlier are on gap-graded soils. The test descriptions in Townsend et al. (1988) indicate fine sand did erode in these tests, so the interpretation of the tests is somewhat complicated and should not be given much weight in assessing the overall behaviour. This leaves few test data for Cu > 3.85, so there is considerable uncertainty in the quantiles in Figure 10.27 beyond Cu = 4.

10.5.3.2 Allan (2018)

Allan (2018) carried out BEP tests at UNSW using a flume which was modelled on the University of Florida tests. Tests were carried out with slope exit similar to that shown in Figure 10.21 and for hole/circle exit, slot (ditch) and plane exit conditions.

The particle size distributions of the soils tested by Allan (2018) are shown in Figure 10.28. These include soils with Cu 6 to Cu 6.5 to supplement the limited data available for high Cu soils.

Figure 10.29 shows the results of Allan (2018) tests superimposed on the Schmertmann (2000) plot. The Allan (2018) tests on soils other than Sydney sand are for circle/hole exit conditions whereas the tests used for the Schmertmann (2000) plot are mostly for slope and plane exit conditions.

Figure 10.30 takes the data from de Wit (1984) for tests ≥ 50% relative density, Townsend and Shiau (1986), Schmertmann (1995), Van Beek et al. (2011, 2015) tests excluding tests in 300 mm flume, and Allan (2018) results corrected to plane and slope exit conditions using the suggested adjustment factors shown in Table 10.8. Shown on the plot are the mean, median and 16th and 84th percentiles for the data.

Figure 10.30 applies to soils with relative density ≥ 50%. Corrections for outlet conditions, relative density and fines content are applied as detailed in Section 10.5.4

10.5.3.3 Recommended Approach to Application of the Schmertmann (2000) Method

It is recommended that the Schmertmann (2000) method be applied as detailed in Section 10.5.3, but instead of using Figure 10.22, Figure 10.30 should be used. Analyses should use the median in Figure 10.30 and check for the 16% and 84% plots.

10.5.3.4 Application of the Schmertmann (2000) Method to Soils Which Have a Uniformity Coefficient Greater than 6

There are two types of soils which may have uniformity coefficients greater than 6:

- Well-graded sand and gravelly sand. Soil mixes 3 and 4 in Figure 10.28 are examples.
- Poorly graded fine to medium sand with silt content typically ≥ about 10%. Soil mixes 6, 7 and 8 are examples.

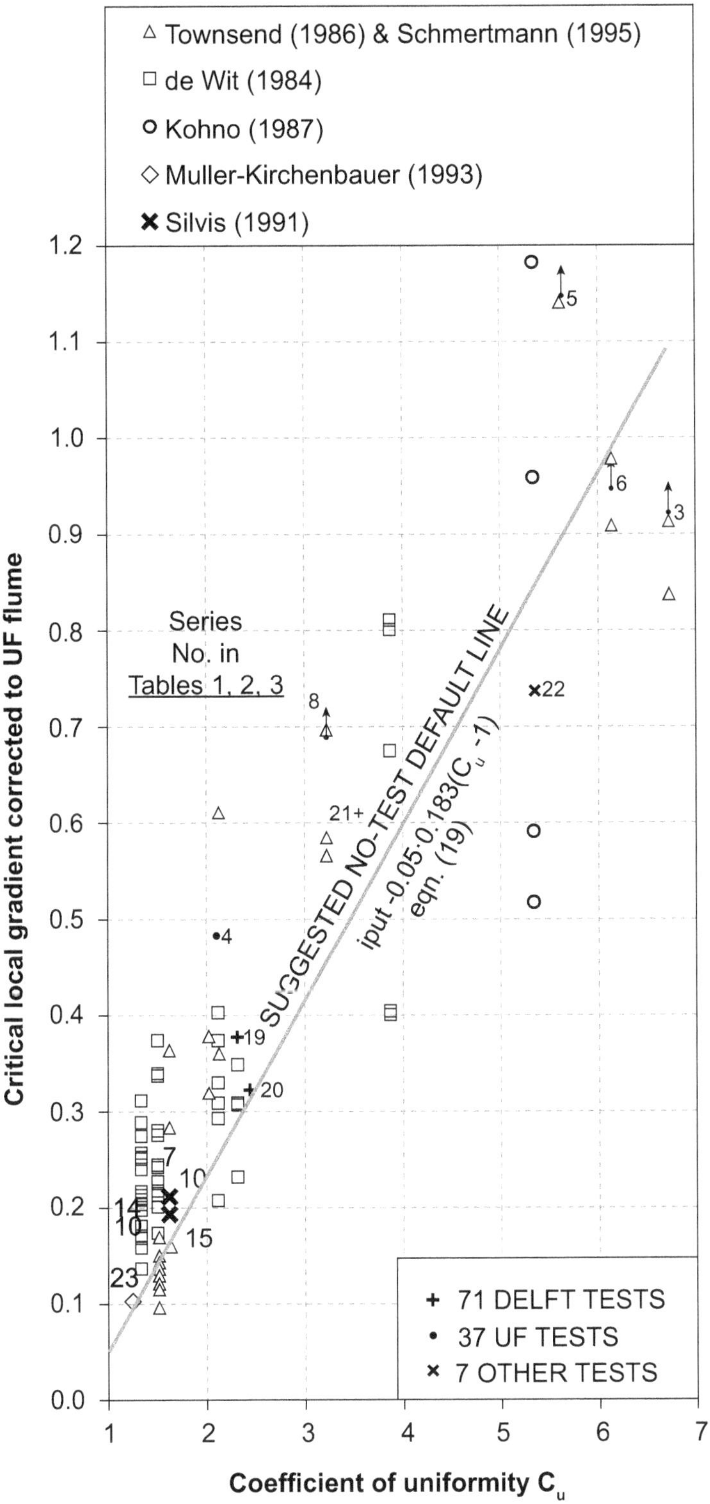

FIGURE 10.27 Schmertmann (2000) plot of local gradient versus Cu but with individual results plotted instead of averages (Allan, 2018).

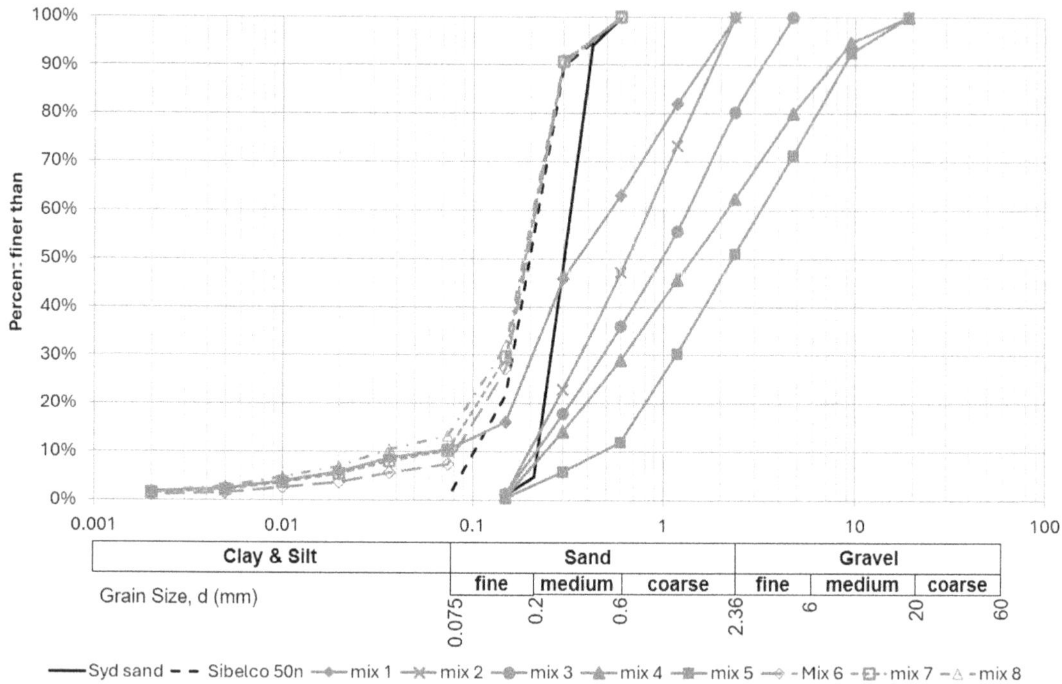

FIGURE 10.28 Particle size distribution of soils tested by Allan (2018).

The suggested approach to considering progression of BEP in these soils is:

Well-Graded Sand and Gravelly Sand
Some factors to consider are as follows:

(a) From Section 10.2.1, soils subject to BEP are typically non-plastic poorly graded fine to medium sand with D_{50} around 0.2 mm to 0.4 mm. The sand particles are readily transported along the BEP channels.
(b) Well-graded sands and gravelly sands have D_{50} around 1.0 mm to 1.5 mm, much coarser and requiring larger hydraulic forces to transport them along BEP channels. As a guide, from Figure 11.3 the Darcy velocity to transport such soils is about double that for soils with D_{50} 0.2 mm to 0.4 mm.
(c) The Allan (2018) BEP test on Mix 3 ($Cu = 6.2$) formed BEP channels and the channels progressed the full test length. The two tests on Mix 4 ($Cu = 8.8$) stalled due to gravel size particles and sharing of pipe paths. They did not progress to break through to the water source upstream. Allan (2018) concluded that sands and gravelly sands with grading as for Mix 4 may be considered not subject to BEP.
(d) Allan (2018) Mixes 3 and 4 are internally unstable according to the Kenney and Lau (1985, 1986) criteria.

It is concluded that non-plastic well-graded sand and gravelly sand with $Cu \geq 6$ are unlikely to be subject to BEP. They are not suited to being assessed by the Schmertmann (2000) or Sellmeijer et al. (2012) methods. They may be subject to suffusion or GBE. Check for this using the Kenney and Lau (1985, 1986) method.

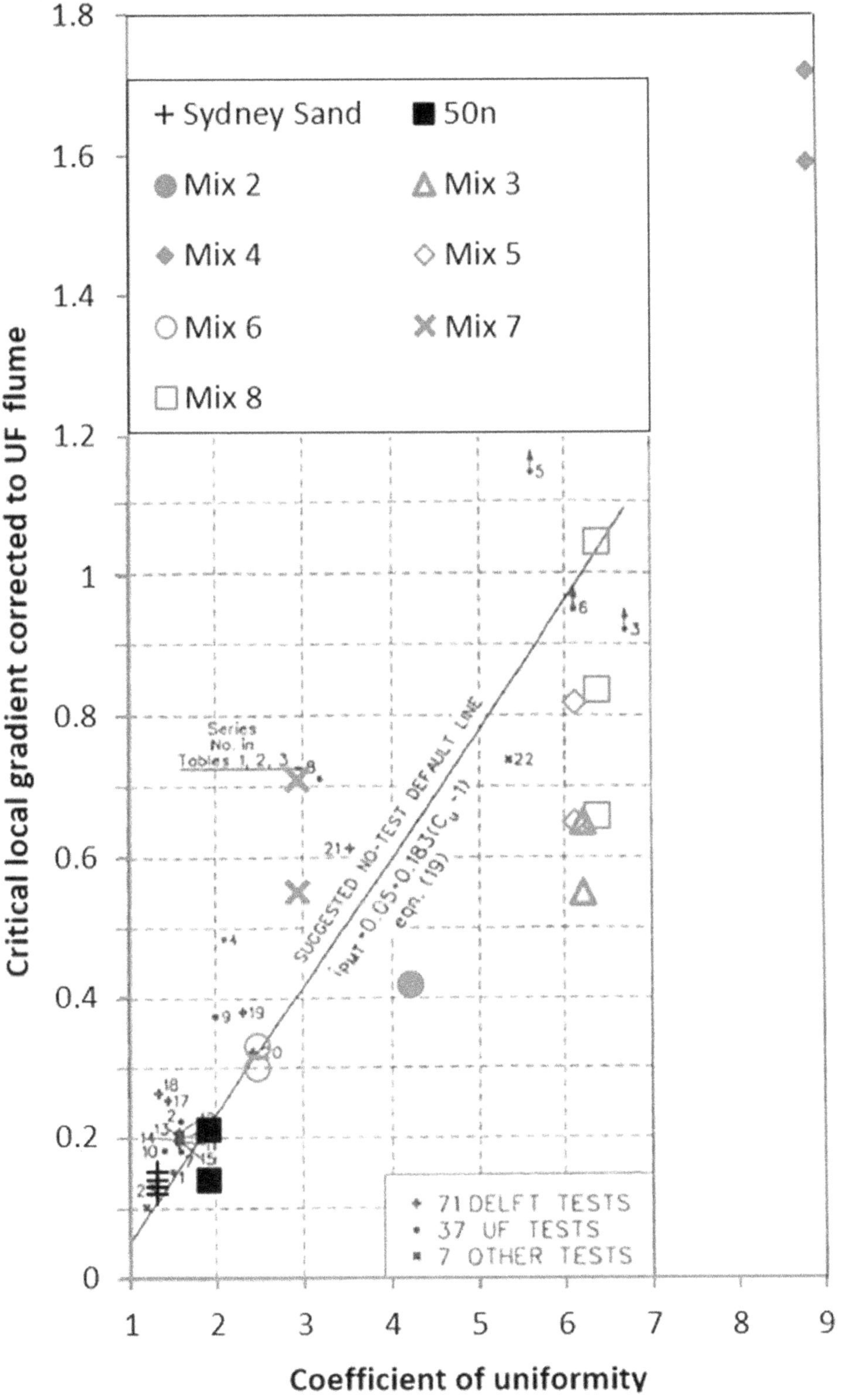

FIGURE 10.29 Results of Allan (2018) backward erosion piping tests superimposed on the Schmertmann (2000) plot. Gradients for Allan (2018) tests corrected to the University of Florida (UF) flume geometry.

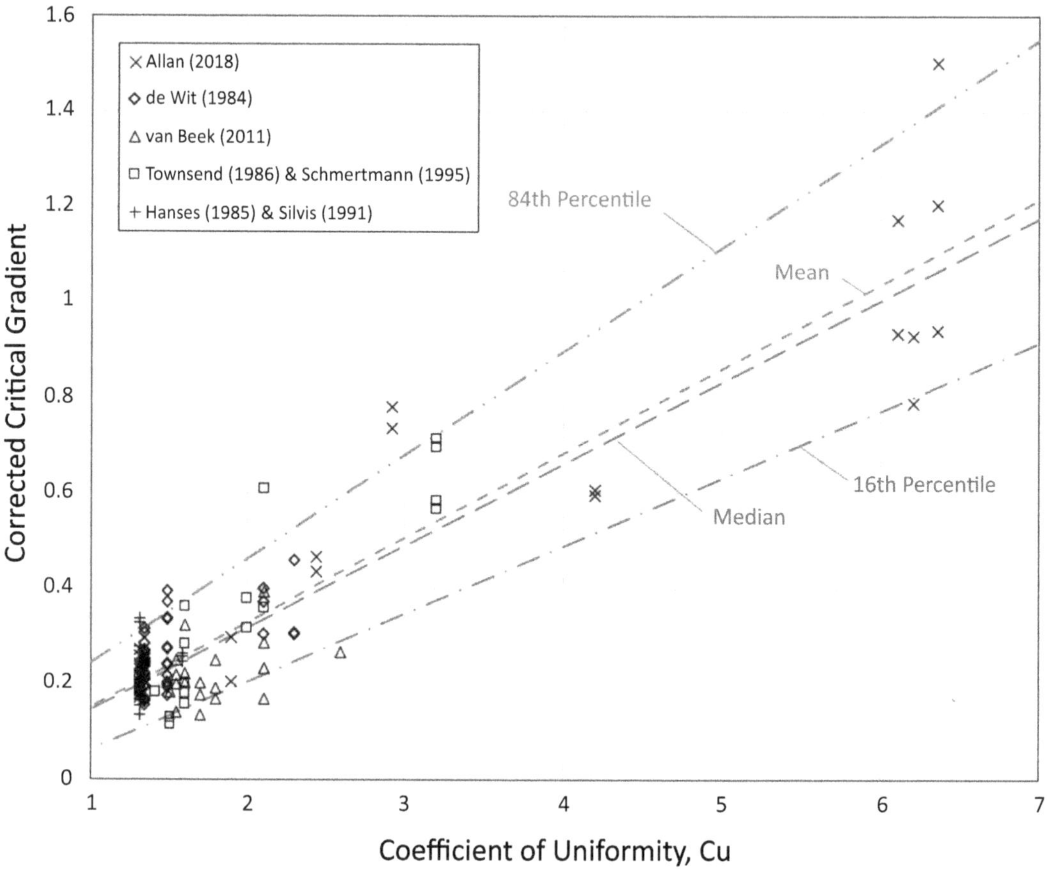

FIGURE 10.30 Coefficient of uniformity (Cu) versus critical local gradient for progression of backward erosion piping using the data from De Wit (1984), Townsend and Shiau (1986), Schmertmann (1995), Van Beek et al. (2011, 2015), and Allan (2018). The plot is for slope and plane exit conditions and for soils with relative density ≥ 50%. (Courtesy of R. Allan.)

Poorly Graded Fine to Medium Sand with Silt Content Typically ≥ about 10%
Some factors to consider are the following:

(a) Dr B. Robbins (Personal communication, 2025) indicates that USACE extrapolates Schmertmann's Cu trend beyond 6 and applies an additional correction factor for fines content.

They calculate the likelihoods of progression for various fines content correction factors and then elicit probabilities of progression from the risk analysis members based on the degree of belief of what is most applicable to the project and failure mode and their confidence in the correction factors.

(b) Dr Robbins makes the further point which the current authors agree with. He says that for these types of soil the main question that needs to be answered is regarding the actual stratigraphy and continuity of strata in the ground. If the risk analysis team believes there is a continuous, high Cu, > 10% fines content layer in the ground (e.g. glacial till deposit), then the probability of progression will be extremely low.

More often than not, the real questions we need answered when encountering these ranges of Cu and fines content are as follows:

- Are the soil index properties representative of the actual soil deposits or are they an artefact of sampling? Has mixing of layered sand and silt resulted in the apparent presence of silty sand?
- What is the probability of having a sand unit subject to BEP (for the given geometry and loading) continuously upstream to downstream given the geomorphology and geology of the area?

(c) The data point for Mix 4 in Figure 10.29 is for a gravelly sand as discussed earlier and should not be considered when extrapolating.

(d) As already discussed in Section 10.5.3, it is recommended that the median line in Figure 10.30 be used for the best estimate instead of the Schmertmann (2000) line in Figure 10.29. The median line better fits the now larger data set. The Schmertmann (2000) line was also a design line, not a line of best fit.

(e) The current authors are reluctant to extrapolate much beyond the data set in Figure 10.30. At Cu = 6 the critical gradient is around 1.0. For most dams or levees founded on such soils the embankment slopes will be ≥ 2H:1V, so the seepage gradient will be about 0.25. That gives a factor of safety of at least 4 on critical gradient from Table 10.5, so the likelihood of progression will be very low.

It is concluded that in general extrapolation should not go beyond the data in Figure 10.30. The default maximum best estimate critical gradient would therefore be = 1.0.

As discussed in Section 10.5.4, it is considered that there is insufficient published data at this time for corrections for fines content to be applied with confidence. Readers should review the published literature for the results of Deltares research and that by others for guidance.

10.5.4 Corrections to the Sellmeijer and Schmertmann Methods for Exit Conditions, Fines Content and Relative Density

10.5.4.1 Effect of Exit Conditions on the Critical Gradient for Progression of Backward Erosion Piping

Allan (2018) carried out tests using Sydney Sand for plane, slope, slot (ditch) and circle (hole) exit conditions. De Wit (1984) carried out tests to assess the effect of exit conditions. Those results are summarised in Appendix A of Van Beek et al. (2015). Van Beek also carried out tests with plane, slope and hole exit conditions. Table 10.8 summarises the results.

Koelewijn and Van Beek (2024) indicated that 3D numerical modelling was confirming that 3D effects were significant.

It is concluded that the exit conditions are an important factor in controlling the critical gradient for progression of backward erosion and that the results from the Sellmeijer, Hoffmans and Schmertmann methods should be adjusted by the suggested factors shown in Table 10.8. All three methods are calibrated against slope and plane laboratory models. If the field exit conditions are represented by those models, no correction is required. If the field exit conditions are best represented by a ditch/slot or hole, then correction factors are required.

TABLE 10.8 Influence of Exit Condition on the Critical Gradient for Progression of Backward Erosion Piping (Fell et al., 2024)

	DE WIT (1984) 1.9 M FLUME		*DE WIT (1984) 5.75 M FLUME*		*VAN BEEK (2015) 1.5 M AND 1.9 M FLUMES*		*ALLAN (2018)*		
EXIT CONDITION	*CRITICAL GRADIENT* i_C	i_C/i_{CPLANE}	*CRITICAL GRADIENT* i_C	i_C/i_{CPLANE}	*CRITICAL GRADIENT* i_C	i_C/i_{CPLANE}	*CRITICAL GRADIENT* i_C	i_C/i_{CPLANE}	*SUGGESTED ADJUSTMENT FACTORS* i_C/i_{CPLANE}
Plane	0.30 to 0.35	1.00	0.15 to 0.185	1.00	n/a	1.00	0.24	1.00	1.00
Slope	n/a	n/a	n/a	n/a	0.24; 0.39[b] 0.14 to 0.26 0.2 av.[c]	1.00	0.25	1.04	1.00
Ditch	0.22 to 0.30	0.8	0.12 to 0.15	0.83			0.18	0.75	0.8
Hole			0.17 to 0.195[a]	1.08[a]	0.15 and 0.25[b] 0.16[c]	0.6 0.8	0.16	0.67	0.7

Notes:
(a) De Wit (1984) hole experiments exited not only from the hole but also the downstream side.
(b) Interbeck 0.33 mm soil.
(c) Baskarp soil.

10.5.4.2 Effect of Fines Content

Lucker and Robbins (2023) carried out BEP tests in a cylindrical test set-up. They tested soils with fines content ranging from 3% to 10%. They concluded that a new factor was required to account for silt content when using the Schmertmann (2000) method but did not quantify it.

Koelewijn and Van Beek (2024) and Deltares (2023) presented some as yet unpublished test data which indicates that soils with higher content of particles finer than 0.016 mm are more resistant to initiation and progression of BEP. The tests are mostly from small-scale (0.3 m long) experiments and have significant scatter in the results. Koelewijn and Van Beek (2024) also briefly describe results from a field scale test.

Allan (2018) tested soils with silt contents 7%, 10% and 12% in a more conventional test set-up, i.e. the Florida flume design. The gradings are shown in Figure 10.28, tests 6, 7 and 8. The results are shown in Figure 10.29. It can be seen that the critical gradients do not show a consistent trend compared to the Schmertmann (2000) recommended Cu versus critical gradient line. They are broadly consistent with the other data. It is noted that the fines content will affect the uniformity coefficient and d_{10} size which are already factors in the Schmertmann (2000) method.

It is concluded that currently there is insufficient evidence to provide guidance on adjustment for fines content. Readers should review the published literature for the results of the Deltares research and that by others.

10.5.4.3 Effect of Porosity and Relative Density on the Critical Gradient for Progression of BEP

Robbins and Van Beek (2015) present Figure 10.31 which shows that the critical gradient for progression of BEP is dependent on the porosity of the sand. These data are for L = 0.34 m, i.e. small-scale tests. The porosity is related to the relative density of the soil.

Appendix A of Van Beek (2015) details her tests and those of De Wit (1984). Figure 10.32 presents the relationship between relative density and the critical gradient for progression. The data has been re-plotted for clarity.

From this it can be seen that for the Dune Sand and Baskarp Sand the trend lines are similar and show a marked dependence of critical gradient on relative density (RD). The effect is less for the Beach Sand.

The ratio of critical gradients for 0.9 and 0.4 relative densities is 1.9 for Dune Sand, 1.45 for Baskarp, and 1.15 for Beach Sand. By comparison the Sellmeijer et al. (2011) method gives a ratio of 1.33 and Schmertmann (2000) gives 1.22, so they appear to underestimate the effect of RD. Sellmeijer et al. (2011) state that their method applies for RD between 50% and 100%.

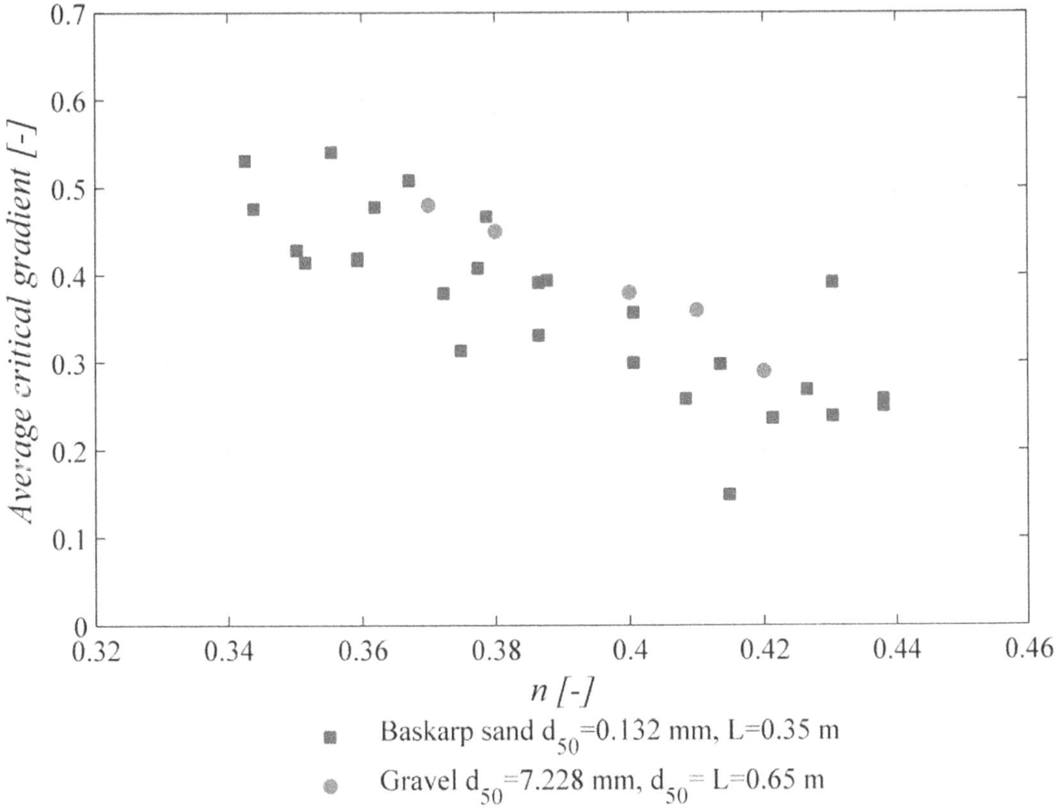

FIGURE 10.31 Influence of porosity on the critical gradient for initiation and progression in experiments with a slope exit (Robbins and Van Beek, 2015; Van Beek et al., 2011).

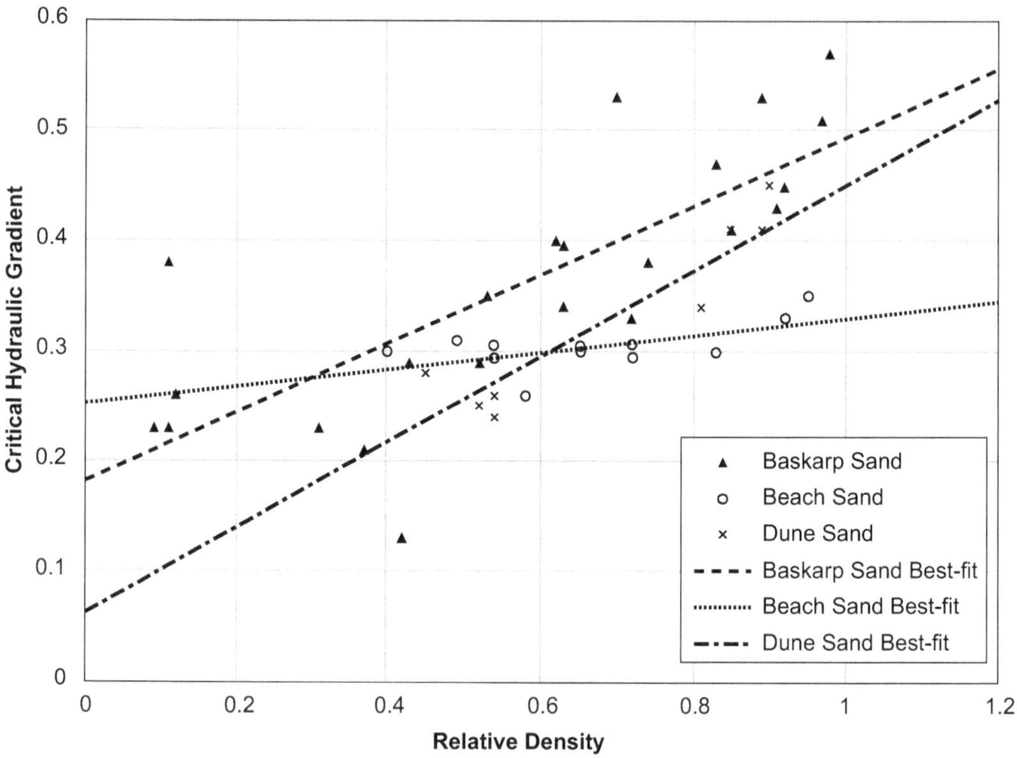

FIGURE 10.32 Relationship between relative density and critical gradient for progression based on tests by De Wit (1984) and Van Beek, reported in Van Beek (2015).

It is concluded that there is a relation between relative density and critical gradient, but the limited data for RD < 50% seems to show less dependence than for RD > 50%.

For the Baskarp Sand for which most data are available, the average critical gradient for RD < 50% is about 0.25, and for > 50% RD about 0.4.

It is recommended therefore that for soils with RD < 50% the critical gradient obtained using the Sellmeijer and Schmertmann methods should be reduced by applying a factor of 0.6. If the RD is just below 50%, a lesser reduction might be applied.

10.5.4.4 Uncertainty in Estimating Average Critical Gradient

See Figure 10.30, which shows the range of data when plotted as corrected critical gradient versus uniformity coefficient Cu.

Figure 10.33 shows the range of data when plotted as critical gradient versus seepage length.

Both these data and the 84% and 16% data are approximated by a range from the median (50%) values of 1.5x for 84% and 0.66x for 16%. These apply where there is sufficient data to determine the range. While the data at field scale is limited, it seems to support the likelihood that uncertainty reduces at large-scale but there are insufficient data points to conclude this. It is therefore suggested that for field scale the same range will apply. This indicates there may be some conservatism in this assumption and hence in the estimated probabilities.

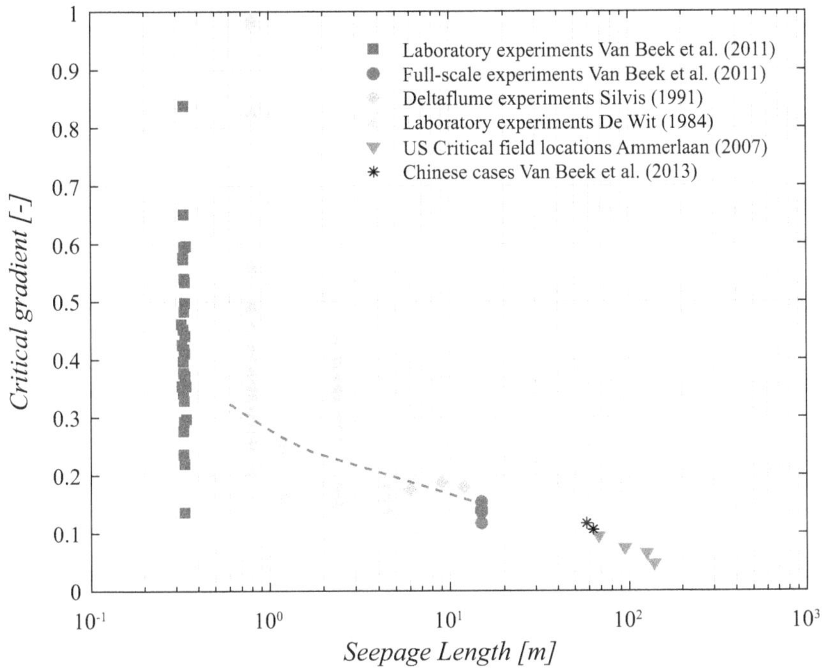

FIGURE 10.33 Critical gradient versus seepage length (L) for laboratory and field data (Van Beek and Hoffmans, 2017). Dashed line is the Sellmeijer et al. (2011) model from Figure 10.34.

There remains uncertainty in the modelling of scale effects and other parameters in the Sellmeijer and Schmertmann models. Figure 10.34 from Van Beek and Hoffmans (2017) illustrates the scale effect for the Sellmeijer et al. (2011) and Hoffmans (2014) methods.

The Sellmeijer et al. (2011) line from Figure 10.34 is plotted on Figure 10.33 and shows a good fit to the median data.

Schmertmann (2000) discusses uncertainty in his method. He estimates that the coefficient of variation is 0.70 allowing for not all factors occurring simultaneously. He assigns a probability of 0.01 for a factor of safety = 2.5 on the critical gradient, and 0.1 for a factor of safety = 1.75.

The suggested probabilities in Table 10.5 take account of Schmertmann (2000) values but allow for the larger scatter in the Cu versus critical gradient in Figure 10.30 compared to what Schmertmann (2000) had in his testing.

Allan (2018) showed that the scatter in laboratory data upon which the Sellmeijer and Schmertmann methods are based was similar. Given this and the uncertainty in scale effects for both methods, it is considered that Table 10.5 may be applied to both methods.

10.5.5 Estimation of Parameters for Use in Estimation of the Critical Gradient for Progression of Backward Erosion

The following is some information provided to assist in the estimation of the parameters required for the estimation of the critical gradient. Apart from geometric parameters these are:

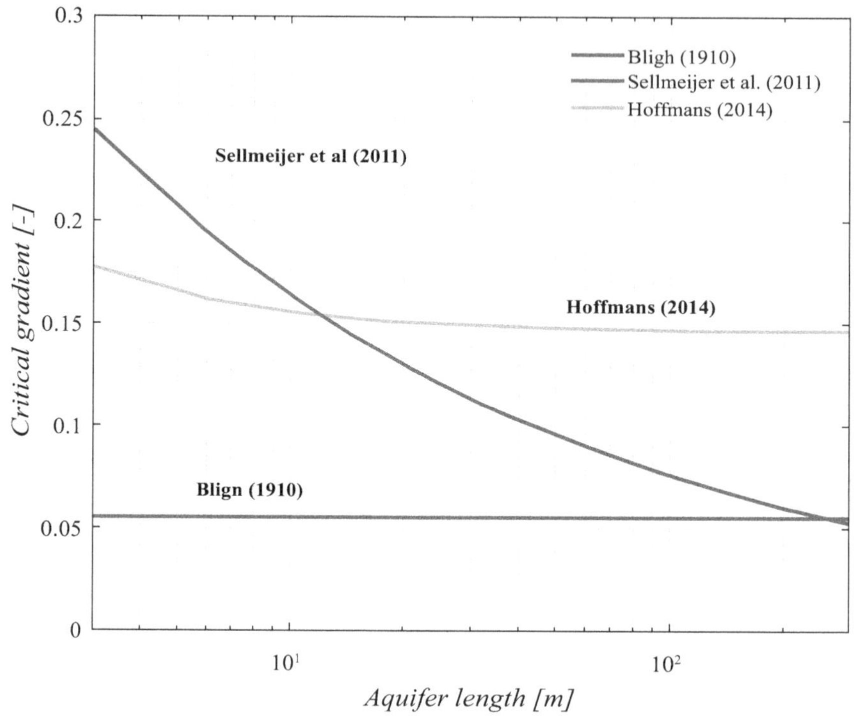

FIGURE 10.34 Illustration of the scale effect (represented by L) in different models given a constant D/L ratio of 0.33 and constant material properties (fine sand) (Van Beek and Hoffmans, 2017).

10.5.5.1 Sellmeijer and Co-Workers at Deltares Method

RD: relative density [%]
U: uniformity coefficient $Cu = d_{60}/ d_{10}$[-]
KAS: roundness [%]
d_{70}: soil particle diameter for which 70% by weight of the soil is finer [m]
γ_p': submerged unit weight of soil particles 9.8(G-1) [kN/m^3]
G: soil particle density [t/m^3]
γ_w: unit weight of water [kN/m^3]
η: Whites drag coefficient [-]
ϑ : bedding angle (angle of repose) of sand [deg]
υ: kinematic viscosity [m^2/sec]
G: gravity [m/sec^2]
K: intrinsic permeability (m^2) which is related to hydraulic conductivity [m/sec] as detailed in Equation 10.13.

10.5.5.2 Schmertmann Method

D_{rf}: relative density [%]
U: uniformity coefficient $Cu = d_{60} / d_{10}$ [-]
d_{60}: particle diameter for which *60%* of the sediment particles is finer (m)
d_{10}: particle diameter for which *10%* of the sediment particles is finer (m)

k_h: horizontal coefficient of permeability, also termed hydraulic conductivity [m/sec]
k_v: vertical coefficient of permeability, also termed hydraulic conductivity [m/sec]
Notes:
(a) The symbols used by the authors of these methods have been retained so readers can relate to the original papers.
(b) In the Sellmeijer method the hydraulic conductivity is not defined in terms of horizontal or vertical permeability. The method would apply to soil with k_h/k_v close to 1.0.

10.5.5.3 Relative Density (Density Index)

The relative density of the sediment should be estimated from SPT and CPT/CPTU tests at the site.

10.5.5.4 Soil Particle Density

Determine from laboratory tests or estimate taking account of the mineralogy of the sediments.

10.5.5.5 d_{10}, d_{15}, d_{50}, d_{60}, d_{70}

Determine from laboratory gradation tests on representative samples of the sediments.

10.5.5.6 Coefficients of Permeability k_h and k_v

Determine from *in situ* permeability tests in the sediments. If these are not available, estimate from the grading of the soils. Figure 10.35 shows the relationship between the horizontal permeability (k_h) and d_{10} size. The permeabilities were determined from *in situ* tests at levee sites along the Mississippi River. USACE (1956) indicates this plot is regarded as more accurate than laboratory tests on remoulded samples. It is noted that the line of fit shown gives permeabilities about 2.5 × the Hazen formula for finer soils and 0.7 × the Hazen formula for coarser soils. The range of permeabilities in the plotted data should be considered when using this figure.

The ratio of k_h/k_v is required to estimate k_v from this data. The ratio should be estimated taking account of any stratification within the sediments. k_v will in all cases be $< k_h$

10.5.5.7 Kinematic Viscosity

The kinematic viscosity of water at 20° centigrade is 1.004×10^{-6} m/sec.

10.5.5.8 Shields Parameter

Hoffmans and van Rijn (2018) indicate that Shields parameter (Ψ) is related to the dimensionless particle size D_* where D_* is determined from:

$$D_* = d_{50}\left(\frac{(\rho_s/\rho - 1)g}{\nu^2}\right)^{1/3} \tag{10.23}$$

Figure 10.36 shows the relationship between D_* and Ψ.

Hoffmans and van Rijn (2018) indicate that the upper envelope of the experimental range, represented by the "Hoffmans, 2014" line, can be seen as representing conditions where there is general bed

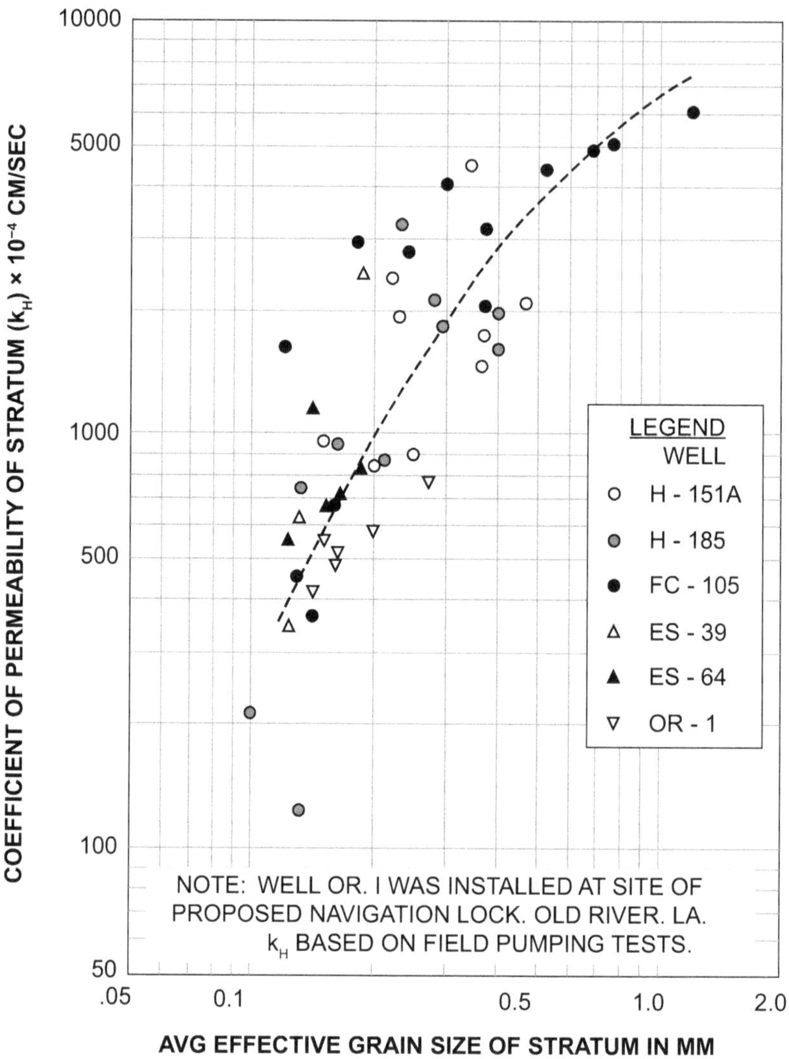

FIGURE 10.35 *In situ* horizontal permeability versus average effective grain size, d_{10} (USACE, 1956).

load transport with particles moving at all locations. For no backward erosion the lower limit of the Shields parameter applies. That may be assumed to be the "Extended Shields (laminar)" line for $D_* > 1$.

10.5.5.9 Whites Drag Coefficient

Sellmeijer (1988) and Sellmeijer et al. (2011) assume that Whites coefficient is 0.25 and a bedding angle of 37° for their database (Sellmeijer et al., 2011). Van Beek et al. (2019) indicate that a value of Whites coefficient of 0.3 gave a better fit to the data. They show that Shields parameter (Ψc) is related to Whites coefficient (η) and the bedding angle (θ) as follows:

$$\Psi c = \eta\ (\pi/6) \tan\theta \qquad (10.24)$$

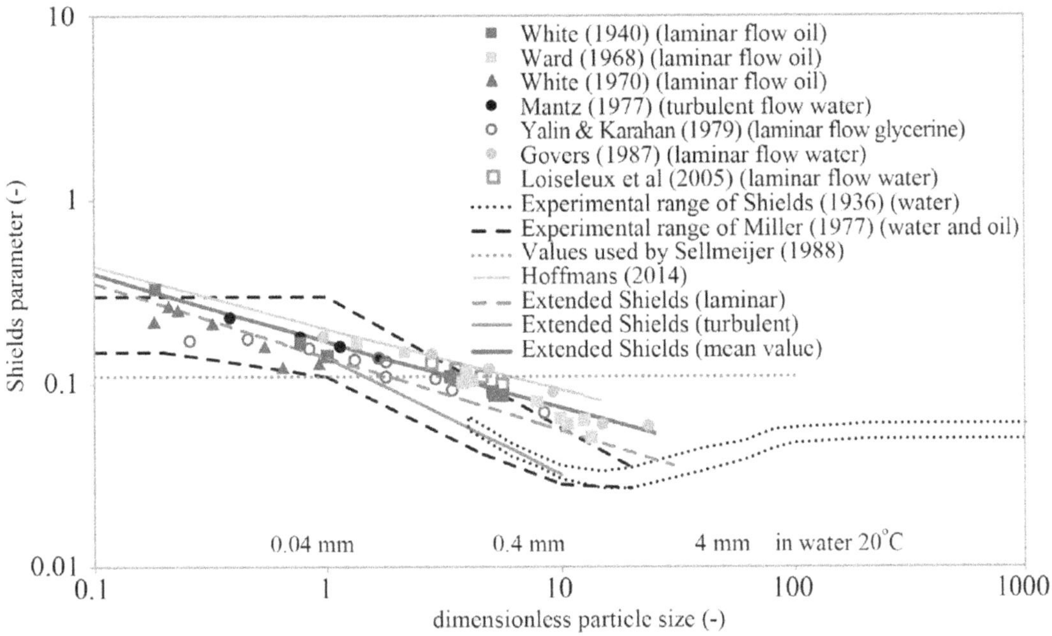

FIGURE 10.36 Shields diagram for laminar and turbulent flows (Hoffmans and van Rijn, 2018; van Rijn, 2014).

For $\theta = 37°$ this gives $\eta \approx 2.5\ \Psi c$. As discussed above Ψc is related to the grain size as shown in Figure 10.36 so Whites coefficient may be estimated using this correlation.

Hoffmans and van Rijn (2018) note that Sellmeijer et al. (1989, 2011), by using a constant value of 0.25 for Whites coefficient (η), effectively use a constant value for $\Psi_{lam,c}$ whereas, as seen in Figure 10.36, it varies with particle size. As a result, the Sellmeijer et al. (1989) approach is satisfactory for fine sand but overestimates η for coarse sand.

10.5.5.10 Bedding Angle

Sellmeijer et al. (1989, 2011) used a constant value of 37° for the bedding angle (θ). Van Beek et al. (2019) show that the bedding angle varies with d_{50} as shown in Figure 10.37.

It is suggested this plot be used with a d_{50} representative of the sediments and that a minimum value of $\theta = 37°$ be adopted.

10.5.5.11 Roundness

The roundness is defined by:

$$Roundness = 4 \cdot \frac{Area}{\pi \cdot (Major\ Axis)^2} \tag{10.25}$$

Sellmeijer et al. (2011) indicate it is a minor factor and may be ignored.

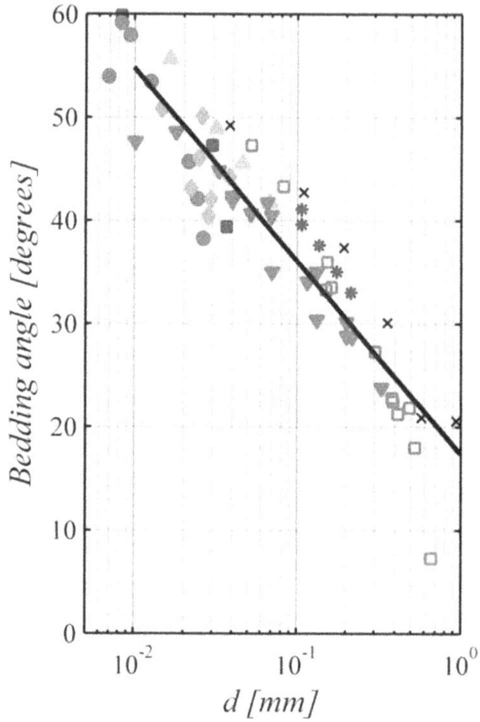

FIGURE 10.37 Relationship between bedding angle (θ) and d_{50} of soil (Van Beek, 2015).

10.6 IM 25—ESTIMATION OF THE PROBABILITY OF INITIATION AND PROGRESSION OF SUFFUSION OR GLOBAL BACKWARD EROSION IN AN EMBANKMENT (IM 25E) OR IN THE FOUNDATION (IM 25F)

10.6.1 Introduction

Douglas et al. (2016, 2019) carried out a number of downflow permeameter tests on silt sand gravel soils with gradings shown in Figure 10.9. These tests showed that there was a gradation of behaviour ranging from internal instability without erosion of particles from the soil, GBE with some finer fraction eroded and suffusion with larger amounts of soil eroded. There was not a clear demarcation between suffusion and GBE.

Wan and Fell (2004c, 2008) also found that suffusive behaviour was not clearly defined and assigned a probability distribution as shown in Figures 10.38 and 10.39.

Kenney and Lau (1985, 1986) developed a method for assessing what they termed internal instability of sand gravel soils. In the terms used in this book their term "internal instability" is equivalent to suffusion. Figures 10.40 and 10.41 show the method. It is recommended this method be used for assessing sand gravel materials typically used for filters. This is discussed further in Chapter 12.

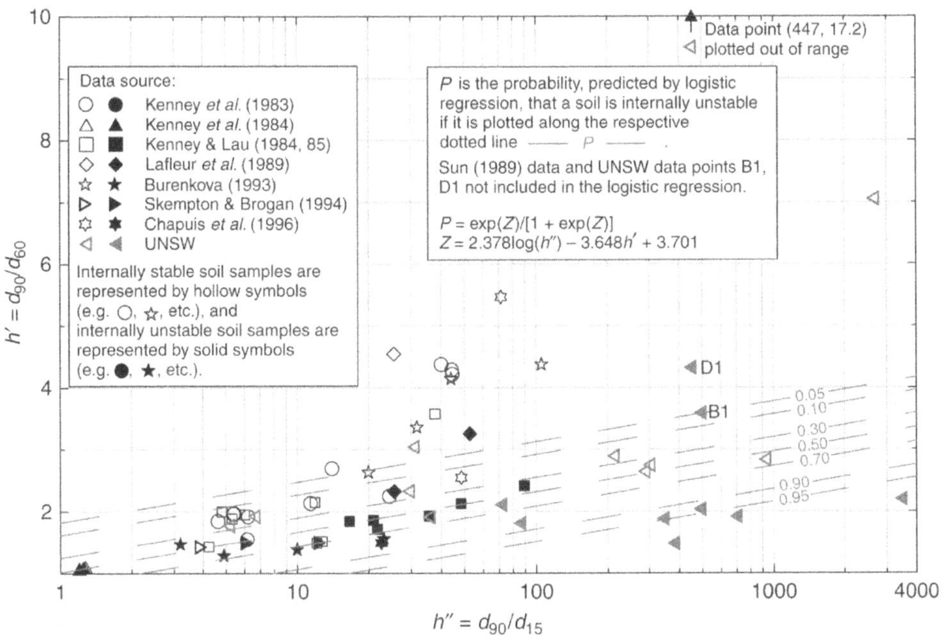

FIGURE 10.38 Contours of the probability of internal instability (potentially suffusive) for silt-sand-gravel soils and clay-silt-sand-gravel soils of limited clay content and plasticity (Wan and Fell, 2004c, 2008). (Used with permission of ASCE conveyed through Copyright Clearance Center Inc.)

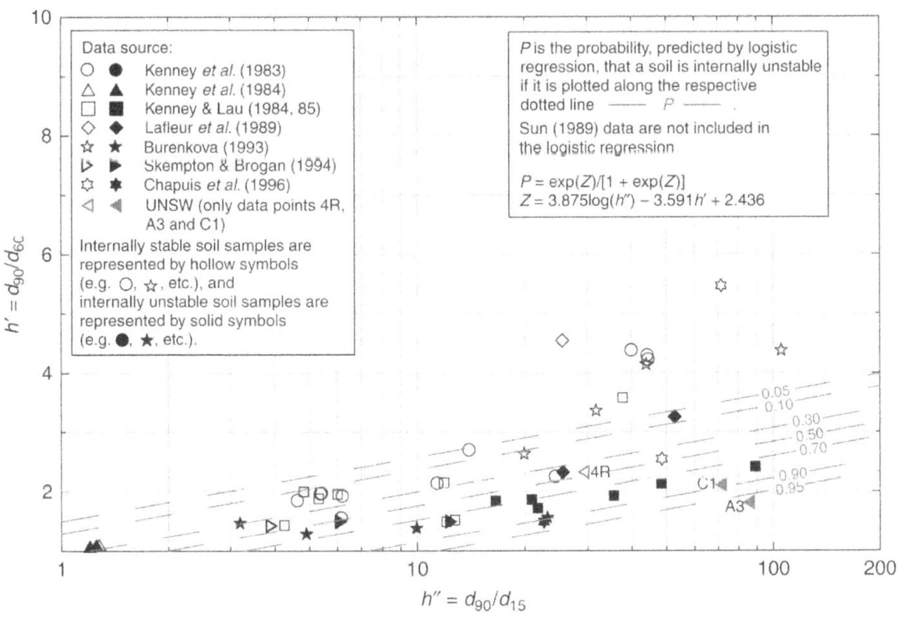

FIGURE 10.39 Contours of the probability of internal instability (potentially suffusive) for sand-gravel soils with less than 10% non-plastic fines passing 0.075 mm (Wan and Fell, 2004c, 2008). (Used with permission of ASCE conveyed through Copyright Clearance Center Inc.)

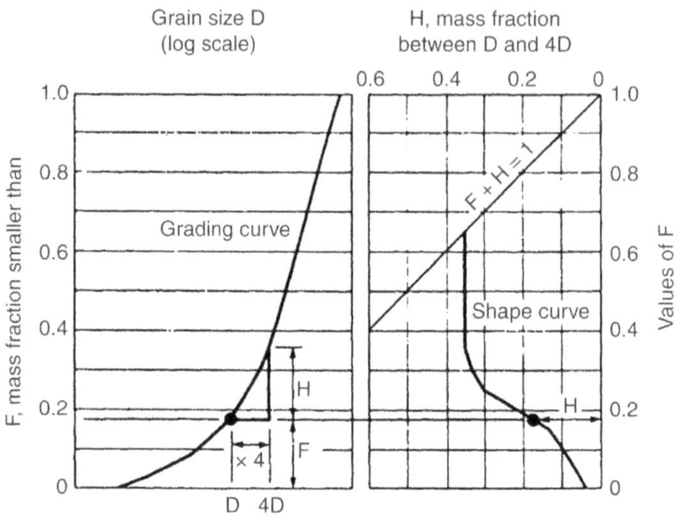

FIGURE 10.40 Method of characterising the shape of the gradation curve (Kenney and Lau, 1985). (Used with permission of *Canadian Geotechnical Journal* conveyed through Copyright Clearance Center Inc.)

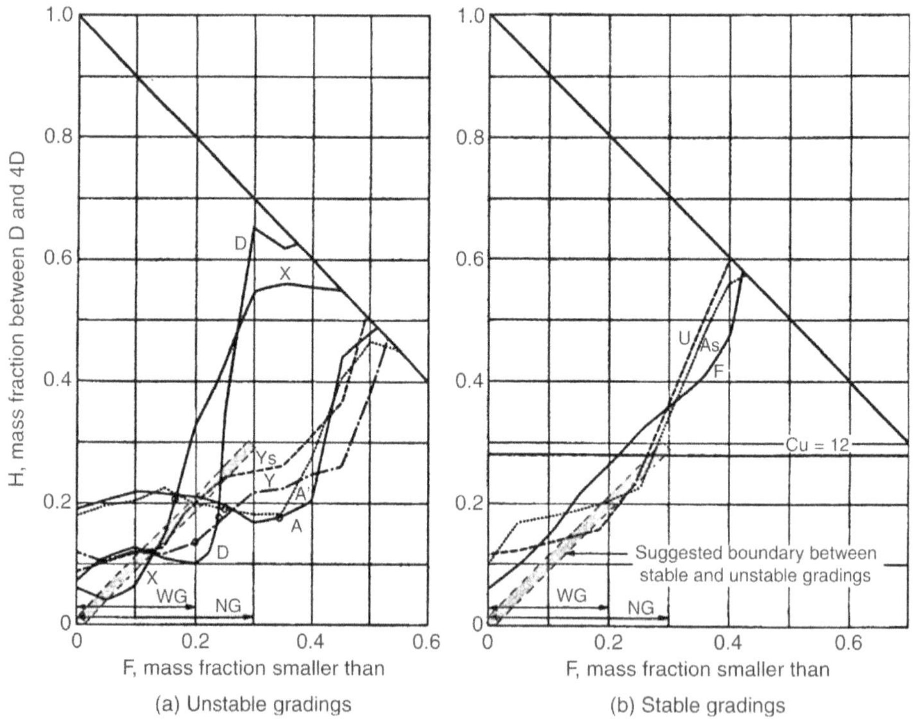

FIGURE 10.41 Method of assessing internal instability (suffusion) by Kenney and Lau (1985), with the revised criteria from Kenney and Lau (1986). LEGEND: WG, limit of internally unstable widely graded soils (Cu > 3). NG, limit of internally unstable soils narrowly graded (Cu < 3). (Used with permission of *Canadian Geotechnical Journal* conveyed through Copyright Clearance Center Inc.)

In this book the Douglas et al. (2016, 2019) method has been preferred because it has been shown to be a reasonable predictor of behaviour not only for the broadly graded silt sand soils they tested but also for those tested by Wan and Fell (2004c, 2008) and Rönnqvist (2015). Douglas et al. (2016, 2019) did not test gap-graded soils and for those soils, the Wan and Fell (2004c, 2008) method is preferred.

10.6.2 Overall Approach to Assessing the Behaviour of Non-Plastic Broadly Graded Silt Sand Gravel Soils

The steps in the assessment are:

First:

Estimate the probability of a continuous layer of soil (P_{cl}) subject to suffusion or GBE.

In the embankment: Is the water retaining zone of the embankment constructed of GW, GP and GM and SW soils with a PI ≤ 4; or are these soils gap graded and PI between 4 and 7?

If so, (P_{cl}) = 1. If not, suffusion and GBE will have a negligible likelihood of occurrence.

In the foundation: Estimate the probability (P_{cl}) there is a continuous layer of non-plastic GW, GP and GM and SW soils in the foundation from upstream of the embankment, to downstream of the embankment. In this context, "non-plastic soils" are non-plastic or Plasticity Index ≤ 4 soils. Soils which are gap graded and have a PI between 4 and 7 may be subject to suffusion or GBE.

Second:

Use the method outlined in Section 10.6.3 to assess the probability the soil will be in the following internal instability categories:

- Suffusion, Major Erosion ($P_{\text{Suffusion Major}}$).
- GBE, Major Erosion ($P_{\text{GBE Major}}$).
- GBE, Minor to Major Erosion ($P_{\text{GBE Major/Minor}}$).
- GBE, Minor, Very Minor Erosion or No Erosion ($P_{\text{GBE Minor}}$).

Allow for the range of soil gradations. Given the uncertainty in the method, and the range of gradations, in most cases there will be some probability of two or more categories.

Note that:

$$\Sigma\,[(P_{\text{suffusion Major}}) + (P_{\text{GBE Major}}) + (P_{\text{GBE Major/Minor}}) + (P_{\text{GBE minor}})] = 1.0.$$

Carry the probability of each relevant internal instability category ($P_{\text{suffusion Major}}$), ($P_{\text{GBE Major}}$) and ($P_{\text{GBE Major/Minor}}$) through the risk analysis because the potential for increased leakage flows is different for each category. $P_{\text{GBE Minor}}$ will have negligibly low probability of a failure mode developing and hence does not need to be carried through the analysis.

Third:

(a) Calculate the seepage gradient within the soil potentially subject to suffusion or GBE (i_{actual}) taking account of the zoning of the embankment and foundation soils, piezometer readings and the reservoir or river levels upstream and downstream.

(b) Estimate the seepage gradient at which suffusion or GBE will initiate (i_{initiate}) using the information in Section 10.6.4.

(c) Estimate the probability of initiation of suffusion or GBE (P_{ig}) from the ratio (i_{actual})/($i_{initiate}$) taking account of the guidance in Section 10.6.4.

Fourth:

Estimate the probability of initiation of internal erosion for each internal erosion category taking account of the probability of a continuous layer (P_{cl}); the probability of each category [($P_{suffusion\ Major}$), ($P_{GBE\ Major}$), ($P_{GBE\ Major/Minor}$)] and the probability of initiation given the seepage gradient (P_{ig}).

Fifth:

Assess the effect of suffusion or GBE on permeability, gradation and settlement behaviour using the information in Sections 10.6.5 and 10.6.7.

10.6.3 Method for Estimating the Probability the Grading of the Soil Is Subject to Suffusion and GBE and for Estimating the Amount of Erosion

The method for estimating the likelihood that a soil is subject to suffusion or GBE is described in Douglas et al. (2016, 2019).

The procedure is:

(a) Obtain representative samples of the soils in the embankment or foundation. Given the coarse nature of the soils large samples are required. These are best obtained from test pits and sampling should take account of segregation which is common in such soils. If excavating test pits is not practical, then sonic drilling using a 150 mm or 200 mm sampler should be used. Samples obtained from small diameter (e.g. 75 mm) boreholes or SPTs will not be representative as they will not sample gravel size particles. See Chapter 4 for details.
(b) Carry out particle size distribution testing using wet sieving. From these select representative gradations.
(c) Calculate the percentage by weight of each soil in the range 0.075 mm to 1.18 mm, and 1.18 mm to 4.75 mm and determine the internal instability category for each sample from Figure 10.42.
(d) To assist in distinguishing between soils which are subject to suffusion or to GBE, estimate the percentage of finer fraction using the method shown in Figure 10.5(a). For soils to be suffusive the finer fraction should be such that the voids between the coarser fraction particles are under-filled. Soils with finer fraction ≤ 30% are likely to be suffusive, and soils with finer fraction ≥ 45% are likely to have over-filled voids and to be subject to GBE or internally unstable but not subject to erosion of particles from the soil. Soils with finer fraction between 30% and 45% are likely to be subject to GBE but may or may not be subject to suffusion.
(e) From this range of results, and taking account of the % finer fraction and the uncertainty in the method as evident from the data in Figure 10.43, use engineering judgement and/or expert elicitation to estimate the probability of each internal instability category.

For important decisions regarding dam or levee safety, laboratory suffusion/GBE tests should be carried out on representative gradations. See Douglas et al. (2016, 2019) for test details. The test cylinder needs to be around 440 mm diameter and 500 mm high. Samples are around 110 kg.

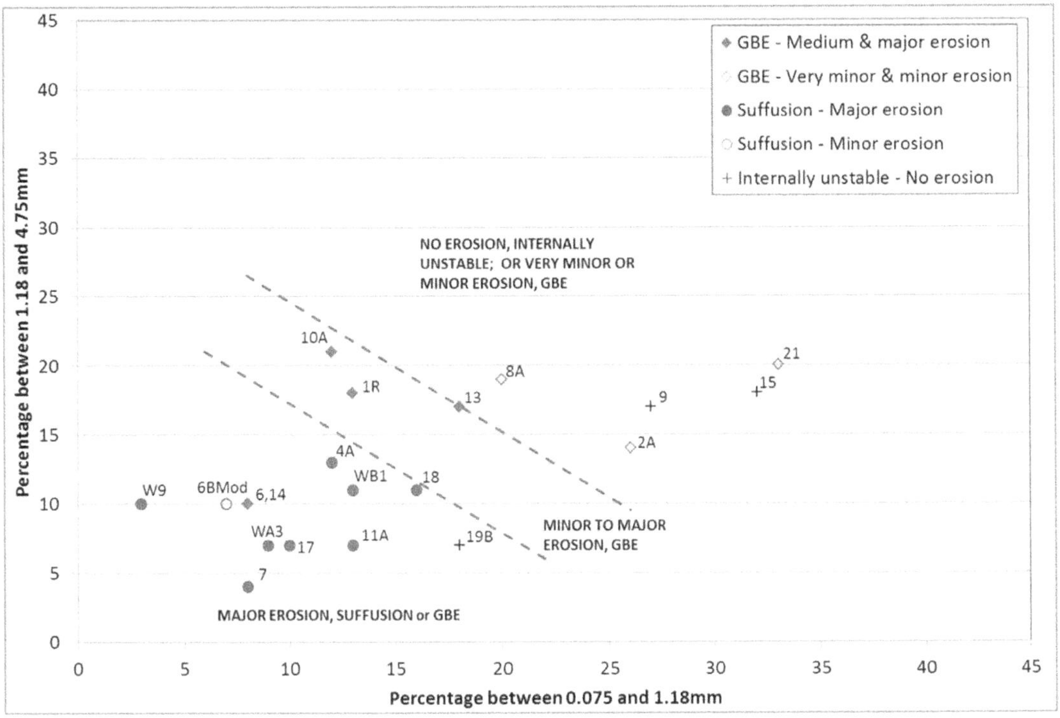

FIGURE 10.42 Suggested method for predicting internal instability category (Douglas et al., 2016, 2019).

In the laboratory testing by Douglas et al. (2016, 2019) the characteristics of the erosion mechanisms are:

Suffusion:

- Erosion throughout the sample, often with most erosion from the top layer.
- Erosion occurs at a gradient of 1 in all cases.
- Erosion occurred rapidly, usually in minutes, occasionally for hours.

Global Backward Erosion:

- Erosion throughout but most on the bottom one or two layers, or only on the bottom one or two layers.
- Erosion generally initiates at gradients 1 and continues throughout the test up to maximum gradients which vary from 2 to 6.
- Erosion occurred slowly, usually in days up to around 100 days in some tests.

Internally Unstable; No Erosion or Very Minor Erosion:

- There is no erosion or < 0.2% of the sample eroded.
- In all cases the manometers and flow rates indicated some movement of particles within the sample altering the permeability, so the soils are internally unstable.
- Depending on the test gradients of 6 or up to 10 were applied and tests had durations generally up to 100 to 200 days.

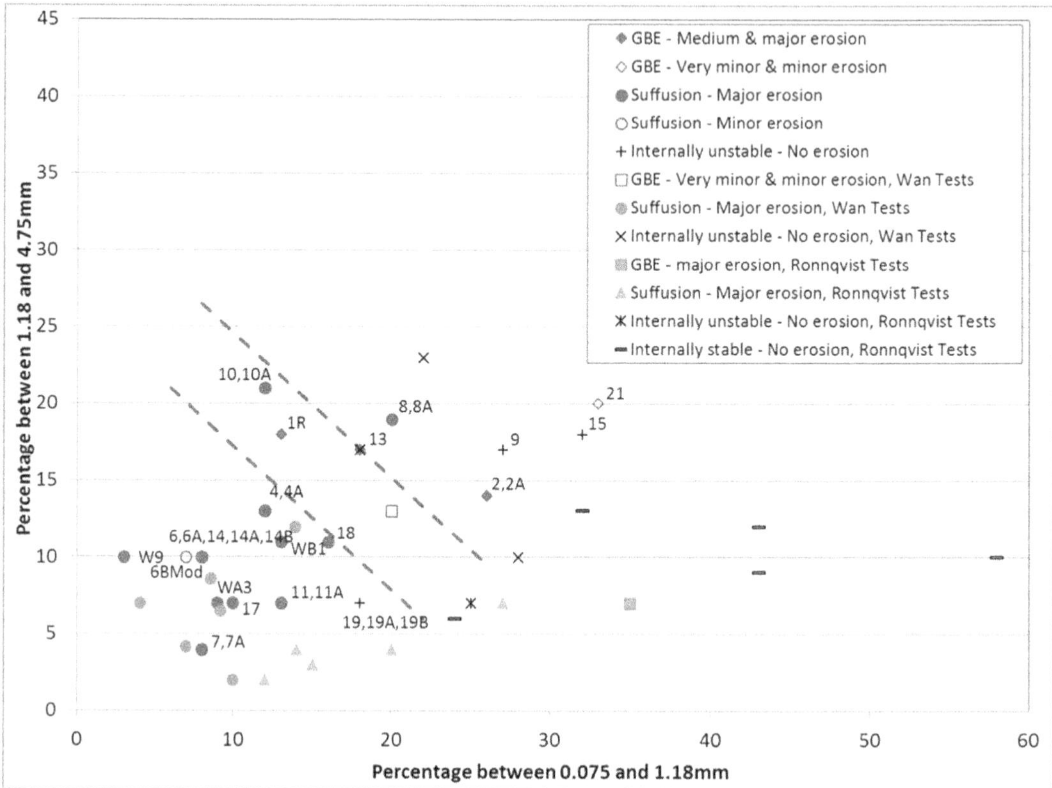

FIGURE 10.43 Suggested method for predicting internal instability categories showing Douglas et al. (2016), Wan and Fell (2004c, 2008) and Rönnqvist (2015) data (Douglas et al., 2016, 2019).

TABLE 10.9 Definition of Erosion Amounts (Douglas et al., 2016, 2019)

EROSION CLASS	*EROSION AMOUNT*
No erosion	No erosion or only a few grams falling from the mesh when first wetted.
Very minor erosion	≤ 0.2% of total sample dry weight
Minor erosion	> 0.2% and < 1% of total sample dry weight
Medium erosion	1% to 5% of total sample dry weight
Major erosion	> 5% of total sample dry weight or 1% to 5% of total sample dry weight but ≥ 10% of the dry weight of the finer fraction

These tests were vertical downward. Samples were compacted dense at around optimum moisture content. The samples were tested in the unsaturated condition.

The amount of erosion has been classified by Douglas et al. (2016, 2019) as shown in Table 10.9.

The boundary between minor erosion and medium erosion equates to the 1 gram/cm^2 limit for Excessive Erosion in the Foster and Fell (2001) classification for filters.

10.6.4 Seepage Gradients at Which Suffusion and Global Backward Erosion Occur

From the laboratory data presented in Section S9.5.3 of Fell et al. (2024), the following can be concluded. The main references upon which the conclusions are reached are shown.

10.6.4.1 For Soils Subject to Suffusion

For downward flows in unsaturated soils: Erosion occurs at a gradient of 1 (unrestricted vertical flow). Erosion occurs rapidly, usually in minutes, occasionally for hours (Douglas et al., 2016, 2019).

For downward flow in saturated soils: The gradient to initiate erosion in the downflow direction may be as low as 0.2 to 0.3 because it is expected the critical gradient will be less than for upward flow.

For upward flows in saturated soils: Particles are carried out of the soil at gradients as low as 0.2 but up to 0.7 in some soils (Wan and Fell, 2004c, 2008; Skempton and Brogan, 1994). The lower gradients are for soils with high void ratios, representing loose soils and the higher gradients for dense soils.

For horizontal flow in saturated soils: Based on the physics of the process it is expected that the gradient to transport the particles horizontally might be less than that to transport them upward vertically. We do not see test data to substantiate this judgement.

For horizontal flow in unsaturated soils: The finer fraction of the soil in the under-filled voids collapses on wetting, and open channels form in the soil in which the water flows and erodes the particles. What gradient is required to transport particles in these open channels for horizontal flow is not known, but it may be similar to that for where flows are in saturated soils as described earlier.

The authors' interpretation is that in the downward flow tests in unsaturated suffusive soils the finer fraction of the soil in the under-filled voids collapses on wetting and open channels form in the soil in which the water flows and erodes the particles. It would be expected that would also apply for horizontal flow.

It is noted that because of the collapse mechanism, the permeability is likely to increase significantly in the unsaturated soils as detailed in Section 10.6.5 even without transport of finer fraction soils from the soil.

10.6.4.2 For Soils Subject to Global Backward Erosion

For downward flow in unsaturated soils: Erosion generally initiates at gradients of 1 and continues throughout the test up to maximum gradients which varied from 2 to 6. Erosion occurs slowly, usually in days up to around 100 days in the tests. There is insufficient data available to distinguish between the behaviour of GBE with Major, Medium and Minor erosion.

For downward flow in saturated soils: It is expected the behaviour would be as for unsaturated soils as the tests in GBE soils effectively become saturated.

For upward flows: Particles begin to be carried out of the soil at gradients > 0.75 and major erosion at gradients > 1.0 which is the heave condition Wan and Fell, 2004c, 2008). However, internal movement of particles is likely to occur at lower gradients.

For horizontal flow: It is expected that behaviour will be similar to that for unsaturated soils.

10.6.4.3 For Internally Unstable Soils, No Erosion or Very Minor Erosion

Erosion initiates at gradients of 1 and continues up to the test gradients of 6 or up to 10 with durations generally up to 100 to 200 days. The behaviour is expected to be similar for all flow directions, and for both unsaturated and saturated soils.

10.6.4.4 For Soils with Some Low Plasticity Fines

There is limited data available. Wan and Fell (2004c, 2008) tested a soil which had about 5% kaolin, and another which had 11% kaolin. These exhibited minor movement of particles out of the soil at gradients around 0.17. These tests had high void ratios. No movement of particles was observed in the other soils with kaolin content.

For important decisions regarding dam or levee safety, laboratory suffusion/GBE tests should be carried out on representative soils, rather than relying on these data.

10.6.5 Effect of Suffusion and Global Backward Erosion on the Properties of the Soil

Suffusion and GBE results in erosion of some of the finer fraction of the soil. How much erosion occurs depends on the grading of the soil, the seepage gradient, duration of seepage and the nature of the exit conditions. If there is a filter or transition zone downstream, erosion may be arrested.

Erosion of some of the finer fraction affects the permeability of the soil. In most cases the permeability will increase, sometimes substantially.

Based on analysis of the laboratory tests carried out by Douglas et al. (2016, 2019) and Rönnqvist (2015) the information in Table 10.10 may be used as a guide to the effect of suffusion and GBE on permeability, where there is either an unfiltered exit to seepage, or the filter or transition is so coarse, or will hold a crack, that continuing erosion conditions exist as defined in Foster and Fell (2001) and discussed in Chapter 12.

Table 10.10 only applies to soils with gradations similar to those tested by Douglas et al. (2016, 2019) as shown in Figure 10.9 and Rönnqvist (2015).

For suffusion within filters with excessive fines, estimate the permeability before and after suffusion using the Sherard et al. (1984a) method (see Equation 9.11).

It is noted that because of the collapse mechanism, the permeability of unsaturated suffusive soils is likely to increase significantly even without transport of finer fraction soils from the soil.

For important decisions regarding dam or levee safety, laboratory suffusion/GBE tests should be carried out on representative soils, rather than relying on Table 10.10.

Douglas et al. (2016) included plots of the particle size distributions for each layer within the tests before and after testing.

Table 17a, b and c in that report summarises the amount of erosion and the % of finer fraction remaining after internal erosion. The following can be observed for the test data. Figure 10.9 of this chapter shows the Particle Size Distribution (PSD) of the test samples referred to below.

(a) The PSDs of the remaining soil are different at the top of the sample to the bottom.
(b) For suffusion, the upper layers erode the most, and for GBE the lowest layer erodes the most.
(c) For suffusion the base mesh size and resulting degree of filtering sometimes determines the PSD of the remaining soil. For example, Test 4 on the 9.5 mm mesh had only 3% finer fraction remaining whereas Test 4A on the 4.75 mm mesh had 20% remaining of the original 30%. However, there was little difference in behaviour of Tests 11 and 11A, and 7 and 7A.
(d) The largest loss of the finer fraction occurs in suffusion. For the most clearly suffusive soils (plotting near the lower part of the test data) the upper layer may lose virtually all the finer fraction; e.g. Tests 4, 7, 7A, 8 and 11.
(e) Soils subject to GBE experience significantly less change in the PSD of the remaining soil.

TABLE 10.10 A Guide to the Effect of Suffusion and Global Backward Erosion on the Permeability of Non-Plastic Gap-Graded and Broadly Graded Silt Sand Gravel Soils

INTERNAL INSTABILITY CATEGORY FROM FIGURE 9.7	*AMOUNT OF SOIL ERODED % OF TOTAL MASS*	*TYPICAL RANGE OF POST EROSION PERMEABILITIES M/SEC*	*RATIO OF POST EROSION PERMEABILITY TO INITIAL PERMEABILITY M/SEC*
Major erosion, Suffusion or GBE	5 to 20	10^{-4} to 10^{-2}	1 to 200 50% likelihood ≥ 50
Minor to Major erosion, GBE	2 to 5	10^{-6} to 10^{-4}	1 to 10
No Erosion Internally Unstable; very Minor or Minor Erosion	0 to 2	10^{-8} to 10^{-6}	1 to 5

Notes:
(a) This information applies to internal erosion for an unfiltered exit.
(b) Values shown are only applicable to soils similar to those tested by Douglas et al. (2016) and Rönnqvist (2015).
(c) More erosion, higher permeabilities and greater ratios are more likely for soils plotting towards the bottom left of Figure 9.7.

(f) Soils 6 and 8 exhibited GBE characteristics on the 4.75 mm base mesh and suffusion on the 9.5 mm base mesh. The resulting % finer fraction was not much different for soil 6, but very different for soil 8.

In Fell et al. (2008) and Fell et al. (2015), it is suggested that the PSD after suffusion can be determined using the method shown in Figure 12.4, which was reasonable for the soils tested by Wan and Fell (2004c, 2008). It assumes half the finer fraction is eroded. Douglas et al. (2016, 2019) conclude that approach is likely to underestimate the amount of finer fraction eroded for the suffusive soils they tested. They suggest that for these suffusive soils it would be better to assume all the finer fraction was eroded.

The authors conclude that for soils which plot in major erosion, suffusion or GBE in Figure 10.42, it would be appropriate to assume that all or most of the finer fraction will erode when considering the post-erosion gradation for filter capability. For soils plotting in the minor to major erosion GBE in Figure 10.42, the method shown in Figure 12.3 may be adopted and is probably on the conservative side.

For important decisions, laboratory tests should be carried out on the actual soil and filter combination in the dam.

10.6.6 Estimation of Seepage Flows Resulting from Suffusion or Global Backward Erosion

It is important to recognise that for unfiltered concentrated leak erosion, erosion will continue, and the resulting pipe will continue to increase in size. However, for suffusion and GBE, erosion will self-limit, leaving behind most of the original soil. The effect of suffusion and GBE is therefore to have increased seepage flows which may or may not result in breach of the embankment.

To estimate the potential seepage flows, the data in Table 10.10, along with data on the as-constructed permeability of the embankment or *in situ* permeability of the foundation, may be used to estimate the permeability of the eroded soil. To estimate flows those carrying out the analyses will need to estimate the area of the embankment or foundation which may be subject to suffusion or GBE, the seepage gradient and the permeability. This is likely to involve judgement, and the results may be uncertain. It may be necessary to carry forward that uncertainty in assessing the safety of the dam and/or the risk analysis.

10.6.7 Estimation of the Amount of Settlement Resulting from Suffusion or Global Backward Erosion

Suffusion or GBE can lead to settlement of the soil material due to loss of the unstable fraction. This can lead to settlement at the crest of an embankment dam and potential PFM if the settlement exceeds the available freeboard (for example during a flood surcharge event). Douglas et al. (2016, 2019) found that there was no consistent trend in the amount of settlement in the laboratory specimens. There was a more consistent trend of the amount of erosion as shown in Table 10.10.

The authors have no field case data for the settlement resulting from suffusion or GBE. The erosion amounts observed in the laboratory specimens after testing shown in Table 10.10 can be used as a guide to assess the potential magnitude of settlement that could develop at the embankment crest if suffusion/GBE occurs within the embankment core zone or within the foundation. This should consider the thickness of the soil unit that is postulated to be subject to suffusion or GBE in the PFM description. Settlement is likely to be less for suffusion because the coarse matrix supports the applied stresses, whereas for GBE the applied stress is carried throughout the soil and therefore settlement is likely to be greater because of erosion of the finer fraction and movement to redistribute stresses.

Engineering judgement and/or expert elicitation can be used to estimate the probability of the estimated crest settlement being sufficient to cause overtopping of the crest and subsequent failure of the embankment.

Consideration should be given to the possibility of voids forming due to formation of a roof near the surface, which could then collapse leading to a rapid settlement mechanism.

Contact Erosion and Other Mechanisms of Erosion into Soil and Rock Foundations

11

11.1 INTRODUCTION

The potential failure modes discussed in this chapter are:

Contact Erosion of the soil in the embankment core resulting from flow of water in a coarse soil such as gravel (IM 28) or in open joints in rock (IM 29) which are in contact with the core.
Contact Erosion of a finer soil stratum in the foundation resulting from flow of water in a coarse soil such as gravel or in open joints in rock which are in contact with the finer soil stratum.
Concentrated Leak Erosion Resulting From Hydraulic Fracture in a narrow cutoff trench with erosion into open joints in the sides of the cutoff trench (IM 30).
Backward Erosion Piping or Suffusion of soil in a cutoff trench with erosion into open joints or coarse soil in the sides of the cutoff trench (IM 31).

These are not common failure modes and for most well-constructed dams have a low likelihood of occurrence.

The text is written around quantitative risk methods. Those carrying out deterministic assessments of dam safety will find that the quantitative methods are not difficult to use and will serve to aid their engineering judgement.

11.2 CONTACT EROSION IN SOILS

Contact erosion occurs where a coarse soil such as gravel is in contact with a fine soil, and flow parallel to the contact in the coarse soil erodes the fine soil as shown in Figure 2.2.

For example, flow through gravel alluvium in the foundations of a dam or dike may erode the base of an overlying silt layer, or erosion of the finer layers of soil in a core may occur into a coarse gravelly layer formed by segregation during construction. Figure 11.1 shows examples.

DOI: 10.1201/9781003542025-11

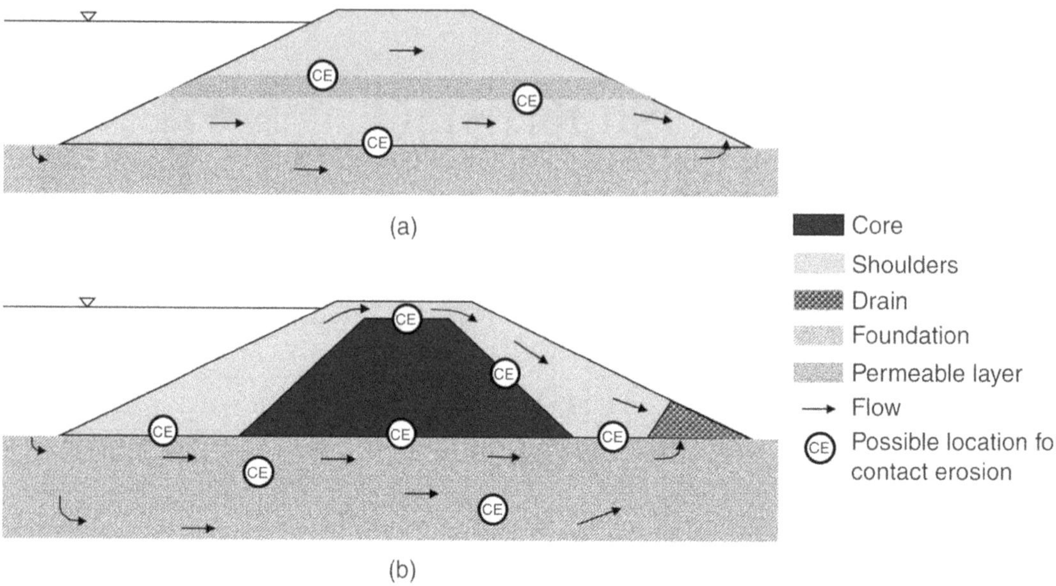

FIGURE 11.1 Examples of where contact erosion may occur. (a) Homogeneous dam with layered fill due to segregation during construction and a coarse foundation soil. (b) Zoned dam with potential for contact erosion at high reservoir levels above the core and for erosion into coarse layers in the foundation (Beguin et al., 2009). (Reproduced with permission of R. Beguin.)

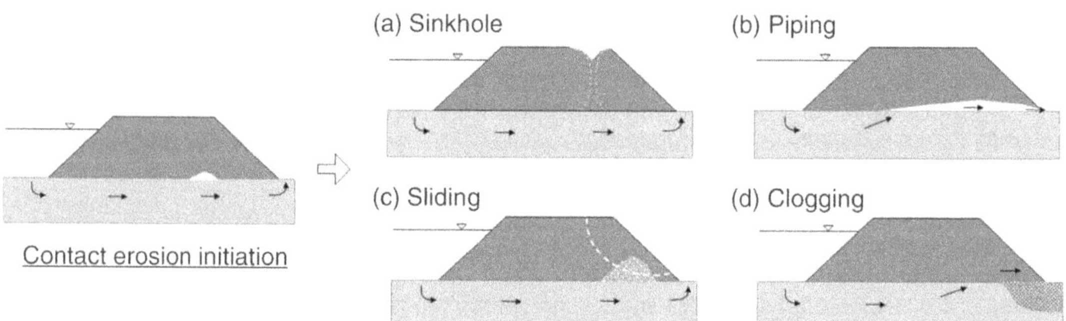

FIGURE 11.2 Potential consequences of contact erosion. Black arrows indicate a groundwater flow through a more permeable layer (light grey) under a less permeable dam (dark grey). (a) sinkhole daylighting, (b) beginning of backward erosion piping, (c) creation of a weaker zone initiating instability and (d) clogging of the permeable layer and increase of pore water pressure (Beguin, 2011). (Reproduced with permission of R. Beguin.)

Figure 11.2 shows some of the potential consequences of contact erosion.

The following is taken from ICOLD (2017):

> Locations with possible high velocities in the coarser layer and high particle size grading contrast between layers are the most likely to experience contact erosion. These characteristics often occur at the interface between the core and a gravelly foundation. The process starts with contact erosion at the interface between the silt or sand and the gravel, often at the contact between the fill and the foundation or in the foundation. This causes a cavity within the fill, and then the pressure drops around that cavity, causing the roof collapse. Those materials fill in the lower part of the cavity and enlarge it at the top; then the roof of the new cavity is decompressed and, step by step, the cavity progresses along

a practically vertical chimney towards the crest or the upstream face [Figure 11.2a]. Leakage flows decrease after the materials fall, and then progressively build up again, carrying away the fallen material until the next occurrence. In most cases experienced in dikes along the Rhone Valley (CFGB, 1997) the sinkhole volume is about 1 to 2 m^3.

Other consequences of contact erosion are possible. First, depending on the mechanical properties of the core, the cavity created by contact erosion may not collapse and can be a beginning for backward erosion piping [Figure 11.2b]. This process has been identified in hydraulic flume tests at scale 1/1, where it appeared that the downstream part of the opening was held open by hydraulic fracture. Second, a weaker zone, less dense because of the development of contact erosion, can cause a loss of stability [Figure 11.2c]. Finally, the eroded fine particles may clog the permeable layer and increase pore water pressure which may result in instability of the downstream slope [Figure 11.2d].

11.3 IM 28 METHOD FOR PREDICTING INITIATION AND PROGRESSION OF CONTACT EROSION IN SOILS

11.3.1 Introduction

Three conditions are needed for initiation of contact erosion in soils:

Presence of a continuous coarse strata in contact with a fine stratum: For contact erosion to occur there needs to be a persistent coarse stratum from upstream to downstream in contact with fine strata.

Geometrical condition: Pores of the coarse layer must be sufficiently large to allow particles from the finer soil to pass through.

Hydraulic condition: Flow velocity in the coarse layer must be sufficient to detach the particles and also to transport them.

If these conditions are satisfied erosion will progress unless arrested by filters which intercept the coarse strata.

11.3.2 Estimation of the Probability (P_{SS}) of the Presence of Continuous Coarse Strata in Contact with Fine Stratum

Estimate the probability (P_{SS}) of the presence of a coarse stratum extending from upstream to downstream in the foundation in contact with a fine stratum in the foundation or in contact with the embankment taking account of the site investigations information, geotechnical model, the depositional environment, and the nature and geometry of the cutoff for the embankment.

11.3.3 Estimation of the Probability ($P_{Contact}$) of the Geometrical Condition for Contact Erosion Being Satisfied

Table 11.1 is derived from Brauns (1985), Wörman and Olafsdottir (1992) and Den Adel et al. (1994). They found that for D_{15}/d_{85} ratios less than in the third column, there is geometrical filtration whatever the hydraulic loading, so contact erosion could not occur. For D_{15}/d_{85} ratios more than in the fifth column, hydraulic loading controls erosion and there is no filtration effect.

The second boundary between "Geometrical and hydraulic condition" and "Hydraulic condition", defines if the coarse layer grading has, or does not have, an influence on the hydraulic criteria for erosion initiation. In practice it seems to not be significant as the hydraulic condition will control.

It is notable that the fifth column ratios are likely to be close to the criteria for continuing erosion $D_{15F/}D_{95B} < 9$ proposed by Foster (1999), Foster and Fell (1999a, 2001). It seems therefore that contact erosion will only occur where the coarse strata can provide limited or no filtering capability against the finer soil.

Beguin et al. (2015) conclude that one condition determining whether a levee may fail by contact erosion is that there is what they termed a continuing erosion condition in which $9\ d_{85core} < d_{15\ Foundation}$. This is the Foster and Fell (1999a, 2001) Excessive Erosion criteria for soils with $D_{95B} > 2$ mm and % passing 0.075 mm < 15%.

The criteria from Table 11.1 indicate that contact erosion may occur at $D_{15}/d_{85} < 7.5$ and 8.1, so somewhat finer than the Excessive Erosion criteria. However, given that the coarse and fine layers are likely to have a range of gradations, it is convenient to assess the likelihood of no, Excessive and Continuing Erosion as detailed in Section 12.2.4 and assume that contact erosion will occur if the Excessive Erosion criteria are met. If the filter characteristics fall just less than those criteria, a probability of contact erosion geometric criteria being satisfied can be assigned by judgement. This is the approach which has been adopted for this document.

For this document it is recommended that the following approach be followed:

(a) Assess the probability of no (P_{NE}), some (P_{SE}), excessive (P_{EE}) and continuing erosion (P_{CE}) between the fine soil and the coarse soil as detailed in Section 12.2.4.
(b) Assume that contact erosion will occur ($P_{Contact} = 1$) if the coarse strata is coarser than Excessive Erosion or Continuing Erosion criteria to the finer soil.
(c) If the filter characteristics between the coarser and finer soil fall in the Some Erosion criteria, assign a probability of contact erosion geometric criteria being satisfied (P_{CSome}) by judgement.
(d) If the filter characteristics fall below the some or No Erosion criteria, assume the probability of contact erosion is negligible ($P_{CNE} = 0$).
(e) Estimate the total probability of geometrical conditions for contact erosion being met from

$$P_{Contact} = [(P_{SE}) \times (P_{CSome})] + [P_{EE}] + [P_{CE}] \tag{11.1}$$

TABLE 11.1 Domain of Geometrical and Hydraulic Influence (Beguin, 2011; reproduced with permission of R. Beguin)

<table>
<tr><td></td><td colspan="5">Grading Ratio D_{15} / d_{85} →</td></tr>
<tr><td>Brauns (1985)
Soil with n = 0.4</td><td rowspan="3">Geometrical condition</td><td>7.5</td><td rowspan="3">Geometrical and hydraulic condition</td><td>25</td><td rowspan="3">Hydraulic condition</td></tr>
<tr><td>Wörman and Olafsdottir (1992)
Soil with $D_{15} = 0.88\ D_H$</td><td></td><td>14.6</td></tr>
<tr><td>Den Adel et al. (1994)
Soil with $d_{85} = d_{50}/0.9$</td><td>8.1</td><td>11.7</td></tr>
</table>

Note: D_{15} is the particle size of the coarser soil for which 15% is finer; d_{85} is the particle size of the finer soil for which 85% is finer; n is porosity.

11.3.4 Estimation of the Probability (P_{HC}) of the Hydraulic Conditions for Contact Erosion Being Satisfied

The following approach is recommended:

For Non-Plastic Soils above and below Coarse Strata

(a) *Estimate the Darcy velocity.*
Estimate the median and range of permeability (hydraulic conductivity) of the coarse strata using the results of *in situ* tests and estimated from the gradation of the soils using the Sherard et al. (1984a) method (see equation 9.11).
Estimate the median and range of the Darcy velocity (= k i where k = hydraulic conductivity, and i = seepage gradient) in the coarse strata for the reservoir level (or river level for levees and dikes) under consideration.

(b) *Estimate the critical Darcy velocity.*
For non-plastic soils beneath the coarse strata use Figure 11.3 to estimate the best estimate and range of critical Darcy velocity from the d_{50} size of the fine soil. Allow for the range of d_{50} sizes and the range in data in Figure 11.3.

For Non-Plastic Silts over the Coarse Strata

Assume that based on the data in Figure 11.3 the critical Darcy velocity is between 0.01 and 0.05 m/sec.

(c) *Estimate the best estimate and range of probabilities of the hydraulic conditions for contact erosion being satisfied (P_{HC})* from the estimated Darcy velocities and critical Darcy velocities.

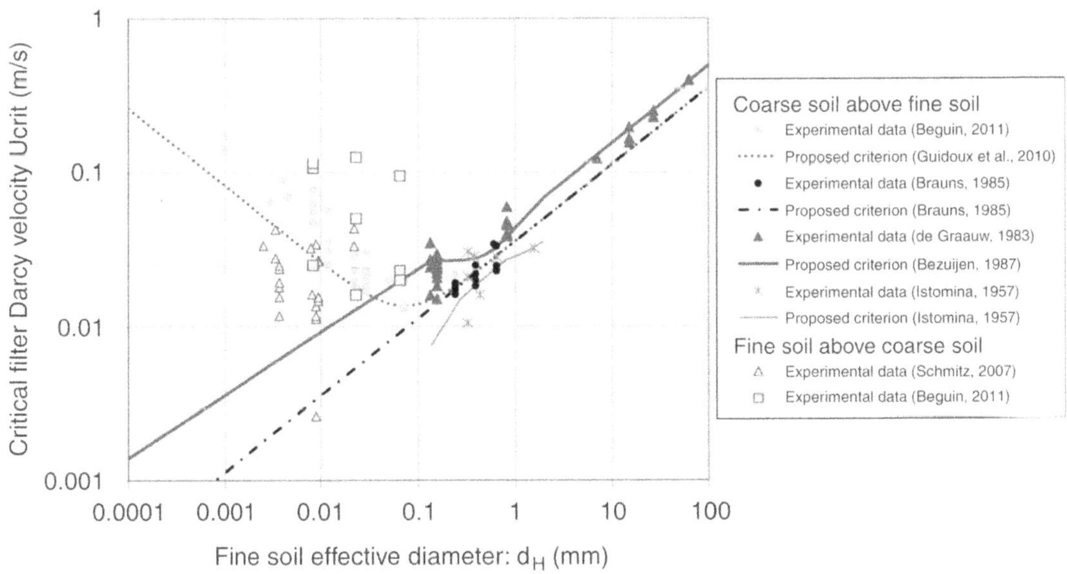

FIGURE 11.3 Critical velocities for contact erosion of soil above and below gravel (Beguin, 2011). (Reproduced with permission of R. Beguin.)

For Plastic Soils above and below the Coarse Strata

(a) *Calculate the hydraulic shear stress on the surface of the fine soil (τ)* from Equation 11.2 (Guidoux et al., 2010):

$$\tau = 0.102\,\rho_w g\, i\, d_H \left(\frac{n_F}{1-n_F}\right) \tag{11.2}$$

where:

ρ_w = unit weight of water
g = acceleration due to gravity = 9.8 kN/m^3
d_H = effective grain diameter of the coarse soil (m)
i = seepage gradient in the coarse soil
n_F = porosity of the coarse soil.

The effective grain size can be estimated from:

$$d_H = \left(\sum_{j=1}^{m} \frac{F_j}{d_j}\right)^{-1} \tag{11.3}$$

where:

F_j = fraction of the total sample mass of the fraction j of the coarse soil
d_j = particle size of the fraction j of the coarse soil.
Note: For uniformly graded coarse soil, d_H approx. = d_{50}.

(b) *Estimate the critical shear stress for the fine soil* as described in Section 9.13.6.
(c) *Estimate the best estimate and range of probabilities of the conditions for contact erosion being satisfied (P_{HC})* from the estimated hydraulic shear stresses and critical shear stresses using engineering judgement.

11.3.5 Estimation of the Probability Contact Erosion in Soil Will Initiate

The probability that contact erosion (P_{CEI}) will initiate is:

$$P_{CEI} = (P_{SS}) \times (P_{Contact}) \times (P_{HC}) \tag{11.4}$$

Alternatively, this may be modelled in an event tree with separate nodes for each conditional probability.

11.4 IM 29 CONTACT EROSION INTO OPEN DEFECTS IN A ROCK FOUNDATION

11.4.1 Introduction

Contact erosion of the core soil into open defects in a rock foundation occurs by water flowing in the open defects applying hydraulic shear stresses on the soil at the contact which exceed the critical shear stress of the soil. The eroded particles are initially transported within the open defects but as erosion progresses a gap or pipe develops in the core at the contact. Contact erosion into open defects in rock is also referred to by some as scour. Figure 11.4 shows the process.

The open defects may be related to joints, bedding partings, shears, faults and solution features. However, in view of the complex nature of solution features such as karst limestone, they are not included in this document. They should be considered on an individual basis by an experienced risk analysis team familiar with karst.

11.4.2 The Steps to Be Followed to Estimate the Probability of Initiation of Contact Erosion into Open Defects in the Foundation

The steps in the assessment are as follows:

(a) Estimate the probability of a persistent pathway of open defects for erosion at the core–foundation contact (P_{CR}) as described in Section 11.4.3.
(b) Estimate the probability of treatment of the core foundation not preventing contact of the core with open defects (P_{TI}) as described in Section 11.4.4.
(c) Estimate the probability of grouting not being effective in cutting off open or infilled defects (P_{GNE}) as described in Section 11.4.5.
(d) Estimate the probability of initiation of erosion of the core material at the core–foundation contact (P_{ic}) using the method for erosion in a crack in the core as described in Section 11.4.6.
(e) Estimate the probability that the geometric condition for contact erosion to progress is met. That is, whether the soil particles will be sufficiently fine to be transported within the open defect (P_{TR}) estimated using the method described in Section 12.2.6.
(f) Estimate the probability of initiation of contact erosion P_{CEI} from:

$$P_{CEI} = (P_{CR}) \times (P_{TI}) \times (P_{GNE}) \times (P_{ic}) \times (P_{TR}). \tag{11.5}$$

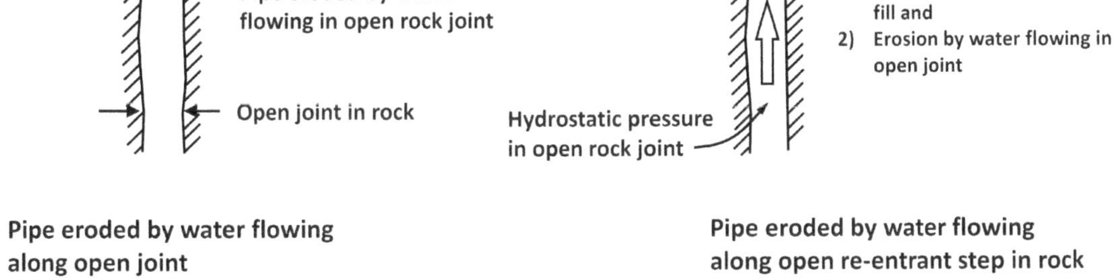

FIGURE 11.4 Contact erosion (or scour) of the core of a dam into an open joint in the foundation (adapted from Independent Panel, 1976).

Do this for each of the potential defect openings and representative soil critical shear stresses and particle size distributions, and take these estimates forward into the assessment of continuation, progression, detection, intervention and repair, and breach.

11.4.3 Estimation of the Probability of a Persistent Pathway of Open Defects in the Rock Foundation (P_{CR})

11.4.3.1 General Approach

For contact erosion to initiate, there must be a persistent open defect or pathway from upstream of the core to downstream of the core.

The probability such a defect or set of defects is present should be assessed from:

(a) The geology and geotechnical model of the foundation. This should include stratigraphy, structure, defect nature, orientation and persistence.
(b) Logs and photographs of cored boreholes and results of water pressure (Lugeon) tests in the foundation.
(c) Mapping and photographs taken during construction and construction reports.
(d) Grouting records, in particular the presence and location of large grout take, which are likely related to open defects.
(e) Records from piezometers in the foundation.
(f) Inspection and geological mapping of exposures of the rock in the river and abutments.

From these determine the spatial distribution of the open defects, the width of opening, persistence and depth. Also assess the presence of infill in the defects. Infill is soil which has been transported into open defects. Having been washed in, it is likely to be readily washed out by water flowing in open parts of the infilled defect.

It should be expected that there will be different conditions in the river section and in each abutment.

In doing this assessment take account of the geological environment and the effects of valley stress relief and monitoring of seepage flows. As described later, valley stress relief features have a major influence on the likelihood of persistent open defects.

The probability of the presence of open defects should be estimated by engineering judgement/expert elicitation with the dam safety assessment team including an experienced engineering geologist and/or geotechnical engineer with dam experience and an understanding of the effects of valley stress relief.

It is best to do this assessment considering different widths of open defects at the core–foundation contact. This will result in a set of probabilities for the different width open defects, e.g. 2 mm to 5 mm, 5 mm to 25 mm, 25 mm to 100 mm, $(PCR)_{2-5}$, $(PCR)_{5-25}$, $(PCR)_{25-100}$ etc.

Those estimating the probabilities must keep in mind that for contact erosion to initiate, the defects must be open and persistent the complete distance from upstream to downstream of the core at the core–foundation contact.

It is useful to consider first if there are potentially persistent defects, and second if they are open. For example, a fault or shear zone may be persistent, over many metres, but it may be seen in all mapping that the defects may be closed. By comparison, joints may be open but not persistent.

If exposures of the foundation show no open cracks or only a few cracks of limited persistence, it is sometimes useful to ask, "Given the lack of evidence of persistent open cracks, what mechanism could result in persistent open cracks the full width of the core foundation?"

For defects such as joints ≤ 5 mm wide, it may be difficult to ascertain if they are present and if so, open and persistent, unless there is very detailed mapping, photography and written records from construction. In many cases such defects may be present but not persistent from upstream to downstream of the core. Wider cracks are more likely to be persistent.

11.4.3.2 Likelihood of Persistent Valley Stress Relief Joints

As shown in Figures 11.5 and 11.6, these defects consisting of open joints in the abutments and open bedding surfaces in the valley floor may be present in the valley slopes and in the valley floor. It is known that interbedded rocks with different stiffnesses such as interbedded shale and sandstone, commonly

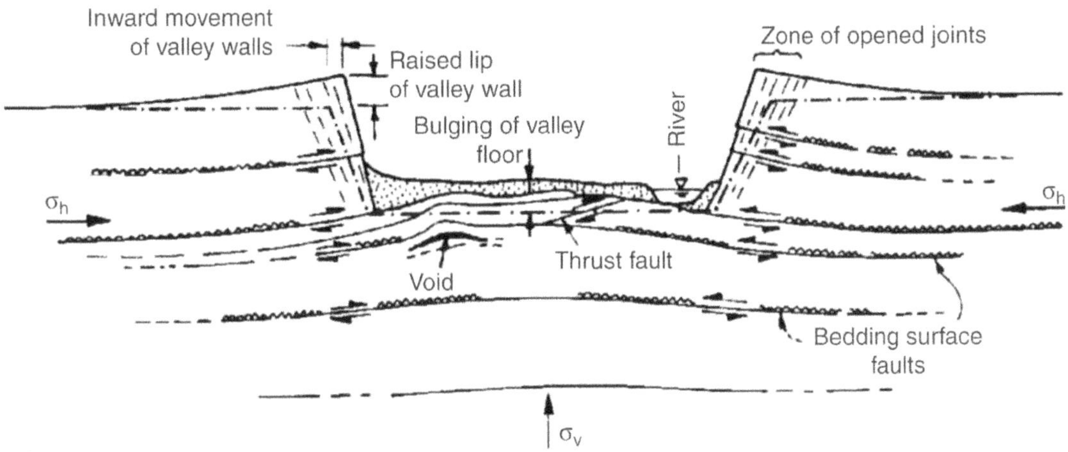

FIGURE 11.5 Valley structures related to stress release in flat-lying rocks (Fell et al., 2015).

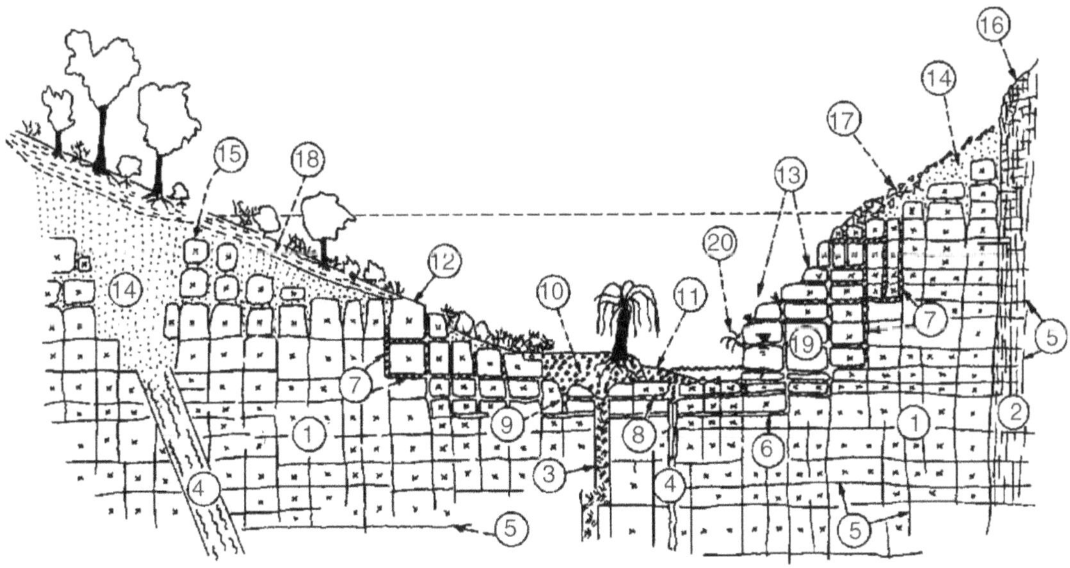

FIGURE 11.6 Diagrammatic cross section through the floor and lowest slopes of a steep-sided river valley, showing the types of features developed by mechanical and chemical weathering processes (Fell et al., 2015, 2024).

LEGEND: Granite (1), microgranite (2), basalt (3) in the form of an intrusive dike, faults (4), joints at depth (5) are mainly tightly closed, but as the valley floor is approached, they are either slightly open (6) due to de-stressing, or else (7) have been open but are now partly or wholly infilled with clay which has migrated down from the chemically weathered zone. In the bedrock just below the river, some joints (8) have become infilled with alluvial silt which has migrated down from the river. (See Fell et al., 2024, Supporting Document Figure S11.12 for explanation of the other numbers.)

result in open joints due to differential strains due to stress relief on valley formation. These strains are often concentrated in bedding surface shears in the weaker rock. Stress relief effects are almost always present, even in dam sites with a wide valley floor and moderate valley wall height and slopes.

Table 11.2 lists some factors to consider in estimation of the likelihood of a persistent infilled or open defect in the rock in the foundation beneath the embankment cutoff.

The dam safety assessment team should estimate the probability for each width defect. The probabilities for each width are independent of each other because all defect sizes may be present. The probabilities may total > 1.0 because they are not conditional probabilities.

TABLE 11.2 Some Geological and Site Investigations and Construction Record Factors Influencing the Likelihood of Persistent Infilled or Open Valley Stress Relief Defects in the Foundation Beneath the Embankment (Fell et al., 2024)

	INFLUENCE ON LIKELIHOOD			
FACTOR	*LESS LIKELY*	*NEUTRAL*	*MORE LIKELY*	*MUCH MORE LIKELY*
Geological environment likely to give open or infilled persistent valley stress relief features	Shale not containing other rocks (sub-horizontal bedded).	Schistose, steeply dipping.	Granite, basalt rhyolite, ignimbrite[d]. Schistose parallel to abutment. Thinly interbedded sedimentary.	Sedimentary—interbedded thin beds of mud rocks and thick beds of sandstone or conglomerate or limestone. Volcanics with interbedded thin beds of tuff.
Spatial continuity of the open or infilled defects[e] in the foundation [c]	Data shows no open or infilled defects aligning spatially in upstream–downstream direction.	Data shows isolated and/or non-persistent open or infilled defects aligning in upstream–downstream direction, or a tortuous path.	Data shows some features or single persistent open or infilled defects partially aligning upstream–downstream direction.	Data shows many or a single open or infilled persistent defect spatially aligning upstream–downstream.
Site investigation and construction data indicating open or infilled defects of this size are/are not present	Good quality site investigation and construction data indicates open or infilled defects of this size are very unlikely to be present.	Site investigation and construction data indicates circumstantial evidence of open or infilled defects of this size are unlikely to be present.	Site investigation and construction data[b] indicates circumstantial evidence open or infilled defects of this size are present.	Good quality site investigation and construction data[a] indicates open or infilled defects of this size are present.

Notes:

(a) For example, records/photographs from excavations, tunnels, downhole imaging, high water losses in drilling, very high grout takes, drill rods dropping.

(b) For example, high water losses in drilling, high leakage rate in dam foundation, very high grout takes, drill rods dropping, low piezometric levels in the abutments.

(c) Data would include the information noted in Notes (a) and (b). Spatial alignment is likely to be able to be related to geologic features such as stress relief joints, shears, etc.

(d) Some basalts (e.g. if interbedded with weak tuff), and some ignimbrites may classify as "Much more likely".

(e) Defects are stress relief joints, sheet joints, and bedding surface partings.

The potential for blasting of the rock foundation for the cutoff or, for example, to form a trench into which the diversion conduit is placed to lead to open defects should be assessed. Construction photographs are the best guide as to whether such features may exist and their likely continuity and opening.

See Fell et al. (2015, 2024) for more information and examples of stress relief features.

11.4.3.3 Likelihood of Persistent Infilled or Open Stress Relief Sheet Joints

Sheet joints occur in the valley sides and sometimes in the valley floor. Figure 11.7 shows the mechanism which leads to their formation.

Sheet joints may be superimposed upon defect sets which have formed from regional stress fields and other valley stress relief defects.

The major variables affecting the likelihood of these being in the dam foundation are:

(a) *Topography.* Sheet joints are most likely to occur in steep sided deep valleys. Table 11.3 is a guide to the influence of topography on the presence of sheet joints.
(b) *Geological environment.* Stress relief defects are common, as are massive rocks such as granite. However, they may occur in other geologic environments.

The site investigation and construction record factors relevant to the likelihood of the presence of infilled or open sheet joints are as detailed in Table 11.2.

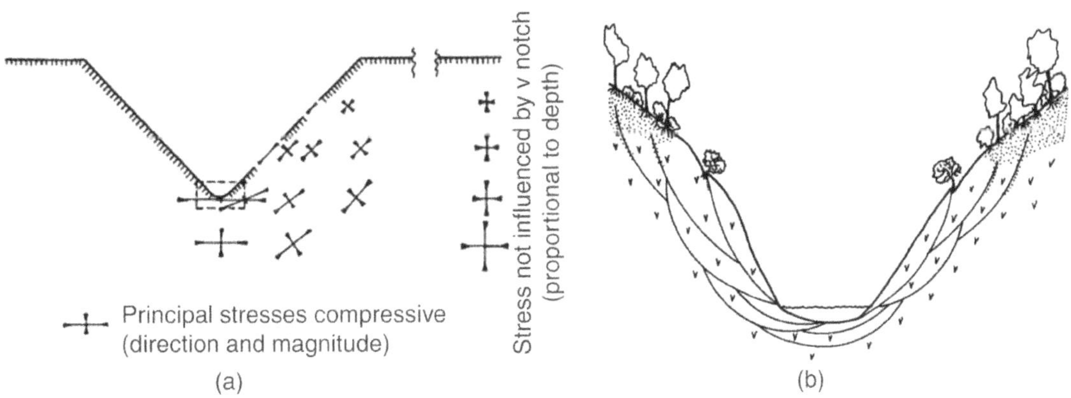

FIGURE 11.7 The mechanism of formation of stress relief sheet joints in valley floor and sides (Fell et al., 2015).

TABLE 11.3 Topographic Factors Influencing the Likelihood of Sheet Joints (Fell et al., 2024)

	AVERAGE ABUTMENT SLOPE (DEGREES)			
VALLEY DEPTH	< 20	20–45	45–60	> 60
< 30 m	LL	LL	N	ML
30–100 m	LL	N	ML	MML
100–300 m	LL	ML	MML	MML
> 300 m	N	ML	MML	MML

Note: LL = less likely; N = neutral; ML = more likely; MML = much more likely.

11.4.4 Estimation of the Probability That Treatment of the Embankment Cutoff Foundation Does Not Prevent Contact of the Core with Open Defects (P_{TI})

If open joints in the core foundation are covered with concrete, shotcrete or slush concrete, that will protect the core from contact erosion. Details of these methods are given in Fell et al. (2005, 2015). How effective this has been depends on the specified requirements and the quality of supervision and inspection during construction. There will be a greater confidence that treatment has been effective if the design and construction organisations are known to have generally followed good practice and there are construction records including photographs and mapping.

The dam safety assessment team should assemble the relevant data and assess the likelihood of treatment not being effective using engineering judgement.

Table 11.4 may be used to assess the probability that foundation treatment, such as treatment of the base and sides of the cutoff trench (or core–foundation contact) by shotcrete, concrete or slush grouting,

TABLE 11.4 Probability (P_{TI}) of Treatment of the Cutoff Foundation Not Preventing Contact of the Core with Open Defects or Solution Features (Fell et al., 2024)

SCENARIOS	*PROBABILITY OF THE TREATMENT NOT PREVENTING CONTACT EROSION (P_{TI})*
Well-documented evidence that there was no treatment of the cutoff foundation.	1.0
No construction records available, the design and construction organisation not known or known but likely to have not paid much attention to inspecting foundations and carrying out surface treatment.	0.1 to 0.5
No construction records available, but knowledge that the practice of the design and construction authority was to inspect foundations and carry out surface treatment.	0.05 to 0.25
Knowledge that the practice of the design and construction authority was to inspect foundations and carry out surface treatment. Recorded evidence that the foundations were mapped, but not in detail. Some evidence that open and/or infilled defects and solution features were covered with concrete or shotcrete.	0.01 to 0.05
Knowledge that the practice of the design and construction authority was to inspect foundations and carry out surface treatment. Well-documented evidence that the foundations were carefully mapped, and all open and infilled defects and solution features were covered with at least 100 mm of concrete or good quality shotcrete, or that they were cleaned out to at least three times the surface width and treated with slush grout.	0.001 to 0.005(2)

Notes:

(1) It is generally more likely that wide open defects, e.g. ≥ 25 mm, will have been treated than narrow open defects, e.g. ≤ 5 mm.

(2) In some cases, it will become apparent that there is little or no likelihood of persistent open defects being present in contact with the core, and the probability of contact erosion into the foundation may be assessed as negligible.

fails to prevent contact of the core with open defects or solution features. Do this for each defect width, for the base and the sides of the cutoff trench, and for the various parts of the foundation, e.g. each abutment and the river section.

11.4.5 Estimation of the Probability of Grouting Not Being Effective in Cutting Off Open or Infilled Defects (P_{GNE})

A well designed and constructed grout cap and grout curtain fills open cracks in the rock and reduces the likelihood of persistent open cracks and the seepage gradient along the cracks.

Fell et al. (2005, 2015) describe good grouting practices. These include:

(a) Orienting grout holes to intersect the open defects (joints, bedding partings, faults and shears).
(b) Adequate grout closure methods, with hole spacing and depth sufficient to allow grouting of the open defects.
(c) Using grout pressures which do not result in hydraulic fracture.

Where there are piezometers in the foundation upstream and downstream of the grout curtain, and these show a large head loss, then this is a guide to the grouting being effective. Conversely if there are no head loss and high pressures downstream of the grout curtain, this is a guide to indicate the grout curtain may not be effective,

Evaluation of the likelihood of grouting not being effective in cutting off the potential open or infilled defect or solution features is as follows:

(a) For open defects, assess the likelihood of grouting not being effective using Table 11.5 and Figure 11.8 to assist in assessing the probability. Note that it may be reasonable to assign different probabilities for defects 2 mm to 5 mm, 5 mm to 25 mm, and 25 mm to 100 mm in width.
(b) For infilled defects, grouting is likely to be ineffective, and it is suggested to assign a probability for grouting not being effective – 1.0.
(c) For no grouting, probability for grouting not being effective = 1.0.

11.4.6 Estimation of the Probability of Initiation of Contact Erosion (P_{IC})

The procedure is:

(a) Estimate the hydraulic shear stress in the open crack assuming conduit flow conditions as described in Section 9.13.2. The hydraulic shear stresses imposed on the core by the water flowing in the open joints is the same as those for an equivalent joint width.

    The hydraulic gradient used in the assessment should be based on the estimated average seepage gradient in the open joint on the core–foundation contact. Account for grouting of the foundation and any piezometric data in estimating the seepage gradients.
(b) Estimate the critical shear stress of the soil in the core at the contact between the core and the foundation as described in Section 9.13.6. For many dams the contact soil is placed at a moisture content wet of standard optimum moisture content and is likely to have further saturated from seepage flow. This should be considered when estimating the critical shear stress. Also take account of the dispersion properties of the soil and the chemistry of the seepage water.

TABLE 11.5 Factors Influencing the Likelihood of Grouting Not Being Effective for Persistent Open Defects and Solution Features (Fell et al., 2024)

FACTOR	*RELATIVE IMPORTANCE FACTOR (RF)*	*LIKELIHOOD FACTOR (LF)* *LESS LIKELY (1)*	*NEUTRAL (2)*	*MORE LIKELY (3)*	*MUCH MORE LIKELY (4)*
Orientation of grout holes compared with the open defects[(a)]	(3)	Grout holes at a wide angle to the dip of the open defect (45 to 90 degrees)	Grout holes at a medium angle to dip of open defects (30 to 45 degrees)	Grout holes at an acute angle to dip of open defects (10 to 30 degrees)	Grout holes parallel or near parallel to dip of open defects (< 10 degrees)
Quality of grouting (closure, number of lines, spacing, grout takes, W/C ratio)	(2)	Three or more lines of grout curtain, with primary holes 6 m or less spacing, secondary, tertiary etc. holes to close to < 10 lugeons or < 25 kg cement/metre take, W/C ratio < 3	Single line grout curtain with primary holes 6 m or less spacing, secondary, tertiary etc. holes to close to < 10 lugeons or < 25 kg cement/metre take, W/C ratio < 3	Single line curtain 5 m to 6 m spacing, no check of closure, high (> 5:1) W/C ratio	Single line curtain, > 10 m spacing, single stage, no secondary holes to check closure, high (> 10:1) W/C ratio
Performance (pore pressures and leakage)[(b)]	(1)	Significant reduction in foundation pore pressures across the grout curtain	Moderate reduction in pore pressures across grout curtain	Minor reduction in pore pressures across grout curtain	No or very little reduction in pore pressures across grout curtain
		$\Delta h_p/h_p > 60\%$	$30\% < \Delta h_p/h_p \leq 60\%$	$10\% < \Delta h_p/h_p \leq 30\%$	$\Delta h_p/h_p \leq 10\%$
		Very low leakage in the foundation	Low leakage in the foundation	Moderate to high leakage in the foundation	High leakage in the foundation OR No performance data available

Notes:

(a) The dip of the open defect will depend on the type of defect:

- For stress relief defects in the valley wall in interbedded sedimentary rock, these defects are likely to be near vertical and parallel to the valley walls (refer to Figure 11.5). For stress relief defects associated with massive rocks, these are likely to be parallel to the valley slopes (refer to Figures 11.6 and 11.7).
- Defects associated with valley bulge are likely to be horizontal or near horizontal.
- Solution features are more likely to develop along the predominant defects sets.

(b) Refer to Figure 11.9 for definition of Δh_p and h_p. When making this assessment take account of the amount and quality of the instrumentation and the duration and range of reservoir levels of the observations.

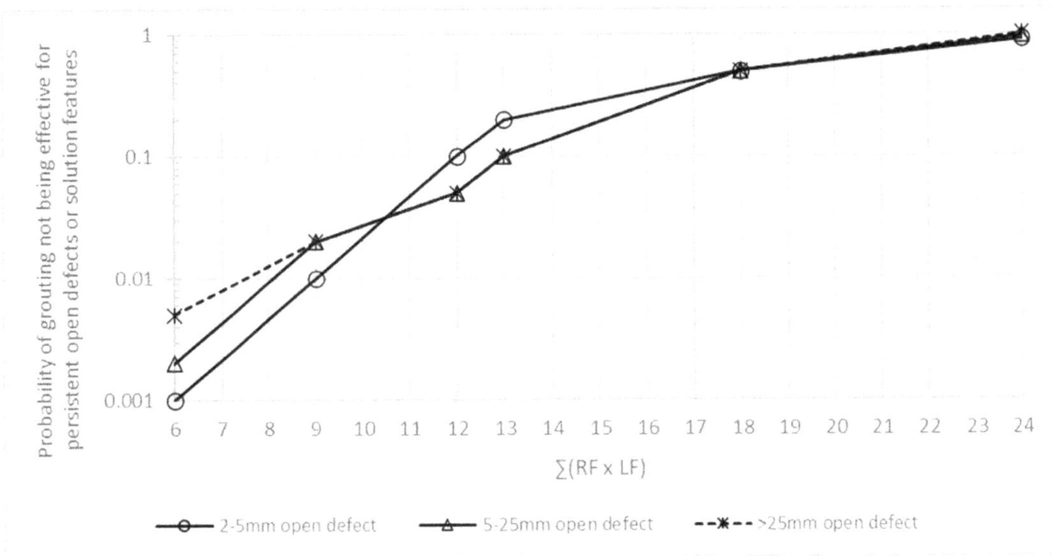

FIGURE 11.8 Probability (P_{GNE}) of grouting not being effective for persistent open defects or solution features versus Σ (Relative importance factor (RF)) × (Likelihood factor (LF)).

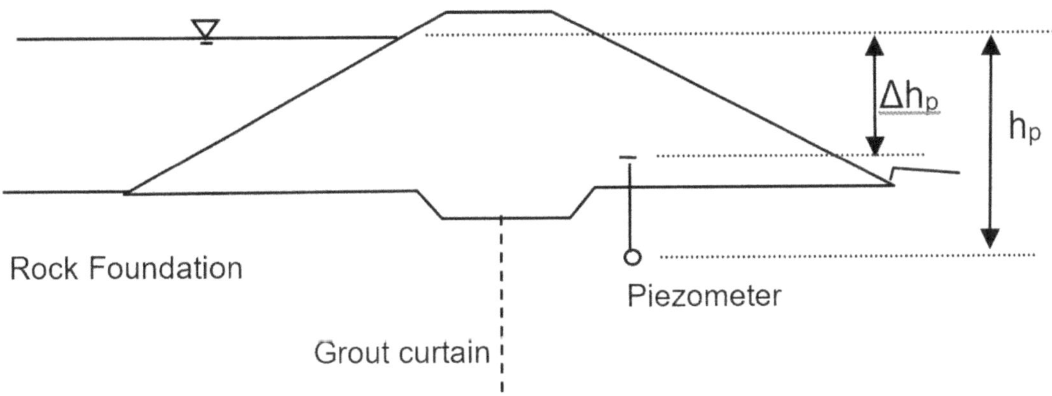

FIGURE 11.9 Definition of Δh_p and h_p (Fell et al., 2024).

(c) Estimate the probability of initiation of contact erosion (P_{IC}) by comparing the hydraulic shear stress to the critical shear stress which will initiate erosion.

These estimates are for each reservoir stage being considered, taking account of the geometry of the core of the embankment and the assumed open joint dimensions and location relative to the reservoir surface.

There is always some uncertainty regarding the input parameters, and this should be allowed for in estimating whether erosion will initiate. There should be checks on the sensitivity of the outcome to the assumptions made.

The prior discussion is modelled around assessing the probability of defects being present in the foundation which are persistent from upstream to downstream of the core and are a constant width. There may be foundations where the information available from construction and site investigations indicates that there will be persistent open defects in the foundation, but they are of varying width. Figure 11.10 shows some examples of this.

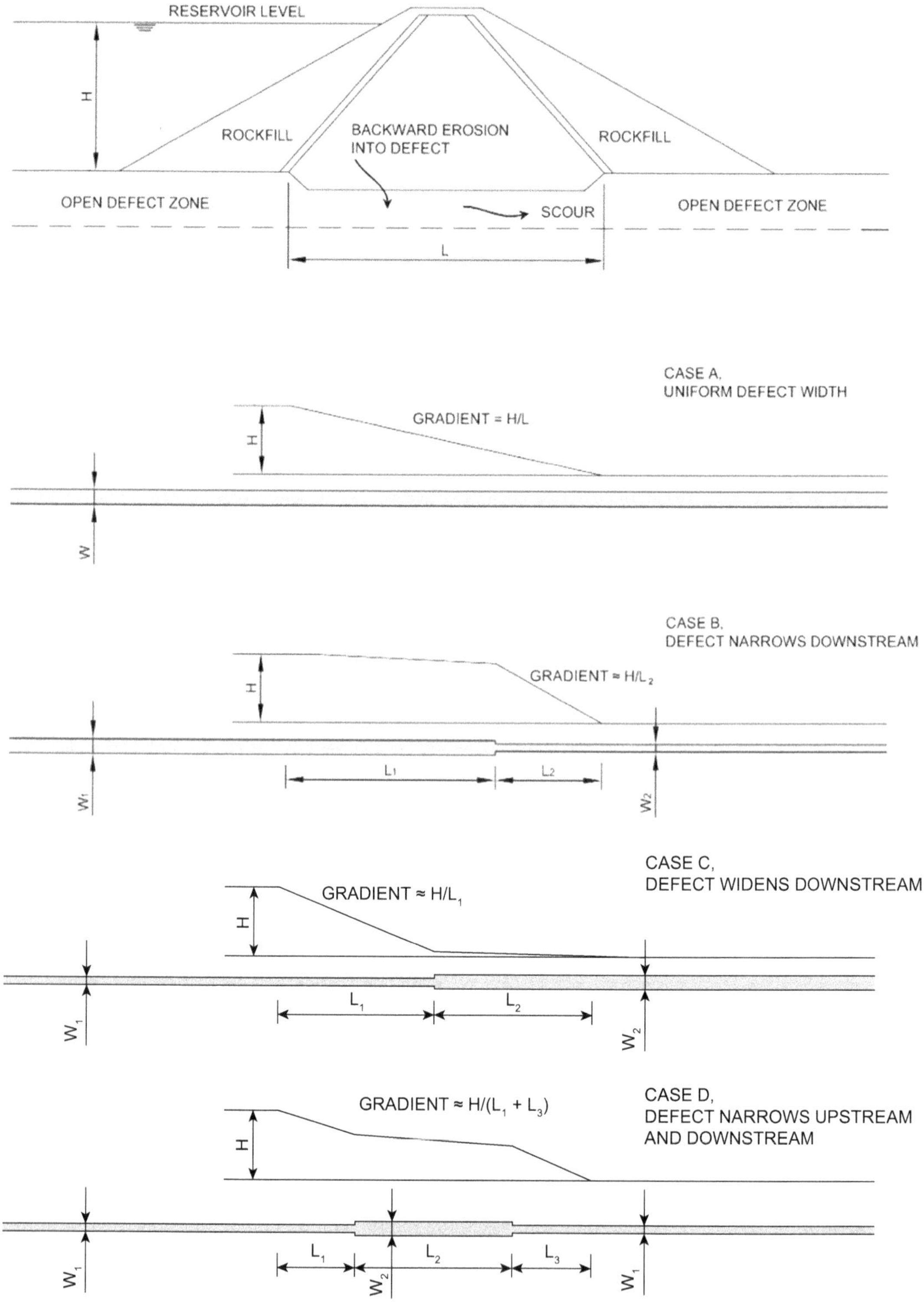

FIGURE 11.10 Examples of foundations with persistent open defects of varying width (Fell et al., 2008).

As shown in Fell et al. (2008), the hydraulic gradient is $\propto (1/\phi^3)$ whereas the imposed hydraulic shear stress is $\propto \phi$, so the gradient is most affected by the defect width or diameter. Hence in the situations shown in Figure 11.10, erosion will initiate in the narrow defects where gradients are high in preference to the wider sections of defect. Within the accuracy of the calculations, it is sufficient to assume that all the head loss occurs in the narrow defects.

11.5 IM 30 CONCENTRATED LEAK EROSION RESULTING FROM HYDRAULIC FRACTURE IN A NARROW CUTOFF TRENCH WITH EROSION INTO OPEN JOINTS IN THE SIDES OF THE CUTOFF TRENCH

11.5.1 The Mechanics of the Process

Figure 11.11 shows a schematic of the process. Arching of the core in the cutoff reduces the total vertical stress in the soil in the cutoff trench. If this is less than the water pressure from the reservoir at that location, hydraulic fracture may occur.

Arching action is more likely to occur if the sides of the cutoff trench are steep, ≥ 60°, the width of the base of the trench ≤ 0.4 trench depth, and the soil in the cutoff trench is not well compacted and is subject to collapse settlement on saturation.

Once hydraulic fracture occurs, concentrated leak erosion may initiate in the opening formed. For the eroded soil to be transported downstream, the hydraulic fracture must exit into an open defect such as a vertical joint.

Hydraulic fracture is more likely to occur on first filling or during large rises in the reservoir level above the pool of record. It is unlikely to occur under normal operating reservoir levels.

The failure of Teton Dam is considered to have occurred because of this failure mode (Independent Panel, 1976).

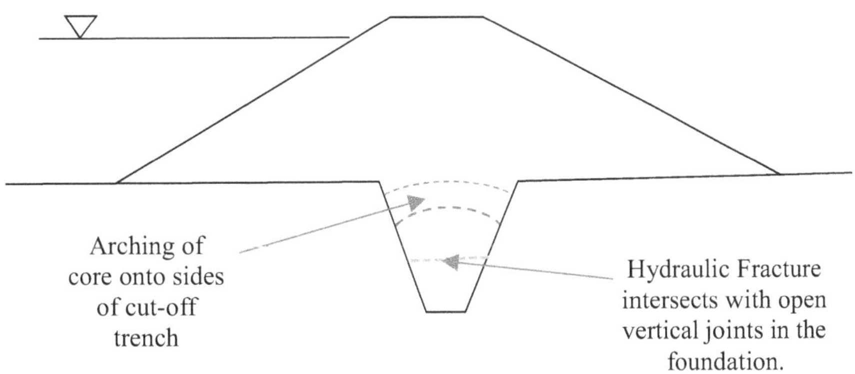

FIGURE 11.11 Schematic of concentrated leak erosion into open joints in the sides of a cutoff trench (Fell et al., 2024).

11.5.2 Estimation of the Probability of Concentrated Leak Erosion in a Narrow Cutoff Trench

The steps in the assessment are as follows:

Estimate the probability and width of open defects that were left untreated at the downstream core–foundation contact in the cutoff trench (P_{path}) from Sections 11.4.3 and 11.4.4. Note that the defects of relevance are above the base of the cutoff so are likely to be wider than the defects at the base of the cutoff. The hydraulic fracture in the cutoff trench needs to coincide with an open joint or other defect to find an unfiltered exit and hence the assessed probability should consider the likelihood of this coincidence. Coincidence is more likely for vertical defects than for horizontal defects.

Estimate the probability of a hydraulic fracture occurring across the cutoff trench (P_{hf}) using Figures 11.12 and 11.13 and Table 11.6.

Estimate the width of the hydraulic fracture taking into consideration the potential for collapse settlement of the soil in the cutoff trench using the methods in Sections 9.5.2 and 9.5.4 or for arching using the method in Section 9.11.2.

Estimate the probability of initiation of erosion through the hydraulic fracture across the cutoff trench (P_{ic}) using the method for erosion in a crack in the core as described in Section 9.13. The hydraulic gradient used in the assessment should be based on the estimated head loss across the cutoff trench.

Estimate the probability that soil particles will be sufficiently fine to be transported within the open defect (P_{TR}) using the method described in Section 12.4.

The probability of initiation of internal erosion by hydraulic fracture across the cutoff trench

$$= P_{path} \times P_{hf} \times P_{ic} \times P_{TR} \tag{11.6}$$

Do this for each of the potential defect openings and soil particle sizes and take these estimates forward into the assessment of continuation, progression, detection, intervention and repair, and breach.

If the embankment is well instrumented, the head loss will be determined from piezometers upstream and downstream of the cutoff trench. Otherwise, the head loss should be estimated using flow nets modelling the embankment and foundation.

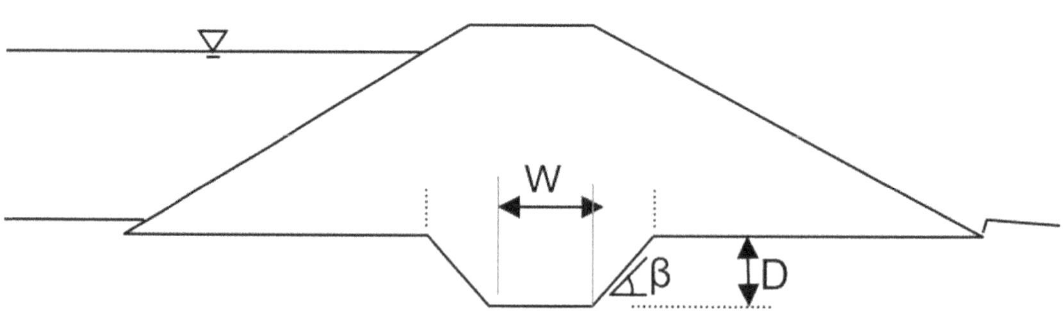

FIGURE 11.12 Definition of terms for arching across a cutoff trench.

TABLE 11.6 Factors Influencing the Likelihood of Hydraulic Fracturing within the Cutoff Trench Due to Arching (IM 30)

FACTOR	*RELATIVE IMPORTANCE FACTOR (RF)*	*LIKELIHOOD FACTOR (LF)*			
		LESS LIKELY (1)	*NEUTRAL (2)*	*MORE LIKELY (3)*	*MUCH MORE LIKELY (4)*
Slope of side slopes of cutoff trench[a]	(3)	$\beta \leq 45°$	$45° < \beta \leq 60°$	$60° < \beta \leq 75°$	$\beta > 75°$
Geometry of cutoff trench[a]	(2)	W/D > 0.75	0.4 < W/D ≤ 0.75	0.25 < W/D ≤ 0.4	W/D ≤ 0.25
Degree of compaction of the soil in the cutoff trench[b]	(2)	Well compacted to e.g. 98% standard MDD, moisture content 1% dry of optimum to 1% wet of standard OWC	Compacted to e.g. 95% standard MDD, moisture content 2% to 3% dry of standard OWC	Layers poorly compacted, dry of standard optimum moisture content e.g. < 93% standard MDD, 2% to 3% dry of standard OWC	Layers very poorly compacted, dry of standard optimum moisture content e.g. < 90% standard MDD, 3% dry of standard OWC
Depth of cutoff trench	(1)	D ≤ 2 m	2 m < D ≤ 4 m	4 m < D ≤ 6 m	> 6 m

Notes:
(a) See Figure 11.12 for definitions of W, D and β.
(b) See Table 9.13 for additional descriptors of the compaction.

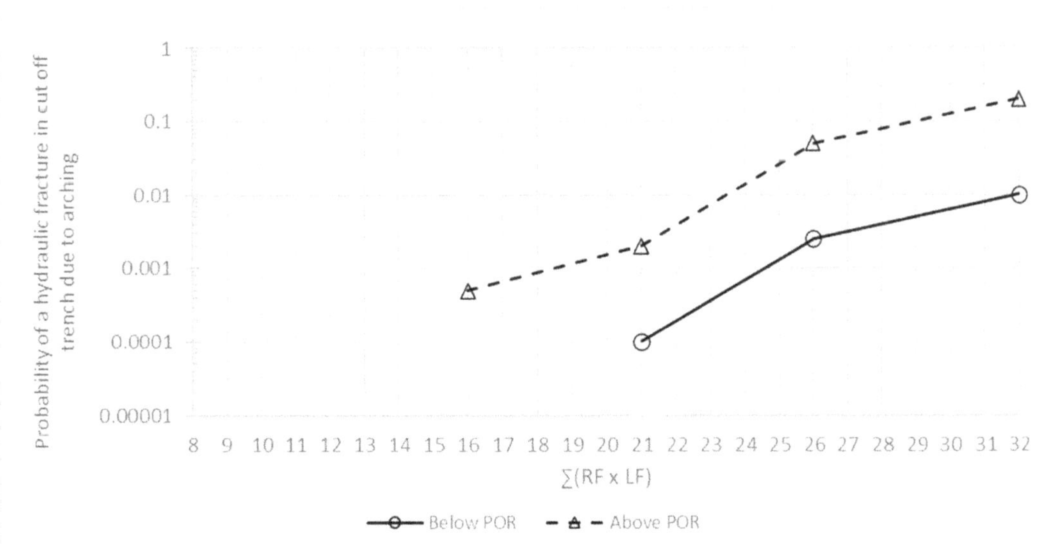

FIGURE 11.13 Probability of hydraulic fracturing in cut off trench due to arching versus Σ (Relative importance factor [RF]) × (Likelihood Factor [LF]).

11.6 IM 31 GLOBAL BACKWARD EROSION OR SUFFUSION OF SOIL IN A CUTOFF TRENCH INTO OPEN JOINTS OR COARSE SOIL IN THE SIDES OF THE CUTOFF TRENCH

11.6.1 Introduction

This mechanism considers the potential of initiation of internal erosion by global backward erosion (GBE) or suffusion due to seepage through the core exiting into an open defect in a rock or a coarse soil foundation. Figure 11.14 is a schematic of the process.

The mechanism only applies to non-plastic silt sand gravel soils. Plastic soils, such as clays, sandy clays and clayey sands commonly used in dam construction are not subject to GBE and suffusion, so this mechanism is not applicable for embankment cores constructed of those soils.

See Chapter 10 for a description of GBE and suffusion and methods to assess whether GBE or suffusion will initiate.

It is important to note that the effects of GBE and suffusion are to erode some of the finer fraction from the silt sand gravel. The effect is to increase the permeability, not to form a pipe. How much soil erodes will be influenced by the width of the open defect or the grading of the coarse soil.

11.6.2 Estimation of the Probability of Global Backward Erosion or Suffusion into Open Defects or Coarse Soil in the Sides of the Cutoff Trench

The steps in the assessment are as follows:

(a) Estimate the probability of the presence of untreated open defects or coarse soil in the sides of the trench which would allow erosion into the foundation (P_{path}) from Sections 11.4.3 and 11.4.4.

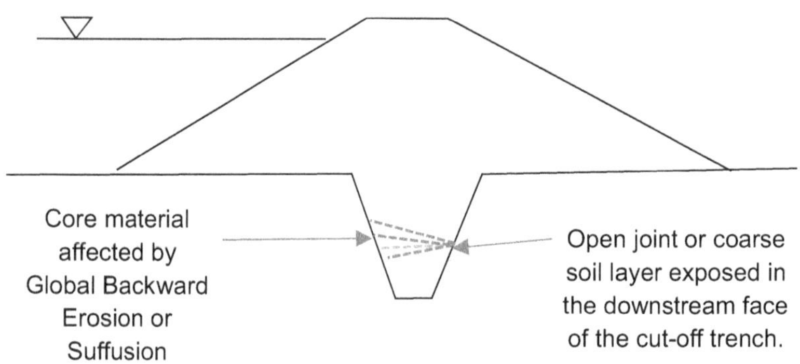

FIGURE 11.14 Schematic global backward erosion or suffusion into open joints or coarse soil in the sides of a cutoff trench.

(b) Estimate the probability that the soil in the core is subject to suffusion or GBE ($P_{SUF,}$ P_{GBE}) using the methods described in Sections 10.6.2 and 10.6.3.
(c) Estimate the probability of initiation of suffusion or GBE using Section 10.6.4 and the actual seepage gradient (P_{IPSUF}, P_{IPGBE}).

The probability of initiation of internal erosion by suffusion = $P_{path} \times P_{SUF} \times P_{IPSUF}$ (11.7)

The probability of initiation of internal erosion by GBE = $P_{path} \times P_{GBE} \times P_{IPGBE}$ (11.8)

(d) Estimate the probability that soil particles will be sufficiently fine to be transported within the open defect or coarse soil (P_{TR}) using the method described in Sections 12.4.
(e) Estimate the effect of GBE or suffusion on the properties of the soil using Sections 10.6.5 and 10.6.6, noting these processes would not lead to an open pipe but would leave a more permeable zone of material along the seepage pathway.

Do this for each of the potential defect openings and coarse soil particle sizes and take these estimates forward into the assessment of continuation, progression, detection, intervention and repair, and breach.

Seepage gradients across the cutoff trench should be estimated allowing for the effects of grouting.

12 Internal Erosion and Piping—Filters and Continuation of Internal Erosion

12.1 GENERAL CONCEPTS AND OVERALL APPROACH

Filters are provided in embankment dams to arrest internal erosion which has initiated in the embankment, foundation and embankment to foundation. That is, the filters are provided to stop erosion continuing.

Many existing dams have no filters or have filters and/or transition zones which are too coarse to satisfy no-erosion filter criteria such as those developed by Sherard and Dunnigan (1989). Foster (1999) and Foster and Fell (1999a, 2001) showed that many such filters, after some erosion of the soil being filtered, will eventually seal with medium size particles eroded from the protected soil so preventing the breach of the dam.

There seems to have been a lack of awareness in the 1950s and 1960s of the need for good filters near the crest of the dam, so the quality control was sometimes relaxed. Some dams even had the filters omitted near the crest (even though sometimes they were shown on drawings). Figure 3.8 shows examples of this.

This chapter describes the methods which are available to allow these scenarios to be assessed. The methods are best used in a risk-based framework in which case the whole of the process—initiation, continuation, progression to form a pipe and breach—is considered. This is discussed in Chapter 14. The text is therefore written around quantitative risk methods. Those carrying out deterministic assessments of dam safety will find that the quantitative methods are not difficult to use and will serve to aid their engineering judgement.

The starting point for assessing existing dams for internal erosion and piping is to determine, as reliably as is practical, the zoning of the embankment, what materials were used to build the dam and their properties, the geology of the foundation and the properties of the materials in the foundation.

The authors' experience is that it is not sufficient to rely on design drawings and the specification and assume those who built the dam followed the specification. Experience with Australian dams constructed in the 1960s and 1970s is that they were often constructed with filters not satisfying the specification. In particular, as-constructed fines contents were often greater than the 5% maximum specified, and sometimes the fines were plastic despite the specification requiring non-plastic materials. These filters are potentially susceptible to holding open a crack and being ineffective as a filter.

It is important therefore to locate and use construction quality control testing rather than relying on specified properties.

DOI: 10.1201/9781003542025-12

Where important decisions on dam safety and upgrades are being made, if that data is not available, test pits or boreholes should be used to obtain samples for testing. Particle size distributions of samples from boreholes may not be properly representative because of fines generated in drilling (e.g. from sonic drilling), or the maximum size particles being limited by sampling methods (e.g. SPT sampler is 35 mm diameter). See Chapter 4 for details of suitable drilling and sampling methods.

12.1.1 Notation

The notation used in this book is described in the following examples:

D_{15F}	Particle size of the filter material for which 15% by weight is finer.
D_{85B}	Particle size of the base material for which 85% by weight is finer.
pp% 2.36 mm	Percent finer than a particle size of 2.36 mm.
Fines content	Percent finer than 0.075 mm = pp% 0.075 mm.

Figure 12.1 shows particle size distributions for two base soils, A and B, and a Zone 2A filter. For broadly graded soils, such as soil B, which have some medium and coarse gravel, USBR (2011), USDA-SCS (1994), Sherard and Dunnigan (1989) and other methods use an adjusted particle size distribution. As shown in Figure 12.1, this involves taking the particle size distribution and adjusting it to what it would be if only the fraction passing the 4.75 mm sieve were used.

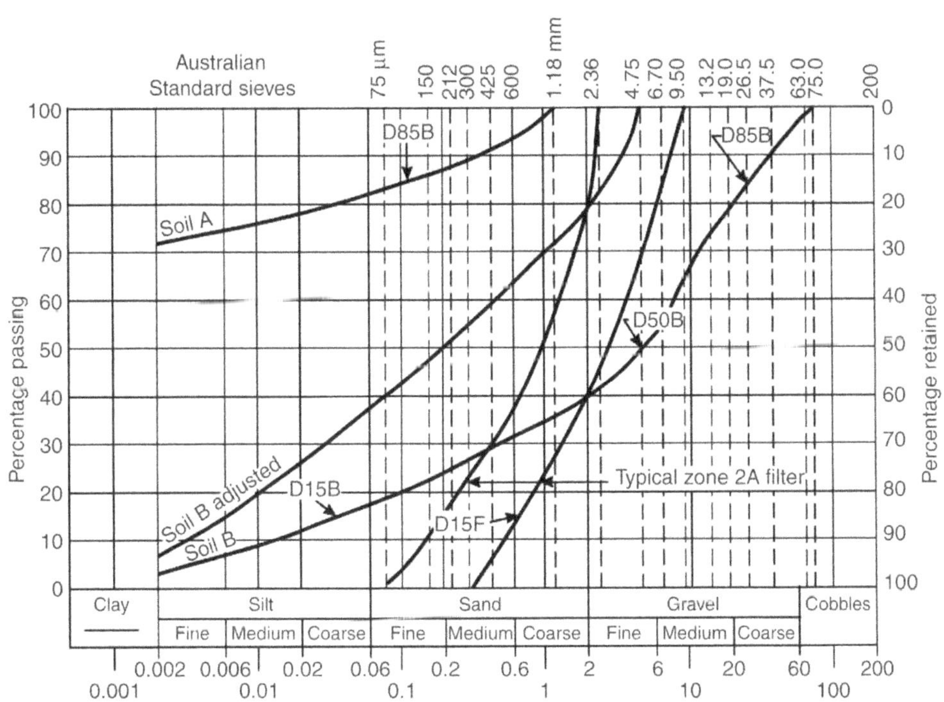

FIGURE 12.1 Filter design notation and adjustment of the particle size distribution for gravelly base soils (Fell et al., 2015).

12.2 PROBABILITY OF CONTINUATION FOR INTERNAL EROSION IN THE EMBANKMENT

12.2.1 Assessment of the Relevant Scenario

Step 1: Assess which of the following four scenarios is most applicable to the dam section and failure mode that is under consideration:

Scenario 1: Homogeneous zoning with no fully intercepting filter or transition.
Scenario 2: Zoned embankment. No filter or transition zone.
Scenario 3: Zoned embankment. Filter or transition zone is present downstream of the core.
Scenario 4: Erosion into an open defect, joint or crack in the foundation, in a wall or conduit.

The example sketches shown in Table 12.1 can be used to help evaluate the most applicable scenario.

Step 2:

For Scenarios 1 and 2: Estimate the probability for a crack persisting through the embankment and giving Continuing Erosion based on the guidance given in the applicable section (Sections 12.2.2 or 12.2.3).

For Scenario 3: A four-way split for filtering behaviour is recommended in the event trees. This is based on Foster and Fell (1999a, 2001):

Seals with No Erosion—the filtering material stops erosion with no or very little erosion of the material it is protecting. The temporary increase in leakage flows is so small that it is unlikely to be detectable. The No Erosion branch on the event tree is a "No Breach" branch.

Seals with Some Erosion—the filtering materials initially allow erosion from the soil it is protecting, but it eventually seals up and stops erosion. Leakage flows due to piping can be up to 100 l/sec but are self-healing.

Seals with Excessive Erosion—the filter material allows erosion from the material it is protecting, and in the process permits large increases in leakage flow (up to 1000 l/sec), but the flows are self-healing.

Continuing Erosion—the filtering material is too coarse to stop erosion of the material it is protecting and continuing erosion is permitted. For concentrated leak erosion unlimited erosion and leakage flows are likely. For suffusion and global backward erosion, flows will be controlled by the permeability of the soil after erosion of the finer fraction.

For Scenario 3: Estimate the probability for Some Erosion, Excessive Erosion and Continuing Erosion based on the guidance given in Section 12.2.4.

For Scenario 4: Estimate the probability for Some Erosion, Excessive Erosion and Continuing Erosion based on the guidance given in Section 12.2.6.

The maximum leakage flows for the various filter erosion categories described earlier are used to assess the potential for unravelling and instability in the breach mechanism phase. See Section 12.2.5 for a description of the basis for the leakage flows.

TABLE 12.1 Guide to Assessing the Probability of Continuation for Internal Erosion in the Embankment

SCENARIO	*EXAMPLES*	*PROBABILITY OF CONTINUATION (P_{NE}, P_{SE}, P_{EE}, P_{CE})*
Scenario 1: Homogeneous zoning with no fully intercepting filter or transition.	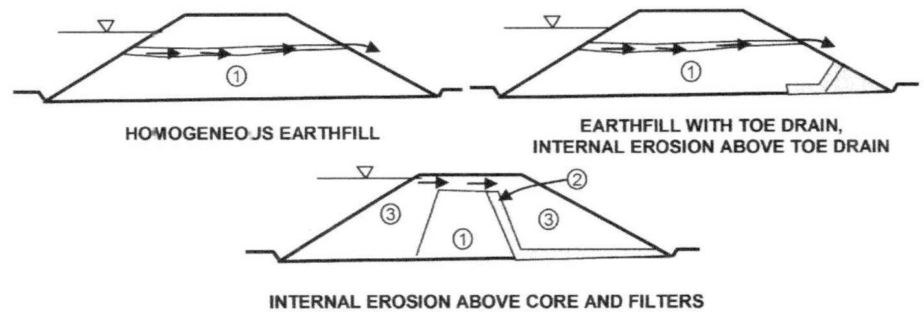	No potential for filtering. Probability for continuing erosion, $P_{CE} = 1.0$ $P_{NE} = P_{SE} = P_{EE} = 0$.
Scenario 2: Zoned embankment, no filter or transition. Scenario 2.1: Downstream shoulder plastic soil, with common cause crack. Scenario 2.2: Downstream shoulder non-plastic granular soil or rockfill. Scenario 2.3: Downstream shoulder plastic soil with no common cause crack. See Section 12.2.3 for more details.	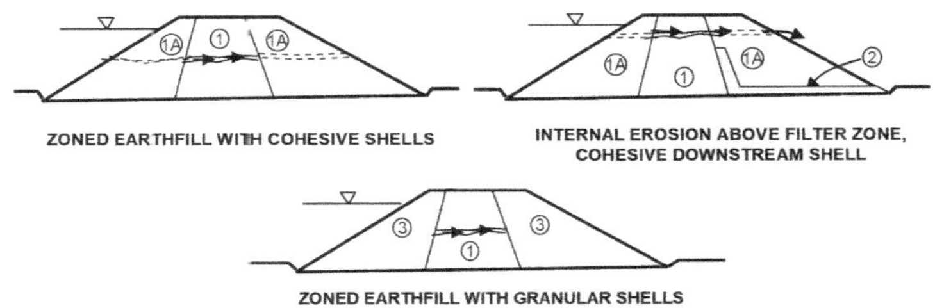	Scenario 2.1: Filtering does not occur if the crack/high permeability zone persists through the downstream shoulder zone. P_{CE} = probability of the crack in the shoulder being present and remaining open, P_{CO}. Calculate $P_{NE} = 1.0 - P_{CE}$. $P_{SE} = P_{EE} = 0$. Scenario 2.2: Follow the procedure outlined in Section 12.2.4 to estimate the probabilities of No Erosion (P_{NE}), Some Erosion (P_{SE}), Excessive Erosion (P_{EE}) and Continuing Erosion (P_{CE}). Scenario 2.3: See Section 12.2.4 Step 3. For continuing erosion an interconnecting pathway must form through the downstream shoulder.

(*Continued*)

TABLE 12.1 (Continued)

SCENARIO	*EXAMPLES*		*PROBABILITY OF CONTINUATION (P_{NE}, P_{SE}, P_{EE}, P_{CE})*
Scenario 3: Filter/transition zone is present downstream of the core or a downstream shoulder zone which is not capable of holding a crack/pipe.	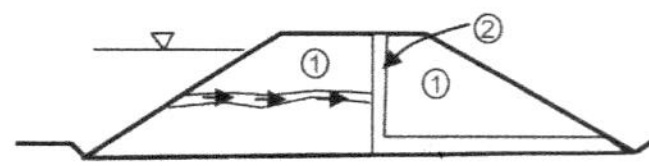 EARTHFILL WITH CHIMNEY FILTER	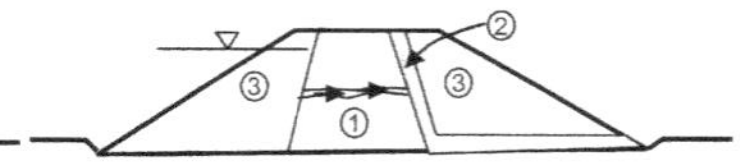 ZONED EARTHFILL WITH CHIMNEY FILTER	Follow the procedure outlined in Section 12.2.4 to estimate the probabilities of No Erosion (P_{NE}), Some Erosion (P_{SE}), Excessive Erosion (P_{EE}) and Continuing Erosion (P_{CE}).
Scenario 4: Piping into an open defect, joint or crack.	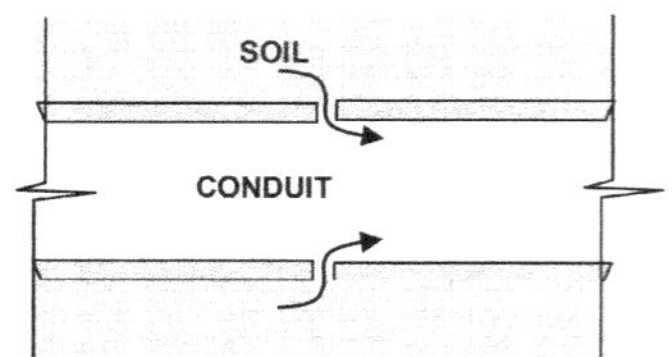 EROSION INTO AN OPEN CRACK OR JOINT IN A CONDUIT OR WALL	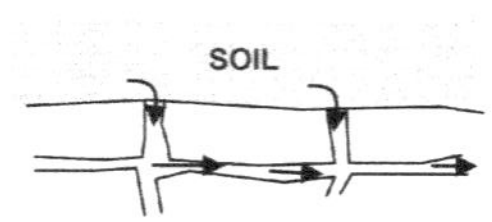 OPEN JOINTED ROCK EROSION INTO OPEN JOINTS IN ROCK FOUNDATION	Step 1: Evaluate the joint/defect opening size that would allow Continuing Erosion (JOS_{CE}) using Table 12.11. Step 2: Estimate the conditional probability for Continuing Erosion (P_{CE}) by estimating the proportion of soils that are coarser than JOS_{CE} and using Table 12.11.

12.2.2 Probability for Continuation—Scenario 1. Homogeneous Zoning

There is no potential for filtering action for this scenario. Adopt a probability of Continuing Erosion $P_{CE} = 1.0$.

12.2.3 Probability for Continuation—Scenario 2. Zoned Embankment. No Filter or Transition

Step 1: Assess the likelihood the crack or gap in the core persists through the downstream shoulder (P_{cpds}).

This depends on the following factors:

The mechanism causing the concentrated leak and whether it also causes cracking in the shoulder. That is whether there is a common cause resulting in cracking of both the core and downstream shoulder.

The material characteristics and width of the downstream shoulder zone, and whether any cracking persists through the shoulder, will determine whether it remains open.

Use engineering judgement and/or expert elicitation to evaluate the probability of a common cause crack through the downstream shoulder (P_{cpds}). In doing this take account of the following:

(a) For cracking resulting from cross-valley differential settlement, whether the foundation profile is the same beneath the full width of the shoulder as for the core.
(b) For cracking resulting from differential settlement in the foundation, whether the foundation conditions beneath the shoulder are the same beneath the full width of shoulder as for the core.
(c) For a crack or gap adjacent to a wall, whether the geometry of the wall and slope below the wall are the same beneath the full width of shoulder as for the core.
(d) For a gap or crack related to a conduit passing through the embankment, whether the geometry of the conduit, the trench into which it is constructed and the compaction of the soil around the conduit are the same beneath the full width of shoulder as for the core.
(e) For desiccation cracking on the crest of a construction shutdown layer, whether the soil in the shoulder is similarly or less susceptible to desiccation cracking as the core.

For (a) and (b), the persistence of slope changes and foundation conditions across the full width of the core and shoulder is critical. If for example a slope change present beneath the core only persists halfway across the shoulder, then it will have a low likelihood that the crack will persist.

Step 2: Assess the likelihood that given the crack or gap persists through the downstream shoulder, it will remain open.

Use Table 12.2 to estimate the probability (P_{CO}) that given a crack or gap is present through the downstream shoulder, it will stay open when water passes through it.

TABLE 12.2 Likelihood of a Crack or Gap in the Downstream Shoulder with Water Flowing in It Remaining Open (P_{CO}) (Fell et al., 2024)

SOIL CLASSIFICATION OF DOWNSTREAM SHOULDER FILL	*FINES CONTENT (PERCENTAGE MINUS 0.075 MM)*	*PLASTICITY OF THE MINUS 0.425 MM FRACTION*	*MOISTURE CONDITION*	*PROBABILITY OF CRACK OR GAP REMAINING OPEN* P_{CO}
Clays, sandy clays (CL, CH, CL-CH)	> 50%	Plastic	Moist or saturated	1.0
Silts (ML or MH)	> 50%	Plastic or non-plastic	Moist or saturated	1.0
Sandy clays, gravelly clays (SC, GC)	15%–50%	Plastic	Moist or saturated	1.0
Silty sands, Silty gravels, Silty sandy gravel (SM, GM)	> 15%	Non-plastic	Moist Saturated	0.7 to 1.0 0.5 to 1.0
Granular soils with some plastic fines (SC, GC)	10% to 15%	Plastic	Moist Saturated	0.2 to 1.0 0.1 to 0.5
Granular soils with some plastic fines (SC, GC)	5% to 10%	Plastic	Moist Saturated	0.02 to 0.2 0.01 to 0.1
Granular soils with some non-plastic fines (SM, SM-SP, SM-SW, GM, GM-GP, GM-GW)	10% to 15%	Non-plastic	Moist Saturated	0.02 to 0.2 0.01 to 0.1
Granular soils with some non-plastic fines (SM, SM-SP, SM-SW, GM, GM-GP, GM-GW)	5% to 10%	Non-plastic	Moist Saturated	0.001 to 0.02 0.0005 to 0.01
Granular soils (SP, SW, GP, GW)	< 5%	Non-plastic Plastic	Moist and saturated Moist and saturated	0.0001 0.001 to 0.01

Notes:
(a) The probabilities are for well compacted materials.
(b) Cemented materials give higher probabilities than indicated in the table. If soils are cemented, use the category that best describes the particular situation.
(c) Saturated conditions only apply where the downstream shoulder is below the tailwater level of the dam.

Step 3: Estimate the likelihood the downstream shoulder will act as a filter.
There are three scenarios:

Scenario 2.1: The downstream shoulder consists of plastic soils or granular soils with > 15% non-plastic fines and there is a common-cause crack through the shoulder. The probability of continuing erosion (P_{CE}) = probability of the crack in the shoulder being present and remaining open = (P_{cpds}) × (P_{CO}).
Scenario 2.2: The downstream shoulder consists of rockfill or non-plastic granular soils or granular soils with some non-plastic fines as described in Table 12.2. The downstream shoulder may act as a filter. Assess the likelihood of continuation of erosion using Section 12.2.4.
Scenario 2.3: The downstream shoulder consists of plastic soils or granular soils with > 15% non-plastic fines but there is no common cause crack in the shoulder.

For this scenario, the concentrated leak through the core would need to find an interconnected pathway through the downstream shoulder zone for a Continuing Erosion condition to develop. The interconnecting pathway through the downstream shoulder fill could be formed by:

- a gap formed by collapse compression of a continuous layer of poorly compacted shoulder fill, or
- a crack through the downstream shoulder formed by hydraulic fracturing.

The probability of an interconnecting pathway (P_{IP}) through the downstream shoulder should be estimated by engineering judgement/expert elicitation. For Scenario 2.3, the probability of continuing erosion (P_{CE}) = (P_{IP}) × (P_{CO}).

If continuing erosion is not predicted, the crack in the core will remain open and continue to allow flow in the crack into the downstream zones. Proceed to consider progression, detection and intervention, and breach, taking account of the potential flow from the open crack. Estimate that flow using the method described in Section 9.13.4.

12.2.4 Probability for Continuation—Scenario 3. Zoned Embankment. Filter or Transition Zone Is Present Downstream of the Core

The method of assessing the probability for continuation depends on the information that is available on the particle size distributions of the core and filter/transition/shoulder materials. The two approaches are as follows:

- If particle size distribution information is available for the core and filter/transition/shoulder materials (either from construction quality control testing and/or borrow area investigations, or from laboratory tests on samples taken from the test pits or boreholes into the filters), then use the approach described in Section 12.2.4.1
- If particle size distribution information is not available for the core and filter/transition/shoulder materials, then use the approximate approach described in Section 12.2.4.2.

The approximate approach described is subject to large uncertainty and the results should not be used to make important decisions regarding dam safety and the requirement for upgrades.

12.2.4.1 *Estimation of the Probability of Continuation, Particle Size Distribution Information Is Available*

The recommended procedure involves the following steps:

Step 1: Adjust and select base soil gradings. Plot the particle size distributions for the base soil, i.e. the core material. If the maximum particle size of the core material is > 4.75 mm, then regrade the core particle size distributions such that the maximum size is 4.75 mm. If the base soil is gap graded, then regrade the base soil grading on the particle size that is missing (i.e. at the finer point of inflection of the grading curve). Select representative gradings of the regraded base soil which are indicative of the finer 5% of the base soil gradings (fine base soil grading), the average grading (average base soil grading) and the coarser 5% of the base soils (coarse base soil grading). Figure 12.2 shows an example.

Step 2: Check for the blow out condition. In cases where there is limited depth of cover over the filter/transition zone, assess the potential for blow out by comparing the seepage pressure head in an open crack at the downstream face of the core/filter interface to the pressure of the soil cover. This is calculated as the ratio of the total stress from the vertical depth of soil (and rockfill) over the crack exit into the filter to the potential reservoir pressure head assuming the filter is effective and seals and full reservoir head is present at the core/filter interface. If the factor of safety is greater than about 0.5, three-dimensional effects will usually be sufficient to prevent blow out. If the factor of safety is less than about 0.1, it should be assumed the filter/transition will not be effective and probability of continuation P_{CE} = 1.0. Between these limits a probability of continuation between 0.1 and 0.9 should be applied. If blow out is not predicted, then follow the proceeding steps.

Step 3: Check if the filter/transition zone will hold an open crack.

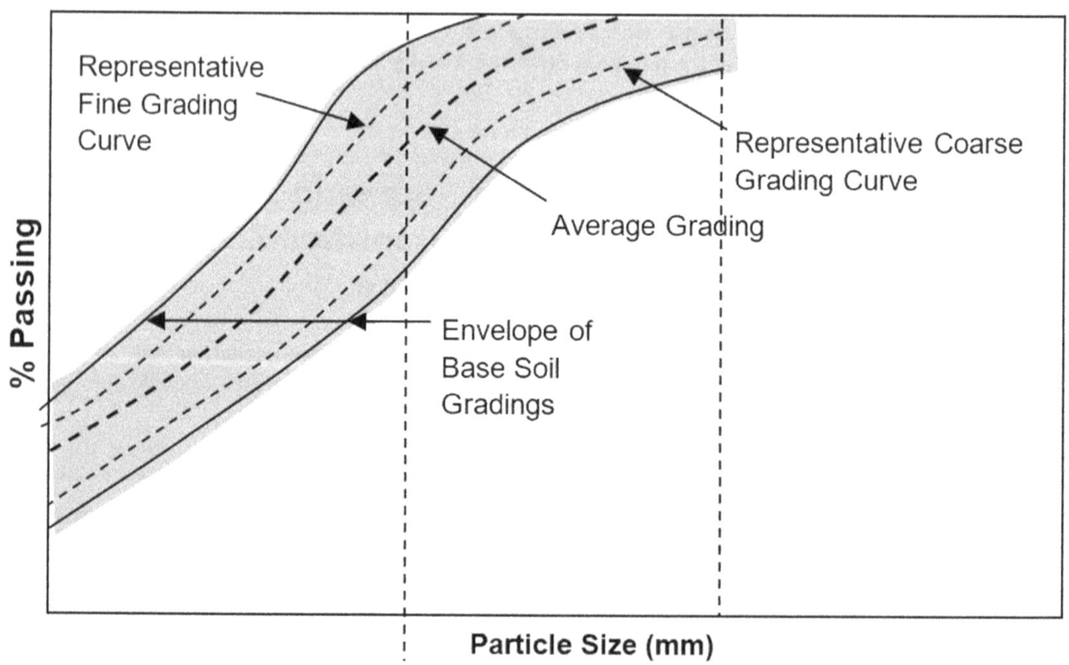

FIGURE 12.2 Example of the selection of representative grading curves (fine, average and coarse) for the assessment of filter compatibility (Fell et al., 2024).

Fell et al. (2024) gathered results from laboratory sandcastle tests using the Vaughan (ICOLD, 1994), Soroush et al. (2012), Rinehart (2012) and Rinehart et al. (2012) methods. They also had results from rectangular tests (Park, 2003), crack box tests (Howard et al., 2015) and observations of filter performance in test pits in existing dams. Based on that data and what is regarded as good filter design practice (FEMA, 2011; USBR, 2011), they concluded the following:

The factors influencing whether a filter will hold a crack and crack remain open:

- Percent fines passing 0.075 mm. High % fines more likely to hold a crack.
- The plasticity of the fines. Plastic fines more likely to hold a crack.
- Whether the filter material is of natural origins or crushed rock. Crushed rock is more likely to hold a crack. Angular natural gravels may behave in a similar manner to crushed rock.
- In-place relative density. Dense filters are more likely to hold a crack than loose filters.
- Uniformity coefficient (Cu). High Cu is more likely to hold a crack.
- The width of the crack. Wide cracks more likely to remain open than narrow cracks.
- As described in Fell et al. (2015), fine filters with carbonate rock such as limestone, marble and dolomite present are likely to hold a crack. Filters with small percentages of sulphide minerals or natural pozzolans may also result in cementation and ability to hold a crack.

Some "anchor points" on probabilities:

(i) Filters constructed from natural (alluvial) materials with ≤ 5% fines passing 0.075 mm and non-plastic fines should be assigned a negligible probability of holding a crack.
  - Based on good filter design practice and good performance of filters satisfying such criteria.
  - Park (2003) concluding, "Conventional filter criteria—no more than 5% fines and fines are non-plastic, are conservative" based on his Vaughan-type sandcastle and square box tests.

Adopted P_{HC} = negligible

(ii) Filters constructed from crushed rock with ≤ 5% fines passing 0.075 mm with non-plastic fines should be assigned a very low probability of holding a crack.
  - Based on Rinehart (2012) testing, crushed rock filters classify marginally worse than natural filters, but all collapsed on full immersion.
  - Based on good filter design practice and good performance of filters satisfying such criteria not differentiating between natural and crushed rock filters.
  - Mejia (2013) Matahina Dam crushed ignimbrite samples with up to 7% non-plastic fines collapsing immediately on full submersion and in test trench.

Adopted P_{HC} = 1.0E-04

(iii) Filters constructed from crushed rock with 5% fines passing 0.075 mm and the fines being plastic (PI ≥ 4) should be assigned a high probability of holding a crack.
  - Based on the Cardinia Dam AECOM (2023) Rinehart-type tests on samples with 4% to 6% fines being Class IV and V.
  - The filter in the sides of Cardinia Dam test pits standing on inundation and losing only the finer fraction.
  - The experience at Ochoco Dam (Redlinger-personal communication) where natural angular gravels with some crushed gravel held a crack.

Adopted $P_{HC} = 0.1$

(iv) Filters constructed from natural (alluvial) materials with 10% fines passing 0.075 mm but the fines being plastic (PI ≥ 4) should be assigned a high probability of holding a crack.

- Based on Park (2003) Sandcastle tests at 10% plastic fines which varied between "collapse in 1 minute" to "did not collapse" depending on moisture content.
- Park (2003) # 34 square test with 10% plastic fines did ultimately collapse and contain the base soil.
- By comparison with (iii) allowing for the filter being natural.

Adopted $P_{HC} = 0.1$

(v) Filters constructed of natural or crushed material with 15% plastic fines (PI ≥ 4) are almost certain to hold a crack.

- AECOM (2023) Greenvale 12% non-plastic fines no-collapse Class VI.
- Park (2003) # 35 square test with 15% fines did not collapse and erosion continued.

Adopted $P_{HC} = 0.9$

(vi) Crushed rock filters are about 10x more likely to hold a crack than natural rounded filters. This is because of the angular crushed particles interlocking and reducing the likelihood of particles detaching from the sides of a crack.

Based on these data it is recommended that the probability that a filter will hold a crack and that erosion will continue be estimated from Table 12.3 taking account of:

TABLE 12.3 Probability of a Filter Holding a Crack (P_{HC}) (Fell et al., 2024)

FINES PLASTICITY	*FINES CONTENT % PASSING 0.075 MM*	*PROBABILITY OF HOLDING A CRACK*	
		NATURAL FILTER	*CRUSHED FILTER*
Non-Plastic and Plasticity Index ≤ 2	2%	Negligible	Negligible
	5%	Negligible	0.0001
	7%	0.001	0.01
	10%	0.01	0.1
	15%	0.1	0.9
Plastic Plasticity Index ≥ 4	2%	0.001	0.01
	5%	0.01	0.1
	7%	0.03	0.3
	10%	0.1	0.9
	15%	0.9	0.99

Notes:

(a) Probabilities are for filters compacted to around 70% relative density. Filters which are not compacted will be less likely to hold a crack. Filters that are heavily compacted (greater than around 90% to 95% relative density) are more likely to hold a crack.

(b) The probabilities are for cracks 2 mm to 25 mm open width.

(c) The probability that a filter will hold a crack and not prevent erosion continuing may be less for cracks narrower than 2 mm and greater for cracks wider than 25 mm. That will depend on filter gradations, filter compaction, filter width and other factors which the risk analysis team may consider relevant.

- The percentage of fines passing 0.075 mm.
- The plasticity of the fines.
- Whether the filters are manufactured from natural sand and gravel, e.g. alluvial sands and gravels, or from crushed rock. Filters with a mix of both should be considered as crushed rock filters.
- The degree of compaction of the filter.
- The estimated width of the crack in the core and filter.

If the fine filter is manufactured from sand and gravel containing carbonate rocks or sulphides, this may result in cementation and ability to hold a crack. Assess the likelihood of holding a crack based on the percentage of carbonate and sulphides and observations of cementation in stockpiles and in the dam.

If the filter or transition holds a crack, erosion will continue and the probability of continuing erosion = P_{HC}.

The probability of the filter not holding a crack = 1 – P_{HC}. Proceed to assessing the filter as detailed in Step 6. The total probability of continuing erosion = P_{HC} + (1–P_{HC}) × (probability of continuing erosion from Step 7).

Step 4: Check if the filter/transition zone is subject to segregation.

Assess the likelihood of segregation of the filter/transition/shoulder materials using Tables 12.4, 12.5 and 12.6. The procedure is based on published data on segregation by Sherard and Dunnigan (1989), Ripley (1986), Kenney and Westland (1993), USDA-SCS (1994), USBR (2011) and USACE (1994) and the expert judgement of the participants in a Fell et al. (2008) workshop.

If a continuous segregated layer is likely to be present, then estimate the grading of the segregated layer. This may be done using the method described in Step 5 but taking into account observations of the gradings of segregated filters in the dam and in stockpiles.

In Table 12.4 a low likelihood of segregation has been assigned to filters placed in layers < 0.45 m thick by spreader box or truck mounted conveyor with care taken to control segregation. These methods are usually satisfactory, but it should be noted that investigations at Cardinia Dam (AECOM, 2023) have shown segregation has occurred in the Zone 2B filter which were placed by spreader box. That material was minus 75 mm angular particles of crushed granite, with a Cu = around 20 to 30, so those factors contributed to the segregation. Lesser segregation was observed in a Zone 2A filter which was minus 40 mm and Cu = around 15 to 20.

Segregation can also occur with filters placed by dumping from a truck and spreading by grader or other equipment, particularly adjacent to walls and around conduits. Figure 12.3 shows an example.

Step 5: Check if the filter/transition zone is subject to suffusion. Evaluate the probability that the filter or transition zone materials are subject to suffusion (P_{IUS}) using Kenney and Lau (1985, 1986) (Figures 10.40, 10.41) for materials with ≤ 15% fines or finer fraction (i.e. the fraction that is finer than the point of maximum inflexion in the PSD curve) and Douglas et al. (2019) (Figure 10.42) for materials with > 15% fines or finer fraction.

Douglas et al. (2019) recommended that if the material is subject to suffusion, that is for materials which are distinctly within the "Major Erosion, Suffusion or GBE" category in Figure 10.42, it is likely that all the finer fraction might be eroded, resulting in a coarser grading. The authors have seen similar outcomes for other tests. It is therefore suggested that in the absence of test data on such filter materials in the embankment, it should be assumed all the finer fraction is eroded.

For materials which fall into the Douglas et al. (2019) "Major to Minor Erosion, GBE" category, less of the finer fraction is likely to erode. From the available data from Wan and Fell (2004c), about half of the finer fraction may erode. Figure 12.4 shows how the gradation may be adjusted in this scenario.

If the effect has a significant effect on the estimated risk, carry out laboratory tests as described in Douglas et al. (2019).

TABLE 12.4 Likelihood of Segregation of Filtering Materials (Fell et al., 2008, 2024)

		LIKELIHOOD FACTOR (LF)			
FACTOR	*RELATIVE IMPORTANCE FACTOR (RF)*	*LESS LIKELY (1)*	*NEUTRAL (2)*	*MORE LIKELY (3)*	*MUCH MORE LIKELY (4)*
Construction practices	(3)	Good construction and stockpiling practices used.	Fair construction and stockpiling practices.	Poor construction or stockpiling practices.	Very poor construction and stockpiling practices with no regard for segregation effects.
		Placed in thin lifts < 0.45 m[(a)] careful control during construction.	End dumping from trucks, spread by dozer in lifts < 0.6 m.	End dumping from trucks, spread by dozer in thick lifts > 0.6 m.	Filters/transition constructed by pushing material over the edge of the core.
Gradation—Comparison to USBR/US Corps filter criteria	(2)	Meets segregation criteria in Table 12.5.	Borderline segregation criteria in Table 12.5.	Fails segregation criteria in Table 12.5	Significant departure from segregation criteria in Table 12.5.
and % sand		> 50% passing 4.75 mm sieve	> 40% passing 4.75 mm sieve	25%–40% passing 4.75 mm sieve	< 25% passing 4.75 mm sieve
Width of zone	(1)	> 6m	3 m–6 m	1.5 m–3 m	< 1.5 m

Note: (a) Placed in spreader boxes or truck mounted conveyors with care to avoid segregation.

TABLE 12.5 Gradation Limits to Prevent Segregation (USDA SCS, 1994; USBR, 1987; USACE, 1994)

MINIMUM D_{10} (MM)	*MAXIMUM D_{90} (MM)*
< 0.5	20
0.5–1.0	25
1.0–2.0	30
2.0–5.0	40
5.0–10	50
10–50	60

TABLE 12.6 Susceptibility of Filter/Transition Zones to Segregation versus Weighted Score (Relative importance factor [RF]) × (Likelihood factor [LF]) (Fell et al., 2008, 2024)

WEIGHTED SCORE FROM TABLE 12.4	*SEGREGATION ASSESSMENT*	*CONSIDERATION OF SEGREGATION EFFECTS FOR FILTER/TRANSITION ASSESSMENT*
6–10	Low potential for segregation	Segregation of filter/transition materials does not need to be considered.
11–17	Moderate potential for segregation	Segregation of filter/transition materials should be considered, unless investigations show otherwise.
18–24	High potential for continuous segregated layers	Segregation should be assumed to be present, unless investigations show otherwise.

Gap-graded soils tested by Wan and Fell (2004c, 2008) in downwards flow at a gradient of 8 lost between 10% and 60% of the finer fraction, with most < 33%. This may be used as a guide to the behaviour of gap-graded filters. For important decisions, carry out laboratory suffusion tests on representative samples of the filter.

Step 6: Evaluate the D_{15F} values for the No, Excessive and Continuing Erosion boundaries.

Use Tables 12.7 and 12.9 for the fine base soil grading, the average base soil grading and the coarse base soil grading. Plot the D_{15F} values for these boundaries on the grading curve limits of the filter/transition material (see Figure 12.6 for an example). Use the adjusted grading curves for the filter/transition zone if required to do so by the preceding steps 4 or 5. Use the Sherard and Dunnigan (1989) criteria as adjusted by Foster and Fell (1999a, 2001) in Table 12.7 for no erosion, and Foster and Fell (1999a, 2001) in Table 12.9 for Excessive and Continuing Erosion criteria.

Step 7: Estimate the probabilities for No Erosion, Some Erosion, Excessive and Continuing Erosion for each representative base soil grading.

Estimate the proportion of the filter/transition gradings that fall into each of the particular erosion categories based on the plot of filter/transition grading curves versus filter erosion boundaries. An example is shown in Figure 12.6.

If there are no filter/transition gradings that fall into the Continuing Erosion category, review the amount and quality of the available data, whether segregation and suffusion have been accounted for and the amount by which the filters gradations are finer than continuing erosion criteria. Test the effects of assuming all the finer fraction of filters is washed out or segregation results in that grading. Then use judgement to either confirm the assessment of zero probability of continuing erosion or assign a small but finite probability.

FIGURE 12.3 Segregation of filter placed against a spillway wall.

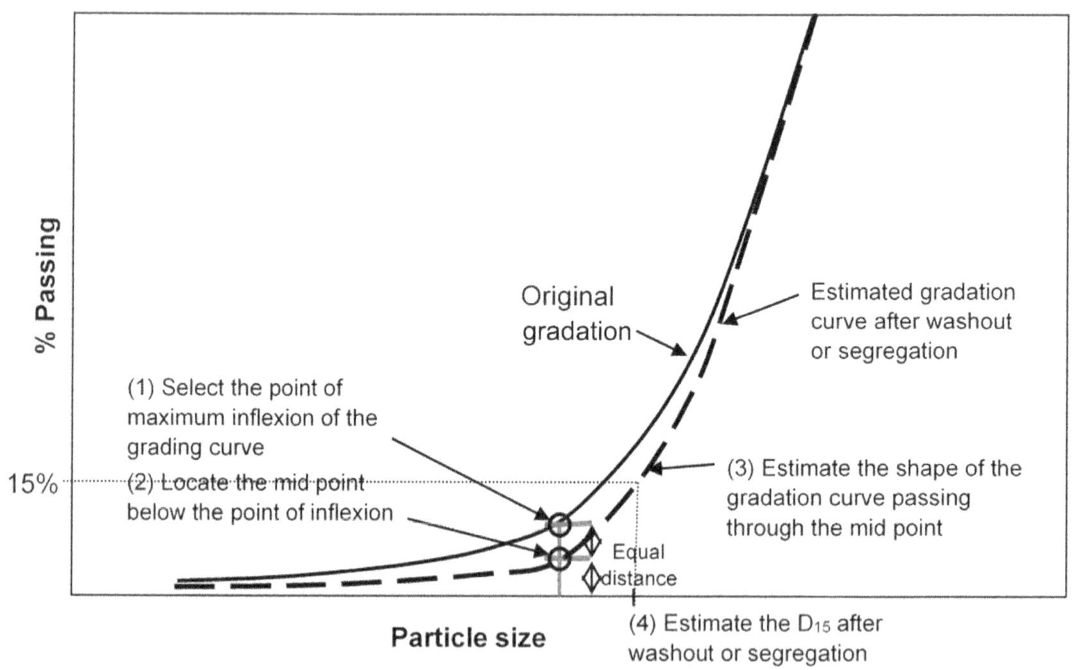

FIGURE 12.4 Approximate method for estimating D_{15F} after washout of the finer fraction from soil subject to major to minor erosion, GBE or soils susceptible to segregation (Fell et al., 2008, 2024).

TABLE 12.7 No-Erosion Filter Criteria by Sherard and Dunnigan (1989) with Adjusted Base Soil Categories by Foster and Fell (1999a, 2001)

BASE SOIL CATEGORY	*FILTERING CRITERIA*
1	$D_{15F} \leq 9 \times$ D85B but not less than 0.2 mm (a)
2A	≤ 0.7 mm (b)
3	$\leq 4 \times D_{85B}$ of base soil after regrading
4A	(35–A) / (35–15) [(4 × D_{85B}) – 0.7 mm] + 0.7 mm. A = % passing 0.075 mm sieve after regrading (If 4 × D_{85B} is less than 0.7 mm, use 0.7 mm)

Note: (a) x ≤ $6D_{85B}$ for dispersive soils. (b) ≤ 0.5 mm for dispersive soils.
Dispersive soils are soils with pinhole classification D1 or D2, or Emerson class 1 or 2.

TABLE 12.8 Base Soil Categories as Adjusted by Foster and Fell (1999a, 2001)

BASE SOIL CATEGORY	*% FINER THAN 0.075 MM (AFTER REGRADING, WHERE APPLICABLE)*	*BASE SOIL DESCRIPTION*
1	> 85	Fine silts and clays
2A	35–85	Silty and clayey sands; sandy clays; and clay, silt, sand, gravel mixes
4A	15–35	Silty and clayey sands and gravel
3	< 15	Sands and gravel

Note: Sherard and Dunnigan (1989) had the boundary between Category 2 and 3 as 40% finer than 0.075 mm. Foster and Fell (1999a, 2001) adjusted this to 35% based on a more extensive range of tests.

TABLE 12.9 Criteria for the Excessive and Continuing Erosion Boundaries (Foster, 1999a; Foster and Fell, 1999a, 2001)

BASE SOIL	*CRITERIA FOR EXCESSIVE EROSION BOUNDARY*	*CRITERIA FOR CONTINUING EROSION BOUNDARY*
Soils with $D_{95B} < 0.3$ mm	$D_{15F} > 9\ D_{95B}$	
Soils with $0.3 < D_{95B} < 2$ mm	$D_{15F} > 9\ D_{90B}$	For all soils: $D_{15F} > 9\ D_{95B}$
Soils with $D_{95B} > 2$ mm and fines content > 35%	Average $D_{15F} > D_{15F}$ which gives an erosion loss of 0.25 g/cm² in the CEF test (0.25 g/cm² contour line in Figure 12.5)	
Soils with $D_{95B} > 2$ mm and fines content < 15%	$D_{15F} > 9\ D_{85B}$	
Soils with $D_{95B} > 2$ mm and fines content 15–35%	$D_{15F} > 2.5\ D_{15F}$ design, where D_{15F} design is given by D_{15F} design = (35–pp% 0.075 mm) (4 D_{85B}–0.7) / 20 + 0.7	

Note: Criteria are directly applicable to soils with DB95 up to 4.75 mm. For soils with coarser particles, determine DB85 and DB95 using grading curves adjusted to give a maximum size of 4.75 mm.

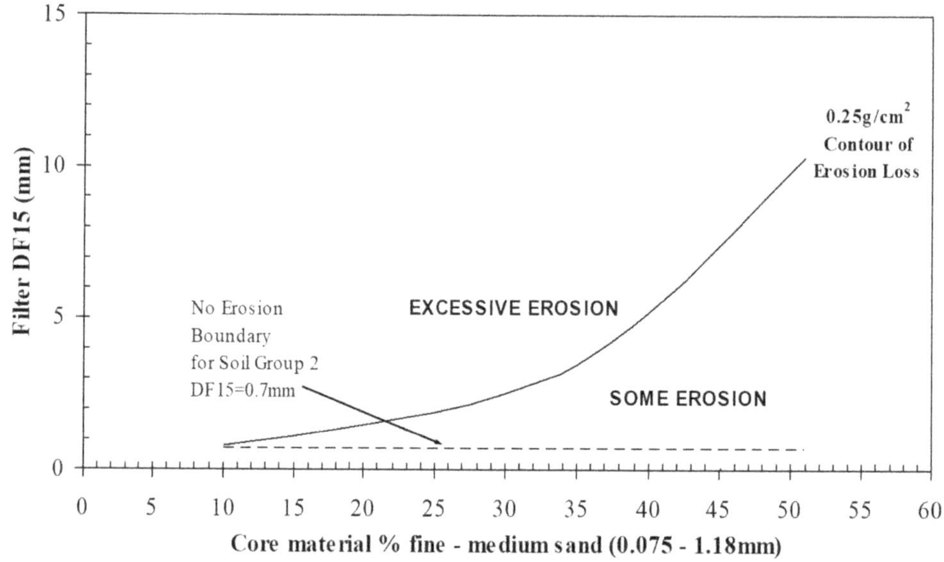

FIGURE 12.5 Criteria for Excessive Erosion boundary (Foster, 1999; Foster and Fell, 1999a, 2001).

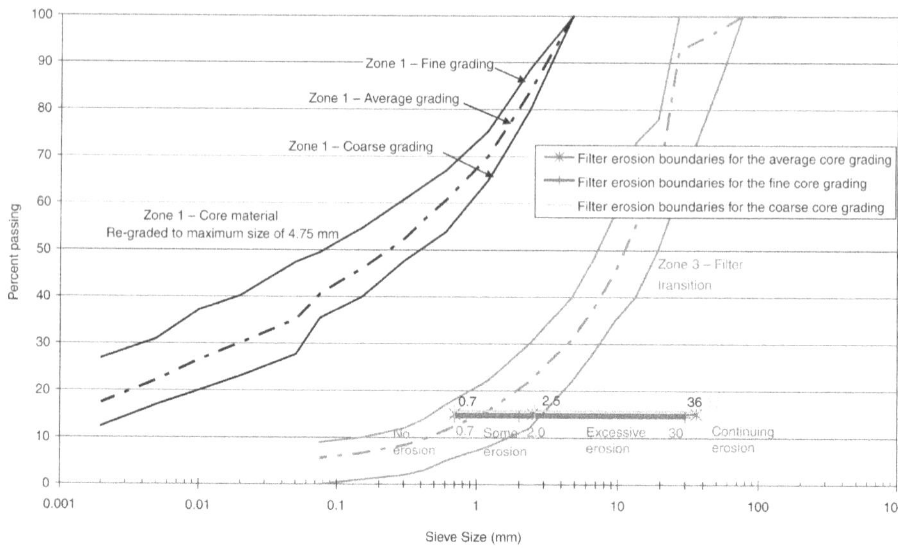

Assessment of zone 1 core against no erosion, excessive erosion and continuing erosion criteria

Core gradation	Base soil sizes (mm)				No erosion	Excessive erosion	Continuing erosion
	DB85 (mm)	DB95 (mm)	% passing 0.075 mm	% fine-medium sand (0.075–1.18 mm)	DF15 (mm)	DF15 (mm)	DF15 (mm)
Fine grading	1.9	3.3	50	25	0.7	2	30
Average	2.4	4	41	29	0.7	2.5	36
Coarse Grading	2.5	4.2	35	30	0.7	2.6	38

FIGURE 12.6 Example of plot showing filter/transition gradings compared to filter erosion boundaries. Evaluate the filter erosion boundaries for the representative fine, average and coarse gradings of the core material (Fell et al., 2008).

The suggested approach is to estimate the proportions for the Some, Excessive and Continuing Erosion categories first (P_{SE}, P_{EE} and P_{CE}) and then calculate $P_{NE} = 1 - (P_{SE} + P_{EE} + P_{CE})$. Do this for each of the representative base soil gradings (fine, average and coarse gradings) as follows:

For the fine base soil grading: P_{NE} fine, P_{SE} fine, P_{EE} fine, and P_{CE} fine.
For the average base soil grading: P_{NE} ave, P_{SE} ave, P_{EE} ave, and P_{CE} ave.
For the coarse base soil grading: P_{NE} coarse, P_{SE} coarse, P_{EE} coarse, and P_{CE} coarse.

Make an initial estimate of the probabilities of the No Erosion, Some Erosion, Excessive Erosion and Continuing Erosion branches by the sum-product of the % of base soil gradings and the % of NE, SE, EE and CE for each representative base soil grading. The calculations are as follows:

$$P_{NE} = (5\% \times P_{NE}\text{ fine}) + (90\% \times P_{NE}\text{ ave}) + (5\% \times P_{NE}\text{ coarse}).$$

$$P_{SE} = (5\% \times P_{SE}\text{ fine}) + (90\% \times P_{SE}\text{ ave}) + (5\% \times P_{SE}\text{ coarse}).$$

$$P_{EE} = (5\% \times P_{EE}\text{ fine}) + (90\% \times P_{EE}\text{ ave}) + (5\% \times P_{EE}\text{ coarse}).$$

$$P_{CE} = (5\% \times P_{CE}\text{ fine}) + (90\% \times P_{CE}\text{ ave}) + (5\% \times P_{CE}\text{ coarse}).$$

The 5% fractions have been selected in the preceding calculations on the basis that these combinations of gradations are representative of the finest and coarsest gradations in contact with the base soil gradation. Selection of the appropriate percentiles for characterising the gradation envelope is a matter of judgement of the risk team and can be adjusted by the risk team based on site specific considerations.

An example of summing of the probabilities is shown in Table 12.10 for the example shown in Figure 12.6.

Use judgement to adjust the calculated percentages to take into account the effects of other factors such as the distribution of the core and filter gradations in the fill, borrow area variability and selective placement of materials.

In cases where there are a limited number of particle size distributions of the base soil and filter, the estimates may be made by judgement from the available data rather than the calculations detailed previously.

12.2.4.2 Approximate Simplified Approach—Particle Size Distribution Information Is Not Available

(a) Estimate the particle size distribution of the core materials based on the likely source of materials. The gradation of the soils may be able to be estimated based on the likely geological origin of the materials (e.g. decomposed granitic soils, residual soils, alluvial fine clays and silts, etc.).
(b) Estimate the particle size distribution of the filter/transition/shoulder materials based on the likely source of the materials and whether they were processed (e.g. run-of-pit alluvial sands and gravels, unprocessed quarry fines or tunnel spoil, processed sand and gravels, etc.). Estimate the likely range of D_{15F} of the filter/transition/downstream shoulder materials.
(c) Evaluate the D_{15F} values for the No, Excessive and Continuing Erosion boundaries for the estimated gradation of the core materials using Tables 12.7 and 12.9.
(d) Estimate the probabilities for No Erosion, Some Erosion, Excessive and Continuing Erosion based on the estimated proportion of the filter/transition gradings that is likely to fall into each of the particular filter erosion categories. The suggested approach is to estimate the probabilities for Some, Excessive and Continuing Erosion (P_{SE}, P_{EE} and P_{CE}) and calculate $P_{NE} = 1 - (P_{SE} + P_{EE} + P_{CE})$.

TABLE 12.10 Example of Estimating Probabilities for No, Some, Excessive and Continuing Erosion for the Example Shown in Figure 12.6 (Fell et al., 2008, 2024)

REPRESENTATIVE BASE SOIL GRADING	*ESTIMATED PROPORTION OF FILTER GRADINGS FALLING INTO EACH FILTER EROSION CATEGORY (FROM FIGURE 10.4)*				
	NO EROSION (NE)	*SOME EROSION (SE)*	*EXCESSIVE EROSION (EE)*	*CONTINUING EROSION (CE)*	*SUM P_{xx}*
Fine Base Soil Grading (represents 5% of finest grading curves)	P_{NE} fine = 20%	P_{SE} fine = 60%	P_{EE} fine = 20%	P_{CE} fine = 0%	100%
Average Base Soil Grading (represents 90% of grading curves)	P_{NE} ave = 20%	P_{SE} ave = 70%	P_{EE} ave = 10%	P_{CE} ave = 0%	100%
Coarse Base Soil Grading (represents 5% of coarsest grading curves)	P_{NE} coarse = 20%	P_{SE} coarse = 70%	P_{EE} coarse = 10%	P_{CE} coarse = 0%	100%
Calculation of Probabilities for No, Some, Excessive and Continuing Erosion (P_{xx})	P_{NE} = (5% × P_{NE} fine) + (90% × P_{NE} ave) + (5% × P_{NE} coarse)	P_{SE} = (5% × P_{SE} fine) + (90% × P_{SE} ave) + (5% × P_{SE} coarse)	P_{EE} = (5% × P_{EE} fine) + (90% × P_{EE} ave) + (5% × P_{EE} coarse)	P_{CE} = (5% × P_{CE} fine) + (90% × P_{CE} ave) + (5% × P_{CE} coarse)	
Calculation Result	20%	69.5%	10.5%	0%	100%
Assigned Probabilities	P_{NE} = 0.20	P_{SE} = 0.69	P_{EE} = 0.11	P_{CE} = 0[(a)]	1.000

Note: (a) See comment in Step 7 regarding assigning zero probability to continuing erosion.

This approach is subject to significant error and should only be used for screening level risk assessments. Under no circumstances should important dam safety decisions be made without actual data on the filter and core particle size distributions and filter plasticity.

12.2.5 Estimate of Potential Leakage Flows for Some, Excessive and Continuing Erosion Filters

Foster et al. (2018) gathered information on leakage flows from case histories where dams have experienced a short period of leakage resulting from internal erosion. This is attributed to the filter or transition being coarser than a no-erosion filter but eventually sealing. In that study allowance was made for adjusting the particle size distributions of the filter or transition to allow for suffusion.

Figure 12.7 summarises the data. This shows that the maximum flows which occur before filters seal is related to the No Erosion, Excessive Erosion and Continuing Erosion boundaries.

Based on these cases, the maximum flows that have been experienced are summarised as follows:

No erosion: No flow (too small to measure).
Some erosion: 3–70 l/sec.
Excessive erosion: 22–560 l/sec.
Continuing erosion: >1000 l/sec and increasing (unlimited unless flow restriction is present).

Foster et al. (2018) concluded that filters coarser than no-erosion filters may be subject to potential maximum leakage flows before the filters seal of:

No-Erosion filters: unlikely to be detectable.
Filters falling into the some erosion category: up to 100 l/sec:
Filters falling into the excessive erosion category: 100 l/sec to 1000 l/sec.
Filters falling into the continuing erosion category: flows of 1000 l/sec and increasing.

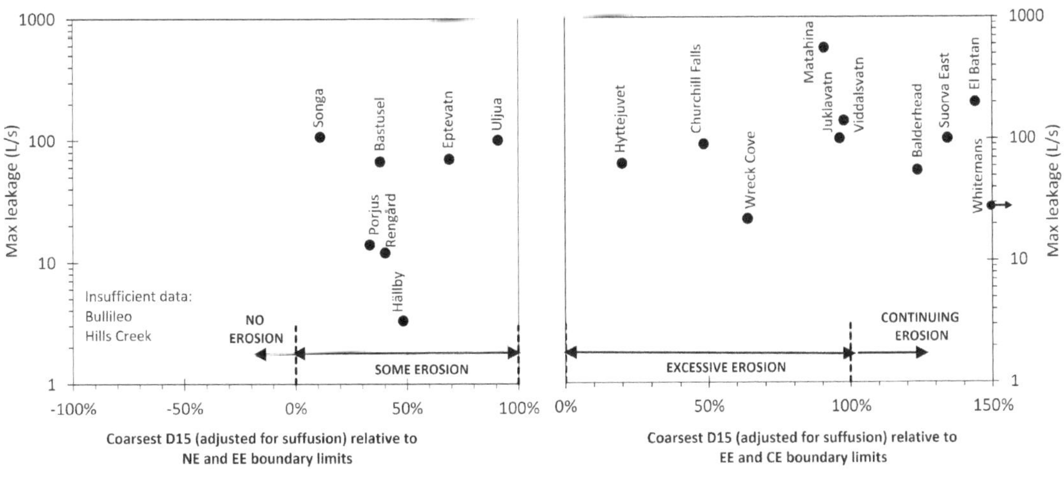

FIGURE 12.7 Maximum leakage rates versus the coarsest D_{15F} of the filter (adjusted for suffusion) relative to the No Erosion, Excessive Erosion and Continuing Erosion boundaries (Foster et al. (2018).

It is recommended that the maximum leakage flows listed earlier be used in the assessment of the probability of a breach mechanism developing. Lower leakage flows are likely if upstream flow limitation occurs, but the factors would need to be carefully considered and justified if they were to be relied upon in the assessment.

12.2.6 Probability for Continuation—Scenario 4. Internal Erosion into an Open Defect, Joint or Crack in the Foundation, in a Wall or Conduit

For erosion to continue through an open defect, the defect needs to be sufficiently open to allow the soil surrounding to penetrate into the defect and be transported within the defect.

One potential limit is the Foster and Fell (1999a, 2001) continuing erosion criteria listed in Table 12.9, which is $D_{15F} > 9D_{95B}$. The filter opening size has been shown by Sherard et al. (1984a) to be $D_{15F} / 9$. This relationship was confirmed by Foster and Fell (1999a, 2001) so the width of an open defect allowing continuing erosion would be the D_{95} of the soil.

A second potential limit is taken from the performance of cement grouts. Laboratory testing by Tjandrajana (1989) showed some cement grout penetrated for openings of $3D_{100}$, but eventually the opening was bridged over by coarser particles. Mitchell (1970) and Karol (1985) indicate that grout will not penetrate if the fracture opening is less than $3D_{100}$. Littlejohn (1985) indicates that grouting is not possible unless the fracture opening is greater than $5D_{100}$.

Based on these, the joint opening size (JOS) which will allow continuing erosion is between:

- D_{95B} of the soil, based on the assumption that crack width is equivalent to the filter opening size of the voids between the particles in a filter.
- $3D_{95B}$ of the soil, based on criteria developed for the opening of joints in rock into which cement grouts will penetrate and flow along the joint.

Table 12.11 has been developed based on these limits. The range of D_{95F} of the soil should be considered in the overall assessment.

Table 12.11 gives best estimate values for erosion into joints or cracks in the foundation and into open joints in conduits and walls with steady flow conditions. For erosion into open joints in conduits or walls where there are dynamic flow conditions, somewhat higher probabilities may apply.

TABLE 12.11 Estimation of Probability for Continuation of Erosion into Open Defects, Joints and Cracks (Fell et al., 2024)

COMPARISON OF JOINT OPENING IN THE DAM (JOS) TO THE D_{95} OF THE SURROUNDING SOIL	
JOINT OPENING IN THE DAM	*PROBABILITY FOR CONTINUING EROSION (P_{CE})*
JOS < 0.4 D_{95}	Negligible
JOS = 0.5 D_{95}	0.0001
JOS = 0.75 D_{95}	0.001
JOS = D_{95}	0.1
JOS = $2D_{95}$	0.5
JOS ≥ $3D_{95}$	0.9

12.3 PROBABILITY OF CONTINUATION FOR INTERNAL EROSION THROUGH THE FOUNDATION

12.3.1 Approach

The probability of continuation of erosion should be estimated by:

(a) Assessing the probability that the exit will be a filtered or unfiltered exit.
  (1) Given the exit is unfiltered the probability of continuation will be 1.0.
  (2) Given the exit is filtered, estimate the probability for Continuing Erosion using the method described in Section 12.2.4.

The probability of continuation will be the product of the probability of an unfiltered exit and the probability assessed considering the filters.

12.3.2 Probability of a Filtered or Unfiltered Exit

The likelihood that the exit will be filtered or unfiltered will depend on the failure mode being considered, the embankment zoning and the details of the foundation geology. The steps are as follows:

Estimate the probability that there will an unfiltered exit (P_{unf}) using Table 12.12 to aid judgement. The conditional probability of continuation for this scenario is equal to (P_{unf}) × 1.0.

Calculate the probability that there will be a filtered exit (P_{fe}) = 1– P_{unf}. Assess the filter materials and materials being eroded in terms of the Continuing Erosion criteria using the procedure in Section 12.2.4.

Table 12.12 has been developed using expert judgement of the Fell et al. (2024) authors taking account of what is good practice and therefore should have low probability, and their experience.

TABLE 12.12 Indicative Probabilities (P_{unf}) of Bypassing the Foundation Filter Resulting in an Unfiltered Exit for Piping through the Foundation (Fell et al., 2024)

SCENARIOS	*EXAMPLES*	*RANGE OF PROBABILITIES OF THE SEEPAGE PATH BYPASSING THE FOUNDATION FILTER*
Foundation filter fully penetrates the foundation material that is erodible. Filter drain has adequate discharge capacity.	Filter trench completely penetrates through the foundation sand/silt layer, infilled defects in rock or desiccated foundation clay and into a non-erodible material (Figure 12.8a).	Negligible Assume probability unfiltered exit = 0.
Foundation filter partially penetrates the foundation material that is erodible. Filter drain has adequate discharge capacity.	Filter trench penetrating through foundation sand/silt layer, infilled defects in rock or desiccated foundation clay but erodible soil or infilled defects in rock remains beneath the filter trench (Figure 12.8b).	0.01 to 0.0001 depending on details.

(Continued)

TABLE 12.12 (Continued)

SCENARIOS	*EXAMPLES*	*RANGE OF PROBABILITIES OF THE SEEPAGE PATH BYPASSING THE FOUNDATION FILTER*
Foundation filter partially penetrates the foundation material that is erodible. Filter drain has inadequate discharge capacity.	Filter trench partially penetrating through foundation sand/silt layer, or desiccated foundation clay (Figure 12.8b).	0.01 to 0.1 depending on the degree to which filter drain capacity is exceeded.
Foundation filter blanket drain directly overlying the foundation material that is erodible, clean-up of the foundation allows seepage into the filter drain.	Filter drain overlying a foundation sand/silt layer, desiccated foundation clay (Figure 12.8c) or infilled defects in rock.	0.02–0.0002 depending on details. If the filter drain is in strips not covering all the foundation (finger drains) the probability will be much higher, e.g. 0.01 to 0.1.
Foundation filter blanket drain directly overlying the foundation material that is erodible, filter drain has inadequate discharge capacity.	Filter drain overlying a foundation sand/silt layer, desiccated foundation clay (Figure 12.8c) or infilled defects in rock.	0.01–0.1 depending on the degree to which filter drain capacity is exceeded.
Low permeability layer is present below the foundation filter drain. Filter drain has adequate discharge capacity.	Low permeability soil layer between the foundation filter/toe drain, and the foundation material that is being eroded (e.g. Figure 12.9). Residual soil left over jointed bedrock below the foundation filter.	0.1–0.5
Scarifying and rolling the foundation below the foundation filter in soils with macrostructure. Filter drain has adequate discharge capacity	Foundation soils or highly weathered rocks containing relict defects, root holes, desiccation cracks or lateritic soils.	0.1–0.5
High permeability soil or rock has an unprotected exit downstream.	Soil layer that can be subject to backward erosion piping or concentrated leak erosion daylights downstream of the dam. Continuous open jointed rock daylights downstream of the dam (Figure 12.10).	0.5–0.9
Blocked or collapsed toe drains	HDPE corrugated pipes, collapsed tile drains, biological growth in pipes.	0.1–0.9
No foundation filter or toe drain system provided, plastic shell materials		1.0

The conditional probabilities should be determined by selecting the most applicable scenario and judging within the range shown taking account of the details of the dam and its foundation. Where there is little or no detailed data available, the default may be to use the upper bound probability for the applicable scenario.

The probability of continuing erosion $P_{CE} = (P_{unf} \times 1.0) + (P_{fe} \times P_{CE})$.

Note that in situations where a foundation filter drain or toe drain system is present, there is still the possibility that seepage paths could bypass the filter drain system due to the geological conditions, compaction of soil foundations below foundation filters, or deteriorated toe drain systems. Examples of these scenarios are included in Table 12.13 and illustrated in Figures 12.8 to 12.10.

It needs to be recognised that low permeability strata beneath horizontal drains may prevent them working effectively. Figure 12.9 shows an alluvial foundation where the lower permeability strata (A and E) will prevent the seepage in the most permeable sand and gravel strata (B and D) from flowing into a filtered exit in the horizontal drain.

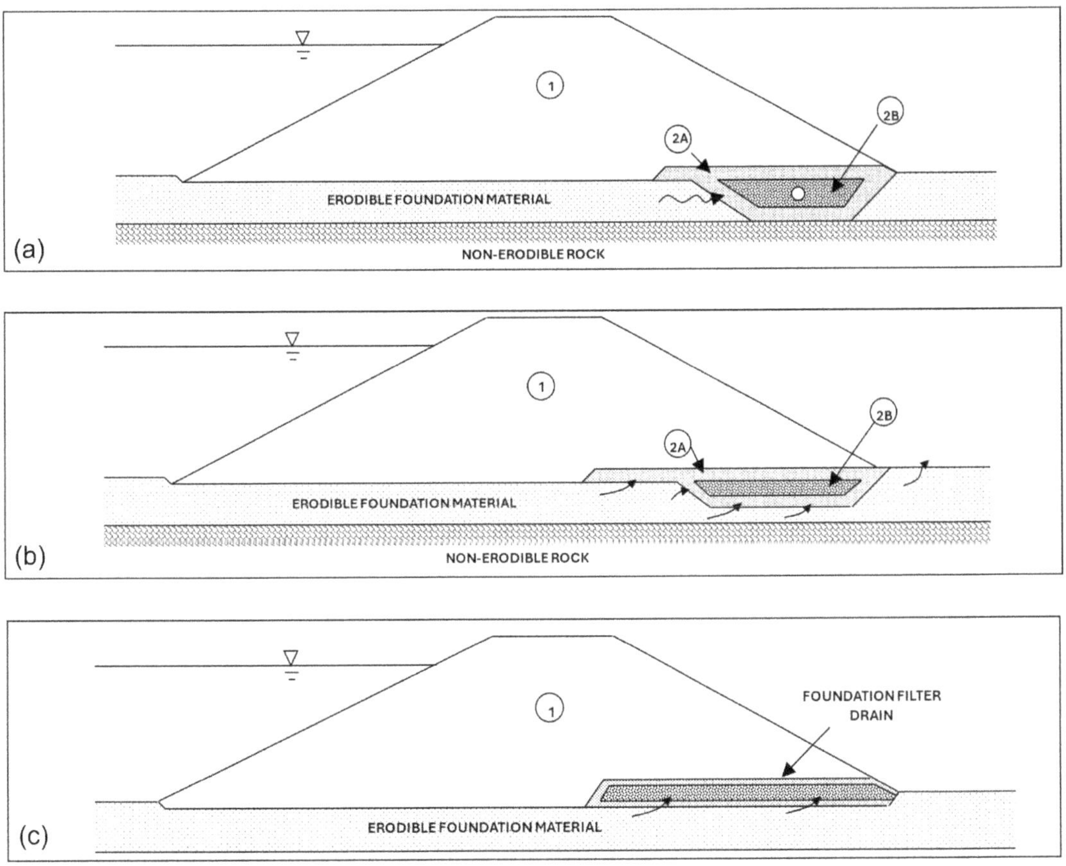

FIGURE 12.8 Examples of scenarios of fully penetrating and partially penetrating foundation filter drains. "Erodible foundation material" may be silt, sand or clay or erodible weathered rock with open defects such as desiccation cracks or open relict joint (Fell et al., 2024).

(a) Example of fully penetrating filter drain.
(b) Example of partially penetrating filter drain.
(c) Example of horizontal filter drain directly overlying erodible foundation material.

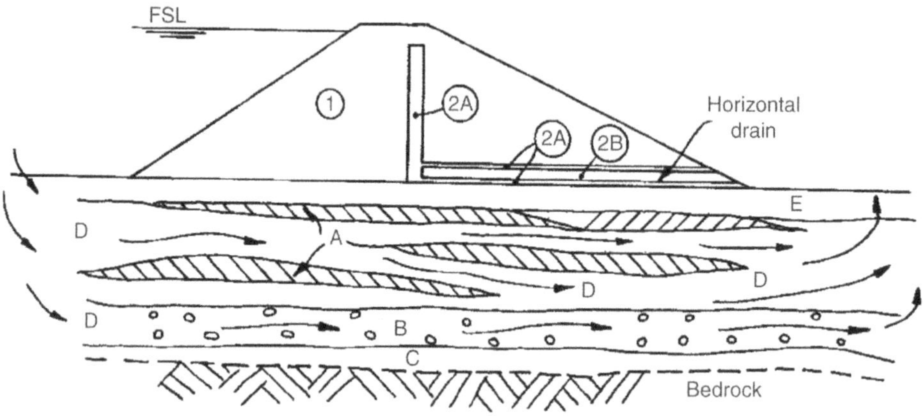

FIGURE 12.9 Example of an embankment where much of the seepage flow will be to an unfiltered exit (Fell et al., 2015).

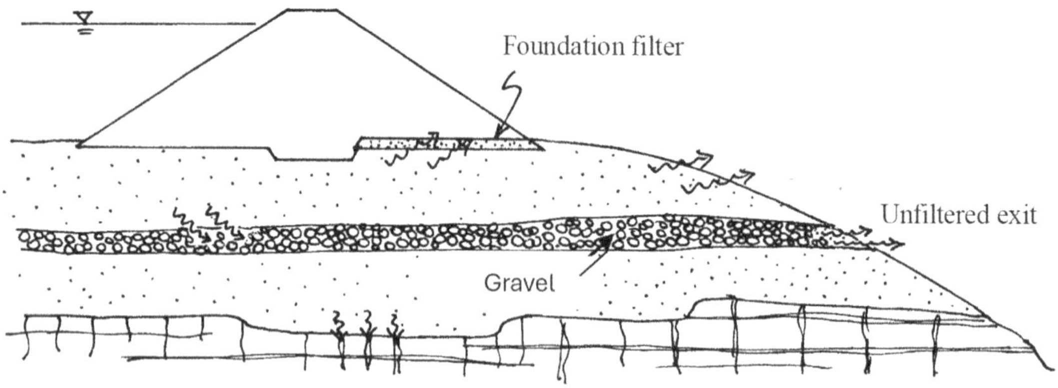

FIGURE 12.10 Example of an embankment where there is an unfiltered exit due to daylighting of the foundation gravel layer downstream of the dam (Fell et al., 2024).

For the situation of backward erosion piping within a foundation sand layer which has an overlying low permeability layer, the likelihood of finding an unfiltered exit is already considered in the assessment of initiation of backward erosion. An unfiltered exit is implicit if a heave condition or a sand boil is present.

12.4 PROBABILITY OF CONTINUATION FOR INTERNAL EROSION OF THE EMBANKMENT AT OR INTO THE FOUNDATION

12.4.1 Erosion into Open Joints in a Rock Foundation

Use the method described in Section 12.4.1 to evaluate the probability of erosion continuing.

12.4.2 Erosion into Coarse-Grained Soil Foundation

Evaluate the probability of the coarse-grained soil foundation layer being sufficiently coarse grained and continuous to allow erosion of the core materials. Evaluate the probability of the foundation soils permitting continuing erosion, excessive erosion and some erosion of the core materials using the methods described in Section 12.4.2.

If the geological conditions of the soil layering are such that the coarse-grained soil layers at the core–foundation contact are unlikely to daylight at an unfiltered exit or need to exit via an interconnected path of another soil layer which is capable of acting as an effective filtering medium, then reduce the probability for continuing erosion using judgement.

13 Internal Erosion and Piping—Progression, Detection, Intervention and Repair, and Breach

13.1 INTRODUCTION

As discussed in Section 3.1.4, for a dam to fail as a result of internal erosion and piping, erosion must progress after initiation and filters fail to arrest the erosion. Then given progression, detection and intervention must fail, and breach initiate.

This chapter provides advice on estimating the probability of the progression, detection, intervention and breach stages of the process for internal erosion in the embankment, foundation and embankment to foundation.

The chapter is written in quantitative risk analysis terms. Those assessing dam safety by deterministic or risk enhanced deterministic methods will find the information useful in making their engineering assessment.

Much of the advice is taken from Fell et al. (2008, 2024) and is based on experience and engineering judgement.

13.2 PROGRESSION OF INTERNAL EROSION

13.2.1 Introduction

Internal erosion progresses if erosion has initiated and filters or a transition have not arrested the erosion. The process depends on the mechanism of internal erosion:

13.2.1.1 Concentrated Leak Erosion

Progression of concentrated leak erosion occurs by enlargement of the eroded crack or gap resulting from the hydraulic shear stresses applied to the walls of the crack or gap. These stresses

DOI: 10.1201/9781003542025-13

increase as the crack or gap enlarges. The steps in the procedure to estimate the probability erosion will progress are:

Step 1: Estimate the probability (P_{ROOF}) that the soil will "form a roof" over a pipe (Section 13.2.2).
Step 2: Estimate the probability (P_{NCF}) that "crack filling" action will not be effective in stopping the erosion process (Section 13.2.3).
Step 3: Estimate the probability (P_{NFL}) that flow in the developing pipe will not be restricted by an upstream zone (or for example a concrete face slab) so the erosion process continues to develop (Section 13.2.4).

The Probability of Progression (P_{PROG}) = (P_{ROOF}) × (P_{NCF}) × (P_{NFL}) (13.1)

13.2.1.2 Contact Erosion

In Non-Plastic Soil Overlying Gravel or an Open Defect in Rock
Progression of contact erosion in non-plastic soil occurs with on-going erosion of the finer soil into the coarse soil or open defect in rock. No pipe is formed so no roof is required. Erosion will progress unless the particles eroding from the soil seal on the filter or transition zone (if present), or the flows are limited by an upstream zone. The likelihood of these happening is estimated using Sections 12.4 and 13.2.4.

In plastic Soil Overlying Gravel or an Open Defect in Rock
The progression process is for the erosion hole in the plastic soil overlying the gravel or open defect to enlarge in a similar way to concentrated leak erosion. The three steps in the estimation of the probability of progression are as for concentrated leak erosion.

13.2.1.3 Backward Erosion Piping (BEP)

Progression of BEP occurs by the channels formed by BEP in the non-plastic soil beneath the plastic soil layer eroding towards the river or reservoir. The method for estimating the probability that progression will reach the river or reservoir is described in Section 10.4.

If the potential BEP pathway requires the pipe to pass through an open defect within a sheet pile wall or concrete element, then flows through the fully developed pipe could be flow limited. Where applicable, this can be evaluated using the guidance provided in Section 13.2.4.

13.2.1.4 Suffusion and Global Backward Erosion

Progression of suffusion and GBE occurs as finer particles are eroded from the soil. As described in Section 10.6.5, the effect is to result in an increase in permeability. No pipe is formed so no roof is required. Erosion will progress unless the particles eroding from the soil seal on the filter or transition zone, or the flows are limited by an upstream zone.

13.2.2 Probability of Forming a Roof (P_{ROOF})

13.2.2.1 Internal Erosion through the Embankment

For internal erosion and piping through the dam or piping from the embankment into a rock foundation by contact erosion, the core must be capable of holding the roof of a pipe. Assess the probability of the soil forming a roof to a pipe using Table 13.1.

13.2.2.2 Internal Erosion through a Soil Foundation

For backward erosion piping through a soil foundation, the roof of a pipe will be formed by layers of soil in the foundation which are plastic or have high fines content, or by the core of the embankment. Other geological conditions which may form a roof within a soil foundation include where, for example, basalts overlie the soil layer.

Assess the probability of the embankment and foundation materials supporting the roof of a pipe in the foundation using Table 13.1.

In most cases the core of the embankment can provide a roof to a developing pipe in the foundation. However, if there are upstream or downstream zones of non-plastic granular material in the embankment that are not capable of supporting a roof of a pipe (e.g. rockfill or gravel shells), then a pipe through the foundation may not be able to fully develop. Figure 13.1b shows an example of this.

13.2.3 Probability of Crack Filling Action Not Being Effective (P_{NCF})

13.2.3.1 Internal Erosion in the Embankment

Estimate the probability (P_{NCF}) of crack filling action not being effective in stopping progression and pipe enlargement by:

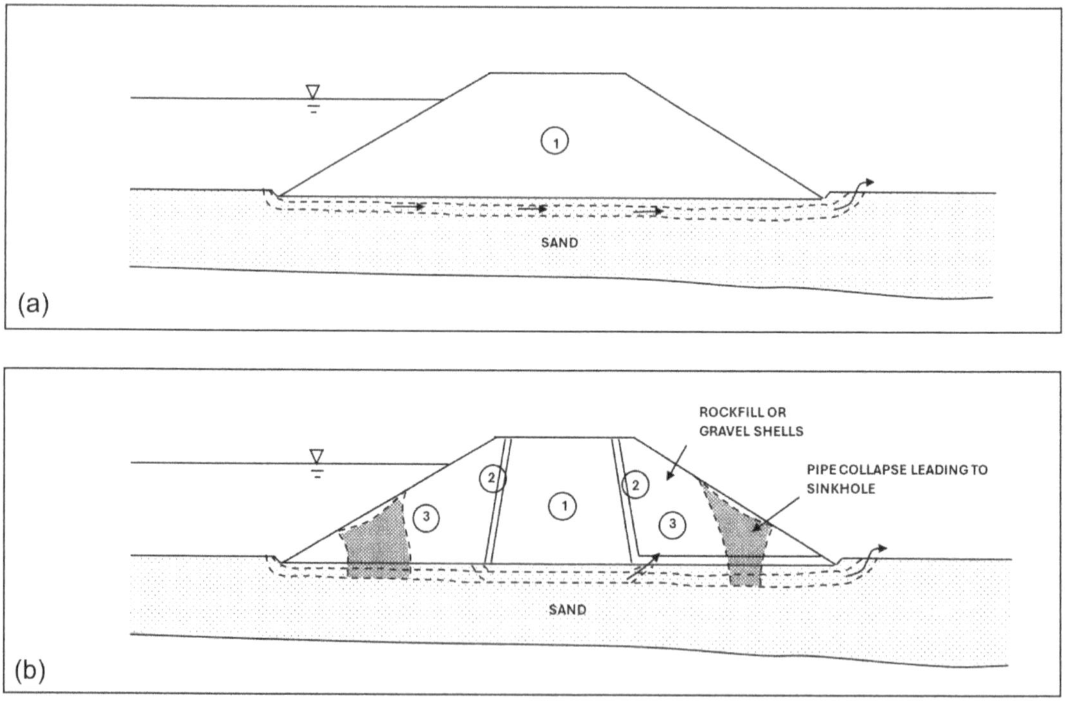

FIGURE 13.1 Scenarios for holding a roof of a pipe for internal erosion through the foundation (Fell et al., 2024).

(a) Homogeneous earthfill dam.
(b) Dam with gravel or rockfill shells.

TABLE 13.1 Probability of a Soil Being Able to Support a Roof to an Erosion Pipe (Fell et al., 2024)

SOIL CLASSIFICATION	*FINES CONTENT (PERCENTAGE MINUS 0.075 MM)*	*PLASTICITY OF THE MINUS 0.425 MM FRACTION*	*MOISTURE CONDITION*	*PROBABILITY OF HOLDING A ROOF*
Clays, sandy clays (CL, CH, CL-CH)	> 50%	Plastic	Moist or saturated	1.0
Silts (ML or MH)	> 50%	Plastic or non-plastic	Moist or saturated	1.0
Sandy clays, Gravelly clays (SC, GC)	15%–50%	Plastic	Moist or Saturated	1.0
Silty sands, Silty gravels, Silty sandy gravel (SM, GM)	> 15%	Non-plastic	Moist Saturated	0.7 to 1.0 0.5 to 1.0
Granular soils with some plastic fines (SC, GC)	10% to 15%	Plastic	Moist Saturated	0.2 to 1.0 0.1 to 0.5
Granular soils with some plastic fines (SC, GC)	5% to 10%	Plastic	Moist Saturated	0.02 to 0.2 0.01 to 0.1
Granular soils with some non-plastic fines (SM, SM-SP, SM-SW, GM, GM-GP, GM-GW)	10% to 15%	Non-plastic	Moist Saturated	0.02 to 0.2 0.01 to 0.1
Granular soils with some non-plastic fines (SM, SM-SP, SM-SW, GM, GM-GP, GM-GW)	5% to 10%	Non-plastic	Moist Saturated	0.001 to 0.02 0.0005 to 0.01
Granular soils (SP, SW, GP, GW)	< 5%	Non-plastic Plastic	Moist and saturated Moist and saturated	Negligible 0.001 to 0.01

Notes:
(a) The probabilities are for well-compacted materials.
(b) Cemented materials give higher probabilities than indicated in the table. If soils are cemented, use the category that best describes the particular situation.
(c) Saturated conditions only apply where the downstream shoulder is below the tailwater level of the dam.
(d) If the granular soils are angular, e.g. crushed rock, higher probabilities may apply. See Table 12.3 for guidance.

Method A

(a) Estimate the particle size distribution of the soil likely to be eroded from the walls of a crack from a zone upstream of the core.
(b) Assess the capability of the filter or transition downstream of the core to act as a filter to these particles using the method described in Section 12.2.4.
(c) Assess whether given the filter or transition is capable of acting as a filter to the eroding particles, whether the filter will now act as a filter to the combination of eroded materials from the core and upstream zones.

Method B

Use Table 13.2 to estimate the probability.

Method A will require some judgement. Method B is approximate and should not be relied upon for important decisions on dam safety.

For piping through the dam, crack filling from an upstream zone could limit the extent of erosion in the core if the materials washed into the crack or pipe can filter against the downstream filter or transition zone. The washed in materials aid in the filtering action against the downstream zone. This will be of greatest benefit in cases where there is poor filter compatibility between the core and downstream filter due to a lack of sand size particles in the core. In these cases, the probability of continuation may be high, but the washed in materials may be capable of filtering against the downstream filter zone, and this reduces the potential for the erosion progressing and the pipe enlarging. There is less benefit where the materials that are washed in are of similar sizes to those already in the core, hence the probabilities for crack filling in Table 13.2 are higher for a well-graded core material compared to those for a core which is deficient in sand sizes. There is very little or no benefit where there is no downstream filter/transition zone.

13.2.3.2 Internal Erosion through the Foundation

The potential for crack filling action for internal erosion in the foundation depends on the zoning of the embankment and the geological features in the foundation.

TABLE 13.2 Probability (P_{NFL}) for Crack Filling Action Not Being Effective in Stopping Pipe Enlargement—Internal Erosion through the Embankment (Fell et al., 2024)

EMBANKMENT ZONING	*UPSTREAM GRANULAR ZONE*	*DOWNSTREAM FILTER OR TRANSITION OR OTHER GRANULAR MATERIAL*	*LIKELIHOOD OF PIPING PROGRESSING—CRACK FILLING ACTION NOT EFFECTIVE*
Homogeneous, earthfill with toe drain, earthfill with horizontal drain, concrete face earthfill, puddle core earthfill, earthfill with core wall, hydraulic fill	None except for rip-rap and filters under these	None, or none effective	1.0
Earthfill with vertical and horizontal drain, zoned earthfill	None	Present	1.0
Central and sloping core earth and rockfill or zoned earthfill with gravel shoulders	Present	Present	0.5 to 1.0 If the core is well graded and has fine to coarse sand sizes (0.075–4.75 mm) already present(1) 0.1 to 0.5 If the core is deficient in sand sized particles and washed in sand material aids in sealing the downstream zone(2)

Notes:

(1) Crack filling is more likely to stop pipe enlargement when the core zone is deficient in sand size particles, and these particles can be provided by washing in from the upstream zone. This aids in sealing of the downstream filter zone. If the core is well graded and has sand sizes present, then the potential benefits of crack filling are less as the sand size particles are already present.

(2) Probability dependent on compatibility of particle sizes of granular soils upstream of the core and in the downstream filter transition.

If the eroding foundation soil layer is located immediately below the embankment, and the embankment forms the roof of the pipe, then evaluate the probability for crack filling action as for erosion in the dam using Table 13.2.

If the eroding foundation soil layer is located further below the embankment, then consider the potential for the overlying soil layers to wash into the developing pipe. There needs to be a filtering material at the downstream end of the flow path for crack filling action to be effective.

The filtering material may be a naturally occurring layer in the foundation, or the embankment filter.

13.2.3.3 Internal Erosion of the Embankment into or at the Foundation

The issues are the same as for internal erosion through the embankment. Estimate the probability of crack filling action not stopping pipe enlargement using Table 13.2.

The authors' experience is that the likelihood of crack filling being effective is difficult to assess and that in many cases it is sufficient to assume it is ineffective.

13.2.4 Probability of Limitation of Flows Not Being Effective (P_{NFL})

13.2.4.1 Flow Limitation by Upstream Zone in the Embankment

Estimate the probability (P_{NFL}) that flow in the developing pipe will not be restricted by an upstream zone in embankment dams using the following procedure:

Upstream Zone Non-Plastic Sand, Gravel or Rockfill

(a) If the zone(s) upstream of the core is (are) non-plastic sand, sandy gravel, gravelly sand with < 5% fines or clean rockfill, upstream flow limitation may be effective. Estimate the flow rate in the eroding crack or gap as it develops as described in Section 9.13 and calculate for this flow what reservoir head will be lost in the upstream zone(s), taking account of the estimated permeability of the upstream zone and the geometry of the crack and upstream zone(s).

(b) When estimating the permeability of the upstream zone(s), take account of the potential for suffusion and GBE as described in Section 10.6.

(c) Calculate the hydraulic shear stresses in the eroding crack or gap accounting for the head loss in the upstream zones and the crack or gap width. If these exceed the critical shear stress, erosion will progress and the crack or gap will become wider as described in Section 9.13.

(d) Repeat steps (a) to (c) to see if an equilibrium condition develops where most reservoir head is lost in the upstream zone and erosion ceases in the core.

(e) From this information estimate the probability that upstream flow limitation will not be effective.

Upstream Zone Other than Non-Plastic Sand, Gravel, Rockfill

(a) For the mechanism causing cracking or flaw in the core also likely to persist into the upstream zone (e.g. common cause cracking):
The probability of upstream flow limitation not being effective = 1.0.

(b) For the mechanism causing cracking or flaw in the core not likely to affect the upstream zone (i.e. no common cause cracking):

- Assess and document the reasons for and against common cause cracking being present. From these, estimate the probability of no common cause cracking (P_{NCCC}).
- Record this and report this outcome. However, unless there is compelling evidence, it should be assumed common cause cracking will be present.
- The risk analysis team may opt to assess the no-common cause scenario in a manner similar to that described previously for non-plastic upstream zones. The authors counsel against doing this and have not done so in practice. The risk analysis team should decide how to model the no-common cause scenario, taking account of the details of the zoning and materials.

For dams with upstream concrete elements, cutoff walls or other structural elements within the embankment or foundation, a similar approach should be adopted. Table 13.3 may be used as a guide to behaviour.

13.2.4.2 Flow into or out of Open Joint in Conduits

Limitation of flows is not applicable to these failure modes and a probability of no flow limitation (P_{NFL}) of 1.0 should be used.

Erosion into open defects in a conduit may lead to the development of a sinkhole on the embankment, and this is considered under the breach node of the event tree (refer to Section 13.4.6).

Leakage out of a pressurised conduit is likely to be limited by the defect in the pipe. Estimate the potential flows out of the defects in the conduit pipe; this is used in Section 13.4.4 to estimate the likelihood for it to cause slope instability.

TABLE 13.3 Probability (P_{NFL}) That Flow in the Developing Pipe Will Not Be Restricted by a Cutoff Wall or a Concrete Element in the Erosion Path (Fell et al., 2024)

CHARACTERISTICS OF CONCRETE ELEMENT OR CUTOFF	*LIKELIHOOD FOR NO FLOW RESTRICTION*
Flow limitation by a concrete element in the embankment	
Concrete slab on upstream slope[(2)]	0.1 to 0.9 depending on the condition of the concrete slab and its joints
Concrete core within embankment (1920s–1930s)[(3)]	0.01 to 0.001
Soil cement wave protection	0.01 to 0.9 depending on the condition of the soil cement wave protection
Partially penetrating concrete core wall in dam (for internal erosion and piping along foundation contact)	0.01 to 0.001 depending on height of the wall
Flow limitation by cutoff walls in the foundation (for internal erosion and piping in the foundation)	
Sheet pile walls	Extruded—0.01 to 0.5 Cold rolled—0.1 to 0.9. If have good piezometers data to show wall is integral may use low end figure or possibly even lower

(Continued)

Table 13.3 (Continued)

CHARACTERISTICS OF CONCRETE ELEMENT OR CUTOFF	*LIKELIHOOD FOR NO FLOW RESTRICTION*
Modern diaphragm walls:	
Cementitious walls—conventional concrete, plastic concrete, cement bentonite.	0.0001 well-constructed, 0.001 serious defects suspected
Non-cementitious walls—soil bentonite[(1)]	0.001 to 0.01
Soil cement bentonite wall	0.01 to 0.1
Column walls—jet grouting, soil mixing	0.01 to 0.1
Open joint, water stop, crack, or other defect in the conduit	0.05 to 0.2

Notes:

(1) For these walls the soil is excavated by excavator or dragline, and the bentonite mixed with the excavated soil using earth moving equipment.

(2) Asphalt concrete facing may provide flow limitation but is likely to be more vulnerable to damage than concrete.

(3) Asphaltic concrete core is likely to provide a similar benefit to a concrete core.

13.2.4.3 Erosion into Jointed Bedrock

The possible scenarios are:

Contact erosion initiating at the core–foundation contact where there is no or a shallow cutoff trench. This is likely to lead to the pipe forming through the embankment. For this case, estimate the probability that flow in the developing pipe will not be restricted by an upstream zone as described for flow limitation by an upstream zone in the embankment.

Erosion initiating within a deep cutoff trench into open joints in a rock foundation. The extent of erosion may become limited by the opening width of the rock defects. This is allowed for in the method for estimating the likelihood of breach due to the flow through the open joints in the rock foundation (Section 13.4.4). The limitation of flows is therefore not applicable to this scenario and a probability of no flow limitation of 1.0 should be used.

13.3 PROBABILITY OF DETECTION, INTERVENTION AND REPAIR

13.3.1 General Principles

The likelihood that a particular failure mode can be detected, and if so, whether it is possible to intervene (e.g. by lowering the reservoir level), or carry out repairs to prevent the dam breaching is usually best considered as two questions:

1. Will this failure mode be detected?
2. Will intervention and repair be successful?

A probability is assigned to each of these questions. The overall probability of successful detection, intervention and repair is the combination of these two probabilities.

The likelihood of detection and successful intervention and repair is dependent on a number of factors, including:

(a) The category of internal erosion and piping, i.e. internal erosion in the embankment, the foundation or embankment to foundation.
(b) The mechanism of initiation of internal erosion—erosion in a crack, suffusion, global backward erosion, backward erosion piping or contact erosion.
(c) The breach mode—gross enlargement of a pipe, instability of the downstream slope, unravelling or sloughing of the downstream slope, settlement of the foundation, sinkhole development.
(d) The nature and the geometry of the materials in the foundation.
(e) The zoning of the embankment and the materials in the embankment.
(f) The reservoir level at the time of the piping incident, and how rapidly it can be drawn down.
(g) The type, location in relation to the piping incident and frequency of monitoring and surveillance at the dam and the training of the staff to recognise a developing internal erosion and piping incident.
(h) The ability to get trained personnel out to the site in the event of a piping incident.
(i) The ability of those responsible to be able to direct emergency release of the reservoir.
(j) The availability of materials and equipment to intervene and carry out repair works.

It is necessary to use judgement to assess these probabilities. The methods described in the following sections should be regarded as aids to judgement.

13.3.2 Some Information on the Rate of Internal Erosion and Piping for Concentrated Leak Erosion

The likelihood of detection and successful intervention or repair depends on the time from when the internal erosion process may be detected to when breach begins.

Fell et al. (2001, 2003) studied case histories of failures and accidents for piping in the embankment, foundation, and embankment to foundation. Based on the case histories and an understanding of the physical processes, they provided guidance on the time for progression beyond when a concentrated leak is first observed and the development of a breach. Table 13.4 to Table 13.7 are based on that study.

Table 13.4 should be used to estimate the approximate likely time to dam failure after a concentrated leak is first observed. Tables 13.1, 13.5, and 13.6 are used in Table 13.4 working from left to right. Upstream flow limitation is assessed as detailed in Section 13.2.4.

Table 13.4 is used by assigning the values to the first four columns and selecting the likely time for progression and breach which best fits the data.

Table 13.5 replaces an original table in Fell et al. (2001, 2003) to assess the likely rate of erosion of the core of the embankment or the soil in the foundation.

In these tables the terms for rates are defined as shown in Table 13.7. Dual descriptors are used to describe intermediate terms, e.g. very rapid—rapid for 6 hours. The terms are applied to part (e.g. progression) or the whole process.

Most of the cases studied were for breach by gross enlargement, so the method is applicable to cases where the mechanism is gross enlargement. It is considered to be reasonably applicable to cases where the final breach is by slope instability, following development of a pipe. It will probably underestimate the time for breach by sloughing. Sloughing is usually a slowly developing breach mode which should take days or weeks to lead to breach.

TABLE 13.4 A Method for the Approximation Estimation of the Time for Progression of Piping and Development of a Breach, for Breach by Gross Enlargement, and Slope Instability Linked to Development of a Pipe (Fell et al., 2001, 2003)

FACTORS INFLUENCING THE TIME FOR PROGRESSION AND BREACH					
ABILITY TO SUPPORT A ROOF FROM TABLE 13.1	*RATE OF EROSION FROM TABLE 13.5*	*UPSTREAM FLOW LIMITER FROM TABLE 13.2*	*BREACH TIME FROM TABLE 13.6*	*APPROXIMATE LIKELY TIME-QUALITATIVE*	*APPROXIMATE LIKELY TIME*
Yes	R or VR	No	VR or R–VR	Very Rapid	< 3 hours
Yes	R	No	R	Very Rapid to Rapid	3–12 hours
Yes	R–M	No	VR	Rapid	12–24 hours
Yes	R	No	R–M		
Yes	R	No	M or S	Rapid to Medium	1–2 days
Yes	R or R–M	No	M or M–S		
Yes	M or R–M	Yes	R or R–M		
Yes	M or R–M	No	S	Medium	2–7 days
Yes	R–M or M	Yes	S		
Yes	M	Yes or No	S	Slow	Weeks—even months or years

Note: VR = Very Rapid; R = Rapid; M = Medium; S = Slow.

TABLE 13.5 Rate of Erosion of the Core or Soil in the Foundation (Fell et al., 2001, 2003)

		TIME FOR EROSION IN THE CORE OF THE EMBANKMENT OR IN THE FOUNDATION	
SOIL CLASSIFICATION	*BEST ESTIMATE EROSION RATE INDEX (I_{HET})*	*GRADIENT ALONG PIPE 0.2*	*GRADIENT ALONG PIPE 0.5*
SM with < 30% fines	< 2	Very Rapid	Very Rapid
SM with ≥ 30% fines	2 to 3	Very Rapid	Very Rapid
SC with < 30% fines	2 to 3	Very Rapid	Very Rapid
SC with ≥ 30% fines	3	Rapid	Very Rapid
ML	2 to 3	Very Rapid to Rapid	Very Rapid
CL-ML	3	Rapid	Very Rapid
CL	3 to 4	Rapid	Very Rapid to Rapid
CL-CH	4	Rapid	Rapid
MH	3 to 4	Rapid	Very Rapid to Rapid
CH with Liquid Limit < 65%	4	Rapid to Medium	Rapid
CH with Liquid Limit ≥ 65%	5	Medium to Slow	Medium

Breach by sinkhole development is potentially a rapid process in the final stages when the sinkhole emerges into the reservoir. It is expected that breach will occur in a small number of hours but there is insufficient case data to support a more refined estimate.

Note that the dispersivity of the soil does not significantly affect the rate of erosion so is not listed as a factor in Table 13.5. For a homogeneous dam the whole of the embankment is the same soil, so in Table 13.5 the soil is considered as the core, and in Table 13.6 as the downstream zone.

TABLE 13.6 Influence of the Material in the Downstream Zone of the Embankment on the Likely Time for Development of a Breach (Fell et al., 2001, 2003)

MATERIAL DESCRIPTION	*LIKELY BREACH TIME*
Coarse-grained rockfill	Slow–medium
Soil of high plasticity (liquid limit ≥ 50%) and high clay size content including clayey gravels	Medium–rapid
Soil of low plasticity (liquid limit < 35%) and low clay size content, all poorly compacted soils, silty sandy gravels	Rapid–very rapid
Sand, silty sand, silt	Very rapid

TABLE 13.7 Qualitative Terms for Times of Development of Internal Erosion, Piping and Breach (Fell et al., 2001, 2003)

QUALITATIVE TERM	*EQUIVALENT TIME*
Slow (S)	Weeks or months, even years
Medium (M)	Days or weeks
Rapid (R)	Hours (> 12 hours) or days
Very Rapid (VR)	< 3 hours

Fell et al. (2001, 2003) show that the method gives a reasonable estimate of the time for progression beyond where a concentrated leak is observed and erosion initiated, and the initial breach and the times are acceptably accurate for the purpose here, which is to assess the likelihood of detection, intervention and repair.

Fell et al. (2001, 2003) caution, however, against over-reliance on these figures for life loss estimates where the estimates are sensitive to the assumed warning times, total breach (or time to failure) times and other factors.

The times estimated in Table 13.4 are only approximate, and hidden or unknown details within a dam or its foundation may give shorter or longer times.

13.3.3 Detection

13.3.3.1 Some General Principles

Detection may be possible in the continuation or early progression phase, or more likely, in the advanced stages of progression and breach formation. Detection is likely to be by:

1. Observation of increased seepage out of the downstream face of the embankment or in the foundation. This may be by visual observation, seepage measurement or more sophisticated methods such as thermal monitoring or geophysical testing of the foundation or the downstream slope.
2. Measured higher pore pressures in the foundation and/or embankment.
3. Settlements, deformation and cracking in the embankment or area downstream of the dam.
4. Walking along the toe of the levee. Heave may be evident by the presence of "blisters" or the ground feeling like walking on a "waterbed".

Whether detection is likely depends on:

1. The rate at which the internal erosion and piping and associated processes, such as instability of the downstream face, occurs.
2. The frequency of inspections and measurement of monitoring equipment.
3. The dam zoning and the location of the concentrated leak and whether the leak will be visible to those doing the inspection.

For example, if a process may go from initiation or first presence of a concentrated leak to breach in say 6 hours, and the dam is only inspected or monitored weekly, it is very unlikely that a piping incident will be detected before breach occurs. However, if the dam is visible by the general population, there is some chance the leak may be detected nonetheless.

The following are some factors which affect the likelihood of successful detection:

(a) Detection early in the internal erosion process is usually difficult, particularly for concentrated leak erosion initiating along a crack or as contact erosion, because the amount of leakage is very small at the start. Fell et al. (2001, 2003) record that most piping incidents are first identified as a concentrated leak in the progression phase.
(b) Backward erosion piping is often first apparent by the presence of sand boils. The presence of conditions potentially leading to heave and backward erosion piping in the foundation may also be detected by piezometers provided they are correctly positioned and read as reservoir levels rise. The progression of BEP beneath a levee may be slow developing, but the authors do not have quantitative information on this. There is also the issue that BEP may only be detected sometime after it initiates.
(c) Suffusion and GBE are more likely to be detected by piezometers or an increase in seepage because the process is slower to develop than for concentrated leak erosion, particularly for GBE.
(d) Visual inspection is a vital tool in detecting internal erosion and piping; whether it is successful is dependent on the factors discussed earlier, but also on such practical issues as the following:
    - Inspections are seldom practical at night, so 30% to 50% of the time (varying throughout the year) detection will not be effective, particularly for rapidly developing piping mechanism. Many dams are not inspected on weekends, further reducing the likelihood of detection.
    - Dense vegetation, runoff from rainfall and snow cover can all hide the presence of a concentrated leak. However, it can be the case that melted snow is a good indicator of areas affected by seepage.
    - For very long embankments, it may not be practical to walk to inspect, so it is less likely small leaks are detected. However, if an all-terrain vehicle is driven slowly along the toe, good inspection can be achieved.

It is known that most internal erosion and piping failures occur at reservoir levels close to or above historic high, and the physical processes are driven by the reservoir water. Hence a good monitoring and surveillance program will have an increased frequency of inspections and reading of critical instruments under such reservoir conditions.

13.3.3.2 Assessing the Probability of Not Detecting Internal Erosion

The probability of not detecting internal erosion is determined by the following:

(a) Assess the probability of not observing the concentrated leak (P_{nol}), allowing for the location of the leak for the failure mode under consideration and factors which may mean the leak cannot be observed. This is done using Table 13.8 and Figure 13.2.
(b) Assess the probability that given the leak is observable ($1 - P_{nol}$), it is not detected (P_{nd}), allowing for the time between the first appearance of the concentrated leak and the frequency of inspections and/or reading of monitoring instruments. This is done using Table 13.9.
(c) The probability of not detecting the internal erosion:

$$= P_{nol} + [(1 - P_{nol}) \times (P_{nd})] \quad (13.2)$$

When using these tables, take account of:

- The location of the potential leak when assessing the probabilities. For example, there might be a dam where leaks in the abutment area may be readily observed because the foundations are low permeability and the vegetation is clear, whereas they may be difficult to observe in

TABLE 13.8 Factors Influencing the Likelihood of Not Observing a Concentrated Leak (P_{nol}) (Fell et al., 2024)

		INFLUENCE ON LIKELIHOOD OF NOT OBSERVING			
FACTOR	*RELATIVE IMPORTANCE*	*LESS LIKELY (1)*	*NEUTRAL (2)*	*MORE LIKELY (3)*	*MUCH MORE LIKELY (4)*
Can a concentrated leak be observed at toe?	(3)	Foundation low permeability soil or rock, so leaks will emerge at the toe. *and* No vegetation or only mown grass at toe, so observation of leakage is easy.	Foundation low permeability soil or rock, so leaks will emerge at the toe. *and/or* Vegetation at toe may preclude observation of seepage.	Foundation medium permeability soil or rock, so leaks may remain in the foundation and not emerge at the toe. *and/or* Dense vegetation at toe makes observation of seepage difficult.	Tail-water drowns part or all of toe or foundation permeable alluvium, so leaks may not emerge at the toe. *and* Dense vegetation at toe makes observation of seepage difficult.
Dam zoning which affects whether leaks emerge on the downstream face of the embankment	(2)	Homogeneous, earthfill with core wall, concrete face earthfill	Earthfill with toe drain, zoned earthfill dam, puddle core	Earthfill with horizontal and chimney drains, zoned earth and rockfill	Central core earth and rockfill dams, concrete face rockfill, rockfill with core wall
Seepage instrumentation weirs, etc.	(1)	Seepage collected to flow to readily observed measuring weir or real-time monitored	Seepage partly collected to flow to measuring weir	Seepage partly collected to flow to measuring weir but masked by rainfall effects	No seepage collection system

the river section because the foundation is high permeability alluvium and the toe overgrown with vegetation.

- The toe of the embankment being drowned out by another reservoir or is in the river, which may make it virtually impossible to detect leaks in this part of the dam.
- The internal erosion mechanism. It may be easier to detect some such mechanisms than others, e.g. backward erosion piping in the foundation because sand boils will form.

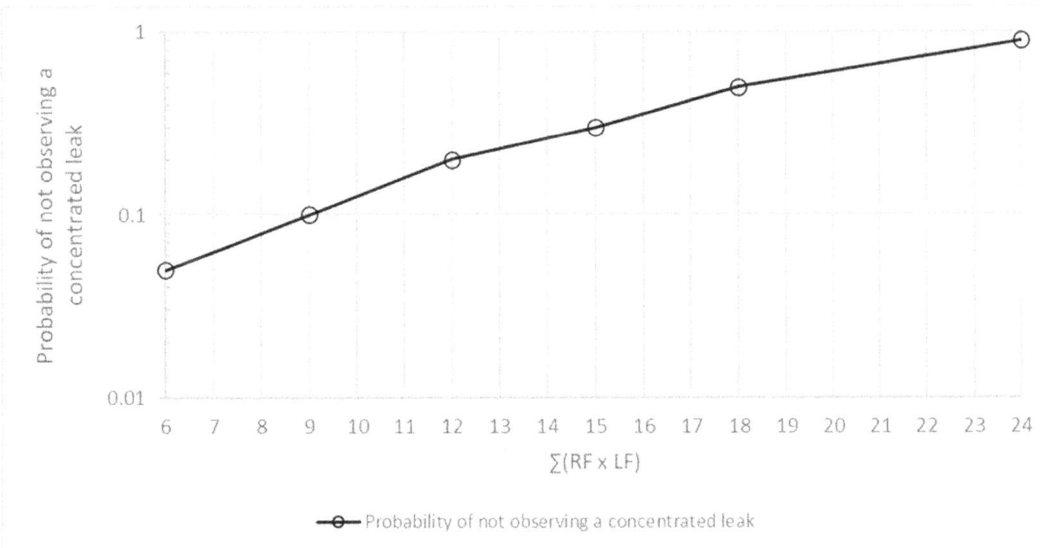

FIGURE 13.2 Probability of not observing a concentrated leak (P_{nol}) versus Σ (Relative importance factor [RF]) × (Likelihood factor [LF]) for internal erosion in an embankment.

TABLE 13.9 Probability That Given the Leak Is Observable It Is Not Detected Given the Time between the First Appearance of the Concentrated Leak and the Initial Breach, and the Frequency of Inspections and/or Reading of Monitoring Instruments (Fell et al., 2024)

FREQUENCY OF INSPECTION AND/OR MONITORING	*PROBABILITY OF NOT DETECTING THE INTERNAL EROSION (P_{nd}) GIVEN THE TIME FOR DEVELOPMENT OF CONCENTRATED LEAK FROM INITIAL APPEARANCE TO INITIAL BREACH FROM TABLE 13.4.*					
	< 3 HOURS	*3 TO 12 HOURS*	*12 TO 24 HOURS*	*1 TO 2 DAYS*	*2 TO 7 DAYS*	*WEEKS OR MONTHS*
Monthly, no public nearby	0.999	0.99	0.95	0.9	0.6	0.1
Monthly, public nearby	0.999	0.8	0.5	0.25	0.1	0.05
Weekly, no public nearby	0.99	0.95	0.9	0.7	0.2	0.1
Weekly, public nearby	0.99	0.75	0.5	0.2	0.1	0.05
Daily, no public nearby	0.9	0.6	0.5	0.1	0.05	0.01
Daily, public nearby	0.8	0.5	0.4	0.1	0.05	0.01
Daily with real time monitoring of leakage	0.2	0.15	0.1	0.1	0.05	0.01

13.3.4 Assessing the Likelihood of Intervention and Repair

Intervention and repair to prevent the progression of internal erosion and piping and breach can take several forms, including:

(i) Drawing down the reservoir level using spillway gates or outlet valves.
(ii) For backward erosion piping, installing pressure relief wells in the foundation of the embankment and/or building reverse filters over "boils" or areas where eroding material is emerging from the foundation. If the developing pipe is not too deep it may be possible to drive sheet piles through the pipe from the dam crest.
(iv) For internal erosion in the embankment, building a weighting berm to reduce the likelihood of heave, slope instability or unravelling.
(v) Constructing a reverse filter at the downstream end of a developing concentrated leak or zone of suffusion and GBE.
(vi) For dams which have a downstream rockfill, transition or filter which is too coarse to arrest erosion, dumping granular material (sand/gravel/rockfill) into the upstream side of sinkholes or the upstream end of a developing concentrated leak to try to block them. For homogeneous dams, dumping plastic soil may be more successful than granular materials.

More than one of these measures may be used together. Which is applicable or feasible will depend on the failure mode and particular circumstances of the dam.

It should be recognised that there may be reluctance on the part of the reservoir owner or operator to release water given the lost revenue that may result or if release of reservoir water is likely to result in property damage and loss of life, for example if levee banks downstream of the dam are likely to be overtopped by the flood resulting from release of the water.

Table 13.10 may be used to assist in assessing the probability that given the concentrated leak is detected, intervention and repair is not successful. This is done for each reservoir level partition. It is not practical to cover all the possible scenarios, and those doing the risk analysis are required to make their assessment within the range of probabilities shown. In making this assessment consideration should be given to the failure mode and location of the developing pipe.

13.3.5 Calculation of the Probability of Not Detecting and Not Intervening

The probability of not detecting and not intervening is calculated as follows:

Probability that intervention fails = [Probability of not observing the concentrated leak because it is not observable] + [Probability leak is observable but not detected] + [Probability leak is observable and detectable, but intervention fails], where:

$$[\text{Probability of not observing the concentrated leak because it is not observable}] = P_{nol} \quad (13.3)$$

$$[\text{Probability leak is observable but not detected}] = (1 - P_{nol}) \times P_{nd} \quad (13.4)$$

$$[\text{Probability leak is observable and detected but intervention fails}] = (1 - P_{nol}) \times (1 - P_{nd}) \times P_{ui} \quad (13.5)$$

or

$$\text{Probability that intervention fails} = P_{nol} + [(1 - P_{nol}) \times P_{nd} + (1 - P_{nol}) \times (1 - P_{nd}) \times P_{ui}] \quad (13.6)$$

TABLE 13.10 Assessment of the Probability That Given the Concentrated Leak Is Detected, Intervention and Repair Are Not Successful (P_{ui}) (Fell et al., 2024)

TIME FOR DEVELOPMENT OF CONCENTRATED LEAK TO INITIAL BREACH	*WHAT CAN BE DONE*	*PROBABILITY OF NOT INTERVENING*
< 3 h	There is too little time to successfully intervene regardless of the failure mode.	0.99
3 to 12 h	In most cases it will be impractical to intervene successfully in this amount of time. Only in cases where there is a straightforward method of intervention and personnel, equipment and materials are available will intervention be successful.	0.9 to 0.99
12 to 24 h	In many cases it will be impractical to intervene successfully in this time. Only in cases where there is a straightforward method of intervention and personnel, equipment and materials are available will intervention be successful; or it is a small storage which can be drawn down to stop the failure mode.	0.85 to 0.99
1 to 2 days	In many cases it will be impractical to intervene successfully in this time. Only in cases where there is a straightforward method of intervention and personnel, equipment and materials are available will intervention be successful; or it is a small storage or medium storage with large gate discharge capacity which can be drawn down to stop the failure mode.	0.7 to 0.95
2 to 7 days	In some cases it will be practical to intervene successfully in this time. In cases where there is a straightforward method of intervention and personnel, equipment and materials are available; or it is a small storage or medium storage with large gate discharge capacity allowing the reservoir to be drawn down to stop the failure mode.	0.2 to 0.9[(a)]
Weeks or months	In some cases it will be practical to intervene successfully in this time. Where there is a straightforward method of intervention, personnel, equipment and materials are available and large resources intervention has a fair chance of being successful; or it is a small or medium storage with large gate discharge capacity allowing the reservoir to be drawn down to stop the failure mode.	0.1 to 0.8[(a)]

Note: (a) Use values less than 0.5 only if there is a high degree of confidence in the assessed value.

This calculation is represented by the sub-event tree structure shown in Figure 13.3.

13.4 PROBABILITY OF BREACH

13.4.1 Overall Approach for Estimating the Probability of Breach

For each general failure mode, including internal erosion in the embankment, in soil or rock foundations and from embankment to foundation:

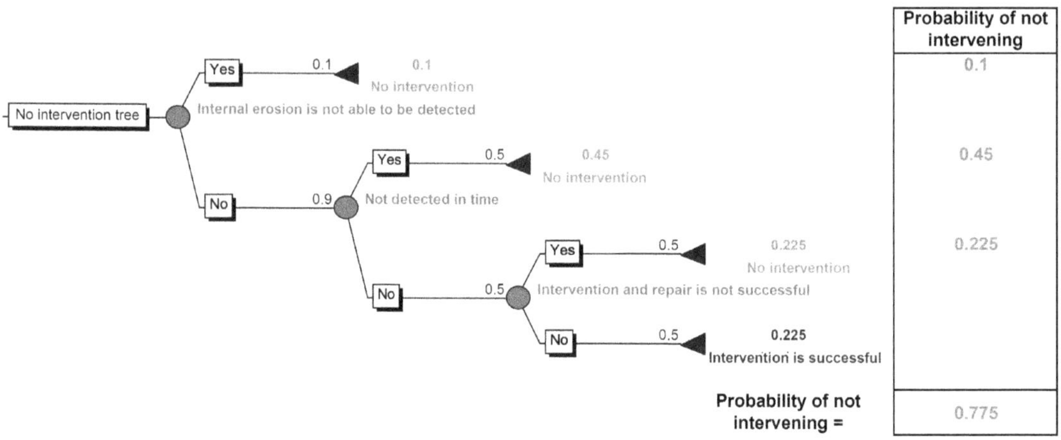

FIGURE 13.3 Sub-event tree for calculating the probability of not intervening.

Step 1: Screen the breach mechanisms depending on the dam zoning type using Section 13.4.2.
Step 2: Estimate the probability of breach by gross enlargement (P_{ge}) of the pipe using Section 13.4.3.
Step 3: Estimate the probability of breach by instability of the downstream slope (P_{si}) using Section 13.4.4. This involves estimating the probability of slope instability occurring and the probability of loss of freeboard given instability.
Step 4: Estimate the probability of breach by unravelling or sloughing of the downstream slope (P_{su}) of the embankment using Section 13.4.5.
Step 5: Estimate the probability of breach by sinkhole development (P_{sd}) using Section 13.4.6.
Step 6: Estimate the overall probability of breach by adding the probabilities for each of the four mechanisms, using the appropriate statistical summation:

$$P_{breach} = 1 - [(1 - P_{ge}) \times (1 - P_{si}) \times (1 - P_{su}) \times (1 - P_{sd})] \tag{13.7}$$

In making these estimates take account of the mechanism of internal erosion, i.e. concentrated leak, backward erosion piping, suffusion, GBE or contact erosion.

In many cases it will be apparent that one of these breach mechanisms is the most likely. It will be sufficient to assume that breach mechanism will apply and not calculate the probability of breach for the other mechanisms.

In most cases the probability of breach will be 1.0 when it is on the continuing erosion branch of the event tree for concentrated leak, backward erosion piping and contact erosion. Exceptions to this will be if the pool level drops below the inlet of the developing pipe before a breach mechanism has time to develop. This is considered in Section 13.4.3.

For these breach mechanisms the probability of breach for concentrated leak erosion is often less than 1.0 when on the some or excessive erosion branches of the event tree. For these cases, the seepage flows that develop through the dam or foundation will be limited by the filter material. The probability of breach will depend on how well the dam can cope with the leakage flows that may occur. Section 12.2.5 provides guidance on the possible magnitude of leakage flows that may develop for the some and excessive erosion conditions.

This will also apply for suffusion and GBE except that as discussed in Section 10.6.5 the flow is likely to be limited by the permeability of the soil after erosion by suffusion and GBE.

13.4.2 Screening of Breach Mechanisms

Figure 13.4 shows the potential breach mechanisms.

For most dam types and failure modes, the likelihood for breach development will be dominated by one or two of the potential breach mechanisms. Breach mechanisms will not necessarily be applicable to some dam zoning types or modes of piping and can be ignored.

Table 13.11 lists those breach mechanisms which should be considered in the assessment for internal erosion through the embankment, internal erosion in soil foundations and from embankment into foundation depending on the dam zoning type. Figure 2.1 in Section 2.1 shows the dam zoning types.

Table 13.11 does not apply to failure modes involving open or infilled defects related to solution features in rock foundations, e.g. karst limestone, because the leakage flows may exceed the capacity of even free draining rockfill.

Internal erosion by the process of suffusion or GBE is very unlikely to lead to the formation of a pipe through the dam or its foundation, and hence the probability of breach by gross enlargement where

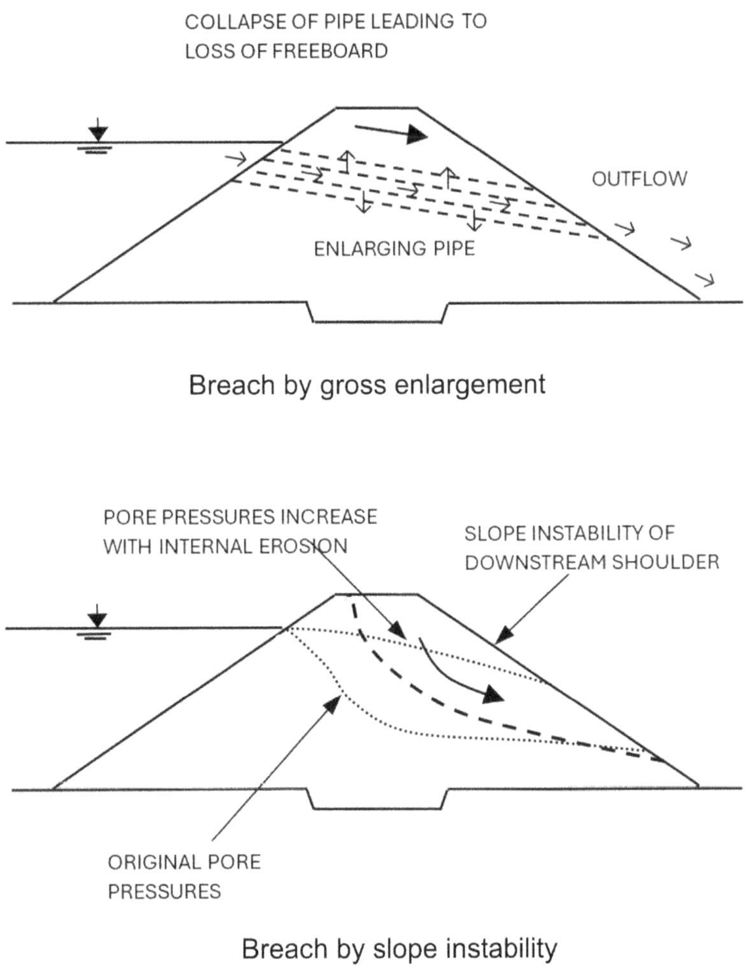

FIGURE 13.4 (A) Potential breach mechanisms—pipe gross enlargement and slope instability.

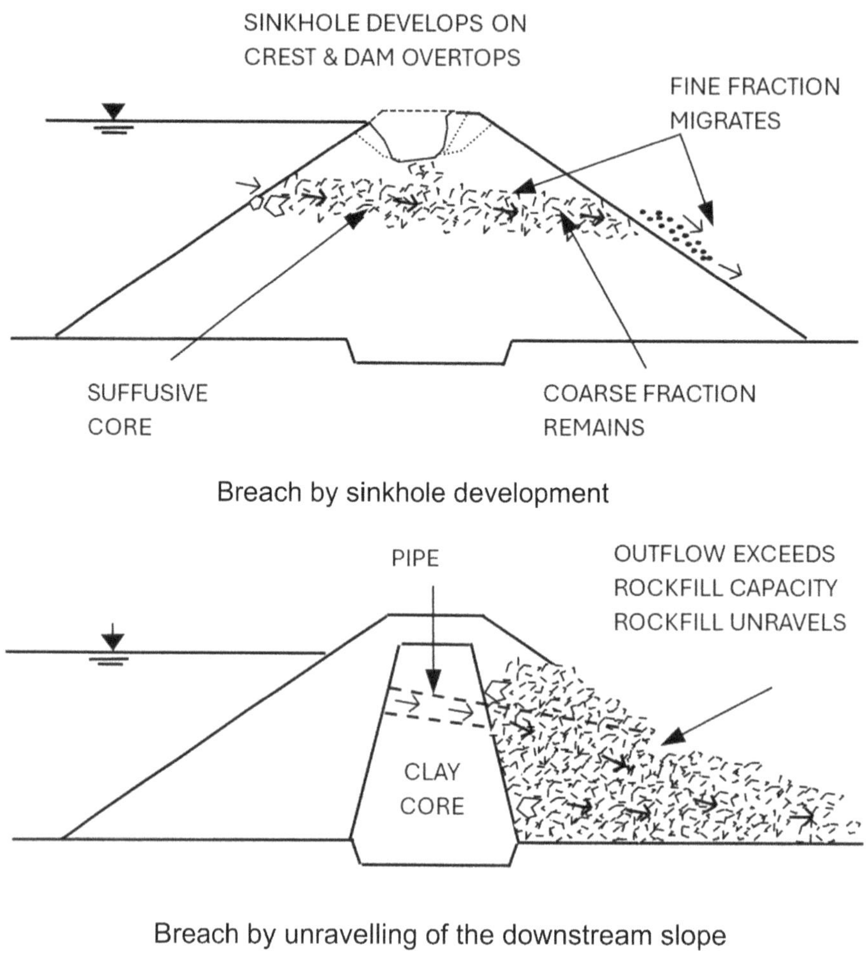

FIGURE 13.4 (B) Potential breach mechanisms—overtopping by sinkhole development and unravelling of the downstream face.

the mode of erosion is suffusion can be excluded. Breach by slope instability or sloughing/unravelling are usually the more critical mechanisms for suffusion and GBE but sinkhole development may also occur. The likelihood of slope instability and unravelling will be dependent on the post-erosion permeability of the soil and extent of the soil in the core or foundation affected.

13.4.3 Estimation of the Probability of Breach by Gross Enlargement

13.4.3.1 Screening for Internal Erosion in the Embankment, Soil Foundation, and Embankment into Foundation

Breach by gross enlargement of the pipe requires a continuing erosion condition. Breach by gross enlargement of the pipe can be considered to be negligible probability when on the "Some Erosion" and "Excessive Erosion" branches of the event tree.

TABLE 13.11 Screening of Breach Mechanisms for Internal Erosion through the Embankment, Internal Erosion in Soil Foundations and from Embankment into Foundation (Fell et al., 2024)

	BREACH MECHANISMS			
DAM ZONING TYPE	*GROSS ENLARGEMENT*	*SLOPE INSTABILITY*	*SLOUGHING OR UNRAVELLING*	*SINKHOLE DEVELOPMENT*
Homogeneous earthfill	✓*	✓	Exclude, except if downstream fill is non-plastic	Exclude except for piping related to conduit through the embankment
Earthfill with filters	✓*	✓	Exclude, except if downstream fill is non-plastic	Exclude except for piping related to conduit through the embankment
Earthfill with rockfill toe	✓*	✓	Exclude, except if downstream fill is non-plastic	Exclude except for piping related to conduit through the embankment
Zoned earthfill	Include, except if downstream fill cannot support a roof	✓*	Exclude, except if downstream fill is non-plastic	Exclude except for piping related to conduit through the embankment
Zoned earthfill and rockfill	Exclude, except if downstream fill can support a roof	✓	✓*	✓
Central core earth and rockfill (or gravel shells)	Exclude, except if downstream fill can support a roof	✓	✓*	✓
Concrete face earthfill	✓	✓*	Exclude, except if downstream fill is non-plastic	Exclude except for piping related to conduit through the embankment
Concrete face rockfill (including gravel fill)	Exclude	Exclude, except if dam is gravel or low permeability	✓*	Exclude
Puddle core earthfill	✓*	✓	Exclude, except if downstream fill is non-plastic	✓
Earthfill with core wall	Exclude	✓*	Exclude, except if downstream fill is non-plastic	Exclude unless there is a defect in the core wall
Rockfill with core wall	Exclude	✓	✓*	✓
Hydraulic fill	Exclude, except if downstream fill can support a roof	✓	✓*	✓

Key:

✓ Breach mechanism should be included for the probability estimate.

✓* Breach mechanism should be included for probability estimate, and usually is the more critical mechanism.

For internal erosion through the embankment, breach by gross enlargement can be considered negligible in cases where the downstream shell is unable to support a roof of a pipe. Use Table 13.12 to assess if this is applicable.

If applicable, estimate the probability of breach by gross enlargement using Table 13.13. This considers whether the reservoir will drop below the level of the pipe before the enlarging pipe develops into a breach. The authors' experience is that these probabilities seldom apply, and if detection and intervention fail, breach is likely in most cases.

13.4.3.2 Screening for Internal Erosion in a Rock Foundation

Breach by gross enlargement is only possible for internal erosion through rock foundations if the infilled defect is sufficiently large such that the embankment may collapse into the void and lowers the crest to below the reservoir level.

Ignore this failure mode unless the size of the infilled defect is very large (e.g. infilled caverns or caves in karst > 2 m width).

13.4.3.3 Assessment of the Probability of Breach by Gross Enlargement for Internal Erosion in the Embankment, Soil Foundation and from Embankment into Foundation

For breach to occur by gross enlargement of a pipe, the pipe must stay open until it is so large that the settlement of the crest due to the pipe or collapse of the embankment into the pipe lowers the crest to below the reservoir level.

If there is no intervention, the process can only stop if one or more of the following occur:

(a) The hydraulic shear stresses in the pipe reach an equilibrium condition with the erosion resistance of the soil. This will not happen unless the reservoir level drops, giving a lower gradient, as the hydraulic shear stress increases with hole diameter for a constant gradient.
(b) The reservoir empties or falls below the entrance of the pipe before a breach mechanism is able to develop. This is a consideration where internal erosion may develop in the upper part of the dam under short duration flood loading conditions.

Table 13.13 provides guidance on the probability of breach by gross enlargement assuming there is no restriction on flows and no intervention to lower the reservoir level. It is applicable to concentrated leak erosion and contact erosion.

TABLE 13.12 Probability of Breach by Gross Enlargement of the Pipe—Ability to Support a Pipe (Screening) (Fell et al., 2024)

DOWNSTREAM SHELL/ZONE	*ABILITY TO SUPPORT A ROOF*	*PROBABILITY OF BREACH BY GROSS ENLARGEMENT*
Downstream shell comprises free draining rockfill, or coarse sandy gravel. Downstream shell comprises sand and gravel, < 5% plastic fines or < 15% non-plastic fines	Very unlikely for piping through dam	Not a likely mode of breach for piping through the dam. Breach by slope instability or unravelling/sloughing are likely to be more critical. Assign probability of breach by gross enlargement $P_{ge} = 0$.
All other cases	Likely	Assess using Table 13.13.

TABLE 13.13 Probability of Breach by Gross Enlargement of the Pipe for Concentrated Leak and Contact Erosion (Fell et al., 2024)

CHARACTERISTICS OF CORE MATERIAL		*TIME FOR RESERVOIR LEVEL TO FALL BELOW THE INVERT OF THE PIPE*	*PROBABILITY OF BREACH BY GROSS ENLARGEMENT*
Soil Classification	Hole Erosion Index I_{HET}		
SM, SC, ML, dispersive soils	≤3 2 to 3		1.0
CL, CL-CH, MH or CH with Liquid Limit < 65%	4 3 to 5	> 2 days 1–2 days < 1 day	0.8 to 0.95 0.6 to 0.8 0.3 to 0.6
CH with Liquid Limit > 65%	5 4 to 6	> 2 weeks 1–2 weeks < 1 week	0.8 to 0.95 0.3 to 0.8 0.1 to 0.3
CH with Liquid Limit < 65% Or CH with Liquid Limit > 65% HET carried out to confirm Hole Erosion Index = 6	6	> 2 weeks 1–2 weeks < 1 week	0.2 to 0.95 0.1 to 0.5 0.05 to 0.1

Notes:
(1) Assuming there is no restriction on flows and no intervention to lower the reservoir level.
(2) I_{HET} from hole erosion tests, not estimated from soil classification.
(3) The basis for the judgemental probabilities is given in the supporting information document in Fell et al. (2024).
(4) Adopt the upper range of breach probability unless testing of I_{HET} and analyses support lower values.

13.4.4 Estimation of the Probability of Breach by Slope Instability

13.4.4.1 Overall Approach

(a) For internal erosion in embankments, soil foundations and from embankment to foundation, estimate the probability of slope instability occurring due to the increased seepage flows (P_{si-i}) using Table 13.14 and Figure 13.5.
(b) For internal erosion in rock foundations, use Tables 13.15 to 13.17 and Figure 13.6.

Estimate the probability of loss of freeboard due to instability (P_{si-lf}) using Table 13.18 and Figure 13.7.

$$\text{The probability of breach by slope instability} = (P_{si-i}) \times (P_{si-lf}) \qquad (13.8)$$

13.4.4.2 Estimation of the Probability Slope Instability Initiates for Internal Erosion in the Embankment, Soil Foundation, and from Embankment into Foundation

The assessment considers whether internal drainage measures in the dam are able to prevent pore pressures rising in the dam and/or foundation and whether the factor of safety of the dam falls below 1.0 if pore pressures do increase.

Estimate the probability of a downstream slide initiating using Table 13.14 and Figure 13.5. It is assumed that seepage in a soil foundation will exit under the embankment.

For important decisions on dam safety or upgrades, estimate the probability of breach using the method described in Chapter 7. For this, estimate the pore pressures in the embankment resulting from

TABLE 13.14 Factors Which Influence Likelihood of Breaching by Instability of the Downstream Slope—Slide Initiates for Internal Erosion in the Embankment, in Soil Foundations, and from Embankment into Foundation (Fell et al., 2024)

		LIKELIHOOD FACTOR (LF)			
FACTOR	*RELATIVE IMPORTANCE FACTOR (RF)*	*LESS LIKELY (1)*	*NEUTRAL (2)*	*MORE LIKELY (3)*	*MUCH MORE LIKELY (4)*
Internal drainage measures in dam	(3)	Good. Filter drains with good discharge capacity. Or free draining rockfill or clean sandy gravel in the downstream zones.	Moderate. Single stage filter zones. Or sandy gravel. Or moderate fines rockfill in the downstream zones.	Limited. Filter drain with excessive fines, poor discharge capacity. Or silty sandy gravel, or high fines content weathered rockfill in the downstream zones.	None. No or limited zoning of materials and no filter drains.
Downstream Slope For dams with an earthfill downstream zone[(a)]	(2)	3H:1V or flatter	2.5H:1V	2H:1V	Steeper than 1.8H:1V
or For dams with a free draining rockfill downstream zone[(b)]		Flatter than 1.75H:1V	1.5H:1V	Steeper than 1.4H:1V	Steeper than 1.3H:1V
Downstream shell materials	(1)	Sandy gravel < 5% fines, coarse-grained, free draining rockfill.	Sandy gravel 5%–20% fines, "dirty" rockfill.	Plastic soils, fine-grained rockfill.	Silty sand, silty sandy gravel, 20%–50% fines.

Notes:

(a) Applies to the following dam types: homogeneous earthfill, earthfill with filter, earthfill with rock toe, zoned earthfill, concrete face earthfill, puddle core earthfill, earthfill with core wall and hydraulic fill.

(b) Applies to the following dam types: zoned earth and rockfill, central core earth and rockfill, concrete face rockfill and rockfill with core wall.

the internal erosion and calculate the factor of safety. Do this for each of the erosion branches. The slope stability analysis should be focused on potential slip surface that may result in a significant loss of freeboard leading to breach. In some dams it will be demonstrable that instability will not occur for some and excessive erosion, and if it does it may take some time for the pore pressures to reach critical conditions, so detection and intervention may be possible.

13.4.4.3 Estimation of the Probability of Slope Instability of the Embankment Initiates for Internal Erosion in a Rock Foundation

(a) Estimate the probability that leakage through the rock foundations exits into the downstream shell P_S using Table 13.15.

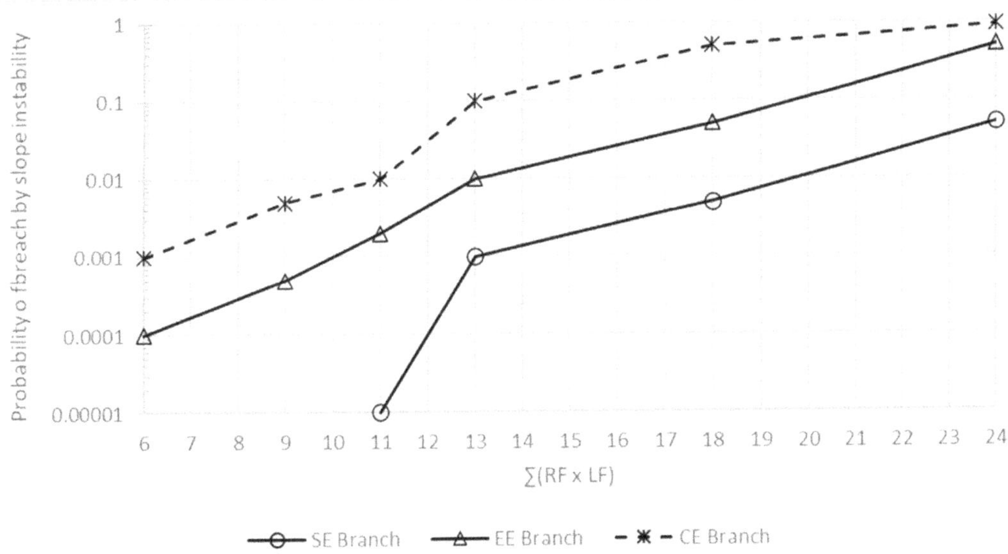

FIGURE 13.5 Estimation of the probability of breach by slope instability—slide initiates for internal erosion in the embankment, in soil foundations, and from embankment into foundation versus Σ (Relative importance Factor [RF]) × (Likelihood Factor [LF]).

Note: Select the probability scale corresponding to the filter erosion condition being considered on the event tree. CE = Continuing Erosion branch, EE = Excessive Erosion branch, and SE = Some Erosion branch.

(b) Estimate the probability that the leak will exceed the drainage capacity of the foundation drains and downstream shell using Table 13.16. The flows in Table 13.16 are taken from Table 9.16. The flow is proportional to crack depth.

(c) Estimate the probability of slope instability initiating due to the increased leakage flows ($P_{si\ i}$) using Table 13.17 and Figure 13.6.

Alternatively and better, assess whether given the leak emerges beneath the embankment, the internal drainage measures in the embankment have sufficient discharge capacity to prevent pore pressures rising in the embankment and/or foundation and whether the factor of safety of the dam falls below 1.0 if pore pressures do increase, taking account of the estimated leakage rate, zoning of the embankment and permeability of the zones.

13.4.4.4 Estimation of the Probability of Loss of Freeboard Due to Slope Instability

The assessment considers whether the resulting sliding deformations are sufficient to result in loss of freeboard, so the reservoir overtops the dam crest. Estimate the probability of loss of freeboard using Table 13.18 and Figure 13.7.

These probabilities do not allow for detection of slope instability and intervention. Those need to be accounted for in Sections 13.3.3 and 13.3.4.

Where practicable carry out stability analyses of the dam to estimate the factor of safety and use the methods described in Chapter 7 to estimate the likelihood of loss of freeboard.

TABLE 13.15 Probability of Seepage Exits from Defects in a Rock Foundation into the Downstream Shell (P_S) (Fell et al., 2024)

SCENARIOS	*PROBABILITY OF THE SEEPAGE PATH EXITING INTO THE DOWNSTREAM SHELL*
The open or infilled defect in the rock foundation daylights downstream of the dam, and the defect is not in direct contact with the downstream shell, and there is very limited interconnectivity through other joint sets	Negligible. Adopt $P_S = 0$
The open or infilled defect in the rock foundation daylights downstream of the dam, and the defect is not in direct contact with the downstream shell, but there is a likely connection into the downstream shell via an interconnected open joint set	0.1 to 0.5
The open or infilled defect in the rock foundation daylights downstream of the dam, and the defect is likely to be in direct contact with the downstream shell	0.5 to 1.0[(a)]
The open or infilled defect in the rock foundation does not daylight downstream of the dam, and the defect is in direct contact with the downstream shell	1.0[(a)]

Note: (a) The geometry of the leakage flow path affects the flow rate. If the leakage path daylights downstream of the dam, it is likely there will be less flow into the downstream shell of the dam.

TABLE 13.16 Assessment of Size of Leak in Defect in a Rock Foundation Relative to Discharge Capacity of Foundation Drains and Downstream Shell (Fell et al., 2024)

	LIKELIHOOD THE LEAK IN THE DEFECT WILL EXCEED THE DISCHARGE CAPACITY OF THE FOUNDATION DRAINS AND DOWNSTREAM SHELL			
WIDTH OF DEFECT OR SOLUTION FEATURE EXITING INTO THE DOWNSTREAM ZONE (FLOW RATE FOR 2 M DEEP CRACK, GRADIENT = 0.15)[b]	*POOR, NO OR LIMITED ZONING OF MATERIALS AND NO FOUNDATION FILTER DRAINS*	*LIMITED FOUNDATION FILTER DRAIN WITH EXCESSIVE FINES, POOR DISCHARGE CAPACITY, OR SILTY SANDY GRAVEL, OR HIGH FINES CONTENT WEATHERED ROCKFILL IN THE DOWNSTREAM ZONES*	*MODERATE SINGLE STAGE FOUNDATION FILTER ZONE, OR SANDY GRAVEL, OR MODERATE FINES ROCKFILL IN THE DOWNSTREAM ZONES*	*GOOD FOUNDATION FILTER DRAINS WITH GOOD DISCHARGE CAPACITY, OR FREE DRAINING ROCKFILL OR CLEAN SANDY GRAVEL IN THE DOWNSTREAM ZONES*
2 mm (2 l/sec)	LL	LL	Negligible	Negligible
5 mm (10 l/sec)	ML	N	LL	Negligible
10 mm (40 l/sec)	MML	ML	N	LL
25 mm (160 l/sec)	MML	MML	MML	ML
50 mm (300 l/sec	MML	MML	MML	MML

Notes:
(a) LL = Less Likely; N = Neutral; ML = More Likely; MML = Much More Likely.
(b) See Table 9.16 for flow rates for other seepage gradients. Flow rate $\propto$ crack depth.

TABLE 13.17 Factors Which Influence Likelihood of Slide Initiating for Internal Erosion in Rock Foundation (Fell et al., 2024)

		LIKELIHOOD FACTOR (LF)			
FACTOR	*RELATIVE IMPORTANCE FACTOR (RF)*	*LESS LIKELY (1)*	*NEUTRAL (2)*	*MORE LIKELY (3)*	*MUCH MORE LIKELY (4)*
Size of leak relative to discharge capacity of the foundation drains and downstream shell (from Table 13.16)	(3)	LL	N	ML	MML
Downstream slope For dams with an earthfill downstream zone (a)	(2)	3H:1V or flatter	2.5H:1V	2H:1V	Steeper than 1.8H:1V
or					
For dams with a free draining rockfill downstream zone (b)		Flatter than 1.75H:1V	1.5H:1V	Steeper than 1.4H:1V	Steeper than 1.3H:1V
Downstream shell materials	(1)	Sandy gravel < 5% fines, coarse grained, free draining rockfill	Sandy gravel 5%–20% fines "dirty" rockfill	Plastic soils, fine-grained rockfill	Silty sand, silty sandy gravel, 20%–50% fines

Notes:
(a) Applies to the following dam types: homogeneous earthfill, earthfill with filter, earthfill with rock toe, zoned earthfill, concrete face earthfill, puddle core earthfill, earthfill with core wall and hydraulic fill.
(b) Applies to the following dam types: zoned earth and rockfill, central core earth and rockfill, concrete face rockfill and rockfill with core wall.

13.4.5 Estimation of the Probability of Breach by Sloughing or Unravelling

For sloughing to be likely to occur, the downstream face generally has to be relatively steep, and the shoulder material a non-plastic soil, probably sandy gravel, or gravelly sand, possibly with some silty fines. The process would have to be allowed to continue until it gradually eroded away the crest and allowed the reservoir to overtop the embankment.

Unravelling usually relates to the progressive removal of individual rocks by fairly large seepage flows flowing through the downstream rockfill.

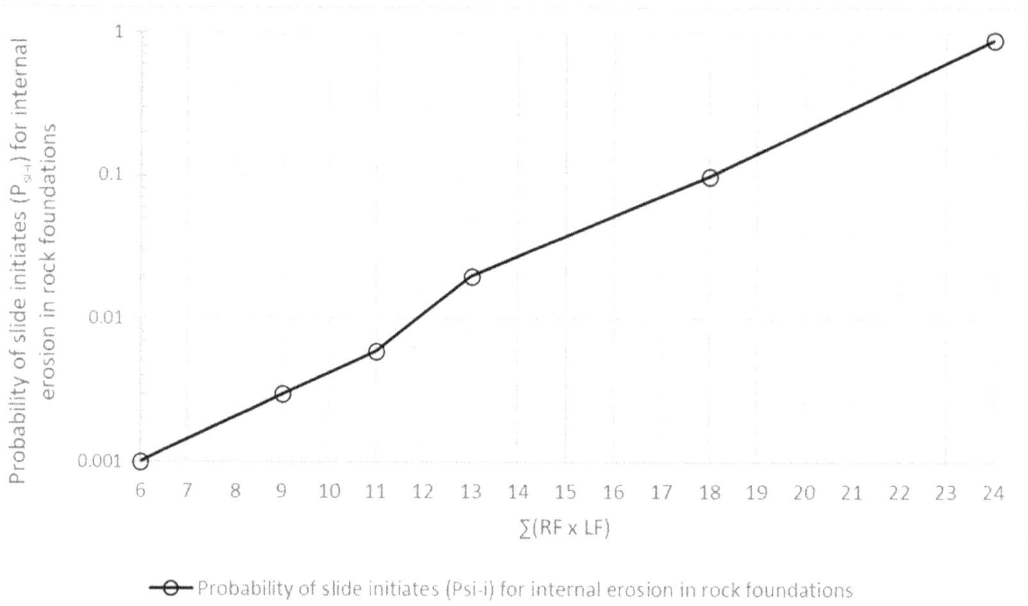

FIGURE 13.6 Estimation of the probability of slide initiates (P_{si-i}) for internal erosion in rock foundations versus ∑ (Relative importance Factor [RF]) × (Likelihood Factor [LF]).

TABLE 13.18 Factors Which Influence Likelihood of Breaching by Instability of the Downstream Slope—Loss of Freeboard (Fell et al., 2024)

		LIKELIHOOD FACTOR (LF)			
FACTOR	*RELATIVE IMPORTANCE FACTOR (RF)*	*LESS LIKELY (1)*	*NEUTRAL (2)*	*MORE LIKELY (3)*	*MUCH MORE LIKELY (4)*
Freeboard compared to dam height at the time of incident	(3)	> 7%	≈ 5%	< 3%	< 1%
Presence of strain weakening soils in the embankment and/or foundation	(2)	Sandy clays, low to medium plasticity, clay size content < 20%, or medium dense to dense dilative non-plastic soils or rockfill.	Clays, sandy clays, clay size content 20% to 40%, or medium dense non-plastic soils.	Clays, sandy clays, high plasticity; clay size content (% passing 0.002 mm) > 40% and/or saturated, very loose sand or loose silty sand contractive on shearing.	As for More Likely, but with very high clay size content or very loose contractive granular soil.
Crest width	(1)	> 9 m	≈ 6 m	< 4 m	≤ 3 m

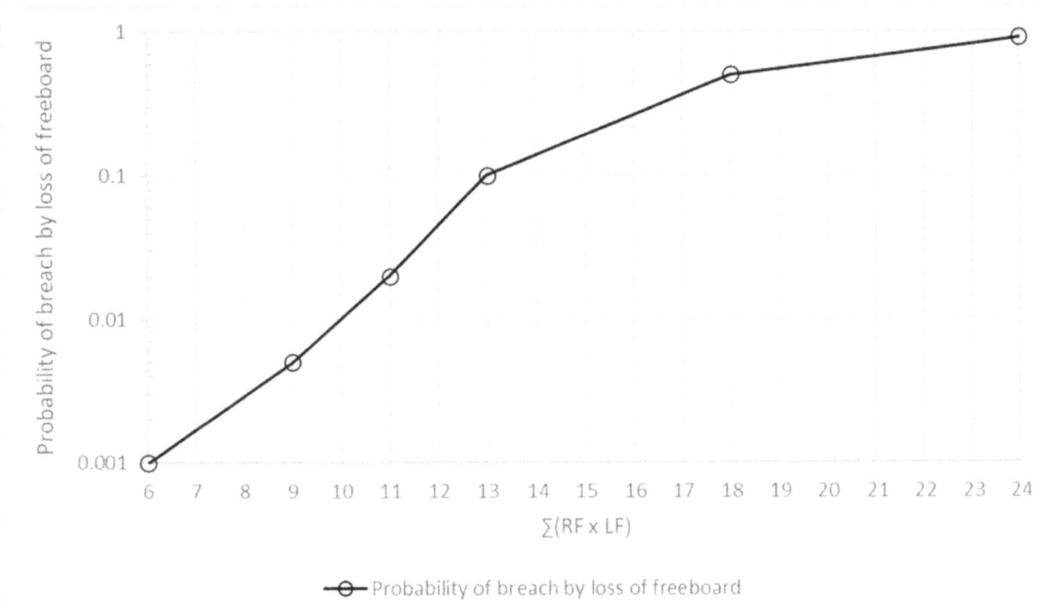

FIGURE 13.7 Estimation of the probability of breach by loss of freeboard versus Σ (Relative importance Factor [RF]) × (Likelihood Factor [LF]).

The approach is as follows:

(a) For internal erosion in rock foundations, estimate the probability that seepage through the rock foundations exits into the downstream shell P_S using Table 13.15.
(b) For internal erosion in the embankment, soil foundation and embankment into foundation, assume that the seepage will emerge into the downstream shell of the embankment, so $P_S = 1.0$.
(c) For dams with a downstream zone of earthfill (i.e. clay, silt, sand or sandy gravel) use Table 13.19 and Figures 13.8 and 13.9 to estimate the probability of breach by sloughing P_{sl}. Table 13.19 applies to the following dam types: homogeneous earthfill, earthfill with filter, earthfill with rock toe, zoned earthfill, concrete face earthfill, puddle core earthfill, earthfill with core wall and hydraulic fill.
(d) For dams with a downstream zone of rockfill, use Table 13.20 and Figures 13.10 and 13.11 to estimate the probability of breach by unravelling P_{un}.
(e) The probability of breach by sloughing or unravelling is equal to $(P_S) \times (P_{sl})$ for dams with a downstream zone of earthfill or $(P_S) \times (P_{un})$ for dams with a rockfill shell.

For cases where a very coarse rockfill toe is present, the probabilities for breach by unravelling can be reduced from those given in Figure 13.10 if calculations indicate that unravelling is very unlikely to occur.

These probabilities do not allow for detection of slope instability and intervention. Those need to be accounted for as described in Sections 13.3.3 and 13.3.4.

Table 13.20 applies to the following dam types: zoned earth and rockfill, central core earth and rockfill, concrete and asphalt face rockfill and rockfill with concrete or asphalt core.

TABLE 13.19 Factors Which Influence Likelihood of Breaching by Unravelling—Dams with an Earthfill Downstream Zone (Fell et al., 2024)

		LIKELIHOOD FACTOR (LF)			
FACTOR	*RELATIVE IMPORTANCE FACTOR (RF)*	*LESS LIKELY (1)*	*NEUTRAL (2)*	*MORE LIKELY (3)*	*MUCH MORE LIKELY (4)*
Material in downstream zone	(3)	Plastic soils (a)	Sandy gravel < 20% non-plastic fines	Silty sand, silty sandy gravel, 20%–50% non-plastic fines	As for more likely, but uncompacted materials
Freeboard at the time of incident	(2)	> 4 m	≈ 3 m	< 2 m	< 1 m
Downstream slope of the embankment	(1)	3H:1V or flatter	2.5H:1V	2H:1V	Steeper than 1.8H:1V

Note: (a) Assign negligible probability. Plastic soils are not subject to unravelling.

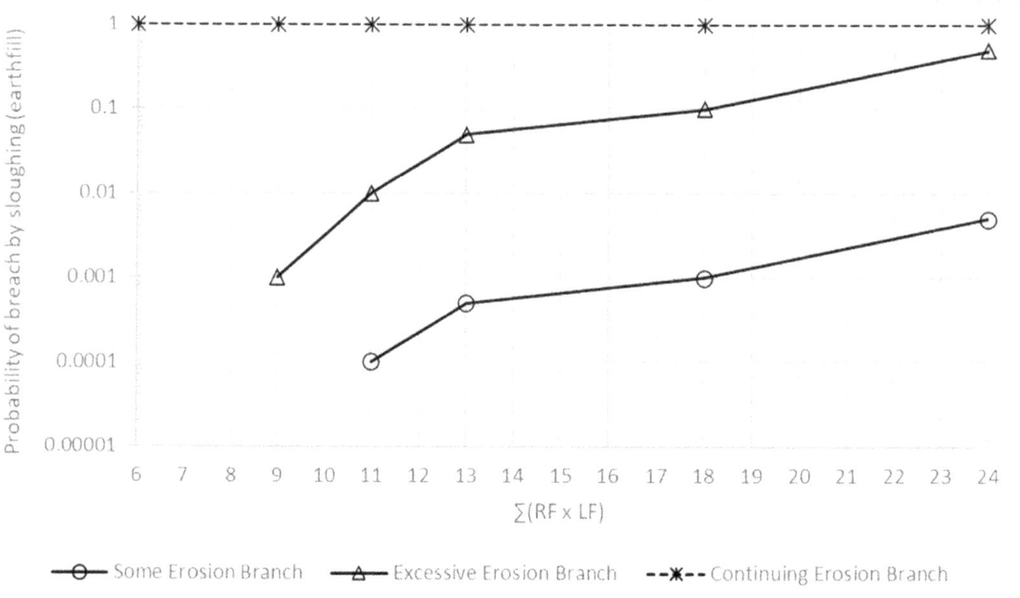

FIGURE 13.8 Estimation of the probability of breach by sloughing of earthfill versus Σ (Relative importance factor [RF]) × (Likelihood factor [LF]) for internal erosion in the embankment, in soil foundations, and from embankment into foundation.

Notes:

(a) Select the probability scale corresponding to the filter erosion condition being considered on the event tree. CE = Continuing Erosion branch, EE = Excessive Erosion branch, and SE = Some Erosion branch.

(b) For CE branch, the dam may fail by slope instability before it might fail by unravelling.

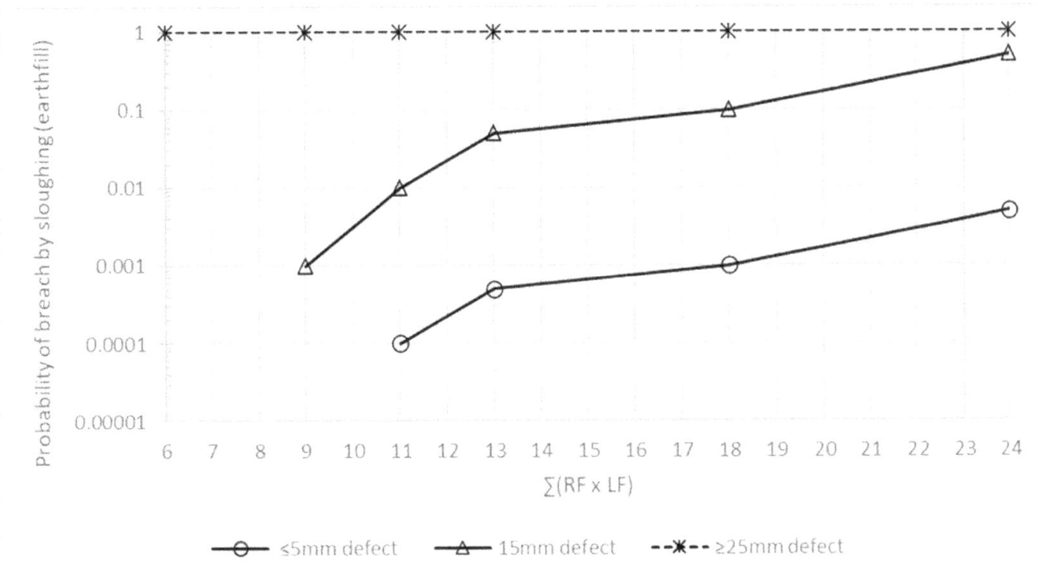

FIGURE 13.9 Estimation of the probability of breach by sloughing in earthfill versus Σ (Relative importance Factor [RF]) × (Likelihood Factor [LF]) for internal erosion in rock foundation

TABLE 13.20 Factors Which Influence Likelihood of Breaching by Unravelling—Dams with a Rockfill Downstream Zone (Fell et al., 2024)

		LIKELIHOOD FACTOR (LF)			
FACTOR	*RELATIVE IMPORTANCE FACTOR (RF)*	*LESS LIKELY (1)*	*NEUTRAL (2)*	*MORE LIKELY (3)*	*MUCH MORE LIKELY (4)*
Material in downstream zone	(3)	Coarse-grained free draining rockfill	Medium-grained "dirty" rockfill	Fine-grained rockfill	As for More Likely, but uncompacted materials
Downstream slope of the embankment	(2)	Flatter than 1.75H:1V	1.5H:1V	Steeper than 1.4H:1V	Steeper than 1.3H:1V
Freeboard at the time of incident	(1)	> 4 m	≈ 3 m	< 2 m	< 1 m

13.4.6 Estimation of the Probability of Breach by Sinkhole Development

13.4.6.1 Overall Approach

Estimate the probability of a sinkhole developing as a result of the internal erosion ($P_{s\text{-}f}$).

Estimate the probability that the sinkhole causes loss of freeboard ($P_{s\text{-}lf}$). Assume the sinkhole develops on the crest unless there is a specific reason to expect it to develop elsewhere on the embankment.

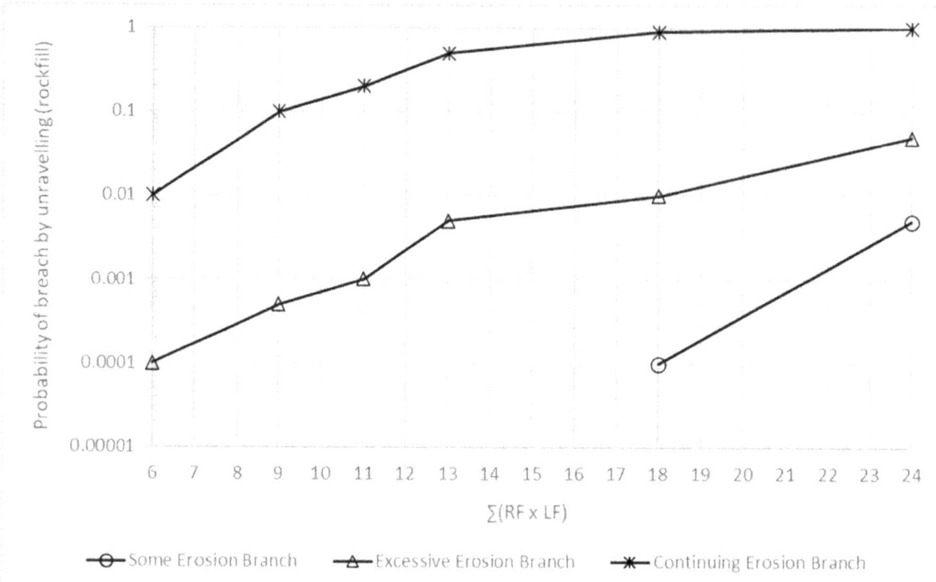

FIGURE 13.10 Estimation of the probability of breach by unravelling in rockfill versus Σ (Relative importance Factor [RF]) × (Likelihood Factor [LF]) for internal erosion in the embankment, in soil foundations, and from embankment into foundation

Note: Select the probability scale corresponding to the filter erosion condition being considered on the event tree. CE = Continuing Erosion branch, EE = Excessive Erosion branch, and SE = Some Erosion branch.

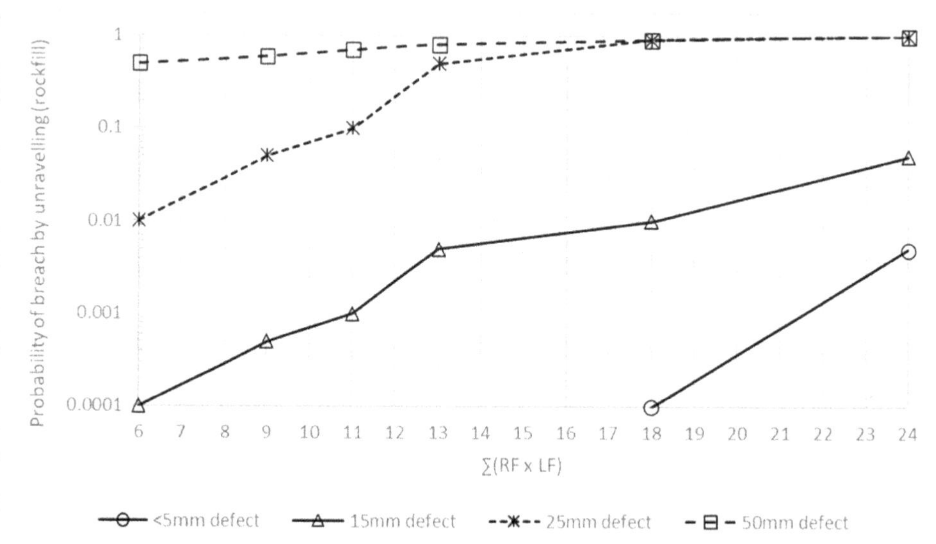

FIGURE 13.11 Estimation of the probability of breach by unravelling of rockfill versus Σ (Relative importance Factor [RF]) × (Likelihood Factor [LF]) for internal erosion in rock foundation.

Note: Select the probability scale corresponding to the width of the open or infilled defect that exits into the downstream shell.

The probability of breach by sinkhole development = probability of sinkhole ($P_{s\text{-}f}$) × probability of loss of freeboard due to the sinkhole ($P_{s\text{-}lf}$).

13.4.6.2 Probability of Sinkhole Formation

Estimate the probability of a sinkhole or crest settlement developing ($P_{s\text{-}f}$) given that internal erosion has initiated and progressed by expert elicitation. Take account of the following information:

(a) Foster and Fell (1999a, 1999b, 2000b) indicate that for their database, sinkhole and localised crest settlements were very common in piping incidents (about 60% of incidents in the embankment, and 30% of incidents in the foundation) but have only led to three failures, for piping in the embankment and around conduits. This compares to 53 accidents, so only about 1 in 20 develop into a breach. Of these, two had been apparent for a long while and should have been fixed before failure. Most sinkholes are relatively shallow and only about half form on the crest.

(b) The authors are not aware of any way of predicting if sinkholes or stopes may develop. The most common occurrence has been in Swedish and Norwegian central core rockfill dams with non-plastic glacial till or moraine cores. The initiating mechanism in these dams is erosion through the core due to concentrated leak erosion or suffusion. Sinkholes have also been observed in embankments with narrow puddle cores.

(c) Sinkholes may take many years to develop and are often found by accident, e.g. by a vehicle wheel dropping into the void.

13.4.6.3 Probability of Loss of Freeboard Due to Sinkhole Formation

For breach to occur by sinkhole development into an erosion pipe in the embankment, the sinkhole or crest settlement would need to be sufficiently large to settle the crest to below reservoir level. For internal erosion in the foundation, loss of freeboard can also occur by excessive settlement of the embankment induced by the loss of foundation materials.

Estimate the probability of loss of freeboard due to sinkhole formation using Table 13.21 and Figure 13.12.

TABLE 13.21 Factors Which Influence Likelihood of Breaching by Sinkhole Development—Loss of Freeboard Given Sinkhole Develops (Fell et al., 2024)

FACTOR	RELATIVE IMPORTANCE FACTOR (RF)	LIKELIHOOD FACTOR (LF) LESS LIKELY (1)	NEUTRAL (2)	MORE LIKELY (3)	MUCH MORE LIKELY (4)
Freeboard at the time of the incident	(3)	> 4 m	≈ 3 m	< 2 m	< 1 m
Width of crest	(2)	> 9 m	≈ 6 m	< 4 m	≤ 3 m
Material in the core of the embankment	(1)	High plasticity clay, well compacted	Low to medium plasticity clays, and sandy clays	Non-plastic, silty sand or silty sandy gravel	As for more likely, poorly compacted/ loose.

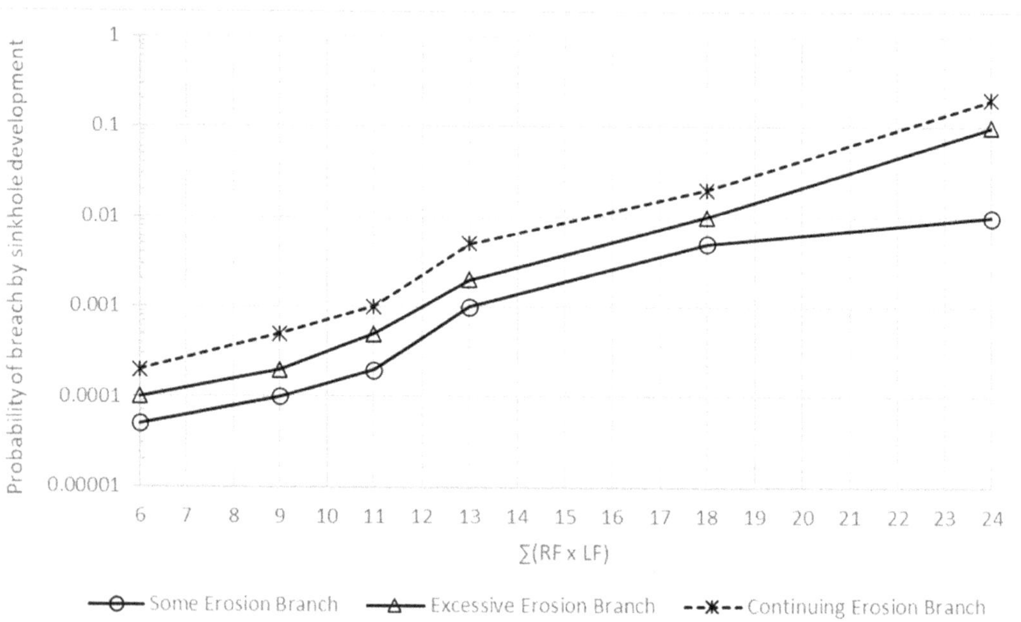

FIGURE 13.12 Estimation of the probability of breach by sinkhole development—loss of freeboard given sinkhole develops versus Σ (Relative importance Factor [RF]) × (Likelihood factor [LF]).

14 Quantitative Risk and Simplified QRA Approaches to Assessing the Safety of a Dam for Internal Erosion and Piping

14.1 INTRODUCTION

Chapters 9 to 13 describe the mechanics of initiation, continuation, progression, detection and repair, and breach for concentrated leak, contact erosion, backward erosion piping, suffusion and global backward erosion.

Those chapters are presented so that readers who are using deterministic approaches to assessing dam safety will find there the information they require. Those who want to use quantitative risk assessments to assess dam safety should use those chapters to understand the mechanics and may then use what is described in those chapters, this chapter and Appendix A to assist in estimating annual probability of failure for use in assessment of the risks as described in Chapter 18.

The guidance in this chapter has been prepared on the basis that the probability estimates are being carried out for quantitative risk assessment and assumes the assessment team is familiar with quantitative risk analysis approaches, including the use of event tree methods. Users should refer to other published guidance for overall quantitative risk analysis methodologies, for example the ANCOLD *Guidelines on Risk Assessment* (ANCOLD, 2022) and the Joint Reclamation and USACE "Best Practices" publication (USACE/Reclamation, 2019) and updates.

In Section 14.10, a Simplified QRA approach is described. This may be used as a guide to the tolerability of risk posed for individual Potential Failure Modes (PFMs) and Initiating Mechanisms (IMs).

The contents of Sections 14.2 to 14.9 of this chapter are largely taken from Fell, R., Foster, M.A., Cyganiewicz, J., Sills, G. and Davidson, R. (2024). *Methods for Estimating the Probability of Failure*

DOI: 10.1201/9781003542025-14

of Embankment Dams by Internal Erosion and Piping. UNICIV Report No. R470, The School of Civil and Environmental Engineering, University of New South Wales, Sydney.

14.2 GENERAL PROCESS

The process for estimating the annual probability of failure by internal erosion and piping failure modes includes the following steps:

Step 1: Assemble and review all information pertinent to internal erosion and piping as detailed in Section 3.2.
Step 2: Identify all potential failure modes and initiating mechanisms associated with internal erosion and piping for:
Internal erosion through the embankment;
Internal erosion through the foundation; and
Internal erosion of the embankment into or at the foundation contact.
Step 3: Screen out those PFMs and IMs that are not credible or have been assessed to have negligible contribution to the annual probability of failure and document the reasons for their exclusion as detailed in Section 14.4.
Step 4: Develop detailed descriptions and sketches of all realistic PFMs and IMs. Guidance on doing this is given in Sections 3.1.4 and 3.4.
Step 5: Decompose each of the PFMs into event trees. Generic event trees have been developed for each PFM. These are presented in Section 14.3.
Step 6: Select the loading partitions and estimate loading probabilities for each of the load conditions (hydrological and seismic) as described in Chapter 5.
Step 7: Estimate the conditional probabilities for each node on the event tree, fully documenting the rationale. Guidance is given for estimating the conditional probabilities for various initiating mechanisms and failure locations in Chapters 9 to 13 and Appendix A.
Step 8: Calculate the annual probability of failure by internal erosion and piping for each PFM and IM and review for consistency.
Step 9: Estimate Potential Life Loss (PLL) for each PFM and IM and loading scenario as discussed in Chapter 18.
Step 10: Use the annual probability of failure and PLL estimates to assess the tolerability of risks as described in Chapter 18.

14.3 EVENT TREES FOR INTERNAL EROSION AND PIPING

Figures 14.1 to 14.4 show example event trees for concentrated leak in the embankment, backward erosion piping and suffusion/global backward erosion in the foundation, and contact erosion. Note that these event trees are not intended to cover all potential failure modes for internal erosion, but the basic structure and principles can be adapted for other internal erosion PFMs.

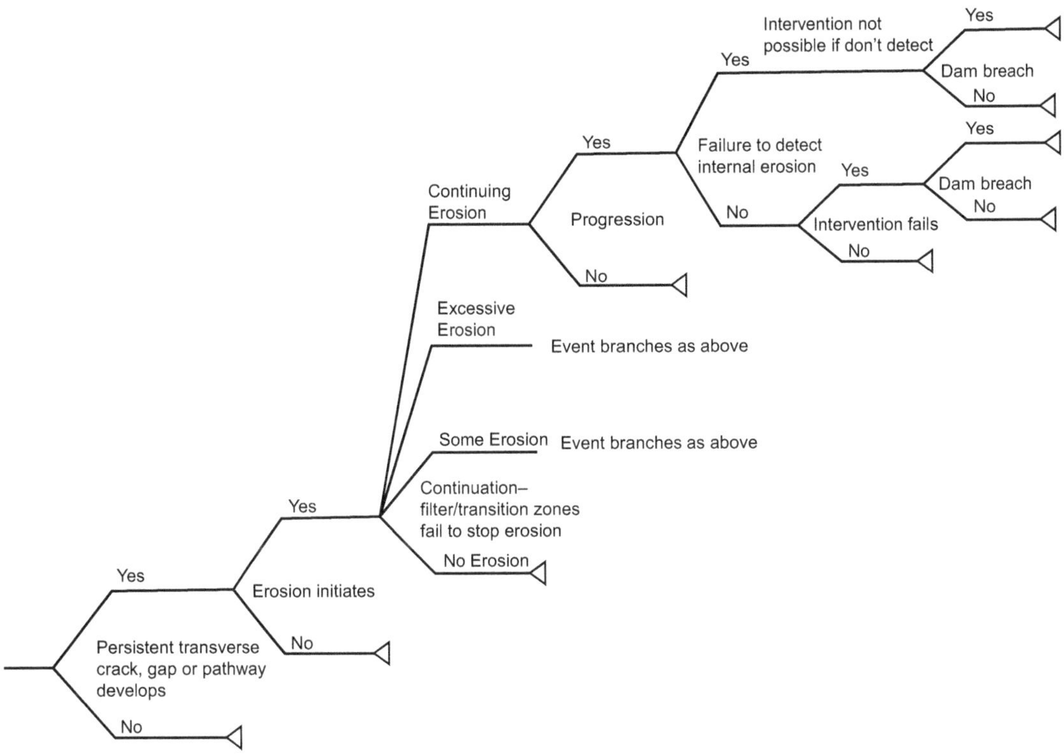

FIGURE 14.1 Event tree for modelling the annual probability of failure for concentrated leak erosion in the embankment.

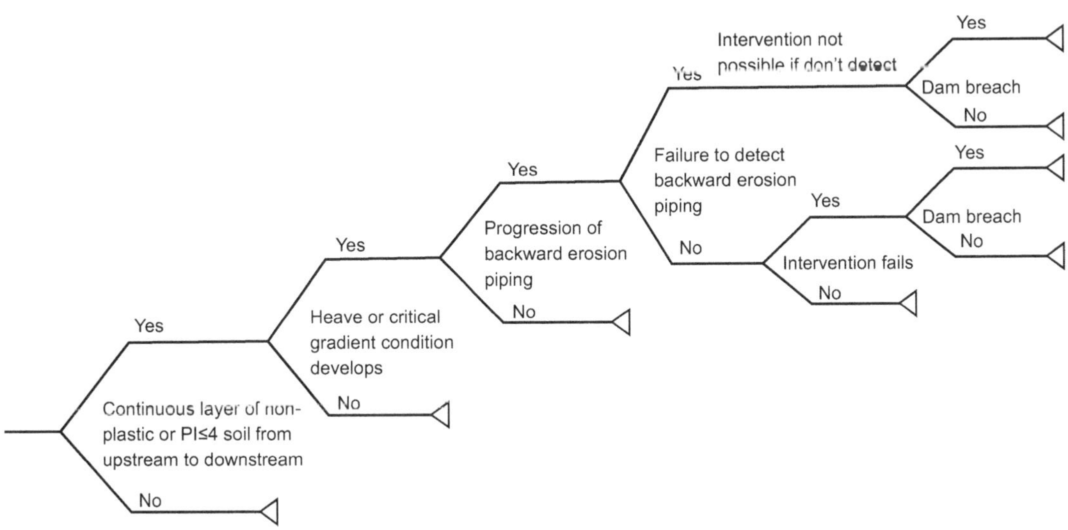

FIGURE 14.2 Event tree for modelling the annual probability of failure for backward erosion piping in the foundation.

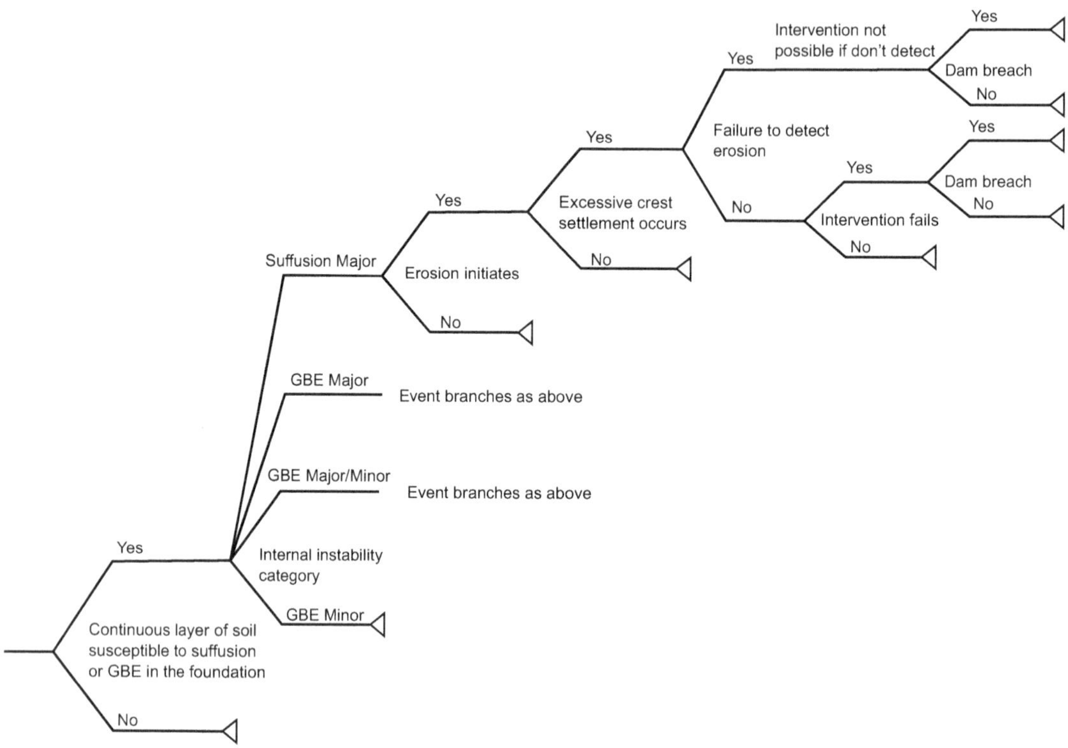

FIGURE 14.3 Event tree for modelling the annual probability of failure for suffusion and global backward erosion in the foundation.

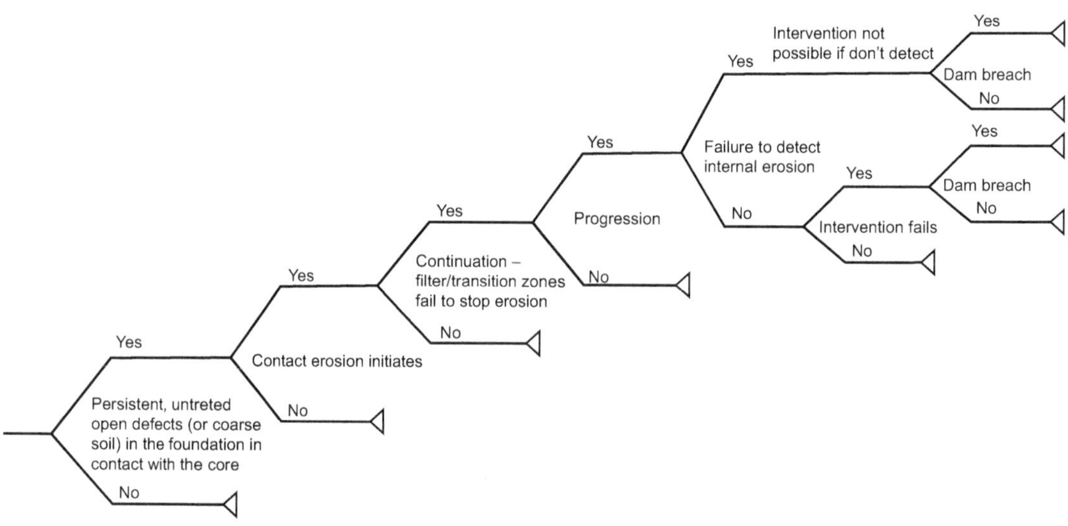

FIGURE 14.4 Event tree for modelling the annual probability of failure for contact erosion of the embankment into the foundation.

14.4 FAILURE MODES ASSESSMENT AND SCREENING

14.4.1 Failure Modes Assessment

See Chapter 3 for guidance on failure modes assessment.

14.4.2 Screening of Failure Modes and Initiating Mechanisms

14.4.2.1 Introduction

Not all potential failure modes and initiating mechanisms for internal erosion will be applicable to the dam being assessed. It is therefore recommended that the PFMs and IMs be screened. The purpose of this process is to systematically review the PFMs and IMs that have been identified and eliminate those from further consideration that are assessed to have negligible contribution to risk.

Each PFM and IM is evaluated by listing the adverse factors that make it "more likely" and the favourable factors that make it "less likely". These are based on the risk analysis team's understanding of the dam, the mechanics of the PFM and available information.

Each PFM and IM is then screened by the analysis team based on the consideration of the list of the more likely and less likely factors. Screening can also be guided by using the screening guidance summarised in Tables 14.1 to 14.6.

These screened PFMs and IMs are not carried forward into the risk analysis. The rationale for the inclusion or exclusion should be fully documented.

The authors have seen examples where risk analysts do not screen out PFMs and IMs in this manner but include them all, often embedded in a spreadsheet. This often results in a waste of time and money by the risk analysis team and potentially to an overestimation of the annual probability of failure because the team feel obliged to assign a probability to every initiating mechanism and even if each is small, they may add up to affect the risk assessment.

14.4.2.2 Internal Erosion through the Embankment

In Table 14.1, the "upper" part of the embankment is the upper 1/3 measured at the maximum height section and continuing across the valley at this level. In Table 14.2 the "lower" part of the embankment is below the top 1/3.

If there is a ridge separating two parts of a dam as shown in Figure 14.5(b), the two parts should be considered as separate embankments.

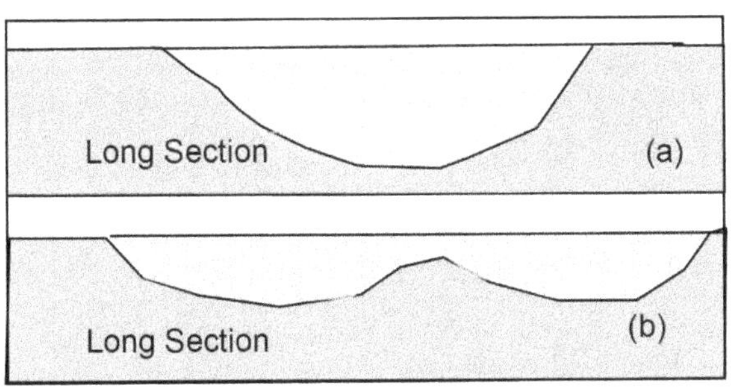

FIGURE 14.5 Longitudinal sections. Section (a): consider as a single embankment. Section (b): consider as two embankments.

TABLE 14.1 Screening of Initiating Mechanisms—PFM Internal Erosion Due to Concentrated Leaks in Transverse Cracks in the Upper Part of Embankment Dams

INITIATING MECHANISM	*SKETCH OF INITIATING MECHANISM*	*EXCLUDE THE INITIATING MECHANISM IF THE FOLLOWING CONDITIONS ARE SATISFIED*	*REFERENCE SECTION AND TABLES*
IM 1—Transverse cracking in the upper embankment resulting from cross-valley differential settlements	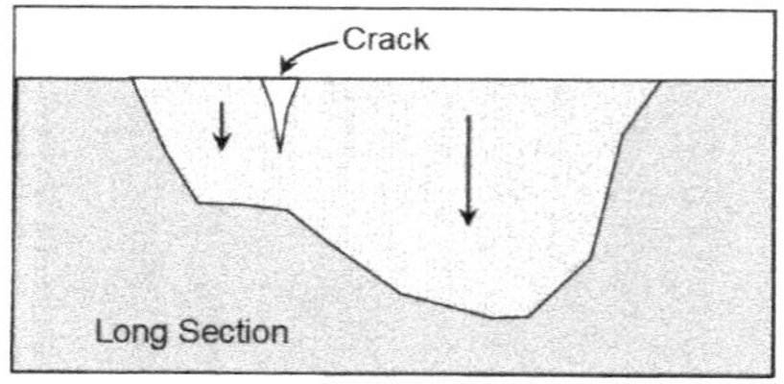	No exclusions apply—always include this initiating mechanism.	Section 9.2.1
IM 2—Transverse crack or gap adjacent to a cliff at the top of the embankment formed by differential settlement	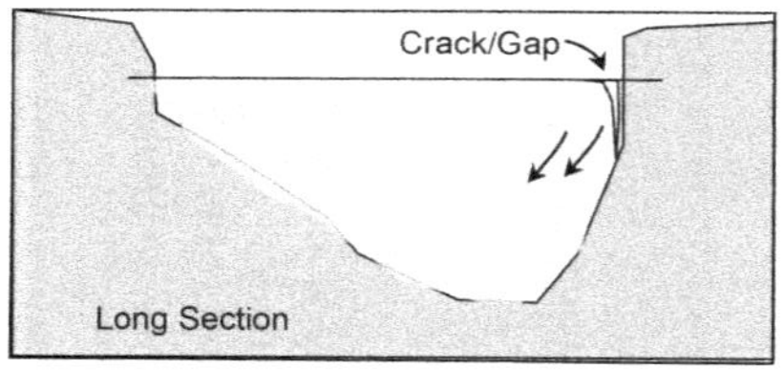	Exclude if: (1) There is no vertical cliff in contact with the embankment, *or* (2) A wide bench is present at the base of the cliff ($W_b/H_w > 2.5$, refer to Figure 9.36), *or* (3) The abutment slope below the cliff is gentle ($\beta_1 \leq 15°$, refer to Figure 9.36).	Section 9.2.9 Table A1 Figure A2
IM 3—Transverse cracking or hydraulic fracture resulting from cross-valley arching	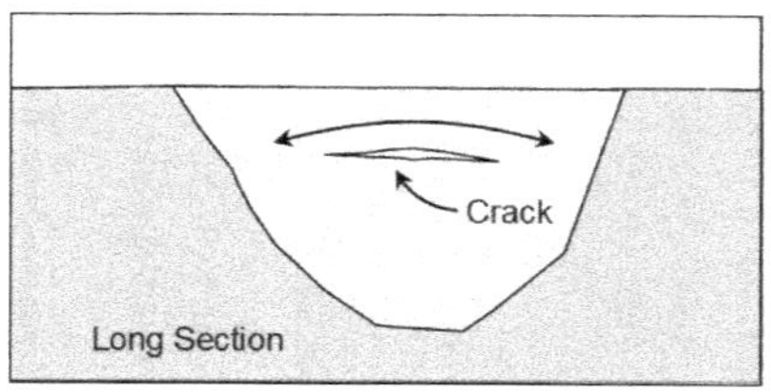	Exclude if: The width of valley to dam height ratio $W_v/H > 2$, (refer to Figure A4).	Section 9.2.2 Table A2 Figures A3, A4
IM 4—Transverse crack resulting from cross section differential settlement due to poorly compacted shoulders	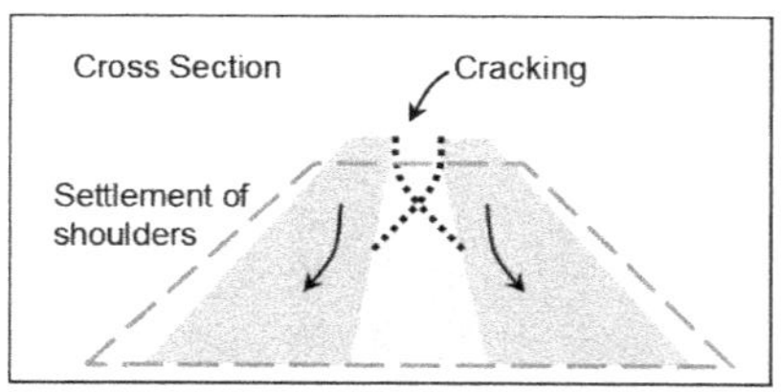	Exclude if: (1) The dam is zoning type homogeneous earthfill, earthfill with filter drains or zoned earthfill, *or for other embankment types* (2) Evidence from relative settlements of core and shoulders that the materials have a similar modulus *or* (3) Wide core, e.g. W/H > with dam height < 15 m.	Section 9.2.6

<table>
<tr><td>IM 5—Transverse cracking in the upper or lower embankment resulting from differential settlements in the foundation beneath the core</td><td>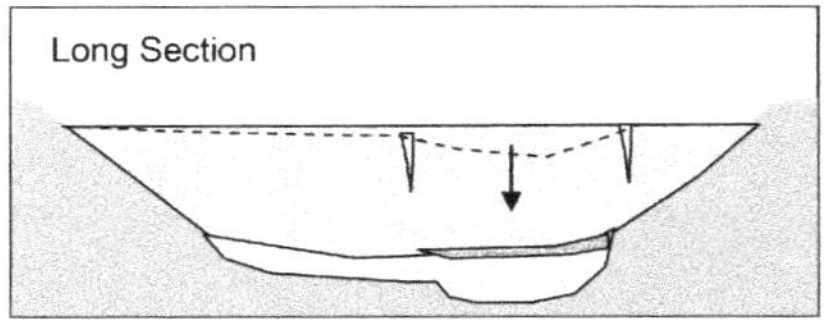</td><td>Exclude if:
There is no compressible soil in the foundation below the core.</td><td>Section 9.2.3
Figures 9.19, A5</td></tr>
<tr><td>IM 6—Transverse cracking due to differential settlements due to embankment staging</td><td>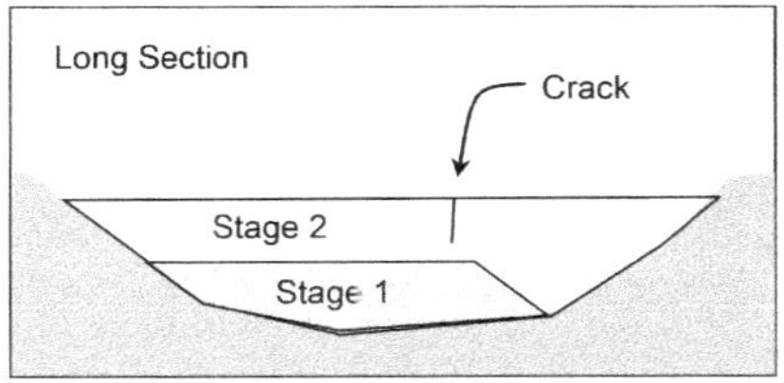</td><td>Exclude if:
The embankment construction was not staged.</td><td>Section 9.2.4
Figure 9.22</td></tr>
<tr><td>IM 7—Desiccation cracking in the crest of the embankment</td><td>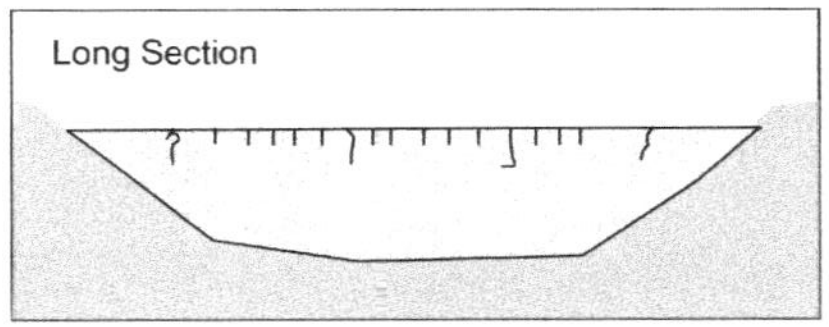</td><td>Exclude if:
Sealed road or concrete pathway > 5 m wide.</td><td>Section 9.8.1
Tables 9.15, A3
Figure A6</td></tr>
<tr><td>IM 8—Desiccation cracking on seasonal shutdown layers and on the surface of staged embankments</td><td>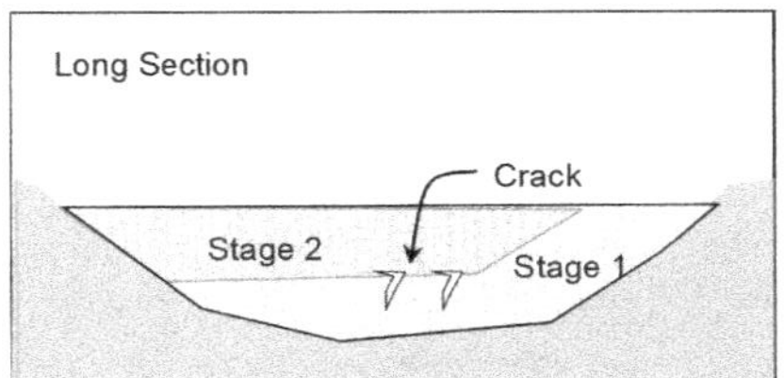</td><td>Exclude if:
There has not been a seasonal shutdown during construction, and the embankment has not been staged,
or
The reservoir stage being considered is below the level of the seasonal shutdown surface,
or
This mechanism only applies above the level of saturation of the core. Below that any desiccation cracks should have swelled and closed,
or
Very good control and clean up practices used—desiccated layers removed from the embankment and replaced with new soil or adequately reworked to specified moisture content.</td><td>See IM 7</td></tr>
</table>

(*Continued*)

TABLE 14.1 (Continued)

INITIATING MECHANISM	*SKETCH OF INITIATING MECHANISM*	*EXCLUDE THE INITIATING MECHANISM IF THE FOLLOWING CONDITIONS ARE SATISFIED*	*REFERENCE SECTION AND TABLES*
IM 13—Transverse cracking in the upper part of the embankment caused by earthquake	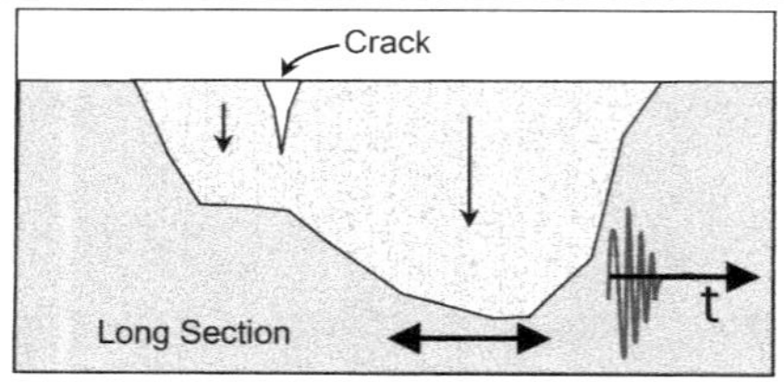	No exclusions apply—always include this failure mode.	Section 9.9

TABLE 14.2 Screening of Initiating Mechanisms—PFM Internal Erosion Due to Concentrated Leaks in Transverse Cracks in the Middle and Lower Parts of Embankment Dams

INITIATING MECHANISM	*SKETCH OF INITIATING MECHANISM*	*EXCLUDE THE INITIATING MECHANISM IF THE FOLLOWING CONDITIONS ARE SATISFIED*	*REFERENCE SECTION AND TABLE*
IM 9—Transverse cracking in the lower/middle embankment resulting from cross-valley differential settlement	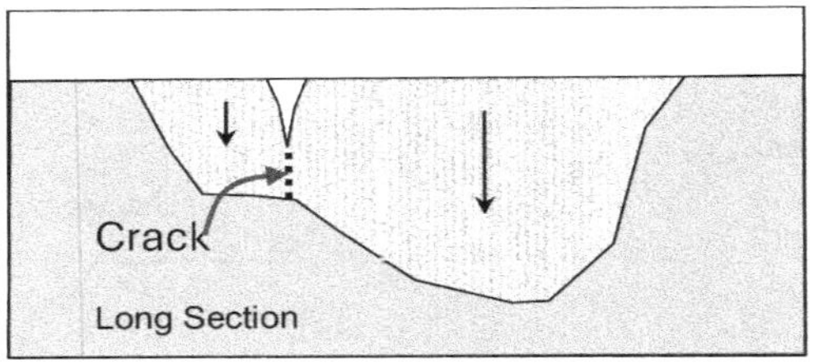	Exclude if: Uniform abutment profile without benches.	Covered in IM 1
IM 10—Hydraulic fracture resulting from arching of the core onto the shoulders of the embankment	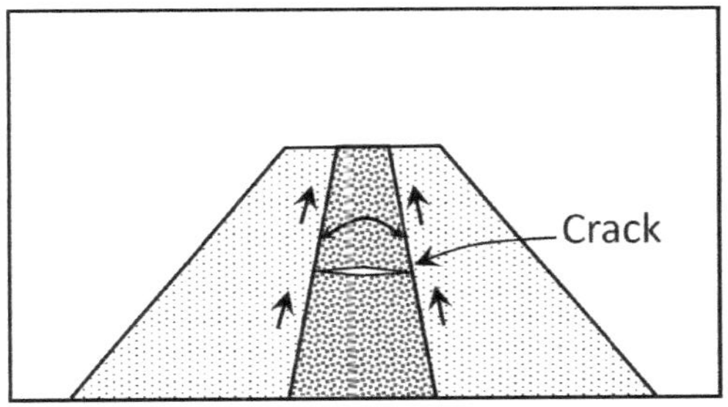	Exclude: (1) For all dam zoning types other than central core earth and rockfill (or gravel shells), and puddle core earthfill dams, *or* (2) Dam has a wide core (W/H>1.0), *or* (3) Core has higher modulus than shells. Shoulders poorly compacted or dumped. Core compacted > 98% SMDD.	Section 9.2.7 Tables A4 Figure A7
IM 11—Transverse cracking in the lower embankment resulting from differential settlement in soil in the foundations under the core	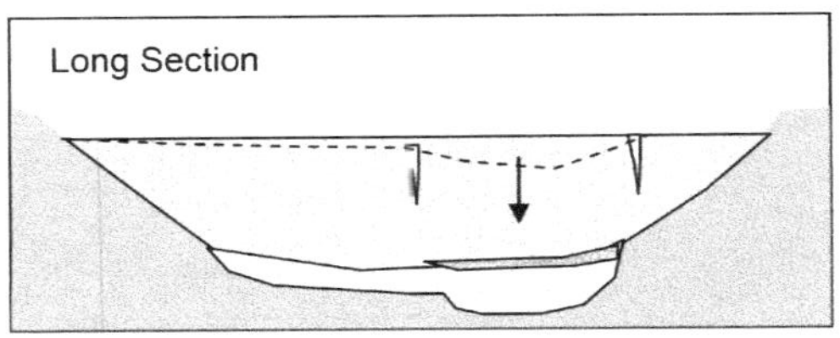	Already considered in upper part of dam.	Covered in IM 5
IM 12—Transverse cracking or hydraulic fracture at the foundation contact due to small-scale irregularities in the foundation profile under the core	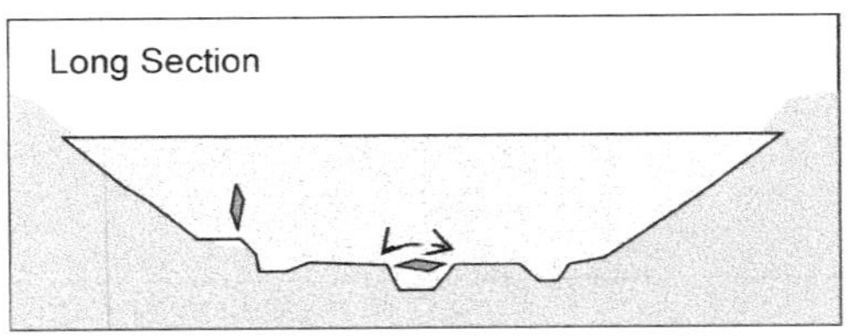	Exclude if: The persistence of the irregularity across the width of the core is less than 50% of the core base width.	Section 9.2.5 Table A5 Figure A8

TABLE 14.3 Screening of Initiating Mechanisms—PFM Internal Erosion Due to Concentrated Leaks in Poorly Compacted Zones in the Embankment, around Conduits or Adjacent to Walls, and Other PFM Related to Walls

INITIATING MECHANISM	*SKETCH OF INITIATING MECHANISM*	*EXCLUDE THE INITIATING MECHANISM IF THE FOLLOWING CONDITIONS ARE SATISFIED*	*REFERENCE SECTION AND TABLE*
IM 14—Transverse crack or continuous gap resulting from poorly compacted zone within the core	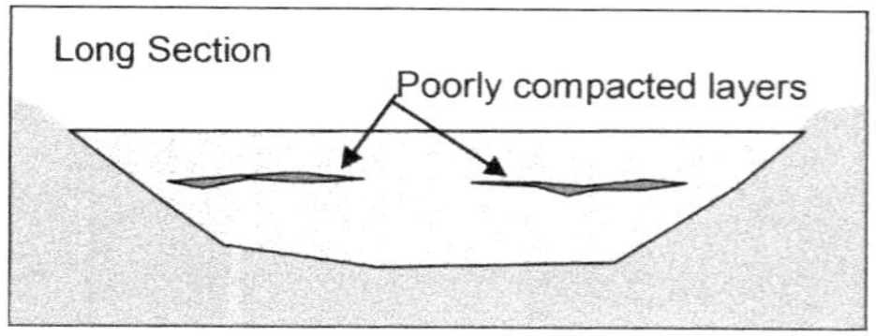	Exclude if: All soils are very well compacted with lift thicknesses less than 200 mm, with good documentation and records: (1) For plastic soils (Plasticity Index > 7), ≥ 98% Standard Maximum Dry Density ratio, moisture content 1% dry to 1% wet of OWC, *or* (2) For non-plastic soils and soils with PI ≤ 7, > 75% Relative Density/Density Index.	Section 9.5 Tables 9.13, 9.14, A6 Figure A9
IM 15—Transverse cracking or continuous flaw resulting from poorly compacted layer on the core–foundation contact	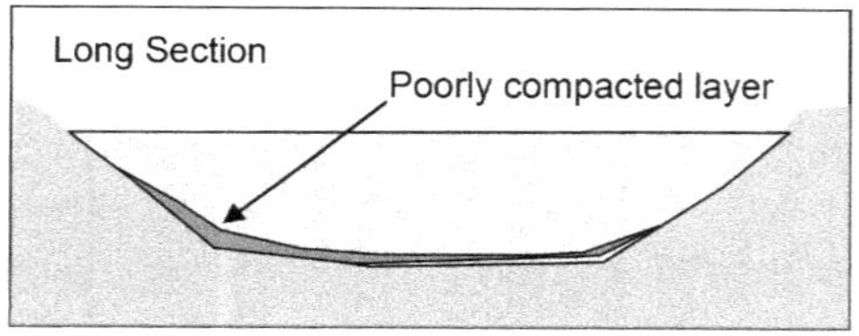	Exclude if: (1) Contact soils are well compacted on a regular foundation surface with good documentation and records, *or* (2) Uniform or regular rock surface or surface treated with shotcrete or concrete to correct slope irregularities, and soils well compacted (contact soil compacted using special compaction methods (e.g. rubber tires, use more plastic material, compaction wet of OWC), *or* (3) Uniform well-compacted soil foundation, with good mixing, bonding and compaction of contact fill. OR (4) Compacted soil foundation	Section 9.5.5 Table A7 Figure A10
IM 18—Continuous crack or gap associated with a conduit through the embankment	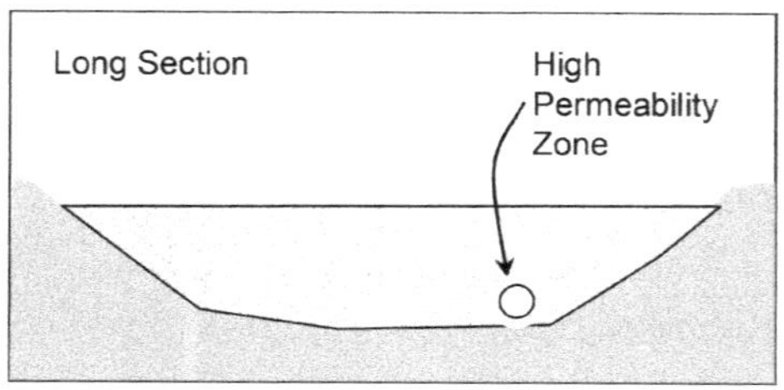	Exclude if: (1) There is no conduit passing through the embankment, *or* (2) The conduit is totally embedded in a trench excavated in non-erodible rock, backfilled to the surface with concrete.	Section 9.6.1 Table A12 Figure A13

IM 19A—Erosion into a (non-pressurised) conduit, and IM 19B—Leakage out of a (non-pressurised) conduit leading to erosion along the conduit	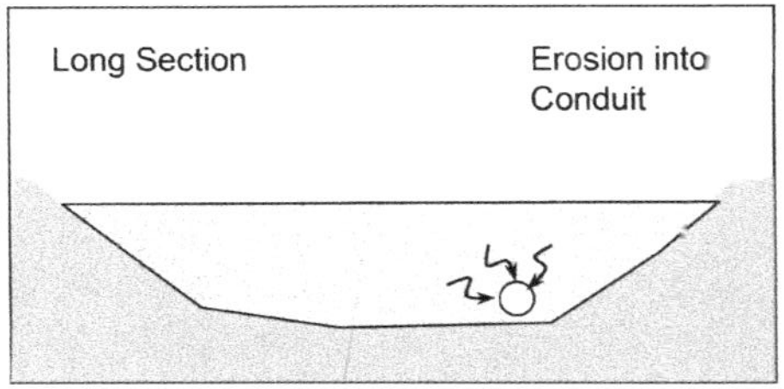	Exclude if: (1) There is no conduit passing through the embankment, *or* (2) Careful internal inspection of conduit showing no evidence of open joints or cracks.	Section 9.7 Table A13 Figure A14
IM 20—Crack or gap resulting from a poorly compacted zone associated with a spillway or abutment wall	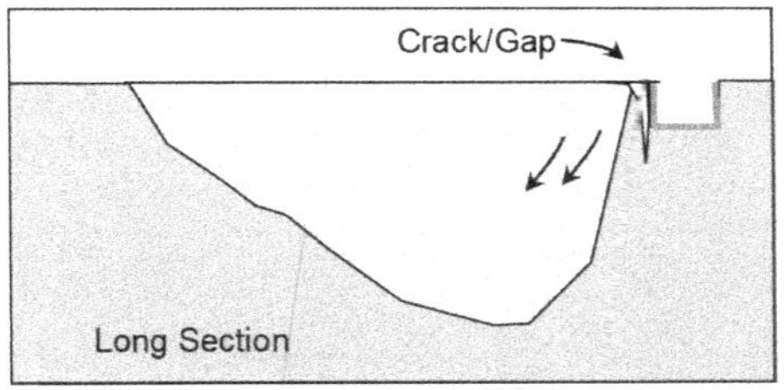	Exclude if: There is no spillway or abutment wall in contact with the embankment.	Section 9.6.2 Table A13 Figure A15
IM 21—Transverse crack or gap adjacent to a spillway or abutment wall resulting from differential settlement adjacent to the wall due to wall profile or foundation profile adjacent to the wall	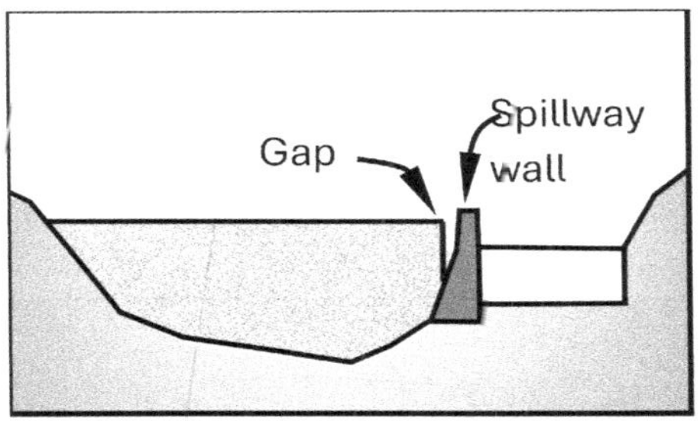	Exclude if: There is no spillway or abutment wall in contact with the embankment.	Section 9.2.8 Table A15 Figures 9.31, A16

(*Continued*)

TABLE 14.3 (Continued)

INITIATING MECHANISM	*SKETCH OF INITIATING MECHANISM*	*EXCLUDE THE INITIATING MECHANISM IF THE FOLLOWING CONDITIONS ARE SATISFIED*	*REFERENCE SECTION AND TABLE*
IM 22—Transverse crack or gap adjacent to a spillway or abutment wall resulting from differential settlement adjacent to the wall	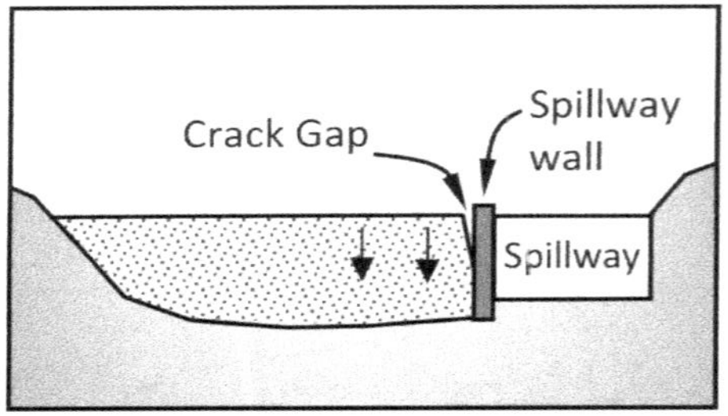	Exclude if: (1) There is no spillway or abutment wall in contact with the embankment, *or* (2) A wide bench is present at the base of the wall (W_b/H_w > 2.5, refer to Figure 9.32, *or* (3) The abutment slope below the wall is gentle (β_1 ≤ 15°, refer to Figure 9.33	Section 9.2.8 Table A16 Figures 9.32, 9.33, A17
IM 23—Persistent crack or gap at the interface of wrap-around connecting embankment to concrete gravity dam	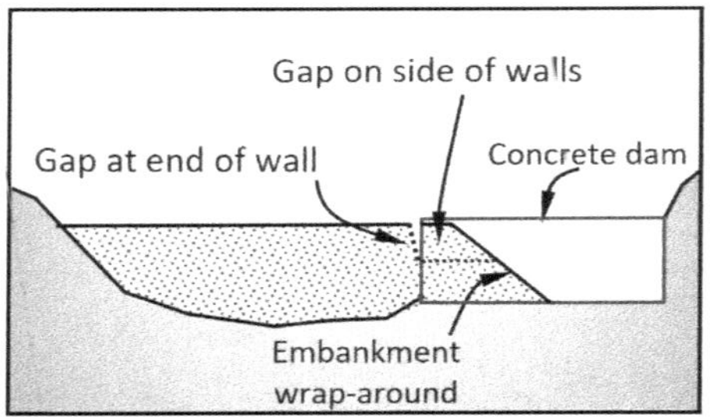	Exclude if: There is no wrap-around connection of an embankment dam to a concrete gravity dam.	Section 9.2.10 Table A17 Figure 9.37

TABLE 14.4 Screening of Initiating Mechanisms—PFM Internal Erosion Due to Concentrated Leaks, Cracks or Gaps Resulting from Freezing

INITIATING MECHANISM	*SKETCH OF INITIATING MECHANISM*	*EXCLUDE THE INITIATING MECHANISM IF THE FOLLOWING CONDITIONS ARE SATISFIED*	*REFERENCE SECTION AND TABLE*
IM 16—Transverse cracking or continuous flaw resulting from freezing at the crest of the embankment	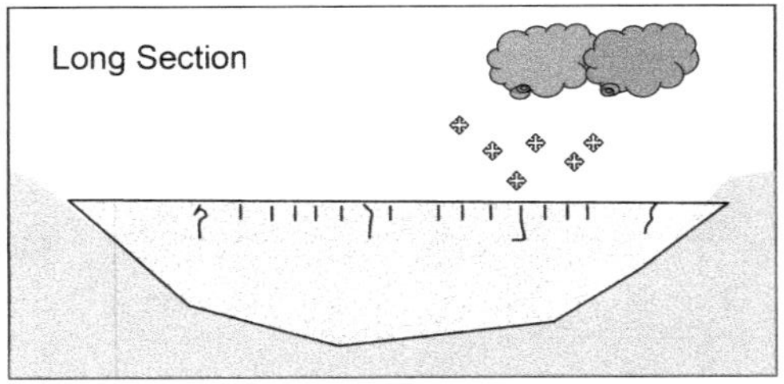	Exclude if: The climate is such that temperatures do not fall below freezing point except possibly overnight or for a day or two.	Section 9.8.2 Table A8 Figure A11
IM 17—Transverse cracking or continuous flaw resulting from freezing on seasonal shutdown layers and on the surface of staged embankments	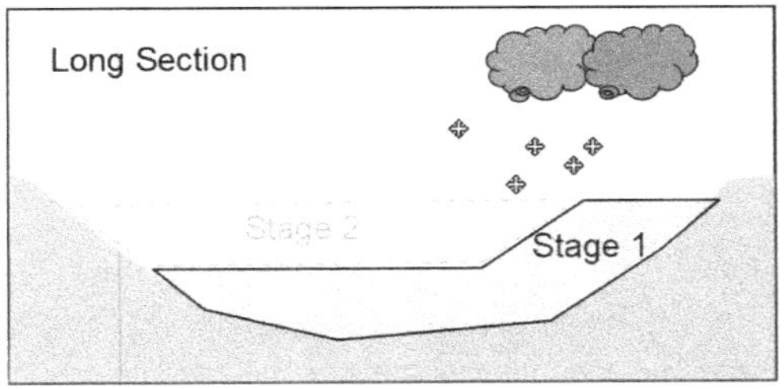	Exclude if: The climate is such that temperatures do not fall below freezing point except possibly overnight or for a day or two, *or* Very good control and clean-up practices were used—frozen layers removed from the embankment and replaced with new soil or adequately reworked to specified moisture content, *or* If the reservoir stage being considered is below the likely depth of cracking.	Section 9.8.1 Table A11 Figure A12

14.4.2.3 Internal Erosion Through the Foundation

TABLE 14.5 Screening of Failure Modes—Internal Erosion through the Foundation

INITIATING MECHANISM	*SKETCH OF INITIATING MECHANISM*	*EXCLUDE THE INITIATING MECHANISM IF THE FOLLOWING CONDITIONS ARE SATISFIED*	*REFERENCE SECTION AND TABLE*
All modes of internal erosion of the foundation (backward erosion piping, suffusion, global backward erosion in a crack)	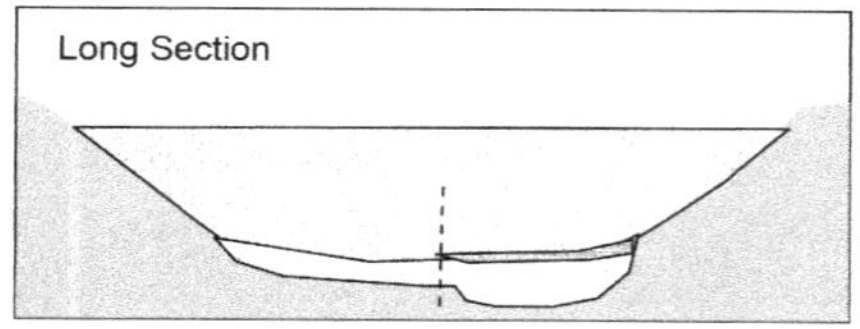	Exclude if: The non-plastic soil layer beneath the dam is not continuous or is intersected by a cutoff trench founded on non-erodible rock. (Note that erosion across the cutoff trench is considered separately in internal erosion of the embankment into or at the foundation).	Section 10.1
IM 24—Backward erosion piping in a non-plastic soil foundation	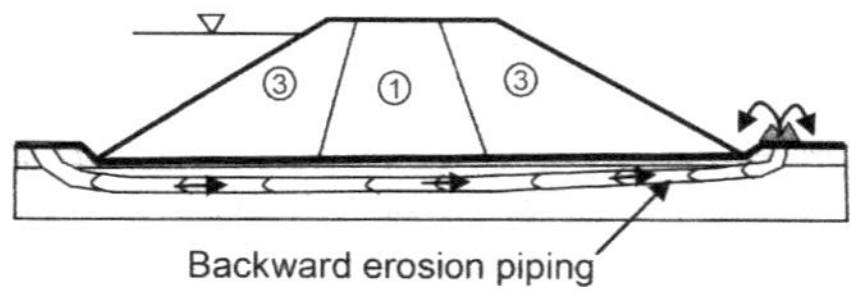	Exclude if: (1) The foundation soil has a Plasticity index > 4, *or* (2) If the non-plastic soil or soil with PI $\leq$ 4 layer is not continuous below the embankment (i.e. it terminates beneath the dam, refer to Figure 9.4), *or* (3) The soil is a silt sand gravel.	Section 10.2
IM 25—Suffusion or global backward erosion in a non-plastic soil in the foundation	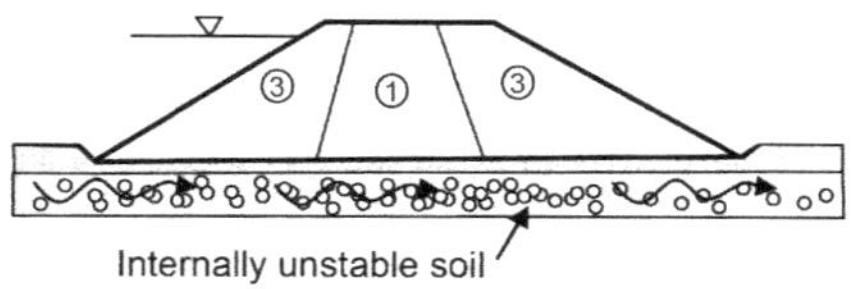	Exclude both suffusion and GBE if: The foundation soil has a Plasticity Index ≥ 7, *and* Exclude suffusion if the proportion of the finer fraction is less than 40% of the total mass of the soil.	Sections 10.2 and 10.6
IM 26—Concentrated leak erosion in a crack in soil or extremely weathered rock in the foundation	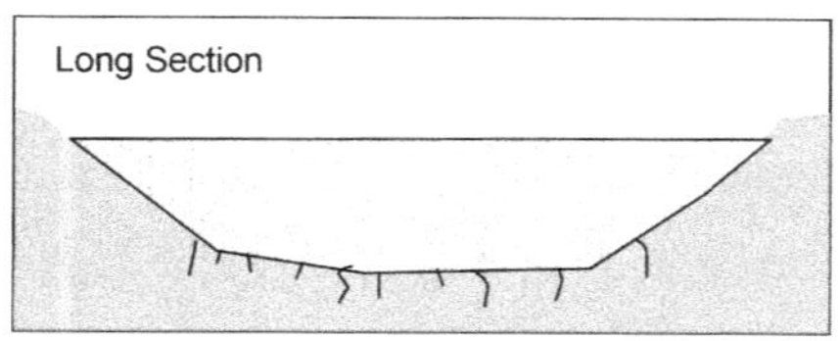	Exclude if: The foundation soil is non-plastic.	Section 9.10
IM 27—Open or infilled defects in a rock foundation potentially leading to gross enlargement or slope instability	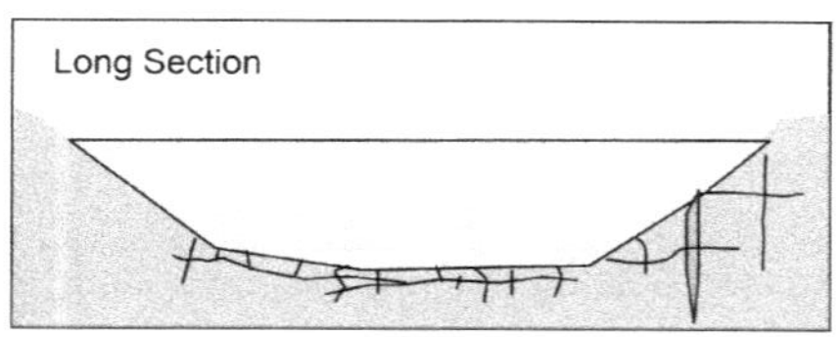	Not considered in this book.	

14.4.2.4 *Internal Erosion of the Embankment into or at the Foundation*

TABLE 14.6 Screening of Initiating Mechanisms—Internal Erosion into or at the Foundation

INITIATING MECHANISM	*SKETCH OF INITIATING MECHANISM*	*EXCLUDE THE INITIATING MECHANISM IF THE FOLLOWING CONDITIONS ARE SATISFIED*	*REFERENCE SECTION AND TABLE*
IM 28—Contact erosion in soils	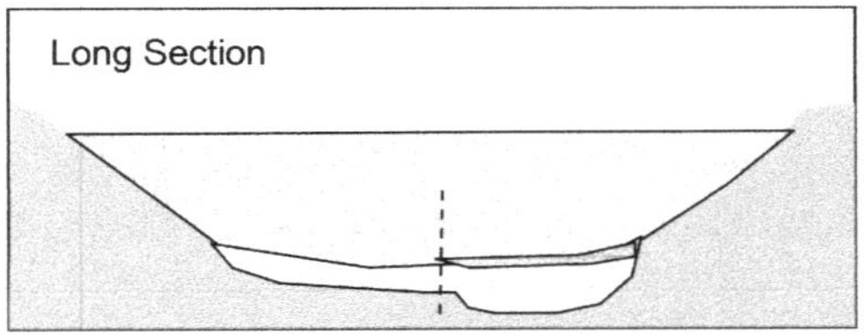	Exclude if: (1) Soil foundation below the core is comprised of fine-grained soils with greater than 12% fines (fraction finer than No 200 sieve [0.075 mm]), *or* (2) Soil foundation below the core is comprised of sands (SP or SW) which are filter compatible with the embankment materials (i.e. satisfy the no erosion criteria, refer to Chapter 12).	Section 11.3
IM 29—Contact erosion into open defects in a rock foundation	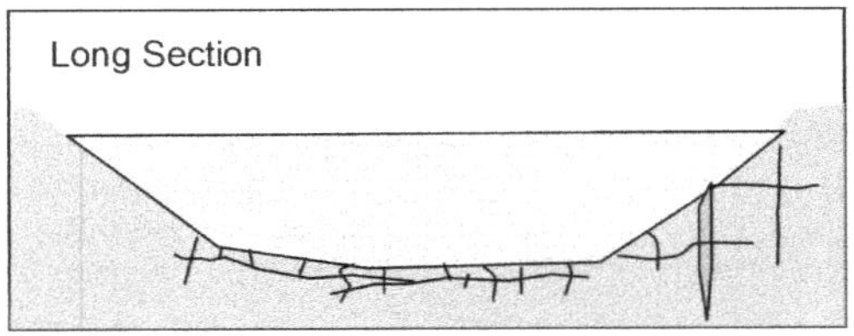 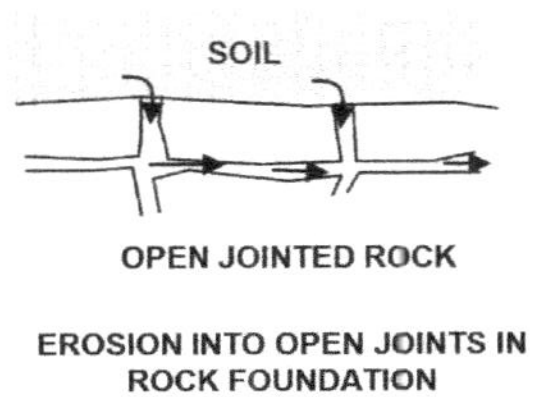	Exclude if: (1) Rock foundation below the core is comprised of rock containing closed rock defects (< 1 mm wide) or defects open less than D_{95} of the fine limit of the core, *or* (2) Rock foundation below the core has been adequately treated (e.g. shotcrete, slush grouting mortar treatment).	Section 11.4

(*Continued*)

TABLE 14.6 (Continued)

INITIATING MECHANISM	*SKETCH OF INITIATING MECHANISM*	*EXCLUDE THE INITIATING MECHANISM IF THE FOLLOWING CONDITIONS ARE SATISFIED*	*REFERENCE SECTION AND TABLE*
IM 30 Hydraulic fracture in a narrow cutoff trench with erosion into open joints in the sides of the cutoff trench	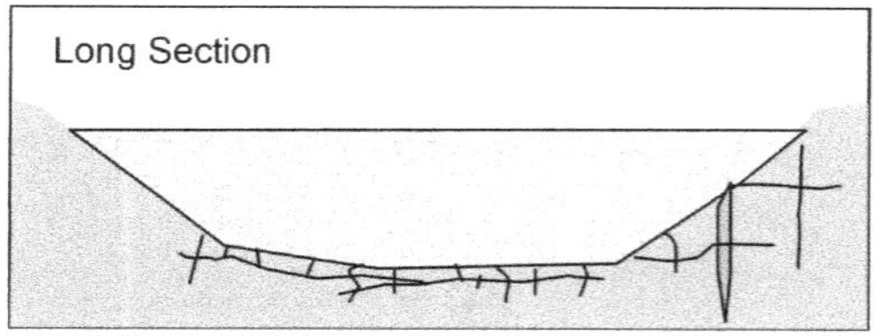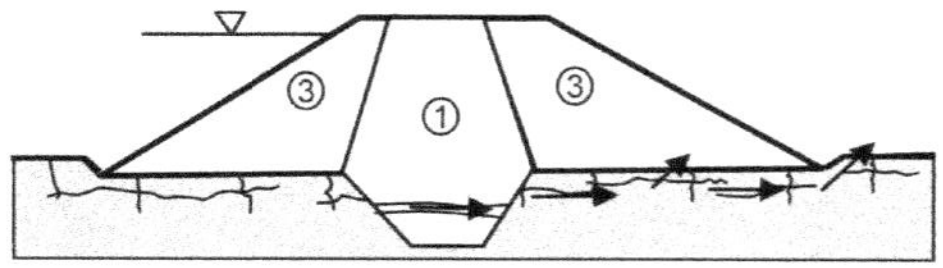Hydraulic fracture in cut-off trench	No trench or wide, shallow cutoff trench	Section 11.5
IM 31—Global backward erosion or suffusion of soil in a cutoff trench into open joints or coarse soil in the sides of the cutoff trench	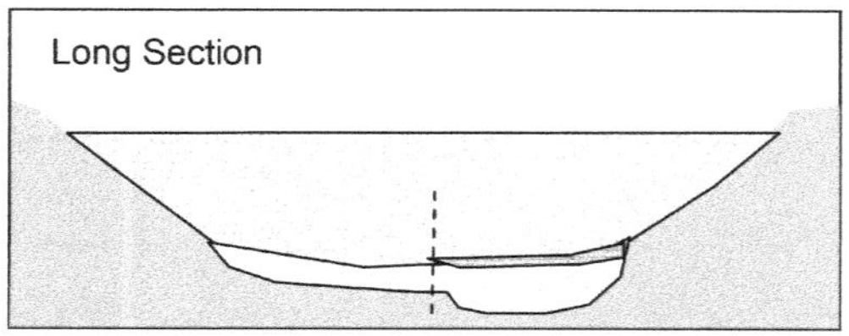	Plastic soils not subject to GBE or suffusion. Sides of cutoff trench treated to cover open joints or coarse soils.	Section 11.6

14.5 PARTITIONING OF RESERVOIR LEVELS AND SEISMIC LOADING

See Chapter 5.

14.6 ASSESSMENT OF CONDITIONAL PROBABILITIES

14.6.1 General Approach

For each failure mode the conditional probabilities at each node in the event tree are estimated for each reservoir and earthquake load partition. They are used in the event trees to calculate probabilities of failure (embankment breach) for each load partition. These probabilities are conditional on the reservoir or earthquake loading and are known as the "system response probabilities".

The estimation of conditional probabilities is covered in Chapters 9 to 13, with Chapters 9, 10 and 11 covering Initiation of Erosion, Chapter 12 Continuation and Chapter 13 Progression, Detection, Intervention and Repair, and Breach.

Tables are presented within the chapters and Appendix A to give guidance on the estimation of conditional probabilities. These tables have been developed to model the physical processes so far as practical. The probabilities have been assessed based on the mechanics of the erosion process and using the experience and expert judgement of the Fell et al. (2024) authors. Where practical these have been anchored to historic data. Anchoring has mainly been possible in the estimation of the probability of initiation of erosion in concentrated leaks and is discussed in the Fell et al. (2024) Supporting Document.

It is essential that those using these methods have a sound knowledge of the mechanics of internal erosion and piping and of risk analysis.

The methods described here and in Appendix A are intended to provide guidance and should not be followed blindly. The risk analysis team should consider whether the outcomes of using this document to estimate conditional probabilities are reasonable and applicable to the potential failure mode under consideration and should be willing to amend the estimates if there are sound reasons for doing so. Such reasons should be well documented.

14.6.2 Estimating Conditional Probabilities

14.6.2.1 Using Relative Importance Factors and Likelihood Factors

A number of the sections for estimation of conditional probabilities are structured so there is a table of factors affecting the likelihood. These tables are in Appendix A and Chapters 12 and 13 and are structured to show:

- The factors
- The relative importance of each factor (RF) with numeric weightings (usually three factors with relative importance weightings of 3, 2 and 1).
- Likelihood factors (LF) for which there are descriptions. Generally, there are four for each factor, with likelihood weightings of 4, 3, 2 and 1.

There is then a figure which links Σ (RF) × (LF) to the conditional probability. The tables have two or more sets of probabilities, the first for "above POR" and the second for "below POR".

The "below POR" figures are for reservoir level stages up to the historic high reservoir pool level (POR). The "above POR" figures are for representative reservoir level stages above this historic high reservoir pool level.

Those carrying out the risk analysis are required to choose which of the descriptors for each factor best reflects conditions at the dam. This assessment should be made on the best available information and using judgement based on geological conditions and experience elsewhere on dams of similar age and design. There may be situations where there is insufficient data to make reliable estimates. In these cases, additional site investigations and/or review of construction data will be needed.

In the tables, the term "negligible" means that the contribution to the probability of failure would be very small indeed—insufficient to affect the risk analysis outcome.

14.6.2.2 Using Scenario Tables

Where the number of factors affecting the estimation of conditional probabilities are few or to be based on limited data, "scenario tables" are used. These were developed by the Fell et al. (2008) Toolbox Development Team based on published information, and the experience of those team members. They were reviewed and revised as considered appropriate by the Fell et al. (2024) authors.

These tables present suggested ranges of conditional probabilities within which the risk analysis team is to select its best estimate based on the details of the dam the team is analysing.

14.7 THE NATURE OF ESTIMATED CONDITIONAL PROBABILITIES AND UNCERTAINTY

14.7.1 The Appendix A Methods Give "Best Estimate" Probabilities of Failure

So far as is practicable the estimates in Appendix A and Chapters 12 and 13 are designed to be defensible in absolute or definitive terms, so they can be used along with probabilities of failure for other potential failure modes such as failure from overtopping. However, as is evident from the Fell et al. (2024) Supporting Document, considerable judgement has been applied in developing the methods, so they are likely to be more reliable in assessing relative probabilities between internal erosion and piping failure modes for a particular dam, and between dams, than assessing absolute values.

The methods provide "best estimates" of the conditional probabilities and hence "best estimate" probabilities of failure.

The estimates are determined by expert judgement based on analyses and laboratory tests modelling the physical processes. They are designed to avoid systematic bias towards conservative or non-conservative probabilities.

Where the risk analysts use expert elicitation to estimate conditional probabilities, they should be best estimate values since the tolerable risk criteria in Chapter 18 are based on median or best estimates of risk.

14.7.2 Adjusting the Appendix A Best Estimates

14.7.2.1 Allowance for Factors Not Included in the Appendix A Methods

It is suggested that the Fell et al. (2024) estimates as presented in Appendix A and Chapters 12 and 13 be adopted except where there are factors not covered in the tables which the risk analysis team believes affect the estimate of the conditional probability for the node in question. These factors may include observations or monitoring data, or physical factors not allowed for in the table. In these cases, the best estimate should be determined by adjusting the estimate to allow for the additional factors. It would not be expected that this would generally result in conditional probabilities greatly different to those estimated by the Fell et al. (2024) method.

The additional factors should be described and the reasoning for the revised estimate provided in the risk analysis report.

14.7.2.2 What to Do if the Appendix A and Chapter 12 and 13 Estimates Seem Incorrect

There may be cases where the Fell et al. (2024) estimates of failure probabilities are significantly different to what the risk analysis team would have expected. This may be due to the following:

- The logic (i.e. construction of the event tree) does not well match the perceived failure mode for the dam being analysed;
- The RF and LF factors do not well represent what the team originally thinks are the key factors or actual observances; or
- The probability value suggested in the toolbox does not well represent what the risk analyst thinks is appropriate.

The suggested steps to follow for each of these situations are:

(a) If the issue appears to be in the logic, look carefully at the failure mode being modelled to identify where the perceived difference has occurred. Experience has shown that the logic presented has stood the test of time and is normally considered appropriate.

Where the analysis team considers the event tree logic is inconsistent with the postulated failure mechanism under consideration, or is not adequately covered, then the failure mode can be developed into its own event tree followed by assigning probabilities using expert elicitation as described in USACE / Reclamation (2019) and ANCOLD (2022).

(a) If there seems to be an issue in the RF and/or LF factors, fully discuss the factors presented to ascertain that the proposed new factors are indeed more important. Be sure that the failure mode is well understood and that the risk analyst has a good understanding of the mechanics and background to the methods as described in the Supporting Document of Fell et al. (2024). If in the end it appears that the factors should be adjusted or changed, make the changes/adjustments and document well what was done and why.

(b) If there seems to be an issue with the probabilities assigned, the risk analyst should be sure they are well aware of the mechanics of internal erosion being assessed and the development of the method's probabilities detailed in the chapters of this book and in Fell et al. (2024) Supporting Document. The adjustment may then be made and fully documented.

14.7.3 Limitations of Appendix A and Chapters 12 and 13 Methods

Fell et al. (2024) is based on the (2024) state of the art on modelling the mechanics of initiation of concentrated leak erosion in cracks and other flaws and by contact erosion, backward erosion piping, global backward erosion and suffusion. These are described in Chapters 9 to 13 of this book and in the Fell et al. (2024) Supporting Document. In some respects, ICOLD (2017) and Fell et al. (2015) are outdated, particularly in regard to estimation of the likelihood of cracks or gaps in poorly compacted layers and the hydraulics of concentrated leak flow in open cracks.

The mechanics of continuation (filter action) are fundamentally simpler, more extensively researched and the methods are less subjective than for initiation, provided the data to do the analyses is available. They have not changed significantly since ICOLD (2017) and Fell et al. (2015) were written.

The modelling of progression, detection and intervention, and breach are more subjective and largely based on case studies and expert judgement. They are not much changed from ICOLD (2017) and Fell et al. (2015).

Most of the logic, modelling of mechanics and analysis, e.g. of the formation of cracks due to differential settlement, laboratory testing and expert elicitation techniques used in this document continue to be actively researched and/or studied. It is likely that these developments in understanding will result in improved methods for assessing probabilities of initiation, continuation, and progression of internal erosion, and breach mechanics. Those using these methods should keep up to date with these developments and assess their practicable application. Unfortunately, some published research is based on a limited number of tests or computer simulations rather than soundly based laboratory testing, so it needs to be treated with caution.

14.7.4 Assessment of Probabilities of Failure for Failure Modes Which Are Not Covered by the Methods Described Here

In some dams there may be failure modes which are not well modelled by what is described here. For these failure modes the risk analysis team should develop an event tree to model the failure mode. Conditional probabilities within the event tree should be estimated using the method described here where the nodes are common to the event tree, and by expert elicitation for the other nodes. Reclamation (2019) and ANCOLD (2022) detail how this should be done.

The Barneich et al. (1996) mapping scheme has proven useful as an aid to judgement. This is presented in Table 14.7.

TABLE 14.7 Mapping Scheme Linking Description of Likelihood to Quantitative Probability (adapted from Barneich et al., 1996)

DESCRIPTION OF CONDITION OR EVENT	*ORDER OF MAGNITUDE OF PROBABILITY ASSIGNED*
Occurrence is virtually certain.	1
Occurrence of the condition or event is observed in the available database.	10^{-1}

(Continued)

Table 14.7 (Continued)

DESCRIPTION OF CONDITION OR EVENT	*ORDER OF MAGNITUDE OF PROBABILITY ASSIGNED*
The occurrence of the condition or event is not observed, or is observed in one isolated instance, in the available database; several potential failure scenarios can be identified.	10^{-2}
The occurrence of the condition or event is not observed in the available database. It is difficult to think about any plausible failure scenario; however, a single scenario could be identified after considerable effort.	10^{-3}
The condition or event has not been observed, and no plausible scenario could be identified, even after considerable effort.	10^{-4}

14.8 SUMMARISING (MAKING THE CASE)

Once a risk estimate is prepared with the use of the methods described, the risk analyst needs to make a summary of the key factors that generated the estimates of probability of failure. An exercise of "making the case" is important so that reviewers and decision makers can quickly focus on the story being told. This can be done relatively easily by reviewing each of the components of the estimate and selecting those that drive most of the final estimate. Once these are determined, they should be reviewed and the main factors that contribute most to the actual estimate of this component should be brought forward into an engineering summary of the failure mode. This engineering summary has traditionally focused on those factors key to the overall estimate. However, in the context of ALARP/SFAIRP considerations, an engineering explanation of PFMs for dams which do not meet modern design standards but with low assessed risks is equally important.

By doing this the redundancy, or lack thereof, in the design can be demonstrated. In the case with much redundancy, i.e. dams with many components that contribute to the overall risk being low, it is important to highlight this. For example, dams with a low chance for concentrated leaks and filtering components meeting modern filter design criteria should be viewed as an overall robust situation with reasonable redundancy. In contrast, a homogeneous dam where the estimated probability of failure is low due almost solely to a low probability of a concentrated leak is a case with little redundancy. It is important to summarise this situation for reviewers and decision makers.

14.9 COMBINING PROBABILITIES

14.9.1 Basic Principles

Risk is computed by finding the product of probabilities and consequences for each path of the event tree.

The principles for combining the probabilities on the event trees and for the different failure modes are as follows:

(a) For mutually exclusive failure modes, as occurs at the end of a specific event tree, the system response probabilities should be added for a specific loading partition.
(b) For failure modes that are not mutually exclusive, the total system response probabilities from the event trees should be calculated for each load partition using De Morgan's rule.

$$P = 1 - (1 - P_{IM1}) \times (1 - P_{IM2}) \times (1 - P_{IM3}) \ldots \text{etc.}$$

The method for presenting the probabilities of failure will depend on which method is intended to be used for computing the risk (e.g. using partitions or system response curves). The risk analyst should refer to other general guidance documents for the combining of probabilities and consequences (e.g. ANCOLD, 2022 and USACE/Reclamation (2019).

14.9.2 Adding Probabilities for Hydrological and Seismic Loads

The following guidelines are provided for computing the annual probabilities of failure for the hydrological and seismic load cases.

14.9.2.1 Annual Probability of Failure for the Hydrologic Load Conditions

The annual probability of failure for the normal and hydrological operating load cases is as follows:

For each hydrologic reservoir level partition, combine the system response probabilities for each failure mode using De Morgan's rule.

The annual probability of failure is then calculated by summing the product of the total system response probability for each load partition by the annual probability of the loading condition.

14.9.2.2 Annual Probability of Failure for the Seismic Load Condition

IM 13 includes transverse cracking in the upper part of embankments resulting from settlement in the embankment (IM 1), in the foundation (IM 5) and adjacent to walls (IM 21 and IM 22). In cases where there are more than one seismic PFM, then the system response probabilities should be combined using De Morgan's rule.

14.9.3 Development of System Response Curves

The following guidance has been developed for users who intend to use system response curves to quantify the risk. The probabilities of failure are estimated for each reservoir level and earthquake loading condition to develop point estimates, and this is used to develop the system response curves. Separate system response curves should be developed for each of the PFMs and IMs identified in the screening process.

The process to develop system response curves for reservoir level loading is summarised as follows:

(a) Select the reservoir level points as described in Chapter 5.
(b) Estimate the system response probabilities for the modes of internal erosion and for each reservoir level.
(c) Select the scale (linear, logarithmic or probabilistic) that best approximates a straight line for the system response probabilities to perform interpolation of intermediate values using the linearisation.
(d) Plot the system response probabilities vs reservoir level on the selected scale. Join the individual point estimates using straight lines. An example is shown in Figure 14.6.

14.10 SIMPLIFIED QRA METHOD FOR ESTIMATING THE ANNUAL PROBABILITY OF FAILURE BY INTERNAL EROSION AND PIPING

14.10.1 General Concepts

In the QRA approach there is an event tree for each Potential Failure Mode (PFM) and Initiation Mechanism (IM) and system response probabilities estimated for each reservoir load partition or

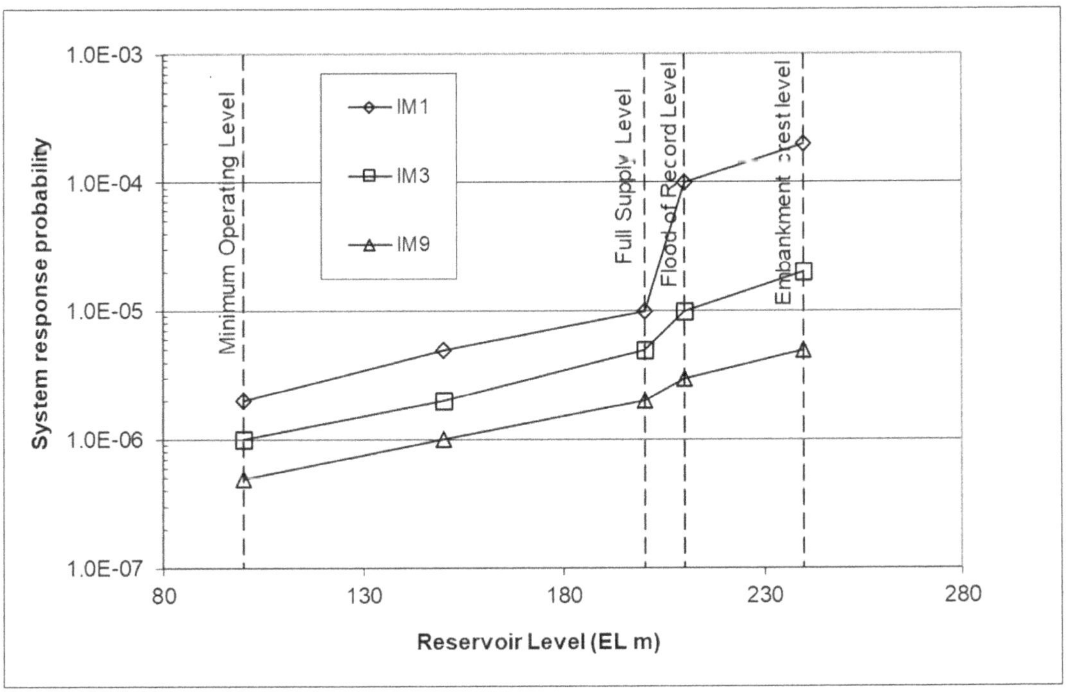

FIGURE 14.6 Example system response curves (Fell et al., 2024).

seismic load partition. There are methods described in Chapters 9 to 13 to estimate the conditional probabilities for initiation, continuation, progression, detection and intervention, and breach.

For the Simplified QRA method, the concept is to provide a simple method based on the QRA approach but assessing the annual probability of failure for only the most important IMs for internal erosion in the embankment, in the foundation and from embankment to foundation.

For each IM, fewer branches of the event tree and only three reservoir levels and three seismic earthquake ground motions are modelled. The annual probability of failure of an IM for each load is estimated.

The PLL for each load case is estimated using the Reclamation Consequence Estimating Methodology (RCEM) method (or other empirical based method) as described in Chapter 18.

The tolerability of the risk for each IM and reservoir level and seismic load is considered separately in the tolerable risk matrix as described in Chapter 18.

The expectation is that the risk estimated in this manner will be estimated within an order of magnitude of what would be estimated by the quantitative risk approach. The method allows for ranking of the risks for each PFM and guides those carrying out the dam safety assessment as to which loading contributes most to the risk.

14.10.2 Simplification of the Event Trees

14.10.2.1 Concentrated Leak Erosion

The branches of the event tree which are included are as follows:

- *Probability of a persistent transverse crack, gap or pathway* for the reservoir partition. This is done using the methods described in Chapters 9 to 11 for each of the IMs listed in Section 14.10.3.
- *Initiation.* Assume that initiation will occur (i.e. probability initiation = 1.0) if the reservoir rises above the base of the crack, gap or continuous pathway. This is somewhat conservative but for unfiltered exits the hydraulic shear stresses are likely to exceed the critical shear stress, so initiation is highly likely to occur. The effect of filters on hydraulic shear stresses is conservatively ignored.
- *Continuation.* This is done using a simplification of the QRA method as described in Section 14.10.5.
- *Progression, detection and intervention, and breach.* For most cases assume the probability of progression, not detecting and failing to intervene and breach is assumed = 1.0.

For most cases progression, detection and intervention, and breach often have much less effect on the outcome, so in the Simplified QRA approach may not need to be considered. Progression should be considered where flow limitation is expected to be an important consideration. An example would be internal erosion through an embankment dam with a concrete core wall. In these cases, refer to the relevant discussion in Chapter 13.

14.10.2.2 Backward Erosion Piping

The branches of the event tree which are included are:

- *The screening test* as described in Section 10.2.
- *Initiation* as described in Section 10.3.

- *Continuation.* Not applicable if BEP has initiated.
- *Progression.* Assume that the probability of progression = 1.
- *Detection, intervention and breach.* Assume the probability of unsuccessful detection and intervention and breach = 1. This is potentially conservative because BEP can often be detected by the presence of sand boils and progression of BEP can be arrested by intervention.

14.10.2.3 Suffusion and Global Backward Erosion

There is no simplified approach for suffusion and GBE. Use QRA approach.

14.10.2.4 Contact Erosion

There is no simplified approach for contact erosion. Use QRA approach.

14.10.3 PFM to Be Considered

The steps are:

(a) Apply the screening tests described in Section 14.4.2.
(b) For those IMs remaining from the screening test, discard any IM which is not in the following list unless there are clear reasons why that IM should be further considered.

14.10.3.1 For Internal Erosion within the Embankment

The authors have identified internal erosion mechanisms which in their experience are more likely to be important for typical embankment dams. The PFMs to be considered may include the following internal erosion initiating mechanisms. The analyst should consider which mechanisms are applicable and most critical to the dam under consideration.

For All Embankments

IM 1 Transverse cracking in the upper embankment resulting from cross-valley differential settlement.
IM 5 Transverse cracking in the upper or lower embankment resulting from differential settlements in soil in the foundation beneath the core.

For Embankments with a Through-Going Conduit

IM 18 Continuous crack or gap resulting from poorly compacted zone around a conduit through the embankment.

For Embankments with a Spillway or Abutment Wall

IM 21 Transverse crack or gap adjacent to a spillway or abutment wall resulting from wall type and profile.
IM 20 Transverse crack or gap resulting from poorly compacted zone associated with a spillway or abutment wall.
IM 22 Transverse crack or gap adjacent to a spillway or abutment wall resulting from settlement of the fill adjacent the wall.

14.10.3.2 For Internal Erosion in the Foundation

Apply the screening test on soil gradation and plasticity as described in Section 10.2. For foundations subject to backward erosion piping or suffusion and global backward erosion:

IM 24 Assessment of the probability of initiation of backward erosion piping in a layer of non-plastic soil or soil with Plasticity Index ≤ 4 in the foundation.
IM 25 Estimation of the probability of initiation of suffusion or global backward erosion in an embankment (IM 25E) or in the foundation (IM 25F).

14.10.3.3 For Erosion of the Core into the Foundation

For embankments where the core is potentially in contact with open jointed rock or gravel soils, apply:

IM 28 Contact erosion in soils.
IM 29 Contact erosion into open defects in a rock foundation.

Those doing the analysis may add PFMs which are important for the dam under consideration and consider them using the principles as described in Section 13.

14.10.4 Loading Cases

14.10.4.1 Reservoir Loading

Consider the following reservoir levels:

(a) POR using the "above POR" conditional probabilities.
(b) Inflow Design Flood (IDF) level.
(c) Level where there is a significant change of probability of initiation, e.g. the level of the base of cracking, or continuation, e.g. the top of filters.

14.10.4.2 Seismic Loading

Consider the following seismic loads:

(a) Earthquake Design Ground Motion (EDGM) which is appropriate for the dam as described in Section 5.1.2.
(b) Earthquake Ground Motion (EGM) which is likely to result in transverse cracking, e.g. Pells and Fell (2002, 2003) Damage Class 2 or higher (see Figures 8.2 and 8.3 and Section 9.9).
(c) EGM at which extensive liquefaction is likely to occur (where applicable).

For dam sites with low to moderate seismic hazard (e.g. 1 in 10,000 AEP peak ground acceleration resulting in Damage Class 0 or 1 in Tables 8.2 and 8.3), seismic-induced internal erosion PFMs are likely to be a small contributor to risk and can be excluded from the simplified approach unless the dam or its foundations are subject to liquefaction.

14.10.5 Assessing the Likelihood of Continuation

There are three scenarios:

Internal erosion pathway is not intercepted by a filter or transition zone. Assume probability of continuation = 1.0 regardless of the type of material downstream of the core.

Internal erosion pathway is intercepted by a filter or transition zone for which as-constructed PSDs of the core and filter available use the method in Chapter 12. Assume the probability of continuation = Σ Probability of excessive and continuing erosion.

Internal erosion pathway is intercepted by a filter or transition zone for which no specified or as-constructed PSDs are available. Assess the likelihood of continuation as described in Table 14.8 and Figure 14.7.

TABLE 14.8 Factors Influencing the Likelihood of Continuation for Dams with a Filter or Transition

		LIKELIHOOD FACTOR (LF)			
FACTOR	*RELATIVE IMPORTANCE FACTOR (RF)*	*LESS LIKELY (1)*	*NEUTRAL (2)*	*MORE LIKELY (3)*	*MUCH MORE LIKELY (4)*
Source of the filter as it affects the likelihood of the filter holding a crack	(3)	Coarse river alluvium, e.g., sandy gravel, GP and GW gravel.	Processed crushed rock, screened and washed. Coarse-grained alluvium, e.g. sand, gravelly sand or glacial soils	Processed crushed rock, screened only, e.g. granite, greywacke, basalt, limestone, *or* Fine-grained alluvium, e.g. silty sand	Ripped rock, e.g. sandstone, siltstone, shale or other weak rock. (b) Unprocessed crushed rock, e.g. quarry run, tunnel spoil, granite, greywacke, basalt, limestone
Filter placement method as it affects the potential for segregation and coarse layers in the filter. All sources except ripped rock.	(2)	Placed in thin lifts < 0.45 m[(a)], careful control during construction	End dumping from trucks, spread by dozer in lifts < 0.6 m	End dumping from trucks, spread by dozer in thick lifts > 0.6 m	Filters/transition constructed by pushing material over the edge of the core
Unified Soil Classification of the core as it affects the D_{85} of the core	(1)	SW, GC, GM, GP, GW	SC, SM	CH, MH, CL, ML With 5% to 15% sand	CH, MH, CL, ML With < 5% sand

Notes:

(a) Placed in spreader boxes or truck mounted conveyors with care to avoid segregation.

(b) Assume P_{CE} = 1.0 regardless of filter placement method and core classification.

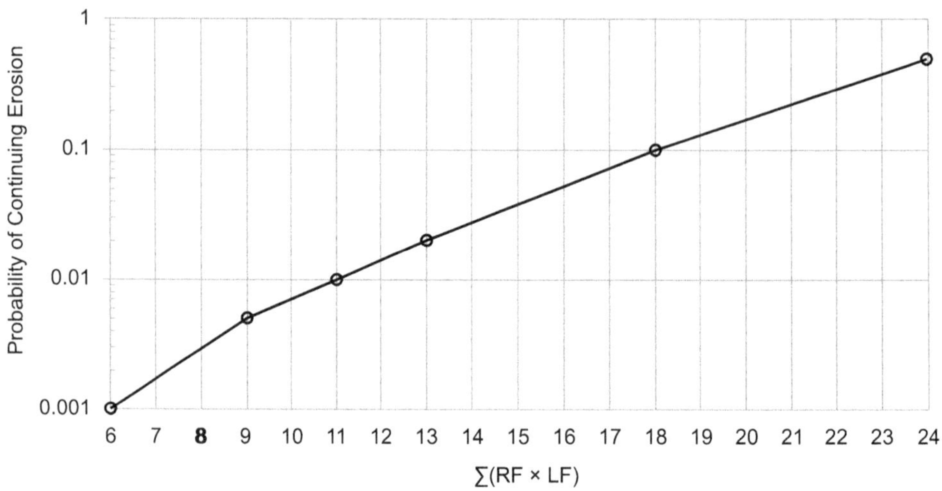

FIGURE 14.7 Probability of continuing erosion versus Σ (Relative importance factor [RF]) × (Likelihood factor [LF])

15 Historical Database Screening Method for Assessing the Annual Probability of Failure Resulting from Internal Erosion and Piping

15.1 CONCEPTUAL FRAMEWORK

Foster et al. (1998, 2000a, 2000b) developed a method for estimating the relative likelihood of failure of embankment dams by internal erosion and piping.

The method involves the adjustment of the historical average probabilities of failure of dams of the three modes of piping to take account of the characteristics of the dam, such as internal zoning, presence of internal filter measures, core properties, compaction and foundation geology, and to take account of the past performance of the dam. These adjustments were made with the use of weighting factors which are multiplied to the average probabilities of failure for each embankment zoning type. The calibration of the weighting factors was determined from the analysis of the characteristics of dam incidents and the dam.

The average probabilities of failure and annual probabilities of failure were determined for the three modes of piping failure from the incident and population statistics taken up to 1986 are summarised in Table 15.1. These are based on the ICOLD (1995) database.

The Database Screening Method presented here is based on the Foster et al. (1998, 2000a, 2000b) method. It:

- Uses the ICOLD (1995) database but adjusts the statistics which allow for under-reporting of failures and accidents. The adjustments are similar to those Foster et al. (1998, 2000a, 2000b) used.

DOI: 10.1201/9781003542025-15

TABLE 15.1 Average Probability of Failure of Embankment Dams by Mode of Piping and Dam Zoning after Foster et al. (1998, 2000a, 2000b)

	EMBANKMENT			*FOUNDATION*			*EMBANKMENT INTO FOUNDATION*		
		AVERAGE ANNUAL P_e (× 10^{-6})			*AVERAGE ANNUAL P_f (× 10^{-6})*			*AVERAGE ANNUAL P_{ef} (× 10^{-6})*	
ZONING CATEGORY	*AVERAGE P_{Te} (× 10^{-3})*	*FIRST 5 YEARS OPERATION*	*AFTER 5 YEARS OPERATION*	*AVERAGE P_{Tf} (× 10^{-3})*	*FIRST 5 YEARS OPERATION*	*AFTER 5 YEARS OPERATION*	*AVERAGE P_{Tef} (× 10^{-3})*	*FIRST 5 YEARS OPERATION*	*AFTER 5 YEARS OPERATION*
Homogeneous earthfill	16	2080	190						
Earthfill with filter	1.5	190	37						
Earthfill with rock toe	8.9	1160	160						
Zoned earthfill	1.2	160	25						
Zoned earth and rockfill	1.2	150	24						
Central core earth and rockfill	(< 1.1)	(< 140)	(< 34)						
Concrete face earthfill	5.3	690	75						
Concrete face rockfill	(< 1)	(< 130)	(< 17)						
Puddle core earthfill	9.3	1200	38						
Earthfill with corewall	(< 1)	(< 130)	(< 8)						
Rockfill with corewall	(< 1)	(< 130)	(< 13)						
Hydraulic fill	(< 1)	(< 130)	(< 5)						
All dams	3.5	450	56	1.7	255	19	0.18	19	4

Notes:
(1) P_{Te}, P_{Tf} and P_{Tef} are the average probabilities of failure over the life of the dam.
(2) P_e, P_f and P_{ef} are the average annual probabilities of failure.

- Adjusts the statistics to allow for information in the updated ICOLD (2019) database. The average dam's age after first fill was 26 years in the ICOLD (1995) database. There were no additional failures and accidents of the ICOLD (1995) dams in the ICOLD (2019) database, so the average age of 58 years in 2018 is used.

Table 15.2 summarises the revised statistics for failures in the embankment. It:

- Uses overall statistics and embankment zoning as a factor rather than using the statistics of individual zoning as was done in Table 15.1.
- Incorporates the development in understanding of the mechanics of internal erosion and piping since 1998 as described in earlier chapters. This results in some different weighting factors than used in the Foster et al. (1998, 2000a, 2000b) method.

The Database Screening Method is to be used only as additional information for the Rsk Enhanced Deterministic Approach to assessing dam safety. It should not be used alone for important dam safety

TABLE 15.2 Historical Statistics of Failures for Internal Erosion and Piping through the Embankment Fell et al. (2024)

CASE	*TOTAL*	*IN DAM*	*AROUND CONDUITS OR WALLS*	*COMMENTS*
No. of internal erosion failures	36	19	15 conduits 2 walls	.
Population of dams	11,192	11,192	5596 conduits 2798 walls	Assumes 50% embankment dams have conduits passing through the dam and that 25% of all dams have wall interfaces.
Historical frequency for internal erosion failures	0.0032	0.0017	0.0027 conduits 0.0007 walls	
Historical frequency for internal erosion failures adjusted for under-reporting	0.0064	0.0034	0.0054 conduits 0.0014 walls	2 × to allow for under-reporting.
Historical frequency First fill failures adjusted for under-reporting	0.00256	0.00136	0.00216 Conduits 0.00056 walls	Assumes 40% occur on first fill.
Historical frequency of failure after first fill adjusted for under-reporting	0.00384			
Historical annual probability of failure after first fill adjusted for under-reporting	66E-06			Assumes 60% after first fill and average age 58 years. Was 56E-06 in Foster et al. (1998, 2000a, 2000b) method.

Note: "First filling" includes the first 5 years of operation.

assessments because it does not model the details of an embankment and monitoring data as well as Simplified QRA and QRA methods.

It is calibrated against the statistics in Table 15.2 so gives similar outcomes (in terms of order of magnitude of the probability estimate) to the updated Piping Toolbox in Fell et al. (2024) as described in Chapters 9–13.

15.2 INTERNAL EROSION AND PIPING IN THE EMBANKMENT

The following summarises the Database Screening Method for internal erosion and piping through the embankment:

For Dams on First Filling and in the First 5 Years of Operation:

The historical average frequency of failure of dams on first filling or the first 5 years of operation = 0.00256.

For the dam being assessed, a weighting factor (w_E) is assessed using Table 15.3.

The weighting factor is obtained by multiplying the individual weighting factors from the table:

$$w_E = w_{E(zoning)} \times w_{E(\mathrm{filt})} \times w_{E(\mathrm{cst})} \times w_{E(\mathrm{cc})} \times w_{E(\mathrm{con})} \times w_{E(\mathrm{ft})} \quad (15.1)$$

The probability of failure on first filling or in the first 5 years = $0.00256 \times w_E$.

For Dams after 5 Years of Operation:

The historical average annual probability of failure of dams after 5 years of operation is 6.6×10^{-5} for all zoning types.

For the dam being assessed, a weighting factor (w_E) is assessed using Table 15.3.

The weighting factor is obtained by multiplying the individual weighting factors from the table:

$$w_E = w_{E(zoning)} \times w_{E(\mathrm{filt})} \times w_{E(\mathrm{cst})} \times w_{E(\mathrm{cc})} \times w_{E(\mathrm{con})} \times w_{E(\mathrm{ft})} \quad (15.2)$$

The annual probability of failure after 5 years of operation = $6.6 \times 10^{-5} \times w_E$ per annum.

15.3 INTERNAL EROSION AND PIPING IN THE FOUNDATION

15.3.1 Introduction

There are only a small number of cases in the database of failures and accidents for internal erosion and piping in the foundation which are for backward erosion piping, suffusion or global backward erosion in the foundation.

TABLE 15.3 Amended Summary of the Weighting Factors for Piping through the Embankment Mode of Failure

	GENERAL FACTORS INFLUENCING LIKELIHOOD OF FAILURE				
FACTOR	*MUCH MORE LIKELY*	*MORE LIKELY*	*NEUTRAL*	*LESS LIKELY*	*MUCH LESS LIKELY*
Embankment Zoning (a)	Homogeneous	Earthfill with foundation filter, earthfill with rock toe, concrete face earthfill, puddle core earthfill.		Earthfill with embankment filter, zoned earthfill, zoned earth and rockfill, earthfill with core wall, hydraulic fill. **[0.3]**	Central core earth and rockfill, rockfill with core wall. (c) **[0.1]**
$w_{E(zoning)}$	**[5]**	**[1.5]**	**[1.0]**		
Embankment Filters (b)		Embankments which do not have embankment filter or transition zones.	Embankment filter or transition present. No data available or poor design and construction. (d)	Embankment filter present—average design and construction. (e)	Embankment filter present—well designed and constructed. (f)
$w_{E(filt)}$		**[2]**	**[1.0]**	**[0.2]**	**[0.015] natural filters. [0.0165] crushed rock filters Cu >20**
Core Soil Type	Dispersive clays. **[5]** Low plasticity silts (ML). **[2.5]**	Clayey and silty sands (SC SM). **[1.2]**	Well graded and poorly graded gravels (GW, GP). **[1.0]**	Clayey and silty gravels (GC, GM). **[0.8]**	High plasticity clays (CH). **[0.3]**
$w_{E(cst)}$	Poorly and well graded sands (SP, SW). **[2]**		High plasticity silts (MH). **[1.0]**	Low plasticity clays (CL). **[0.8]**	
Core Compaction (g)	No formal compaction.	Rolled, modest control.	Puddle, hydraulic fill.		Rolled, good control.
$w_{E(cc)}$	**[6]**	**[1.2]**	**[1.0]**		**[0.2]**
Conduits (h)	Conduit through the embankment—many poor details.	Conduit through the embankment—some poor details.	Conduit through embankment—typical USBR practice.	Conduit through embankment—good detailing including downstream filters.	No conduit through the embankment.
$w_{E(con)}$	**[6]**	**[2]**	**[1.0]**	**[0.1]**	**[0.5]**

(Continued)

TABLE 15.3 (Continued)

	GENERAL FACTORS INFLUENCING LIKELIHOOD OF FAILURE				
FACTOR	*MUCH MORE LIKELY*	*MORE LIKELY*	*NEUTRAL*	*LESS LIKELY*	*MUCH LESS LIKELY*
Foundation Profile $w_{E(ft)}$	**[20]**	**[4]**	**[1]**	**[0.25]**	**[0.05]**
With step	> 50°, > 1.75% settlement. (i)	35° to 50°, 2.0% to 0.75%	25° to 35°, 1.5% to 0.4% 35° to 50°, 0.4% to 0.1%	15° to 25°, > 2% to 0.5% 25° to 35°, 0.25% to ≤ 0.1%	15° to 25°, < 0.75% to ≤ 0.1%
Without step	> 50°, > 2.5% settlement	35° to 50°, 2.5% to 1%	25° to 35°, 1.0% to 0.5% 35° to 50°, 0.5% to < 0.1%	15° to 25°, > 1% to 0.5% 25° to 35°, 0.5% to ≤ 0.1%	15° to 25°, < 0.5% to ≤ 0.1%
With compressible soil	> 60°, > 2.0% settlement	25° to 60°, 2.0% to 0.5%	15° to 60°, 0.5% to 0.2%	15° to 60°, 0.2% to ≤ 0.1%	15° to 60°, ≤ 0.1%

Notes:

(a) Embankment zoning shown in Figure 2.1.
(b) See Figure 2.1 (1) for definition of "embankment filter". Also referred to as "chimney filter".
(c) Concrete face rockfill dams not susceptible to internal erosion and piping in the same sense as other zoning so are not covered by this method. See Chapter 16.
(d) Poor design may relate to filters not extending above full supply level (FSL) and/or taken from sources likely to result in excessive fines or gap grading and segregation and gradation so coarse that continuing erosion is possible.
(e) Average design, filters taken above FSL but not to dam crest, filters designed to satisfy Terzaghi filter criteria or modern criteria, but limited construction control testing.
(f) Filter taken to near crest level, designed and constructed with good control. Likely to satisfy no-erosion or some-erosion criteria.
(g) The methods of compaction are:
No formal compaction—fill materials in the core were dumped in place and either no compaction, compaction by animal hooves or by travel of construction equipment only.
Rolled, modest control—core materials were rolled but with poor control of moisture content (e.g. varying greater than ± 2% of optimum water content) and/or compacted in relatively thick layers.
Rolled, good control—core materials compacted in thin layers, good control of moisture content within ±2% of optimum water content and greater than 95% of standard compaction.
(h) Conduit descriptions. See Appendix C for details.
Conduit through the embankment—many poor details: No downstream filter and three or more of the poor detailing characteristics described in Appendix C.
Conduit through the embankment—some poor details: No downstream filter and two poor detailing characteristics.
Conduit through the embankment—typical USBR practice: Details of conduits typical of USBR practice in the 1950s–1970s which are described in Appendix C.
Conduit through the embankment—good detailing including downstream filters: Good detailing and the downstream end of the conduit surrounded by a well designed and constructed filter zone.
No conduit through the embankment: Outlet tunnel in the abutment, no outlet conduit, e.g. flood control dams, or outlet through a concrete section of the dam or embedded in the foundation as shown in Figure 9.54.
(i) Post-construction settlement for 10 years after construction.

It is therefore not possible to develop a data-based approach to assessing the likelihood of failure for these PFMs.

15.3.2 Recommended Method

The recommended method is as follows.

15.3.2.1 General

Identify cases where the likelihood of failure from these PFMs is negligible based on the lack of continuity and/or classification of the soil strata, or because the strata has been fully intercepted by a cutoff trench, or by a sheet pile wall, or other type of cutoff. See later for screening on soil classification.

15.3.2.2 Backward Erosion Piping

Identify cases where the likelihood of failure from BEP is significant because of the presence of foundation soil subject to BEP with evidence of sand boils.

For other cases with foundation soil subject to BEP, carry out an assessment of the likelihood of initiation of BEP based on available information and seepage flow nets for reservoir at full supply level and at dam crest level. See Section 10.3 for the method for this assessment.

Assume that given BEP initiates the probability of progression is one.

Allow for the likelihood of detection and intervention as described in Chapter 13. Allow for the fact that few BEP failures occur in levees despite there being many instances of sand boils because the boils are detected and measures are taken to stop the progression of BEP.

15.3.2.3 Suffusion and Global Backward Erosion

Identify cases where the likelihood of suffusion or GBE is significant based on the presence of non-plastic silt sand gravel soils in the foundation having gradations making them susceptible to suffusion or GBE (see later), and sinkholes in the floor of the reservoir upstream of the dam.

Taking account of the performance of the dam, assess the likelihood that this could lead to failure. Historically, suffusion and GBE are more likely to experience an accident than failure. This is supported by the laboratory experiments described in Section 10.6 which show the result of suffusion and GBE is to erode only some of the finer fraction, giving an increase in permeability.

For other cases where there are soils subject to suffusion or GBE but there has been no history of evidence of erosion, the probability of failure by suffusion or GBE is likely to be very low or negligible, taking account of the likelihood of detection and intervention.

15.3.2.4 Screening of Soils Subject to BEP, Suffusion or GBE Based on Classification

The following is based on Fell et al. (2024) and is as presented in Section 10.2.3.

If the soil is non-plastic or has a Plasticity Index $\leq$ 4, then follow the procedures in Section 10.3 to assess the probability of initiation of backward erosion piping (BEP) for SP, SM and ML soils, and Section 10.4 for initiation of suffusion and global backwards erosion (GBE) for GW, GP and GM and SW soils.

If the soil is plastic with a Plasticity Index > 7, then the likelihood of BEP and suffusion and GBE is negligible under the seepage gradients which occur in foundations and the embankment of earth and earth and rockfill dams. For such soils assess if concentrated leak erosion may occur through a crack in the soil using the procedures detailed in Sections 9, 11, 12 and 13.

If the soil is low plasticity with a Plasticity Index between 4 and 7, the likelihood of BEP is negligible. Suffusion or GBE may occur in gap-graded GW, GP and GM and SW soils including for seepage gradients ≤ 4.

15.4 EROSION OF FOUNDATION SOIL INTO OPEN DEFECTS IN ROCK OR INTO COARSE GRAVEL

15.4.1 Description of the Potential Failure Mode

Figures 15.1 and 15.2 show the scenarios covered by this PFM.

The historic case data for piping in the foundation is dominated by erosion of soil in the foundation into open joints and erosion into coarse gravel underlying the soil. For that reason, these PFMs are considered in this screening method.

For erosion into open joints, the scenario is that the rock with open joints is overlain by soil, and the downstream shoulder of the dam is founded on this soil. The soil erodes into the open joints and

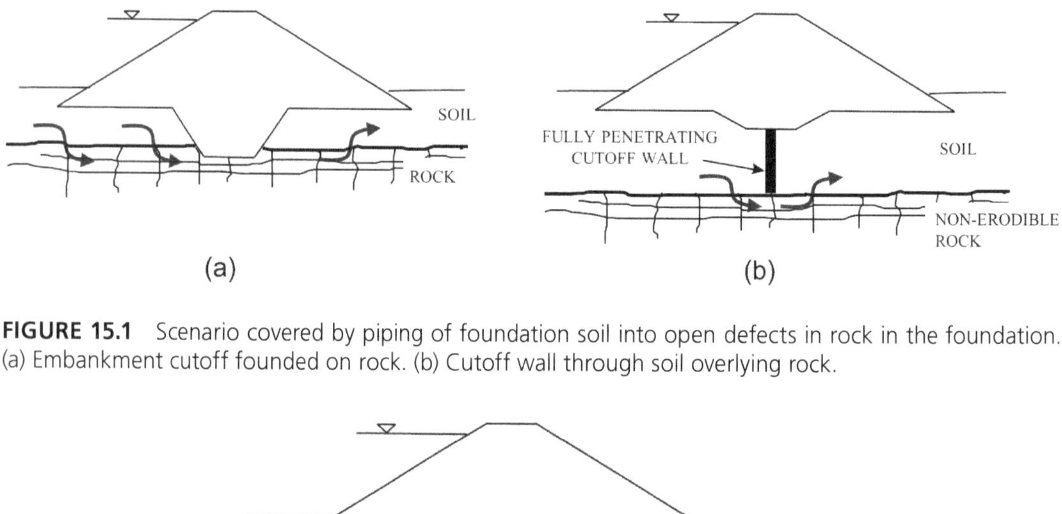

FIGURE 15.1 Scenario covered by piping of foundation soil into open defects in rock in the foundation. (a) Embankment cutoff founded on rock. (b) Cutoff wall through soil overlying rock.

FIGURE 15.2 Scenario covered by piping of foundation soil into coarse gravel in the foundation.

results in settlement of the embankment, potentially resulting in loss of freeboard, slope instability or piping through the embankment.

For erosion into coarse gravel, the scenario is that the coarse gravel is overlain by soil, and the downstream shoulder of the dam is founded on this soil. The soil erodes into the coarse gravel and results in settlement of the embankment, potentially resulting in loss of freeboard, slope instability or piping through the embankment.

The internal mechanism of erosion of the soil into the open defects or gravel is contact erosion.

For this PFM the requirements are:

- Persistent open defects or continuous gravel layer in the foundation beneath the downstream shoulder of the embankment.
- Soil overlying the open defects or gravel sufficiently fine to be eroded into and transported in the open defect or gravel (geometric condition).
- Seepage flow in the open defects sufficient to erode overlying soil (hydraulic condition). This is potentially affected by grouting for rock foundations and construction of cutoff walls for gravel and rock foundations.

Erosion related to karst limestone and soluble rock such as gypsum and salt in the foundation is too site dependent to generalise and cannot be considered in a historic method. They must be considered on a site-specific approach. They are, however, likely to be problematic for internal erosion and piping and leakage from the reservoir.

15.4.2 Recommended Method

For Dams on First Filling and in the First 5 Years of Operation

The historical average frequency of failure of dams by piping of foundation soil into open defects or coarse gravel on first filling or the first 5 years of operation adjusted for under-reporting = 0.0025.

For the dam being assessed, a weighting factor (w_F) is assessed using Table 15.4.

The weighting factor is obtained by multiplying the individual weighting factors from the table:

$$w_F = w_{F(zoning)} \times w_{F(rgs)} \times w_{F(cts)} \times w_{F(sg)} \times w_{F(obs)} \times w_{F(mon)} \text{ for soil foundation} \quad (15.3)$$

and:

$$w_F = w_{F(zoning)} \times w_{F(rgr)} \times w_{F(ctr)} \times w_{F(rg)} \times w_{F(obs)} \times w_{F(mon)} \text{ for rock foundation.} \quad (15.4)$$

The probability of failure on first filling or in the first 5 years = 0.0025 × w_F.

For Dams after 5 Years of Operation

The historical average annual probability of failure of dams after 5 years of operation is 1.7×10^{-5} for all zoning types.

For the dam being assessed, a weighting factor (w_F) is assessed using Table 15.4.

The weighting factor is obtained by multiplying the individual weighting factors from the table.

The annual probability of failure after 5 years of operation = $1.7 \times 10^{-5} \times w_F$.

TABLE 15.4 Summary of Weighting Factors—Piping of Foundation Soil into Open Defects in Rock or into Coarse Gravel

	GENERAL FACTORS INFLUENCING LIKELIHOOD OF FAILURE				
FACTOR	*MUCH MORE LIKELY*	*MORE LIKELY*	*NEUTRAL*	*LESS LIKELY*	*MUCH LESS LIKELY*
Zoning					
$W_{E(zoning)}$			**[1.0]** Same for all zoning types.		
Geological Environment (a) (f) *Rock in foundation*	Sedimentary—interbedded thin beds of mud rocks within thick beds of sandstone or conglomerate or limestone. Volcanics with interbedded thin beds of tuff.	Granite, gneiss, basalt rhyolite, ignimbrite. (b)	Schistose parallel to abutment. Thinly interbedded sedimentary.	Sandstone. Schistose, steeply dipping.	Shale, siltstone, claystone (not interbedded with other rocks, sub-horizontal bedded).
$W_{F(rgr)}$	**[5]**	**[3]**	**[1.0]**	**[0.5]**	**[0.2]**
Alluvium or colluvium in the foundation	Fast-flowing river alluvium, glacial moraine, resulting in uniform coarse gravel (GP) or cobbles.	River alluvium, landslide debris, glacial moraine resulting in silt sand gravel soils (GM) subject to suffusion. (c)	Well-graded gravel (GW), sandy gravel (GW).	Well-graded sand (SW).	All plastic soils (SC, GC, CL, CL-CH, CH). Silty sand (SM), Silt (ML, MH).
$W_{F(rgs)}$	**[5]**	**[3]**	**[1.0]**	**[0.2]**	**[0.01]**
Cutoff Type $W_{F(cts)}$ or $W_{F(ctr)}$	No cutoff in soil overlying rock or gravel.	Sheet pile wall. Diaphragm wall cutoff in the soil overlying rock or gravel.	Cutoff trench through soil overlying rock or gravel. No grouting of rock.		Cutoff trench through soil overlying rock. Grouting of rock.
	[2]	**[1.5]**	**[1.0]**		**[0.2]**
Soil Overlying Rock or Gravel (d)	Silt (ML), silty sand (SM), sand (SP).	Silt sand gravel soils (GM) subject to suffusion. (c) Dispersive clays.	Clay (CL), (CL-CH).	Clayey gravel (GC), clayey sands (SC).	Sandy gravel (GW) internally stable, or subject to GBE. Clay (CH).
$W_{F(sg)}$	**[5]**	**[2]**	**[1.0]**	**[0.5]**	**[0.2]**
Observations of Seepage	Muddy seepage, sudden increases in leakage.	Seepage gradually increasing, clear, sinkholes, sand boils.	Seepage steady, clear or not observed.		Seepage measured none or very small.
$W_{F(obs)}$	**[up to 10]**	**[2]**	**[1.0]**		**[0.5]**

and/or Observations of Pore Pressures in Foundation	Sudden fluctuations in pressures.	Gradually increasing pressures in foundation	High pressures measured in foundation under downstream shoulder and/or at the toe. (e)		Low pore pressures in foundation under downstream shoulder and/or at the toe. (e)
$w_{F(obp)}$	**[up to 10]**	**[2]**	**[1.0]**		**[0.8]**
Monitoring and Surveillance	Inspections annually.	Inspections monthly.	Irregular seepage observations, inspections weekly.	Weekly—monthly seepage monitoring, weekly inspections.	Daily monitoring of seepage, daily inspections.
$w_{F(mon)}$	**[2]**	**[1.2]**	**[1.0]**	**[0.8]**	**[0.5]**

Notes:

(a) Likely to result in persistent open defects in rock or persistent coarse alluvium or colluvium.
(b) Some basalts (e.g. if interbedded with weak tuff, and some ignimbrites may classify as "Much more likely".
(c) See Section 10.6 for details of assessment of suffusion.
(d) For erosion of the soil into the open defect or gravel, the soil must be finer than the opening of the defect or the voids between the gravel particles. See Chapter 11 for details.
(e) Compared to ground surface level at the toe of the dam.
(f) Take into consideration knowledge of site geology and geotechnical conditions from original site investigations, construction records and investigations of the dam post construction. Also account for exposure of foundation soils and rock upstream and downstream of the dam.

15.5 INTERNAL EROSION AND PIPING FROM THE EMBANKMENT TO THE FOUNDATION

The following outlines the amended method. It is only marginally different to the Foster et al. (1998, 2000a, 2000b) method.

For Dams on First Filling and in the First 5 Years of Operation

The frequency of failure on first filling or in the first 5 years = 0.00018 allowing for under-reporting. See Appendix C for details.

For the dam being assessed, a weighting factor (w_{EF}) is assessed using Table 15.5.

The weighting factor is obtained by multiplying the individual weighting factors from the table:

$$W_{EF} = w_{E(zoning)} \times w_{EF(rgs)} \times w_{EF(cot)} \times w_{F(ecm)} \times w_{EF(cst)} \times w_{EF(obs)} \times w_{EF(mon)} \text{ for soil foundation} \quad (15.5)$$

and:

$$W_{EF} = w_{E(zoning)} \times w_{EF(rgr)} \times w_{EF(cot)} \times w_{F(ecm)} \times w_{EF(gr)} \times w_{EF(obs)} \times w_{EF(mon)} \text{ for rock foundation} \quad (15.6)$$

The probability of failure on first filling or in the first 5 years = $0.00018 \times w_{EF}$.

For Dams after 5 Years of Operation

The historical average annual probability of failure of dams by erosion of the embankment into the foundation after 5 years of operation is 3.6×10^{-6} for all zoning types allowing for under-reporting. See Appendix for details.

For the dam being assessed, a weighting factor (w_{EF}) is assessed using Table 15.5.

The weighting factor is obtained by multiplying the individual weighting factors from the table.

The annual probability of failure after 5 years of operation = $3.6 \times 10^{-6} \times w_{EF}$.

TABLE 15.5 Summary of Weighting Factors—Piping of Embankment into Open Defects in Rock or into Coarse Gravel.

FACTOR	GENERAL FACTORS INFLUENCING LIKELIHOOD OF FAILURE				
	MUCH MORE LIKELY	*MORE LIKELY*	*NEUTRAL*	*LESS LIKELY*	*MUCH LESS LIKELY*
Zoning $W_{E\ (zoning)}$	**[1.0]** Same for all zoning types.				
Geological Environment (a, g) Rock at the core–foundation interface	Sedimentary—interbedded thin beds of mud rocks and thick beds of sandstone or conglomerate or limestone. Volcanics with interbedded thin beds of tuff.	Granite, gneiss, basalt rhyolite, ignimbrite. (b)	Schistose parallel to abutment. Thinly interbedded sedimentary.	Sandstone. Schistose, steeply dipping.	Shale, siltstone, claystone (not interbedded with other rocks, sub-horizontal bedded).
$w_{EF(rgr)}$	**[5]**	**[3]**	**[1.0]**	**[0.5]**	**[0.2]**
Alluvium or colluvium at the core–foundation interface	Fast-flowing river alluvium, glacial moraine, resulting in uniform coarse gravel (GP) or cobbles.	River alluvium, landslide debris, glacial moraine resulting in silt sand gravel soils (GM) subject to suffusion. (c)	Well-graded gravel (GW), Sandy gravel (GW).	Well-graded sand (SW).	All plastic soils, (SC, GC, CL, CL-CH, CH). Silty sand (SM), silt (ML, MH).
$W_{EF(rgs)}$	**[5]**	**[3]**	**[1.0]**	**[0.2]**	**[0.01]**
Foundation Cutoff Trench (d)	Deep and narrow cutoff trench.		Average cutoff trench width and depth.	Shallow or no cutoff trench.	
$w_{EF(cot)}$	**[1.5]**		**[1.0]**	**[0.8]**	
Erosion Control Measures of Core Foundation (e)	No erosion control measures, open jointed bedrock or open work gravels.	No erosion control measures, average foundation conditions.	No erosion control measures, good foundation conditions.	Erosion control measures present, poor foundations.	Good to very good erosion control measures present and good foundation.
$w_{EF(ecm)}$	**[up to 5]**	**[1.2]**	**[1.0]**	**[0.5]**	**[0.3–0.1]**
Grouting of Foundations $w_{EF(gr)}$		No grouting on rock foundations. **[1.3]**	Soil foundation only—not applicable. **[1.0]**	Rock foundations grouted. **[0.8]**	

(Continued)

TABLE 15.5 (Continued)

FACTOR	GENERAL FACTORS INFLUENCING LIKELIHOOD OF FAILURE				
	MUCH MORE LIKELY	*MORE LIKELY*	*NEUTRAL*	*LESS LIKELY*	*MUCH LESS LIKELY*
Core Soil Type (f) $W_{EF(cst)}$	Dispersive clays. **[5]** Low plasticity silts (ML). **[2.5]** Poorly and well-graded sands (SP, SW). **[2]**	Clayey and silty sands (SC, SM). **[1.2]**	Well-graded and poorly graded gravels (GW, GP). **[1.0]** High plasticity silts (MH). **[1.0]**	Clayey and silty gravels (GC, GM). **[0.8]** Low plasticity clays (CL). **[0.8]**	High plasticity clays (CH). **[0.3]**
Observations of Seepage $W_{EF(obs)}$	Muddy leakage, Sudden increases in leakage. **[up to 10]**	Leakage gradually increasing, clear, Sinkholes. **[2]**	Leakage steady, clear or not monitored. **[1.0]**	Minor leakage. **[0.7]**	Leakage measured none or very small. **[0.5]**
Monitoring and Surveillance $W_{F(mon)}$	Inspections annually. **[2]**	Inspections monthly. **[1.2]**	Irregular seepage observations, inspections weekly. **[1.0]**	Weekly—monthly seepage monitoring, weekly inspections. **[0.8]**	Daily monitoring of seepage, daily inspections. **[0.5]**

Notes:

(a) Likely to result in persistent open defects in rock or persistent coarse alluvium or colluvium. Values adopted are as for piping in the foundation.
(b) Some basalts (e.g. if interbedded with weak tuff, and some ignimbrites may classify as "Much more likely".
(c) See Section 10.6 for details of assessment of suffusion.
(d) The cutoff trench would be considered to be "deep" if the trench is > 3–5 m deep from the general foundation level and "narrow" if the width to depth ratio (W/D) is < about 1.0, where the width is measured at the top of the cutoff trench. A cutoff trench would be considered to be "shallow" if it is < 2–3 m, and "average" cutoff trench width and depth is depth 2–5 m and W/D > 1.0.
(e) Erosion control measures are the design features used to protect the core materials from within the cutoff trench being eroded into the foundation and these can include slush concrete or shotcrete on rock foundations and filters located on the downstream side of the cutoff trench for soil or rock foundations.
(f) For erosion of the soil into the open defect or gravel, the soil must be finer than the opening of the defect or the voids between the gravel particles. See Chapter 11 for details.
(g) Take into consideration knowledge of site geology and geotechnical conditions from original site investigations, construction records and investigations of the dam post construction. Also account for exposure of foundation soils and rock upstream and downstream of the dam.

Assessment of the Safety of Concrete Face Rockfill and Asphalt Face Embankment Dams

16

16.1 INTRODUCTION

This chapter presents a description of Concrete Face Rockfill Dams (CFRD) and Asphalt Face Dams (AFD).

Methods are described for how to assess the safety of these dams for overtopping, slope instability, seismic loads and slope instability and loss of freeboard resulting from leaks in the face slab, using deterministic and risk-based approaches. Where applicable reference is made to the methods for earth and rockfill dams described in earlier chapters

16.2 CONCRETE FACE ROCKFILL DAMS (CFRD)

16.2.1 Historic Development of Concrete Face Rockfill Dams

The development of concrete face rockfill dams (CFRD) has been described by ICOLD (1989, 2010), Cooke (1984, 1993, 1999, 2000), and Cruz et al. (2009).

The first rockfill dam to have a concrete face was constructed in California in 1895. This followed on from construction of timber faced dumped rockfill dams beginning in the 1850s.

These early CFRD often had steep (0.54H:1V to 0.75H:1V) slopes, with a facing of hand-placed rocks to stop the face from ravelling. Then dams of greater height, up to 100 m high (Salt Springs Dam, California), were constructed with concrete faces and rockfill dumped in thick layers, often greater than 20 m or 35 m, and placed without compaction other than sluicing.

The design of CFRD was mainly empirical, based on experience and judgement and often driven by constructability. The typical features of designs of CFRD up until the late 1950s are shown in

DOI: 10.1201/9781003542025-16

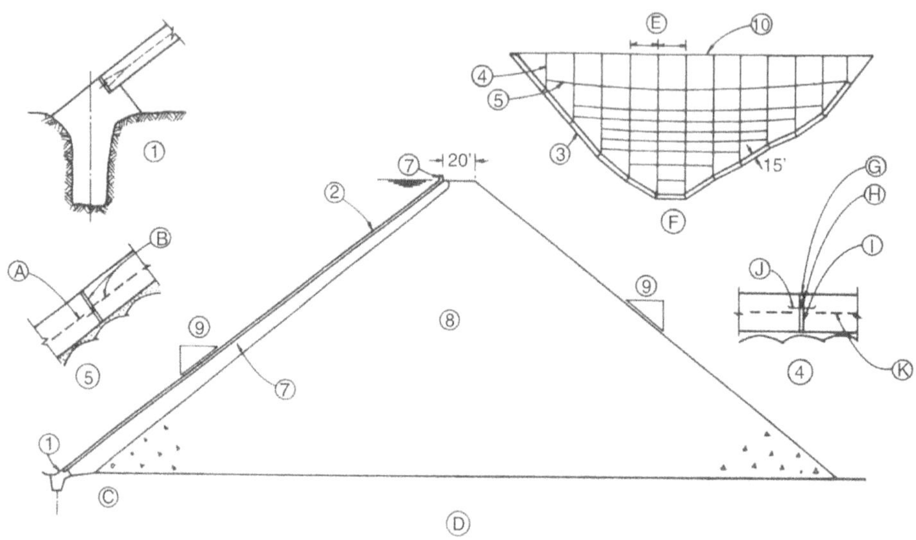

LEGEND: (1) Cutoff trench, (2) Concrete face, (3) Plinth, (4) Vertical joint, (5) Horizontal joint, (6) Parapet, (7) Crane-placed large rock, (8) Dumped rockfill, (9) Slope, (10), Curved axis. (A) Reinforcement, (B) 1.9 cm redwood filler and Z water stop, (C) Grout curtain, (D) Cross section of dam, (E) 18 m, (F) Elevation of face, (G) Mastic, (H) Pre-moulded asphalt, (I) Compressible joint filler, (J) U copper, (K) Reinforcement.

FIGURE 16.1 Features of early concrete face rockfill dam design (ICOLD, 1989).

Figure 16.1. Some design features such as the materials used to support the face slab persisted into the 1960s.

Many of these dams performed well. However, several higher dams leaked excessively due to deformation of the concrete face, with resulting opening of joints and cracking. This could be attributed to the low modulus of the dumped rockfill, and to the detailing of joints, which allowed compression of horizontal and vertical joint fillers in the central part of the dam face (which is under compression) and resulted in increased opening of other joints, including the perimetric joint. The leakage did not endanger the dam stability but in some cases was unacceptably high for operating reasons.

Over the period 1955 to 1965 there was a general adoption of compaction of rockfill. This change was brought about by a realisation that dumping and sluicing of rockfill led to significant segregation, with the accumulation of larger rock at the base of the layer, leaving large voids and being particularly compressible. There was also a realisation that weaker rocks tended to lose strength on saturation, leading to settlement if placed as dumped rockfill.

The change from a deep cutoff trench in rock as shown in Figure 16.1, to the adoption of the use of a plinth, or toe slab as shown in Figure 16.2, was done on the grounds that excavation for the trench could loosen and fracture the rock, making it more permeable, and that the plinth could be an adequate cutoff if founded on suitable rock, grouted and anchored to the rock with steel bars. Experience is that is correct.

16.2.2 CFRD General Arrangement as Described in ICOLD (1989, 2010)

Figure 16.2 shows the general arrangement and zoning of CFRD constructed of sound, free draining rockfill, on a strong rock foundation.

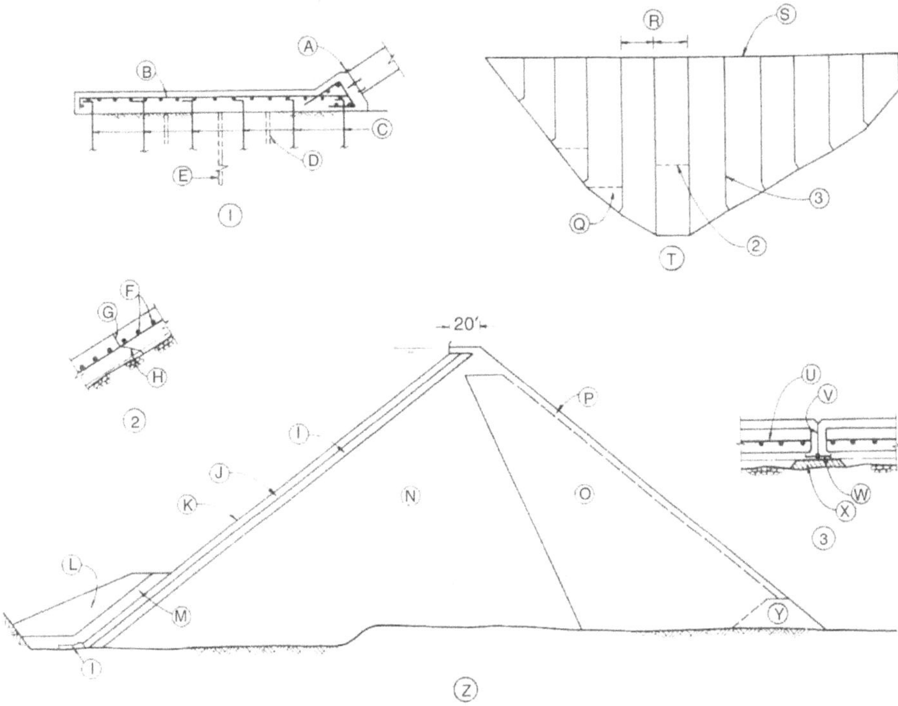

LEGEND: (1) Plinth, (2) Horizontal joint, (3) Vertical joint. (A) Perimetric joint, (B) Steel reinforcement, (c) Anchor bars, (D) Consolidation grout holes, (E) Grout curtain, (F) Horizontal reinforcement, (G) Width form, (H) Broom joint, (I) Zone 2E, selected small rock placed in the same layer thickness as Zone 2D, (J) Zone 2D, processed small rock, (K) Concrete face, (L) Zone 1B, random, (M) Zone 1A, impervious soil, (N) Zone 3A, quarry run rockfill or gravel fill, about 1.0 m layers, (O) Zone 3B, quarry run rockfill or gravel fill, about 1.5 m to 2.0 m layers, (P) Available large size rock dozed to face, (Q) Starter slab, (R) 18 m, (S) Straight axis, (T) Elevation of face, (U) Horizontal reinforcement, (V) Surface painted with asphalt, (W) Copper water stop, (X) Mortar pad, (Y) Zone 3D, plus 0.3 m rockfill, (Z) Section of dam.

FIGURE 16.2 General arrangement for zoning of CFRD constructed of sound rockfill on a strong rock foundation (adapted from ICOLD, 1989).

The embankment consists of:

Plinth. Reinforced concrete slab cast on sound non-erodible low permeability rock to join the face slab to the foundation.

Face slab. Reinforced concrete, between 0.25 and 0.6 m thick, with vertical, some horizontal and perimetric joints to accommodate deformation which occurs during construction and when the water load is applied.

Zone 2D. Transition rockfill, processed rockfill or alluvium, grading from silt to cobble size or, in more recent dams, from silt to coarse gravel size. The transition provides uniform support for the face slab and acts as a semi-impervious layer to restrict flow through the dam in the event that cracking of the face slab or opening of joints occurs.

Zone 2E. Fine rockfill, selected fine rock which acts as a filter transition between Zone 2D and Zone 3A in the event of leakage through the face slab.

Zone 3A. Rockfill, quarry run, free draining rockfill placed in layers about 0.9 m thick, except adjacent to Zone 2E where it is 0.45 m. This zone provides the main support for the face slab and is compacted to a high modulus to limit settlement of the face slab.

Zone 3B. Coarse rockfill, quarry run, free draining rockfill placed in layers about 1.5 to 2.0 m thick. Larger rock may be pushed to the downstream face. This zone is less affected by the water load than Zone 3B, so a lower modulus is acceptable. The thicker layers allow placement of larger rock.

The boundary between Zone 3A and 3B may vary from what is shown in Figure 16.2 to the dam centreline and even upstream of the centreline. Marulanda and Pinto (2000) warn that, at least for high dams, this may lead to larger settlements.

The zone designation discussed previously is used throughout the book. The transition Zones 2D and 2E have been so designated to differentiate from Zones 2A, 2B and 2C used in earth and rockfill dams. ICOLD (1989, 2010) uses different zoning terms, including naming what is Zones 2D here as Zone 2B and what is Zone 2E here as Zone 3A. There is no consistent terminology in published papers and books on CFRD.

16.2.3 Discussion on CFRD Zoning

The major differences between earlier designs and the more modern designs are as follows:

(a) The rockfill in earlier designs was not compacted so settlements were large, resulting in a higher likelihood of face slab joints damage or perimetric joint opening.
(b) Earlier dams had a high permeability zone supporting the face slab. More modern design provides Zone 2D, which is a silt sand gravel material with relatively low permeability to control leakage if it occurs. Section 16.2.4 discusses this in more detail.
(c) Modern face slabs only have vertical joints, not both vertical and horizontal.
(d) The plinth design has changed as described earlier.

Many variations of the zoning shown in Figure 16.2 are adopted to meet site conditions and the quality of construction materials available.

Cruz et al. (2009) present 28 typical cross sections of dams constructed in the period 1970 to 2008. Most of the dams given as examples in Cruz et al. (2009) have the equivalent of a Zone 2D.

Kangaroo Creek Dam is an example of a dam which followed the earlier design and had a high permeability zone (Zone 1) supporting the face slab. Figure 16.3 shows the cross section. The dam was constructed in 1968–1969 and the rockfill was well compacted. It was raised with a reinforced concrete U beam on the crest in the early 1980s.

The zoning is as follows:

Zone 1. Face Slab Support. Rockfill 300 mm maximum, 10 mm lower limit, compacted in 600 mm layers.

Zone 2. Drainage Zone. Rockfill 900 mm maximum, not more than 20% passing 25 mm sieve, compacted in 900 mm layers.

Zone 3. Bulk Rockfill. Rockfill 900 mm maximum, not more than 30% passing 25 mm sieve, compacted in 900 mm layers.

Zone 4. Horizontal Drainage Zone. Rockfill 1200 mm maximum, not more than 20% passing 25 mm sieve, compacted in 1200 mm layers.

Zone 5. *In situ* river gravel in the river valley.

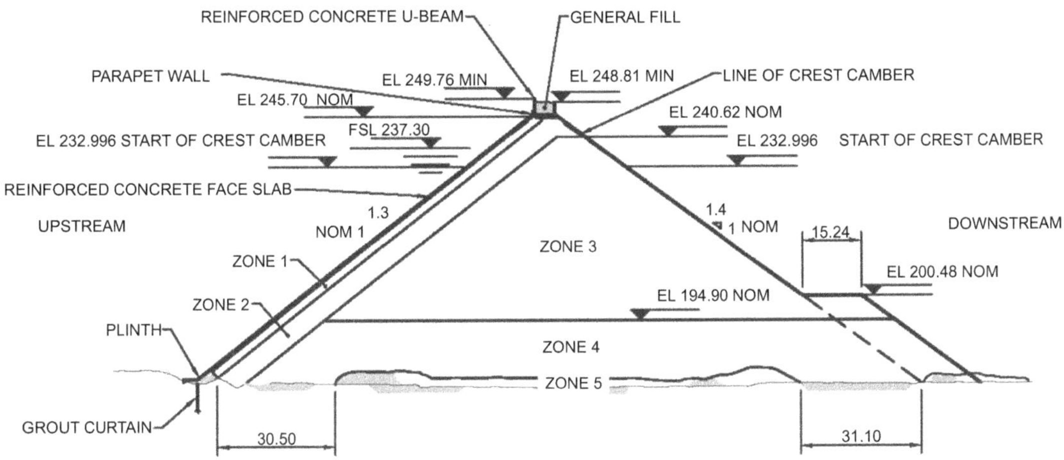

FIGURE 16.3 Kangaroo Creek Dam, South Australia. Typical cross section of the original embankment showing the zoning (Buchanan et al., 2016).

The dam is an extreme consequence classification. As discussed in Buchanan et al. (2016), a QRA based safety review indicated that under seismic loading the face slab joint and perimetric joint water stops may rupture, and the resulting leakage might not be arrested by Zone 1 and Zone 2.

Zones 3 and 4 rockfill were constructed of schist and gneiss rockfill and are not high permeability, and the leakage might result in saturation of the rockfill and instability of the downstream slope.

The embankment required raising for flood reasons, and the decision was made to construct a berm as shown in Figure 16.4 to improve stability in the event of a major leak.

A two-stage filter drain system Zone 6B and 6C was placed between the original rockfill and the new rockfill berm (Zone 7A). Coarse free draining rockfill Zone 7C at the downstream toe was designed to control potential unravelling in the case of a leakage incident. Water stops were upgraded as described in Maisano et al. (2016).

16.2.4 Zone 2D Properties

16.2.4.1 Historic Development of Zone 2D

Zone 2D has an important role in CFRD safety because it is designed to control leakage in the event the face slab, the perimetric joint or vertical joints are damaged.

In CFRD constructed in the 1960s, the function of Zone 2D was seen as providing uniform support for the face slab. Specifications required that the Zone 2D was screened to remove all materials less than 25 or 50 mm. This was to ensure that, in the event of a leak in the face slab, there would be no fines to be washed away which might lead to loss of support of the face slab. However, with this grading Zone 2D was very permeable and rocks were easily dislodged during construction, and a rough, porous surface led to excess concrete being required for the face slab.

Beginning with the 110 m high Cethana Dam in 1971 (Wilkins et al., 1973), there was a change in design approach which resulted in Zone 2D being specified as crusher run or quarry run (after passing through a grizzly) rockfill passing 150 to 225 mm (up to 300 mm). This resulted in a material with a lower permeability and a smoother and more stable surface on which to construct the face slab. Many dams have been successfully constructed in this way, and this was the established practice until about

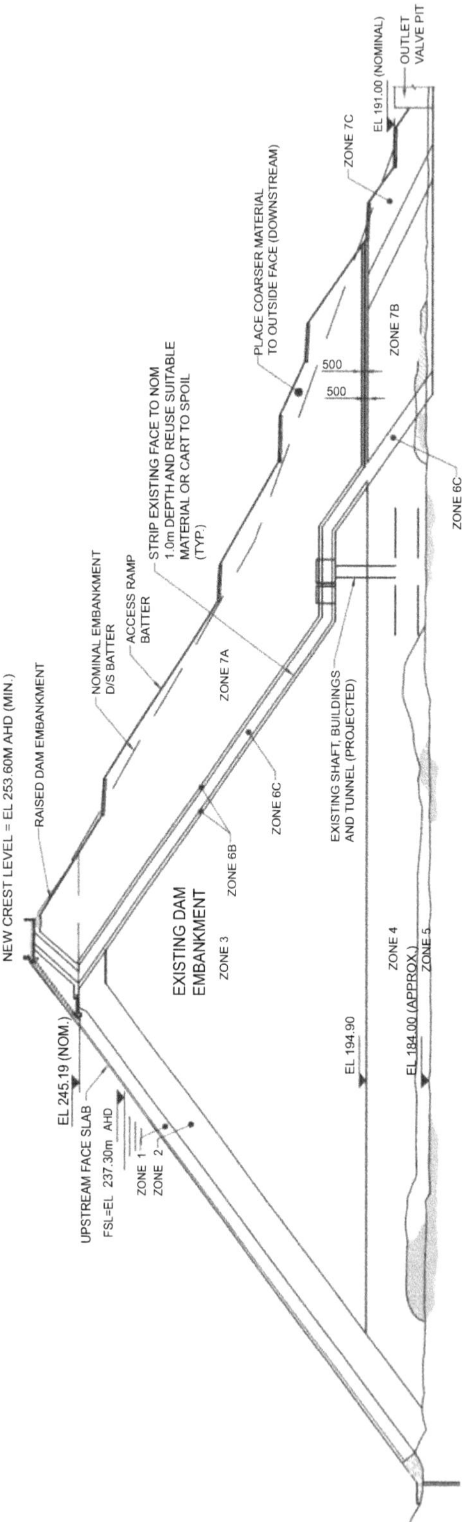

FIGURE 16.4 Kangaroo Creek Dam, South Australia. Typical cross section of the upgraded dam (Buchanan et al., 2016).

the mid-1980s. Table 16.1 shows the specified gradations for Cethana and two South American dams constructed according to these principles.

Sherard (1985) points out that quite large leakage had occurred at some dams constructed with crusher run Zone 2D including 1800 litres/second at the Alto Anchicaya Dam on first filling. Sherard notes that, for such wide graded materials with a small percentage of sand size material, segregation was a problem giving a high permeability, and this could in part explain why such large leakage had occurred.

Sherard (1985b) suggests that about 40% of sand size (passing 4.76 mm) is required to avoid the segregation and the gradation should be as shown in Figure 16.5.

Sherard also shows that leakage through the dam, resulting from a crack in the face slab, is controlled more by the permeability of Zone 2D than by the face slab crack aperture.

Sherard (1985) indicates that a grading for Zone 2D as shown in Figure 16.5 will be "stable", i.e. internally stable, not susceptible to washing out of the fines, and that a permeability of 10^{-6} m/sec would be achieved. Earlier Zone 2D equivalents are not internally stable and are susceptible to the finer particles washing from the zone, leading to a higher permeability and possibly some settlement of the face slab as the fines are eroded.

ICOLD (1989a) recommended a Zone 2D grading virtually the same as that suggested by Sherard (1985). They indicated that a maximum of 10% to 12%, passing 0.075 mm is desirable, while giving 15% as the upper limit.

ICOLD (2010) modified these gradations to specify more completely the finer materials but in particular to reduce the % fines Sherard (1985b) and ICOLD (1989) had recommended to a maximum of 7%. This was to reduce the likelihood that the Zone 2D might hold a crack. They specify the fines passing 0.075 mm must be non-plastic.

Sherard (1985b) and ICOLD (1989) indicate that Zone 2D should be 4 m to 5 m wide measured horizontally, possibly wider for dams higher than 150 m. ICOLD (2010) recommends 2 m to 4 m depending on height. If a dam has significantly narrower Zone 2D, the seepage gradients will be higher and there will be a higher seepage gradient and a greater likelihood of segregation resulting in coarse gradation material persisting across the zone. This may result in a higher likelihood that Zone 2D may not perform its function to limit leakage flows.

TABLE 16.1 Zone 2D Gradation Specifications (Sherard, 1985)

	CETHANA	*ALTO ANCHICAYA*	*FOZ DO AREIA*
Year	1971	1974	1980
Size % finer			
300 mm		100	
225 mm	100		
150 mm	80–100		100
100 mm			
75 mm	51–100	50–100	55–100
40 mm			
20 mm	15–63	25–60	18–65
5 mm	0–40	0–30	5–22
0.5 mm	0–17	0–5	0–8
0.075 mm	0–5		0–1

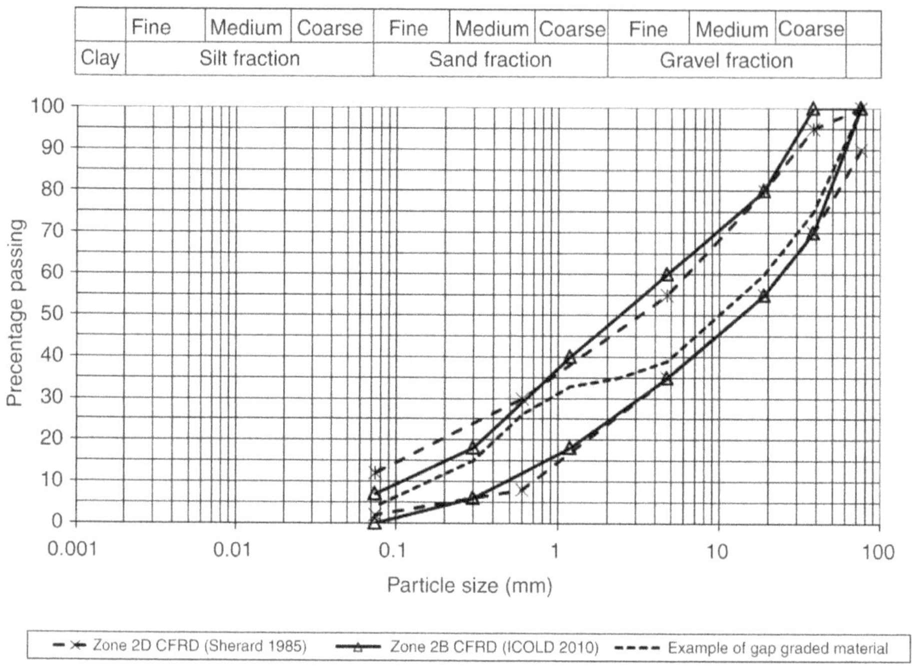

FIGURE 16.5 Sherard (1985) and ICOLD (2010) Zone 2D Gradation Limits (example of gap-graded material also shown).

Fitzpatrick et al. (1985) and Sherard (1985b) and Cooke (2000) point out that an added advantage of having Zone 2D graded as shown in Figure 16.5 is that, if a crack forms in the face slab, it can be sealed by spreading silty sand over the crack and having the silty sand wash into it, with further erosion controlled by Zone 2D acting as a filter. This method has been used to reduce leakage on several dams including Shiroro Dam in Nigeria (Bodtman and Wyatt, 1985), Khao Laem Dam in Thailand (Watakeekul et al., 1985) and Bailey Dam in the USA (Cooke, 2000).

16.2.4.2 Assessment of Zone 2D in Existing CFRD

Internal Instability

Douglas et al. (2016, 2019) tested materials with gradations to simulate the fine and coarse limits of Zone 2D as recommended by ICOLD (2010). The gradations are shown in Figure 16.5. Table 16.2 summarises the performance of these tests.

It can be seen that:

(a) Both fine and coarse limits were internally unstable where the mesh was too coarse to act as an effective filter, but the fine limit soil eroded only to a very minor degree.
(b) The coarse limit soil experienced medium/minor erosion, but this was probably related to the use of the coarser base mesh giving mesh/EOSCE = 2.4.
(c) The permeability of the two soils was not greatly affected by the erosion, but the permeabilities were significantly higher than Sherard (1985b) expected to achieve (10^{-6} m/sec).
(d) The gradients at which erosion occurs are likely to be experienced in a CFRD.

TABLE 16.2 Summary of Results of Permeameter Tests of Soils Representing Zone 2D of Concrete Face Rockfill Dams Based on ICOLD (2010) Gradations (Douglas et al., 2016, 2019)

TEST NUMBER	*REPRESENTING*	*MESH/ EOSCE(a, b, c)*	*GRADIENT AT WHICH EROSION OCCURS*	*EROSION % OF TOTAL*	*AVERAGE PERMEABILITY (M/SEC)*		*TYPE OF EROSION*
					START OF TEST	*FINISH OF TEST*	
13	ICOLD Coarse limit	2.4	2	1.2 Medium/Minor	7×10^{-5}	1.4×10^{-4}	GBE
21	ICOLD Fine limit	1.2	4	0.1 Very Minor	6×10^{-6}	4×10^{-6}	GBE

Notes:
(a) "Mesh" = opening size in mm of the mesh supporting the sample in the test.
(b) EOSCE = Equivalent opening size for continuing erosion using the Foster and Fell (2001) criteria.
(c) Mesh/EOSCE is the ratio between the mesh size used in the permeameter test and the EOSCE of the Zone 2D gradation.

What is important to consider is that the performance of Zone 2D is related to the gradation of the zone downstream of it. Tests by Douglas et al. (2016, 2019) on other silt sand gravel materials showed that if the material downstream is sufficiently coarse, major rapid erosion may occur even for soils which did not erode for a continuing erosion condition.

If the dam has a well designed and constructed Zone 2E finer rockfill placed in reduced thickness layers downstream of Zone 2D as described later, this will not occur.

If, however, there is only a single zone, Zone 2D placed against Zone 3A rockfill, the conditions modelled in Douglas et al. (2016, 2019) may be met, and in the event of a leak in the face slab, major erosion of the finer fraction of Zone 2D may occur, resulting in a much higher permeability, and potentially settlement resulting in additional opening of joints in the face slab.

To avoid this, good design practice has required Zone 2E downstream of Zone 2D to be specified and constructed as follows:

- The layer thickness should be reduced to 0.40 to 0.45 m.
- Maximum size 400 mm to 450 mm.
- A coarse limit for the rockfill should be specified with a $D_{15} \leq 40$ mm.

By doing this, segregation of Zone 3A is less likely to occur and the resulting rockfill is more likely to act as a filter to Zone 2E.

If the dam has Zone 2D and Zone 2E but Zone 2E is very coarse or segregated, leaving Zone 2E not acting as a filter to Zone 2D, this could also replicate the conditions of the tests by Douglas et al. (2016, 2019).

For existing dams those assessing dam safety need to consider this potential issue, taking account of the specification and actual construction of Zones 2E and 3A. Photographs during construction will often assist in this assessment.

Another potential issue is that if the Zone 2D material is gap graded, as shown for example in Figure 16.5, and therefore potentially subject to suffusion, the finer fraction is likely to erode, leaving only the coarse fraction which may be too permeable to limit leakage through a defect in the concrete face to tolerable values.

The Potential for Zone 2D Holding a Crack and Not Acting to Limit Leakage Flows
If the criteria in Table 12.3 for assessing the likelihood a filter or transition will hold a crack are applied, a Zone 2D with 12% non-plastic fines would have a probability of holding a crack (P_C) = 0.05 if Zone 2D was constructed of natural rounded alluvium, and P_C = 0.5 if constructed of crushed rock. The corresponding values for 7% fines are (P_C) = 0.001 if Zone 2D was constructed of natural rounded alluvium, and P_C = 0.01 if constructed of crushed rock.

These should be considered in relation to the possibility of common cause cracking, e.g. from joints near the abutments opening resulting in cracking of Zone 2D as described in Section 16.3.5 and Figure 16.29.

ICOLD (2010) (pp. 255–257) discusses problems of two CFRD with deep cracking of the Zone 2D material with 10% to 15% non-plastic fines during construction and operation. During construction, cracks were observed in the Zone 2D up to 60–100 mm wide and 3 m deep. This was caused by differential settlements of the rockfill zone. It would appear that in one of the dams (Xingo), the cracks in the Zone 2D were believed to have reopened during operation and this adversely affected its ability to limit flows through leaks in the face slab (a complete seal was not achieved when they dumped silty sand on the face). These seem to be important case histories and possibly a reason why ICOLD modified its recommended curves to limit the fines to less than 7%.

In view of this, when assessing the safety of existing dams with high fines content in Zone 2D, there should be consideration of the possibility of the Zone 2D holding a crack taking account of the actual fines content. The method for assessing this is given in Section 12.2.4.

16.2.5 Use of Gravel as Rockfill

Cooke (1984) discusses the use of gravel (really sandy gravel or sandy gravel with some silt) instead of rockfill and maintains that, if available, this material can be very suitable. He points out the following:

- Compacted gravels commonly have high modulus of compressibility, up to 5 to 10 times that of some compacted rockfill. In his experience since face slab movements vary roughly inversely with the modulus, and directly with the square of height of the dam, gravel fills are desirable for higher dams.
- If significant fines content is allowed, then chimney drains, abutment drains, filters and intermediate drainage layers will be required.

Amaya and Marulanda (1985) describe the use of gravel for the 125 m Golillas Dam in Colombia. In this dam the bulk of the fill (i.e. Zone 2 in Figure 16.6) was natural river gravel. The gravels were washed for drainage zones (Zone 3) in the dam.

Sierra et al. (1985) describe the use of gravels in Salvajina Dam in Colombia. The 148 m high dam had its "Zone 3A" constructed of gravels from gold mining dredging for Zone 2 as designated in Table 16.1.

The Hydro-Electric Commission of Tasmania used gravels to construct the 82 m high Crotty Dam in 1991 (Figure 16.7). Zone 2A is the equivalent of Zone 2D. Zone 2B is free draining rockfill or gravel, higher permeability than the gravels in Zone 3A.

Gouhou Dam in China (Chen, 1993), Chen and Zhang (2006), was constructed of sandy gravel with a finer grading envelope than the earlier dams (Table 16.3).

The zoning of this dam is shown in Figure 16.8. As constructed, there was little difference in grading between Zones II, III and IV. The dam failed by slope instability when leakage occurred through

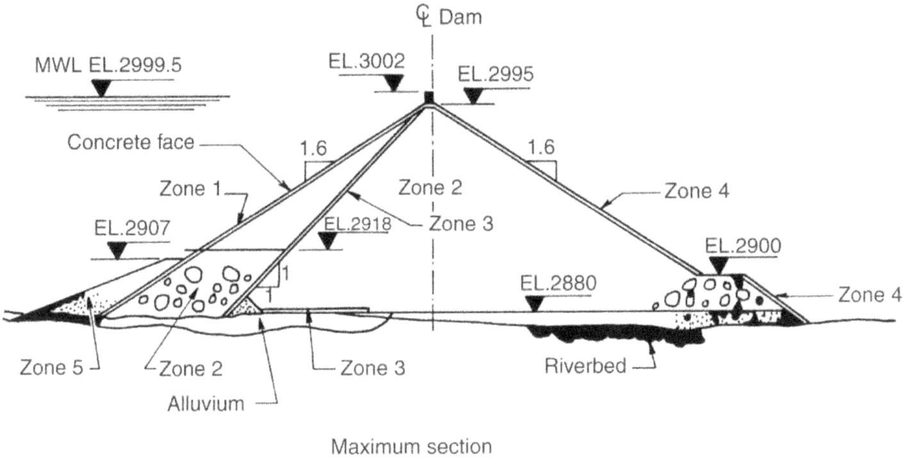

Zone	Description
1	Processed dirty gravels
2	Unprocessed dirty gravels
3	Processed clean gravels
4	Oversize from zone 2
5	Silty clays

FIGURE 16.6 Golillas Dam (Amaya and Marulanda, 1985). (Used with permission of ASCE conveyed through Copyright Clearance Center Inc.)

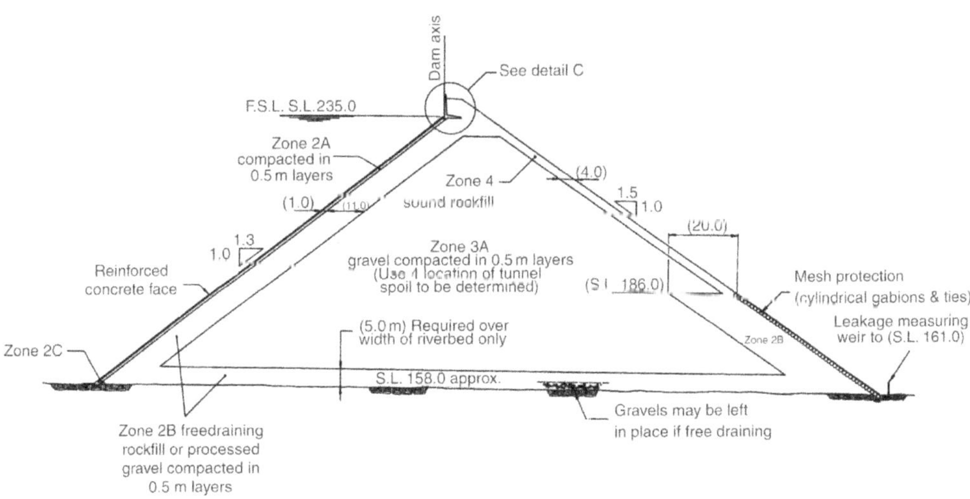

FIGURE 16.7 Crotty Dam, Tasmania, Australia (Fell et al., 2015; courtesy of Hydro Tasmania).

joints in the face, particularly at the join of the face slab and the crest wall which was below the reservoir normal FSL. The relatively fine fill, with no higher permeability zones such as those built into other gravel fill CFRD dams, contributed to the failure. Figure 16.9 shows the failed dam and the lack of a coarse gravel free draining zone.

TABLE 16.3 Grading of River Gravels Used as "Rockfill" in CFRD (Fell et al., 2015)

DAM	*CROTTY*	*GOLILLAS*	*SALVIJINA*	*GOUHOU*
YEAR	*1971*	*1978*	*1984*	*1992*
Sieve Size	% Passing	% Passing	% Passing	% Passing
300 mm	100	100	100	100
150 mm	85–100 (98)(b)	60–100	65–100 (98)	100
75 mm	34–98 (85)	35–80	35–100 (80)	60–100 (90)
25 mm	26–70 (55)	20–50	10–80 (40)	55–90 (65)
4.76 mm	6–42 (16)	10–30	0–30 (8)	25–65 (42)
1.18 mm	2–22 (10)	4–22	0–25 (6)	15–45 (28)
0.075 mm	0–4 (2)	0–12 (a)	0–12 (3)	5–10 (6)

Notes:
(a) Average 8% for one source, 12% for second.
(b) Average in brackets.

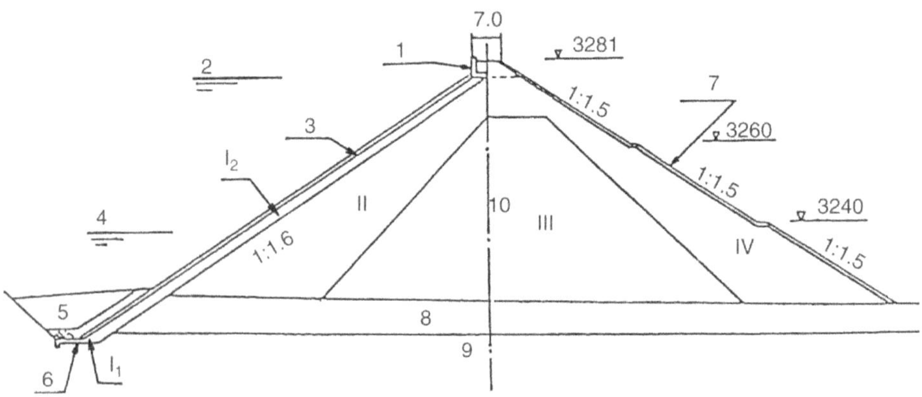

LEGEND: (1) Crest wall, (2) Normal water level, (3) Face slab, (4) Dead water level, (5) Random fill, (6) Clay, (7) Masonry, (8) Alluvium, (9) Dam axis.

FIGURE 16.8 Gouhou Dam (Chen, 1993).

16.2.6 Use of Low Permeability Rockfill

When the available rockfill breaks down under compaction to form a layer on the surface of each layer of rockfill with a large proportion of sand and silt/clay size particles, the resulting fill may have adequate modulus but may not be free draining. Rocks which are likely to do this are sandstones, siltstones, shale, schists and phyllites.

CFRD have been constructed using such rockfill but required zoning to provide flow limitation as provided by Zone 2D and high permeability drainage layers within the embankment and on the foundation.

Mangrove Creek Dam is an example of this design and is discussed in detail in Mackenzie and McDonald (1985). Figure 16.10 shows the zoning. Zones 2C and 2CD are high permeability drainage zones.

FIGURE 16.9 Gouhou Dam. Photograph of failed dam showing sandy gravel fill with no high permeability drainage zone. (Photograph courtesy of Professor Chen Zuyu.)

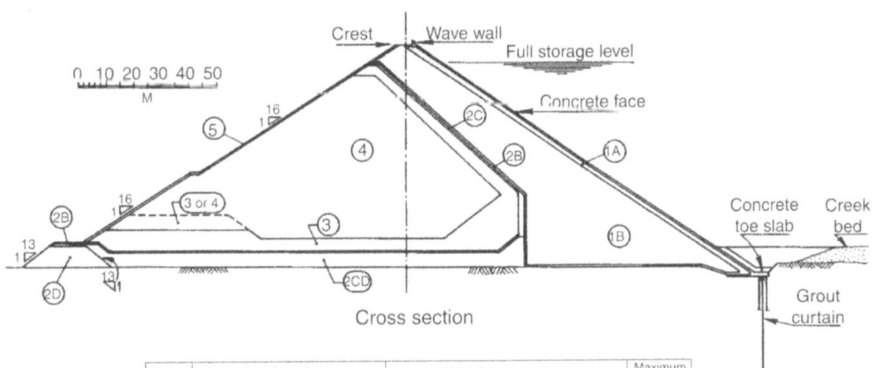

Zone	Function	Description	Maximum size mm
1A	Semi-pervious zone	Fresh siltstone	150
1B		Fresh sandstone and siltstone	450
2A	Filters		20
2B		Fresh basalt	75
2C	Chimney drain		300
2CD	Bottom drain		600
2D	Toe drain	Fresh basalt from 100mm to 1200m	1200
3	Free draining layer	Fresh sandstone and/or siltstone	600
4	Random zone	Fresh or weathered sandstone and siltstone	450
5	Downstream slope protection	Fresh sandstone from 600mm to 1200mm	1200

FIGURE 16.10 Mangrove Creek Dam (Mackenzie and McDonald, 1985). (Used with permission of ASCE conveyed through Copyright Clearance Center Inc.)

16.2.7 Side Slopes and Analysis of Slope Stability

For CFRD constructed of hard, free draining rockfill, the upstream and downstream slopes are usually 1.3H to 1V or 1.4H to 1V, which corresponds approximately to the angle of repose of loose dumped rockfill. With slopes within this range ravelling of the faces should not occur.

When gravel is used for the dam "rockfill" zones, flatter slopes are needed to prevent ravelling of the face. Usually 1.5H:1V has been adopted in these cases, although 1.6H:1V has also been used.

When low permeability rockfill has been used, flatter slopes have sometimes been adopted, e.g. 1.5H:1V for Mangrove Creek Dam (Mackenzie and McDonald, 1985).

If foundation strengths dictate, flatter slopes may also be required, e.g. 2.2H to 1V was used for Winneke Dam (Casinader and Watt, 1985).

Haul roads may have been constructed on downstream slopes, or defined berms may have been incorporated in the face. In these cases, steeper slopes between the "berm" located by the haul road may have been used, e.g. 1.25H:1V was used for Foz do Areia Dam (Pinto et al., 1985).

In some cases, the upper 10 m to 15 m of the dam may have been steepened to as much as 1.25H to 1V to provide the camber of the crest (Fitzpatrick et al., 1985).

As pointed out by Sherard and Cooke (1987) and Fitzpatrick et al. (1985), the stability of the slopes in the dam is not usually analysed. This is in recognition of the fact that well designed and constructed CFRD founded on rock foundations with no weak zones such as bedding shears in sedimentary rock have no pore pressures in the rockfill and will remain stable under static loads when constructed to the slopes described earlier. However, existing dams with non free draining rockfill may experience some degree of saturation of the rockfill if the face slab is damaged, e.g. in an earthquake, especially if there is no low permeability zone (Zone 2D) beneath the face to limit seepage into the embankment. For these dams, estimates must be made of potential pore pressures in the embankment allowing for the zoning and nature of the rockfill, including anisotropy of permeability, and analyses of stability carried out.

If stability analysis is to be carried out, knowledge of the shear strength properties is required. The determination of rockfill properties is discussed in Section 7.6. It is important to model the nonlinearity in the rockfill shear strength.

If the foundation may control stability, the analyses should model the foundation.

The deformations of CFRD under earthquake loading is discussed in Section 16.3.3.

16.2.8 Crest Wall

Many CFRD have a reinforced concrete retaining wall ("wave-wall", "crest wall" or "parapet wall") at the crest of the CFRD to reduce the volume of rockfill. Walls up to 3 m to 5 m high have been used. Figure 16.11 shows details of one example.

The base of the wall is usually above full supply level so not subject to water load except in flood conditions. The wall is joined to the face slab with a flexible joint, such as that shown in Figure 16.11. The joint should be vertical, not normal to the plane of the slab, so that differential settlement can be accommodated. The actual joint detail and whether there is evidence of differential movement and opening of the joint is important when assessing whether the joint may leak when the reservoir rises above the joint.

Some dams have the parapet wall cast integrally with the face slab. These are commonly no higher than 1.5 m to 3 m high. The detailing of water stops in the construction joints is an important factor in considering whether leakage may occur under high reservoir conditions.

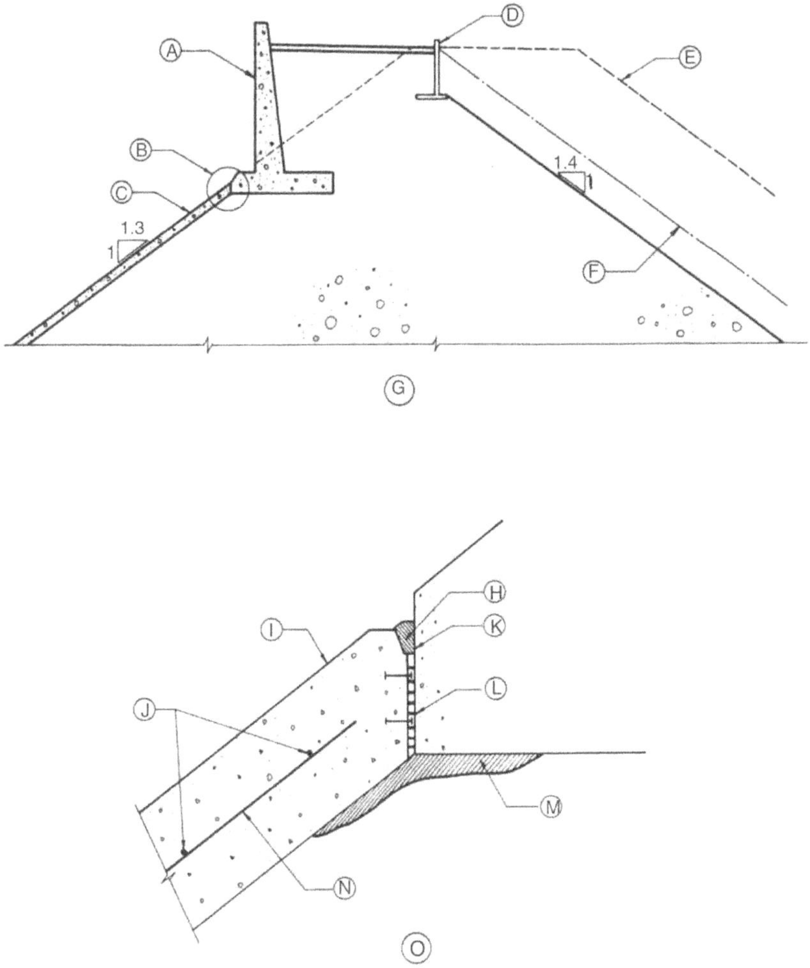

LEGEND: (A) Upstream parapet wall, (B) See detail, (C) Concrete face, (D) Downstream parapet wall, (E) Additional rockfill if both parapet walls are omitted, (F) Additional rockfill if downstream parapet wall is omitted, (G) Crest details, (H) IGAS, (I) Concrete face, (J) Horizontal reinforcement, (K) IGAS primer, (L) Pre-moulded filler, (M) Mortar pad, (N) Inclined reinforcement, (O) Detail: concrete face and upstream parapet wall joint.

FIGURE 16.11 Crest details (ICOLD, 1989).

16.2.9 Plinth

The principal purpose of the plinth (or "toe-slab") is to provide a "watertight" connection between the face slab and the dam foundation. Figure 16.12 shows several types of plinth designed to accommodate particular conditions.

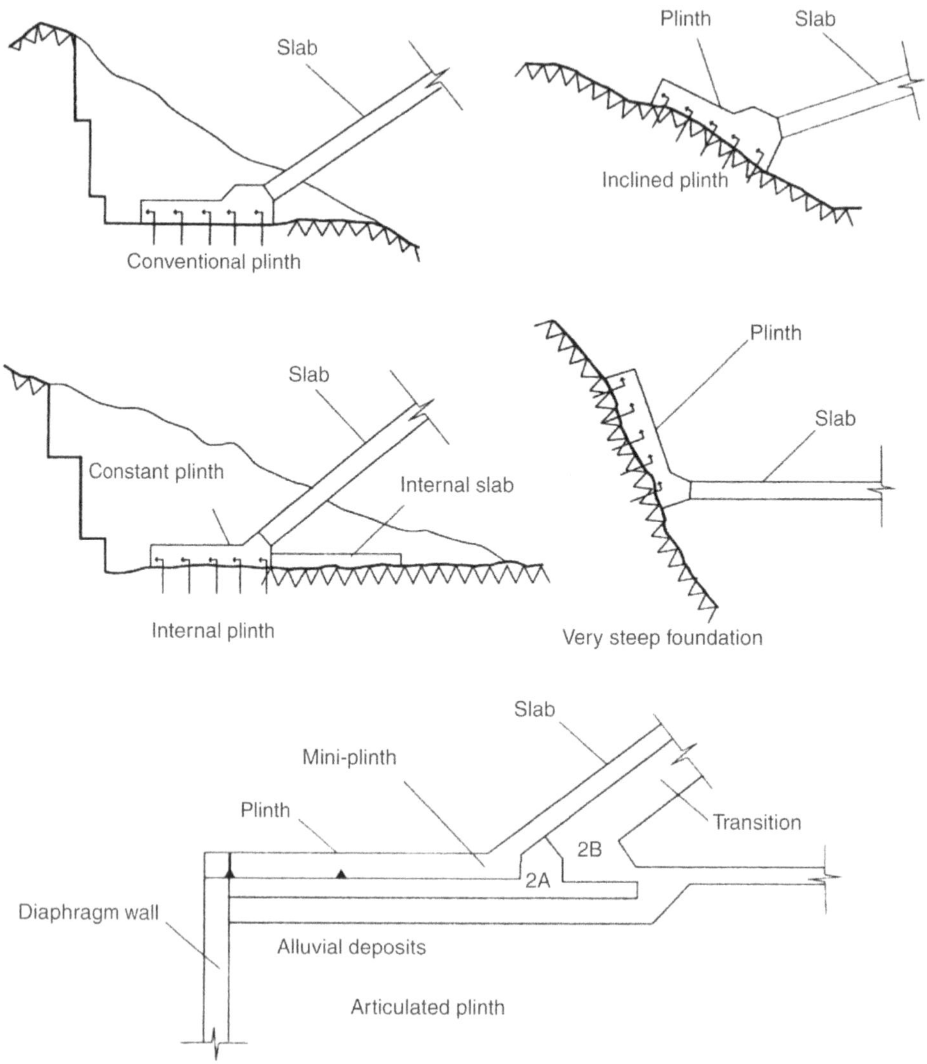

FIGURE 16.12 Types of plinth (Cruz et al., 2009). (Reprinted with permission of GA Educacao LTDA, Rua Ernesto Alves, 150, Floresta District, Porto Alegre, RS-Brazil)

16.2.9.1 Plinth Width

The plinth is usually founded on strong, non-erodible rock which is grouted and has been carefully excavated and cleaned up with a water jet to facilitate a low permeability cutoff. For these conditions, the plinth width is of the order of 1/20 to 1/25 of the water depth (ICOLD, 1989; Cooke and Sherard, 1987). ICOLD (2010) recommends 1/15 to 1/20 of the water depth for competent groutable non-erodible rock. Recommendations are provided in ICOLD (2010) for less competent rock.

When assessing existing dams, consider the width in relation to these criteria and the potential for erosion of the rock beneath the plinth, taking account of the orientation and persistence of open defects or features such as clay seams or other erodible material, the erosion properties of the features and what measures have been taken to provide a filter exit for such seepage. For erosion to occur there must be a persistent open defect within or adjacent to the erodible material. Well designed and executed grouting should have sealed such defects.

Cooke (2000), Marulanda and Pinto (2000) and Cruz et al. (2009) indicate that a recent evolution at that time was to reduce the length of the plinth to say 3 m to 5 m and maintain an overall seepage gradient beneath the plinth by using an un-dowelled reinforced slab under the rockfill as shown in the "internal plinth" example in Figure 16.12 (labelled "internal slab"). This can give economies in rock excavation. ICOLD (2010) provides an example.

16.2.9.2 Plinth Thickness

The minimum plinth thickness is usually between 0.3 m and 0.4 m and 0.6 m for medium height dams, but Cruz et al. indicate it may be up to 1.00 m for the lower plinths of high dams (> 120 m). ICOLD (2010) suggests the thickness should = 0.3 m + 0.003 H where H = reservoir head in metres.

The actual thickness is usually more because of the need to fill over-excavation and to make up for irregularities in the topography. Where the height of the extra concrete is large, it is common to construct the plinth in two stages; the first stage is to fill the irregularities.

Where there is extra concrete there is potential for additional settlement of the face slab and damage to the perimetric joint.

16.2.9.3 Plinth Stability

It is necessary to ensure that the plinth is stable against sliding and overturning under the imposed forces. For a plinth of normal thickness there is adequate friction resistance on the base, unless there are unfavourably oriented low-strength bedding planes, joints or shears in the foundations. The plinth is usually anchored to the rock with grouted dowels, which are generally 25 mm to 35 mm diameter, reinforcing steel bars, 3 m to 5 m long and are installed at 1.0 m to 1.5 m spacing.

For plinths in existing dams which are thicker than normal, due to over-break or irregularities in the foundation, the stability of the plinths should be analysed assuming the uplift pressure under the slab is zero at the downstream toe and varies linearly to full reservoir head at the upstream toe. No support should be assumed from the face slab on the understanding that the perimetric joint may have opened, and no support from the rockfill should be allowed for, since significant displacement into the rockfill would be necessary to mobilise the resistance.

16.2.10 Face Slab

Figures 16.13 and 16.14 show examples of the face slab joints and joint details.

16.2.10.1 Face Slab Deformations

Figure 16.15 shows the zones of compression and tension within the face slab as the water load from the reservoir is applied. Also shown are the movements at the perimetric joint.

These are accounted for in the design of the joints in the face slab and the perimetric joint.

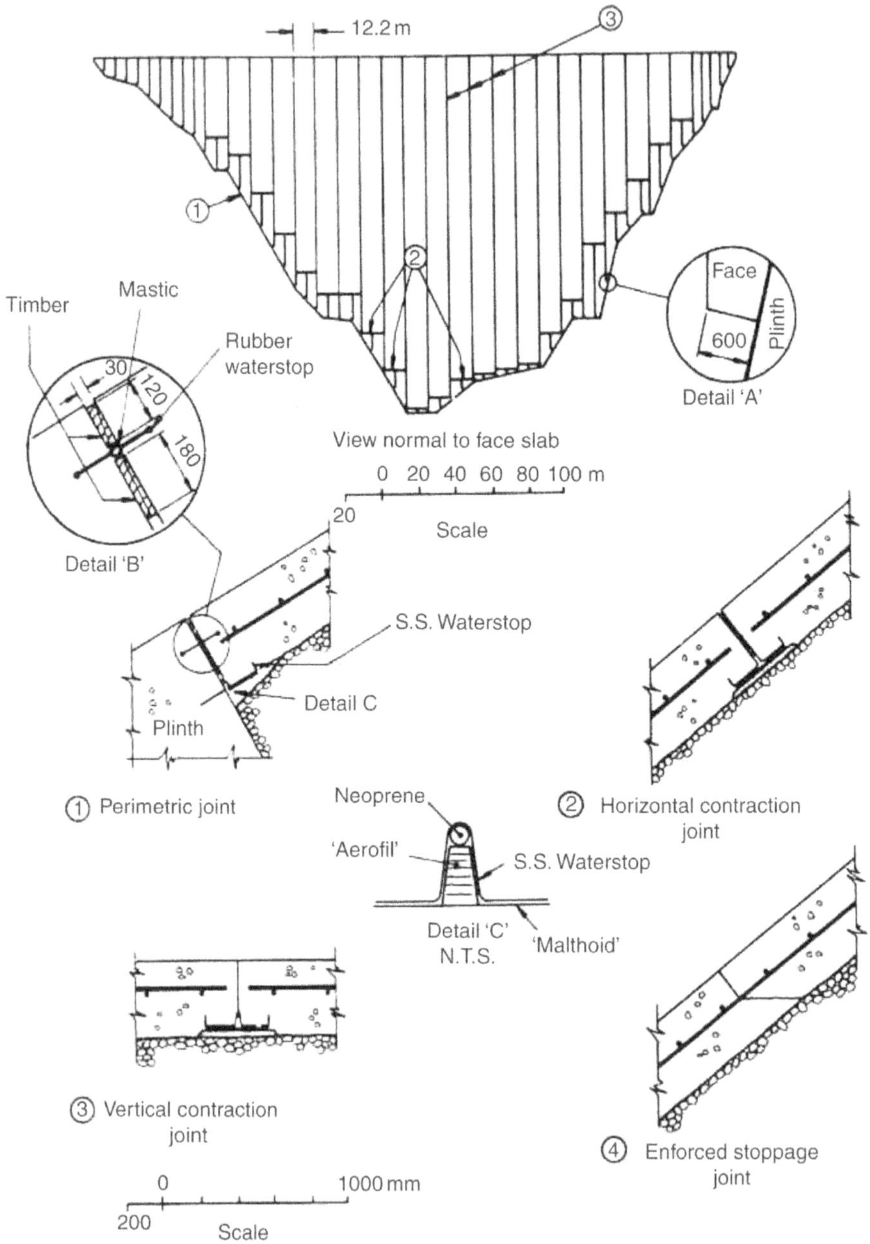

FIGURE 16.13 Joint layout and details for Reece Dam (Fitzpatrick et al., 1985). (Used with permission of ASCE conveyed through Copyright Clearance Center Inc.)

16.2.10.2 Face Slab Thickness

The face slab thickness has generally been determined from past experience. ICOLD (1989a), Cooke and Sherard (1987) and Cooke (2000) recommend the following:

- For dams of low and moderate height (up to 100 m): Use constant thickness = 0.25 m or 0.30 m.
- For high and/or very important dams: Use thickness = 0.3 m + 0.002 H where H = water head in metres.

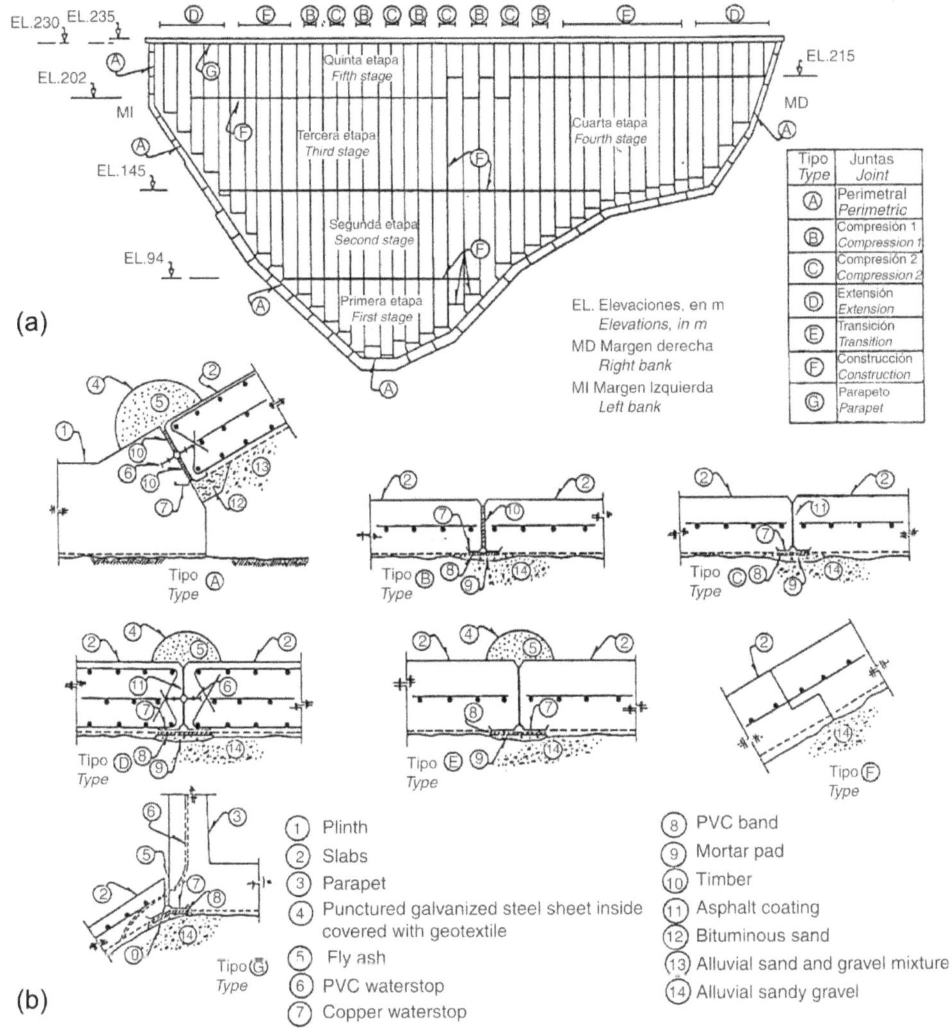

FIGURE 16.14 Joint details for Aguamilpa dam: (a) elevation showing joint layout and (b) joint details (Valencia and Sandoval, 1997).

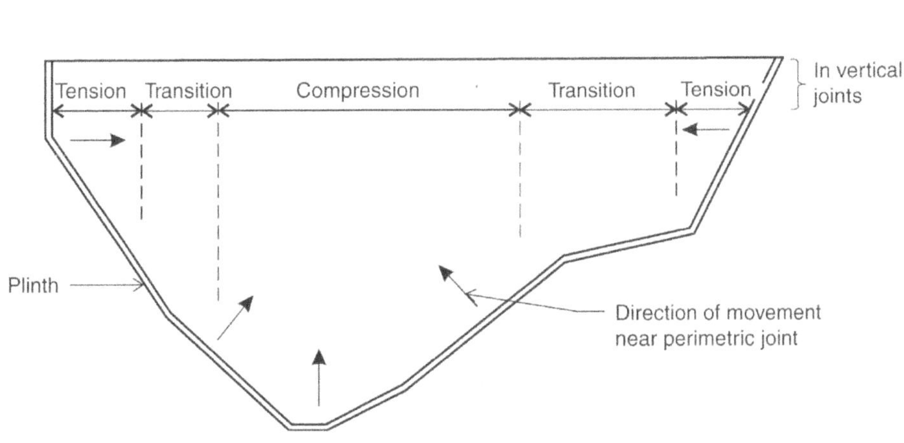

FIGURE 16.15 Face slab elevation showing movement directions and stress zones (Fell et al., 2015).

ICOLD (2010) indicate that practice is to use 0.3 m + 0.002 H to 0.3 m + 0.004 H for dams > 100 m high.

These are minimum thicknesses; average thicknesses will be greater due to irregularities in the compacted Zone 2D. They assume a well designed and constructed Zone 2D as detailed earlier.

16.2.10.3 Face Slab Reinforcement

Steel reinforcement is provided to control cracking due to temperature and shrinkage.

ICOLD (1989a, 2010) and Cooke and Sherard (1987) recommend the use of 0.4% reinforcing steel in each direction, with possible reduction to 0.3% or 0.35% in areas of the slab which will definitely be in compression, while retaining 0.4% within about 15 m of the perimeter. Cooke (1997, 1999, 2000), recommends the use of 0.3% horizontal, 0.4% vertical and 0.4% both ways within 15 m of the perimetric joint. He points out that several experiences of cracks within about 20 m of the perimeter and near changes in the plinth slope, from relatively steep to flat upper slope, may justify 0.5% each way in these areas. Cooke (2000) recommends 0.5% in areas of major change in plinth slope.

The reinforcing steel is placed as a single mat at or just above the centreline. The area of steel is calculated on the theoretical minimum thickness (Cooke and Sherard, 1987; ICOLD, 1989a).

Reinforcement is provided in the face and plinth at perimetric joints, and in some cases across vertical joints which are likely to be under compression, to control spalling. Examples are shown in Figures 16.16 and 16.17.

For assessing existing dams, there may be a higher likelihood of cracking of the face slab if it is significantly thinner and has less reinforcement than described earlier.

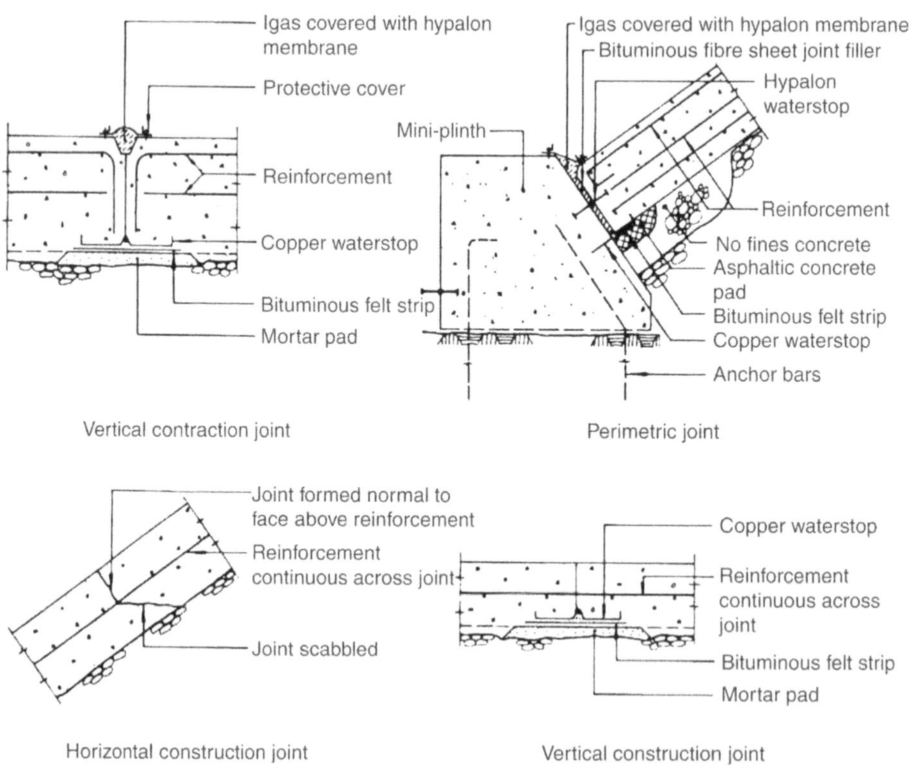

FIGURE 16.16 Joint details for Khao Laem Dam (Watakeekul et al., 1985). (Used with permission of ASCE conveyed through Copyright Clearance Center Inc.)

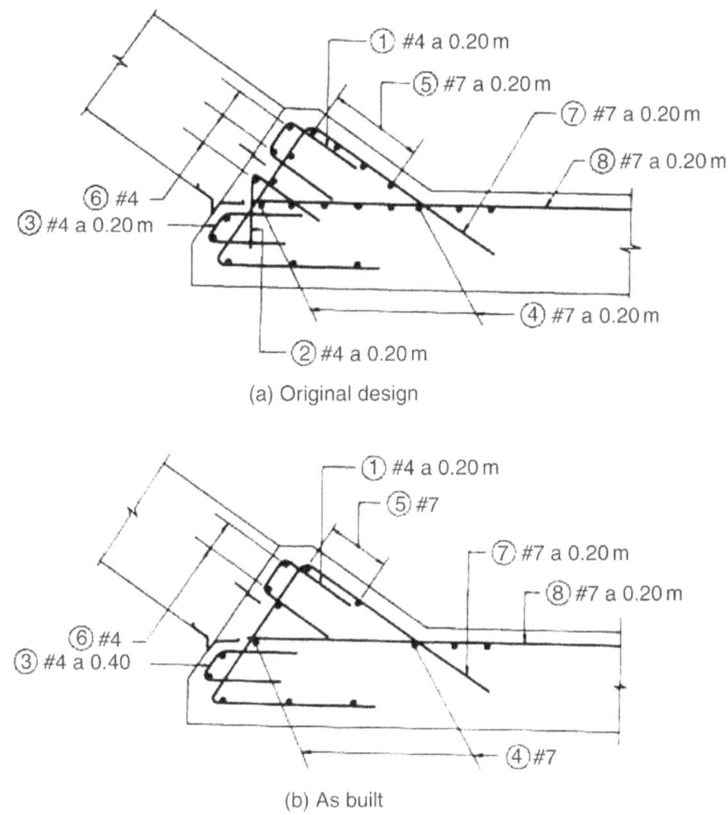

FIGURE 16.17 Anti-spalling steel in Salvajina Dam (Hacelas and Ramirez, 1987). (Used with permission of ASCE conveyed through Copyright Clearance Center Inc.)

16.2.10.4 Horizontal Joints

ICOLD (1989a) indicates that design practice at that time did not include horizontal joints, except for construction joints in which the reinforcing steel is carried through the joint without water stops. This approach was adopted because, when horizontal joints with water stops were used, it was difficult to obtain good quality concrete around the water stops and some joints experienced spalling under compression and mild rotation.

Cruz et al. (2009) and ICOLD (2010) indicated that was still current practice at that time. Details of such joints are shown in Figure 16.13 and Figure 16.14.

16.2.10.5 Vertical Joints

Vertical joints are generally provided at 12, 15, 16 or 18 m spacing depending on construction factors. For smaller dams, narrower spacing is common, e.g. 6 m.

Most CFRD have been constructed with each of the vertical joints being a construction joint with a copper water stop at the base as shown in Figure 16.16 and Figures 16.13 and 16.14. The joints are generally painted with asphalt, not filled with a compressible filler, as these have been shown to compress under load and cause opening of the perimetric joint. The exception is very high CFRD. ICOLD (2010) and Cruz et al. (2009) report that following the damage to the face slab on Barra Grande, Campos

Novos and Mohale dams, a compressive filler of wood or equivalent material has been used to absorb and distribute compressive stresses on several high CFRD and CFRD in narrow valleys.

Cooke and Sherard (1987) advocated carrying the horizontal reinforcement through all but a few joints near the abutments, in the manner shown in Figure 16.16. This method gives a construction cost advantage and a reduction in potential leakage, by potentially eliminating water stops and anti-spalling steel.

For Aguamilpa Dam and many more recent dams, different details were used, depending on whether the joints were expected to be in tension or compression as shown in Figure 16.14.

16.2.10.6 Perimetric Joint

Instrumentation of CFRD has indicated that compressive strains develop in more than 90% of the face slab due to settlement of the rockfill.

When the reservoir is filled there is further displacement of the face slab, which leads to closing of vertical joints over most of the slab and opening of the perimetric joint and those joints near the abutments. The face slab also pulls away from the plinth, and offsets normal to the face slab and parallel to the joint due to shear movement of the face. The joint is a common cause of leakage if not well designed, constructed and inspected.

To accommodate these movements, joints with multiple water stops are provided. Figures 16.18 to 16.20 show some examples.

The design shown in Figure 16.13 for Reece Dam is typical of practice up until around the mid-1980s. The joint includes two water stops:

- Primary—copper or stainless steel “W” or “F” shaped.
- Secondary—central “bulb” water stop made of rubber, hypalon or PVC.

Cooke and Sherard (1987) and ICOLD (2010) indicate that this arrangement has performed adequately on dams up to 75 m high, where perimetric joint movements are generally relatively small. Fitzpatrick et al. (1985) indicate that in two dams, 39 m and 26 m high, only the primary water stop was used because very small movements were expected.

More recently, particularly for higher dams, a third water stop has been included in the form of mastic or fly ash filler covered with a PVC or hypalon sheet. These are shown in Figures 16.18, 16.19 and 16.20.

Cooke and Sherard (1987) argued that, because of the difficulty of having concrete around the central “bulb” water stop, a joint with the copper (or stainless steel) water stop underlain by asphalt impregnated sand or concrete mortar and the mastic type stop is the preferable detail.

It will be noted that in the designs shown, a wood plank approximately 12.5 mm to 20 mm thick, or some other compressible filler of similar thickness, is placed between the face slab and plinth to prevent concentration of stresses in the joint during construction and before reservoir filling.

ICOLD (2010) provides detailed information on water stop design and construction.

16.2.11 Post-Construction Settlements

Post-construction settlements of the embankment are important because they influence the likelihood of damage to the face slab and opening of vertical and perimetric joints. Hunter et al. (2003), Fell et al. (2015) and Hunter (2003) discuss settlement and lateral deformation in detail. What follows is a summary of that information.

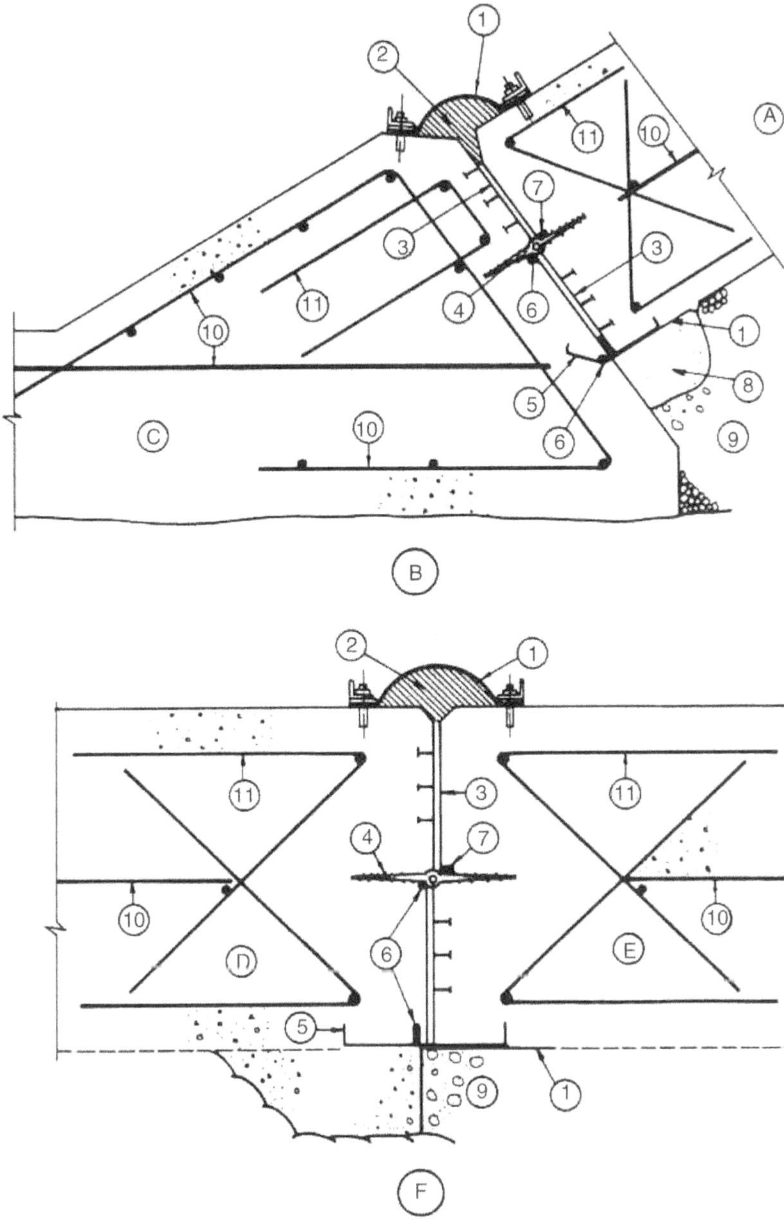

LEGEND: (1) Hypalon band, (2) Mastic filler, (3) Compressible wood filler, (4) PVC waterstop, (5) Copper waterstop, (6) Neoprene cylinder, (7) Styrofoam filler, (8) Sand-asphalt mixture, (9) Zone 2, (10) Steel reinforcement, (11) Steel reinforcement to protect concrete against crushing and to protect waterstop. (A) Face slab, (B) Perimetric joint, (C) Plinth, (D) Plinth, (E) Face slab, (F) Abutments.

FIGURE 16.18 Salvajina Dam perimeter and near abutment vertical joints (ICOLD, 1989a).

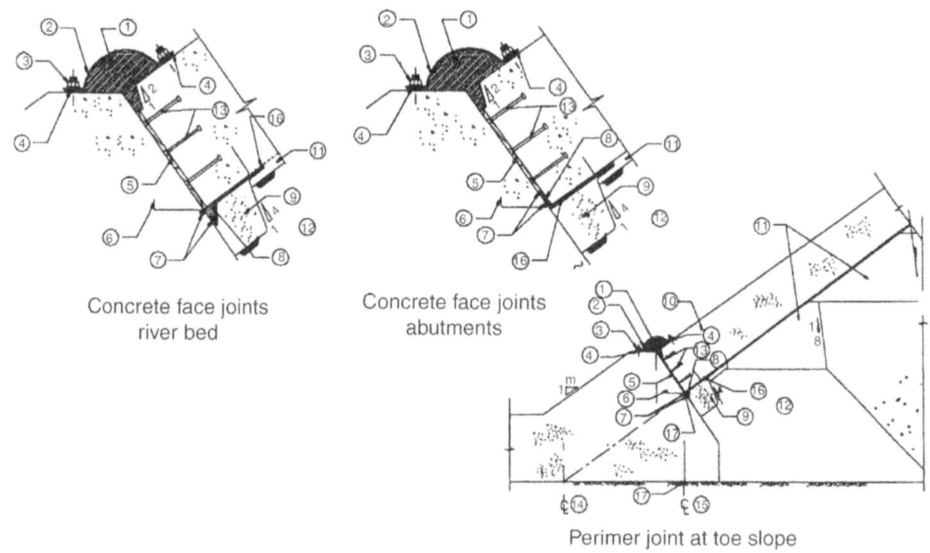

LEGEND: (1) Fly-ash, (2) PVC band, (3) Anchor bolts, (4) Epoxy glue, (5) Abarco wood, (6) Copper seal, (7) Circular neoprene, (8) Styrofoam, (9) Sand/asphalt mix, (10) Slab, (11) Concrete curb, (12) Zone 2A, (13) Steel bolts, (14) Plinth excavation reference line, (15) Plinth construction reference line, (16) PVC strip.

FIGURE 16.19 Antamina Dam joint details (Amaya and Marulanda, 2000).

16.2.11.1 *Effect of Compaction and Rock Strength*

Table 16.4 summarises the range of total post-construction settlement at 10 and 30 years and the long-term creep rate determined from the data set of CFRD. Figures 16.21 to 16.23 show plots of settlement versus time.

The general trend of the vertical crest settlement versus log time plot is for rapid rates of crest settlement during first filling. The steep portions of the curves in Figures 16.21 to 16.23 are typical for CFRD where first filling started more than about 0.5 years after the end of the main rockfill construction. The steeper portion of the curve is to be expected given that stress levels in the rockfill exceed those previously experienced. This is followed by long-term movements at a near constant creep strain rate (on log scale). The amount of deformation that occurs during first filling comprises a significant proportion of the total crest settlement, greater than 50% in a number of cases.

For the dumped and sluiced rockfill, Figure 16.21 indicates:

- Significantly greater magnitude of settlement for the dumped and sluiced rockfill compared with compacted rockfill. The rates of settlement (on log scale) post first filling (time-dependent deformation) are also significantly greater for the dumped rockfill.
- The poorly sluiced rockfill (Salt Springs and Dix River) have higher long-term creep rates (strain per log cycle of time) than the well-sluiced rockfill.

For well-compacted rockfill the post-construction crest deformation records presented in Figure 16.22 and Figure 16.23 indicate the following:

- For embankments constructed of rockfill of medium to high intact strength rock sources, the total magnitude of settlement at 10 years is on average approximately twice that of quarried,

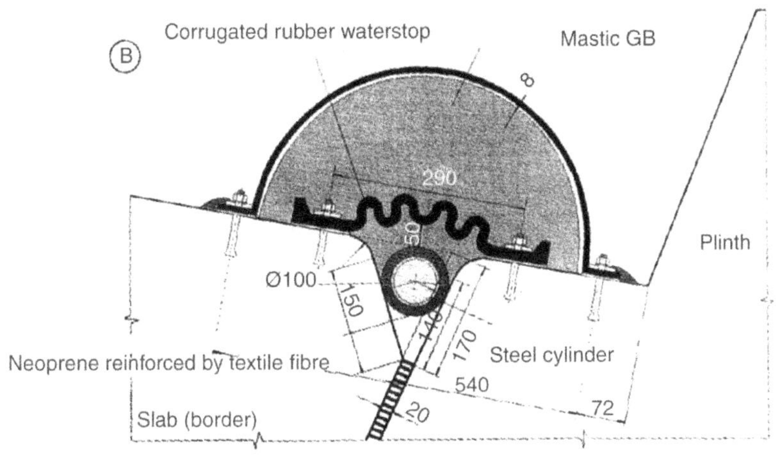

(a)

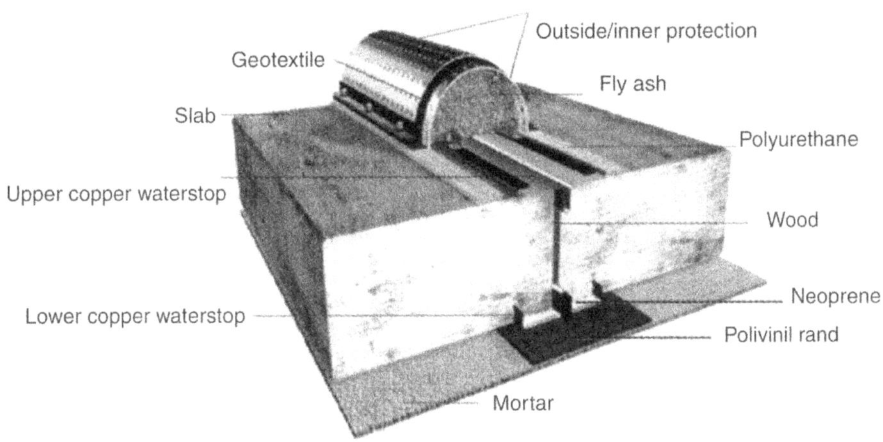

(b)

FIGURE 16.20 Perimeter joint details for (a) Mazar Dam; (b) El Cajon Dam (Cruz et al., 2009). (Reprinted with permission of GA Educacao LTDA, Rua Ernesto Alves, 150, Floresta District, Porto Alegre, RS-Brazil)

TABLE 16.4 Post-Construction Total Crest Settlement and Long-Term Creep Rate for CFRD (Hunter, 2003; Hunter et al., 2003)

	TOTAL POST-CONSTRUCTION SETTLEMENT (% OF DAM HEIGHT)		*LONG-TERM CREEP RATE (%/LOG CYCLE)*
ROCKFILL CLASSIFICATION	*10 YEARS*	*30 YEARS*	
Dumped rockfill	*0.6 TO 1.0*	*1.0 TO 1.5*	*0.3 TO 1.5*
Well-compacted rockfill:			
Medium to high strength	0.15 to 0.4	-	0.05 to 0.25
Very high strength, quarried	0.06 to 0.2	-	0.02 to 0.10
Gravel rockfill	< 0.05 to 0.2	-	< 0.10

Note: (1) Rock substance unconfined compressive strength medium 6–20 MPa, high 20–70 MPa, very high 70–240 MPa.

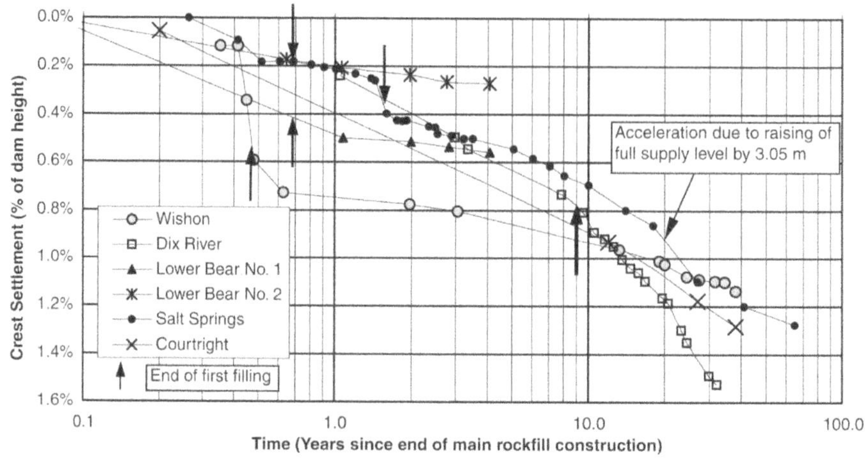

FIGURE 16.21 Post-construction crest settlement versus time for dumped sluiced rockfill CFRD (Hunter, 2003; Hunter et al., 2003).

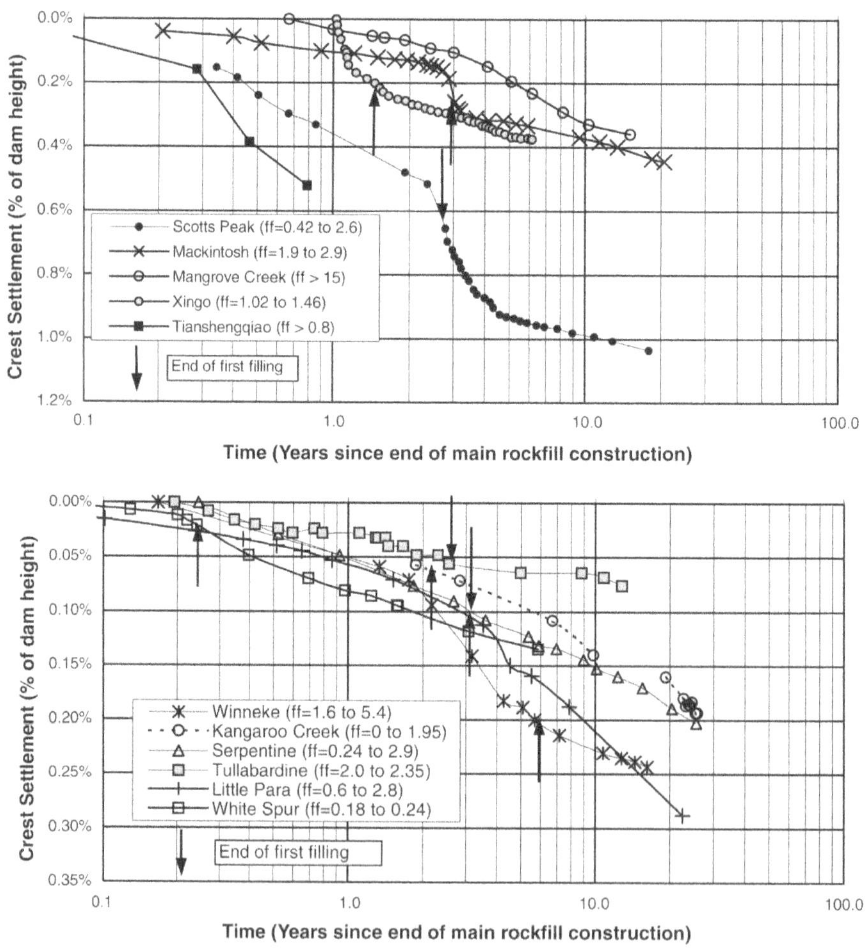

FIGURE 16.22 Post-construction crest settlement versus time for CFRD constructed of well-compacted quarried rockfill of medium to high intact strength (Hunter, 2003; Hunter et al., 2003).

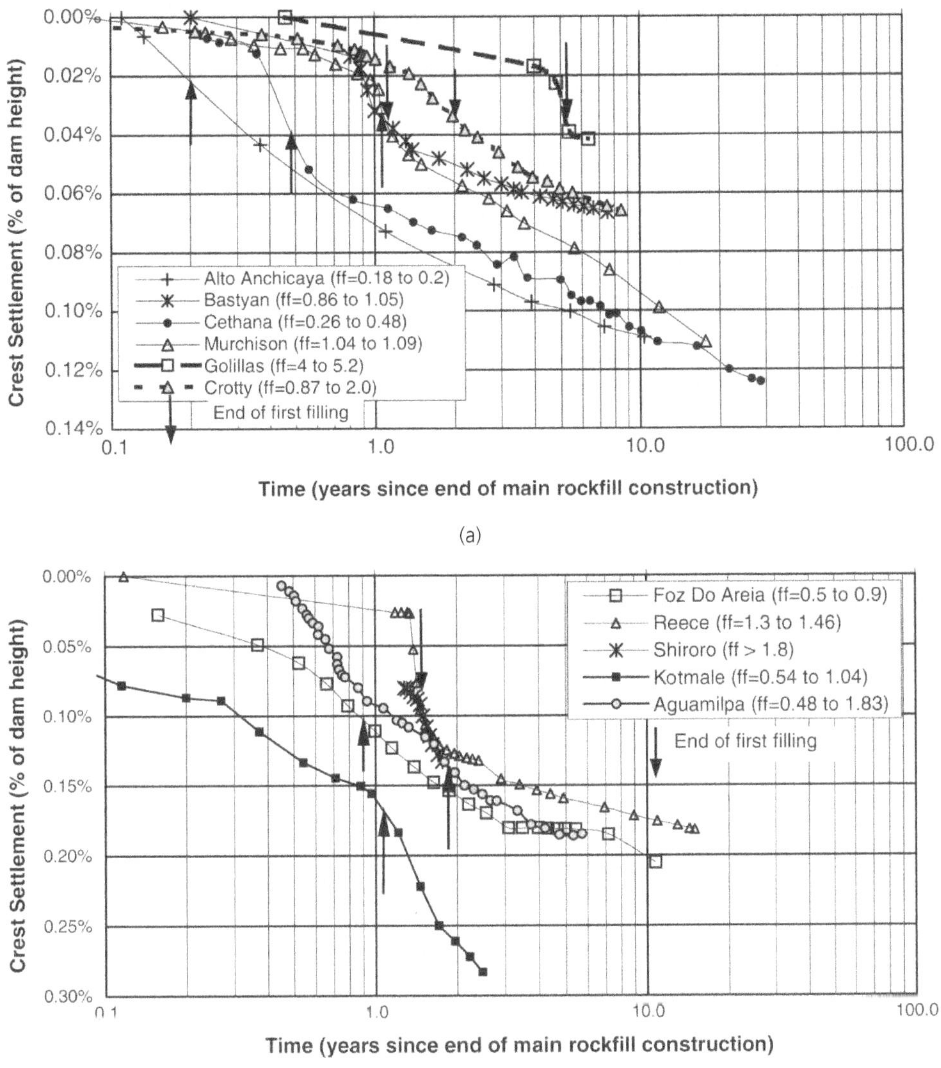

FIGURE 16.23 Post-construction crest settlement versus time for CFRD constructed of well-compacted quarried rockfill of very high intact strength and of well-compacted gravels (Hunter, 2003; Hunter et al., 2003).

very high strength rockfill. The long-term creep rate is also significantly higher for the weaker rockfill, 2 to 10 times that of quarried, very high strength rockfill.

- For the well compacted sandy gravels (Crotty and Golillas) the post-construction deformation is less than that for the well-compacted, very high strength quarried rockfill. This is likely to be due to several reasons but is considered to be mainly a result of the rounded shape of the gravels. The point area of contact between particles of rounded gravel will be significantly greater than that of angular quarried rockfill. Hence, the contact stresses will be significantly less, resulting in less particle breakage.

When assessing the likelihood of open perimetric joints being present, the observed settlement and creep rate should be compared to those in Table 16.4. Dams with higher than these are more likely to have the perimetric joint open.

16.2.11.2 Performance of CFRD under Large Seismic Loading

The 156 m high Zipingpu CFRD dam was impacted by a Magnitude 8 Wenchuan earthquake in May 2008. The dam is sited 17 km from the epicentre and recorded horizontal peak ground accelerations of over 2 g at the crest with a bedrock PGA of 0.5 g. The damage to the dam is described in Wieland (2009), Cruz et al. (2009), Chen (2008a, 2008b) and Xu (2008).

The most significant damage was to vertical joints which suffered compressive stress failure and offset to a horizontal joint between the second and third stage face slab. Figures 16.24 and 16.25 show the damage. There was also lateral deformation at the crest as shown in Figure 16.26.

The maximum settlement at the dam crest due to this massive earthquake loading was 735 mm (0.5% of dam height) and 180 mm downstream. The water level was low (54 m below crest level) in the dam at the time of the earthquake. Maximum leakage went from 10 l/sec to 18 l/sec.

This is an important case study because it shows that the face slabs of CFRD are vulnerable to damage under large seismic loads because the settlements induced by the earthquake may be far larger than experienced under normal operating conditions (see Table 16.4).

ICOLD (2010) and Wieland (2009) indicate that they expect the performance of well-designed CFRD under seismic loading to be generally good. Wieland (2009) points out that cross canyon excitation under large earthquakes may result in local deformations of joints, rupturing of water stops and cracking of face slabs. Stiff crest elements may behave differently to the remainder of the embankment and result in separation of the element from the face slab.

How well the embankment performs will therefore be dependent on the joint details, and whether there is a well designed and constructed Zone 2D to control leakage in the event face slab and joints are damaged.

To assess the likelihood of cracking, estimate the settlement under seismic load as described in Section 16.3.3 and compare to the observed settlement to date. If there are potentially large additional settlements, cracking might be expected.

16.2.12 Joint Displacements and Leakage Rates

Hunter et al. (2003), Fell et al. (2015) and ICOLD (2010) present tables of measured perimetric joint displacements and leakage rates for a number of dams. These may be used as a guide to performance for existing dams.

16.2.13 Additional Information

This section has set out to summarise the history of development, features and performance of CFRD.

Hunter et al. (2003) and Fell et al. (2015) have more detail than the prior summary, particularly regarding performance of CFRD. The former can be freely downloaded.

Those seeking more detail should also refer to the following volumes:

J. Barry Cooke Volume, Concrete Face Rockfill dams. Chinese Committee on Large Dams, Beijing, September 2000.

Proceedings, International Symposium on Concrete Face Rockfill Dams, September 2000, International Commission on Large Dams, Paris.

Proceedings, Second Symposium on Concrete Face Rockfill Dams, Comite Brasileiro de Barragens, Rio de Janeiro, October 1999.

Concrete Face Rockfill Dams, P.T. Cruz, B. Materon and M. Freitas. CRC Press, 2009.

FIGURE 16.24 Damage of vertical joint of face of Zipingpu CFRD looking upwards towards the crest: (a) damage, (b) preparation for repairs (Wieland, 2009; photographs courtesy of Xu Zeping, China). (Reproduced courtesy of M. Wieland.)

FIGURE 16.25 Damage of horizontal joint between Stages 2 and 3 of the concrete face of Zipingpu CFRD (Wieland, 2009, photographs courtesy Xu Zeping, China). (Reproduced courtesy of M. Wieland.)

FIGURE 16.26 Damage at downstream crest of Zipingpu CFRD (Wieland, 2009, photograph courtesy Xu Zeping, China). (Reproduced courtesy of M. Wieland.)

ICOLD (1989) *Rockfill dams with concrete facing*. State-of-the-art. International Committee on Large Dams, Paris. Bulletin 70.

ICOLD (2010) *Bulletin 141, Concrete Face Rockfill Dams,* Concepts for Design and Construction. ICOLD, Paris.

16.3 ASSESSING THE SAFETY OF CONCRETE FACE ROCKFILL DAMS

16.3.1 Introduction

Potential failure modes for concrete face rockfill dams (CFRD) are the following:

(a) Flood overtopping leading to slope instability or unravelling leading to loss of freeboard.
(b) Deformations under seismic loads leading to loss of freeboard.
(c) Slope instability leading to loss of freeboard. This is most likely initiated by a leak forming in the face slab or joints under flood or seismic loading.
(d) Slope instability and loss of freeboard in the upper part of the embankment where a vertical joint in the face slab opens, with common cause cracking of Zone 2D at the joint, potentially leading to internal erosion of Zone 2D and increased leakage.

CFRD inherently have a low likelihood of failure provided that:

(a) They are designed and constructed with adequate freeboard for floods.
(b) Have good face slab and perimetric joint details.
(c) Have a modern Zone 2D to limit leakage flows.
(d) Have high permeability downstream zones capable of discharging leakage flows without build-up of pore pressure or unravelling, well treated foundation beneath the plinth and adequate free draining high permeability zones taken to crest level to accommodate possible leakage flows.
(e) Have well designed plinth founded on a non-erodible rock foundation with good foundation treatment beneath the plinth.

The low likelihood of failure applies particularly for reservoir levels up to the pool of record and for small seismic hazard loads. Higher likelihoods may apply for reservoir levels above the POR and large seismic loads.

The concrete of the face slab, and concrete surrounding the perimetric and other joints or cracks which may form in the face slab, is essentially non-erodible, so leakage should not increase rapidly. For well managed dams, detection and intervention to limit leakage flows should be possible.

However, in many existing dams the Zone 2D and 2E materials are not present or not designed to modern standards and are high permeability, often poorly graded and potentially subject to suffusion, with the finer particles potentially being washed out and permeability increased. This can lead to settlement, further opening of joints or cracks, and leakage flows.

Left unchecked, the process may not be self-limiting, and very large leakage, leading to a breach mechanism by unravelling or instability of the downstream slope, may result. At what stage this may occur is dependent on the zoning and the nature of the rockfill, e.g. gravel fill, and low strength weathered rockfill, is likely to have lower permeability and be more likely to reach an unstable condition than other rockfill.

The one failure of a modern CFRD (Gouhou dam, Chen, 1993) occurred at apparently relatively small leakage at the junction between the parapet wall and face slab, but the zoning of that dam was poor, with the fill being silty sandy gravel without high permeability drainage zones.

16.3.2 Flood Overtopping

16.3.2.1 Deterministic and Risk Enhanced Deterministic Assessment

As for other embankment dams. See Sections 6.7.2 and 6.5.5.

16.3.2.2 Quantitative and Simplified QRA

Assess as described in Sections 6.7.4, 6.7.5 and 6.7.6.

16.3.3 Deformations under Seismic Loads Leading to Loss of Freeboard

This considers deformations induced by an earthquake. It does not include potential for slope instability post earthquake if the face slab or joints are damaged.

Assess as described in Section 8.12.1.

16.3.4 Slope Instability Leading to Loss of Freeboard—General Approach

As discussed in Section 16.2.7, the stability of the slopes in CFRD is not usually analysed. This is in recognition of the fact that well designed and constructed CFRD have no pore pressures in the rockfill and will remain stable under static loads.

However, some existing dams with non free draining rockfill or with gravel instead of rockfill may experience some degree of saturation and build-up of pore pressure in the rockfill if the face slab is damaged, especially if there is no low permeability zone (Zone 2D) beneath the face to control seepage into the embankment.

For these dams, estimates should be made of potential pore pressures in the embankment allowing for cracking of the face slab and opening of joints and the resulting flows through the crack or open joint, the zoning of the embankment and its ability to limit leakage flows, and the nature of the rockfill and its ability to discharge leakage flows without build-up of pore pressure and initiation of failure by slope instability or unravelling.

For these dams:

(a) Assess the likelihood of a crack in the face slab or joints as described in Section 16.3.5.
(b) Assess the location, width and length of cracks or open joints.
(c) Estimate leakage flows in the crack or open joint as described in Section 16.3.6.
(d) Develop flow nets of the embankment and foundation to model pore pressures in the embankment resulting from the leak(s). Use rockfill permeabilities estimated as described in Section 6.4.3.
(e) Analyse stability based on the embankment zoning, pore pressures and shear strengths.
(f) Take account of low strength strata in the foundation which may influence stability.

Based on these analyses, assess the likelihood of instability and loss of freeboard as described in Section 7.14 for deterministic and risk-based assessments.

When making this assessment care should be taken to account for differing zoning in the upper part of the dam, such as for Mangrove Creek Dam, Figure 16.10. The drainage zone for that dam does not extend above full storage level, so the potential for build-up of pore pressure in the upper part of the dam is greater for reservoir levels above FSL than below.

For dams with a crest wall (or parapet wall), the joint between the wall and the face slab (see Figure 16.11) may be vulnerable to opening and allowing leakage. If this detail is combined with the drainage zone not extending high into the embankment, build-up of pore pressures and reduction in factor of safety may result.

16.3.5 Leak in the Face Slab or Joints, Leading to Slope Instability and Loss of Freeboard under Flood and Seismic Loads

Figure 16.27 shows the failure path for a CFRD.

As discussed in Southcott et al. (2003) and Fell et al. (2015), the likelihood of failure can be assessed qualitatively or quantitatively in this approach. The sequence of events to be modelled are as follows:

- Initiation—start of a leak through an open crack in the face slab or damaged joint, or though the foundation below the plinth.
- Continuation—transition zones such as Zone 2D fail to control leakage and Zone 2E fails to prevent erosion of finer fraction from Zone 2D.
- Progression—erosion of finer fraction from Zone 2D results in settlement of the face slab, further opening of the leaking joint.
- Breach initiates. Leakage increases, resulting in increased pore pressures and slope instability or unravelling initiates.
- Detection and intervention fail.
- Breach—uncontrolled flow over or through the embankment.

Figure 16.28 shows the event tree for this potential failure mode.

The failure paths should be considered for flood and seismic loads. For deterministic and risk enhanced deterministic assessments the flood will be the reservoir level for the Inflow Design Flood

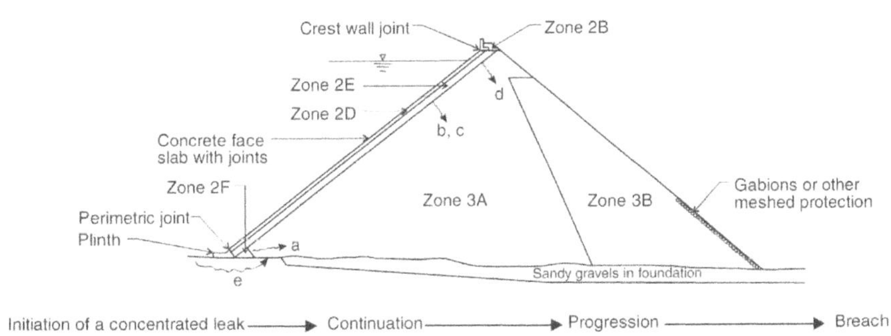

FIGURE 16.27 Failure paths for concrete face rockfill dam (adapted from Fell et al., 2015).

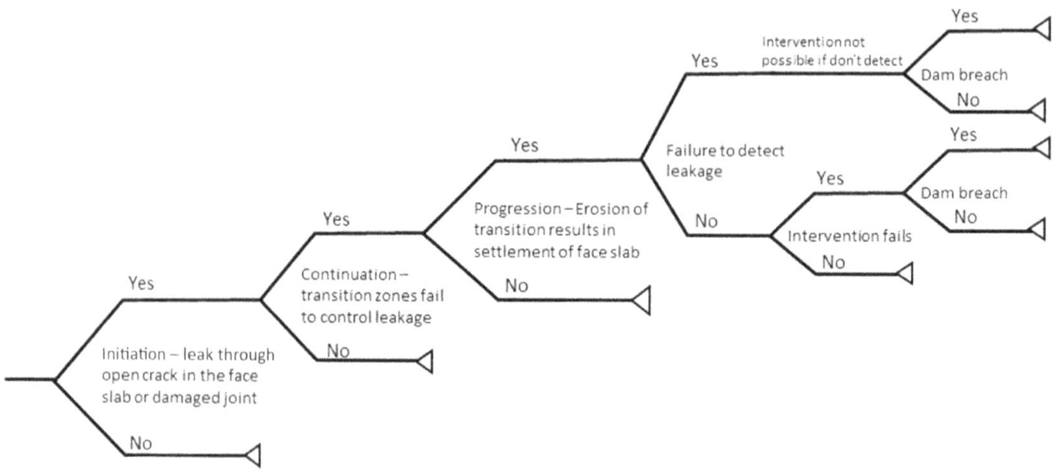

FIGURE 16.28 Event tree for CFRD for PFM leak in face slab or joints, leading to slope instability and loss of freeboard.

and seismic loads will be from the Earthquake Design Ground Motion for the dam Consequence Classification.

For quantitative and simplified QRAs, the flood reservoir partitions up to the dam crest level, and all earthquake ground motion partitions should be modelled.

16.3.5.1 Assessment of the Likelihood of a Leak in the Face Slab, Joints or Foundation

Table 16.5 summarises the factors to be considered for assessing the likelihood of initiation of leakage under normal operating and flood (to dam crest level) conditions.

Also detailed are suggestions on the likelihood of initiation of a leak. The term "compared to population" suggests the measurable value, e.g. slab displacement, can be compared to the data in Section 16.2 and described in Hunter et al. (2003), Fell et al. (2015) and ICOLD (2010). For example, if the measured value is large compared to the population, a higher likelihood might be assigned.

It will be apparent that the estimate will be subjective and difficult to quantify. Some points that should be kept in mind are as follows:

- It is very unlikely a leak will form under normal operating conditions (water levels within annual levels), unless for example the face slab or joints have already been damaged, a perimetric joint is right at the limit of its displacement or there is a deteriorating condition of joint filler or water stop measures.
- It is more likely that a leak would form under reservoir water levels higher than the historic high (because the loads will be greater than the dam has experienced before) and under large seismic loading.

Tables showing suggested values for the conditional probabilities of initiation are included in Southcott et al. (2003). The authors have found these to be conservative and recommend that the conditional probabilities be instead estimated by expert elicitation/engineering judgement.

TABLE 16.5 Factors to Be Considered and Methods for Assisting in the Estimation of the Likelihood of Initiation of a Concentrated Leak in CFRD (adapted from Southcott et al., 2003)

INITIATOR	*FACTORS TO CONSIDER*	*METHODS FOR ASSISTING IN THE ESTIMATION OF THE LIKELIHOOD OF CONCENTRATED LEAK*
(A) Factors applicable to all load conditions		
Perimetric joint opens	Crest and face slab deformation, first fill and continuing.	Compared to population and degree of variation across dam.
	Joint meter and face slab inclinometer displacement.	Compared to design displacement and to population.
	Deep plinth, and local deep fill at plinth in over-break and steep plinth sections.	Calculate face slab and joint displacement from depth, construction modulus, for Zone 3A or modulus of Zone 2F, allowing for water loading (depth).
	Leakage flow rates vs reservoir level and time. If there is increasing leakage, or 10 to 20 l/sec may be a leak, > 100 l/sec certainly a leak.	If no leakage measurements are available, use a range of values of likelihood to reflect the uncertainty.
	Water stop deterioration, accounting for water stop type, acidity of water, displacement.	Inspect when reservoir levels are low.
Vertical joints open	Crest settlement, horizontal displacement and face slab deformation first fill and continuing.	Compared to population and degree of variation across dam.
	Joint meter and face slab inclinometer displacements.	Compared to design displacement and to population.
	Valley profile—e.g. steep valley sides leading to large differential settlements.	Estimated total settlement and surface profile.
	Opening of vertical crest wall joints.	Inspect when reservoir levels are low.
	Leakage flow rates vs reservoir level and time. If there is increasing leakage or > 10 to 20 l/sec may be a leak, > 100 l/sec certainly a leak.	If no leakage measurements use a range of values.
Face slab cracking	Crest and face slab deformation, first fill and continuing.	Compared to population.
	Staged rockfill and/or staging of face slab construction effects, with potentially different modulus for lower and upper part.	Construction data, calculation from rockfill depth, construction modulus, allowing for water loading (depth) and link to total deformation.
	Deep plinth, and local deep fill at plinth in over break and steep plinth sections.	Construction data, from rockfill depth, construction modulus, allowing for water loading (depth) calculation and link to total deformation.
	Location of and nature of difficult construction joints	
	Leakage flow rates vs reservoir level and time. Increasing leakage or > 10 to 20 l/sec may be a leak, >100 l/sec certainly a leak	If no leakage measurements are available, use a range of values of likelihood to reflect the uncertainty.
	Crack mapping of face slab	Inspect when reservoir levels are low.
	Sudden changes in slab thickness	From design drawings.
(A) Factors applicable to all load conditions.		
Crest wall joint opens	Visual inspection/displacement on joint.	Joint should be above reservoir FSL. If not, inspect when reservoir level is low.
	Design detail.	See Figures 16.11 and 16.14 for good details.

(Continued)

TABLE 16.5 (Continued)

INITIATOR	*FACTORS TO CONSIDER*	*METHODS FOR ASSISTING IN THE ESTIMATION OF THE LIKELIHOOD OF CONCENTRATED LEAK*
Internal erosion under plinth	Part or all of plinth foundation erodible, e.g. soils, completely weathered rock, clay infill joints.	Check construction geology mapping and photographs.
	Hydraulic gradient.	Calculate from plinth width.
	Grouting—curtain and consolidation.	Check construction records for grout details and takes.
(B) Additional factors applicable to flood loading.		
All initiators.	AEP of flood levels.	Hydrological analysis.
	Historic high reservoir level.	Likelihoods much less for flood levels below historic high than above.
Crest wall joint opens	Increased face slab deformations, joint openings, under flood loading.	Calculate from dam geometry, properties, existing deformations and openings allowing for added water load.
	AEP of flood levels which reach the parapet crest wall joint.	Hydrological analysis. Detail of joint and its condition.
(C) Additional factors applicable to seismic loading.		
All initiators	Earthquake Magnitude, PGA, vs AEP	From seismic hazard study.
	Additional settlement induced by earthquake.	Pells and Fell (2002, 2003) and simplified methods.
	Crest wall geometry + geometry of the perimetric joint where the slab abutments a spillway wall.	From design drawings and inspection.
	Are there liquefiable (not necessarily flow liquefaction) zones in the foundation? What is the AEP of liquefaction, and will freeboard be lost if liquefaction occurs?	Chapter 8, Sections 8.6 to 8.12.

16.3.5.2 Assessment of the Likelihood of Continuation of a Leak

The procedure is as follows:

(a) Given a crack or opening of a joint is likely, estimate the location, width and length of the crack.
(b) Estimate the leakage flow using the method described in Section 16.3.6. Take account of the ability of Zone 2D to limit flows.
(c) Assess whether Zone 2E will act as a filter to Zone 2D. If other than a no-erosion filter, what is the expected grading of Zone 2D and what would the likely flow rate be for this condition?

Table 16.6 summarises the factors to be considered and methods for assisting in estimation of the likelihood of continuation of a concentrated leak.

16.3.5.3 Will Zone 2E Act as a Filter to Zone 2D and Arrest Erosion?

In this assessment, the particle size distribution of Zones 2D versus 2E, and 2E versus 3A, are assessed against the no erosion, excessive erosion and continuing erosion criteria in Foster and Fell (2001) as described in Section 12.2.4.

TABLE 16.6 Factors to Be Considered and Methods for Assisting in the Estimation of the Likelihood of Initiation of Continuation of a Leak in CFRD (adapted from Southcott et al., 2003)

INITIATOR	*FACTORS TO CONSIDER*	*METHODS FOR ASSISTING IN THE ESTIMATION OF LIKELIHOOD OF CONTINUATION*
Continuation. Zones 2D and 2E fail to control leakage.		
Perimetric joint opens, leak initiates	As-constructed particle size distributions, susceptibility to segregation, suffusion and GBE of Zone 2D, and filtering characteristics of zones 2D, 2E, 3A.	Assess Zone 2D for suffusion and GBE using Douglas et al. (2017, 2019); see Section 10.6 Assess the likelihood of Zone 2D holding a crack as described in Section 12.2.4. Assess Zone 2E versus Zone 2D against no-erosion, excessive erosion and continuing erosion criteria. See Section 12.2.4. Assess Zone 3A versus Zone E filtering.
Vertical joint opens, face slab cracks, crest wall joint opens, leak initiates	As above.	As above.
Internal erosion under plinth	Presence of extent of filter(s) under rockfill downstream of plinth. If present, particle size of filter and erodible material in the foundation.	Assess filter versus potentially eroded foundation against no-erosion, excessive erosion and continuing erosion criteria. See Section 12.2.4.
All joints, cracks	Displacement/disruption or cracking of Zone 2D by seismic loading.	Judged from estimate of overall settlement using Pells and Fell (2002, 2003). See Chapter 8. Section 8.3.

Note: In Southcott et al. (2003) it was assumed that erosion of Zone 2D would be concentrated leak in cracks. That is now considered less likely than suffusion and GBE of Zone 2D. The table has been amended accordingly.

For that assessment the Zone 2D gradation is regraded on the 4.75 mm sieve. It may also be possible to estimate the gradation of the soil eroding from Zone 2D based on data in Douglas et al. (2016, 2019) and check the filter capability of Zone 2E to arrest erosion of those particles.

In many dams there will be no gradations available for Zones 2D and 2E, and those assessing the dam safety will have to rely on specified gradations, construction photographs and their experience to estimate gradations.

16.3.5.4 Will Zone 2D Hold a Crack and Be Ineffective as a Filter?

Assess whether the fines content of Zone 2D makes it susceptible to holding a crack. Use the method described in Section 12.2.4, Table 12.3 to make this assessment.

As shown in Figure 16.15, most of the face slab is in compression. In those areas Zone 2D will also likely be in compression so it should not be cracked.

The joints on the edges of the face slab are potentially in tension and may open. It is likely that Zone 2D at that location will also be in tension. If Zone 2D is susceptible to holding a crack, it is possible that a crack or more likely several cracks may form in Zone 2D. If the open joint and a crack in Zone 2D coincide, a concentrated leak scenario could develop. The likelihood of this occurring is likely to be very low and would probably require some related common cause effect.

An example of an exception to this is shown in Figure 16.29. Here the left abutment upper face slab joint is where the face joins to the concrete abutment. The upper right abutment has a similar detail but

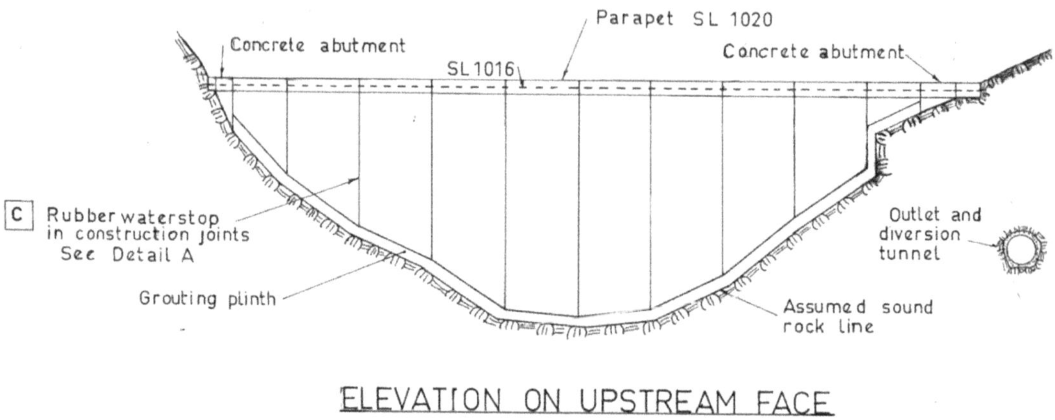

FIGURE 16.29 Elevation on the upstream face of a CFRD where concrete gravity structures were used on the upper abutments.

also has the second joint above a near vertical change in foundation profile. At these locations, cross-valley differential settlement potentially results in a crack in Zone 2D and also opening of the face slab joint.

16.3.5.5 Assessment of the Likelihood of Progression

In many existing dams, the Zones 2D and 2E will be potentially subject to suffusion or Global Backward Erosion (GBE). The effect will be to erode some of the finer fraction of these zones, resulting in a higher permeability and some additional settlement which may cause further opening of cracks or joints.

Whether Zone 2D is subject to suffusion or GBE should be assessed using the method of Douglas et al. (2016, 2019), which is described in Section 10.6. It can be expected that many dams constructed before the 1970s will have the earlier type of Zone 2D so will be likely to experience this erosion. As discussed in Section 16.2.4, even dams with the more modern Zone 2D may experience such erosion if the Zone downstream of Zone 2D is very coarse.

Table 16.7 lists the factors to be considered. They are similar to those relating to continuation.

In the case of CFRD, it is unlikely a hole or pipe will form as would happen in an earthfill core. Rather, what is likely to happen is a continuing erosion of the finer fraction from the zones in the dam, including the rockfill. The erosion of the finer material will cause settlement, a likely greater opening of the joint or crack which initiated leakage, and so on unless there is intervention.

16.3.5.6 Detection and Intervention

Table 16.8 lists the factors to be considered. What is important to any assessment of successful detection and intervention to prevent failure is the frequency of inspections, whether leakage of the dam can be measured or observed, and if so whether it is real time monitoring or only weekly or longer measurements.

Intervention depends on the ability of the owner of the dam to respond to the leakage incident, by lowering the water in the reservoir to below the cracked zone, or by relieving the head driving the seepage and/or to seal the crack or opening by, for example, placing sand onto the leak and have it wash into the crack, and be prevented from eroding further by filter action on Zone 2D, or Zone 2F at the perimetric joint. The erosion process for a well designed and constructed CFRD is not likely to be so

TABLE 16.7 Factors to Be Considered and Methods for Assisting in the Estimation of the Likelihood of Progression of Leakage in CFRD (adapted from Southcott et al., 2003)

INITIATOR	*FACTORS TO CONSIDER*	*METHODS*
All initiators	Zoning and particle size distribution of each zone, and potential for internal erosion in Zones 2D, 2E, 3A, 3B leading to washing out of fines and increased permeability.	Assess Zone 2D and Zone 2E for suffusion and GBE and resulting permeability using Douglas et al. (2016, 2019) in Section 10.6.
	Flow rate and location of the initial leak in the face slab or joints.	Judged from the initiating mechanism, permeability of Zones 2D and 2E, and precedent in other CFRD.
	Potential for downstream zones to intercept eroding material, leading to slowing down of internal erosion.	Assess as a filter against the eroding fines for no-erosion, excessive erosion and continuing erosion criteria. Section 12.2.4.
	Presence of Zones 1A and 1B upstream of the plinth as shown in Figure 16.2, and whether it will act to seal on Zones 2F, 2D as shown in Figure 16.27.	As above.

TABLE 16.8 Factors to Be Considered and Methods for Assisting in the Estimation of the Likelihood of Detection, Intervention and Formation of a Breach in CFRD

BREACH AND MODE	*FACTORS TO CONSIDER*	*METHODS FOR ASSISTING IN THE ESTIMATION OF LIKELIHOOD OF DETECTION, INTERVENTION AND FORMATION OF A BREACH*
All breach modes	Inspection regime for the dam.	Account for the inspection regime under normal operating, flood and earthquake conditions.
	Seepage monitoring.	Is seepage from the dam monitored, can it separate seepage through the foundation, face slab, and from rainfall onto the downstream rockfill? Is seepage monitored automatically or only daily, weekly, monthly, etc.? Is the toe of the dam drowned out by tailwater or covered by disposed fill?
	Ability to intervene, draw down the reservoir.	Account for location and capacity of outlet works, reservoir volume, river inflows.
	Ability to control leakage by placing sand and/or gravel onto the leak area and have it seal on Zone 2D.	Location of the leak, accessibility of site, availability of barges, personnel, whether the joint or opening is covered by debris, Zone 1A etc.

rapid as for earth and earth and rockfill dams, but may still progress in days or weeks, so detection and intervention need to be done soon after the significant leak develops if intervention is to be successful.

16.3.5.7 Assessment of the Likelihood of a Breach Forming

Table 16.9 lists the factors to be considered and the methods which can be used to calculate whether slope instability or unravelling will occur.

The potential for slope instability is greater for dams with low permeability rockfill or gravel fill instead of rockfill. The drainage capacity of drainage zones needs to be checked against potential leakage flows resulting from cracks in the face slab and/or damage to joints.

TABLE 16.9 Factors to Be Considered and Methods for Assisting in the Estimation of the Likelihood of Formation of a Breach in CFRD

BREACH AND MODE	*FACTORS TO CONSIDER*	*METHODS FOR ASSISTING IN THE ESTIMATION OF LIKELIHOOD OF DETECTION, INTERVENTION AND FORMATION OF A BREACH*
Slope instability and unravelling	Estimated likely flow rate of the leak.	Initiator mechanism, continuation and progression factors, historic precedents for modern and older CFRD. Estimate leakage flow rates through the concrete face by considering potential crack or joint openings and estimate flows through the crack as described in Section 16.3.6. Estimate the flow limiting effects of the Zone 2D using flow nets as described in Sherard (1985) and described in Section 16.3.6.
Unravelling	Discharge capacity of rockfill zones, compared to estimated leakage flow rates, giving gradients through the dam.	Estimate discharge capacity from zoning, particle size distribution, presence of steel meshing on downstream face and methods for assessing unravelling of rockfill using the methods described in Section 6.7.
Slope instability	Discharge capacity of rockfill zones compared to estimated leakage flow rates, giving pore pressure rise in the dam.	As above. Calculate factors of safety allowing for the permeability of the rockfill (Section 6.4.3) and shear strength of the rockfill (Section 7.6). Assess as described in Section 6.7.

16.3.5.8 Concluding Remarks

There is no deterministic approach to assess this potential failure mode. It is necessary to use risk-based methods and expert elicitation.

It will be apparent that what is described previously is only a framework, and quantification will be difficult and subject to considerable uncertainty. It may be sufficient in many cases only to use qualitative terms. The process of formally considering all failure initiators and failure paths is a valuable one, even if the estimates of likelihood are approximate.

16.3.6 Estimating Leakage Flows in Cracks in the Face Slab or Open Joints

Estimates of leakage flows require estimates of:

- The location, length and width of the crack or joint opening.
- Whether the flows may be limited by the Zone (e.g. Zone 2D) supporting the face slab allowing for potential for suffusion, GBE or cracking of that Zone as described earlier.

The leakage flow estimates can then be compared to the ability of the downstream rockfill zone to pass the leakage flows without unravelling or slope instability as described in Section 6.7.

16.3.6.1 Leakage Through Cracks with No Downstream Zone to Limit Flows

For leakage in cracks and joint opening which have no limitation by the downstream zone, e.g. where coarse rockfill supports the face slab or a gap has formed beneath the slab, the following method can be used.

As shown next, the leakage rate is determined by the Darcy–Weisbach friction factor which is related to the surface roughness of the crack in the face slab or the open joint. Cracks in concrete are rough because the sides of the crack will be formed in sand and gravel aggregates in the concrete. Joints, including vertical and perimetric joints, are formed as the slab is constructed and are relatively smooth.

Figure 16.30 shows the relationship between the Darcy–Weisbach friction factor (*f*) and Reynolds number (R) for ratios of (concrete surface roughness height, Ks) / (crack width, b) between 0.0001 and 2.0, and for smooth joints. Also shown are the Wahl and Mortensen (2024) data discussed later, which are for rough joints in cracked concrete.

This figure has been developed by Dr W. Peirson, who provided the following information.

Darcy friction factors *f* as a function of crack Reynolds number *R* presented in the Moody diagram format (see Wahl and Mortensen, 2024, Figure 3, *J. Hydraul. Eng.*, 151(1): 06024009) where *b* is the crack width, *U* is the mean velocity in the crack, *ν* is the water kinematic viscosity, *ρ* is the water density and *τ* is the wall stress. Wahl and Mortensen's measurements for cracked concrete are shown as a thick grey line. The short dashed line is the classic Lamb (1945, *Hydrodynamics*. Dover. ISBN 9780521458689) crack analytical result.

The theoretical curves are derived from Schlichting and Gersten (2000, *Boundary-Layer Theory*, Springer, ISBN 3-540-66270-7), Chapter 17, applicable to cracks. The Peirson and Fell, 2024

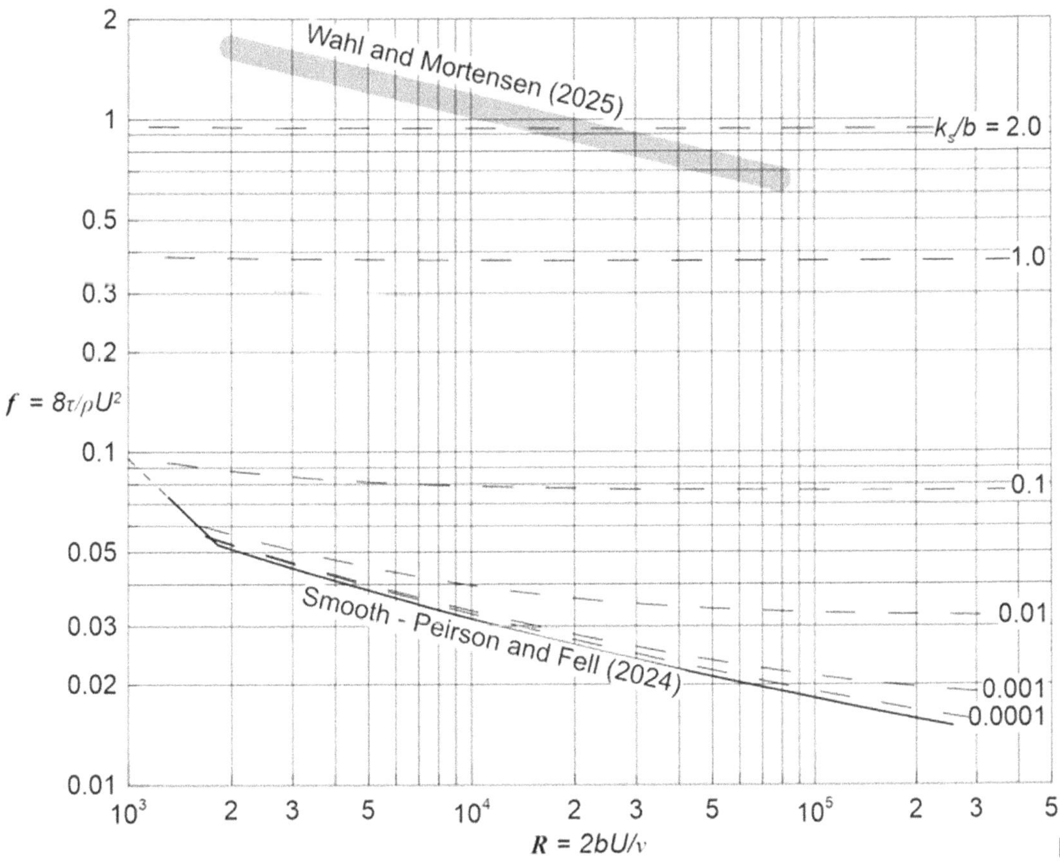

FIGURE 16.30 Darcy friction factors (*f*) as a function of Reynolds number (R) for ratios of surface roughness (K_s) / crack width (b) from 0.0001 to 1.0 (Peirson, *pers. comm.* 2025).

(*Euro. J. Env. Civil Eng.* DOI 10.1080/19648189.2024.2428253) smooth crack result is shown as a solid line. Families of curves for equivalent wall roughness height k_s are shown as long dashed lines. Note that k_s is not necessarily the height of the wall roughness but a measure of their effect on the flow equivalent to a flat surface covered in sand grains. Concrete is conventionally regarded as having a characteristic roughness between 0.3 and 3 mm, depending on the quality of finish (Peirson, *pers. comm.* 2025).

Flow in Cracks in the Face Slab

The method described by Wahl and Mortensen (2024) may be used. In summary this is as follows:

Estimate the average flow velocity in the crack using Equation 16.1:

$$\frac{P}{\gamma}+z=\frac{V^2}{2g}+f\frac{L}{4R_h}\frac{V^2}{2g} \tag{16.1}$$

where:

P/γ = pressure head at the level of the crack (m)
z = thickness of the face slab at the crack (m)
V = average velocity of flow through the crack (m/sec)
f = Darcy–Weisbach friction factor
L = flow path length through the crack = z (m)
R_h = hydraulic radius = cross section area divided by its wetted perimeter = z (m)
g = acceleration due to gravity = 9.81 m^2/sec.

Table 16.10 is a summary of Darcy–Weisbach *f* values estimated by Wahl and Mortensen (2024) for the tests carried out by Kanitkar et al. (2011). These may be used as an initial estimate in Equation 16.1. These are also shown in Figure 16.30.

From this initial estimate of flow velocity V, calculate the Reynolds number using Equation 16.2 and re-estimate the Darcy–Weisbach *f* value using Equation 16.3. Use this to re-estimate V from Equation 16.1. Repeat until convergence is achieved.

$$R = 4R\ V / \nu \tag{16.2}$$

Where ν = kinematic viscosity of water = 1.0×10^{-6} m^2/sec at 20° C

TABLE 16.10 Summary of Darcy–Weisbach Friction Factors Estimated by Wahl and Mortensen (2024) for Laboratory Tests Carried Out by Kanitkar et al. (2011)

CRACK WIDTH (MM)	*DARCY–WEISBACH FRICTION FACTOR f*	
	RANGE	*MEDIAN*
1	1.35–2.0	1.8
2	0.95–1.13	1.05
3	0.85–1.0	0.95
5	0.80–0.97	0.85
6.5	0.69–0.90	0.80
10	0.68–0.72	0.70

$$f = 11R^{-0.25} \text{ (for rough joints)} \quad (16.3)$$

$$\text{The total flow rate (m}^3\text{/sec) through the crack} = V\ z\ CL \quad (16.4)$$

Where CL = crack length (m).

The authors have developed Figure 16.31 for cracks with rough walls using the Wahl and Mortensen (2024) data.

Flow in Open Vertical and Perimetric Joints

For flow in open vertical and perimetric joints the same equations apply except that instead of Equation 16.3 the Darcy–Weisbach friction factor will be smaller because the joints are relatively smooth and may be estimated from Figure 16.30.

The authors have developed Figure 16.32 assuming open joints with smooth side walls. The assumptions are:

- No downstream flow control exists.
- Entrance and exit losses ignored.
- Sides of the joint are assumed to be "hydraulically smooth".
- Friction factor as shown for smooth surfaces which are equivalent to $f = 0.32R^{-0.25}$ where R is the Reynolds number.

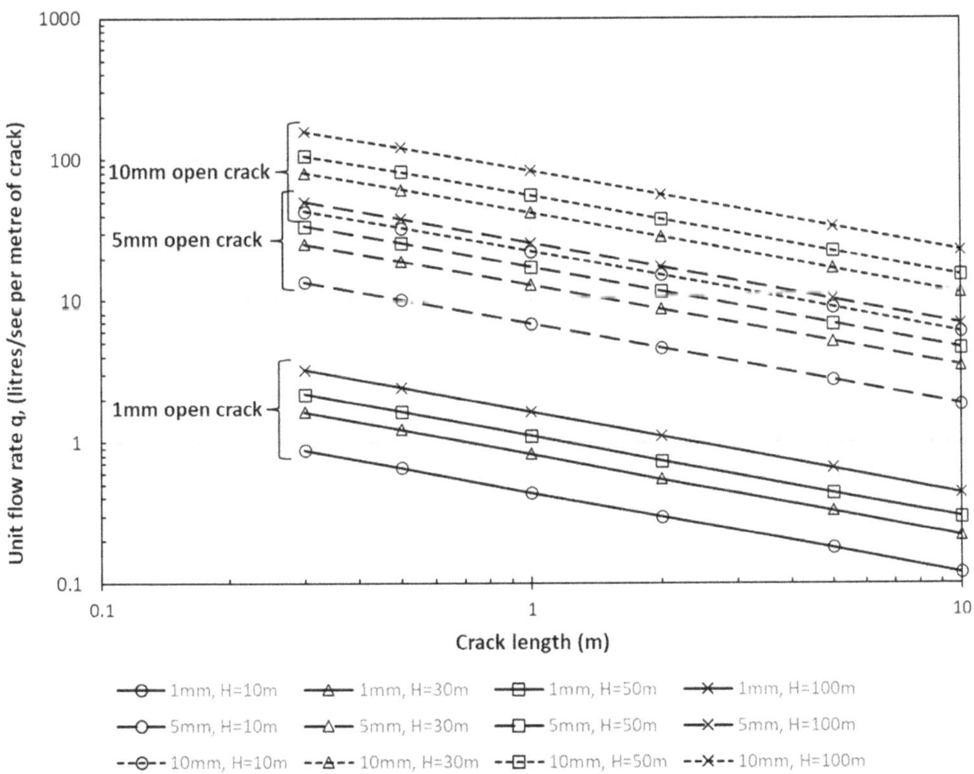

FIGURE 16.31 Estimated unit flow rate (litres/sec per metre length of crack) through rough open cracks, 1 mm, 5 mm and 10 mm opening width.

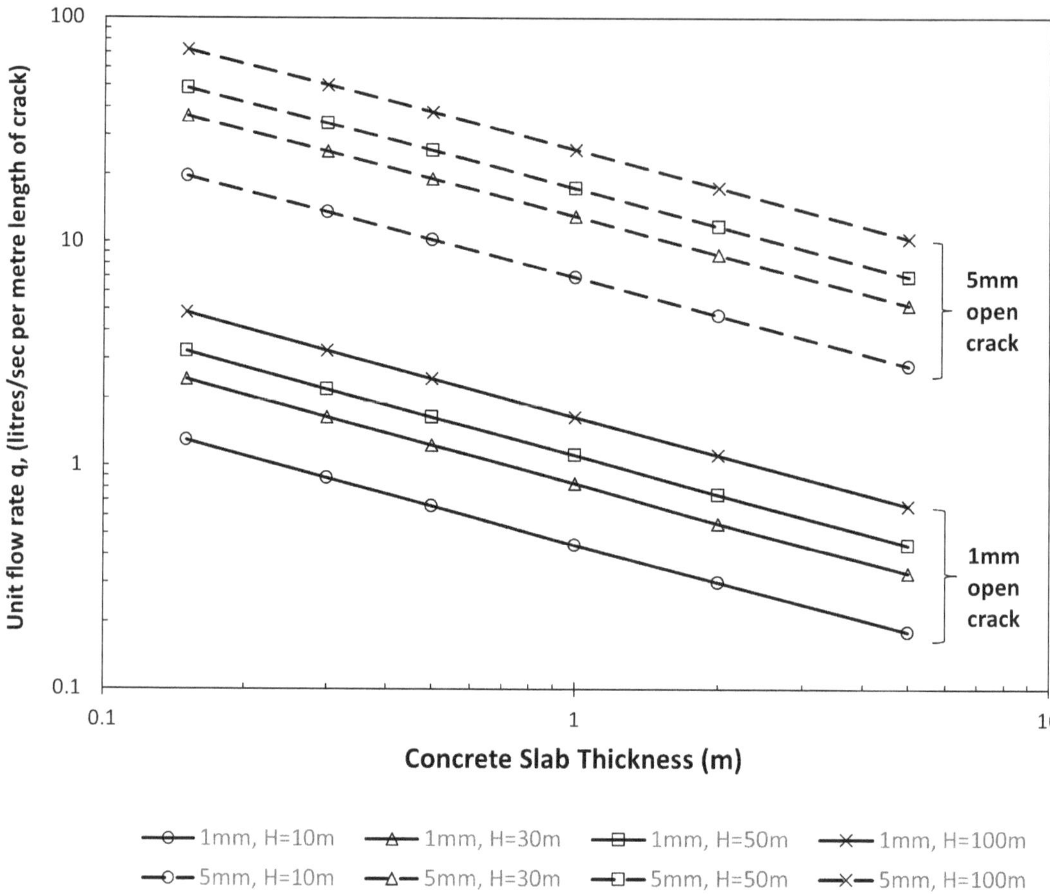

FIGURE 16.32 (a) Estimated unit flow rate (litres/sec per metre length of crack) through smooth open joints, 1 mm and 5 mm opening width.

Comparing these results with those for rough open cracks calculated as described earlier indicates the estimated flow rate for smooth gaps is three to eight times higher for narrow gaps (1–5 mm opening) and 1.5 to 5 times higher for wide gaps (10–25 mm opening).

16.3.6.2 Leakage through Cracks with Flows Limited by a Downstream Zone, e.g., by Zone 2D Designed According to ICOLD (2010)

Sherard (1985) estimated flows in cracks in CFRD using flow nets assuming that all the head loss was in the zone supporting the face slab. Table 16.11 summarises the results.

This method gives significantly higher estimated flows than those in Figure 16.31 for high permeability supporting zones e.g. $k = 10^{-1}$ m/sec and 10^{-2} m/sec. In these situations, head losses in the crack would control flows. If in doubt as to whether the downstream zone will limit flows, estimate the flows with and without downstream control and adopt the lower of the estimates.

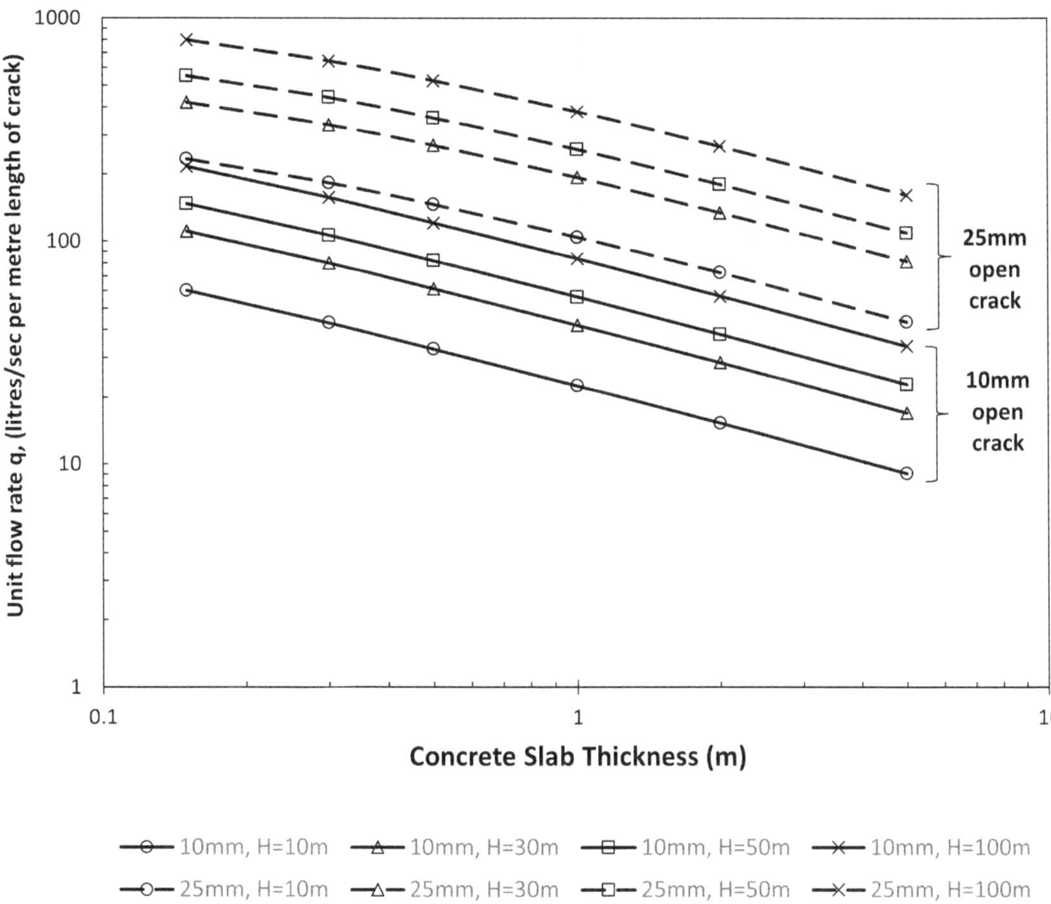

FIGURE 16.32 (b) Estimated unit flow rate (litres/sec per metre length of crack) through smooth open joints, 10 mm and 25 mm opening width.

TABLE 16.11 Estimated Flow Rate for Cracks with Flows Controlled by the Downstream Zone (Sherard, 1985)

CRACK WIDTH (MM)	*FLOW RATE M^3/SEC/M CRACK LENGTH*
1	0.33 kH
10	0.44 kH
100	0.64 kH

LEGEND:
k = permeability of zone supporting the face slab (m/sec). H = pressure head at the crack (m).

16.4 ASPHALT FACE DAMS (AFD)

16.4.1 Zoning and Design Features

The term "asphalt" has been used for consistency with asphalt core dams. ICOLD uses the term "bituminous concrete".

The following is a summary of what is described in ICOLD (1999) *Bulletin 114, Embankment Dams with Bituminous Concrete Facing.*

Figures 16.33 and 16.34 show the components of an asphalt face dam (AFD) and types of facing systems.

16.4.1.1 Facing System

Type A Double Lining Facing

The upper layer of asphalt is the actual water barrier.

The drainage layer collects leakage through the upper asphalt layer and takes the leakage to where it can be monitored and discharged without build-up of pressure and potential saturation of the embankment. See Figure 16.35 for an example.

The second asphalt layer which is beneath the drainage layer acts to limit the leakage into the embankment from the drainage layer.

The binder layer is created by spraying bitumen onto the upstream face of the embankment.

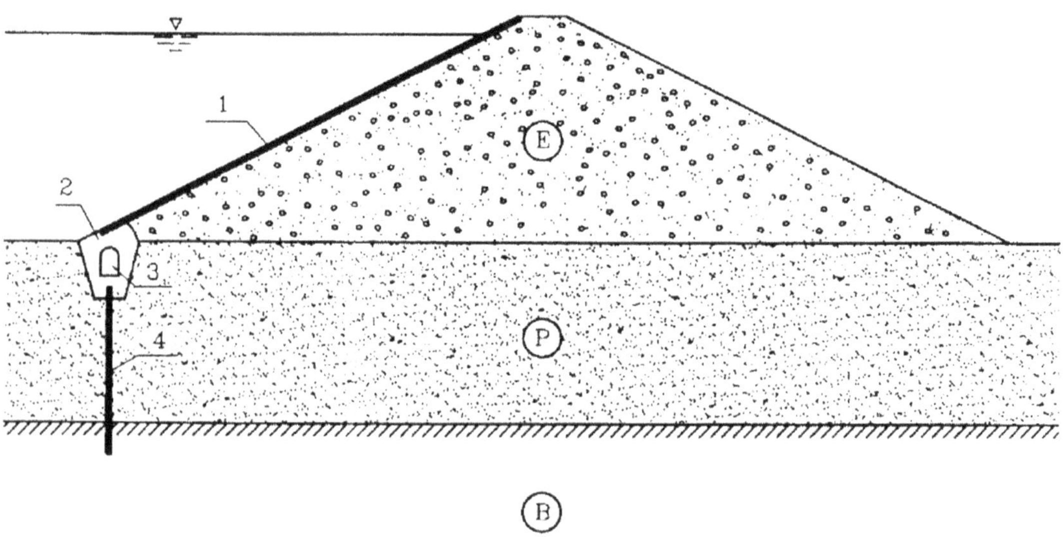

LEGEND:

E Embankment.
B Bedrock.
P Pervious foundation soil.
1 Facing system.
2 Plinth.
3 Gallery (optional).
4 Diaphragm cutoff wall.

FIGURE 16.33 Embankment dam with upstream asphalt facing founded on a soil foundation (ICOLD, 1999).

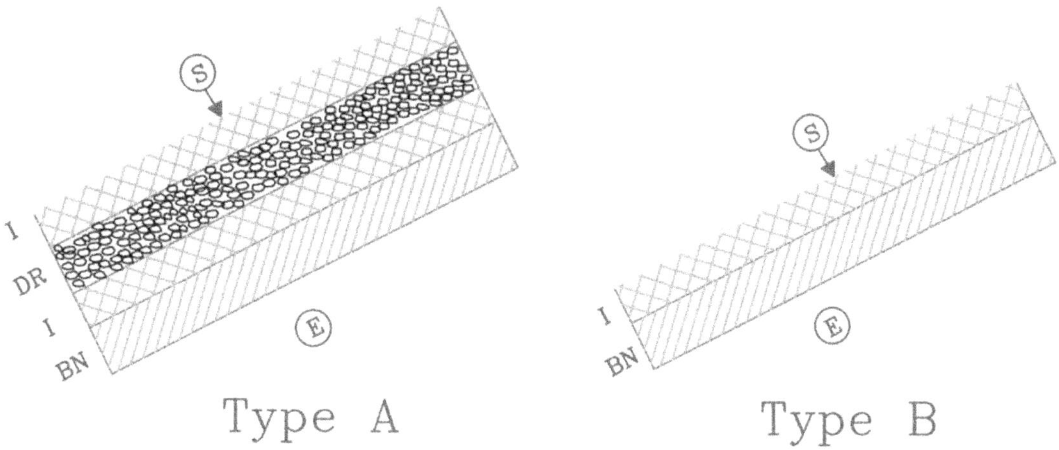

LEGEND:

A Double lining facing.	I Impervious asphalt layer.	E Embankment.
B Single lining facing.	DR Drainage layer.	S Seal coat.
	BN Binder layer.	

FIGURE 16.34 Types of facing AFD system (ICOLD, 1999).

Type B Single Lining Facing

There is no drainage layer so no ability to collect and monitor leakage through the asphalt face.

ICOLD (1999) indicates that the asphalt face layer is commonly 60 mm to 80 mm thick but may be up to 150 mm thick for high dams.

16.4.1.2 Asphalt Layer

See ICOLD (1999) for details of the design and construction of the asphalt layers.

16.4.1.3 Drainage Layer

The drainage layer is a porous bituminous aggregate mix with 2% to 5% bitumen content. The aggregate is from 5 mm to 8 mm size. The drainage layer thickness is commonly 50 mm to 150 mm. ICOLD (1999) indicates 80 mm minimum thickness is desirable and that the permeability of the drainage layer should be at least 10^{-4} m/sec.

16.4.1.4 Embankment

ICOLD (1999) indicates that the preferred material is well graded free draining rockfill for the upstream shoulder. The permeability should ideally be between 10^{-4} to 10^{-6} m/sec but lower permeability earthfill may be used, provided an internal drainage system is provided.

ICOLD (1999) emphasises that it is important that the fill is well compacted, so settlements of the face are small.

ICOLD (1999) does not comment further on zoning. The importance of zoning and the ability to control pore pressures and internal erosion and piping are discussed in Section 16.4.2.

16.4.1.5 Plinth and Connection of the Face to the Plinth

The plinth is concrete which may be keyed into a trench or placed on the rock surface. The plinth may be anchored, and the rock should be grouted. Figures 16.35, 16.36 and 16.37 show examples.

As the embankment settles differential movement between the facing and the plinth occurs. ICOLD (1999) indicates that the facing should overlap the plinth, so the connecting joint is parallel rather than perpendicular to the face. The facing should be bonded to the plinth. Water stops are not preferred. The thickness of the face should be increased at the plinth.

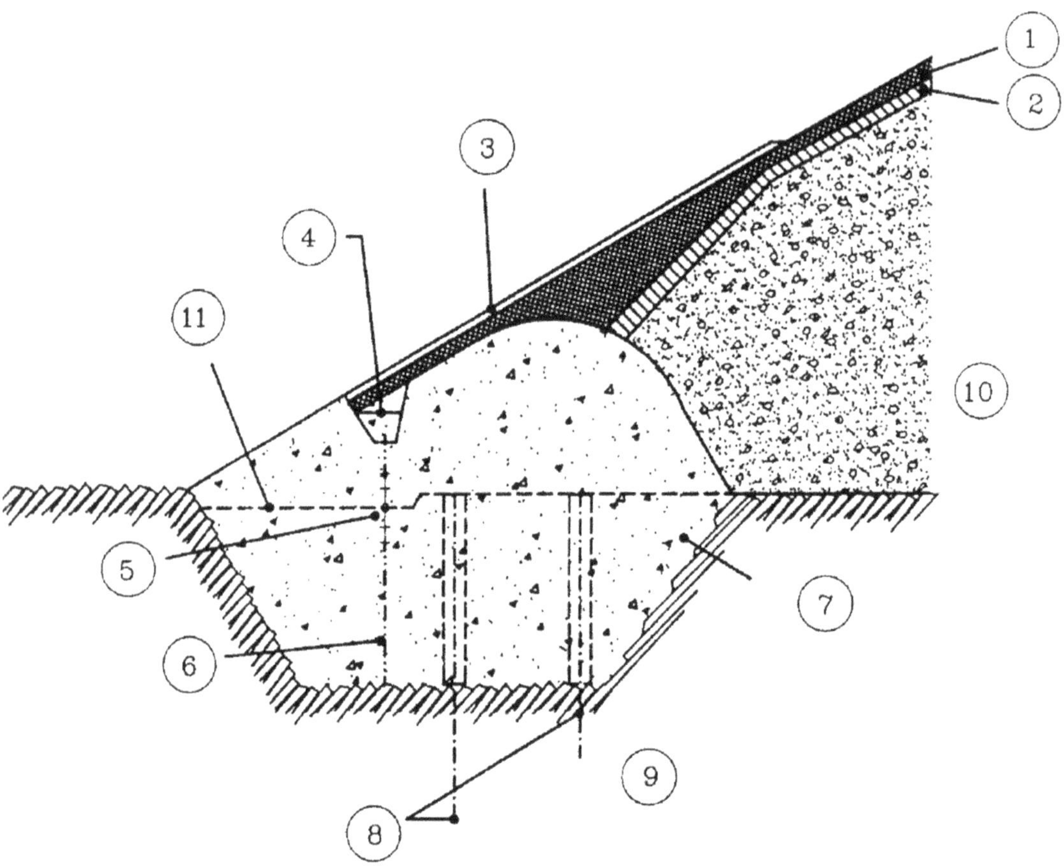

LEGEND:

1 Asphalt impervious layer.
2 Binder layer.
3 Asphalt protective layer.
4 Sand mastic bituminous filling.
5 Horizontal waterstop.
6 Vertical waterstop.
7 Plinth.
8 Grouting.
9 Rock.
10 Embankment.
11 Structural joint.

FIGURE 16.35 Example of connection of a single lining facing (Type B) to a plinth founded on rock (ICOLD, 1999).

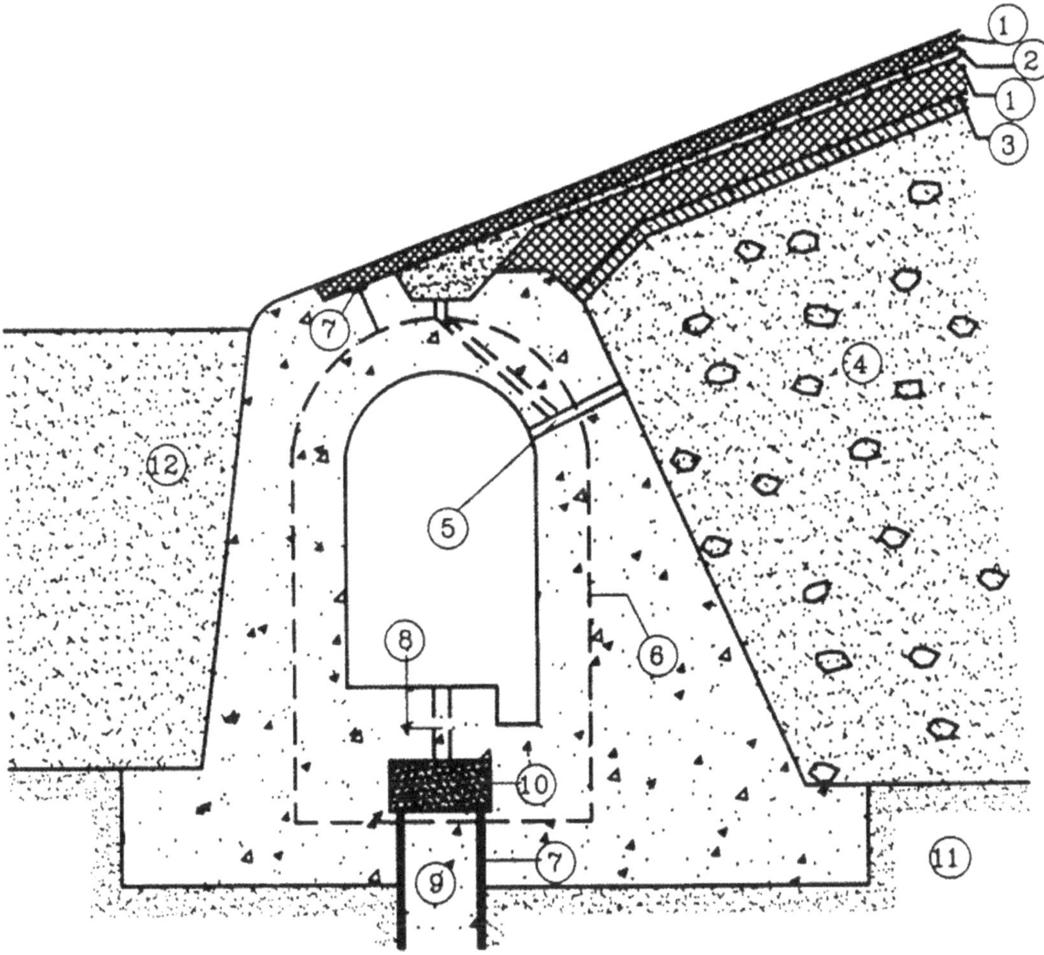

LEGEND:

1 Asphalt layer.
2 Drainage layer.
3 Binder layer.
4 Embankment.
5 Leakage collectors.
6 Waterstop.
7 Waterstop embedded in bituminous mastic.
8 Vertical drain.
9 Concrete diaphragm cutoff wall.
10 Expanded polystyrene.
11 Soil foundation.
12 Earthfill.

FIGURE 16.36 Example of connection of a double lining (Type A) face to a plinth gallery founded on pervious soil (ICOLD, 1999).

16.4.1.6 Connection to Ancillary Structures

Connection of the face to a rigid structure is a point of potential cracking due to settlement of the face. The design should be shaped so that hydrostatic forces will tend to close the joint. Figure 16.38 shows an example.

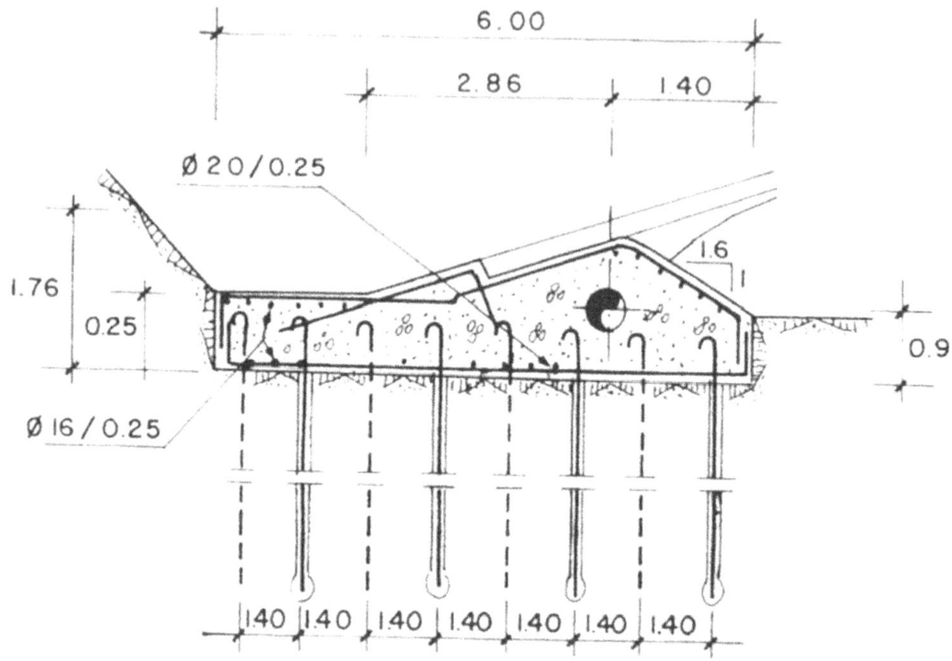

FIGURE 16.37 Slab type plinth (Justo, 1990). (Reproduced with permission of the Earthquake Engineering Research Institute.)

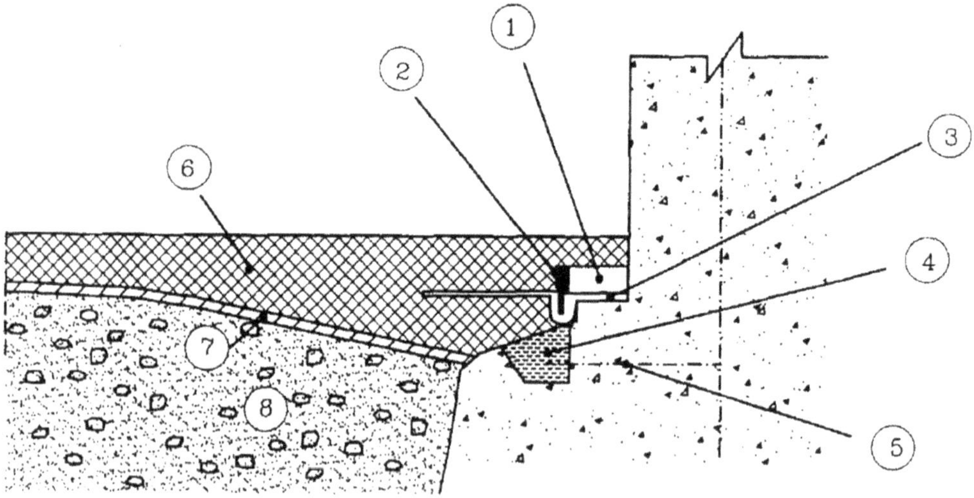

LEGEND:

1 Concrete block glued-in.
2 Bituminous sealing compound.
3 Copper foil with unfilled loop.
4 Bituminous mastic.
5 Waterstop.
6 Asphalt facing.
7 Bituminous binder.
8 Embankment.

FIGURE 16.38 Example of connection of the face to a rigid structure (ICOLD, 1999).

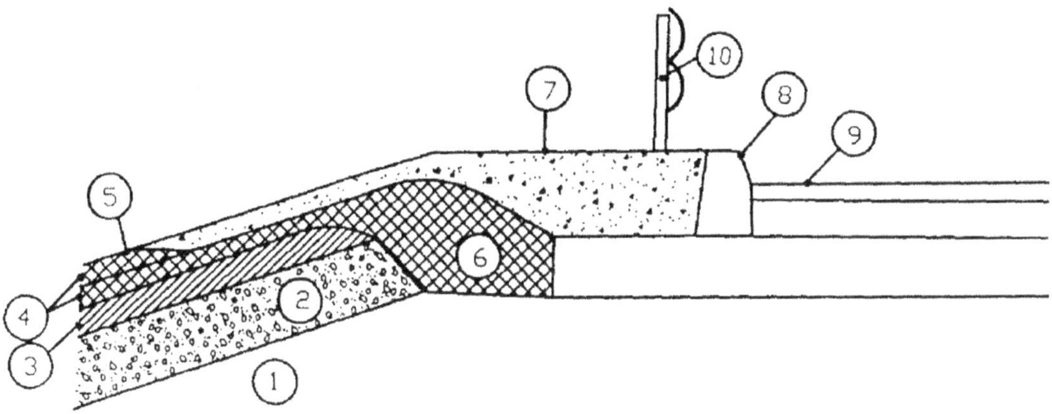

LEGEND:

1 Embankment.
2 Subgrade.
3 Binder layer.
4 Asphalt in two courses.
5 Seal coat.
6 Impervious asphalt.
7 Sidewalk/footpath.
8 Curb.
9 Roadway on crest of dam.
10 Guard rail.

FIGURE 16.39 Example of the connection of the face to the crest of the embankment (ICOLD, 1999).

16.4.1.7 Crest Detail

Figure 16.39 shows an example of the connection of the face to the crest of the embankment.

16.4.2 Design Principles

ICOLD (1999) lists the design requirements for an asphalt facing. These include:

- Low permeability throughout its lifetime.
- Water tightness of connections between the face and plinth, plinth and cutoff, and other structures such as the spillway and intake structure.
- Sufficient flexibility to tolerate without cracking the displacements resulting from the deformations of the supporting embankment including where the face joins the plinth, crest and spillway or intake structures.
- Good underdrainage of the facing system to control water pressure if leakage occurs.
- Designed to cope with wave action and seismic loads.

ICOLD (1999) also requires:

- Sound design of the dam with asphalt face should make the dam safe even in the event that the face loses its integrity.
- This requires consideration of hypotheses regarding the size of probable damage, an estimate of resulting pore pressures and quantity of flow through the embankment and the design of the embankment to withstand such conditions.

These are important criteria which should take into account the zoning of the embankment, the properties of each zone in regard to stability of the downstream slope and internal erosion and piping.

We note that these design principles and the design features discussed in Section 16.4.1 may not all have been followed in existing dams.

16.4.3 Performance of Asphalt Face Dams

Justo (1990) presents a summary of the performance of Asphalt Face Dams (AFD) up to 1991. This includes tables of data for 43 dams. Readers are encouraged to access this paper which can be downloaded from the publishers for a modest fee. The data shows:

(a) There are two what Justo (1990) terms "failures". These are:
   - Mulungushi Dam in Zambia, a 46 m high dumped friable schist rockfill, which settled 1.4 m or 3.1%. A crack formed parallel to the abutment slope, with leakage up to 7.1 m^3/sec. The dam did not breach. It was remediated by dumping clayey soil upstream of the face.
   - Scotts Peak Dam, Australia. Cracks formed during first filling more than 50 mm wide in places mainly at the contact between stiff gravel fill used for the lower 1/3 of the upstream part of the embankment, and more compressible argillite rockfill. Leakage rates were up to 125 l/sec. The dam did not breach. It was remediated by filling leaks with gravel and sand and applying butyl rubber patches on them. A silty sand gravel blanket was dumped into water to cover the affected area. Figures 16.40 and 16.41 show the details. Fitzpatrick (1976) provides details including Figure 16.42 showing the extent of cracking and the relation to the top of the gravel fill.

(b) Several dams had cracks in the asphalt face at the join with the plinth (Grosse, Licheyu, Hochwurten) and culverts (Alesani, Winscar). Major repairs were required for Grosse and Hochwurten.

(c) One had plinth itself rupture resulting in 15 l/sec (El Siberio). The plinth was not anchored and rotated under water load.

(d) Dumped rockfill and rock placed by directional blasting from the dam abutment without compaction experienced relatively large settlements 3.1%, 0.54%, 0.26%, 0.4% and 0.45%.

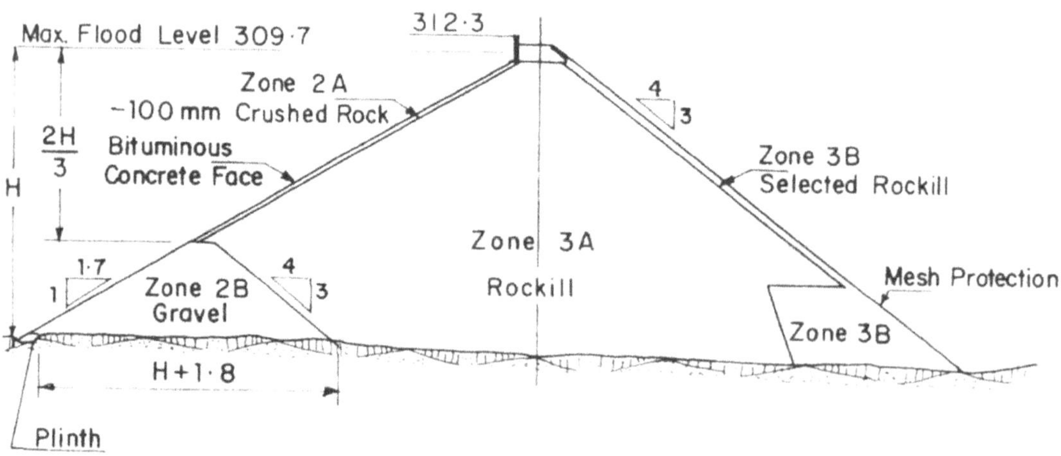

FIGURE 16.40 Scotts Peak Dam (Fitzpatrick, 1976; Cole and Fone, 1979).

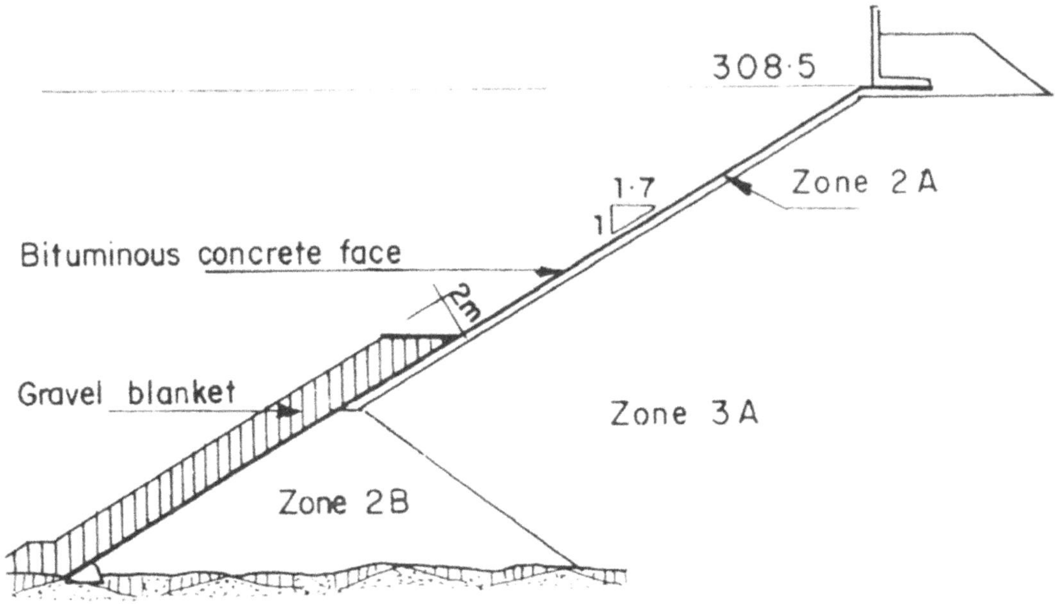

FIGURE 16.41 Scotts Peak Dam repairs (Fitzpatrick, 1976; Cole and Fone, 1979).

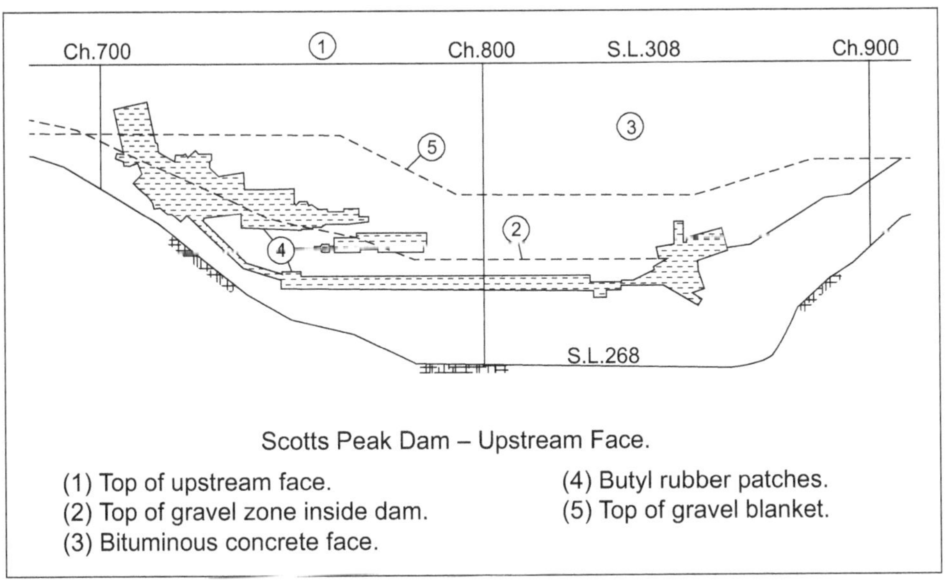

FIGURE 16.42 Scotts Peak Dam butyl rubber patching (Fitzpatrick, 1976; Cole and Fone, 1979).

(e) Dams with ≥ 0.5% settlement were:

- Ghrib, 25 l/sec, 0.54% crest settlement. Concrete slab slope protection slid down.
- Huesna, 0.3 l/sec, 0.5% crest settlement. No damage.
- Lichyu, leaked, no rate. crack at plinth, displacement up to 0.15 m. 0.5% crest settlement. Cracks at plinth, up to 150 mm relative deflection.

- Mulungushi, 7100 l/sec, 3.1% crest settlement. See earlier.
- Prims, 10 l/sec, 0.8% crest settlement. No damage.
- Scotts Peak, 125 l/sec, 0.84% crest settlement. See earlier.
- Shibianyu Leaked, no rate, 0.45% crest settlement. "Caves and cracks".
- Steinbach no leak rate, 0.78% crest settlement. No damage.

From ICOLD (1982) and ICOLD (1999), there were probably about 175 AFD when Justo (1990) was written.

Damwatch (personal communication) gathered case data on AFD which had experienced earthquake loading. This data is summarised in Table 16.12.

This is not a large database, but it indicates that cracking of the asphalt facing may occur for Pells and Fell (2002, 2003) damage class 1 and greater. The cracking is not only a result of settlement induced by the earthquake but is a result of loading on the facing during the earthquake.

16.5 ASSESSING THE SAFETY OF ASPHALT FACE DAMS

16.5.1 Introduction

Potential failure modes for AFD are as follows:

(a) Overtopping.
(b) Slope instability leading to loss of freeboard.
(c) Deformations under seismic loads leading to loss of freeboard.
(d) Development of a leak in the asphalt face leading to slope instability and loss of freeboard under flood and seismic loads.
(e) Development of a leak in the asphalt face resulting in internal erosion and piping. This applies to asphalt face earthfill embankments, not to asphalt face rockfill embankments.

16.5.2 Overtopping Failure

16.5.2.1 Deterministic Assessment

As for other embankment dams. For asphalt face earthfill embankments see Section 6.5. For asphalt face rockfill embankments see Section 6.7.

16.5.2.2 Quantitative and Simplified QRA Risk Assessment

For asphalt face earthfill embankments assess as described in Section 6.5.5.

For asphalt face rockfill embankments assess as described in Sections 6.7.4, 6.7.5 and 6.7.6.

For both, account for detection, intervention and breach as described in Section 6.10.

16.5.3 Slope Instability Leading to Loss of Freeboard

The safety of a dam from slope instability leading to loss of freeboard depends on the likelihood of leaks in the asphalt face, the face drainage system and the zoning and properties of the embankment.

TABLE 16.12 Summary of Asphalt Facing Dam Case Histories under Earthquake Loading (courtesy of Damwatch Engineering)

DAM	*COMPLETED*	*HEIGHT (M)*	*U/S SLOPE H/V*	*EARTHQUAKE MAGNITUDE MW*	*FOUNDATION PGA*	*DAMAGE CLASS*[b]	*OBSERVED CRACKING*
River Valley							
Tataragi Terada et al., (1997)	1973	64	1.8	7.2 70 km	0.07 g	0/1	None
Yashio Tsukada et al. (2013)	1992	90	2.0	9.0 300 km	0.05 g	2	Transverse at both abutments
Minamikawa Saddle Yamaguchi et al. (2012)	1987	20	2.0	9.0 182 km	0.26 g	3	Transverse up to 3 mm wide
Rectangular Shape							
Higashi-Fuji Nakamura et al. (2002)	1971	22	2.5	5.8 15 km	0.087 g	0	Cracks in reservoir floor and embankment up to 10 mm wide
Numappara	1973	38	2.5	9.0 300 km	0.21 g	3	Multiple cracks 1 to 2 mm wide
Van Norman LAD[a] Davis and Bardet (1996), Davis and Sakado (1994)	1977	47	3.5	6.7 11.3 km Uplift across site 27–30 cm	0.32 g	1	Multiple cracks compressible up to 25 mm, shear 6 to 13 mm wide
Van Norman North Dike[a]		36	3.0			1	Multiple cracks and transverse crack extending full height of downstream and upstream slope, up to 1.2 m deep in test pits

Notes:
(a) Van Norman Dams are two parts of a rectangular reservoir. The embankments are earthfill with embankment filter drain. Cracking extended from crest to at least reservoir level which was 7 m below the crest.
(b) Damage class from Pells and Fell (2002, 2003) based on earthquake magnitude and PGA at the dam foundation. See Section 8.3.2.

If a crack forms in the asphalt facing and there is no drainage system or the drainage system cannot discharge the resulting leakage flow, full reservoir pressure will potentially occur on the upstream slope beneath the asphalt.

16.5.3.1 Embankments Constructed of Free Draining Rockfill

Rockfill embankments will be able to discharge some leakage flows with little or no build-up of pore pressures. However, unless there is the equivalent of Zone 2D for CFRD, the leakage flow may increase as the finer fraction of the rockfill is eroded resulting in additional settlement and opening of cracks in the facing and increasing leakage flows.

For these dams assess the likelihood, location and leakage flow which may occur as described later and assess the likclihood of failure as described for CFRD in Section 16.3.5.

16.5.3.2 Embankments Constructed of Non-Plastic Silt Sand Gravel

Leaks in the face of embankments constructed of sand and gravel, sand or silty sand will lead to build-up of pore pressures in the embankment potentially leading to slope instability. How high the pore pressures are depends on zoning of the embankment, permeability of the zones and size and location of the leak(s).

Silt sand gravel materials may be subject to suffusion or GBE. This should be assessed as described in Section 10.6, and the resulting permeability used in the seepage analyses. If a significant amount of the finer fraction erodes this may result in settlement which may result in widening of cracks in the facing and increased leakage.

16.5.3.3 Embankments Constructed of Plastic Soils

Leaks in the face of embankments constructed of clay, sandy clay, clayey sand or clayey gravel should be limited by the low permeability of the soil but may result in increased pore pressures in the embankment. These will depend on zoning and whether drainage zones are provided, such as embankment and horizontal filter drains.

For these soil embankments assess the safety as follows:

(a) Assess the likelihood of a crack in the face as described in Section 16.5.6.
(b) Assess the location, width and length of cracks.
(c) Estimate the leakage flows in the crack(s) as described in Section 16.3.6.
(d) Develop flow nets of the embankment and foundation to model pore pressures in the embankment resulting from the leak(s).
(e) Analyse stability based on the embankment zoning, pore pressures and shear strengths.

Based on these analyses, assess the safety of the dam as described in Section 7.14 for deterministic and risk-based assessments.

16.5.4 Deformations under Seismic Loads Leading to Loss of Freeboard

This considers deformations induced by an earthquake. It does not include potential for slope instability post earthquake if the face slab or joints are damaged.

Assess as described in Section 8.12 for deterministic and risk-based assessments.

16.5.5 Failure Resulting from Internal Erosion and Piping

Given the asphalt face is cracked, the embankment is potentially subject to failure by internal erosion and piping. This should be assessed by the following:

(a) Estimate the annual probability of cracking (P_{CA}) and the location, width and length of cracking in the asphalt face as described in Section 16.5.6.
(b) Estimate the leakage which may result from the cracking using the method described in Section 16.3.6.
(c) Estimate the probability that given cracking and resulting leakage, the face drainage system will not be able to collect all the resulting seepage (P_{NCC}).
(d) Assume that as a result full reservoir pressure is applied to the upstream face of the embankment beneath the asphalt facing.
(e) Determine which PFM of internal erosion and piping apply and use the methods in Chapters 9, 12, 13 and 15 to estimate the conditional probability of failure (P_{Piping}). Make allowance for upstream flow limitation that the asphalt face may provide.

Do this for each partition of flood and seismic loading.

$$\text{The annual probability of failure} = \Sigma\, P_{CA} \times P_{NCC} \times P_{Piping} \qquad (16.5)$$

For deterministic assessment use the IDF and EDGM applicable to the dam.

Assess the potential for internal erosion and piping in the foundation as detailed in Chapters 9 12, 13 and 15.

16.5.6 Assessment of the Likelihood of Cracks in the Asphalt Face

16.5.6.1 Normal Operating and Flood Loading

From ICOLD (1999) and Justo (1990), it is apparent that the likelihood of cracking is related to post-construction settlement, details of the join between the face and the plinth and with culverts, spillway walls etc. Type A facing systems should be less susceptible to through-going cracks than Type B.

The authors are not aware of any detailed database of AFD failures, so it is difficult to quantify estimates of the likelihood of cracking. Tables 16.13 and 16.14 provide some guidance to engineering judgement and expert elicitation.

Other factors to consider include:

(a) The drainage layer in Type A face has limited discharge capacity. The main benefit is to collect leakage so it can be detected and intervention initiated.
(b) Asphalt facing is potentially inherently more susceptible to cracking than CFRD, as there are no joints or reinforcement and it relies on creep properties to accommodate strains.
(c) Unlike CFRD, designs mostly do not include the equivalent of Zone 2D to limit leakage.

16.5.6.2 Cracking under Seismic Loading

ICOLD (1999) does not discuss the effects of seismic loads on AFD. It is suggested that the information in Table 16.12 may be used as a guide to the likelihood of cracking under seismic loads.

TABLE 16.13 Factors Affecting the Likelihood of the Presence of Open Cracks and Leakage in the Asphalt Face of AFD

		LIKELIHOOD FACTOR (LF)			
FACTOR	*RELATIVE IMPORTANCE FACTOR (RF)*	*LESS LIKELY (1)*	*NEUTRAL (2)*	*MORE LIKELY (3)*	*MUCH MORE LIKELY (4)*
Post-construction crest settlement[(a)] or Embankment material and degree of compaction	(3)	< 0.2% Well compacted Sandy Gravel Earthfill ≥ 98% SMDD	0.2 to 0.5% Compacted Rockfill Earthfill 95% SMDD	0.5 to 1.0% Dumped Rockfill Different stiffness fill[(f)] Earthfill 93% SMDD	> 1.0% Dumped Rockfill subject to collapse settlement Earthfill ≤ 90% SMDD
Join of asphalt face to plinth and associated structures	(2)	As for ICOLD (1999)[(b)]	Joins parallel to face, fair details	Joins normal to face, fair details	Joins normal to face, poor details
Asphalt lining type[(c)]	(2)	Type A with thick high permeability drainage layer	Type A with thin low permeability drainage layer	Type B Thickness 60 to 80 mm	Type B Thickness ≤ 50 mm
Leakage[(d)]	(2)	Measured with gallery or other reliable system. Very small or no leakage.	Measured Small rate, consistent with minor cracking	Not measured or inadequate or unreliable measuring system, irregular leakage rates.	Measured High rate, consistent with open crack(s)[(e)]

Notes:
(a) First 10 years.
(b) See description in Section 16.4.1.
(c) See Figure 16.34 for description of lining types.
(d) Those carrying out the safety assessment to use their experience and judgement. As a guide 5 l/sec is a small rate, a high rate is ≥ 25 l/sec.
(e) Probability of cracking = 1 for this condition.
(f) As in Scotts Peak Dam.

TABLE 16.14 Likelihood of the Presence of Open Cracks and Leakage in the Asphalt Face versus Σ (Relative importance Factor [RF]) × (Likelihood Factor [LF])

NEGLIGIBLE	*VERY LOW*	*LOW*	*MEDIUM*	*HIGH*	*VERY HIGH*
9	**12**	**18**	**24**	**30**	**36 Σ (RF × LF)**

Numerical modelling of the performance of AFD under seismic loading is an active research topic, e.g. ETH Faculty of Geomechanics and Geosystems Engineering have a current (2025) project, details of which are on the internet.

For important decisions on safety of AFD, it may be necessary to consider using up-to-date numerical modelling.

Assessing the Safety of Concrete and Asphaltic Concrete Core Embankment Dams

17

17.1 CONCRETE CORE EARTHFILL DAMS (CCED)

CCED are typically earthfill dams with a concrete core constructed ahead of placing the earthfill. As shown in Figure 17.1, the concrete core was usually excavated into the foundation and keyed into rock.

ANCOLD (2000) indicates that in Australia these dams were constructed in the years from 1895 to 1939.

Davidson and Hansen (2015) indicate that the earliest CCED in the USA was constructed in 1865, but most USA CCED were constructed from 1890 to 1931.

The concrete core was usually tapered with increasing width with depth below the crest. For one 12 m high Australian dam the concrete core was only 0.2 m to 0.5 m wide, but more typically they were more like Hume Embankment No. 1 for which the concrete core was tapered from 0.6 m wide at the top to 1.8 m at the base. Hume No. 1 Embankment is 39 m high.

The concrete core usually has either nominal or no reinforcement.

A number of CCED in Australia were constructed with a puddle clay zone upstream of the core. Puddle clay is clay compacted at a moisture content well above the plastic limit and compacted by the workers with shovels or with rammers to knead the puddle clay, or in one case, by sheep's feet!

The puddle clay acted with the concrete core to control seepage.

In later CCED such as Hume No. 1 Embankment shown in Figure 17.2, a "selected clay" zone was provided upstream and downstream of the concrete core. This was more plastic soil than the shoulders and placed at a higher moisture content and in thinner layers.

These dams were constructed before formal compaction of earthfill by rollers was introduced. The earthfill was typically placed by horse and dray and the only compaction was by the passage of the horses. Figure 17.3 shows an example where horse drawn scoops were used to transport and place the earthfill near the concrete core in thin layers, and placement in the outer shoulder was by tipping from rail carts.

DOI: 10.1201/9781003542025-17

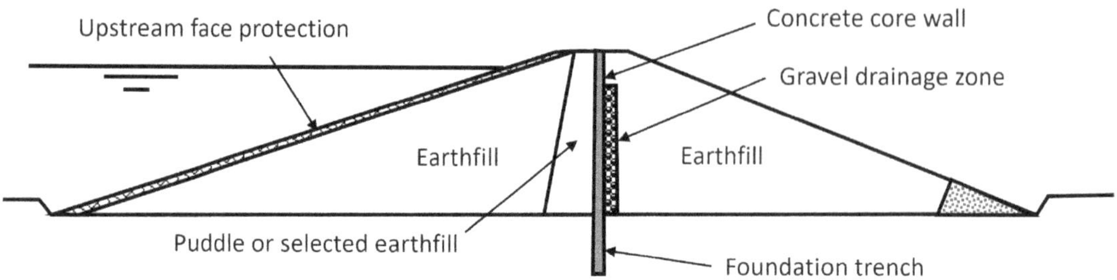

FIGURE 17.1 Typical concrete core earthfill dam.

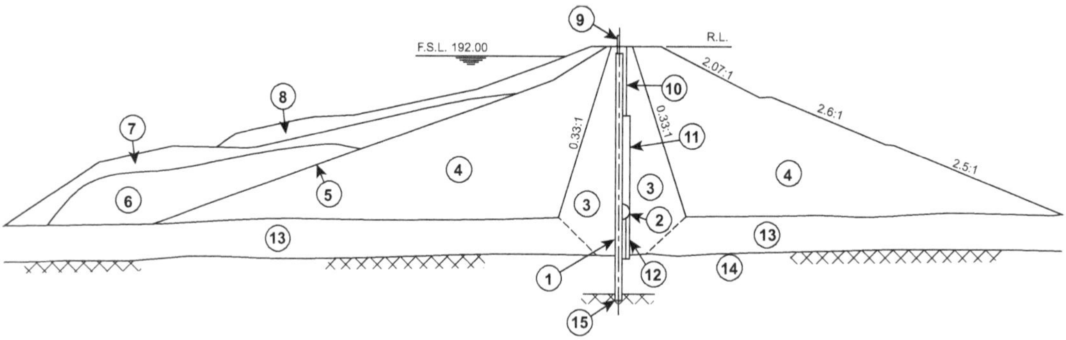

LEGEND:

1. Concrete core wall.
2. Blister Gallery.
3. Selected clayey soil.
4. Sandy clay–clayey sand.
5. Concrete slabs on original upstream face.
6. Rockfill placed in 1939 after drawdown-induced failure.
7. Rockfill placed in 1941 to improve stability.
8. Rockfill placed in 1965 for the raising of full supply level.
9. Parapet wall.
10. "U"-shaped concrete drains.
11. Gravel drainage layer.
12. Concrete buttresses under the blister gallery.
13. Alluvium.
14. Weathered granite gneiss.
15. Core wall 1.5 m into "solid granite".

FIGURE 17.2 Cross section of Hume Embankment No. 1 prior to upgrading (Cooper et al., 1997).

As described in Cooper et al. (1997), where the poorly compacted outer shoulder earthfill is saturated by seepage through the foundation and/or through cracks in the concrete core, the soil behaves as a contractive soil in shear and the available strength is the undrained strength, so significantly lower than assuming peak effective strengths apply. For Hume No. 1 Embankment, the resulting factor of safety near an abutment was approaching 1.0 under higher reservoir levels and deformations accelerated each time the reservoir rose within about 3 m of the FSL. A full height filter and berm have been constructed to achieve adequate factors of safety.

In some CCED, a gravel drainage zone was provided downstream of the core to intercept any seepage through the core. Several CCED in Hong Kong have a sand drainage zone. Figure 17.2 shows this for Hume No. 1 Embankment. For that dam the gravel drains into a gallery which was constructed as part of the core wall. For the upper part of the embankment U-shaped concrete drains were used instead of the gravel.

The concrete cores on several Australian CCED have displaced downstream with time. The top of the concrete core has moved as much as 1 m in one dam, and that resulted in cracking of the concrete core at the foundation level with seepage resulting in saturation of the downstream fill. The resulting low undrained strength of the saturated earthfill necessitated construction of a full height filter and stabilising berm.

The downstream displacement occurred mostly in the first 2 years after construction and has resulted in a wedge of the upstream embankment forming with a downwards displacement of about 0.3 m.

At Hume Embankment No. 1, the deformation caused cracking of the concrete core where the core joined the Southern Training Wall of the spillway. That junction can be seen in the distance in Figure 17.3, and Figure 17.4 shows the detail. The cracking extended about 10 m from the junction. The concrete "nib" in the tower block cracked under the load applied by the core wall.

Remedial works have included reinforcing the nib and construction of a full-height filter and stabilising berm, partly required to account for potential liquefaction of alluvium in the foundation under seismic loads. It has also necessitated construction of a controlled low strength material (CLSM), i.e. low strength; 2MPa; bentonite-cement concrete) wall and soil-bentonite cutoff walls as shown in

FIGURE 17.3 Hume No. 1 Embankment concrete core wall under construction. Rail trucks carrying earthfill to the embankment where it was side dumped and spread by horses and dray. (Photograph courtesy of MDBA and WaterNSW.)

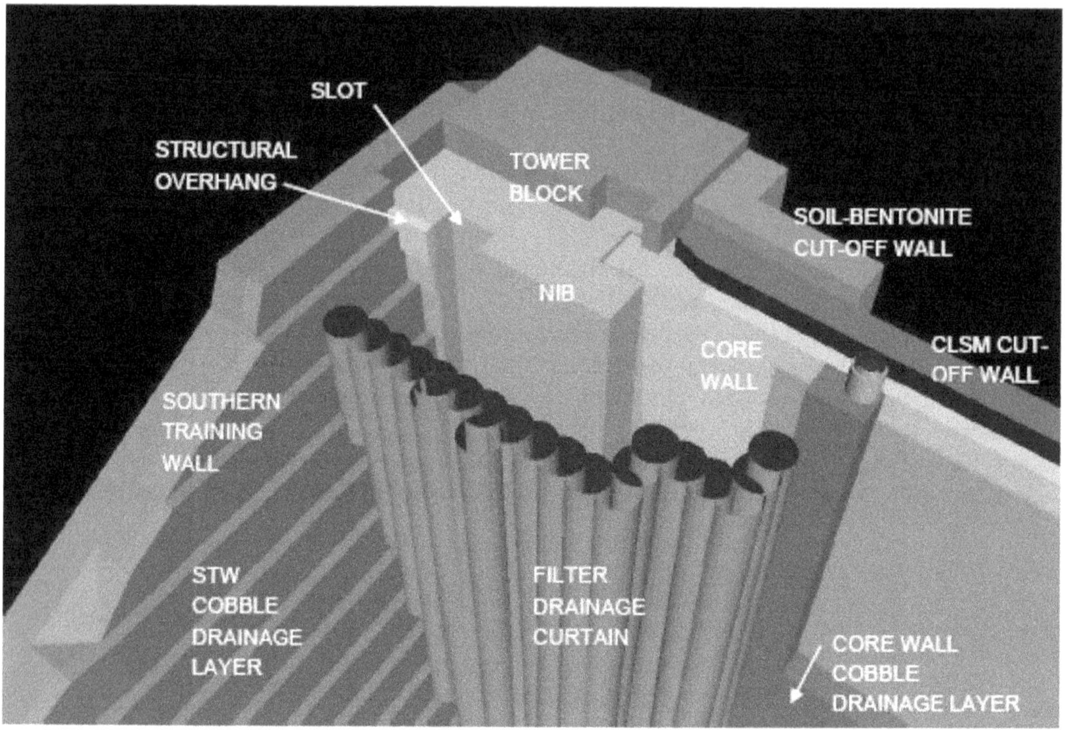

FIGURE 17.4 Hume No. 1 Embankment remedial works constructed 1996 to 2003 (Westmore et al., 2011).

Figure 17.4 to reduce the likelihood of major leak around the junction. Later, filter columns were constructed as shown in Figure 17.5 to control erosion and limit leakage flows in the event the nib fails under flood or seismic loads.

17.2 ASPHALTIC CONCRETE CORE EMBANKMENT DAMS (ACED)

ICOLD (2018) *Bulletin 179, Asphalt Concrete Cores for Embankment Dams*, Høeg (1993) and Høeg and Wang (2025) describe typical ACED dam features. Figure 17.6 shows the general layout and definition of zoning according to ICOLD (2018). Actual zoning depends on site conditions and available construction materials. ICOLD (2018) gives examples.

17.2.1 Asphalt Concrete Core

ICOLD (2018) indicates that under normal and good conditions:

Core width (m) $\geq$ 0.07 $\times$ hydraulic head (m) with a minimum of 0.50 m.

ICOLD (2018) indicates that the Norwegian Regulatory Authority adopted:

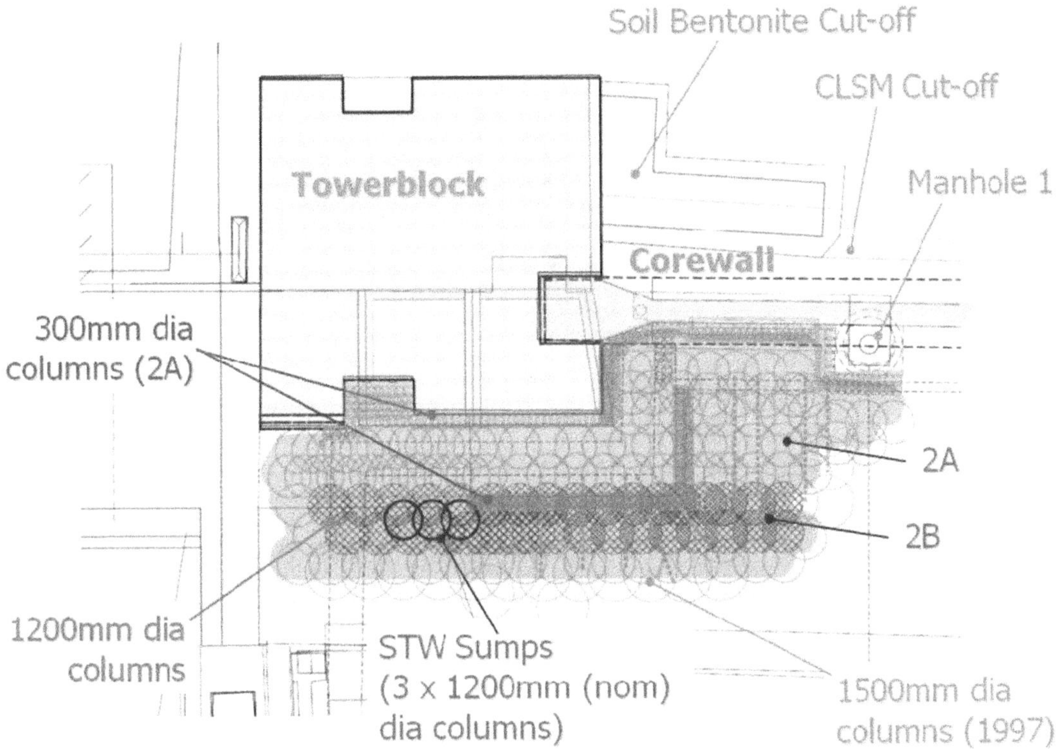

FIGURE 17.5 Hume No. 1 Embankment secant column filter and drainage system (Westmore et al., 2011).

Core width (m) = 0.50 m + [(0.7/100) × (H-50)] where H = dam height in metres. The minimum width = 0.50 m.

ICOLD (2018) indicates that:

(a) Towards the abutments and for the connection to the plinth, the core width should be increased to double core width.
(b) The core is generally vertical and centrally located but may be somewhat upstream to connect to a wave wall or parapet.
(c) In very high dams the upper part of the core may be inclined downstream as shown in Figure 17.6.

The asphaltic concrete mix is sand and gravel-sized concrete aggregates up to 16 mm or 18 mm size, a "filler" of silt-sized particles of crushed rock with or without some cement, and bitumen.

The sand and gravel aggregates are graded to give a maximum density according to a Fuller curve as shown in Figure 17.7. Høeg (1993) indicates that the bitumen content is typically 5.5% to 6.5% and is a little higher than theoretically required to fill the voids between the aggregates.

The core is compacted to air porosity by volume < 3% which gives an extremely low permeability.

Høeg (1993), ICOLD (2018) and Høeg and Wang (2025) provide details of mix design and specification, construction equipment and experiments designed to assess the ability of the asphaltic concrete to accommodate deformations without cracking.

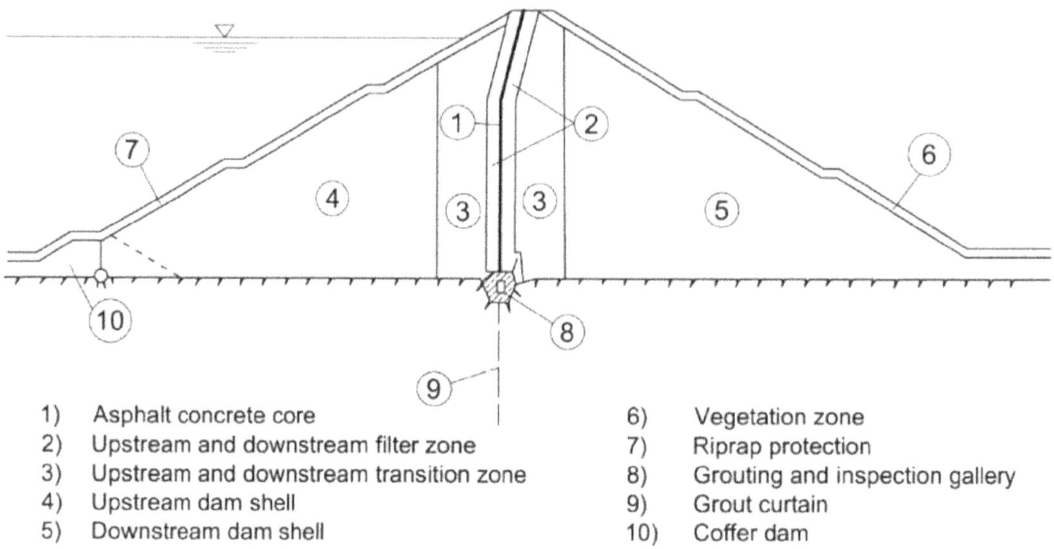

FIGURE 17.6 General layout and definition of zoning for Asphalt Concrete Embankment Dams (ICOLD, 2018).

17.2.2 Filter Zones

The filter zones provide immediate support for the core and act as filters in the event the core is cracked. The asphalt core is not susceptible to erosion in cracks, so filters are not required to arrest erosion of the core in a well designed and constructed core. If the core is cracked, the principle is that the upstream filter will erode into the crack but be prevented from eroding into the transition or rockfill by the downstream filter. The filter material that erodes into the crack will help to limit leakage.

Høeg (1993) and Høeg and Wang (2025) indicate that the filters are placed concurrent with the core and should be not less than 1.0 m wide.

ICOLD (2018) gives the following requirements for the filters:

(a) Well graded crushed aggregates or natural gravels.
(b) It is important that there is a proportion of crushed or angular gravel to give good support for the core.
(c) The grading should "preferably comply" with:

- Maximum size of 63 mm
- $D_{50F} \geq 10$ mm
- $D_{15F} \leq 10$ mm
- Percentage fines passing 0.074 mm ≤ 5%

Høeg (1993) indicates that some designers recommend the addition of fine-grained materials to the upstream filter. The principle is that in the event the core cracks, fine particles from the upstream filter are transported into the crack and limit flows until the crack heals by viscous flow.

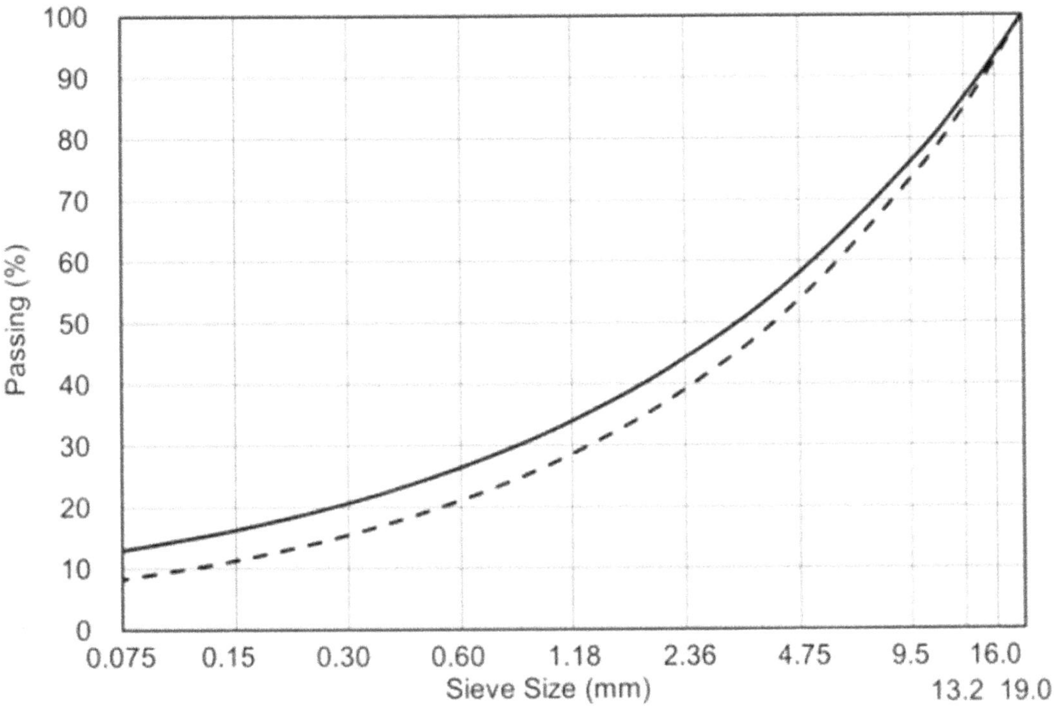

FIGURE 17.7 Fuller gradation (dashed line n = 0.45, F = 8.3%) and modified Fuller gradation (solid line, n = 0.41, F = 13%) with a maximum grain size of 19 mm (Høeg and Wang, 2025). (Copyright © 2025 Taylor & Francis Group, reprinted by permission of Informa UK Limited, trading as Taylor & Francis Ltd, https://www.tandfonline.com.)

17.2.3 Transition Zones

The transition zones are designed to transition in filtering terms and stiffness between the filter and rockfill shells. Høeg (1993) describes the material in the transition zones as crushed rock up to 200 mm size. The width may vary from relatively narrow as shown in Figure 17.7 to wider as in Figure 17.9.

ICOLD (2018) gives the following requirement for the transition:

$$D_{100F} \geq 0.25 D_{100TRANS}$$

where:

D_{100F} = the particle size for which 100% of the filter is finer

$D_{100TRANS}$ = the particle size for which 100% of the transition is finer.

Note that ICOLD (2018) sometimes uses the term "transition" to describe the filter.

17.2.4 Rockfill Shells

Høeg (1993) describes two zones, Zone 4A, quarried rock < 400 mm diameter and Zone 4B, quarried rock < 800 mm diameter. Generally good quality strong fresh or SW rock is preferred but Høeg (1993) indicates lesser quality rock has been used.

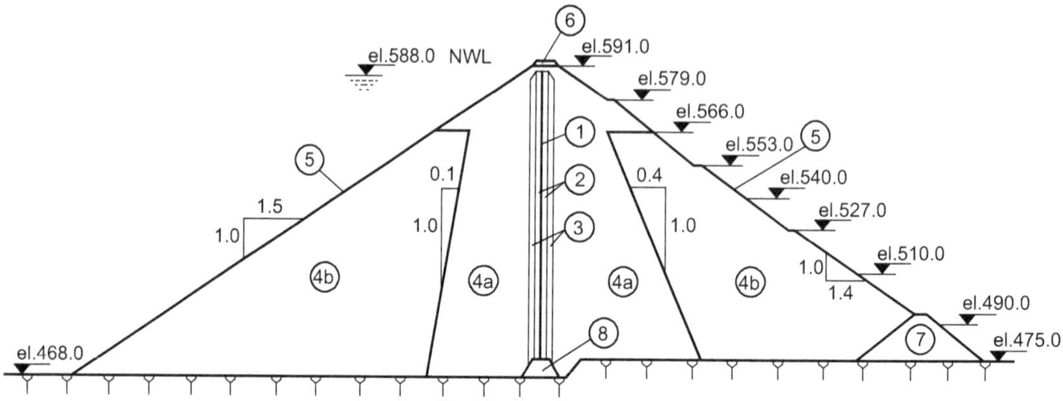

LEGEND:

Zone 1. Asphalt concrete core.
Zone 2. Filter.
Zone 3. Transition.
Zone 4a. Shoulder rockfill.
Zone 4b. Shoulder rockfill.

Zone 5. Rip-rap.
Zone 6. Rockfill crest protection.
Zone 7. Rockfill toe.
Zone 8. Concrete sill.

FIGURE 17.8 Cross section of Storglomvatn Dam showing zoning (Høeg and Wang, 2025). (Copyright © 2025 Taylor & Francis Group, reprinted by permission of Informa UK Limited, trading as Taylor & Francis Ltd, https://www.tandfonline.com.)

Figure 17.8 shows the zoning of Storglomvatn Dam, which is typical of Norwegian practice. This has a narrower transition Zone 3 and uses two rockfill zones, Zone 4a adjacent to the transition, and Zone 4b on the outer shoulders. Table 17.1 adapted from Høeg (1993) describes the materials in each zone and the compaction requirements.

Høeg and Wang (2025) stress the importance of having an especially well-compacted downstream rockfill shoulder zone (Zone 4a and 4b) to prevent the core from undergoing excessive lateral displacements during impounding and earthquake loading.

The overall approach is to have very stiff filters, transitions and shells so settlement during and post construction and deformations under reservoir load are small, reducing the likelihood of deformation and cracking of the core.

17.2.5 Plinth

Shown as "grouting and inspection gallery" in Figure 17.6 and as "concrete sill" in Figure 17.8.

The main purpose is a cap for the grouting and as the contact between the core and foundation. It is anchored to the foundation by rock bolts. ICOLD (2018) indicates that the plinth should be designed so the slope up the abutments is not steeper than 45°, but if it must be steeper, it should be inclined upstream to achieve a higher stress contact between the core and plinth.

A gallery may be incorporated into the plinth to allow grouting during or post construction if required.

Table 17.1 Description of the Embankment Zones in ACED (developed from descriptions in Høeg, 1993)

ZONE				*COMPACTION*	
NO.	*DESCRIPTION*	*MATERIAL*	*LAYER THICKNESS (M)*	*VIBRATORY ROLLER MIN. WEIGHT (TONNES)*	*NUMBER OF PASSES*
1	Asphalt concrete core	Bitumen 180 aggregate (0–18 mm) filler (< 0.075 mm)	0.2	Approx 0.25 to 0.5 (trials on site approx. 0.5 m)	Until air porosity < 3%
2	Filter	Crushed rock or natural gravel 0–60 mm	0.2	1.5	3–6
3	Transition	Crushed rock 0–200 mm	0.4	15 + water sluicing	4
4a	Shoulder rockfill shell	Quarried rock 0–400 mm	0.8	15 + water sluicing	8
4b	Shoulder rockfill shell	Quarried rock 0–800 mm	1.6	15	6
5	Rip-rap slope protection	Selected large blocks of rock > 0.5 m^3	Individually placed by backhoe	n/a	n/a

17.3 FAILURE BY OVERTOPPING

Use the methods described in Chapter 6.

A detailed assessment of the effect of the 2 m wide concrete core in a 30 m high earthfill dam in Australia showed that the concrete core has little effect on the rate of erosion or width and depth of a breach following overtopping by flood or subsequent to liquefaction in the foundation.

It is therefore suggested that the effect of concrete or asphalt concrete cores be regarded as being small when considering failure by overtopping.

17.4 FAILURE BY SLOPE INSTABILITY RESULTING FROM DEFECTS IN THE CORE UNDER NORMAL OPERATING AND FLOOD CONDITIONS

17.4.1 Concrete Core Earthfill Dams

The following is a sequence of events which may lead to failure of a CCED by slope instability:

(a) A defect forms in the concrete core or at a contact between the core and a rigid structure. Alternatively, a preferential seepage pathway exists in the foundation below the concrete core.
(b) Reservoir water flows through the defect or foundation seepage pathway.
(c) Flow through the defect in the concrete core or seepage occurring through the foundation results in saturation of earthfill downstream of the core.

(d) Under high reservoir levels, the factor of safety becomes less than 1.0.
(e) The resulting deformation further opens defects in the core, further saturating the downstream earthfill.
(f) Detection and intervention, e.g. by lowering of the reservoir, fails.
(g) Breach initiates.

When assessing the likelihood of leakage through a defect in the concrete core take account of:

- The deformation of the concrete core and resulting potential to crack.
- Details of connection between the concrete core, the foundation and rigid structures.
- The nature of the earthfill upstream of the concrete core, and whether it will limit flows through the defect in the concrete core.
- Whether drainage measures downstream of the concrete core will be sufficient to avoid saturation of the downstream earthfill.
- The nature of the fill downstream of the concrete core, e.g. whether it is plastic soil, and if so well or poorly compacted, and if non-plastic, the drainage capacity.
- Seepage and deformation monitoring.

When assessing the factor of safety take account of:

- Pore pressures downstream of the concrete core likely to result from leakage through defects in the concrete core.
- Stability analyses allowing for undrained strength behaviour in saturated poorly compacted earthfill.
- Reservoir levels and downstream tailwater levels.
- Deformation monitoring, in particular whether deformations accelerate under higher reservoir levels.

Assess the safety of the embankment as described in:

Deterministic. Section 18.5.2.
Risk Enhanced Deterministic. Section 18.5.2 and related sections in Chapter 7.
Quantitative and Semi Quantitative Risk Assessment. Section 7.14.

17.4.2 Asphaltic Concrete Core Embankment Dams

Høeg (1993) and ICOLD (2018) present information which shows the good performance of over 200 ACED, some of which are up to 50 years old. Dams up to 150 m high have been constructed. Most are on rock foundations, but some have been constructed partly on rock, partly on soil foundation.

The following sequence of events is likely to be required for an ACED to fail by slope instability:

(a) A defect forms in the asphalt core.
(b) Reservoir water flows through the defect. The upstream filter flows into the defect.
(c) The downstream filters and transition fail to arrest the erosion of the upstream filter.
(d) The defect in the core widens as a result of loss of support from the downstream filter. Flows through the defect increase.
(e) Flow through the defect in the core results in pore pressures in the downstream fill and instability or unravelling of the downstream shell.
(f) Detection and intervention, e.g. by lowering of the reservoir, fails.
(g) Breach initiates.

The sequence of events is similar to that for internal erosion and piping failure (by concentrated leak) in earth and earth and rockfill dams.

It is recommended that the safety of an existing ACED be assessed as described in Chapter 9 taking account of the following:

A defect forms in the asphalt core or at the interface of the core with the plinth resulting from:

(a) Faulty construction of the core, e.g. layers not properly overlapping because of poor construction control.
(b) Cross valley differential settlements.
(c) Differential settlement resulting from part of the foundation being more compressible than the rest, e.g. part on a soil foundation.
(d) Differential settlement adjacent to a spillway wall or other structure.
(e) Cracking or rupture of the core if it joins a rigid structure as described in Section 17.1 for the concrete core wall of Hume Dam.
(f) Arching of the asphalt core onto the stiff filters and transition resulting in hydraulic fracture. See Wang et al. (2010) and Wang and Høeg, 2010) for some observed performance. These show that the asphalt core settled more than the shoulders for the dams they describe but the measured strains in the asphalt core are compressive.
(g) Collapse settlement of not so well compacted upstream rockfill resulting in a wedge of the upper part of the dam moving upstream.
(h) Concentrated leakage and internal erosion through defects in the rock mass below the plinth (similar to the plinth of CFRD).

17.4.2.1 Likelihood of a Defect

(a) The probability of cracking may be estimated using the information in Chapter 9 because the mechanics are essentially the same as for earth and earth and rockfill dams. If the ACED is designed and constructed with well compacted zones the post-construction settlement will be small, so the resulting extensional strains will be small and the likelihood of cracking small.
    The high tensile strength of the asphalt compared to soil may be offset by the asphalt core being narrow compared to earthfill cores, so any crack is likely to be persistent through the core.
(b) Assess whether the viscous properties of the asphalt will close the crack taking account of the literature on that subject.
(c) The core of a well designed and constructed ACED is non-erodible. Narrow cracks are unlikely to widen under the forces imposed by water flowing through a crack. If however the filter and/or transition do not provide adequate support for the core, the core may be damaged, and it is possible that pieces of the core may break away resulting in widening of the crack.

17.4.2.2 Ability of Filter and Transition to Arrest Erosion

(d) Assess whether the downstream filter is a no erosion, some erosion, excessive erosion or continuing erosion filter to the upstream filter. Take account of the potential for segregation and suffusion of the filter.
(e) Assess whether the transition is a no erosion, some erosion, excessive erosion or continuing erosion filter to the downstream filter, taking account of the as-constructed gradation of the aggregates in the filter and the transition. Take account of the potential for segregation and suffusion of the transition.
(f) The ICOLD (2018) requirement for the transition is

$$D_{100F} \geq 0.25D_{100TRANS}$$

which would not necessarily satisfy no-erosion filter criteria.

17.4.2.3 Flow Through the Core Results in Instability or Unravelling of the Downstream Shell

(g) Estimate the flow rate through the defect in the core using the method described in Section 16.3.6. Allow for the potential for the downstream filter to act as a downstream control for the flow. If erosion of the asphalt core does initiate, then the size of enlargement of the defect and resulting flow will be dependent on the filter compatibility between the asphalt core and filter zone.

(h) Assess the likelihood of downstream instability or unravelling as described in Section 6.8 given the estimated flow rate and the geometry and permeability of the downstream zones.

17.4.2.4 Detection and Intervention, e.g. by Lowering of the Reservoir Fails

Assess the likelihood of successful detection and intervention as described in Section 7.14.

17.4.2.5 Summary

An ACED designed and constructed according to Høeg (1993), ICOLD (2018) and Høeg and Wang (2025) as summarised earlier will have an extremely low likelihood of failure.

A dam with less well constructed asphalt core, large post-construction settlements potentially resulting in cracking, a downstream filter and/or transition which is too coarse by design and/or as a result of segregation during construction, and downstream rockfill which is constructed of rock which breaks down during compaction to give a low permeability will have a higher likelihood of failure. As for CFRD, if lower permeability rockfill or gravel fill is used for the downstream shoulder there should be a drainage zone to control pore pressures. If absent, the likelihood of instability will be higher than for a well designed and constructed AECD.

The authors' experience in regards to other types of dams is that such shortcomings are not uncommon, and we suggest that the commentary on AECD in ICOLD (1992, 2018) on the success of AECD may be applicable to well designed and constructed dams, but it should not be assumed that all AECD are as well designed and constructed.

Zhang et al. (2016) describe leakage in an ACED. The dam is 61 m high and experienced up to 24 l/sec leakage. Seepage emerged in a 1.5 m thick zone 32 m below the dam crest. Investigations showed that there was a defect in the core, probably resulting from a change in core width. Repairs of the core reduced the leakage to 1.5 l/sec.

17.5 FAILURE BY SLOPE INSTABILITY RESULTING FROM DEFECTS IN THE CORE UNDER SEISMIC LOADS

17.5.1 Concrete Core Dams

Seismic loading may result in additional deformations and cracking of the concrete core and increase the likelihood of rupture of the connection between the concrete core and rigid structures.

Additional settlement induced by the earthquake may also result in opening of joints in the concrete core. This may be assessed using the approach described in Section 9.9 as a guide to performance.

The likelihood of failure should be assessed as described in Section 17.4.1 taking these factors into account.

17.5.2 Asphaltic Concrete Core Embankment Dams (AECD)

ICOLD (2018) discusses the effect of seismic loads on AECDs. These may be summarised as:

(a) No existing AECD has been subject to major seismic loading. Wang et al. (2010) discuss the effects of earthquake on Chinese dams, but none were close to the earthquake source.
(b) The central location of the core is favourable regarding deformations of the core which may occur.
(c) Laboratory testing by Wang and Høeg (2010) and others show the asphaltic concrete core can withstand severe seismic loads without significant material degradation and cracking.
(d) Dynamic nonlinear elasto-plastic numerical analyses can be used to estimate the response of an AECD to seismic loads.
(e) The largest dynamic amplifications occur in the upper 20% of the dam and cracking of the core may result.

The current authors note that the emphasis of these analyses is for cross section analyses. We believe that the potential for transverse cracking resulting from cross-valley differential settlements should be considered and may be a larger contributor to risk than cracking resulting directly from cross section deformations.

We recommend that the likelihood of cracking be assessed as for other embankment dams as described in Section 9.9 allowing for the presence of the core.

The likelihood of failure resulting from this PFM should be assessed as described in Section 17.4.2.

ICOLD (2018) describes design measures which may be taken if analyses indicate that the upper part of the core may experience excessive deformations and cracking under the safety evaluation earthquake:

- Use high quality, well graded, well compacted rockfill/gravel in the top part of the dam.
- Flatten the exterior slopes of the top part of the dam.
- Use horizontal reinforcement, e.g. geogrids in the top part of the dam.
- Increase the core width to allow for deeper tension cracks or larger shear displacements.
- Use an asphalt concrete mix with softer bitumen type and/or higher bitumen content to increase the core flexibility and ductility.
- Design the downstream filter as a wide filter to arrest any erosion from the upstream filter zone through the cracks so the eroded material in the upstream filter acts as a crack stopper.

An existing dam which has most or all of these measures will be less likely to fail or experience an incident under seismic loading than one with few or none of these features.

17.6 FAILURE BY INTERNAL EROSION AND PIPING

17.6.1 Deterministic

Assess as described in Section 18.5.3.

17.6.2 Quantitative and Semi-Quantitative

Assess the likelihood of failure from internal erosion and piping for both CCED and AECDs using the method described in Chapters 9, 12 and 13, taking account of:

(a) The likelihood, location and width of cracking assessed as described earlier.
(b) For CCED, whether there is common cause cracking in earthfill zones upstream and downstream of the concrete core.
(c) For CCED, whether there is a drainage system downstream of the concrete core and, if so, its drainage capacity relative to possible leakage flows.
(d) For ACED, the zoning of the embankment, in particular whether there is a filter and transition downstream of the asphalt core and, if so, how effective it will be in limiting leakage.
(e) The ability of the core to act as a flow limiter and for an ACED to self-heal.

Dam Safety Assessment

18

18.1 INTRODUCTION

Four approaches to safety assessment of dams are described in this book:

- Deterministic, described as Standards based by some.
- Risk Enhanced Deterministic.
- Quantitative Risk.
- Simplified QRA.

We have adopted the term "deterministic" rather than "standards based" because there are no internationally accepted standards for dam safety assessment for overtopping, safety under seismic loading, or internal erosion and piping.

The principles relating to each approach are provided as follows. Further discussion on application of these approaches is provided in Sections 18.5, 18.6 and 18.7.

18.1.1 Deterministic

18.1.1.1 Overtopping and Wave Overtopping

Estimate the maximum reservoir level and wave run-up for the Inflow Design Flood. Deterministic criteria are satisfied if the embankment crest is not overtopped.

18.1.1.2 Slope Instability Leading to Loss of Freeboard

Assess the factor of safety against slope instability under normal operating and flood loading. Deterministic criteria are satisfied if the factor of safety is greater than commonly accepted factors of safety.

18.1.1.3 Seismic Loading Leading to Loss of Freeboard

Embankments not subject to liquefaction: Estimate crest settlement under the Earthquake Design Ground Motion. Compare to the freeboard with the reservoir at Full Supply Level. Deterministic criteria are satisfied if the embankment crest is not overtopped.

DOI: 10.1201/9781003542025-18

Embankments subject to liquefaction: Estimate the post-earthquake factor of safety. If greater than an acceptable factor of safety, estimate crest settlement under the EDGM using the methods for embankments not subject to liquefaction. Compare to the freeboard with the reservoir at Full Supply Level. Deterministic criteria are satisfied for low and significant classification dams if the embankment crest is not overtopped.

If post-earthquake factors of safety are less than acceptable for low and significant classification dams and for all or high, very high and extreme classification dams, use QRA methods.

18.1.1.4 Internal Erosion and Piping

Assess whether the embankment zoning and the filters are sufficient to arrest erosion should it initiate in the embankment or foundation or from embankment to foundation. See Section 18.5.3 for more details.

18.1.1.5 Defensive Design Principles for Earthquake Design of Embankment Dams

Fell et al. (2015) describe the concept of "defensive design" of embankment dams for earthquake.

The general philosophy is to apply logical, commonsense measures to the design of the dam, to take account of the cracking, settlement and displacements which may occur as the result of an earthquake. These measures are at least as important (probably more so) as attempting to calculate accurately the stability during earthquake or the likely deformations. The most important measures which can be taken are as follows:

(a) Provide ample freeboard, above normal operating levels, to allow for settlement or slumping or fault movements which displace the crest.
(b) Use well designed and constructed filters downstream of the earthfill core (and correctly graded rockfill zones downstream of a concrete face for concrete face rockfill dams) to control erosion of the core (or face), should such occur in the earthquake. Filters should be taken up to the dam crest level, so they will be effective in the event of large crest settlements, which are likely to be associated with transverse cracking. For larger dams, full width filters (2.5 m to 3 m) might be adopted instead of narrower (say 1.5 m) filters placed by spreader boxes. This would give greater security in the event of large crest deformations.
(c) Provide ample drainage zones to allow for discharge of flow through possible cracks in the core. For example, ensure that at least part of the downstream zone is free draining or that extra discharge capacity is provided in the vertical and horizontal drains for an earthfill dam with such drains. In this regard some embankment dam types are inherently more earthquake resistant than others. In general, the following would be in order of decreasing resistance:
    - Concrete face rockfill.
    - Sloping upstream core earth and rockfill.
    - Central core earth and rockfill.
    - Earthfill with chimney and horizontal drains.
    - Zoned earth-earth rockfill.
    - Homogeneous earthfill.
(d) Avoid, densify, drain (to be non-saturated) or remove potentially liquefiable materials in the foundation or in the embankment. Filters, rockfill and other granular materials in the embankment should be well compacted if they are likely to become saturated, so they will be dilatant and not liquefy or build up pore pressure under cyclic loading.

(e) Avoid founding the dam on strain weakening clay soils, completely weathered rock or weak rock with the potential to strain weaken. In many older, smaller dams, this was not done. Post-earthquake stability can be an issue if the earthquake causes even relatively minor movements which can take the foundation strength from peak to a strength close to or at residual.
(f) The foundation under the core should so far as is practicable be shaped to avoid sharp changes in profile across the valley. These are likely to make the core more susceptible to cracking due to differential settlement under earthquake (and normal) loading. In practice it is seldom practicable to achieve much by slope modification, and the potential effects of cracking have to be managed by providing filters and drainage capacity as described earlier.

When assessing an existing dam, the use of these "defensive design" measures is seldom practical (except in remedial works). However, it is useful to gauge the safety of the existing dam by comparing it with this list. Where the dam fails to meet many or most of these features, particularly (a) to (e), this may be a better guide to the fact that the dam may not be very secure against earthquake than a lot of analysis.

Dams which have well designed and constructed filters, have adequate stability against normal loads, and do not have liquefiable or strain weakening zones or foundations will be able to withstand the loading from very large earthquakes.

18.1.2 Risk Enhanced Deterministic

18.1.2.1 Overtopping and Wave Overtopping

Assess as for Deterministic approach.

For Type Z1 and Z4 earthfill embankments, if the IDF plus wave run-up results in overtopping of the crest, use the database method described in Section 6.5.2 to provide additional information on the likelihood of failure to assist in making the dam safety assessment.

There is no Risk Enhanced Approach for other embankment types (Type Z3, Z5 and Z6).

18.1.2.2 Slope Instability Leading to Loss of Freeboard

Assess as for Deterministic approach.

Use the Modified Silva et al. (2008) method described in Section 7.14.2 to estimate the probability of initiation of instability and estimate the likelihood of loss of freeboard as described in Section 7.14.3. Use these to provide additional information to assist in making the assessment of dam safety.

18.1.2.3 Safety under Seismic Loading

Embankments not subject to liquefaction:

There is no Risk Enhanced Deterministic Approach.

18.1.2.4 Embankments and Foundations Subject to Liquefaction

As for Deterministic approach.

Use the Modified Silva et al. (2008) method described in Section 7.14.2 to estimate the probability of initiation of post-earthquake instability and estimate the likelihood of loss of freeboard as described in Section 7.14.3. Use these to provide additional information to assist in making the assessment of dam safety.

Internal erosion and piping:

As for Deterministic Approach.

Use the historical database screening method in Chapter 15 to assess the annual probability of failure and use this to assist in making the assessment of dam safety.

18.1.3 Quantitative Risk Assessment

In the Quantitative Risk Approach there is an event tree for each Potential Failure Mode (PFM) and Initiating Mechanism (IM) and conditional probabilities estimated for each reservoir load partition or seismic load partition. There are methods in Chapters 6 to 13, 16 and 17 to estimate the conditional probabilities for initiation, continuation, progression, detection and intervention, and breach. From these the annual probability of failure is estimated for each PFM and IM.

Estimates of Potential Life Loss (PLL) are made for each PFM.

The combined annual probabilities of failure and corresponding PLL for all PFMs are compared to tolerable societal and individual risk criteria described in Section 18.6.

18.1.4 Simplified QRA

For the Simplified QRA method, the concept is to provide a simpler method based on the QRA but assessing the annual probability of breach for only the most important PFMs. For these PFMs, fewer branches of the event tree and fewer reservoir/seismic load partitions are modelled, and for some PFMs simplified methods are used to estimate conditional probabilities of failure. The annual probabilities of failure and related PLL for each PFM are considered individually in a semi-quantitative risk matrix described in Section 18.6.

The results of engineering analyses are considered in all risk assessment approaches.

18.1.5 Summaries

Table 18.1 summarises where the methods for assessing deterministic, risk enhanced deterministic and quantitative risk analyses for each PFM are to be found.

An estimate of the Potential Life Loss is required. An overview of available methods is provided in Sections 18.2 and 18.3. Section 18.4 describes how individual risk may be estimated.

Section 18.5 presents a detailed summary of Deterministic and Risk Enhanced Deterministic methods.

Section 18.6 presents details of tolerable risk criteria for QRA methods.

Section 18.7 presents semi-quantitative risk matrices for Risk Enhanced and Simplified QRA.

18.2 ESTIMATION OF POTENTIAL LIFE LOSS

The following sections are edited versions of the commentary on this topic in ANCOLD (2022).

TABLE 18.1 Location of Deterministic Methods and Annual Probability of Failure Estimates for Quantitative and Risk Enhanced Dam Safety Assessments

TYPE OF ANALYSIS	OVERTOPPING AND WAVE OVERTOPPING	SLOPE INSTABILITY	SEISMIC LOADING	INTERNAL EROSION AND PIPING		
				EMBANKMENT	FOUNDATION	EMBANKMENT TO FOUNDATION
Deterministic	Whether IDF can be passed through spillway with adequate freeboard 6.5.5, 6.6.6, 6.7.2.	Factor of safety compared to adjusted baseline factor of safety 7.12.	*Non-liquefied*: Estimate crest settlements using database and simplified methods 8.3.2, 8.3.3, 8.12.1. *Liquefied* post-earthquake F of S ≥ 1.3 8.11.1, 8.12.2.	Zoning and Filter properties 18.5.3.	Foundation requirements and treatment 18.5.3.	Foundation requirements and treatment 18.5.3.
Risk Enhanced Deterministic	As for deterministic and database method 6.5.2.	As for deterministic and modified Silva method 7.14.2 and assess loss of freeboard 7.14.3.	*Non-liquefied*: As for deterministic. *Liquefied*: As for deterministic and modified Silva method 7.14.2 and assess loss of freeboard 7.14.3.	As for deterministic and database screening 15.2.	As for deterministic and database screening 15.3.	As for deterministic and database screening 15.4.
Quantitative Risk	*Earthfill* time for initiation of erosion and for head cut compared to flood hydrograph 6.5.3. May also use physically based numerical modelling 6.5.4. *Earthfill with rip-rap*: Probability of loss of rip-rap and breach of remaining earthfill 6.6.6. *Rockfill and earth and rockfill*: Probability based on factor of safety against displacement of rock 6.7.3, 6.7.4, 6.7.5, 6.7.6.	Probability of initiation of instability and loss of freeboard, detection intervention and breach 7 14.1 to 7.14.6.	*Non-liquefied*: Estimate P_{fail} using database and simplified methods 8.3.2, 8.3.3, 8.12.1. *Liquefied*: Post-earthquake F of S and deformations 8.12.2.	QRA all PFM and IM Chapters 9, 12 and 13.	QRA all PFM and IM Chapters 10, 12 and 13.	QRA all PFM and IM Chapters 11, 12 and 13.

(Continued)

TABLE 18.1 (Continued)

TYPE OF ANALYSIS	*OVERTOPPING AND WAVE OVERTOPPING*	*SLOPE INSTABILITY*	*SEISMIC LOADING*	*INTERNAL EROSION AND PIPING*		
				EMBANKMENT	*FOUNDATION*	*EMBANKMENT TO FOUNDATION*
Simplified QRA	*Earthfill*: Database 6.5.2 and/or as for quantitative 6.5.3. *Earthfill with rip-rap*: As for quantitative 6.6.6. *Rockfill and earth and rockfill*: As for quantitative 6.6.6.	Probability of initiation of instability and loss of freeboard 7.14.1, 7.14.2 and 7.14.3.	As for quantitative.	QRA selected PFM 15.10.	QRA selected PFM 15.10.	QRA selected PFM 15.10.

18.2.1 Approaches to Estimating PLL

There is a range of relatively simple empirical methods for estimating PLL which have been derived from databases of historic dam failures.

Of the available empirical models, the RCEM (USBR, 2015) is considered the most suitable for applying to large dams based upon the extensive database, the availability of guidance documentation and the industry experience in its application. Information is available on the development, sample applications and a guideline for application (USBR, 2014, 2015).

When estimating life loss using fatality rates from empirical methods such as RCEM, it is important to remember that these are derived from historic dam failure and natural flood events. Characteristics specific to that particular event, location and population are likely to have contributed to the life loss outcome, and such characteristics are then implicit in the fatality rates. This is a potential limitation when applying RCEM to scenarios which are not well represented by the case histories such as large urban populations and close proximity of the population at risk (PAR) to the dam.

Simulation models may be used to estimate the PLL in situations which are not well represented by the case histories used to develop the empirical approaches such as where evacuation may be constrained or where different warning and evacuation scenarios need to be modelled.

Examples of simulation models are as follows:

LifeSim—www.rmc.usace.army.mil/Software/LifeSim/

The model explicitly simulates the warning and mobilisation of the PAR and predicts the spatial distribution of fatalities across the structures and transport networks expected to be inundated. This capability provides additional insights that can be used to better understand and take measures to reduce the public safety risks. Uncertainties associated with the warning of dam failure and mobilisation of PAR are also addressed by LifeSim, and each component can be sampled in the Monte Carlo framework.

HEC-FIA—www.hec.usace.army.mil/software/hec-fia/

The Hydrologic Engineering Centre—Flood Impacts Assessment (HEC-FIA) model is a spatially distributed dynamic simulation model for assessing the public safety and economic consequences of natural and dam-break flooding. Loss of life is estimated in HEC-FIA using a simplified version of LIFESim.

Life Safety Model LSM—www.lifesafetymodel.net/

Developed by HR Wallingford, UK. Differences between the LSM and LIFESim models are that LIFESim is based on case studies of catastrophic flooding whereas the LSM was designed to use first principles and physically based relationships wherever possible to estimate life safety consequences. Because the LSM relies on a first principles approach, the model is parameter and data intensive.

The most appropriate approach will depend upon the level of required detail and the characteristics of the dam and downstream population. Table 18.2 provides a guide as to the applicable approaches for different levels of assessment. The method needs to be sufficient to categorise the consequences for selecting the appropriate dam classification.

Simulation models are required for assessing the effectiveness of early warning systems for evaluating non-structural risk reduction options.

A major issue in the use of these simulation models is the selection of key parameters and predicting the performance of emergency preparedness. The authors' experience is that even consultants with expertise and extensive experience in dam break modelling, estimating PLL and quantitative risk analysis may need assistance from those who developed the methods when they are first using the model. Sensitivity testing of major inputs is required.

TABLE 18.2 Suggested Method for Estimating Potential Life Loss Related to Dam Safety Assessment Method and Dam Classification

DAM SAFETY ASSESSMENT METHOD	DAM CLASSIFICATION	SUGGESTED MODELS FOR ESTIMATING PLL BASED ON ANCOLD (2022)		
		ANCOLD RISK ASSESSMENT LEVEL	APPROACH	MODELS
Deterministic	Low Significant. High, Very High and Extreme.	Basic or Moderate	PLL estimated by engineering judgement based on PAR. or Estimate PLL from PAR, warning times using empirical methods.	N/A RCEM
Risk Enhanced Deterministic	Low Significant. High, Very High and Extreme.	Basic or Moderate	PLL estimated by engineering judgement based on PAR. or Estimate PLL from PAR, warning times using empirical methods	N/A RCEM
Quantitative Risk	Low, Significant, High and Very High.	Detailed	Estimate PLL from PAR, warning times etc. using detailed application of empirical method, or use simulation model.	RCEM LifeSim HEC-FIA
Quantitative Risk	Extreme.	Very Detailed	Estimate PLL from PAR, warning times etc. using simulation model, judgement and assessment of PAR evacuation routes and emergency management procedures.	LifeSim Life Safety Model
Simplified QRA	All	Moderate	Estimate PLL from PAR, warning times using empirical methods.	RCEM

18.2.2 Incremental Consequences

The tolerable risk criteria discussed in Sections 18.6.2 and 18.6.3 apply to incremental PLL.

Incremental consequences are those over and above any losses which might have occurred for the same natural event or conditions, had the dam not failed.

The baseline for calculating the incremental consequences should be no failure of the dam rather than no dam as the consequences downstream of the dam may be increased or decreased because of the dam's existence.

18.2.3 Estimation of Population at Risk (PAR)

18.2.3.1 General Population

The PAR within the dam-break inundation extent is a key input to estimates of PLL. The PAR is dependent on many factors, including the type of property inundated and whether the dam-break occurs during the day or night. The PAR assessment involves estimating for each inundated location with PAR:

- The potential occupancy (i.e. number of people who could be on the property at any one time).
- The extent to which that capacity is used (e.g. hotels will not always be full).
- The proportion of total time that those using the property are likely to be there (e.g. even when booked into a hotel, people are unlikely to be there all day).

The PAR is the number of people occupying the dam failure inundation area prior to the issuance of any warning—those who would get their feet wet if they took no action to evacuate.

The PAR can be based on the number and types of buildings inundated and the number of occupants expected to be within those buildings before dam failure warnings are issued. For each building, the type and PAR can be assessed using a combination of aerial photography, census data, Google maps and Street View, GIS datasets of land use and planning zones, and internet searches. Field visits are required for more detailed assessments.

Estimates of PAR need to consider both the building's capacity and the extent to which different types of buildings are likely to be occupied during the day and night. Identifying buildings with large PAR (e.g. schools and hospitals) is particularly important.

18.2.3.2 Population at Risk on Roads

For many applications, fatalities on roads do not need to be estimated separately. However, there are some cases where the fatality on roads may represent a significant proportion of the estimated PLL, e.g. a rural setting with limited or no buildings but a road is inundated, or the road, road bridge or culvert washed away, or where there is a major road carrying a large volume of traffic through the inundation extent.

For estimating PLL from flooding of roads, the PAR should include those persons in vehicles which are within the resulting dam break flood at the time the flood reaches the road and those who may drive into the flooded area. It should not be assumed that vehicles will continue to drive into the flooded area, although one or more may drive into the flooded area depending on site distance, vehicle speed, weather conditions and day or night time.

The estimates of PLL should take account of the traffic volumes, numbers of passengers in each vehicle and likely traffic speed in the affected area as this affects the exposure time vehicles are in the affected area.

Campbell et al. (2013) describe a method for assessing the PLL in this situation.

Where a culvert or bridge has been washed away in the flood, the PLL should be estimated taking account of the traffic volumes, numbers of passengers in each vehicle, site distance and the likelihood vehicles may be able to stop before driving into the void formed by the culvert or bridge failing, and the likelihood of passenger fatality in the event the vehicle drives into the void.

RTA (2010) provide guidance on assessing risks in this situation.

18.2.3.3 Itinerant Populations

For some dams in rural locations the PLL is mainly related to itinerant persons within the dam-break area. These may be persons fishing or camping. The PLL can be estimated using the same models as discussed earlier, but allowance must be made for the exposure time of the PAR as discussed later. Consideration should be given to the probability that itinerants may leave the area prior to any dam break event because of flooding as a result of recent rainfall in the area.

18.2.4 Warning Time

A key input to assessing the potential loss of life from dam failure is the estimated time required to identify the threat, and to warn and evacuate the PAR.

The warning times are closely linked to the failure mode; for example, failures triggered by flood and those by earthquake will have different warning times.

There are multiple contributors to the available warning time. First, the potential flooding scenario needs to be identified, and this will depend upon the failure mode. Next, information regarding the flooding scenario (i.e. a warning) must be communicated to the PAR. The PAR then needs to hear the warning and make a decision to take protective or evasive action. For the evacuation to be successful, the protective action has to begin in time for the PAR to reach safety before the danger arrives (e.g. dam-break flooding).

For RCEM, estimates of PLL warning times are estimated judgementally, with sensitivity testing for the impact of variations on the PLL estimates. For simulation models, the uncertainties in each of the key components of the warning time can be incorporated by using distributions of each component rather than single fixed values.

When estimating warning time, consideration should be given to factors where the threat may not be detected until it is too late to issue a warning. This may occur if the dam sits directly above buildings or in situations where in a flood or following an earthquake, dam staff may not be able to get to the site to detect a threat. Instruments transmitting data to a control centre may not be reliable in this situation or be in the wrong place to detect a threat.

18.2.5 Exposure Scenarios and Factors

Overall dam failure scenarios comprise:

- Exposure scenario at the dam—dam element that fails, flood/earthquake or normal operating conditions, failure mode.
- Exposure factors scenario in the dam-break affected zone—according to time of day, day of week and season of year.

PAR and their vulnerability vary according to time of day, day of week and season of the year.

Exposure factors can be highly variable, as illustrated by the following:

- Residents of nursing home in dam-break zone—1.0 (there continuously).
- Residents who live in dam-break zone, but work elsewhere—0.70 (away from home 10 hours per day for 5 days a week, 52 weeks a year).
- Schoolchildren who live outside dam-break zone, the school being in the zone—0.14 (at school 6 hours a day, 5 days a week, 42 weeks a year).
- Tourists who visit a tourist park in the dam-break zone for two hours a month—0.0027.

Such exposure factors directly affect the risk imposed on particular individuals (*individual risk*) but only affect the estimation of PLL if they change the PAR at particular times.

18.3 RCEM METHOD FOR ESTIMATING PLL

18.3.1 Summary of the Method

The following summarises the USBR (2014, 2015) RCEM method. Readers must refer to the original documents before applying the method. The references are freely available for download.

Figures 18.1 and 18.2 show fatality rate versus depth × velocity. Dam break analyses are required to allow these to be estimated in relation to the location of the PAR. There are two graphs, depending on the warning time.

USBR (2014, 2015) provide the following information:

18.3.1.1 Warning Time

The amount of warning that the PAR would be expected to receive in the event of dam failure. Specifically, the amount of time between receiving the warning and the advent of threatening flows. There are two different warning categories considered in RCEM described as follows.

***Adequate Warning*:**
An undefined amount of time that would allow most of the PAR to understand the threat posed by dam failure, to take reasonable actions to leave the inundation plain and to successfully move to a safe location. However, even if given adequate warning, there are a multitude of reasons that people may chose not to leave or are unable to leave. "Adequate" cannot be defined as an exact amount of time because adequate warning is very dependent on site specific conditions. For example, 30 minutes may be an adequate warning for residents of a small town to evacuate, but it may take many hours of warning to enable residents of a large city to evacuate.

***Little or no Warning*:**
A limited (but undefined) amount of time that essentially results in most or much of the PAR receiving an inadequate notification (reflecting quality and timeliness of the warning) of an impending failure and a resulting inability to get out of the inundation plain (or seek adequate shelter from flooding).

Fiedler et al. (2014) suggest little to no warning is typically less than an hour, and adequate warning is "typically more than an hour, although there could be situations with a dense population where hours of warning are not adequate".

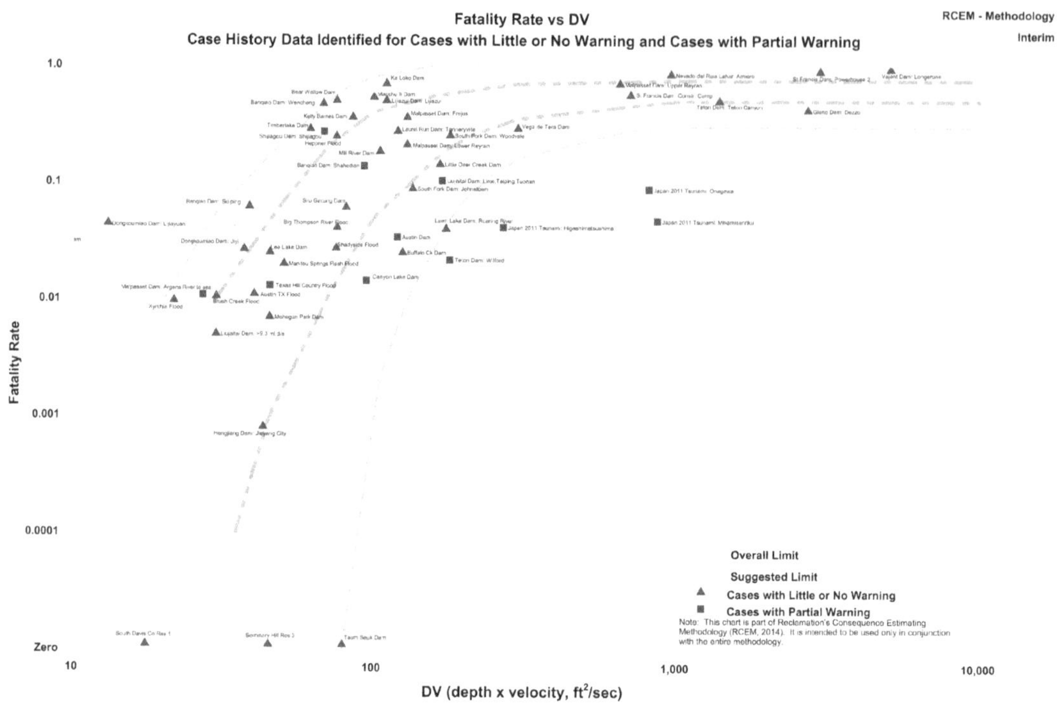

FIGURE 18.1 Fatality Rate vs DV—Case History Data identified for cases with Little or No Warning and Cases with Partial Warning (USACE / Bureau Reclamation, 2015).

Note: 1 ft^2/sec = 0.093 m^2/sec.

USBR (2014, 2015) indicate that the "suggested" limits in Figures 18.1 and 18.2 were selected visually based on the most representative case history data points for each warning time scenario. The "overall" limits were also established visually and are intended to represent the upper and lower bounds of fatality rates.

USBR (2014, 2015) recommend that when using RCEM, users should compare the dam under consideration to the case histories on which the method was developed. If there are similarities with a case history that might be used as a starting point for estimating PLL, taking account of differences in warning time and time of day and year.

In the absence of a similar case history, it is suggested that the geometric mean of the Suggested Limit curves on the RCEM plots be used as the best estimate. That is half-way between the two suggested limit curves in Figures 18.1 or 18.2, whichever is applicable. This is preferable to using the arithmetic mean, that is average of the two PLL limit values from the plot.

18.3.2 An Approximate Application of the RCEM Method for Use in Risk Enhanced and Deterministic Assessments

Where results of dam-break studies are not available, PLL may be estimated as described next. Note that the RCEM method is in itself subject to large uncertainty, and the uncertainty will be even greater using this approximation.

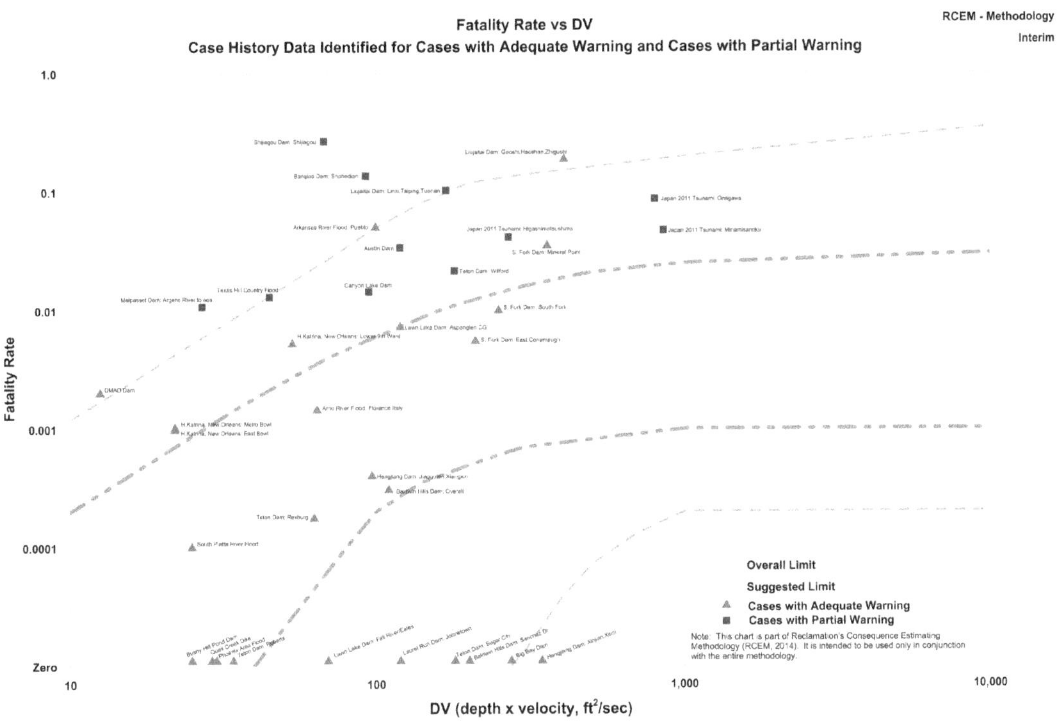

Legend:

Constant Fatality Rate	
1.0	
0.1	
0.01	
0.001	
0.0001	

FIGURE 18.2 Fatality Rate vs DV—Case History Data identified for cases with Adequate Warning and Cases with Partial Warning (USACE / Bureau Reclamation, 2015).

Note: 1 ft²/sec = 0.093 m²/sec.

From Figures 18.1 and 18.2 it can be seen that:

(a) Fatality rate > 0.01 for all cases with little or no warning. Of these 21/34 or about 60% of case data are between 0.1 and 1.
(b) Fatality rate is between 0.01 and 0.1 for partial warning.
(c) Fatality rate is between 0.0001 and 0.01 for adequate warning, but about half the case data had zero life loss.

Galic (2023) re-plotted the case data as shown in Figure 18.3.

It is suggested that an approximate method to estimate PLL would be to:

- Estimate PAR as best can be done from the available information.
- Assess the likely warning category for the PAR.
- Estimate the likely fatality rate for this warning category from the prior data.
- Estimate the PLL from Figure 18.3 using the PAR and likely fatality rate.
- Assess and report a likely range of PLL.

18.3.2.1 Example

For a PAR = 100, little or no warning, 60% probability fatality rate between 0.1 and 1, 40% probability fatality rate between 0.01 and 0.1; gives 60% probability PLL between 10 and 100; 40% probability PLL between 1 and 10.

Likely PLL between 5 and 50 but given approximate estimate, adopt PLL between 10 and 100 in risk matrix.

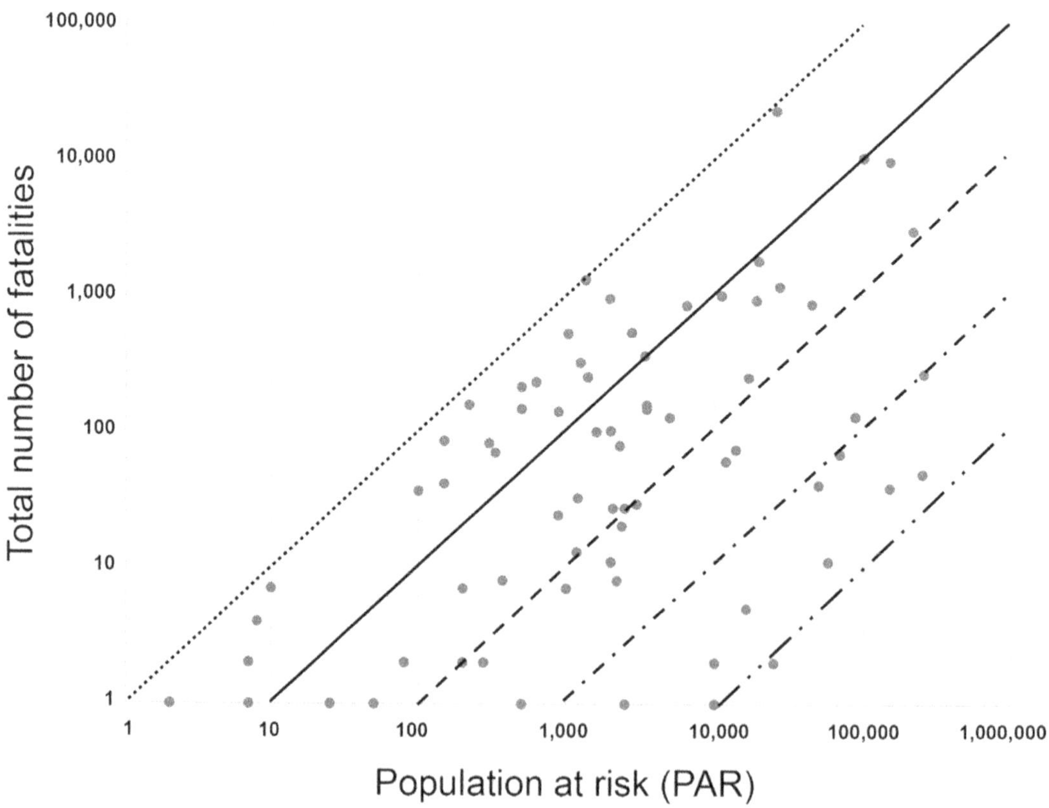

FIGURE 18.3 Non-zero RCEM case history data points replotted to show the number of fatalities as a function of the size of the population at risk (Galic, 2023).

18.4 ESTIMATION OF INDIVIDUAL RISK

Maslin et al. (2012) provide the following method for assessing the risk to the individual most at risk.

The total risk to the individual for a given dam is described as follows:

$$IR = \Sigma IR_{fs}$$

where:

IR_{fs} = individual risk for each failure scenario.

$$IR_{fs} = (f \times EF_{day} \times TF_{day} \times WEF_{day} \times FF) + (f \times EF_{night} \times TF_{night} \times WEF_{night} \times FF) \quad (18.1)$$

where:

f = the annual probability of failure for a particular failure scenario. It should be noted that the failure probability must account correctly for the relevant load partitions.

EF = exposure factor—the proportion of time that the individual most at risk is at the structure or location most at risk. Note that in many cases a dam break inundation extent affects both residential areas and industrial or commercial areas. Therefore, the total exposure time for the person most at risk should include the time they occupy the residential structure and also while present at school or their workplace if these are located within the same inundation extents.

TF = time of day factor—proportion of time a person will be asleep or awake.

WEF = warning and evacuation factor—the probability that the person or group most at risk does not receive sufficient warning to allow for safe evacuation or ignores the warning and does not evacuate. This factor is dependent on the time of day.

FF = fatality factor—the conditional probability of fatality given the person has not evacuated.

Tables 18.3–18.7 provide guidance on these factors.

Methods for estimating fatality rates for the PAR (e.g. RCEM) are not applicable to estimating individual risk. As explained in Maslin et al. (2012), these fatality rates should not be used for individual risk calculations because those fatality rates are based on large-scale averages, and they do not specifically allow for consideration of potential evacuation, provision of shelter or the vulnerability of the individual most at risk.

18.5 DETERMINISTIC AND RISK ENHANCED DETERMINISTIC APPROACHES TO ASSESSING DAM SAFETY

18.5.1 Overtopping and Wave Overtopping Failure of the Embankment

18.5.1.1 Deterministic—All Embankment Types

Determine the maximum reservoir level for the inflow design flood (IDF) by routing the flood through the spillway. Assume that the reservoir is at full supply level or normal operating level at the beginning of the flood.

TABLE 18.3 Exposure Factors EF (Maslin et al., 2012)

HYPOTHETICAL PERSONS	*ASSUMPTION DAY*	*EF DAY*	*ASSUMPTION NIGHT*	*EF NIGHT*
Children at school within dam break inundation extent	6 h/16 h day, 5 days/week	0.3	0 h/8 h day, 5 days/week	0
Permanent resident of nursing home or hospital within dam break inundation extent	16 h/16 h day, 7 days/week	1	8 h/8 h, night, 7 days per/week	1
Resident who lives in the dam break inundation extent but works outside	6 h/16 h day, 5 days/week—weekdays 12 h/16 h day, 2 days/week—weekends	0.5	8 h/8 h, night, 5 days/week—weekdays 8 h/8 h, night, 2 days/week—weekends	1
Retired resident who lives in the dam break inundation extent	12 h/16 h per day 7 days/week	0.75	8 h/8 h per night 7 days/week	1
Resident who lives and works (or attends school) in dam break inundation extent	12 h/16 h per day 7 days/week	0.75	8 h/8 h per night 7 days/week	1
Worker who lives outside the dam break inundation extent but works inside	10 h/16 h day, 5 days/week	0.45	0 h/8 h night, 5 days/week	0

TABLE 18.4 Time of Day Factor TF (Maslin et al., 2012)

DESCRIPTION	*ASSUMPTION*	*TF*
Daylight	16 h/day	0.67
Night time (dark)	8 h/day	0.33

(a) If the maximum reservoir level exceeds the dam crest level, the dam fails to satisfy deterministic safety criteria.
(b) If the maximum reservoir level is below embankment crest level but the estimated wave run-up for the 2% waves ($R_{u2\%}$) from Section 6.3.6 exceeds the freeboard, the dam fails to satisfy deterministic safety criteria.
(c) If the freeboard exceeds the run-up for the 2% waves, there will be at most very minor overtopping flows, the likelihood of failure by overtopping may be assumed to be negligible and the dam satisfies deterministic safety criteria.

Use the actual crest level allowing for settlement since construction, residual camber and repairs to the crest.

18.5.1.2 Risk Enhanced Deterministic

Type Z1 Embankments:

Determine the maximum reservoir level for the IDF by routing the flood through the spillway.

Then assess as described for deterministic assessment, but if the IDF plus wave run-up results in overtopping of the crest, use the database method described in Section 6.5.2 to provide additional information on the likelihood of failure to assist in making the dam safety assessment.

TABLE 18.5 WEF for a Range of Scenarios (Maslin et al., 2012)

WARNING TIME	*WARNING EFFECTIVENESS*[1]	*TIME OF DAY*	*WEF (LOW AND MEDIUM SEVERITY FLOODING)*	*WEF (HIGH SEVERITY FLOODING, COMPLETE INUNDATION)*
< 0.25 h	Not applicable	Day	0.1	0.8
		Night	1	1
0.25–1 h	Not effective[2]	Day	0.08	0.4
	Not effective	Night	0.2	0.7
	Effective[3]	Day	0.04	0.2
	Effective	Night	0.1	0.4
1–2 h	Not effective	Day	0.04	0.08
	Not effective	Night	0.08	0.15
	Effective	Day	0.02	0.04
	Effective	Night	0.04	0.07
2–12 h	Not effective	Day	0.02	0.04
	Not effective	Night	0.04	0.06
	Effective	Day	0.01	0.02
	Effective	Night	0.02	0.03
> 12 h	Not applicable	Not applicable	0.001	0.005

Notes:

(1) The warning effectiveness relates to a number of factors including clarity, accuracy, forcefulness and urgency (Graham, 2008).

(2) The warning would be considered not effective if it is issued prior to dam failure, or if the person issuing the warning has not observed the dam failure.

(3) The warning would be considered effective if dam failure is in process or the dam has failed and the magnitude of downstream flooding has been observed.

TABLE 18.6 Fatality Factors FF for Persons within Structures (Maslin et al., 2012)

DESCRIPTION OF STRUCTURE	*DESCRIPTION OF FLOODING*	*FF*
Permanent residential structure (house)	High severity flooding, DV > 15 m^2/s	0.9
	Medium severity flooding, partial inundation, DV > 4.6 m^2/s *and* D > 3 m	0.3
	Low severity flooding, partial inundation, DV < 4.6 m^2/s	0.1
Temporary residential structure (caravan)	High severity flooding, DV > 15 m^2/s	1
	Medium severity flooding, partial inundation, DV > 4.6 m^2/s *and* D > 3 m	0.9
	Low severity flooding, partial inundation, DV < 4.6 m^2/s	0.3
Permanent residential structure (nursing home or aged care facility)	High severity flooding, DV > 15 m^2/s	1
	Medium severity flooding, partial inundation, DV > 4.6 m^2/s *and* D > 3 m	0.5
	Low severity flooding, partial inundation, DV < 4.6 m^2/s	0.3
School	High severity flooding, DV > 15 m^2/s	1
	Medium severity flooding, partial inundation, DV > 4.6 m^2/s *and* D > 3 m	0.5
	Low severity flooding, partial inundation, DV < 4.6 m^2/s	0.3

Type Z2, Z3, Z5 and Z6 Embankments:

There is no risk enhanced deterministic method for these embankments.

Type Z4 Embankments:

Treat as a variant of Type Z1.

TABLE 18.7 Fatality Factor FF for Individual Most at Risk with No Provision for Shelter (Maslin et al., 2012)

DESCRIPTION	DESCRIPTION OF FLOODING	FF
Adult	DV < 0.6 m^2/s	0.01
	DV 0.6–0.8 m^2/s	0.5
	DV > 0.8 m^2/s	1
Frail or older person and infants	DV < 0.4 m^2/s	0.1
	DV > 0.4 m^2/s	1
Child	DV < 0.4 m^2/s	0.01
	DV > 0.4 m^2/s	1

Include consideration of flow concentration effects (e.g. due to uneven crest level, or flow concentration along the abutment groin).

18.5.2 Slope Instability Leading to Overtopping Failure

18.5.2.1 Deterministic

For the deterministic approach estimate the factor of safety against slope instability of the downstream slope as described in Chapter 7.

If the estimated factor of safety ≥ the adjusted baseline factor of safety from Tables 7.3 and 7.4 and if the crest deformations are monitored and show normal behaviour as discussed in Section 7.13, the likelihood of failure leading to loss of freeboard may be assumed to be extremely low or negligible and the dam considered as satisfying deterministic acceptance criteria.

If the factor of safety is less than the adjusted baseline factor of safety, it is necessary for those carrying out the dam safety assessment to use their judgement to consider if the dam is satisfactory regarding slope stability of the embankment. The risk enhanced approach is provided as an aid to that judgement.

18.5.2.2 Risk Enhanced Deterministic

For a risk enhanced approach, use the modified Silva et al. (2008, 2025) method to estimate the conditional probability of failure for the IDF reservoir level as described in Sections 7.14.2 and 7.14.3 and assess the annual probability of failure against semi-quantitative tolerable risk criteria described in Section 18.7.

18.5.3 Failure Resulting from Internal Erosion and Piping

As discussed in Chapters 9, 10 and 11, failure may result from internal erosion and piping in the embankment, foundation and embankment to foundation.

There are no commonly applied deterministic criteria for assessing the safety of an embankment for internal erosion and piping. The following are the authors' suggested criteria.

18.5.3.1 Deterministic

A dam which satisfies deterministic criteria would have all of the following characteristics:

Internal Erosion and Piping in the Embankment

Zoning:

To satisfy deterministic criteria there should be a filtered exit to all potential seepage or flow in cracks in the embankment.

As can be seen in Figure 18.4, this is achieved with the following types of embankments:

- Earthfill with vertical and horizontal filter drain.
- Central core earth and rockfill.
- Sloping core earth and rockfill.

Zoned earthfill embankments may also satisfy these criteria if the downstream zone is non-plastic sandy gravel or gravelly sand which acts as a filter.

To satisfy deterministic criteria, the filter must extend from the foundation to the maximum reservoir level for the IDF to control the continuation of internal erosion. In practice that means for most dams the filter should extend to the crest and extend the full length of the embankment.

Many dams have filters, but these do not extend to the crest of the embankment. In many dams the top of the filters is at FSL. In others the top of the filters is where the designer's seepage flow net emerged on the downstream slope of the core with the reservoir at FSL. That was consistent with the understanding of internal erosion and piping at the time, that erosion initiated from seepage within the core and that flow nets would not have time to develop under flood conditions. It is now recognised that concentrated leak erosion initiates in transverse cracks in the core and the flow in the cracks occurs as the reservoir rises.

Our experience is also that many as-constructed or work-as-executed drawings are not accurate. Crest details are often not as designed with filters not taken as high as designed or constructed narrower than shown in design drawings because of room constraints at the top of the embankment.

Many saddle dams are homogeneous embankments with no filters, even though there is a high potential for life loss if they fail.

Filter design and construction:

To satisfy deterministic criteria, the filter as designed and constructed should satisfy no-erosion filter criteria and have ≤ 5% non plastic fines passing 0.075 mm. Documentation of particle size distribution testing during construction is essential to be confident this requirement is met.

We note that many existing dams do not satisfy these requirements.

Many dams have filters which are coarser than required to satisfy no erosion filter criteria and/or have fines content > 5% so may hold a crack and be ineffective as a filter.

Internal Erosion and Piping in the Foundation

To satisfy deterministic criteria the following is required:

Rock foundations:

- Cutoff for the core founded on non-erodible rock.
- Cutoff foundation taken below the level of wide open cracks, and below the level where there are infilled joints.
- Foundation grouted to suitable closure.

See Fell et al. (2015) for details of cutoff and grouting requirements.

Plastic soil foundations:

- Cutoff founded below roots and open cracks.
- A no-erosion horizontal filter drain on the foundation downstream of the cutoff as shown in Figure 18.4 for the earthfill embankment with vertical and horizontal filter drains.

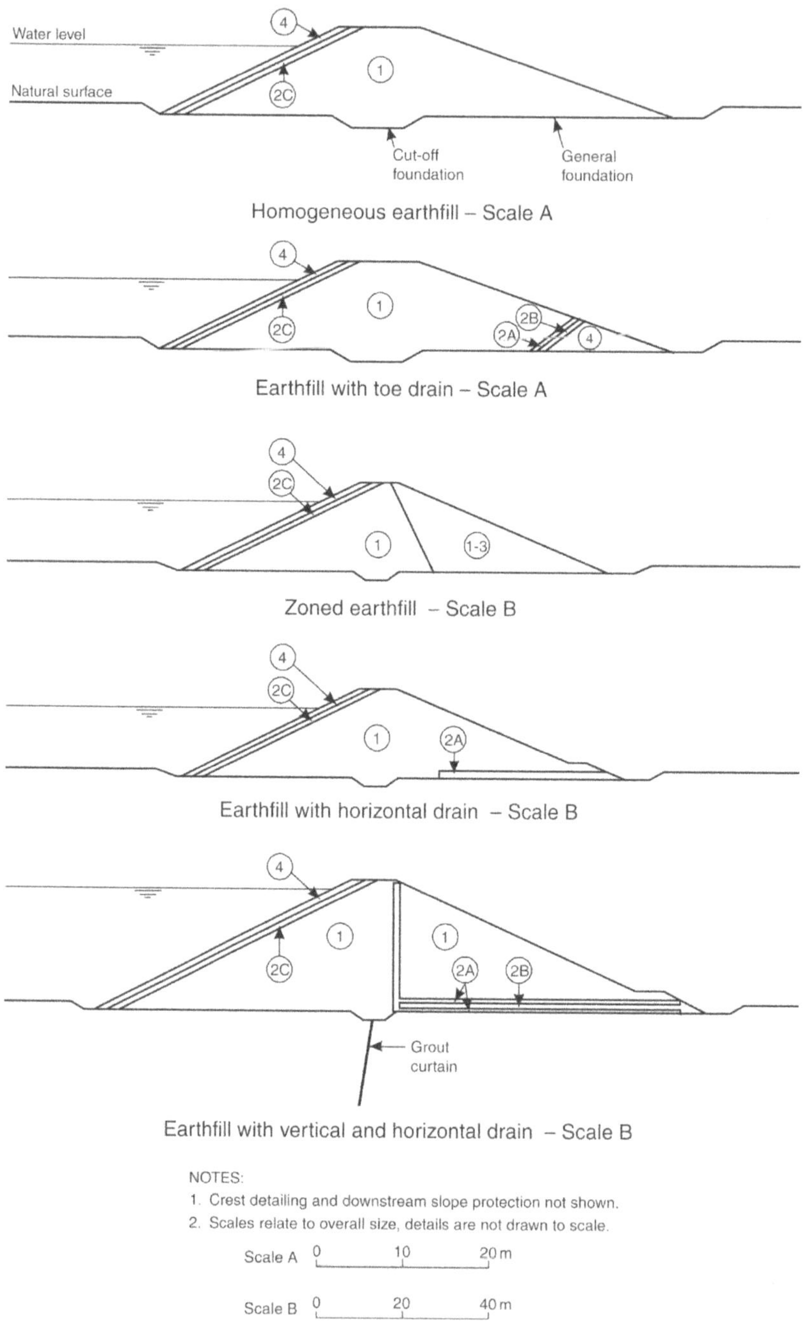

FIGURE 18.4 Typical zoning of embankment dams (Fell et al., 2015).

Non-plastic soil foundations:

- A factor of safety > 1.6 against heave or critical exit gradient for initiation of backward erosion piping at the toe of the embankment for the IDF reservoir level using conservative assumptions and/or monitoring of pore pressures,

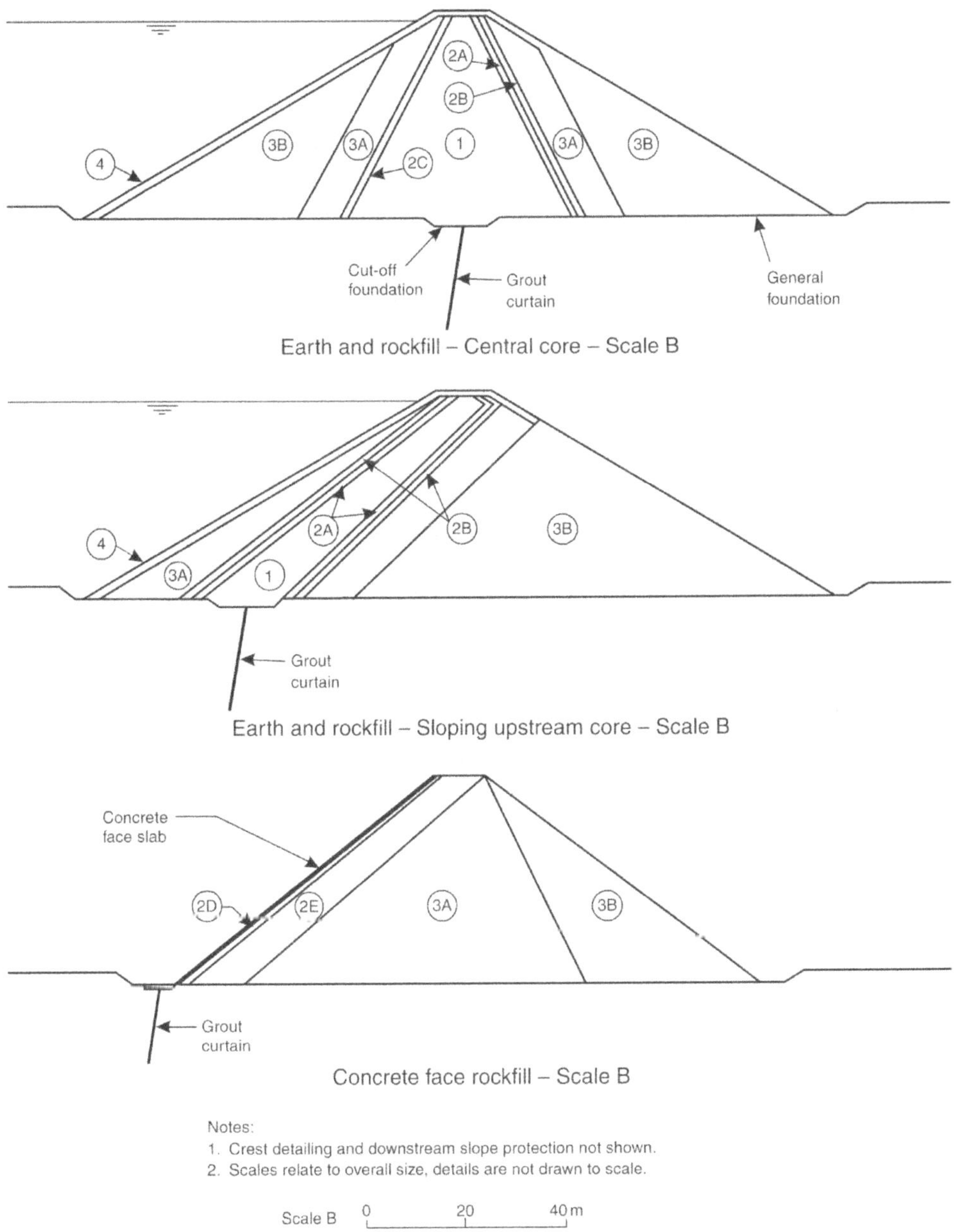

FIGURE 18.4 (Continued)

or

a fully penetrating cutoff through the foundation soils subject to backward erosion piping.

- No-erosion filter drain on the foundation downstream of the cutoff as shown in Figure 18.4 for the earthfill embankment with vertical and horizontal filter drains.
- If there is silt sand gravel in the foundation, it should not be subject to suffusion or contact erosion.

Erosion of the Embankment into the Foundation
To satisfy deterministic criteria the following is required:
Rock foundation:

- Cutoff foundation treated with shotcrete and/or slush grout to fill open defects if they are present in the base and downstream side of the cutoff trench. Not all foundations have open defects, so not all dams require shotcrete and/or slush grout.

Soil foundation:

- Cutoff foundation treated with shotcrete and/or slush grout to fill open defects in the downstream side of the cutoff trench.
  or
- A no-erosion filter provided on the downstream side of the cutoff trench.

Discussion
These criteria are based on the dam being high, very high or extreme consequence of failure class.

Those carrying out the dam safety assessment will have to use their judgement to consider if the dam satisfies these criteria, and whether a significant or low consequence of failure dam can be assessed as satisfactory even though it does not have all these characteristics.

18.5.3.2 Risk Enhanced Deterministic

As for deterministic plus, apply the historical database approach as described in Chapter 15 to assess the likelihood of failure and assessed against the semi-quantitative risk criteria described in Section 18.7.

18.5.4 Failure Resulting from Loss of Freeboard under Seismic Loads

18.5.4.1 Embankments Not Subject to Liquefaction

Deterministic approach:

(a) Use one or more of the database methods in Section 8.3.2 to estimate the deformation for the earthquake design ground motion. If the estimates of crest settlements are less than 10% of the available freeboard before the earthquake with reservoir level at FSL or normal operating level, accept that the likelihood of failure by overtopping is negligible for that loading and the embankment satisfies deterministic criteria.
(b) If the estimated crest settlements by the database methods are greater than this, use one or more of the simplified methods in Section 8.3.3 to estimate deformations. If the estimated deformations from both database and simplified methods are less than 20% the available freeboard, accept that the likelihood or failure by overtopping is negligible for that loading and the embankment satisfies deterministic criteria.
(c) If the crest settlements are greater than these, the dam does not satisfy deterministic criteria.

Risk enhanced deterministic approach:
There is no risk enhanced approach.

18.5.4.2 Embankments Subject to Liquefaction

Deterministic approach:

Estimate the post-earthquake factor of safety for the earthquake design ground motion (EDGM) using:

- Representative geometry of the slope and foundations. Pay particular attention to the extent and continuity of liquefied strata for the EDGM.
- Reasonable lower bound estimates of shear strength parameter for liquefied zones, reasonable but conservative shear strengths for non-liquefied zones.
- Conservative estimates of pore pressure build-up and strain weakening in other zones.

Model the whole dam and obtain factors of safety for downstream and upstream failure surfaces. Take account of the guidance in Section 7.12.

If the factor of safety is ≥ 1.3, accept that the likelihood of failure will be small and probably tolerable subject to the quality and extent of information available. This has been the authors' practice for some years and is consistent with what FEMA (2005a) recommends.

Deterministic criteria may be satisfied for low and significant consequence of failure classification. For high, very high and extreme classification dams, use risk-based methods to assess dam safety.

Risk enhanced deterministic approach:

(a) Estimate the post-earthquake factor of safety for the earthquake design ground motion using:
  - Representative geometry of the slope and foundations. Pay particular attention to the extent and continuity of liquefied strata.
  - Best estimates of shear strength parameters including liquefied shear strength.
  - Best estimates of pore pressure build-up and strain weakening in other zones.
  - Model the whole dam and foundation and obtain factors of safety for downstream and upstream failure surfaces. Take account of the guidance in Section 7.12.

(b) Use the modified Silva et al. (2008, 2025) method described in Section 7.14 to estimate the conditional probability of initiation of instability as discussed for static loading. Assume Category III to allow for the uncertainty in liquefied strength and undrained strength of other plastic soils.

(c) Estimate the likelihood of loss of freeboard given instability initiates using Tables 7.10 and 7.11. Assume that the reservoir is at full supply level.

Use engineering judgement to assess the dam safety, taking account of these results and the dam consequence of failure classification. If in doubt use the QRA method.

It is important to note that the assessment is based on deformations. The pseudo-static factor of safety method should not be used for the reasons described in Section 8.5.

18.6 TOLERABLE RISK CRITERIA FOR QUANTITATIVE RISK ASSESSMENT APPROACH TO ASSESSING DAM SAFETY

18.6.1 Introduction

In the QRA approach the estimates of annual probability of failure and consequences of failure for all PFMs and load partitions are compared to tolerable risk criteria. See ANCOLD (2022) for details.

ANCOLD (2022) and other guidelines such as USACE (2014) and FERC (2016) require that three conditions are to be satisfied for risks to be considered tolerable:

- The societal risk.
- The individual risk.
- The as-low-as-reasonably-practicable (ALARP) requirement.

ANCOLD (2022) defines individual and societal risk as:

Individual risk	The increment of risk imposed on a particular individual by the existence of a hazardous facility. This increment of risk is an addition to the background risk to life, which the person would live with on a daily basis if the facility did not exist.
Societal risk	The risk of multiple fatalities resulting from a failure. This special term is a subset of societal concerns as defined above, though societal concerns is also often used to mean all such concerns other than societal risk.

USACE (2014) and FERC (2016) have similar definitions.

The following sections provide details of ANCOLD (2022), USACE (2014) and FERC (2016) tolerable risk criteria. It will be seen that they are similar but with important differences. Readers may decide which criteria they adopt or may have guidelines from their country to follow.

18.6.2 Tolerable Societal Risk Criteria

18.6.2.1 ANCOLD (2022) Risk Criteria

ANCOLD (2022) proposed the following guidelines for the tolerability of public safety risks for the general community:

- For existing dams, a societal risk that is higher than the limit line, shown on Figure 18.5, is unacceptable, except in exceptional circumstances. For new dams or major augmentations such as raising the full supply/normal operating reservoir level, the limit of tolerability is an order of magnitude lower.
- The guideline applies to "best estimates" of the risks.
- Consider what possibilities exist for further risk reduction in order that the decision-maker can decide whether risks are ALARP.

These criteria are applied to the combined risk from all potential failure modes and IMs. An example is shown in Figure 18.6.

ANCOLD (2022) provides the following additional information:

> The qualification "except in exceptional circumstances" requires explanation. In the view of ANCOLD, the discretion to decide that circumstances are exceptional should not reside with the owner but should be a matter for government. The justification for tolerating such high risks is in the "wider interests of society". Risks, which would normally be unacceptable, can be tolerated on account of the special benefits, which the dam brings to society.
>
> If a dam owner is going to use the except in exceptional circumstances provision, the ALARP test should first be applied to demonstrate that all things reasonable have been done or planned to be done, before relying on the except in exceptional circumstances provision.

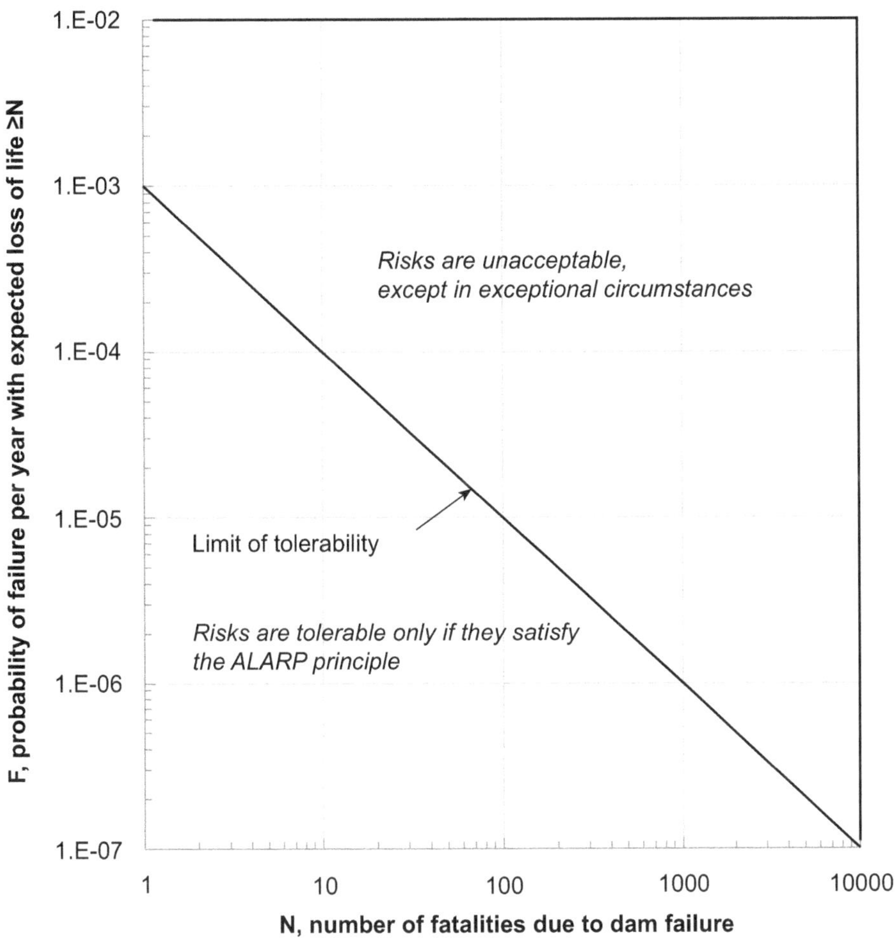

FIGURE 18.5 ANCOLD (2022) Societal Risk Guideline: Existing Dams.

ANCOLD (2022) provides detail on how to assess ALARP. In practice it is a difficult concept to apply, and readers should inform themselves and then consider ALARP in the context of their country's regulatory and legal framework.

18.6.2.2 United States Army Corp of Engineers (USACE, 2014) Tolerable Risk Criteria

Figure 18.7 shows the USACE (2014) guidelines for societal and individual risk.

USACE (2021) says the societal guideline applies to all potential failure modes and population exposure scenarios. The individual guideline is for the person most at risk for the sum of contributions from all potential failure modes.

USACE (2021) explains the low probability–high consequence part of the societal risk plot as follows:

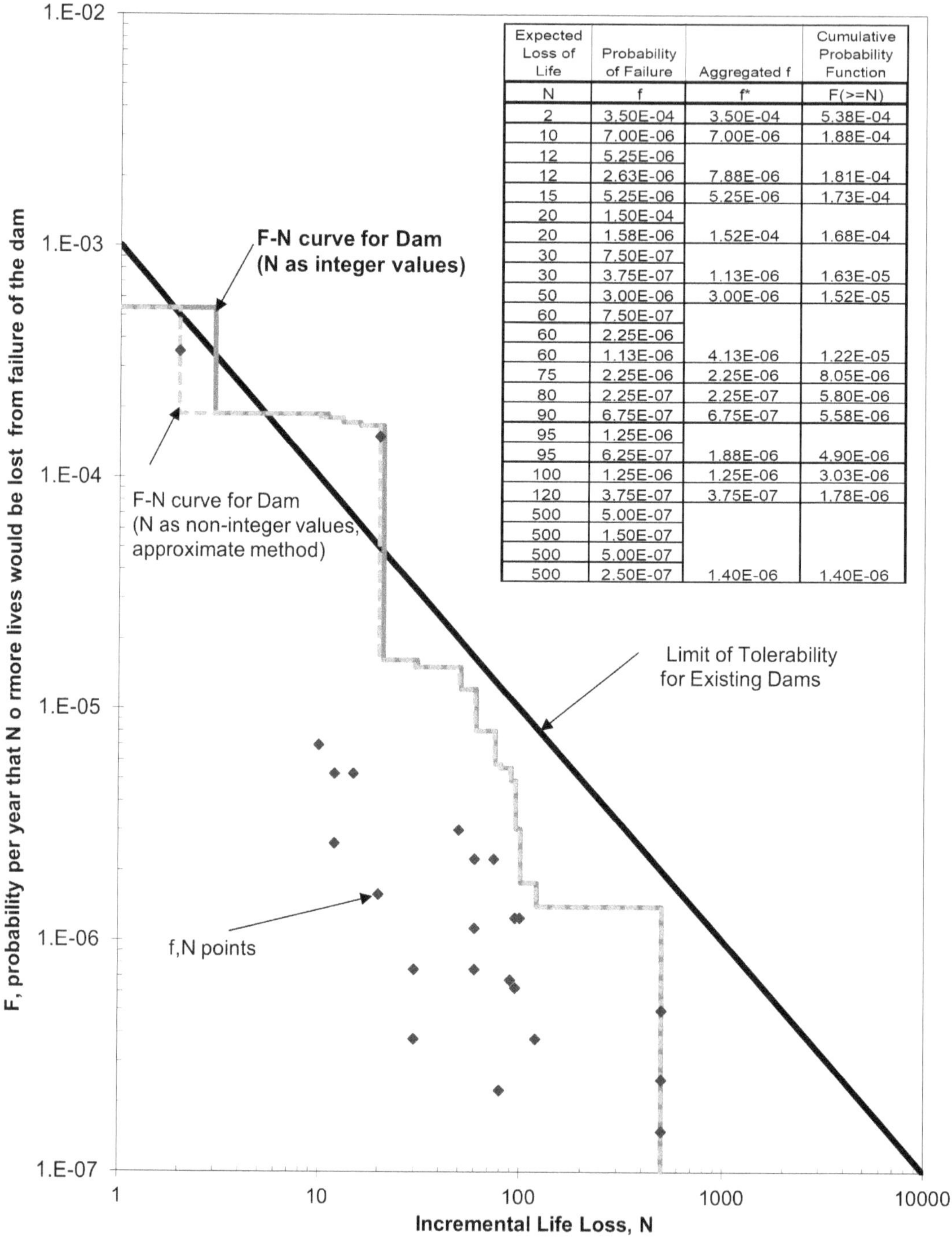

FIGURE 18.6 Example Plot of F-N Line and f,-N pairs (ANCOLD (2022).

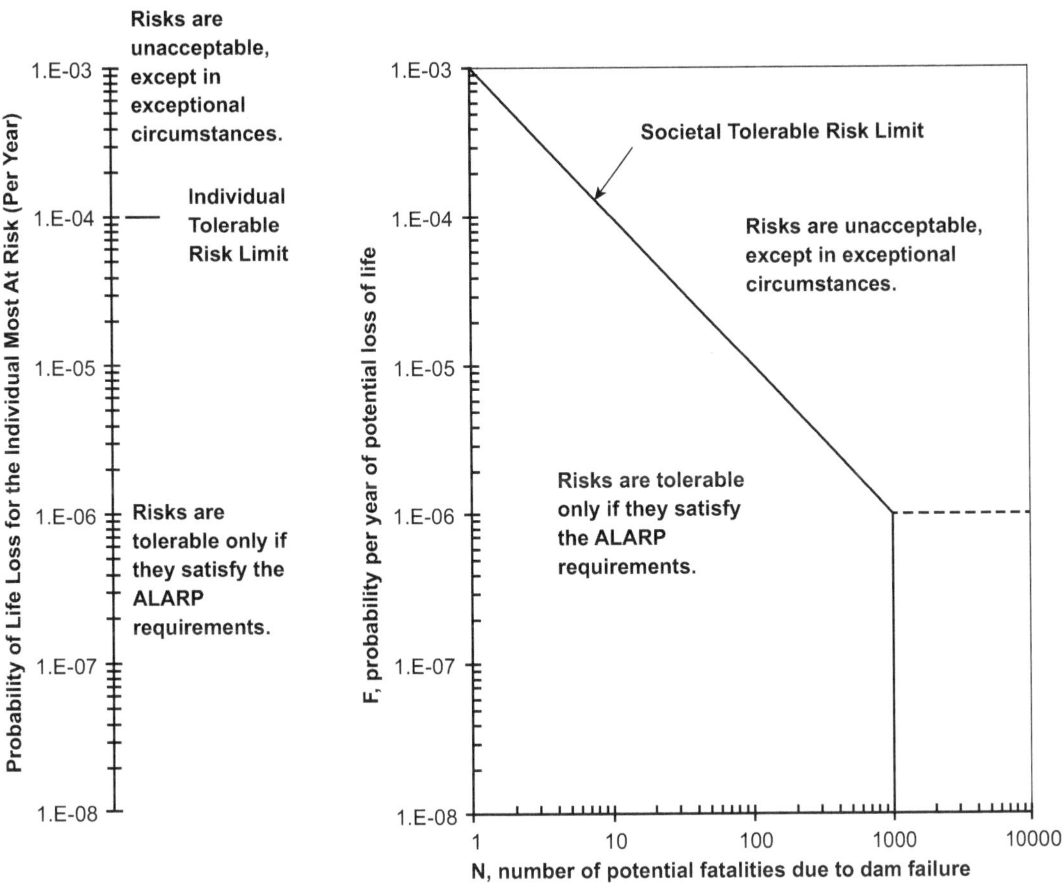

FIGURE 18.7 Guidelines for (a) Individual and (b) Societal Risk (USACE, 2014, 2021).

"As a minimum, dams that can cause more than 1000 fatalities in a single event should:

- Have the highest expectations for the standard of care.
- Have the most up-to-date hazard and response evaluations including:
  - Site specific probabilistic seismic hazard analysis.
  - Site specific hydrologic analysis.
  - State-of-the-art supporting analysis.
- Have a quantitative risk analysis to support the tolerability decision.
- Consider increasing the priority of safety related Operation and Maintenance (O&M) activities.
- Have a robust surveillance and observation plan.
- Have a study outlining potential ALARP actions which have been considered".

18.6.2.3 US Federal Energy Regulatory Commission (FERC, 2016)

The FERC (2016) societal risk guideline is shown in Figure 18.8. This is as for USACE (2014, 2021) but with a second diagonal line two orders of magnitude below the tolerable risk reference line.

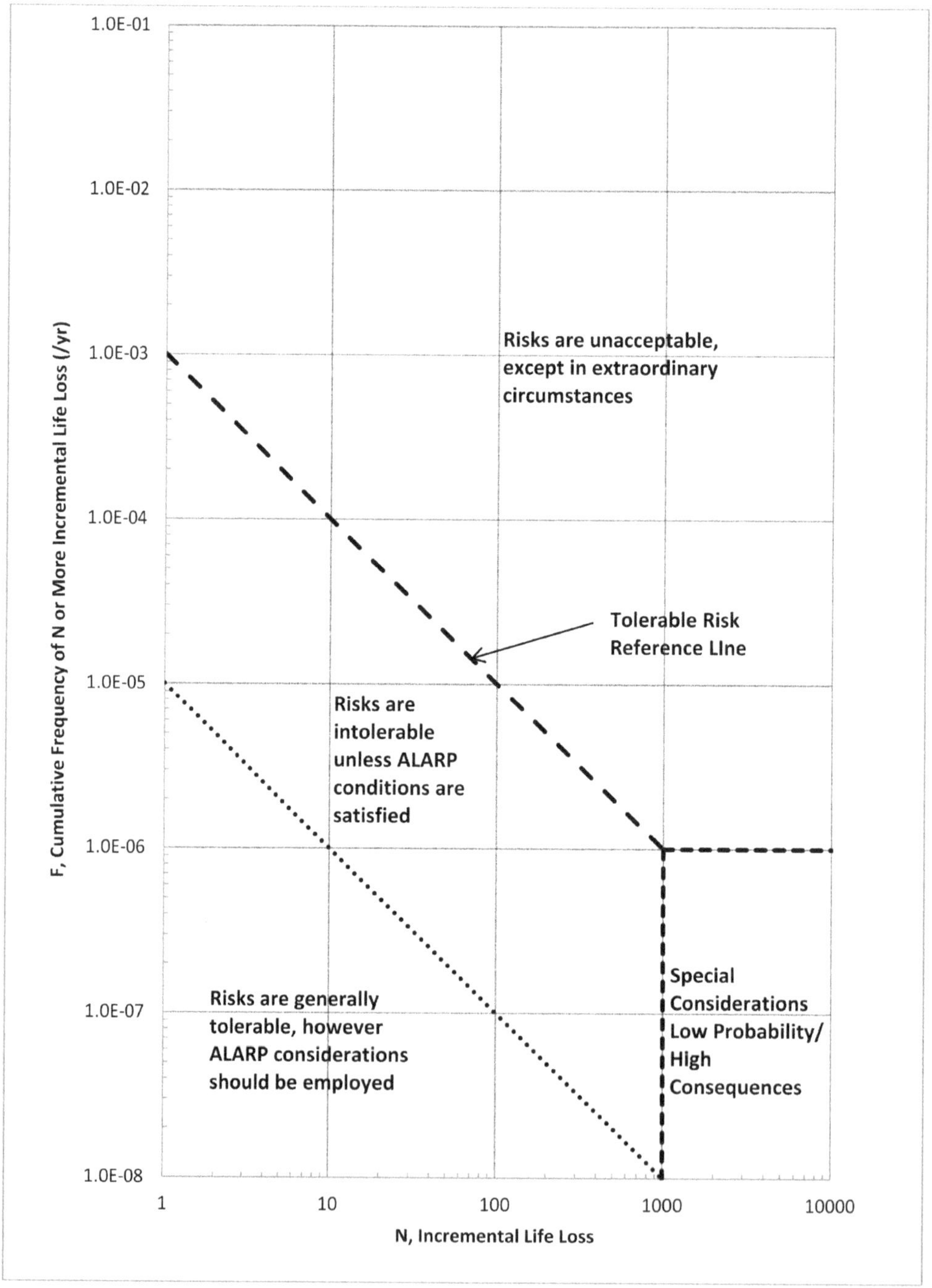

FIGURE 18.8 FERC (2016) Societal Risk Guideline for Incremental Risk (F-N).

FERC (2016) also requires consideration of Average Annual Life Loss (AALL) criteria. Readers should refer to the reference for details.

18.6.2.4 US Bureau of Reclamation Tolerable Risk Criteria

The USBR does not have tolerable societal risk criteria. It uses risk analyses to prioritise dam safety improvements within its large portfolio of dams.

The approach is described in Reclamation (2022). Figure 18.9 shows the risk portrayal chart. These criteria are applied to risks from internal erosion and piping, failure resulting from seismic loads, and flood overtopping separately, so different to the ANCOLD (2022), USACE (2014, 2021) and FERC (2016) criteria which apply to the summation of risk from all PFMs.

The diagonal part of the plot is the same as ANCOLD (2022) and USACE (2014, 2021) limit line.

The upper horizontal is explained to be analogous to the tolerable individual risk guideline.

The lower horizontal line is explained as a truncation of the diagonal line. Reclamation (2022) says, "otherwise that facility with an estimated AEP lower than one in a million would be competing for funding (for upgrade) with those with a much higher risk of failure and more straightforward risk reduction options".

18.6.3 Tolerable Individual Risk Criteria

ANCOLD (2022) proposes the following guideline for the tolerability of public safety risks for the general community:

- For existing dams, an individual risk to the person, who is most at risk, that is higher than 10^{-4} per annum is unacceptable, except in exceptional circumstances.
- For new dams or major augmentations of existing dams, an individual risk to the person, who is most at risk, that is higher than 10^{-5} per annum is unacceptable, except in exceptional circumstances.
- Use "best estimates" of the risks for comparison with the public safety policies and criteria.

ANCOLD (2022) says that for evaluating the tolerability of individual risks, those individual risks higher than 10^{-5} per annum should only be regarded as tolerable after careful consideration of such factors as:

- The age of those at risk, which influences the average background risk levels. This recognises the limit value of 10^{-4} would be a little more than double the minimum average background risk, which is that of females aged 12,
- The benefit which the dam provides in the wider interests of society and
- The ALARP principle.

The emphasis in demonstrating whether risks for existing dams are tolerable should therefore be on reducing risks below 10^{-5} per annum as opposed to the limit value of 10^{-4} per annum.

As shown in Figure 18.7a, USACE (2014) has tolerable individual risk of 10^{-4} per annum subject to ALARP. Figure 18.10 shows the FERC (2016) individual risk criteria.

These criteria are therefore similar. USACE (2014, 2021) and FERC (2016) have a generally tolerable category—below 10^{-6} per annum.

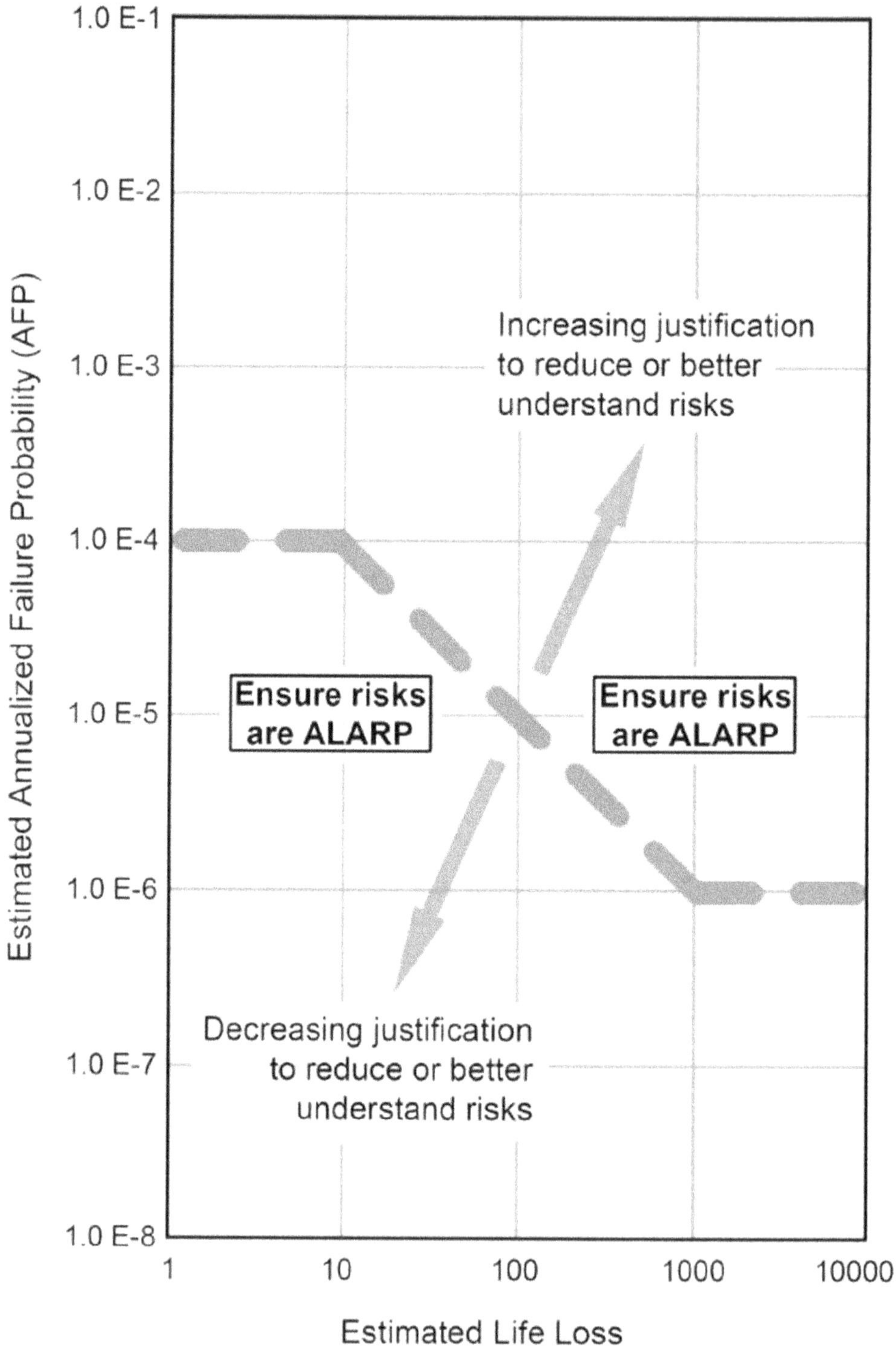

FIGURE 18.9 Annotated reclamation risk portrayal chart. Justification to reduce or better understand risks would generally be considered to increase above the visual guideline and decrease below it. The ALARP concept is considered to apply everywhere on the chart (Reclamation, 2022).

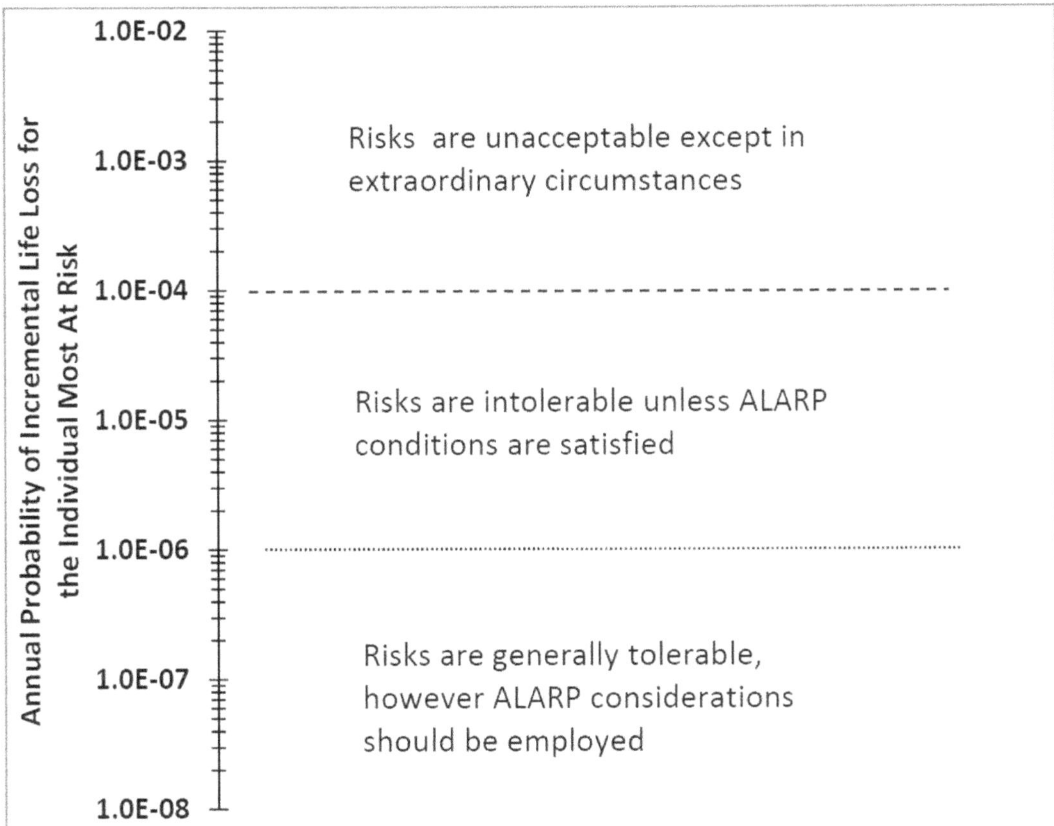

FIGURE 18.10 FERC (2016) Individual Life Safety Risk Guideline for Incremental Risk.

18.7 SEMI-QUANTITATIVE TOLERABLE RISK CRITERIA

Tables 18.8 and 18.9 are two suggested semi-quantitative risk classification matrices. Owners may choose which approach best suits their needs.

The consequence classification is based on the Canadian Dams Association CDA (2007) classification shown in Table 5.1. It relies on estimating the potential life loss which may be done using the RCEM method (USBR, 2014, 2015).

The annual probability of failure classification is calculated as described in Sections 18.1 and 18.5.

The risk classification terms in Table 18.8 are based on ANCOLD (2022) *Guidelines on Risk Assessment* societal risk criteria. For consequences controlled by environmental and cultural values, or infrastructure and economics, the consequence classification from the CDA classification should be used.

Semi-quantitative risk matrix criteria may be applied to individual PFMs using database estimates of failure from overtopping, the modified Silva et al. (2008) method for failure from slope instability, empirical database and simplified methods of failure for seismic loads, and database method for internal erosion and piping.

Some may choose to use these criteria for QRAs rather than using the F-N type plots discussed in Section 18.6.

TABLE 18.8 Semi-Quantitative Matrix for Risk Classification of Dams—With Terms Guiding on the Acceptability of the Societal Risk

LIKELIHOOD OF FAILURE / ANNUAL PROBABILITY OF FAILURE		*CONSEQUENCE CLASSIFICATION/POTENTIAL LOSS OF LIFE (PLL)*				
		LOW	*SIGNIFICANT*	*HIGH*	*VERY HIGH*	*EXTREME*
		NONE	*≤ 1*	*≤ 10*	*≤ 100*	*>100*
Very High	1.0E-02 to 1.0E-03	Tolerable subject to ALARP	Unacceptable	Unacceptable	Unacceptable	Unacceptable
High	1.0E-03 to 1.0E-04	Tolerable subject to ALARP	Tolerable subject to ALARP	Unacceptable	Unacceptable	Unacceptable
Moderate	1.0E-04 to 1.0E-05	Tolerable	Tolerable subject to ALARP	Tolerable subject to ALARP	Unacceptable	Unacceptable
Low	1.0E-05 to 1.0E-06	Tolerable	Tolerable	Tolerable subject to ALARP	Tolerable subject to ALARP	Unacceptable
Very Low	1.0E-06	Tolerable	Tolerable	Tolerable	Tolerable subject to ALARP	Tolerable subject to ALARP

TABLE 18.9 Semi-Quantitative Matrix for Risk Classification of Dams—With a Number Risk Classification System

LIKELIHOOD OF FAILURE / ANNUAL PROBABILITY OF FAILURE		*CONSEQUENCE CLASSIFICATION/POTENTIAL LOSS OF LIFE (PLL)*				
		LOW	*SIGNIFICANT*	*HIGH*	*VERY HIGH*	*EXTREME*
		NONE	*≤ 1*	*≤ 10*	*≤ 100*	*>100*
Very High	1.0E-02 to 1.0E03	4	5	6	7	8
High	1.0E-03 to 1.0E-04	3	4	5	6	7
Moderate	1.0E-04 to 1.0E-05	2	3	4	5	6
Low	1.0E-05 to 1.0E-06	1	2	3	4	5
Very Low	≤ 1.0E-06	1	1	2	3	4

Notes:
(a) Classification 1 is the lowest risk and Classification 8 the highest risk.
(b) Classifications 1 and 2 risks are likely to represent generally acceptable risks to the public downstream of the dam.
(c) Classifications 5, 6, 7 and 8 are likely to represent unacceptable risks to the public downstream of the dam.
(d) Classifications 3 and 4 may or may not represent tolerable risks to the public downstream. More detailed risk analysis is required to better define the risks and/or check ALARP.

19 Measures for Upgrading Dams to Reduce Risks

19.1 INTRODUCTION

This chapter describes methods for upgrading existing embankment dams which do not meet deterministic and/or risk-based dam safety criteria.

This includes upgrades to reduce risks from:

(a) Flood overtopping.
(b) Slope instability and deformations leading to loss of freeboard.
(c) Deformations resulting from liquefaction of the foundations or embankment leading to loss of freeboard.
(d) Internal erosion in the embankment.
(e) Internal erosion in the foundation.

Upgrades are commonly required to address deficiencies in more than one of these categories, e.g. to address flood overtopping and internal erosion.

The principles of upgrading are discussed and some examples in which the authors have been involved are described.

Some common issues which arise in the design and construction of dam upgrades are also discussed.

19.2 RISK REDUCTION STRATEGY

Risk reduction measures may be designed to:

(a) Meet deterministic design criteria for individual PFMs, e.g. flood overtopping, or
(b) Reduce risks from individual PFMs to satisfy tolerable quantitative risk criteria, or
(c) Reduce risks from all PFMs contributing significantly to the risk to meet tolerable quantitative risk criteria, or
(d) As for (c) and satisfying ALARP principles.

DOI: 10.1201/9781003542025-19

Which approach is taken is for the dam owner to determine in consultation with regulators.

Many dam owners have a number of dams which do not meet dam safety criteria. The most effective path to reducing risks to the public from their portfolio of dams may be to stage upgrades, addressing the PFM or PFMs which contribute most to the risk rather than meeting deterministic criteria.

An important principle which the ANCOLD (2022) guidelines propose is that risks to the public during dam upgrades should not be greater than the risks of the dam before the upgrade. This is discussed further in Section 19.8.1. If adopted, this principle can influence the design of an upgrade. For example, it may be desirable to install full height filters on a zoned earthfill dam, but to do so may increase risks during construction if part of the downstream shoulder has to be removed to allow construction of the filter. In these cases, the filters may only be designed and constructed in the upper part of the dam to the level the reservoir can be safely lowered during construction.

19.3 REDUCING RISKS FROM FLOOD OVERTOPPING

19.3.1 The General Principles

It is not uncommon for existing spillways having insufficient discharge capacity to pass design floods and as a result the dam not satisfying deterministic and/or risk-based safety criteria for flood overtopping. This often occurs when revisions of flood hydrology are carried out resulting in larger inflow design floods than the spillway was designed for, and/or the population at risk has increased since the dam was constructed, resulting in a higher consequence of failure classification and required Inflow Design Flood (IDF).

There are several ways the risk from flood overtopping can be reduced:

(a) Increase the spillway discharge capacity by widening the existing spillway, effectively widening by installing a labyrinth spillway crest, or lowering the spillway crest level and installing gates.
(b) Lowering the spillway crest level with a resultant loss of storage.
(c) Constructing a secondary spillway.
(d) Raising the embankment crest level and passing an increased depth of flow and discharge flow through the existing spillway.

For low-height weirs and retarding basin embankments which are drowned out in larger floods, the likelihood of failure can be reduced by providing protection to the crest and downstream slope against erosion with Roller Compacted Concrete (RCC) or proprietary blockwork. FEMA (2014) provides details. Davidson et al. (2003) give an example. These methods are not suited to higher embankment dams.

The following section discusses the principles and provides examples of raising of existing embankment dams as part of flood overtopping upgrades.

Spillway design is not covered in this book. Some guidance may be obtained from:

FCERM (2022) Spillway Design Guide. Flood and Coastal Erosion Risk Management Research and Development Programme Research Report, FRS 21238/R1. United Kingdom government Environment Agency.

USBR (1987) Design of Small Dams, 3rd Edition. U.S. Department of the Interior, Bureau of Reclamation.

USBR (2014) Design Standards No. 14. Appurtenant Structures for Dams, Spillways and Outlet Works. Chapter 3. General Spillway Design Considerations. U.S. Department of the Interior, Bureau of Reclamation.

ICOLD (2016) ICOLD Bulletin 172. Technical Advancements in Spillway Design, Progress and Innovations from 1985 to 2015. International Commission on Large Dams. Paris.

19.3.2 Raising of Embankment Dams to Reduce Risks from Overtopping

19.3.2.1 The Most Common Methods Available

As discussed in USACE (2004), the principal methods for raising embankment dams include:

(a) Parapet walls with minor raising of the embankment. Figure 19.1 shows the principles for an arrangement where the raising of the embankment is sequenced so that traffic flow can be maintained during construction. That is not always a requirement.
Parapet walls are commonly the most cost effective for raises of up to about 4.5 m. The parapet wall is usually a maximum height of around 2.7 m. Whilst higher walls may be possible, USACE indicates a limitation is interference with visual inspection of the upstream shoulder from a vehicle on the crest.
The walls may be cast *in situ* concrete or prefabricated. Prefabricated walls are limited in height to what can be transported and installed. For Buffalo Dam (Figure 19.8), the prefabricated walls were placed onto strips of concrete to provide a level base, and the gap below the base grouted with cement grout.
Other methods of construction include casting the concrete base *in situ* with pre-fabricated vertical wall, and all cast *in situ*.
Parapet wall detailing is important. If they are only subject to wave run-up, the detail may be designed as a wave wall. However, if the walls are required to hold the reservoir, they need to be designed to be stable in sliding and overturning, and to not be subject to internal erosion and piping beneath the wall. This is discussed further later.
(b) Crest raising by steepening the upper slopes of the embankment as shown in Figure 19.2. This is usually limited to less than 2 m or 3 m. The core is extended as shown, and no-erosion filters constructed to the top of the core. The filters must be connected to the existing filter or transition, often requiring excavation into the existing embankment as shown.
(c) Crest raising by constructing downstream as shown in Figure 19.3. Downstream construction allows larger raises of the embankment crest. It requires temporary lowering of the crest of the existing embankment so the new core and no-erosion filters can be joined to the existing core and filter/transition.

For deterministic design, the top of the core should be at least to the reservoir level for the Inflow Design Flood with freeboard for wave run-up, plus camber to allow for post-construction settlement. The crest level should be designed so that the likelihood of wave overtopping is negligible as discussed in Section 6.5.5.

For risk-based upgrades, the levels will be as required to satisfy tolerable risk criteria.

Filters should extend to the top of the core, so they are able to arrest concentrated leak erosion in the core in the event the core is cracked by desiccation or on-going settlement.

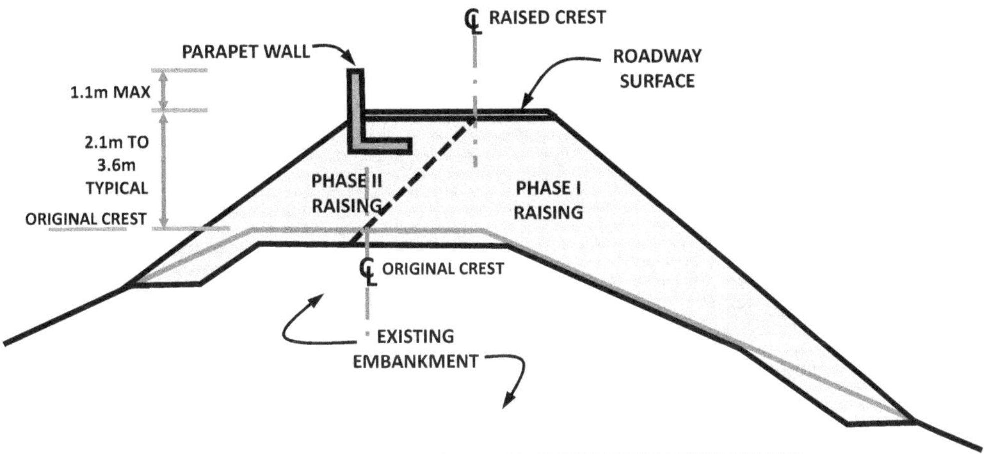

NOTES:

1) PHASE CREST RAISE WITH PARAPET TO MAINTAIN TRAFFIC FLOW ON CREST DURING CONSTRUCTION.
2) WALL DETAILS TO BE ADAPTED TO EXISTING EMBANKMENT GEOMETRY AND ZONING. IMPERVIOUS, FILTER, SLOPE PROTECTION, AND OTHER CRITICAL ELEMENTS TO BE EXTENDED UPWARD AS NECESSARY.

NOT TO SCALE

FIGURE 19.1 Embankment raising by parapet wall and minor raising of the embankment. (Adapted from USACE, 2004.)

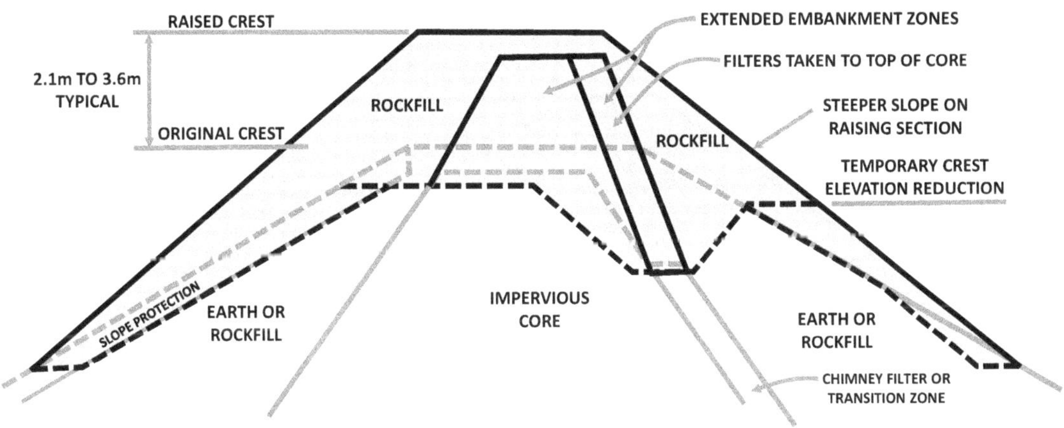

1) TEMPORARY REMOVAL OF TOP OF CURRENT DAM TO EXPOSE INTERNAL ZONES. LIMIT LENGTH OF REMOVAL BASED ON ABILITY TO CLOSE IN CASE OF FLOOD EVENT.
2) EXTEND INTERNAL EMBANKMENT ZONES TO HEIGHT AND WIDTH NECESSARY.
3) UPPER SLOPES TYPICALLY STEEPENED USING HIGH STRENGTH OUTER SHELL MATERIAL SUCH AS ROCKFILL.
4) ADD UPSTREAM SLOPE PROTECTION AS NECESSARY

NOT TO SCALE

FIGURE 19.2 Embankment raising by steepening the upper slopes of the embankment. (Adapted from USACE, 2004.)

19.3.2.2 Parapet Wall Design Principles Where the Reservoir May Rise Up against the Parapet Wall

Figure 19.4 shows schematically good practice for parapet wall detailing where the reservoir level may rise up against the wall under flood conditions.

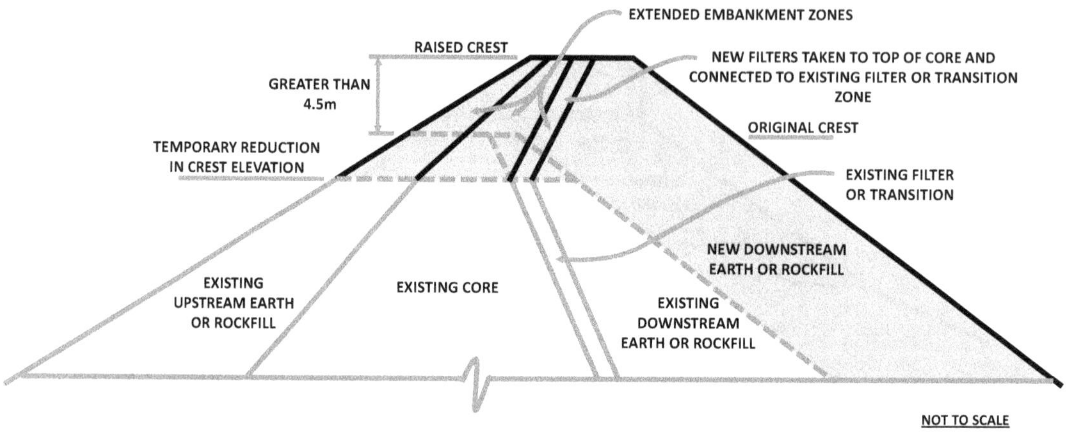

FIGURE 19.3 Embankment raising by downstream construction. (Adapted from USACE, 2004.)

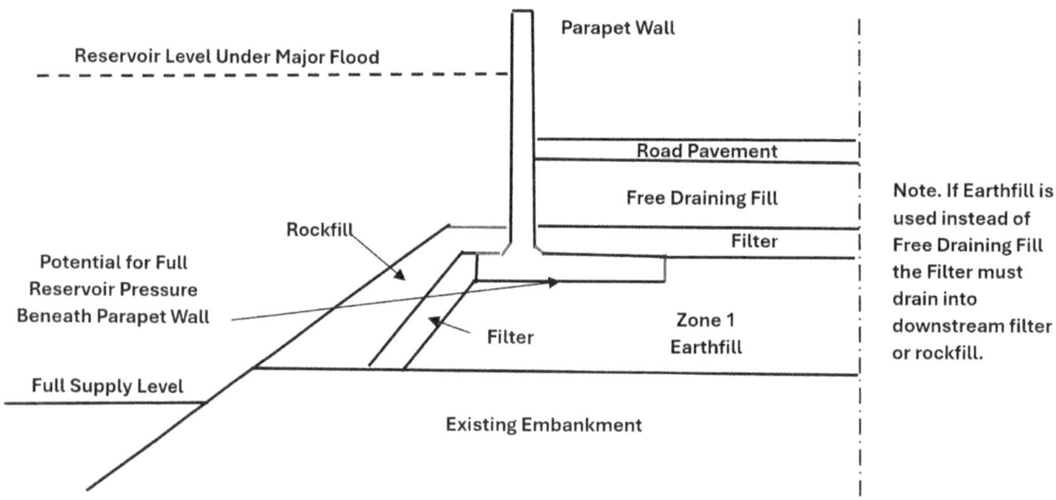

FIGURE 19.4 Parapet wall detailing.

The design principles are:

(a) The parapet wall is to be designed such that it is integral with the Zone 1 earthfill. This is to ensure the parapet deforms with the earthfill core and is not affected by differential deformations associated with settlement of other zones. Good practice is to embed much of the parapet base within earthfill core and incorporate a width of core downstream of the parapet toe, and for the global stability to meet the criteria described in Section 7.12.

(b) As the reservoir rises against the parapet wall, it will tend to rotate about the downstream toe of the wall. A gap may form beneath the wall and full reservoir pressure develop as uplift on the base of the parapet wall if the wall is embedded in earthfill. Figure 19.5(a) shows the free body diagrams for that case and Figure 19.5(b) for where the wall is in free draining fill.

(c) Design should incorporate adequate factor of safety against buoyancy from full uplift under the new Dam Crest Flood.

(d) A filter is placed over the wall to arrest concentrated leak erosion in the gap if it were to initiate. The filter must drain into the overlying free draining fill, or if earthfill is used instead of free draining fill, into the downstream filter. Figure 19.20 shows an example.
(e) The combined weight of the filter, free draining fill and road pavement is sufficient to be greater than the reservoir pressure over the full base length of the parapet wall so heave will not occur.
(f) The filter, free draining fill and pavement provide an adequate factor of safety against horizontal sliding of the parapet wall under the major flood reservoir level.
(g) Provide adequate bearing capacity for the parapet wall on to the earthfill core for a range of load conditions.
(h) Consider the design for joints between lengths of the parapet wall to accommodate forecast future embankment deformation. For large dams, this would include consideration of compression deformations within the central valley section and extension deformation over the mid- to upper abutments.

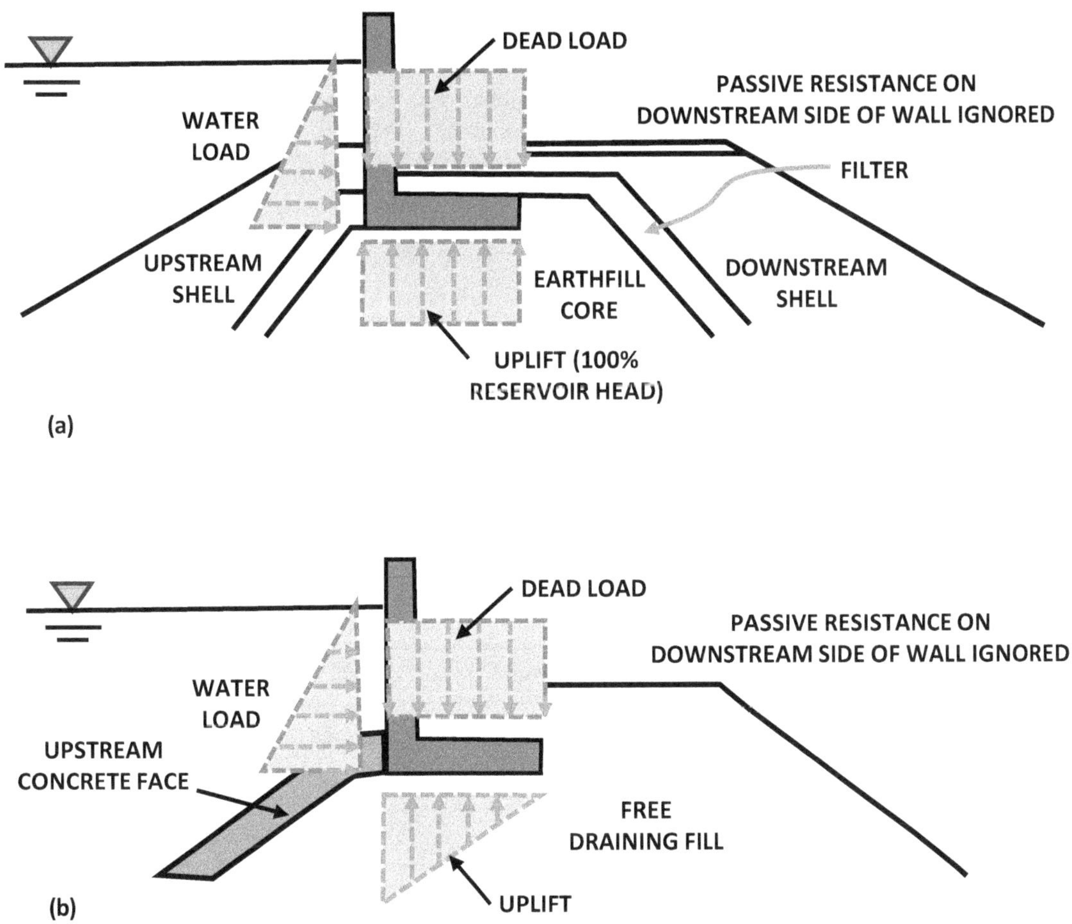

FIGURE 19.5 Free body diagrams for (a) where the parapet wall is embedded in earthfill; (b) where the parapet wall is founded on free draining fill.

(i) Consider details for connection of the parapet wall to the foundation or structure on the abutments.

19.3.2.3 A Case History of Crest Raising with a Parapet Wall

Figures 19.6 and 19.7 show Buffalo Dam, Australia, as constructed in 1965. Figure 19.8 shows the design of the main embankment upgrade carried out in 2003. The upgrade addressed risks resulting from internal erosion and piping in the upper part of the embankment and flood overtopping. Details are provided in Newman and Foster (2006).

The upgrade was an interim risk-based upgrade and included:

(a) Increasing the flood capacity to pass the 1 in 50,000 AEP flood by raising the crest level of the embankments.
(b) A 2.2 m composite earthfill and parapet wall raise of the main and secondary embankments. The use of a parapet wall reduced the total volume of additional fill while maintaining the required width of road on the dam crest.
(c) Reconstruction of the upper part of the main embankment with no-erosion filters and rockfill to address the risk posed by concentrated leak erosion resulting from cross-valley and cross section differential settlement. The original Zone 2C filter transition has excessive fines and potentially held a crack, making it ineffective as a filter. The potential for cracking below the filters was assessed to be low and the residual risks less than the tolerable risks using ANCOLD (2003) *Guidelines on Risk Management.*

LEGEND

Zone	Purpose	Description
1	Earthfill Core	Sandy clay, high to medium plasticity. Well compacted
2C	Filter Transition	Silty sand gravel, with 5% to 25% fines passing 0.075mm (mostly > 10% fines). Fines have a Plasicity Index of about 15. Compaction variable
3A	Fine Rockfill Transition	Silty sandy gravel, loose dumped or trafficked only.
3B	Rockfill	Sandy gravelly boulders, with 10% to 15% fines passing 0.075mm, loose dumped

FIGURE 19.6 Lake Buffalo Dam. Main dam cross section as constructed in 1965 (Newman and Foster, 2006).

FIGURE 19.7 Lake Buffalo Dam immediately after construction in 1965 (Newman and Foster, 2006).

19.4 UPGRADE MEASURES TO REDUCE THE RISK FROM SLOPE INSTABILITY AND DEFORMATIONS LEADING TO LOSS OF FREEBOARD

19.4.1 Introduction

The reasons for upgrading an embankment for potential slope instability and loss of freeboard are usually:

(a) Abnormal deformations detected by monitoring of crest settlement and lateral displacements.
(b) Stability analyses carried out as part of a dam safety assessment indicating that the factor of safety under normal operating or flood loading does not meet deterministic dam safety criteria. This may be because pore pressures are higher than the dam designers assumed but more commonly because the design assumed that conventional effective stress analyses were applicable, whereas undrained strengths in poorly compacted saturated earthfill in the embankment should be used.
(c) Site investigations carried out as part of dam safety assessments or other investigations, e.g. for potential liquefaction in the embankment or foundation, show low strength zones within the embankment or foundation.

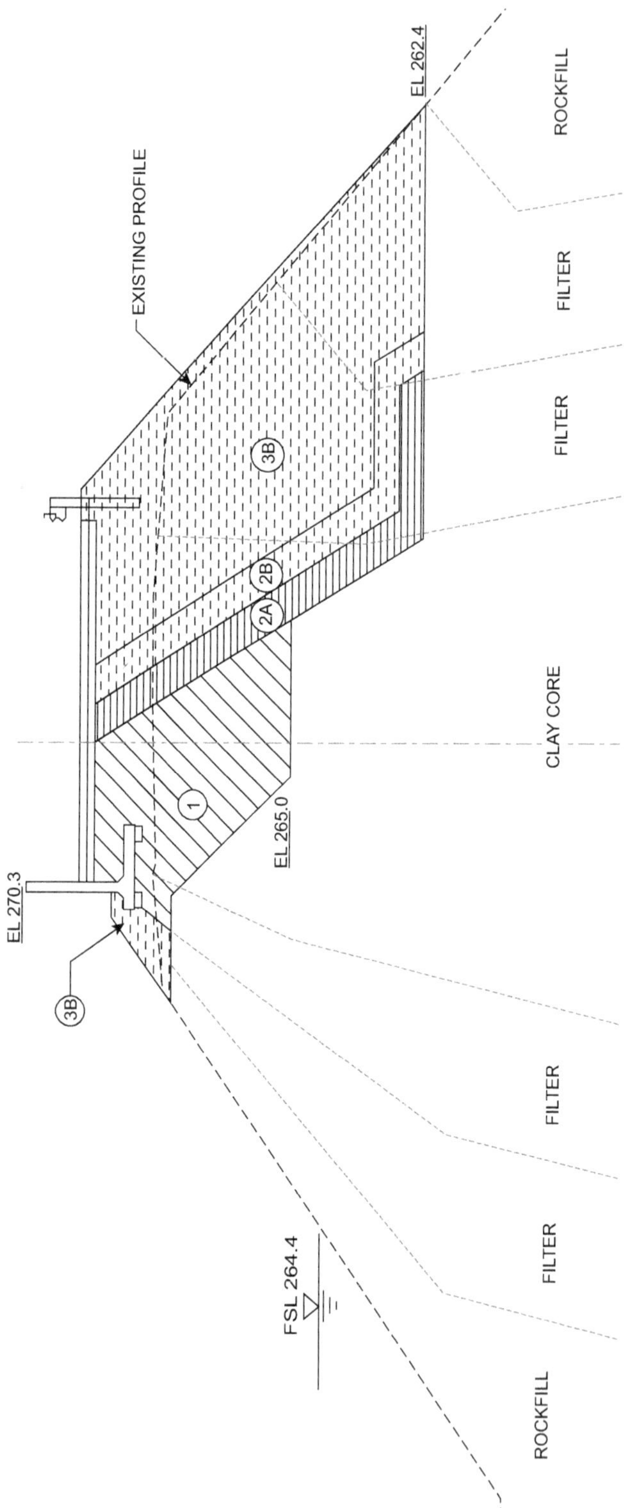

FIGURE 19.8 Buffalo Dam. Cross section of main dam upgrade (Newman and Foster, 2006).

This section only deals with downstream instability. The authors' experience is that remedial works for upstream instability are only carried out when drawdown has resulted in slope failure. We do not know of an example where an upstream upgrade has been carried out because estimated factors of safety were below commonly accepted values, or deformations were somewhat greater than normal. Such situations are usually managed by controlling the rate of reservoir drawdown.

Upgrades for potential upstream slope failure resulting from liquefaction under seismic loads are covered in Section 19.5.3.

19.4.2 Some General Principles

Slope instability upgrades usually involve construction of a full height or partial height berm downstream of the existing embankment. The design should involve the following:

(a) The post-upgrade factor of safety should meet the criteria described in Section 7.12.
(b) This should include potential failure surfaces in the embankment and foundation.
(c) The analyses should be based on adequate shear strength information. If available data is inadequate, consider carrying out additional site investigations and laboratory testing.
(d) If, as is commonly the case, the embankment does not have No Erosion filters, they should be incorporated into the upgrade so far as is practicable.
(e) Foundation preparation for the berm should be to remove low strength strata consistent with the assumptions for stability analyses. Preparation should be as described in Fell et al. (2015).
(f) For soil foundations, incorporate a filter drain on the foundation below the new berm.
(g) Specify the classification, quality and compaction of the material in the berm and the foundation to meet modern standards as described in Fell et al. (2005, 2015).
(h) If the design involves removing some existing fill prior to building the berm, check the factors of safety during this excavation and develop procedures to maintain tolerable risks during construction.

19.4.3 A Case History—Cairn Curran Dam Upgrade

Cairn Curran Dam is located in Victoria, Australia. It is a 44 m high zoned earthfill embankment constructed between 1947 and 1956. There are two embankments, one each side of a gated spillway.

Figure 19.9 shows the general cross section of the embankments. The foundation rock level is higher adjacent to the spillway and slopes downstream. Site investigations showed that the impervious core and semi-impervious zones were constructed of sandy clay, CL/CH and that the sandy gravel zone was in fact silty clayey sand.

A quantitative risk analysis indicated that the likelihood of a concentrated leak was high at the spillway interface due to a discontinuity formed by the wall and geometry of the wall interface. There was also evidence of movement within the embankment fill, spillway approach walls and the road kerb had moved relative to the spillway wall. It was determined that an upgrade was required.

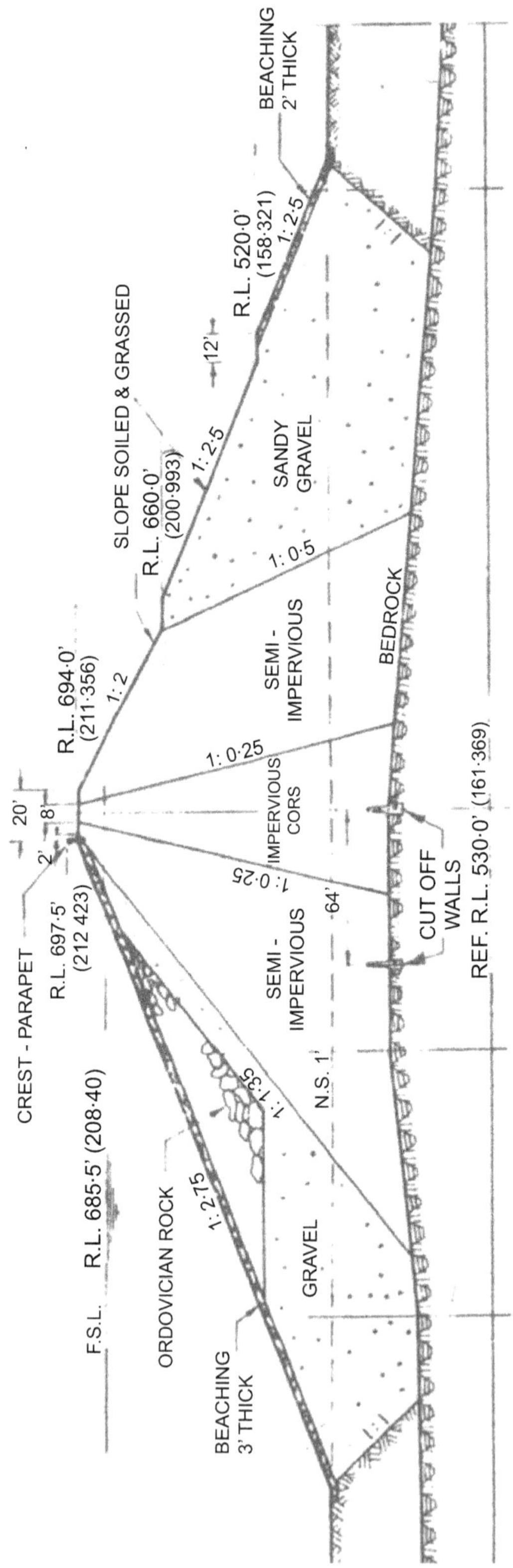

FIGURE 19.9 Cairn Curran Dam; cross section as designed (Swindon et al., 2003).

The upgrade involved constructing full height filters for the parts of the embankments adjacent to the spillway.

Site investigations for the upgrade design determined that there were softened zones at the embankment foundation interface and at about mid height dipping downstream. The upper softened layer was at the level of the phreatic surface.

Figure 19.10 shows the geotechnical model for a cross section 10 m from the spillway wall. The softened zones had undrained shear strength around 10 kPa to 25 kPa.

As shown in Figure 19.11, stability analyses for this cross section showed low factors of safety, less than 1.0 for failure through the softened zone at the embankment foundation contact.

Site investigations showed that the softening of the fill at a section 25 m from the wall was significantly less than at 10 m from the wall. Figure 19.11 shows the geotechnical model at that section. Analyses of stability at that section gave factors of safety of 3.2 for the upper surface and 1.34 for the lower surface. It was therefore assessed that 3D effects were acting to keep the slope from failing. 3D limit equilibrium and numerical analyses confirmed 3D effects. The average of the factors of safety for the lower softened surface at 10 m and 25 m from the wall was 1.01 consistent with the deformations observed.

The indications were that the whole embankment adjacent to the wall was moving downstream parallel to the foundation in that area.

The excavation and backfilling for the upgrade were staged to maintain tolerable factors of safety during construction of the upgrade. The first stage of the upgrade was to remove the wedge of "upstream

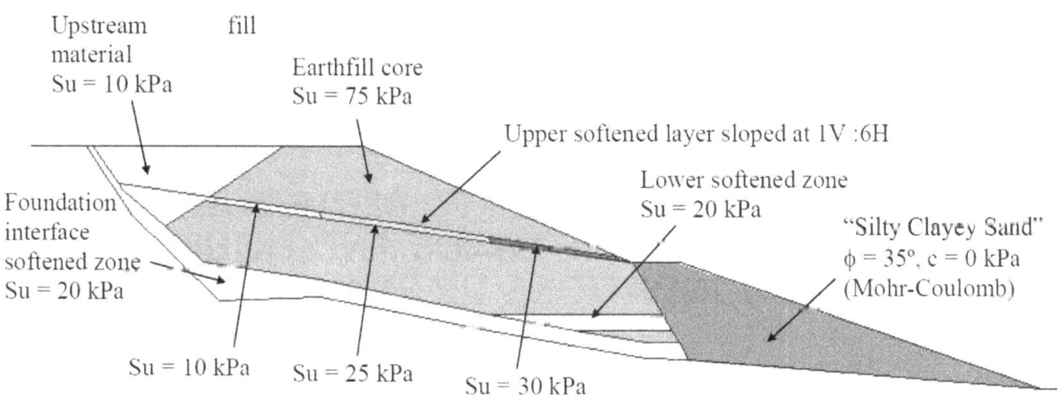

FIGURE 19.10 Geotechnical model; left embankment 10 m from the spillway wall (Swindon et al., 2003).

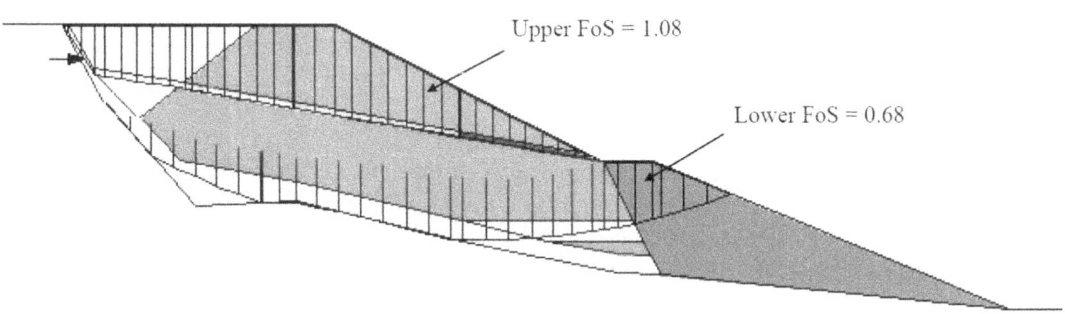

FIGURE 19.11 Results of analysis of stability; section 10 m from the spillway wall (Swindon et al., 2003).

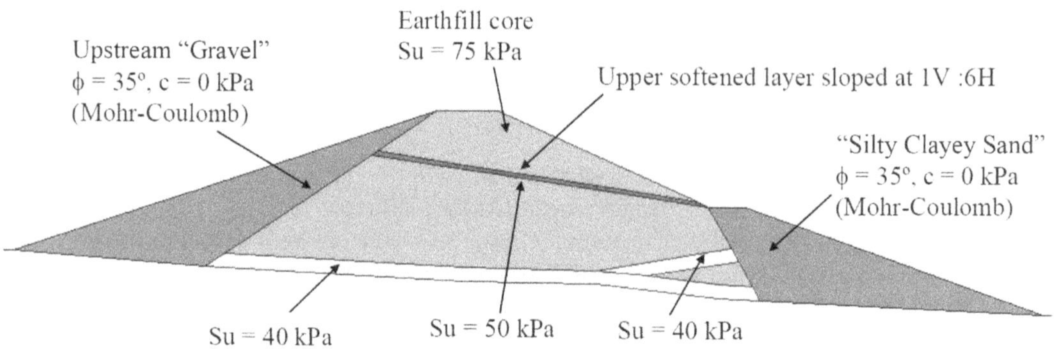

FIGURE 19.12 Geotechnical model; 25 m left of the spillway wall (Swindon et al., 2003).

material" behind the spillway training wall. The lower part of the berm had been constructed before the remainder. Figures 19.13 and 19.14 show the design of the berm and staging of construction.

Post this 2002 upgrade which addressed the instability adjacent to the spillway, the 2D factor of safety at the section 10 m from the wall with the berm was 1.35, and 1.82 at 25 m from the wall. The average of these is 1.52, which was considered satisfactory.

The berm has a Zone 2A fine filter, designed as a No Erosion filter to the existing fill, and Zone 2B between Zone 2A and the rockfill.

Full-height filters and crest raise for the remainder of the embankment were completed in 2007.

19.5 UPGRADE MEASURES TO REDUCE THE RISK RESULTING FROM SEISMIC LOADS AND LIQUEFACTION OF THE FOUNDATIONS OR EMBANKMENT

19.5.1 General Approaches

The method or methods suitable for upgrading a dam and its foundation for seismic loads will be site specific and will require a thorough understanding of the liquefaction mechanics, where liquefaction is potentially of concern, the extent of liquefied foundation and embankment, and the consequences for the dam. It will also depend on the objectives, probably measured in residual risk terms.

The remedial measures may consist of one or more of the following:

(a) Do nothing and accept the potential damage and risks.
(b) Adding filters and possibly raising the crest level of the embankment so the risks posed by deformations and related transverse cracking are reduced to tolerable levels.
(c) Modifying the liquefiable soil in the foundation (and the embankment if applicable), and/or constructing stabilising berms to limit deformations to tolerable levels.

It is not uncommon to use a combination of methods, e.g. to carry out treatment on the downstream but do nothing upstream and tolerate the risks posed by upstream deformations.

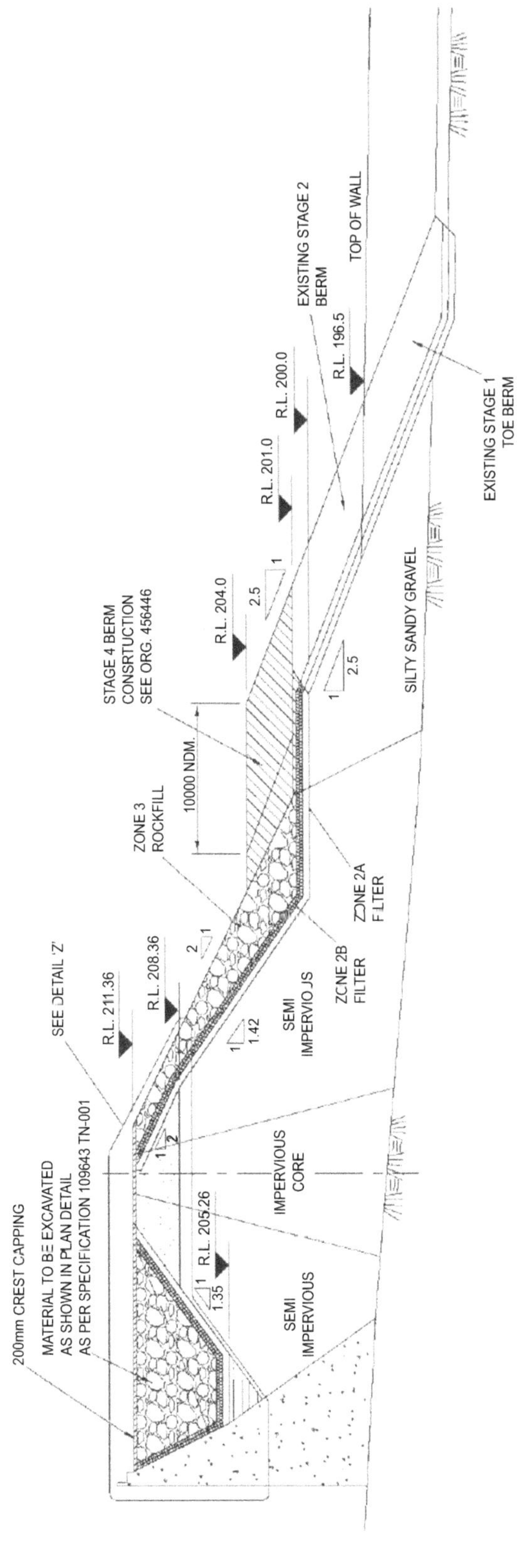

FIGURE 19.13 Cross section showing the left embankment upgrade (Swindon et al., 2003).

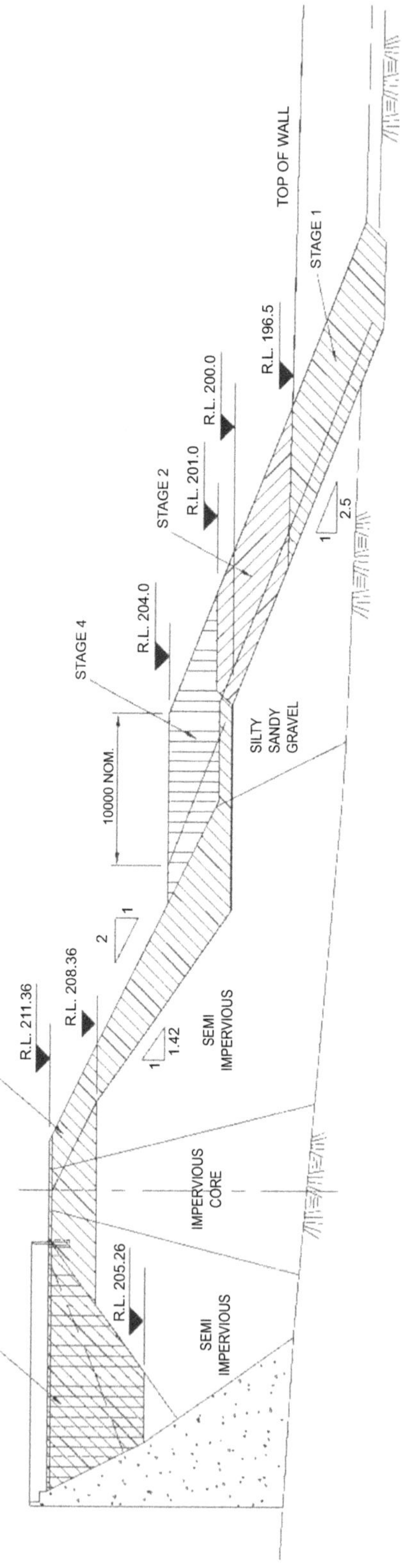

FIGURE 19.14 Staging of the left embankment upgrade (Swindon et al., 2003).

19.5.2 Upgrading of Embankment Dams Not Subject to Liquefaction

The following are the most common remedial measures which may be required:

(a) Provision of filters or upgrade of existing filters in the upper part of the dam. This is usually done to reduce risks of internal erosion and piping to tolerable levels for flood and normal loading, and the upgrade for earthquake loads is achieved at the same time.
(b) For the few dams where the existing freeboard is insufficient to cope with earthquake-induced deformations, the embankment may have to be raised, often along with raising for increased flood capacity.
(c) Where there are strain weakening soils such as high clay content, high plasticity overconsolidated clays in which strains may localise under earthquake deformations, it may be necessary to add a stabilising berm.

19.5.3 Upgrading of Embankment Dams Subject to Liquefaction

The following are the most common remedial measures which may be required:

(a) Construction of a stabilising berm, most commonly founded upon the potentially liquefiable soil after it has been treated by ground improvement as shown in Figure 19.15(c). Alternatively, remove the potentially liquefiable soil from the foundation of the berm.
(b) The purpose of the ground improvement is to densify the soil so it will no longer liquefy and/or to provide additional strength to the potentially liquefiable soil strata. Stone columns and other vertical drainage measures have the added role of draining excess pore pressures developed by the cyclic loading of the earthquake so reducing the likelihood of liquefaction.
(c) Provision of filters or upgrade of existing filters in the upper part of the dam to reduce the risks from internal erosion and piping in transverse cracks induced by the earthquake to tolerable levels.
(d) Where freeboard is insufficient, the embankment may have to be raised.

These works may be required only on the downstream of the embankment, or on both upstream and downstream. Quite commonly only part of the length of the embankment may require remedial works, e.g. if only part is founded upon liquefiable soil.

The ground improvement method used will depend on the nature and depth of the liquefiable soils, the techniques available and cost. Mitchell (2008), Idriss and Boulanger (2008) and Seed et al. (2003) discuss the various ground improvement methods and their limitations. These are summarised in Table 19.1.

The most widely used methods in the authors' experience have been "stone" columns, deep soil mixing, and removal and replacement. Examples are described in Davidson et al. (2003), Toose et al. (2007), Mejia (2005), Dise et al. (1994) and Luehring et al. (2001).

There are detailed design matters which need to be considered when designing "stone" columns. These include:

(a) If the columns are situated on the downstream side of the embankment and there is no cutoff through the alluvium upon which the dam is constructed, the columns may act as drains for seepage water. In this case they should be backfilled with sandy gravel designed to act as a filter to the surrounding soil or backward erosion piping may initiate into the columns. This

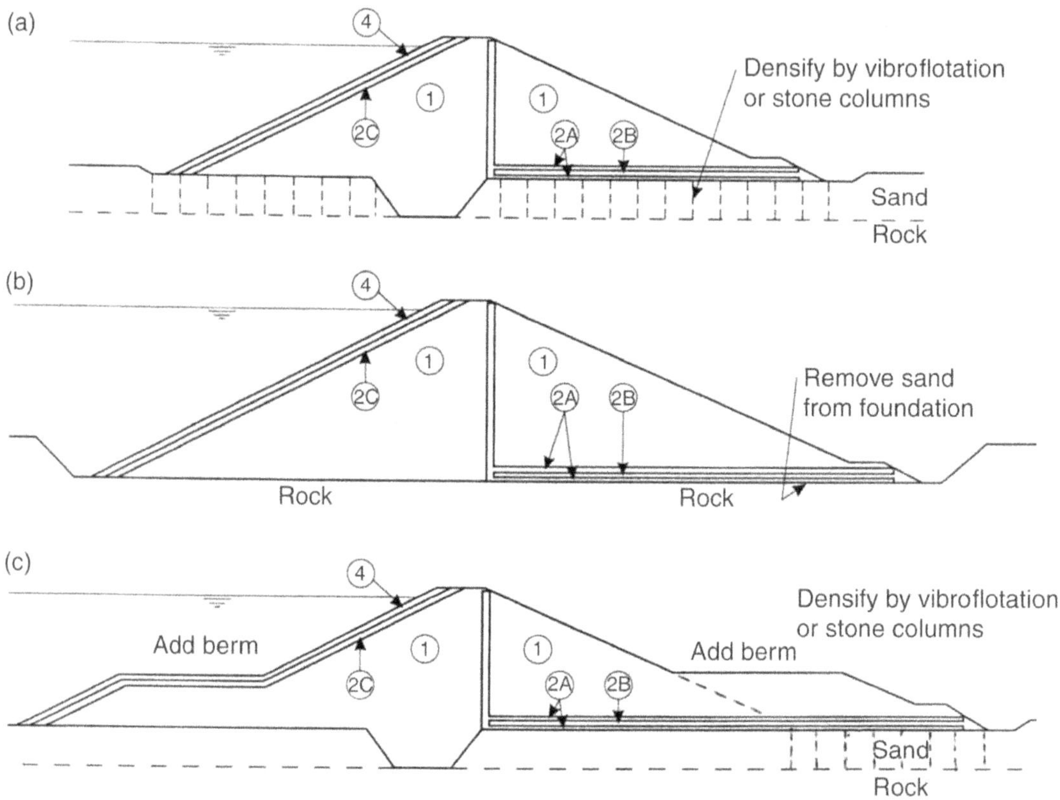

FIGURE 19.15 Principles of ground improvement or treatment to reduce the likelihood of liquefaction in dam foundations: (a) densification by vibroflotation or stone columns; (b) remove potentially liquefiable soils; (c) densification by vibroflotation or stone columns under a berm for an existing dam (Fell et al., 2015).

TABLE 19.1 Summary of Ground Improvement Methods and Their Application for Remedial Works for Liquefiable Foundations (Fell et al., 2015)

METHOD	*METHOD OF IMPROVEMENT*	*SOILS FOR WHICH THE METHOD IS APPLICABLE*	*LIMITATIONS AND COMMENTS*
Vibroflotation	Densification, increased lateral stresses	Sand, sand with some silt, gravelly sand	Ineffective if silt content > 20% approx. May not penetrate gravelly strata.
Vibro-replacement e.g. stone columns	Densification increased lateral stresses, reinforcement, increased drainage	Sand, silt, clay	Ineffective densification if silt content > 20% approx. May not penetrate gravelly strata. Columns may be constructed by driving casing, removing soil from within casing and compacting "stone" as casing is withdrawn. Treatment must penetrate through liquefiable soil or treatment will be ineffective at the base.
Dynamic compaction	Densification increased lateral stresses	Sand, silty sand	Ineffective if silt content > 20% approx. Effective only in the upper 10 m. Quality control difficult. Vibration of adjacent structures.

(*Continued*)

TABLE 19.1 (Continued)

METHOD	*METHOD OF IMPROVEMENT*	*SOILS FOR WHICH THE METHOD IS APPLICABLE*	*LIMITATIONS AND COMMENTS*
Compaction grouting	Densification increased lateral stresses, reinforcement	Sand, silt, clay	Ineffective at depths < 6 m. Augers must penetrate through liquefiable soil or treatment will be ineffective at the base. Quality control may be difficult.
Deep soil mixing	Reinforcement, reduce earthquake-induced strains and pore pressures	Sand, silt, clay	Augers must penetrate through liquefiable soil or treatment will be ineffective at the base. Quality control may be difficult. Soil-cement elements may be brittle.
Jet grouting	Reinforcement, reduce earthquake-induced strains and pore pressures	Sand, silt, clay	Uses very high pressures so should not be used where hydraulic fracture may damage the embankment core. Difficult to control diameter of treatment.
Vertical drainage	Drainage	Sand, sand with some silt	Ineffective in silty sands and silts.
Permeation grouting	Reinforcement	Sand, gravel, sandy gravel, gravelly sand	Particulate grouts, e.g. microfine cement, will not penetrate fine sands because the sand acts as a filter. Chemical grouts expensive and some toxic. Quality control difficult.
Removal and replacement	Removes liquefiable soil	All soils	Dewatering, stability during construction may be problems. Gives high degree of confidence in final product.

can create problems with the construction of the columns as the backfill is low permeability. For some projects the backfill has been designed as a Some Erosion filter, using the Foster and Fell (2001) method which limits the potential for erosion and results in a higher permeability than a no-erosion filter.

The authors are aware of sand boils forming in stone columns used to support a structure as silt surrounding the stone columns eroded into the columns. The columns were constructed of gravel, with no consideration for filtering.

(b) The composite strength of the columns and the surrounding soil is estimated allowing for the relative areas of the columns and the surrounding soil. The strength of the surrounding soil should be taken as the liquefied strength unless it can be demonstrated that the soil between the columns is densified so that it is no longer liquefiable or the soil is sufficiently permeable to allow dissipation of pore pressures built up during the cyclic loading. The latter can be assessed by the method of Seed and Booker (1977). Idriss and Boulanger (2008) indicate that method has been refined by Onoue (1988), Iai and Koizumi (1986) and Pestana et al. (1997), but they caution that uncertainties in the hydraulic conductivities of the surrounding soil (particularly as it is disturbed by construction of the columns) and the columns themselves make it difficult to have confidence of efficient drainage except in relatively high permeability soils.

(c) Seed et al. (2003) discuss the effect of the stiffer stone columns attracting the cyclic shear stresses and potentially reducing the build-up of pore pressure in the liquefiable soil. They caution against relying upon this because the columns still flex if they are longer than three times their diameter.

19.5.4 Examples of Treatment of Liquefiable Foundations

19.5.4.1 Yarrawonga Weir–Stone Columns

Yarrawonga Weir is a major irrigation diversion weir on the Murray River in Australia. It has a gated spillway structure and 7 m high earthfill embankment founded on loose alluvial sand interbedded with silty clay. It was constructed in the 1930s.

Investigations and analyses carried out in 2002 (Davidson et al., 2003) showed that the loose sand in the foundation of the earthfill embankment was susceptible to liquefaction even under the OBE. Deformations during and post-earthquake were estimated as > 4 m with loss of freeboard and breach of the reservoir. It was decided that an upgrade with stone columns in the foundation was required to maintain earthquake-induced deformations to tolerable levels.

Figure 19.16 shows a cross section of the embankment with the upgrade works with stone columns upstream and downstream. The upgrade also included construction of a downstream stabilising berm and protection of the downstream slope from erosion during flood overtopping with a proprietary blockwork slope protection.

As described in Davidson et al. (2003), stone columns were selected because of their ability to drain excess pore pressures during the earthquake and to increase the density of the loose foundation soils, making the soil less susceptible to liquefaction and having a higher liquefied strength. Post upgrade, the required post-earthquake factor of safety was ≥ 1.3 and deformations < 200 mm.

The stone columns were constructed of minus 35 mm sandy gravel. The grading is shown in Figure 19.17. The columns were designed to act as a some erosion filter to the foundation soils. If the columns had been constructed of clean gravel as is common for roadworks or building foundations, it would have allowed erosion of the foundation soils into the columns. Backfilled with sandy gravel designed as a No Erosion filter, the columns would have been too low permeability to act as drains and would be difficult to compact.

The columns were constructed on a 2.4 m triangular grid spacing by driving a 600 mm diameter steel column with gravel plug to bedrock. The sandy gravel column was placed in layers and compacted as the steel column was withdrawn.

19.5.4.2 Lance Creek Dam—Deep Soil Mixing

Lance Creek Dam, Victoria, Australia, is a 17 m high zoned earthfill embankment originally constructed in 1911 and raised in 1986. Investigations in 2002 and later showed the presence of potentially liquefiable silty sand in the foundation. Estimated post-earthquake factors of safety were less than 1.0 and large deformations were predicted.

As described in Toose et al. (2007), it was determined that to limit deformations to levels at which tolerable risk criteria would be achieved, ground treatment was required to increase the post-earthquake shear strength in the liquefiable silty sand. The foundation under both upstream and downstream shoulders required treatment. Stone columns were an option, but the silty nature of the sand meant that close column spacing was required for them to be effective. That and the difficulty of constructing stone columns in the upstream foundation through the embankment meant that deep soil mixing (DSM) was the preferred option.

Figures 19.18 and 19.19 show the layout of the deep soil mixing. The treatment zones are located where they provide optimal contribution to the post-earthquake stability. The reservoir level was drawn down to allow construction of a working platform on the upstream shoulder of the embankment. The specified fully cured undrained strength of the grouted DSM columns was 500 kPa. Within the treated area the columns were 30% of the area.

Details of the design and construction are provided in Toose et al. (2007).

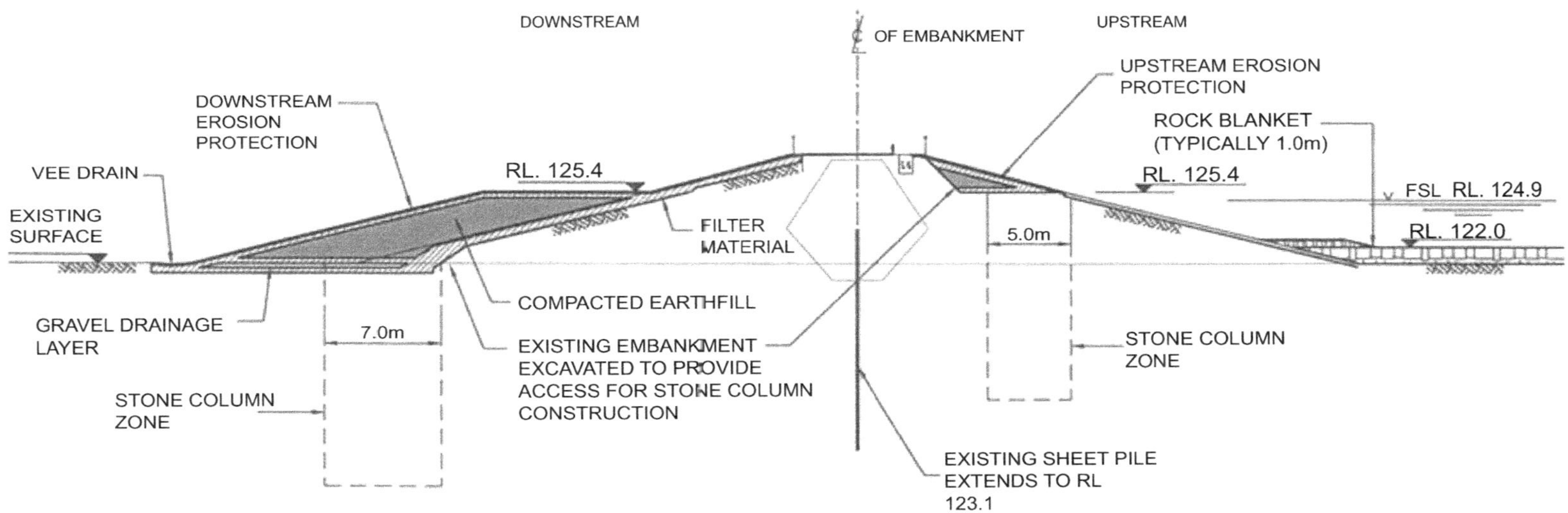

FIGURE 19.16 Yarrawonga Weir. Stone column remedial works to address post-earthquake instability and deformations (Davidson et al., 2003).

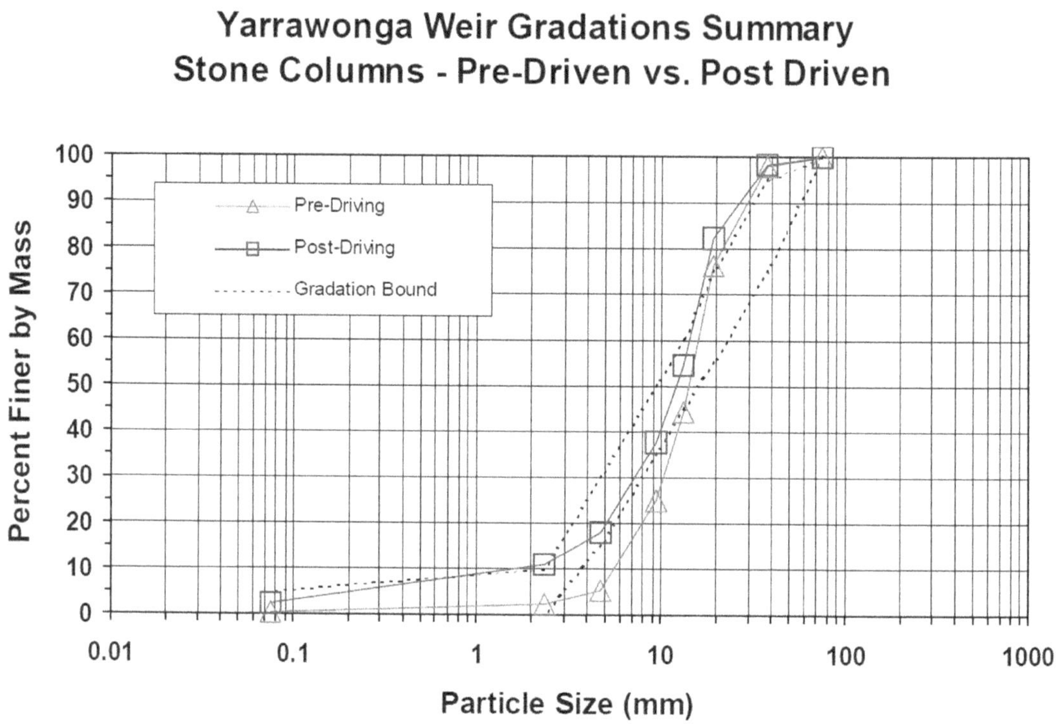

FIGURE 19.17 Yarrawonga Weir. Grading of the stone column material (Davidson et al., 2003).

19.5.4.3 Toorourrong Dam—Grouted Stone Columns

Toorourrong Dam, Victoria, Australia, is a puddle core earthfill dam constructed in 1885. It experienced several piping incidents in the embankment in the 1970s. It is founded on a very complex alluvial deposit, some of which was assessed to be susceptible to liquefaction under large earthquakes. The dam also had insufficient spillway capacity to pass the design floods. Details are given in Johnson and Arnold (2011).

The upgrade shown in Figure 19.20 consisted of raising the dam with a concrete parapet wall and embankment raise; ground improvement of the liquefiable foundations by construction of grouted stone columns; and provision of filters and drains to control internal erosion in the embankment and the foundation.

The stone columns can potentially act as drains and because the foundation soils were very fine, they would have required a fine filter backfill to prevent internal erosion of the foundation soil into the columns. This was too fine to be constructed using the equipment available, so grouted stone columns were used. These provided the additional shear strength required in the foundation but did not allow erosion into the columns.

The very wide berm for cross section (a) was required to achieve the required factor of safety for the post-earthquake liquefied soil condition. That section did not have stone column treatment.

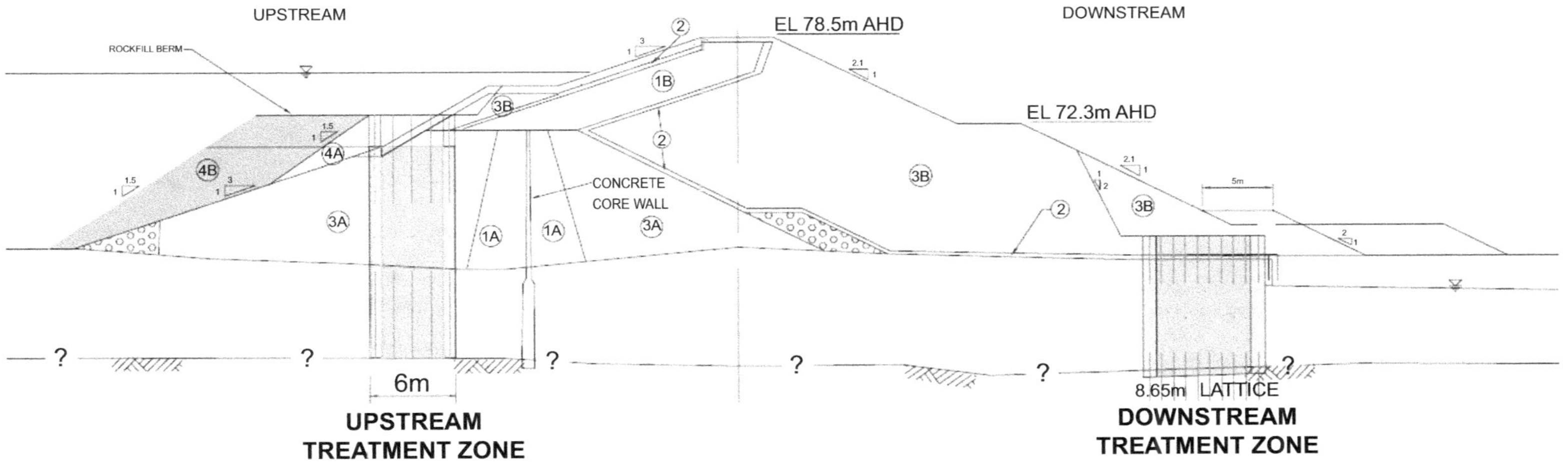

FIGURE 19.18 Cross section of Lance Creek Dam showing the deep soil mixing treatment zones (Toose et al., 2007).

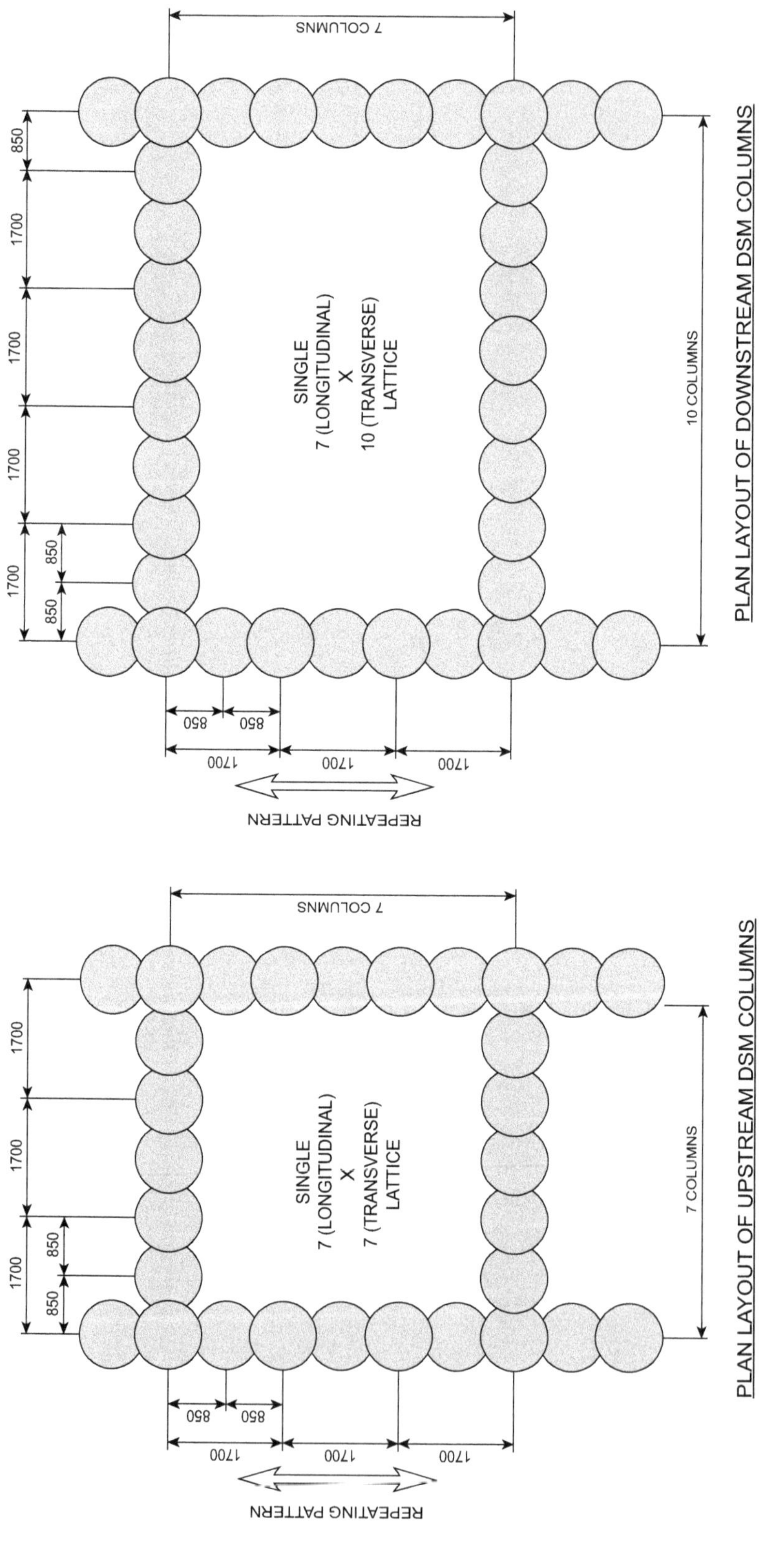

FIGURE 19.19 Lance Creek Dam. Plan layout of deep soil mixing columns (Toose et al., 2007).

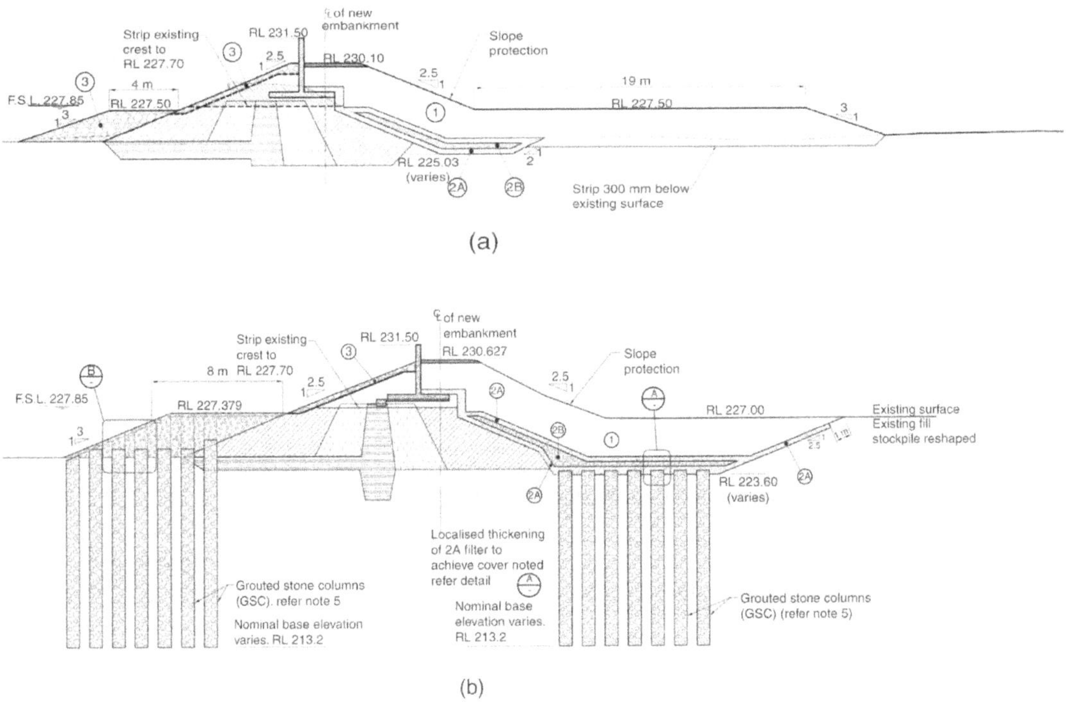

FIGURE 19.20 Upgrade of Toorourrong Dam for flood, seismic loads and internal erosion and piping. (a) Typical section; (b) section where there were more extensive potentially liquefiable soils (Fell et al., 2015 from Johnson and Arnold, 2011).

19.6 EXAMPLES OF DAM UPGRADES TO ADDRESS DEFICIENCIES IN INTERNAL EROSION AND PIPING CONTROL

19.6.1 Upgrades to Reduce the Likelihood of Continuation of Erosion by Providing Filters and Cutoffs

Figures 19.21 to 19.28 show examples of dam upgrades which were carried out to address deficiencies in the control of internal erosion and piping by upgrading the filters.

In some cases, the upgrades also address flood and potential liquefaction in the dam foundation.

The following are some details of these upgrades:

19.6.1.1 Wingecarribee Dam (Figure 19.21)

The upgrade for this 1970s dam consisted of the excavation of a 3 m deep × 1.2 m wide trench to intercept the existing sloping chimney filter which had been designed and constructed only to reservoir full supply level. This was common practice up to and in a few cases beyond that time. Many dam engineers at the time did not appreciate the fact that internal erosion in plastic soils such as used for this core would erode in cracks formed by differential settlement and/or desiccation, and erosion could initiate in hours or days if the reservoir rose under flood conditions. They had assumed that piping occurred as a result of seepage and that there was insufficient time under flood conditions to establish a flow net above FSL.

The new chimney filter extension was provided with transverse outlet trenches at 50 metre intervals. The outlets have adequate discharge capacity because if internal erosion occurs the erosion will cause the eroded soil to form a "cake" on the filter and the flow of water in the crack will be almost fully stopped.

The filter was compacted in the trench by a vibrating plate attachment to an excavator.

19.6.1.2 Laanecoorie Dam (Figure 19.22)

Laanecoorie Dam is a puddle core earthfill dam constructed from 1889 to 1891. It breached by overtopping during a flood in 1909 and was reconstructed.

During major floods in 2010 and 2011, the embankment cracked and deformed largely because the downstream slopes were partly inundated by high tailwater levels causing collapse settlement of the shoulder fill.

Investigations found evidence of cracking and root holes through the core. The dam was upgraded in 2013 to address potential instability and internal erosion and piping. Hunter et al. (2014) provide details.

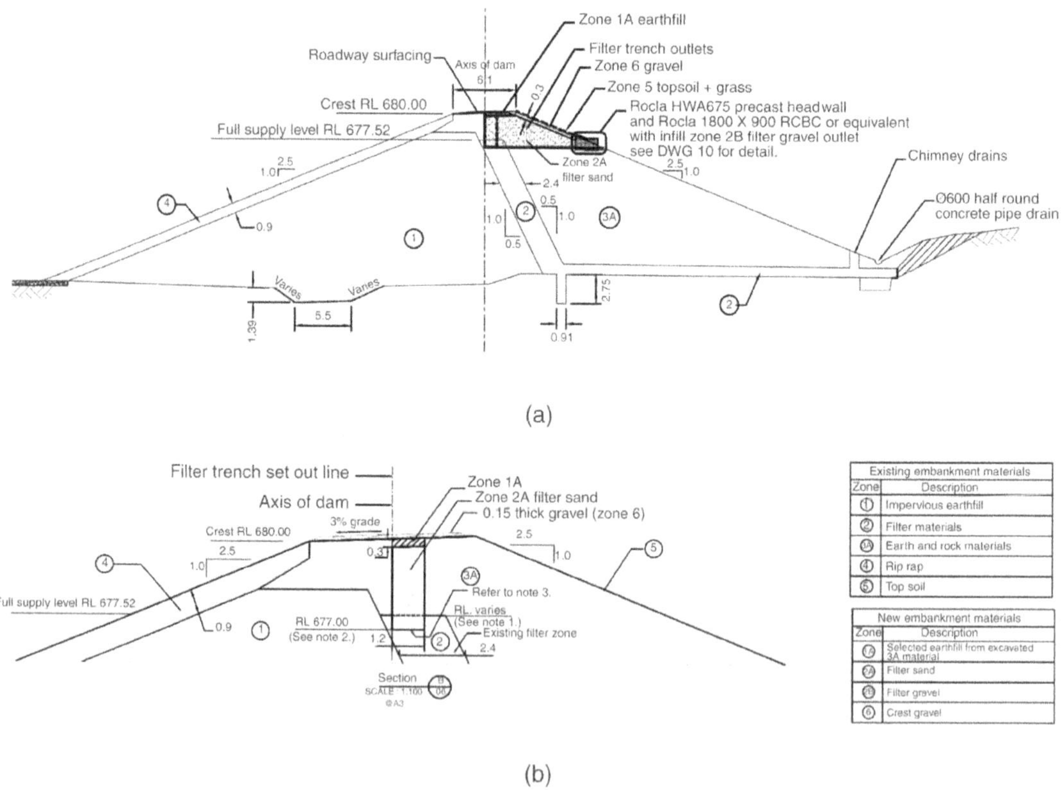

FIGURE 19.21 Extension of chimney filter to dam crest, Wingecarribee Dam: (a) typical section; (b) details of filter extension (Fell et al., 2015, courtesy of WaterNSW and AECOM).

The dam was buttressed with a rockfill berm with a two stage filter between the existing dam fill and the rockfill. Filters were also provided to control internal erosion of the foundation into the rockfill.

In the dam abutments a vertical trench up to 3 metres deep was excavated and backfilled with fine filter. This was provided to further control internal erosion in the foundation particularly in the upper 2 m to 3 m where roots from trees had been found in test pits.

It will be noted that the filters were constructed at a minimum practicable width of 0.75 m because of the high cost of filter materials. The filters were placed by excavator bucket progressively as the fill was raised. On other projects "Flo-Con" side cast conveyor equipment has been used. This narrow width can only be achieved with very strict construction supervision.

On some projects segregation of the coarse filter has been a problem requiring very careful placement, and to accommodate the coarser grading a minimum width of 1.0 metre or more has been adopted.

The use of rockfill rather than earthfill has been found to be the best solution in a number of dam upgrades because it is easier to place the rockfill in the narrow widths required, and placement is less affected by wet weather.

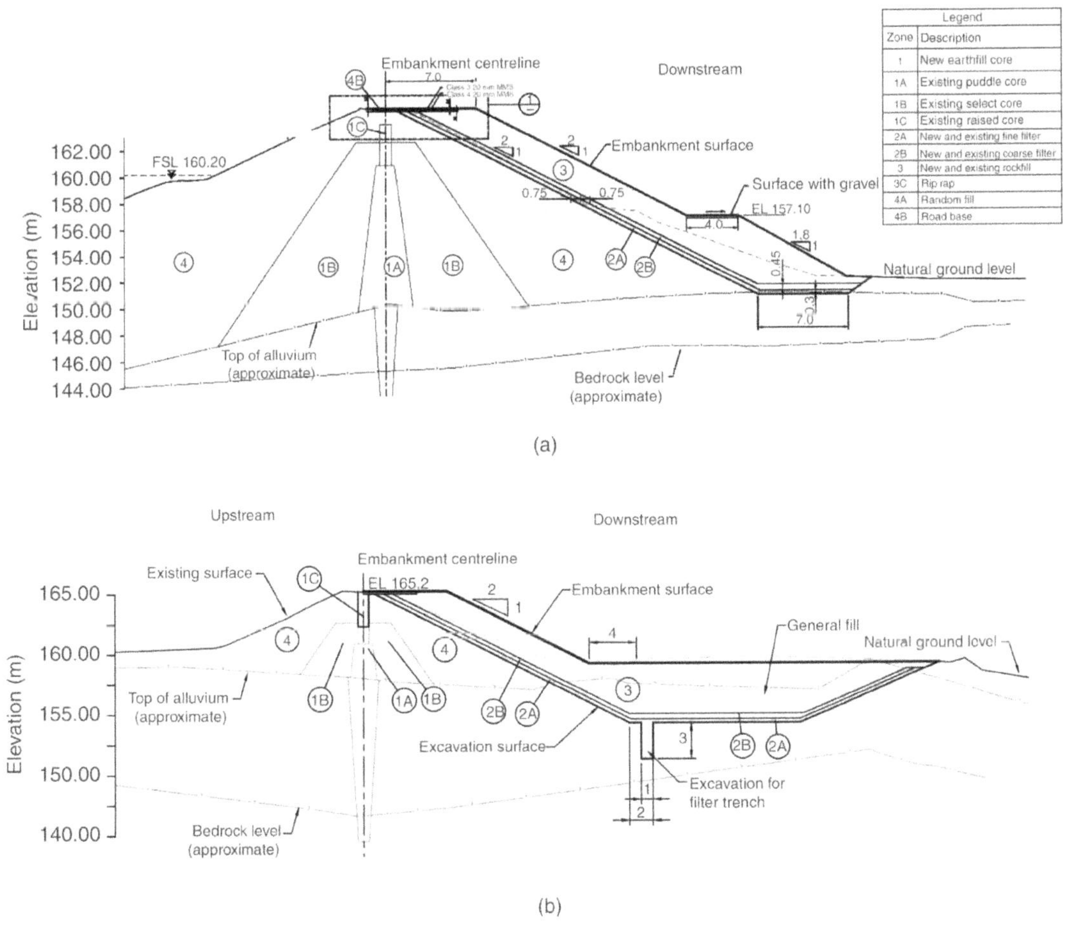

FIGURE 19.22 Upgrade of Laanecoorie Dam for internal erosion and piping: (a) typical section in river bed; (b) typical section in abutment (Fell et al., 2015, courtesy of Goulburn-Murray Water and AECOM).

19.6.1.3 Tullaroop Dam

Tullaroop Dam is a 43 m high zoned earth and rockfill dam in Victoria, Australia, constructed in the late 1950s. Figure 19.23 shows the cross section of the original dam. The earthfill is well compacted but the rockfill was not compacted.

There had been a history of cracking dating from 1987. As discussed in Section 9.2.1, in 2004 cracks up to 60 mm wide developed in the left abutment downstream of the crest over a two week period. Significant factors contributing to the crack formation were the presence of pre-existing shear planes in the foundation not removed as part of the embankment construction, and extensional strains resulting from the sharp change in foundation slope on the left abutment from excavation of the temporary diversion channel. The timing of the crack in 2004 was assessed as due to a combination of sustained drawdown and the extended drought period. The extended drawdown resulted in reduction in pore pressures to historic lows and increased effective stresses to historic highs post first filling within the central to upstream portion of the broad earthfill core, resulting in settlement in the mid to lower zone of the earthfill core as evidenced within the vertical settlement gauge.

As described in Hunter et al. (2017), the reservoir was subject to prolonged drought conditions remaining at low levels for almost 10 years.

In September 2010 the reservoir filled rapidly (in days) and experienced a historic high level 1.5 m above FSL and another historic high level in January 2011 2.7 m above FSL. In May 2011 three piezometers within the earthfill core recorded very rapid rises in pore water pressure equivalent to 12 to 13 m pressure head. The levels were close to and just above the embankment surface. The piezometers then showed a steady fall returning to pre-rises levels in 4 to 6 weeks.

Extensive investigations indicated that:

(a) The pore pressure rises were real, not due to instrument malfunction.
(b) Water had probably entered the upstream slope in longitudinal cracks similar to what had been observed on the downstream berm during the prolonged low reservoir levels.
(c) Piezometers installed in the core in 1985 had been installed in boreholes drilled by wash boring methods. Problems had been recorded during this work including loss of drill water circulation consistent with hydraulic fracture within the embankment core.
(d) The water from the longitudinal cracks may have penetrated into the fractures formed by hydraulic fracture during the drilling and caused the pore pressure rises. Figure 19.24 shows the postulated seepage pathways.
(e) Not all the evidence supported this concept, but no viable alternative was found to be the cause of the pore pressure rises.
(f) It was concluded that it could not be ruled out that there were interconnecting defects through the core in which concentrated leak erosion could initiate under high reservoir levels.

A near full height filter berm was constructed in 2015/2016. Cowan et al. (2016) provide some details. Figure 19.25 shows the cross section.

Key aspects of the upgrade works included:

- A buttress comprising fine and coarse filters and an outer shoulder of rockfill.
- Construction of an intermediate filter trench from RL 213 m local to the maximum section to manage dam safety risk from concentrated leak erosion during construction prior to removal of the existing rockfill shoulder below RL 213 m.
- Construction of a filter trench from a bench at RL 209 m for a depth of 6 m across the full length of the embankment designed to intercept potential seepage and concentrated leak pathways in the lower portion of the embankment. It was not practicable to excavate to foundation level in this area because of construction slope instability risks.

- Construction of a coarse rockfill berm at the toe of the buttress with high discharge capacity to reduce the residual risk. The existing Zone 2 transition gravel was too coarse to be a No Erosion filter but should eventually act as a filter in the event concentrated leak erosion initiated. The rockfill berm was designed on "Swedish Berm" principles described in Section 19.6.2.
- Connection of the new filter buttress to that constructed in 2005 in the left abutment.
- Provision of filters around the outlet conduit through the embankment.

19.6.1.4 Upper Yarra Dam

Upper Yarra Dam is a 90 m high earth and rockfill dam in Victoria, Australia. It was constructed in the 1950s. Figure 19.26 shows the cross section and zoning. The dam is an extreme consequence of failure classification.

Quantitative risk assessments indicated that upgrade works were required to reduce the risks from internal erosion and piping. The main contributors to the risk were the potential for concentrated leak

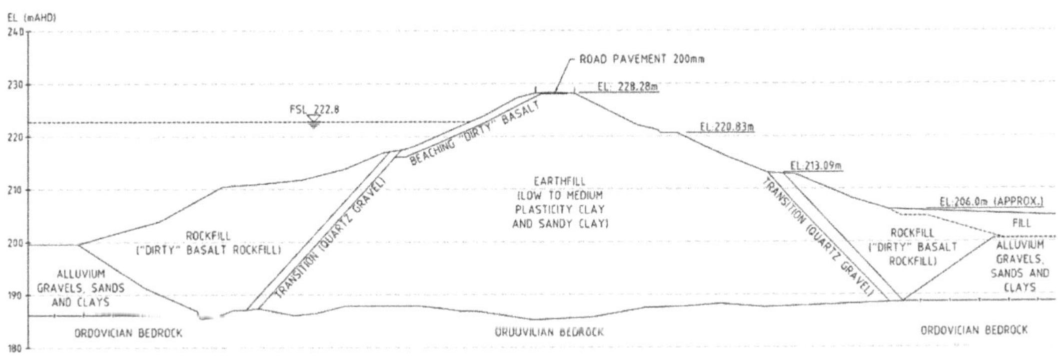

FIGURE 19.23 Tullaroop Dam; original dam cross section (Cowan et al., 2016).

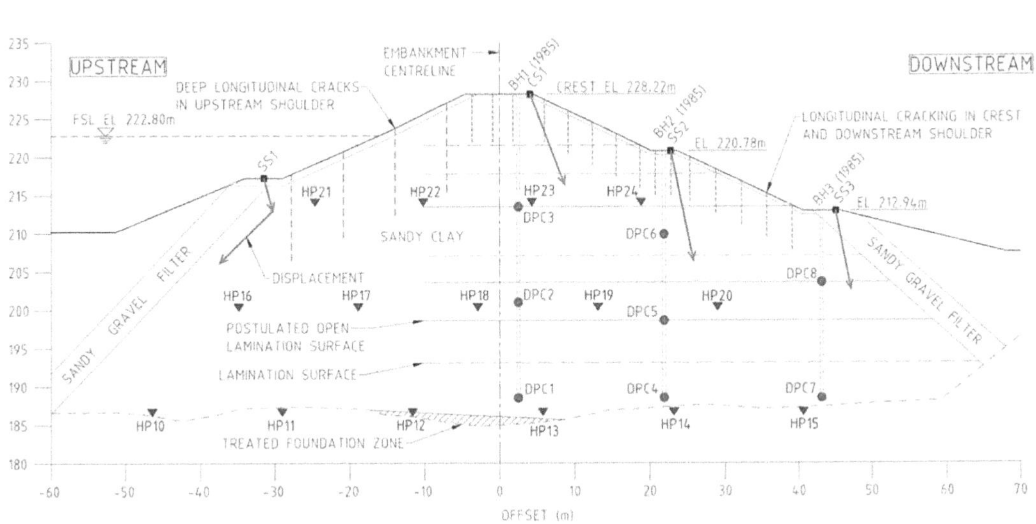

FIGURE 19.24 Tullaroop Dam; postulated seepage pathways (Hunter et al., 2017).

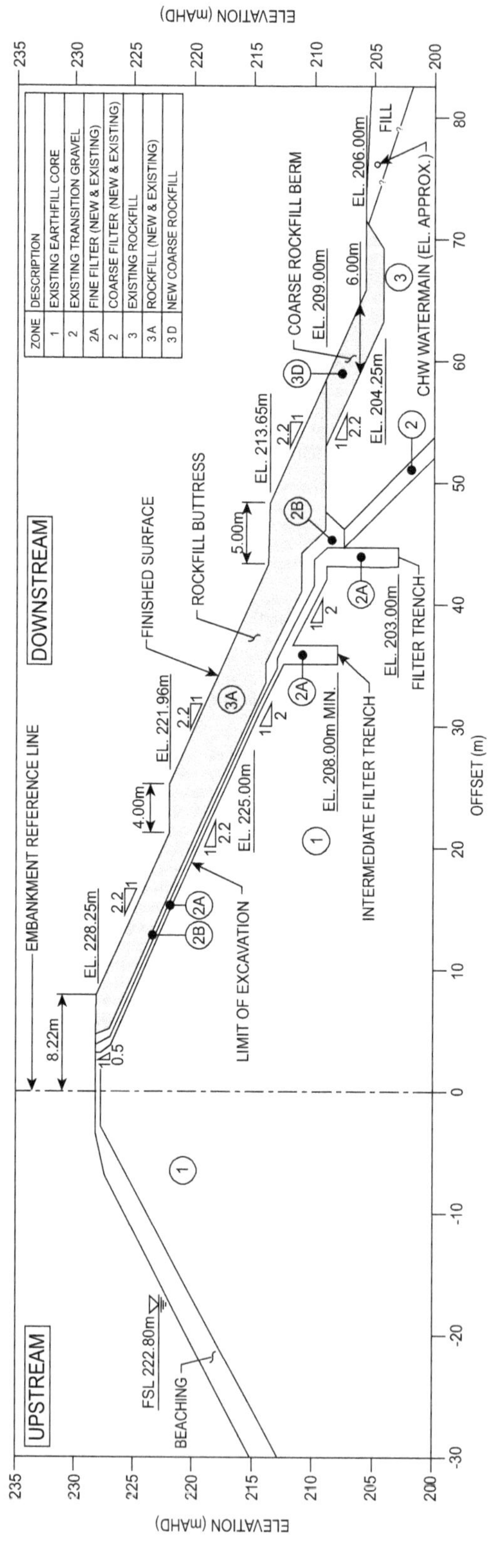

FIGURE 19.25 Tullaroop Dam; cross section showing 2015 upgrade berm and filters (Cowan et al., 2016).

erosion in transverse cracks in the right and left abutments and in a potential shear surface in the upper central part of the embankment described in Section 9.2.6 and Figure 9.27.

Extensive site investigations, analyses and engineering assessments were carried out. These are described in Hunter et al. (2022) and Chia et al. (2022). The numerical analyses are summarised in Section 9.3.2, Figure 9.42 and Table 9.8.

Several options to reduce the risk were considered. These are summarised in Hunter et al. (2022). The chosen option was to excavate 10 m from the upper part of the embankment across the full length of the embankment. In the left and right abutments, the excavation was taken to 15 m below the original crest level as shown in Figures 19.27 and 19.28.

At that level secant filter columns were installed to the level of the base of potential transverse cracking under flood and seismic loading indicated by the numerical analyses. The maximum depth of these columns was up to 31 m below embankment crest level. The columns were 1.2 m diameter overlapping 0.4 m and up to 15.6 m deep.

After the filter columns were installed the embankment core and two stage filters and rockfill were constructed to bring the embankment up to crest level.

Construction works and the reservoir level were managed to maintain the risks to the public downstream to at or below the risks before the upgrade.

19.6.2 Upgrades to Reduce the Likelihood of Breach

The likelihood of breach by unravelling of the downstream rockfill for dams, which have large leakage flows before the filters seal, can be reduced by providing a berm constructed of large diameter rockfill.

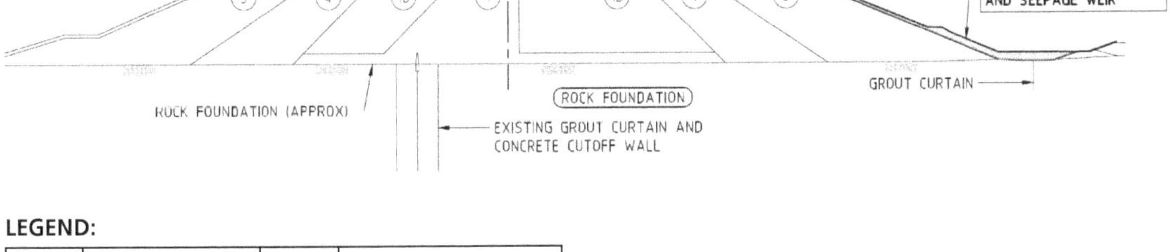

LEGEND:

Zone		Zone	
1	Earthfill Core	3	Rockfill
1A	Select Earthfill	3C	Rip-rap
1B	Soft Rock	4	Indifferent Rockfill

FIGURE 19.26 Upper Yarra Dam. Cross section showing zoning of the original embankment and the outline of the embankment upgrade in the central part of the embankment. (Figure courtesy of Melbourne Water and AECOM.)

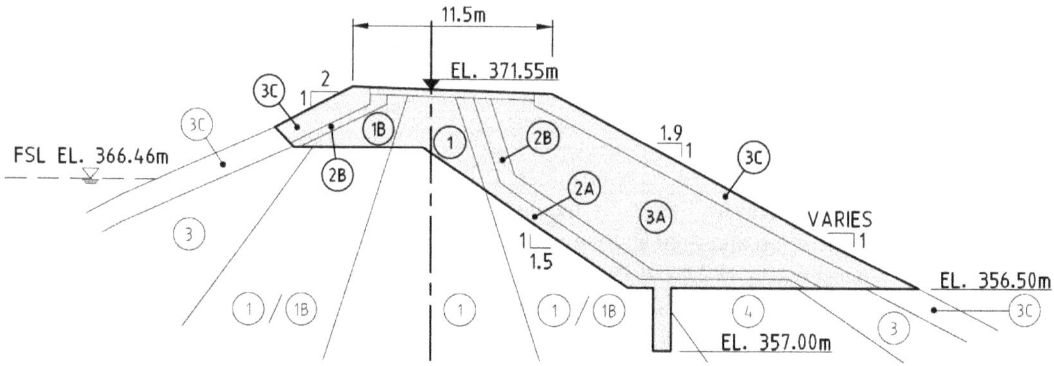

LEGEND:

Zone		Zone	
1	Earthfill Core	2B	Coarse Filter
1B	Soft Rock	3A	Rockfill
2A	Fine Filter	3C	Rip-rap

FIGURE 19.27 Upper Yarra Dam. Cross section showing the upgrade on the left abutment. (Figure courtesy of Melbourne Water and AECOM.)

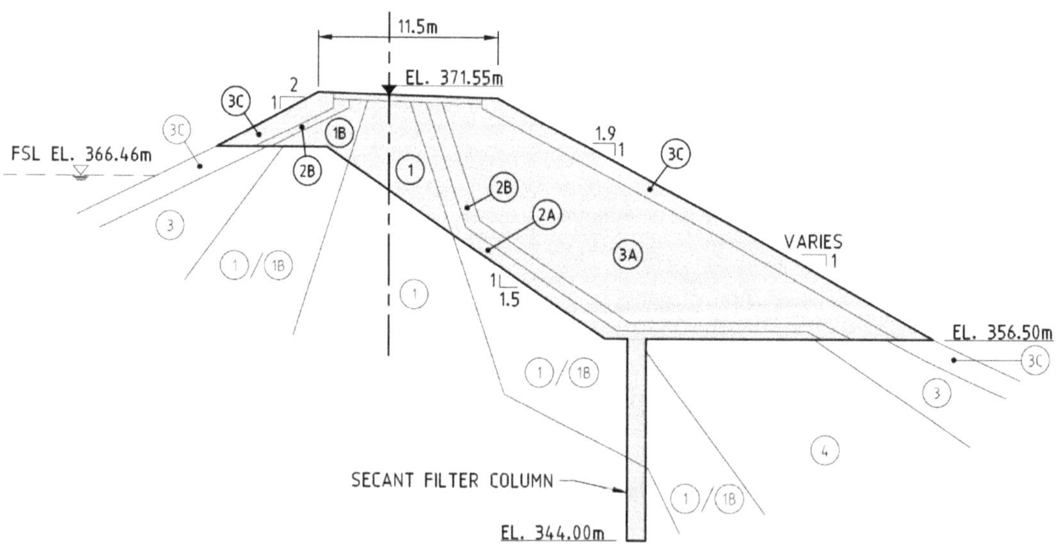

FIGURE 19.28 Upper Yarra Dam. Cross section showing the upgrade on the right abutment. (Figure Courtesy of Melbourne Water and AECOM.)

This has been used in a number of dams, particularly in Sweden. These dams are usually central core earth and rockfill dams. The cores are glacial till, and the downstream filter was often constructed of tunnel spoil or spoil from underground power stations.

As discussed in Nilsson (2007a, 2007b, 2009) and Rönnqvist (2010), the filters were often too coarse to satisfy No Erosion criteria and fell into the some or excessive erosion range. Many of the dams had experienced internal erosion and piping incidents with significant leakage flows before the filters sealed.

To address this problem, the concept of the "Swedish berm", a rockfill berm on the dam's downstream rockfill shoulder, was introduced. This rockfill berm was designed to reduce the likelihood of failure of the dam by unravelling of the downstream slope resulting from the potentially large flow through the dam that might result from substantial erosion of the core. The flows through the dam are estimated assuming the core is eroded such that the permeability of the damaged core is the same as the rockfill and the flow through the dam is controlled by the upstream and downstream rockfill zones.

Figure 19.29 shows the assumed seepage conditions. The permeability of the downstream and upstream rockfill is determined from the gradation of the rockfill. From these the flow rate for which the berm is designed to prevent unravelling of the downstream slope can be calculated. Details of the design method are given in EBL (2005).

In Norway the regulator has some minimum requirements for the throughflow for which the berm should be designed. These are described in NVE (2012). In summary the requirements are that a total design leakage of 5 m^3/sec or a flow per metre length of dam of 0.5 m^3/sec is required. The size of the rockfill in the berm is determined by the methods outlined in Section 6.7.4.

The principle of the berms is that they will control the breach process until the filters seal with material transported by the leakage flow from upstream zones of the dam as has happened in many incidents.

As shown in Figures 19.30 and 19.31, the berm may be full height or only at the toe depending on where internal erosion and piping may be expected to occur and the nature of the existing rockfill.

Nilsson (2013) advises that a transition layer is provided between the existing fill and the rockfill berm. This is designed to allow the fines from the core and the existing rockfill to pass so there is not a build-up of pressure on the interface. This will result in higher permeability existing rockfill which needs to be allowed for in designing the new rockfill in the berm.

If sealing on the transition is likely, then the berm must be designed accordingly to resist higher pore pressures.

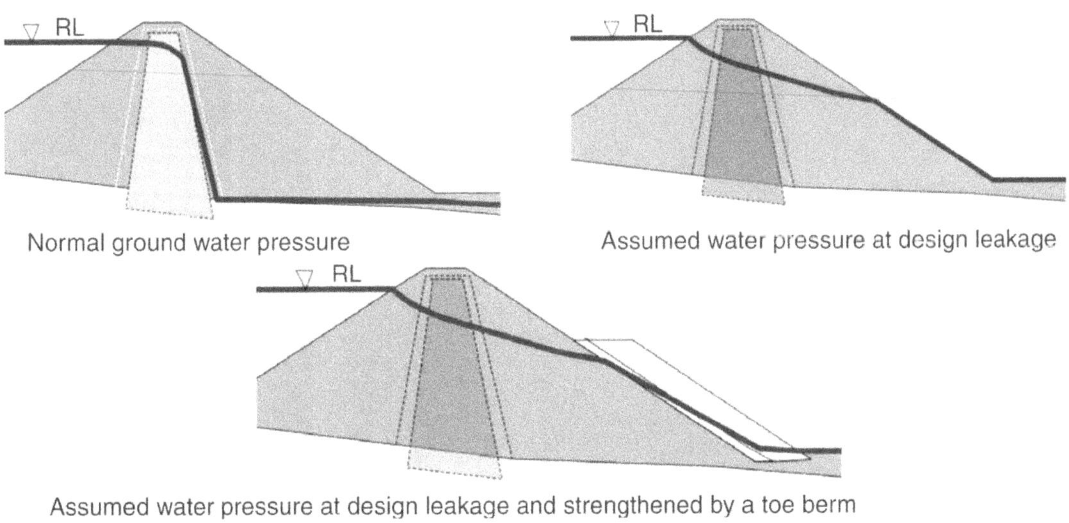

FIGURE 19.29 Assumed phreatic surfaces within the embankment for the design of a "Swedish berm" (Nilsson, 2009).

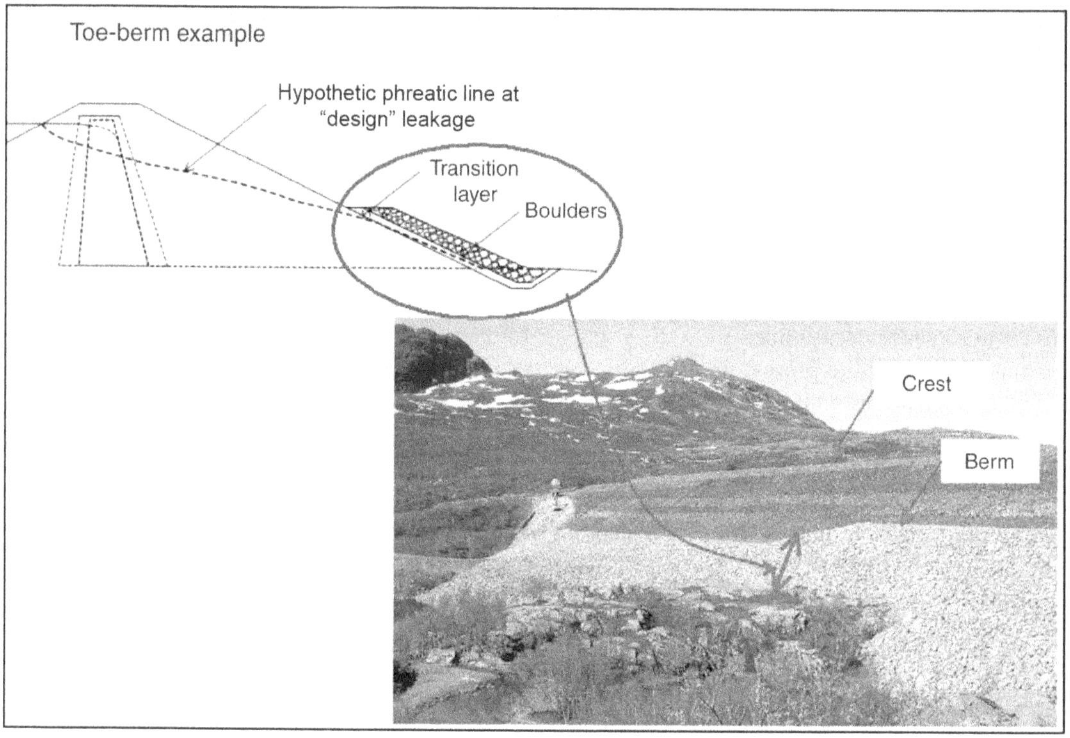

FIGURE 19.30 Example of a part height "Swedish berm" (Nilsson, 2009).

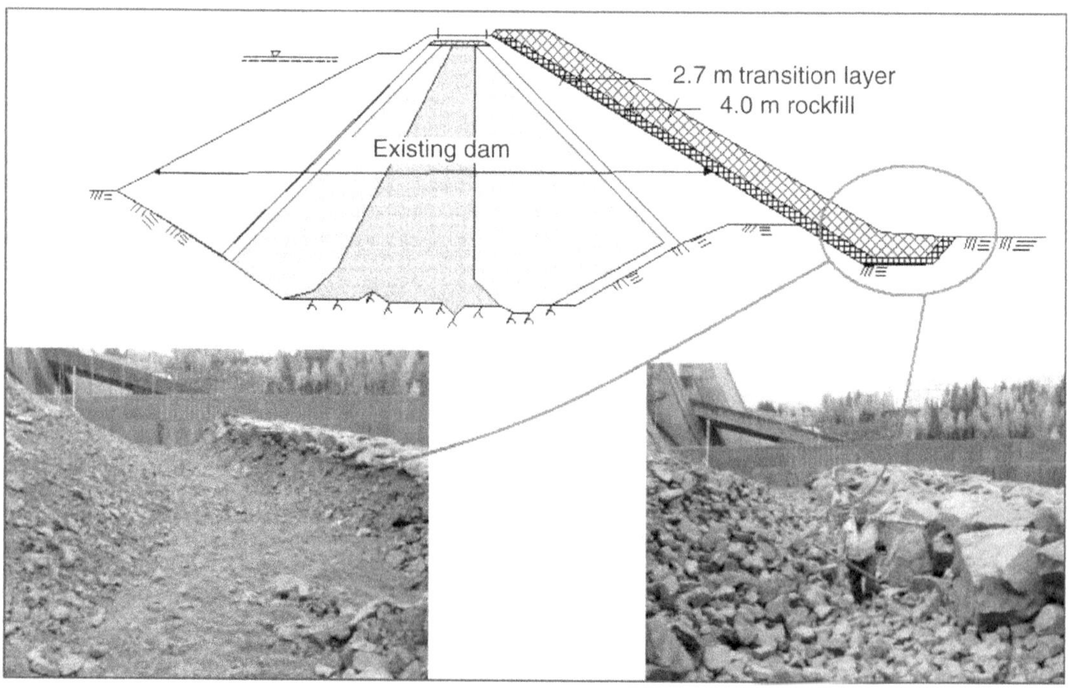

FIGURE 19.31 Example of a full-height "Swedish berm" (Nilsson, 2009).

19.7 MEASURES TO MITIGATE THE LIKELIHOOD OF BACKWARD EROSION PIPING (BEP)

19.7.1 "Flood Fighting"

Flood fighting is a term used in the USA and is measures taken before and during a flood to reduce the likelihood of failure of a dam or levee from the progression of BEP.

The USACE has flood fighting manuals which describe the measures and procedures that may be taken and organisational matters. The manuals cover overtopping, BEP and internal erosion in levee embankments and around pipes and other structures through the levee. USACE (2017a) Appendix K and USACE (2015) provide valuable information which will be applicable elsewhere.

During a flood the most common measure taken to arrest the progression of BEP is to ring sand boils. Figure 19.32 shows an example of a sand boil and Figure 19.33 shows an example of ringing the sand boil with sandbags. The principle is to raise the water level in the sand boil and so reduce the gradient from upstream to downstream in the BEP channel beneath the levee, so sand is no longer transported in the channel and progression of BEP is stopped. Water but not sand continues to flow in the boil.

FIGURE 19.32 Sand boil on the land side of a levee.

FIGURE 19.33 Ringing a single sand boil with sandbags to raise the water level downstream (Fell et al., 2015, courtesy of George Sills).

19.7.2 Long Term Measures to Reduce the Progression of Backward Erosion Piping

The most commonly adopted longer term mitigation measures are:

- *Pressure Relief Wells*:
  Installing pressure relief wells downstream of the levee can reduce the hydraulic head and gradients at the toe of the structure, thereby lowering the upwards seepage gradient at the toe of the dam or levee, and reducing the likelihood of sand being transported in sand boils or into drainage ditches penetrating the sand.
- *Cutoff Walls*:
  A partial or complete cutoff wall can be installed in the foundation to restrict seepage gradients and prevent the initiation and progression of BEP.
- *Berms*:
  Adding a downstream berm can increase the path length for seepage, which reduces the uplift pressures and gradients at the exit point, making erosion less likely. The added vertical pressure from the berm at the toe of the dam or levee makes it less likely heave will occur.

19.7.2.1 Pressure Relief Wells and Berms

In the event that pore pressures are calculated, or known from piezometers, to be too high, and as a result the factor of safety against heave is too low, the factor of safety can be increased by adding a weighting berm (see Figure 19.34) and/or by providing pressure relief wells.

The weighting berm is effective by adding to the vertical stress and provides a working platform to install and maintain the pressure relief wells. The pressure relief drains are effective by reducing the pore pressure.

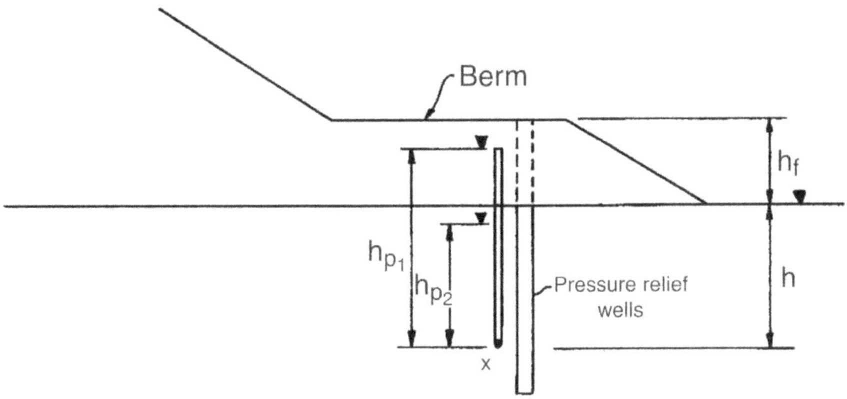

FIGURE 19.34 Control of "heave" by pressure relief wells and weighting berm (Fell et al., 2015)

If the original pore pressure head at x was hp_1, the original factor of safety would have been:

$$F_{UT} = \frac{\gamma_{sat} h}{\gamma_w hp_1} \tag{19.1}$$

If a berm of height h_f and unit weight γ_f was constructed, then:

$$F_{UB} = \frac{\gamma_{sat} h + \gamma_f h_f}{\gamma_w hp_1} \tag{19.2}$$

If no berm was provided but instead pressure relief wells were constructed, resulting in a reduction of pore pressure to γ_w hp_2, then:

$$F_{UW} = \frac{\gamma_{sat} h}{\gamma_w hp_2} \tag{19.3}$$

If both the berm and pressure relief wells were constructed:

$$F_{UWB} = \frac{\gamma_{sat} h + \gamma_f h_f}{\gamma_w hp_2} \tag{19.4}$$

The following should be noted:

- It is difficult to precisely predict pore pressures near the toe of a dam or levee, because they are greatly affected by local variations in permeability, assumptions on the ratio of horizontal to vertical permeability, and the permeability of the strata overlying the sand. These factors should be considered in carrying out the finite element analyses of seepage and the sensitivity of the pore pressures to assumptions checked. It is necessary to be conservative and to provide piezometers to check the assumptions.
- The berm will be most effective if it is free draining, at least for the situation shown in Figure 19.34.
- Pressure relief wells are designed to relieve artesian pressures and are usually constructed to be self-draining at the original ground surface or below if they drain into drainage pipes installed below ground surface level.

- The pressure relief wells will provide the greatest benefit at each well and least benefit midway between wells. For conservatism the design should be based on the pore pressures midway between wells. Piezometers are required to check the effectiveness of the wells.

19.7.2.2 Design of Pressure Relief Wells

USACE (2000, 2025) provides useful information on the design of pressure relief wells. Some of the important points are as follows:

(a) The wells should penetrate through the permeable strata. Partially penetrating wells may not be effective.
(b) The spacing of wells depends on the depth, stratification and permeability of the foundation soils; distance to the source of seepage; characteristics of the strata overlying the high permeability strata; and the degree of pressure relief desired.
(c) Wells are potentially subject to corrosion and bacterial clogging or carbonate encrustation and must be designed so the screens can be cleaned by flushing and/or be replaced. Therefore, access to the top of the wells is required.
(d) The water collected by the wells may have water quality issues and may need to be collected and treated.

Figure 19.35 shows a well design from USACE (2000). They indicate this design has been used widely and has given good service. They make the following points:

(a) *Riser pipe and screen*. The well screen normally extends from just below the top of the permeable stratum to the bottom of the well, with a solid riser pipe installed from the top of the permeable strata to the surface. In zones of very fine sand or silt interbedded with the permeable strata the screen is replaced by unperforated (blank) pipe. The screen and riser are most commonly stainless steel or plastic because they are corrosion resistant. However, plastic risers should be considered with caution as they are susceptible to damage.
(b) *Filter*. The filter should be designed according to the methods outlined in Chapter 12. USACE (2000) uses the Sherard and Dunnigan (1989) approach. The filter has to control erosion in the soil surrounding the well and not erode into the well screen.
(c) *Well installation*. Proper methods of drilling, backfilling and development of relief wells must be used, or the well will be of little use. USACE (2000) gives details.

Sills and Harder (2016) point out pressure relief wells are unreliable as a means of remediating sand boil activity in much damaged foundations because flows are localised in pipes in the foundation and the wells may or may not intersect the pipes. As discussed next, Sills and Harder (2016) recommend constructing fully penetrating cutoff walls in these circumstances. They say that in some situations a cutoff wall penetrating 5 m to 8 m below the top of the sandy strata coupled with pressure relief wells may be satisfactory.

19.7.2.3 Cutoff Walls

Sills and Harder (2016) recommend that for levees with severely damaged foundations from BEP, remediation may take two forms:

(a) Fully penetrating cutoff wall through the aquifer and past any piping damage. Figure 19.36 shows the principle.
(b) Partially penetrating cutoff wall penetrating at least 15 ft to 25 ft (4.5 m to 7.5 m) below the bottom of the top stratum as shown in Figure 19.37. They point out that a partially penetrating wall

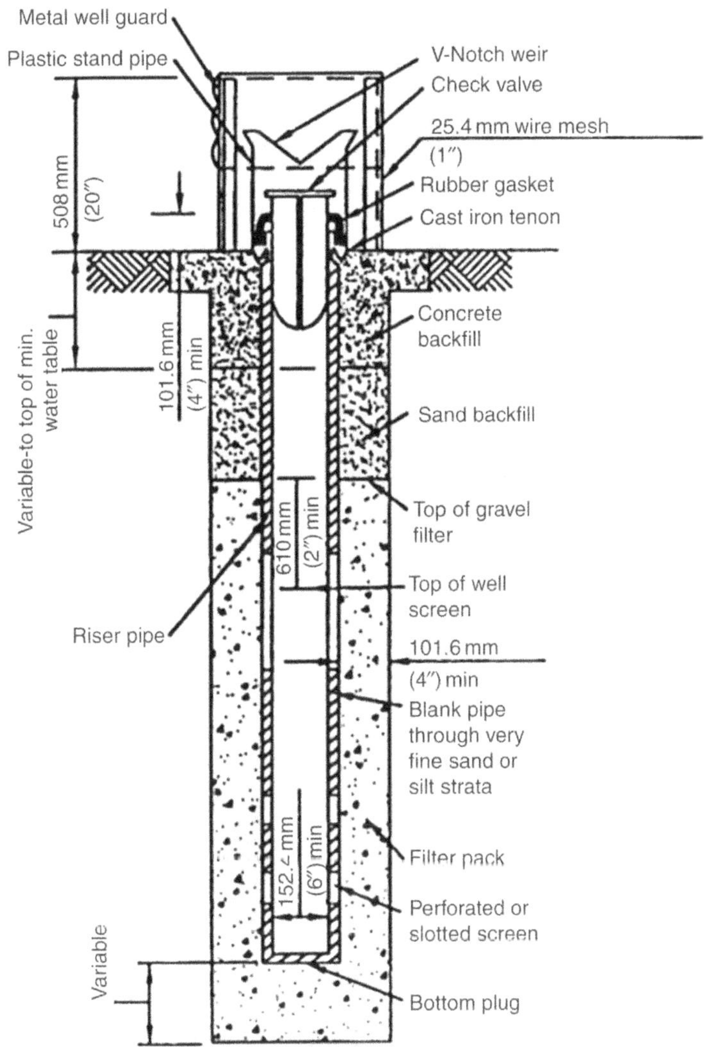

FIGURE 19.35 Typical USACE relief well (USACE, 2000).

may not reduce the pressures which resulted in the initiation of sand boils. Sills and Harder (2016) caution using this approach if the top stratum contains continuous lenses of pervious soil because pressures nearing full river level may occur upstream of the cutoff wall and this could lead to stability problems. It may be necessary to construct a weighting berm as shown in Figure 19.37.

Both these options are addressing the progression of BEP, cutting off the BEP channels.

19.7.2.4 Filter Cutoffs

Koelewijn and Van Beek (2024) indicate that Deltares are investigating several methods of mitigating BEP. This includes construction of a coarse sand filter "barrier" as shown in Figure 19.38. The principle is that the sand in the barrier will be stable under the potential BEP hydraulic gradient and arrest the progression of BEP. The sand barrier is installed in supported trenches from the ground surface. This method is being investigated with large scale flume experiments and pilot field applications.

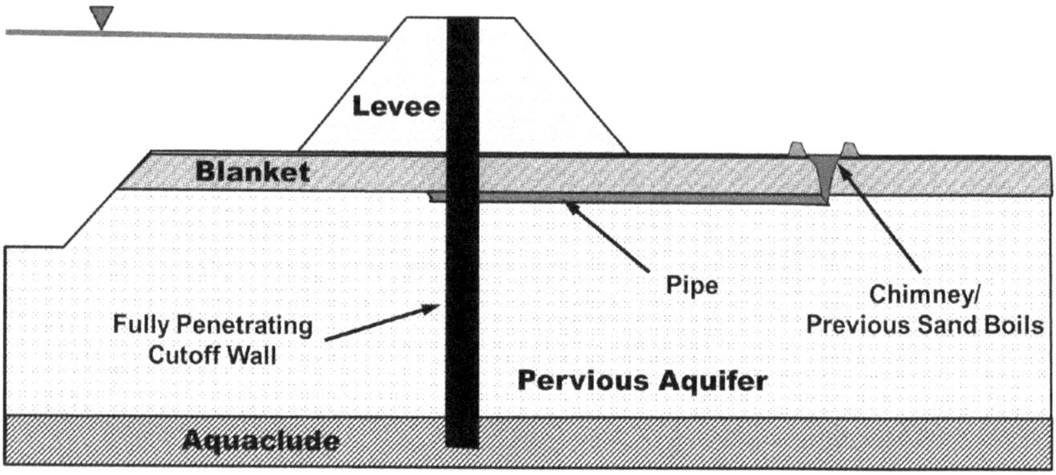

FIGURE 19.36 Recommended remediation approach using a fully penetrating cutoff wall through a levee to address both past piping damage and original under-seepage problems (Sills and Harder, 2016).

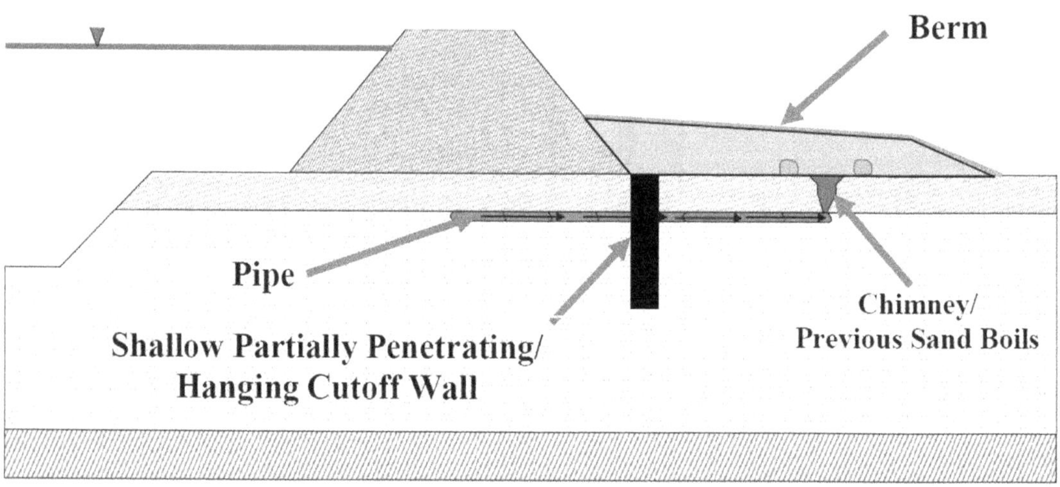

FIGURE 19.37 Recommended remediation approach using a partially penetrating cutoff wall and a seepage berm for levee foundations damaged by piping during previous flood events (Sills and Harder, 2016).

19.8 SOME ISSUES WHICH ARISE IN CONSTRUCTING DAM UPGRADES

19.8.1 Managing Dam Safety Risks during Construction

The following is a brief description of some of the issues which arise in construction of upgrades for flood overtopping and internal erosion and piping.

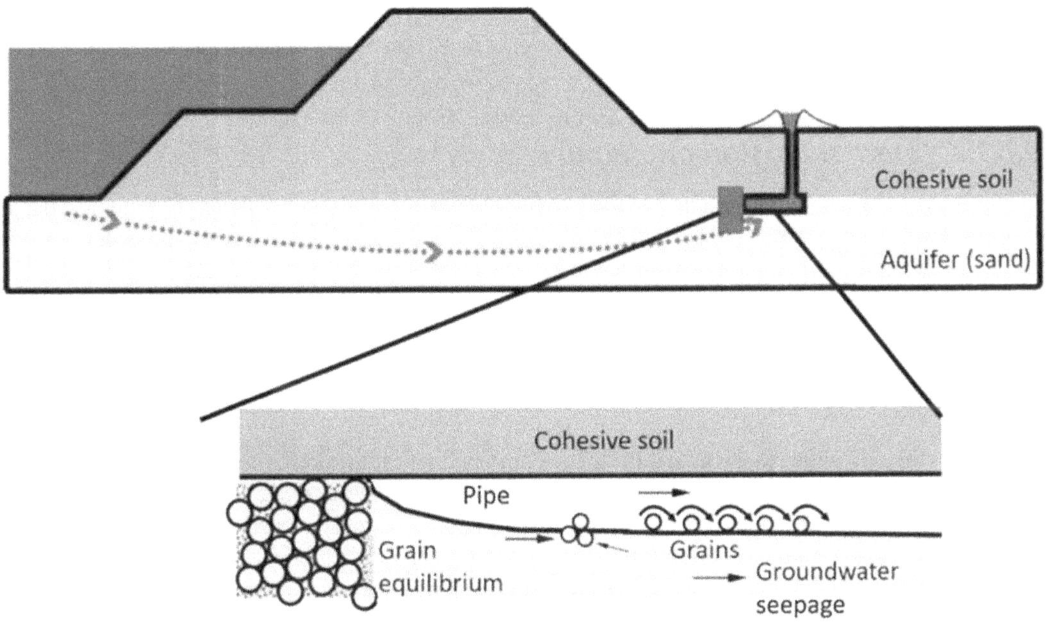

FIGURE 19.38 The principle of using a sand filter barrier to mitigate the progression of backward erosion piping (Koelewijn and van Beek, 2024).

19.8.1.1 Flood Overtopping of Cut-Down Crest

It is common when constructing upgrades to have to remove the upper part of the embankment so the new core and filters can be connected to the existing core and filters. Figures 19.27 and 19.28 for Upper Yarra Dam show an example.

The principles involved in managing risks from the crest lowering are as follows:

- So far as is practicable, design and construct the works so the risks during construction are no higher than those for the dam prior to the upgrade. ANCOLD (2022) discusses how this is done in quantitative risk terms. In a risk enhanced deterministic approach, the annual probability of the reservoir reaching the cutdown crest level should be no higher than for the existing dam. The annual probability of reaching the cutdown crest level should allow for reservoir drawdown, seasonality and duration of the construction works.
- Timing the work to be during the season when flooding is less likely.
- Draw down the reservoir level prior to commencing excavation of the crest.
- Staging the work so only part of the crest is lowered at any time, and materials and equipment are available to re-instate the crest to the original level in the event flooding is forecast.
- Working extended hours, e.g. 24/7, so the time of construction is kept to a minimum.
- Prepare an addendum to the Emergency Action Plan so that there is a clear understanding of responsibilities and trigger levels for emergency backfilling of excavations.
- If appropriate (e.g. for on-stream storages), include a regime of forecasting reservoir level based on forecast rainfall and use this as part of the decision-making on emergency backfill works within the addendum to the Emergency Action Plan.

Hunter et al. (2015) and Robinson et al. (2015) describe these measures for the Rowallan Dam Upgrade. Davidson and Dann (2011) outline the staging of works and Dam Safety Management Plan for managing the crest excavation into Hinze Dam.

19.8.1.2 Internal Erosion and Piping

When it is necessary to excavate some of the existing downstream fill and deficient filter materials, as for Upper Yarra Dam (Figures 19.27 and 19.28) and Tullaroop Dam (Figure 19.25), the benefit of that material in acting as a filter, albeit not necessarily very effectively, may increase the conditional probability of initiation and continuation of concentrated leak erosion.

This risk is managed by:

- Timing the works to be during the season when flooding and high reservoir levels are less likely.
- Drawing down the reservoir level prior to commencing excavation of the crest.
- Staging the work so only part of the downstream slope is exposed at any time.
- Enhanced surveillance including inspection and, where appropriate, instrumentation. Examples would be inspection for wet areas/seepage and then monitoring of seepage.
- Working extended hours, e.g. 24/7, so the time of construction is kept to a minimum.
- Preparing an addendum to the Emergency Action Plan so that there is a clear understanding of responsibilities and trigger levels for emergency works.
- Having contingency plans in place to place a filter and weighting fill over any crack in which water is flowing and there is a potential for concentrated leak erosion to initiate.

Examples of risk management are given in Hunter and Toose (2009) and Davidson and Dann (2011) for the Hinze Dam Raising Project, Hunter et al. (2015) and Robinson et al. (2015) for the Rowallan Dam Upgrade, and Arnold et al. (2016) for the Greenvale Dam.

19.8.1.3 Stability of Cut-Down Slope

The stability of the cut-down slope during construction of upgrades is an important issue.

The factors controlling the design of the slope are as follows:

(a) The nature and shear strength of the materials in the cut slope. Well compacted clay soils can be cut steeper than non-plastic soils because the stability is controlled by the undrained strength which is increased compared to drained strengths by partial saturation and overconsolidation of the soil during compaction.
(b) The height of the slope.
(c) The consequences of instability of the slope.
(d) Potential for pre-existing cracks within the core zone, e.g. longitudinal cracks may exist in the outer part of the core zone due to cross sectional differential settlement. Figure 19.39 shows an example.

In most cases the requirement will be that the factor of safety satisfies the criteria in Tables 7.3 and 7.4 for existing dams in operation.

Examples of satisfactory cut slopes are in Figure 19.8, which shows the 0.5H:1V slope at the Buffalo Dam upgrade; Figure 9.26 which shows the 0.5H:1V slope at Eppalock Dam. For the deep excavation at Upper Yarra Dam the slope was 1.5H:1V as shown in Figures 19.27 and 19.28.

FIGURE 19.39 Example of instability of cut slope due to pre-existing longitudinal cracking in core zone, 0.5H:1V batter slope.

19.8.1.4 Protection of Cut Slopes

The clay in cut slopes will desiccate from exposure to the sun and potentially experience erosion from rainfall. For Eppalock Dam, a bitumen spray coat was used as shown in Figure 9.26. This was successful and the bitumen was stripped off before the new filters were placed.

Protection by laying filter cloth on the slope has been used with reasonable success but rainfall runoff tends to flow under the cloth and erode the soil beneath. It can also be blown off the slope by wind.

At Upper Yarra Dam (Toose, *pers. comm.*), to overcome this, plastic sheeting was used with geotextile that was anchored in a trench at the crest and pinned to the slope with additional protection provided by strips of wire mesh for longer duration protection where required. Figure 19.40 shows the system used. The coverings were effective in reducing the drying effects and retaining moisture in the earthfill, thereby reducing the amount of final trimming required to prepare the batter. The earthfill "sweated" under the plastic and remained moist for several weeks. Some trimming was always required but mostly surficial.

An approach which has been used successfully at the Tullaroop Dam Upgrade and the Greenvale Dam Upgrade was to excavate the slope to within about 300 mm of the final slope, accepting it will be damaged by desiccation and rainfall, and strip it to final level progressively as the new fill is placed

FIGURE 19.40 Upper Yarra Dam. Protection of the excavated slope during construction of the upgrade. (a) Geotextile with strips of wire mesh. (b) Final clean up prior to placing filters. (Photographs courtesy of J. Toose.)

19.8.2 Filters: Manufacture, Placement and Compaction

19.8.2.1 Too High Fines Content

As discussed in Section 12.2.4, filters with excessive fines content are susceptible to holding a crack and being ineffective as a filter. Table 12.3 provides a method for estimating the likelihood a filter will hold a crack.

It is common to have contractual difficulties relating to the gradation and durability of filters. These occur at the supplier and contractor levels.

Almost invariably suppliers and contractors will provide samples of proposed filter materials for approval which do not satisfy the requirement for limiting fines, or they have trouble meeting specification requirements during construction. This is due to several causes:

(a) Some countries like Australia have standardised concrete sand aggregates that require more than 5% fines content, sometimes up to 15%. Hence suppliers are not commonly required to supply washed sand products containing low fines contents.
(b) They may not recognise the importance of these requirements and have not understood that they will be enforced.
(c) The main contractor does not provide the supplier with the details from the technical specification and therefore important requirements are overlooked.
(d) It often costs more to process the filters to reduce the fines content to specified level and may require additional steps in the process plant, such as washing in a cyclone, which were not allowed for in contract bid prices. This can result in major delays in supply of filters and may affect the program for the upgrade.
(e) They may not have allowed for potential breakdown of the filters on handling, placement and compaction. This breakdown could result in generating additional fines but can also be breakdown of the sand particle structure changing the grading (Cowan et al., 2016).
(f) Dry sieving may have been used in the supplier's tests when wet sieving should be used. As a result, fines contents were underestimated.

It is necessary to spell out clearly in specifications the requirements and advise the tenderers that the specification will be enforced.

One approach used is to require the contractor to supply and stockpile the filters well ahead of when they are required to be placed in the embankment so any problems can be sorted out in a timely manner. Another approach is to have a separate contract for supply of filters so they can be supplied early. Another approach is to undertake inspection and testing of filters as part of detailed design and include pre-approved quarry sources within the technical specification.

It is good practice to specify ≤ 2% or 3% non-plastic fines for the as-supplied filter, anticipating that handling and compaction during placement in the embankment may result in some breakdown so in place the fines content should not exceed 5%, which is the normal specified maximum in place in the embankment.

19.8.2.2 Use of Crushed Rock for Fine Filters

Many suppliers and contractors will attempt to manufacture fine filters by crushing rock. This has the advantage that one source and plant can supply coarse and fine filters. The problems with this are as follows:

(a) Almost invariably the crushing process results in a deficiency in coarse sand size material and a surplus of fines. The first reaction from the contractor is sometimes to want to alter the specification, which is usually not possible. When the specification is enforced, there is often a lot of wastage which the contractor claims was unexpected, so claims for the additional costs result.
(b) In attempting to wash out the fines in the washing plant, this can also lead to removal of the finer sand fraction pushing the grading toward the coarse envelope at the bottom end.
(c) The crusher run product is often gap graded so the finer fraction may be eroded by seepage flows because the filter is internally unstable and so potentially outside the coarse limit. This

is particularly a problem when rock with fine grain crystals is used, such as basalt, dolerite and some greywacke.

(d) Crushed filters are potentially susceptible to segregation during placement. Particle shape can also be an issue with specific type of crushers and rock types.

(e) Crushed filters have a greater potential to hold a crack because of the greater interlock of the angular particles. This may require further reducing the limit on fines content in the filter.

Part of the answer to this is to warn tenderers about the problems in the contract documents. The other important thing to do is to try to ensure that the supplier and/or contractor has a skilled crushing plant team and the correct equipment. In addition, assessing commercial quarries during the detailed design phase can assist in ruling in/out quarries. If sources of natural sand and gravel are available, they are preferred because they avoid these issues, but they may present difficulties with fines content.

19.8.2.3 Possible Effects If Carbonate Rocks Are Used as Filter Materials

Rocks or natural aggregates containing significant proportions of carbonate minerals (e.g. limestone, dolomite, marble, gypsum or calcite) are likely to be unsuitable for use in filter zones, because they could cause a filter zone to become cemented and be capable of sustaining an open crack.

The authors consider that carbonate materials should not be adopted for use in a filter zone, particularly fine filters.

Readers should be aware that carbonate may leach into other rock. In one case carbonates in wind aeolian soil overlying basalt resulted in significant percentage of carbonate in the filter made by crushing the basalt.

19.8.2.4 Filter Placement and Compaction

Figures 19.41 and 19.42 show two methods for placing filters. For small extensions of filters such as shown in Figure 19.22(b) for Laanecoorie Dam, the method shown in Figure 19.42 is used. A trench is excavated through the existing fill and the filter placed in the trench. This may be adopted for trench depths up to about 3 m provided the trench remains stable. The filter is placed in thin layers, e.g. 200 to 300 mm after compaction and compacted by a vibrating plate.

For most upgrades, filters are placed against the excavated slope as shown in Figure 19.42. The fine filter is placed by front end loader or excavator laying the filter against the slope. The coarse filter is then placed up against the fine filter. Care must be taken to avoid segregation.

Filters are placed in layers from 400 mm to 600 mm thickness after compaction and compacted together by for example a 10 tonne or 12 tonne smooth drum vibrating roller. For narrow filters, e.g. 0.75 m wide for each filter, a smaller roller with thinner lift thickness is required. Cowan et al. (2016) indicate that for the Eppalock Dam Upgrade, a 2.4 tonne smooth drum vibrating roller was used

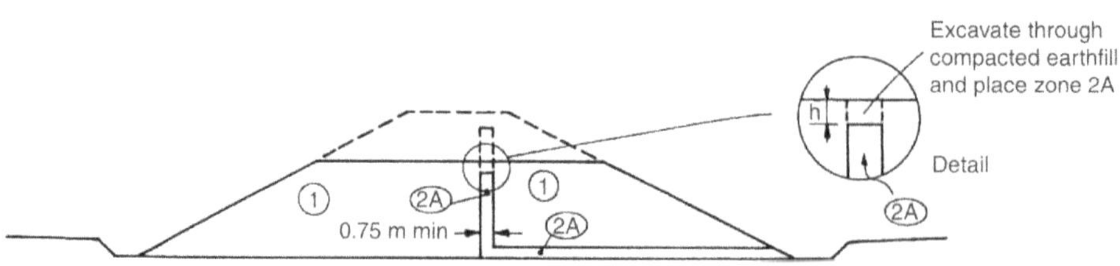

FIGURE 19.41 Construction of vertical chimney drain by excavation through earthfill (Fell et al., 2015).

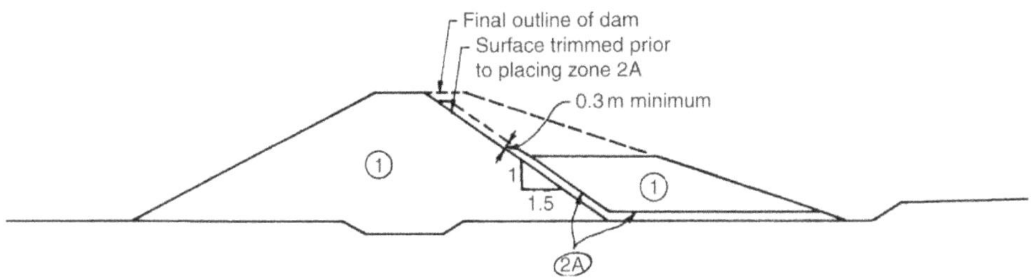

FIGURE 19.42 Construction of inclined chimney drain by placement on downstream slope of earthfill (Fell et al., 2015).

with 300 mm lift thickness. Water is added to achieve the required degree of compaction and reduce segregation.

The requirement for compaction of filters is for the filter to be dense and dilative on shear so as to not be subject to liquefaction if saturated and subject to seismic loading. Typically, a relative density ≥ 65% to 70% is specified for fine (Zone 2A) filters, but as Cowan et al. (2016) describe this may be difficult to achieve. It may be necessary in some cases to accept relative densities less than 65% to 70%, e.g. 55% or 60%, for chimney filters which are less likely to be saturated.

This is offset though by avoiding over-compaction of filters, e.g. to 85% or 90% relative density, which can increase the potential of the filter to hold a crack or not collapse, in addition to issues with potentially generating higher fines due to breakdown.

Coarse filters (Zone 2B) have gravel particles in them which makes carrying out the maximum and minimum densities required to calculate relative densities impractical. These are commonly specified with a methods specification, e.g. placed in 600 mm layer thickness before compaction and rolled with three passes of a 10 tonne or 12 tonne smooth drum vibrating roller.

Trials of compaction methods before beginning placement in the embankment are required to develop satisfactory procedures. Trials should be undertaken for the range of filter placement methods and arrangements to be employed; for example, in filter trenches, where there are variations in horizontal width and to test layer thickness and compaction equipment type. For layer thickness, it is important to test the full lift thickness, and this may require separately testing the upper and lower part of the lift to ensure adequate density is achieved throughout. The equipment used for the trial should be the same equipment to be employed on the embankment.

Cowan et al. (2016) describe in detail the issues which may arise even when those involved are well experienced in design and construction.

Appendix A
Methods for Estimating Conditional Probabilities of Transverse Cracking Leading to Concentrated Leak Erosion

This appendix presents methods for estimating conditional probabilities of transverse cracking which potentially may lead to initiation of concentrated leak erosion.

These are taken from Fell, R., Foster, M.A., Cyganiewicz, J., Sills, G. and Davidson, R. (2024). *Methods for Estimating the Probability of Failure of Embankment Dams by Internal Erosion and Piping*. UNICIV Report No. R470, The School of Civil and Environmental Engineering, University of New South Wales, Sydney.

Chapter 9 summarises the technical background to these methods. More information is provided in the Supporting Document of Fell et al. (2024).

IM 1 TRANSVERSE CRACKING IN THE UPPER EMBANKMENT RESULTING FROM CROSS-VALLEY DIFFERENTIAL SETTLEMENT

The recommended method is as follows:

(a) Determine the as-constructed cross-valley foundation profile beneath the core of the dam allowing for the excavation carried out to form the foundation for the core. Do not rely upon design drawings, as these often fail to represent the actual constructed profile.

(b) Determine the abutment slope (θ) and for valley profiles with a step, h_1 / H as defined in Figure 9.13.

(c) Determine the maximum total post-construction settlement of the dam crest as a percentage of the height of the dam at the location of maximum settlement. This should be calculated from monitoring data. If monitoring only began some years after construction, estimate the total settlement using the methods described in Section 9.2.1.

(d) In the event no data or survey is available, use a default settlement of 0.25% for modern dams and 0.5% for dams built before 1950.

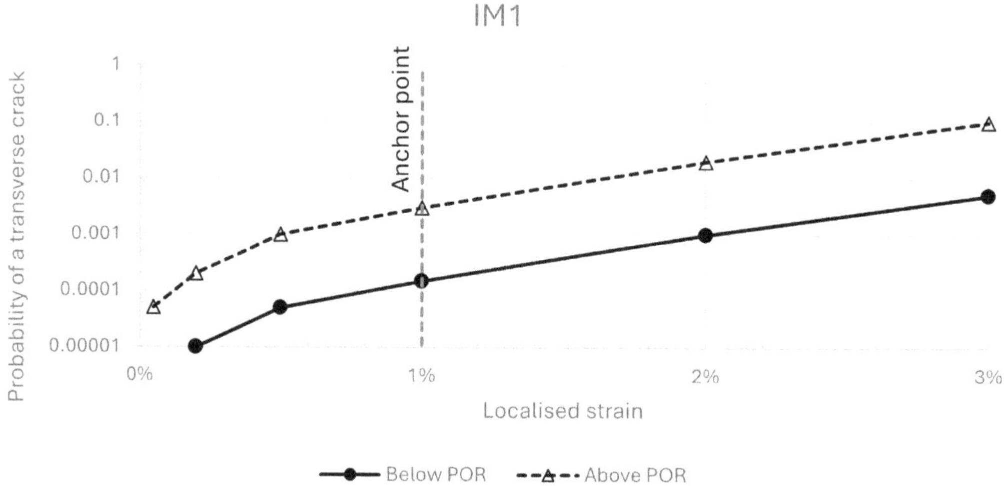

FIGURE A1 IM 1. Transverse cracking resulting from cross-valley differential settlement. Probability of persistent open cracking in the upper part of the embankment versus localised tensile strain at the crest %.

Note: "POR" refers to the Pool of Record level.

(e) Estimate the localised strain using Figure 9.11 for valley profiles without a step and Figure 9.12 for valley profiles with a step.
(f) If there is good quality survey monitoring of post-construction longitudinal deformations, consider whether these are consistent with the estimated localised strains, keeping in mind that the survey marks are often widely spaced and provide averaged strains.
(g) Use this localised strain to estimate the probability of a crack being present using Figure A1.

For risk assessments that support important dam safety decisions, the use of cross-valley numerical analysis methods is recommended to evaluate the magnitude and extent of localised tensile strains. Refer to He et al. (2017, 2019) for guidance.

For dams for which cross-valley numerical analyses have been carried out, use the calculated maximum localised tensile strain at the crest of the embankment to estimate the probability of a persistent open crack from Figure A1.

IM 2 TRANSVERSE CRACK OR GAP ADJACENT TO A CLIFF AT THE TOP OF THE EMBANKMENT FORMED BY DIFFERENTIAL SETTLEMENT

Use Table A1 and Figure A2 to estimate the probability of a persistent gap being present.

TABLE A1 IM 2—Factors Influencing the Likelihood of a Transverse Crack or Gap Adjacent to a Cliff at the Upper Part of the Embankment Formed by Settlement of the Embankment

		LIKELIHOOD FACTOR (LF)			
FACTOR	*RELATIVE IMPORTANCE FACTOR (RF)*	*LESS LIKELY (1)*	*NEUTRAL (2)*	*MORE LIKELY (3)*	*MUCH MORE LIKELY (4)*
Cross-valley profile under embankment core[(a) (b)]	(3)	$W_b/H_w > 2.5$ Note: If this condition is present, probability = zero	$1.0 < W_b/H_w \leq 2.5$	$0.25 < W_b/H_w \leq 1.0$	$.W_b/H_w \leq 0.25$
Post-construction settlement of the embankment as a % of maximum embankment height	(2)	s/H < 0.25%	0.25% < s/H < 0.75%	0.75% < s/H < 2.0%	s/H ≥2.0%
Slope of the abutments under embankment core[(a)]	(1)	$15° > \beta_1 \leq 25°$ Note: If $\beta_1 < 15°$ probability = zero	$25° < \beta_1 \leq 45°$	$45° < \beta_1 \leq 60°$	$\beta_1 > 60°$

Notes:
(a) See Figure 9.33 for definitions of W_b, H_w, β_1.
(b) This mechanism only applies for $W_b/H_w \leq 2.5$ and $\beta_1 \geq 15°$.

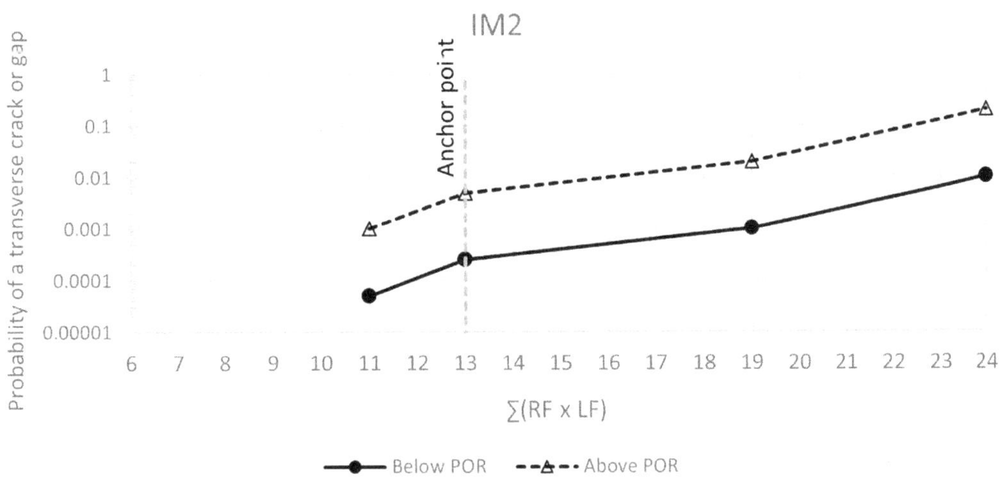

FIGURE A2 IM 2. Probability of a persistent transverse crack or gap in the upper part of the embankment adjacent a cliff versus Σ (Relative importance factor [RF]) × (Likelihood factor [LF]).

Note: "POR" refers to the Pool of Record level.

IM 3 TRANSVERSE CRACKING OR HYDRAULIC FRACTURE RESULTING FROM CROSS-VALLEY ARCHING

Use Tables A2 and Figure A3 to estimate the probability of a persistent crack or hydraulic fracture being present.

TABLE A2 IM 3—Factors Influencing the Likelihood of Cracking or Hydraulic Fracturing in Embankment Resulting from Cross-Valley Arching

FACTOR	*RELATIVE IMPORTANCE FACTOR (RF)*	*LIKELIHOOD FACTOR (LF)*			
		LESS LIKELY (1)	*NEUTRAL (2)*	*MORE LIKELY (3)*	*MUCH MORE LIKELY (4)*
Slope of abutments under embankment core[(a)]	(3)	Moderate abutment slope β_1, $\beta_2 \leq 45°$	Moderately steep abutment slopes $45° < \beta_1, \beta_2 \leq 60°$	Steep abutments $60° < \beta_1, \beta_2 \leq 75°$	Very steep abutments, $\beta_1, \beta_2 > 75°$ or β_1 near vertical, $\beta_2 > 60°$
Cross-valley geometry under embankment core[(a)]	(2)	$W_v/H > 0.75$ Note: If $W_v/H > 2$, exclude this failure mode.	$0.4 < W_v/H \leq 0.75$	Narrow deep valley $0.25 < W_v/H \leq 0.4$	Very narrow deep valley $W_v/H < 0.25$
Height of embankment	(1)	Less than 15 m	15 m to 30 m	30 m to 60 m	> 60 m

Notes:

(a) See Figure A4 for definitions of W_v, H, β_1 and β_2.
(b) If the soil in the lower part of the core is poorly compacted or subject to collapse compression on saturation, and the upper part is not, increase the weighted factor Σ (LF × RF) by 1 or 2.
(c) See 9.2.2 for examples of numerical modelling which show where hydraulic fracture is likely to occur. In the absence of other information, assume hydraulic fracture may occur at all levels within the core. For important decisions, carry out dam specific numerical analyses.

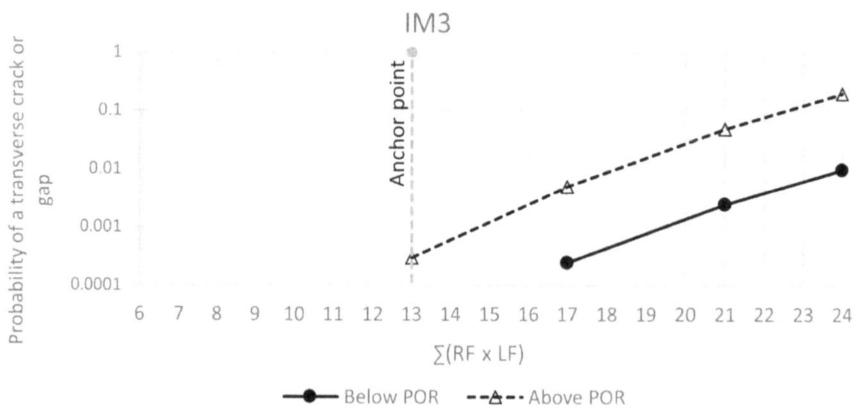

FIGURE A3 IM 3. Probability of persistent cracking or hydraulic fracturing in an embankment due to cross-valley arching versus Σ (Relative importance factor [RF]) × (Likelihood factor [LF]).

Note: "POR" refers to the Pool of Record level.

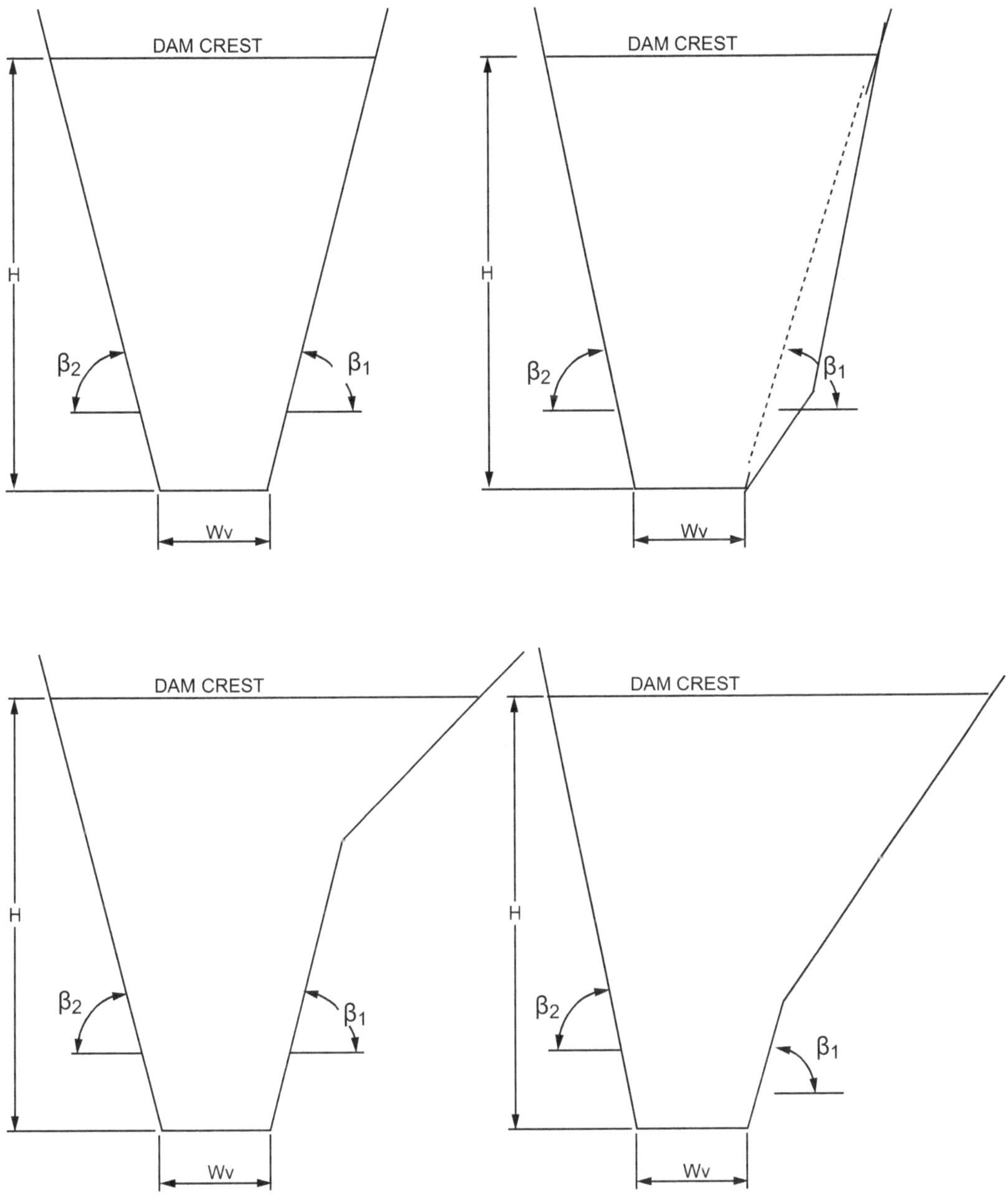

FIGURE A4 Longitudinal profiles of the dam showing the definition of terms for cross-valley arching.

IM 4 TRANSVERSE CRACKING RESULTING FROM CROSS SECTION DIFFERENTIAL SETTLEMENT DUE TO POORLY COMPACTED SHOULDERS

Use engineering judgement and/or expert elicitation to estimate the probability of a through-going shear.

IM 5 TRANSVERSE CRACKING IN THE UPPER AND LOWER EMBANKMENT RESULTING FROM DIFFERENTIAL SETTLEMENTS IN THE SOIL IN THE FOUNDATION UNDER THE CORE

The method to estimate the presence of transverse cracking is as follows:

(a) Determine the cross-valley stratigraphy beneath the core of the dam. From this determine the profile of foundation rock and the compressible soil units remaining beneath the core after foundation preparation for the dam construction. Use all available data and do not rely solely upon design drawings, as these often fail to represent the actual stratigraphy and foundation excavation.
(b) Determine the slope of the interface of the foundation (α), as defined in Figure 9.19, and the thicknesses of the soil units. Depending on which scenario is applicable, α may be α_1, α_2 or α_3.
(c) Estimate the compressibility properties of the soil in the foundation using soil mechanics principles.
(d) From the thickness of the soil units and compressibility properties, estimate the settlement within the soil units across the valley resulting from the weight of the embankment on the soils. From this determine the maximum settlement in the foundation soils imposed on the embankment.
(e) If there is monitoring data, for example from settlement plates installed at the top of the soil foundation prior to dam construction, use this data instead of or in combination with estimated settlements to determine the maximum settlement.

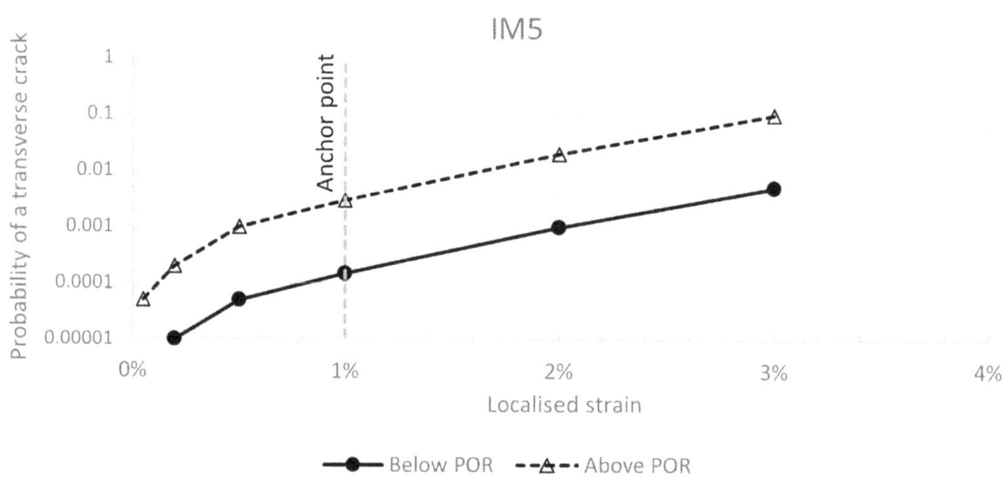

FIGURE A5 IM 5. Transverse cracking resulting from differential settlements in soil in the foundation beneath the core. Probability of persistent open cracking in the upper part of the embankment versus adjusted localised strain %.

Note: "POR" refers to the Pool of Record level.

(f) Express the maximum settlement as a percentage of the height of the embankment above the upper surface of the soil foundation at the point where maximum settlements are calculated to occur.
(g) Estimate the localised strain using Figure 9.21.
(h) Adjust the localised strain for settlement in the soil during construction using Table 9.2.
(i) Use this localised strain to estimate the probability of a crack being present using Table A6.

For risk assessments that support important dam safety decisions, cross-valley numerical analysis methods are recommended to evaluate the magnitude and extent of localised tensile strains. Refer to He et al. (2017, 2019) for guidance.

IM 6 TRANSVERSE CRACKING DUE TO DIFFERENTIAL SETTLEMENTS RESULTING FROM EMBANKMENT STAGING

No additional information.

IM 7 DESICCATION CRACKING IN THE CREST OF THE EMBANKMENT

Overall Approach

First. Obtain the average annual rainfall and average annual pan evaporation for the dam site. In Australia, the rainfall can be obtained from the Bureau of Meteorology website. Select the rain gauge location best representing conditions at the dam. The pan evaporation is available as contour maps of average annual pan evaporation. Estimate the evaporation at the dam site from the contours.

Calculate the Average Annual Net Evaporation = (Average Annual Pan Evaporation) – (Average Annual Rainfall).

Second. Determine the Unified Soil Classification of the dam core from available information, and the thickness, width and nature of the road pavement and/or other zones over the core on the dam crest.

Third. Use Table 9.15 to estimate the maximum likely depth of desiccation cracking. If the reservoir stage being considered is below the maximum depth of desiccation cracking, the probability of a crack being present due to desiccation can be assumed to be zero.

Fourth. For cases where the reservoir stage is above the base of potential desiccation cracking, follow the next procedure.

Estimating the Probability of a Persistent Desiccation Crack

Use Tables A3 and Figure A6 to estimate the probability (PPCracking) of a desiccation crack being present and persisting across the width of the core.

TABLE A3 IM 7—Factors Influencing the Likelihood of Desiccation Cracking (PCracking) in the Upper Part of Embankment Dams

		LIKELIHOOD FACTOR (LF)			
FACTOR	*RELATIVE IMPORTANCE FACTOR (RF)*	*LESS LIKELY (1)*	*NEUTRAL (2)*	*MORE LIKELY (3)*	*MUCH MORE LIKELY (4)*
Crest zoning and surface layer over core	(3)	Sealed road or concrete pathway ≥ 5 m wide[(1)] *or* ≥ 0.5 m thickness of granular fill over the core[1]	Gravel road ≥ 0.2 m thick ≥ 3 m wide *or* Concrete or sealed pathway ≥ 2 m wide *or* ≥ 0.2 m Granular fill over the top of the core	Gravel road or pathway ≥ 3 m wide and 0.05 m to 0.2 m thick *or* Sealed pathway < 2 m wide *or* 0.05 m to 0.2 m Granular fill over the top of the core	No road or pathway, or < 0.05 m thick gravel or granular fill
Average annual net evaporation	(2)	175 mm to 425 mm	450 mm to 900 mm	925 mm to 1500 mm	1500 mm to 2000 mm
Unified Soil Classification of the core	(1)	Silty sand, clayey sand, clayey silt SM-SC-ML	Silty clay and sandy clay CL LL < 35%	Clay and sandy clay CL-CH LL 35% to 50% and MH	Clay CH LL > 50%

Note: (1) If either of these conditions are present, then the probability that desiccation cracking will be present and persist across the core is negligible.

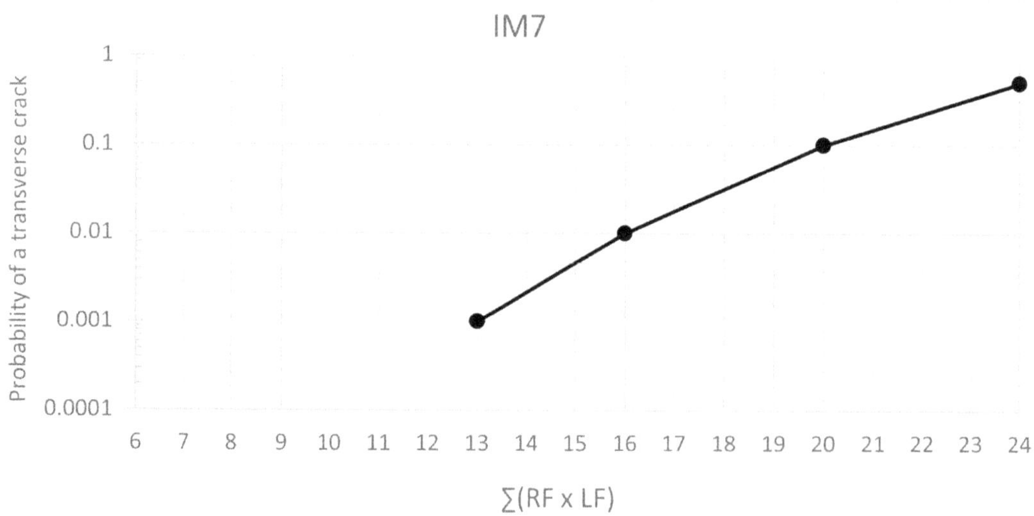

FIGURE A6 IM 7. Probability of a persistent transverse crack in the core due to desiccation versus Σ (Relative importance factor [RF]) × (Likelihood factor [LF]).

Estimating the Depth and Width of Desiccation Cracking

Use Table 9.15 to estimate the depth and width of desiccation cracking. Where a range of depth and width are shown, use the smaller depth/width for the bottom of the range of average annual net evaporation and the larger depth/width for the top of the range of net evaporation, and interpolate in between.

For important dam safety decisions, test pit excavations at the crest are recommended to be carried out to help evaluate the potential for the desiccation cracking to be present and to evaluate the characteristics of the cracks if present (width, depth and persistence across the crest).

If test pits have been excavated in the dam crest and mapped for the depth, width and persistence of cracking, take account of this information allowing for the rainfall and evaporation leading up to when the test pits were excavated.

IM 8 DESICCATION CRACKING ON SEASONAL SHUTDOWN LAYERS AND ON THE SURFACE OF STAGED EMBANKMENTS

Use the method for IM 7 to estimate the likelihood of desiccation cracking on the shutdown or staging surface. Then use expert elicitation to assess the probability these cracks were left in the shutdown or staging surface prior to placing new fill.

For seasonal shutdowns, e.g. in winter when evaporation is lower than in summer, use net evaporation for the period of the shutdown.

IM 9 CRACKING IN THE LOWER/MIDDLE EMBANKMENT RESULTING FROM CROSS-VALLEY DIFFERENTIAL SETTLEMENT

IM 1 includes this potential for cracking.

IM 10 HYDRAULIC FRACTURE RESULTING FROM ARCHING OF THE CORE ONTO THE SHOULDERS OF THE EMBANKMENT

This mode is applicable to central core earth and rockfill (or gravel shells) dams and puddle core earthfill dams. It is not applicable to other dam types. The method is shown in Table A4 and Figure A7.

TABLE A4 IM 10—Factors Influencing the Likelihood of Cracking or Hydraulic Fracture Resulting from Arching of the Core onto the Shoulders of the Embankment

		LIKELIHOOD FACTOR (LF)			
FACTOR	*RELATIVE IMPORTANCE FACTOR (RF)*	*LESS LIKELY (1)*	*NEUTRAL (2)*	*MORE LIKELY (3)*	*MUCH MORE LIKELY (4)*
Core geometry Width (W) / Height (H)	(3)	W/H > 1.0[(c)]	0.5 < W/H ≤ 1.0	0.25 < W/H ≤ 0.5	W/H ≤ 0.25[(g)]
Relative stiffness of core and shells	(2)	Core has higher modulus than shells. Shoulders poorly compacted or dumped.	Modulus of core same or marginally lower than shoulders. Shoulders well-compacted and medium to high modulus.	Core significantly lower modulus than shoulders, e.g. 0.2x to 0.5x. Shoulders well-compacted and high modulus.	Core much lower modulus than shoulders, e.g. 0.1x, or subject to collapse compression as detailed in Table 9.17. Shoulders well-compacted and high modulus.
Height of embankment	(1)	Less than 15 m	15 m to 30 m	30 m to 60 m	> 60 m

Notes:
(a) The most likely location for arching to occur is in the upper to middle part of the dam.
(b) Core materials subject to collapse settlement as detailed in Table 9.17 may result in arching and low stresses. Collapse settlement may also lead to softened zones and a crack in the vicinity of the contact between the saturated and unsaturated parts of the core, i.e. at the phreatic surface.
(c) The likelihood of arching is negligible for embankments with these characteristics.
(d) The mechanism is considered to be significant only for reservoir levels above the pool of record.
(e) W = width of the core at depth H below the crest of the embankment.
(f) Finite element analyses which properly model the history of the dam and its properties including collapse on saturation may be used to assess the likelihood of cracking or hydraulic fracture.
(g) Puddle clay core dams with W/H ≤ 0.17 are particularly susceptible to hydraulic fracture (Skempton, 1990). Consider adopting a higher probability of hydraulic fracture for such dams.

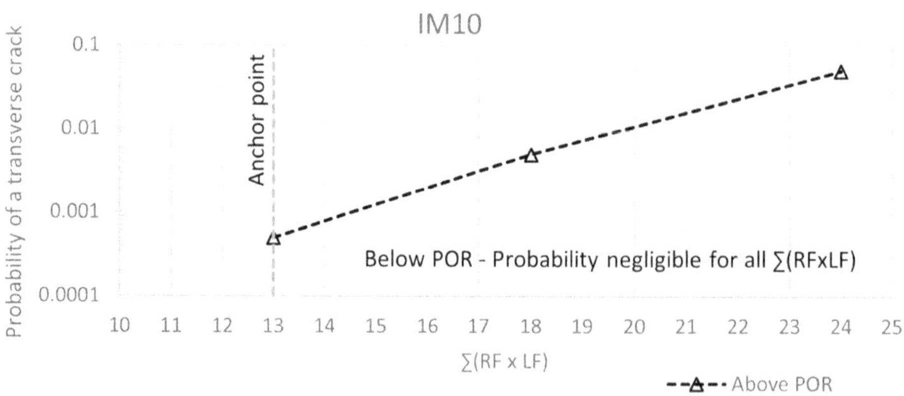

FIGURE A7 IM 10. Probability of a persistent transverse crack or hydraulic fracture resulting from arching of the core onto the shoulders of the embankment versus Σ (Relative importance factor [RF]) × (Likelihood factor [LF]).

Note: "POR" refers to the Pool of Record level. The probability for this initiating mechanism is assumed to be negligible for the below POR cases.

IM 11 TRANSVERSE CRACKING IN THE LOWER EMBANKMENT RESULTING FROM DIFFERENTIAL SETTLEMENT IN SOIL IN THE FOUNDATIONS UNDER THE CORE

IM 5 includes potential cracking low in the embankment.

IM 12 TRANSVERSE CRACKING OR HYDRAULIC FRACTURE OVER SMALL-SCALE IRREGULARITIES IN THE FOUNDATION PROFILE UNDER THE CORE

Empirical Method. This method is calibrated against historic data of failures and accidents. It is approximate and should not be relied upon for important decisions. The method is:

TABLE A5 IM 12—Factors Influencing the Likelihood of Cracking or Hydraulic Fracturing in Embankment Dams Due to Small-Scale Irregularities in the Foundation Profile under the Core

		LIKELIHOOD FACTOR (LF)			
FACTOR	*RELATIVE IMPORTANCE FACTOR (RF)*	*LESS LIKELY (1)*	*NEUTRAL (2)*	*MORE LIKELY (3)*	*MUCH MORE LIKELY (4)*
Persistence of the irregularity across the core	(3)	Persistent < 75%[(b)]	75% ≤ Persistent < 90%	90% ≤ Persistent < 100%	Persistent 100%
Small-scale irregularities in abutment profile	(2)	Uniform abutment profile, or irregularities treated by slope modification[(c)]	Steps, benches, depressions in rock foundations less than 5% of the embankment height	Steps, benches, depressions in rock foundations 5% to 10% of the embankment height	Steps, benches, depressions in rock foundation 10% to 20% of the embankment height[(a)]
Post-construction settlement as a percentage of dam height s/H	(1)	s/H ≤ 0.5%	0.5% < s/H ≤ 1.0%	1.0% < s/H ≤ 2.0%	s/H > 2.0%

Notes:

(a) Larger irregularities are covered in Section IM 1.

(b) An irregularity with less than 75% persistence across the core is assumed to have a negligible contribution to the probability of a transverse crack.

(c) If these conditions are met the probability of a persistent transverse crack or hydraulic fracture may be negligible.

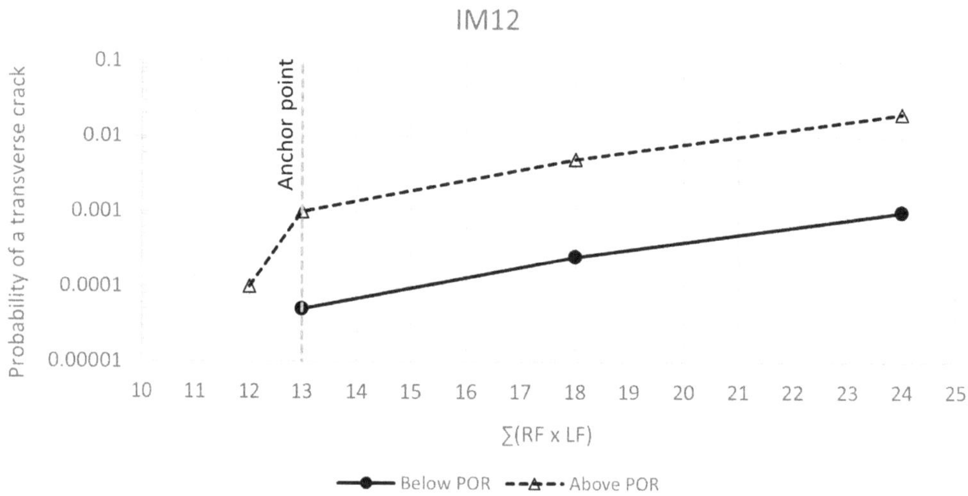

FIGURE A8 IM 12—Probability of a persistent transverse crack or hydraulic fracture in an embankment dam due to small-scale irregularities in the foundation profile under the core versus Σ (Relative importance factor [RF]) × (Likelihood factor [LF]).

Notes:

(1) If the reservoir stage represents a large increase above the POR larger probabilities may be appropriate for above POR Σ (RF × LF) = 24.

(2) "POR" refers to the Pool of Record level.

IM 13 TRANSVERSE CRACKING IN THE UPPER PART OF THE EMBANKMENT CAUSED BY EARTHQUAKE

No additional information; see Section 9.9.

IM 14 TRANSVERSE CRACK OR CONTINUOUS GAP RESULTING FROM POORLY COMPACTED ZONE WITHIN THE CORE

Probability of the Presence and Characteristics of a Continuous Poorly Compacted Layer within the Core and Subject to Collapse Settlement

To estimate the probability of a continuous poorly compacted layer or layers use either or both of the methods in Table A6 to estimate Σ RF × LF and then use that estimate in Figure A9. Use all available information to make the estimates.

TABLE A6 IM 14—Factors Influencing the Likelihood of a Poorly Compacted Layer(s) in the Embankment

FACTOR	RELATIVE IMPORTANCE FACTOR (RF)	LIKELIHOOD FACTOR (LF) LESS LIKELY (1)	NEUTRAL (2)	MORE LIKELY (3)	MUCH MORE LIKELY (4)
Method based on compaction equipment, layer thickness and moisture content.					
Compaction equipment	(3) (a)	As for "neutral" but with good documentation and records (b)	Soil compacted by suitable rollers, disking to blend material and compatible layer thickness for the equipment being used	Soil placed and compacted by bulldozer, no compaction by rollers, no disking, or rolled in thick layers beyond the capability of the roller	Soil placed with no formal compaction (e.g. by horse and cart in old dams, or by pushing into place by excavator or bulldozer or in very thick layers)
and Layer thickness			Layer thickness 150 mm to 250 mm after compaction	Layer thickness at or beyond the limit of compaction equipment (e.g. > 300 mm to 450 mm after compaction)	No control on layer thickness, often > 450 mm to 600 mm loose
and moisture content			Around standard compaction optimum moisture content	Dry of standard compaction optimum moisture content, e.g. 2% to 3% dry.	Well dry of standard compaction optimum moisture content. E.g. ≥ 4% dry
Method based on measured compaction density ratio and moisture contents					
AND/OR Measured or estimated compaction density ratio and moisture content	(3) (a)	All very well compacted to e.g. ≥ 98% standard MDD, moisture content 1% dry of optimum to 1% wet of standard OMC (b)	Well compacted to e.g. 95%–98% standard MDD, moisture content 2% dry of optimum to 1% wet of standard OMC	Layers poorly compacted, dry of standard optimum moisture content e.g. < 93% standard MDD, 2% to 3% dry of standard OMC	Layers very poorly compacted, dry of standard optimum moisture content e.g. < 90% standard MDD, ≥4 % dry of standard OMC
Factors applying to both alternatives					
Core geometry Width (W)/Height (H)	(2)	W/H > 2.0 *or* W > 15 m	0.75 < W/H ≤ 2.0 *or* 15 m < W ≤ 9 m (	0.25 < W/H ≤ 0.75 *or* 9 m < W ≤ 4.5 m	W/H ≤ 0.25 *or* W < 4.5 m
Borrow area variability	(1)	Uniform soil and moisture contents in the borrow areas,	Uniform or minor variability of soil and moisture contents in the borrow areas,	Variable soil and moisture contents in the borrow areas,	Very variable soil and moisture contents in borrow areas,
And site supervision		Good site supervision, fully documented.	Good site supervision but limited documentation available.	Moderate site supervision, limited documentation.	Poor site supervision, no documentation.

Notes:
(a) Make an assessment based on a combination of the available data.
(b) If there are good records of construction method and results from construction to demonstrate soils were compacted to this standard, this mechanism does not apply.

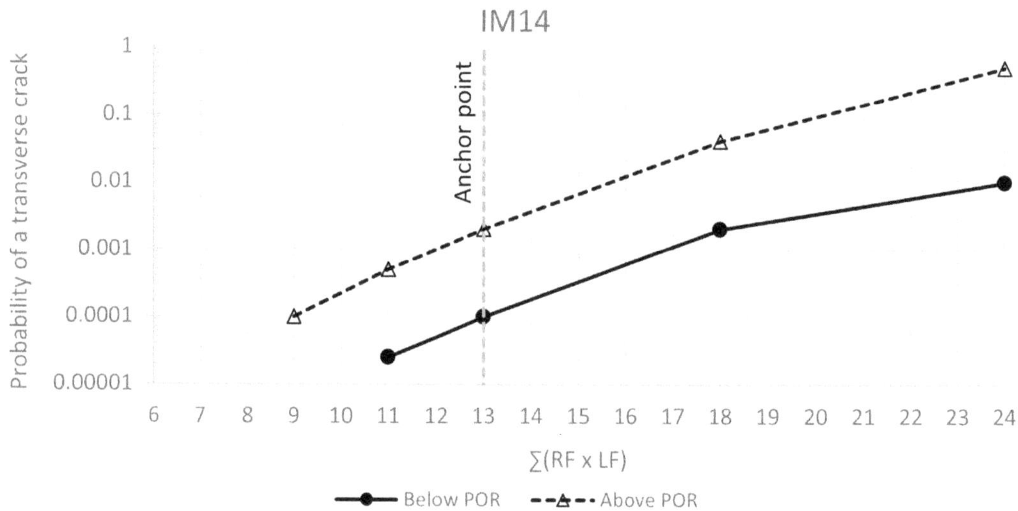

FIGURE A9 IM 14. Probability of a poorly compacted layer(s) in the embankment versus Σ (Relative importance factor [RF]) × (Likelihood factor [LF]) (applies to upper, middle and lower parts of the embankment).

Note: "POR" refers to the Pool of Record level.

Record the following:

- The estimated probability of a poorly compacted zone.
- The estimated thickness and depth below the dam crest of the poorly compacted layer. This may be one layer significantly thicker than the specified layer thickness but may be several layers or even the whole embankment.
- The estimated likely best estimate and range of standard compaction relative compaction and moisture content relative to standard compaction optimum moisture content (OWC) in the poorly compacted zone.

Estimation of the Total Settlement in the Layer(s) of Soil Subject to Collapse Settlement

The procedure is as follows:

(a) *For poorly compacted layers at or above the Pool of Record*:

Assess the total thickness of the layers of soil which are poorly compacted (T_p). There may be a single layer or several layers. The minimum poorly compacted layer thickness adopted should be 300 mm.

Assess the likely degree of compaction in the poorly compacted layer, taking into account the factors in Table A13.

Estimate the amount by which the poorly compacted layer may collapse (C_F) from C_S in Table 9.13 and T_p.

$C_F = C_S \times T_p$.

Record C_F and the location and depth below the dam crest of the poorly compacted layers.

(b) *For poorly compacted layers below a reservoir level which has been reached several times and for a total of months*:

Settlement = 0.1 C_F

This may often occur at the Full Supply Level.

(c) *For poorly compacted layers between (a) and (b)*:

Estimate settlement between C_F and 0.1 C_F taking account of the number of times and duration the reservoir has reached this level.

Table 9.13 has been developed from laboratory data on plastic soils. In the absence of better information, it may be applied to non-plastic soils. However, caution should be applied in doing this, and for important decisions laboratory tests on the soil should be carried out.

Estimate, Given the Total Settlement, the Probability of a Continuous Gap or Crack through the Embankment Core for Each of the Four Mechanisms

Mechanism A. The soil behaves as a series of clods with openings between the clods through which water passes.

Estimate the probability of the soil within the poorly compacted layer behaving as a series of clods using Table 9.14. Take account of any observations from site investigations into the poorly compacted layer(s) and of the soil in undisturbed bulk samples in the laboratory.

The probability of a continuous gap for this mechanism is equal to (Probability of a poorly compacted layer being present, from Table A13) × Probability of a continuous pathway within the poorly compacted layer, from Table 9.14).

Poorly compacted plastic soils may behave as a series of clods with openings between the clods. Non-plastic soils are unlikely to form clods but may settle locally on saturation resulting in continuous pathways for water to flow. In the absence of other information Table 9.14 may be applied to non-plastic soils.

Mechanism B. The soil surrounding a conduit or adjacent a wall collapses on saturation forming a gap at the contact of the soil with a conduit or wall.

This mechanism is covered in IM 18 for conduits and IM 20 for walls.

Mechanism C. The soil collapses on saturation causing settlement of the embankment core which results in arching of the core onto the filters and/or shoulders, resulting in a gap at the top of the saturated soil.

This mechanism is covered in IM 10. Consider the effect of poorly compacted layers when evaluating IM 10.

Mechanism D. Additional settlement of the embankment resulting from collapse settlement in the poorly compacted layer causes transverse cracking.

For valley cross sections such as 1–1 in Figure 9.52, the effect of collapse settlement will be to result in tensile strains over the ends of the poorly compacted layer. That is at A and B.

Use IM 1 with the collapse settlement in the poorly compacted layer assumed to be additional post-construction settlement unless it is already included in the settlement assessment for IM 1.

For valley cross sections as 2–2 in Figure 9.52, the effect of collapse settlement will be to result in tensile strains over the ends of the poorly compacted layer. That is at C and D.

Use IM 5 with the collapse settlement assumed to be equivalent to the same amount of post-construction settlement in the foundation.

IM 15 TRANSVERSE CRACKING OR CONTINUOUS FLAW RESULTING FROM POORLY COMPACTED LAYER ON THE CORE–FOUNDATION CONTACT

Estimation of the Probability of the Presence of and Characteristics of a Continuous Poorly Compacted Layer at the Core–Foundation Contact

Use Table A7 and Figure A10 to estimate the probability of a continuous poorly compacted layer. Use all available information to make the estimates and be prepared to use expert elicitation to adjust the outcomes.

Record the following:

- The estimated probability of a continuous poorly compacted layer.
- The location of and estimated thickness and depth below the dam crest of the poorly compacted layer.
- The estimated likely best estimate and range of standard compaction relative compaction and moisture content relative to standard compaction optimum moisture content in the poorly compacted layer.

IM 16 TRANSVERSE CRACKING OR CONTINUOUS FLAW RESULTING FROM FREEZING AT THE CREST OF THE EMBANKMENT

Probability of Transverse Cracking or Flaw

Freezing conditions can result in frost heave and formation of ice lenses in the crest of dams. When the ice thaws, loosened and/or cracked soil may be present in which internal erosion may initiate if the reservoir rises sufficiently high. The procedure is:

First—Consider the maximum likely depth of freezing for the soil in the core of the dam, and the climate using Table A9. If the reservoir stage being considered is below the likely depth of freezing, the probability of a crack or flaw due to freezing can be assumed = zero.

TABLE A7 IM 15—Factors Influencing the Likelihood of Poorly Compacted Zones on the Core–Foundation/Abutment Contact

FACTOR	*RELATIVE IMPORTANCE FACTOR (RF)*	*LIKELIHOOD FACTOR (LF)*			
		LESS LIKELY (1)	*NEUTRAL (2)*	*MORE LIKELY (3)*	*MUCH MORE LIKELY (4)*
Either Foundation preparation below the core, rock foundations,	(3)	Uniform rock surface or surface treated with extensive shotcrete, or concrete to correct slope irregularities(c)	Regular rock surface, or rock surface treated with some shotcrete or concrete to correct slope irregularities(b)	Irregular rock surface, with minimal slope correction or treatment	Very irregular rock surface, overhangs with no slope correction, shotcrete or concrete treatment
or Foundation preparation below the core, soil foundations	(3)	Uniform well-compacted soil foundation(c)	Compacted soil foundation(b)	Irregular or benched soil with no compaction	Poor stripping of soil foundation leading to poor compaction of first lift
Compaction methods for contact zone	(2)	As for "neutral" but with good documentation and records(d)	Soil compacted using special compaction methods (rubber tyres, use more plastic materials, compaction wet of OMC)	Soil placed and compacted by bulldozer, no compaction by rollers. Layer thickness at the limit of compaction equipment (e.g. > 300 mm or 450 mm after compaction)(a)	Poor compaction methods used, soil poorly compacted or allowing segregation against foundation surface. Thick layer thickness, often > 450 mm or 600 mm loose.
Continuity of features, Core geometry Width (W) /Height (H)	(1)	W/H > 2.0 *or* W > 15 m	0.75 < W/H ≤ 2.0 *or* 15 m < W ≤ 9 m	0.25 < W/H ≤ 0.75 *or* 9 m < W ≤ 4.5 m	W/H ≤ 0.25 *or* W < 4.5 m

Notes:
(a) Smaller thicknesses are for plastic soil, larger thicknesses for non-plastic soil.
(b) For most cases this mechanism may be judged as of negligible likelihood and assigned zero probability.
(c) For this foundation preparation the probability of this mechanism is negligible.
(d) Even situations where soil is well compacted can soften or loosen if the contact is irregular.
(e) For homogeneous earthfill dams, assess foundation preparation and compaction methods for the central portion of the section.

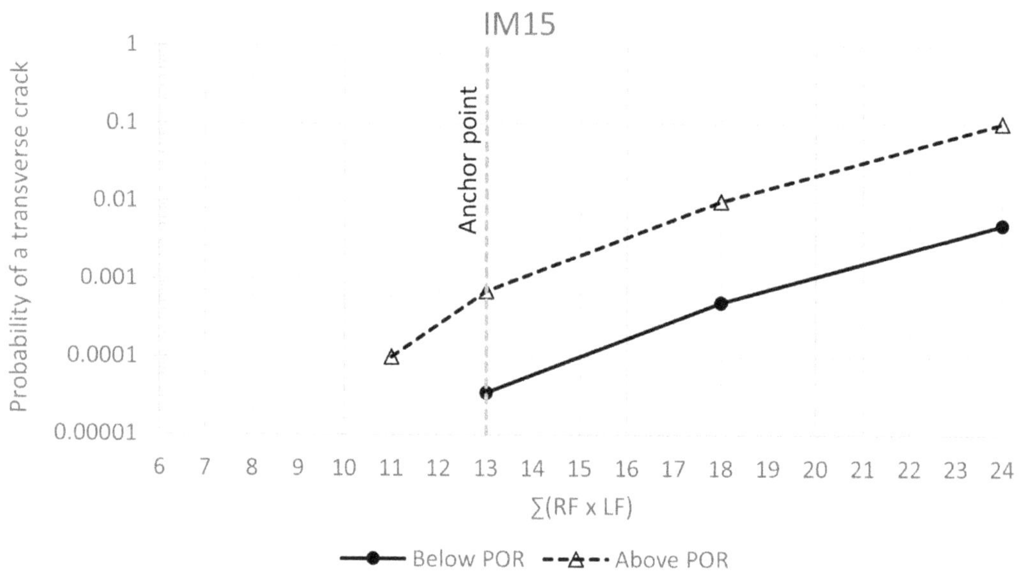

FIGURE A10 IM 15. Probability of a poorly compacted layer on the core foundation contact versus ∑ (Relative importance factor [RF]) × (Likelihood factor [LF]).

Note: "POR" refers to the Pool of Record level.

Second—For cases where the reservoir stage is above the base of potential freezing, follow the next procedure to estimate the probability of frost action affecting the dam and the depth of the frost action.

Third—Estimate the likelihood of transverse cracking or flaw in the upper part of the embankment using Table A8 and Figure A11.

For climates where temperatures do not fall below freezing point except possibly overnight or for a day or two, the probability of a persistent crack or flaw resulting from freezing = zero.

Fourth—Estimate the width of crack or flaw which may result from the frost action using Table A10. Assume that the flaw width may persist to the base of the effects of frost in Table A9.

Estimation of the Width of a Flaw Resulting from Freezing

The effects of frost action are complex and include formation of cracks due to heave and formation of ice lenses. The latter may melt in summer months and leave pathways in which erosion may initiate. Where there is specific information about frost effects for the dam being assessed, that information should be used by the risk analysis team to assess likely defect widths. In the absence of such data Table A10 should be used. These widths may be applied to the full depth of penetration of frost from Table A10, as the evidence is that the ice lenses may be as thick at the base of freezing as at the surface.

TABLE A8 IM 16—Factors Influencing the Likelihood of Transverse Cracking or Flaw in the Upper Part of Embankment Dams Due to Freezing

		LIKELIHOOD FACTOR (LF)			
FACTOR	*RELATIVE IMPORTANCE FACTOR (RF)*	*LESS LIKELY (1)*	*NEUTRAL (2)*	*MORE LIKELY (3)*	*MUCH MORE LIKELY (4)*
Climate	(3)	Other climates where temperatures do not fall below freezing point except possibly overnight or for a day or two[(a)]	Temperate climate where temperatures may remain below freezing point for up to 1 month	Subarctic or alpine climates where temperatures remain below freezing point for 1 to 3 months	Subarctic or alpine climates where temperatures remain below freezing point for 3 months or more
Classification of core material	(2)	Clean gravel (GP, GW) and sand (SP, SW), less than 3% finer than 0.02 mm	Gravelly (GP, GW) and sandy (SP, SW) soils with between 3% and 6% finer than 0.02 mm High plasticity clays (CH)	Silty gravelly (GM, GW-GM, GP-GM) and silty sandy soils (SM, SW-SM, SP-SM) with 6% to 15% finer than 0.02 mm Clayey sands and gravels (SC, GC) and clays with plasticity index < 12	Silts (ML, MH), silty sands (SM) with > 15% finer than 0.02 mm, and clayey silts (ML-CL)
Crest zoning-surface layer over core	(1)	Greater than 2 m of rockfill over the core	Gravelly material or rockfill 1 m to 2 m thick over the core	Gravel material or rockfill 0.45 m to 1 m thick over the core	No surface layer with dam core extending to crest level or thin (less than 75 mm) road pavement or gravelly material

Note: (a) If climatic conditions are as described here, the probability of a persistent crack or flaw resulting from freezing = zero.

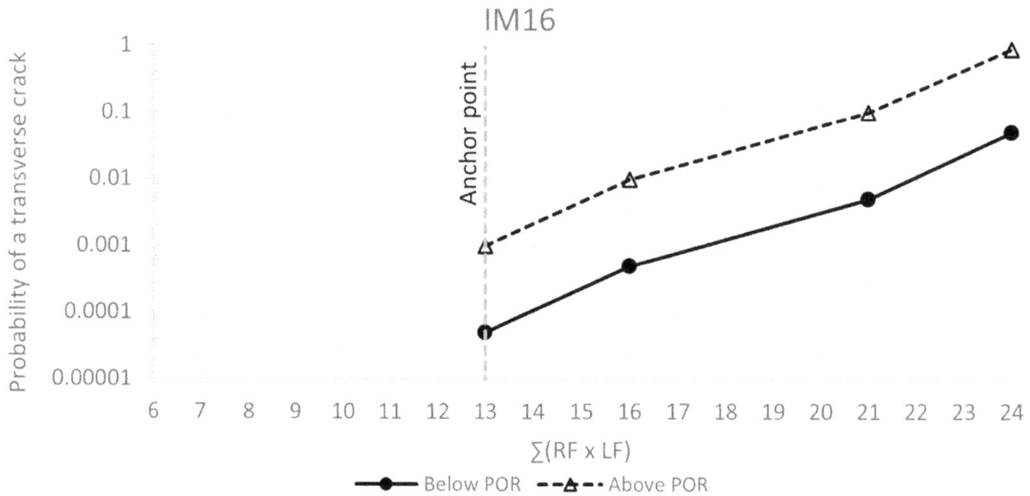

FIGURE A11 IM 16. Probability of persistent cracking or flaw in the crest due to freezing versus Σ (Relative importance factor [RF]) × (Likelihood factor [LF]).

Note: "POR" refers to the Pool of Record level.

TABLE A9 Maximum Likely Depth of Freezing Induced Flaws Based on Climate

CLIMATE	*MAXIMUM LIKELY DEPTH (METRES)*
Subarctic or alpine climates where temperatures remain below freezing point for 3 months or more	1.8
Subarctic or alpine climates where temperatures remain below freezing point for 1 to 3 months	1.3
Temperate climate where temperatures may remain below freezing point for up to 1 month	0.6
Other climates where temperatures do not fall below freezing point except possibly overnight or for a day or two	0

TABLE A10 Width of Frost Induced Crack or Flaw Versus Σ (RF × LF)

FOR IM 16 Σ (RF × LF) FROM TABLE A8	*FOR IM 17 Σ (RF × LF) FROM TABLE A11*	*WIDTH OF CRACK OR FLAW (MM)*
24	24	20
21	21	10
16	16	5
13	13	2
12	12	Zero

IM 17 TRANSVERSE CRACKING OR CONTINUOUS FLAW RESULTING FROM FREEZING ON SEASONAL SHUTDOWN LAYERS AND ON THE SURFACE OF STAGED EMBANKMENTS

The procedure to estimate the likelihood of a persistent crack or flaw is as follows:

First—This mechanism only applies where there has been a seasonal shutdown during construction, or the embankment has been staged with the stage being subject to freezing conditions during the shutdown.

Second—Where the mechanism applies, estimate the probability of transverse cracking or flaws (P_{CF}) using Table A11 and Figure A12.

Third—Assess the probability the cracking or flaw will be continuous across the core (P_{cont}). This likelihood is to be assessed by the risk analysis team based on the information available and their judgement.

Fourth—The probability of persistent crack or flaw = (P_{CF}) × (P_{cont}).

Fifth—Estimate the likely depth of the crack or flaw below the seasonal shutdown from Table A9, and the width of the crack or flaw from Table A10.

IM 18 CONTINUOUS CRACK OR GAP ASSOCIATED WITH A CONDUIT THROUGH THE EMBANKMENT

Estimation of the Probability of a Continuous Crack or Gap along a Conduit

Assess the probability of a continuous crack or gap around the conduit using Table A12 and Figure A13.

In Figure A13, for reservoir stages between POR and POR + 20%, interpolate between the probabilities for POR and POR + 20%.

"First fill" probabilities apply where the reservoir rises above the conduit level for the first time.

Estimating the Width of Crack or Gap in Poorly Compacted Soil around a Conduit

The procedure is as follows:

- Assess the thickness of the layer of soil which is poorly compacted (T). There may be a single layer or several layers. To do this take account of the dimensions of the conduit and the trench in which it is placed.

TABLE A11 IM 17—Factors Influencing the Likelihood of Transverse Cracking or Flaws on Seasonal Shutdown Layers during Construction and Staged Construction Surfaces Due to Freezing

		LIKELIHOOD FACTOR (LF)			
FACTOR	*RELATIVE IMPORTANCE FACTOR (RF)*	*LESS LIKELY (1)*	*NEUTRAL (2)*	*MORE LIKELY (3)*	*MUCH MORE LIKELY (4)*
Construction practices regarding clean-up of frozen layers after construction shutdowns or the surface of the earlier stage of the dam	(3)	Very good control and clean up practices. Frozen layers removed from embankment and replaced with new soil or adequately reworked to specified moisture content. If this condition is present, probability = zero.	Good control and practices, surfaces scarified, moisture adjusted to specified range, surface re-compacted.	Moderate control. Attempts to scarify frozen layers, but depth of scarifying insufficient or difficulties with moisture control on re-compacting the soil.	Poor control. No attempt to scarify or remove frozen layers, poor moisture control on re-compacting the soil.
Climate	(2)	Other climates where temperatures do not fall below freezing point except possibly overnight or for a day or two.	Temperate climate where temperatures may remain below freezing point for up to 1 month.	Subarctic or alpine climates where temperatures remain below freezing point for 1 to 3 months.	Subarctic or alpine climates where temperatures remain below freezing point for 3 months or more.
Classification of core material	(1)	Clean gravel (GP, GW) and sand (SP, SW), less than 3% finer than 0.02 mm.	Gravelly (GP, GW) and sandy (SP, SW) soils with between 3% and 6% finer than 0.02 mm. High plasticity clays (CH)	Silty gravelly (GM, GW-GM, GP-GM) and silty sandy soils (SM, SW-SM, SP-SM) with 6% to 15% finer than 0.02 mm. Clayey sands and gravels (SC, GC) and clays with Plasticity Index < 12.	Silts (ML, MH), silty sands (SM) with > 15% finer than 0.02 mm, and clayey silts (ML-CL).

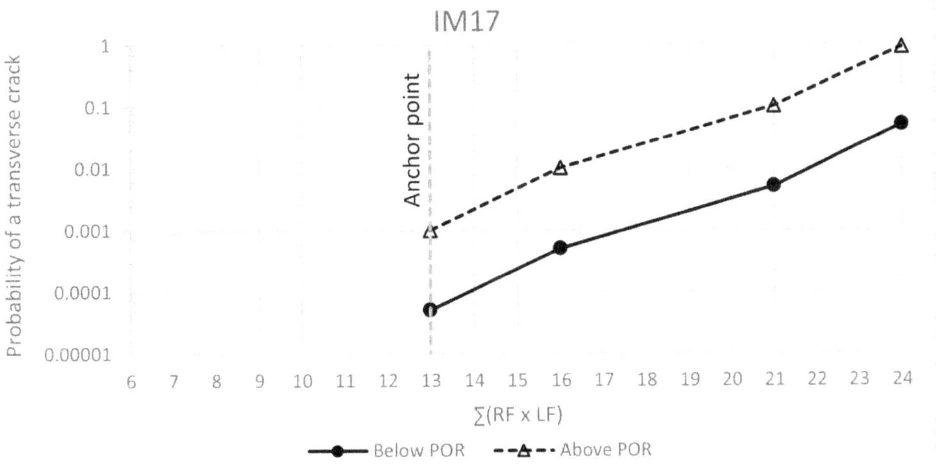

FIGURE A12 IM 17. Probability of transverse cracking or flaws on seasonal shutdown layers during construction and staged construction surfaces due to freezing versus Σ (Relative importance factor [RF]) × (Likelihood factor [LF]).

Note: "POR" refers to the Pool of Record level.

TABLE A12 IM 18—Factors Influencing the Likelihood of Poorly Compacted Zones along the Outside of a Conduit

		LIKELIHOOD FACTOR (LF)			
FACTOR	*RELATIVE IMPORTANCE FACTOR (RF)*	*LESS LIKELY (1)*	*NEUTRAL (2)*	*MORE LIKELY (3)*	*MUCH MORE LIKELY (4)*
Conduit type and surround	(3)	Concrete encased round pipe, concrete precast or cast *in situ*, sloping sides.	Concrete encased round pipe, concrete precast or cast *in situ*, vertical sides.	Masonry, brick, vertical sides.	Any round pipe (including corrugated metal pipe), not concrete encased.
Cutoff collars	(2)	No cutoff collars.	Well detailed cutoff collars, USBR (1987) design.	Poorly detailed cutoff collars, widely spaced.	Poorly detailed cut off collars, close spacing.
Compaction of earthfill around the conduit	(2)	Compaction by mechanical equipment to ≥ 98% standard MDD at OMC−1% to OMC+2%.	Compaction by hand and mechanical equipment to ≥ 95% standard MDD at OMC−2% to OMC+1%.	Compaction by hand equipment, thick layers, ≥ 2% to 3% dry of OMC.	No formal compaction, or poor compaction practices used adjacent to conduits (e.g. thick layers inappropriate for equipment).
Conduit trench details	(2)	Trench totally in non-erodible rock, backfilled to the surface with concrete. (a)	Wide, slopes flatter than 1H:1V, base width not less than conduit width plus 2 m either side.	Medium width, depth, and slope.	Narrow, deep, near vertical sides in soil or rock, backfilled with soil.

Note: (a) If the conduit is totally embedded in a trench totally in non-erodible rock, backfilled to the surface with concrete, it should not be considered as a conduit through the embankment (probability = zero).

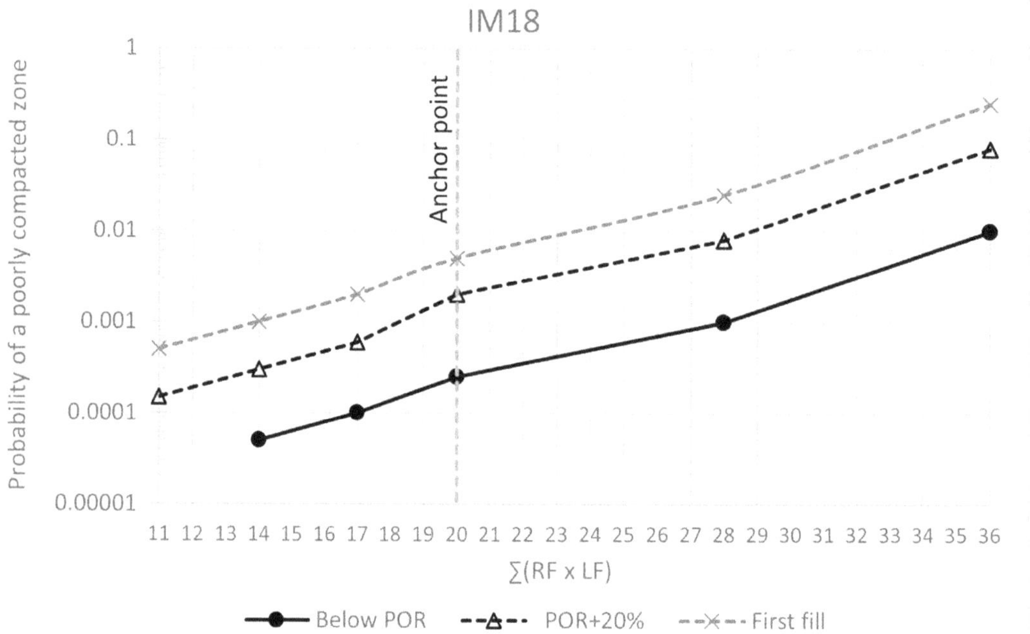

FIGURE A13 IM 18. Probability of a poorly compacted zone associated with a conduit versus Σ (Relative importance factor [RF]) × (Likelihood factor [LF]).

Notes:
(1) "POR" refers to the Pool of Record level.
(2) POR + 20% probabilities apply where the reservoir rise is greater than 20% increase in previously recorded hydraulic head (e.g. for flood protection dams).

- Estimate the amount by which the layer may collapse (C_S) using Table 9.13 as a guide. Table 9.13 requires an overburden stress to select C_S. For this mechanism, use the reservoir pressure at the level of the conduit as the stress causing collapse.
- Estimate the width or height of the gap (G) which is formed from G = (T) × (C_S).

IM 19A EROSION INTO A NON-PRESSURISED CONDUIT

Overall Approach

There are two overall approaches. Both require engineering judgement and/or expert elicitation.

If the internal condition of the conduit is regularly inspected and there is good documentation of the inspections:

(a) Assess the location, probability and width of a crack or an open joint ("a defect") which persists through the conduit based on those inspections (P_{CC}).
(b) Given a defect is present, estimate the probability of the soil surrounding the conduit (P_{EC}) eroding into the conduit, taking account of the width of the defect and particle size

distribution of the soil around the conduit, observations of erosion of soil and/or seepage into the conduit, the external water pressure on the conduit, and the flow conditions in the conduit under operating conditions.

(c) Estimate the probability that given erosion into the conduit has occurred, a sinkhole will develop and erosion will progress into erosion along the conduit (P_{PE}).

The probability of erosion into the conduit progressing to erosion along the conduit = $P_{CC} \times P_{EC} \times P_{PE}$.

If the internal condition of the conduit is not known:

(a) Assess the probability of an open defect being present (P_{CC}) using Table A13 and Figure A14.
(b) Assume that given there is an open defect, the probability of the soil surrounding the conduit (P_{EC}) eroding into the conduit = 1.
(c) Estimate the probability that given erosion into the conduit has occurred, it will progress into erosion along the conduit (P_{PE}).

The probability of erosion into the conduit progressing to erosion along the conduit = $P_{CC} \times P_{EC} \times P_{PE}$.

It should be noted that this mechanism can apply for erosion into the conduit from soil in the foundation and in the embankment.

Estimation of the Probability of and Width of an Open Defect (P_{CC})

If the internal condition of the conduit is regularly inspected and there is good documentation of the inspections, then estimate the probability and width of a crack which persists through the conduit walls and the probability and width of an open joint based on those inspections. Make allowance for the potential for cracks or open joints not persisting through the full width of the conduit walls or through joints.

If the internal condition of the conduit is not known, use Table A12 and Figure A14.

The width is not predictable, so it is suggested that it be assumed that the defects are wide enough to allow erosion of the soil surrounding the conduit to erode into the conduit.

For important decisions, the insides of the conduit should be inspected by remote camera. The conduit may need to be dewatered to allow inspection by camera or to allow persons inside to inspect for defects.

Given a Defect Is Present, Estimation of the Probability of the Soil Surrounding the Conduit Eroding into the Conduit (P_{EC})

If the internal condition of the conduit is regularly inspected:

(a) If there is evidence of soil eroded into the conduit, e.g. soil deposited into the conduit, or evidence of a cavity outside the conduit, assume $P_{EC} = 1$.
(b) If there is no evidence of eroded soil, that may only reflect the likelihood eroded soil would be carried away by water flowing in the conduit. Estimate the probability soil will be eroded into the open defect using the method described in Section 12.2.6.

If the internal condition of the conduit is not known:

In the absence of data regarding the width of any defect assume $P_{EC} = 1$.

TABLE A13 IM 19A—Factors Influencing the Probability of an Open Defect in a Non-Pressurised Conduit When the Internal Condition Is Not Known

		LIKELIHOOD FACTOR (LF)			
FACTOR	*RELATIVE IMPORTANCE FACTOR (RF)*	*LESS LIKELY (1)*	*NEUTRAL (2)*	*MORE LIKELY (3)*	*MUCH MORE LIKELY (4)*
Concrete, masonry, or brick Conduits	(3)	Cast *in situ* concrete. Precast concrete pipes with high quality joints with water stops.	Concrete pipe with bell and spigot joints.	Precast concrete culverts. Cast *in situ* concrete culvert with poor or no water stops.	Masonry, brick.
OR Steel/Cast Iron Pipe Conduits	(3)	New steel or cast iron with extensive corrosion protection. Concrete encased steel or iron.	"New" cast iron or steel pipe ("New" < 20 years cast iron, 10 years steel).	"Old" cast iron or steel pipe, not encased ("old" > 60 years cast iron, 30 years steel). Corrugated metal pipe < 5 years.	"Old" corrugated metal pipe (> 10 years).
Settlement of the conduit	(2)	Rock foundation or measured small settlements.	Shallow compressible foundation soils, calculated small settlements.	Deep compressible foundation soils calculated large settlements.	Measured significant settlement and differential settlement.
Conduit operation	(1)	Wet conduit, with low velocity (< 1.5 metres/sec) flows.	Wet conduit, medium to high velocity flows.	Wet conduit, high velocity (> 5 metres/sec).	Wet conduit, surging flows, medium or high velocity.

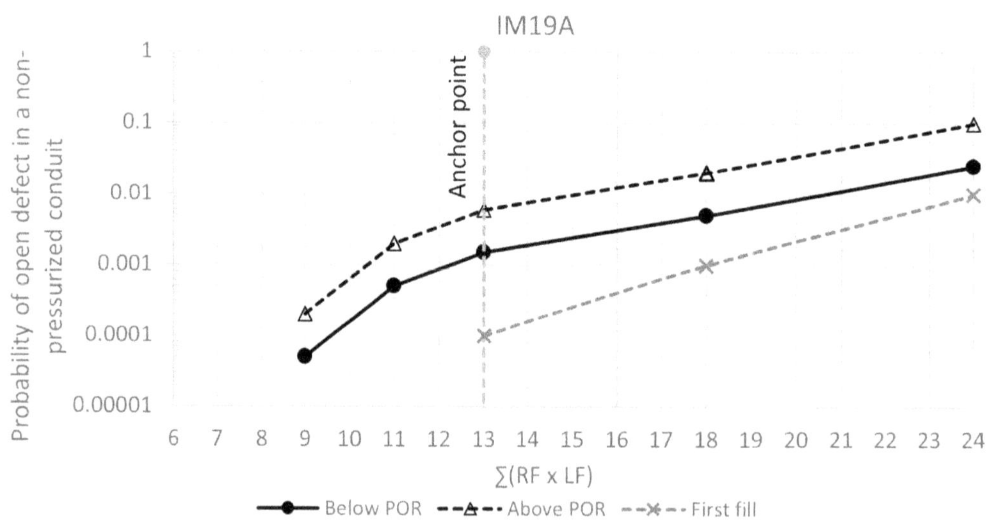

FIGURE A14 IM 19A. Probability of an open defect in a non-pressurised conduit versus Σ (Relative importance factor [RF]) × (Likelihood factor [LF]).

Note: "POR" refers to the Pool of Record level. First fill is the first time the conduit has water in it.

Estimate the Probability That Given Erosion into the Conduit Has Occurred, It Will Progress into Erosion along the Conduit (PPE)

First, estimate using expert elicitation the probability that given erosion into the conduit has occurred, a sinkhole will form, and if so, where.

The following factors should be taken into account in making this estimate:

(a) Non plastic soils are more likely to erode into a defect than plastic soils.
(b) Many sinkholes have formed in silt sand gravel non plastic soils, e.g. fill constructed from glacial moraine and till.
(c) Soils which are suffusive will most likely not form a sinkhole for the full height of the fill but be subject to the finer fraction eroding, leaving the highly permeable coarser fraction.
(d) A sinkhole is more likely to occur if the defect is wide open compared to the continuing erosion criteria for the soil.

Second, estimate the probability that given a sinkhole forms, piping will develop along the conduit.

Assume that the mechanism of erosion is hydraulic fracture and use the method described in Section 9.2.7, taking account of the location of the sinkhole, the details of the conduit downstream of the sinkhole and the reservoir pressure at the sinkhole.

IM 19B LEAKAGE OUT OF A PRESSURISED CONDUIT LEADING TO EROSION ALONG THE CONDUIT

The overall approach is as follows:

(a) Assess the location, probability of a crack or an open joint ("a defect") which persists through the conduit using the method described for IM 19A.
(b) Estimate the potential pressure on the outside of the defect resulting from the water pressure in the conduit.
(c) Given a defect is present, and the pressure estimated in (b), estimate the probability of formation of a pipe along the outside of the conduit downstream of the defect using the method for IM 18.
(d) Given a pipe forms downstream, estimate the probability the pipe head cuts upstream to the reservoir or inter-connects with a pre-existing defect through the upstream fill.

The risk analysts will have to use expert elicitation and there will be considerable uncertainty in the outcomes because the mechanisms are not readily quantified.

IM 20 TRANSVERSE CRACK OR GAP ASSOCIATED WITH A SPILLWAY OR ABUTMENT WALL RESULTING FROM POOR COMPACTION ADJACENT TO THE WALL

Use Tables A14 and Figure A15 to estimate the probability of a continuous poorly compacted layer. Use all available information to make the estimates and be prepared to use expert elicitation to adjust the outcomes.

Record the following:

- The estimated probability of a continuous poorly compacted layer or layers.
- The location of and estimated thickness and depth below the dam crest of the poorly compacted layer(s).
- The estimated likely best estimate and range of standard compaction relative compaction, and moisture content relative to standard compaction optimum moisture content (SOWC) in the poorly compacted layer(s).

TABLE A14 IM 20—Factors Influencing the Likelihood of a Poorly Compacted Layer Associated with a Spillway or Abutment Wall

		LIKELIHOOD FACTOR (LF)			
FACTOR	*RELATIVE IMPORTANCE FACTOR (RF)*	*LESS LIKELY (1)*	*NEUTRAL (2)*	*MORE LIKELY (3)*	*MUCH MORE LIKELY (4)*
Compaction of earthfill adjacent to the wall	(3)	Compaction by mechanical equipment to ≥ 98% standard MDD at OMC −1% to OMC +2%	Compaction by hand and mechanical equipment to ≥ 95% standard MDD at OMC −2% to OMC+1%	Compaction by hand equipment, thick layers ≥ 2% to 3% dry of standard OMC	No formal compaction, or poor compaction practices, e.g. placed in very thick lifts and no moisture content control
Slope and finish on wall	(2)	> 0.25H:1V Smooth, planar	0.1 ≤ H/V < 0.25 Smooth, planar	Vertical, rough and irregular	Vertical and horizontal steps (e.g. masonry or brick walls)
Concrete buttresses	(1)	None	Single, slopes flatter than 0.1H:1V	Single, with poor details such as vertical sides	Several close together preventing good compaction

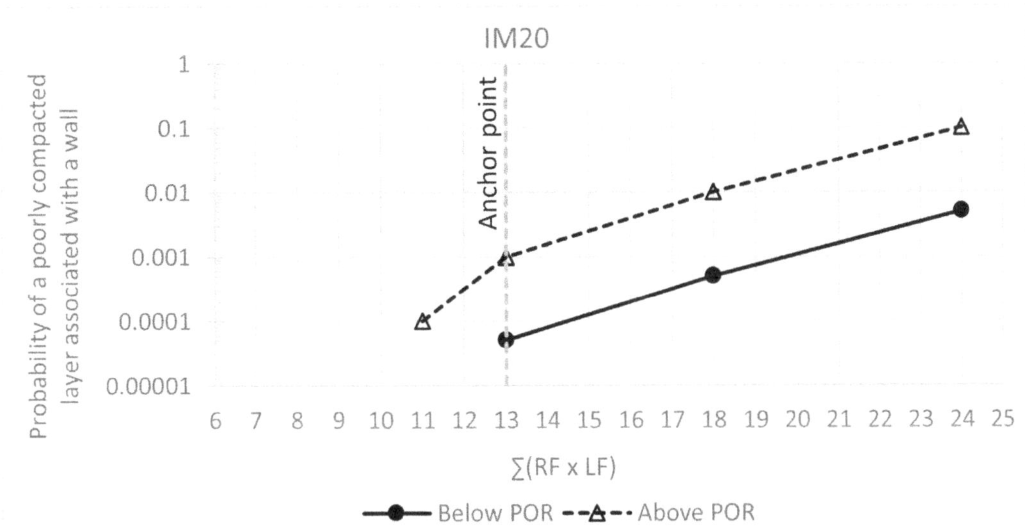

FIGURE A15 IM 20. Probability of a poorly compacted layer associated with a spillway or abutment wall versus Σ (Relative importance factor [RF]) × (Likelihood factor [LF]).

Note: "POR" refers to the Pool of Record level.

IM 21 TRANSVERSE CRACK OR GAP ADJACENT TO A SPILLWAY OR ABUTMENT WALL RESULTING FROM DIFFERENTIAL SETTLEMENT ADJACENT THE WALL DUE TO WALL PROFILE OR FOUNDATION PROFILE ADJACENT THE WALL

Figure 9.31 shows a situation where settlement of the dam fill may result in a gap forming between the fill and the upper, steeper part of the wall.

A gap may also form if the retaining wall is slender and has marginal factor of safety against stability.

Use Table A15 and Figure A16 to estimate the likelihood of a gap or crack being present.

IM 22 TRANSVERSE CRACK OR GAP ADJACENT TO A SPILLWAY OR ABUTMENT WALL RESULTING FROM DIFFERENTIAL SETTLEMENT ADJACENT TO THE WALL

Figure 9.32 shows a situation where settlement of the dam fill may result in a gap forming between the fill and the upper, steeper part of the wall.

Use Table A16 and Figure A17 to estimate the likelihood of a transverse crack or gap.

TABLE A15 IM 21—Factors Influencing the Likelihood of a Crack or Gap Adjacent to a Spillway or Abutment Wall

FACTOR	*RELATIVE IMPORTANCE FACTOR (RF)*	*LIKELIHOOD FACTOR (LF)* *LESS LIKELY (1)*	*NEUTRAL (2)*	*MORE LIKELY (3)*	*MUCH MORE LIKELY (4)*
Slope of wall	3)	Sloping, flatter than 0.5H to 1V	Sloping, 0.1H to 1V to 0.5H to 1V	Vertical or steep with flatter slope on lower part	Overhanging over the core width
Wall type, wall stiffness	(2)	Very stiff gravity wall or counterfort wall	Thin gravity wall or counterfort wall	Cantilever wall not normally subjected to cyclic reservoir level conditions	Cantilever wall, subject to cyclic reservoir level conditions. Poor freeze/thaw details.
Concrete buttresses	(1)	Single or multiple with good details and proven good compaction around the buttress	Single with probably good to reasonable compaction around the buttress	None or single with poor details such as vertical sides, little evidence of good compaction	Several close together preventing good compaction and little evidence of good compaction

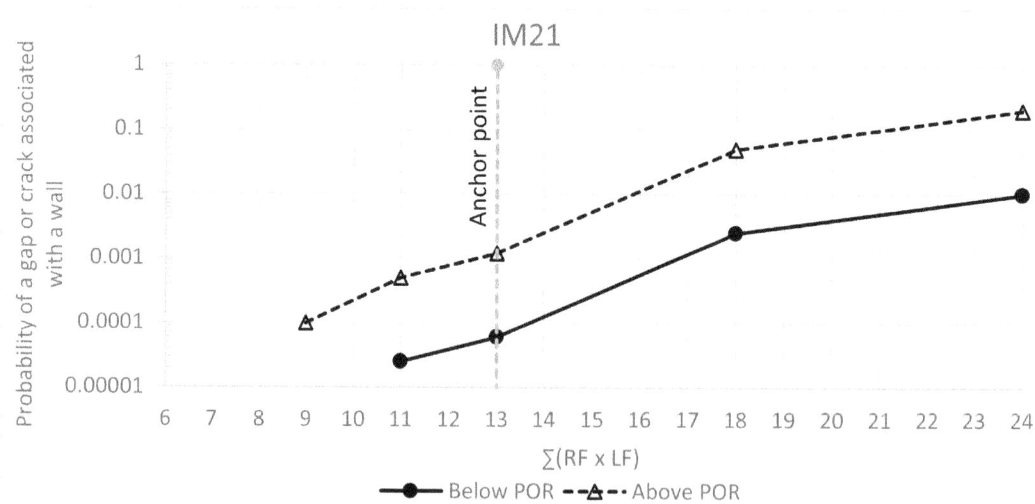

FIGURE A16 IM 21. Probability of a persistent gap or crack associated with a spillway or abutment wall versus Σ (Relative importance factor [RF]) × (Likelihood factor [LF]).

Note: "POR" refers to the Pool of Record level.

TABLE A16 IM 22—Factors Influencing the Likelihood of a Transverse Crack or a Gap Resulting from Settlement of Fill Adjacent to the Wall

FACTOR	*RELATIVE IMPORTANCE FACTOR (RF)*	*LIKELIHOOD FACTOR (LF)*			
		LESS LIKELY (1)	*NEUTRAL (2)*	*MORE LIKELY (3)*	*MUCH MORE LIKELY (4)*
Cross-valley profile under embankment core[(a) (b)]	(3)	$W_b/H_w > 2.5$ Note: If this condition is present, probability = zero	$1.0 < W_b/H_w \leq 2.5$	$0.25 < W_b/H_w \leq 1.0$	$W_b/H_w \leq 0.25$
Post-construction settlement of the embankment as a % of maximum embankment height	(2)	s/H < 0.25%	0.25% < s/H < 0.75%	0.75% < s/H < 2.0%	s/H ≥2.0%
Slope of the abutments under embankment core[(a)]	(1)	$15° > \beta_1 \leq 25°$ Note: If $\beta_1 < 15°$ probability=zero	$25° < \beta_1 \leq 45°$	$45° < \beta_1 \leq 60°$	$\beta_1 > 60°$

Notes:
(a) See Figure 9.33 for definitions of W_b, H_w, β_1.
(b) This mechanism only applies for $W_b/H_w \leq 2.5$ and $\beta_1 \geq 15°$.

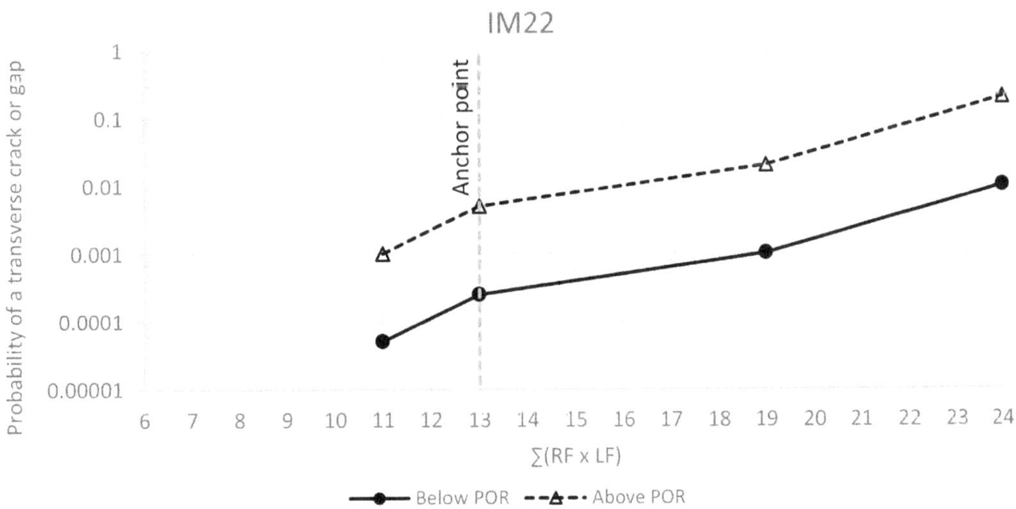

FIGURE A17 IM 22. Probability of a persistent transverse crack or a gap resulting from settlement of embankment adjacent a cliff versus Σ (Relative importance factor [RF]) × (Likelihood factor [LF]).

Note: "POR" refers to the Pool of Record level.

IM 23 PERSISTENT CRACK OR GAP AT THE INTERFACE OF WRAP-AROUND CONNECTING EMBANKMENT TO A CONCRETE GRAVITY DAM

Figure 9.37 shows typical details of a wrap-around connection of an embankment dam to a concrete gravity dam.

There is a potential for the embankment to move away from the concrete dam at PQ and particularly at RS due to settlement of the embankment dam during and after construction. It is recommended that this situation be assessed as follows:

The primary control on seepage and initiation of erosion is considered to be along QR. The likelihood of a crack or gap being present, and if so, the probability erosion will initiate should be assessed as for an embankment abutting a very stiff retaining wall (refer to IM 21).

When assessing the seepage gradient along QR, the likelihood there will be gaps or poorly compacted soil along PQ and RS should be assessed, taking account of the factors in Table A17. For the worst scenarios, there will be a gap and no head loss in the seepage path on PQ and RS if there is a gap there which can be seen on inspection.

IM3, IM 10, AND IM 12 WIDTHS OF CRACKING AND HYDRAULIC FRACTURE RESULTING FROM ARCHING

For IM 3, IM 10 and IM 12, crack widths should be estimated as shown in Table A18.

For IM 3 and IM 10, these widths apply for all reservoir stages. For IM 12, these widths should be considered applicable to reservoir stages where the increase in reservoir pressure is 100 kPa or more

TABLE A17 Factors to Be Considered in Assessing Seepage Gradients on the Wrap-Around

FACTORS WHICH MAKE IT LIKELY THERE IS A GOOD SEEPAGE CONTACT ALONG PQ AND RS, AND GRADIENTS WILL BE LOWER	*FACTORS WHICH MAKE IT LIKELY THERE IS A POOR SEEPAGE CONTACT ALONG PQ AND RS, AND GRADIENTS WILL BE HIGHER*
Uniform slope with no overhangs.	Overhangs in the concrete.
Uniform concrete slopes with at least 0.1H to 1V slope.	Change in slope of concrete such as shown in Figure 9.29 allowing a gap to form as the embankment settles.
Well compacted embankment shoulders or zoning with all materials having a similar and high modulus.	Poorly compacted shoulders leading to large settlements during and post construction.
Low embankments and/or post-construction settlements < 0.25% of dam height adjacent to the interface.	High embankment and/or post-construction settlements ≥ 1.0% of dam height adjacent to the interface.
Reservoir level at least 6 m below dam crest level so there is a lesser likelihood a crack will persist to reservoir level.	Reservoir level approaching dam crest level.

TABLE A18 Crack Widths for IM 3, IM 10 and IM 12

	CRACK WIDTHS IN MM FOR Σ (RF × LF) VALUE FROM TABLES A2, A4 AND A5			
FAILURE MODE	*13*	*17*	*21*	*24*
IM 3	1	20	40	50
IM 10	1	20	40	50
IM 12	1	10	20	25

above Pool of Record level. For increases in reservoir pressure from 0 kPa to 100 kPa, the crack widths should be scaled according to the pressure increases, so for 50 kPa, the crack widths would be half those for 100 kPa.

These widths are approximate and given only as a guide.

Appendix B
Guidelines for Backfilling Investigation Drill Holes, Holes Resulting from Cone Penetration Tests, and Test Pits (ANCOLD, 2020)

B1 OVERALL PLANNING

It is important at the planning stage of investigations to consider the backfilling methods. The key objectives for backfilling investigation holes within the embankments should include the following:

(a) Backfilling to a condition similar to the existing condition prior to investigations. This includes material types, internal zoning and compaction.
(b) The dam safety risks for the condition after backfilling should be equal to or, where possible, less than the existing condition prior to investigations.

It is good practice in the planning phase of investigations to consider potential scenarios that could impact dam safety during the investigations. It should be asked in the planning phase (and revisited whilst on site): *Has this test pit/borehole the potential to expose the embankment to an increased dam safety risk?* If the answer is yes, then mitigation measures should be implemented to manage this risk. If these risks cannot be mitigated, then the test pit/borehole should not be undertaken.

Mitigation measures can include:

(a) Timing the investigations to periods when the reservoir is at a lowered level or temporary lowering of the reservoir level.
(b) Review of the weather forecast in the lead up to the investigations and delaying works if significant wet weather is forecast.
(c) Having materials on site for emergency backfilling. For test pits this might include filters and/or rockfill in the event that a seep is exposed and either seepage rates increase or are

turbid. It is good practice to have materials (such as filters) on standby and close to the investigations for this potential occurrence (even if low risk) and discuss this with the owner. The owner will often be willing to provide support during the works (e.g. access to a truck or front end loader that is present on site with an experienced operator).

(d) Ensuring investigation holes are safely closed before the end of the working day (this could include temporary backfilling of a test pit with filter material if necessary).

These dam safety measures are in addition to health and safety considerations.

B2 BACKFILLING TEST PITS

Test pits that extend either partially or fully through the embankment core can leave a transverse defect that exposes the embankment to an increased likelihood of internal erosion and piping. Because of this test pits deeper than approximately 1.0 to 1.5 m excavated in a transverse direction across the full width of the embankment core should not be undertaken. The alternative of staggering test pits should be considered whereby one pit is excavated in the downstream portion of the core and a separate offset test pit excavated in the upstream portion of the core.

Poor compaction of earthfill on backfilling will result in settlement of the test pit. This has dam safety impacts (as indicated earlier) and also presents a safety concern for operators and the public. When wetted, the poorly compacted earthfill will soften and settle, leaving depressions (further exacerbating wetting by allowing water to pool) and development of soft areas on the surface that present a health and safety risk. Backfilling with compaction only by the bucket of the excavator or backhoe is not acceptable practice.

Guidelines for backfilling test pitting include:

(a) Stockpiling separately the different materials during excavation (including capping materials and internal zoning) to minimise contamination.

(b) Ensuring the earthfill core (and other cohesive zones) is well compacted. This includes:

  (i) Placement and compaction in thin lifts (approximately 100 mm compacted but depends on the compaction plant and material being compacted).

  (ii) Moisture conditioning the earthfill if required (in the case of earthfill that has dried in the upper or outer embankment).

  (iii) Compaction using plant that can access all areas of the layer especially against vertical faces. This might require more than one piece of compaction plant (e.g. a small roller and a hand operated impact compactor). The use of roller wheels and vibrating plate attachments to excavators can be useful for compaction but are not suited to all earthfill types.

  (iv) Scarifying smooth surface of each layer prior to placing the next layer.

Figures B1 and B2 show good practice for backfilling of test pits.

(c) Using filter materials downstream of the earthfill core. For homogeneous or zoned earthfill embankments, a filter material can also be used for backfill. The filter should extend longitudinally across the full width and over the full height of the test pit and be continuous between layers. The use of a properly installed filter within a test pit will more than offset the potential for a defect created by the excavation of that pit, and filter materials are more readily compacted than earthfill.

Where filters are present in an embankment, these will likely be contaminated during test pitting and will unlikely be suitable for reuse as backfill. The use of imported filters is recommended for backfilling to replace filter zones. It is important to connect into the existing filter in the base and sides of the test pit when backfilling.

B3 BACKFILLING OF BOREHOLES

It is good practice to backfill all boreholes. This will avoid the potential for creation of a void within the core and interconnection of potential defect pathways.

Widespread practice is to backfill the borehole with a non-shrink bentonite cement grout. Consultants generally have a standard mix design that they use. Otherwise, reference can be made to Mikkelsen and Green (2003) and Contreras et al. (2007). Backfilling with drill cuttings is not acceptable for backfilling the core of an embankment.

FIGURE B1 Backfilling the large test pits at Rowallan Dam.

FIGURE B2 Backfilling test pits with vibrating plate and pneumatic impact rammer.

Good practice for grouting boreholes includes:

(a) Testing the grout mix prior to use. The properties of bentonite can vary, and this may affect the mix design. Testing the mix design is recommended to:
   (i) Confirm the mix design is non-shrink on curing and that the fluid mix does not bleed.
   (ii) Test the viscosity of the mix and the ability to tremie the grout into the borehole.
   (iii) Test the set strength of the mix and its compatibility with the strength properties of the earthfill core.
(b) Using a driller that is well equipped. This includes the use of a high-speed shear mixer for mixing the grout, having measuring scales, tremie pipes, etc.
(c) Hydrating the bentonite well before addition of cement. Adding the cement will tend to thicken the mix significantly as it cures and should be done only just before using a tremie to place into the base of the hole.
(d) Using tremie pipes that extend to the base of the borehole. During grouting, the tremie should always remain within the grout.
(e) Delivering grout into the top of the tremie pipe by gravity (i.e. don't connect the pump line to the tremie pipe) so as to not potentially hydraulically fracture the earthfill core. This can be done either by pumping the grout to a funnel at the top of the tremie or pouring the grout in by bucket.
(f) Stage grouting of earthfill core to reduce the risk of hydraulic fracture. An example is to use grout stages of 5 m and waiting between stages for the grout to partially set. This is usually on the order of 30 to 60 minutes. This can be checked as part of the initial trial grout. It is important to note that grout set times will increase significantly in cold weather or if cold water is used to prepare the mix.

The engineering geologist or geotechnical engineer supervising the drilling should keep account of the volume of grout poured into a hole and check this against the expected volume. Checks on the grout level at stages during grouting should also be done (especially in the case of grouting in puddle core earthfill as it is more susceptible to hydraulic fracture from applied pressure), as this may identify if there is a problem with grout loss into the embankment.

Where the borehole crosses internal zones within an embankment, it is important to ensure the backfilling is compatible with the internal zoning. Where filter zones are intersected, they should be reinstated. Consideration needs to be given to the compatibility of the sand for backfill compared to the filter and, if necessary, multiple filter backfill materials used (if this is required, then a larger diameter borehole is likely also required).

B4 BACKFILLING BOREHOLES FOR INSTRUMENTS

Several considerations for grouting of instrument installations in boreholes are as follows:

(a) For sand pack installed vibrating wire piezometers, the use of bentonite zones above and below the sand zone are recommended for sealing. The bentonite is in granular form placed by tremie pipe. It swells on hydration forming the seal. Where grout is used between stages, the compatibility between the grout and sand zone can be checked by a column test (Figure B3).
(b) For Casagrande piezometers, ensure there is enough space between the standpipe and the borehole annulus for installing the sand pack, bentonite seal and tremie pipe for grouting.

FIGURE B3 Column tests to check compatibility between grout and sand surrounding piezometers.

(c) For inclinometer installations, grout mix strengths closer to the core stiffness should be used (rather than soft set grouts) to avoid the potential for buckling of the inclinometer casing post installation.
(d) For vibrating wire piezometers, there are methods where the piezometers are grouted in (i.e. no sand zones). Refer to Mikkelsen and Green (2003) and Contreras et al. (2007, 2011, 2012) for details.

Consideration needs to be given to the layering and permeability variation in earthfill to the type of piezometer installation (i.e. a broader sand pack zone may be a better alternative to capture pore pressure within higher permeability layers rather than a discrete point for grouted in piezometer).

B5 BACKFILLING OF HOLES RESULTING FROM CONE PENETRATION TESTS (CPT, CPTU)

The practices for grouting boreholes apply to grouting up holes resulting from CPT and CPTU. Often the grout mix needs to be thinned in order for it to flow within the smaller diameter tremie pipe inserted in the CPT hole.

Good practice for grouting holes resulting from CPT and CPTU includes the following:

(a) Pre-drilling through any capping layer and backfilling this with compacted earthfill. This is a useful practice to stop cave-in from pavement layers blocking the CPT/CPTU hole.
(b) On removal of the CPT/CPTU rods, immediately insert a small diameter tremie pipe whilst the rig is still set up over the hole. If the tremie gets to the base of the CPT, then staged grouting can begin and the rig can move off the hole.
(c) If the tremie pipe cannot reach the base of the CPT/CPTU hole, then the CPT/CPTU rods are re-inserted into the hole (with a dummy tip) and back grouting is then undertaken through the CPT/CPTU rods. The grouting is done by retracting the CPT/CPTU rods from the hole base 1 to 2 m and grouting, then retracting a further 1 to 2 m and grouting until the CPT/CPTU hole is fully grouted. Grouting through the CPT/CPTU rods is a slow process but is required to adequately grout up holes.
(d) CPT/CPTU holes within puddle cores (soft to firm clay earthfill) will close on extraction. This is considered acceptable, and no grouting is required in some cases but should be reviewed on a case-by-case basis. It would not be acceptable where there are known leaks/defects within the puddle core for example. Where not acceptable, grouting will need to be undertaken through the CPT/CPTU rods (i.e. remove instrumented cone and install a dummy tip cone and push back down the original hole). Even where assessed as acceptable, the top of the hole should be grouted, as it is likely to be open to depths up to 5 to 6 m.

REFERENCES

Contreras, I.A., Grosser, A.T. and Ver-Strate, R.H. (2007). The use of the fully-grouted method for piezometer installation. *Proc. 7th Int. Symp. On Field Measurement in Geomechanics*, Boston, MA, ASCE Geotechnical Publication.

Contreras, I.A., Grosser, A.T. and Ver Strate, R.H. (2011). Practical aspects of the fully-grouted method for piezometer installation. *Proc. Eighth International Symposium on Field Measurements in Geomechanics.*

Contreras, I.A., Grosser, A.T. and Ver Strate, R.H. (2012). Update of the fully-grouted method for piezometer installation. *Geotechnical News*, June, 20–25.

Mikkelsen, P.E. and Green, G.E. (2003). Piezometers in fully grouted boreholes. *Proc. 6th Int. Symp. on Field Measurement in Geomechanics*, Oslo, Norway.

Appendix C

Development of the Historical Database Screening Method for Assessing the Annual Probability of Failure Resulting from Internal Erosion and Piping

C1 INTERNAL EROSION AND PIPING IN THE EMBANKMENT

C1.1 Introduction

The following outlines the amendments to the Foster et al. (1998, 2000a, 2000b) method taking account of improved understanding of the mechanics of internal erosion since the UNSW method was developed. The basic structure of the Foster et al. (1998, 2000a, 2000b) method is maintained.

Historical Probability of Failure, All Zoning Types

Table C1 summarises the historical data. This includes failures and accidents.

Table C2 is for failures and is based on the data in Table C1 and allows for under-reporting of failures as described in Fell et al. (2024).

The historical frequencies for initiation of internal erosion derived in Table C1 underestimates the probability of an incident for the following reasons:

- The accidents and failures cases include both cracks and poorly compacted zones in which erosion initiates. There would be more cases where a crack or poorly compacted zone was present, but it was not detected either because erosion did not initiate in the crack, or erosion did initiate but was stopped by filters.

TABLE C1 Historical Frequencies of Internal Erosion, Large Dams up to 1986 (Fell et al., 2024)

CASE	*TOTAL*	*IN DAM*	*AROUND CONDUITS OR WALLS*	*COMMENTS*
No. of internal erosion failures	36	19	15 conduits 2 walls	.
No. of internal erosion accidents	73	52	1 conduit 4 walls	
No. of seepage accidents	36	30	6 conduits 0 walls	
Total No. of failures and accidents	145	101	38 conduits 6 walls	.
Population of dams	11,192 1866 pre-1950 = 16.6%	11,192	5,596 conduits 2,798 walls	Assumes 50% embankment dams have conduits passing through the dam and that 25% of all dams have wall interfaces.
Historical frequency for internal erosion failures and accidents	0.013	0.009	0.0067 conduits 0.002 walls	
Historical frequency for first fill: From database: Adjusted for under-reporting:		0.0036 0.0108	0.0027 conduits 0.0008 walls 0.008 conduits 0.0024 walls	Assumes 40% occur on first fill, 60% after first fill. See Table S4.2 in Fell et al. (2024).
Historical frequency after first fill: From database: Adjusted for under-reporting:		0.0062 0.019	0.0040 conduits 0.0012 walls 0.012 conduits 0.0036 walls	
Historical annual frequency after first fill adjusted for under-reporting as described in Appendix C		3.3×10^{-4}/annum	2.1×10^{-4}/annum conduits 0.6×10^{-4}/annum walls	Average dam age after first fill of 26 years in ICOLD (1995) database. In ICOLD (2019) the data was to 2018 so the average age was 58 years. There were no additional failures and accidents of the ICOLD (1995) dams in ICOLD (2019) so the age in 2018 is used.

Note: "First filling" includes the first 5 years of operation.

- Not all failures and accidents would have been reported in the ICOLD (1995, 2019) studies. Possible reasons are legal issues, cultural differences in reporting failures, poor inspection and poor ICOLD representation. The authors' experience in Australian dams is that more historic accident cases are found with gathering of data for safety reviews for dams.
- Internal erosion incidents that were detected but not recognised as being an incident, and incidents that have been lost in the passage of time.
- Cracking which is present in dams but above historic high reservoir levels.

Fell et al. (2024) estimated the proportion of under-representation of cracks, and poorly compacted zone cases by assessing the various reasons separately, as shown in Table C3.

TABLE C2 Historical Statistics of Failures in the Embankment Based on Data in Fell et al. (2024)

CASE	*TOTAL*	*IN DAM*	*AROUND CONDUITS OR WALLS*	*COMMENTS*
No. of internal erosion failures	36	19	15 conduits 2 walls	.
Population of dams	11,192	11,192	5596 conduits 2798 walls	Assumes 50% embankment dams have conduits passing through the dam and that 25% of all dams have wall interfaces.
Historical frequency for internal erosion failures	0.0032	0.0017	0.0027 conduits 0.0007 walls	
Historical frequency for internal erosion failures adjusted for under-reporting	0.0064	0.0034	0.0054 conduits 0.0014 walls	2 × to allow for under-reporting. See Table C3.
Historical frequency First fill failures adjusted for under-reporting	0.00256	0.00136	0.00216 Conduits 0.00056 walls	Assumes 40% occur on first fill.
Historical annual frequency of failure after first fill adjusted for under-reporting	0.00384			
Historical annual probability of failure after first fill adjusted for under-reporting	66E-06			Assumes 60% after first fill and average age 58 years. Was 56E-06 in Foster et al. (1998, 2000a, 2000b) method.

Note: "First filling" includes the first 5 years of operation.

There is some useful information relevant to the prior calculation in the ICOLD Bulletin 99 update:

(a) From ICOLD (2019) Table 10.1, the total number of internal erosion failures up to 1986 is 72. That is double what was assumed in the 2008 Toolbox, so failures are under-reported in ICOLD (1995). Many of the additional failures are from the USA pre-1986 and were not included in the Foster et al. (1998, 2000) database.

(b) Based on this it seems likely accidents are also under-reported by at least a similar amount, say 2 X rather than 1.67 X as assumed for Fell et al. (2008).

(c) None of the additional failures were in the accident category in the Foster et al. (1998, 2000) database.

(d) In the Foster et al. (1998, 2000) database the average age of the dams was 26 years (to 1986). They would be average age 26 + (2018 − 1986) = 58 years or about double what was used in Fell et al. (2008).

C1.2 Dam Zoning

In the Foster et al. (1998, 2000a, 2000b) method zoning was accounted for by using the historical annual probabilities of failure for each zoning type.

TABLE C3 Estimated Proportion of Under Representation of Crack, and Poorly Compacted Zone Cases

REASON FOR UNDER-REPRESENTATION	*ESTIMATED PROPORTION*	*JUSTIFICATION*
Internal erosion failures not all reported	2.0 X	ICOLD (2019) shows failures were under-reported in the 1995 database by a factor of 2.
Accidents not reported	2.0 X	Based on a review of Australian dam incidents. 6 cases are reported in ICOLD database, but a questionnaire to Australian dam owners revealed a further 4 cases that had not been reported (10 / 6 = 1.67 X). However, based on under-reporting of failures demonstrated by ICOLD (2019), this is increased to 2.0 X.
Crack developed, but not wide enough/soil sufficiently erosion resistant for erosion to initiate, or crack swelling/crack closing	2.0 X	Estimate 50% of dams in the population have core materials which would be capable of resisting erosion for seepage through a 1–2 mm wide crack (refer to Table S4.6 in Fell et al., 2024). factor = 100 / 50 = 2 X.
Erosion initiated but stopped by filters	1.33 X	An estimated 40% of dams in the population have a downstream zone capable of stopping erosion (refer to Table S4.7 in Fell et al., 2024). Factor is 100 / 60 = 1.67 X. However, there is likely to be some double counting with case 3 above and hence halve the factor (= 1.33 X). That is a dam may have both narrow cracks and filters.
TOTAL	3.0 X	See Fell et al. (2024) Supporting Document for details.

For the Database Screening Method, the zoning is a factor to be used to amend the updated annual probability of failure from Table C2.

Table C4 summarises the data from Table 15.1 from the Foster et al. (1998, 2000a, 2000b) method in columns 2, 3 and 4 and analyses of the data to determine the likelihood class for each zoning category.

C1.3 Embankment Filters

The approach taken is to first estimate the likelihood of failure of dams for which the specified grading of filters is known and meets USBR (1980, 1987) *Earth Manual* or modern filter criteria. This is based on a selection of case data for dams which the authors have had involvement. Table C5 summarises the data.

This is taken as the multiplying factor for the "Much Less Likely" category in Table 15.3. The "Neutral" category is assigned a multiplying factor = 1.0.

The % of dams in these categories are assessed from the data in Foster et al. (1998, 2000a, 2000b) and the authors' experience.

Based on this data it has been assessed that:

For Dams Where the Specified Grading of Embankment Filters Is Known and Meets USBR (1977, 1987) or Modern Criteria:

Assume probability of filter failure = 0.015, or 1.5%. This is controlled by the likelihood of the filter having excess fines and potentially holding a crack.

For dams with filters are constructed of crushed rock with Cu ≥ 20, so likely to be segregated assume probability of filter failure = 0.0165 or 1.65%.

TABLE C4 Assessment of the Relative Likelihood of Failure in the Embankment for the Zoning Classes Based on Foster et al. (1998, 2000a, 2000b)

ZONING CATEGORY	*AVERAGE P_{Te}*	*AVERAGE ANNUAL P_e (× X 10^{-6})*		*ANALYSIS OF DATA*				
	(× 10^{-3})	*FIRST 5 YEARS OPERATION*	*AFTER 5 YEARS OPERATION*	*RELATIVE LIKELIHOOD(3)*	*ADOPT RELATIVE LIKELIHOOD(4)*	*LIKELIHOOD CLASS*	*% DAMS*	*RELATIVE LIKELIHOOD X %*
Homogeneous earthfill	16	2080	190	4.6	5	MML	9.5	0.47
Earthfill with filter	1.5	190	37	0.4	n/a		15	
Earthfill with embankment filter					0.25	LL	9	0.02
Earthfill with foundation filter					1.0	ML	6	0.06
Earthfill with rock toe	8.9	1160	160	2.6	2.0	ML	6.1	0.12
Zoned earthfill	1.2	160	25	0.36	0.35	LL	35.9	0.12
Zoned earth and rockfill	1.2	150	24	0.33	0.35	LL	9.3	0.03
Central core earth and rockfill	(< 1.1)	(< 140)	(< 34)	< 0.3	0.2	MLL	8.4	0.02
Concrete face earthfill	5.3	690	75	1.5	1.5	ML	4.1	0.06
Concrete face rockfill	(< 1)	(< 130)	(< 17)	< 0.3	0.1	MLL	2.8	n/a(7)
Puddle core earthfill	9.3	1200	33	2.7	2.0	ML	4.7	0.09
Earthfill with corewall	(< 1)	(< 130)	(< 8)	< 0.3	0.25	LL	2.4	0.006
Rockfill with corewall	(< 1)	(< 130)	(< 13)	< 0.3	0.1	MLL	0.9	0.001
Hydraulic fill	(< 1)	(< 130)	(< 5)	< 0.3	0.25	LL	0.9	0.001
ALL DAMS	**3.5**	**450**	**55**	**1.0**				**1.0**

Notes:

(1) P_{Te} is the average probability of failure over the life of the dam.
(2) P_e is the average annual probability of failure.
(3) "Relative likelihood" is calculated value based on (P_{Te} for zoning type) / (P_{Te} for All Dams).
(4) "Adopt relative likelihoods" based on average P_{Te} adjusted to take account of the relative vulnerability of embankment zonings for initiation and continuation of piping and to achieve overall neutral outcome.
(5) Earthfill with filter zoning split into two categories. Embankment with foundation filter does not have embankment filter. Embankment with embankment filter may have foundation filter.
(6) *Relative likelihoods to be applied to updated overall statistics including 2x allowance for under-reporting as shown in Table C2.*
(7) CFRD dams not susceptible to internal erosion and piping in the same sense as other dams. See Chapter 16.

Assume that 10% of all embankment dams in the database are in this category based on experience in Australia.

For Other Dams with Embankment Filter or Transition Zones:

Some factors to consider include:

(a) This includes dams which have filters which may be designed and constructed to USBR (1977, 1987) or modern criteria, but the specifications and construction data are not available. These may be considered the "average" dam.

Assume these represent 40% of all dams in the database. The factor = 0.2 for these based on 10x the preceding value.

(b) Many dams have filters which do not reach FSL being based on flow nets. Most did not extend above FSL.

(c) Control of widths and location was often poor, so gaps occur.

(d) The understanding in earlier dams was just to transition from fine to coarse. Gradations and plasticity were not regarded as important.

Assume these represent 10% of all dams in the database. The factor = 1.0 based on relativity with the other filter categories.

For Dams with No Embankment Filters:

Based on Foster et al. (1998, 2000a, 2000b) it is assumed that 40% of dams in the database have no filters. The likelihood factor of 2 is assessed to give an overall neutral outcome.

Table C6 summarises this assessment.

C1.4 Core Soil Type and Geological Origin

It is concluded that no change is required from the Foster et al. (1998, 2000a, 2000b) method for core soil type.

It is questionable if soil geological origin should be considered since soil type should be sufficient. It has been deleted.

C1.5 Compaction

From Foster et al. (1998, 2000a, 2000b):

The methods of compaction are briefly described as follows to give guidance on the application of the method.

- *No formal compaction*—fill materials in the core were dumped in place and either no compaction, compaction by animal hooves or by travel of construction equipment only.
- *Rolled, modest control*—core materials were rolled but with poor control of moisture content (e.g. varying greater than 2% of optimum water content) and/or compacted in relatively thick layers.
- *Rolled, good control*—core materials compacted in thin layers, good control of moisture content within 2% of optimum water content and greater than 95% of standard compaction.

The failure and population statistics of core compaction were used to derive the weighting factors, and the results are shown in Table C7. Hydraulic fill and puddle methods of compaction were assigned a factor of 1.0 as they are taken into account by the hydraulic fill and puddle core zoning categories.

TABLE C5 Calibration of Likelihood of Embankment Filter Failure for Dams Where the Specified Grading of Filters Is Known and Meets USBR (1977, 1987) or Modern Criteria

	GENERAL FACTORS INFLUENCING LIKELIHOOD OF FILTER FAILURE				
FACTOR	*MUCH MORE LIKELY*	*MORE LIKELY*	*NEUTRAL*	*LESS LIKELY*	*MUCH LESS LIKELY*
Filter category as-constructed	Continuing Erosion or Excessive Erosion filter with low permeability downstream zone.	Excessive Erosion filter with high permeability downstream zone.	No Erosion or Some Erosion filter. 10% non plastic or 5% plastic fines.	No Erosion or some Erosion filter. 7% non-plastic or 2% plastic fines.	No-erosion filter ≤ 5% non-plastic fines
Proportion of the dam population which has filters in this category (P_{FC}) (a)	0.0 to 0.01 **[0.01] No Seg.** **[0.01] Seg.** Dam C 1A/2A Nil Dam C 1A/2A seg.0.005 Dam C 1A/2B Nil Dam C 1A/2B seg.0.78 Dam M 0.01 Dam G A n/a Dam G B n/a Dam O Nil	0.0 **[0.01] No Seg.** **[0.01] Seg.** Dam C 1A/2A Nil Dam C 1A/2A seg. n/a Dam C 1A/2B Nil Dam C 1A/2B seg. n/a Dam M n/a Dam G A n/a Dam G B n/a Dam O Nil	0.0 to 0.30 **[0.10] No Seg.** **[0.10] Seg.** Dam C 1A/2A Nil Dam C 1A/2A seg. n/a Dam C 1A/2B Nil Dam C 1A/2B seg. n/a Dam M 0.10 Dam G A 0.05 Dam G B 0.30 Dam O 0.10	0.025 to 0.68 **[0.40] No Seg.** **[0.60] Seg.** Dam C 1A/2A 0.025 Dam C 1A/2A seg. (0.945) Dam C 1A/2B 0.04 Dam C 1A/2B seg. (0.12) Dam M 0.25 Dam G A 0.50 Dam G B 0.68 Dam O 0.40	0.45 to 0.975 **[0.48] No Seg.** **[0.28] Seg.** Dam C 1A/2A 0.975 Dam C 1A/2A seg.0.05 Dam C 1A/2B 0.96 Dam C 1A/2B seg.0.10 Dam M 0.65 Dam G A 0.45 Dam G B 0.02 Dam O 0.50
Probability of being a Continuing Erosion filter (P_{CE}) (b) (c)	0.5 to 1.0 [0.75]	[Say 0.1]	0.01 to 0.1 [0.05]	0.001 to 0.01 [0.005]	0.0001 [0.0001]
$(P_{FC}) \times (P_{CE})$ No Seg. $\Sigma = 0.0155$	0.0075	0.001	0.005	0.002	0.000048
$(P_{FC}) \times (P_{CE})$ Seg. $\Sigma = 0.0165$	0.0075	0.001	0.005	0.003	0.000028

Notes:
(a) Based on experience of investigating existing dams, most constructed in the 1960s and 1970s.
(b) Based on the filter category and the probability of holding a crack given the fines content and plasticity, based on Table 12.3.
(c) Or an Excessive Erosion filter with seepage flows resulting in failure of the dam.

TABLE C6 Summary of Adopted Likelihood Factors

FILTER DESCRIPTION	*PROPORTION OF DATABASE POPULATION*	*LIKELIHOOD FACTOR*	*PROPORTION × LF*
Embankment filter present—well designed and constructed.	0.10	0.02	0.002
Embankment filter present—average design and construction.	0.40	0.2	0.08
Embankment filter or transition present. No data available or poor design and construction.	0.10	1	0.1
Embankment types which do not have filters.	0.40	2	0.8
TOTALS	1.00		0.982

TABLE C7 Weighting Factors for Core Compaction—Piping through the Embankment, $\boldsymbol{w_{E(cc)}}$ (Foster et al., 1998)

METHOD OF COMPACTION OF CORE MATERIALS	*FAILURES*		*ACCIDENTS*			*ORIGINAL WEIGHTING FACTORS*			
	NO.	*%*	*NO.*	*%*	*POPULATION %*	*%FAIL / %POP*	*%ACC / %POP*	*ASSIGNED FACTOR $W_{E(CC)}$*	*REVISED FACTOR*
Hydraulic fill	0	-	5	-	-	-	-	**1.0**	**1.0**
Puddle	5	-	14	-	-	-	-	**1.0**	**1.0**
No formal compaction	11	37.9	8	16.7	8	4.7	2.1	**5.0**	**6.0**
Rolled, modest control	13	44.8	25	52.1	32	1.4	1.6	**1.4**	**1.0**
Rolled, good control	5	17.2	15	31.3	60	0.3	0.5	**0.3**	**0.2**
Unknown	18		35						
No. of cases	52	100	102	100	100	Sum of Factor × Pop% = 1.03			0.98

From Chapter 9:
See Table 9.14.
Comment:

Based on Table C7 and Table 9.14, there should be greater difference between no formal compaction and modest control, and rolled good control should be lower. However, cracking due to differential settlement occurs even for well compacted soil, so any change should be limited. Amended as shown in Table C7.

C1.6 Conduits

The categories used to describe the degree of detailing incorporated into the design of conduits located through the embankment are described as follows:

- *Conduit through the embankment—many poor details*: No downstream filter and three or more of the poor detailing characteristics described later.
- *Conduit through the embankment—some poor details*: No downstream filter and two poor detailing characteristics.
- *Conduit through the embankment—typical USBR practice*: Details of conduits typical of USBR practice in the 1950s–1970s which are described later.
- *Conduit through the embankment—good detailing including downstream filters*: Good detailing and the downstream end of the conduit surrounded by a well designed and constructed filter zone.
- *No conduit through the embankment*: outlet tunnel in the abutment, no outlet conduit, e.g. flood control dams, or outlet through a concrete section of the dam or embedded in the foundation as shown in Figure 9.54a.

Conduits through the embankment include both conduits above the level of the general foundation of the dam and conduits located in trenches excavated through the foundation of the dam.

Poor details of outlet conduits can include any of the features described in the following:

1. Outlet conduit located in a deep and narrow trench as shown in Figure 9.54b.
2. Outlet conduit located in a narrow trench in soil or erodible rock with potential for desiccation cracking as shown in Figure 9.53c.
3. Culvert or pipe with vertical sides and in a narrow trench with steep cut slopes making compaction of the soil adjacent the culvert or pipe difficult. Figure 9.55 shows good practice.
4. Poor conduit geometry such as overhangs or circular pipe with no support as shown in Figure 9.53b.
5. Poorly designed seepage cutoff collars or other features that make compaction of the backfill around the conduit difficult.
6. Corrugated metal form work used for concrete surround as shown in Figure 9.53c.
7. Old cast iron or other types of metal pipes in badly deteriorated condition or of unknown condition.

Typical USBR practice in the 1950s to 1970s for the detailing of conduits includes the following features (USBR, 1977, 1987):

1. No downstream filter surrounding the outlet conduit.
2. Special compaction around the outlet conduit with special materials and hand tampers.
3. Outlet conduits were typically concrete formed in place with rectangular or horseshoe shaped sections.
4. Concrete cutoff collars spaced at 4 m to 5 m.
5. Trench slopes excavated at 1V:1H.

The weighting factors could not be determined directly from the failure statistics as whilst a particular dam which failed by piping was known to have a conduit through the dam, piping may have occurred elsewhere in the body of the dam and be totally unrelated to the conduit. All that can be inferred from the incident statistics is that the presence of conduits through the dam appears to be an influential factor affecting the initiation of piping and likely to have a high weighting factor.

The estimated percentage of dams in the population that fall into each of the conduit descriptions and the assigned weighting factors are shown in Table C8. To calibrate the weighting factors, a conduit with many poor details was considered to be equivalent to a continuous zone of poor compaction. Accordingly, an upper bound weighting of 5 was adopted broadly consistent with the weightings

TABLE C8 Weighting Factors for Conduits through the Embankment—Piping through the Embankment, $w_{E(con)}$

DESCRIPTION OF CONDUIT DETAILING	*ESTIMATED PERCENTAGE OF POPULATION*	*ASSIGNED WEIGHTING FACTOR*	*FACTOR × POPULATION %*
Conduit through embankment—many poor details	5%	5	0.25
Conduit through embankment—some poor details	10%	2	0.20
Conduit through embankment—typical USBR practice	25%	1.0	0.25
Conduit through embankment—good detailing including downstream filter	10%	0.6	0.06
No conduit through embankment	50%	0.5	0.25
Total	100%	(Check Sum = 1.0):	1.01

from core compaction as a baseline. The lower bound weighting factor for dams with no outlet conduit through the embankment was assigned a factor of 0.5 assuming the historical probability of failure by piping may have been halved if the dams which failed by piping around the conduit had no conduit. This is consistent with the data for failures in Table C2. The weighting factors of the intermediate categories were selected such that when they are applied to the population, the result is neutral. The weighting factor for "conduit through embankment—good detailing including downstream filter" has been reduced from 0.8 to 0.6 consistent with the revised description.

C1.7 Foundation Profile

As discussed in Fell et al. (2024), most transverse cracking is a result of post-construction settlement with abutments with steps in the profile, and dams with compressible soil in part of the foundation.

Tables C9 and C10 present a summary of data from Fell et al. (2024). The factors in Column 1 are used in Table 15.3.

C1.8 Observations of Seepage

Most seepage collection systems measure seepage through the dam and in the foundation combined. For most dams, seepage through the foundation will be greater than seepage through the dam, often much greater.

In view of this and the difficulty in quantifying the third, fourth and fifth terms for observation of seepage in the Foster et al. (1998, 2000a, 2000b) method, this factor has been excluded in the amended method.

C1.9 Monitoring and Surveillance

The experience in QRA is that the estimated probability of successful detection and intervention for piping in the embankment is often small because the time from the initiation of piping to breaching of the dam is often short (e.g. often less than 6 hours from the initial signs of muddy leakage to breaching) so not much affected by the inspection frequency.

In view of this, and that this is not an inherent property of the dam, this factor has not been included in the amended method.

TABLE C9 Slope Geometry and % Post-Construction Settlement vs Probability of Transverse Cracking, Cross-Valley Differential Settlement, Foundation Profile without and with a Step and Showing Adopted Factors in Table 15.3

LIKELIHOOD OF FAILURE [ADOPTED FACTORS ADJUSTED TO GIVE NEUTRAL OUTCOME]	*PROBABILITY OF PERSISTENT CRACKING (RATIO VS NEUTRAL)*	*LOCALISED STRAIN %*	*SLOPE AND POST-CONSTRUCTION SETTLEMENT*	
			FOUNDATION WITHOUT A STEP	*FOUNDATION WITH A STEP*
Much More Likely [20]	0.1 (30x)	3% (0.05) (a) 0.01 (b) (0.0005) (c) 1% (d)	> 50° > 2.5%	> 50° > 1.75%
More Likely [4]	0.02 (6x)	2% (0.30) (a) 0.09–0.18 (0.05) (c) 10% (d)	35° to 50° 2.5% to 1.00%	35° to 50° 2.0% to 0.75%
Neutral [1]	0.003	1% (0.30) (a) 0.59 (b) (0.50) (a) (0.15) (a) 0.23 (b) (0.10) (c) 40% (d)	35° to 50° 0.5% to ≤ 0.1% 25° to 35° 1.0% to 0.5%	35° to 50° 0.4% to 0.1% 25° to 35° 1.5% to 0.4%
Less Likely [0.25]	0.001 (0.3x)	0.5% (0.50) (a) 0.59 (b) (0.15) (a) 0.23 (b) (0.035) (c) 30% (d)	25° to 35° 0.5% to ≤ 0.1% 15° to 25° 1.0% to 0.5%	25° to 35° 0.25% to ≤ 0.1% 15° to 25° > 2% to 0.25%
Much Less Likely [0.05]	0.0002 (0.066x)	0.2% (0.15) (a) 0.59 (b) (0.10) (c) 20% (d)	< 15° to 25° < 0.50% to ≤ 0.1%	<15° to 25° <0.75% to ≤ 0.1%

Notes:
(a) Estimated proportion of foundations in this slope range.
(b) Settlement % from Figure 7.39 summarised in Table C10.
(c) (a) × (b).
(d) Adopted % in this "Likelihood of Failure" class.

C1.10 Limitations of the Method

The following points discuss the limitations of the proposed method.

(a) As the weighting factors are often based on low numbers of accident and failure cases, some of the factors and the baseline annual probabilities of failure for the zoning categories are sensitive to the occurrence of only one or two piping failures for dams with a particular zoning category or some other characteristic. This may tend to either underestimate or

TABLE C10 Slope of the Soil-Rock Interface and % Post-Construction Settlement vs Probability of Persistent Transverse Cracking, Embankments with Soil in the Foundation beneath the Core (Data from Figure 7.39 and Table C11)

LIKELIHOOD OF FAILURE	*PROBABILITY OF PERSISTENT CRACKING*	*LOCALISED STRAIN %*	*SLOPE AND POST-CONSTRUCTION SETTLEMENT*
Much More Likely	0.1	3	> 60° > 2.0%
More Likely	0.02	2	25° to 60° 2.0% to 0.5%
Neutral	0.003	1	15° to 60° 0.5% to 0.2%
Less Likely	0.001	0.5	15° to 60° 0.2% to ≤ 0.1%
Much Less Likely	0.0002	0.2	< 15° to 60° ≤ 0.1%

TABLE C11 Frequency of Percentage Settlement at 10 Years after Construction from Figure 7.39

PERCENT SETTLEMENT AT 10 YEARS	*NUMBER OF DAMS*	*PERCENTAGE OF DAMS*
≤ 0.1	7	9
0.1 to 0.25	19	24
0.25 to 0.5	28	35
0.5 to 0.75	11	14
0.75 to 1.0	7	9
1.0 to 1.5	4	5
1.5 to 2	3	4
> 2	0	0
Total	79	100

overestimate the influence of these factors. However, attempts were made in the analysis of the weighting factors to highlight these cases and to check the reasonableness of the factors based on the expected susceptibility of the particular conditions for piping failure.

(b) The probabilities of failure are based on large dams (> 15 m height) and so the method may tend to underestimate the probability of failure of piping if applied to smaller dams.

(c) The analysis of the weighting factors assumes the factors to be independent of each other. However, it is probable there are interdependencies between some of the factors. Therefore, when the weightings are multiplied together, some "doubling-up" of the weighting factors may occur, and this may tend to over or under emphasise some factors. Any obvious cases of this "doubling-up" of factors were accounted for in the analysis and any remaining cases are probably not significant.

(d) Since the method is based on accidents and failures, it is inherently difficult to get low weighting factors using this approach. Dams with favourable characteristics often have no failures on which to base a weighting factor on and so it must be based on some other basis. This was particularly evident for the influence of embankment filters on piping through the embankment where a value judgement was required to assess the weighting factor for well designed and constructed filters. Therefore, care should be taken in applying the probability of failure obtained by this method to acceptable risk guidelines particularly at very low probabilities of failure.

C2 INTERNAL EROSION AND PIPING IN THE FOUNDATION

C2.1 Development of the Database Screening Method for Piping of Foundation Soil into Open Defects in Rock or into Coarse Gravel

C2.1.1 Introduction

The Foster et al. (1998, 2000a, 2000b) method has been amended, taking account of improved understanding of the mechanics of internal erosion since the method was developed. The basic structure of the Foster et al. (1998, 2000a, 2000b) method has been maintained.

C2.1.2 Historical Probability of Failure, All Zoning Types

From Table 17.1 the average frequency of failure in the foundation = 1.7×10^{-3}. Adjusting this by 2x for under-reporting gives 3.4×10^{-3}.

From Foster et al. (1998, 2000a, 2000b), the probability of failure due to piping through the foundation is significantly higher early in the operation of the dam, with 25% of the failures occurring on first filling of the dam and 75% occurring within the first 5 years of operation.

Therefore, the average frequency of failure occurring on first filling, or the first 5 years adjusted for under-reporting = $3.4 \times 10^{-3} \times 0.75 = 2.5 \times 10^{-3}$.

From Table C1 the average annual probability of failure after 5 years = 1.9×10^{-6}.Adjusting this by 2x for under-reporting and 26/58 for the longer period of record gives average annual probability of failure = 1.7×10^{-5} / annum.

C2.1.3 Development of Table 15.4

Table 15.4 is based on the Foster et al. (1998, 2000a, 2000b) method table with minor adjustments based on the authors' experience and current understanding of mechanics.

C3 PIPING FROM THE EMBANKMENT INTO THE FOUNDATION

C3.1 Introduction

The Foster et al. (1998, 2000a, 2000b) method has been amended taking account of improved understanding of the mechanics of internal erosion since the method was developed. The basic structure of the Foster et al. (1998, 2000a, 2000b) method has been maintained.

C3.2 Historic Frequency of Failure by Piping of the Embankment into the Foundation

From Table C.1 from Foster et al. (2000a, 2000b), the average frequency of failure for piping of the embankment into the foundation = 1.8×10^{-4}.

Adjusting this by 2x for under-reporting gives 3.6×10^{-4}.

From Table 8.9 of Foster et al. (1998) all failures (four of them) occurred on first filling or in the first 5 years, but accidents (31 of them) were split 50% between first filling and in the first 5 years, and after 5 years. However, in view of the small number of failures, Foster et al. (1998, 2000a, 2000b) assumed 50% of failures occur on first filling and the first 5 years.

Assuming 50% occur on first filling or the first 5 years, the average frequency of failure = $0.5 \times 3.6 \times 10^{-4} = 1.8 \times 10^{-4}$ in that period.

The average annual probability of failure after 5 years = 4.0×10^{-4}. Adjusting this by 2x for under-reporting and 26/58 for the longer period of record gives average annual probability of failure = 3.6×10^{-6}.

C3.2 Development of Table 15.5

Table 15.5 is based on the Foster et al. (1998, 2000a, 2000b) method table with minor adjustments based on the authors' experience and current understanding of mechanics.

REFERENCES

Fell, R., Foster, M.A., Cyganiewicz, J., Sills, G. and Davidson, R. (2024). *Methods for Estimating the Probability of Failure of Embankment Dams by Internal Erosion and Piping*. UNICIV Report No R470, The School of Civil and Environmental Engineering, University of New South Wales, Sydney.

Fell, R., Foster, M.A., Cyganiewicz, J., Sills, G., Vroman, N. and Davidson, R. (2008). *A Unified Method for Estimating Probabilities of Failure of Embankment Dams by Internal erosion and Piping*. UNICIV Report No R 446, The School of Civil and Environmental Engineering, University of New South Wales, Sydney.

Foster, M., Fell, R. and Spannagle, M. (2000a). The statistics of embankment dam failures and accidents. *Canadian Geotechnical Journal*, Vol. 37, No. 5, National Research Council Canada, Ottawa, 1000–1024, ISSN: 0008–3674.

Foster, M., Fell, R. and Spannagle, M. (2000b). A method for assessing the relative likelihood of failure of embankment dams by piping. *Canadian Geotechnical Journal*, Vol. 37, No. 5, National Research Council Canada, Ottawa, 1025–1062, ISSN: 0008–3674.

Foster, M., Spannagle, M. and Fell, R. (1998). *Report on the Analysis of Embankment Dam Incidents*. UNICIV Report No R374, School of Civil and Environmental Engineering, University of New South Wales ISBN: 85841 349 3; ISSN 0077–880X.

ICOLD (1995). *Bulletin 99-Dam failure Statistical Analysis*. International Commission on Large Dams, Paris.

ICOLD (2019). *Bulletin 99 Update. Statistical Analysis of Dam Failures*. International Commission on Large Dams, Paris.

USBR (1977, 1987). *Design of Small Dams*. United States Department of the Interior, Bureau of Reclamation, Denver.

USBR (1980, 1985). *Earth Manual*. US Dept. Interior Water Resources Technical Publication, Denver.

References

Aas, G., Lacasse, S., Lunne, T. and Höeg, K. (1986). Use of in-situ tests for foundation design in clay. *Use of in-situ Tests in Geotechnical Engineering, Geotechnical Special Publications*, No. 6, ASCE, 1–30.

Abt, S.R., Thornton, C.I., Scholl, B.A. and Bender, T.R. (2013). Evaluation of overtopping riprap design relationships. *Journal of American Water Resources Association*, Vol. 49, No. 4, 923–937.

AECOM (2017). *Upper Yarra Dam-Numerical Models for Risk Workshop*. Unpublished Memorandum dated 8th August 2017.

AECOM (2023). *Cardinia Dam DSU: Phase 1 and Phase 2 Geotechnical Investigations Factual Report*. 6th June 2023. (Unpublished report).

Akroyd, T.N.N. (1957, 1975). *Laboratory Testing in Soil Engineering*. G.T.Foulis for Soil Mechanics Ltd., London.

Allan, R.J. (2018). *Backward Erosion Piping*. PhD thesis, School of Civil and Environmental Engineering, University of New South Wales.

Amaya, F. and Marulanda, A. (1985). Golilas Dam—design, construction and performance. In *Concrete Face Rockfill Dams, Design, Construction and Performance*, Cooke, J.B. and Sherard, J.L. (eds.). ASCE Geotechnical Engineering Division, 98–120.

Amaya, F. and Marulanda, A. (2000). *Colombian Experience in the Design and Construction of CFRD Dams*. J. Barry Cooke Volume, Concrete Face Rockfill dams. Chinese Committee on Large Dams.

ANCOLD (2000). *Dam Technology in Australia, 1850–1999*. Cole, B. (ed.). Australian National Committee on Large Dams. www.ancold.org.au.

ANCOLD (2003). *Guidelines on Risk Assessment*. Australian National Committee on Large Dams. www.ancold.org.au.

ANCOLD (2019). *Guidelines for Design of Dams and Appurtenant Structures for Earthquake*. Australian National Committee on Large Dams. www.ancold.org.au.

ANCOLD (2020). *Guidelines on Geotechnical Investigations of Dams, Foundations and Appurtenant Structures*. Australian National Committee on Large Dams. www.ancold.org.au.

ANCOLD (2022). *Guidelines on Risk Assessment*. Australian National Committee on Large Dams. www.ancold.org.au.

Andrus, R.D. and Stokoe, K.H., II (2000). Liquefaction resistance of soils from shear-wave velocity. *Journal of Geotechnical and Geoenvironmental Engineering*, ASCE, Vol. 126, No. 11, 1015–1025.

Anon (1972). The preparation of maps and plans in terms of engineering geology. *Quarterly Journal of Engineering Geology and Hydrogeology, London*, Vol. 5, 293–381.

Anon (1981). Recommended symbols for engineering geological mapping. *Bulletin IAEG*, Vol. 24, 227–234.

Arnold, M., Hunter, G. and Foster, M. (2016). Dam safety management for upgrade of an extreme consequence storage. *Proceedings ANCOLD 2016 Conference*. Australian National Committee on Large Dams. www.ancold.org.au.

Arroyo, M. and Gens, A. (2022). *Computational analyses of Dam I Failure at the Corrego de Feijao Mine in Brumadinho. final report for VALE S.A., August 2021*. CIMNE International Center for Numerical Methods in Engineering.

Arulanandan, K. and Perry, E.B. (1983). Erosion in relation to filter design criteria in earth dams. *Journal of the Geotechnical Engineering Division*, ASCE, Vol. 109, No. GT5, 682–698.

AS (1726–2017). *Geotechnical Site Investigations*. Standards Australia.

Azzouz, A.S., Baligh, M.M. and Ladd, C.C. (1983). Corrected field vane strength for embankment design. *Journal of the Geotechnical Engineering Division*, ASCE, Vol. 109, No. GT5, 730–734.

Baecher, G.B., Pate, M.E. and de Neufville, R. (1980). Dam failure in benefit cost analysis. *Journal of the Geotechnical Engineering Division*, ASCE, Vol. 106, No. 1, 101–105.

Barden, L., McGown, A. and Collins, K. (1973). The collapse mechanism in partially saturated soils. *Engineering Geology*, Vol. 7, 49–60.

Bardet, J.P., Ichii, K. and Lin, C.H. (2000). *EERA: A Computer Program for Equivalent-Linear Earthquake Site Response Analyses of Layered Soil Deposits*. University of Southern California, Los Angeles.

Barneich, J., Majors, D., Moriwaki, Y., Kulkarni, R. and Davidson, R., (1996). Application of reliability analysis in the environmental impact report (EIR) and design of a major dam project. *Proceedings of Uncertainty 1996, Geotechnical Engineering Division*, ASCE.

Barton, N.R. (1982). Shear strength investigations for surface mining. In *Stability in Surface Mining. Proceedings of the 3rd International Conference on Stability in Surface Mining*, Brawner, C.O. (ed.), 1–3 June 1981. Vancouver, B.C. Society of Mining Engineering of the American Institute of Mining. Metallurgy and Petroleum Engineering, New York, 171–196.

Barton, N.R. and Bandis, S. (1982). Effects of block size on the shear behaviour of jointed rock. Keynote Lecture, *23rd US Symposium on Rock Mechanics*, Berkeley, CA.

Barton, N.R. and Bandis, S. (1991). *Review of Predictive Capabilities of JRC-JCS Model in Engineering Practice*. Publ. No. 182. Norwegian Geotechnical Institute, Oslo, Norway.

Baziar, M.H. and Dobry, R. (1995). Residual strength and large scale deformation potential of loose silty sands. *Journal of the Geotechnical Engineering Division*, ASCE, Vol. 121, No. 12, 896–906.

Beaty, M.H. and Byrne, P.M. (2008). Liquefaction and deformation analysis using a total stress approach. *Journal of Geotechnical and Geoenvironmental Engineering*, ASCE, Vol. 134, No. 8, 1059–1072.

Beaty, M.H. and Byrne, P.M. (2011). *UBCs and Constitutive Model*. Itasca UDM Website 904aR.

Been, K. (1999). The critical state line and its application to soil liquefaction. In *Physics and Mechanics of Soil Liquefaction*, P.V. Lade and J.A. Yamamuro (eds.). Balkema, 195–204.

Beguin, R. (2011). *Etude multi-échelle de l'érosion de contact dans les ouvrages hydrauliques*. Thèse de l'Université Joseph Fourrier Grenoble.

Beguin, R., Fry, J.J., Courivaud, J.R., Picault, C. and Duchesne, L (2015). Modes of failures of zoned dams with pervious downstream shell. In *Proceedings, XVI Dam Monitoring International Conference*, Wintera, J. and Wity, A. (eds.). TKZ2015, Wierchomla, Poland, 29 September.

Beguin, R., Guidoux, C., Faure, Y.H. and Ho, C.C. (2009). Stability conditions and erosion rates at the contact between two soils, subjected to a water flow parallel to the interface. *Presentation at ICOLD European Working Group on Internal Erosion*, St Petersburg, April.

Benahmed, N. and Bonelli, S. (2012). Investigating piping erosional behaviour of cohesive soils. *European Journal of Environmental and Civil Engineering*, Vol. 16, No. 1, 43–58.

Bendahmane, F., Marot, D., Rosquoët, F. and Alexis, A. (2008). Experimental parametric study of suffusion and backward erosion. *International Journal of Geotechnical and Geoenvironmental Engineering*, Vol. 134, No. 1, 57–67.

Bishop, A.W., Hutchison, J.N., Penman, A.D.M. and Evans, H.E. (1969). *Geotechnical Investigation into the Causes and Circumstances of the Disaster of 21 October 1966. Aberfan Tribunal*, H.M.S.O. Welsh Office, London, 1–80.

Bodtman, W.L. and Wyatt, J.D. (1985). Design and performance of Shiroro rockfill dam. In *Proceedings of the Symposium on Concrete Face Rockfill Dams—Design, Construction and Performance*, Cooke and Sherard (ed.). ASCE New York, Detroit, MI, 231–251.

Bolton, A., Brandon, T.L., Duncan, J.M., Mitchell, J.K. and Tiwari, B. (2006). *Literature Review, Depth of Cracking in Dams Worldwide*. Prepared for USBR No. 05PG810038, Virginia Polytechnic Institute and State University.

Bombardelli, F.A., Palermo, M. and Pagliara, S. (2018). Temporal evolution of jet induced scour depth in cohesionless granular beds and the phenomenological theory of turbulence. *Physics of Fluids*, Vol. 30, No. 8, 085109.

Bonelli, S. and Brivois. O. (2008). The scaling law in the hole erosion test with a constant pressure drop. *International Journal Numerical Methods in Engineering*, 1573–1595.

Bonelli, S., Fell, R. and Benahmed, N. (2013). Concentrated leak erosion (Chapter 4). In *Erosion in Geomechanics Applied to Dams and Levees*, Bonelli S. (ed.). Wiley, Hoboken and ISTE, London, 271–342.

Bonelli, S., Olivier, B. and Damien, L. (2007). The scaling law of piping erosion. *Proceedings 18th Congress Francais Mechanique*, Grenoble.

Boulanger, R.W. (2010). *A Sand Plasticity Model for Earthquake Engineering Applications*. Report No. UCD/CGM-10-01, Center for Geotechnical Modelling, Department of Civil and Environmental Engineering, University of California, Davis, CA.

Boulanger, R.W. and Idriss, I.M. (2006). Liquefaction susceptibility criteria for silts and clays. *Journal of Geotechnical and Geoenvironmental Engineering*, ASCE, Vol. 132, No. 11, 1413–1426.

Boulanger, R.W. and Idriss, I.M. (2012). Probabilistic standard penetration test–based liquefaction–triggering procedure. *Journal of Geotechnical and Geoenvironmental Engineering*, ASCE, Vol. 138, No. 10, 1185–1195.

Boulanger, R.W. and Idriss, I.M. (2014). *CPT and SPT Based Liquefaction Triggering Procedures*. Report No. UCD/CGM-14-01. Center for Geotechnical Modelling, Department of Civil and Environmental Engineering. University of California at Davis.

Boulanger, R.W. and Idriss, I.M. (2015a). CPT-based liquefaction triggering procedure. *Journal of Geotechnical and Geoenvironmental Engineering*, ASCE, 04015065. https://doi.org/10.1061/(ASCE)GT.1943-5606.0001388.

Boulanger, R.W. and Idriss, I.M. (2015b). Magnitude scaling factors in liquefaction triggering procedures. *Soil Dynamics and Earthquake Engineering*, Vol. 79, 296–303. https://doi.org/10.1016/j.soildyn.2015.01.004.

Boulanger, R.W., Wilson, D.W. and Idriss, I.M. (2012). Examination and re-evaluation of SPT-based liquefaction triggering case histories. *Journal of Geotechnical and Geoenvironmental Engineering*, ASCE, Vol. 138, No. 8, 898–909.

Boulanger, R.W., Wilson, D.W. and Idriss, I.M. (2013). Closure to examination and re-evaluation of SPT-based liquefaction triggering case histories. *Journal of Geotechnical and Geoenvironmental Engineering*, ASCE, Vol. 138, No. 8, 2000–2001.

Boulanger, R.W. and Ziotopoulou, K. (2012). *PM4Sand (Version 2): A Sand Plasticity Model for Earthquake Engineering Applications. Center for Geotechnical Modelling, Report No UCD.CGM-12-03*. Department of Civil and Environmental Engineering, University of California, Davis, CA.

Boulanger, R.W. and Ziotopoulou, K. (2013). Formulation of a sand plasticity plane-strain model for earthquake engineering applications. *Journal of Soil Dynamics and Earthquake Engineering*, Elsevier, Vol. 53, 254–267. https://doi.org/10.1016/j.soildyn.2013.07.006.

Boulanger, R.W. and Ziotopoulou, K. (2018). *PM4Silt (Version 1): A Silt Plasticity Model for Earthquake Engineering Applications. Report No. UCD/CGM-18/01*, Center for Geotechnical Modelling, Department of Civil and Environmental Engineering, University of California, Davis, CA.

Boulanger, R.W. and Ziotopoulou, K. (2023). *PM4Silt (Version 2.1): A Silt Plasticity Model for Earthquake Engineering Applications. Report No. UCD/CGM-23/02*, Center for Geotechnical Modelling, Department of Civil and Environmental Engineering, University of California, Davis, CA. [February 2023; revised June 2023].

Bowles, D.S., Xu, Y., Ehasz, J.L., Briaud, J. and Ey, J.C. (2014). Obtaining reasonable breach hydrographs based on breach parameter estimates from regression equation. *ASDSO Conference Proceedings*.

Bowles, J.E. (1978). *Engineering Properties of Soils and Their Measurement*. McGraw-Hill.

Brauns, J. (1985). Erosionsverhalten geschichteten Bodens bei horizontaler Durchstromung. *Wasserwirtschaft*, Vol. 75, 448–453.

Bray, J.D., Macedo, J. and Travasarou, T. (2018). Simplified procedure for estimating seismic slope displacements for subduction zone earthquakes. *Journal of Geotechnical and Geoenvironmental. Engineering*, ASCE, Vol. 144, No. 3, 04017124.

Bray, J.D. and Sancio, R.B. (2006). Assessment of the liquefaction susceptibility of fine-grained soils. *Journal of Geotechnical and Geoenvironmental Engineering*, ASCE, Vol. 132, No. 9, 1165–1177. https://doi.org/10.1061/(ASCE)1090.0241(2006)132.9(1165).

Bray, J.D. and Sancio, R.B. (2008). Closure to assessment of the liquefaction susceptibility of fine-grained soils. *Journal of Geotechnical and Geoenvironmental Engineering*, ASCE, Vol. 134, No. 7, 1031–1034. https://doi.org/10.1061/(ASCE)1090.0241(2008)134.7(1031).

Bray, J.D. and Travasarou, T. (2007). Simplified procedure for estimating earthquake induced deviatoric slope displacements. *Journal of Geotechnical and Geoenvironmental Engineering*, ASCE, Vol. 133, No. 4, 381–392.

BRE (1999). *An Engineering Guide to the Safety of Embankment Dams in the United Kingdom*. Building Research Establishment, Garston.

Briaud, J.-L. (2008). Case histories in soil and rock erosion. *Journal of Geotechnical and Geoenvironmental Engineering*, ASCE, Vol. 134, No. 10, 1425–1447.

Briaud, J.-L., Govindasamy, A.V. and Shafii, I. (2017). Erosion charts for selected geomaterials. *Journal of Geotechnical and Geoenvironmental Engineering*, ASCE, Vol. 143, No. 10. https://doi.org/10.1061/(ASCE)GT.1943-5606.0001771.

Brunsden, D., Doornkamp, J.C., Fookes, P.G., Jones, D.K.C. and Kelly, G.M.H. (1975). Large scale geomorphological mapping and highway engineering design. *Quarterly Journal of Engineering Geology and Hydrogeology, London*, Vol. 8, 227–253.

Buchanan, P., Barker, M., Maisano., P. and Jonker, M. (2016). Kangaroo Creek Dam embankment raising and stabilization-balancing competing objectives. *Proceedings ANCOLD Conference*. Australian National Committee on Large Dams. www.ancold.org.au.

Bui, H., Fell, R. and Song, C. (2004). *Two and Three Dimensional Numerical Modelling of the Potential for Cracking of Embankment Dams during Construction*. UNICIV Report No. 426, The School of Civil and Environmental Engineering, The University of New South Wales, Sydney. www.engineering.unsw.edu.au/civil-engineering/uniciv-reports.

Bui, H., Tandrijana, V., Fell, R. Song, C. and Khalili, N. (2005). *Two and Three Dimensional Numerical Modelling of the Potential for Cracking of Embankment Dams-Supplementary Report*. UNICIV Report No. 438, The School of Civil and Environmental Engineering, The University of New South Wales, Sydney. www.engineering.unsw.edu.au/civil-engineering/uniciv-reports.

Campbell, J., Barker, G., Southcott, P. and Wallis, M. (2013). Flooded cars: Estimating the consequences to itinerants exposed to dam break floods on roads. *Proceedings Joint NZSOLD and ANCOLD Conference on Dams*. Australian National Committee on Large Dams. www.ancold.org.au.

Casinader, R.J. and Watt, R.E. (1985). Concrete face rockfill dams of the Winneke Project. In *Concrete Face Rockfill Dams, Design, Construction and Performance*, Cooke, J.B. and Sherard, J.L. (eds.). ASCE Geotechnical Engineering Division, 140–162.

Castro, G. (1969). *Liquefaction of Sands*. PhD thesis, Harvard Soil Mechanics Series, No. 81, Harvard University.

CDA (2013). *Dam Safety Guidelines 2007, 2013 Edition*. Canadian Dam Safety Association.

Cetin, K.O., Seed, R.B., Der Kiureghian, A., Tokimatsu, K., Harder, L.F., Kayen, R.E. and Moss, R.E.S. (2004). Standard penetration test-based probabilistic and deterministic assessment of seismic soil liquefaction potential. *Journal of Geotechnical and Geoenvironmental Engineering*, ASCE, Vol. 130, No. 12, 1314–1340.

CFGB. (1997). *Internal Erosion: Typology, Detection and Repairs*. Barrages & Réservoirs N° 6 – French Committee on Large Dams.

Chang, H.H. (1998). Riprap stability on steep slopes. *International Journal of Sediment Research*, Vol. 13.

Chapius, R.P. (1986a). *Use of Rotational Device on Cohesive Soils*. Transportation Research record 1089.

Chapius, R.P. (1986b). Quantitative measurement of the scour resistance of natural soil clays. *Canadian Geotechnical Journal*, Vol. 23, 132–141.

Chapius, R.P. and Gatien, T. (1986). An improved rotating cylinder technique for quantitative measurements of scour resistance of clays. *Canadian Geotechnical Journal*, Vol. 23, No. 1, 83–87.

Charles, J.A. and Skinner, H.D. (2001, July). Compressibility of foundation fill. *Geotechnical Engineering*, Vol. 149, No. 3, 145–157, Proceedings of the Institution of Civil Engineers.

Chen, H. (2008a). Consideration of dam safety after Wenchuan earthquake in China. *Proc. 14th World Conf. on Earthquake Engineering*, Special Session S13, Seismic aspects of large embankment and concrete dams, Beijing.

Chen, Q. and Zhang, L.M. (2006). Three-dimensional analysis of water infiltration into the Gouhou rockfill dam using saturated–unsaturated seepage theory. *Canadian Geotechnical Journal*, Vol. 43, 449–461.

Chen, S. (2008b). Impact of the "5.12" Wenchuan earthquake on Zipingpu concrete face rockfill dam and its analysis. *Proc. Int. Conf. on Dam Safety Management*, Nanjing.

Chen, Z. (1993). Breaching of the Gouhou concrete face sand and gravel dam. *Symposium on High Rockfill Dams*, Beijing, October.

Chia, S., He, K. and Hunter, G. (2022). Strain and crack propagation analysis to inform on depth of filtering: Upper Yarra Dam safety upgrade. *Proceedings ANCOLD Conference, Sydney*. Australian National Committee on Large Dams. www.ancold.org.au.

Christian, J.T., Ladd, C.C. and Baecher, G.B. (1992). Invited lecture: Reliability and probability in stability analysis; in Stability and Performance of Slopes and Embankments—11. *ASCE Geotechnical Special Publication*, No. 31, 1071–1111.

CIRIA (1987). *Design of Reinforced Grass Waterways (R116)*.

Cole, B.A. and Fone, P.J.E. (1979). Repair of Scotts Peak Dam, Tasmania. *ICOLD Congress Q49, R15*. International Commission on Large Dams, Paris.

Cooke, J.B. (1984). Progress in rockfill dams (18th Terzaghi lecture). *Journal of Geotechnical Engineering*, ASCE, Vol. 110, No. 10, 1383–1414.

Cooke, J.B. (1993). Rockfill and the rockfill dam. In *Proceedings of the International Symposium on High Earth-Rockfill Dams*, Jiang, Zhang and Qin (ed.). Beijing, 1–24.

Cooke, J.B. (1997). The concrete face rockfill dam. In *Seventeenth Annual USCOLD Lecture Series: Non-Soil Water Barriers for Embankment Dams*. San Diego, 117–132.

Cooke, J.B. (1999). The development of today's CFRD dam. *Proceedings of the Second Symposium on Concrete Face Rockfill Dams*, Florianopolis, Brazil, Brazilian Committee on Dams, 3–11.

Cooke, J.B. (2000). The high CFRD. *Journal of Barry Cooke Volume Concrete Face Rockfill Dams*, 20th ICOLD Congress Beijing 2000, 1–4.

Cooke, J.B. and Sherard, J.L. (1987). Concrete-face rockfill dams: II design. *Journal of Geotechnical Engineering*, ASCE, Vol. 113, No. 10, 1113–1133.

Cooper, B., Khalili, N. and Fell, R. (1997). Large deformations due to undrained strain weakening slope instability at Hume Dam No. 1 embankment. *Nineteenth International Congress on Large Dams, Florence*, ICOLD Vol. 2, (Q73 R46), 797–818, International Commission on Large Dams, Paris.

Cornforth, D.H., Worth, E.G. and Wright, W.L. (1974). Observations and analysis of a flow slide in sand fill. In *Proc. of the Symposium on Field Instrumentation in Geotechnical Engineering*, British Geotechnical Society (ed.). England, Butterworths, London, 136–151.

Cowan, J., Kelly, C. and Hunter, G. (2016). Tullaroop Dam: Dealing with filter breakdown and compaction during construction. *Proceedings ANCOLD 2016 Conference*. Australian National Committee on Large Dams. www.ancold.org.au.

Cruz, P.T., Materon, B. and Freitas, M. (2009). *Concrete Face Rockfill Dams*. CRC Press, Leiden.

Cubrinovski, M. and Ishihara, K. (2000). Flow potential of sandy soils with different grain compositions. *Soils and Foundations*, Vol. 40, No. 4, 103–119.

Dafalias, Y.F. and Manzari, M.T. (2004). Simple plasticity sand model accounting for fabric change effects. *Journal of Engineering Mechanics*, ASCE, Vol. 130, No. 6, 622–634.

Davidson, R.R. and Dann, C. (2011). Hinze Dam stage 3: Dam safety delivered through innovation. *ASDSO Conference*, Washington, DC, 25028 September.

Davidson, R.R., Fox, S. and Cooper, B. (2003). *Seismic Upgrade of Yarrawonga Weir, ICOLD Congress*. International Committee on Large Dams, Paris.

Davidson, R.R. and Hansen, K. (2015). Revisiting a forgotten dam building technology: Concrete core walls in embankment dams. *ANCOLD Conference Proceedings*. Australian National Committee on Large Dams. www.ancold.org.au.

Davidson, R.R., Vreugdenhil, R. and Foster, M. (2000). The dam safety upgrade at Lake Eppalock. *ANCOLD Conference on Dams*. Australian National Committee on Large Dams, Hobart. www.ancold.org.au.

Davies, M., McRoberts, E. and Martin, T. (2002). Static liquefaction of tailings: Fundamentals and case histories. *ASDSO Conference Tailings Dams 2002*. Association of State Dam Safety Officials.

Davis, B.K. (2015). Innovations in electronic core orientation. *AUSIMM Bulletin*. https://www.ausimmbulletin.com/feature/innovations-in-electronic-core-orientation/.

Davis, C.A. and Bardet, J.P. (1996). Performance of two reservoirs during 1994 Northridge Earthquake. *Journal Geotechnical Engineering*, ASCE, Vol. 122, No. 8. https://doi.org/10.1061/(ASCE)0733-9410(1996)122:8(613).

Davis, C.A. and Sakado, M.M. (1994). Response of the Van Norman complex to the Northridge Earthquake. *11th Conference, Association of State Dam Safety Officials*, Boston, September, Proceedings, 241–255.

Dawson, R.F. (1994). *Mine Waste Geotechnics*. PhD thesis, Department of Civil Engineering, University of Alberta, Edmonton, Alberta.

Dawson, R.F., Morgenstern, N.R. and Stokes, A.W. (1998). Liquefaction flow slides in Rocky Mountain coal mine waste dumps. *Canadian Geotechnical Journal*, Vol. 35, 328–343.

Dearman, W.R. (1991). *Engineering Geological Mapping*. Elsevier.

Deltares (2023). *Eindrapport Piping in Getijdenzand. Hedwigeproject*. (Final Report Piping in Tidal Sand, Research into additional resistance to Piping) Documentnummer1220-164143.R08 v 3.0.

De Wit, J.M. (1984). *Research report on sand boil model tests*. Delft, Netherlands.

De Wit, J.M., Sellmeijer, J.B. and Penning, A. (1981). Laboratory tests on piping. *Proc. 10th Int. Conf. Soil Mechanics and Foundation Engineering*, Stockholm, June. Balkema, Rotterdam, 517–520.

Den Adel, H., Koender, M.A. and Bakker, K.J. (1994). The analysis of relaxed criteria for erosion-control filters. *Canadian Geotechnical Journal*, Vol. 31, No. 6, 829–840.

Dezert, T., Kiplesund, G.H. and Sigtryggsdóttir, F.G. (2022). Riprap protection exposed to overtopping phenomena: A review of laboratory experimental models. *Water*, Vol. 14, 2722. https://doi.org./10.3390/w14172722.

Di Nardi, J., Palermo, M. and Pagliara, S. (2021). A critical analysis of jet-induced scour formulas. *International Junior Researcher and Engineer Workshop in Hydraulic Structures*. Utah State University, 6.

Dise, K., Stevens, M.G. and Von Thun, J.L. (1994). USBR experiences using DDC at 3 dams for mitigation of liquefaction risk. *ASCE Geotechnical Special Proceedings*, No. 45, 1–23.

Douglas, K.D. (2003). *The Shear Strength of Rock Masses*. PhD Thesis, School of Civil and Environmental Engineering, The University of New South Wales, Sydney.

Douglas, K.J., Fell, R., Peirson, W.L. and Studholme, H. (2016). *Experimental Investigation of Global Backward Erosion and Suffusion*. UNICIV Report No. R465. School of Civil and Environmental Engineering, UNSW. ISBN: 0-7334-3719-2.

Douglas, K.J., Fell, R., Peirson, W.L. and Studholme, H. (2019). Experimental investigation of global backward erosion and suffusion. *Canadian Geotechnical Journal*, Vol. 56, 789–807.

Druyts, F.H. and Van Den Berg (1988). The failure of the Tygerpoort and Spitkop Dams. *The Civil Engineer in South Africa*, 277–283.

Duncan, J.B. (1992). State of the art—static stability and deformation analysis in stability and performance of slopes and embankments. *Geotechnical Special Publication*, ASCE, No. 31, 222–266.

Duncan, J.B. (1996). *Soil Slope Stability Analysis, in Landslides Investigation and Mitigation.* Transport Research Board, N.R.C. Spec. Report 247, Washington, DC, 337–371.

Duncan, J.M., Wright, S.G. and Wong, K.S. (1990). Slope stability during rapid drawdown. *Proc. H. Bolton Seed Memorial Symposium*, Berkeley, CA, Bitech Publishers, Vancouver, 253–272.

EBL (2005). *Stability and Breaching of Embankment Dams.* Report on Sub-Project 2: Stability of Downstream Shell and Dam Toe during Large through-Flow. EBL Kompetanse AS, Publication No: 186-2005, ISBN 82-436-0521-5.

Eckerlin, R.D. (1992). Mud Mountain Dam concrete cut off wall: A case history. *Bulletin of the Association of Engineering Geologists*, Vol. 29, No. 1, 11–32.

Eckerlin, R.D. (1993). *Mud Mountain Dam Concrete Cutoff Wall: Geotechnical Practice in Dam Rehabilitation*, Anderson, I.R. (ed.). ASCE Special Publication, 284–298.

Eckersley, J.D. (1986). *The Initiation and Development of Slope Failures with Particular Reference to Flow Slides.* PhD thesis, Department of Civil and Systems Engineering, James Cook University of North Queensland, Australia.

Eckersley, J.D. (1990). Instrumented laboratory flow slides. *Geotechnique*, Vol. 40, No. 3, 489–502.

El Howayek, A., Huang, P.T., Bisnett, R. and Santagata, M.C. (2011). *Identification and Behaviour of Collapse Soils.* Publication FWHA/IN/JTRP-2011/a2. Joint Transportation Research Program, Indiana Department of Transportation and Purdue University, West Lafayette, Indians. https://doi.org/10.5703/1288284314625.

Eurotop (2018). *Manual on Wave Overtopping of Sea Defences and Related Structures: An Overtopping Manual Largely Based on European Research, But for Worldwide Application*, 2nd ed. www.overtopping-manual.com.

Fell, R., Foster, M.A., Cyganiewicz, J., Sills, G. and Davidson, R. (2024). *Methods for Estimating the Probability of Failure of Embankment Dams by Internal Erosion and Piping.* UNICIV Report No R470, The School of Civil and Environmental Engineering, University of New South Wales, Sydney.

Fell, R., Foster, M.A., Davidson, R., Cyganiewicz, J., Sills, G. and Vroman, N. (2008). *A Unified Method for Estimating Probabilities of Failure of Embankment Dams by Internal Erosion and Piping.* UNICIV Report R 446, The School of Civil and Environmental Engineering, University of New South Wales, Sydney, Australia 2052. ISBN:85841 413 9. www.engineering.unsw.edu.au/civil-engineering/uniciv-reports.

Fell, R. and Fry, J.J. (2007). The state of the art of assessing the likelihood of internal erosion of embankment dams, water retaining structures and their foundations. In *Internal Erosion of Dams and Their Foundations*, Fell, R. and Fry, J.J. (eds.). Taylor and Francis, London, 1–24.

Fell, R. and Fry, J.J. (2013). State of the art on the likelihood of internal erosion of dams and levees by means of testing. Chapter 1 In *Erosion in Geomechanics Applied to Dams and Levees*, Bonelli, S. (ed.). Wiley, Hoboken, and ISTE, London, 1–100.

Fell, R., Macgregor, J.P., Stapledon, D.H. and Bell, G. (2005). *Geotechnical Engineering of Dams.* Balkema, Leiden.

Fell, R., MacGregor, J.P., Stapledon, D.H., Bell, G. and Foster, M. (2015). *Geotechnical Engineering of Dams*, 2nd ed. Taylor Francis Group, London.

Fell, R., Wan, C.F., Cyganiewicz, J. and Foster, M. (2001). *The Time for Development and Detectability of Internal Erosion and Piping on Embankment Dams and Their Foundations.* UNICIV Report No. R-399. ISBN: 84841 366 3. School of Civil and Environmental Engineering, The University of New South Wales.

Fell, R., Wan, C.F., Cyganiewicz, J. and Foster, M. (2003). Time for development of internal erosion and piping in embankment dams. *Journal Geotechnical and GeoEnvironmental Engineering*, ASCE, Vol. 129, No. 4, 307–314.

FEMA (2005a). *Conduits through Embankment Dams, Best Practices for Design, Construction, Problem Identification and Evaluation, Inspection, Maintenance, Renovation and Repair*, L-266, Federal Emergency Management Agency.

FEMA (2005b). *FEMA P-65: Earthquake Analyses and Design of Dams.* Federal Emergency Management Agency, USA.

FEMA (2011). *Filters for Embankment Dams, Best Practices for Design and Construction.* US Federal Emergency Management Agency. http://www.fema.gov.

FEMA (2014). *Technical Manual: Overtopping Protection for Dams.* FEMA P-1015. Federal Emergency Management Agency. US Department of Homeland Security.

FERC (2016). *Risk-Informed Decision Making Guidelines.* Chapter 3, Risk Assessment. Federal Energy Regulatory Commission.

Ferkh, Z. and Fell, R. (1994). Design of embankments of soft clay. *XIII Int. Conf. on Soil Mech. and Foundation Engineering*, New Delhi, India, 733–738.

Fiedler, W., Osmun, D., Engemoen, W. and Feinberg, B. (2014). Reclamation's new life loss estimating methodology. *34th USSD Annual Meeting and Conference, Dams and Extreme Events*, 1383–1396.

Fitzpatrick, M.D. (1976). Scotts Peak Dam: Cracking of bituminous concrete face. *ICOLD Congress Q41*. International Commission on Large Dams, Paris, 54–60.

Fitzpatrick, M.D., Cole, B.A., Kinstler, F.L. and Knoop, B.P. (1985). Design of concrete faced rockfill dams. In *Concrete Face Rockfill Dams, Design, Construction and Performance*, Cooke, J.B. and Sherard, J.L. (eds.). ASCE Geotechnical Engineering Division, 410–434.

Flynn, S., Zamanian, S., Vahedifard, F., Shafieezadeh, A. and Schaaf, D. (2021). Data-driven model for estimating the probability of riverine levee breach due to overtopping. *Journal of Geotechnical and Geoenvironmental Engineering*, ASCE, Vol. 148, No. 3.

Forster, I.R. and MacDonald, R.B. (1998). Post-earthquake response procedures for embankment dams—lessons from the Loma Prieta Earthquake. *ANCOLD Bulletin, No. 109, 46–64*. Australian National Committee on Large Dams. www.ancold.org.au.

Foster, M.A. (1999). *The Probability of Failure of Embankment Dams by Internal Erosion and Piping*. PhD Thesis, School of Civil and Environmental Engineering, The University of New South Wales, Sydney.

Foster, M.A. and Fell, R. (1999a). *Assessing Embankment Dam Filters Which Do Not Satisfy Design Criteria*. UNICIV Report No. R-376, School of Civil and Environmental Engineering, University of New South Wales. ISBN: 85841 343 4, ISSN 0077-880X.

Foster, M.A. and Fell, R. (1999b). *A Framework for Estimating the Probability of Failure of Embankment Dams by Piping Using Event Tree Methods*. UNICIV Report No. R-377. School of Civil and Environmental Engineering, The University of New South Wales. ISBN 85841 343 4. www.engineering.unsw.edu.au/civil-engineering/uniciv-reports.

Foster, M.A. and Fell, R. (2001). Assessing embankment dams, filters who do not satisfy design criteria. *Journal Geotechnical and Geoenvironmental Engineering*, ASCE, Vol. 127, No. 4, 398–407.

Foster, M.A., Fell, R. and Spannagle, M. (2000a). The statistics of embankment dam failures and accidents. *Canadian Geotechnical Journal*, Vol. 37, No. 5, National Research Council Canada, Ottawa, 1000–1024. ISSN: 0008-3674.

Foster, M.A., Fell, R. and Spannagle, M. (2000b). A method for assessing the relative likelihood of failure of embankment dams by piping. *Canadian Geotechnical Journal*, Vol. 37, No. 5, National Research Council Canada, Ottawa, 1025–1062, ISSN: 0008-3674.

Foster, M.A., Rönnqvist, H. and Fell, R. (2018). The performance of embankment dams with filters coarser than no-erosion design criteria. *International Journal Hydropower and Dams*, Vol. 20, No. 4, 64–71.

Foster, M.A., Spannagle, M. and Fell, R. (1998). *Report on the Analysis of Embankment Dam Incidents*. UNICIV Report No R374, School of Civil and Environmental Engineering, University of New South Wales ISBN: 85841 349 3; ISSN 0077-880X.

Fox, S. and Fyfe, G.E. (2001). It's more fun when you do it yourself: Constructing the Lake Eppalock main embankment remedial works project. *Proceedings ANCOLD Conference on Dams*. Australian National Committee on Large Dams. www.ancold.org.au.

France, J.W. (2024). *Lesson Learned from Dam Incidents and Failures*. Static Liquefaction Should Be Considered as a Potential Failure Mode for Dams That Have Loose Sands or Silts in Their Embankments or Foundations. Association of State Dam Officials (ASDSO).

France, J.W., Alvi, I.A., Miller, A.C., Williams, J.L. and Higinbotham, S. (2022). *Independent Forensic Team Final Report*. Investigation of Failures of Edenville and Sanford Dams.

Fredlund, D.G., Morgenstern, N.R. and Widger, R.A. (1978). The shear strength of unsaturated soils. *Canadian Geotechnical Journal*, Vol. 15, No. 3, 313–321.

Fredlund, D.G. and Rahardjo, H. (1993). *Soil Mechanics for Unsaturated Soils*. Wiley, New York.

Frizell, K.H., Ruff, J.F. and Mishra, S. (1998). Simplified guidelines for riprap subject to overtopping flow. *Proceedings of the Annual Association of State Dam Officials* (ASDSO), Las Vegas, Nevada.

Galic, D. (2023). RCEM at 10: Why Reclamation is still using an empirical approach to consequence estimation. *ANCOLD Conference 2023*, Australian National Committee on Large Dams, Cairns. www.ancold.org.au.

Gilbert, P. and Miller, P. (1991). *A Study of Embankment Performance during Overtopping*. Technical Report GL-91-23, US Army Corps of Engineers, Mississippi.

Gillon, M.D. (2007). Re-evaluation of internal erosion incidents at Matahina Dam, New Zealand. In *Internal Erosion of Dams and Their Foundations*, Fell, R. and Fry, J.-J. (eds.). Taylor Francis, 115–132.

Glynn, M.E. and Kusmaul, J. (2004). *Prediction of Piping Erosion Along Middle Mississippi River Levees—an Empirical Model.* US Army Corps of Engineers, Engineering Research Center, Geotechnical and Structural Laboratory, ERDC/GSL TR-04-12.

Gordon, J.L. and Duguid, D.R. (1970). Experiences with cracking at Duncan Dam. *Tenth Int. Congress on Large Dam, Montreal.* Question R29. International Commission on Large Dams, Paris, 469–486.

Graham, W. (2008). *A Procedure for Estimating Vehicle-Related Fatalities from Dam Failure Flooding.* Unpublished memo.

Graybeal, K.D. (1988). Seepage control at Mud Mountain Dam. *Sixteenth International Congress on Large Dams*, San Francisco, ICOLD. Q61. R.29, 539–555.

Graybeal, K.D. and Levallois, J. (1991). Construction of a cut off wall with the hydrofraise through the core of Mountain Dam (USA). *Proceedings 16th ICOLD Congress on Large Dams*, Vienna. International Commission on Large Dams, Paris, 879–901.

Guidoux, C., Faure, Y.H., Beguin, R. and Ho, C.C. (2010). Contact erosion at the interface between granular filter and various base-soils with tangential flow. *Journal of Geotechnical and Geoenvironmental Engineering*, ASCE, Vol. 136, No. 5, 655–775.

Hacelas, J.E. and Ramirez, C.A. (1987). Closure by authors on design features of Salvajina Dam. *Journal Geotechnical Engineering*, ASCE, Vol. 113, No. 10, 1175–1180.

Hanson, G.J. (1990). Surface erodibility of earthen channels at high stresses: Part II, developing an in-situ device. *Transactions American Society of Agricultural Engineers*, Vol. 33, No. 1, 132–137.

Hanson, G.J. (1991). Development of a jet index to characterize erosion resistance of soils in earthen spillways. *Transactions American Society of Agricultural Engineers*, Vol. 36, No. 5, 2015–2352.

Hanson, G.J. and Cook, K.R. (2004). Apparatus test procedure and analytical methods to measure soil erodibility in-situ. *Applied Engineering in Agriculture*, Vol. 20, No. 4, 455–462.

Hanson, G.J., Cook, K.R., Hahn, W. and Britton, S.L. (2003). *Observed Erosion Processes during Embankment Overtopping Tests.* ASAE Paper No. 032066, American Society Agricultural Engineers, St Joseph, MI.

Harder, L.F. Jr. and Boulanger, R.W. (1997). Application of K_α correction factors. *Proceedings NCEER Workshop on Evaluation of Liquefaction Resistance of Soils*, Nat. Center. For Earthquake Engineering Research, State Univ. of New York at Buffalo, 167–190.

Harder, L.F. Jr., Bray, J.D., Volpe, R.L. and Rodda, K.V. (1991). Performance of Earth Dams during the Loma Prieta Earthquake Proceedings. *Second International Conference on Recent Advances in Geotechnical Earthquake Engineering and Soil Dynamics*, Paper No. LP05, March.

Hartung, D.C. and Scheuerlein, H. (1970). Design of overflow rockfill dams. *Proceedings of the 10th International Congress on Large Dams*, Vol. 1, 587–598.

He, K. (2016). *Numerical Modelling of Cracking in Embankment Dams.* PhD thesis, School of Civil and Environmental Engineering, University New South Wales, Sydney, Australia.

He, K., Fell, R. and Song, C. (2017). *Stresses and Strains in Embankment Dams Resulting from Differential; Settlements.* UNICIV Report No. R-466, School of Civil and Environmental Engineering, University of New South Wales, Sydney, Australia. ISBN: 978-0-7334-3773-1. www.engineering.unsw.edu.au/civil-engineering/uniciv-reports.

He, K., Fell, R. and Song, C. (2019). Transverse cracking in embankment dams resulting from cross valley differential settlements. *European Journal of Environmental and Civil Engineering.* https://doi.org/10.1080/19648189.2019.1691663.

He, K., Song, C. and Fell, R. (2021). Numerical modelling of transverse cracking in embankment dams. *Computers and Geotechnics.* https://doi.org/10.1016/j.compgeo.2021.104028.

Head, K.H. (1980, 81, 85). *Manual of Soil Laboratory Testing*, Vols 1, 2 and 3. Pentech Press.

Hencher, S. and Richards, L.R. (2014). Assessing the shear strength of rock discontinuities at laboratory and field scales. *Rock Mechanics and Rock Engineering*, Vol. 48, No. 3. https://doi.org/10.1007/s00603-014-0633-6.

Hewlett, H., Boorman, L., and Bramley, M. (1987). *Design of Reinforced Grass Waterways.* CIRIA Report 116, Construction Industry Research and Information Association, London.

HKIE (1998). *Soil Nails in Loose Fill: A Preliminary Study.* Hong Kong Institution of Engineers, Geotechnical Division.

Høeg, K. (1993). *Asphaltic Concrete Cores for Embankment Dams: Experience and Practice.* Stadkraft Veidekke, Norwegian Geotechnical Institute.

Høeg, K. and Wang, W. (2025). *Asphalt Core Embankment Dams.* CRC Press, Balkema, ISBN 978-1-138-03763-2. https://doi.org/10.1201/9781315177786.

Hoek, E. and Bray, J.W. (1981). *Rock Slope Engineering.* Institution of Mining and Metallurgy, London.

Hoek, E. and Brown, E.T. (1997). Practical estimation of rock mass strength. *International Journal of Rock Mechanics and Mining Sciences*, Vol. 34, No. 8, 1165–1186.

Hoek, E., Carranza-Torres, C. and Corkum, B. (2002). Hoek-Brown failure criterion-2002 edition. *Proceedings NARMS-TAC Conference*, Toronto, 1, 267–273.

Hoffmans, G.J.C.M. (2009). Closure problem to jet scour. *Journal of Hydraulic Research*, Vol. 47, No. 1, 100–109.

Hoffmans, G.J.C.M. (2014). An overview of piping models. In *Scour and Erosion: Proceedings of the 7th International Conference on Scour and Erosion, Perth, Australia*, Cheng, L., Draper, S. and An, H. (eds.). CRC Press, Boca Raton, FL, 3–18.

Hoffmans, G.J.C.M. and Van Rijn, L. (2018). Hydraulic approach for predicting piping in dikes. *Journal of Hydraulic Research*, Vol. 56, No. 2, 268–281. https://doi.org/10.1080/00221686.2017.1315747.

Holtz, R.D. and Kovacs, W.D. (1981). *Introduction to Geotechnical Engineering*. Prentice Hall.

Howard, T., Rudkin, C. and Irey, P. (2015). Large scale filter performance tests. *ASDSO Conference*, New Orleans.

Hunter, G. (2003). *The Deformation Behaviour of Embankment Dams and Landslides in Natural and Constructed Soil Slopes*. PhD Thesis, School of Civil and Environmental Engineering, The University of New South Wales, Sydney.

Hunter, G. and Fell, R. (2003a). Mechanics of failure of soil slopes leading to "rapid" failure. *Proceedings International Conference on Fast Slope Movement, Prediction and Prevention for Risk Mitigation*, Naples, Editor Picarelli, L. Patron editore, Bologna, 283–296.

Hunter, G. and Fell, R. (2003b). Travel distance for rapid landslides in constructed and natural slopes. *Canadian Geotechnical Journal*, Vol. 40, No. 6, 1123–1141.

Hunter, G. and Fell, R. (2003c). *The Deformation Behaviour of Embankment Dams*. UNICIV Report No. R-416, ISBN 0077-880X, School of Civil and Environmental Engineering, The University of New South Wales, Sydney. www.engineering.unsw.edu.au/civil-engineering/uniciv-reports.

Hunter, G., Fell, R. and McGrath, S. (2007). Tullaroop dam: Embankment cracking and risk based assessment of remedial works. *ANCOLD Bulletin* No 135, 16–37. Australian National Committee on Large Dams. www.ancold.org.au.

Hunter, G., Foster, M. and Arnold, M. (2022). Risk and ALARP assessment for Upper Yarra embankment upgrade selection. *Proceedings ANCOLD 2022 Conference*. Australian National Committee on Large Dams. www.ancold.org.au.

Hunter, G., Glastonbury, J., Ang, D. and Fell, R. (2003). *The Performance of Concrete Face Rockfill Dams*. UNICIV Report No. 413. www.engineering.unsw.edu.au/civil-engineering/uniciv-reports.

Hunter, G., Jeffery, D. and Chia, S. (2017). Tullaroop Dam: Ongoing cracking and unusual pore pressure response in a 60 year old earth embankment. *Proceedings ANCOLD Conference*. Australian National Committee on Large Dams. www.ancold.org.au.

Hunter, G., Jeffery, D. and Kelly, C. (2014). Cracking and dam safety of a 100 year old puddle core embankment. *Proceedings ANCOLD 2014 Conference*. Australian National Committee on Large Dams. www.ancold.org.au.

Hunter, G., Pattle, A. and Foster, M. (2015). Partial reconstruction of Rowallan Dam to repair a piping incident. *Proceedings ANCOLD Conference 2015*. Australian National Committee on Large Dams. www.ancold.org.au.

Hunter, G. and Toose, J. (2009). Embankment design and construction risk management for the right abutment at Hinze Dam. *Proceedings ANCOLD Conference*, Adelaide, Australian National Committee on Large Dams. www.ancold.org.au.

Hutchison, J.N. (1988). General report: Morphological and geotechnical parameters of landslides in relation to geology and hydrogeology. *Proc. 5th Int. Symp. on Landslides, Lausanne*, Bonnard, C. (ed.). Balkema, 3–36.

Hutchison, J.N. (2001). Reading the ground: Morphology and Geology in Site Appraisal. *The 4th Glossop Lecture, Quarterly Journal of Engineering Geology and Hydrogeology*, Vol. 34, 8–50.

Hynes, M.E. and Olsen, R.S. (1999). Influence of confining stress on liquefaction resistance. *Proc., Int. Workshop on Phys. And Mech. of Soil Liquefaction*, Balkema, Rotterdam, The Netherlands, 145–152.

Iai, S. and Koizumi, K. (1986). Estimation of earthquake induced excess pore water pressure for gravel drains. *Proceedings, 7th Japan Earthquake Engineering Symposium*, 679–684.

ICOLD (1974). *Lessons from Dam Incidents*. Bulletin 99. International Commission on Large Dams, Paris.

ICOLD (1982). *Bituminous Concrete Facings for Earth and Rockfill Dams*. ICOLD Bulletin 32a. International Committee on Large Dams, Paris.

ICOLD (1989). *Rockfill Dams with Concrete Facing*. Bulletin 70. State-of-the-art. International Committee on Large Dams, Paris.

ICOLD (1992). *Bulletin 84, Bituminous Cores for Fill Dams.* International Commission on Large Dams, Paris.

ICOLD (1994). *Embankment Dams Granular Filters and Drains.* Bulletin 95. International Commission on Large Dams, Paris.

ICOLD (1995). *Bulletin 99-Dam Failure Statistical Analysis.* International Commission on Large Dams, Paris.

ICOLD (1999). *Embankment Dams with Bituminous Concrete Facing.* ICOLD Bulletin 114. International Committee on Large Dams, Paris.

ICOLD (2010). *Concrete Face Rockfill Dams: Concepts for Design and Construction.* Bulletin 141. International Commission on Large Dams, Paris.

ICOLD (2012). *Selecting Seismic Parameters for Large Dams—Guidelines (Revision of Bulletin 72).* Bulletin 148. International Commission on Large Dams, Paris.

ICOLD (2017). *Bulletin 164 on Internal Erosion in Existing Dams, Dikes and Levees and Their Foundations.* International Commission on Large Dams, Paris.

ICOLD (2018). *Bulletin 179, Asphalt Concrete Cores for Embankment Dams.* International Commission on Large Dams, Paris.

ICOLD (2019). *Bulletin 99 Update: Statistical Analysis of Dam Failures.* International Commission on Large Dams, Paris.

Idriss, I.M. and Boulanger, R.W. (2008). *Soil Liquefaction during Earthquakes.* Monograph MNO-12, Earthquake Engineering Research Institute, Oakland, CA.

Idriss, I.M. and Boulanger, R.W. (2010). *SPT–Based Liquefaction Triggering Procedures.* Report No UCD/CGM-10/02, Center for Geotechnical Modelling. University of California at Davis.

Idriss, I.M. and Boulanger, R.W. (2015). 2nd Ishihara Lecture: SPT and CPT based relationships for the residual shear strength of liquefied soils. *Soil Dynamics and Earthquake Engineering*, Vol. 68, 57–68.

Independent Panel (1976). *Failure of Teton Dam, Independent Review Panel to Review the Cause of Teton Dam Failure.* Report to U.S. Department of the Interior and State of Idaho, December.

Ishihara, K. (1993). Liquefaction and flow failures during earthquakes. *Geotechnique*, Vol. 33, No. 3, 351–415.

Jacobs (2016). *Retarding Basin Failure Modes. Guidance Note.* Unpublished Report Jacobs Consulting, Melbourne.

Jamiolkowski, M., Ladd, C., Germaine, J. and Lancellotta, R. (1985). New developments in field and laboratory testing. *11th International Conference Soil Mechanics and Foundation Engineering*, Vol. 1, Balkema, 57–153.

Jefferies, M.G. and Been, K. (2006). *Soil Liquefaction, a Critical State Approach.* Taylor and Francis, London.

Johnson, C. and Arnold (2011). Toorourrong Reservoir—small dam, big problems. *Proceedings ANCOLD 2011 Conference.* Australian National Committee on Large Dams. www.ancold.org.au.

Justo, J.L. (1990). *Asphaltic Concrete Face Dams.* Chapter 10 in "*Advances in Rockfill Structures*". Kluwer Academic Press.

Kanitkar, R., Wei Liu, W., Houlahan, G. and Kianoush, R. (2011). Water flow through cracks in thick concrete sections. *Concrete International*, April.

Karol, R.H. (1985). Grout penetrability. In *Issues in Dam Grouting.* ASCE Geotechnical Engineering, 1–26.

Kenney, T.C., Chanal, R., Chin, E., Ofoeghu, G.I., Omange, G.N. and Ume, C.A. (1985). Controlling construction size of granular filters. *Canadian Geotechnical Journal*, Vol. 22, 32–43.

Kenney, T.C. and Lau, D. (1985). Internal stability of granular filters. *Canadian Geotechnical Journal*, Vol. 22, 215–225.

Kenney, T.C. and Lau, D. (1986). Closure to: Internal stability of granular filters. *Canadian Geotechnical Journal*, Vol. 23, 420–423.

Kenney, T.C. and Westland, J. (1993). Laboratory study of segregation of granular materials. In *Filters in Geotechnical and Hydraulic Engineers*, Brauns, Heibaum and Schuler (eds.), Balkema, Rotterdam, 313–319.

Khalili, N., Fell, R. and Tai, K.S. (1996). A simplified method for estimating failure induced deformation. In *Proc. Seventh Int. Symp. On Landslides*, Senesett, K. (ed.). Balkema, Rotterdam, 1263–1268.

Khan, D. and Ahmad, Z. (2011). Stabilization of angular-shaped riprap under overtopping flows. *World Academy of Science, Engineering and Technology*, Vol. 59, 1153–1157.

Koelewijn, A. and Van Beek, V. (2024). Powerpoint presentation to *ICOLD Technical Committee E.* ICOLD Conference, New Delhi.

Koenders, M.A. and Sellmeijer, J.B. (1992). Mathematical model for piping. *Journal of Geotechnical. Engineering*, ASCE, Vol. 118, No. 6, 943–946.

Kohno, I., Nishigaki, M. and Takeshita, Y. (1987). Levee failure caused by seepage and preventative measures. *Natural Disaster Science*, Vol. 9, No. 2, 55–76.

Koppejan, A.W., Wamelan, B.M. and Weinberg, L.J.H. (1948). Coastal flow slides in the Dutch province of Zeeland. *Proceedings of the 2nd International Conference on Soil Mechanics and Foundation Engineering*, Rotterdam, Vol. 5, 89–96.

Kramer, S.L. (1988). Triggering of liquefaction flow slides in coastal soil deposits. *Engineering Geology*, Vol. 26, 17–31.

Kramer, S.L. (1996). *Geotechnical Earthquake Engineering*. Prentice Hall.

Kulhawy, F.H. (1992). On the evaluations of static soil properties, in Stability and Performance of Slopes and Embankments. *Geotechnical Special Publication*, ASCE, No. 31, 95–115.

Lade, P.W. (1986). *Advanced triaxial testing of soils*. University of Sydney, School of Civil and Mining Engineering.

Lawrence, P. (2002). *Summary of Outcomes of "Workshop on Cracking of Homogeneous Earth Dams in Semi-Arid Environments*". Maricopa County, USA.

Lawton, E.C., Fragaszy, R.J. and Hardcastle, J.H. (1989). Collapse of compacted clayey sand. *Journal of Geotechnical Engineering*, ASCE, Vol. 115, No. 9, 1252–1267.

Lawton, E.C., Fragaszy, R.J. and Hetherington, M.D. (1992). Review of wetting induced collapse in compacted soil. *Journal of Geotechnical Engineering*, ASCE, Vol. 118, No. 9, 1376–1394.

Leonards, G.A. and Narain, M. (1963). Flexibility of clay and cracking of earth dams. *Journal Soil Mechanics and Foundation Engineering*, ASCE, Vol. 89, No. SM2, 47–97.

Lim, S.S. (2006). *Experimental Investigation of Erosion in Variably Saturated Clay Soils*. PhD thesis, School of Civil and Environmental Engineering, University of New South Wales, Sydney, Australia.

Lim, S.S. and Khalili, N.K. (2010). Laboratory measurement of erosion of clay soils using rotating cylinder test. *Geotechnical Testing Journal*, Vol. 32, No. 3.

Littlejohn, G.S. (1985). Chemical grouting 1, 2 and 3. *Ground Engineering*, March, April, May.

Lucker, S.L. and Robbins, B.A. (2023). Backwards erosion piping through silty sand mixtures. *Symposium "Management for Safe Dams*", ICOLD Annual Meeting, Gothenburg.

Luehring, R., Snortland, N., Stevens, M. and Mejia, L. (2001). Liquefaction mitigation of a silty dam foundation using vibro-stone columns and drainage wicks: A case history at Salmon Lake Dam. *Bureau of Reclamation Water Operation and Maintenance Bulletin*, No. 198, 1–15.

Lunne, T., Robertson, P.K. and Powell, J.J.M. (1997). *Cone Penetration Testing in Geotechnical Practice*. Blackie Academic and Professional, London.

Lupini, J.F., Skinner, A.E. and Vaughan, P.R. (1981). The drained residual strength of cohesive soils. *Geotechnique*, Vol. 31, No. 2, 181–213.

Mackenzie, P.R. and McDonald, L.A. (1985). Mangrove Creek Dam: Use of soft rock for rockfill. In *Concrete Face Rockfill Dams, Design, Construction and Performance*, Cooke, J.B. and Sherard, J.L. (eds.). ASCE Geotechnical Engineering Division, 208–230.

Maisano, P., Buchanan, P. and Schmidt, T. (2016). Omega-type external waterstops on a CFRD-An Australian first. *Proceedings ANCOLD Conference*. Australian National Committee on Large Dams. www.ancold.org.au.

Makdisi, F.I. and Seed, H.B. (1978). A simplified procedure for estimating dam and earthquake induced deformations. *Journal Geotechnical Engineering*, ASCE, Vol. 1, 105, No. GT7, 849–867.

Marcuson, W.F., Hynes, M.E. and Franklin, A.G. (1990). Evaluation and use of residual strength in seismic safety analysis of embankments. *Earthquake Spectra*, Vol. 6, No. 3, 529–572.

Marjoribanks, R. (2011). *Geological Methods in Mineral Exploration and Mining*, 2nd ed. Springer.

Marot, D., Alexis, A. and Bendahmane, F. (2007). A specific triaxial device for the study of internal erosion in cohesive soils. *Proceedings of Workshop on internal; erosion and piping of dams*, Aussois France May. Balkema, 159–166.

Marsal, R.J. (1973). Mechanical properties of rockfill. In *Embankment Dam Engineering J.* Wiley and Sons, NY, 109–201.

Martin, G.R. (1988). Geotechnical Aspects of Earthquake Engineering. State of the Art Report. Australia-New Zealand Conference on Geomechanics, Sydney.

Martins, R. (1981). *Hydraulics of Overflow Rockfill Dams*. Lisboa Laboratorio Nacional de Engenharia Civil.

Marulanda, A. and Pinto, N.L. (2000). Recent experience on design, construction, and performance of CFRD dams. In *Concrete Face Rockfill Dams*, J. Barry Cooke Volume, Mori, S., Dijkstra, G. and Borgatti (eds.). Chinese Committee on Large Dams, Beijing, 279–315.

Maslin, K., Foster, M. and McDonald, L. (2012). Estimating individual risk: A pragmatic and defensible approach. *Proceedings ANCOLD Conference*, Perth, Australian National Committee on Large Dams. www.ancold.org.au.

Maslin, K. and Rodd, R. (2016). Estimating the probability of breach for overtopping of low to medium height homogenous earthfill embankments. *Proceedings ANCOLD/NZSOLD 2016 Conference*, Adelaide, Australian National Committee on Large Dams. www.ancold.org.au.

Mason, P.J. and Arumugam, K. (1985). Free jet scour below dams and flip buckets. *Journal of Hydraulic Engineering*, Vol. 111, No. 2, 220–235; Vol. 113, No. 9, 1192–1205.

Mejia, L.H. (2005). Seismic retrofit of embankment dams to mitigate liquefaction: Selected case histories. *Earthquake Engineering and Soil Dynamics*, 1–14.

Mejia, L.H. (2013). Field testing of crushed ignimbrite for dam filter material. *Proceedings ISSD Conference*, Phoenix, Arizona.

Mishra, S.K. (1998). *Riprap Design of Overtopped Embankments*. PhD dissertation, Department of Civil Engineering, Colorado State University, Fort Collins, CO.

Mitchell, J.K. (1970). In place treatment of foundation soils. *Journal Soil Mechanics and Foundation Engineering*, ASCE, Vol. 92, No. SM1, 73–100.

Mitchell, J.K. (1976). *Fundamentals of Soil Behaviour*. John Wiley & Sons.

Mitchell, J.K. (2008). Mitigation of liquefaction potential of silty sands. In *From Research to Practice in Geotechnical Engineering*, Laier, J.E., Crapps, D.K. and Hussein, M.H. (eds.). Geotechnical Special Publication 180, ASCE, Reston, VA.

Moffat, R.M. and Fannin, R.J. (2011). A hydromechanical relation governing the internal stability of cohesionless soil. *Canadian Geotechnical Journal*, Vol. 48, No. 3, 413–424.

Moffat, R.M., Fannin, R.J. and Garner, S.J. (2011). Spatial and temporal progression of internal erosion in cohesionless soil. *Canadian Geotechnical Journal*, Vol. 48, No. 3, 399–412.

Montgomery, J., Boulanger, R.W. and Harder, L.F., Jr. (2014). Examination of the K_s overburden correction factor on liquefaction resistance. *Journal of Geotechnical and Geoenvironmental Engineering*, ASCE, 04014066. https://dx.doi.org/10.1061/(ASCE)GT.1943-5606.0001172.

Moon, A.T. (1984). Effective shear strength parameters for stiff fissured clays. *Proceedings Fourth Australia New Zealand Conf. on Geomechanics*, 107–111.

Moriwaki, Y. and Mitchell, J.K. (1977). *The Role of Dispersion in the Slaking of Intact Clay*. ASTM STP 623, 287–302.

Morris, M.W., Hassan, M., Escarameia, M. and van Hoven, A. (2012). Assessing the performance of grass cover on flood embankments. In *Comprehensive Flood Risk Management*, Klijn, F. and Schweckendiek, T. (eds.). Taylor & Francis. https://doi.org/10.1201/b13715-54.

Moss, R.E.S., Seed, R.B., Kayen, R.E., Stewart, J.P., Der Kiureghian, A. and Cetin, K.O. (2006). CPT-based probabilistic and deterministic assessment of in situ seismic soil liquefaction potential. *Journal of Geotechnical and Geoenvironmental Engineering,* ASCE, Vol. 132, No. 8, 1032–1051.

Moye, D. G. (1967). Diamond drilling for foundation exploration. *Civil Engineering Transactions. J. I.E Aust.*

Muller-Kirchenbauer, H., Rankl, M. and Schlotzer, C. (1993). Mechanism for regressive erosion beneath dams and barrages. In *Filters in Geotechnical and Hydraulic Engineering*, Brauns, J., Heibaum, M. and Schuler, U. (eds.). Balkema, 360–376.

Nakamura, Y., Ohne, Y., Narita, K., Okumura, T., Nomura, K., Shimazaki, M. and Mizuno, T. (2002). *Earthquake Damages and Remedial Works for an Earth Dam with Asphalt Facing*. Aico-cc.co.jp

NCHRP (2006). *Riprap Design Criteria, Recommended Specifications, and Quality Control*. National Cooperative Highway Research Program Report 568, Transportation Research Board.

Newman, S. and Foster, M. (2006). *Lake Buffalo Dam Risk Reduction Upgrade*. ANCOLD Bulletin No 132, 45–56. Australian National Committee on Large Dams, Brisbane. www.ancold.org.au.

Newmark, N.M. (1965). Effects of earthquakes on dams and embankments. *5th Rankine Lecture, Geotechnique*, Vol. 15, No. 2, 139–160.

Nilsson, A. (2007a). The susceptibility of internal erosion in the Suova Dam. In *Internal Erosion of Dams and their Foundations*, Fell, R. and Fry, J.J. (eds.). Taylor & Francis, 167–172.

Nilsson, A. (2007b). Filters and internal erosion in Swedish dams. In *Internal Erosion of Dams and their Foundations*, Fell, R. and Fry, J.J. (eds.). Taylor & Francis, 173–178.

Nilsson, A. (2009). Upgrading by a downstream berm to improve the defense from internal erosion—Swedish application. *A Prepared Contribution, ICOLD*, Brasilia, May.

Nilsson, A. (2013). Personal communication.

Norwegian Geotechnical Institute. (1992). *Rockfill Dams, Design and Construction*, Kjaernsli, B. Valstad, T. and Höeg, K. (eds.). Norwegian Geotechnical Institute, Norwegian Institute of Technology, Oslo.

NVE (2012). *Veileder for fyllingsdammer*. Norges vassdrags-og energidirektorat Veileder nr. 4.

Olsen, S.M., Castro, G. and Stark, T.D. (2000). 1907 static liquefaction flow failure of the north dike of Wachusett Dam. *Journal of Geotechnical and Geoenvironmental Engineering*, ASCE, Vol. 126, No. 12, 1184–1193.

Olsen, S.M. and Stark, T.D. (2002). Liquefied strength ratio from liquefaction flow failure case histories. *Canadian Geotechnical Journal*, Vol. 39, No. 3, 629–647.

Olson, S.M. and Stark, T.D. (2003). Yield strength ratio and liquefaction analysis of slopes and embankments. *Journal of Geotechnical and Geoenvironmental Engineering*, ASCE, Vol. 129, No. 8, 727–737.

Onoue, A. (1988). Diagrams considering well resistance for designing spacing ratio of gravel drains. *Soils and Foundations*, Vol. 28, No. 3, 160–168.

Ostrovsky, N. and Wan, C.-F. (2023). A method for assessing the likelihood of overtopping failure for earthfill embankment dams. *Proceedings of the ANCOLD Conference*, Cairns, Australian National Committee on Large Dams. www.ancold.org.au.

Park, Y. (2003). *Investigation of the Ability of Filters to Stop Erosion through Cracks in Dams*. PhD thesis, Department of Civil and Environmental Engineering. Virginia Institute of Technology.

Patton, F.D. (1966). Multiple modes of shear failure in rock. *Proceedings 1st International Congress of Rock Mechanics, Lisbon*, Vol. 10, 509–513.

Patton, F.D. and Hendron, A.J. (1972). General report on mass movements. *Proceedings 2nd International Congress of International Association of Engineering Geology*, V-GR1-V-GR57.

PEER (2023). *PEER Workshop on Liquefaction Susceptibility*. Pacific Earthquake Engineering Center. Report 2023/02.

Peirson, W.L. and Fell, R. (2024). Pressurised flow in cracks in embankment dam cores. *European Journal of Environmental and Civil Engineering*. https://doi.org/10.1080/19648189.2024.2428253.

Peirson, W.L. and Fell, R. (2025). Downstream filter control of flow in cracks in embankment dam cores. *Journal of Environmental and Civil Engineering*. https://doi.org/10.1080/19648189.2025.2547252.

Peirson, W.L., Figlus, J., Pells, S.E. and Cox, R.J. (2008). Placed rock as protection against erosion by flow down steep slopes. *Journal of Hydraulic Engineering*, ASCE, Vol. 134, No. 9, 1370–1375.

Peirson, W.L., Prapakaran, V., Fell, R. and Douglas, K. (2022). On the hydraulics of flow in cracks in embankment dam cores. *European Journal of Environmental and Civil Engineering*. https://doi.org/10.1080/19648189.2022.20811993.

Pells, S. and Fell, R. (2002). *Damage and Cracking of Embankment Dams by Earthquakes, and the Implications for Internal Erosion and Piping*. UNICIV Report R 406, The University of New South Wales, ISBN 85841 3752.

Pells, S. and Fell, R. (2003). Damage and cracking of embankment dams by earthquake and the implications for internal erosion and piping. *Proceedings 21st International Congress on Large Dams, Montreal*. ICOLD, Paris Q83–R17, International Commission on Large Dams, Paris.

Pells Sullivan Meynink (2017). *Upper Yarra Dam Crack Propagation Analysis*. Unpublished Report dated 18th August.

Pestana, J.M., Hunt, C.E. and Goughnour, R.R. (1997). *FEQDrain: A Finite Element Computer Program for the Analysis of the Earthquake Generation and Dissipation of Pore Water Pressure in Layered Sand Deposits with Vertical Drains*. Report No. EERC 97-17, Earthquake Engineering Res. Ctr., University of California at Berkeley.

Pinto, N.L.S., Filho Marques, P.L. and Maurer, E. (1985). Foz do Areia dam—design, construction and behaviour. *Proceedings of the Symposium on Concrete Face Rockfill Dams—Design, Construction and Performance*, Cooke and Sherard (eds.). Detroit, Michigan, ASCE, New York, 173–191.

Powledge, G.R., Ralston, D.C., Miller, P., Chen, Y.H., Clopper, P.E. and Temple, D.M. (1989). Mechanics of overflow erosion on embankments, II: Hydraulic and design considerations. *Journal of Hydraulic Engineering*, ASCE, Vol. 115, No. 8, 1056–1075.

Purss, C. (2021). *Extension Strains in Embankment Dams and the Potential for Development of Transverse Cracks*. Master of Civil Engineering, School of Civil and Environmental Engineering, UNSW Sydney.

Purss, C. and Hunter, G. (2022). Extensional strains in embankment dams and the potential for development of transverse cracks. *Proceedings ANCOLD Conference 2022*. Australian National Committee on Large Dams. www.ancold.org.au.

Reclamation (2017). *Evaluation of Numerical Models for Simulating Embankment Dam Erosion and Breach Processes*. DS-2017-02. US Bureau of Reclamation, Denver Co.

Reclamation (2022). *Public Protection Guidelines: A Risk Informed Framework to Support Dam Safety Decision Making*. Bureau of Reclamation, Denver, CO.

Rinehart, R. (2012). *Development of a Test to Determine Cementation Potential of Embankment Dam Granular Filter Material—Results of Phase III Research*. Bureau of Reclamation Report DSO-12-01.

Rinehart, R., Pabst, M., Ngozi-Bullock, K. and Jackson, B. (2012). A test for cementation potential of granular filter material. *International Conference on Scour and Erosion*, (ICSE6), Paris, 104–112.

Ripley, C.F. (1986). Discussion of: Internal stability of granular filters. *Canadian Geotechnical Journal*, Vol. 23, 255–258.

Robbins, B.A. and Corcoran, M.K. (2022). *Calculation of Levee-Breach Widening Rates*. USACE Report ERDC/GSL TR-22-8.

Robbins, B.A. and Sharp, M.K. (2016). Incorporating uncertainty into backward erosion piping risk assessments. *Flood Risk 2016–3rd European Conference on Flood Risk Management*. https://doi.org/10.1051/e3sconf/20160703007.

Robbins, B.A. and Van Beek, V. (2015). Backward erosion piping: A historical review and discussion of influential factors. *ASDSO Conference*.

Robertson, P.K. (1990). Soil classification using CPT. *Canadian Geotechnical Journal*, Ottawa, Vol. 27, No. 1, 151–158.

Robertson, P.K. (1994). Suggested terminology for liquefaction. *Proceeding 47th Canadian Geotechnical Conference*, Halifax, Nova Scotia, 277–286.

Robertson, P.K. (2010). Evaluation of flow liquefaction and liquefied strength using the cone penetration test. *Journal of Geotechnical and Geoenvironmental Engineering*, ASCE, Vol. 136, No. 6, 842–853.

Robertson, P.K. (2016). Cone Penetration Test (CPT)—based soil behaviour type (SBT) classification system—an update. *Canadian Geotechnical Journal*. https://dx.doi.org/10.1139/cgj-2016-044.

Robertson, P.K. (2022). Evaluation of flow liquefaction and liquefied strength using the cone penetration test: An update. *Canadian Geotechnical Journal*, Vol. 59, No. 4, 620–624. https://doi.org/10.1139/cgj-2020-0657.

Robertson, P.K. and Campanella, R.G. (1984). *Guidelines for Use, Interpretation and Application of the CPT and CPTU*. Soil Mechanics Series No. 105, Civil Eng. Dept., The University of British Columbia, Vancouver, BC, Canada.

Robertson, P.K., de Melo, L., Williams, D.J. and Wilson, G.W. (2019). *Report of the Expert Panel on the Technical Causes of the Failure of Feijão Dam I*.

Robertson, P.K. and Wride, C.E. (1998). Evaluating cyclic liquefaction potential using the cone penetration test. *Canadian Geotechnical Journal*, Vol. 35, No. 3, 442–459.

Robinson, K., Pattle, A. and Shurvell, T. (2015). Rowallan Dam: Management of flood risk during a major dam upgrade. *Proceedings ANCOLD Conference 2015*. Australian National Committee on Large Dams. www.ancold.org.au.

Rollins, K.M., Roy, J., Athanasipoulos-Zekkos, A., Zekkos, D., Amaroso, S., Cao, Z., Milana, G., Vassallo, M. and Di Giulio, G. (2022). A new V_S—based liquefaction triggering procedure for gravelly soils. *Journal of Geotechnical and Geoenvironmental Engineering*, ASCE, Vol. 148, No. 6, 04022040.

Rönnqvist, H. (2010). *Predicting Surfacing Internal Erosion in Moraine Core Dams*. KTH. TRITA-LWR Lic Thesis 2050.

Rönnqvist, H. (2015). *On the Assessment of Internal Erosion of Dam Cores of Glacial Till*. Doctoral Thesis, Luleå University of Technology, Department of Civil, Environmental and Natural Resources Engineering, Division of Mining and Geotechnical Engineering, Sweden.

RTA (2010). *Culvert Risk Assessment Guideline*. New South Wales Roads and Traffic Authority.

Sail, Y., Marot, D., Sibille, L. and Alexis, A. (2011). Suffusion tests on cohesionless granular matter. *European Journal of Environmental and Civil Engineering*, Vol. 15, No. 5.

Sasitharan, S., Robertson, P.K., Sego, D.C. and Morgenstern, N.R. (1993). Collapse behaviour of sand. *Canadian Geotechnical Journal*, Vol. 30, No. 4, 569–577.

Savage, S., Douglas, K., Fell, R., Peirson, W. and Berndt, R. (2019). Modelling the erosion and swelling of the sides of transverse cracks in embankment dams. *Journal of Geotechnical and Geoenvironmental Eng*, ASCE, Vol. 145, No. 5, 04019015.

Schmertmann, J.H. (1978). *Guidelines for Cone Penetration Test, Performance and Design*. US Federal Highway Administration FWHA TS-78-209.

Schmertmann, J.H. (1995). *Report on Flume Tests for Piping and Scour of HL-2 Sands*. University of Florida.

Schmertmann, J.H. (2000). The non-filter factor of safety against piping through sands. In *ASCE Geotechnical Special Publication No. 111, Judgment and Innovation*, Silva, F. and Kavazanjian, E. (eds.). ASCE, Reston.

Seed, H.B. (1979). Consideration in the earthquake resistant design of earth and rockfill dams. *Geotechnique*, Vol. 29, 215–263.

Seed, H.B. (1983). Earthquake-resistant design of earth dams. *Proc., Symp. Seismic Design of Earth Dams and Caverns*. ASCE, New York, 41–64.

Seed, H.B. (1987). Design problems in soil liquefaction. *Journal of Geotechnical Engineering*, ASCE, Vol. 113, No. GT8, 827–845.

Seed, H.B. and Booker, J.R. (1977). Stabilization of potentially liquefiable sand deposits using gravel drains. *Journal of Geotechnical Engineering*, ASCE, Vol. 103, No. GT7, 757–768.

Seed, H.B., Makdisi, F.I. and De Alba, P. (1978). Performance of earth dams during earthquakes. *Journal of Geotechnical Engineering*, ASCE, Vol. 104, No. GT7.

Seed, H.B., Tokimatsu, K., Harder, L.F. and Chung, R.M. (1985). The influence of SPT procedures in soil liquefaction resistance evaluations. *Journal of Geotechnical Engineering*, ASCE, Vol. 111, No. 12, 1425–1445.

Seed, H.B., Woodward, R.J. and Lundgren, R. (1962). Prediction of swelling potential for compacted clays. *Journal of Soil Mechanics and Foundation Division*, ASCE, Vol. 88, No. SM4, 107–131.

Seed, R.B. (2010). *Technical Review and Comments: 2008 EERI Monograph Titled Soil Liquefaction during Earthquakes*. Geotechnical Report No, UCB/GT-2010/01, University of California, Berkeley.

Seed, R.B., Cetin, K.O., Moss, R.E.S., Kammerer, A., Wu, J., Pestana, J., Riemer, M., Sancio, R.B., Bray, J.D., Kayen, R. and Faris, A. (2003). *Recent Advances in Soil Liquefaction Engineering: A Unified and Consistent Framework*. Earthquake Engineering Research Center Report No EERC 2003-06.

Seed, R.B. and Harder, L.F. (1990). SPT-based analysis of cyclic pore pressure generation and undrained residual strength. In *Proceedings, Seed Memorial Symposium*, Duncan, J.M. (ed.). BiTech Publishers, Vancouver, British Columbia, 351–376.

Sellmeijer, H., López de la Cruz, J., Van Beek, V. and Knoeff, H. (2012). Fine-tuning of the backward erosion piping model through small-scale, medium-scale and IJkdijk experiments. *European Journal of Environmental and Civil Engineering*, Vol. 15, No. 8, 2011, 1139–1154.

Sellmeijer, J.B. (1988). *On the Mechanism of Piping under Impervious Structures*. Delft University of Technology.

Sellmeijer, J.B., Calle, E. and Sip, J.W. (1989). Influence of aquifer thickness on piping below dikes and dams. *International Symposia. Analytical Evaluation of Dam Related Safety Problems*, Vol. 21, 357. Copenhagen. Danish National Committee on Large Dams.

Sellmeijer, J.B. and Koenders, M.A. (1991). A mathematical model for piping. *Applied Mathematical Modelling*, Vol. 115, 646–661. Butterworth-Heinemann.

Sherard, J.L. (1953). *Influence of Soil Properties and Construction Methods on the Performance of Homogeneous Earth Dams*. Ph.D Thesis, Harvard University.

Sherard, J.L. (1967). Earthquake considerations in earth dam design. *Journal of Soil Mechanics and Foundations Division*, ASCE, Vol. 93, No. SM4, 377–401.

Sherard, J.L. (1973). Embankment dam cracking. In *Embankment Dam Engineering*, Hirschfeld, R.C. and Poulos, S.J. (eds.). John Wiley & Sons, 308–312.

Sherard, J.L. (1985). The upstream zone in concrete faced rockfill dams. In *Concrete Face Rockfill Dams Design, Construction and Performance*, Cooke, J.B. and Sherard, J.L. (eds.). ASCE Geotechnical Engineering Division, 618–641.

Sherard, J.L. (1986). Hydraulic fracturing in embankment dams. *Journal of Geotechnical Engineering*, ASCE, Vol. 112, No. 10, 905–927.

Sherard, J.L., Cluff, L.S. and Allen, C.R. (1974). Potentially active faults in dam foundations. *Geotechnique*, Vol. 24, No. 3, 367–428.

Sherard, J.L. and Cooke, J.B. (1987). Concrete-face rockfill dam: 1. Assessment. *Journal of Geotechnical Engineering*, ASCE, Vol. 113, No. 10, 1096–1112.

Sherard, J.L., Decker, R.S. and Ryker, N.L. (1972). Hydraulic fracturing in low dams of dispersive clay. *Proceedings, Specialty Conference on Performance of Earth and Earth-Supported Structures*, ASCE, Vol. 1, Part 1, 653–689.

Sherard, J.L. and Dunnigan, L.P. (1989). Critical filters for impervious soils. *Journal of Geotechnical Engineering*, ASCE, Vol. 115, No. 7, 927–947.

Sherard, J.L., Dunnigan, L.P. and Talbot, R.T. (1984). Basic properties of sand and gravel filters. *Journal of Geotechnical. Engineering*, ASCE, Vol. 110, No. 6.

Sierra, J.M., Ramirez, C.A. and Hacelas, J.E. (1985). Design features of salvajina dam. In *Concrete Face Rockfill Dams, Design, Construction and Performance*, Cooke, J.B. and Sherard, J.L. (eds.). ASCE Geotechnical Engineering Division, 266–285.

Sills, G.L. and Harder, L.F. (2016). Decades of foundation deterioration threatening public safety. *Proceedings of the 36th Annual Conference, United States Society on Dams* (USSD), Denver, CO, April 11–15, 883–907.

Sills, G.L. and Vroman, N. (2007). A review of Corps of Engineers levee seepage practices in the United States. In *Internal Erosion of Dams and their Foundations*, Fell, R. and Fry, J.J. (eds.). Taylor & Francis, 209–218.

Silva, F., Lambe, T.W. and Marr, W.A. (2008). Probability and risk of slope failure. *Journal of Geotechnical and Geoenvironmental Engineering*, ASCE, Vol. 134, No. 12, 1691–1699.

Silva, F., Lambe, T.W. and Marr, W.A. (2010). Probability and risk of slope failure: Discussions and closures. *Journal of Geotechnical and Geoenvironmental Engineering*, ASCE, Vol. 136, No. 5, 764–771.

Silva, F., Lambe, T.W. and Marr, W.A. (2025). *Probability and Risk of Slope Failure*. Personal Communication.

Silvis, F. (1991). *Verificatie Piping Model; Proeven in de Deltagoot. Evaluatierapport*. Rapport Grondmechanica Delft, CO 317710/7.

Skempton, A.W. (1985). Residual strength of clays in landslides, folded strata and the laboratory. *Geotechnique*, Vol. 35, No. 1, 3–18.

Skempton, A.W. (1986). Standard penetration test procedures and the effects in sands of overburden pressure, relative density, particle size, aging and over-consolidation. *Geotechnique*, Vol. 36, No. 3, 425–447.

Skempton, A.W. (1990). Historical development of British embankment dams to 1960. In *Clay Barriers for Embankment Dams*. Thomas Telford, London, 15–52.

Skempton, A.W. and Brogan, J.M. (1994). Experiments on piping in sandy gravels. *Geotechnique*, Vol. 44, No. 3, 449–460.

Sladen, J.A. and Hewitt, K.J. (1989). Influence of placement method on the in-situ density of hydraulic sand fills. *Canadian Geotechnical Journal*, Vol. 26, 453–466.

Solvik, Ø. (1981). Throughflow and stability problems in rockfill dams exposed to exceptional loads. *Sixteenth International Congress on Large Dams, Q.67 R.20, ICOLD*, International Commission on Large Dams, Paris, 333–343.

Soroush, A., Shourijeh, P.T., Aghajani, H.F. and Mohammadinia, A. (2012). A review of the sand castle test for assessing collapsibility of filter in dams. *ASTM Geotechnical Testing Journal*, Vol. 35, No. 4, 1–14.

Southcott, P.H., Herweynen, R. and Fell, R. (2003). *Assessing the Likelihood of Concentrated Leaks Leading to Failure in Concrete Face Rockfill Dams*. ANCOLD Bulletin Issue No.124, 17–28. Australian National Committee on Large Dams. www.ancold.org.au.

Stark, T.D. and Duncan, J.M. (1987). *Mechanisms of Strength Loss in Stiff Clays*. Virginia Polytechnic Institute and State University, Dept. of Civil Engineering, Geotechnical Engineering.

Stark, T.D. and Duncan, J.M. (1991). Mechanisms of strength loss in stiff clays. *Journal of Geotechnical Engineering*, ASCE, Vol. 117, No. 1, 139–155.

Swaisgood, J.R. (1998). Seismically-induced deformation of embankment dams. *6th US National Conference on Earthquake Engineering*, Seattle, Washington.

Swaisgood, J.R. (2003). Embankment dam deformations caused by earthquakes. *Pacific Conference on Earthquake Engineering, Paper No 014*.

Swindon, A., Griggs, T., Herweynen, R. and Fell, R. (2003). Cairn Curran upgrade—investigations and geotechnical modelling. *ANCOLD 2003 Conference Proceedings*. Australian National Committee on Large Dams. www.ancold.org.au.

Talbot, J.R. (1994). The mechanics of cracking in embankment dams. *Proceeding of the Fracture Mechanics Applied to Geotechnical Engineering*, ASCE, 118–130.

Terada, M., Hayakawa, T. and Nakamura, I. (1997). Dynamic behaviour and analysis of rockfill dam with asphalt facing during Hyogo-Ken-Nanbu Earthquake. *Proceedings Earthquake Engineering for Dams and Reservoirs*, Hinks, J. (ed.).

Terzahgi, K and Peck, R.B. (1967) Soil Mechanics in Engineering Practice. Wiley, New York.

Thorne, C.P. (1984). Strength assessment and stability analysis for fissured clay. *Geotechnique*, Vol. 34, No. 3, 305–322.

Tjandrajana, R. (1989). *Penetration of Cement Grout into Fractures*. M.Eng.Sc. Thesis. School of Civil Engineering, The University of New South Wales.

Toose, J., Hunter, G. and Watkins, G. (2007). Design and construction of the Lance Creek Reservoir seismic upgrade. *Proceedings ANCOLD Conference 2007*. Australian National Committee on Large Dams. www.ancold.org.au.

Townsend, F.C., Bloomquist, D., Shiau, J-M., Martinez, R. and Rubin, H. (1988). *Evaluation of Filter Criteria and Thickness for Migrating Piping in Sands*. Department of Civil Engineering, University of Florida, Report for U.S. Bureau of Reclamation.

Townsend, F.C. and Shiau, J.M. (1986). *Analytical and Experimental Evaluation of Piping and Filter Design for Sands*. University of Florida, Gainsville, FL.

Tsukada, T., Uchita, Y., Yamamoto, H., Shimada, Y. and Takasawa, K. (2013). Study on behaviour of AFRD during earthquake and the conducted reinforcement. *ICOLD 2013 International Symposium*, Seattle.

USACE (1956). *Investigation of Under-Seepage and Its Control, Lower Mississippi River Levees*. USACE Waterways Experimental Station, Vicksburg, Technical Memorandum No 3-424.

USACE (1984). *EM 1110-3-138, Engineering and Design—Pavement Criteria for Seasonal Frost Conditions—Mobilization Construction*. US Army Corps of Engineers.

USACE (1994). *EM1110-2-2300 Earth and Rock-Fill Dams—General Design and Construction Considerations.* US Army Corps of Engineers.

USACE (2000). *Engineering Manual EM 1110-2-1913, Design and Construction of Levees.* US Army Corps Engineers.

USACE (2002, 2011). *Coastal Engineering Manual, EM 1110-2-1100.* CECW-EW US Army Corps Engineers.

USACE (2003). *Slope Stability, Engineering Manual 1110-2-1902 Department of the Army, Corps of Engineers.* Washington, DC. www.usace.army.mil/inet/usacoe-docs/eng-manuals/em1110-2-1902/entire.pdf

USACE (2004). *EM 1110-202300. General Design and Construction Considerations for Earth and Rockfill Dams.* U. S. Army Corps of Engineers, Washington, DC.

USACE (2005). *Engineering Technical Letter ETL 1110-2-569. Engineering Design—Design Guidance for Levee Under-Seepage.* US Army Corps of Engineers.

USACE (2014). *Safety of Dams—Policy and Procedures.* ER 1110-2-1156.

USACE (2015). *Flood Fight Handbook—Preparing for a Flood.* U.S. Army Corps of Engineers-Walla Walla District.

USACE (2017a). *Best Practices Chapter C-1. Consequences of Dam and Levee Failure.* US Army Corp of Engineers.

USACE (2017b). *EM 1110-2-1913, Appendix K, Emergency Flood Fighting and Flood Protection.* US Army Corps Engineers.

USACE (2019). *Best Practices Chapter D-1. Erosion of Soil and Rock.* US Army Corp of Engineers.

USACE (2021). *Rationale Behind the US Army Corps of Engineers Tolerable Risk Guidelines from 2010 to 2021.* RMC-TR-2021-05.

USACE (2022). *Evaluation, Design and Construction of Levees.* USACE Engineering Design EM 1110-2-1913. US Army Corps Engineers.

USACE (2025). *EM 1110-2-1914, Design, Construction and Maintenance of Relief Wells.* US Army Corps Engineers, Washington, DC.

USACE/Reclamation (2019). *Best Practices in Dam and Levee Safety Risk Analysis.* US Army Corps Engineers, Washington, DC.

U.S. Army Corps of Engineers (1939). *Report on the Slide of a Portion of the Upstream Face of the Fort Peck Dam, Fort Peck, Montana.* United States Government Printing Office.

USBR (1987). *Design Standard No.13, Embankment Dams, Chapter 4, Static Stability Analysis.* US Bureau of Reclamation, Denver, CO.

USBR (1998). *Earth Manual.* US Bureau of Reclamation, Denver, CO.

USBR (2011). *Design Standards. Embankment Dams, No. 13, Chapter 5. Protective Filters.* US Bureau of Reclamation, Denver, CO.

USBR (2012). *Design Standards No. 13, Embankment Dams, Chapter 6, Freeboard.* US Department of the Interior, US Bureau of Reclamation, Denver, CO.

USBR (2013). *Design Standards No. 13, Embankment Dams, Chapter 4, Static.* US Department of the Interior, US Bureau of Reclamation, Denver, CO.

USBR (2014). *RCEM—Reclamation Consequence Estimating Methodology: Dam Failure and Flood Event Case History Compilation.* Interim draft, February. United States Bureau of Reclamation, Denver, CO.

USBR (2015). *RCEM—Reclamation Consequence Estimating Methodology: Guidelines for Estimating Life Loss for Dam Safety Risk Analysis.* Interim, July. United States Bureau of Reclamation, Denver, CO.

US Bureau of Reclamation (2013). *Red Willow Dam Incident Report.* Technical Memorandum No RW-8311-2013-12.

US Bureau of Reclamation (2014). *Red Willow Dam, Hydraulic Slug Testing.* Technical Memorandum No 86-68320-14-1.

USCOLD (1988). *Lessons from Dam Incidents USA-II.* ASCE.

US Army Corps of Engineers (1970). *Engineering and Design—Stability of Earth and Rock Fill Dams.* Engr. Manual EM 1110-2-1902, Dept. of the Army, Corp of Engrs., Office of the Chief of Engineers.

USDA-SCS (1994). *Gradation Design of Sand and Gravel Filters.* United States Department of Agriculture, Soil Conservation Service, Part 633, Chapter 26, National Engineering Handbook.

USSD (2022). Analysis of seismic deformations of embankment dams. *Prepared by the USSD Committee on Earthquakes Sub-Committee on Seismic Deformation Analysis of Embankment Dams.* U.S. Society on Dams, August.

Valencia, G.F. and Sandoval, E.M. (1997). Aguamilpa Dam behaviour. *Proceedings, Seventeenth Annual USCOLD Lecture Series, Non-Soil Water Barriers for Embankment Dams*, San Diego, CA, 133–147.

Van Beek, V.M. (2015). *Backward Erosion Piping: Initiation and Progression*. PhD Thesis, Delft University of Technology, Delft.

Van Beek, V.M., Bezuijen, B. and Sellmeijer, H. (2013). Backward erosion piping. Chapter 3 In *Erosion in Geomechanics Applied to Dams and Levees*, Bonelli, S. (ed.). Wiley, Hoboken, and ISTE, London, 193–269.

Van Beek, V.M. and Hoffmans, G.J.C.M. (2017). Evaluation of Dutch backward erosion piping models and a future perspective. *Proceedings 25th Meeting of the European Working Group on Internal Erosion*.

Van Beek, V.M., Knoeff, H., Förster, U. and Koelewijn, A. (2010). Observations of piping at the IJKdijk test site in the Netherlands. *EWG Internal Erosion Workshop*, Granada, April 13.

Van Beek, V.M., Knoeff, H. and Sellmeijer, J.B. (2011). Observations on the process of backwards erosion piping in small-medium and full scale experiments. *European Journal of Environmental and Civil Engineering*, Vol. 15, No. 8, 1115–1137.

Van Beek, V.M., Knoeff, J.G. and Sellmeijer, J.B. (2012a). Observations on the process of piping by under-seepage in cohesionless soils in small, medium and full scale experiments. *European Journal of Environmental and Civil Engineering*, Vol. 15, No. 8, 1115–1137.

Van Beek, V.M., Qiuling, Y.A., Meindert, V. and Barends, F. (2012b). Validation of Sellmeijer's model for backward piping under dikes on multiple sand layers. *ICSE6 Paris*, August 27–31, 543–550.

Van Beek, V.M., Robbins, B.A., Hoffmans, G.J.C.M., Bezuijen, A. and van Rijn, L.C. (2019). Use of incipient motion data for backward erosion piping models. *International Journal of Sediment Research*, Vol. 34, 401–408.

Van Beek, V.M., Van Essen, H.M., Vandenboer, K. and Bezuijen, A. (2015). Developments in modelling of backward erosion piping. *Geotechnique*, Vol. 65, No. 9, 740–754.

Van Rijn, L.C. (2014). *Review of Piping Processes and Models*. Report Prepared for Deltares. www.leovanrijn-sediment.com.

Varnes, D.J. (1978). Movement types and processes. In *Landslides, Analysis and Control*, Schuster, R.L. and Krizek, R.J. (eds.). National Academy of Sciences, 11–33.

Vrymoed, J. and Lam, W. (2006). *Earthquake Performance of Austrian Dam, California during the Loma Prieta Earthquake*. Division of Safety of Dams, Department of Water Resources, ASDSO Conference, Boston MA. http://damsafety.water.ca.gov/seismic_austrian.cfm.

Vuola, P., Konrad, J.M. and Bartsch, M. (2007). *Effects of Frost and Thaw on Dams: In Assessment of the Risk of Internal Erosion of Water Retaining Structures: Dams, Dykes and Levees*. Intermediate Report of the European Working Group of ICOLD, Aufleger, M., Fry, J.J., Goltz, M. and Perzlmaier, S. (eds.).

Wahl, T.L. and Mortensen, J.L. (2024). Friction factors for flow through cracks in concrete hydraulic structures. *Journal of Hydraulic Engineering*, ASCE, Vol. 151, No. 1, 06024009.

Wahl, T.L., Regazzoni, P. and Erdogan, Z. (2008). *Determining Erosion Indices of Cohesive Soils with the Hole Erosion Test and Jet Erosion Test*. US Department of Interior, Bureau of Reclamation Report DSO-08-05.

Wair, B.R., De Jong, J.T. and Shantz, T. (2012). *Guidelines for Estimation of Shear Wave Velocity Profiles*. PEER Report 2012/08. Pacific Earthquake Engineering Research Center.

Walker, B.F., Blong, R.J. and MacGregor, J.P. (1987). Landslide classification, geomorphology and site investigations. In *Soil Slope Instability and Stabilisation*, Walker, B.F. and Fell, R. (eds.). Balkema, 1–52.

Wallace, M. (1987). How to prevent frost heave. *Concrete Construction*, April, 369–372.

Wan, C.F. (2006). *Experimental Investigation of Piping Erosion and Suffusion of Soils in Embankment Dams and Their Foundations*. PhD thesis, School of Civil and Environmental Engineering, The University of New South Wales, Sydney, Australia.

Wan, C.F. and Fell, R. (2002). *Investigation of Internal Erosion and Piping of Soils in Embankment Dams by the Slot Erosion Test and the Hole Erosion Test*. UNICIV Report No. R-412, ISBN 85841 379 5, School of Civil and Environmental Engineering, The University of New South Wales, Sydney. www.engineering.unsw.edu.au/civil-engineering/uniciv-reports.

Wan, C.F. and Fell, R. (2004a). Investigation of rate of erosion of soils in embankment dams. *Journal of Geotechnical and GeoEnvironmental Engineering*, ASCE, Vol. 130, No. 4, 373–380.

Wan, C.F. and Fell, R. (2004b). Laboratory tests on the rate of piping erosion of soils in embankment dams. *Geotechnical Testing Journal*, Vol. 27, No. 3, 295–303.

Wan, C.F. and Fell, R. (2004c). *Experimental Investigation of Internal Iinstability of Soils in Embankment Dams and Their Foundations*. UNICIV Report No.429, School of Civil and Environmental Engineering, The University of New South Wales, Sydney, ISBN 85841 396 5. www.engineering.unsw.edu.au/civil-engineering/uniciv-reports.

Wan, C.F. and Fell, R. (2008). Assessing the potential of internal erosion and suffusion in embankment dams and their foundations. *Journal of Geotechnical and Geoenvironmental Engineering*, ASCE, Vol. 134, No. 3, 410–407.

Wang, W. and Høeg, K. (2010). The asphalt core embankment dam: A very competitive alternative. *Proceedings of 1st the International Symposium on Rockfill Dams*, Chengdu.

Wang, W., Høeg, K. and Zhang, Y. (2010). Design and performance of the Yele asphalt-core rockfill dam. *Canadian Geotechnical Journal*, Vol. 47, 1365–1381. https://doi.org/10.1139/T10-028.

Wang, Z.L. and Makdisi, F.I. (1999). Implementing a bounding surface hypo plasticity model for sand into the FLAC program. In *FLAC and Numerical Modelling in Geomechanics*. A.A. Balkema, Netherlands, 483–490.

Watakeekul, S., Roberts, G.J. and Coles, A.J. (1985). Khao Laem: A concrete face rockfill dam on karst. In *Concrete Face Rockfill Dams, Design, Construction and Performance*, Cooke, J.B. and Sherard, J.L. (eds.). ASCE Geotechnical Engineering Division, 336–361.

Weijers, J.B.A. and Sellmeijer, J.B. (1993). A new model to deal with the piping mechanism. In *Filters in Geotechnical and Hydraulic Engineering*. Brauns, Herbaum and Schuler (eds.). Balkema, Rotterdam.

Westmore, R., George, A. and Wilson, R. (2011). Hume Dam spillway Southern junction filter and drainage works. *ANCOLD Conference Proceedings 2011*. Australian National Committee on Large Dams. www.ancold.org.au.

Whitman, R.V. (1984). Evaluating calculated risk in geotechnical engineering: Seventeenth terzaghi lecture. *Journal Geotechnical Engineering*, ASCE, Vol. 110, No. 2, 143–188.

Wieland, M. (2009). Concrete face rockfill dams in highly seismic regions. *Proceedings 1st International Symposium on Rockfill Dams, Chengdu.* Chinese National Committee on Large Dams, Beijing.

Wilkins, J.K., Mitchell, W.R., Fitzpatrick, M.D. and Liggins, T. (1973). The design of Cethana concrete face rockfill dam. *Proceedings, 11th ICOLD Congress*, Rio de Janeiro (Q.42 R.3). International Commission on Large Dams, Paris, 25–43.

Win, S.S. (2006). *Tensile Strength of Compacted Soils Subject to Wetting and Drying*. Master of Engineering thesis, School of Civil and Environmental Engineering, The University of New South Wales, Sydney, Australia.

Winckler, C., Davidson, R., Yenne, L. and Pilz, J. (2014). CPTu based characterization of tailings liquefaction susceptibility. *Proceedings 34th USSD Annual Conference*.

Wolff, T.F. (2002). *Performance of Levee Under-Seepage Controls: A Critical Review*. US Army Corps of Engineers, Engineering Research Center, Geotechnical and Structural Laboratory, ERDC/GSL TR-02-19.

Woodward Clyde (1999). *Lake Eppalock Main Embankment Remedial Work*. Unpublished Design Report, for Goulburn Murray Water.

Wörman, A. and Olafsdottir, R. (1992). Erosion in a granular medium interface. *Journal of Hydraulic Research*, Vol. 30, No. 5, 639–655.

Wroth, C.P. and Houlsby, G.T. (1985). Soil mechanics—property characterization and analysis procedures. *11th ICSFME*, Vol. 1, 1–55.

Xu, Z. (2008). Performance of Zipingpu CFRD during the strong earthquake. *Proc. 10th Int. Conf. on Landslides and Engineered Slopes*, Xian, China.

Yamaguchi, Y., Kondo, M. and Kobori, T. (2012). Safety inspections and seismic behaviour of embankment dams during the 2011 off the Pacific Coast of Tohoku earthquake. *Soils and Foundations*, Vol. 52, No. 5, 945–955.

Youd, T.L., Idriss, I.M., Andrus, R.D., Arango, I., Castro, G., Christian, J.T., Dobry, R., Finn, W.D.L., Harder, L.F., Hynes, M.E., Ishihara, K., Koester, J.P., Liao, S.S.C., Marcuson, W.F., Martin, G.R., Mitchell, J.K., Moriwaki, Y., Power, M.S., Robertson, P.K., Seed, R.B. and Stokoe, K.H. (2001). Liquefaction resistance of soils: Summary report from the 1996 NCEER and 1998 NCEER/NSF Workshops on evaluation of liquefaction resistance of soils. *Journal of Geotechnical and Geoenvironmental Engineering*, ASCE, Vol. 127, No. 10, 817–834.

Youd, T.L. and Perkins, D.M. (1978). Mapping of liquefaction-induced ground failure potential. *Journal of Geotechnical Engineering*, ASCE, Vol. 104, No. 4, 433–446.

Zhang, J., Wang, J. and Cui, H. (2016). Causes of the abnormal seepage field in a dam with asphaltic concrete core. *Journal of Earth Science*, Vol. 27, No. 1, 74–82.

Zhong, S. (2023). Learnings from deep test pits investigations into the filters for the Cardinia Dam safety upgrade. *Proceedings ANCOLD Conference, Cairns*. Australian National Committee on Large Dams. www.ancold.org.au.

Zhu, F., Zhang, W. and Puzrin, A.M. (2024). The slip surface mechanism of delayed failure of the Brumadinho tailings dam in 2019. *Communications Earth & Environment*, Vol. 5, No. 33.

Ziotopoulou, K. and Boulanger, R.W. (2012). Constitutive modelling of duration and overburden effects in liquefaction evaluations. *Proc. Second Int. Conf. on Performance-Based Design in Earthquake Geotechnical Engineering*, Taormina, Italy.

Ziotopoulou, K. and Boulanger, R.W. (2013). Calibration and implementation of a sand plasticity plane-strain model for earthquake engineering applications. *Journal of Soil Dynamics and Earthquake Engineering*, Vol. 53, 268–280. https://dx.doi.org/10.1016/j.soildyn.2013.07.009.

Index

A

B

C

D

E

T

U

V

W

Z

For Product Safety Concerns and Information please contact our EU
representative GPSR@taylorandfrancis.com
Taylor & Francis Verlag GmbH, Kaufingerstraße 24, 80331 München, Germany

www.ingramcontent.com/pod-product-compliance
Lightning Source LLC
LaVergne TN
LVHW081517300926
849161LV00006B/1203

* 9 7 8 1 0 3 2 8 8 9 0 4 7 *